COHERENCE AND ULTRASHORT PULSE LASER EMISSION

Edited by **Dr. F. J. Duarte**

Coherence and Ultrashort Pulse Laser Emission
Edited by Dr. F. J. Duarte

Published by InTech
Janeza Trdine 9, 51000 Rijeka, Croatia

Publishing Process Manager Jelena Marusic

Technical Editor Teodora Smiljanic

Cover Designer InTech Design Team

Image Copyright Kostyantyn Ivanyshen, 2010. Used under license from Shutterstock.com

Additional hard copies can be obtained from orders@intechopen.com

Coherence and Ultrashort Pulse Laser Emission, Edited by Dr. F. J. Duarte
p. cm.
ISBN 978-953-307-242-5

For EU product safety concerns:
IN TECH d.o.o., Prolaz Marije Krucifikse Kozulić 3, 51000 Rijeka, Croatia,
info@intechopen.com or visit our website at intechopen.com.

Meet the editor

F.J. Duarte is a laser physicist based in Western New York, USA. He is the author and editor of several well-known books on tunable lasers including Dye Laser Principles (Academic, New York, 1990) and Tunable Laser Optics (Elsevier Academic, New York, 2003). His most recent edited work is Tunable Laser Applications, 2nd Edition (CRC, New York, 2009).

Dr. Duarte has made key experimental and theoretical contributions to the field of narrow-linewidth tunable laser oscillators. These include original oscillator architectures and the generalized multiple-prism grating dispersion theory. He has also pioneered the use of Dirac's quantum notation in the description of generalized N-slit interference and classical optics phenomena. Currently, his research focuses on further developments of dispersive narrow-linewidth laser oscillators and very large N-slit laser interferometers. Dr. Duarte's contributions are cited in some 130 laser and optics books including several classics. He received the Engineering Excellence Award from the Optical Society of America, is a Fellow of the Australian Institute of Physics, and a Fellow of the Optical Society of America.

Contents

Preface

In an optimized pulsed laser source, the coherence of its emission (or linewidth) is intimately related to the duration of the emission pulse, by one of the most beautiful expressions of quantum optics

$$\Delta\nu\Delta t \approx 1$$

which is an alternative version of Heisenberg's uncertainty principle

$$\Delta p\Delta x \approx h$$

In this volume, recent contributions on coherence provide a useful perspective on the diversity of various coherent sources of emission and coherent related phenomena of current interest. These papers provide a preamble for a larger collection of contributions on ultrashort pulse laser generation and ultrashort pulse laser phenomena. Papers on ultrashort pulse phenomena include works on few cycle pulses, high-power generation, propagation in various media, and a variety of applications of current interest. Undoubtedly, *Coherence and Ultrashort Pulse Emission* offers a rich and practical perspective on this rapidly evolving field.

F. J. Duarte

Rochester
New York, USA

Part 1

Coherence and Quantum Phenomena

1

Electrically-Pumped Organic-Semiconductor Coherent Emission: A Review

F. J. Duarte
Interferometric Optics, Rochester, New York,
Department of Electrical and Computer Engineering, University of New Mexico, New Mexico,
USA

1. Introduction

Organic lasers came into existence via the introduction of the pulsed optically-pumped liquid organic dye laser by Sorokin and Lankard (1966) and Schäfer et al. (1966). An additional momentous contribution was the discovery of the continuous wave (CW) liquid organic dye laser by Peterson et al. (1970) which opened the way for the development of narrow-linewidth tunability in the CW regime plus the eventual introduction of femtosecond lasers (see, for example, Dietel et al. (1983) and Diels, (1990)). The narrow-linewidth tunable pulsed dye laser was demonstrated by Hänsch (1972) and improved by Shoshan et al. (1977), Littman and Metcalf (1978), Duarte and Piper (1980, 1981). All these developments in practical organic tunable lasers, spanning the visible spectrum, "created a renaissance in diverse applied fields such as medicine, remote sensing, isotope separation, spectroscopy, photochemistry, and other analytical tasks" (Duarte et al. (1992)).

An early development, in the field of tunable lasers, was also the discovery of solid-state pulsed optically-pumped organic dye lasers by Soffer and McFarland (1967) and Peterson and Snavely (1968). However, it was not until the 1990s that, due to improvements in the dye-doped polymer gain media, this class of lasers would again be the focus of research attention (see, for example, Duarte (1994), Maslyukov et al. (1995), Costela et al. (2003)). An additional effort in optically-pumped tunable laser research is the work on organic semiconductor lasers based on thin-film conjugated polymers (see, for example, Holzer et al. (2002)).

All this activity has been conducted on optically pumped organic lasers although researchers from the onset have also been interested on the direct electronic excitation of tunable organic lasers (Steyer and Schäfer, 1974; Marowsky et al., 1976). Some recent reviews mentioning efforts towards realizing coherent emission from direct electrical excitation of organic semiconductors, include Kranzelbinder and Leising (2000), Baldo et al. (2002), Samuel and Turnbull (2007), and Karnutsch (2007). Most of these reviews give ample attention to conjugated polymer gain media.

In this chapter, experimental results demonstrating coherent emission from electrically-excited pulsed dye-doped organic semiconductors, in microcavity configurations, are reviewed. The reported emission is single-transverse-mode, and given the 300 nm cavity length, also single-longitudinal mode. In the spectral domain the emission is *indistinguishable* from broadband dye laser radiation. The radiation is generated from a tandem semiconductor structure where the emission medium are regions of coumarin 545 tetramethyl dye-doped Alq_3. An alternative description for the emission medium would be a laser-dye-doped tandem organic light emitting diode (OLED).

This work came to light in 2005 when researchers working on electrically-pumped tandem organic semiconductors reported on highly-directional coherent emission in the pulsed regime (Duarte et al., 2005). This pulsed coherent emission was characterized by a nearly diffraction limited beam and an interferometrically estimated linewidth of $\Delta\lambda \approx 10.5$ nm (Duarte et al., 2005; Duarte, 2007). In 2008 a detailed analysis of the measured emission characteristics led to the conclusion that the observed radiation was *indistinguishable* from broadband dye laser emission (Duarte, 2008). This coherent emission was generated in a sub-micrometer asymmetrical cavity comprised of a high reflector and a low reflectivity output coupler (Duarte et al., 2005; Duarte, 2007, 2008). This sub micrometer cavity was collinearly confined within an interferometric configuration which selects a single-transverse mode. The emission medium is the laser dye coumarin 545 tetramethyl.

Subsequently, using a tetramethyl dye emitting in the red, and a similar experimental arrangement, Liu et al. (2009) also reported on coherent emission in the pulsed regime. More recently, however, a paper by Samuel et al. (2009) has formulated several criticisms to the work reported by Liu et al. (2009) and interrogates their laser interpretation. Here, central aspects of Liu et al. (2009) and Samuel et al. (2009) are also reviewed and discussed in light of well-known laser, and amplified spontaneous emission (ASE), literature standards. Furthermore, the results and interpretation disclosed by Duarte et al. (2005) and Duarte (2007, 2008) are reexamined, again leading to the conclusion that the emission from the interferometric emitter is *indistinguishable* from broadband dye laser emission.

2. Coherent emission from electrically excited organic semiconductors

For completeness the salient features of the experiments discussed by Duarte et al. (2005) and Duarte (2007, 2008) are reiterated here. These experiments involve pulsed electrical excitation of organic semiconductors integrating two emitter regions in series. The active medium in each region is a coumarin 545 tetramethyl (C 545 T) dye-doped Alq_3 matrix. The structure of this class of high-brightness tandem organic semiconductors has been described in detail elsewhere (Duarte et al., 2005; Liao et al., 2004). The dye C 545 T has also been demonstrated to be a high-gain and efficient laser dye under pulsed optical excitation, by Duarte et al. (2006). Using a simple grating-mirror cavity the tuning range of this lasers is 501-574 nm. Maximum emission is observed at $\lambda \approx 555$ nm and the laser grating-narrowed linewidth is $\Delta\lambda \approx 3$ nm (Duarte et al., 2006). These results are presented in detail in Section 3.

Using the double stack electrically-excited organic light emitting diode (OLED) structure configured within an asymmetrical sub microcavity, and collinearly confined within a double interferometric structure, Duarte et al. (2005) reported on a nearly diffraction limited beam with a near-Gaussian profile and high visibility interferograms. The experimental arrangement is shown in Figure 1. The sub microcavity has a high reflectivity back mirror, that is also the cathode, and a low reflectivity output coupler, which is also the anode. This output coupler is configured by a layer of ITO and the glass interface. The external surface of the glass output coupler is antireflection coated with MgF_2 to avoid intra-glass interference. A detail description of the semiconductor structure is given in Duarte et al. (2005).

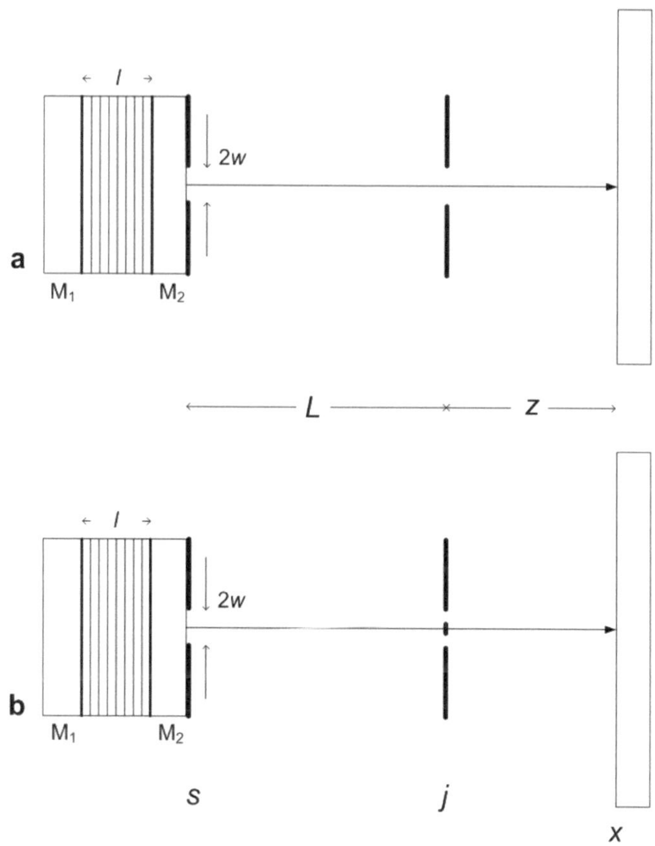

Fig. 1. (a) Electrically-pumped organic semiconductor interferometric emitter depicting the sub micrometer cavity with $l \approx 300$ nm. M_1 is a total reflector and M_2 is a low reflectivity output coupler (see text). (b) Double-slit interferometric configuration used to determine the coherence of the emission. The slits are 50 μm wide separated by 50 μm. The distance to the interferometric plane is z (from Duarte (2008)).

The overall length of this asymmetrical sub micrometer cavity is 300 nm. This interferometric emitter has been described as a *doubly interferometrically confined organic semiconductor* (DICOS) emitter where the emission medium is a laser-dye-doped Alq_3 matrix (Duarte, 2007). As described by Duarte et al. (2005) the DICOS emitter is excited with high-voltage pulses, at 100V, with ns rise times. This interferometric emitter works in the following manner: the first 150 μm aperture allows the propagation of a highly divergent, multiple-transverse-mode beam. The second $2w = 150$ μm aperture, positioned along the optical axis at $L \approx 130$ mm from the first aperture, allows propagation of a single-transverse mode exclusively. The optimum value of L is a function of wavelength and aperture dimensions (Duarte, 1993). That emission, the emission precisely along the optical axis, corresponds to a single transverse mode, with a near-Gaussian profile (Figure 2), and exhibits a divergence near the diffraction limit as defined by the dimensions of the aperture (Duarte et al., 2005; Duarte, 2007, 2008). The digital profile of this near-Gaussian beam is shown in Figure 3.

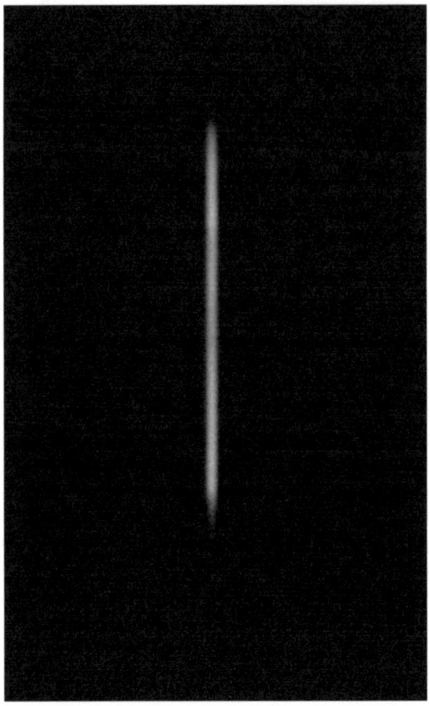

Fig. 2. Black and white silver halide photograph of the emission beam recorded at z = 340 mm. As shown in Duarte et al. (2005) the spatial profile of this single-transverse-mode emission is near-Gaussian and the beam divergence is ~ 1.1 times its diffraction limit (from Duarte et al. (2005)).

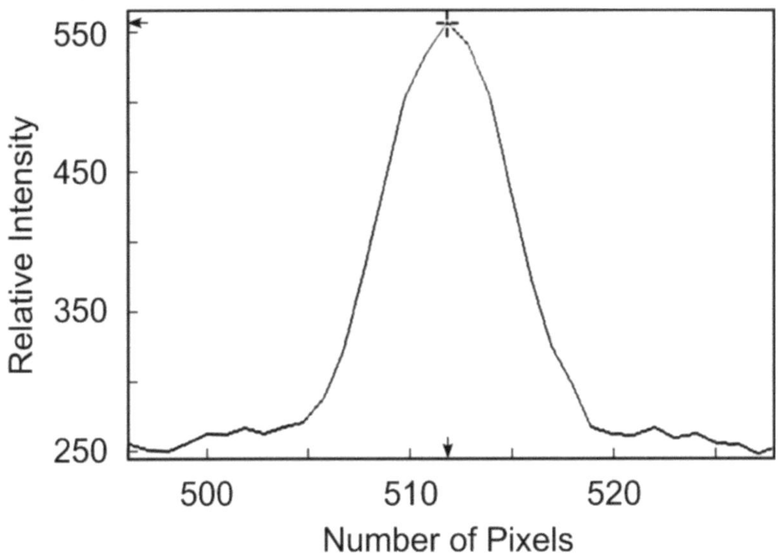

Fig. 3. Digital profile of the near-Gaussian emission beam, with a measured divergence ~ 1.1 times its diffraction limit, recorded at z = 340 mm. Each pixel is 25 μm wide (from Duarte et al. (2005)).

Considering the uncertainty in the measurement plus the uncertainty in the dimensions of the aperture a convergence towards the diffraction limit might be possible. In essence, the function of the double interferometric array is analogous to the highly discriminatory function of a multiple-prism grating configuration in narrow-linewidth laser oscillators (Duarte, 1999).

As mentioned the emission beam profile is near Gaussian and exhibits a divergence of $\Delta\theta = 2.53 \pm 0.13$ mrad which is ~ 1.1 times the diffraction limit as defined by the $2w \approx 150$ μm dimensions of the apertures (Duarte et al., 2005). The emission is also characterized by high visibility double-slit interferograms with $\mathcal{V} \approx 0.9$ (see Figure 4) which approaches the visibility regime observed from interferograms generated with the $\lambda \approx 543.30$ nm transition of a He-Ne laser with $\mathcal{V} \approx 0.95$ (see Figure 5) (Duarte, 2007). The interferometrically determined linewidth of the electrically-excited dye emission is $\Delta\lambda \approx 10.5$ nm (Duarte, 2007, 2008). Given the extremely short length of the cavity ($l \approx 300$ nm), this linewidth is consistent with single-longitudinal-mode emission since the free-spectral range is $\delta\lambda \approx 486$ nm. Pulsed output power is in the nW regime (Duarte et al., 2005) and results are summarized in Table 1. As an

$\Delta\theta$ (mrad)	$\Delta\lambda$ (nm)	$\mathcal{V}$	λ (nm)	Threshold A/cm^2
2.53[a]	~10.5[b]	0.9	~540	~ 0.8

[a] This $\Delta\theta$ corresponds to ~ 1.1 times the diffraction limit
[b] This $\Delta\lambda$ was determined using the interferometric method described in Duarte (2007, 2008).

Table 1. Emission parameters of the organic semiconductor interferometric emitter

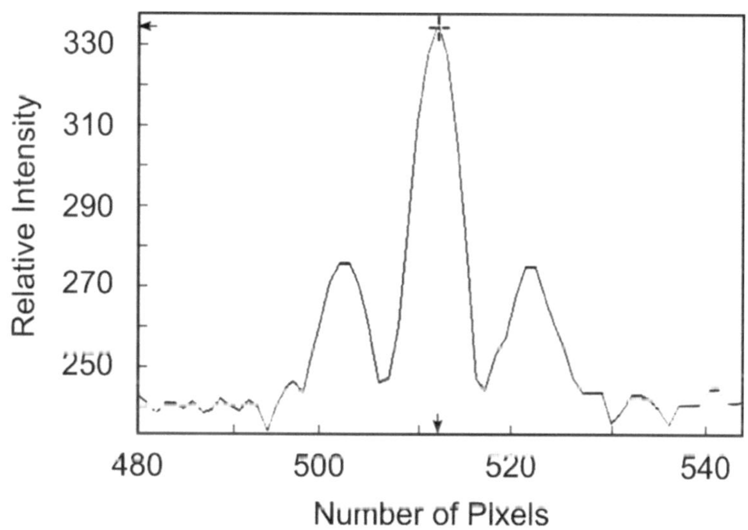

Fig. 4. Double-slit interferogram of the emission from the interferometric emitter using the configuration depicted in Figure 1b. The visibility recorded at $z = 50$ mm is $\mathcal{V} \approx 0.9$ leading to an interferometrically determined linewidth of $\Delta\lambda \approx 10.5$ nm. Each pixel is 25 μm wide (from Duarte et al. (2005)).

explanatory note it should be mentioned that since both axial apertures can be physically represented as an array of a large number of sub apertures they can be considered as interferometric arrays. Indeed, interferometry of the emission is performed by replacing the second aperture by a double-slit arrangement also known as a Young-slit configuration.
Furthermore, absence of the second aperture causes the emission to be, as previously mentioned, highly divergent and multi transverse mode. The similarities of the interferograms corresponding to the electrically excited DICOS emitter and the narrow-linewidth green He-Ne laser (Figures 4 and 5) are self evident. It should be indicated that the noise in the interferogram depicted in Figure 4 is mainly detector noise given the much lower intensity levels and the fact that the digital detector was not cooled.

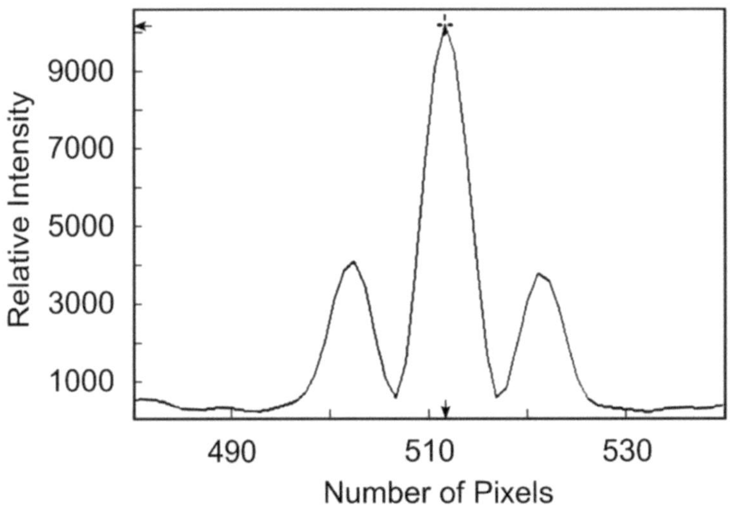

Fig. 5. Double-slit interferogram of the emission from the $\lambda \approx 543.30$ nm narrow-linewidth He-Ne laser using the interferometric configuration depicted in Figure 1b. The visibility recorded at $z = 50$ mm is $\mathcal{V} \approx 0.95$ while the measure laser linewidth is $\Delta\lambda \approx 0.001$ nm. Each pixel is 25 μm wide (from Duarte et al. (2005)).

3. Optically-pumped coumarin 545 tetramethyl tunable laser

As the experiments reported by Duarte et al. (2005) began it became apparent that the dye used in the tandem organic semiconductor, or tandem OLED, that is coumarin 545 tetramethyl (or C 545 T) had not been reported in the literature as a laser dye. The molecular structure of C 545 T is depicted in Figure 6. It is well known that many dyes with good to strong fluorescence characteristics might not necessarily become laser dyes. Thus, an standard laser experiment was designed to investigate the emission properties of C 545 T. If this dye was not capable of emitting coherent emission in its optically pumped version then the likelihood of observing coherent emission in the electrically-pumped regime would be infinitesimally small.
The experiment consisted in using a 3 mM solution of C 545 T in ethanol in a wedged optical cell deployed in a straight forward tunable optical cavity as depicted in Figure 7. The excitation laser is a Nitrogen laser ($\lambda \approx 337$ nm) yielding approximately 7 mJ/pulse in pulses with a duration of ~ 10 ns (FWHM).

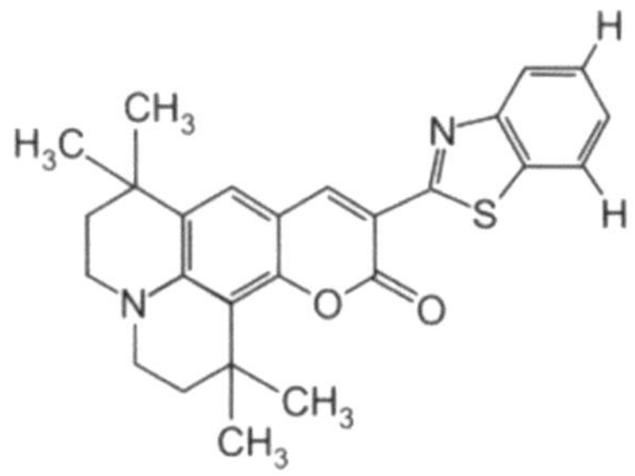

Fig. 6. Molecular structure of the laser dye coumarin 545 tetramethyl (C 545 T) (from Duarte et al., 2006).

These experiments demonstrated that C 545 T not only lased but lased extremely well under pulsed optical excitation. The measured laser efficiency was found to be ~ 14%, with a nearly diffraction limited beam divergence of $\Delta\theta \approx 1.2$ mrad., laser linewidth of $\Delta\lambda \approx 3$ nm, and an exceptional tuning range of $501 \le \lambda \le 574$ nm (see Figure 8) (Duarte et al., 2006). Thus, C 545 T adds to the excellent laser performance of the family of coumarin tetramethyl laser dyes (Chen et al., 1988; Duarte, 1989). Table 2 summarizes the performance of this optically-pumped C 545 T tunable laser.

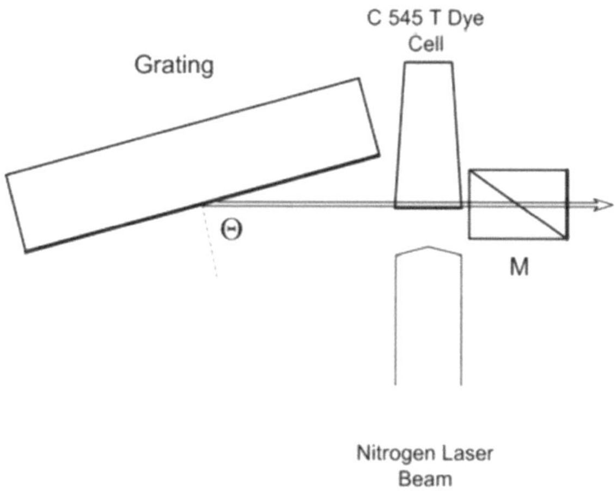

Fig. 7. Transversely-excited coumarin 545 tetramethyl dye laser. The tuning-narrowing diffraction grating has 3000 lines/mm and the output coupler-mirror is configured with a Glan-Thompson polarizer to yield laser emission polarized parallel to the plane of propagation (from Duarte et al. (2006)).

$\Delta\theta$ (mrad)	$\Delta\lambda$ (nm)	Tuning range (nm)	Efficiency %	Dye concentration mM
1.2	~3.0	$501 \le \lambda \le 574$	~14	3.0

Table 2. Performance of the grating-tuned optically-pumped C 545 T dye laser (from Duarte et al. (2006)).

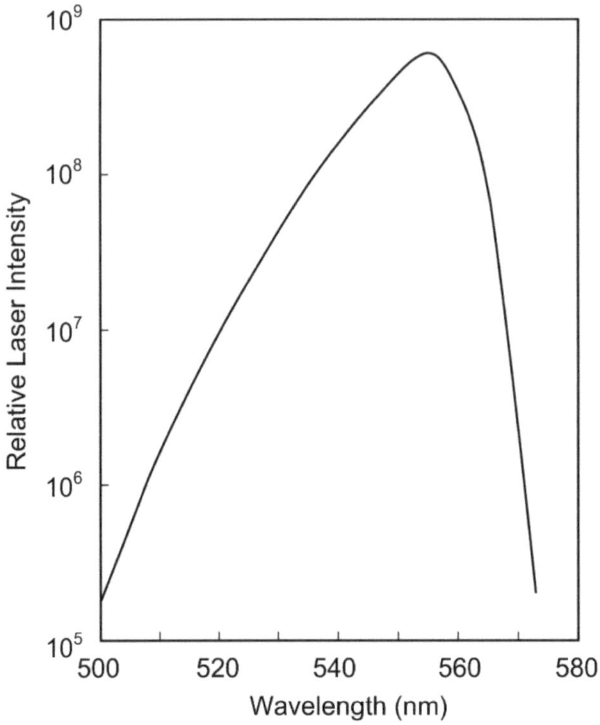

Fig. 8. Tuning curve of the transversely-excited coumarin 545 tetramethyl dye laser. The tuning range of the emission is $501 \leq \lambda \leq 574$ nm and the dynamic range of its output intensity span approximately four orders of magnitude (from Duarte et al. (2006)).

4. Microcavity emission in the red

Using an experimental configuration partially similar to that disclosed by Duarte et al. (2005) and Duarte (2007, 2008) (without the second aperture), Liu et al. (2009) reported on laser emission at 621.7 nm using a red emitting tetramethyl dye-doped active medium. A summary of this report includes a linewidth of $\Delta\lambda \approx 1.95$ nm, a beam divergence of $\Delta\theta \approx 32$ mrad, interferometric visibility of $\mathcal{V} \approx 0.89$, and a current threshold of 0.86 A/cm.[2] However, in a recent paper Samuel et al. (2009) interrogate various aspects of Liu et al. (2009), including:

1. The linewidth reduction from 2.62 nm, below threshold, to 1.95 nm, above threshold, is deemed as insufficient evidence of lasing. An analogous comment is made in reference to beam divergence (Samuel et al., 2009).
2. The threshold current density of 0.86 A/cm^2 is said to be "five orders of magnitude smaller" than expected (Samuel et al., 2009).

   Thus, the output emission reported in Liu et al. (2009) is not classified by Samuel et al. (2009) as corresponding to laser emission. In a more general context Samuel et al. (2009) highlight the importance of polarization in organic laser emission and formulate further assertions including:
3. "Interference effects can be observed perfectly well using a lamp and a pair of slits... The observation of interference phenomena is intriguing, but the source is small and

therefore capable of giving high-visibility fringes when illuminating double slits" (Samuel et al., 2009).

4. "The typical linewidth of ASE in an organic semiconductor is 10 nm" (Samuel et al., 2009).

Here, the comments and assertions of Samuel et al. (2009) are examined in light of the published dye laser literature in Sections 5, and 7-9.

5. Microcavity lasers and thresholds

Experiments published on optically-pumped liquid dye lasers with a cavity length $l \leq \lambda$ indicate that a near "zero-threshold-laser" emission is observed for $l \approx \lambda / 2$ (De Martini et al., 1988). These results were obtained with a multiple-transverse-mode emission beam (De Martini et al., 1988). Albeit the beam waist, on focus at the gain region, of the excitation beam is not given by these authors, it can be shown that even with an emission beam waist of only a few micrometers there would be an enormous number of transverse modes due to the $l \approx \lambda / 2$ length of the cavity. The relevant fact here is that De Martini et al. (1988) did observe a near "zero-threshold-laser" emission under multiple-transverse-mode conditions. These experiments provide very persuasive evidence in support of threshold behavior at very low excitation densities. In other words, the experiments of De Martini et al. (1988) strongly suggest that with a suitable gain medium, and microcavity configurations, high threshold energy densities are not required. Subsequently, with a suitable gain medium, and microcavity, current density thresholds in the 10-100 kA/cm^2 range, as mentioned by Samuel et al. (2009), should not be necessary. Therefore, low threshold behavior at 0.8–0.9 A/cm^2 (Duarte et al., 2005; Duarte, 2007, 2008; Liu et al., 2009) is consistent with what would be expected in a sub micrometer cavity where the conditions $l < \lambda$ do apply. It should also be mentioned that in our own experiments rapid destruction of the laser dye-doped semiconductor medium was observed at peak excitation voltages, in the nanosecond regime, approaching 10 kV corresponding to current densities of only ~ 190 A/cm^2 (Duarte, 2008). The emission beam profile under such extreme excitation conditions is shown in Figure 9.

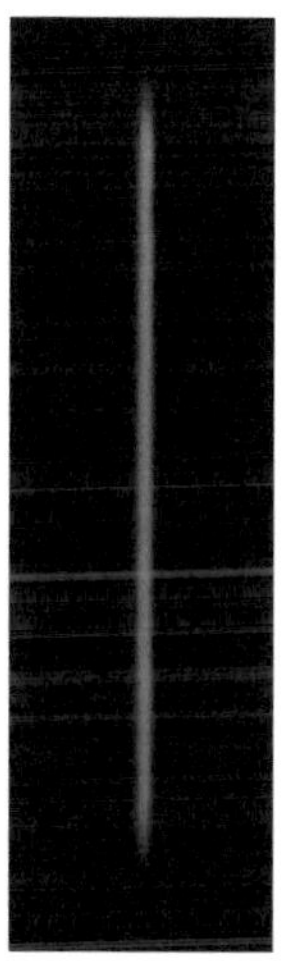

Fig. 9. Black and white silver halide photograph of the near-Gaussian emission beam, from the interferometric emitter (DICOS), recorder under nanosecond pulsed excitation at ~ 10 kV per pulse. The corresponding excitation current density is ~ 190 A/cm.^{2} (from Duarte (2008)).

6. Beam divergence and transverse mode structure

The ideal *diffraction limited* beam divergence, derived from the uncertainty principle $\Delta x \Delta p \approx h$ (Dirac, 1978), is given by Duarte (2003)

$$\Delta\theta \approx (\lambda / \pi w) \tag{1}$$

where λ is the emission wavelength and w the emission beam waist. However, the beam divergence from a cavity can be augmented by geometrical factors included in the expression (Duarte, 1990a)

$$\Delta\theta \approx (\lambda / \pi w) \left(1 + (L_R / B)^2 + (L_R A \;/ B\;)^2\;\right)^{1/2} \tag{2}$$

where

$$L_R = (\pi w^2 / \lambda)$$

is the Rayleigh length and A and B are propagation terms from propagation matrix theory (Duarte, 1989, 1990a). In well-designed narrow-linewidth laser oscillator cavities the beam divergence often approaches the diffraction limit as the term in parenthesis approaches unity (Duarte, 1990a, 1999). For a complete matrix treatment of tunable laser resonators the reader should refer to Duarte (1989, 1992, 2003).

For a simple mirror-mirror resonator, in the absence of intracavity beam expansion $A \approx 1$ and B becomes the intra cavity length l (see Figure 1), so that $B = l$, and

$$\Delta\theta \approx (\lambda / \pi w) \left(1 + 2(L_R / l)^2\right)^{1/2} \tag{3}$$

For a microcavity the condition $l \leq \lambda$ applies, and $L_R >> l$, so that

$$\Delta\theta >> (\lambda / \pi w) \tag{4}$$

Thus, large beam divergences are inherent to resonators where the cavity length is in the sub micrometer, or nanometer, regime as in the case of what is understood for sub micrometer cavities where the condition $l \leq \lambda$ applies. Thus the use of a secondary aperture along the propagation axis, as depicted in Figure 1, is necessary if near single-transverse-mode emission is desired.

A more accurate description of the diffraction limited divergence as given in (Duarte, 2008)

$$\Delta\theta \approx (\lambda \pm \Delta\lambda) / (\pi w) \tag{5}$$

where $\Delta\lambda$ is the usual linewidth of emission. In this regard, a cumulative spatial detector (such as a silver-halide photographic plate) registers

$$\Delta\theta \approx (\lambda + \Delta\lambda) / (\pi w) \tag{6}$$

which illustrates to a first approximation that the narrowing in the spectral width, as the emission transitions from ASE to lasing, leads to a decrease in the beam divergence.

The conditions outlined in Equation (4) lead to an emission characterized by a very large number of transverse modes as can be quantified via the interferometric equation (Duarte, 1993, 2007, 2008)

$$|\langle x|s\rangle|^2 = \sum_{j=1}^{N} \Psi(r_j)^2 + 2\sum_{j=1}^{N} \Psi(r_j) \left(\sum_{m=j+1}^{N} \Psi(r_m)\cos(\Omega_m - \Omega_j) \right) \tag{7}$$

where the phase term in parenthesis relates directly to the exact geometry of the cavity and the wavelength of emission (Duarte, 2003). Moreover, because the cavity is so short the free spectral range, given by

$$\delta\lambda \approx \lambda^2 / (2l) \tag{8}$$

is very wide ($\delta\lambda \approx 486$ nm). Thus for emission linewidths in the 10 nm range (as in Duarte et al. (2005) and Duarte (2007, 2008)), only single-longitudinal-mode emission for each transverse mode is allowed.

Therefore, for $l \approx 300$ nm, and $2w \approx 150$ μm, the experimenter using a conventional configuration is confronted with large beam divergences and a multitude of transverse modes. One way to overcome this compounded problem is to position, at a distance determined by Equation (7), a second slit to filter out all the unwanted modes and allow only a single-transverse-mode that, as discussed previously, under appropriate linewidth conditions, would only allow a single-longitudinal mode. This concept led to the doubly interferometrically confined organic semiconductor (DICOS) emitter described in (Duarte et al., 2005; Duarte, 2007, 2008) and depicted in Figure 1.

Equation (7) is also used to generate a series of numerically based interferograms, using the emission wavelength and interferometric parameters applicable to the interferometric emitter, which are used in a graphical technique to estimate the approximate linewidth of the emission (see Duarte (2007, 2008)). This interferometrically determined linewidth yields a conservative upper limit for the emission linewidth, in this case $\Delta\lambda \approx 10.5$ nm (Duarte, 2007, 2008).

7. Visibility of double-slit interferograms

It is well-known that "interference effects can be observed perfectly well using a lamp and a pair of slits" (Samuel et al., 2009). Indeed, in their classic experiment Thompson and Wolf (1957) observed beautiful fringes with $\mathcal{V} \approx 0.593$ from a simple mercury lamp. Nevertheless, the literature indicates that interferograms with high visibility, $\mathcal{V} \geq 0.9$, are coherently speaking more special. In a directly relevant experiment while using a similar geometry as (Duarte et al., 2005; Duarte, 2007, 2008), and a small organic semiconductor source, Saxena et al. (2006) have reported on visibilities of $\mathcal{V} \approx 0.4$ which is related to a measured linewidth of $\Delta\lambda \approx 100$ nm. Experiments on the measurement of interferometric visibility for dye ASE sources give a visibility of $\mathcal{V} \approx 0.65$ which is related to a measured linewidth of $\Delta\lambda \approx 17$ nm (Dharmadhikari et al., 2005).

The use of double-slit interference techniques, to characterize the coherence of laser emission, is a well-known and accepted practice documented in the laser literature (Nelson and Collins, 1961; Shimkaveg et al., 1992; Trebes et al. 1992; Ditmire et al., 1996; Lucianetti et al., 2004). Indeed, double-slit interferometric measurement have been used to characterize laser emission since the dawn of the laser age (Nelson and Collins, 1961). In this regard, laser emission is associated with visibilities in the approximate $0.85 \leq \mathcal{V} \leq 1.0$ range (Trebes et al., 1992; Lucianetti et al., 2004). In our own experiments we have measured a visibility of $\mathcal{V} \approx 0.90$ using the configuration depicted in Figure 1b and $\mathcal{V} \approx 0.95$ for the equivalent interferometric configuration while using illumination from the narrow-linewidth $3s_2$–$2p_{10}$ transition ($\Delta\lambda \approx 0.001$ nm) of the He-Ne laser at $\lambda \approx 543.3$ nm (Duarte et al., 2005; Duarte, 2007; 2008).

Thus the existing literature documents that the use of double-slit interferograms is a well accepted practice in determining the coherence of laser sources since 1961. It is further well established that for typical ASE and non laser emission the measured visibilities are below $\mathcal{V} \approx 0.65$. The published literature also indicates that emission with visibility in the range of $0.85 \leq \mathcal{V} \leq 1.0$ is highly coherent and therefore associated with laser emission. Besides citing the refereed literature, and in addition to the comparison with the coherence from the green He-Ne laser, we now refer to a direct experiment that registered the interferograms from the optically-pumped high-power C 545 T dye laser and the electrically-pumped C 545 T dye-doped interferometric emitter. The interferograms were recorded directly on black and white silver-halide film thus leaving a permanent photographic record of the experiment (Duarte, et al., 2005; Duarte, 2008).

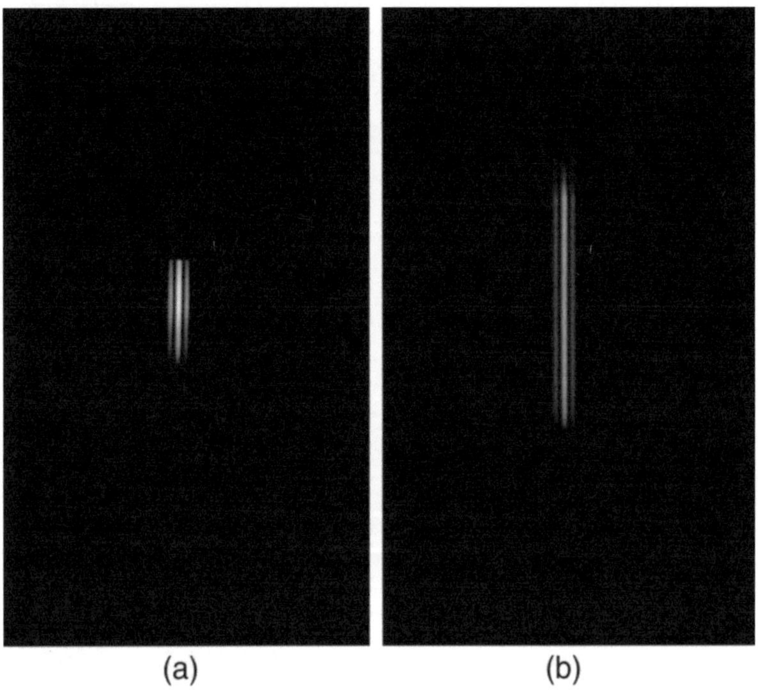

Fig. 10. Back and white silver halide photographs of double-slit interferograms produced by (a) laser emission from a grating-narrowed C 545 T dye laser (Duarte et al., 2006) and (b) from the C 545 T dye-doped electrically-excited organic semiconductor interferometric emitter as depicted in Figure 1b. In both cases the slits are 50 μm wide, separated by 50 μm, and z = 175 mm (from Duarte et al. (2005)).

Figure 10 shows the interferograms recorded under identical interferometric conditions, for the optically-pumped C 545 T dye laser (with $\Delta\lambda \approx 3$ nm) (Duarte et al., 2006), and the electrically-excited interferometric emitter (depicted in Figure 1b). Although these interferograms, at first glance, appear to be quite similar, examination under magnification reveals slightly broader characteristics from the interferometric emitter that are quite consistent with the interferometric estimate (Duarte, 2007, 2008) of $\Delta\lambda \approx 10.5$ nm.

8. Emission bandwidth of ASE versus laser emission linewidth

The presence of ASE, in narrow-linewidth dye lasers, is a physics and engineering challenge that has been a subject of sustained interest among researchers for quite a while (Duarte, 1990, 2003). In particular, optimized oscillator cavities yielding tunable narrow-linewidth laser emission with extremely low levels of ASE have been successfully developed and optimized (Duarte and Piper, 1980; Duarte et al., 1990). In this effort it has been learned that ASE in dye lasers can be as wide as 50-60 nm (Dujardin and Flamant, 1978; Duarte and Piper, 1980; Bor, 1981; McKee et al., 1982) and can be successfully suppressed, in optimized multiple-prism grating cavities, to levels as low as

$$(\rho_{ASE} / \rho_{laser}) \approx 10^{-9}$$

for highly-coherent, highly-polarized, laser emission with $\Delta\nu \approx 360$ MHz at 590 nm (or $\Delta\lambda \approx 0.00042$ nm) (Duarte et al., 1990). Here it should be noted that the highly polarized emission (parallel to the plane of propagation) is almost entirely a by-product of the cavity architecture (Duarte et al., 1990; Duarte, 1990a).

On the other hand, dye lasers are also capable of emitting high-power emission in a broadband mode while using basic *mirror-mirror* resonators. Iconic examples of such lasers are the dye laser reported by Schäfer et al. (1966), with a bandwidth of $\Delta\lambda \approx 10$ nm, and the dye lasers reported by Spaeth and Bortfeld (1966) with emission bandwidths in the $4.5 < \Delta\lambda \leq 7$ nm range.

In summary, pulsed laser dye ASE can be as wide as 50-60 nm (Dujardin and Flamant, 1978; Duarte and Piper, 1980; Bor, 1981; McKee et al., 1982), and high-power dye lasers can be designed to either deliver broadband laser emission in the approximate $4.5 \leq \Delta\lambda \leq 10$ nm range (Schäfer et al., 1966; Spaeth and Bortfeld, 1966; Baltakov et al., 1973), or highly coherent emission with laser linewidths as narrow as $\Delta\lambda \approx 0.0004$ nm at 590 nm (Duarte, 1999). The characteristics of the emission depend on various parameters including the excitation conditions, laser dye concentration, and very importantly, the resonator architecture. Again, dye lasers can emit in the ASE regime (pre-laser emission), the broadband laser regime, and the narrow-linewidth laser regime.

9. Polarization in organic dye lasers

Shäfer (1990) and Duarte (1990b) provide detailed discussions on the polarization characteristics of laser and ASE emission in dye lasers. It is beautifully illustrated, for instance, that the intrinsic polarization orientation of the emission from a particular dye can be controlled by choosing the orientation of the polarization of the excitation laser relative to the propagation plane of the dye emitter (Duarte, 1990b). In the case of the copper-laser-pumped rhodamine 590 dye laser these polarization alternatives enable the designer to engineer a broadband dye laser with either an emission polarization perpendicular to the plane of propagation or nearly undefined (that is, nearly unpolarized). Furthermore, the overall final polarization for narrow-linewidth laser emission depends critically on various other factors. Thus, for laser-pumped dye lasers the final outcome is a function of the orientation of the polarization of the pump laser, the polarization preference of the grating, the orientation of the windows of the active region, the degree of polarization selection of the beam expander, and the presence or absence of an intracavity polarizer (Duarte, 1990b). Therefore, one avenue to optimize the overall conversion efficiency of a laser oscillator consists in carefully matching the polarization preference of all the above mentioned parameters to yield efficient, highly polarized, narrow-linewidth laser emission (Duarte and Piper, 1984). One recent example of such optimization is the narrow-linewidth solid-state multiple-prism grating laser oscillator shown in Figure 11. The rhodamine 6G dye doped polymer matrix is configured to favor polarization parallel to the plane of propagation, this polarization is then reinforced by the multiple-prism beam expander, the 3300 line/mm diffraction grating, and the Glan-Thompson polarizer output coupler (Duarte, 1999). This dispersive oscillator yields single-transverse-mode, single-longitudinal-mode, highly-polarized laser emission (parallel to the plane of propagation) with a linewidth of $\Delta\nu \approx 350\,\text{MHz}$ (see Figure 12) and a pulse duration of $\Delta t \approx 3$ ns which means that the emission is nearly limited by Heisenberg's uncertainty principle.

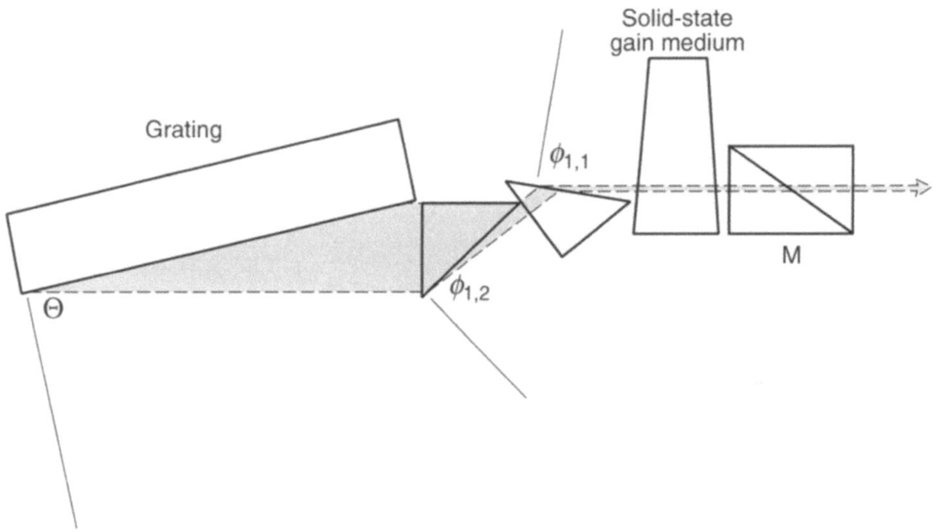

Fig. 11. Multiple-prism grating solid-state dye laser oscillator yielding single-transverse-mode and single-longitudinal-mode laser emission at a linewidth of $\Delta\nu \approx 350\,\text{MHz}$ ($\Delta\lambda \approx 0.0004$ nm at 590 nm) at extremely low levels of ASE. The temporal pulse (~ 3 ns at FWHM) is nearly Gaussian and its tuning range is $550 \leq \lambda \leq 603\,\text{nm}$. Due to the polarization preference of the grating, the multiple-prism beam expander, and the Glan-Thompson polarizer output coupler, the emission is highly polarized parallel to the plane of propagation. The gain medium is rhodamine 6 G dye-doped PMMA at a concentration of 0. 5 mM (from Duarte (1999)).

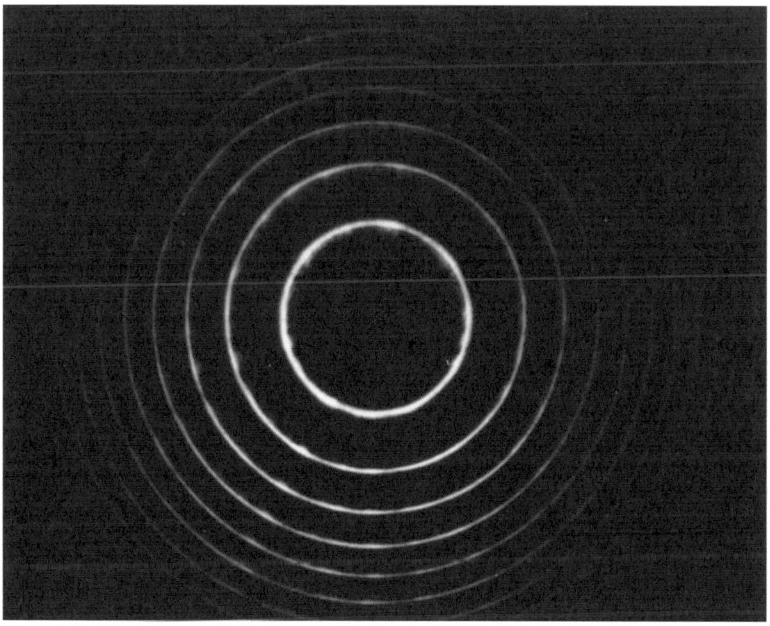

Fig. 12. Fabry-Perot interferogram of the single-longitudinal-mode emission, at $\Delta\nu \approx 350$ MHz, from the optimized solid-state multiple-prism grating laser oscillator (from Duarte (1999)).

The ASE level of this highly polarized single-longitudinal-mode emission is quoted as $\sim 10^{-6}$. The polarization discussion in this section is aimed at reminding the readership that, in organic dye lasers, polarization of the emission depends principally on the cavity architecture.

10. Additional comments on broadband laser emission

One of the characteristics of dye gain media is its high gain. The original reason to introduce trapezoidal dye optical cells (Duarte, 1980) was to successfully *avoid broadband lasing* due to the resonator cavity effect provided by the uncoated windows. With the trapezoidal geometry the excited dye region emits only ASE along the optical axis. This ASE can then be guided through the intracavity dispersive optics so that, following proper alignment of the cavity, low noise tunable narrow-linewidth emission can be achieved (Duarte, 1980, 1984, 1999).

A subtle point emerges here that is worth further attention. When broadband lasing occurs, due to the low reflectivity of parallel uncoated windows, and despite the absence of external mirrors, it is because a resonator cavity has been configured by the uncoated windows. Indeed, it is improper to label this broadband lasing, contained in a directional laser beam, as ASE. Certainly, the same effect occurs in solid-state dye lasers leading to the use of dye-doped polymer matrices configured in trapezoidal geometries (Duarte, 1994, 1999). As a footnote it should be mentioned that once broadband lasing occurs, due to the low-reflectivity effect mentioned here, it is not possible to bring the emission under control with the external optics shown in Figure 11. Hence, it is imperative to suppress broadband lasing prior to achieving controllable low noise narrow-linewidth emission.

11. Additional comments on the microcavity emission in the red

As already indicated by Samuel et al. (2009), Liu et al. (2009) adopted an experimental approach similar to that of Duarte et al. (2005). In particular, they use double slit interference to determine that the emission interferograms yield a visibility of $\mathcal{V} \approx 0.51$ prior to threshold and a $\mathcal{V} \approx 0.89$ post threshold (Liu et al., 2009). This post-threshold visibility is close to the value reported by Duarte et al. (2005) and Duarte (2007, 2009) of $\mathcal{V} \approx 0.9$. The interferogram published by Liu et al. (2009) is perfectly symmetrical and devoid of noise. If this represents a series of averaged interferograms it is not mentioned by the authors (Liu et al., 2009), however, in our experience a perfectly-symmetrical noiseless interferogram would be an extremely rare occurrence (see Figures 4 and 5).

The magnitude of the change in interferometric visibility reported in (Liu et al., 2009), from $\mathcal{V} \approx 0.51$ to $\mathcal{V} \approx 0.89$, is not consistent with the relatively small transition in linewidth, from 2.36 nm to 1.95 nm. From an alternative perspective a bandwidth of 2.26 nm is unusually narrow to be associated with dye ASE as discussed in the previous section. Moreover, the reported pre threshold visibility of $\mathcal{V} \approx 0.51$ is too low to be associated to a linewidth of 2.26 nm.

12. Discussion

Recently, Newton's ring interferometry has been used on standard C 545 T dye-doped OLED devices yielding measured linewidths in the $40 \leq \Delta\lambda \leq 110$ nm range with the authors attributing the narrower linewidth ($\Delta\lambda \approx 40$ nm) to "microcavity effects" (Tsia et al., 2010). These results are consistent with those of Saxena et al. (2006) who reported $\Delta\lambda \approx 100$ nm and $\mathcal{V} \approx 0.4$ for their organic semiconductor device. These completely independent results illustrate that the high degree of coherence reported in Duarte et al. (2005) and Duarte (2007, 2008), is not observed in the standard OLED devices of Saxena et al. (2006) or Tsia et al. (2010). For emission originating from a laser dye medium, such as C 545 T, it is experimentally justified to expect, according to the literature, a wide spectral coverage for ASE (50-60 nm) (Dujardin and Flamant, 1978; Duarte and Piper, 1980; Bor, 1981; McKee et al., 1982) and relatively broad laser linewidths, in the approximate $4.5 \leq \Delta\lambda \leq 10$ nm range, for simple mirror-mirror cavities (Schäfer et al., 1966; Spaeth and Bortfeld, 1966; Baltakov et al., 1973). By contrast, Samuel et al. (2009) mention that "typical linewidth of ASE in an organic semiconductor is 10 nm." Although this statement might be relevant for conjugated polymer gain media it does not apply to laser-dye based gain media (Dujardin and Flamant, 1978; Duarte and Piper, 1980; Bor, 1981; McKee et al., 1982).

The visibility of the interferograms $\mathcal{V} \approx 0.90$ (Duarte et al., 2005; Duarte, 2007, 2008) is considerably higher than the visibilities associated with standard OLED devices (Saxena et al., 2006; Tsia et al., 2010), higher than the visibility associated with ASE (Dharmadhikari et al., 2005), and consistent with the interferometric visibilities of laser emission (Trebes et al., 1992; Lucianetti et al., 2004). Further, the interferometrically determined linewidth $\Delta\lambda \approx 10.5$ nm is consistent with the linewidth reported for broadband dye laser emission in the iconic paper of Schäfer et al. (1966). As a result of the double interferometric confinement (Figure 1b) the emission radiation corresponds to a single-transverse-mode (Figures 2 and 3) and a single-longitudinal-mode as discussed in Section 6. As suggested in (Duarte, 2007) the likely origin for the observed spectral coherence is the resonance established in the asymmetrical sub microcavity, or nanocavity. In this regard, under excitation with a fast rise time pulse (Duarte et al., 2005; Duarte, 2007) it would only take a tiny fraction of a nanosecond for the emission to execute an enormous number of intracavity return passes since $l \approx 300$ nm. Previous studies on optically-pumped solid-state narrow-linewidth dye laser oscillators, using asymmetrical cavity configurations, established that only a few intracavity return passes are necessary to reach laser threshold (Duarte, 2003).

The lack of access, via mirror controls, associated with the class of sub microcavity configuration used in these experiments, does deny the opportunity to observe the emission as a function of cavity alignment. As previously indicated (Duarte, 2008) this might require the construction of a series of devices with various output coupler reflectivities in addition to changing, by design, the angle of incidence, relative to the optical axis, at the output coupler mirror. Since mirror dependence of the emission was not demonstrated an explicit claim of a conventional laser device, or "an electrically-pumped dye laser," was not made in Duarte et al. (2005) or Duarte (2007, 2008). Nevertheless, the experimental evidence clearly shows that the coherent emission reported in these papers (Duarte et al., 2005; Duarte, 2007, 2008) is, spatially and spectrally, indistinguishable from broadband dye laser emission.

13. Conclusion

In this chapter experimental results demonstrating spatially coherent, and interferometrically coherent, emission from pulsed electrically-pumped coumarin 545 tetramethyl laser-dye-doped organic semiconductors have been reviewed. This review also takes into account interpretational issues brought forward in recent publications. In particular, this review re-examines concepts and parameters relevant to amplified spontaneous emission, linewidth, and polarization in tunable organic dye lasers. This has been done by revisiting published results on dye amplified spontaneous emission, broadband dye lasers, and narrow-linewidth tunable laser oscillators using laser dye gain media. This has been performed using seminal and historically significant references from the peer-reviewed literature. This exercise leads us to reaffirm the previous interpretation of Duarte et al. (2005) and Duarte (2007, 2008) that the emission observed from the electrically-excited doubly interferometrically confined organic semiconductor emitter is, coherently speaking, *indistinguishable from broadband dye laser emission.*

14. References

Baldo, M. A., Holmes, R. J., and Forrest, S. R. (2002). Prospects for electrically pumped organic lasers, *Phys. Rev. B* 66, 035321.

Baltakov, F. N., Barikhin, B. A., Kornilov, V. G., Mikhnov, S. A., Rubinov, A. N., and L. V. Sukhanov (1973). 110-J pulsed laser using a solution of rhodamine 6G in ethyl alcohol, *Sov. Phys. Tech. Phys.* 17, 1161-1163.

Bor, Z. (1981). Amplified spontaneous emission from N_2 laser pumped dye lasers, *Opt. Commun.* 39, 383-386.

Chen, C. H., Fox, J. L., Duarte, F. J., and Ehrlich, J. J. (1988). Lasing characteristics of new coumarin-analog dyes: broadband and narrow-linewidth performance, *Appl. Opt.* 27, 443-445.

Costela, A., García Moreno, I., and Sastre, R. (2003). Polymeric solid-state dye lasers: recent developments, *Phys. Chem. Chem. Phys.* 5, 4745-4763.

De Martini, F., and Jakobovitz, J. R. (1988). Anomalous spontaneous-emission-decay phase transition and zero-threshold laser action in a microscopic cavity, *Phys. Rev. Lett.* 60, 1711-1714.

Diels, J. C. (1990). Femtosecond dye lasers, In: *Dye Laser Principles*, Duarte F. J., Hillman, L. W., (Eds.), pp. 41-132, Academic, New York.

Dietel, W., Fontaine, J. J., and Diels, J. C. (1983). Intracavity pulse compression with glass: a new method of generating pulses shorter than 60 fs, *Opt. Lett.* 8, 4-6.

Ditmire, T., Gumbrell, E. T., Smith, R. A., Tisch, J. W. G., Meyerhofer, D. D., and Hutchison, M. H. R. (1996). Spatial coherence of soft x-ray radiation produced by high order harmonic generation, *Phys. Rev. Lett.* 77, 4756-4759.

Dirac. P. A. M. (1978). *The Principles of Quantum Mechanics*, Oxford University, London.

Dharmadhikari, J. A., Dharmadhikari, A. K., and Kumar, G. R. (2005). High-contrast interference pattern of amplified spontaneous emission from dyes under transient grating excitation *Opt. Lett.* 30, 765-767.

Duarte, F. J. (1989). Ray transfer matrix analysis of multiple-prism dye laser oscillators, *Opt. Quantum. Electron.* 21, 47-54.

Duarte, F. J. (1990a). Narrow-linewidth pulsed dye laser oscillators, In: *Dye Laser Principles*, Duarte F. J., Hillman, L. W., (Eds.), pp. 133-183, Academic, New York.

Duarte, F. J. (1990b). Technology of pulsed dye lasers, In: *Dye Laser Principles*, Duarte F. J., Hillman, L. W., (Eds.), pp. 239-285, Academic, New York.

Duarte, F. J. (1992). Multiple-prism dispersion and 4x4 ray transfer matrices, *Opt. Quantum Electron.* 24, 49-53.

Duarte, F. J. (1993). On a generalized interference equation and interferometric measurements, *Opt. Commun.* 103, 8-14.

Duarte, F. J. (1994). Solid-state multiple-prism grating dye laser oscillators, *Appl. Opt.* 33, 3857-3860.

Duarte, F. J. (1999). Solid-state multiple-prism grating oscillator: optimized cavity architecture, *Appl. Opt.* 38, 6347-6349.

Duarte, F. J. (2003). *Tunable Laser Optics*, Elsevier-Academic, New York.

Duarte, F. J. (2007). Coherent electrically excited organic semiconductors: visibility of interferograms and emission linewidth, *Opt. Lett.* 32, 412-414.

Duarte, F. J. (2008). Coherent electrically excited semiconductors: coherent or laser emission? *Appl. Phys. B.* 90, 101-108.

Duarte, F. J., Ehrlich, J. J., Davenport, W. E., and Taylor, T. S. (1990). Flashlamp pumped narrow-linewidth dispersive dye laser oscillators: very low amplified spontaneous emission levels and reduction of linewidth instabilities, *Appl. Opt.* 29, 3176-3179.

Duarte, F. J., Liao, L. S., and Vaeth, K. M. (2005). Coherent characteristics of electrically excited tandem organic light emitting diodes, *Opt. Lett.* 30, 3072-3074.

Duarte, F. J. , Liao, L. S., Vaeth, K. M., and Miller, A. M. (2006)., Widely tunable green laser emission using the coumarin 545 tetramethyl dye as the gain medium, *J. Opt. A: Pure Appl. Opt.* 8, 172-174.

Duarte, F. J., Paisner, J. A., and Penzkofer, A. (1992). Dye lasers: introduction by the feature editors, *App. Opt.* 31, 6977-6978.

Duarte, F. J., and Piper, J. A. (1980). A double-prism beam expander for pulsed dye lasers, *Opt. Commun.* 35, 100-104 (1980).

Duarte, F. J., and Piper, J. A. (1981). Prism preexpanded grazing-incidence grating cavity for pulsed dye lasers, *Appl. Opt.* 20, 2113-2116.
Duarte, F. J., and Piper, J. A. (1984). Narrow linewidth high prf copper laser-pumped dye-laser oscillators, *Appl. Opt.* 23, 1391-1394.
Dujardin, G., and Flamant, P. (1978). Amplified spontaneous emission and spatial dependence of gain in dye amplifiers, *Opt. Commun.* 24, 243-247.
Hänsch, T. W. (1972). Repetitively pulsed tunable dye laser for high resolution spectroscopy, *Appl. Opt.* 11, 895-898.
Holzer, W., Penzkofer, A., Pertsch, T., Danz, N., Bräuer, A., Kley, E. B., Tillman, H., Bader, C., Hörhold, H. H. (2002). Corrugated neat thin-film conjugated polymer distributed-feedback lasers, *Appl. Phys. B.* 74, 333-342.
Karnutsch, C. (2007). *Low Threshold Organic Thin Film Devices*, Cuvillier, Göttingen.
Kranzelbinder, G., and Leising, G. (2000). Organic solid-state lasers, *Rep. Prog. Phys.* 63, 729-762.
Liao, L. S., Klubek, K. P., and Tang, C. W. (2004). High-efficiency tandem organic light-emitting diodes, Appl. Phys. Lett. 84, 167-169.
Littman, M. G., and Metcalf, H. J. (1978). Spectrally narrow pulsed dye laser without beam expander, *Appl. Opt.* 17, 2224-2227.
Liu, X., Li, H., Song, C., Liao, Y., and Tian, M. (2009). Microcavity organic laser device under electrical pumping, *Opt. Lett.* 34, 503-505.
Lucianetti, A., Janulewicz, K. A., Kroemer, R., Priebe, G., Tümmler, J., Sandner, W., Nickless, P. V., Redkorechev, V. I (2004). Transverse spatial coherence of a transient nickel like silver soft-x-ray laser pumped by a single picosecond laser pulse, *Opt. Lett.* 29, 881-883.
Maslyukov, A., Solokov, S., Kailova, M., Nyholm, K., and Popov, S. (1995). Solid state dye laser with modified poly(methyl methacrylate)-doped active elements, *Appl. Opt.* 34, 1516-1518.
McKee, T. J., Lobin, J., and Young, W. A. (1982). Dye laser spectral purity, *Appl. Opt.* 21, 725-728.
Nelson, D. F., and Collins, R. J. (1961). Spatial coherence in the output of an optical maser *J. Appl. Phys.* 32, 739-740.
Peterson, O. G., Tuccio, S. A., and Snavely, B. B. (1970). CW operation of an organic dye solution laser, *Appl. Phys. Lett.* 17, 245-247.
Peterson, O. G., and Snavely, B. B. (1968). Stimulated emission from flashlamp-excited organic dyes in polymethyl methacrylate, *Appl. Phys. Lett.* 12, 238-240.
Samuel, I. D. W., Namdas, E. B., and Turnbull, G. A. (2009). How to recognize lasing, *Nat. Photon.* 3, 546-549.
Samuel, I. D. W., and Turnbull, G. A. (2007). Organic semiconductor lasers, *Chem. Rev.* 107, 1272-1295.
Saxena, K., Mehta, D. S., Srivastava, R., and Kamalasanan, M. N. (2006). Spatial coherence properties of electroluminescence from Alq_3-based organic light emitting diodes, *Appl. Phys. Lett.* 89, 061124.
Schäfer, F. P., Schmidt, W., and Volze, J. (1966). Organic dye solution laser, *Appl Phys. Lett.* 9, 306-309.
Schäfer, F. P. (1990). Principles of dye laser operation, In: *Dye Lasers*, Schäfer, F. P. (Ed.), pp. 1-89, Springer, Berlin.

Shimkaveg, G. M., Carter, M. R., Walling, R. S., Ticehurst, J. M., Koch, J. A., Mrowka, S., Trebes, J. E., MacGowan, B. J., Da Silva, L. B, Mathews, D. L., London, R. A., Stewart, R. E. (1992). X-ray laser coherence experiments in neon like yttrium, In: *Proceedings of The International Conference on Lasers' 91*, Duarte F. J., and Harris, D. G. (Eds.), pp. 84-92, STS: McLean, VA.

Shoshan, I., Danon, N. N., and Oppenheim, U. P. (1977). Narrowband operation of a pulsed dye laser without intracavity beam expansion, *J. Appl. Phys.* 48, 4495-4497.

Soffer, B. H., and McFarland, B. B. (1967). Continuously tunable narrow-band organic dye lasers, *Appl. Phys. Lett.* 10, 266-267.

Sorokin, P. P. and Lankard, J. R. (1966). Stimulated emission observed from an organic dye, chloro-aluminum phthalocyanine, *IBM J. Res. Develop.* 10, 162-163.

Spaeth, M. L., and Bortfeld, D. P., Stimulated emission from polymetine dyes, *Appl. Phys. Lett.* 9, 179-181 (1966).

Thompson, B. J., and Wolf, E. Two-beam interference with partially coherent light*J. Opt. Soc. Am.* 47, 895-902 (1957).

Trebes, J. E., Nugent, K. A., Mrowka, S., London, R. A., Barbee, T. W., Carter, M. R., Koch, J. A., MacGowan, B. J., Matthews, D. L., Da Silva, L. B., Stone, G. F., and Feit, M. D. (1992). Measurement of the spatial coherence of a soft x-ray laser, *Phys. Rev. Lett.* 68, 588-591.

Tsai, C-H., Tien, K-C., Chen, M-C., Chang, K-M., Lin, M-S., Cheng, H-C., Lin, Y-H., Chang, H-W., Lin, H-W., Lin, C-L., and Wu, C-C. (2010). Characterizing coherence lengths of organic light-emitting devices using Newton's rings apparatus, *Org. Electron.* 11, 439-444.

2

Coherence of XUV Laser Sources

Sebastian Roling and Helmut Zacharias
Westfälische Wilhelms-Universität Münster
Germany

1. Introduction

This article reviews recent developments in research on coherence properties of femtosecond XUV and x-ray lasers. Coherence is one of the most conspicuous features of laser sources. It describes how an electromagnetic wave field is correlated at different times and at different points in space. Coherence is a property that enables stationary interferences and thereby enables holographic and diffractive or lens less imaging.

While a large number of different laser systems radiating in the infrared, visible and UV spectral range have been developed in the past five decades since the invention of the laser, the generation of intense and coherent XUV and x-ray radiation is still a formidable goal of basic research. Developed first in the mid and far infrared spectral region free electron lasers (FEL) provide a new means to generate widely tunable coherent radiation [Madey, 1971; Deacon et al., 1977; Oepts et al., 1995]. In the XUV and soft x-ray spectral region the present lack of appropriate mirrors prevent the construction of optical resonators, essential for a laser with well-defined mode properties. Therefore in this spectral region free-electron lasers have to rely on the principle of self-amplified spontaneous emission (SASE), where in a single path enough gain is accumulated to evolve a pulse from shot noise. These machines then provide pulsed XUV and x-ray radiation which shows only partial coherence (Kondratenko & Saldin, 1980; Bonifacio et al., 1984; Murphy & Pellegrini, 1985). A characterisation of the coherence properties of SASE FEL radiation is not only important for the understanding and improvement of the SASE process. With the concomitant development of coherent diffraction imaging it is also of great practical importance in a broad field of fundamental scientific applications, like holographic imaging of artificial magnetic structures or single pulse imaging of large biomolecules. The theoretical description of the coherence properties of SASE FEL has made tremendous progress in the recent years. In this chapter we will briefly review measurement methods and description of partially coherent optical fields. The considerations will be applied to characterize the spatial and temporal coherence of pulses from the SASE free electron laser in Hamburg (FLASH) and of high harmonic radiation.

2. Basic definitions

Coherence is a measure of the correlation properties of different electromagnetic wave fields. For a more general comprehension some basic definitions are reviewed following the books of Mandel and Wolf (Mandel & Wolf, 1995) and Goodman (Goodman, 2000). The mutual correlation function Γ describes the coherence of an electromagnetic field E between two positions $\mathbf{r}_1$ and $\mathbf{r}_2$ at different times t and $(t + \tau)$:

$$\Gamma(\mathbf{r}_1,\mathbf{r}_2,\tau)=E(\mathbf{r}_1,t)E^*(\mathbf{r}_2,t+\tau) \quad (1)$$

For a fixed distance ($\mathbf{r}_i$ - $\mathbf{r}_j$) this simplifies to the mutual time correlation function:

$$\Gamma_{ij}(\mathbf{r}_i-\mathbf{r}_j,\tau)= E_i(t)E_j^*(t+\tau) \quad (2)$$

The absolute value of the normalised correlation function $\gamma_{12}(\tau)$

$$|\gamma_{12}(\tau)|=\Gamma_{12}(\tau)/\sqrt{\Gamma_{11}(0)\Gamma_{22}(0)}\,. \quad (3)$$

can be measured as the visibility V of the interference fringes of two interfering partial beams. The visibility V is connected to the normalised correlation $\gamma_{12}(\tau)$ function via

$$V=\gamma_{12}(\tau)\sqrt{I_1I_2}\,/\,(I_1+I_2) = (I_{max}-I_{min})\,/\,(I_{max}+I_{min}), \quad (4)$$

where I_1 and I_2 are the intensities of the interfering partial beams and I_{max} and I_{min} are the maximum and minimum intensities of the interference fringes, respectively. The coherence time τ_c can be defined as the half width at half maximum [HWHM] of $\gamma_{12}(\tau)$. A more general definition for arbitrary functions is given by

$$\tau_c = \int_{-\infty}^{\infty}|\gamma_{12}(\tau)|^2\, d\tau. \quad (5)$$

From the mutual correlation function Γ the cross spectral density $W(\mathbf{r}_1, \mathbf{r}_2, \omega)$ is then defined as

$$W(\mathbf{r}_1,\mathbf{r}_2,\omega)= \int_{-\infty}^{\infty}\Gamma(\mathbf{r}_1,\mathbf{r}_2,\tau)\mathbf{e}^{i\omega\tau}\,d\tau\,. \quad (6)$$

From this expression one can readily recognize that it forms a Fourier pair with the mutual correlation function Γ. The cross spectral density constitutes a measure between the spectral amplitudes of any particular frequency component ω of the electromagnetic field at the spatial points $\mathbf{r}_1$ and $\mathbf{r}_2$. When the cross spectral density W is treated for one point $\mathbf{r}_1 = \mathbf{r}_2 = \mathbf{r}$ it equals the power spectrum $S(\mathbf{r}, \omega)$ of the light

$$S(\mathbf{r},\omega)=W(\mathbf{r},\mathbf{r},\omega) \quad (7)$$

Analogous to the mutual coherence function Γ also the cross spectral density W can be normalised

$$\mu(\mathbf{r}_1,\mathbf{r}_2,\omega)=W(\mathbf{r}_1,\mathbf{r}_2,\omega)\,/\,\sqrt{S(\mathbf{r}_1,\omega)}\sqrt{S(\mathbf{r}_2,\omega)} \quad (8)$$

The normalised cross spectral density $\mu(\mathbf{r}_1,\mathbf{r}_2,\omega)$ is usually referred to as the spectral degree of coherence. In analogy to the coherence time of the light pulse also the degree of transverse coherence ζ is defined as

$$\zeta = \iint|\gamma(\mathbf{r}_1,\mathbf{r}_2)|^2 I(\mathbf{r}_1)I(\mathbf{r}_2)d\mathbf{r}_1 d\mathbf{r}_2 \,/\left[\int I(\mathbf{r})d\mathbf{r}\right]^2. \quad (9)$$

Here $I(\mathbf{r})$ denotes the intensity distribution of the electromagnetic field. An experimentally easily accessible value for the transverse coherence is given by the (r.m.s) transverse coherence length $\xi_{x,y}$. The effective transverse coherence length at a distance z from the source is $\Xi_{x,y}(z) = \xi_{x,y} \cdot \Sigma_{x,y}(z) / \sigma_{sx,y}$, where $\Sigma_{x,y}(z)$ denotes the r.m.s. beam size at the position z and σ_s is the r.m.s source size. This article reviews recent theoretical and experimental results concerning the coherence time t_c, the degree of transverse coherence ζ and accordingly the transverse coherence length $\xi_{x,y}$ of free electron lasers and high-harmonic generation light sources in the XUV and x-ray spectral regime.

3. The SASE FEL process

The generation of coherent radiation in a Self-Amplified Spontaneous Emission (SASE) process by a relativistic electron beam passing through a series of regularly spaced magnetic fields with opposite direction - a so-called undulator that is shown in Fig. 1a - was first proposed by Kondratenko and Saldin in 1980 (Kondratenko & Saldin, 1980). In 1985 Murphy and Pellegrini (Murphy & Pellegrini, 1985) pursued this idea of a single-pass amplifier starting from noise for soft x-ray radiation. Since in the x-ray spectral region optical cavities would feature significant losses present machines operate on a single-pass through an undulator. Therefore a SASE FEL consists of an electron accelerator and a sufficiently long undulator.

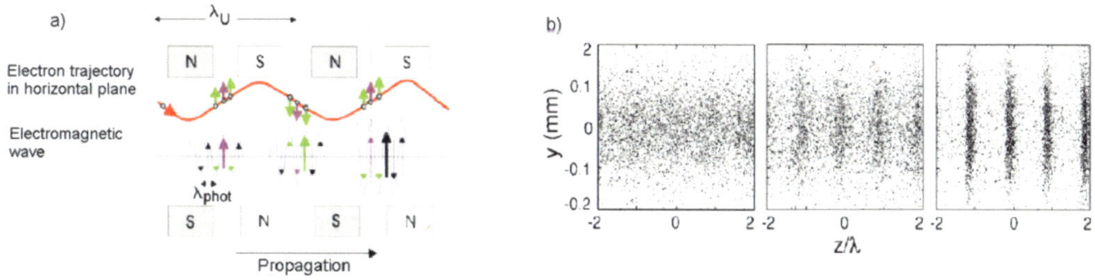

Fig. 1. (a) The electron orbit in a periodic undulator field (undulator period λ_U). Three electrons and the field of an electromagnetic wave (wavelength λ_{phot}) are shown at three positions along the undulator. The phase slip of the electrons with respect to the light field by exactly one wavelength from one undulator period to the next is discernible. (b) The density modulation of the electron bunch as it develops along the undulator, at the beginning of the undulator, in the middle and at the exit. The pictures are taken from reference (Feldhaus, Arthur & Hastings, 2005).

When the quality of the electron beam in terms of charge density, emittance and energy spread is high enough, the interaction between the electrons oscillating in the undulator magnetic field and the synchrotron radiation that is spontaneously emitted, leads to a density modulation of the electron bunch in the longitudinal direction, see Fig. 1b.
The period length of this modulation equals the wavelength of the undulator radiation (microbunching). Figure 1a shows a bunch of three electrons and the field of the electromagnetic wave at three positions along the undulator. An electron is accelerated or decelerated by the electric field of the radiation depending on its relative phase with respect to the light wave. This leads to a velocity modulation which then transforms into a longitudinal density modulation if the undulator is sufficiently long. The microbunched electrons will then radiate in phase with each other because of their longitudinal grouping

being exactly one optical wavelength. Thereby, the microbunching is sharpened further which in turn enhances the intensity of the electromagnetic field. This effect is referred to as Self-Amplified Spontaneous Emission (SASE). The radiated intensity of such a FEL is therefore proportional to the square of the number of particles involved and presently reaches values of the order of 1 mJ per pulse (Feldhaus, Arthur & Hastings, 2005; Ackermann et al., 2007; Emma et al., 2010). A list of some essential parameters of already operating and proposed SASE FEL can be found in Table I.

	FLASH	LCLS	XFEL
Minimum Wavelength (fundamental) [nm]	4.45	0.12	0.1
Maximum Photon energy (fundamental) [eV]	276	8270	12400
Pulse duration	29 fs at 24 nm	500 to < 10 fs	100 fs
Coherence time	6 fs at 24 nm	< 1 fs	0.22 fs
Pulse energy [mJ]	0.3	1.3 - 3	30

Table I. Main parameters of FLASH (Mitzner, 2008; Mitzner, 2009; DESY Homepage, 2010), LCLS (McNeil, 2009; Emma et al., 2010), XFEL (Altarelli et al., 2006).

4. Theoretical simulations

Bonifacio et al. (Bonifacio et al., 1994a; 1994b) presented a systematic theoretical study of the temporal structure, the linewidth, noise, fluctuations and the spectrum of a SASE FEL. The latter is directly related to the temporal coherence (see equation 8). The results of the 1D time-dependent numerical model show that the temporal as well as the spectral structure consist of several spikes. Saldin et al. report a systematic approach for an analytical description of SASE FEL in the linear mode of amplification where SASE does not yet reach full saturation (Saldin et al., 2000b). In addition to the average radiated power and the spectral envelope also the angular distribution of the radiation intensity in the far field and the degree of transverse coherence was calculated. This general result was applied to the special case of an electron beam with a Gaussian profile and a Gaussian energy distribution. Applying the three dimensional time-dependent numerical code FAST (Saldin et al., 1999) an investigation of the amplification process in a SASE FEL was implemented. The fluctuations of the current density in the electron bunch are uncorrelated not only in time but also in space. In contrast to the 1D model this fact is taken into account within the 3D model. The results were compared with analytical results showing that in the high-gain linear limit there is good agreement between both approaches. Importantly, it was found that even after completion of the transverse mode selection process the degree of transverse coherence of the radiation from a SASE FEL significantly differs from the ideal situation of complete coherence. This is a consequence of the interdependence of the longitudinal and transverse coherence. The SASE FEL shows poor longitudinal coherence which develops only slowly with increasing undulator length and thus prevents a full transverse coherence. For sake of generality the undulator length z is replaced by the normalized undulator length

$$\hat{z} = \Gamma \cdot z \,, \qquad (10)$$

where $\Gamma = \left[\pi j_0 K_1^2 / I_A \lambda_U \gamma\right]^{1/3}$ denotes the gain parameter. Here j_0 is the beam current density, λ_U is the undulator period, $I_A = mc^3/e \approx 17$ kA is the Alfven current and $\omega = 2\pi c/\lambda$ the frequency of the electromagnetic wave. The coupling factor K_n is given as $K_n = K(-1)^{(n-1)/2} \cdot \left[J_{n-1/2}(Q) - J_{n+1/2}(Q)\right]$, where $Q = n K^2 / \left[2\left(1+K^2\right)\right]$. With the FEL efficiency parameter $\rho = c\gamma_z^2 \Gamma / \omega$ the normalized coherence time is given as

$$\hat{\tau}_c = \rho \omega \tau_c . \tag{11}$$

The detuning C is given as $C = k_U - \omega / 2c\gamma_z^2$, where k_U denotes the wave number of the undulator. The temporal structure of the radiation pulse at the undulator length $\hat{z} = 15$ and a spectrum of the radiation from a SASE FEL calculated with the linear simulation code FAST are shown in Fig. 2. One can recognize that several temporal and spectral modes are present in the FEL radiation. Due to the connection between the frequency and time domain via a Fourier transformation a poor longitudinal coherence can readily be deduced from Fig. 2b. It had been found that the coherence time achieves its maximal value near the saturation point and then decreases drastically.

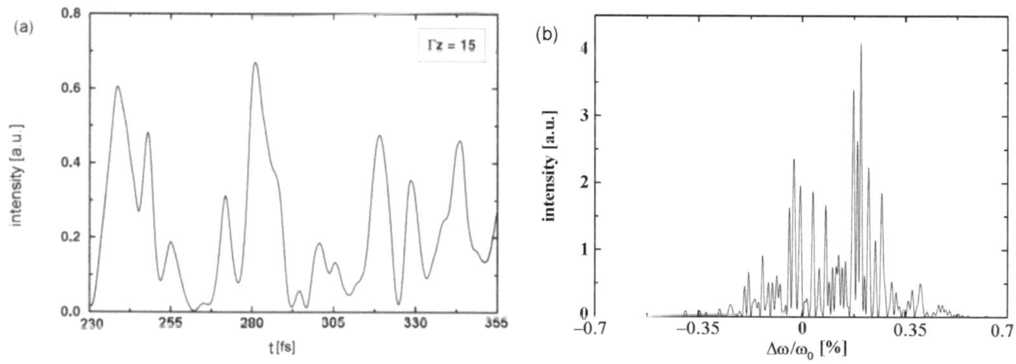

Fig. 2. (a) Temporal structure of the radiation pulse ($\lambda = 13$ nm) at the undulator length $\hat{z} = 15$. (b) Typical spectrum of the radiation from the FEL amplifier starting from shot noise at the undulator length $\hat{z} = 15$. Both calculations had been performed with the linear simulation code FAST. After reference (Saldin et al., 2000b).

A simulation of the degree of transverse coherence of the radiation of a SASE FEL plotted versus the normalized undulator length is shown in Fig. 3. The transverse degree of coherence ζ was found to approach unity asymptotically as $(1-\zeta) \propto 1/z$ at $\Gamma z \approx 15$ (Saldin et al., 2000b). At the beginning of the SASE process a large number M of transverse radiation modes are excited when the electron beam enters the undulator. At that stage the fluctuations of the current density in the electron bunch are still uncorrelated not only in time but also spatially. Since these different modes have different spatial overlap with the electron beam the amplification they experience will be different. Therefore, the number of modes decreases upon further amplification. As the fundamental mode has the best overlap with the electron beam the total field amplitude is dominated by the fundamental mode, which contributes according to this simulation close to 99.9 % to the total radiation power

(Saldin et al., 2001). Transverse coherence establishes quickly at an early stage of amplification due to this transverse mode selection. One could mistakenly deduce that the latter effect proceeds further at higher z and the degree of transverse coherence should approach unity exponentially but not as $(1-\zeta) \propto 1/z$ as it is found in (Saldin et al., 2000b). However, at this point, one should take a closer look at the spiky temporal and spectral structure of the FEL pulses shown in Fig. 2. In the high-gain linear regime the radiation of the SASE FEL consists of a spatial fundamental mode, however, at many different frequencies. The transverse distributions of the radiation field of spatially fundamental modes are also slightly different for different frequencies. As a result of the spatial interference of these longitudinal modes full transverse coherence is not achieved during the SASE process. To conclude, the interdependence between longitudinal coherence and transverse coherence is accountable for the fact that full transverse coherence is not achieved even after completition the mode selection process.

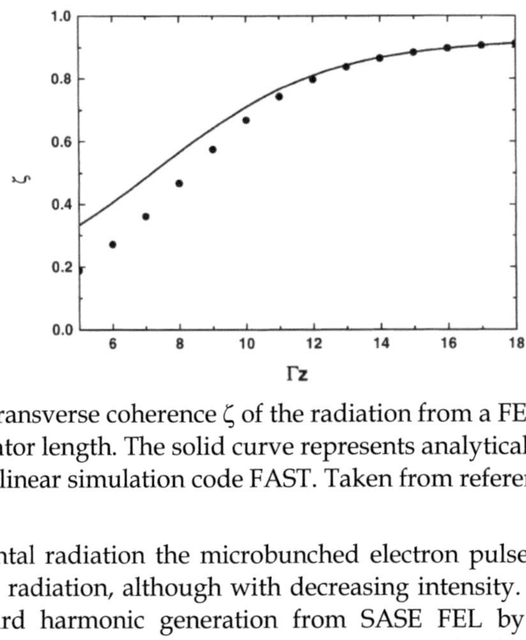

Fig. 3. The degree of transverse coherence ζ of the radiation from a FEL amplifier as a function of the undulator length. The solid curve represents analytical results and the circles are obtained with the linear simulation code FAST. Taken from reference (Saldin et al., 2000b).

Besides the fundamental radiation the microbunched electron pulse also supports gain at odd harmonics of the radiation, although with decreasing intensity. Saldin et al. evaluated the properties of third harmonic generation from SASE FEL by employing the time-dependent FEL simulation code GINGER (Saldin et al., 2006). The maximal value of the coherence time depends on the saturation length and, therefore, on the number of cooperating electrons N_c. For the coherence time $\hat{\tau}_c$ of the fundamental harmonic they arrived at

$$(\hat{\tau}_c)_{max} \approx \sqrt{\pi \ln N_c / 18}\,. \tag{12}$$

The evolution of the longitudinal coherence for higher harmonics with increasing undulator length is shown in Fig. 4. It is obvious that the normalized coherence time $\hat{\tau}_c$ evolves in three different stages. Initially (up to $\hat{z}$ = 7 - 8, see Fig. 4) the coherence time increases linearly as it should occur for the spontaneous emission of the radiation from the undulator. When the process of nonlinear harmonic generation begins to dominate the spontaneous emission the coherence time drops sharply. Around $\hat{z} = 10 - 11$ a plateau is obtained where the ratio of the coherence time of the nth harmonic to that of the first harmonic scales with $1/\sqrt{n}$. The structure of the third harmonic radiation indicates a similar spiky characteristic both in time as well as in the spectral distribution (Saldin et al., 2006), as shown for the fundamental in Fig. 2. It was found that a sharpening of the peaks in the intensity distribution leads to the decrease of the coherence time for the higher harmonics. When the amplification process enters the nonlinear regime at a normalized undulator length of $\hat{z} \geq 11$, the relative sharpening of the intensity peaks of higher harmonics becomes stronger, and the coherence time decreases even further. In fact, the coherence time at the saturation point of $\hat{z} = 13$ develops for higher harmonics approximately inversely proportional to the harmonic order n, as is shown in Fig.4.

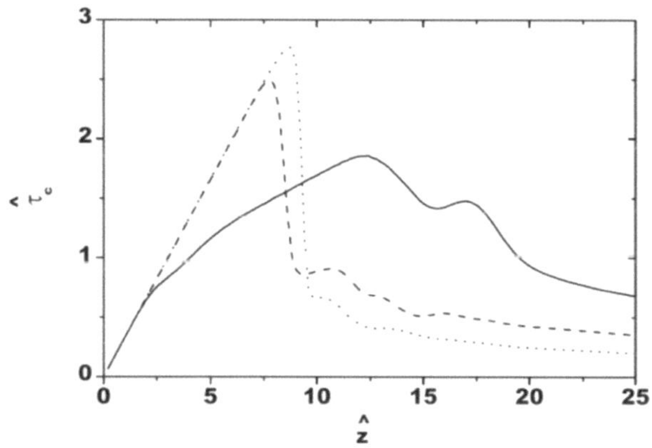

Fig. 4. Normalized coherence time of a SASE FEL as a function of normalized undulator length. Solid, dashed, and dotted lines correspond to the fundamental, 3rd, and 5th harmonic, respectively. Taken from reference (Saldin et al., 2006).

5. Experimental results

The experimental investigation of the spatial and temporal coherence properties of SASE FELs is of crucial importance, both from a fundamental point of view to assess different theoretical models, as well as for envisioned applications such as coherent imaging and holography in the x-ray spectral regime (Eisebitt et al., 2004; Chapman et al., 2006).

5.1 Transverse coherence

The first measurements of the transverse coherence of a SASE FEL operating in the VUV spectral regime have been performed by Ischebeck and co-workers in a Youngs double slit experiment (Ischebeck et al., 2003). Double slits with different separations, crossed slits and circular apertures were used to study the transverse coherence of the VUV light of the TESLA Test Facility, the predecessor of FLASH at DESY. The set-up is shown schematically in Fig. 5. It basically consisted of a Ce:YAG scintillator to convert the VUV light with wavelengths between λ = 80 nm and λ = 120 nm into visible light and a camera. The spatial distribution of the FEL radiation as well as the interference patterns from the different insertable apertures were observed. With a grating spectrometer equipped with an intensified CCD camera the spectral distribution of the light has been measured, although consecutively to the coherence measurements. To measure the transverse coherence properties double slits (width d=100 μm and d=200 μm) with different slit separations between 0.5 mm and 3 mm were inserted at a distance of 11,84 m downstream from the undulator exit.

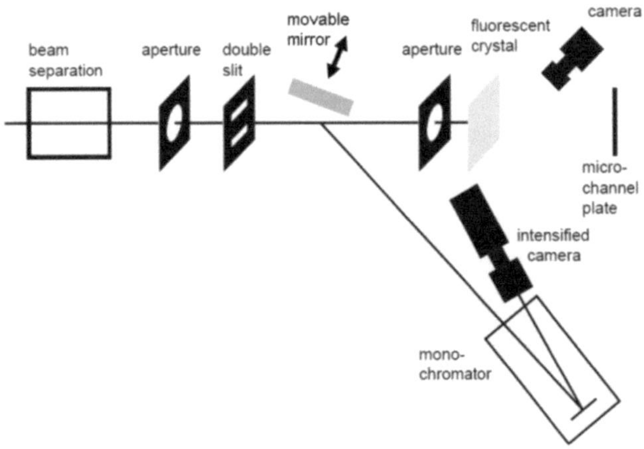

Fig. 5. The experimental set-up used for the first measurements of the transverse coherence at the TESLA Test Facility, after (Ischebeck et al., 2003).

Figure 6 shows the measured visibility (eqn. 4, circles) decreasing from V = 0.83 for 0.5 mm slit separation to V = 0.23 for 3.0 mm slit separation. The measured beam profile (asterisks) with a Gaussian fit (solid line) and a simulated beam profile taken from Saldin et al. (Saldin, Schneidmiller & Yurkov, 1999) are also shown. The experimental results yield a high degree of transverse coherence of up to V = 0.83 in the center of the beam as it is expected from FEL theory presented above. The spatial coherence then decreases rapidly with increasing radial distance from the center of the beam.

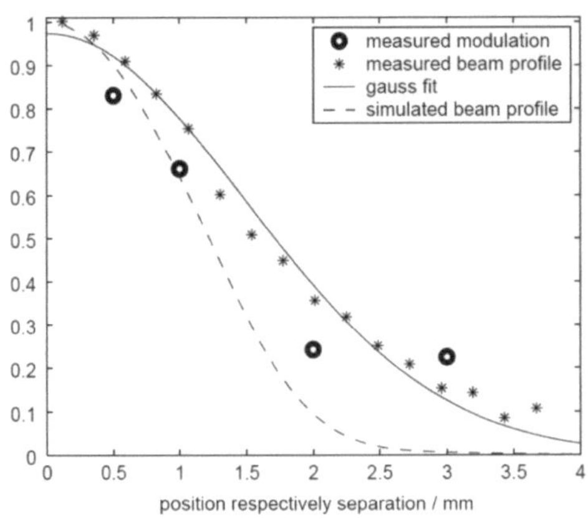

Fig. 6. Visibility in the center of the interference pattern as a function of the slit separation. For comparison, the beam profile measured with a MCP detector as a function of the position is shown. Taken from reference (Ischebeck et al., 2003).

Singer et al. used Youngs double slit to measure the transverse coherence properties of FLASH operated at $\lambda = 13.7$ nm (Singer et al., 2008). The experimental results are compared to results from a phenomenological modeling in terms of the Gaussian-Shell model (GSM). For the measurements a Young's double slit experiment with slit separations of 150, 300 and 600 µm was used. The double slit was positioned 20 m downstream from the undulator. Figure 7 shows the experimentally obtained absolute value of the complex degree of coherence with a Lorentzian fit yielding a transverse coherence length of $\xi_x = 300 \pm 15$ µm for the horizontal direction and $\xi_y = 250 \pm 13$ µm for the vertical direction. The degree of coherence $|\gamma|$ decreases faster with the beam radius then the intensity. It is clear that the radiation field is not fully coherent, but rather partially coherent as it is expected from the measurements shown in reference (Ischebeck et al., 2003).

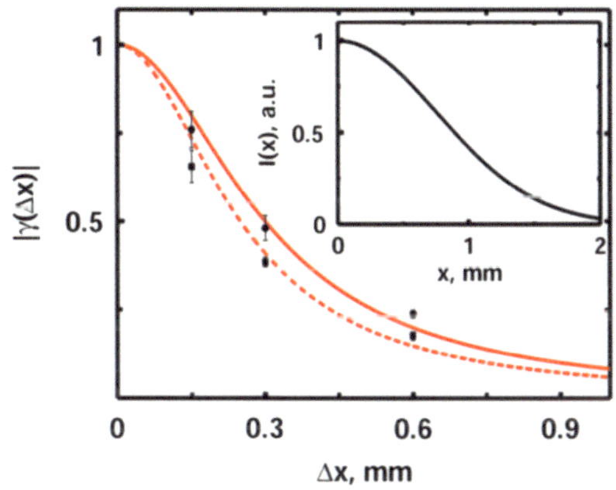

Fig. 7. Absolute value of the complex degree of coherence $|\gamma|$ at $\lambda = 13.7$ nm as a function of slit separation in the horizontal (circles) and vertical (squares) directions as a result of the fit to experimental data. A Lorentzian fit to the data of $|\gamma(\Delta x)|$ is shown by solid (horizontal direction) and dashed (vertical) lines. The intensity distribution is shown in the inset. Taken from reference (Singer et al., 2008).

For an analysis of the coherence properties of FLASH the experimental results are compared to the results from a general theoretical approach which is based on a decomposition of statistical fields into a sum of independently propagating transverse modes. This method is known as the Gaussian-Schell model (GSM). Thereby the cross-spectral density $W(\mathbf{r}_1, \mathbf{r}_2)$ (see eqn. 6) is represented as a sum of independent coherent modes $\Psi(\mathbf{r}_1, \mathbf{r}_2)$

$$W(\mathbf{r}_1, \mathbf{r}_2) = \sum_{j=0} \beta_j \Psi_j^*(\mathbf{r}_1) \Psi_j(\mathbf{r}_2). \tag{10}$$

Here Ψ_j and β_j are the amplitudes and coefficients, respectively. To evaluate the coherence properties of the FEL, the real FEL source was substituted by an equivalent model planar source described by a Gaussian Schell-model (GSM) cross spectral density

$$W(\mathbf{r}_1, \mathbf{r}_2,) = A_0^2 \cdot \exp[-\frac{\left(\mathbf{r}_1^2 + \mathbf{r}_2^2\right)}{4\sigma_I^2} - \frac{\left(\mathbf{r}_1 + \mathbf{r}_2\right)^2}{2\sigma_\mu^2}] \tag{11}$$

Here σ_I and σ_μ are the standard deviations of the spectral density and spectral degree of coherence at the source plane, respectively. The coordinates $\mathbf{r}_1$ and $\mathbf{r}_2$ are located in the plane of the source and A_0^2 is the normalization constant. The analysis shows that six transverse modes are contributing to the total radiation field of FLASH. From Fig. 8a one can easily see that the contribution of the modes falls exponentially with the mode number j. Thereby the contribution of the first mode is about 59% and the contribution of the second mode is 24% of the total radiation power. The contribution of the sixth mode is more than two orders of magnitude smaller. From Fig. 8b a transverse coherence length of $\xi_x = 715$ µm can be obtained. Thus the experimentally measured values were considerably lower than those calculated in terms of the GSM. The apparent source size corresponding to the measured values for the coherence length was calculated making use of the GSM. The resulting value of σ_I = 180 µm was in good agreement with the source size observed in wave front measurements (Kuhlmann et al., 2006), but 2.5 times larger than considered in the theoretical modeling.

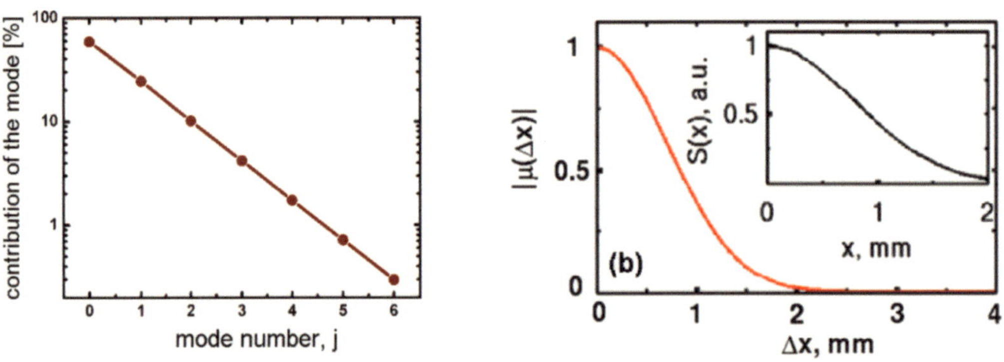

Fig. 8. (a) Contribution of individual modes in the GSM to the cross spectral density. (b) Absolute value of the spectral degree of coherence taken along a line through the middle of the beam profile. In the inset the spectral density S(x) is shown. Calculations for the FEL operating in saturation are performed in the frame of a Gaussian-Schell model 20 m downstream from the source at a wavelength λ = 13.7 nm; after reference (Singer et al., 2008).

Recently Vartanyants & Singer employed the GSM to evaluate the transverse coherence properties of the proposed (Altarelli et al., 2006) SASE 1 undulator of the European XFEL, scheduled to begin operation in 2014. Simulations were made for a GSM source with an rms source size σ_s = 27.9 µm and a transverse coherence length ξ_s = 48.3 µm at the source for a wavelength of λ = 0.1 nm (corresponding to $h \cdot \nu$ = 12 keV), taken from the XFEL technical design report (Altarelli et al., 2006). Figure 9 shows the evolution of the beam size $\Sigma(z)$ and the transverse coherence length $\Xi(z)$ with the distance of propagation z. At a distance z = 500 m from the source a transverse coherence length of $\Xi(z) = 348$ µm and a beam size of $\Sigma(z) = 214$ µm is obtained for the XFEL. Thus the coherence decreases less rapid than the spatial intensity of the beam. At present this prediction differs significantly from the experimental results for both TTF and FLASH. Distinct from a synchrotron source the coherence properties of the radiation field from the XFEL is of the same order of magnitude both for the vertical and the horizontal direction.

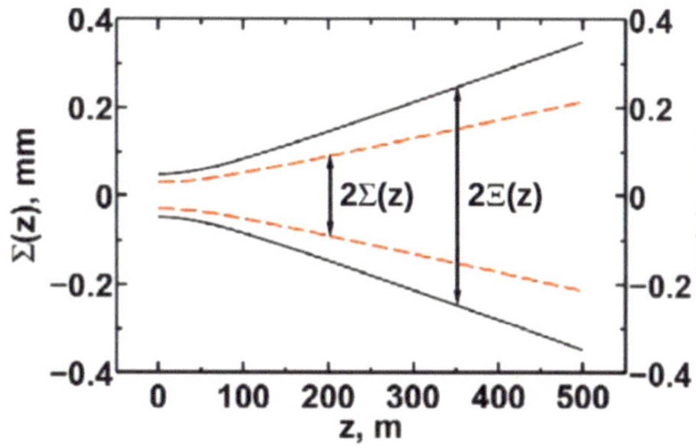

Fig. 9. Beam size $\Sigma(z)$ (dashed line) and the transverse coherence length $\Xi(z)$ (solid line) at different distances z from the SASE 1 undulator of the European XFEL, from (Vartanyants & Singer, 2010).

Earlier, results of numerical time dependent simulations for the coherence properties of LCLS XFEL at SLAC in Stanford have been reported (Reiche, 2006). For a wavelength of λ = 0.15 nm (corresponding to $h \cdot \nu$ = 8.27 keV) the effective coherence area within which the field amplitude and phase have a significant correlation to each other amounts to 0.32 mm². That is about five times larger than the spot size with a value of 0.044 mm² when evaluated at the first experimental location 115 m downstream from the undulator.

5.2 Temporal coherence

For an experimental measurement of the temporal coherence ideally time-delayed amplitude replicas of the FEL pulses should be brought to interference. However, the lack of amplitude splitting optical elements in the x-ray regime permits only the use of wavefront splitting mirrors in grazing incidence. These elements can then, however, be applied in a broad spectral region. Such a beam splitter and delay unit is shown schematically in Fig. 10(a).

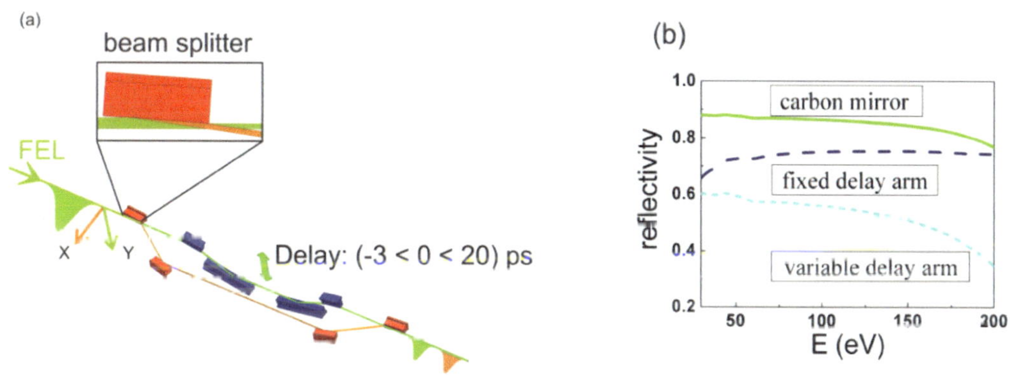

Fig. 10. (a) Schematic drawing of the layout of the autocorrelator. Grazing angles of 3° and 6°for the fixed and variable delay arms, respectively, are employed to ensure a high reflectivity of the soft x-ray radiation. (b) Calculated reflectivity for amorphous carbon coated silicon mirrors for $h\nu$ = 30 to 200 eV. The full green line shows the reflectivity of a single mirror for a grazing angle of 6°.

Based on geometrical wave front beam splitting by a sharp mirror edge and grazing incident angles the autocorrelator covers the fundamental energy range of FLASH (20 - 200 eV) with an efficiency of better than 50%, see Fig. 10 (b). Grazing angles of 3° and 6° for the fixed and variable delay arms, respectively, are employed to ensure a high reflectivity of the soft x-ray radiation. Looking in the propagation direction the beam splitter with a sharp edge reflects the left part of the incoming FEL pulse horizontally into a fixed beam path. The other part of the beam passes this beam splitting mirror unaffected and is then reflected vertically by the second mirror into a variable delay line. A variable time delay between -5 ps and +20 ps with respect to the fixed beam path can be achieved with a nominal step size of 40 as. The seventh and eighth mirror reflect the partial beams into their original direction. Alternatively, small angles can be introduced to achieve and vary a spatial overlap of the partial beams. Mitzner et al. investigated the temporal coherence properties of soft x-ray pulses at FLASH at $\lambda = 23.9$ nm by interfering two time-delayed partial beams directly on a CCD camera (Mitzner et al., 2008). Fig. 11 shows two interferograms at zero and 50 fs delay, respectively. The overlap of the two partial beams is $\Delta x \approx 1.2$ mm which corresponds ~44% of the beam diameter in this case where an 1 mm aperture is set 65 m in front of the detector near the center of the beam profile. In these particular cases the contrast of the interference fringes yields via equation (4) a visibility of V = 0.82 and V = 0.07, respectively.

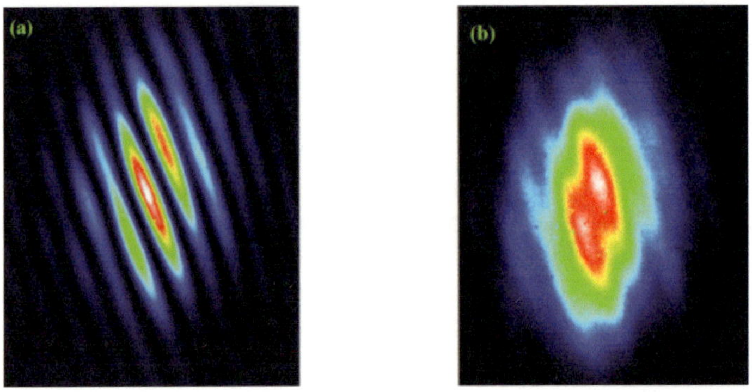

Fig. 11. Single exposure interference fringes at $\lambda = 24$ nm (a) at zero and (b) at 55 fs delay between both partial beams. The crossing angle of the partial beams is $\alpha = 60$ µrad.

Scanning now the delay between the two pulses and calculating at each time step the visibility of the interference fringes (applying Equation 4) the temporal coherence properties of FLASH pulses are investigated. Figure 12 shows the time delay dependence of the average visibility observed for two different wavelengths, $\lambda = 23.9$ nm and $\lambda = 8$ nm. Each data point (red dots) is the average of the visibility of ten single exposure interference pictures. In Fig. 12(a) the (averaged) visibility of V = 0.63 at zero time delay rapidly decreases as the time delay is increased. The central maximum of the correlation can be described by a Gaussian function (green line) with a width of 12 fs (FWHM). Then a coherence time corresponding to half of the full width of $\tau_{coh} = 6$ fs is obtained. Remarkably, the visibility, i.e. the mutual coherence, is not a monotonic function of the delay time between both partial beams. Instead, a minimum at about 7.4 fs after the main maximum and a secondary maximum at about 12.3 fs appear, symmetrically on both sides of the main maximum. In addition, a small but discernible increase of the visibility occurs at a delay around 40 fs.

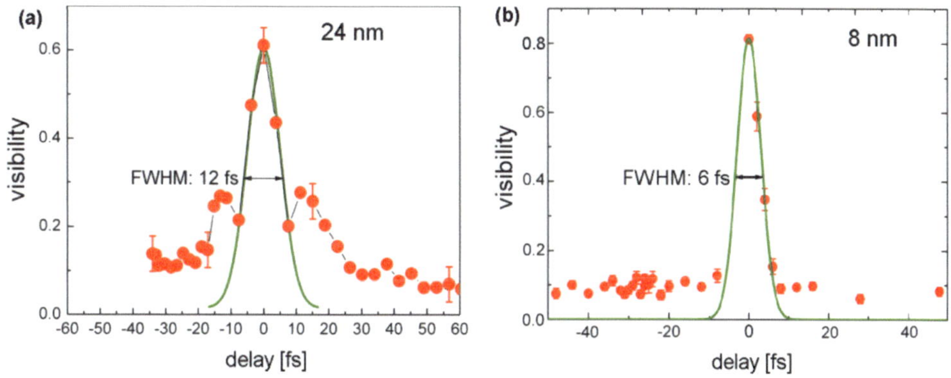

Fig. 12. Observed visibility (experimental data points: red dots) as a function of time delay for (a) λ = 24 nm and (b) λ = 8 nm. The green line depicts a Gaussian function with a coherence time of (a) τ_{coh} = 6 fs and (b) τ_{coh} = 3 fs, representing a single Fourier transform limited pulses.

Since the interferences were measured for independent single pulses of the FEL and then their visibilities averaged, this behavior of the temporal coherence function reflects an intrinsic feature of the FEL pulses at the time of the measurements. The radiation of SASE FELs consists of independently radiating transverse and longitudinal modes. In the time domain the radiation is emitted in short bursts with random phase relationship between the bursts. Time domain and spectral domain are related to each other via a Fourier transformation which leads to narrow spikes within the bandwidth of the undulator in the spectral domain, see also the calculated spectrum of a SASE FEL shown in Fig. 2. In the linear autocorrelation experiments shown in Fig. 12a (Mitzner et al., 2008) these independent modes can interact at longer time delays as a cross correlation. This behavior was found to be accountable for the non-monotonous decay of the visibility. A second sub-pulse at Δt = 12 fs and a weak third one at Δt = 40 fs can be stated as a reason for this behavior. Figure 12 (b) shows the result from an analogous measurement at λ = 8 nm. From a Gaussian fit with a FWHM of 6 fs a coherence time of τ_c = 3 fs is obtained. The non-monotonous decay that was discussed before for the 24 nm measurement is not apparent here.

Recently, a similar measurement also utilizing an autocorrelator that employs wave front beam splitters was performed for FLASH radiating at λ = 9.1 nm and λ = 33.2 nm (Schlotter et. al, 2010). These data were compared to Fourier transformed spectral bandwidth measurements obtained in the frequency domain by single-shot spectra. A good agreement with the measurements in the time domain was found. In addition to single shot exposures the temporal coherence was measured in the 15-pulse-per-train mode. Figure 13 shows the time delay dependence of the average visibility observed for two different wavelengths, λ = 33.2 nm (single shot: black triangles; 15 bunches: red triangles) and λ = 9.6 nm (single shot: black squares; 15 bunches: red dots). In order to plot the data of both wavelengths into one graph the abscissa is given in cτ/λ which represents the number of periods of the lightwave. For the 15 bunch per train data a clearly lower coherence at longer timescales is observed than for the single shot data.

To explain this behavior we should take a look at a single point in the interference pattern. If a maximum of the intensity appears at this point for zero delay and for a path length differences $n \cdot \lambda$, minima will appear for $\frac{n}{2} \cdot \lambda$. The wavelength of the FEL radiation shows

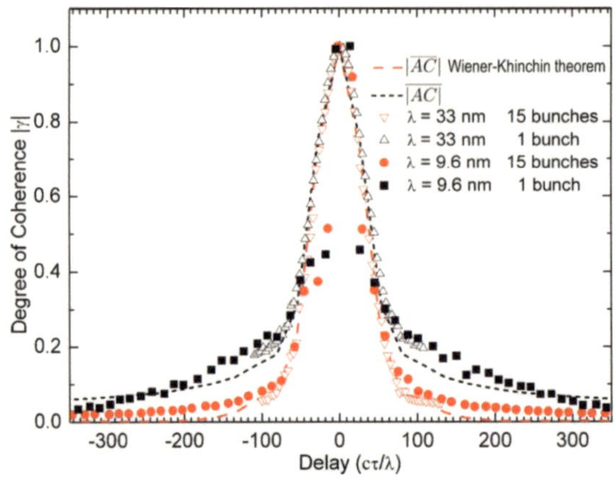

Fig. 13. The normalized degree of coherence $|\gamma|$ plotted versus the delay given in units of the wavelength. The dashed curve was calculated from spectral measurements at 33.2 nm. Taken from reference (Schlotter et al., 2010).

small shot-to-shot fluctuations. Therefore for longer path length differences (n ~ 50) a π phase difference occurs for different wavelengths λ_k. At the same point of the detector the interference pattern corresponding to a wavelength λ_1 may now show a maximum while the interference pattern corresponding to a wavelength λ_2 shows a minimum. Thus, the visibility appears blurred, when k = 15 bunches with slightly different wavelengths form interference patterns before the read-out of the detector.

5.3 Coherence enhancement through seeding

An essential drawback of SASE FEL starting from shot noise is the limited temporal coherence. Therefore, the improvement of the temporal coherence is of great practical importance. One idea to overcome this problem was presented by Feldhaus et al. (Feldhaus et al., 1997). The FEL described consists of two undulators and an X-ray monochromator located between them (see Fig. 14). The first undulator operates in the linear regime of amplification and starts from noise. The radiation output has the usual SASE properties with significant shot-to-shot fluctuations. After the first undulator the electron beam is guided through a by-pass, where it is demodulated. The light pulse on the other hand is monochromatized by a grating. At the entrance of the second undulator the monochromatic X-ray beam is recombined with the demodulated electron beam, thereby acting as a seed for the second undulator. For this purpose, the electron micro-bunching induced in the first undulator must be destroyed, because this electron micro-bunching from the first undulator corresponds to shot noise that was amplified. The degree of micro-bunching can thus be characterized by the power of shot noise which has the same order of magnitude as the output power of the FEL. When the radiation now passes the monochromator only a narrow bandwidth and thus only a small amount of the energy is transmitted. Thus at the entrance of the second undulator a radiation-signal to shot-noise ratio much larger than unity has to be provided. This can be achieved because of the finite value of the natural energy spread in the beam and by applying a special design of the electron by-pass.
At the entrance of the second undulator the radiation power from the monochromator then dominates over the shot noise and the residual electron bunching, such that the second stage of the FEL amplifier will operate in the steady-state regime when the input signal

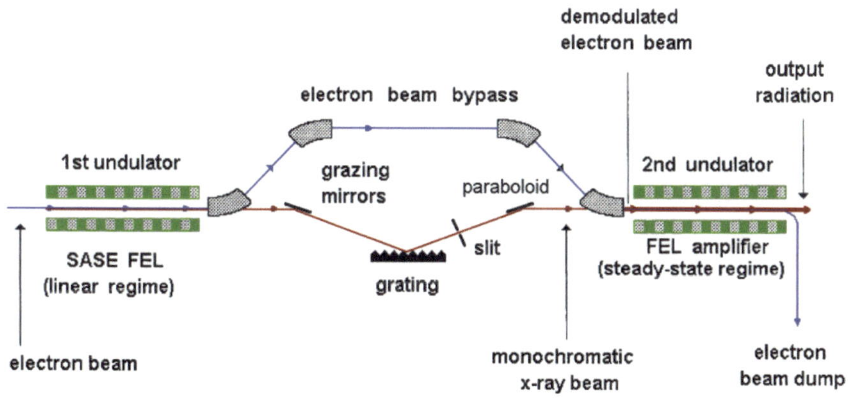

Fig. 14. Principal scheme of a single-pass two-stage SASE X-ray FEL with internal monochromator; after (Saldin et al., 2000a).

bandwidth is small with respect to the FEL amplifier bandwidth. The second undulator will thus amplify the seed radiation. The additional benefits derived from this configuration are superior stability, control of the central wavelength, narrower bandwidth, and much smaller energy fluctuations than SASE. Further, it is tunable over a wide photon energy range, determined only by the FEL and the grating.

An alternative approach is based on seeding with a laser, see ref. (Yu et al., 1991, 2000). Such a scheme has been applied at the Deep Ultraviolet FEL (DUV FEL) at the National Synchrotron Light Source (NSLS) of Brookhaven National Laboratory (BNL) (Yu et al., 2003). The set-up is shown in Fig. 15. In high-gain harmonic generation (HGHG) a small energy modulation is imposed on the electron beam by its interaction with a seed laser (1) in a short undulator (8) (the modulator) tuned to the seed wavelength λ. The laser seed introduces an energy modulation to the electron bunch. In a dispersive three-dipole magnetic chicane (9) this energy modulation is then converted into a coherent longitudinal density modulation. In a second long undulator (10) (the radiator), which is tuned to the *n*th odd harmonic of the seed frequency, the microbunched electron beam emits coherent radiation at the harmonic frequency $n\lambda$, which is then amplified in the radiator until saturation is reached. The modulator (resonant at λ = 800 nm) of the DUV FEL is seeded by an 800 nm CPA Ti:sapphire laser (pulse duration: 9 ps). This laser drives also the rf gun of the photocathode producing an electron bunch of 1 ps duration. In this way an inherent synchronization between the electron bunches and the seeding pulses is achieved. The output properties of the HGHG FEL directly maps those of the seed laser which can show a high degree of temporal coherence. In the present case the output HGHG radiation shows a bandwidth of 0.23 nm FWHM (corresponding to ~0.1%), an energy fluctuation of only 7% and a pulse length of 1 ps (equal to the electron bunch length) when the undulator is seeded with an input seed power of P_{in} = 30 MW. The bandwidth within a 1 ps slice of the chirped seed is 0.8 nm (corresponding to 0.1% bandwidth) and the chirp in the HGHG output is expected to be the same, i.e., 0.1% · 266 nm = 0.26 nm. This is consistent with a bandwidth of Δλ = 0.23 nm [FWHM] experimentally observed. A Fourier-transform limited flat-top 1 ps pulse would have a bandwidth of Δλ = 0.23 nm and a 1 ps (FWHM) Gaussian pulse would have a bandwidth of Δλ = 0.1 nm. Besides the high degree of temporal coherence a further advantage compared to a SASE FEL is the reduced shot-to-shot fluctuations of the output radiation if the second undulator operates in saturation. A similar scheme will also be applied at the FERMI FEL at Elettra that will start operation in 2011 providing wavelengths down to λ = 3 nm for the fundamental and λ = 1 nm for the third harmonic (Allaria et al., 2010).

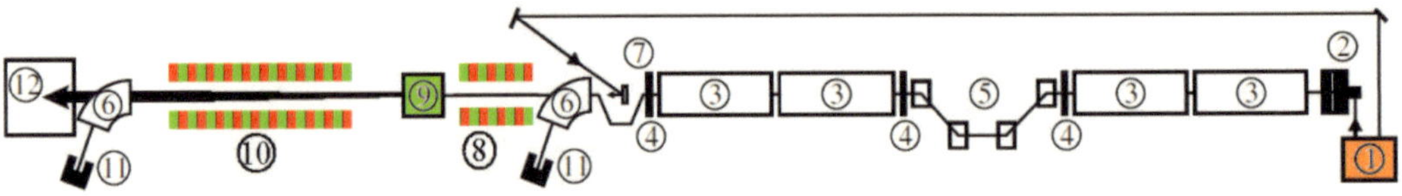

Fig. 15. The NSLS DUV FEL layout. 1: gun and seed laser system; 2: rf gun; 3: linac tanks; 4: focusing triplets; 5: magnetic chicane; 6: spectrometer dipoles; 7: seed laser mirror; 8: modulator; 9: dispersive section; 10: radiator; 11: beam dumps; 12: FEL radiation measurements area. After reference (Yu et al., 2003).

Another possibility to generate coherent radiation from an FEL amplifier is seeding with high harmonics (HH) generated by an ultrafast laser source whose beam properties are simple to manipulate, see reference (Sheehy et al., 2006; Lambert et al., 2008). In this way extremely short XUV pulses are obtained, down to a few femtoseconds. Such a scheme was applied at the Spring-8 compact SASE source (Lambert et al., 2008) and is depicted schematically in Fig. 16.

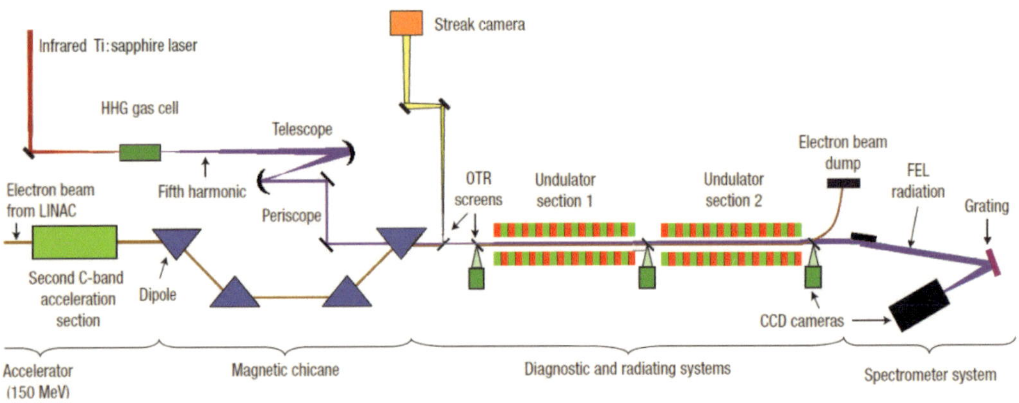

Fig. 16. Experimental setup for HHG seeding of the Spring-8 Compact SASE source, after (Lambert et al., 2008).

A Ti:sapphire laser (800 nm, 20 mJ, 100 fs FWHM, 10 Hz) that is locked to the highly stable 476 MHz clock of the accelerator passes a delay line that is necessary to synchronize the HHG seed with the electron bunches. For this purpose a streak camera observes the 800 nm laser light and the electron bunch signal from an optical transition radiation (OTR) screen. The beam is then focused into a xenon gas cell in order to produce high harmonics. Using a telescope and periscope optics the HHG seed beam is spectrally selected, refocused and spatially and temporally overlapped with the electron bunch (150 MeV, 1 ps FWHM, 10 Hz) in the two consecutive undulator sections 1 and 2. Both undulators are tuned to λ = 160 nm, corresponding to the fifth harmonic of the laser. The beam position is monitored on optical transition radiation (OTR) screens. The output radiation is characterized with an imaging spectrometer for different seeding pulse energies between 0.53 nJ and 4.3 nJ per pulse.

Figure 17 shows the spectra of the unseeded undulator emission (purple, enlarged 35 times), the HHG seed (yellow, enlarged 72 times) and the seeded radiation output (green) for a seed pulse energy of 4.3 nJ. A spectral narrowing for the seeded output radiation and a significant shift to longer wavelengths compared with the seed radiation is obvious. The measured relative spectral widths of the seeded FEL are reduced compared to the unseeded one from 0.54% to 0.46 % (0.53 nJ seed) and from 0.88% to 0.44% (4.3 nJ seed). A similar narrowing is observed for the spectra of the third (λ = 53.55 nm) and fifth harmonic (λ = 32.1 nm).

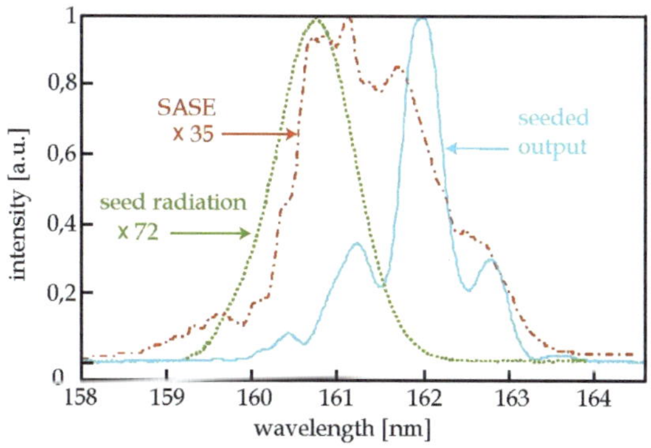

Fig. 17. Experimentally obtained spectra of the FEL fundamental emission (λ = 160 nm). SASE (red), seed radiation (green) and seeded output (blue), after (Lambert et al., 2008).

For a fully coherent seed pulse the seeded FEL should also show a high temporal coherence which, however, is not yet experimentally confirmed. The pulse should then also show a duration close to the Fourier transform limit. From the measured spectral width of $\Delta\lambda$ = 0.74 nm (for 0.53 nJ seed) one might conclude a Fourier transform limited duration of 57 fs. Currently several facilities using HHG as a seed source are proposed or under construction e.g. references (McNeil et al., 2007; Miltchev et al., 2009).

6. Temporal coherence of high-order harmonic generation sources

The generation of high-order harmonics of a short laser pulse in a gas jet has attracted a lot of attention since the first discovery in the late 1980s (McPherson et al., 1987; Ferray et al., 1988; Li et al., 1989). High harmonic radiation has become a useful short-pulse coherent light source in the XUV spectral regime (Haarlammert & Zacharias, 2009; Nisoli & Sansone, 2009). By focussing an intense femtosecond laser pulse into rare gases odd order high harmonics of the original laser frequency can be generated.

This can be explained in terms of the three step model (Corkum, 1993; Kulander et al., 1993; Lewenstein et al., 1994). The focused pumping laser beam typically has intensities of more than 10^{13} W/cm^2, which is in the order of the atomic potential. This leads to a disturbance of the atomic potential of the target atoms allowing the electron to tunnel through the remaining barrier, see Fig. 18a. Figure 18b shows how the electron is then accelerated away from the atom core by the electric field of the driving laser lightwave. After half an optical cycle the direction of the driving laser field reverses and the electron is forced to turn back to the core. There, a small fraction of the electrons recombine with the ion, and the energy that was gained in the accelerating processes before plus the ionization energy I_P is emitted as light, see Fig. 18c. When the electrons turn back to the core they can basically follow two

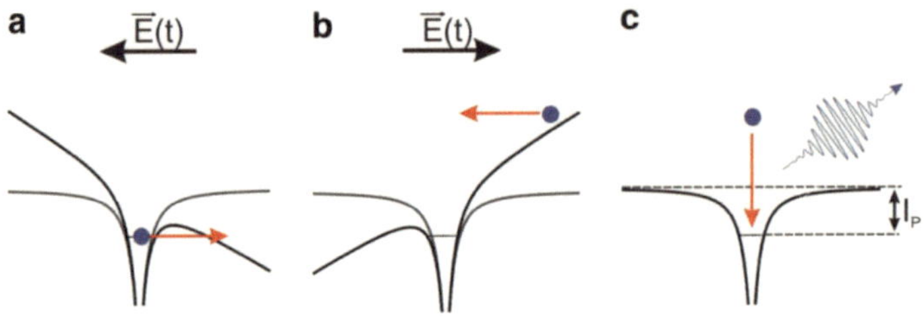

Fig. 18. Illustration of the three step model for high harmonic generation. (a) deformation of the atomic potential and tunnel ionization of the target atoms (b) acceleration of the free electrons in the laser electric field (c) recombination and photon emission.

different trajectories, a short one and a long one, respectively. The short trajectory shows an excursion time close to half an optical cycle, whereas the long trajectory takes slightly less than the whole optical period. Both of them show different phase properties with respect to the dipole moment of the particular harmonic. The phase of the short trajectory does not significantly vary with the laser intensity as opposed to the phase of the long trajectory that varies rapidly with the laser intensity (Lewenstein et al., 1995; Mairesse et al., 2003). The energy acquired by the electron in the light field corresponds to the ponderomotive energy U_P

$$U_p = e^2 E_o^2 / 4m_e\omega^2 . \tag{12}$$

Here E_0 denotes the electric field strength, e the elementary charge, m_e the electron mass and ω the angular frequency. The maximum photon energy emitted, the cut-off energy, is given by

$$E_{cutoff} = 3.17 \cdot U_{pon} + I_p , \tag{13}$$

where I_p denotes the ionization potential of the atom.
A theoretical study of the coherence properties of high order harmonics generated by an intense short-pulse low-frequency laser is presented particularly for the 45th harmonic of a 825 nm wavelength laser (Salières, L'Huillier & Lewenstein, 1995). First, the generation of the radiation by a single atom is calculated by means of a semi-classical model (Lewenstein et al., 1994). Phase and amplitude of each harmonic frequency are evaluated and then in a second step propagated in terms of the slowly varying amplitude approximation (L'Huillier et al., 1992). Harmonic generation is optimized when the phase-difference between the electromagnetic field of the driving laser and the electromagnetic field of the output radiation is minimized over the length of the medium. At this point phase-matching is achieved. It is shown that the coherence properties and consequently the output of the harmonics can be controlled and optimized by adjusting the position of the laser focus relative to the nonlinear medium.
Bellini et al. investigated experimentally the temporal coherence of high-order harmonics up to the 15th order produced by focusing 100 fs laser pulses into an argon gas jet (Bellini et al., 1998; Lyngå et al., 1999). The visibility of the interference fringes produced when two spatially separated harmonic sources interfere in the far-field was measured as a function of time delay between the two sources. The possibility to create two phase-locked HHG sources that are able to form an interference pattern in the far-field had been demonstrated earlier (Zerne et. al., 1997). A high transverse coherence that is necessary for two beams to interfere under an angle had been proven by a Youngs double-slit set-up (Ditmire et al., 1996). The experimental set-up used for the coherence measurements is shown in Fig. 19.

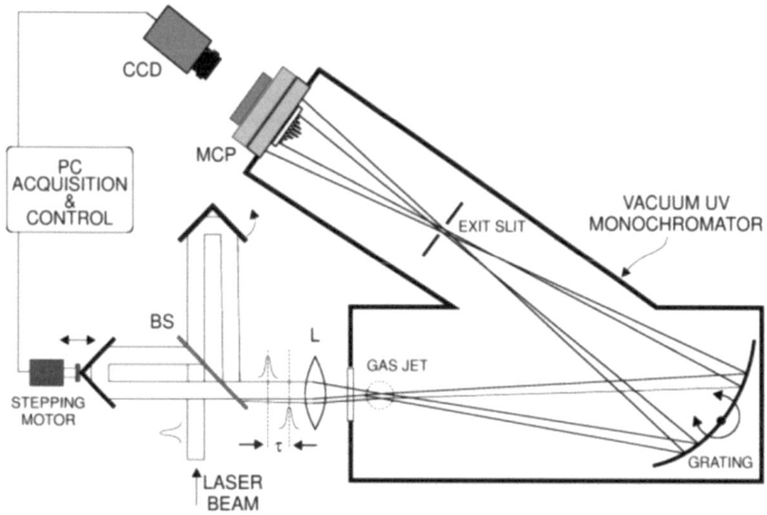

Fig. 19. Experimental setup for the measurement of the temporal coherence of high-order harmonics. BS is a broadband 50% beam splitter for 800 nm. L is the lens used to focus the two pulses, separated by the time delay τ, into the gas jet. Taken from reference (Bellini et al., 1998).

The laser used was an amplified Ti:sapphire system delivering 100 fs pulses with 14 nm spectral width centered around 790 nm at a 1 kHz repetition rate and with an energy up to 0.7 mJ. A Michelson interferometer placed in the path of the laser beam produced pairs of near infrared pump pulses which had equal intensities and whose relative delay could be accurately adjusted by means of a computer controlled stepping motor. The beams were then apertured down and focused into a pulsed argon gas jet. In order to avoid interference effects in the focal zone and to prevent perturbations of the medium induced by the first pulse one arm of the interferometer was slightly misaligned. Thus the paths of the two pulses were not perfectly parallel to each other and formed a focus in two separate positions of the jet. Both pulses then interacted with different Ar ensembles and produced harmonics as two separate and independent sources that may interfere in the far field. Behind the exit slit of a monochromator spatially overlapped beams were detected on the sensitive surface of a MCP detector coupled to a phosphor screen and a CCD camera.

To determine the temporal coherence of the high-order harmonics the time delay between both generating pulses was varied in steps of 5 or 10 fs and successive recordings of the interference patterns were taken. The fringe visibility V is calculated according to equation 4 for the different delays Δt and for different points in the interference pattern in order to analyze the temporal coherence properties spatially for inner and outer regions of the beam. The coherence time was obtained as the half width at half maximum of the curve shown in Fig. 20.

The coherence times measured at the center of the spatial profile varied from 20 to 40 fs, relatively independently of the harmonic order. In the outer region a much shorter coherence time is observed. This can be explained when the different behavior of the phases of the long and the short trajectory due to the laser intensity is taken into account. In a simulation the contributions of these trajectories are examined. Because the long trajectory shows a rapid variation of the dipole phase that leads to a strong curvature of the phase

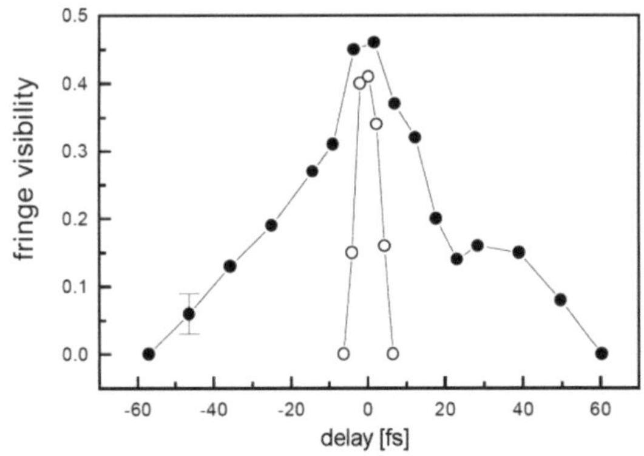

Fig. 20. Visibility curves as a function of the delay for the 15th harmonic, for the inner (full symbols) and outer (open symbols) regions. Taken from (Bellini et. al., 1998).

front, the radiation emitted from this process has a short coherence time due to the chirp caused by the rapid variation of the phase during the pulse. As opposed, the short trajectory shows a long coherence time. Since the radiation emerging from the short trajectory is much more collimated than that from the long trajectory its contribution to the outer part of the observed interference pattern is much lower than that of the latter. At this point it is necessary to emphasize that in this experiment the temporal coherence of two phase-locked HHG sources is evaluated, where the time delay is introduced between the partial beams of the driving Ti:sapphire laser. Therefore the two XUV pulses have to be assumed to be identical.

Hemmers and Pretzler presented an interferometric set-up operating in the XUV spectral range (Hemmers & Pretzler, 2009). The interferometer consisted of a combination of a double pinhole (similar to Young's double slit) and a transmission grating. In the case of a light source consisting of discrete spectral lines, it allows to record interferograms for multiple colors simultaneously. The experimental setup is shown in Fig. 21.

The pinholes were mounted such that a defined rotation around the beam axis was possible. A transmission grating placed behind the pinholes dispersed the radiation spectrally. Spectra were recorded by a CCD camera placed at a distance of L = 135 cm from the pinholes. This set-up is suitable to be used as a spectrometer with the double pinhole as a slit. The spectral resolution is determined by the pinhole diffraction, which creates Airy spots in the far-field. With the described geometry this leads to a spectral resolution in the range of $\Delta\lambda$ = 0.3 nm at a wavelength of λ = 20 nm, sufficient to separate individual odd harmonics with spectral separation of about 1 nm in that spectral range. Furthermore, the combination of a rotatable double pinhole and a transmission grating acts as a spectrally resolved Young's double slit interferometer with variable slit spacing. The time delay between the partial beams was realized in the following manner: a varying path difference between the two interfering beams was achieved by rotating the double pinhole around the grating normal. As illustrated in Fig. 21, the path difference in the beams diffracted into first order by the grating varies as

$$\Delta s = D * \sin\beta * \sin\gamma = D * \sin\beta * N\lambda \,. \tag{12}$$

Here γ denotes the diffraction angle and β the rotation angle of the pinholes with respect to the grating.

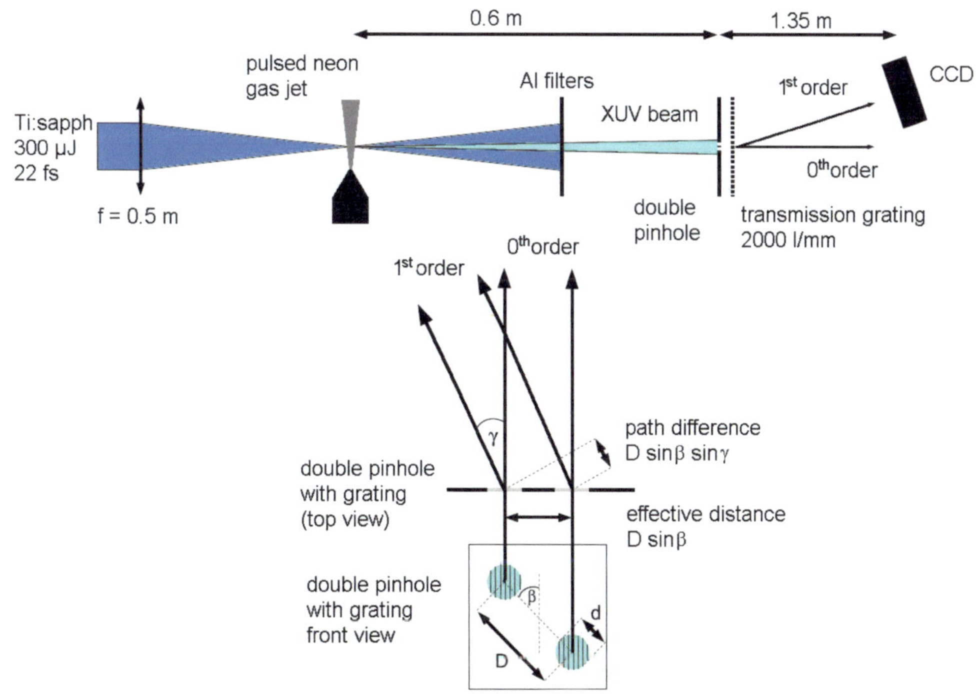

Fig. 21. Experimental set-up of high harmonics generation and the rotatable pinhole interferometer; after (Hemmers & Pretzler, 2009)

This allowed the variation of the path length difference $|\Delta s|$ between zero ($\beta = 0$: pinholes perpendicular to the dispersion direction) and $200 \cdot \lambda = 16{,}7$ fs at $\lambda = 25$ nm ($\beta = \pm\pi/2$) with respect to the given geometrical parameters. When the two diffracted Airy disks overlap partially an interference pattern occurs on the detector for each single harmonic if the light is sufficiently coherent. The visibility V for different delays was then calculated according to equation 4.

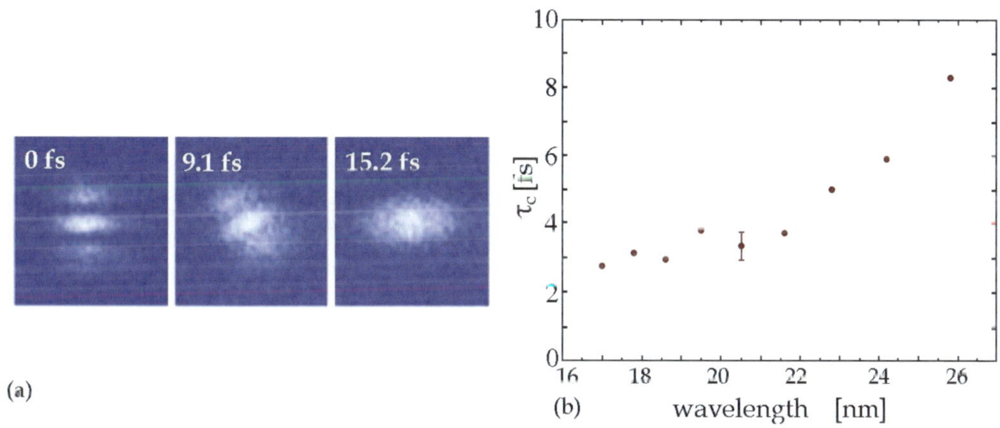

Fig. 22. (a) Interference patterns for different time delays. (b) Coherence times τ_c for the harmonics H17 - H 26; after reference (Hemmers & Pretzler, 2009).

The temporal coherence lengths of the single harmonics (H47 - H31, λ = 17 - 25 nm) generated by a Ti:sapphire laser focused into a neon gas jet were determined. The results are shown in Fig. 22. Interference patterns for different path length differences show that the high visibility that occurs for zero delay degrades for Δs = 2.74 µm and then vanishes at Δs = 4.57 µm (Fig. 22a). An observed decrease of the coherence length from 2.5 µm at λ = 25 nm to 0.8 µm at λ = 17 nm corresponds to coherence times of t_c = 8.3 fs and t_c = 2.7 fs, respectively.

7. Summary

In this article recent developments in research on coherence properties of free electron lasers and high-harmonic sources in the xuv and soft x-ray spectral regime were reviewed. Theoretical results applying 1D- and 3D-numerical simulations for SASE FEL yield good spatial coherence but rather poor temporal coherence which is confirmed by experimental studies. Promising seeding methods for the improvement of the coherence properties and the stability of the output radiation power of free electron lasers were discussed. Several FEL user facilities are nowadays proposed or under construction applying such sophisticated techniques. The first FEL utilizing a HHG scheme, SCSS at Spring-8 in Japan, has succesfully proven this principle in the x-ray spectral regime. A scheme making use of a HHG light source for the seeding of an undulator is now applied at S-FLASH. The outstanding coherence properties of such light sources discussed in this article predestine HHG for the seeding of FEL. Nevertheless FEL based on SASE are established light sources in the XUV and soft x-ray spectral regime. This technique will also be employed for upcoming projects such as the European XFEL. For the latter also a pulse splitting and delay unit (autocorrelator) similar to that one at FLASH presented in this article is now under construction. With this device a measurement of the coherence properties as well as jitter free hard x-ray pump - x-ray probe experiments will be possible.

8. References

Ackermann, W. et al. (2007). Operation of a free-electron laser from the extreme ultraviolet to the water window. *Nature Photonics*, 1, (2007) 336-342

Allaria, E.; Callegari, C.; Cocco, D.; Fawly, W.M.; Kiskinova, M.; Masciovecchio, C. & Parmigiani, F. (2010). The FERMI@Elettra free-electron-laser source for coherent x-ray physics: photon properties, beam transport system and applications. *New Journal of Physics*, 12, (2010) 075022

Altarelli, M.; Brinkmann, R.; Chergui, M.; Decking, W.; Dobson, B.; Düsterer, S.; Grübel, G.; Greaff, G.; Graafsma, H.; Haidu, J.; Marangos, J.; Pflüger, J.; Redlin, H.; Riley, D.; Robinson, I.; Rossbach, J.; Schwarz, A.; Tiedtke, K.; Tschentscher, T.; Vartaniants, I.; Wabnitz, H.; Weise, H.; Wichmann, R.; Witte, K.; Wolf, A.; Wulff, M. & Yurkov, M. (2006). *The European X-Ray Free-Electron Laser Technical Design Report*. DESY XFEL Project Group, European XFEL Project Team, Deutsches Elektronen-Synchrotron, ISBN 978-3-935702-17-1, Hamburg

Bellini, M.; Lyngå, M.; Tozzi, A.; Gaarde, M.B.; Hänsch, T.W.; L'Huillier, A. & C.-G. Wahlström (1998). Temporal Coherence of Ultrashort High-Order Harmonic Pulses. *Physical Review Letters*, 81, (1998) 297-300

Bonifacio, R.; Pellegrini, C. & Narducci, L.M. (1984). Collective instabilities and high-gain regime in a free electron laser. *Optics Communications*, 50, (1984) 373-378

Bonifacio, R.; De Salvo, L.; Pierini, P.; Pivvella, N.; Pellegrini , C. (1994a). A study of linewidth, noise and fluctuations in a FEL operating in SASE. *Nuclear Instruments and methods in Physics Research A,* 341, (1994) 181-185

Bonifacio, R.; De Salvo, L.; Pierini, P.; Pivvella, N.; Pellegrini , C. (1994b). Spektrum, Temporal Structure, and Fluctuations in a High-Gain Free-Electron Laser starting from Noise. *Physical Review Letters,* 73, (1994) 70-73

Chapman, H.N.; Barty, A.; Bogan,M. J.; Boutet, S.; Frank, M.; Hau-Riege, S. P.; Marchesini, S.; Woods, B.W.; Bajt, S.; Benner, W.H.; London, R.A.; Plönjes, E.; Kuhlmann, M.; Treusch, R.; Düsterer, S.; Tschentscher, T.; Schneider, J. R.; Spiller, E.; Möller, T.; Bostedt, C.; Hoener, M.; Shapiro, D.A.; Hodgson, K. O.; van der Spoel, D.; Burmeister, F.; Bergh, M.; Caleman, C.; Huldt, G.; Seibert, M.M.; Maia, F. R. N. C.; Lee, R. W.; Szöke, A.; Timneanu, N. & Hajdu J. (2006). Femtosecond diffractive imaging with a soft-X-ray free-electron laser. *Nature Physics,* 2, (2006) 839 - 843

Corkum P. (1993). Plasma perspective on strong-field multiphoton ionisation. *Physical Review Letters,* 71, (1993) 1994-1997

Deacon, D.A.G.; Elias, L.R.; Madey, J.M.J.; Ramian, G.J.; Schwettman, H.A. & Smith, T.I. (1977). First Operation of a Free-Electron Laser. *Physical Review Letters,* 38, (1977), 892-894

Deutsches Elektronen-Synchrotron Homepage (2010). Record wavelength at FLASH - First lasing below 4.5 nanometres at DESY´s free-electron laser, Hamburg (June 2010), http://zms.desy.de/news/desy_news/2010/flash_record/index_eng.html

Ditmire, T.; Gumbrell, E. T.; Smith, R. A.; Tisch, J. W. G.; Meyerhofer, D. D. & Hutchinson M.H.R. (1996). Spatial Coherence Measurement of Soft X-Ray Radiation Produced by High Order Harmonic Generation. *Physical Review Letters,* 77,(1996) 4756-4758

Eisebitt S.; Lüning, J.; Schlotter, W. F.; Lörgen, M.; Hellwig, O.W.; Eberhardt W. & Stöhr J. (2004). Lensless imaging of magnetic nanostructures by X-ray spectro-holography. *Nature,* 432, (2004) 885-888

Emma, P.; Akre, R.; Arthur, J.; Bionta, R.; Bostedt, C; Bozek, J.; Brachmann, A.; Bucksbaum, P.;Coffee, R.; Decker, F. J.; Ding, Y.; Dowell, D.; Edstrom, S.; Fisher, A., Frisch, J.; Gilevich, S.; Hastings, J.; Hays, G.; Hering, Ph.; Huang, Z. ; Iverson, R.; Loos, H.; Messerschmidt, M.; Miahnahri1, A.; Moeller, S.; Nuhn, H.-D. ; Pile, G.; Ratner, D.; Rzepiela, J.; Schultz, D.; Smith, T.; Stefan, P.; Tompkins, H.; Turner, J.; Welch, J.; White, W.; Wu, J.; Yocky, G. & Galayda, J. (2010). First lasing and operation of an ångstrom-wavelength free-electron laser. *Nature photonics,* 4, (2010) 641-647

Feldhaus, J.; Saldin, E.L.; Schneider, J.R.; Schneidmiller, E.A.; M.V. Yurkov (1997). Possible application of X-ray optical elements for reducing the spectral bandwidth of an X-ray SASE FEL. *Optics Communications,* 140, (1997) 341-352

Feldhaus, J.; Arthur, J & Hastings J. B. (2005). X-ray free electron lasers. *Journal of Physics B: Atomic, Molecular and Optical Physics,* 38, (2005) 799-899

Ferray, M; L'Huillier, A; Li, X; Lompre, L; Mainfray, G & Manus, C. (1988). Multiple-harmonic conversion of 1064-nm radiation in rare-gases, *Journal of Physics B: Atomic, Molecular and Optical Physics,* 21, (1988) L31-L35

Geloni, G.; Saldin, E.; Schneidmiller E.; Yurkov, M. (2008). Transverse coherence properties of X-ray beams in third-generation synchrotron radiation sources. *Nuclear instruments and Methods in Physics Research A: Accelerators, Spectrometers, Detectors and Associated Equipment,* 588, (2008) 463-493

Goodman, J.W. (2000). *Statistical Optics*. Wiley & Sons, ISBN 0 471 39916 7, New York

Haarlammert, T. & Zacharias, H. (2009). Application of high-harmonic radiation in surface science. *Current Opinion in Solid State and Materials Science*, 13, (2009) 13-27

Hemmers, D. & Pretzler, G. (2009). Multi-color XUV Interferometry Using High-Order Harmonics. *Applied Physics B: Lasers and Optics*, 95, (2009) 667-674

Ischebeck, R.; Feldhaus, J.; Gerth, Ch.; Saldin, E.; Schmüser, P.; Schneidmiller, E.; Steeg, B.; Tiedtke, K. ; Tonutti, M.; Treusch, R. & Yurkov M. (2003). Study of the transverse coherence at the TTF free electron laser, *Nuclear Instruments and Methods in Physics Research A*, 507, (2003) 175–180

Kondratenko, A.M. & Saldin, E.L. (1980). Generation of coherent radiation by a relativistic electron beam in an ondulator. *Particle Accelerators*, 10, (1980) 207-216

Kuhlmann, M.; Caliebe, W.; Drube, W. & Schneider, J. R. (2006). *HASYLAB Annual Report*, Part I, (2006) 213

Kulander, K; Schafer, K & Krause, J. (1993). *Super-intense laser-atom physics*. Plenum Press, New York and London

Lambert, G.; Hara, T.; Garzella, D.; Tanikawa, T.; Labat, M.; Carre, B.; Kitamura, H.; Shintake, T.; Bougeard, M.; Inoue, S.; Tanaka, Y.; Salieres P., Merdju, H.; Chubar, O., Gobert, O.; Tahara, K. & Couprie M.-E. (2008). Injection of harmonics generated in gas in a free-electron laser providing intense and coherent extreme-ultraviolet light. *Nature Physics*, 4, (2008) 296-300

Lewenstein, M.; Balcou, P.; Ivanov, M.; L'Huillier, A. & Corkum P. (1994). Theory of high-harmonic generation by low frequency laser fields. *Physical Review A*, 49, (1994), 2117-2132

L'Huillier, A.; Balcou, P.; Candel, S.; Schafer, K.J. & Kulander, K.C. (1992). Calculations of high-order harmonic-generation processes in xenon at 1064 nm. *Physical Review A*, 46, (1992) 2778 - 2793

Li, X; L'Huillier, A.; Ferray, M.; Lompre, L.; Manfray, G. & Manus, C. (1989); Multiple-harmonic generation in rare gases at high laser intensity, *Physical Review A*, 39 (1989) 5751-5761

Lyngå, C.; Gaarde, M.B.; Delfin, C.; Bellini, M.; Hänsch, T.W.; L'Huillier, A. & Wahlström, C.-G. (1999). Temporal coherence of high-order harmonics, *Physical Review A*, 60, (1999) 4823-4830

Mairesse, Y.; de Bohan, A.; Frasinski, L. J.; Merdji, H.; Dinu, L. C.; Monchicourt, P.; Breger, P.; Kovačev, M.; Taïeb, R.; Carré, B.; Muller, H. G.; Agostini, P. & Salières, P. (2003). Attosecond Synchronization of High-Harmonic Soft X-rays, *Science*, 302, (2003) 1540-1543

Madey, J.M.J. (1971). Stimulated Emission of Bremsstrahlung in a Periodic Magnetic Field. Journal of Applied Physics, 42, (1971) 1906-1930

Mandel, L. & Wolf, E. (1995). *Optical coherence and quantum optics,* Cambridge university press, ISBN 0 521 41711 2, Cambridge

McNeil, B.W.J.; Clarke, J.A., Dunning D.J.; Hirst, G.J.; Owen, H.L.; Thompson N.R.; Sheehy, B. & Williams, P.H. (2007). An XUV-FEL amplifier seeded using high harmonic generation. *New Journal of Physics*, 9, (2007) 1367-2630

McNeil, B. (2009); First light from hard X-ray laser, *Nature Photonics*, 3, (2009) 375-377

McPherson, A; Gibson, G; Jara, H; Johann, U; Luk, T; McIntyre, I; Boyer, K. & Rhodes, C.K. (1987). Studies of multiphoton production of vacuum ultraviolet-radiation in rare-gases. *Journal of the Optical Society of America B*, 4, (1987) 595-601

Miltchev, V.; Azima, A.; Bödewadt, J.; Curbis, F.; Drescher, M.; Delsim-Hashemi, H.; Maltezopoulos, T.; Mittenzwey, A.; Rossbach, J.; Tarkeshian, R.; Wieland, M., Düsterer, S.; Feldhaus, J.; Laarmann, T.; Schlarb, H.; Meseck, A.; Khan, S. & Ischebeck, R. (2009). Technical design of the XUV seeding experiment at FLASH. *Proceedings of FEL*, pp. 503-506, Liverpool, UK

Mitzner, R.; Siemer, B.; Neeb, M.; Noll, T.; Siewert, F.; Roling, S.; Rutkowski, M.; Sorokin, A. A.; Richter, M.; Juranic, P.; Tiedtke, K.; Feldhaus, J.; Eberhardt, W. & Zacharias H.(2008), Spatio-temporal coherence of free electron laser pulses in the soft x-ray regime, Optics Express, 16, (2008) 19909-19919

Mitzner, R.; Sorokin, A.A.; Siemer, B.; Roling, S.; Rotkowski, M.; Zacharias, H.; Neeb, M.; Noll, T.; Siewert, F.; Eberhardt, W.; Richter, M.; Juranic, P.; Tiedtke, K. & Feldhaus, J. (2009). Direct autocorrelation of soft-x-ray free-electron-laser pulses by time-resolved two-photon double ionization of He. *Physical Review A*, 80, (2009), 025402

Murphy, J.B. & Pellegrini, C (1985). Generation of high-intensity coherent radiation in the soft x-ray and vacuum-ultraviolet region. *Journal of the Optical Society of America* B, 2, (1985) 259-264

Nisoli, M. & Sansone, G. (2009). New frontiers in attosecond science. *Progress in Quantum Electronics*, 33, (2009) 17-59

Oepts, D.; van der Meer, A.F.G.; van Amersfoort, P.W. (1995). The Free-Electron-Laser user facility FELIX. *Infrared Physics Technology*, 1, (1995) 297-308

Reiche, S. (2006). Transverse coherence properties of the LCLS x-ray beam. *Proceedings of FEL*, pp. 126-129, Bessy, Berlin, Germany

Saldin, E.L.; Schneidmiller, E.A. & Yurkov, M.V. (1999). FAST: a three-dimensional time-dependent FEL simulation code, *Nuclear Instruments and Methods in Physics Research A*, 429, (199), 233-237

Saldin, E.L.; Schneidmiller, E.A. & Yurkov, M.V. (2000a). Optimization of a seeding option for the VUV free electron laser at DESY, *Nuclear Instruments and methods in Physics research A*, 445, (2000) 178-182

Saldin, E.L.; Schneidmiller, E.A. & Yurkov, M.V. (2000b). Diffraction effects in the self-amplified spontaneous emission FEL, *Optics Communications*, 186, (2000) 185-209

Saldin, E.L.; Schneidmiller, E.A. & Yurkov, M.V. (2001). Limitations of the transverse coherence in the self-amplified spontaneous emission FEL, *Nuclear Instruments and Methods in Physics Research A*, 475, (2001) 92-96

Saldin, E.L.; Schneidmiller, E.A. & Yurkov, M.V. (2006). Properties of the third harmonic of the radiation from self-amplified spontaneous emission free electron laser, *Physical review special topics – accelerators and beams*, 9, (2006) 030702

Salières, P.; L'Huillier, A. & Lewenstein M. (1995). Coherence Control of High-Order Harmonics, *Physical Review Letters*, 74, (1995) 3776-3779

Schlotter, W.; Sorgenfrei, F.; Beeck, T.; Beye, M.; Gieschen, S.; Meyer, H., Nagasono, M.; Föhlisch, A. & Wurth, W. (2010). Longitudinal coherene measurements of an extreme-ultraviolet free-electron laser, *Optics Letters*, 35, (2010) 372-374

Sheehy, B.; Clarke, J.A.; Dunning, J.; Thompson N.R. & McNeil B.W.J. (2006). High Harmonic Seeding and the 4GLS XUV-FEL. *Proceedings of FLS*, pp. 36-38, Hamburg, Germany

Singer, A.; Vartanyants, I. A.; Kuhlmann, M.; Duesterer, S.; Treusch, R & Feldhaus, J. (2008). Transverse-Coherence Properties of the Free-Electron-Laser FLASH at DESY. *Physical Review Letters*, 101, (2008) 254801

Vartanyants, I.A. & Singer, A. (2010). Coherence properties of hard x-ray synchrotron sources and x-ray free-electron lasers, *New Journal of Physics* 12, (2010) 035004

Yu, L.H. (1991). Generation of intense UV radiation by subharmonically seeded single-pass free-electron lasers, *Physical Review Letters*, 44, (1991) 5178–5193

Yu, L.H.; Babzien, M.; Ben-Zvi, I.; DiMauro, L. F.; Doyuran, A.; Graves, W.; Johnson, E.; Krinsky, S.; Malone, R.; Pogorelsky, I.; Skaritka, J.; Rakowsky, G.; Solomon, L.; Wang, X. J.; Woodle, M.; Yakimenko, V.; Biedron, S. G.; Galayda, J. N.; Gluskin, E.; Jagger, J. ; Sajaev V. & Vasserman I. (2000). First lasing of a high-gain harmonic generation free-electron laser experiment. *Nuclear Instruments and Methods in Physics Research A*, 445, (2000) 301-306

Yu, L.H.; DiMauro, L.; Doyuran, A.; Grave, W.S.; Johnson, E.D.; Heese, R.; Krinski, S.; Loos, H.; Murphy, J.B.; Rabowsky, G.; Rose, J.; Shaftan, T.; Sheehy, B.; Skaritka, J.; Wang, X.J. & Zu, W. (2003). First Ultraviolet High-Gain Harmonic-Generation Free-Electron Laser. *Physical Review Letters*, 91, (2003) 074801

Zerne, R.; Altucci, C.; Bellini, M.; Gaarde, M. B.; Hänsch, T. W.; L'Huillier, A.; Lyngå, C.; Wahlström, C.-G. (1997). Phase-Locked High-Order Harmonic Sources, *Physical Review Letters*, 79, (1997) 1006-1009

3

Laser Technology for Compact, Narrow-bandwidth Gamma-ray Sources

Miroslav Shverdin, Felicie Albert, David Gibson, Mike Messerly, Fred Hartemann, Craig Siders, Chris Barty
Lawrence Livermore National Lab
USA

1. Introduction

Compton-scattering is a well-known process, observed and described by Arthur H. Compton in 1924, where the energy of an incident photon is modified by an inelastic scatter with matter (Compton, 1923). In 1948, Feenberg and Primakoff presented a theory of Compton backscattering, where photons can gain energy through collisions with energetic electrons (Feenberg & Primakoff, 1948). In 1963, Milburn and Arutyanyan and Tumanyan developed a concept for Compton-scattering light sources based on colliding an accelerated, relativistic electron beam and a laser (Arutyunyan & Tumanyan, 1964; Milburn, 1963).
When an infrared photon scatters off a relativistic electron beam, its wavelength can be Doppler-upshifted to X-rays. Under properly designed conditions, we can generate high brightness, high flux, MeV-scale photons by colliding an intense laser pulse with a high quality, electron beam accelerated in a Linac. Despite being incoherent, the Compton-generated gamma-rays share many of the laser light characteristics: low divergence, high flux, narrow-bandwidth, and polarizability. Traditional laser sources operate in a $0.1-10$ eV range, overlapping most of the molecular and atomic transitions. Transitions inside the nucleus have energies greater than 0.1 MeV. By matching the gamma-ray energy to a particular nuclear transition, we can target a specific isotope, akin to using a laser to excite a particular atomic or molecular transition.
Narrow-bandwidth gamma-ray sources enable high impact technological and scientific missions such as isotope-specific nuclear resonance fluorescence (NRF) (Bertozzi & Ledoux, 2005; Pruet et al., 2006), radiography of low density materials (Albert et al., 2010), precision nuclear spectroscopy (Pietralla et al., 2002), medical imaging and treatment (Carroll et al., 2003; Bech et al., 2009), and tests of quantum chromodynamics (Titov et al., 2006). In traditional X-ray radiography, the target must have a higher atomic number than the surrounding material. Hence, one could conceal an object by shielding with a higher Z-number material. In NRF gamma-ray imaging, the MeV class photons have very long absorption lengths and will transmit through meter lengths of material unless resonantly absorbed by a specific isotope. Some applications of NRF tuned gamma-rays include nuclear waste imaging and assay, monitoring of special nuclear material for homeland security, and tumor detection for medical treatment.
Compton-based sources are attractive in the 100 keV and higher energy regime because they are highly compact and can be more than 15 orders of magnitude brighter than alternative methods for producing photons in this energy regime: Bremsstrahlung radiation

or synchrotron sources. A major challenge for Compton-scattering based gamma-ray generation is its inefficiency, caused by the low Thomson scattering cross-section. In a typical Compton interaction, only 1 in 10^{10} of the incident laser photons is converted to gamma-rays. Producing a high gamma-ray flux requires a highly energetic, intense, short-pulse laser, precisely synchronized to the electron beam with sub-picosecond precision. Production of high brightness gamma-rays also requires a high quality, low emittance, and low energy spread electron beam. The electron beam is typically generated in a Linac, by accelerating electron bunches produced by the photoelectric effect at a photocathode RF gun. For optimum, high charge electron generation, the few picosecond duration photocathode laser must operate above the photocathode material work function (3.7 eV or 265 nm for copper), and have a flattop spatial and temporal shape.

While various Compton-scattering based sources have been in existence since 1970s, they suffered from low brightness, low flux, and wide bandwidth. Recent advances in laser and accelerator technology have enabled production of high-flux, narrow-bandwidth gamma-ray sources with a highly compact footprint machine. For example, at Lawrence Livermore National Lab, we have recently demonstrated a 2nd generation monoenergetic gamma-ray (MEGa-Ray) source termed T-REX (Thomson-Radiated Extreme X-rays) with a record peak brilliance of $1.5\mathrm{x}10^{15}$ photons/mm^2/mrad^2/s/0.1% bandwidth at 478 keV (Albert et al., 2010). In this chapter, we will give a brief overview of Compton-scattering and describe the laser-technology for MEGa-ray sources with emphasis on a recently commissioned T-REX machine, at LLNL (Gibson et al., 2010). We will review basic concepts, such as Chirped Pulse Amplification (CPA), pulse dispersion and compression, and nonlinear frequency conversion in the context of compact Compton sources. We will also describe some of the novel CPA developments such as hyper-dispersion stretching and compression, and narrowband CPA with Nd:YAG amplifiers that have recently been demonstrated in our group (Shverdin, Albert, Anderson, Betts, Gibson, Messerly, Hartemann, Siders & Barty, 2010). We will conclude with a brief overview of areas for future laser research for the continuing improvement of source size, brightness, flux, efficiency, and cost.

2. Overview of Compton scattering

2.1 Basic properties

Compton scattering sources, which have been widely studied over the past decades (Esarey et al., 1993; Hartemann & Kerman, 1996; Leemans et al., 1997), rely on energy-momentum conservation, before and after scattering. The energy of the scattered photons, E_x, depends on several electron and laser beam parameters:

$$E_x = \frac{\gamma - \sqrt{\gamma^2-1}\cos\phi}{\gamma - \sqrt{\gamma^2-1}\cos\theta + \bar{\lambda}k_0(1-\cos\theta\cos\phi + \cos\psi\sin\theta\sin\phi)} E_L \tag{1}$$

where γ is the electron relativistic factor, ϕ is the angle between the incident laser and electron beams, θ is the angle between the scattered photon and incident electron, ψ the angle between the incident and scattered photon, $k_0 = 2\pi/\bar{\lambda}_c$ is the laser wavenumber (reduced Compton wavelength $\bar{\lambda}_c = 3.8616 \times 10^{-13}$ m), and finally E_L is the laser energy. Here, we assume that $\beta = v/c \simeq 1$, where v is the electron velocity. For a head-on collision ($\phi = 180°$) and on-axis observation in the plane defined by the incident electron and laser beams ($\theta = 0°$ and $\psi = 0$), the scattered energy scales as $4\gamma^2 E_L$. In our experiments, where $\gamma \simeq 200$, $k_0 \simeq 10^7$, we can

ignore the electron recoil term, $4\gamma k_0 \bar{\lambda}_c$. One can obtain high-energy (MeV) scattered photons with relatively modest electron beam energies in a compact footprint machine.
When the laser and the electron focal spots, w_0, are matched, the number of the generated γ-rays is approximately $N_x = (\sigma/\pi w_0^2) N_L N_e$, where N_L and N_e are respectively the number of laser photons and electrons at the interaction point, and $\sigma = 6.65 \times 10^{-25}$ cm^2 is the Thomson cross-section. In a linear accelerator, the electron beam focal spot size scales with electron energy as $1/\gamma$ and the γ-ray yield scales as γ^2. Compton scattering sources become more efficient at higher energies and produce a highly collimated beam with good energy-angle correlation.

2.2 Modelling

To properly model the Compton scattering spectrum and the broadening effects, we need to consider electrons with a particular phase space distribution interacting with a Gaussian-paraxial electromagnetic wave. For simplification, we can neglect laser's wavefront curvature for interaction geometries with a slow focus ($f/\# > 10$). For a detailed description of our formalism, see (Hartemann, 2002). We first calculate the Compton scattering frequency using the energy-momentum conservation law:

$$\kappa - \lambda = \bar{\lambda}_c (k_\mu q^\mu), \tag{2}$$

where κ and λ are the incident and scattered light cone variables, $k_\mu = (k, 0, 0, k)$ is the incident laser pulse 4 wavevector, and q_μ is the scattered 4-wavevector. Solving this equation for q, we obtain q_c, the wavenumber for the scattered radiation in spherical coordinates (θ, ϕ):

$$q_c = \frac{k(\gamma - u_z)}{\gamma - u_z \cos\theta + k\bar{\lambda}_c (1 - u_x \sin\theta \cos\phi - u_y \sin\theta \sin\phi - u_z \cos\theta)}, \tag{3}$$

where $u_\mu = (\gamma, u_x, u_y, u_z)$ is the electron 4-velocity. For a counter-propagating geometry, $\theta = \pi$ and $\phi = 0$. Next, we generate a random normal distribution of particles with velocities u_x, u_y, relativistic factor γ and with respective standard deviations $\Delta u_x = j\epsilon_x/\sigma_x$, $\Delta u_y = j\epsilon_y/\sigma_y$ and Δ_γ. The quantities ϵ and υ refer to the electron beam normalized emittance and spot size. j is the jitter: e. g. $j = 1$ in the absence of jitter. Electron beam jitter and emittance are the biggest contributors to the spectral broadening. An example of a normalized spectrum calculated with the method outlined above is shown in Fig. 1. An experimental spectrum from our T-REX source, measured with a germanium detector collecting photons scattered at 48 degrees from an aluminium plate is shown in Fig. 2 (Albert et al., 2010). The γ-ray beam profile is shown in Fig. 2 inset.
From Fig. 2, the spectrum has several distinctive features. The tail after 400 keV is mainly due to the high energy Bremsstrahlung and to pile-up (multiple-photon events) in the detector. The main peak has a maximum at 365 keV, which corresponds to an incident energy of 478 keV. The full width at half maximum of this peak is 55$\pm$5 keV, which corresponds to a relative bandwidth of 15%. The bandwidth of the γ-ray spectrum increases as the square of the normalized emittance - critical for the design of a narrow bandwidth Compton scattering source. The energy spread and jitter of the laser are negligible, while those of the electron beam are 0.2% and $j = 2$, respectively. From Eq. 1, the γ-ray spectrum is broadened by 0.4% by the electron beam energy spread. The electron beam divergence, $\epsilon/\gamma r$, where r is the electron beam radius, broadens the γ-ray spectrum by 4.5% At energies below 250 keV, we record a broad continuum with two additonal peaks at 80 keV and 110 keV. These features are artifacts of several processes during beam detection and can be reproduced by a

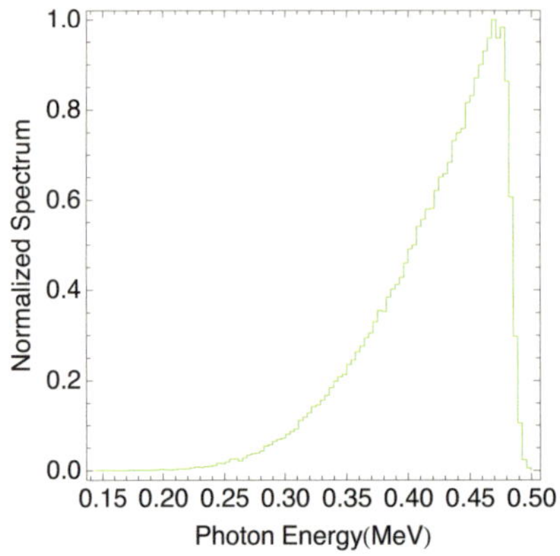

Fig. 1. Example of a spectrum simulated with Mathematica, using 10^5 particles and 100 bins for an electron beam energy of 116 MeV and a laser wavelength of 532 nm. The other parameters are $j = 2$, $\epsilon_x = 5$ mm mrad, $\epsilon_y = 6$ mm mrad, $\sigma_x = 35$ μm and $\sigma_y = 40$ μm.

Monte-Carlo simulation. We utilized MCNP5 code (Forster et al., 2004), with modifications to include Compton scattering of linearly polarized photons (G.Matt et al., 1996) to describe the realistic experimental set-up. Fig. 2 shows the simulated pulse height spectrum expected for single-photon counting. The continuum below 250 keV is caused by incomplete energy absorption (elastic Compton scattering) in the detector itself. The broad peak at 110 keV arises from double Compton scattering off the Al plate and the adjacent wall, followed by photoabsorption in the Ge Detector. Since the detector is shielded with lead, the first peak is due to X-rays coming from the lead K_α and K_β lines, respectively, at 72.8 keV, 75 keV, 84.9 keV and 87.3 keV.

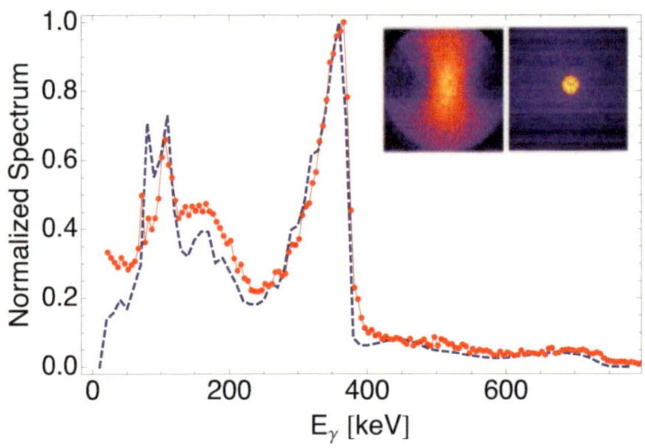

Fig. 2. On axis spectrum recorded after scattering off the Al plate and corresponding Monte Carlo simulation. The images correspond to the full beam and the signal transmitted through the collimator, respectively.

3. Laser technology

3.1 Laser system overview

We describe laser technology underlying high brightness, Compton-scattering based monoenergetic gamma-ray sources, focusing on a recently commissioned T-REX (Gibson et al., 2010) and a 3rd generation Velociraptor, currently under construction at Lawrence

Livermore National Lab. The two main systems of these machines consist of a low emittance, low energy spread, high charge electron beam and a high intensity, narrow bandwidth, counter-propagating laser focused into the interaction region to scatter off the accelerated electrons. A simplified schematic of the Compton-scattering source is shown in Fig. 3.

A state of the art ultrashort laser facilitates generation of a high charge, low emittance electron beam. A photogun laser delivers spatially and temporally shaped UV pulses to RF photocathode to generate electrons with a desired phase-space distribution by the photoelectric effect. Precisely synchronized to the RF phase of the linac, the generated electrons are then accelerated to relativistic velocities. The arrival of the electrons at the interaction point is timed to the arrival of the interaction laser photons. A common oscillator, operating at 40 MHz repetition rate, serves as a reference clock for the GHz-scale RF system of the Linac, seeds both laser systems, and facilitates subpicosecond synchronization between the interaction laser, the photogun laser, and the RF Linac phase. Typical parameters for the laser system are listed in Table 1.

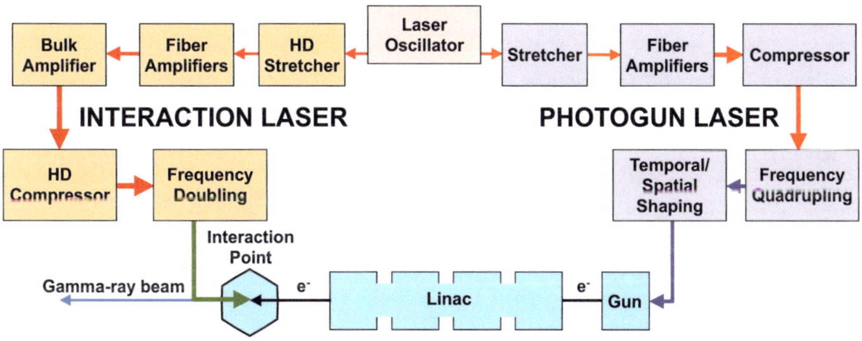

Fig. 3. Block diagram of the Velociraptor compton source with details of the laser systems.

Parameters	Oscillator	Photogun Laser	Interaction Laser
Repetition Rate	40.8 MHz	10-120 Hz	10-120 Hz
Wavelength	1 μm	263 nm	532 nm or 1 μm
Energy	100 pJ	30-50 μJ	150 mJ - 800 mJ
Pulse Duration	1 ps	2 or 15 ps	10 ps
Spot size on target	n/a	1-2 mm dia	20-40 μm RMS

Table 1. Summary of key laser parameters for T-REX and Velociraptor compton source

Because the oscillator produces ultrashort pulses with energies in the pico-Joule to nano-Joule range, and the final required energies are in the milli-joule to Joule range, pulses are amplified by 70-100 dB in a series of amplifiers. Chirped pulse amplification (CPA) enables generation of highly energetic picosecond and femtosecond pulses (Strickland & Mourou, 1985). The key concept behind CPA is to increase the pulse duration prior to amplification, thereby reducing the peak intensity during the amplification process. The peak intensity of the pulse determines the onset of various nonlinear processes, such as self-focusing, self-phase modulation, multi-photon ionization that lead to pulse brake-up and damage to the medium. Nonlinear phase accumulation, or B-integral, is guided by Eq. 4:

$$\phi_{NL} = \frac{2\pi}{\lambda} \int_{-\infty}^{\infty} n_2 I(z) dz \quad (4)$$

where, $I(z)$ is a position dependent pulse intensity and n_2 is the material dependent nonlinear refractive index ($n = n_0 + n_2 I$). ϕ_{NL} should normally be kept below 2-3 to avoid significant pulse and beam distortion. For a desired final pulse intensity, we minimize the value of the accumulated B-integral by increasing the beam diameter and the chirped pulse duration. After amplification, the stretched pulse is recompressed to its near transform limited duration. The four stages of the CPA process are illustrated in Fig. 4. The stretcher and the compressor impart the largest pulse dispersion. In our laser systems, the chirped pulse duration is between 1.5 and 3 ns to maximize the final pulse energy.

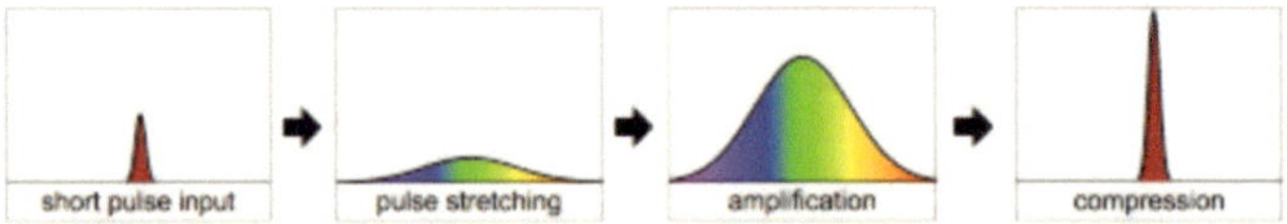

Fig. 4. Basic CPA scheme: ultrashort seed pulse is stretched, amplified, and recompressed.

The photogun laser consists of a mode-locked and phase-locked Yb:doped fiber oscillator, an all-reflective pulse stretcher, and a series of fiber amplifiers producing over 1 mJ/pulse at 10 kHz at the output. The pulses are then re-compressed to near transform limited (200 fs) duration in a compact multi-layer-dielectric grating compressor. Next, two nonlinear crystals generate the 4th harmonic of the fundamental frequency in the UV. Temporally and spatially shaped pulses are finally delivered at the RF photogun.
The interaction laser system (ILS) parameters are optimized to generate high flux, narrowband γ rays. The interaction laser produces joule-class, $\approx$ 10 ps pulses at 1064 nm, at up to 120 Hz repetition rate, which are then frequency doubled to 532 nm. Separate fiber amplifiers pre-amplify the pulses prior to injection into a bulk power amplifier chain. The pulses are then recompressed, frequency doubled, and focused into the interaction region to scatter off the electrons.
Picosecond pulse duration is desirable to both minimize peak laser intensity, which leads to nonlinear scattering effects, and decrease the laser linewidth, $\Delta\omega_{laser}$, which broadens the generated γ-ray bandwidth, $\Delta\omega_\gamma$, as $\Delta\omega_\gamma = \sqrt{\Delta\omega_{res}^2 + \Delta\omega_{laser}^2}$. Here, $\Delta\omega_{res}$ is the contribution to γ-ray bandwidth from all other effects, such as e$^-$ beam emittance, energy spread, focusing geometry, etc. Due to narrow laser pulse bandwidths, we utilize a novel type of high dispersion pulse stretcher and pulse compressor. The ILS pulses are finally frequency doubled to increase the generated γ-ray energy.
Several alternate types of Compton-scattering γ-ray sources, not covered here, exist (see (D'Angelo et al., 2000)). Some utilize different laser technologies such as Q-switched, nanosecond duration interaction laser pulses (Kawase et al., 2008). Other sources are based on Free Electron Laser (Litvinenko et al., 1997) or CO_2 technologies (Yakimenko & Pogorelsky, 2006). Several sources based on low charge, high repetition rate electron bunches and low energy, high repetition rate laser pulses coupled to high finesse resonant cavities have been proposed (Graves et al., 2009) and demonstrated (Bech et al., 2009).

3.1.1 Dispersion management

The action of various dispersive elements is best described in the spectral domain. Given a time dependent electric field, $E(t) = A(t)e^{i\omega_0 t}$, where we factor out the carrier-frequency term, its frequency domain is the Fourier transform, $\mathcal{A}(\omega) = \int_{-\infty}^{\infty} A(t)e^{-i\omega t}dt = \sqrt{I_s(\omega)}e^{i\phi(\omega)}$. Here, we explicitly separate the real spectral amplitude, $\sqrt{I_s(\omega)}$, and phase, $\phi(\omega)$. We define spectral intensity $I_s(\omega) \equiv |\mathcal{A}(\omega)|^2$, and temporal intensity $I(t) \equiv |A(t)|^2$ We can Taylor expand the phase of the pulse envelope $\mathcal{A}(\omega)$ as

$$\phi(\omega) = \sum_{n=0}^{\infty} \frac{1}{n!}\phi^{(n)}(0)\omega^n \tag{5}$$

where $\phi^{(n)}(0) \equiv \frac{d^n\phi(\omega)}{d\phi^n}$ evaluated at $\omega = 0$. Term n=1 is the group delay, corresponding to the time shift of the pulse; terms $n \geq 2$ are responsible for pulse dispersion. Terms $\phi^{(n)}(0)$, where n=2 and 3, are defined as, respectively, group delay dispersion (GDD) and third order dispersion (TOD). For a gaussian, unchirped, pulse, $E(t) = E_0 e^{-\frac{t^2}{\tau_0^2}}$, we can analytically calculate its chirped duration, τ_f assuming a purely quadratic dispersion ($\phi^{(n)}(0) = 0$ for $n > 2$).

$$\tau_f = \tau_0\sqrt{1 + 16\frac{GDD^2}{\tau_0^4}} \tag{6}$$

The transform limited pulse duration is inversely proportional to the pulse bandwidth. For a transform limited gaussian, the time-bandwidth product (FWHM intensity duration [sec] x FWHM spectral intensity bandwidth [Hz]) is $\frac{2}{\pi}\log 2 \approx 0.44$. From Eq. (6), for large stretch ratios, the amount of chirp, GDD, needed to stretch from the transform limit, τ_0, to the final duration, τ_f, is proportional to τ_0.

In our CPA system, near transform limited pulse are chirped from 200 fs (photogun laser) and 10 ps (interaction laser) to few nanosecond durations. Options for such large dispersion stretchers include chirped fiber bragg gratings (CFBG), chirped volume Bragg gratings (CVBG), and bulk grating based stretchers. Prism based stretchers do not provide sufficient dispersion for our bandwidths. The main attraction of CFBG and CVBG is their extremely compact size and alignment insensitivity. Both CFBG and CVBG have shown great recent promise but still have unresolved issues relating to group delay ripple that affect recompressed pulse fidelity (Sumetsky et al., 2002). High pulse contrast systems typically utilize reflective grating stretchers and compressors which provide a smooth dispersion profile. The grating stretcher and compressor designs are driven by the grating equation which relates the angle of incidence ψ and the angle of diffraction ϕ measured with respect to grating normal for a ray at wavelength λ,

$$\sin(\psi) + \sin(\phi) = \frac{m\lambda}{d} \tag{7}$$

where d is the groove spacing and m is an integer specifying the diffraction order. Grating stretchers and compressors achieve large optical path differences versus wavelength due to their high angular resolution, $\frac{d\phi}{d\lambda}$ (Treacy, 1969; Martinez, 1987).

A stretcher imparts a positive pulse chirp (longer wavelengths lead the shorter wavelengths in time) and a compressor imparts a negative pulse chirp. The sign of the chirp is important, because other materials in the system, such as transport fibers, lenses, and amplifiers, have positive dispersion. A pulse with a negative initial chirp could become partially recompressed during amplification and damage the gain medium.

We need to dispersion balance the CPA chain, to recompress the pulse to its transform limit. The total group delay, GD_{total} vs wavelength for the system must be constant at the output, or $GD_{total}(\lambda) = GD_{stretcher}(\lambda) + GD_{compressor}(\lambda) + GD_{fiber}(\lambda) + GD_{oscillator}(\lambda) + GD_{amplifier}(\lambda) + GD_{material}(\lambda) = C_1$, where C_1 is an arbitrary constant. From Eq. (5), $\phi(\omega) = \int_{-\infty}^{\infty} GD(\omega)d\omega$. Any frequency dependence in the total group delay will degrade pulse fidelity.

Several approaches to achieving dispersion balance exist. In the Taylor's expansion (Eq. (5)) of the phase of a well behaved dispersive element, when $n < m$ the contribution of a term n to the total phase is much larger than of a term m, or $|\phi^{(n)}(0)\Delta\omega^n/n!| \gg |\phi^{(m)}(0)\Delta\omega^m/m!|$, where $\Delta\omega$ is the pulse bandwidth. One approach to achieve dispersion balance is to calculate the total GDD, TOD, and higher order terms for the system and attempt to zero them. For example, when terms up to the 4th order are zero, the system is quintic limited. This term-by-term cancellation approach can become problematic when successive terms in the Taylor's expansion do not decrease rapidly. It is often preferable to minimize the residual group delay, GD_{RMS}, over the pulse bandwidth (Eq. 8).

$$GD_{RMS} = \sqrt{\frac{\int_{-\infty}^{\infty} \left(GD(\omega) - \bar{GD}\right)^2 |\mathcal{E}(\omega)|^2 d\omega}{\int_{\infty}^{\infty} |\mathcal{E}(\omega)|^2 d\omega}} \tag{8}$$

where $\bar{GD}$ is the mean group delay, $\bar{GD} \equiv \frac{\int_{-\infty}^{\infty} GD(\omega)|\mathcal{E}(\omega)|^2 d\omega}{\int_{-\infty}^{\infty} |\mathcal{E}(\omega)|^2 d\omega}$.

A sample pulse stretcher is shown in Fig. 5. The stretcher consists of a single grating, a lens of focal length f, a retro-reflecting folding mirror placed f away from the lens and a vertical roof mirror. The folding mirror simplifies stretcher configuration and alignment by eliminating a second grating and lens pair. A vertical roof mirror double passes the beam through the stretcher and takes out the spatial chirp. The total dispersion of the stretcher is determined by the distance from the grating to the lens, L_f. When $L_f = f$, the path lengths at all wavelengths are equal and the net dispersion is zero. When $L_f > f$, the chirp becomes negative, same as in the compressor. In a stretcher, $L_f < f$ producing a positive chirp. We can calculate the dispersion, or $GD(\omega) = n_\omega L(\omega)/c$, where $L(\omega)$ is the frequency dependent propagation distance and n_ω is the frequency dependent refractive index, using various techniques. Here, we give a compact equation for $GDD(\omega) = \partial GD(\omega)/\partial\omega$, from which all other dispersion terms can be determined. Assuming an aberration free stretcher in the thin lens approximate, $GDD(\omega)$ is then given by Eq. 9.

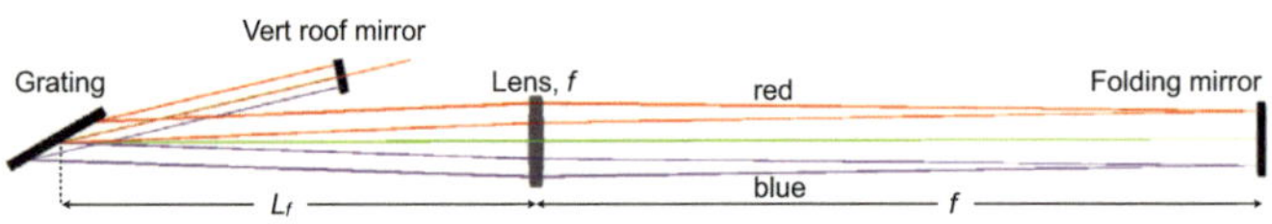

Fig. 5. Grating and lens based stretcher is folded with a flat mirror. The distance from the grating to the focal length of the lens determine the sign and magnitude of the chirp.

$$GDD(\omega) = 2(f - L_f)\frac{\lambda}{\pi c^2}\frac{[\sin(\phi) + \sin(\psi)]^2}{\cos^2(\phi)} \tag{9}$$

where the diffraction angle, ϕ is a function of ω. The aberration free approximation is important, because a real lens introduces both chromatic and geometric beam aberrations which modify higher order dispersion terms from those derived from Eq. 9.

A compressor, which is typically a final component in the CPA system undoes all of the accumulated dispersion. The compressor used on the photogun laser is shown in Fig. 6. Compared to the stretcher, the lens is absent. The magnitude of the negative chirp is now controlled by the distance from the horizontal roof to the grating. The horizontal roof, here folds the compressor geometry and eliminates a second grating. As in the stretcher of Fig. 5, the compressor's vertical roof mirror double passes the beam and removes the spatial chirp. Because a compressor has no curved optics, it introduces no geometric or chromatic aberrations. The dispersion of a real compressor is very precisely described by the Treacy formula or its equivalent given by Eq. 10.

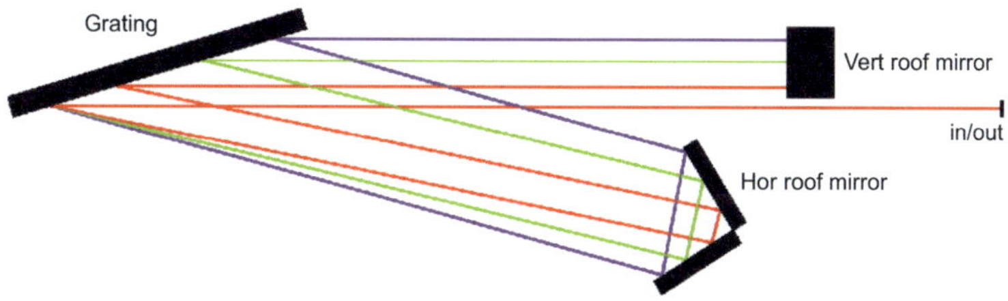

Fig. 6. Folded grating compressor. The distance from the horizontal roof mirror to the grating determines the magnitude of the negative chirp.

$$GDD(\omega) = -L_1 \frac{\lambda}{\pi c^2} \frac{\left[\sin(\phi) + \sin(\psi)\right]^2}{\cos^2(\phi)} \tag{10}$$

The actual stretcher design for Velociraptor photogun laser is shown in Fig. 7. The all-reflective Offner design (Cheriaux et al., 1996) uses a convex and a concave mirror to form an imaging telescope with magnification of -1 and relay planes at the radius of curvature (ROC) center of the concave mirror. The grating is placed inside the ROC to impart a positive chirp. A vertical roof mirror folds the stretcher geometry and eliminates the second grating. The Offner stretcher is compact and has low chromatic and geometric aberrations. Consequently, its dispersion profile is nearly aberration free, meaning the *GDD* and higher order dispersion terms are closely predicted by Eq. 9.

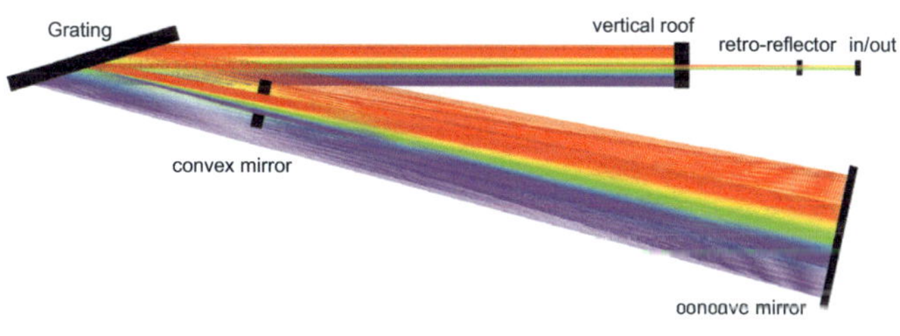

Fig. 7. Raytraced design of the Offner stretcher for the photogun laser on Velociraptor.

3.2 Fiber oscillator and amplifiers

Essential to building a compact, highly stable laser system is a fiber front-end. Fibers lasers are highly portable, reliable, relatively insensitive to external perturbations, provide long term hands free operation, and have been scaled to average powers above 1 kW in a diffraction limited beam (Jeong et al., 2004). When the required pulse energy is above a mJ, a fiber oscillator and amplifiers can serve as a front end in a multi-amplifier-stage system. We employ a 10 mW Yb:doped mode-locked oscillator which, when compressed, produces 250 pJ, sub 100 fs, near transform limited pulses at 40.8 MHz repetition rate with a full bandwidth from 1035 nm to 1085 nm. The oscillator, based on a design developed at Cornell (Ilday et al., 2003), fits on a rack mounted, 12"x12" breadboard (see Fig. 8).

On T-REX and Velociraptor, a single fiber oscillator seeds both, the photogun laser and the interaction laser systems. An experimental oscillator spectrum showing the bandpass for the photogun and the interaction lasers is shown in Fig. 9. Two different chains of fiber Yb:doped fiber amplifiers tuned for peak gain at 1053 nm for PDL and 1064 nm for ILS, boost the seed pulse to $\approx$100 μJ/pulse. The seed is chirped to a few nanosecond duration prior to amplification. Each fiber amplifier stage provides 20 dB of gain. The repetition rate is gradually stepped down to 10 kHz with acousto-optic modulators (AOM), inserted in between fiber amplifiers. The AOMs also remove out of band amplified spontaneous emission (ASE), preserving pulse fidelity.

The total energy from the fiber amplifier is limited by the total stored energy and by the accumulated nonlinear phase (Eq. 4). Several initial amplifier stages consist of telecom-type, 6.6 μm core polarization maintaining (PM) fibers. The output from this core sized fiber is limited to $\approx 1\mu$J. Next, a series of large mode field diameter, 29 μm core photonic band-gap (PBG) fibers accommodate pulse energies up to 100 μJ. We are currently developing even larger core (80 μm) PBG fiber rod based amplifiers to attain over 1 mJ per pulse at the output. A major challenge with PBG and other large core fibers is careful control of the refractive index uniformity to prevent generation of higher order spatial modes, which degrade beam quality.

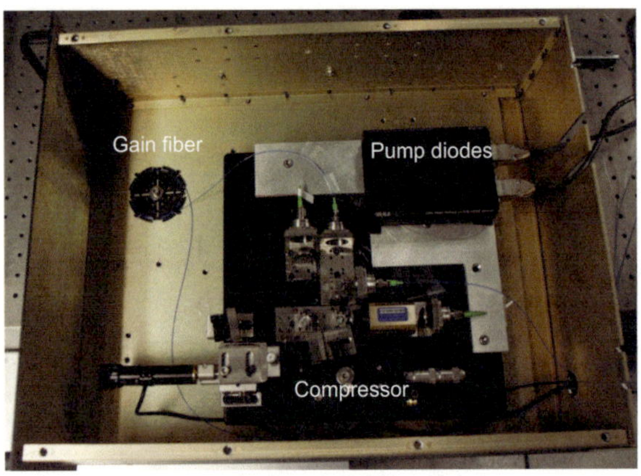

Fig. 8. The ultrashort Yb^+ doped fiber oscillator fits on a 12"x12" breadboard. When recompressed, the anomalous dispersion oscillator generates 100 fs pulses.

4. Amplification

The action of the amplifier is illustrated in Fig. 10. An input pulse with intensity I_{in} is amplified to output intensity I_{out}. An amplifier can be chacterized by its small signal (undepleted) gain, G_0, saturation fluence, J_{sat}, and its frequency dependent lineshape function, $g(\nu)$. In general, we need to solve the laser rate equations to determine the output pulse shape, spectral distribution, and energy. For a homogeneous gain medium, when (1) the temporal pulse variation and the population inversion is sufficiently small (compared to T_2 coherence time) to justify rate equation analysis, and (2) transient effects relating to spontaneous emission and pumping occur on a time scale much longer than the pulse

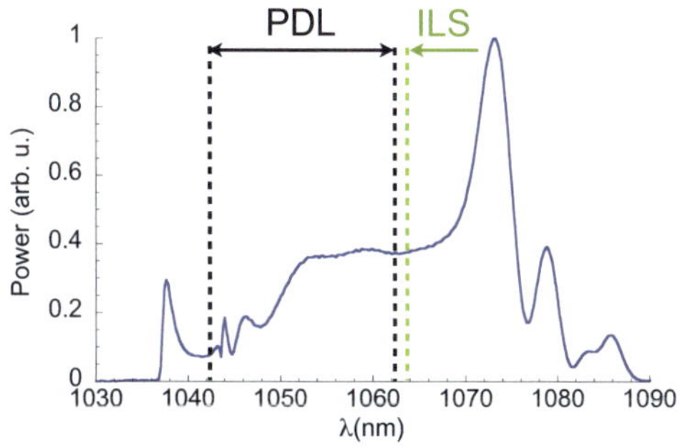

Fig. 9. Experimental pulse spectrum of the Velociraptor oscillator of Fig. 8. The same oscillator seeds both the interaction and the photogun lasers.

duration, an analytical solution can be obtained using a Frantz-Nodvik approach (Frantz & Nodvik, 1963). Eq. (11) is derived for a monochromatic input pulse (Planchon et al., 2005).

$$I_{out}(t) = I_{in}(t) \times \left[1 - \left(1 - G_0^{-1}\right) e^{-J_{in}(t)/J_{sat}(t)}\right]^{-1} \tag{11}$$

Here, $J_{in}(t) \equiv \int_0^t I_{in}(t')dt'$ is the instantaneous fluence. Eq. (11) can be modified to describe amplification of a chirped pulse. This involves determining the frequency dependence of the small gain, $G_0(\nu)$ and the saturation fluence, $J_{sat}(\nu)$, both of which depend on the emission cross-section $\sigma_e(\nu)$. For a four-level system, $J_{sat}(\nu) = h\nu/\sigma_e(\nu)$, and $G_0 = \exp\left[N\sigma_e(\nu)\right]$, where N is the population inversion. If the instantaneous pulse frequency, $\bar{\nu} = \frac{1}{2\pi}\frac{d\phi(t)}{dt}$ is a monotonic function of t, we can define time as a function of frequency, $t(\nu)$. In this instantaneous frequency approach, we can rewrite Eq. (11) as a function of ν. Various aspects of chirped pulse amplification, such as total output energy, gain narrowing, square pulse distortion, and spectral sculpting can be analytically calculated using the modified form of Eq. (11).

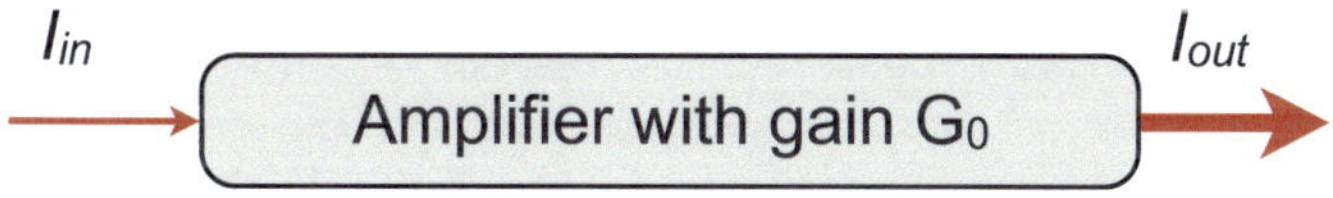

Fig. 10. Basic scheme for pulse amplification. A gain medium with stored energy coherently amplifies the input pulse.

4.1 Chirped pulse amplification with narrowband pulses

In CPA, the gain bandwidth of the amplifying medium is typically broad enough to minimize gain narrowing of the seed pulse. On the photogun laser, Yb doped glass has a wide bandwidth spanning from 1000 nm to 1150 nm, and is well suited for amplifying 100 fs pulses. On the interaction laser, however, bulk amplification is in Nd:YAG, which has a narrow bandwidth of 120 GHz. The seed pulses gain narrow to sub nanometer bandwidths after amplification. Traditional two grating stretchers and compressors cannot provide adequate dispersion in a table-top footprint. In this section, we will describe novel hyper-dispersion technology that we developed for CPA with sub-nanometer bandwidth pulses (Shverdin, Albert, Anderson, Betts, Gibson, Messerly, Hartemann, Siders & Barty, 2010). The meter-scale stretcher and compressor pair achieve 10x greater dispersion compared to standard two-grating designs. Previously, D. Fittinghoff, *et. al.* suggested hyper-dispersion

compressor arrangements (Fittinghoff et al., 2004). F. J. Duarte described a conceptually similar hyper-dispersion arrangements for a prism-based compressor (Duarte, 1987).
We utilize commercial Nd:YAG amplifiers for two reasons: (1) Nd:YAG technology is extremely mature, relatively inexpensive, and provides high signal gain; (2) nominally 10 ps transform limited laser pulses are well-suited for narrowband γ-ray generation. Employing a hyper-dispersion stretcher and compressor pair, we generated 750 mJ pulses at 1064 nm with 0.2 nm bandwidth, compressed to near transform limited duration of 8 ps.
The nearly unfolded version of the hyper-dispersion compressor is shown in Fig. 11, with a retro-mirror replacing gratings 5-8. Compared to standard Treacy design, this compressor contains two additional gratings (G2 and G3). The orientation of G2 is anti-parallel to G1: the rays dispersed at G1, are further dispersed at G2. This anti-parallel arrangement enables angular dispersion, $\frac{d\theta}{d\lambda}$, which is greater than possible with a single grating. The orientation of G3 and G4 is parallel to, respectively, G2 and G1. G3 undoes the angular dispersion of G2 and G4 undoes the angular dispersion of G1 producing a collimated, spatially chirped beam at the retro-mirror. After retro-reflection, the spatial chirp is removed after four more grating reflections. The number of grating reflections (8), is twice that in a Treacy compressor. High diffraction efficiency gratings are essential for high throughput efficiency. We utilize multi-layer-dielectric (MLD) gratings developed at LLNL with achievable diffraction efficiency >99% (Perry et al., 1995). The magnitude of the negative chirp is controlled by varying L_1, the optical path length of the central ray between G1 and G2 and L_2, the optical path length between G2 and G3.
We derive an analytical formula (Eq. 12) for group delay dispersion (GDD) as a function of wavelength for the compressor using Kostenbauder formalism (Kostenbauder, 1990; Lin et al., 1993).

$$GDD = -\frac{\lambda}{\pi c^2}\frac{[\sin(\phi)+\sin(\psi)]^2}{\cos^4(\phi)} \times \left[2L_1\cos^2(\phi) + L_2[\cos(\phi)+\cos(\psi)]^2\right] \qquad (12)$$

Here ψ is the angle of incidence and ϕ is the angle of diffraction of the central ray of wavelength λ at the first grating measured with respect to grating normal. We assume that the groove density of gratings 1 and 2 is the same. Note that the expression for GDD reduces to that of the standard two-grating compressor when L_2=0. Higher order dispersion terms can be derived by noting that ϕ is a function of wavelength.

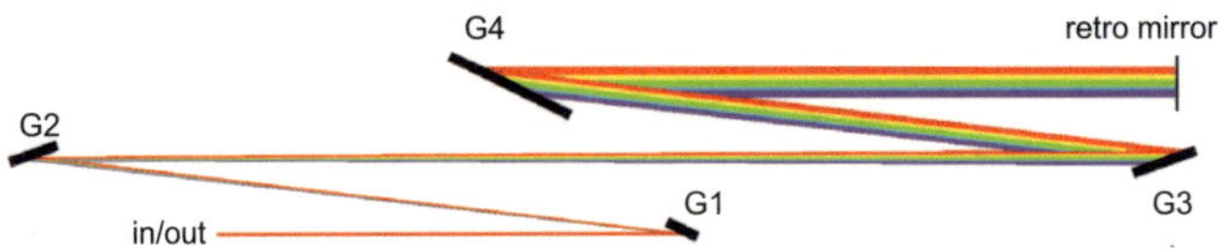

Fig. 11. Unfolded version of the hyper-dispersion compressor with anti-parallel gratings.

Compressor design can be folded to reduce the total number of gratings and simplify compressor alignment. The compressor shown in Fig. 12 has been designed for Velociraptor and is similar to the experimental design on T-REX. The compressor consists of two 40x20 cm multi-layer dielectric (MLD) gratings arranged anti-parallel to each other, a vertical roof mirror (RM), and a series of two periscopes and a horizontal roof mirror. These six mirrors set the height and the position of the reflected beam on the gratings and inverts the beam in the plane of diffraction. The beam is incident at Littrow-3 degrees (64.8^o) on the 1740 grooves/mm gratings. The vertical roof mirror here, is equivalent to the retro-mirror in Fig. 11. The beam undergoes a total of 8 grating reflections and 16 mirror reflections in the compressor. The total beam path of the central ray is 22 m. On T-REX, the MLD gratings had a diffraction efficiency above 97%, enabling an overall compressor throughput efficiency of 60%. Here, the magnitude of the chirp is tuned by translating the horizontal roof mirror. The compressor, with its relatively compact footprint of 3x0.7 m provides GDD=-4300 ps^2/rad, or a pulse dispersion of 7000 ps/nm. This chirps the incident 0.2 nm bandwidth pulse to 3 ns. The two grating separation in a standard Treacy compressor with the same dispersion would be 32 m.

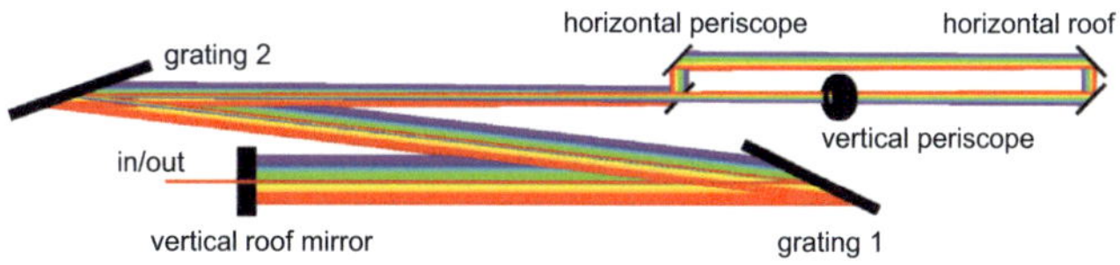

Fig. 12. The compact hyper-dispersion compressor consisting of two 1740 grooves/mm gratings has a footprint of 3 x 0.7 m and group delay dispersion of -4300 ps^2/rad.

The hyper-dispersion stretcher design is conceptually similar. The unfolded version is shown in Fig. 13. Compared to the standard Martinez stretcher, the hyper-dispersion design contain two extra mirrors G1 and G4, arranged anti-parallel to G2 and G3. We define L_f as the path length from G2 to the first lens. Then the total ray path length from G1 to the first lens $(L_1 + L_f)$ must be smaller than the lens focal length, f to produce a positive chirp. The magnitude of the chirp is controlled by varying the value of $L_1 + L_f - f$.
We modify Eq. 12 to derive the GDD formula for the aberration free hyper-dispersion stretcher shown in Fig. 13:

$$GDD = -\frac{\lambda}{\pi c^2}\frac{[\sin(\phi)+\sin(\psi)]^2}{\cos^4(\phi)} \times \left[L_1\cos^2(\phi) + (L_f - f)\left[\cos(\phi)+\cos(\psi)\right]^2\right] \tag{13}$$

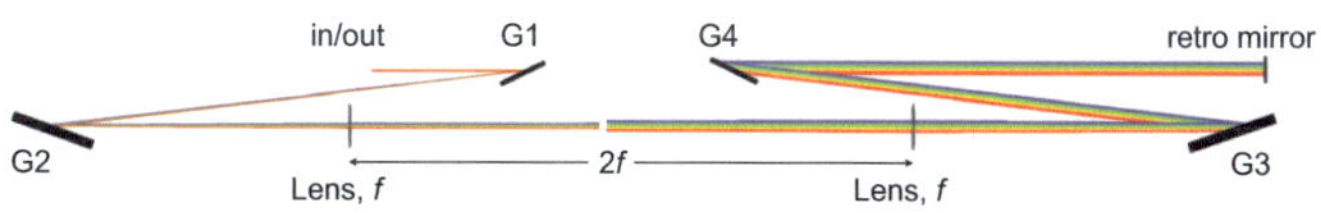

Fig. 13. Unfolded hyper-dispersion stretcher utilizes four gratings, as opposed to two gratings in the standard Martinez design.

The folded CAD version of the stretcher built for T-REX is shown in Fig. 14. For high fidelity pulse recompression, the stretcher is designed with a nearly equal and opposite chirp compared to that of the compressor. The small difference accounts for other dispersive elements in the system. We again use two large 1740 grooves/mm MLD gratings, with footprints of 20x10 cm and 35x15 cm. The beam is incident at the same Littrow 3° angle as in the compressor. A large, 175 mm diameter, f = 3099 mm lens accommodates the large footprint of the spatially chirped beam. A folding retro-mirror is placed f away from the lens, forming a 2-f telescope seen in the unfolded version of Fig. 13. The beam height changes through off-center incidence on the lens. After 4 grating reflections, the beam is incident on the vertical roof mirror, which is equivalent to the retro-mirror in Fig. 13. The two 45^0

mirrors fold the beam path, rendering a more compact footprint. After 8 grating reflections, the compressed pulse arrives and 2nd pass retro mirror. The beam is then retro-reflected through the stretcher, undergoing a total of 16 grating reflections. We double pass through the stretcher to double the total pulse chirp. Beam clipping on the lens prevents reducing the lens to G2 distance to match compressor dispersion in a single pass.
Chromatic and geometric lens aberrations modify higher order dispersion terms in the stretcher, requiring raytracing for more accurate computation. We use a commercial ray-tracing software (FRED by Photon Engineering, LLC) to compute ray paths in the stretcher and in the compressor. From raytrace analysis, the GDD for the stretcher is 4300 ps^2/rad, and the TOD/GDD ratio is -115 fs, at the 1064 nm central wavelength; for the compressor, the TOD/GDD ratio is -84 fs. The TOD mismatch would result in a 3% reduction in the temporal Strehl ratio of the compressed pulse. We can match the GDD and the TOD of the stretcher/compressor pair by a 1° increase of the angle of incidence on gratings 1 and 2 in the stretcher.

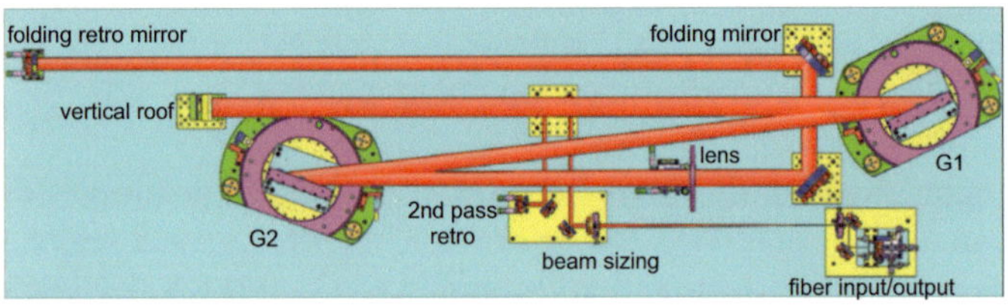

Fig. 14. The compact hyper-dispersion stretcher matches the GDD of the compressor.

We employed the hyper-dispersion stretcher-compressor pair in our interaction laser. Commercial Q-switched bulk Nd:YAG laser heads amplified stretched pulses from the fiber chain to 1.3 J, with 800 mJ remaining after pulse recompression. We characterize the compressed pulse temporal profile using multi-shot second harmonic generation (SHG) frequency resolved optical gating (FROG) (Kane & Trebino, 1993; Kane et al., 1994) technique. FROG is a commonly used method for measurement of ultrashort pulses. Compared to autocorrelation which cannot measure the actual pulse intensity, FROG measures both the intensity and the phase of the recompressed pulse. In its multi-shot, SHG implementation, a FROG measurement consists of measuring the frequency spectrum of the auto-correlation signal at each relative delay between the two pulse replicas. Mathematically, we determine $I_{FROG}^{SHG}(\omega,\tau) = |\int_{-\infty}^{\infty} E(t)E(t-\tau)\exp(-i\omega t)dt|^2$, which is a 2D spectrogram plotting delay, τ vs frequency (Trebino, 2000). Several inversion algorithms exist to process the FROG data and extract pulse intensity and phase.
In the measurement, we use a 0.01 nm resolution 1 m spectrometer (McPherson Model 2061) to resolve the narrow bandwidth pulse spectrum at the output of a background free SHG auto-correlator. The measured field of the FROG spectrogram, $\sqrt{I(\omega,\tau)}$ is shown in Fig. 15(a). Numerical processing then symmetrizes the trace and removes spurious background and noise. The FROG algorithm converges to the spectrogram shown in Fig. 15(b). The FROG algorithm discretizes the measurement into a 512x512 array. The FROG error between the measured and the converged calculated profile, defined as $E_{FROG} \equiv \sqrt{\frac{1}{N^2}\sum_{i,j=1}^{N}\left[I_{FROG}^{meas}(\omega_i,\tau_j) - I_{FROG}^{calculated}(\omega_i,\tau_j)\right]^2}$, where N=512 is the array dimension, is $5.3\text{x}10^{-3}$.
The intensity profile corresponding to FROG spectrum of Fig. 15(b) is shown in Fig. 16(a). The

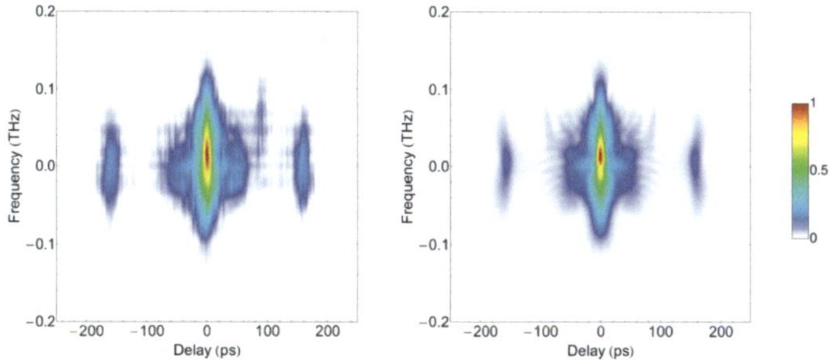

(a) (b)

Fig. 15. Experimental measurement of the 800 mJ pulse duation. (a) Field of the experimentally measured FROG spectrogram. (b) Lowest error field obtained by a FROG algorithm using a 512×512 discretization grid.

pulse is slightly asymmetric and contains a small pre-pulse, caused by a small residual TOD mismatch. We calculate that the FWHM is 8.3 ps, with 84% of the pulse energy contained in the 20 ps wide bin indicated by the dashed box, and the temporal strehl ratio is 0.78. The temporal strehl is the ratio of the peak measured intensity, and the peak transform limited intensity for the measured spectral profile. The temporal waveform on the logarithmic scale (Fig. 16(b)) shows a post pulse at 160 ps, 100 dB lower than the main pulse. This post-pulse causes the satellite wings and the fringing in the FROG measurement. Frequency doubling will further improve the pulse contrast.
FROG also measures the pulse spectrum (Fig. 17). Comparing the FROG measured spectrum with the direct IR spectral measurement performed with an f=1 m spectrometer indicates good agreement.

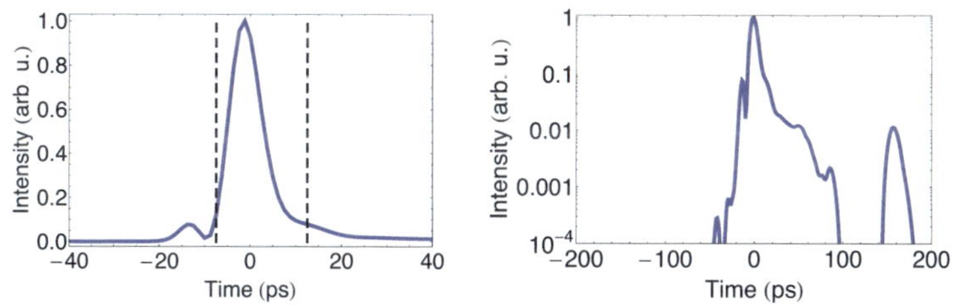

(a) (b)

Fig. 16. Temporal pulse intensity obtained by analyzing a numerically processed FROG spectrogram on the linear scale (a) and log scale (b). FWHM of the pulse duration is 8.3 ps, and 84% of the energy is contained in the 20 ps bin (dashed box).

5. Frequency conversion

High power, high energy laser technology is well developed in the 800 nm to 1.1 μm wavelength range. One can generate other wavelengths through a nonlinear conversion process. Fundamentally, the response of a dielectric medium to an applied electric field can be described by an induced polarization, $\boldsymbol{P} = \epsilon_0 \chi^{(1)} \boldsymbol{E} + \epsilon_0 \chi^{(2)} \boldsymbol{E}\boldsymbol{E} + ...$, where ϵ_0 is the free space permittivity, $\chi^{(n)}$ is the nth order susceptibility, and $\boldsymbol{E}$ is the vector electric field. Because

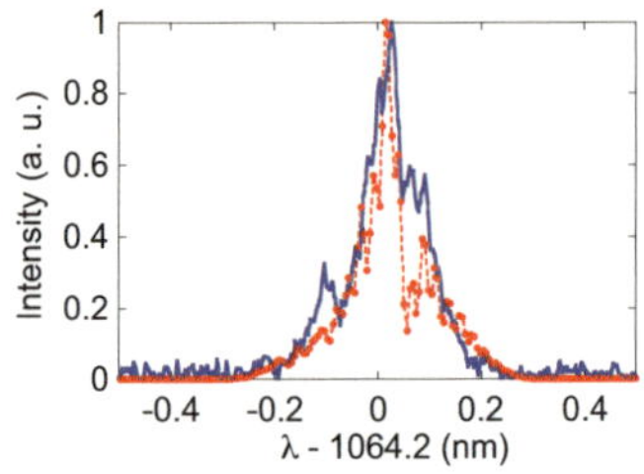

Fig. 17. Pulse spectrum from FROG (red dots) and spectrometer (solid line) measurements.

$\chi^{(1)} \gg \chi^{(2)}$ for an off-resonant medium, higher order terms become important only when the applied electric field is sufficiently high. In a $\chi^{(2)}$ process, $\boldsymbol{EE}$ term produces excitation at twice the fundamental frequency. Let $\boldsymbol{E} = \boldsymbol{A}\cos(\omega t)$, then $\boldsymbol{E}^2 = \frac{\boldsymbol{A}^2}{2}(1 + \cos(2\omega t))$. The magnitude of the nonlinear susceptibility varies with the applied frequency and depends on the electronic level structure of the material. Under well-optimized conditions harmonic efficiencies can exceed 80%. When selecting an appropriate nonlinear crystal, we consider various application dependent factors such as the magnitude of the nonlinear coefficient, acceptance bandwidth, absorption, thermal acceptance, thermal conductivity, walk-off angle, damage threshold, and maximum clear aperture. For pulse durations in the 200 fs to 10 ps range and for the fundamental wavelength $\approx$1 μm, beta barium borate (BBO) is an excellent candidate for 2ω, 3ω, and 4ω generation. The main draw back, is that the largest clear crystal aperture is $\approx$20 mm which limits its use to low pulse energies ($< 10 - 100$ mJ). For higher pulse energies, deuterated and non-deuterated potassium dihydrogen phosphate (DKDP and KDP), lithium triborate (LBO) and yttrium calcium oxyborate (YCOB) can be grown to much larger apertures. YCOB is particularly attractive for its high average power handling, high damage threshold, and large effective nonlinearity (Liao et al., 2006). For frequency doubling, typical required laser intensities are in the 100 MW/cm^2 to 10 GW/cm^2 range.

The crystal must be cut along an appropriate plane to allow phase matching and to maximize the effective nonlinear coefficient, d_{eff}, which is related to $\chi^{(2)}$ and the crystal orientation. The interacting waves at ω and 2ω acquire different phases, $\phi(\omega) = k_\omega z = n_\omega \omega z/c$ and $\phi(2\omega) = k_{2\omega} z = 2n_{2\omega}\omega z/c$ as they propagate along the crystal in z direction. An interaction is phase matched when $k_{2\omega} = 2k_\omega$. A uniaxial crystal contains two polarization eigenvectors, one parallel to the optic axis (the axis of rotation symmetry) and one perpendicular to it. An electric field inside the crystal contains a component perpendicular to the optic axis (ordinary polarization) and a component in the plane defined by the optic axis and the direction of propagation (extraordinary polarization), as illustrated in Fig. (18). The refractive index of the extraordinary polarization, n^e, varies with θ, the angle between the direction of propagation and the optic axis; the ordinary refractive index, n^o has no angular dependence. In the example shown in Fig. (18), the crystal is rotated along the y-axis until $n^e_{2\omega}(\theta) = n^o_\omega$. The illustrated phase matching condition, where both incident photons have the same polarization is known as type I phase matching. In Type II phase matching, the incident field has both an ordinary and an extraordinary polarization component.

Coupled Eqs. (14-15), given in SI units, describe Type I, 2ω generation process relevant for 200 fs - 10 ps duration pulses. Here, we make a plane wave approximation, justified when we are not focusing into the crystal, and when the crystal is sufficiently thin to ignore beam walk-off effects. We also ignore pulse dispersion in the crystal, justified for our pulse bandwidth and crystal thickness. We can account for two-photon absorption, which becomes important for 4ω generation in BBO, by adding $\beta|A_{2\omega}|^2 A_{2\omega}$ term to the left hand side of Eq. (15).

$$\frac{\partial A_{\omega}}{\partial z} + \frac{1}{v_{g,\omega}}\frac{\partial A_{\omega}}{\partial t} + \frac{\alpha_{\omega}}{2}A_{\omega} = i\frac{2\omega}{n(\omega)c}d_{eff}A_{\omega}^{*}A_{2\omega}\exp(-i\Delta kz) \tag{14}$$

$$\frac{\partial A_{2\omega}}{\partial z} + \frac{1}{v_{g,2\omega}}\frac{\partial A_{2\omega}}{\partial t} + \frac{\alpha_{2\omega}}{2}A_{2\omega} = i\frac{2\omega}{n(2\omega)c}d_{eff}A_{\omega}^{2}\exp(i\Delta kz) \tag{15}$$

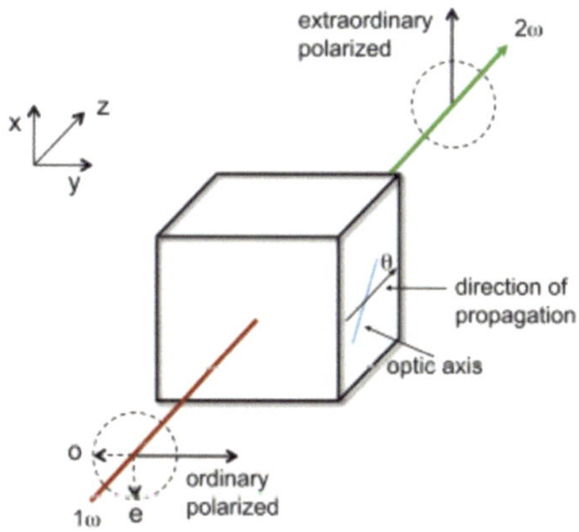

Fig. 18. Frequency doubling with OOE type phasematching in a uniaxial nonlinear crystal. θ is the phasematching angle between the optic axis and the propagation direction.

We obtain an analytical solution assuming qausi-CW pulse duration, which eliminates the time dependent terms, and a low conversion efficiency, or a constant A_{ω}. The efficiency of 2ω harmonic generation, $\eta_{2\omega} = I_{2\omega}/I_{\omega}$, reduces to Eq. (16):

$$\eta_{2\omega} = \frac{8\pi^2 d_{eff}^2 L^2 I_{\omega}}{\epsilon_0 n_{\omega}^2 n_{2\omega} c\lambda_{\omega}^2}\frac{\sin^2(\Delta kL/2)}{(\Delta kL/2)^2} \tag{16}$$

In our laser systems, we implement frequency conversion on both, the photogun and the interaction laser systems (Gibson et al., 2010). On T-REX, we generate the 4th harmonic of the fundamental frequency by cascading two BBO crystals. The first, 1 mm thick crystal cut for Type I phase matching, frequency doubles the incident pulse from 1053 nm to 527 nm. The second 0.45 mm thick BBO crystal cut for Type I phase matching,frequency doubles 527 nm pusle to 263 nm. The overall conversion efficiency from IR to UV is 10%, yielding 100 μJ at 263 nm. Here, frequency conversion is primarily limited by two-photon absorption in the UV and the group velocity mismatch (GVM) between the 2ω and the 4ω pulses. GVM results in temporal walk-off of the pulse envelopes and, in the frequency domain, is equivalent to the acceptance bandwidth.

On the interaction laser, we frequency double the high energy pulses to increase the final γ-ray energy. On T-REX we use a large aperture (30x30 mm) 6 mm thick DKDP crystal to frequency double 800 mJ pulse from 1064 nm to 532 nm with up to 40% conversion efficiency. Here, the pulse bandwidth is relatively narrow ($\approx$0.2 nm) and group velocity walk-off is insignificant. The conversion efficiency is primarily limited by beam quality and temporal pulse shape. Generated 532 nm pulse energy is plotted vs the compressed input pulse energy in Fig. 19.

At maximum IR energy, the conversion efficiency unexpectedly decreases. This may indicate onset of crystal damage, degradation in pulse quality, or an increase in phase mismatch.

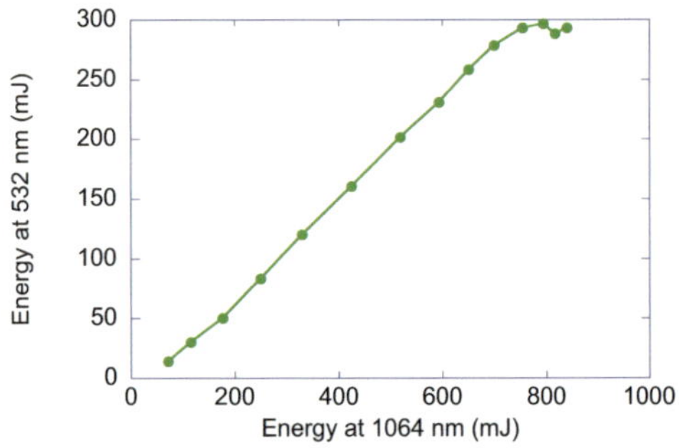

Fig. 19. Frequency doubling of the 10 ps T-REX ILS laser with a peak efficiency of 40%.

6. Pulse shaping

A uniform elliptical laser shape in space and time minimizes the contribution of space-charge force on the electron beam emittance (Kapchinskij & Vladimirskij, 1959; Li & Lewellen, 2008). Due to technical difficulties of generating this shape, simulations suggest that a uniform cylindrical pulse in space and time is an alternative (Cornacchia & *et. al.*, 1998). We generate the cylindrical shape by time-stacking the UV laser pulse in a Hyper-Michelson interferometer (Siders et al., 1998) and clipping the UV beam with a hard edge aperture. It is also possible to mode-convert the UV beam from a Gaussian to a flat-top with a commercial refractive beam shaper based on aspheric lenses.

The pulse shaper is a Michelson-based ultrafast multiplexing device having nearly 100% throughput and designed for high energy shaped pulse generation. The pulse train generates a train of replicas of the input pulse delayed with respect to each other with femtosecond precision. The pulse stacker used in T-REX (Gibson et al., 2010) consists of 4 stages, stacking 16 pulses. The laser pulse passes through a series of beam splitters, each time being recombined following an adjustable delay path. The built pulse stacker is shown in Fig. 20. At the output of the pulse stacker, two orthogonally polarized pulse trains are recombined and interleaved in time at a polarizing beamsplitter. The individual pulses must be sufficiently delayed to avoid high frequency intensity modulation that results from interferences of same-polarized pulses.

An example of a stacked flat-top pulse from the pulse stacker is shown in Fig. 21. The figure shows a cross-correlation of the stacked UV pulse obtained on T-REX. The 15 ps stacked UV (4ω) pulse is cross-correlated with the $\approx$0.75 ps IR (1ω) pulse (blue line). The predicted cross-correlation on the basis of the pulse energies of each of the 16 stacked pulses is shown by the dashed line. Cross-correlations between the IR and the 16 individual UV pulses are shown below the main pulse.

7. High energy laser pulse recirculation

In this section we describe a novel technique for recirculating high power, high energy, picosecond laser pulses, akin to the interaction laser pulses on T-REX. The motivation for laser recirculation for compton-scattering sources is two-fold. First, a major fundamental limitation of these sources is the extremely small Thomson scattering cross-section, $\sigma_T = \frac{8\pi}{3} r_e^2 = 6.65 \times 10^{-25}\text{cm}^2$, where r_e classical electron radius, which leads to low conversion

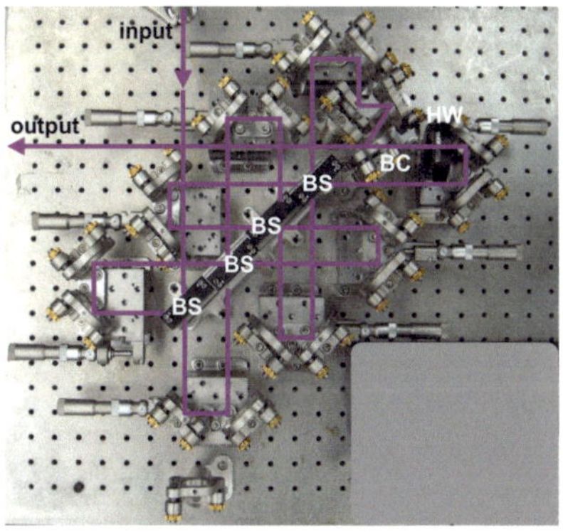

Fig. 20. Photograph of the pulse stacker on T-REX photogun laser.

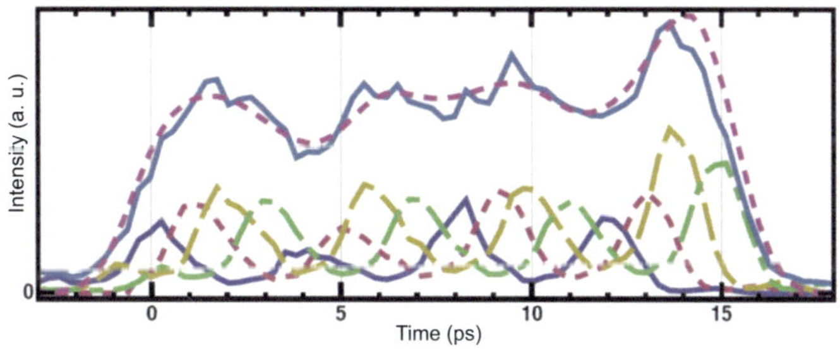

Fig. 21. Cross-correlation measurement of the temporally shaped UV pulse from T-REX photogun laser indicates a 15 ps FWHM pulse duration.

efficiency from laser photons to γ rays. Only 1 in 10^{10} of laser photons is doppler upshifted to γ-ray energy. The overall efciency of the compton-scattering source could be increased by reusing the laser photons after each interaction with the electron bunch. Second, the joule-class, short pulse lasers operate at a few Hz to 100 Hz type of repetition rates. Linacs can operate at kHz and higher repetition rates. Increasing the repetition rate of the interaction laser would increase the average brightness of the γ-ray source.

The pulse recirculation scheme that we have developed is general and could be applied to various other phenomena that involve high intensity lasers interacting with an optically thin medium such as cavity ring down spectroscopy, high-harmonic generation in short gas jets, or laser based plasma diagnostics. The pulse recirculation scheme is based on injection and trapping a single laser pulse inside a passive optical cavity. A thin nonlinear crystal acts as an optical switch, trapping the frequency converted light. This technique, termed recirculation injection by nonlinear gating (RING) is compatible with joule class, 100s of Watts of average power, picosecond laser pulses. In the simplest implementation of this technique, the incident laser pulse at the fundamental frequency enters the resonator and is efficiently frequency doubled. The resonator mirrors are dichroic, coated to transmit the 1ω light and reflect at 2ω (see Fig. 22). The upconverted 2ω pulse becomes trapped inside the cavity. After many roundtrips, the laser pulse decays primarily due to Fresnel losses at the crystal faces and cavity mirrors. The crystal thickness is optimized for high conversion efficiency.

Current pulse recirculation schemes are based on either resonant cavity coupling (Gohle et al., 2005; Jones et al., 2005) or active (electro-optic or acousto-optic) pulse switching (Yu & Stuart,

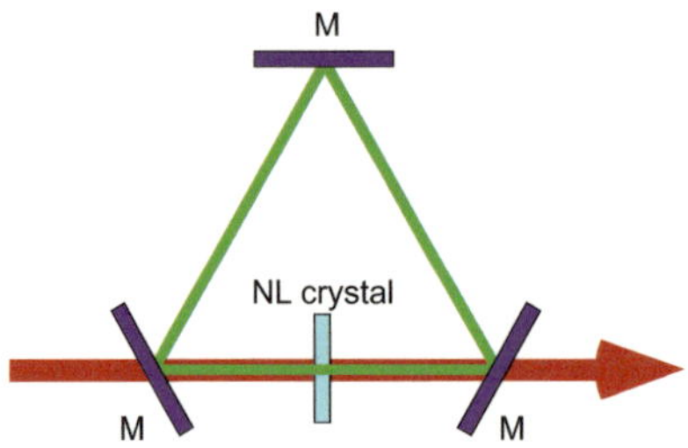

Fig. 22. Conceptual design of the RING picosecond pulse recirculation cavity.

1997; Mohamed et al., 2002) into and out of the resonator. Active pulse switching schemes are suitable for low intensity, nanosecond duration pulses (Meng et al., 2007). Resonant cavity coupling requires interferometric cavity alignment and MHz and higher repetition rates. To date, researchers have attained up to 100x enhancement for 1 W average power, $\approx$50 fs duration incident pulses with per pulse energy $<$1 μJ. Compared to active pulse switching schemes, the main advantage of RING is an order of magnitude reduction of the accumulated nonlinear phase with each roundtrip. Here, the crystal thickness is a few mm, compared to a few cm thick crystal inside a pockels cell. Compared to resonant cavity coupling, RING increases the intracavity repetition rate, while maintaining nearly the same peak pulse power. A resonant cavity increases the peak pulse power while decreasing the intracavity repetition rate.

We describe a low energy, millijoule-scale and a high energy, joule-scale pulse recirculation experiments (Shverdin, Jovanovic, V. A. Semenov, Brown, Gibson, Shuttlesworth, Hartemann, Siders & Barty, 2010). We chose a Fabry-Perot configuration for our RING cavity to maximize the cavity finesse (Fig. 23). In this arrangement, the concave spherical mirrors are spaced $\approx 2f$ apart, where $f = 375$ mm is the focal length of each of the cavity mirrors. The mirrors' multi-layer dielectric coating on the surface internal to the cavity is 99.8% reflective at 527 nm and 98% transmissive at 1053 nm. The mirrors' flat surface external to the cavity is anti-reflection coated at both wavelengths. Any plane inside the cavity is relay imaged back onto itself, which minimizes diffraction losses and supports an arbitrary incident spatial profile. The difference in the RING cavity designs for the two experiments involves the aperture size of the optical components and the choice of the doubling crystal.

The CAD experimental design is shown in Fig. 24. The cavity is contained inside two interconnected vacuum chambers. One of the vacuum chambers contains the nonlinear crystal and one of the spherical mirrors. The second chamber contains the other spherical mirror. Concave spherical mirrors act as negative lenses in transmission. A positive lens prior to the cavity adjusts the beam curvature to produce a collimated wavefront for the input 1ω (IR) beam inside the cavity. The 2ω beam is collimated when traveling from right to left, and focuses in the middle when traveling in the opposite direction. The cavity contains an internal focus to simulate an electron beam interaction region. Vacuum compatible actuators control the tip/tilt of the mirrors, the phase matching angle of the crystal, and the total cavity length. The chambers are pumped down to 10^{-3} Torr range to minimize nonlinear phase accumulation and prevent air breakdown at the focus.

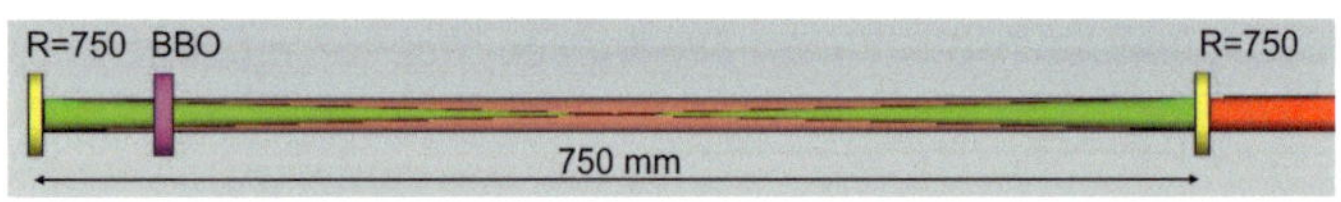

Fig. 23. Ray trace of the designed confocal resonator-type RING cavity.

In the low energy experiment, we inject a 10 Hz, 1.7 mJ, 10 nm bandwidth pulse centered at 1053 nm and chirped to 2.25 ps through the right 1" diameter cavity mirror, which is a negative lens in transmission. Upstream beam sizing optics produce a collimated w_0=4 mm gaussian inside the cavity. The nonlinear crystal is a 10x10x1 mm Type I SHG BBO crystal. The crystal has a single layer MgF_2 antireflection (AR) coating at both wavelengths (0.3% loss per surface). The choice of BBO is motivated by its excellent thermomechanical properties, broad spectral and thermal acceptance, and a relatively low nonlinear index to effective nonlinear coefficient ratio, n_2/d_{eff}. We measure 270 μJ at 527 nm after the crystal, corresponding to peak intensity ≈ 0.8 GW/cm^2. The majority of the residual IR is coupled out of the cavity through the end mirror.

The precise cavity length, L_{cav}, is slightly longer than $2f$ to account for refraction through the crystal. $L_{cav} = 2f + \Delta L$, where ΔL=L_c(1-1/n_c). Here, L_c is the crystal's thickness and n_c is its refractive index at 2ω. ΔL=2 mm for the 1 mm thick BBO.

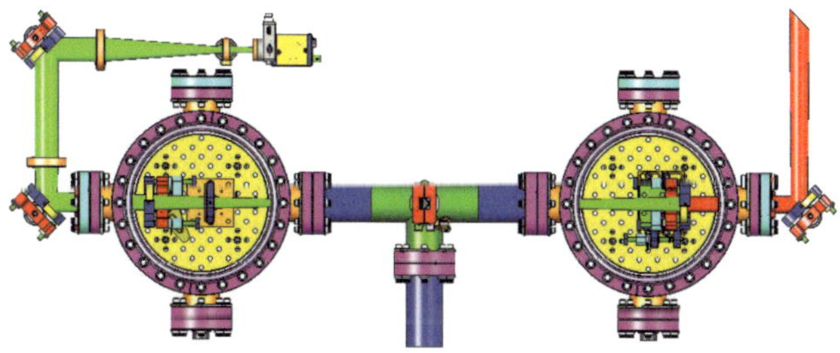

Fig. 24. CAD of the RING cavity design: two interconnected vacuum chambers contain the nonlinear crystal and dichroic mirrors.

We measure cavity enhancement at 2ω, by recording the leakage 527 nm light that passes through the end mirror, on a 1.2 GHz Si photodiode and 15 GHz digital signal oscilloscope and analyzing the resulting cavity ring-down signal. We define enhancement as $enh \equiv \sum_{n=0}^{N} I_n/I_0$, where I_n is pulse power after n roundtrips, and N is the total number of roundtrips. Dichroic mirrors and green bandpass filters scrape off any residual IR from the detected beam. The measured ring-down signal (Fig. 25) shows approximately 170 pulses spaced by the 5 ns, cavity roundtrip time. Impedance mismatch between the photodiode and the oscilloscope introduces some ringing in the recorded waveform.

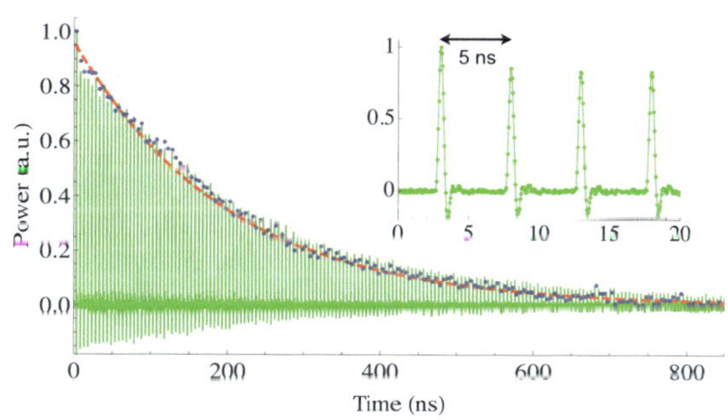

Fig. 25. Low energy cavity ring-down signal indicates 40x signal enhancement. The red dashed line is an exponential fit to the pulse power at each roundtrip (blue circles). The total cavity length sets the 5 ns pulse spacing (inset figure).

The circles in Fig. 25 correspond to the normalized power in each pulse, determined by integrating the voltage signal. The expected power in each successive pulse is $(1-\alpha)$ times the power of the previous pulse, or $I_n=I_0(1-\alpha)^n$, where I_0 is the initial signal power and α is the loss per roundtrip. The total cavity enhancement is then $[1-(1-\alpha)^n]/\alpha$. The dashed line in the figure is a fit to pulse power with α = 0.025, corresponding to *enh*=40. Contributions above the 1.8% Fresnel losses, could be attributed to hard edge diffraction and scattering in the nonlinear crystal. In the measured trace, the ratio of the powers in the first to the second pulses is larger than for subsequent pulses; an effect caused by additional diffraction losses during the first pass. The waveform also exhibits "picket fence" effect, where the power of many pulses is higher than of the adjacent pulses. This is likely caused by a slight cavity misalignment. We explicitly calculate cavity enhancement by summing over all of the observed pulses, obtaining *enh*=36. We estimate the total accumulated nonlinear phase, $\phi_{NL}=\frac{2\pi n_2 L_c I_{peak}}{\lambda}\frac{1-(1-\alpha)^N}{\alpha}$=0.7 rad, where, n_2 is the nonlinear refractive index (8.8x10^{-16} cm^2/W).

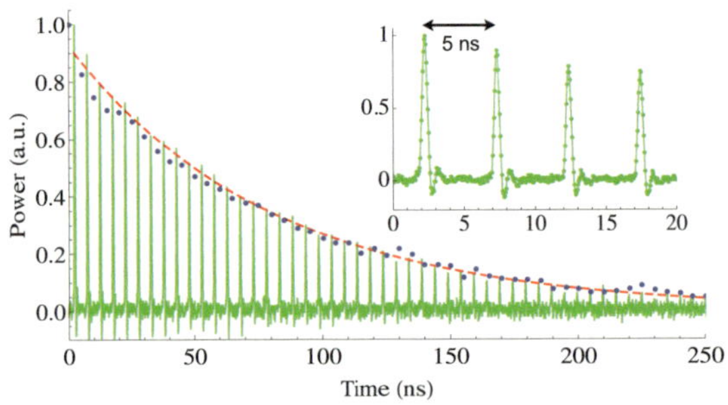

Fig. 26. High energy (177 mJ) cavity ring-down signal indicates 17x signal enhancement.

In the high energy experiment, both the spatial and the temporal profiles exhibited significant aberrations. We injected 10 Hz, 677 mJ, 0.25 nm bandwidth pulses at 1064 nm with a FWHM pulse duration of 20 ps. Autocorrelation measurements of the IR pulse indicate that 70% of the 677 mJ is contained in a wide 400 ps pulse pedestal. Compared to the low energy experiment, the main changes to the RING cavity involve replacing 1″ diameter cavity mirrors with 2″ diameter mirrors and replacing the small aperture BBO crystal with 30x30x6 mm deuterated potassium dihydrogen phosphate (DKDP) cut for Type II phase matching. After frequency doubling, we generate 177 mJ at 532 nm in an slightly elliptical 12x15 mm FWHM beam. Computer simulations indicate that the pulse at 532 nm is 16 ps FWHM and 50% of total energy is contained in the wide pedestal. We estimate the resulting peak pulse intensity at 4 GW/cm^2.

The high energy ring-down signal is shown in Fig. 26. We observe pulses over ≈50 roundtrips. Here, the loss coefficient, α=0.06, resulting in cavity enhancement, *enh*=17, the same value as obtained by explicitly summing the powers in each pulse. The estimated total nonlinear phase is 2.8 rad. We attribute the significant degradation in cavity enhancement to the poor spatial beam quality of the high energy 1ω laser. The near field spatial profile suffers from high frequency intensity modulation causing high hard edge diffraction losses at the crystal. When we replaced the DKDP crystal with a smaller aperture 20x20x1.2 mm BBO crystal, higher diffraction losses reduced cavity enhancement to 11.

The peak power scaling of the RING cavity is primarily limited by the nonlinear phase accumulation in the crystal. For a gaussian pulse, the bandwidth doubles for $\phi_{NL}\approx$2.4.

In monoenergetic gamma-ray generation, this increases the bandwidth of the generated photons (Albert et al., 2010). Other deleterious effects include whole beam self-focusing and modulation instability growth. Increasing beam size and correspondingly the aperture of the cavity optics mitigates nonlinear phase accumulation. DKDP crystals are available in sizes up to 40x40 cm, potentially enabling recirculation of 100 J, 10 ps pulses.

RING cavity design is scalable to very high peak and average power recirculation. Linear absorption in the crystal is the primary limitation to the maximum sustainable average power inside the cavity. LBO, YCOB, and BBO crystals are particularly attractive candidates for high average power operation. For a simple edge cooling scheme, finite element simulation of the thermal profile inside the crystal indicate that LBO and YCOB support up to 1.2 kW of total recirculating power. The peak recirculating power limit could be significantly increased with surface cooling schemes.

Deploying RING on a Compton scattering light source could lead to more than an order of magnitude increase in average source brightness of the generated γ-ray flux. The RING cavity would be integrated into the γ-ray source after a relatively simple modification of the interaction point architecture.

8. Conclusion

We provided a brief overview of Compton-scattering based compact monoenergetic γ-ray sources, emphasizing the recently commissioned 2nd generation T-REX, and the currently under construction, 3rd generation Velociraptor machine at LLNL. We detailed the underlying laser technology and described several technological breakthroughs which enable development of γ-ray sources with the highest peak brightness in a compact footprint.

We expect continued improvement in future Compton-based light sources. In the laser realm, some of the fruitful future research will include: (1) boosting the laser repetition rates to develop joule-class, kW and MW average power systems, to increase the average γ-ray source brightness; (2) developing tunable and multi-resonant joule-class, picosecond optical parametric amplifiers, to either simultaneously generate multiple γ-ray energies or rapidly tune the output energy; (3) preparing proper laser pulse and beam formats to further reduce the γ-ray bandwidth; (4) demonstrating appropriate beam pointing and scanning technology to enable γ-ray beam rastering.

This work was performed under the auspices of the U.S. Department of Energy by University of California, Lawrence Livermore National Laboratory under Contract W-7405-ENG- 48. We also acknowledge support of DOE/NA-22.

9. References

Albert, F., Anderson, S. G., Anderson, G. A., Betts, S. M., Gibson, D. J., Hagmann, C. A., Hall, J., Johnson, M. S., Messerly, M. J., Semenov, V. A., Shverdin, M. Y., Tremaine, A. M., Hartemann, F. V., Siders, C. W., McNabb, D. P. & Barty, C. P. J. (2010). Isotope-specific detection of low-density materials with laser-based monoenergetic gamma-rays, *Opt. Lett.* 35(3): 354–356.

Arutyunyan, F. R. & Tumanyan, V. A. (1964). Quasi-monochromatic and polarized high-energy gamma rays, *Soviet Physics Uspekhi* 7(3): 339.

Bech, M., Bunk, O., David, C., Ruth, R., Rifkin, J., Loewen, R., Feidenhans, R. & Pfeiffer, F. (2009). Hard x-ray phase-contrast imaging with the compact light source based on inverse compton x-rays, *J. Synchrotron Radiat.* 16: 43–47.

Bertozzi, W. & Ledoux, R. J. (2005). Nuclear resonance fluorescence imaging in non-intrusive cargo inspection, *Nucl. Instrum. Meth. B* 241: 820–825.

Carroll, F. E., Mendenhall, M. H., Traeger, R. H., Brau, C. & Waters, J. W. (2003). Pulsed tunable monochromatic x-ray beams from a compact source: New opportunities, *Am. J. Roentgenol.* 181: 1197–1202.

Cheriaux, G., Rousseau, P., Salin, F., Chambaret, J. P., Walker, B. & Dimauro, L. F. (1996). Aberration-free stretcher design for ultrashort-pulse amplification, *Opt. Lett.* 21: 414–416.

Compton, A. H. (1923). A quantum theory of the scattering of x-rays by light elements, *Phys. Rev.* 21(5): 483–502.

Cornacchia, M. & *et. al.* (1998). Linac coherent light source (lcls) design study report, *Technical report*, SLAC-R-521, Stanford Linear Accelerator Center, Stanford, CA.

D'Angelo, A., Bartalini, O., Bellini, V., Sandri, P. L., Moricciani, D., Nicoletti, L. & Zucchiatti, A. (2000). Generation of compton backscattering [gamma]-ray beams, *Nucl. Instrum. Meth. A* 455(1): 1–6.

Duarte, F. J. (1987). Generalized multiple-prims dispersion theory for pulse compression in ultrafast dye lasers, *Opt. Quant. Electron.* 19: 223–229.

Esarey, E., Ride, S. K. & Sprangle, P. (1993). Nonlinear thomson scattering of intense laser pulses from beams and plasmas, *Phys. Rev. E* 48(4): 3003–3021.

Feenberg, E. & Primakoff, H. (1948). Interaction of cosmic-ray primaries with sunloght and starlight, *Phys. Rev.* 73(5): 449–469.

Fittinghoff, D. N., Molander, W. A. & Barty, C. P. J. (2004). Hyperdispersion grating arrangements for compact pulse compressors and expanders, *Frontiers in Optics*, Optical Society of America, Rochester, NY, p. FThL5.

Forster, R. A., Cox, L. J., Barrett, R. F., Booth, T. E., Briesmeister, J. F., Brown, F. B., Bull, J. S., Geisler, G. C., Goorley, J. T., Mosteller, R. D., Post, S. E., Prael, R. E., Selcow, E. C. & Sood, A. (2004). Mcnp(tm) version 5, *Nucl. Instrum. Meth. B* 213: 82–86.

Frantz, L. M. & Nodvik, J. S. (1963). Theory of pulse propagation in a laser amplifier, *J. Appl. Phys.* 34: 2346–2349.

Gibson, D. J., Albert, F., Anderson, S. G., Betts, S. M., Messerly, M. J., Phan, H. H., Semenov, V. A., Shverdin, M. Y., Tremaine, A. M., Hartemann, F. V., Siders, C. W., McNabb, D. P. & Barty, C. P. J. (2010). Design and operation of a tuna ble mev-level compton-scattering-based γ-ray source, *PRST-AB* vol. 13, page 070703 (2010).

G.Matt, Feroci, M., Rapisarda, M. & Costa, E. (1996). Treatment of compton scattering of linearly polarized photons in monte carlo codes, *Radiat. Phys. Chem.* 48(4): 403–411.

Gohle, C., Udem, T., Herrmann, M., Rauschenberger, J., Holzwarth, R., Schuessler, H., Krausz, F. & Hänsch, T. (2005). A frequency comb in the extreme ultraviolet, *Nature* 436: 234–237.

Graves, W. S., Brown, W., Kartner, F. X. & Moncton, D. E. (2009). Mit inverse compton source concept, *Nucl. Instrum. Meth. A* 608(1, Supplement 1): S103–S105.

Hartemann, F. (2002). *High Field Electrodynamics*, CRC Press, Boca Raton, FL.

Hartemann, F. V. & Kerman, A. K. (1996). Classical theory of nonlinear compton scattering, *Phys. Rev. Lett.* 76(4): 624–627.

Ilday, F., Buckley, J., Lim, H., Wise, F. & Clark, W. (2003). Generation of 50-fs, 5-nj pulses at 1.03 μm from wave-breaking free fiber laser, *Opt. Lett.* 28: 1365–1367.

Jeong, Y., Sahu, J. K. & Nilsson, J. (2004). Ytterbium-doped large-core fiber laser with 1.36 kw continuous-wave output power, *Opt. Express* 12: 6088–6092.

Jones, R. J., Moll, K. D., Thorpe, M. J. & Ye, J. (2005). Phase-coherent frequency combs in the vacuum ultraviolet via high-harmonic generation inside a femtosecond enhancement cavity, *Phys. Rev. Lett.* 94: 193201.

Kane, D. J., Taylor, A. J., Trebino, R. & DeLong, K. W. (1994). Single-shot measurement of the intensity and phase of a femtosecond uv laser pulse with frequency-resolved optical gating, *Opt. Lett.* 19: 1061–1063.

Kane, D. J. & Trebino, R. (1993). Single-shot measurement of the intensity and phase of an arbitrary ultrashort pulse by using frequency-resolved optical gating, *Opt. Lett.* 18: 823–825.

Kapchinskij, I. M. & Vladimirskij, V. V. (1959). Limitations of proton beam current in a strong focusing linear accelerator associated with beam space charge, *in* L. Kowarski (ed.), *Conference on High Energy Accelerators and Instrumentation*, CERN, p. 274.

Kawase, K., Kando, M., Hayakawa, T., Daito, I., Kondo, S., Homma, T., Kameshima, T., Kotaki, H., Chen, L.-M., Fukuda, Y., Faenov, A., Shizuma, T., Fujiwara, M., Bulanov, S. V., Kimura, T. & Tajima, T. (2008). Sub-mev tunably polarized x-ray production with laser thomson backscattering, *Rev. Sci. Instrum.* 79: 053302.

Kostenbauder, A. G. (1990). Ray-pulse matrices: A rational treatment for dispersive optical systems, *IEEE J. Quantum Elect.* 26: 1148–1157.

Leemans, W., Schoenlein, R., Volfbeyn, P., Chin, A., Glover, T., Balling, P., Zolotorev, M., Kim, K.-J., Chattopadhyay, S. & Shank, C. (1997). Interaction of relativistic electrons with ultrashort laser pulses: generation of femtosecond x-rays and microprobing of electron beams, *IEEE J. Quantum Elect.* 33(11): 1925–1934.

Li, Y. & Lewellen, J. (2008). Generating a quasiellipsoidal electron beam by 3d laser-pulse shaping, *Phys. Rev. Lett.* 100: 074801.

Liao, Z. M., Jovanovic, I. & Ebbers, C. A. (2006). Energy and average power scalable optical parametric chirped-pulse amplification in yttrium calcium oxyborate, *Opt. Lett.* 31(9): 1277–1279.

Lin, Q., Wang, S., Alda, J. & Bernabeu, E. (1993). Transformation of pulsed nonideal beams in a four-dimension domain, *Opt. Lett.* 18: 669–671.

Litvinenko, V. N., Burnham, B., Emamian, M., Hower, N., Madey, J. M. J., Morcombe, P., O'Shea, P. G., Park, S. H., Sachtschale, R., Straub, K. D., Swift, G., Wang, P., Wu, Y., Canon, R. S., Howell, C. R., Roberson, N. R., Schreiber, E. C., Spraker, M., Tornow, W., Weller, H. R., Pinayev, I. V., Gavrilov, N. G., Fedotov, M. G., Kulipanov, G. N., Kurkin, G. Y., Mikhailov, S. F. & Popik, V. M. (1997). Gamma-ray production in a storage ring free-electron laser, *Phys. Rev. Lett.* 78: 4569–4572.

Martinez, O. E. (1987). 3000 times grating compressor with positive group velocity dispersion: Application to fiber compensation in 1.3-1.6 μm region, *IEEE J. Quantum Elect.* 23: 59–64.

Meng, D., Sakamoto, F., Yamamoto, T., Dobashi, K., Uesaka, M., Nose, H., Ishida, D., Kaneko, N. & Sakai, Y. (2007). High power laser pulse circulation experiment for compact quasi-monochromatic tunable x-ray source, *Nucl. Instrum. Meth. B* 261: 52.

Milburn, R. H. (1963). Electron scattering by an intense polarized photon field, *Phys. Rev. Lett.* 10(3): 75–77.

Mohamed, T., Andler, G. & Schuch, R. (2002). Development of an electro-optical device for storage of high power laser pulses, *Opt. Commun.* 214: 291–295.

Perry, M. D., Boyd, R. D., Britten, J. A., Decker, D., Shore, B. W., Shannon, C. & Shults, E. (1995). High-eff iciency multilayer dielectric diffraction gratings, *Opt. Lett.* 20: 940–942.

Pietralla, N., Berant, Z., Litvinenko, V. N., Hartman, S., Mikhailov, F. F., Pinayev, I. V., Swift, G., Ahmed, M. W., Kelley, J. H., Nelson, S. O., Prior, R., Sabourov, K., Tonchev, A. P. & Weller, H. R. (2002). Parity measurements of nuclear levels using a free-electron-laser generated γ-ray beam, *Phys. Rev. Lett.* 88(1): 012502.

Planchon, T. A., Burgy, F., Rousseau, J.-P. & Chambaret, J.-P. (2005). 3d modeling of amplification processes in cpa laser amplifiers, *Appl. Phys. B* 80: 661–667.

Pruet, J., McNabb, D. P., Hagmann, C. A., Hartemann, F. V. & Barty, C. P. J. (2006). Detecting clandestine material with nuclear resonance fluorescence, *J. Appl. Phys.* 99: 123102.

Shverdin, M. Y., Albert, F., Anderson, S. G., Betts, S. M., Gibson, D. J., Messerly, M. J., Hartemann, F., Siders, C. W. & Barty, C. P. J. (2010). Chirped pulse amplification with narrowband pulses, *Opt. Lett.* 35: 2478–2480.

Shverdin, M. Y., Jovanovic, I., V. A. Semenov, S. M. B., Brown, C., Gibson, D. J., Shuttlesworth, R. M., Hartemann, F. V., Siders, C. W. & Barty, C. P. J. (2010). High-power picosecond laser pulse recirculation, *Opt. Lett.* 35: 2224–2226.

Siders, C. W., Siders, J. L. W., Taylor, A. J., Park, S.-G. & Weiner, A. M. (1998). Efficient high-energy pulse-train generation using a 2 n-pulse michelson interferometer, *App. Opt.* 37: 5302–5305.

Strickland, D. & Mourou, G. (1985). Isotope-specific detection of low density materials with laser-based mono-energetic gamma-rays, *Opt. Commun.* 56(219-221).

Sumetsky, M., Eggleton, B. & de Sterke, C. (2002). Theory of group delay ripple generated by chirped fiber gratings, *Opt. Express* 10: 332–340.

Titov, A. I., Fujiwara, M. & Kawase, K. (2006). Parity non-conservation in nuclear excitation by circularly polarized photon beam, *J. Phys. G: Nucl. Part. Phys.* 32: 1097–1103.

Treacy, E. B. (1969). Optical pulse compression with diffraction gratings, *IEEE J. Quantum Elect.* 5(454-458).

Trebino, R. (2000). *Frequency-Resolved Optical Gating: The Measurement of Ultrashort Laser Pulses*, Kluwer Academic Publishers, Boston, MA.

Yakimenko, V. & Pogorelsky, I. V. (2006). Polarized γ source based on compton backscattering in a laser cavity, *Phys. Rev. Spec. Top.-AC* 9: 091001.

Yu, D. & Stuart, B. (1997). A laser pulse trapper for compton backscattering applications, *Particle Accelerator Conference*, PAC'97, Vacouver, DC, Canada.

4

Quantum Manipulations of Single Trapped-Ions Beyond the Lamb-Dicke Limit

M. Zhang and L. F. Wei
Quantum Optoelectronics Laboratory and School of Physics and Technology, Southwest Jiaotong University, Chengdu 610031 China

1. Introduction

An ion can be trapped in a small region in three dimensional space by the electromagnetic trap, and can be effectively manipulated by the applied laser pulses (1; 2; 3). The system of trapped ions driven by laser pulses provides a powerful platform to engineer quantum states (4; 5; 6; 7) and implement quantum information processings (QIPs) (8; 9; 10). Typically, manipulations and detections quantum information can be utilized to test certain fundamental principles (such as EPR and Shrödinger cat "paradox") in quantum mechanics and implement quantum computation, quantum telegraph, and quantum cryptography, etc.(see, e.g., (11; 12; 13; 14)).
Up to now, several kinds of physical systems, e.g., cavity-QEDs (15), superconducting Josephson junctions (16), nuclear magnetic resonances (NMRs) (17), and coupled quantum dots (18), etc., have been proposed to physically implement the QIPs. The system of trapped ions is currently one of the most advanced models for implementing the QIPs, due to its relatively-long coherence time. Indeed, the coherent manipulations up to eight trapped ions had already been experimentally demonstrated (14).
However, most of the quantum manipulations with the trapped cold ions are within the Lamb-Dicke (LD) approximation, e.g., the famous Cirac-Zoller model (8). Such an approximation requires that the coupling between the quantum bit (encoded by two atomic levels of a trapped ion) and the data bus (the collective vibration mode of the ions) should be sufficiently weak. Sometimes, the LD limits could be not rigorously satisfied for typical single trapped-ion system, and thus higher-order powers of the LD parameter must be taken into account (19). Alternatively, the laser-ion interaction beyond the LD approximation might be helpful to reduce the noise in the ion-trap and improve the cooling rate(see, e.g., (20; 21; 22)), and thus could be utilized to high-efficiently realize QIPs.
In this chapter, we summarize our works (23; 24; 25; 26; 27; 28; 29; 30; 31) on how to design proper laser pulses for the desirable quantum-state engineerings with single trapped-ions, including the preparations of various typical quantum states of the data bus and the implementations of quantum logic gates beyond the LD approximation. The chapter is organized as: In Sec. 2 we derivate the dynamical evolutions of a single trapped ion (driven by a classical laser beam) with and beyond the LD approximation. Based on the quantum dynamics beyond the LD approximation, in Sec. 3, we discuss how to use a series of laser pulses to generate various vibrational quantum states of the trapped ion, e.g., coherence states, squeezed coherent states, squeezed odd/even coherent states and squeezed vacuum states, etc. We also present the approach (by properly setting the laser pulses) to realize control-NOT

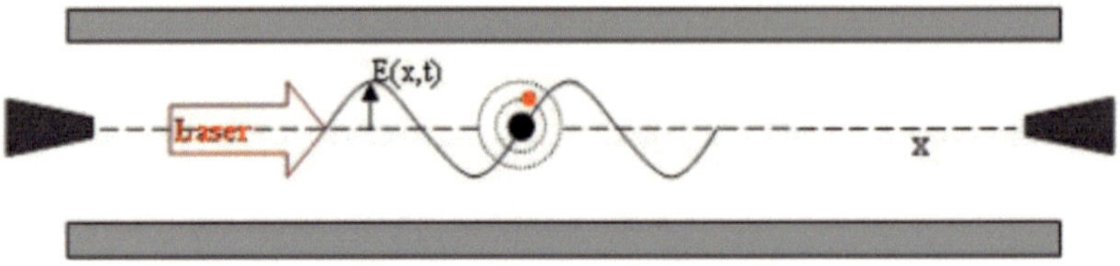

Fig. 1. (Color online) A sketch of an two-level ion trapped in one-dimensional electromagnetic trap (Only the vibrational motion along the x-direction is considered) and driven by a classical laser field propagating along the x direction.

(CNOT) gates (between the internal and external freedoms of a single trapped ion) beyond the LD approximation. In Sec. 4 we give a brief conclusion.

2. Dynamics of single trapped ions driven by laser beams

An electromagnetic-trap can provides a three-dimensional potential to trap an ion (with mass m and charge e). The potential function takes the form (3)

$$U(x,y,z) \approx \frac{1}{2}(\alpha x^2 + \beta y^2 + \gamma z^2), \tag{1}$$

near the trap-center ($x, y, z = 0$), with α, β, and γ bing the related experimental parameters. Therefore, a trapped ion (see, Fig.1) has two degrees of freedom: the vibrational motion around the trap-center and the internal atomic levels of the ion. For simplicity, we consider that the trap provides a pseudopotential whose frequencies satisfy the condition $v_x = v \ll v_y, v_z$. This implies that only the quantized vibrational motion along the x-direction (i.e., the principal axes of the trap) is considered. For the internal degree of freedom of the trapped ion, we consider only two selected levels, e.g., the ground state $|g\rangle$ and excited state $|e\rangle$. The Hamiltonian describing the above two uncoupled degrees of freedom can be written as

$$\hat{H}_0 = \hbar\nu(\hat{a}^\dagger\hat{a} + \frac{1}{2}) + \frac{\hbar}{2}\omega_a\hat{\sigma}_z, \tag{2}$$

where, $\hbar$ is the Planck's constant divided by 2π, $\hat{a}^\dagger$ and $\hat{a}$ are the bosonic creation and annihilation operators of the external vibrational quanta of the cooled ion with frequency ν. $\hat{\sigma}_z = |e\rangle\langle e| - |g\rangle\langle g|$ is the Pauli operator. The transition frequency ω_a is defined by $\omega_a = (E_e - E_g)/\hbar$ with E_g and E_e being the corresponding energies of the two selected levels, respectively.
In order to couple the above two uncoupled degrees of freedom of the trapped ion, we now apply a classical laser beam $E(x,t) = A\cos(k_l x - \omega_l t - \phi_l)$ (propagating along the x direction) to the trapped ion (see Fig. 1), where A, k_l, ω_l, and ϕ_l are its amplitude, wave-vector, frequency, and initial phase, respectively. This yields the laser-ion interaction

$$V = erE(x,t). \tag{3}$$

Certainly, $x = \sqrt{\hbar/2m\nu}(\hat{a} + \hat{a}^\dagger)$ and $r = \langle g|r|e\rangle(|g\rangle\langle e| + |e\rangle\langle g|)$, and thus the above interaction can be further written as

$$\hat{H}_{\rm int} = \frac{\hbar\Omega}{2}\hat{\sigma}_x\left(e^{i\eta(\hat{a}+\hat{a}^\dagger)-i\omega_l t-i\vartheta_l} + e^{-i\eta(\hat{a}+\hat{a}^\dagger)+i\omega_l t+i\vartheta_l}\right), \tag{4}$$

where, $\hat{\sigma}_x = \hat{\sigma}_- + \hat{\sigma}_+$ with $\hat{\sigma}_- = |g\rangle\langle e|$ and $\hat{\sigma}_- = |e\rangle\langle g|$ being the usual raising and lowering operators, respectively. The Rabi frequency $\Omega = eA\langle g|r|e\rangle/\hbar$ describes the strength of coupling between the applied laser field and the trapped ion. Finally, $\eta = k_l\sqrt{\hbar/2m\nu}$ is the so-called LD parameter, which describes the strength of coupling between external and internal degrees of freedom of the trapped ion.
Obviously, the total Hamiltonian, describing the trapped ion driven by a classical laser field, reads

$$\hat{H} = \hat{H}_0 + \hat{H}_{\text{int}} = \hbar\nu(\hat{a}^\dagger\hat{a} + \frac{1}{2}) + \frac{\hbar}{2}\omega_a\hat{\sigma}_z + \frac{\hbar\Omega}{2}\hat{\sigma}_x\left(e^{i\eta(\hat{a}+\hat{a}^\dagger)-i\omega_l t-i\vartheta_l} + e^{-i\eta(\hat{a}+\hat{a}^\dagger)+i\omega_l t+i\vartheta_l}\right). \quad (5)$$

This Hamiltonian yields a fundamental dynamics in this chapter for engineering the quantum states of trapped cold ion.

2.1 Within Lamb-Dicke approximation

Up to now, most of the experiments (see, e.g., (10)) for engineering the quantum states of trapped ions are operated under the LD approximation, i.e., supposing the LD parameters are sufficiently small ($\eta \ll 1$) such that one can make the following approximation:

$$e^{\pm i\eta(\hat{a}+\hat{a}^\dagger)} \approx 1 \pm i\eta(\hat{a} + \hat{a}^\dagger). \quad (6)$$

Under such an approximation, the Hamiltonian (5) can be reduced to

$$\hat{H}_{\text{LDA}} = \hat{H}_0 + \frac{\hbar\Omega}{2}\hat{\sigma}_x\left[e^{-i\omega_l t-i\vartheta_l} + e^{i\omega_l t+i\vartheta_l} + i\eta e^{-i\omega_l t-i\vartheta_l}(\hat{a}+\hat{a}^\dagger) - i\eta e^{i\omega_l t+i\vartheta_l}(\hat{a}+\hat{a}^\dagger)\right], \quad (7)$$

and can be further written as

$$\begin{aligned}\hat{H}'_{\text{LDA}} &= \tfrac{\hbar\Omega}{2}\left(e^{-i\omega_a t}\hat{\sigma}_- + \hat{\sigma}_+ e^{i\omega_a t}\right)\\ &\times\left[e^{-i\omega_l t-i\vartheta_l} + e^{i\omega_l t+i\vartheta_l} + i\eta e^{-i\omega_l t-i\vartheta_l}(e^{-i\nu t}\hat{a} + e^{i\nu t}\hat{a}^\dagger) - i\eta e^{i\omega_l t+i\vartheta_l}(e^{-i\nu t}\hat{a} + e^{i\nu t}\hat{a}^\dagger)\right]\end{aligned} \quad (8)$$

in the interaction picture defined by the unity operator $\hat{U} = \exp(-it\hat{H}_0/\hbar)$. Note that, to obtain the interacting Hamiltonian (8), we have used the following relations

$$\begin{cases} e^{-i\alpha\hat{\sigma}_z}\hat{\sigma}_z e^{i\alpha\hat{\sigma}_z} = \hat{\sigma}_z \\ e^{-i\alpha\hat{\sigma}_z}\hat{\sigma}_+ e^{i\alpha\hat{\sigma}_z} = e^{i2\alpha}\hat{\sigma}_+ \\ e^{-i\alpha\hat{\sigma}_z}\hat{\sigma}_- e^{i\alpha\hat{\sigma}_z} = e^{-i2\alpha}\hat{\sigma}_- \\ e^{-i\alpha\hat{a}^\dagger\hat{a}} f(\hat{a}^\dagger,\hat{a}) e^{i\alpha\hat{a}^\dagger\hat{a}} = f(e^{i\alpha}\hat{a}^\dagger, e^{-i\alpha}\hat{a}). \end{cases} \quad (9)$$

Above, α is an arbitrary parameter and $f(\hat{a}^\dagger,\hat{a})$ is an arbitrary function of operators $\hat{a}^\dagger$ and $\hat{a}$. Now, we assume that the frequencies of the applied laser fields are set as $\omega_l = \omega_0 + K\nu$, with $K = 0, \pm 1$ corresponding to the usual resonance ($K = 0$), the first blue- ($K = 1$), and red- ($K = -1$) sideband excitations (4), respectively. As a consequence, the Hamiltonian (8) can be specifically simplified to (30)

$$\hat{H}^0_{\text{LDA}} = \frac{\hbar\Omega}{2}\left(e^{i\vartheta_l}\hat{\sigma}_- + e^{-i\vartheta_l}\hat{\sigma}_+\right) \qquad \text{for } K = 0, \quad (10)$$

$$\hat{H}^{\text{r}}_{\text{LDA}} = \frac{\hbar\Omega}{2}i\eta\left(e^{-i\vartheta_l}\hat{\sigma}_+\hat{a} - e^{i\vartheta_l}\hat{\sigma}_-\hat{a}^\dagger\right) \quad \text{for } K = -1, \quad (11)$$

$$\hat{H}^{\mathrm{b}}_{\mathrm{LDA}} = \frac{\hbar\Omega}{2} i\eta \left(e^{-i\vartheta_l} \hat{\sigma}_+ \hat{a}^\dagger - e^{i\vartheta_l} \hat{\sigma}_- \hat{a} \right) \quad \text{for } K = -1, \tag{12}$$

under the rotating-wave approximation (i.e., neglected the rapidly-oscillating terms in Eq. (8)). Obviously, the Hamiltonian $\hat{H}^0_{\mathrm{LDA}}$ describes a Rabi oscillation and corresponds to a one-qubit operation, and Hamiltonians $\hat{H}^{\mathrm{r}}_{\mathrm{LDA}}$ and $\hat{H}^{\mathrm{b}}_{\mathrm{LDA}}$ correspond to the Jaynes-Cummings (JC) model (JCM) and anti-JCM (4), respectively. It is well-known that the JCM describes the basic interaction of a two-level atom and a quantized electromagnetic field (15). Here, the quantized electromagnetic field is replaced by the quantized vibration of the trapped ion, and thus the entanglement between the external motional states and the internal atomic states of the ions can be induced. Certainly, these JC interactions take the central role in the current trap-ion experiments for implementing the QIPs (10).

2.2 Beyond the LD approximation

In principle, quantum motion of the trapped ions beyond the usual LD limit ($\eta \ll 1$) is also possible. Utilizing the laser-ion interaction outside the LD regime might be helpful to reduce the noise in the ion-trap and improve the cooling rate, and thus could be utilized to efficiently realize QIPs. Indeed, several approaches have been proposed to coherently operate trapped ions beyond the LD limit (see, e.g., (19)).
Because $[\hat{a}, \hat{a}^\dagger] = 1$, the term $\exp[\pm i\eta(\hat{a} + \hat{a}^\dagger)]$ in Eq. (5) can be expanded as

$$e^{\pm i\eta(\hat{a}+\hat{a}^\dagger)} = e^{-\eta^2/2} e^{\pm i\eta\hat{a}^\dagger} e^{\pm i\eta\hat{a}} = e^{-\eta^2/2} \sum_{n,m}^{\infty} \frac{(\pm i\eta\hat{a}^\dagger)^n (\pm i\eta\hat{a})^m}{n!m!}. \tag{13}$$

Comparing to the expansion Eq. (6), the above expansion is exactly, and the LD parameter can be taken as an arbitrary value. With the help of relation (9) and (13), the Hamiltonian (5) can be written as:

$$\begin{aligned} \hat{H}'_{\mathrm{NLDA}} = \tfrac{\hbar\Omega}{2}\hat{\sigma}_+ \Big(& e^{i(\omega_a+\omega_l)t} e^{i\vartheta_l} e^{-\eta^2/2} \textstyle\sum_{n,m}^{\infty} \frac{(-i\eta\hat{a}^\dagger e^{ivt})^n (-i\eta\hat{a} e^{-ivt})^m}{n!m!} \\ & + e^{i(\omega_a-\omega_l)t} e^{-i\vartheta_l} e^{-\eta^2/2} \textstyle\sum_{n,m}^{\infty} \frac{(i\eta\hat{a}^\dagger e^{ivt})^n (i\eta\hat{a} e^{-ivt})^m}{n!m!} \Big) + H.c \end{aligned} \tag{14}$$

in the interaction picture defined by unity operator $\hat{U} = \exp(-it\hat{H}_0/\hbar)$.
For the so-called sideband excitations $\omega_l = \omega_0 + K\nu$ with K being arbitrary integer, the above Hamiltonian can be reduced to (23; 26)

$$\hat{H}^r_{\mathrm{NLDA}} = \frac{\hbar\Omega}{2} e^{-\eta^2/2} \left[e^{-i\vartheta_l} (i\eta)^k \hat{\sigma}_+ \sum_{j=0}^{\infty} \frac{(i\eta)^{2j} (\hat{a}^\dagger)^j \hat{a}^{j+k}}{j!(j+k)!} + H.c \right] \quad \text{for } K \le 0 \tag{15}$$

$$\hat{H}^b_{\mathrm{NLDA}} = \frac{\hbar\Omega}{2} e^{-\eta^2/2} \left[e^{-i\vartheta_l} (i\eta)^k \hat{\sigma}_+ \sum_{j=0}^{\infty} \frac{(i\eta)^{2j} (\hat{a}^\dagger)^{j+k} \hat{a}^{j}}{j!(j+k)!} + H.c \right] \quad \text{for } K \ge 0 \tag{16}$$

under the rotating wave approximation. Above, $K = 0$, $K < 0$, and $K > 0$ correspond to the resonance, the kth blue-, and red-sideband excitations (with $k = |K|$), respectively. Comparing with Hamiltonian (10)-(12) with LD approximation, the above Hamiltonian (without performing the LD approximation) can describes various multi-phonon transitions (i.e., $k > 1$) (19). This is due to the contribution from the high order effects of LD parameter.

2.3 Time evolutions of quantum states

The time evolution of the system governed by the above Hamiltonian can be solved by the time-evolution operator $\hat{T} = \exp(-it\hat{H}/\hbar)$, with $\hat{H}$ taking $\hat{H}^0_{\rm LDA}$, $\hat{H}^r_{\rm LDA}$, $\hat{H}^b_{\rm LDA}$, $\hat{H}^r_{\rm NLDA}$, and $\hat{H}^b_{\rm NLDA}$, respectively. For an arbitrary initial state $|\varphi(0)\rangle$, the evolving state at time t reads

$$|\varphi(t)\rangle = \hat{T}|\varphi(0)\rangle = \sum_{n=0}^{\infty} \frac{1}{n!}\left(\frac{-it}{\hbar}\right)^n \hat{H}^n\,|\varphi(0)\rangle. \tag{17}$$

Typically, if the external vibrational state of the ion is initially in a Fock state $|m\rangle$ and the internal atomic state is initially in the atomic ground sate $|g\rangle$ or excited one $|e\rangle$, then the above dynamical evolutions can be summarized as follows (23; 26):

i) For the resonance or red-sideband excitations $K \leq 0$:

$$\begin{cases} |m\rangle|g\rangle \longrightarrow |m\rangle|g\rangle,\ m < k, \\ |m\rangle|g\rangle \longrightarrow \cos(\Omega_{m-k,k}t)|m\rangle|g\rangle + i^{k-1}e^{-i\vartheta_l}\sin(\Omega_{m-k,k}t)|m-k\rangle|e\rangle;\ m \geq k, \\ |m\rangle|e\rangle \longrightarrow \cos(\Omega_{m,k}t)|m\rangle|e\rangle - (-i)^{k-1}e^{i\vartheta_l}\sin(\Omega_{m,k}t)|m+k\rangle|g\rangle \end{cases} \tag{18}$$

ii) For the resonance or blue-sideband excitations $K \geq 0$:

$$\begin{cases} |m\rangle|g\rangle \longrightarrow \cos(\Omega_{m,k}t)|m\rangle|g\rangle + i^{k-1}e^{-i\vartheta_l}\sin(\Omega_{m,k}t)|m+k\rangle|e\rangle, \\ |m\rangle|e\rangle \longrightarrow |m\rangle|e\rangle,\ m < k, \\ |m\rangle|e\rangle \longrightarrow \cos(\Omega_{m-k,k}t)|m\rangle|e\rangle - (-i)^{k-1}e^{i\vartheta_l}\sin(\Omega_{m-k,k}t)|m-k\rangle|g\rangle,\ m \geq k. \end{cases} \tag{19}$$

The so-called effective Rabi frequency introduced above reads

$$\Omega_{m,k} = \begin{cases} \Omega^L_{m,k} = \frac{\Omega}{2}\eta^k\sqrt{\frac{(m+k)!}{m!}}, \quad k = 0, 1, \\ \\ \Omega^N_{m,k} = \frac{\Omega}{2}\eta^k e^{-\eta^2/2}\sqrt{\frac{(m+k)!}{m!}}\sum_{j=0}^{m}\frac{(i\eta)^{2j}m!}{(j+k)!j!(m-j)!}, \quad k = 0, 1, 2, 3, \ldots. \end{cases} \tag{20}$$

The above derivations show that the dynamics either within or beyond the LD approximation has the same form (see, Eqs. (18) and (19)), only the differences between them is represented by the specifical Rabi frequencies $\Omega^L_{m,k}$ and $\Omega^N_{m,k}$. Certainly, the dynamical evolutions without LD approximation are more closed to the practical situations of the physical processes. Furthermore, comparing to dynamics with LD approximation (where $k = 0, 1$), the dynamics without LD approximation (where $k = 0, 1, 2, 3, \ldots$) can describes various multi-phonon transitions.

Certainly, when the LD parameters are sufficiently small, i.e., $\eta \ll 1$, the rate

$$\gamma = \frac{\Omega^L_{m,k}}{\Omega^N_{m,k}} = \frac{e^{\eta^2/2}}{\sum_{j=0}^{m}\frac{(i\eta)^{2j}m!}{(j+k)!j!(m-j)!}} \sim 1, \tag{21}$$

and thus the dynamics within LD approximation works well. Whereas, if the LD parameter are sufficiently large, the quantum dynamics beyond the LD limit must be considered.

3. Quantum-state engineerings beyond the LD limit

3.1 Preparations of various motional states of the single ions

The engineering of quantum states has attracted considerable attention in last two decades. This is in order to test fundamental quantum concepts such as non-locality, and for implementing various potential applications, including sensitive detections and quantum information processings. Laser-cooled single trapped ions are the good candidates for various quantum-state engineering processes due to their relatively-long decoherence times. Indeed, various engineered quantum states of trapped cold ions have been studied. The thermal, Fock, coherent, squeezed, and arbitrary quantum superposition states of motion of a harmonically bound ion have been investigated (4; 5; 6; 7). However, most of the related experiments are operated under the LD approximation. In this section we study the engineerings of various typical vibrational states of single trapped ions beyond the LD limit.

There are serval typical quantum states in quantum optics: coherence states, odd/even coherent states and squeezed states, etc.. They might show highly nonclassical properties, such as squeezing, anti-bunching and sub-Poissonian photon statistics. In the Fock space, coherence state $|\alpha\rangle$ can be regarded as a displaced vacuum state, i.e.,

$$|\alpha\rangle = D(\alpha)|0\rangle = e^{-|\alpha|^2/2}\sum_{n=0}^{\infty}\frac{\alpha^n}{\sqrt{n!}}|n\rangle, \tag{22}$$

where $D(\alpha) = \exp(\alpha\hat{a}^\dagger - \alpha^*\hat{a})$ is the so-called displace operator, with α being a complex parameter and describing the strength of displace. Similarly, the squeezed states $|\psi_s\rangle$ are generated by applying the squeeze operator to a quantum state

$$|\psi_s\rangle = \hat{S}(\xi)|\psi\rangle, \tag{23}$$

where $\hat{S} = \exp(\xi^*\hat{a}^2/2 - \xi\hat{a}^{\dagger 2}/2)$, and ξ is a complex parameter describing the strength of the squeezes. On the other hand, the odd/even coherent states are the superposed states of two coherent states with different phases

$$|\alpha_o\rangle = C_o(|\alpha\rangle - |-\alpha\rangle), \tag{24}$$

$$|\alpha_e\rangle = C_e(|\alpha\rangle + |-\alpha\rangle). \tag{25}$$

Above, $C_o = [2 - 2\exp(-2|\alpha|^2)]^{-1/2}$ and $C_e = [2 + 2\exp(-2|\alpha|^2)]^{-1/2}$ are the normalized coefficients.

According to the above definitions: Eqs. (22)-(25), one can obtain the squeezed coherent state:

$$|\alpha_s\rangle = \hat{S}(\xi)|\alpha\rangle = \sum_{n=0}^{\infty} G_n(\alpha,\xi)|n\rangle, \tag{26}$$

squeezed vacuum state:

$$|0_s\rangle = \hat{S}(\xi)|\alpha = 0\rangle = \sum_{n=0}^{\infty} G_n(0,\xi)|n\rangle, \tag{27}$$

squeezed odd state:

$$|\alpha_{o,s}\rangle = \hat{S}(\xi)|\alpha_o\rangle = \sum_{n=0}^{\infty} O_n(\alpha,\xi)|n\rangle, \tag{28}$$

and squeezed even state:

$$|\alpha_{e,s}\rangle = \hat{S}(\xi)|\alpha_e\rangle = \sum_{n=0}^{\infty} E_n(\alpha,\xi)|n\rangle. \tag{29}$$

respectively. Above,

$$G_n(\alpha,\xi) = \frac{1}{\sqrt{\cosh(r)n!}}\left(\frac{e^{i\theta}\sinh(r)}{2\cosh(r)}\right)^{n/2}\exp[\frac{e^{-i\theta}\sinh(r)\alpha^2}{2\cosh(r)} - \frac{|\alpha|^2}{2}]H_n(\frac{\alpha}{\sqrt{2e^{i\theta}\sinh(r)\cosh(r)}}). \tag{30}$$

with $\xi = r\exp(i\theta)$, and $H_n(x)$ being the hermitian polynomials. Finally,

$$O_n(\alpha,\xi) = C_o\left(G_n(\alpha,\xi) - G_n(-\alpha,\xi)\right) \tag{31}$$

and

$$E_n(\alpha,\xi) = C_e\left(G_n(\alpha,\xi) + G_n(-\alpha,\xi)\right). \tag{32}$$

Based on the quantum dynamics of laser-ion interaction beyond the LD limit, in what follows we discuss how to use a series of laser pulses to generate the above vibrational quantum states of the trapped ions. Assuming the trapped ion is initialized in the state $|0\rangle|g\rangle$, to generate the desirable quantum states we use the following sequential laser pulses to drive the trapped ion:

1) Firstly, a pulse with frequency $\omega_l = \omega_a$, initial phase ϑ_1, and duration t_1 is applied on the trapped ion to yield the evolution:

$$|0\rangle|g\rangle \to |\psi_1\rangle = \cos(\Omega_{0,0}t_1)|0\rangle|g\rangle + i^{-1}e^{-i\vartheta_1}\sin(\Omega_{0,0}t_1)|0\rangle|e\rangle \tag{33}$$

2) Secondly, a pulse with frequency $\omega_l = \omega_a - \nu$, initial phase ϑ_2, and duration t_2 is applied on the ion to generate:

$$\begin{aligned}|\psi_1\rangle \to |\psi_2\rangle = &\cos(\Omega_{0,0}t_1)|0\rangle|g\rangle + ie^{i(\vartheta_2-\vartheta_1)}\sin(\Omega_{0,0}t_1)\sin(\Omega_{0,1}t_2)|1\rangle|g\rangle \\ &+i^{-1}e^{-i\vartheta_1}\sin(\Omega_{0,0}t_1)\cos(\Omega_{0,1}t_2)|0\rangle|e\rangle\end{aligned} \tag{34}$$

3) Thirdly, a pulse with frequency $\omega_l = \omega_a - 2\nu$, initial phase ϑ_3, and duration t_3 is applied on the ion to generate:

$$\begin{aligned}|\psi_2\rangle \to |\psi_3\rangle = &\cos(\Omega_{0,0}t_1)|0\rangle|g\rangle + ie^{i(\vartheta_2-\vartheta_1)}\sin(\Omega_{0,0}t_1)\sin(\Omega_{0,1}t_2)|1\rangle|g\rangle \\ &+e^{i(\vartheta_3-\vartheta_1)}\sin(\Omega_{0,0}t_1)\cos(\Omega_{0,1}t_2)\sin(\Omega_{0,2}t_3)|2\rangle|g\rangle \\ &+i^{-1}e^{-i\vartheta_1}\sin(\Omega_{0,0}t_1)\cos(\Omega_{0,1}t_2)\cos(\Omega_{0,2}t_3)|0\rangle|e\rangle\end{aligned} \tag{35}$$

4) Successively, after the Nth pulse (with frequency $\omega_l = \omega_a - (N-1)\nu$, initial phase ϑ_N, and duration t_N) is applied on the ion, we have the following superposition state:

$$|\psi_{N-1}\rangle \to |\psi_N\rangle = \sum_{n=0}^{N-1} C_n|n\rangle|g\rangle + B_N|0\rangle|e\rangle \tag{36}$$

with

$$C_n = \begin{cases} \cos(\Omega_{0,0}t_1), & n = 0, \\ ie^{i(\vartheta_2-\vartheta_1)}\sin(\Omega_{0,0}t_1)\sin(\Omega_{0,1}t_2), & n = 1, \\ (-1)^{n-1}i^n e^{i(\vartheta_{n+1}-\vartheta_1)}\sin(\Omega_{0,0}t_1)\sin(\Omega_{0,n}t_{n+1})\prod_{j=2}^{n}\cos(\Omega_{0,j-1}t_j), & n > 1, \end{cases} \tag{37}$$

and

$$B_N = \begin{cases} i^{-1}e^{-i\vartheta_1}\sin(\Omega_{0,0}t_1), & N=1, \\ i^{-1}e^{-i\vartheta_1}\sin(\Omega_{0,0}t_1)\prod_{j=2}^{N}\cos(\Omega_{0,j-1}t_j), & N>1 \end{cases} \tag{38}$$

If the duration of Nth ($N>1$) laser pulse is set to satisfy the condition $\cos(\Omega_{0,N-1}t_N)=0$, then $B_N=0$. This implies that the internal state of the trapped ion is returned to the ground state $|g\rangle$ and its external vibration is prepared on the superposition state $|\psi_N^{\text{ex}}\rangle = \sum_{n=0}^{N-1} C_n|n\rangle$. Because the durations t_N and initial phases ϑ_N are experimentally controllable, we can set $C_n = G_n(\alpha,\xi)$, $O_n(\alpha,\xi)$, or $E_n(\alpha,\xi)$ with $n \le N-2$ by selecting the proper t_N and ϑ_N. As a consequence, we have the vibrational superposition state:

$$|\psi_N^G\rangle = \sum_{n=0}^{N-2} G_n(\alpha,\xi)|n\rangle + C_{N-1}^G|n-1\rangle, \tag{39}$$

$$|\psi_N^O\rangle = \sum_{n=0}^{N-2} O_n(\alpha,\xi)|n\rangle + C_{N-1}^O|n-1\rangle, \tag{40}$$

$$|\psi_N^E\rangle = \sum_{n=0}^{N-2} E_n(\alpha,\xi)|n\rangle + C_{N-1}^E|n-1\rangle, \tag{41}$$

and

$$|C_{N-1}^G|^2 = 1 - \sum_{n=0}^{N-2} |G_n(\alpha,\xi)|^2, \tag{42}$$

$$|C_{N-1}^O|^2 = 1 - \sum_{n=0}^{N-2} |O_n(\alpha,\xi)|^2, \tag{43}$$

$$|C_{N-1}^E|^2 = 1 - \sum_{n=0}^{N-2} |E_n(\alpha,\xi)|^2. \tag{44}$$

If the number of the laser pulses, i.e., N, are sufficiently large, the superposition states (39), (40), and (41) can well approach the desirable squeezed coherent state, squeezed odd and even coherent states, respectively. Specifically, when $\xi=0$, the usual coherent state, odd/even coherent states are generated.

Table 1 presents some numerical results for generating squeezed coherent states, where the typical values of parameters α and ξ, e.g., $\alpha=2$ and $\xi=0.5$, are considered. When $\alpha=0$, the squeezed coherent states correspond to the squeezed vacuum states. On the other hand, some results for generating squeezed odd/even coherent states are shown in table 2, and

α	ξ	N	$\Omega t_1(\vartheta_1)$	$\Omega t_2(\vartheta_2)$	$\Omega t_3(\vartheta_3)$	$\Omega t_4(\vartheta_4)$	$\Omega t_5(\vartheta_5)$	$\Omega t_6(\vartheta_6)$	F
0	0.5	5	0.7080(0)	0(0)	53.9174(π)	0(0)	4065.1(π)		0.9983
0	0.8	5	1.0859(0)	0(0)	43.9507(π)	0(0)	4065.1(π)		0.9816
1	0.0	5	1.8966(0)	7.1614($3\pi/2$)	46.0741(0)	343.5682($\pi/2$)	4065.1(0)		0.9981
2	0.5	5	2.5666(0)	5.3267($3\pi/2$)	43.7265(0)	371.8479($\pi/2$)	4065.1(π)		0.9882
2	0.5	6	2.5666(0)	5.3267($3\pi/2$)	43.7265(0)	371.8479($\pi/2$)	1826.3(π)	3635.9(0)	0.9910
2	0.8	6	2.2975(0)	6.8280($3\pi/2$)	43.9627(0)	59.8973($\pi/2$)	1877.7(π)	3635.9(0)	0.9821

Table 1. Parameters for generating squeezed coherent states, with the typical $\eta = 0.25$

odd/even	α	ξ	N	$\Omega t_1(\vartheta_1)$	$\Omega t_2(\vartheta_2)$	$\Omega t_3(\vartheta_3)$	$\Omega t_4(\vartheta_4)$	$\Omega t_5(\vartheta_5)$	$\Omega t_6(\vartheta_6)$	F
odd	1	0.0	5	3.2413(0)	9.6943($3\pi/2$)	0(0)	436.5962($\pi/2$)	4065.1(0)		0.9927
odd	2	0.5	6	3.2413(0)	7.7328($3\pi/2$)	0(0)	438.6774($\pi/2$)	0(0)	3635.9(0)	0.9974
odd	2	0.8	6	3.2413(0)	9.9520($3\pi/2$)	0(0)	85.3471($\pi/2$)	0(0)	3635.9(0)	0.9763
even	1	0.0	5	1.3106(0)	0(0)	59.9889(0)	0(0)	4065.1(0)		0.9995
even	2	0.5	5	2.2687(0)	0(0)	61.3354(0)	0(0)	4065.1(0)		0.9870
even	2	0.8	5	1.8498(0)	0(0)	51.1086(0)	0(0)	4065.1(0)		0.9900

Table 2. Parameters for generating squeezed odd/even coherent states, with the typical $\eta = 0.25$

where, $\xi = 0$ corresponds to the usual odd/even coherent states. By properly setting the durations t_N and phases ϑ_N of each laser pulse, the desirable squeezed states of the trapped ion could be generated with high fidelities (e.g., 99%), via few steps of laser pulse operations (e.g., $N = 6$). Here, the fidelities $F = \langle\psi_{\text{generated}}|\psi_{\text{expected}}\rangle$ are defined by the overlaps between the generated quantum states $|\psi_{\text{generated}}\rangle$ (being $|\psi_N^G\rangle$, $|\psi_N^O\rangle$, or $|\psi_N^E\rangle$) and the expected states $|\psi_{\text{expected}}\rangle$.

Above, the frequencies, durations, phases, and powers of the laser pulses should be set sufficiently exact for realizing the high fidelity quantum state operations. In fact, it is not difficult to generate the desired laser pulse with current experimental technology. It has been experimentally demonstrated that, a Fock state up to $|n = 16\rangle$ can be generated from the vibrational ground state $|0\rangle$, by sequentially applying laser pulses (with exact frequencies, durations, and phases, etc.) to the trapped ion (2; 4). This means that here the generation of the superposed Fock states, being coherent state, even/odd coherent states, or squeezed quantum states, etc., is also experimentally realizability.

On the other hand, the generation of the superposed Fock states is limited in practice by the existing decays of the vibrational and atomic states. Experimentally, the lifetime of the atomic excited states $|e\rangle$ reaches 1s and the coherence superposition of vibrational states $|0\rangle$ and $|1\rangle$ can be maintained for up to 1ms (32; 33). Therefore, roughly say, preparing the above superposed vibrational Fock states is experimentally possible, as the durations of quantum operation are sufficiently short, e.g., < 0.1 ms. To realize a short operation time, the Rabi frequencies Ω (describing the strength of the laser-ion interaction) should be set sufficiently large, e.g., when $\Omega = 10^5$ kHz, the total duration $t = t_1 + t_2 + t_3 \approx 0.04$ ms for generating squeezed vacuum state (see, table 1).

Certainly, by increasing the Rabi frequency $\Omega \propto \sqrt{P}$ via enhancing the power P of the applied laser beams, the operational durations can be further shorten. For example, if the Rabi frequency Ω increases ten times, then the total durations presented above for generating the desired quantum states could be shortened ten times. Therefore, the desired quantum states could be realized more efficiently. Indeed, the power of the laser applied to drive the trapped cold ion is generally controllable, e.g., it can be adjusted in the range from a few microwatts to a few hundred milliwatts (32).

3.2 Implementations of fundamental quantum logic gates

There is much interest in constructing quantum computer, which uses the principles of quantum mechanics to solve certain problems that could never been solved by any classical computer. An quantum computer is made up of many interacting units called quantum bits or qubits, which usually consist of two quantum states, e.g., $|0\rangle$ and $|1\rangle$. Unlike the classical bit, which hold either 0 or 1, qubit can hold $|0\rangle$, $|1\rangle$, or any quantum superposition of these states. Therefor, an quantum computer can solve certain problems much faster than any of our current classical computers, such as factoring large numbers (34) and searching large databases (35).

It has been shown that a quantum computer can be built using a series of one-qubit operations and two-qubit controlled-NOT (CNOT) gates, because any computation can be decomposed into a sequence of these basic logic operations (36; 37; 38). Therefore, a precondition work for realizing quantum computation is to effectively implement these fundamental logic gates. Indeed, since the first experiment of demonstrating quantum gates (10), with a single trapped cold ion, much attention has been paid to implement the elementary quantum gates in the systems of trapped cold ions, due to its relatively-long coherence times. However, most of previous approaches are work in LD limit. Following, we study the generations of CNOT gates beyond the LD limit.

A CNOT gate of a single trapped ion can be expressed as (10)

$$\begin{array}{ccc} \text{Input state} & & \text{Output state} \\ |0\rangle|g\rangle & \longrightarrow & |0\rangle|g\rangle \\ |0\rangle|e\rangle & \longrightarrow & |0\rangle|g\rangle \\ |1\rangle|g\rangle & \longrightarrow & |1\rangle|e\rangle \\ |1\rangle|e\rangle & \longrightarrow & |1\rangle|g\rangle \end{array} \tag{45}$$

Above, two lowest occupation number states $|0\rangle$ and $|1\rangle$ of the external vibration of the ion encode the control qubit and selected two atomic levels $|g\rangle$ and $|e\rangle$ of the internal degree of freedom of the ion define the target qubit. It is seen that, if the control qubit is in $|0\rangle$ the target qubit unchanged, whereas, if the control qubit is in $|1\rangle$ the target qubit is changed: $|g\rangle \rightleftarrows |e\rangle$, and thus a CNOT operation is implemented.

Without preforming the LD approximation, the CNOT gate could be implemented by sequentially applying three pulses of the laser beams:

1) A resonance pulse with frequency $\omega_l = \omega_a$, initial phase ϑ_1, and the duration t_1 is applied to implement the operation:

$$\begin{array}{l} |0\rangle|g\rangle \longrightarrow |\psi_a^{(1)}\rangle = \alpha_{11}^{(1)}|0\rangle|g\rangle + \alpha_{12}^{(1)}|0\rangle|e\rangle \\ |0\rangle|e\rangle \longrightarrow |\psi_b^{(1)}\rangle = \alpha_{21}^{(1)}|0\rangle|e\rangle + \alpha_{22}^{(1)}|0\rangle|g\rangle \\ |1\rangle|g\rangle \longrightarrow |\psi_c^{(1)}\rangle = \alpha_{31}^{(1)}|1\rangle|g\rangle + \alpha_{32}^{(1)}|1\rangle|e\rangle \\ |1\rangle|e\rangle \longrightarrow |\psi_d^{(1)}\rangle = \alpha_{41}^{(1)}|1\rangle|e\rangle + \alpha_{42}^{(1)}|1\rangle|g\rangle \end{array} \tag{46}$$

2) A red sideband pulse with the frequency $\omega_l = \omega_a - \nu$, initial phase ϑ_2, and duration t_2 is applied to implement the evolution

$$\begin{array}{l} |\psi_a^{(1)}\rangle \longrightarrow |\psi_a^{(2)}\rangle = \alpha_{11}^{(2)}|0\rangle|g\rangle + \alpha_{12}^{(2)}|0\rangle|e\rangle + \alpha_{13}^{(2)}|0\rangle|g\rangle \\ |\psi_b^{(1)}\rangle \longrightarrow |\psi_b^{(2)}\rangle = \alpha_{21}^{(2)}|0\rangle|g\rangle + \alpha_{22}^{(2)}|0\rangle|e\rangle + \alpha_{23}^{(2)}|1\rangle|g\rangle \\ |\psi_c^{(1)}\rangle \longrightarrow |\psi_c^{(2)}\rangle = \alpha_{31}^{(2)}|0\rangle|e\rangle + \alpha_{32}^{(2)}|1\rangle|g\rangle + \alpha_{33}^{(2)}|1\rangle|e\rangle + \alpha_{34}^{(2)}|2\rangle|g\rangle \\ |\psi_d^{(1)}\rangle \longrightarrow |\psi_d^{(2)}\rangle = \alpha_{41}^{(2)}|0\rangle|e\rangle + \alpha_{42}^{(2)}|1\rangle|g\rangle + \alpha_{43}^{(2)}|1\rangle|e\rangle + \alpha_{44}^{(2)}|2\rangle|g\rangle \end{array} \tag{47}$$

3) A resonance pulse with the initial phase $\vartheta_3 = -\pi/2$ and duration $t_3 = \pi/(4\Omega^L_{0,0})$ is again applied to implement the following operation

$$\begin{aligned}
&|\psi_a^{(2)}\rangle \longrightarrow |\psi_a^{(3)}\rangle = \alpha_{11}^{(3)}|0\rangle|g\rangle + \alpha_{12}^{(3)}|0\rangle|e\rangle + \alpha_{13}^{(3)}|1\rangle|g\rangle + \alpha_{14}^{(3)}|1\rangle|e\rangle \\
&|\psi_b^{(2)}\rangle \longrightarrow |\psi_b^{(3)}\rangle = \alpha_{21}^{(3)}|0\rangle|g\rangle + \alpha_{22}^{(3)}|0\rangle|e\rangle + \alpha_{23}^{(3)}|1\rangle|g\rangle + \alpha_{24}^{(3)}|1\rangle|e\rangle \\
&|\psi_c^{(2)}\rangle \longrightarrow |\psi_c^{(3)}\rangle = \alpha_{31}^{(3)}|0\rangle|g\rangle + \alpha_{32}^{(3)}|0\rangle|e\rangle + \alpha_{33}^{(3)}|1\rangle|g\rangle + \alpha_{34}^{(3)}|1\rangle|e\rangle + \alpha_{35}^{(3)}|2\rangle|g\rangle + \alpha_{36}^{(3)}|2\rangle|e\rangle \\
&|\psi_d^{(2)}\rangle \longrightarrow |\psi_d^{(3)}\rangle = \alpha_{41}^{(3)}|0\rangle|g\rangle + \alpha_{42}^{(3)}|0\rangle|e\rangle + \alpha_{43}^{(3)}|1\rangle|g\rangle + \alpha_{44}^{(3)}|1\rangle|e\rangle + \alpha_{45}^{(3)}|2\rangle|g\rangle + \alpha_{46}^{(3)}|2\rangle|e\rangle
\end{aligned} \tag{48}$$

Above, the coefficients $\alpha_{11}^{(1)}, \alpha_{12}^{(1)}$, and $\alpha_{13}^{(1)}$ et. al. are derived in APPENDIX A, basing on the dynamical evolution Eqs. (18) and (19).
With the above sequential three operations, the system undergoes the following evolution:

$$\begin{array}{ll}
\text{Input state} & \qquad\qquad \text{Output state} \\
|0\rangle|g\rangle \longrightarrow |\psi_a^{(1)}\rangle & \longrightarrow |\psi_a^{(2)}\rangle \longrightarrow |\psi_a^{(3)}\rangle \\
|0\rangle|e\rangle \longrightarrow |\psi_b^{(1)}\rangle & \longrightarrow |\psi_b^{(2)}\rangle \longrightarrow |\psi_b^{(3)}\rangle \\
|1\rangle|g\rangle \longrightarrow |\psi_c^{(1)}\rangle & \longrightarrow |\psi_c^{(2)}\rangle \longrightarrow |\psi_c^{(3)}\rangle \\
|1\rangle|e\rangle \longrightarrow |\psi_d^{(1)}\rangle & \longrightarrow |\psi_d^{(2)}\rangle \longrightarrow |\psi_d^{(3)}\rangle
\end{array} \tag{49}$$

Obviously, if $|\psi_a^{(3)}\rangle = |0\rangle|g\rangle$, $|\psi_b^{(3)}\rangle = |0\rangle|e\rangle$, $|\psi_c^{(3)}\rangle = |1\rangle|e\rangle$, and $|\psi_d^{(3)}\rangle = |1\rangle|g\rangle$, the CNOT gate (45) is realized. This means that the coefficients should satisfy the condition (see, Eq. (48)):

$$\alpha_{11}^{(3)} = \alpha_{22}^{(3)} = \alpha_{34}^{(3)} = \alpha_{43}^{(3)} = 1, \tag{50}$$

i.e.,(see, APPENDIX A):

$$\begin{cases}
1 = \cos(\Omega_{0,0}t_1)\cos(\Omega_{0,0}t_3) - e^{i(\vartheta_3-\vartheta_1)}\sin(\Omega_{0,0}t_1)\cos(\Omega_{0,1}t_2)\sin(\Omega_{0,0}t_3), \\
1 = -e^{i(\vartheta_1-\vartheta_3)}\sin(\Omega_{0,0}t_1)\sin(\Omega_{0,0}t_3) + \cos(\Omega_{0,0}t_1)\cos(\Omega_{0,1}t_2)\cos(\Omega_{0,0}t_3), \\
1 = -ie^{-i\vartheta_3}\cos(\Omega_{1,0}t_1)\cos(\Omega_{0,1}t_2)\sin(\Omega_{1,0}t_3) - ie^{-i\vartheta_1}\sin(\Omega_{1,0}t_1)\cos(\Omega_{1,1}t_2)\cos(\Omega_{1,0}t_3), \\
1 = -ie^{i\vartheta_3}\cos(\Omega_{1,0}t_1)\cos(\Omega_{1,1}t_2)\sin(\Omega_{1,0}t_3) - ie^{i\vartheta_1}\sin(\Omega_{1,0}t_1)\cos(\Omega_{0,1}t_2)\cos(\Omega_{1,0}t_3).
\end{cases} \tag{51}$$

In the previous approaches, within the LD limit, for implementing quantum gate, the $k\pi$ pulses $k = 1, 2, 3, ...$ of the laser beams are usually applied to drive the ions. The relevant LD parameters should be set sufficiently small such that the LD approximation could be satisfied. Here, in order to realize the desirable quantum gate with arbitrarily LD parameters (i.e., beyond the LD limit), we assume that the durations of the applied laser beams could be arbitrarily set (instead of the so-called $k\pi$ in the previous approaches). Indeed, the desired quantum gate could be implemented for arbitrarily experimental LD parameters, as long as the durations t_N and initial phases ϑ_N of the applied three laser beams ($N = 1, 2, 3$) are accurately set such that Eq. (50) is satisfied.
For equation (50), there maybe exist many solutions for any selected LD parameter. However, the practical existence of decoherence imposes the limit that the total duration of the present three-step pulse should be shorter than the decoherence times of both the atomic and motional states. This implies that exact solutions to Eq. (50) do not exist for certain experimental LD parameters. Certainly, for these parameters approximative solutions to Eq. (50) still exist and the CNOT gate could be approximately generated. The fidelity F for such an approximated realization is defined as the minimum among the probability amplitudes: $\alpha_{11}^{(3)}, \alpha_{22}^{(3)}, \alpha_{34}^{(3)}$, and $\alpha_{43}^{(3)}$, i.e. (27),

η	$\Omega t_1 = \Omega t_3$	Ωt_2	F
0.17	192.01	673.50	0.9935
0.18	292.03	637.85	0.9984
0.20	1.6690	255.07	0.9935
0.25	1.7287	207.94	0.9986
0.30	9.0200	438.00	0.9998
0.31	31.000	552.89	1.0000
0.35	177.00	1107.0	1.0000
0.40	2.0260	170.07	0.9999
0.50	97.304	57.310	0.9975
0.60	61.710	75.480	1.0000
0.70	66.900	160.85	0.9969
0.75	175.99	665.98	0.9998
0.85	73.110	105.91	0.9985
0.90	12.400	460.43	0.9992
0.95	25.300	415.21	0.9973

Table 3. Parameters for generating high-fidelity CNOT gates with arbitrarily selected LD parameters from 0.17 to 0.95 (27). Here $\vartheta_1 = -\vartheta_3 = \pi/2$, and $t_1 = t_3$.

$$F = \min\left\{\alpha_{11}^{(3)}, \alpha_{22}^{(3)}, \alpha_{34}^{(3)}, \alpha_{43}^{(3)}\right\} \tag{52}$$

It has been shown that the lifetime of the atomic excited state $|e\rangle$ reaches 1 s and the coherence superposition of vibrational states $|0\rangle$ and $|1\rangle$ can be maintained for up to 1 ms (32; 33). These allow us to perform the required three-step sequential laser manipulations; e.g., for $\Omega/(2\pi) \approx$ 500 kHz and $\eta = 0.20$ (4) the total duration $t = t_1 + t_2 + t_3 \approx 0.08$ ms. Table 3 provides some numerical results for arbitrarily selected LD parameters (not limited within the LD regime requiring $\eta \ll 1$) from 0.17 to 0.95. Here, the durations of the applied laser beam pulses are defined by the parameters Ωt_i with $i = 1, 2, 3$. It is seen that the fidelities of implementing the desirable CNOT gate for these LD parameters are significantly high (all of them larger than 99%).

3.3 Simplified approaches for implementing quantum logic gates

In fact, the above approach (with three laser pulses) for generating CNOT gate can be simplified, by using only two laser pulses (28). With the sequential two operations 1) and 2), the system undergoes the following evolutions:

$$\begin{array}{ccc} \text{Input state} & & \text{Output state} \\ |0\rangle|g\rangle \longrightarrow |\psi_a^{(1)}\rangle & \longrightarrow & |\psi_a^{(2)}\rangle \\ |0\rangle|e\rangle \longrightarrow |\psi_b^{(1)}\rangle & \longrightarrow & |\psi_b^{(2)}\rangle \\ |1\rangle|g\rangle \longrightarrow |\psi_c^{(1)}\rangle & \longrightarrow & |\psi_c^{(2)}\rangle \\ |1\rangle|e\rangle \longrightarrow |\psi_d^{(1)}\rangle & \longrightarrow & |\psi_d^{(2)}\rangle \end{array} \tag{53}$$

If

$$\alpha_{11}^{(2)} = \alpha_{22}^{(2)} = \alpha_{33}^{(2)} = \alpha_{42}^{(2)} = 1, \tag{54}$$

i.e.,(see, APPENDIX A):

$$\begin{cases} 1 = \cos(\Omega_{0,0}t_1), \\ 1 = \cos(\Omega_{0,0}t_1)\cos(\Omega_{0,1}t_2), \\ 1 = i^{-1}e^{-i\vartheta_1}\sin(\Omega_{1,0}t_1)\cos(\Omega_{1,1}t_2), \\ 1 = i^{-1}e^{i\vartheta_1}\sin(\Omega_{1,0}t_1)\cos(\Omega_{0,1}t_2), \end{cases} \tag{55}$$

then $|\psi_a^{(2)}\rangle = |0\rangle|g\rangle$, $|\psi_b^{(2)}\rangle = |0\rangle|e\rangle$, $|\psi_c^{(2)}\rangle = |1\rangle|e\rangle$, and $|\psi_d^{(2)}\rangle = |1\rangle|g\rangle$. Thus, the desirable CNOT gate (45) is realized.

The above condition could be achieved by properly setting the relevant experimental parameters, t_1, t_2, ϑ_1, and η as

$$t_1 = \frac{2p\pi}{\Omega_{0,0}}, t_2 = \frac{2m\pi}{\Omega_{0,1}}, \theta_1 = (-1)^{q+1}\frac{\pi}{2}, \tag{56}$$

with $p, q, m = 1, 2, 3...$, and

$$\eta^2 = 1 \quad \frac{q-0.5}{2p} = 2 \quad \frac{\sqrt{2}(n-0.5)}{m}, \tag{57}$$

with $n = 1, 2, 3, ...$ This means that the CNOT gate (45) could be implemented, since all the experimental parameters related above are accurately controllable. This implies that the required laser pulses for implementing a CNOT gate could be really reduced to two ones. Due to decoherence(i.e., the limit of durations $t_1 + t_2$), the integers p, q, m, n could not take arbitrary large values to let (57) be satisfied exactly, although their approximated solutions

η	$t_1\Omega$	$t_2\Omega$	$\theta_1 = \pm\frac{\pi}{2}$	F
0.18	689.6666	993.3470	+	0.9907
0.20	1525.600	1410.200	+	0.9958
0.22	1261.700	1463.000	+	0.9984
0.25	1504.000	1192.800	-	0.9978
0.28	457.4064	326.7189	+	0.9967
0.30	328.6193	438.1591	-	0.9993
0.38	351.1877	284.3625	+	0.9994
0.40	490.0675	170.1623	+	0.9975
0.44	304.5596	283.1649	-	0.9980
0.53	347.0706	190.9979	+	0.9970
0.57	443.4884	492.7649	+	0.9993
0.64	385.5613	481.9516	-	0.9967
0.68	316.7005	395.8756	-	0.9950
0.73	377.2728	292.1111	-	0.9953
0.80	276.8880	281.2143	-	0.9980
0.86	436.5392	486.4535	+	0.9999
0.90	471.0198	460.5527	-	0.9971
0.96	318.7519	394.2894	+	0.9969

Table 4. CNOT gates implemented with sufficiently high fidelities for arbitrarily selected parameters (28).

η	$t_1\Omega$	$\alpha_{11}^{(1)}=\alpha_{21}^{(1)}$	$\alpha_{12}^{(1)}=\alpha_{22}^{(1)}$	$\alpha_{31}^{(1)}=\alpha_{41}^{(1)}$	$\alpha_{32}^{(1)}=\alpha_{42}^{(1)}$
0.18	267.75	0.97520	-0.22135	-0.21954	0.97560
0.20	243.65	0.99948	0.03218	0.03193	0.99949
0.22	179.76	0.97284	-0.23146	-0.23053	0.97306
0.24	168.13	1.00000	0.00000	0.00000	1.00000
0.26	129.49	0.97165	-0.23640	-0.24006	0.97076
0.28	130.92	0.99376	0.11153	0.11041	0.99389
0.30	104.95	0.99505	-0.09935	-0.09778	0.99521
0.32	92.35	0.99374	-0.11172	-0.10790	0.99416
0.34	79.49	0.98271	-0.18517	-0.18852	0.98207
0.36	80.64	0.99586	0.09088	0.09407	0.99557
0.38	67.33	0.99541	-0.09572	-0.09377	0.99559
0.40	68.44	0.98505	0.17225	0.16794	0.98580
0.42	54.57	0.98859	-0.15063	-0.15383	0.98810
0.44	55.56	0.99646	0.08411	0.08530	0.99636
0.46	111.30	0.97957	-0.20111	-0.19843	0.98012
0.48	41.83	0.97796	-0.20881	-0.20607	0.97854
0.50	42.72	1.00000	0.00000	0.00000	1.00000
0.52	43.67	0.97497	0.22234	0.21915	0.97570
0.54	87.23	1.00000	0.00000	0.00000	1.00000
0.56	132.93	0.96361	0.26731	0.26592	0.96400
0.58	118.42	0.97544	-0.22027	-0.22025	0.97544
0.60	29.81	0.99320	-0.11640	-0.11358	0.99353

Table 5. CNOT gates implemented with high fidelities for arbitrarily selected parameters(from 0.18 to 0.60) (29).

are still exists. This implies that the desirable CNOT gate is approximately implemented with a fidelity $F < 1$. Similar to that of Sec. 3.2, here the fidelity F of implementing the desirable gate is defined as the minimum among the values of $\alpha_{11}^{(2)}$, $\alpha_{22}^{(2)}$, $\alpha_{33}^{(2)}$ and $\alpha_{42}^{(2)}$, i.e., $F = \min\{\alpha_{11}^{(2)}, \alpha_{22}^{(2)}, \alpha_{33}^{(2)}, \alpha_{42}^{(2)}\}$.

In table 4, we present some experimental parameters arbitrarily selected to implement the expected CNOT gate (45), with sufficiently high fidelities $F > 99\%$. For example, we have $F = 99.58\%$ for the typical LD parameter $\eta = 0.2$. It is also seen from the table that the present proposal still works for certain large LD parameters, e.g., 0.90, 0.96, etc.. For the experimental parameters $\eta = 0.2$ and $\Omega/(2\pi) \approx 500$ kHz, the total duration $t_1 + t_2$ for this implementation is about 0.9 ms. By increasing the Rabi frequency Ω via enhancing the power of the applied laser beams, the durations can be further shorten, and thus the CNOT gates could be realized more efficiently.

In fact, a CNOT gate (*apart from certain phase factors*) (39) could be still realized by a single laser pulse. By setting $\vartheta_1 = \pi/2$ and

$$\cos(\Omega_{0,0}t_1) = \sin(\Omega_{1,0}t_1) = 1, \tag{58}$$

η	$t_1\Omega$	$\alpha_{11}^{(1)} = \alpha_{21}^{(1)}$	$\alpha_{12}^{(1)} = \alpha_{22}^{(1)}$	$\alpha_{31}^{(1)} = \alpha_{41}^{(1)}$	$\alpha_{32}^{(1)} = \alpha_{42}^{(1)}$
0.62	30.64	0.99720	0.07473	0.07201	0.99740
0.64	31.52	0.96241	0.27158	0.26906	0.96312
0.66	62.42	0.99952	-0.03096	-0.03027	0.99954
0.68	95.26	0.99509	0.09898	0.09985	0.99500
0.70	353.53	0.99224	0.12437	0.12458	0.99221
0.72	178.61	0.97983	-0.19981	-0.20082	0.97963
0.74	82.49	0.99876	-0.04974	-0.05286	0.99860
0.76	50.12	0.99715	-0.07548	-0.07608	0.99710
0.78	186.67	0.96719	-0.25406	-0.25835	0.96605
0.80	155.88	0.99888	0.04737	0.04576	0.99895
0.82	175.54	0.99258	-0.12162	-0.12311	0.99239
0.84	17.27	0.97693	-0.21356	-0.21395	0.97685
0.86	18.04	0.99867	-0.05149	-0.05191	0.99865
0.88	18.88	0.99205	0.12582	0.12453	0.99221
0.90	19.79	0.95031	0.31129	0.31157	0.95022
0.92	153.85	0.99321	0.11610	0.11507	0.99336
0.94	39.40	0.99517	0.09817	0.09647	0.99534
0.96	60.04	0.99627	0.08632	0.08611	0.99629
0.98	122.12	0.99709	0.07617	0.07482	0.99720

Table 6. CNOT gates implemented with high fidelities for arbitrarily selected parameters(from 0.62 to 0.98) (29).

the operation (46) can realize the following two-qubit quantum operation (see, APPENDIX A)

$$\begin{array}{l} |0\rangle|g\rangle \longrightarrow |0\rangle|g\rangle \\ |0\rangle|e\rangle \longrightarrow |0\rangle|e\rangle \\ |1\rangle|g\rangle \longrightarrow -|1\rangle|e\rangle \\ |1\rangle|e\rangle \longrightarrow |1\rangle|g\rangle \end{array} \tag{59}$$

which is equivalent to the standard CNOT gate (45) between the external and internal states of the ion, apart from the phase factors -1.

Obviously, the condition (58) can be satisfied by properly setting the relevant experimental parameters: t_1 and η, as

$$t_1 = \frac{2n\pi}{\Omega_{0,0}}, \eta^2 = 1 - \frac{m - \frac{3}{4}}{n}, \ n, m = 1, 2, 3...., \tag{60}$$

with n and m being arbitrary positive integers. Because of the practical existence of decoherence, as we discussed above, the duration of the present pulse should be shorter than the decoherence times of both the atomic and motional states of the ion. This limits that the integers n could not take arbitrary large values to let Eq. (60) be exactly satisfied.

In tables 5 and 6 we present some numerical results for setting proper experimental parameters Ωt_1 (all of them $\lesssim 0.1$ ms for the experimental Rabi frequency $\Omega/(2\pi) \approx 500$ KHz), to implement quantum operation (59) for the arbitrarily selected LD parameters (not limited within the LD regime requiring $\eta \ll 1$) from 0.18 to 0.98. It is seen that, the probability amplitudes $\alpha_{11}^{(1)} = \alpha_{21}^{(1)}$ and $\alpha_{32}^{(1)} = \alpha_{42}^{(1)}$ are desirably large, most of them could reach to 0.99. While, unwanted probability amplitudes $\alpha_{12}^{(1)} = \alpha_{22}^{(1)}$ and $\alpha_{31}^{(1)} = \alpha_{41}^{(1)}$ are really significantly small; all of them is less than 0.32. This implies that the lowest fidelity $F = \min\{\alpha_{11}^{(1)}, \alpha_{21}^{(1)}, \alpha_{32}^{(1)}, \alpha_{42}^{(1)}\}$ for implementing the quantum operation (59) is larger than 95%. Certainly, the above approximated solutions could be further improved by either relaxing the limit from the decoherence time or increasing Rabi frequency Ω (via increasing the powers of the applied laser beams) to shorten the operational time. Designing the applied laser pulse with so short duration is not a great difficulty for the current experimental technology, e.g., the femto-second (10^{-15}s) laser technique. Also, our numerical calculations show that the influence of the possibly-existing fluctuations of the applied durations is really weak. For example, for the Rabi frequency $\Omega/(2\pi) \approx 500$ kHz, the fluctuation $\delta t \approx 0.1\,\mu$s of the duration lowers the desirable probability amplitudes, i.e., $\alpha_{11}^{(1)}$ and $\alpha_{32}^{(1)}$ presented in tables 5 and 6, just about 5%. Thus, even consider the imprecision of the durations, the amplitude of the desirable elements, $\alpha_{11}^{(1)}$ and $\alpha_{32}^{(1)}$, are still sufficiently large, e.g., up to about 0.95. Therefore, the approach proposed here to implement the desirable quantum operation (59) for arbitrary LD parameters should be experimentally feasible.

Finally, we consider how to generate the standard CNOT gate (45) from the quantum operation (59) produced above. This could be achieved by just eliminating the unwanted phase factors in (59) via introducing another off-resonant laser pulse (10). Indeed, a first blue-sideband pulse (of frequency $\omega_L = \omega_{ea} + \nu$ and initial phase ϑ_2) induces the following evolution

$$|1\rangle|e\rangle \longrightarrow \cos(\Omega_{0,1}t_2)|1\rangle|e\rangle - e^{i\vartheta_2}\sin(\Omega_{0,1}t_2)|0\rangle|a\rangle, \tag{61}$$

but does not evolve the states $|0\rangle|g\rangle$, $|1\rangle|g\rangle$ and $|0\rangle|e\rangle$. Above, $|a\rangle$ is an auxiliary atomic level, and ω_{ea} being the transition frequency between it and the excited state $|e\rangle$. Obviously, a "π-pulse" defined by $\Omega_{0,1}t_2 = \pi$ generates a so-called controlled-Z logic operation (10)

$$\begin{aligned} |0\rangle|g\rangle &\longrightarrow |0\rangle|g\rangle \\ |0\rangle|e\rangle &\longrightarrow |0\rangle|e\rangle \\ |1\rangle|g\rangle &\longrightarrow |1\rangle|g\rangle \\ |1\rangle|e\rangle &\longrightarrow -|1\rangle|e\rangle \end{aligned} \tag{62}$$

For the LD parameters from 0.18 to 0.98, and $\Omega/(2\pi) \approx 500$ kHz, the durations for this implementation are numerically estimated as $3.3 \times 10^{-3} \sim 1.2 \times 10^{-2}$ ms. Therefore, the standard CNOT gate (45) with a single trapped ion could be implemented by only two sequential operations demonstrated above.

4. Conclusions

In this chapter, we have summarized our works on designing properly laser pulses for preparing typical motional quantum states and implementing quantum logic gates with single trapped ions beyond the LD approximation. To generate the typical vibrational states (e.g., the coherence states, squeezed coherent states, odd/even coherent states and squeezed states) and implement CNOT gates of a single trapped ion, serval sequential laser pules are proposed to be applied to the trapped ion. It is shown that by properly set the frequency, duration, and phase, etc., of the each laser pulse the desired vibrational states and CNOT gates of a single trapped ion could be well generated with high fidelities.

5. APPENDIX A:

$$\begin{aligned}
\alpha_{11}^{(1)} &= \alpha_{21}^{(1)} = \cos(\Omega_{0,0}t_1)\\
\alpha_{12}^{(1)} &= i^{-1}e^{-i\vartheta_1}\sin(\Omega_{0,0}t_1)\\
\alpha_{22}^{(1)} &= i^{-1}e^{i\vartheta_1}\sin(\Omega_{0,0}t_1)\\
\alpha_{31}^{(1)} &= \alpha_{41}^{(1)} = \cos(\Omega_{1,0}t_1)\\
\alpha_{32}^{(1)} &= i^{-1}e^{-i\vartheta_1}\sin(\Omega_{1,0}t_1)\\
\alpha_{42}^{(1)} &= i^{-1}e^{i\vartheta_1}\sin(\Omega_{1,0}t_1)
\end{aligned} \tag{63}$$

$$\begin{aligned}
\alpha_{11}^{(2)} &= \cos(\Omega_{0,0}t_1)\\
\alpha_{12}^{(2)} &= i^{-1}e^{-i\vartheta_1}\sin(\Omega_{0,0}t_1)\cos(\Omega_{0,1}t_2)\\
\alpha_{13}^{(2)} &= -i^{-1}e^{i(\vartheta_2-\vartheta_1)}\sin(\Omega_{0,0}t_1)\sin(\Omega_{0,1}t_2)\\
\alpha_{21}^{(2)} &= i^{-1}e^{i\vartheta_1}\sin(\Omega_{0,0}t_1)\\
\alpha_{22}^{(2)} &= \cos(\Omega_{0,0}t_1)\cos(\Omega_{0,1}t_2)\\
\alpha_{23}^{(2)} &= -e^{-i\vartheta_2}\cos(\Omega_{0,0}t_1)\sin(\Omega_{0,1}t_2)
\end{aligned} \tag{64}$$

$$\begin{aligned}
\alpha_{31}^{(2)} &= e^{-i\vartheta_2}\cos(\Omega_{1,0}t_1)\sin(\Omega_{0,1}t_2)\\
\alpha_{32}^{(2)} &= \cos(\Omega_{1,0}t_1)\cos(\Omega_{0,1}t_2)\\
\alpha_{33}^{(2)} &= i^{-1}e^{-i\vartheta_1}\sin(\Omega_{1,0}t_1)\cos(\Omega_{1,1}t_2)\\
\alpha_{34}^{(2)} &= -i^{-1}e^{i(\vartheta_2-\vartheta_1)}\sin(\Omega_{1,0}t_1)\sin(\Omega_{1,1}t_2)
\end{aligned} \tag{65}$$

$$\begin{aligned}
\alpha_{41}^{(2)} &= i^{-1}e^{i(\vartheta_1-\vartheta_2)}\sin(\Omega_{1,0}t_1)\sin(\Omega_{0,1}t_2)\\
\alpha_{42}^{(2)} &= i^{-1}e^{i\vartheta_1}\sin(\Omega_{1,0}t_1)\cos(\Omega_{0,1}t_2)\\
\alpha_{43}^{(2)} &= \cos(\Omega_{1,0}t_1)\cos(\Omega_{1,1}t_2)\\
\alpha_{44}^{(2)} &= -e^{i\vartheta_2}\cos(\Omega_{1,0}t_1)\sin(\Omega_{1,1}t_2)
\end{aligned} \tag{66}$$

$$
\begin{aligned}
\alpha_{11}^{(3)} &= \cos(\Omega_{0,0}t_1)\cos(\Omega_{0,0}t_3) - e^{i(\vartheta_3-\vartheta_1)}\sin(\Omega_{0,0}t_1)\cos(\Omega_{0,1}t_2)\sin(\Omega_{0,0}t_3) \\
\alpha_{12}^{(3)} &= -ie^{-i\vartheta_3}\cos(\Omega_{0,0}t_1)\sin(\Omega_{0,0}t_3) - ie^{-i\vartheta_1}\sin(\Omega_{0,0}t_1)\cos(\Omega_{0,1}t_2)\cos(\Omega_{0,0}t_3) \\
\alpha_{13}^{(3)} &= ie^{i(\vartheta_2-\vartheta_1)}\sin(\Omega_{0,0}t_1)\sin(\Omega_{0,1}t_2)\cos(\Omega_{0,1}t_3) \\
\alpha_{14}^{(3)} &= ie^{i(\vartheta_2-\vartheta_1-\vartheta_3)}\sin(\Omega_{0,0}t_1)\sin(\Omega_{0,1}t_2)\sin(\Omega_{0,1}t_3)
\end{aligned} \tag{67}
$$

$$
\begin{aligned}
\alpha_{21}^{(3)} &= -ie^{i\vartheta_1}\sin(\Omega_{0,0}t_1)\cos(\Omega_{0,0}t_3) - ie^{i\vartheta_3}\cos(\Omega_{0,0}t_1)\cos(\Omega_{0,1}t_2)\sin(\Omega_{0,0}t_3) \\
\alpha_{22}^{(3)} &= -e^{i(\vartheta_1-\vartheta_3)}\sin(\Omega_{0,0}t_1)\sin(\Omega_{0,0}t_3) + \cos(\Omega_{0,0}t_1)\cos(\Omega_{0,1}t_2)\cos(\Omega_{0,0}t_3) \\
\alpha_{23}^{(3)} &= -e^{i\vartheta_2}\cos(\Omega_{0,0}t_1)\sin(\Omega_{0,1}t_2)\cos(\Omega_{1,0}t_3) \\
\alpha_{24}^{(3)} &= -e^{i(\vartheta_2-\vartheta_3)}\cos(\Omega_{0,0}t_1)\sin(\Omega_{0,1}t_2)\sin(\Omega_{1,0}t_3)
\end{aligned} \tag{68}
$$

$$
\begin{aligned}
\alpha_{31}^{(3)} &= -ie^{i(\vartheta_3-\vartheta_2)}\cos(\Omega_{1,0}t_1)\sin(\Omega_{0,1}t_2)\sin(\Omega_{0,0}t_3) \\
\alpha_{32}^{(3)} &= e^{-i\vartheta_2}\cos(\Omega_{1,0}t_1)\sin(\Omega_{0,1}t_2)\cos(\Omega_{0,0}t_3) \\
\alpha_{33}^{(3)} &= \cos(\Omega_{1,0}t_1)\cos(\Omega_{0,1}t_2)\cos(\Omega_{1,0}t_3) - e^{i(\vartheta_3-\vartheta_1)}\sin(\Omega_{1,0}t_1)\cos(\Omega_{1,1}t_2)\sin(\Omega_{1,0}t_3) \\
\alpha_{34}^{(3)} &= -ie^{-i\vartheta_3}\cos(\Omega_{1,0}t_1)\cos(\Omega_{0,1}t_2)\sin(\Omega_{1,0}t_3) - ie^{-i\vartheta_1}\sin(\Omega_{1,0}t_1)\cos(\Omega_{1,1}t_2)\cos(\Omega_{1,0}t_3) \\
\alpha_{35}^{(3)} &= ie^{i(\vartheta_2-\vartheta_1)}\sin(\Omega_{1,0}t_1)\sin(\Omega_{1,1}t_2)\cos(\Omega_{2,0}t_3) \\
\alpha_{36}^{(3)} &= e^{i(\vartheta_2-\vartheta_1-\vartheta_3)}\sin(\Omega_{1,0}t_1)\sin(\Omega_{1,1}t_2)\sin(\Omega_{2,0}t_3)
\end{aligned} \tag{69}
$$

$$
\begin{aligned}
\alpha_{41}^{(3)} &= -e^{i(\vartheta_1-\vartheta_2+\vartheta_3)}\sin(\Omega_{1,0}t_1)\sin(\Omega_{0,1}t_2)\sin(\Omega_{0,0}t_3) \\
\alpha_{42}^{(3)} &= -ie^{i(\vartheta_1-\vartheta_2)}\sin(\Omega_{1,0}t_1)\sin(\Omega_{0,1}t_2)\cos(\Omega_{0,0}t_3) \\
\alpha_{43}^{(3)} &= -ie^{i\vartheta_3}\cos(\Omega_{1,0}t_1)\cos(\Omega_{1,1}t_2)\sin(\Omega_{1,0}t_3) - ie^{i\vartheta_1}\sin(\Omega_{1,0}t_1)\cos(\Omega_{0,1}t_2)\cos(\Omega_{1,0}t_3)
\end{aligned}
$$

$$\alpha_{44}^{(3)} = \cos(\Omega_{1,0}t_1)\cos(\Omega_{1,1}t_2)\cos(\Omega_{1,0}t_3) - e^{i(\vartheta_1-\vartheta_3)}\sin(\Omega_{1,0}t_1)\cos(\Omega_{0,1}t_2)\sin(\Omega_{1,0}t_3)$$

$$\alpha_{45}^{(3)} = -ie^{i\vartheta_2}\cos(\Omega_{1,0}t_1)\sin(\Omega_{1,1}t_2)\cos(\Omega_{2,0}t_3)$$

$$\alpha_{46}^{(3)} = ie^{i(\vartheta_2-\vartheta_3)}\cos(\Omega_{1,0}t_1)\sin(\Omega_{1,1}t_2)\sin(\Omega_{2,0}t_3) \tag{70}$$

6. References

[1] W. Paul, *Rev. Mod. Phys.* 62, 531 (1990).

[2] D. J. Wineland, C. Monroe, W. M. Itano, D. Leibfried, B. E. King, and D. M. Meekhof, *J. Res. Natl. Inst. Stand. Technol.* 103, 259 (1998).

[3] D. Leibfried, R. Blatt, C. Monroe, and D. Wineland, *Rev. Mod. Phys.* 75, 281 (2003).

[4] D.M. Meekhof, C. Monroe, B.E. King, W.M. Itano and D.J. Wineland, *Phys. Rev. Lett.* 76, 1796 (1996).

[5] D. Leibfried, D. M. Meekhof, B. E. King, C. R. Monroe, W. M. Itano, and D. J. Wineland, *Phys. Rev. Lett.* 77, 4281 (1996).

[6] D. Leibfried, D. M. Meekhof, B. E. King, C. R. Monroe, W. M. Itano, and D. J. Wineland, *Phys. Rev. Lett.* 89, 247901 (2002).

[7] R. L. de Matos Filho and W. Vogel, *Phys. Rev. Lett.* 76, 608 (1996).

[8] J. I. Cirac, and P. Zoller, *Phys. Rev. Lett.* 74, 4091 (1995).

[9] K. Mølmer and A. Sørensen, *Phys. Rev. Lett.* 82, 1835 (1999).

[10] C. Monroe, D.M. Meekhof, B.E. King, W.M. Itano and D.J. Wineland, *Phys. Rev. Lett.* 75, 4714 (1995).

[11] C. A. Sackett, D. Kielpinski, B. E. King, C. Langer, V. Meyer, C. J. Myatt, M. Rowe, Q. A. Turchette, W. M. Itano, D. J. Wineland, and C. Monroe, *Nature (London)* 404, 256 (2000).

[12] F. Schmidt-Kaler, H. Häffner, M. Riebe, S. Gulde, G.P.T. Lancaster, T. Deuschle, C. Becher, C.F. Roos, J. Eschner, and R. Blatt, *Nature (London)* 422, 408 (2003).

[13] D. Leibfried, E. Knill, S. Seidelin, J. Britton, R. B. Blakestad, J. Chiaverini, D. B. Hume, W. M. Itano, J. D. Jost, C. Langer, R. Ozeri, R. Reichle, and D. J. Wineland, *Nature (London)* 438, 639 (2005).

[14] H. Häffner, W. Hänsel, C. F. Roos, J. Benhelm, D. Chek-al-kar, M. Chwalla, T. Körber, U. D. Rapol, M. Riebe, P. O. Schmidt, C. Becher, O. Ghühne, W. Dür, and R. Blatt, *Nature (London)* 438, 643 (2005).

[15] J. M. Raimond, M. Brune, and S. Haroche, *Rev. Mod. Phys.* 73, 565 (2001).

[16] Y. Makhlin, G. Schon, and A. Shnirman, *Rev. Mod. Phys.* 73, 357 (2001).

[17] N. A. Gershenfeld and I. L. Chuang, *Science* 275, 350 (1997).

[18] D. Loss and D. P. DiVincenzo, *Phys. Rev. A* 57, 120 (1998).

[19] W. Vogel and R. L. de Matos Filho, *Phys. Rev. A* 52, 4214 (1995).

[20] D. Stevens, J. Brochard, and A. M. Steane, *Phys. Rev. A* 58, 2750 (1998).

[21] A. Steane, *Appl. Phys. B: Lasers Opt.* B64, 623 (1997).

[22] G. Morigi, J. I. Cirac, M. Lewenstein, and P. Zoller, *Europhys. Lett.* 39, 13 (1997).

[23] L. F. Wei, Y.X. Liu and F. Nori, *Phys. Rev. A* 70, 063801 (2004).

[24] Q. Y. Yang, L. F. Wei, and L. E. Ding, *J. Opt.* B7, 5 (2005).

[25] M. Zhang, H. Y. Jia, X. H. Ji, K. Si, and L. F. Wei, *Acta Phys. Sin.*, 57, 7650 (2008).

[26] L. F. Wei, S.Y. Liu and X.L. Lei, *Phys. Rev. A* 65, 062316 (2002).

[27] L. F. Wei, M. Zhang, H.Y. Jia, and Y. Zhao, *Phys. Rev. A* 78, 014306 (2008).
[28] M. Zhang, X. H. Ji, H. Y. Jia, and L. F. Wei, *J. Phys.* B42, 035501 (2009).
[29] M. Zhang, H. Y. Jia, and L. F. Wei, *Opt. Communications.* 282, 1948 (2009).
[30] H. J. Lan, M. Zhang, and L. F. Wei, *Chin. Phys. Lett.* 27, 010304 (2010).
[31] M. Zhang, *"quantum-state manipulation with trapped ions and electrons on the liquid Helium"*, Southwest Jiaotong University Doctor Degree Dissertation, 2010.
[32] P. A. Barton, C. J. S. Donald, D. M. Lucas, D. A. Stevens, A. M. Steane and D. N. Stacey, *Phys. Rev. A* 62, 032503 (2000).
[33] Ch. Roos, Th. Zeiger, H. Rohde, H. C. Näerl, J. Eschner, D. Leibfried, F. Schmidt-Kaler, and R. Blatt, *Phys. Rev. Lett.* 83, 4713 (1999).
[34] P. Shor, in Proceedings of the 35th Annual Symposium on the Foundations of Computer Science, edited by Shafi Goldwasser (IEEE Computer Society Press, New York, 1994), p. 124.
[35] L. K. Grover, *Phys. Rev. Lett.* 79, 325 (1997).
[36] T. Sleator and H. Weinfurter, *Phys. Rev. Lett.* 74, 4087 (1995).
[37] D. P. DiVincenzo, *Phys. Rev. A* 51, 1015 (1995).
[38] A. Barenco, D. Deutsch, A. Ekert, and R. Jozsa, *Phys. Rev. Lett.* 74, 4083 (1995).
[39] C. Monroe, D. Leibfried, B.E. King, D.M. Meekhof, W.M. Itano, and D.J. Wineland, *Phys. Rev. A* 55, R2489 (1997).

5

Coherent Optical Phonons in Bismuth Crystal

Davide Boschetto and Antoine Rousse
Laboratoire d'Optique Appliquée, ENSTA ParisTech/Ecole Polytechnique/CNRS, Chemin de la Hunière, 91761 Palaiseau cedex
France

1. Introduction

This chapter is devoted to the study of the lattice vibrations within a crystal structure using laser-based time resolved spectroscopic methods. We have used ultrafast pulsed laser systems to extend the analysis of atomic displacements into the time domain with femtosecond resolution. This field has gained a lot of attention from a large scientific community during the past decades. Actually, the knowledge of all the properties in a crystal depends on two key factors: the way in which atoms arrange themselves to give rise to the crystal structure, and the movement of both electrons and atoms within this structure. Many physical parameters, such as the transport properties, strongly depend on the dynamical behavior of both electrons and atoms. Besides, the mutual interaction between them determines the pathway of chemical reactions and phase transitions. As an example, one challenge in solid state physics is the understanding and control of high temperature superconductivity, in which the dynamical aspect may play a crucial role. If there is a lattice displacement responsible for the Cooper pairs formation, its selective excitation could be exploited to reach a superconductivity phase even at temperature well above the known transition temperature. The general question is how can we modify the phase of a material using photons. This type of studies belong to a broader research area, known as the science of photoinduced phase transition. An important point is that high frequency lattice vibrations are the first response of ions to the external pumping laser pulse. Therefore, the study of such movements could clarify the energy relaxation channels of the excess energy stored into the electrons subsystem from the pumping laser pulse.

Another important application of the study of coherent phonon concerns nanostructures. When reducing the size down to the nanometric scale, the ratio between the atomic displacement and the structure size increases considerably. Together with the confinement effect, this produces major modifications of the physical properties.

The general movement of the atoms within the crystal structure can be seen as an elastic deformation, which can be described as a sum of several normal modes of vibration of the lattice. The quantum of energy associated to a normal mode of vibration of the lattice is called phonon. In order to study each of these elementary vibrations, there is a need to excite and detect selectively the desired phonon mode. This is possible only if we can excite and detect the phonon mode coherently, i.e. the atomic vibration has a given phase which is kept in a time window, called the dephasing time or damping time.

Coherent phonons were observed in many different materials, and their extensive study in novel materials represents a large part of the activity centered on the femtosecond laser

applications. Here, we will focus our attention on bismuth crystal mainly because it is the simplest case for which the description approaches the ideal situation of a single harmonic oscillator. This will allow an understanding of the key points in coherent phonons studies.
This chapter is structured as follows. First, we will define the main properties of coherent phonons, highlighting the difference between coherent and incoherent phonons. We will then address the question of how to excite and detect a coherent phonon, focussing on the different optical properties. The dependence of phonon parameters in bismuth crystals on several factors such like excitation fluence and lattice temperature will be discussed.

2. Coherent and incoherent phonons

It is well known that atoms in the lattice vibrate around their equilibrium positions, with a vibration amplitude that increases with temperature. Such vibrations can be decomposed in a sum of normal modes, which depend on the symmetry of the lattice. The quantum of energy associated to the vibration of a given normal mode is called phonon (Ashcroft and Mermin, 1976).
The temperature effect results in an incoherent movement of the atoms in the lattice, in which each atom vibrates independently. In this case there is no phase relationship between the vibration of each atom. We call these incoherent phonons. Instead, in the case of a coherent phonon, there is a well defined phase relationship, and the generic ith atom moves following a sinusoidal function $sin(\omega t - \vec{k} \cdot \vec{x_i} + \phi)$, where t is the time, ω and $\vec{k}$ are the vibration frequency and wavevector, respectively, ϕ is the initial phase and $\vec{x_i}$ is the position of the ith atom in the lattice.
The difference between a coherent and an incoherent phonon can be visualized thinking of the sea surface. Without perturbations, the surface of the sea is modulated by several waves, which sum to each other. This gives rise to the well known sea surface fluctuations, where each part of the surface can vibrate independently from their neighbors. This is an incoherent movement at the surface of the sea, which could be associated to the incoherent phonons. On the other hand, we all know that when a boat comes across the sea, it generates a wave form modulation of the sea surface, whose spatial frequency and amplitude depends for example on the speed and the size of the boat. This wave generates a modulation of the sea surface which propagates as a coherent wave, which can be associated to coherent phonons. We will see in the following that the laser pulse propagating into the crystal plays the same role than the boat on the surface of the sea.
Only coherent optical phonons will be discussed in this chapter. First, optical phonons are the first response of the lattice to the external laser pulse excitation, mainly because they are generated much more rapidly due to their much higher vibration frequency near the center of the Brillouin zone. Moreover, the optical phonons represent the movement of the atoms within the elementary cell, and can be regarded as the elementary atomic movement.
In the most general way, a coherent optical phonon displacement must be described as the following

$$Q(t) = A_{ph}e^{-t/\tau} sin\left(\omega t - \vec{k} \cdot \vec{x_i} + \phi\right) \tag{1}$$

where A_{ph} is the phonon amplitude and τ is the phonon damping time. As it will be shown, laser pulse excitation produces phonons whose wavevector is always very close to zero, therefore we will neglect it hereafter. Instead, the parameters of interest in the study of coherent phonon are the amplitude, the frequency and the damping time. On the other hand, the initial phase is often used to discriminate between the different excitation mechanisms.

3. Excitation and detection of coherent optical phonons

In order to study the dynamical behavior of a given phonon mode two key factors are required. We must be able to excite coherently the phonon we are interested in, and we must be able to detect the corresponding atomic displacements. The pump probe technique using the very high time resolution of modern laser systems is a unique tool for this purpose. The excitation and detection mechanisms are independent, so it is possible to excite the coherent phonon using one laser pulse and detect it using any other pulse of the desired wavelength, from hard X-ray to infrared spectral range. We will first describe the excitation of coherent optical phonon by a laser pulse in visible and near infrared range, and then show how the detection can be done from reflectivity measurements.

3.1 Excitation mechanisms

The excitation of a coherent optical phonon can be viewed in a classical frame by considering the case of a forced harmonic oscillator, which follows the equation

$$\frac{dQ^2(t)}{dt^2} + \frac{2}{\tau}\frac{dQ(t)}{dt} + \omega^2 Q(t) = \frac{F(t)}{m} \tag{2}$$

where the applied external force $F(t)$ is due to the pumping laser pulse. All the existing mechanisms used to describe the excitation of coherent phonon are focussed on the expression of the driving force. We can distinguish mainly three mechanisms, the impulsive stimulated Raman scattering (ISRS) (Yan, 1985), the theory of displacive excitation of coherent phonon (DECP) (Zeiger, 1992) and the temperature gradient theory (TGT) (Garl, 2008 , a). Before going into the summary description of each of these models, we must recall some fundamental experimental results. In the case of a transparent material, the phonon excitation and detection follows the Raman selection rules, therefore providing an experimental evidence that the excitation mechanism is probably a pure stimulated Raman process. Instead, in the case of opaque materials, only the completely symmetric optical phonons are observed, namely the A_{1g} mode (Zeiger, 1992), if they exist, even when the Raman cross section associated with the A_{1g} phonon is much lower than the other modes with different symmetry. The case of superconductor iron pnictide is a great example of this (Mansart, 2009). In pump probe experiments on opaque materials, phonons modes other than A_{1g} were detected only at low temperature (Ischioka, 2006). The only exception is the case of graphite (Ischioka, 2008), in which the shearing mode with E_{1g} symmetry can be excited with higher efficiency with respect to the A_{1g} mode.

3.1.1 Impulsive stimulated Raman scattering

Let's suppose that the crystal under study is transparent and has a Raman active mode of frequency ω, and that we let propagate into the crystal two laser pulses with frequency and wavevector $(\omega_1, \vec{k_1})$ and $(\omega_2, \vec{k_2})$, respectively. If the following relation is satisfied

$$\omega_1 - \omega_2 = \omega \tag{3}$$

the phonon with frequency ω will be excited into the crystal, with a wavevector given by the phase matching condition

$$\vec{k} = \vec{k_1} - \vec{k_2} \tag{4}$$

As a laser pulse is usually in the visible or near infrared range, the corresponding wavevector is orders of magnitude smaller than the edge of the Brillouin zone, therefore the excited phonon wavevector is close to the Γ point and can be approximated by zero.

In the case of a short laser pulse, the spectral shape can be large enough so that the two frequencies ω_1 and ω_2 are within the same pulse, and therefore the phonon can be excited by only one laser pulse.

The key hypothesis of the ISRS theory (Yan, 1985; Merlin, 1997) is that the polarizability α is not constant, but rather depends on the relative distance between the atoms and therefore on the phonon displacement (Boyd, 2003)

$$\alpha(t) = \alpha_0 + \left(\frac{\partial\alpha(t)}{\partial Q}\right)_0 \cdot Q(t) \tag{5}$$

where α_0 is the polarizability corresponding to the equilibrium position. The external force acting in equation 2 can be written as (Yan, 1985)

$$F(t) = \frac{1}{2}N\left(\frac{\partial\alpha(t)}{\partial Q}\right)_0 : \vec{E}\cdot\vec{E} \tag{6}$$

where $\vec{E}$ is the optical electric field and N is the volume density of oscillators. If $\tau >> \omega$, the solution of the equation 2 is

$$Q(z>0,t>0) = Q_0 e^{-\frac{1}{\tau}\left(t-\frac{zn}{c}\right)} sin\left[\omega\left(t-\frac{zn}{c}\right)\right] \tag{7}$$

where z is the propagation direction of the laser. It is possible to show that the phonon amplitude Q_0 is proportional to $e^{-\omega^2\tau_L^2/4}$ where τ_L is the laser pulse duration. Therefore, the maximum phonon amplitude is reached when the following condition is satisfied

$$\tau_L << \frac{2\pi}{\omega} \tag{8}$$

This condition is called the *impulsive limit*. Obviously, when $\tau_L >> 2\pi/\omega$, $Q_0 \sim 0$.

We point out that this theory is in perfect agreement with all the existing experimental results on transparent materials. Instead, for opaque materials, an extension of the ISRS was made by (Stevens, 2002) by proving that the stimulated Raman scattering is defined rather by two different tensors instead of one, having the same real parts but distinct imaginary parts in the absorbing region. This model was successfully applied to the case of *Sb*.

3.1.2 Theory of displacive excitation of coherent phonon

The theory of displacive excitation of coherent phonon (DECP) (Zeiger, 1992) was developed to address the case of absorbing material, in which only the A_{1g} mode was observed, regardless to the value of the Raman tensor coefficients. The key point of this theory is that the laser pulse, affecting both the density of electrons in the conduction band and their temperature, results in an abrupt change in the equilibrium position of the atoms within the elementary cell and produces an atomic displacement that sets up the oscillations of the atoms around their new equilibrium positions. By assuming that the main effect of the laser pulse is the excitation of electrons from the valence to the conduction band, the external force in equation 2 is

$$F(t) = \omega^2 \kappa j(t) \tag{9}$$

where κ is a constant and $j(t)$ is the density of electrons in the conduction band.
If $\tau >> \frac{2\pi}{\omega}$ and $\tau_L << \frac{2\pi}{\omega}$, the solution of equation 2 is

$$Q(t > 0) = Q_0 \left[e^{-\beta t} - e^{-\frac{t}{\tau}} cos\left(\omega t + \phi\right) \right] \tag{10}$$

where β is the electrons decay rate. This theory explains the unique excitation of A_{1g} mode on the basis of thermodynamics arguments.

3.1.3 Temperature gradient theory

This theory approaches the excitation of coherent phonon from an analysis of all the forces acting into the crystal when the interaction with the laser pulse occurs (Garl, 2008 , b). Three types of forces can be recognized, namely the ponderomotive force, the polarization force analog to Raman scattering process, and the thermal force, produced by the thermal gradients of both of electrons and lattice. Therefore

$$F(t) = F^{pond} + F^{pol} + F^{grad} \tag{11}$$

The key point is that a quantitative analysis shows that the force due to gradient temperature is the largest, and therefore excites the coherent phonons.

3.2 Electrons and lattice temperature in photoexcited absorbing crystal

In absorbing material, the interaction with the laser pulse produces several effects, which could be described looking at the time scale on which these effects take place, as shown in figure 1 (Boschetto, 2010 , a). The laser pulse energy is first stored into the electrons, causing changes in free carriers density as well as in their temperature. We must point out here that the word *temperature* must be taken carefully. Actually, we usually talk about temperature only if the distribution of kinetic energy follows a maxwellian distribution. This happens once the system reaches an equilibrium condition, which could be stable, i.e. not changing in time, or metastable, therefore evolving in time. If the collision time between the particles is short enough in comparison to the time scale under study, the particles subsystem, in our case electrons or phonons, are in equilibrium and therefore we could define a temperature for each subsystem. What happens when the distribution in kinetic energy is not maxwellian goes beyond the scope of this chapter. The pump pulse is exponentially absorbed at the surface of the crystal, generating therefore a gradient in electrons temperature along the direction of propagation of the light. At this stage, we can therefore define two temperatures independently. On one hand, the electrons temperature, and on the other hand the lattice temperature. On a time scale shorter than the electron phonon coupling time, the electrons did not have enough time to exchange their energy with the lattice.
Therefore, the lattice is at room temperature whereas the electrons reach very high temperature because of their low heat capacity. Once the electrons are excited, they start sharing their energy with the lattice through electron phonon coupling. The time required for the electrons and phonons subsystem to equilibrate can be calculated by using the two temperatures model (Anisimov, 1975). In this model, we consider that the energy exchange between electrons and phonons scales linearly with the temperature difference between these two subsystems. The two coupled differential equations describing the evolution of the electrons and lattice temperature, namely T_e and T_l, in both space and time are (Anisimov, 1975)

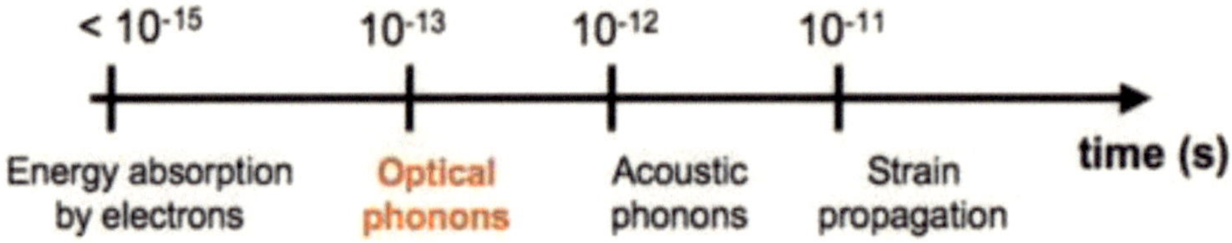

Fig. 1. Time scales of several processes taking place when a femtosecond pulse excites a crystal, reprinted by permission of the publisher (Taylor & Francis Ltd, http://www.tandf.co.uk/journals) from (Boschetto, 2010 , a).

$$C_e \frac{\partial T_e}{\partial t} = \frac{\partial}{\partial z}\left(\kappa_e \frac{\partial T_e}{\partial z}\right) - g(T_e - T_l) + P(z,t)$$
$$C_l \frac{\partial T_l}{\partial t} = g(T_e - T_l) \tag{12}$$

where C_e and C_l are the heat capacities of electrons and lattice, respectively, κ_e is the electrons thermal conductivity, g is the electrons phonons coupling constant and $P(z,t)$ is the absorbed energy density. Here we neglected the lattice thermal diffusion, which takes place on a much longer time scale with respect to the temporal ranges presented in this chapter.

If the electron diffusion is a very slow process with respect to the electrons phonons equilibration time, the final lattice temperature in the skin depth can be calculated simply by the requirement of energy conservation. In the opposite case, the excess energy stored in the electrons subsystem will escape from the skin depth, therefore resulting in a lower lattice temperature.

3.3 Transient reflectivity in absorbing crystal

In the previous paragraphs we have seen that the interaction of the femtosecond pump laser pulse with absorbing crystal will generate a modification in electrons and lattice temperature, as well as will excite one or more coherent phonon displacements. If the dielectric constant is modified by these changes, the signature of each of these effects should show up in the transient reflectivity. A general approach to the description of the transient reflectivity in absorbing materials, regardless to the mechanism invoked for coherent phonon excitation, was developed in reference (Boschetto, 2008 , a). We will give here the major outlines.

The reflectivity depends on the real and imaginary part of the dielectric constant, namely ϵ_{Re} and ϵ_{Im}, through the well known Fresnel formula. The point is now to evaluate the way in which the dielectric constant changes when a coherent phonon mode is set up into the crystal, as well as its dependence on electrons and lattice temperature. For sake of simplicity, let's assume that the dielectric constant of the crystal under study can be well described by the Drude model (Ashcroft and Mermin, 1976)

$$\epsilon_{Re} = 1 - \frac{\omega_p^2}{\omega_l^2 + \nu_{e-ph}^2}$$
$$\epsilon_{Im} = \frac{\omega_p^2}{\omega_l^2 + \nu_{e-ph}^2} \frac{\nu_{e-ph}}{\omega_l} \tag{13}$$

where ω_l and ω_p are the laser frequency and the plasma frequency, respectively, whereas ν_{e-ph} is the electron phonon collision frequency. The plasma frequency is defined as a function of the effective electron mass m_e, the electron charge e and the free carriers density $n_e(t)$ as the following

$$\omega_p^2 = \frac{4\pi e^2 n_e(t)}{m_e^*} \tag{14}$$

The key point of this approach is that the electron phonon collision frequency can be expressed in the frame of the kinematical theory as (Ziman, 2004)

$$\nu_{e-ph} = \sigma_{e-ph} n_{ph} v_e \tag{15}$$

where σ_{e-ph} is the electron phonon scattering cross section, n_{ph} is the phonon density and v_e is the electron velocity. Knowing that

$$n_{ph} \cong \frac{n_a T_L(t)}{T_D} \tag{16}$$

where n_a is the atoms density in the crystal and T_D is the Debay temperature, and assuming a circular scattering cross section, the change in the electron phonon collision frequency due to the coherent phonon can be expressed as

$$\frac{\Delta\nu_{e-ph}}{\nu^0_{e-ph}} = \frac{\Delta T_L}{T_0} + 2\frac{\Delta Q(t)}{Q_0} \tag{17}$$

where T_0 and Q_0 are the lattice temperature and phonon displacement at equilibrium before the pump pulse interact with the crystal. By taking

$$\Delta n_e(t) \propto T_e(t) \tag{18}$$

we can write the changes in transient reflectivity as the following (Boschetto, 2008 , a)

$$\frac{\Delta R}{R} = A_e T_e(t) + A_L T_L(t) + A_{ph} Q(t) \tag{19}$$

where A_e, A_L and A_{ph} are constant which depend on the value of the partial derivatives of the reflectivity with respect to the real and imaginary part of the dielectric function as well as on the probe pulse wavelength. The algebraic sign of these constants can be different, implying a competition in the induced reflectivity changes produced by the electrons and lattice temperature. After the pump pulse arrival, the electrons temperature reaches the maximum, whereas the lattice is still cold. As described in equation 12, for longer time delay the electrons temperature decreases whereas the lattice temperature increases, until they have the same temperature, and the equilibrium is reached. Therefore, we must expect a transient behavior in the reflectivity while the two temperatures are changing, and we expect a plateau in the reflectivity when the equilibrium is reached.

3.4 Set up for coherent optical phonon study

The study of coherent optical phonon in time domain requires the use of ultrafast laser pulses, typical of 50 fs or less, depending on the phonon frequency under investigation. Obviously, the higher the phonon frequency the higher time resolution is required. Such a study is usually performed in a pump probe set up, summarized in figure 2.

The basic idea is that one pulse is used to excite the sample, whereas a second pulse is used to probe it. A controlled delay stage between the two beams allows to probe the sample at any given time delay from the pump arrival on the sample. The pump and probe wavelength can be different (Papalazarou, 2008). If the probe is in the visible range, we usually measure

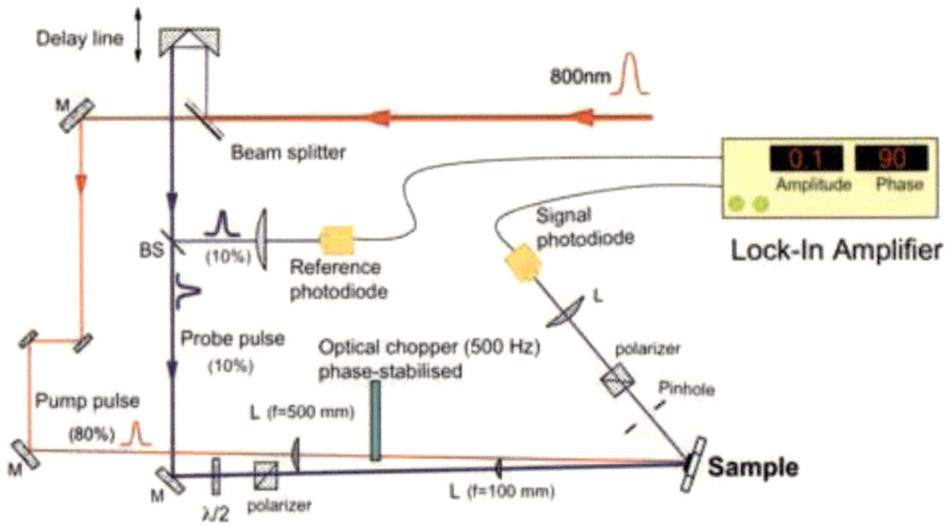

Fig. 2. Example of experimental set up for pump and probe measurement in reflectivity.

either the induced changes in reflectivity or in transmission of the sample. If the probing wavelength is in the hard X-ray range, a diffraction pattern could be recorded and analyzed at different time delay (Sokolowski, 2003; Beaud, 2007; Rousse, 2001). Here, we will focus mainly on pump probe experiments using one laser pulse at a given wavelength for recovering the transient reflectivity (Boschetto, 2008 , a). In this case, a laser beam is split into two parts, one is used as pump pulse and the other as probe pulse. Eventually, the relative polarisation is changed to match the Raman selection rule for the phonon under study. The excitation of coherent optical phonon requires some attention with respect to the fluency used in the experiment. Typically, they show up only in a certain range of pump fluency, depending on the sample as well as on the pump wavelength. However, the most sensible part of the experiment is the signal detection. Coherent phonon displacement gives usually rise to very small changes in reflectivity, as they are only a tiny perturbation with respect to the equilibrium configuration in the crystal. Therefore, how to measure the signal is here the key point. The main point to extract the phonon signal from the reflectivity is to get rid of all sources of noise. Laser fluctuations can be accounted for by using a reference photodiode. We then measure the difference between the signal photodiode and the reference photodiode. In order to have a very high signal to noise ratio, a differential measurement coupled to a spectral filtering of the signal is required. This can be accomplished by using a chopper on the path of pump beam, which will therefore excite the sample at frequency lower then the probing pulse frequency. For example, for 1 KHz repetition rate laser, we use 500 Hz chopping frequency on the pump pulse. The difference between the signal and reference photodiodes is then analyzed by a lock-in amplifier at the pump chopping frequency. The lock-in amplifier produces a spectral filtering of the input signal, giving as an output signal only the changes in the reflectivity induced by the pump pulse. This is depicted in figure 3. This method results in a very high sensitivity. At 1 kHz laser repetition rate, we reach a signal to noise ratio of 10^5, which is today the state of the art at this laser repetition rate (Boschetto, 2008 , a). Moreover, using 80 MHz repetition rate it was possible to approach the shot noise limit of 10^{-8}, which allowed the detection of the vibrations of only two atomic thin layers of graphene (Boschetto, 2010 , b).

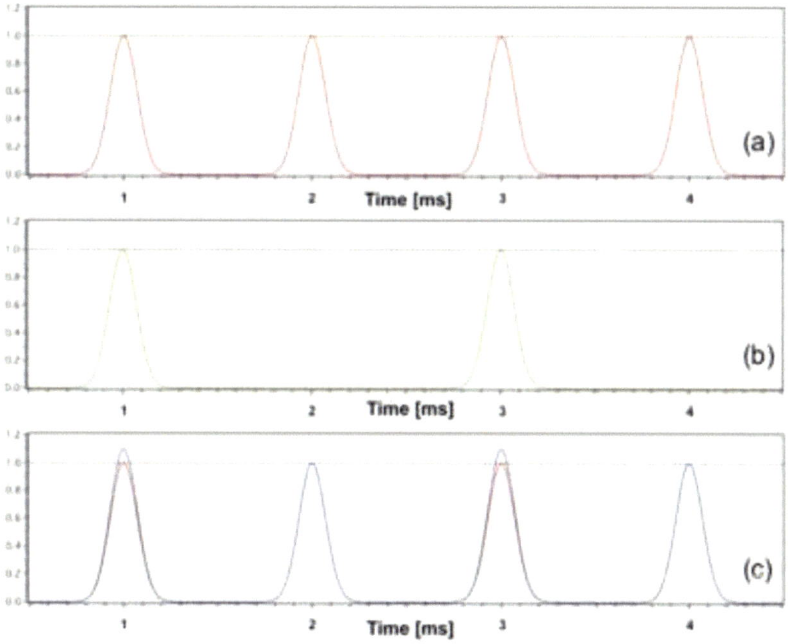

Fig. 3. Probe (a), pump (b) and pump perturbed probe (c) using a chopper at half the laser repetition rate.

4. Bismuth crystal

The study of optical phonons in bismuth crystal is probably the most important from a pedagogical point of view. Actually, the elementary cell of bismuth contains only two atoms, displaced along the main diagonal of the cell (Boschetto, 2010 , a). Therefore, the study of the vibrational movement approaches a standard bookcase on harmonic oscillator, which is easier to understand. Within the bismuth structure, only two normal modes of vibration exist, namely the A_{1g} and the E_g mode. The A_{1g}, also called *breathing mode*, corresponds to the displacement of the atoms along the diagonal of the elementary cell, and remind the bookcase of standard harmonic oscillator. This mode of vibrations is also called "completely symmetric" because it preserves all the symmetry operations of the elementary cell. We note also that this displacement does not change the size of the elementary cell. The frequency of this mode, measured by Raman scattering process, is 2.92 THz, i.e. the vibration period is 342 fs (Lannin, 1975). This is a rather low frequency for optical phonon, basically because of the high atomic mass of bismuth, which is 209 u. The E_g mode corresponds to the displacement of the atoms in the plane perpendicular to the principal diagonal of the cell, and is therefore degenerate in the plane. This mode has an even lower frequency of 2.22 THz, corresponding to a vibrational period of 450 fs (Lannin, 1975). Any atomic movement within the elementary cell can be seen as superposition of these two normal modes. In contrast with its simple structure, bismuth has a very complex band structure and unique electronic properties (Edelman, 1977), which have intrigued the curiosity of scientists for decades. According to both experimental and theoretical results, valence and conduction band overlap only around a small region of the

Brillouin zone, allowing only a very small part of the electrons to be in the conduction band, which results in a low conductivity. However, it is interesting to note that electrons in bismuth can propagate over a long distance before experiencing any collision, giving rise to a quite large mean free path also at room temperature (Pippard, 1952).

5. Optical phonon in bismuth single crystal

Figure 4 shows a typical transient reflectivity in bismuth single crystal (Boschetto, 2008 , a), oriented along the [111] direction with respect to the trigonal elementary cell, corresponding to an excitation fluence of 2.7 mJ/cm^2. The oscillations due to the coherent A_{1g} optical phonon can clearly be seen, superimposed to a exponential like relaxation signal until a plateau is reached. The reflectivity recovers in around 4 ns. In figure 4, three different dynamics can be highlighted. Just after the excitation pulse arrival, the reflectivity increases because of the increase in electrons temperature. Besides, the coherent optical phonon oscillation is set up by this abrupt modification in electrons energy. While the lattice oscillates, the reflectivity decreases and passes through the zero in going toward negative value of the signal. This is due to a competition between the first two terms in equation 19. Indeed, for bismuth probed at 800 nm wavelength, reflectivity derivative calculations show that $A_e > 0$ whereas $A_L < 0$. For positive time delay, while the electrons temperature decreases from its maximum value reached at $t \sim 0$, the lattice temperature increases due to electron phonon collisions. Therefore, the first two terms in equation 19 compete until the equilibrium is reached, at around 20 ps in our case, which is characterized by the plateau. For longer time delay, ranging in the 100 ps to ns range, the thermal diffusion is responsible for heat transfer out of the skin depth, resulting in a decrease in the crystal temperature at the surface. The red solid line in figure 4 correspond to a fit by the equation 19. Such a very good agreement was found for all the pump fluences ranging from 1 to 25 mJ/cm^2, which corresponds to the damage threshold we found experimentally.

The optical phonon parameters are strongly dependent on the pumping fluence, and this dependence gives important information on the crystal properties. We will focus our attention on the amplitude, frequency and damping time of the phonon. The amplitude generally increases linearly with the fluence. This is due to the fact that increasing the fluence results in both a larger number of free carriers as well as a stronger electrons temperature gradient, implying a bigger initial atomic displacement. Figure 5 shows an example of the phonon frequency in bismuth as a function of the pumping fluence. A small departure from the linear behavior is observed for fluences around 15 mJ/cm^2, close to the crystal damage threshold. This non linearity indicates that in this regime the atoms may displace in a stronger anharmonic regime.

On the other hand, increasing the fluence results in a red shift of the phonon frequency, as it can be observed in figure 6, in agreement with previous findings (DeCamp, 2001; Murray, 2005).

This can be explained again by the fact that increasing the fluence, the density of electrons excited into the conduction band increases as well, resulting in a softening of the interatomic bonding. This is a general observation in many materials, with the only exception so far known of V_2O_3 (Mansart, 2010 , a) and graphite (Ischioka, 2008), in which the bonding are strengthen by the electrons excitation.

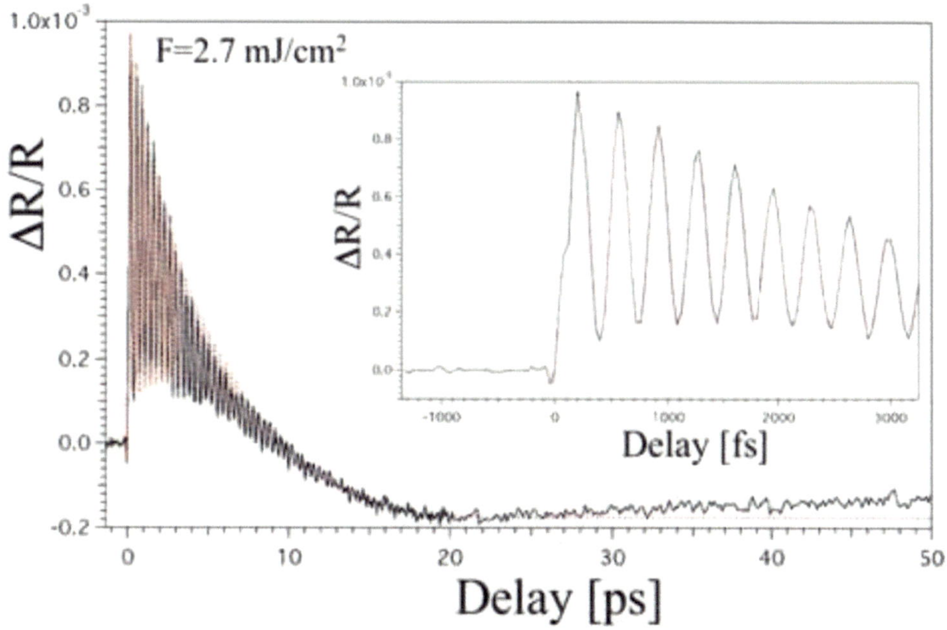

Fig. 4. Transient bismuth reflectivity at excitation fluence of 2.7 mJ/cm^2, reprinted with permission from (Boschetto, 2008 , a) (http://prl.aps.org/abstract/PRL/v100/i2/e027404), ©(2008) by the American Physical Society*.

A closer look into the phonon frequency behavior shows a more sophisticated response. Actually, a more perfect adjustment of the phonon behavior highlights that the frequency of the oscillations is not constant and it rather depends on time delay (Hase, 2002). Therefore, a better insight into the change of phonon frequency versus fluence can be obtained by measuring the phonon frequency versus the time delay, as depicted in figure 7. At time close to zero, the phonon frequency undergoes the largest red shift. Then, it relaxes back at its unperturbed value on a time scale that depends on the excitation fluence. This can be well understood by the fact that after the excitation by the pump pulse, the electrons relax back, therefore causing again the hardening of the atomic bonding. Figure 7 shows the time dependent evolution of phonon frequency for various pumping fluences. The initial softening followed by exponential relaxation is observed for any excitation fluence above 4 mJ/cm^2. For smaller fluence, the changes in frequency are within the error bars, and therefore cannot be investigated. Time resolved X-ray diffraction was successfully applied to the study of A_{1g} phonon in bismuth (Fritz, 2007), proving that the frequency red shift is mainly due to electronic softening of the interatomic potential.

The phonon damping constant τ^{-1} also has a linear dependence versus the pumping fluence (Boschetto, 2008 , b), as shown in figure 8. The phonon damping time could be originated from both a decrease in phonon population or a dephasing in the oscillation. The coherent pump probe measurement does not allow to discriminate between these two possibles effects. A pure dephasing implies simply a lost of coherence, even at constant phonons population. Instead, phonons population decay can have three main origins. First, the phonon phonon

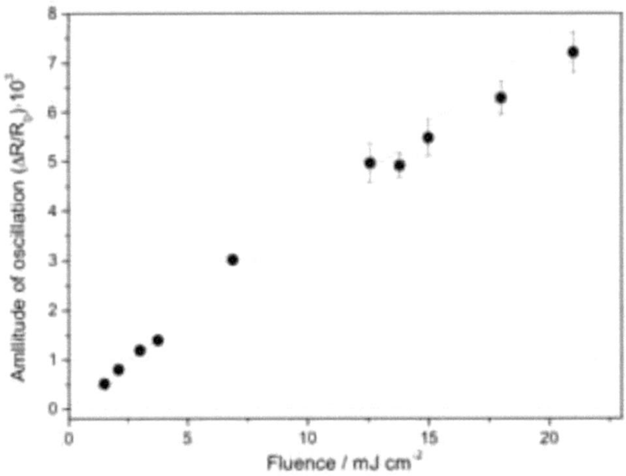

Fig. 5. Phonon amplitude versus pumping fluence (Garl, 2008 , a).

scattering, in which one phonon decays into two phonons, whose wavevectors and energy match both the energy and momentum conservation laws. This phonon-phonon transition is due to anharmonic effects, which introduces off diagonal terms in the phono- phonon interaction matrix, which would be pure diagonal in a perfect harmonic crystal. Of course, the strongest the anharmonicity, the higher is the transition probability between two different eigen modes. For example, in our case one optical phonon at the Γ point can be annihilated to produce two acoustical phonons outside the center of the Brillouin zone. Because of the momentum conservation, we cannot measure those acoustical phonons, and therefore we cannot monitor the phonon energy relaxation channel. The second possible reason for phonon damping is the electron phonon interaction. Actually, the electron phonon collisions implies energy and momentum exchange in both directions, from electrons to phonons and vice versa. Therefore, the excess energy stored in coherent phonons oscillation can be also lost in the interaction with the electrons. A third mechanism for phonon damping is scattering from impurities, which could also change the phase of oscillations.

5.1 Temperature dependence of optical phonon

The initial crystal temperature plays an important role in the measured phonon properties (Garl, 2008 , a;b). Figure 9 (a) shows the transient reflectivity for several initial crystal temperatures from 50 K up to 510 K, which is close to the melting point. Two main features can be highlighted. The phonon amplitude decreases whereas the phonon frequency undergoes a red shift when increasing the initial crystal temperature, as shown in the Fourier transform of the signal depicted in figure 9 (b).

The strong dependence of the coherent phonon amplitude on initial crystal temperature supports the scenario of excitation mechanism by DECP or TGM rather then a pure ISRS. This dependence could be explained as the following. The coherent optical phonon is generated by the increase in the free carriers density and/or the changes in electrons temperature, because of the gradient set up by the pump pulse. At very low temperature, the density of thermally excited electrons as well as their initial energy distribution is very low. The arrival of the pump pulse produces a larger modification with respect to the electrons initial conditions,

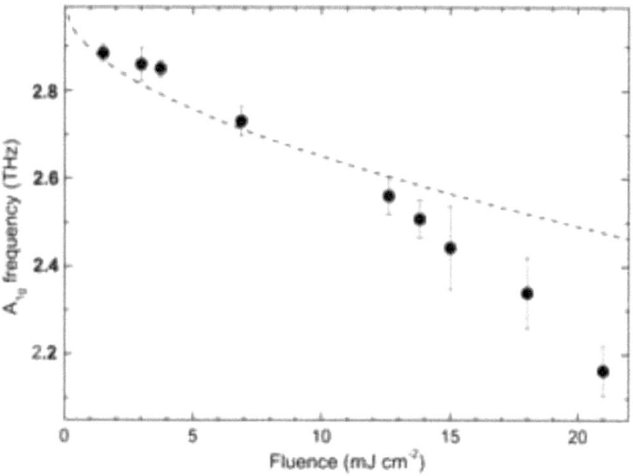

Fig. 6. Phonon frequency versus pumping fluence, reprinted with permission from (Garl, 2008 , b) (http://prb.aps.org/abstract/PRB/v78/i13/e134302), ©(2008) by the American Physical Society*.

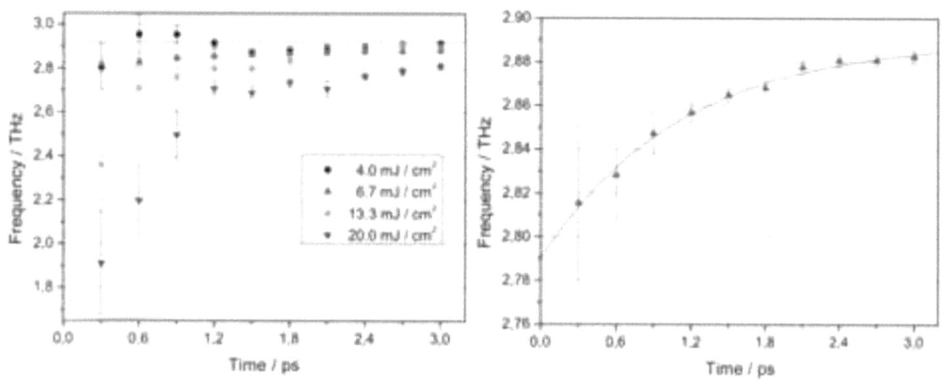

Fig. 7. Phonon frequency versus time delay (Garl, 2008 , a).

producing a larger coherent phonon amplitude. Therefore, at low temperature more energy can be stored in the coherent phonon displacement.

The behavior of the phonon frequency versus the temperature reinforces the scenario of bond softening by electrons excitations (Fritz, 2007). Actually, when increasing the temperature, the free carriers density increases, resulting in a softening of both atomic bonds and phonon frequency. This is in agreement with the results shown in the previous paragraph. The Fourier transform at 50 K shows also the existence of a secondary peak at 2.12 THz, which results in amplitude modulations of the A_{1g} optical phonon. This peak corresponds to the E_g optical mode, which can be observed only at very low temperature, as reported also by other teams (Ischioka, 2006). This mode is excited with a very low efficiency with respect to the A_{1g} mode

5.2 Electrons and lattice temperatures in photoexcited bismuth crystal

In the previous paragraph we have explained the transient reflectivity in bismuth crystal by the competition of changes in both electrons and lattice temperatures superimposed to the

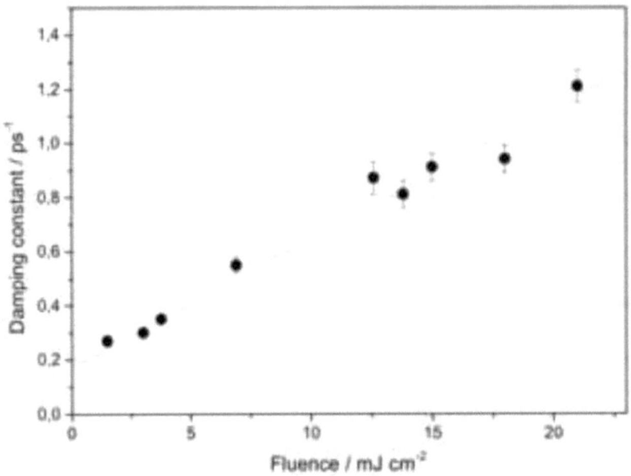

Fig. 8. Phonon damping constant versus fluence, reprinted with permission from (Garl, 2008 , b) (http://prb.aps.org/abstract/PRB/v78/i13/e134302), ©(2008) by the American Physical Society*.

oscillations due to coherent optical phonon. The interesting question concerns the quantitative knowledge of both temperatures and their evolution in time. The final temperature at equilibrium, when the plateau is reached at around 20 *ps*, can be calculated by energy conservation neglecting the electrons diffusion. Knowing that the laser penetration depth in bismuth at 800 *nm* wavelength is 30 *nm*, the calculated final equilibrium temperature is then 1300 *K*, which must be compared to the melting temperature of 544 *K*. Obviously, this would mean that the plateau in reflectivity is rather due to either a transition of the crystal to liquid phase or a mixture of solid and liquid phase, instead of simple heating of the crystal as described by equation 19. In order to discriminate the formation of any liquid phase, we performed a double pump experiment (Boschetto, 2010 , a), with the second pump pulse arriving around 25 ps after the first pump pulse, shown in figure 10.
It can be clearly seen that the behavior of the reflectivity after the arrival of the second pump pulse is exactly the same as after the arrival of the first pump pulse. Moreover, the coherent phonon amplitude, frequency and damping time excited by the two pump pulses are nearly the same. This is a strong indication that the there is no transition to liquid phase, neither any mixture of phase, otherwise the phonon parameters could not be the same.

6. Dielectric function measurement in bismuth single crystal

Reflectivity measurement gives only partial information about the transient state of the crystal, mainly because there is no direct link with the crystal structure and electronic configuration. Instead, we can gain a deeper understanding on the crystal dynamics under femtosecond photoexcitation by recovering the transient real and imaginary part of the dielectric function. There are mainly two ways to perform such a measurement, either the use of white light pulse combined to a spectrometer (Kudryashov, 2007), or using a double probe pulse at two known angles (Uteza, 2004). The former method gives access to a broad range response of the dielectric function, but it has a lower signal to noise ratio. Instead, by using a two probes set up, we can recover the dielectric function only for one wavelength at time, but

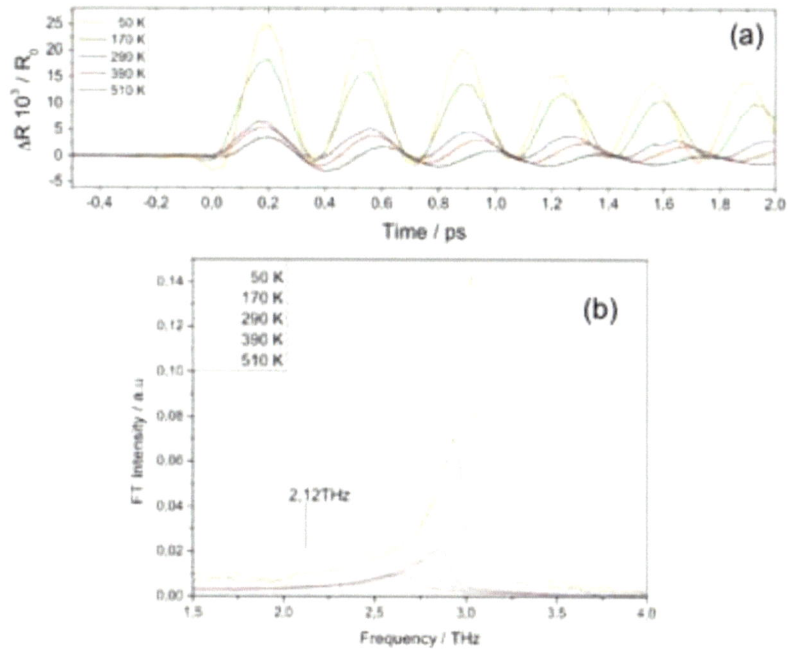

Fig. 9. Transient reflectivity (a) and its Fourier transform (b) for several initial crystal temperatures (Garl, 2008 , a).

the signal to noise ratio is much higher. The changes in reflectivity in photoexcited bismuth being very small, we had to use the second method. The real and imaginary part of dielectric constant are then recovered by using the Fresnel formulas. Figure 11 shows the transient behavior of the real and imaginary part of the dielectric constant at 800 nm in photoexcited bismuth (Garl, 2008 , a).

After the pump pulse arrival, we observe simultaneously an increase of the imaginary part and a decrease of the real part. This is consistent with the excitation of electrons in the conduction band, which enhances the conductivity. The coherent oscillations in both real and imaginary part shows that the electronic band structure is modulated by the coherent phonon displacement. This is not surprising because the excited phonon is at Γ point, and therefore the induced changes in interatomic distance concern the skin depth as whole. As the band structure depends also on the mean interatomic distance, its modulation will change periodically the electrons band structure as well. Instead, the relaxation behavior rises some question. When the plateau is reached, both the real and imaginary part are significantly different from the liquid phase value, as shown in the table 1 (Boschetto, 2010 , a). This confirms the aforementioned statement that when the equilibrium is reached, the skin depth is still in the solid state. This clearly indicates that the reached equilibrium temperature is well below the melting temperature, although the pumping energy density is higher than the enthalpy of melting.

This can be justified by the electrons transport out of the excited region. Actually, the very strong gradient set up by the pump pulse can be responsible of a fast electrons transport. Two main mechanisms could arise such a fast transport, namely the ballistic electrons transport and the electrons diffusion. The former takes place when the electron mean free path is longer then the skin depth. As the electrons Fermi velocity in bismuth reaches 10^8 cm/s (Landolt,

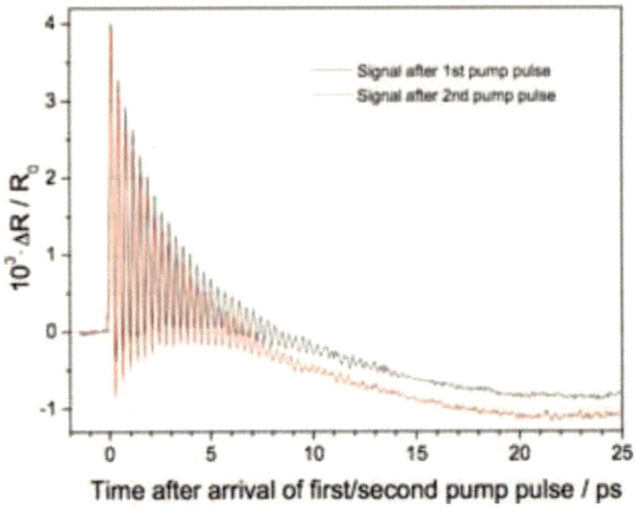

Fig. 10. Transient reflectivity of bismuth corresponding to double pump experiment, with the second pump arriving 25 *ps* after the first pump pulse, reprinted by permission of the publisher (Taylor & Francis Ltd, http://www.tandf.co.uk/journals) from (Boschetto, 2010 , a).

	ϵ_{Re}	ϵ_{Im}	$\lvert\epsilon\rvert$
Solid	-16.25	15.40	22.39
Liquid	-11.0	28.9	30.92
Transient	-13.80	11.30	17.84

Table 1. Real and imaginary part of the dielectric function at 800 nm for the solid and liquid phase, as well as for the plateau of the transient reflectivity, reprinted by permission of the publisher (Taylor & Francis Ltd, http://www.tandf.co.uk/journals) from (Boschetto, 2010 , a).

2006), the excited electrons would leave the skin depth on a time scale comparable with the pulse duration. This would imply that the skin depth temperature does not change, in contrast with previous observations. Instead, diffusive transport is a much slower process, which takes place when the electrons mean free path is small in comparison to the skin depth. Using the equations 12 and taking into account the electrons thermal conductivity, we found at equilibrium a temperature increase of only 20 K. The final lattice temperature can also be calculated by using the known changes in reflectivity toward the crystal temperature (Wu, 2007), as well as by using the coherent optical phonon parameters dependence on initial crystal temperature (Garl, 2008 , a). All these methods give around the same value of the temperature rise.
The scenario of fast electrons transport is further supported by the comparison with recent experiments by time resolved electrons diffraction on 30 nm bismuth thin film (Sciaini, 2009), in which a pump fluence of 1.3 mJ/cm^2 was enough to produce a transition to the liquid phase. Actually, in thin film the electrons cannot propagate in the direction of the temperature gradient, and their confinement results in a larger increase in the lattice temperature. Instead, in our case we used bulk crystal, in which there is no confinement of electrons.

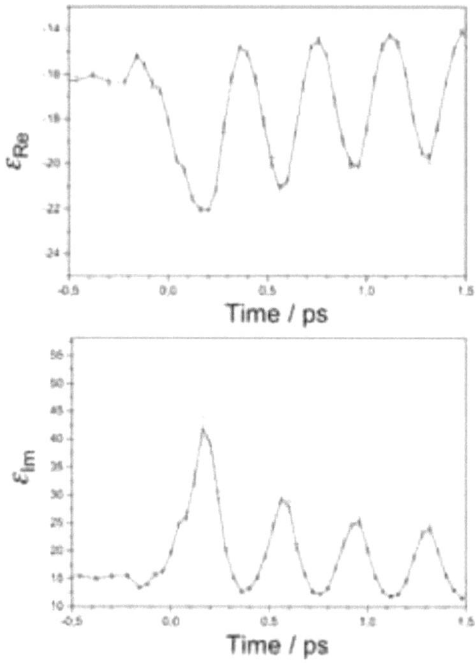

Fig. 11. Transient dielectric function of photoexcited bismuth (Garl, 2008 , a).

In order to further investigate the dynamics of electrons and lattice in photoexcited bismuth, the dielectric function should be recovered in a larger spectral range. This would also gives access to the modifications in the electrons band structure and its correlation with the coherent phonon mode.

7. Coherent optical phonon in strongly correlated materials

A great interest in studying coherent optical phonons is obviously in those materials in which electrons and lattice are strongly coupled, as well as in materials where the lattice displacements may play a role in the phase transitions. The strongly correlated materials are a large class of materials that often meet these conditions and in which the interplay between all the degrees of freedom of the crystal such as spin, charge and lattice gives rise to exotic phase diagrams. This subtle equilibrium can be strongly perturbed by the external laser pulse, leading sometimes to new phases. For example, in a manganite crystal it has even been shown that a phase transition can be induced by a coherent phonon in absence of resonant absorption (Rini, 2007). Among the strongly correlated electrons materials, the V_2O_3 crystal is of primary importance, as it is the prototype of Mott transition. Time resolved measurements on this compound have shown the excitation of both coherent acoustic and optical phonons (Mansart, 2010 , a). Interestingly, the laser pulse excitation produces a stiffening of the coherent A_{1g} mode. This indicates that the electrons excitation by the pump laser pulse gives rise to a hardening of the atomic bonding, in contrast with the founding in other materials like bismuth. Recently, the attention of a large part of the international community working on superconductors has been focussed on iron pnictides, which are new compounds showing interesting phase diagram and high superconductivity transition temperature. The coherent phonon study in the pnictide $Ba(Fe_{1-x}Co_x)_2As_2$ (Mansart, 2009) has suggested that

the A_{1g} mode does not participate to the phase transition. These results have also allowed the evaluation of the electrons phonon coupling constant (Mansart, 2010 , b), as well as to invalidate the Bardeen-Cooper-Schrieffer theory as origin of the superconductivity in this material. We point out here that the coherent phonon spectroscopy is the key approach to determine the electron phonon coupling constant of a given phonon mode.
The extension of coherent phonon studies to many other processes can be reached also by the development of tunable sources in a large spectral range. Especially, the advance in both femtosecond X-ray sources and in THz sources will allows a deeper insight in the correlations between the phonons and the physical properties in many materials.

8. Conclusions

In conclusions, in this chapter we have suggested how to approach the study of coherent optical phonon, focussing our attention on the pedagogical case of bismuth. We have shown that it is possible to control selectively the atomic displacement corresponding to one phonon mode. The study of the A_{1g} mode in bismuth has revealed some general properties of the coherent optical phonon as function of the pump pulse excitation as well as of the initial crystal temperature. As the changes in reflectivity gives only partial information on the electrons and phonon dynamics, we have shown the use of double probe pulse to recover the transient behavior of the real and imaginary part of the dielectric function. This study has demonstrated that the excess energy brought by the pump pulse is transported away from the skin depth by fast electrons diffusion, preventing any formation of liquid phase. We have discussed some examples of coherent phonon studies in strongly correlated electrons materials and shown that investigating coherent phonon dynamics will allow to gain fundamental knowledges on the physical properties of many materials.

9. References

Anisimov, S. I. et al. (1975). Electron emission from metal surfaces exposed to ultrashort laser pulses. *Sov. Phys. JETP*, Vol. 39, No. 2, August 1974, 375-377.

Ashcroft, N. W. Mermin, N. D. (1976). *Solid State Physics*, Saunders College, ISBN 0-03-083993-9, New York, New York, United States of America.

Beaud, P. et al. (2007). Spatiotemporal Stability of a Femtosecond HardX-Ray Undulator Source Studied by Control of Coherent Optical Phonons. *Physical Review Letters*, Vol. 99, October 2007, 174801.

Boschetto, D. et al (2008). Small Atomic Displacements Recorded in Bismuth by the Optical Reflectivity of Femtosecond Laser-Pulse Excitations. *Physical Review Letters*, Vol. 100, January 2008, 027404.

Boschetto, D. et al (2008). Lifetime of optical phonons in fs-laser excited bismuth. *Applied Physics A*, Vol. 92, May 2008, 873-876.

Boschetto, D. et al (2010). Ultrafast dielectric function dynamics in bismuth. *Journal of Modern Optics*, Vol. 57, Issue No. 11, 20 June 2010, 953-958.

Boschetto, D. et al (2010). Coherent interlayer vibrations in bilayer and few-layer graphene. *Submitted.*

Boyd, R. W. (2003). *Nonlinear Optics*, Academic Press, ISBN 0-12-121682-9, San Diego, California, United States of America.

DeCamp, M. F. et al. (2001). Dynamics and coherent control of high amplitude optical phonons in bismuth. *Physical Review B*, Vol. 64, August 2001, 092301.

Edelman, V. S. (1977). Properties of electrons in bismuth. *Sov. Phys. Usp.*, Vol. 20, October 1977, 819-835.

Fritz, D. M. et al. (2007). Ultrafast bond softening in bismuth: mapping a solid's interatomic potential with X-rays. *Science*, Vol. 315, February 2007, 633-636.

Garl, T. (2008). Ultrafast Dynamics of Coherent Optical Phonons in Bismuth, *PhD thesis*, July 2008, Ecole Polytechnique.

Garl, T. et al. (2008). Birth and decay of coherent optical phonons in femtosecond-laser-excited bismuth, *Physical Review B*, Vol. 78, October 2008, 134302.

Hase, M. et al. (2002). Dynamics of Coherent Anharmonic Phonons in Bismuth Using High Density Photoexcitation. *Physical Review Letters*, Vol. 88, No. 6, January 2002, 067401.

Ishioka, K. et al. (2006). Temperature dependence of coherent A_{1g} and E_g phonons of bismuth. *Journal of Applied Physics*, Vol. 100, November 2006, 093501.

Ishioka, K. et al. (2008). Ultrafast electron-phonon decoupling in graphite. *Physical Review B*, Vol. 77, March 2008, 121402.

Kudryashov, S. I. et al. (2007). Intraband and interband optical deformation potentials in femtosecond-laser-excited alpha-Te. *Physical Review B*, Vol. 75, February 2007, 085207.

Landolt-Börnstein (2006). Numerical Data and Functional Relationships in Science and Technology. *Edited by O. Madelung, U. Rössler, and M. Schulz, Landolt-Börnstein*, New Series, Group III, Vol. 41C (Springer-Verlag, Berlin, 2006).

Lannin, J. S. et al. (1975). Second order Raman scattering in the group V_b semimetal Bi Sb and As. *Physical Review B*, Vol. 12, No. 2, July 1975, 585-593.

Mansart, B. el al. (2009). Observation of a coherent optical phonon in the iron pnictide superconductor $Ba(Fe_{1-x}Co_x)_2As_2$ (x = 0.06 and 0.08). *Physical Review B*, Vol. 80, (November 2009), 172504.

Mansart, B. el al. (2010). Ultrafast dynamical response of strongly correlated oxides: role of coherent optical and acoustic oscillations. *Journal of Modern Optics*, Vol. 57, June 2010, 959-966.

Mansart, B. et al. (2010). Ultrafast transient response and electron-phonon coupling in the iron-pnictide superconductor $Ba(Fe_{1-x}Co_x)_2As_2$. *Physical Review B*, Vol. 82, July 2010, 024513.

Merlin, R. (1997). Generating coherent THz phonons with light pulses. *Solid State Communications*, Vol. 102, No. 2-3,1997, 207-220.

Murray, E. D. et al. (2005). Effect of lattice anharmonicity on high-amplitude phonon dynamics in photoexcited bismuth. *Physical Review B*, Vol. 72, August 2005, 060301.

Papalazarou, E. et al. (2008). Probing coherently excited optical phonons by extreme ultraviolet radiation with femtosecond time resolution, *Applied Physics Letters*, Vol. 93, July 2008, 041114.

Pippard, A. B., et al. (1952). The Mean Free Path of Conduction Electrons in Bismuth. *Proceedings of Royal Society A*, Vol. 65, August 1952, 955-956.

Rini, M. et al. (2007). Control of the electronic phase of a manganite by mode-selective vibrational excitation. *Nature*, Vol. 449, September 2007, 72-74.

Rousse, A., et al. (2001). Non-thermal melting in semiconductors measured at femtosecond resolution. *Nature*, Vol. 410, March 2001, 65-68.

Sciaini, G. et al. (2009). Electronic acceleration of atomic motions and disordering in bismuth. *Nature*, Vol. 458, March 2009, 56-59.

Sokolowski-Tinten, K. et al. (2003). Femtosecond X-ray measurement of coherent lattice vibrations near the Lindemann stability limit. *Nature*, Vol. 422, March 2003, 287-289.

Stevens, T. E. et al. (2002). Coherent phonon generation and the two stimulated Raman tensors. *Physical Review B*, Vol. 65, March 2002, 144304.
Uteza, O. P. et al. (2004). Gallium transformation under femtosecond laser excitation: Phase coexistenceand incomplete melting. *Physical Review B*, Vol. 70, August 2004, 054108.
Yan, Y.-X. et al. (1985). Impulsive stimulated Raman scattering: General importance in femtosecond laser pulse interaction with matter, and spectroscopy applications. *Journal of Chemical Physics*, Vol. 83, 1985, 5391-5399.
Wu, A. Q. et al. (2007). Coupling of ultrafast laser energy to coherent phonons in bismuth. *Applied Physics Letters*, Vol. 90, June 2007, 251111.
Zeiger, H.J. et al. (1992). Theory for displacive excitation of coherent phonons. *Physical Review B*, Vol. 45, January 1992, 768-778.
Ziman, J. M. (2004). *Electrons and Phonons*, ISBN 0-19-850779-8, Oxford University Press, New York, United States of America.

6

Quantum Interference Signal from an Inhomogeneously Broadened System Excited by an Optically Phase-Controlled Laser-Pulse Pair

Shin-ichiro Sato and Takayuki Kiba
Division of Biotechnology and Macromolecular Chemistry,
Graduate School of Engineering, Hokkaido University,
Sapporo 060-8628
Japan

1. Introduction

Control of quantum interference (QI) of molecular wavefunctions excited by a pair of femtosecond laser pulses that have a definite optical phase is one of the basic schemes for the control of versatile quantum systems including chemical reactions. The QI technique with the pulse pair, or the double pulse, has been applied to several atomic, molecular systems in gas phase (Scherer et al., 1991; Scherer et al., 1991; Ohmori et al., 2006) and condensed phases (Bonadeo et al., 1998; Mitsumori et al., 1998; Htoon et al., 2002; Sato et al., 2003; Fushitani et al., 2005). A basic theory of the double-pulse QI experiment for a two-level molecular system in gas phase has been given in the original paper by Scherer et al.(Scherer et al., 1991; Scherer et al., 1991). In their beautiful work, they derived the expression for the QI signal from a two-level system including a molecular vibration. However, the effect of inhomogeneous broadening, which is not very significant in the gas phase, has not been taken into account.

Although the overwhelming majority of chemical reactions take place in solution, there have been very few experimental studies on the coherent reaction control of polyatomic molecules in condensed media, due to rapid decoherence of wavefunctions. Electronic dephasing times of polyatomic molecules in solution, which have been mainly measured by photon-echo measurements, are reported to be < 100 fs at room temperature(Fujiwara et al., 1985; Bardeen &Shank, 1993; Nagasawa et al., 2003). These fast quantum-phase relaxations are considered to be caused by solute-solvent interactions such as elastic collisions or inertial (librational) motions (Cho &Fleming, 1993). Thus, understanding the role for the solvent molecules in dephasing mechanism and dynamics is strongly required.

Here, we (1) derive a compact and useful expression for the QI signal for an inhomogeneously broadened two-level system in condensed phases, when the system was excited by an optically phase-controlled laser-pulse pair (Sato, 2007), and (2) introduce our experimental results on the electronic decoherence moderation of perylene molecule in the γ–cyclodextrin (γ-CD) nanocavity (Kiba et al., 2008).

2. Theory

In general, the homogeneous broadening gives a Lorentz profile:

$$S_L(\omega) = \frac{1}{2\pi} \frac{\gamma_l}{(\omega_0 - \omega)^2 + (\gamma_l / 2)^2}. \tag{1}$$

On the other hand, the inhomogeneous broadening gives a Gauss profile:

$$S_G(\omega) = \frac{1}{\sqrt{\pi}\gamma_g} e^{-(\omega-\omega_0)^2/\gamma_g^2}. \tag{2}$$

When both the homogeneous and inhomogeneous broadening exist, the spectral profiles are given by a convolution of $S_L(\omega)$ with $S_G(\omega)$, namely, Voigt profile:

$$S_V(\omega) = \int_{-\infty}^{\infty} d\omega' S_L(\omega') S_G(\omega + \omega_0 - \omega'). \tag{3}$$

As pointed out by Scherer *et al.*, the QI signal is the free-induction decay and the Fourier transform of the optical spectral profile. According to the convolution theorem in the Fourier transform, the expression for the QI signal should have the form in principle:

$$QI(t_d) = FT[S_V(\omega)] = FT[S_L(\omega)] \cdot FT[S_G(\omega)] \propto \cos[\omega_0 t_d] \exp\left[-\frac{\gamma_l t_d}{2}\right] \exp\left[-\frac{\gamma_g^2 t_d^2}{4}\right], \tag{4}$$

where t_d is a time delay between the laser-pulse pair. However, in the above discussion, the laser pulse is assumed to be impulsive, that is, the effects of a finite time width or a spectral width of the actual laser pulse is not taken into accounts. The purpose of this paper is to derive the expression for the QI signal that includes the effects of non-impulsive laser pulses. The procedure for derivation is two steps; first, we derive the expression for the homogeneously broadened two-level system, and then we obtain the expression for the inhomogeneously broadened system by integrating the result of the homogeneously broadened system weighted by the inhomogeneous spectral distribution function.

2.1 homogeneously broadened two-level system

Let us consider a two-level electronic system interacting with a phase-controlled femtosecond-laser pulse pair (Figure 1). When the ground-state energy is assumed to be zero, that is the system is referenced to the molecular frame, the electronic Hamiltonian for the two-level system with the homogeneous broadening is given by

$$\hat{H} = (\varepsilon - i\gamma_l / 2)|e\rangle\langle e|, \tag{5}$$

where γ_l is a homogeneous relaxation constant that stands for a radiative or a non-radiative decay constant. An electronic transition dipole operator is expressed as

$$\hat{\mu} = \mu_{eg}\left(|e\rangle\langle g| + |g\rangle\langle e|\right). \tag{6}$$

The interaction Hamiltonian between the system and a photon field is given by

$$\hat{V} = -\hat{\mu}\, E(t)\,, \tag{7}$$

where photoelectric field $E(t)$ in the double-pulse QI experiments is given by the sum of E_1 and E_2, each of which has a Gauss profile：

$$E(t) = E_1(t) + E_2(t)\,, \tag{8}$$

$$E_1(t) = E_0 \exp\left[-t^2 / 2\tau^2\right]\cos(\Omega t)\,, \tag{9}$$

$$E_2(t) = E_0 \exp\left[-(t-t_d)^2 / 2\tau^2\right]\cos\left[\Omega(t-t_d)\right], \tag{10}$$

where τ is a standard deviation of an each laser pulse in time domain, and related to a standard deviation Γ of the each laser pulse in frequency domain by $\tau = 1/\Gamma$, and Ω is a common carrier frequency of the laser pulses. The phase shift of the photon field is defined as delay-time (Xu et al., 1996): the delay-time t_d between double pulses is finely controlled with attoseconds order in the optical phase-controlled experiments. This definition is natural in the optical phase-shift experiments (Albrecht et al., 1999).

To derive the expression for the QI signal, we divide the time region into the free-evolution regions and the interaction regions. (Fig. 2) Then, the time evolution of the system from the initial electronic state $|\psi(t=0)\rangle = |g\rangle$ is given by the equation:

$$|\psi(t)\rangle = \hat{U}(t-t_d-\delta)\,\hat{W}_2\,\hat{U}(t_d-2\delta)\,\hat{W}_1\,|g\rangle\,, \tag{11}$$

where the time evolution operator in the absence of the photon field is defined by

$$\hat{U}(t,t') = \exp\left[-i\hat{H}(t-t')/\hbar\right], \tag{12a}$$

or by replacing as $\Delta t = t - t'$

$$\hat{U}(\Delta t) = \exp\left[-i\hat{H}\Delta t/\hbar\right]. \tag{12b}$$

Within the framework of the first order perturbation theory, (Louisell, 1973) the time evolution operator $\hat{W}_j\ (j=1,2)$ in the presence of the photon field is given by

$$\begin{aligned} W_j &= \hat{U}(t_j+\delta, t_j-\delta)\left\{1 - \frac{1}{i\hbar}\int_{t_j-\delta}^{t_j+\delta} dt'\, \hat{U}(t_j-t')\,\hat{\mu}\,E_j(t')\,\hat{U}(t'-t_j)\right\} \\ &= \hat{U}(2\delta)\left\{1 - \frac{1}{i\hbar}\int_{t_j-\delta}^{t_j+\delta} dt'\, \hat{U}(t_j-t')\,\hat{\mu}\,E_j(t')\,\hat{U}(t'-t_j)\right\} \\ &\equiv \hat{U}(2\delta)\,(1 - \frac{1}{i\hbar}\hat{F}_j)\,. \end{aligned} \tag{13}$$

The substitution of Eq. (9) into Eq. (8) yields

$$|\psi(t)\rangle = \hat{U}(\delta)\left(\hat{U}(t) + \frac{i}{\hbar}\hat{U}(t)\hat{F}_1 + \frac{i}{\hbar}\hat{U}(t-t_d)\,\hat{F}_2\,\hat{U}(t_d)\right)|g\rangle\,, \tag{14}$$

where $\hat{F}$ is defined as an electronic transition operator, and $\hat{U}(\delta)$ a global phase factor, which will be neglected hereafter, because it does not affects final results in the state density matrix. The projection of Eq. (10) onto the excited state $|e\rangle$ gives

$$\begin{aligned}\langle e|\psi(t)\rangle &= \langle e|\left(\hat{U}(t)+\frac{i}{\hbar}\hat{U}(t)\hat{F}_1+\frac{i}{\hbar}\hat{U}(t-t_d)\hat{F}_2\,\hat{U}(t_d)\right)|g\rangle \\ &= \frac{i}{\hbar}\langle e|\left(\hat{U}(t)\hat{F}_1+\hat{U}(t-t_d)\hat{F}_2\,\hat{U}(t_d)\right)|g\rangle \\ &= \frac{i}{\hbar}\left(\exp\left[(-i\omega_0-\gamma_l/2)\,t\right]\langle e|\hat{F}_1|g\rangle+\exp\left[(-i\omega_0-\gamma_l/2)(t-t_d)\right]\langle e|\hat{F}_2|g\rangle\right)\end{aligned} \quad , \tag{15}$$

where $\omega_0=\varepsilon/\hbar$. The matrix element of an electronic transition operator $\hat{F}_j$ is calculated as

$$\begin{aligned}\langle e|\hat{F}_j|g\rangle &= \langle e|\int_{t_j-\delta}^{t_j+\delta}dt'\,\hat{U}(t_j-t')\,\hat{\mu}\,\hat{U}(t'-t_j)|g\rangle \\ &= \mu_{eg}\langle e|\int_{t_j-\delta}^{t_j+\delta}dt'\,\hat{U}(t_j-t')\left(\,|g\rangle\langle e|+|e\rangle\langle g|\,\right)\hat{U}(t'-t_j)|g\rangle\,E_j(t') \\ &= \mu_{eg}\int_{t_j-\delta}^{t_j+\delta}dt'\,\langle e|U(t_j-t')|e\rangle\langle g|U(t'-t_j)|g\rangle\,E_j(t') \\ &= \mu_{eg}\int_{t_j-\delta}^{t_j+\delta}dt'\,e^{(-i\omega_0-\gamma_l/2)(t_j-t')}E_j(t').\end{aligned} \tag{16}$$

Using a rotating-wave approximation, the matrix element is further calculated as

$$\begin{aligned}\langle e|\hat{F}_j|g\rangle &= \frac{1}{2}E_0\mu_{eg}\int_{-\infty}^{+\infty}dx e^{-i(\omega_0-\Omega)x}e^{-\gamma_l x/2-x^2/(2\tau^2)} \\ &= \sqrt{\frac{\pi}{2}}E_0\mu_{eg}\tau\exp\left[-\frac{(\omega_0-\Omega)^2\tau^2}{2}\right]\equiv F\exp\left[-\frac{(\omega_0-\Omega)^2\tau^2}{2}\right].\end{aligned} \tag{17}$$

The substitution of Eq. (13) into Eq. (11) yields

$$\langle e|\psi(t,t_d)\rangle = \frac{iF}{\hbar}\exp\left[(-i\omega_0-\gamma_l/2)t\right]\left\{\,1+\exp\left[(i\omega_0+\gamma_l/2)\,t_d\,\right]\right\}e^{-(\omega_0-\Omega)^2/(2\Gamma^2)}. \tag{18}$$

The absolute square of Eq. (14) gives the density matrix element $\rho_{ee}(t,t_d,\omega_0,\Omega)$ for the excited state

$$\begin{aligned}\rho_{ee}(t,t_d,\omega_0,\Omega) &= \langle e|\psi(t,t_d)\rangle\langle\psi(t,t_d)|e\rangle \\ &= \frac{2F^2}{\hbar^2}e^{-(\omega_0-\Omega)^2/\Gamma^2}\left\{e^{-\gamma_l t}+e^{-\gamma_l(t-t_d)}+e^{-\gamma_l t/2}e^{-\gamma_l(t-t_d)/2}\cos\left[\omega_0 t_d\right]\right\}.\end{aligned} \tag{19}$$

The first and second term give population decays of the excited state created by the first and second pulses, respectively. The third term is the interference term that is the product of coherence decays and an oscillating term.

2.2 inhomogeneously broadened system

In the previous section, the inhomogeneous broadening was not taken into consideration. The effects of inhomogeneous decay can be taken into account by summing up ρ_{ee} that originates from inhomogeneously broadened spectral components (Allen &Eberly, 1975). When the inhomogeneous spectrum function is given by a Gauss function in Eq. (2), the expectation value of the excited-state density function can be written as:

$$\begin{aligned}\langle \rho_{ee}(t_d,\omega_0,\Omega)\rangle &= \int_{-\infty}^{+\infty} d\omega\, S_G(\omega,\omega_0)\,\rho_{ee}(t_d,\omega,\Omega)\\ &= \frac{F^2}{\hbar^2}\int_{-\infty}^{+\infty} d\omega\, S_G(\omega,\omega_0)\, e^{-(\omega-\Omega)^2/\Gamma^2}\left\{e^{-\gamma_l t}+e^{-\gamma_l(t-t_d)}+e^{-\gamma_l t/2}e^{-\gamma_l(t-t_d)/2}\cos(\omega t_d)\right\}\end{aligned} \tag{20}$$

In the above equation, the two-center Gaussian functions can be rewritten as a one-center Gaussian function;

$$\begin{aligned}S_G(\omega,\omega_0)\,e^{-(\omega-\Omega)^2/\Gamma^2} &= \frac{1}{\sqrt{\pi}\gamma_g} e^{-(\omega-\omega_0)^2/\gamma_g{}^2} e^{-(\omega-\Omega)^2/\Gamma^2}\\ &= \frac{1}{\sqrt{\pi}\gamma_g}\exp\left[-\frac{(\omega_0-\Omega)^2}{\gamma_g^2+\Gamma^2}\right]\exp\left[-\frac{\gamma_g^2+\Gamma^2}{\gamma_g^2\Gamma^2}\left(\omega-\frac{\Gamma^2\omega_0+\gamma_g^2\Omega}{\gamma_g^2+\Gamma^2}\right)^2\right].\end{aligned} \tag{21}$$

By defining a reduced decay constant γ_a and a reduced frequency ω_a;

$$\frac{1}{\gamma_a^2} \equiv \frac{\gamma_g^2+\Gamma^2}{\gamma_g^2\Gamma^2}, \quad \omega_a \equiv \frac{\Gamma^2\omega_0+\gamma_g^2\Omega}{\gamma_g^2+\Gamma^2} \tag{22}$$

Eq. (21) becomes a simple form:

$$S_G(\omega,\omega_0)\,e^{-(\omega-\Omega)^2/\Gamma^2} = \frac{1}{\sqrt{\pi}\gamma_g}\exp\left[-\frac{(\omega_0-\Omega)^2}{\gamma_g^2+\Gamma^2}\right]\exp\left[-\frac{(\omega-\omega_a)^2}{\gamma_a^2}\right] \tag{23}$$

By carrying out the Gauss integral and the Fourier integral of the Gaussian function, the final form of Eq. (20) becomes:

$$\langle \rho_{ee}(t,t_d,\omega_0,\Omega)\rangle = \frac{\gamma_a}{\gamma_g}\frac{F^2}{\hbar^2}\exp\left[-\frac{(\omega_0-\Omega)^2}{\gamma_g^2+\Gamma^2}\right]\left\{e^{-\gamma_l t}+e^{-\gamma_l(t-t_d)}+2\cos[\omega_a t_d]e^{-\gamma_l t/2}e^{-\gamma_l(t-t_d)/2}e^{-\frac{\gamma_a^2 t_d^2}{4}}\right\} \tag{24}$$

In the conventional QI experiments, the QI signal is obtained as total fluorescence integrated over time. Thus, the QI signal is calculated from Eq. (24) as following:

$$\begin{aligned}QI(t_d) &= -\int_{t_d}^{\infty} dt \frac{d\langle \rho_{ee}(t,t_d,\omega_0,\Omega)\rangle}{dt} dt = \langle \rho_{ee}(t=t_d,t_d,\omega_0,\Omega)\rangle\\ &= \frac{\gamma_a}{\gamma_g}\frac{F^2}{\hbar^2}\exp\left[-\frac{(\omega_0-\Omega)^2}{\gamma_g^2+\Gamma^2}\right]\left\{1+e^{-\gamma_l t_d}+2\cos[\omega_a t_d]\,e^{\frac{-\gamma_l t_d}{2}}e^{-\frac{\gamma_a^2 t_d^2}{4}}\right\}\end{aligned} \tag{25}$$

In the above derivation, the pure dephasing was not taken into account and a transverse relaxation time constant T_2 and a longitudinal relaxation constant T_1 is related by

$$\frac{1}{T_2} = \frac{\gamma_l}{2} = \frac{1}{2T_1}. \tag{26}$$

However, in general, there also exists a pure dephasing γ^* that is brought about from elastic solute-solvent collisions. (Louisell, 1973) Thus, the transverse relaxation time constant should be rewritten as:

$$\frac{1}{T_2}=\frac{\gamma_l}{2}+\gamma^*=\frac{1}{2T_1}+\frac{1}{T_2^*}$$

The final expression for the QI signal is given by

$$QI(t_d)=\frac{\gamma_a}{\gamma_g}\frac{F^2}{\hbar^2}\exp\left[-\frac{(\omega_0-\Omega)^2}{\gamma_g^2+\Gamma^2}\right]\left\{1+e^{-\frac{t_d}{T_1}}+2\cos[\omega_a t_d]e^{-\frac{t_d}{T_2}}e^{-\frac{\gamma_a^2 t_d^2}{4}}\right\}. \tag{27}$$

By comparing the third term in Eq. (27) with Eq. (4), we obtain

$$QI(t_d)\propto\cos[\omega_a t_d]\exp\left(-\frac{t_d}{T_2}\right)\exp\left(-\frac{\gamma_a^2 t_d^2}{4}\right). \tag{28}$$

We notice that ω_0 and γ_g in the impulsive excitation are replaced by ω_a and γ_a, respectively, in the non-impulsive excitation. These reduced constants, of course, approaches ω_0 and γ_g in the limiting case of impulsive laser pulses; that is, when $\Gamma >> \gamma_g$, the following relations can be deduced.

$$\omega_a \cong \omega_0\,,\ \gamma_a \cong \gamma_g\,.$$

In the reverse limiting case of $\gamma_g >> \Gamma$, that is, in the case of quasi continuum wave (CW) laser, we notice that

$$\omega_a \cong \Omega\,,\ \gamma_a \cong \Gamma\,.$$

Under this condition, if we further assume that $\frac{1}{T_2} >> \Gamma$, the QI signal can be approximately written as

$$QI(t_d)=\frac{\Gamma}{\gamma_g}\frac{F^2}{\hbar^2}\exp\left[-\frac{(\omega_0-\Omega)^2}{\gamma_g^2}\right]\left\{1+e^{-\frac{t_d}{T_1}}+2\cos[\Omega t_d]e^{-\frac{t_d}{T_2}}\right\}. \tag{28}$$

This result may be the time-domain expression for the hole-burning experiments. These two extreme situations are schematically drawn in Fig. 3. Figure 3 infers that the overlap of the laser-pulse spectrum with the absorption spectrum plays a role of the effective spectral width for the system excited by the non-impulsive laser pulse.

Figure 4 shows the interference term of QI signals calculated for intermediate cases. The red sinusoidal curve of the QI signal was calculated for $\gamma_g = 100\ cm^{-1}$ and $\Gamma = 200\ cm^{-1}$, while the blue one was calculated for $\gamma_g = 200\ cm^{-1}$ and $\Gamma = 100\ cm^{-1}$. All the other parameters were common for the two calculations. The frequency of the QI signal is altered by the ratio of γ_g to Γ for the cases of non-zero detuning (e.g $\omega_0 - \Omega \neq 0$).

3. Cyclodextrin nanocavity caging effect on electronic dephasing of perylene in γ-CD

It is obvious that the inhibition or the moderation of dephasing is quite important subject for the development of coherent control techniques for more general reactions. In another word, protection of molecular wavefunctions from the surrounding environment becomes important issue for realization of quantum control techniques in condensed phases. For that purpose, we aimed for the protection of the quantum phase of a guest molecule using the size-fit nano-space in a cyclodextrin nanocavity (Kiba et al., 2008).

Cyclodextrins (α-, β-, or γ-CD), which are oligosaccharides with the hydrophobic interior and the hydrophilic exterior, are used as nanocavities because of their unique structures and the fact that six(α-), seven(β-), or eight(γ-) D-glucopyranose units determine the sizes whose diameters are ~5.7, 8.5, and 9.5 Å, respectively. The ability of CDs to encapsulate organic and inorganic molecules in aqueous solution has led to intensive studies of their inclusion complexes.(Douhal, 2004) We intuitively imagined that the confinement of a guest molecule within the CD nanocavity will reduce perturbations from the surrounding environment which causes decoherence. Several studies on CD complexes with aromatic compounds using steady-state and ultrafast time-resolved spectroscopy have been reported (Hamai, 1991; Vajda et al., 1995; Chachisvilis et al., 1998; Matsushita et al., 2004; Pistolis &Malliaris, 2004; Sato et al., 2006). However, there were no experiments, to our knowledge, which interrogate the effect of CD inclusion on the inhibition of decoherence.

3.1 Sample preparation

Perylene (Sigma Chemical Co.), γ-CD (Kanto Kagaku), and tetrahydrofuran (JUNSEI) was used without further purification. A Milli-Q water purification system (Millipore) was used for purification of water. Perylene / γ-CD aqueous solution for measurements was prepared by the following procedure; perylene was deposited by evaporation from saturated ethanol solution into an inner surface of a beaker, and then 10^{-2} M aqueous solution of γ-CD was added into that. The stock solution was sonicated for 5 minutes and stirred 12 hours, and then filtered in order to remove the aggregates of unsolubilized perylene. The concentration of perylene was 5×10^{-7} M that was determined from the absorption spectrum. Steady-state fluorescence and fluorescence-excitation spectra were measured with an F-4500 fluorescence spectrometer (Hitachi) at room temperature.

3.2 Quantum interference measurement using an optical-phase-controlled pulse pair

Experimental setup for the QI measurement is schematically drawn in Figure 5. The optical-phase-controlled pulse pair was generated by splitting femtosecond pulses (844 nm, ~ 40 fs, 80 MHz) from a Ti: sapphire laser (Tsunami, Spectra physics) into two equal parts by means of a Michelson interferometer.(Sato et al., 2003) A delay time t_d of pulse pair was determined by the difference in an optical path length of the two arms of the interferometer. A coarse delay was varied by a stepper-motor-driven mechanical stage on the one arm. A relative optical phase angle of two pulses was controlled with a fine delay produced by a liquid-crystal modulator (LCM, SLM-256, CRI), which can vary an optical delay with tens of atto seconds precision (approximately λ/100 of the laser wavelength). A dual-frequency (f_1 and f_2) mechanical chopper was used to modulate the laser field. The cross-beam fluorescence component that was proportional to E_1E_2 was picked up through lock-in amplifier (NF5610B) referenced to the differential frequency $f_1 - f_2$. A group velocity dispersion (GVD) of the laser output from the interferometer was compensated by a prism pair. The pulse pair from the interferometer was frequency-doubled by a BBO crystal. The frequency-doubled pulse pair was reflected by a dichroic mirror (DM) and used to excite a sample molecule, while the fundamental pulse pair transmitted through the DM was used to measure laser-fringe intensity. The fringe intensity measured here was used to determine the relative optical phase angle of two beams. The fluorescence dispersed by a monochromator (P250, Nikon) was detected by a photomultiplier tube (R106, Hamamatsu). The excitation wavelength in this measurement was fixed at 422 nm to minimize the effects of change in laser pulse shape. Fluorescence was measured at the 0-0 peak that was located at 440 nm for bulk solvent and at 450 nm for γ-CD, respectively. The typical pulse duration was obtained to be 47 fs fwhm at the sample point, assuming a Gaussian pulse. All the spectral measurements were performed using a 10 mm cuvette at room temperature (293 K).

3.3 The spectrum narrowing of steady-state fluorescence and fluorescence-excitation spectra of perylene in γ-CD

Steady-State fluorescence and fluorescence-excitation spectra of perylene in a γ-CD aqueous solution and in THF solution are shown in Figure 6. Each excitation wavelength of the fluorescence spectra was 420 nm for γ-CD and 409 nm for THF, respectively. The excitation spectra were measured by monitoring at 480 nm for γ-CD and 470 nm for bulk solvent, respectively. The stoichiometry of perylene/γ-CD complex was confirmed by measuring a pH dependence of their fluorescence spectra. The fluorescence of perylene disappeared with addition of 0.2 M NaOH to the solution. This is because the deprotonation of a neutral γ-CD molecule gives rise to form an anion in alkaline solution; thus the 1:2 complex will be dissociated owing to electronic repulsion forces between two associating γ-CD molecules which have negatively charged hydroxyl groups. This result is consistent with the behavior of 1:2 complex previously reported (Pistolis &Malliaris, 2004).

Quite interesting point in Figure 6 is that the each band in γ-CD were narrowed in comparison with that in bulk solvent, and the vibrational structure due to the ν_{15} mode (in-plane stretching motion of the center ring between the two naphthalene moieties) became clear in γ-CD. This spectral narrowing of perylene in γ-CD was comparable to that measured in MTHF at 77 K (Figure 7). Because the spectral broadening is generally caused by solute-solvent interactions, the spectral narrowing of the guest in γ-CD at room temperature is likely to be caused by the isolation of the guest from the solvent. If perylene molecules were not encapsulated by γ-CD, the broad vibrational structure like those observed in bulk solvents such as n-hexane or THF, which are shown in Figure 7 for comparison, would be observed due to the direct interaction with water molecules in solution. The fluorescence excitation spectra were fitted to a sum of ν_7 and ν_{15} vibronic bands, each of which has a Voigt lineshape. The contributions of the ν_7 and ν_{15} vibrational modes were taken into account in this fitting. The FWHM of the lowest energy vibronic band ($v' = 0$ for both ν_7 and ν_{15} mode) in several solvents are shown in Table. 1. It is remarkable that the linewidth of the vibronic band of perylene in γ-CD is narrowed even compared to that in a non-polar solvent such as *n*-hexane.

Generally speaking, origins of the spectral narrowing for perylene in γ-CD would be brought about from static and dynamic factors. The spectral width observed for molecules in bulk solvents arises from different local environments (static effect) and/or velocity distribution of solvent molecules colliding with solute molecules (dynamical effect). These effects lead to a Gaussian distribution of electronic energy gaps, that is, an inhomogeneously broadened (ensemble-averaged) spectrum. The interaction with surrounding environment plays a major role for both homogeneous and inhomogeneous broadenings. We intuitively imagine that the spectral narrowing is originated from the isolation of a guest molecule from the surrounding environment. However, the situation is not so simple, since this phenomenon was not observed for every combination of other host/guest CD complexes. For example, the excitation spectra of perylene/γ-CD and anthracene/β-CD complex were shown in Figure 4 in order to compare the spectral linewidth. The significant spectral narrowing was observed for the case of perylene/γ-CD (Figure 8(a)), whereas almost no narrowing was observed for anthracene/β-CD complex (Figure 8(b)). Since both guest molecules are non-polar, the relative extent (size) of guest molecule relative to the CD cavity sizes was a key factor of the spectral narrowing. In aqueous solution of CD inclusion complex, it is well known that the CD nanocavity contains some solvent water molecules accompanied with the guest molecule.(Douhal, 2004) Therefore, the guest molecules don't suffer from the solvent relaxations that would bring about the spectral broadening. Actually, a Stokes-shift in γ-CD was very small (30 cm^{-1}). As shown in Figure 6a, the 0-0 transition bands of fluorescence and excitation spectra in γ-CD were almost overlapped. This spectral feature indicates that there is no space inside the CD cavity for solvent reorientation between photo absorption and emission. On the other hand, for the case of anthracene / β-CD complex, where the cavity size is larger than the guest molecule, the water molecules are loosely captured in CD cavity, in which the water molecules have the degree of freedom to affect the spectral properties of guest molecule. Therefore, the spectral changes (i.e. narrowing and nearly zero Stoke's Shift) are likely to appear only when the size of guest molecule is just-fitted to the interior size of CD cavity.

There is an issue that whether the size-fit effect within CD cavity contributes to homo- or inhomogeneously to the spectral changes. Generally, the homogeneous broadening gives a Lorentz profile, and the inhomogeneous broadening gives a Gauss profile. In condensed phases, the spectral lineshape contains both homo- and inhomogeneous contributions, and is described by Voigtian which is the convolution of a Lorentzian with a Gaussian as described in the theoretical section. In principle, it is possible to separate a homogeneous component from an inhomogeneous component in the steady-state electronic spectra, by fitting each peak to a Voigt function.(Srajer &Champion, 1991) However, this method includes ambiguity since the deconvolution is necessary, and it is troublesome to determine each parameter uniquely. On the other hand, the QI time profile is the Fourier transform of the steady-state spectrum as previously mentioned. This means the QI time profile is the product of homogeneous dephasing (exponential decay) and inhomogeneous dephasing (Gauss-type decay). Thus, the separation of the two components is much easier in time-domain. In the next section, we discuss the distinction between homogeneous and inhomogeneous broadenings from the result of time-domain QI measurements.

3.4 The QI signals of perylene in γ-CD

The QI signals of perylene (solid line) in γ-CD and in THF solution are shown together with the fringe signals (dotted line) in Figure 9. The intensity of the QI signal was plotted as a function of fine time delay which was defined by the liquid-crystal phase-shifter. The QI signal oscillated with the frequency almost twice that of the fringe, since the fringe signal was measured for the fundamental laser light. A QI signal observed in γ-CD survives at 180 fs, although the signal in THF solution almost diminishes at the same delay time. The QI signal should be enhanced or depreciated according to the phase relation of the molecular wave function, and should oscillate with the period corresponding to the energy interval between electronic ground and excited states while quantum phase of molecular wavefunction created by the first pulse is preserved as shown in Eq.(27). Intramolecular vibrational relaxations and/or solute-solvent interactions disturb the quantum phase of a molecular wavefunction created by the first pulse, and induce decoherence of the wavefunctions. Decoherence reduces the amplitude of QI signal decays as the delay time increases. Therefore, a decay curve of the amplitude of the QI signal represents the electronic dephasing of the sample molecules. It should be noted that the dephasing includes homogeneous and inhomogeneous contributions as described in the previous section. Figure 10 displays the envelope function of QI Signal, in which the absolute square root of the QI signal is plotted as a function of the delay time after t_d = 100 fs. We abandoned the data before t_d = 100 fs because the overlapping of the laser-pulse pair deforms the QI time profile. At a glance, the electronic dephasing of perylene in γ-CD is slower than that in THF solution in Fig. 10.

The envelope of QI signal was fitted to the Eq. (28) in order to estimate the homogeneous dephasing time T_2. The QI signal fitting was carried out together with the fitting of steady-state electronic spectra to Voigt functions, simultaneously, in order to eliminate the ambiguity which arises from estimation of the homogeneous dephasing time and the inhomogeneous linewidth value. In this fitting, we used the following procedures; the initial estimated value of homogeneous and inhomogeneous linewidth were obtained from the rough fitting of vibronic bands to the sum of Voigtian. The QI signal was fitted to the Eq. (28) using obtained inhomogeneous linewidth value (γ_g) in order to estimate the homogeneous dephasing time (T_2). We used the vibronic bandwidth value which was overlapped with laser spectrum. Average value of two vibronic band weighted with an area intensity, (v_7=1, v_{15}=0), (v_7=1, v_{15}=1), was used in the case of γ-CD aqueous solution, and (v_7=0, v_{15}=3), (v_7=1, v_{15}=1) was used in the case of THF solution. The QI signal of the pure THF solvent and the 10^{-2} M γ-CD aqueous solution were used as an instrumental response function, and the spectral linewidth value of laser pulse (Γ) was calculated from the pulse duration of the instrumental response function. The steady-state spectrum was fitted again by using T_2 obtained from QI signal fitting, and estimated the inhomogeneous linewidth value. The QI signal fitting was carried out again by using the obtained γ_g. Fitting of steady-state spectra and QI signal was iterated until the fitting parameters T_2 and γ_g were converged, and we found the best parameter set which can reconstruct the steady-state spectra and the time profile consistently. The estimated dephasing time constant (T_2), the homogeneous (γ_l) and inhomogeneous (γ_g) linewidth values obtained from QI signal analysis was summarized in Table 2. From the analysis of dephasing curve, the homogeneous electronic dephasing time (T_2) of perylene in THF and γ-CD nanocavity were estimated to be T_2 = 23 ± 3 fs and T_2 = 42 ± 5 fs, respectively. It was found that the encapsulation of perylene molecule into CD nanocavities brings about the lengthening of T_2. The same excitation wavelength (422 nm) was used for the QI measurements in γ-CD and in THF, in order to avoid the influence caused by the change in laser pulse shape. The same excitation energy caused the situation that a vibrational excess energy above S_1 origin is different for the two measurements since the absorption spectrum of perylene in CD nanocavities are red-shifted from that in THF solution. The excess energies are approximately 1500 cm^{-1} in the γ-CD and 900 cm^{-1} in THF, respectively. In the photon-echo studies, it was found that the excitation with large vibrational excess energy accelerates the electronic dephasing in large molecules such as cresyl violet; the acceleration was attributed to intramolecular vibrational relaxations. Therefore, the faster electronic dephasing time would be expected for the measurement of CD inclusion complex if only the difference in excess energy were taken into consideration in our experimental condition. However, our experimental finding was opposite; the dephasing time of perylene in γ-CD was longer than that in THF even with the higher excess energy. Therefore, we can conclude that the CD caging effect brings the lengthening of dephasing time, which overcomes the shortening of the dephasing time due to the increased excess energy. The longer dephasing time should be expected when the excitation laser wavelength is located around 0-0 transition of the absorption spectrum.

4. Conclusion

We have shown that the decays of QI signal obtained by the non-impulsive excitation can be written as the product of exponential decay and Gauss decay, and the Gauss decay constant is given by

$$\frac{1}{\gamma_a^2} \equiv \frac{\gamma_g^2 + \Gamma^2}{\gamma_g^2 \Gamma^2}.$$

We have also shown that the frequency of QI signal is given by

$$\omega_a \equiv \frac{\Gamma^2 \omega_0 + \gamma_g^2 \Omega}{\gamma_g^2 + \Gamma^2}.$$

In general, it is often difficult to fit the optical absorption spectrum with Voigt functions in the frequency domain, because the Voigt function includes the convolution integral, and one often finds several parameter sets of the least-squared fits. This situation makes it difficult to separate homogeneous components from inhomogeneous components in the frequency-domain spectrum. In contrast, the fitting procedure is rather easier in the QI experiment, once the expression that includes the effect of laser-pulse width is given. This is because the homogeneous and inhomogeneous components are the simple product in the QI experiment. By analyzing the frequency-domain spectrum and the time-domain QI profile simultaneously (e.g. global fit), the reliable determination of homogeneous and inhomogeneous components of relaxations becomes possible.

	Linewidth[a]	$\nu_{exc\ max}$	$\nu_{fluo\ max}$	Stokes-Shift
γ-CD / water	278 cm^{-1}	22245 cm^{-1}	22215 cm^{-1}	30 cm^{-1}
THF	373 cm^{-1}	22837 cm^{-1}	22717 cm^{-1}	120 cm^{-1}
n-hexane	314 cm^{-1}	22959 cm^{-1}	22894 cm^{-1}	65 cm^{-1}

Table 1. Comparison of the electronic spectra of perylene in solutions at room temperature.

a. FWHM of lowest energy vibronic band (v′ = 0 for both ν_7 and ν_{15} mode) resolved from the fluorescence excitation spectra.

	T_2	□g
γ-CD / water	42 ± 5 fs	180 ± 20 cm^{-1}
THF	23 ± 3 fs	270 ± 35 cm^{-1}

Table 2. Best-fit parameter set (homogeneous dephasing time and inhomogeneous linewidth value (FWHM)) obtained from QI signal and steady-state spectra.

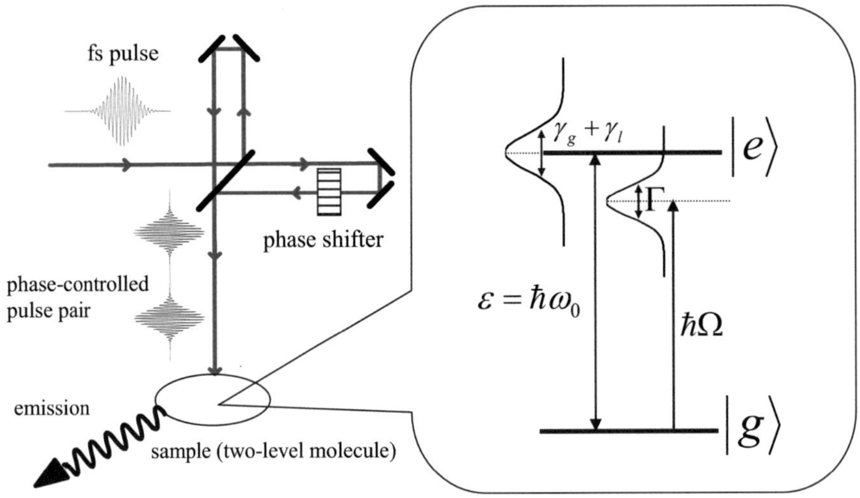

Fig. 1. Schematic drawing of the QI experiment with a phase-controlled laser-pulse pair.

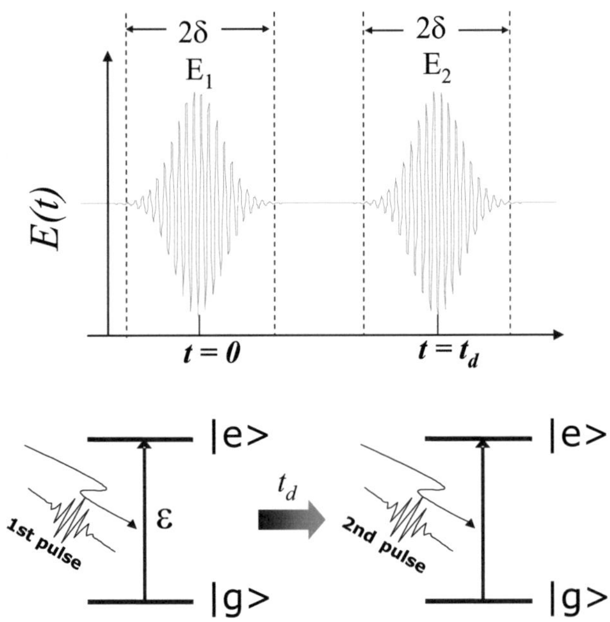

Fig. 2. Time domains: free evolution and interaction with laser pulses.

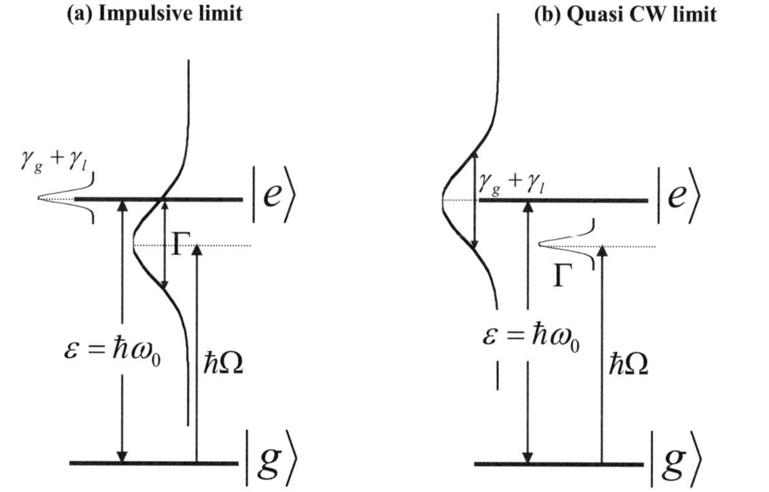

$$\psi_1(r,t) \propto \varphi(r)\, e^{-\omega_0 t} \qquad \psi_1(r,t) \propto \varphi(r)\, e^{-\Omega t}$$

Fig. 3. Limiting cases: impulsive laser (left) and quasi CW laser (right).

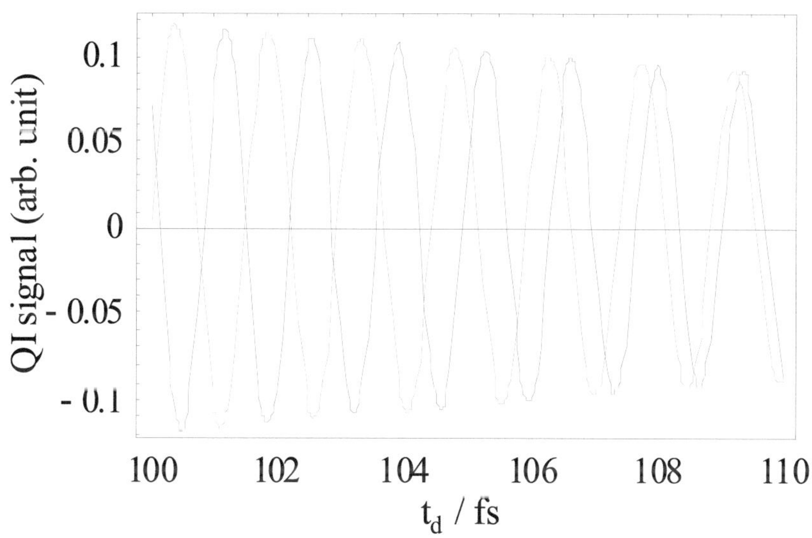

Fig. 4. QI signals simulated for the intermediate cases. The parameters specific for each curve were: $\gamma_g = 100\ cm^{-1}$ and $\Gamma = 200\ cm^{-1}$ (for the red curve), $\gamma_g = 200\ cm^{-1}$ and $\Gamma = 100\ cm^{-1}$ (for the blue curve). The common parameters for the two curves were $\omega_0 = 25000\ cm^{-1}$, $\Omega = 22000\ cm^{-1}$, $\gamma_l = 100 cm^{-1}$, and $\gamma^* = 25\ cm^{-1}$.

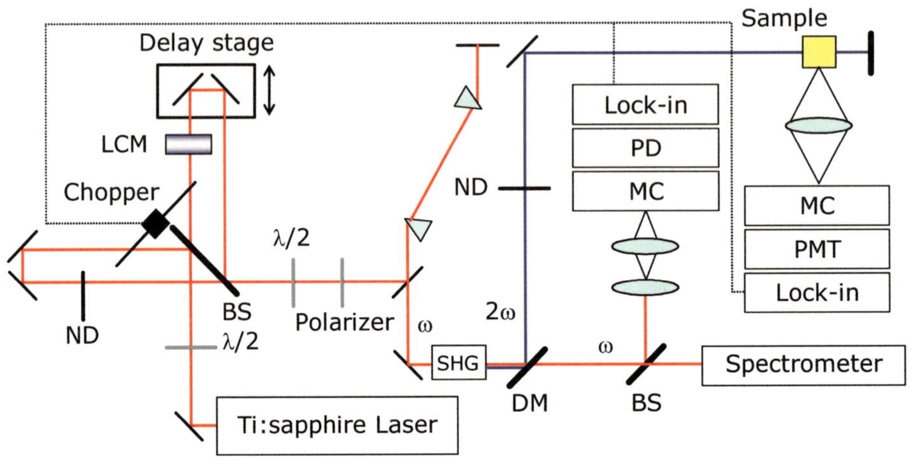

Fig. 5. The experimental setup for the quantum wavepacket interferometry. Abbreviations in the schematic diagram are used for optical beam splitter (BS), the second harmonic generator (SHG), the dichroic mirror (DM), the liquid crystal modulator (LCM), the neutral density filter (ND), monochromator (MC), the photomultiplier (PMT), and the photo diode (PD).

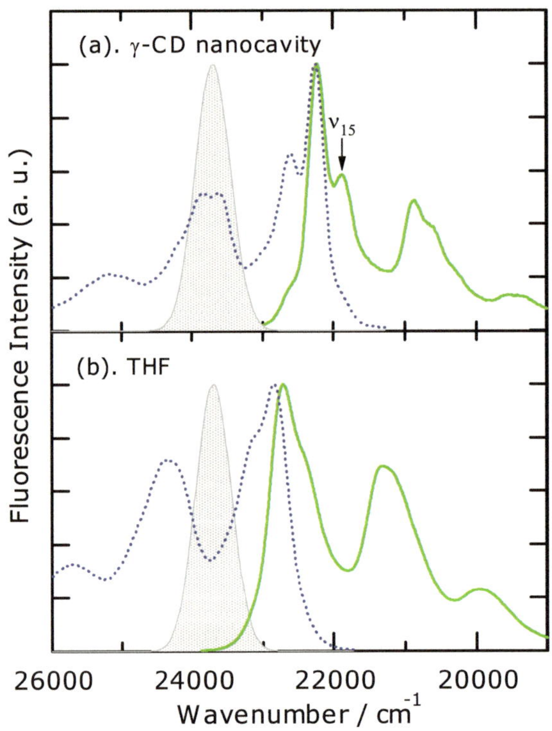

Fig. 6. Steady-state fluorescence (solid line) and fluorescence-excitation (dotted line) spectra of perylene (a) in γ-CD nanocavity and (b) in THF. Spectrum of the excitation pulse used in the quantum interference measurement is also shown for comparison (shaded area).

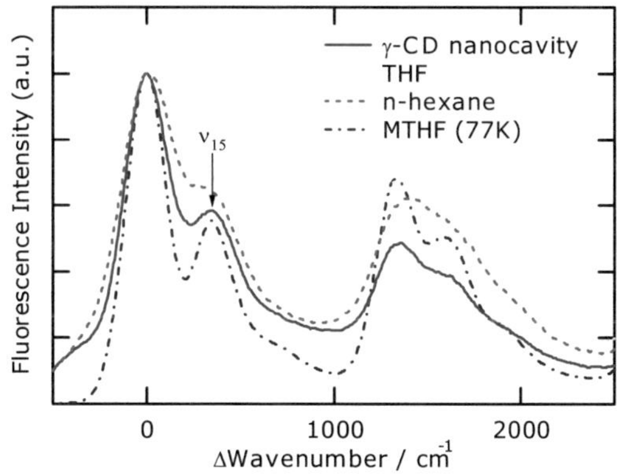

Fig. 7. Steady-state fluorescence spectra of perylene in γ-CD nanocavity (solid line), in THF (dotted line), in n-hexane (dashed line) and in MTHF at 77 K (dash-and-dotted line). The spectra are displayed as wavenumber shift from 0-0 transition in order to compare the spectral line-shapes.

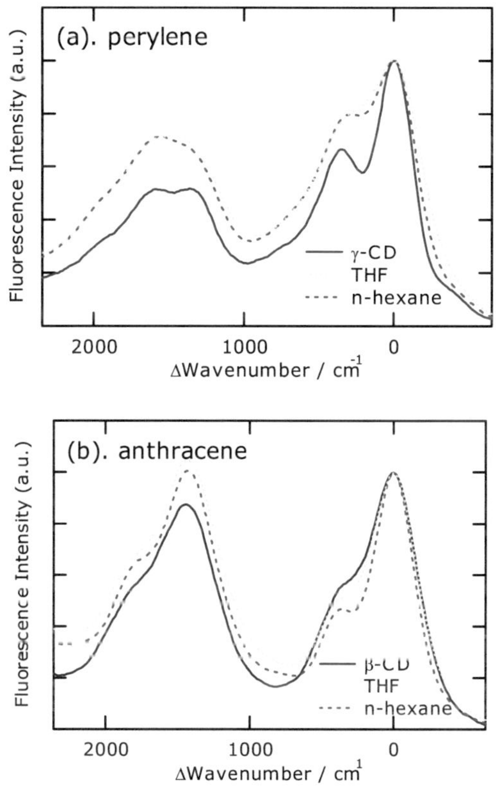

Fig. 8. Steady-state fluorescence-excitation spectra of (a) perylene in γ-CD (solid line), in THF (dotted line) and in n-hexane (dashed line), and (b) anthracene in β-CD (solid line), in THF (dotted line) and in n-hexane (dashed line). The spectra are displayed as wavenumber shift from 0-0 transition in order to compare the spectral line-shapes.

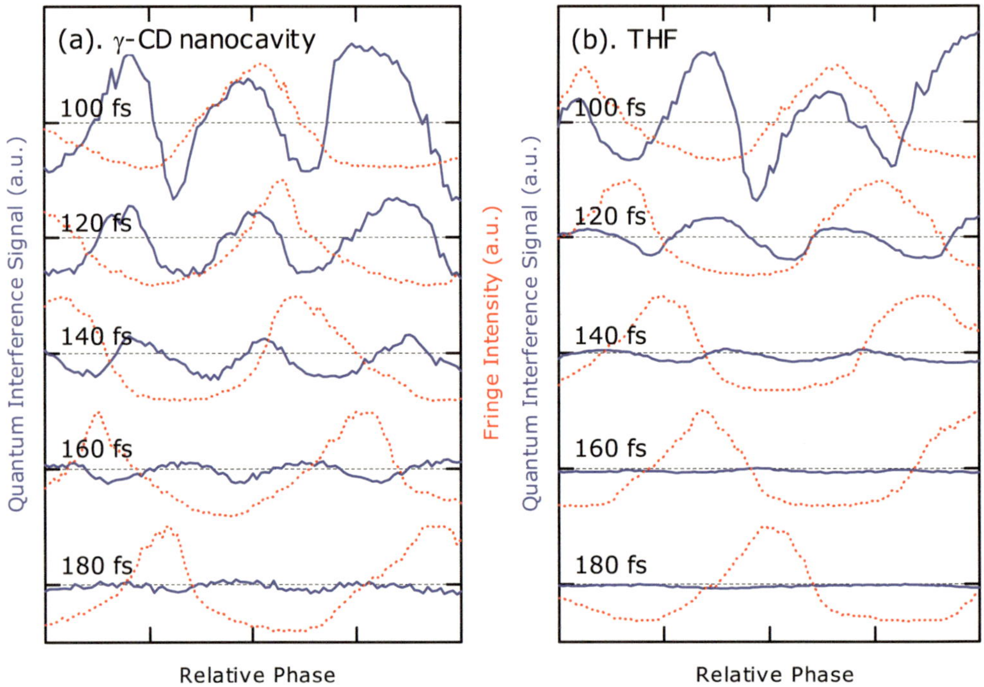

Fig. 9. Quantum interference signals (solid line) of perylene in (a) γ-CD nanocavity and in (b) THF. Fringe signals (dashed line) are also shown as a measure of relative optical phase.

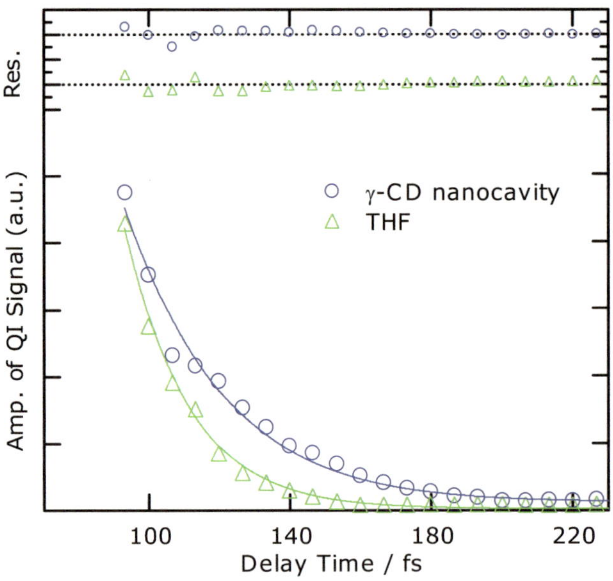

Fig. 10. Electronic dephasing curves of perylene in γ-CD nanocavity (open circles) and in THF (open triangles), where the oscillating amplitude of the QI signal is plotted as a function of the delay time. The dephasing curves were fitted to a theoretical equation. Solid lines are fits of experimental data.

5. References

Albrecht, A. W., Hybl, J. D., Faeder, S. M. G. and Jonas, D. M. (1999). Experimental distinction between phase shifts and time delays: Implications for femtosecond spectroscopy and coherent control of chemical reactions, *Journal of Chemical Physics*, Vol.111, 10934-10956.

Allen, L. and Eberly, J. H. (1975). Optical Resonance and Two-Level Atoms. New York, Dover.

Bardeen, C. J. and Shank, C. V. (1993). Femtosecond electronic dephasing in large molecules in solution using mode suppression, *Chemical Physics Letters*, Vol.203(5,6), 535-539.

Bonadeo, N. H., Erland, J., Gammon, D., Park, D., Katzer, D. S. and Steel, D. G. (1998). Coherent Optical Control of the Quantum State of a Single Quantum Dot, *Science*, Vol.282, 1473-1476.

Chachisvilis, M., Garcia-Ochoa, I., Douhal, A. and Zewail, A. H. (1998). Femtochemistry in nanocavities: dissociation, recombination and vibrational cooling of iodine in cyclodextrin *Chemical Physics Letters*, Vol.293(3-4), 153-159.

Cho, M. and Fleming, G. R. (1993). Photon-echo measurements in liquids: numerical calculations with model systems *Journal of Chemical Physics*, Vol.98(4), 2848-2859.

Douhal, A. (2004). Ultrafast Guest Dynamics in Cyclodextrin Nanocavities, *Chemical Review*, Vol.104, 1955-1976.

Fujiwara, M., Kuroda, R. and Nakatsuka, H. (1985). Measurement of ultrafast dephasing time of Cresyl Fast Violet in cellulose by photon echcues with incoherent light, *Journal of Optical Society of America B*, Vol.2, 1634-1639.

Fushitani, M., Bargheer, M., hr, M. G. and Schwentner, N. (2005). Pump–probe spectroscopy with phase-locked pulses in the condensed phase: decoherence and control of vibrational wavepackets, *Physical Chemistry Chemical Physics*, Vol.7, 3143-3149.

Hamai, S. (1991). Exciplex Formaition between Perylene and N,N-Simethylaniline in a Ternary Inclusion Compound with γ-Cyclodextrin in H_2O-Ethanol (7:3) Mixture, *Bulletin of the Chemical Society of Japan*, Vol.64, 431-438.

Htoon, H., Takagahara, T., Kulik, D., Baklenov, O., A. L. Holmes, J. and Shih, C. K. (2002). Interplay of Rabi Oscillations and Quantum Interference in Semiconductor Quantum Dots, *Physical Review Letters*, Vol.88, 087401-1-4.

Kiba, T., Kasajima, T., Nishimura, Y. and Sato, S.-i. (2008). Cyclodextrin Nanocavity Caging Effect on Electronic Dephasing, *ChemPhysChem*(9), 241-244.

Louisell, W. H. (1973). Quantum Statistical Properties of Radiation. New York, John Wiley & Sons, Inc.

Matsushita, Y., Suzuki, T., Ichimura, T. and Hikida, T. (2004). Cavity Size Effect on the Excited State Dynamics of Methyl 4-(Dimethylamino)benzoate-Cyclodextrin Complexes, *Journal of Physical Chemistry A*, Vol.108, 7490-7496.

Mitsumori, Y., Mizuno, M., Tanji, S., Kuroda, T. and Minami, F. (1998). Coherent transients involving 2P excitons in ZnSe, *Journal of Luminescence*, Vol.76&77, 113-115.

Nagasawa, Y., Seike, K., Muromoto, T. and Okada, T. (2003). Two-Dimensional Analysis of Integrated Three-Pulse Photon Echo Signals of Nile Blue Doped in PMMA, *Journal of Physical Chemistry A*, Vol.107(14), 2431-2441.

Ohmori, K., Katsuki, H., Chiba, H., Honda, M., Hagihara, Y., Fujiwara, K., Sato, Y. and Ueda, K. (2006). Real-Time Observation of Phase-Controlled Molecular Wave-Packet Interference, *Physical Review Letters*, Vol.96, 093002.

Pistolis, G. and Malliaris, A. (2004). Evidence for Highly Selective Supramolecular Formation between Perylene/ç-CD and Pyrene/ç-CD Complexes in Water, *Journal of Physical Chemistry B*, Vol.108, 2846-2850.

Sato, S.-i. (2007). Theory of the Quantum Interference Signal from an Inhomogeneously Broadened Two-Level System Excited by an Optically Phase-Controlled Laser-Pulse Pair, *Journal of Chemical Theory and Computation*, Vol.3, 1158-1162.

Sato, S.-i., Nakamura, T., Nitobe, S., Kiba, T., Hosokawa, K., Kasajima, T., Otsuka, I., Akimoto, S. and Kakuchi, T. (2006). Structure and Excitation Relaxation Dynamics of Dimethylanthracene Dimer in a γ-Cyclodextrin Nanocavity in Aqueous Solution, *Journal of Physical Chemistry B*, Vol.110, 21444-21449.

Sato, S.-i., Nishimura, Y., Sakata, Y. and Yamazaki, I. (2003). Coherent Control of Oscillatory Excitation Transfer in Dithia-1,5[3,3]anthracenophane by a Phase-Locked Femtosecond Pulse Pair, *Journal of Physical Chemistry A*, Vol.107, 10019-10025.

Scherer, N. F., Carlson, R. J., Matro, A., Du, M., Ruggiero, A. J., Romero-Rochin, V., Cina, J. A., Fleming, G. R. and Rice, S. A. (1991). Fluorescence-detected wave packet interferometry: Time resolved molecular spectroscopy with sequences of femtosecond phase-locked pulses, *Journal of Chemical Physics*, Vol.95, 1487-1511.

Scherer, N. F., Matro, A., Ziegler, L. D., Du, M., Carlson, R. J., Cina, J. A. and Fleming, G. R. (1991). Fluorescence-detected wave packet interferometry. II. Role of rotations and determination of the susceptibility, *Journal of Chemical Physics*, Vol.96, 4180-4194.

Srajer, V. and Champion, P. M. (1991). Investigations of Optical Line Shapes and Kinetic Hole Burning in Myoglobin, *Biochemistry*, Vol.30, 7390-7402.

Vajda, S., Jimenez, R., Rosenthal, S. J., Fidler, V., Fleming, G. R. and Castner, E. W., Jr. (1995). Femtosecond to nanosecond solvation dynamics in pure water and inside the g-cyclodextrin cavity., *Journal of the Chemical Society, Faraday Transactions*, Vol.91, 867-873.

Xu, L., Spielmann, C., Poppe, A., Brabec, T., Krausz, F. and Hänsch, T. W. (1996). Route to phase control of ultrashort light pulses, *Optics Letters*, Vol.21, 2008-2010.

7

Quantum Control of Laser-driven Chiral Molecular Motors

Masahiro Yamaki[1], Sheng H. Lin[1,2], Kunihiko Hoki[3] and Yuichi Fujimura[3]
[1]*Department of Applied Chemistry, National Chiao Tung University, Hsinchu, Taiwan 30010*
[2]*Institute of Atomic and Molecular Sciences, Academia Sinica, Taipei, Taiwan 10617*
[3]*Department of Chemistry, Graduate School of Science, Tohoku University, Sendai, Japan 980-8578*

1. Introduction

The design and control of functional molecular machines and devices is one of the fascinating and challenging research targets in molecular science (Feringa et al., 2000; Kinbara & Aida, 2005; Kay et al., 2007). They were originally inspired from biological machines such as ATP synthases (Boyer, 1993; Abrahams et al., 1994) and myosin and kinesin (Jülicher et al., 1997). They now include various kinds of artificial molecular machines such as transmitters, shuttles, nanocars and logic gates (Balzani et al., 2008), which can be driven by external forces at the molecular level. Some of them are not simply sized-down versions of macroscopic machines and are controlled at the quantum level (Roncaglia & Tsironis, 1998).

Lasers are energy sources over a wide range of wave lengths from mid-infrared to ultraviolet, which make it possible to drive various sizes of molecular machines without any direct contact. Lasers are expected to play an important role as a source of external forces for controlling molecular machines because lasers have various controlling-parameters such as central frequencies, pulse shapes, photon polarizations and time differences between two pulses (Assion et al., 1998; Gouliemakis et al., 2004).

Based on coherent control theory (Kosloff et al. 1989; Shi & Rabitz, 1990; Shapiro & Brumer, 2000), laser pulses can be designed to produce the maximum desired target with minimum laser energy (Assion et al., 1998; Rice & Zhao, 2000; Gordon & Fujimura, 2002; Bandrauk et al., 2002). Molecular machines can be controlled through coherent interactions between lasers and molecules at a quantum level (Hoki et al., 2003). The procedures are sometimes called "quantum ignition" for driving molecular motors (Fujimura et al., 2004). The time evolution is obtained by solving the time-dependent Schrödinger equation or the Liouville equation (Sugawara & Fujimura, 1994; Ohtsuki et al., 1999; Hoki et al., 2001). Application of coherent control theory enables extraction of key factors for driving molecular motors with a unidirectional motion, though we have to wait for further experimental progress to carry out coherent control experiments on artificial molecular machines. In this chapter, we present fundamental principles for unidirectional motions of chiral molecular motors driven by linearly polarized laser pulses having no photon helicity.

In Sec. 2, we first clarify the role of molecular chirality. We discuss the mechanism of unidirectional motions of chiral motors. For this purpose, it is instructive to mention mechanisms of unidirectional motions of bio-motors. In bio-motors, the unidirectional motions are explained in terms of so-called nonequilibrium fluctuations of a Brownian motion with a saw-toothed ratchet potential (Astumian & Hänggi, 2002; Reimann, 2002). Molecular chirality is an essential factor for the unidirectional motion of a rotary motor driven by a linearly polarized laser pulse. The motion is basically determined by both the asymmetric potential originating from its molecular chirality and a periodic perturbation of laser-molecule interactions (Hoki et al., 2003).

We next present the results of a quantum dynamics simulation of simple, real chiral molecules to clarify the mechanism of unidirectional motions. The directional motion is determined by molecular chirality, not by the external laser field, which gives periodic perturbations to motors. We treat molecular motors in a heat bath by using the Liouville equation to demonstrate dephasing effects due to interactions with the heat bath. Here, we consider the case in which the magnitude of fluctuation forces causing dephasing is weak compared with that of laser-motor interactions. The situation is common for the treatment of molecule-laser interactions but different from the case of bio-motors, the dynamics of which can be explained in terms of nonequilibrium fluctuations of bath modes. We also present results of unidirectional motions obtained by using a pump-dump laser-ignition method (Hoki et al., 2004). We finally present a practical method for detecting quantum dynamics of molecular motors in real time.

In Sec. 3, we briefly present results of optimal control for unidirectional motions of chiral molecular motors. A local and global control methods were applied to chiral molecular motors (Yamaki, 2005; Yamaki, 2008).

In Sec. 4, we treat a simple molecular machine consisting of two internal rotors, one of which is a propeller and the other of which is a motor driven by laser pulses. We discuss the mechanism of energy transmission from the motor to the propeller.

In Sec. 5, we present some future research subjects in laser-driven molecular motors after a summary of the chapter.

2. Unidirectional rotations of molecular motors

First of all, we define the equation of motion and the Hamiltonian of the light-driven molecular motors, which are used throughout this chapter. The system Hamiltonian of a molecular motor in the presence of an electric field of light $\boldsymbol{E}(t)$ within the long-wave approximation is written as

$$\hat{H}_S^{\Gamma}(t) = \hat{H}_0^{\Gamma} - \hat{\boldsymbol{\mu}}^{\Gamma} \cdot \boldsymbol{E}(t)\,, \tag{1}$$

where $\hat{H}_0^{\Gamma}$ is a molecular Hamiltonian that consists of a kinetic energy operator, $\hat{T}$ and a potential energy operator, $\hat{V}_0^{\Gamma}$; $\hat{\boldsymbol{\mu}}^{\Gamma}$ is a dipole moment vector; $\Gamma(=S, R)$ indicates molecular chirality, which is sometimes omitted if it is not necessary in the later sections. Explicit forms of $\hat{H}_S^{\Gamma}(t)$ will be defined later. The total Hamiltonian including surrounding effects is written as

$$\hat{H}_S^{\Gamma}(t) = \hat{H}_{\boldsymbol{S}}^{\Gamma}(t) + \hat{H}_B + \hat{H}_{\boldsymbol{S}B}^{\Gamma} \tag{2}$$

with the bath Hamiltonian $\hat{H}_B$ and the motor-bath interaction Hamiltonian $\hat{H}_{SB}^{\Gamma}$. We can obtain results of interactions between the motor and laser pulses as a special case. The quantum dynamics of the molecular motors can be generally described by the time-dependent Liouville equation given as

$$i\hbar\frac{\partial}{\partial t}\hat{\rho}^{\Gamma}(t)=\hat{L}^{\Gamma}(t)\hat{\rho}^{\Gamma}(t), \tag{3}$$

where $\hat{\rho}^{\Gamma}(t)$ denotes the density operator of molecular motors in the heat bath, and $\hat{L}^{\Gamma}(t)=[\hat{H}^{\Gamma}(t),\ ]$ with commutator $[\ ,\ ]$. The time-dependent coupled equation (3) is solved numerically with an initial condition $\hat{\rho}^{\Gamma}(0)$ given by the Boltzmann distribution at temperature T.

2.1 Molecular chirality and periodically perturbed chiral molecular motors

One of the ideas for creating unidirectional motions is shown in the upper panel of Fig. 1. Consider a saw-toothed ratchet potential $V_0(\phi)$. The potential $V_0(\phi)$ is characterized by the L periodicity as $V_0(\phi) = V_0(\phi+L)$, and by the broken spatial symmetry that is expressed as $V_0(\phi-c) \neq V_0(-\phi+c)$ for any c. The asymmetric static potential energy $V_0(\phi)$ is not sufficient to create unidirectional motion of the system, and one of the typical ratchet systems introduces a time-correlated tilting force $f(t)$, which is known as a nonequilibrium fluctuation of a Brownian motion. As shown in the upper-left panel of Fig. 1, when the force is negative, a mass point falls down to the left side. On the other hand, when the force is positive, the mass point falls down to the right side. In this way, by combining static asymmetric potential $V_0(\phi)$ and unbiased force such that the time average of $f(t)$ equals zero, a unidirectional motion in the system can be obtained. Here, the direction is determined by the shape of the static asymmetric potential energy $V_0(\phi)$, and the mass point moves in the intuitive direction in which the slope to climb is gentle.

A rotary motion of a chiral molecule can also be characterized by asymmetric potential energy. Consider an idealized chiral molecule with two rigid groups, A and B, as shown in Fig. 2. Here, the mass of A is set to be heavy compared with that of B, and A is taken as the body and B is taken as the rotating group. These two groups are connected to each other by a single bond, and the rotary motion is described in terms of coordinate ϕ. We call a pair of the chiral molecules (S)- and (R)-motors in this paper. The potential energy of rotation $V^{\Gamma}(\phi)$ with $\Gamma = S$ or R is characterized by an asymmetric potential, and the potential energy satisfies $V^S(\phi) = V^R(-\phi)$ since the two motors (S) and (R) have mirror image. The rigid group B has plus and minus extremities to create an electric dipole moment. Therefore, the lowest order interaction energy between the dipole moment and an oscillating electric field with a linear photon polarization can be written as $\cos(\phi)\,f(t)$. We note that a linearly polarized light has no photon helicity, i.e., no photon angular momentum.

The time-dependent effective potential energy of the rotary motion under an oscillating electric field is schematically shown in the lower panel of Fig. 1. The major difference from the above-mentioned tilting ratchet is that the chiral molecule is periodically perturbed, and the entire effective potential has L periodicity as $V^{\Gamma}(\phi,t) = V^{\Gamma}(\phi + L,t)$. In contrast to the case of the tilting ratchet, it is not always obvious whether the system creates a unidirectional motion or not, although the same rotational direction as that in the upper panel in Fig. 1 is expected. In the next subsection, we examine the direction of the rotary motion by using a real model molecule.

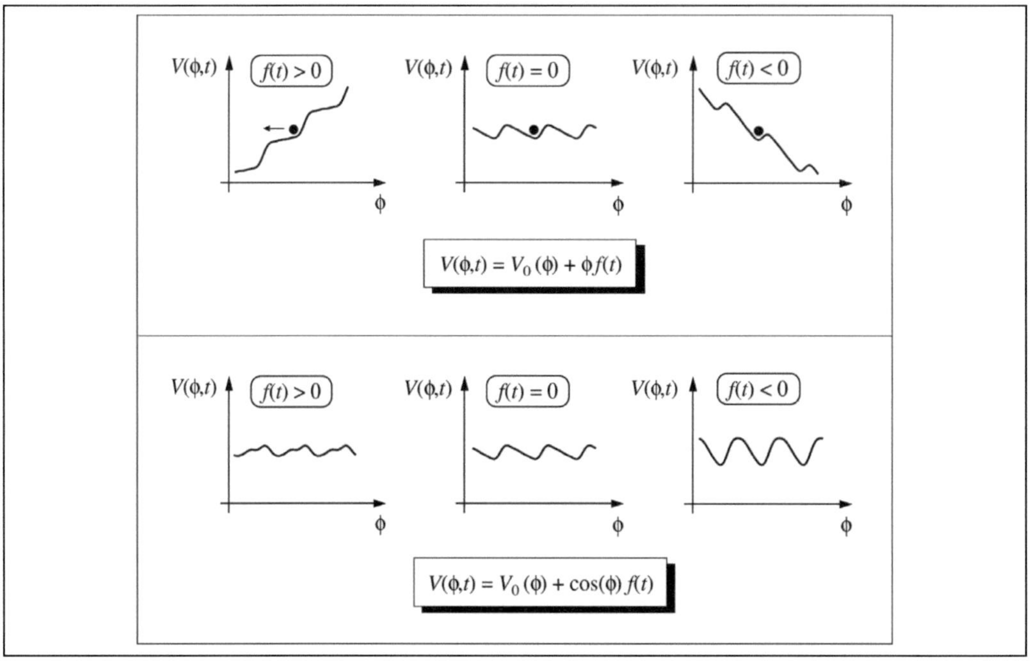

Fig. 1. Upper panel: a model for unidirectional Brownian motions. Lower panel: a model for unidirectional motions of a chiral molecular motor induced by a linearly polarized laser pulse.

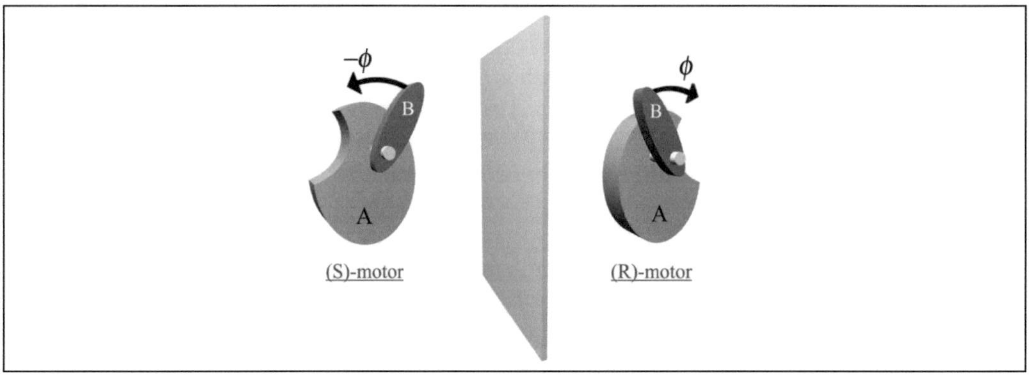

Fig. 2. A simplified model for chiral molecular motors, (*S*)- and (*R*)-motors, which are mirror images of each other. They have two rigid groups, A and B. The parameter ϕ denotes the rotational angle between A and B.

2.2 Quantum dynamics simulation of a chiral molecular motor

We chose (*R*)-2-chloro-5-methylcyclopenta-2,4-dienecarbaldehyde shown in Fig. 3a as a model system. Note that the molecule has a chiral center at the C^3 atom. Consider the internal rotation of the CHO group around the C^2-C^3 bond, and the coordinate of the rotation ϕ as the dihedral angle O^1-C^2-C^3-H^4. In the molecular frame, the *z*-axis is chosen along the C^2-C^3 bond, and the *x*-axis is set toward the H^4 atom at a right angle to the *z*-axis.

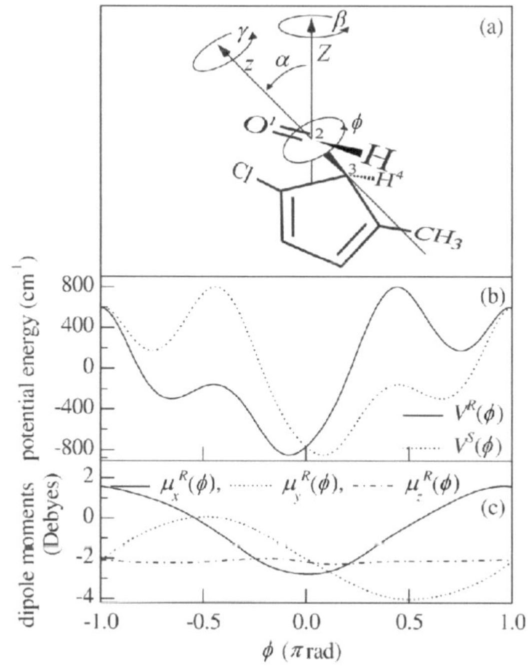

Fig. 3. (a) (*R*)-2-chloro-5-methylcyclopenta-2,4-dienecarbaldehyde as a chiral molecular motor. The aldehyde (-CHO) group is an motor driven by a laser pulse. (b) The ground state potential of (*R*)- and (*S*)- motors as a function of dihedral angle ϕ. (c) Cartesian components of the dipole moment vector in the molecular frame (*x*,*y*,*z*). Reproduced with permission from *J. Chem. Phys.*, 119, 12393 (2003).

Consider a quantum dynamics of the motor in the electronic ground state since the frequency of the laser used is in the far-IR range. The model system has the following essential properties: (i) owing to molecular chirality, the potential energy function $V^{\Gamma}(\phi)$ is asymmetric such that $V^S(\phi - c) \neq V^S(-\phi + c)$ for any *c*, (ii) there is angle-ϕ dependency of the dipole moment vector because of the sufficient electronegativity of the O^1 atom, and (iii) because the rotational constants of the molecule are large enough, the effects of the entire rotation of the molecule can be ignored. Therefore, the molecule interacting with a linearly polarized electric-field component of laser pulses can be a realistic model of the periodically perturbed ratchet system.

Molecular parameters were calculated by using the GAUSSIAN 98 package of programs (Frisch et al., 1998) with the 6-31G* basis set and the MP2 method. The moment of inertia *I* is estimated as 17.6 amu Å^2. The rotational constants are 1.97, 1.16 and 0.79 GHz, the time scale of which is sufficiently large to safely ignore the rotational effects of the entire molecule, or treat them classically. The potential energy curves for (*R*)- and (*S*)-motors shown in Fig. 3b were evaluated at each dihedral angle ϕ, where all the other structural parameters were optimized. It can be seen that these potential curves are asymmetric with respect to ϕ and are mirror images of each other. As can be seen from the optimized structure for the (*R*)-motor, the rotation in the negative direction of ϕ has a gentle slope compared to that of the positive direction and vice versa in the relation for the (*S*)-motor. These are well-known properties of a chiral molecule.

Figure 3c shows three computed Cartesian components of the dipole moment of the (*R*)-motor, μ_x^R, μ_y^R and μ_z^R as a function of ϕ. If a rigid dipole moment rotates in *xy*-plane, then these functions can be expressed in the form of sinusoidal functions as

$$\boldsymbol{\mu}^{\Gamma}(\phi) = \mu^{\Gamma}\cos(\phi)\boldsymbol{e}_x + \mu^{\Gamma}\sin(\phi)\boldsymbol{e}_y + \boldsymbol{\mu}_0^{\Gamma}, \tag{4}$$

where μ^{Γ} and $\boldsymbol{\mu}_0^{\Gamma}$ are constants. It can be seen from Fig. 3c that this is a fairly good approximation and μ^R is about −2.0 Debye. Therefore, the interaction energy with electric fields of a few $\mathrm{GVm^{-1}}$ is comparable to the internal rotational barrier height, ΔV=1650 $\mathrm{cm^{-1}}$. Assuming an electric field of a *z*-polarized laser pulse as

$$\boldsymbol{E}(\mathrm{t}) = f(\mathrm{t})\cos(\omega \mathrm{t})\boldsymbol{e}_Z, \tag{5}$$

the effective potential in Eq. (1) becomes

$$\mathrm{V}^{\Gamma}(\phi) - \boldsymbol{\mu}^{\Gamma}(\phi)\cdot\boldsymbol{E}(\mathrm{t}) = \mathrm{V}^{\Gamma}(\phi) - f(\mathrm{t})\mu^{\Gamma}\sin(\alpha)\cos(\phi+\gamma)\cos\omega \mathrm{t} - \boldsymbol{\mu}_0^{\Gamma}\cdot\boldsymbol{E}(\mathrm{t}). \tag{6}$$

Here, α and γ are the Euler angles between the molecular frame *xyz* and the laboratory frame *XYZ*. The last term is a constant with respect to the variable ϕ and does not make any contribution to the dynamics of the molecular motor. The second term is the time-correlated periodic function of ϕ, which rocks the potential energy function $V^{\Gamma}(\phi)$ as shown in the lower panels of Fig. 1.

Let us now consider the rotational dynamics including motor-bath interactions. Equation (3) can be reduced to a so-called Lindblad-type equation (Lindblad, 1976) by assuming the bath consists of harmonic oscillators and the motor-bath interaction is given in the first order of displacements for both bath motor coordinates as

$$i\hbar\frac{\partial}{\partial t}\hat{\rho}^{\Gamma}(t) = \left[\hat{H}_S^{\Gamma}(t), \hat{\rho}^{\Gamma}(t)\right] + \frac{i}{2}\sum_n\left\{\left[\hat{A}_n^{\Gamma}\hat{\rho}^{\Gamma}(t), \hat{A}_n^{\Gamma\dagger}\right] + \left[\hat{A}_n^{\Gamma}, \hat{\rho}^{\Gamma}(t)\hat{A}_n^{\Gamma\dagger}\right]\right\}, \tag{7}$$

where $\hat{\rho}^{\Gamma}(t)$ is the reduced density operator of the motor and the system–bath coupling is phenomenologically introduced by the interaction operator $\hat{A}_n^{\Gamma}$.

We set the interaction operator $\hat{A}_n^{\Gamma}$ to $|n+1\rangle^{\Gamma} a_n^{\Gamma}\langle n|^{\Gamma} + |n\rangle^{\Gamma} b_n^{\Gamma}\langle n+1|^{\Gamma}$. That is, only inelastic scattering processes were taken into account and elastic interactions were ignored. To semi-quantitatively discuss relaxation effects, both the interaction parameters a_n and b_n were set to

$$\frac{a_n^2}{b_n^2} = \exp\left(-\frac{E_{n+1}-E_n}{kT}\right), \tag{8a}$$

$$a_n^2 + b_n^2 = \frac{\hbar}{\tau_0}\frac{E_{n+1}-E_0}{E_1-E_0}, \tag{8b}$$

where E_n denotes the *n*th eigenstate of the motor in the ground state, and τ_0 is relaxation time from n = 1 to 0. Condition (8a) assures detailed balance of the system at temperature *T*, and condition (8b) specifies the relaxation time such that the more the system is excited, the faster the energy relaxes. The quantum master equation, Eq. (7), was solved by means of the split operator and a finite difference method with a fast Fourier transform algorithm using 256 grids for ϕ.

In order to discuss the dynamics of molecular motors in a quantum mechanical way, we now introduce an angular momentum operator for the internal rotation $-i\hbar(\partial/\partial\phi)$. An expectation value of the angular momentum operator at time t in the coordinate representation is defined as

$$\ell^{\Gamma}(t) = \int_{-\pi}^{\pi} d\phi' \int_{-\pi}^{\pi} d\phi \left\{ -i\hbar \frac{\partial \delta(\phi' - \phi)}{\partial \phi} \rho^{\Gamma}(\phi, \phi'; t) \right\}. \tag{9}$$

The instantaneous angular momentum of an ensemble of randomly oriented motors, averaged over all of the Euler angles $\Omega(\alpha,\beta,\gamma)$, $\left\langle \ell^{\Gamma}(t) \right\rangle_{\Omega}$,

$$\left\langle \ell^{\Gamma}(t) \right\rangle_{\Omega} = \frac{1}{8\pi^2} \int \ell^{\Gamma}(t) d\Omega, \tag{10}$$

was calculated as an index of rotations of motors. The above integration was computed by the trapezoidal rule with 16 grid points for α and 64 grid points for γ.

Let us apply a linearly polarized electric field, Eq. (5) with the envelop function $f(t)$,

$$f(t) = \begin{cases} E_0 \sin^2 \dfrac{\pi t}{t_p} & (0 \le t < t_p) \\ 0 & (t < 0,\ t_p \le t) \end{cases}. \tag{11}$$

Here, the pulse length t_p is 30 ps and the central frequency ω is 124 cm^{-1}. The frequency is related to the frequency of a normal vibration which localizes to the torsion of the aldehyde group. The maximum amplitude of the electric field E_0 is set to 3.4 GVm^{-1}. Thus, the maximum difference of the dipole interaction energy, which is $2\mu^{\Gamma}E_0$, is larger than the barrier height of the potential energy $\Delta V \approx 1{,}600$ cm^{-1}, and the wave packet of the motor can overcome the barrier even if the frequency of the laser is non-resonant to anharmonic potential energy function $V^{\Gamma}(\phi)$.

Figure 4 shows the time evolution of instantaneous angular momentum of randomly oriented molecular motors with initial density being set to the Boltzmann distribution at $T = 150$ K. The axis on the left-hand side denotes angular momentum in units of $\hbar$, and the axis on the right-hand side denotes angular frequency in units of Hz. In Fig. 4, we obtained $\left\langle \ell^{R}(t) \right\rangle_{\Omega} \approx 1.6\hbar$ for (R)-motors and $\left\langle \ell^{S}(t) \right\rangle_{\Omega} \approx 1.6\hbar$ for (S)-motors after application of the laser pulse. In the absence of a relaxation process (τ_0=∞) as a special case, $\left\langle \ell^{\Gamma}(t) \right\rangle_{\Omega}$ reaches the maximum value at around the pulse peak and maintains a certain value even after the pulse is turned off. This means that the motion created is not a pendulum one but a unidirectional rotation across the potential barrier. It should also be noted that rotational directions of randomly oriented motors are opposite in (R)- and (S)-motors. If molecules do not have chirality, the angular momentum becomes zero. Here, an achiral motor is obtained by substituting a methyl group of the motor with a chlorine atom and its potential becomes symmetric as $V^{R}(\phi) = V^{S}(\phi)$.

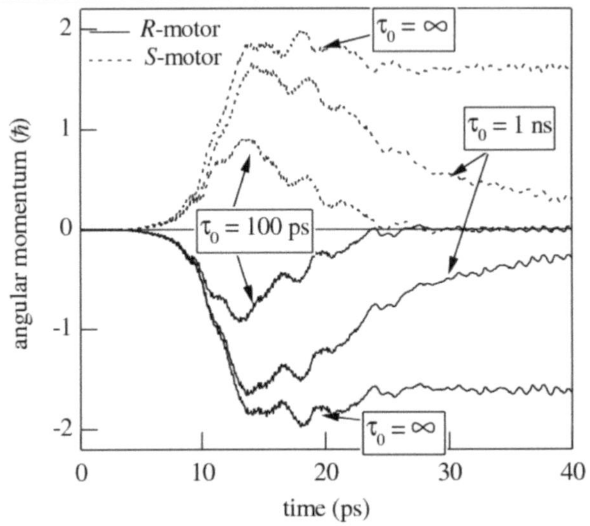

Fig. 4. Temporal behaviours of the instantaneous angular momentum of randomly oriented (*R*)-motors (solid lines) and those of (*S*)-motors (dotted lines) at T =150 K. τ_0 denotes the relaxation time the first excited vibrational state to the ground state. Reproduced with permission from *J. Chem. Phys.*, 119, 12393(2003).

Figures 4 also shows the relaxation effects on $\left\langle \ell^{\Gamma}(t) \right\rangle_{\Omega}$. The relaxation time τ_0 from the first excited vibrational state to the ground state is taken as a relaxation parameter. In the simulation, we assumed the system is weakly fluctuating by neglecting ultrafast inertial effects. Such a condition can be realized considering that the molecules are surrounded by solvent cage molecules or imbedded in rigid solvents under low temperature conditions. Strong τ_0 dependence can be seen in Fig. 4, which is mainly due to the assumption that effective relaxation times were taken to be proportional to ΔV, being shorter than that of τ_0 by about two orders. One of the possibilities to drive rotation continuously is to supply laser pulses repeatedly to recover the loss of energies and accelerate the rotation.

Figure 5 shows the effective potential and rotational wave packets of the (*R*)-motor at several specific Euler configurations to qualitatively understand its unidirectional motion. The Euler angle α was fixed at 0.5π to give the maximum interaction between the dipole moment vector and the photon polarization vector. The amplitude of the electric field was the same as that in Fig. 4. The rotational wave packet was initially localized around the stable configuration, $\phi \approx 0$. It can be seen from Fig. 5 that the initial rotational wave packet at the configuration $\gamma = 0$ at $\omega t = \pi$ or $\gamma = \pi$ at $\omega t = 0$ moves toward the left-hand side with a gentle slope when the wave packet is rocked by a strong pulsed laser. On the other hand, Figs. 5c and 5d show that the rotational wave packet rocked by a pulsed laser at the configuration $\gamma = \pm 0.5\pi$ cannot obtain sufficient angular momentum to cross the potential barrier after the laser pulse is turned off. The origin of the unidirectional motion is the asymmetry of the rotational potential. The linearly polarized laser pulse acts as a kick to the molecular motor. The initial angular momentum is large enough to determine the rotation toward the gentle slope of the rotational potential after the application of laser pulse.

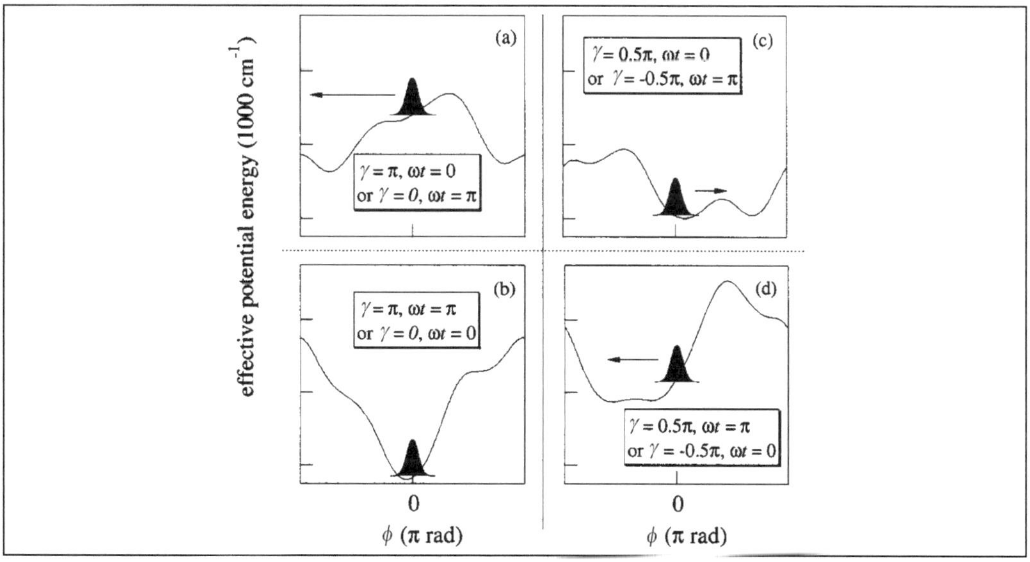

Fig. 5. A schematic view of a localized torsional wave packet on the effective potential of an (*R*)-motor fixed at a specific Euler angle γ and time ωt. The effective potential is constructed from the torsional potential and time-dependent periodic interaction between the chiral motor and laser pulse. Reproduced with permission from *J. Chem. Phys.*, 119, 12393(2003).

2.3 Femtosecond pump–dump laser-ignition method

In this subsection, we describe another method called the pump-dump laser-ignition method for driving unidirectional motions of chiral molecular motors (Hoki et al., 2004). Figure 6 shows the control scheme. A femtosecond vis/UV pump pulse denoted by (1) creates a wave packet in an electronic excited state V_g^{Γ}. When the wave packet moves along the potential energy surface (PES) and reaches an appropriate position, it is dumped to the electronic ground state V_g^{Γ} by applying the second pulse (2). In the dumping process, the kinetic energy of the rotary motion is conserved before and after the dumping process as a result of the Franck–Condon principle. Therefore, the pump–dump laser excitation can ignite a unidirectional motion of the rotational wave packet because the wave packet has a certain kinetic energy to overcome its activation barrier height in the ground state V_g^{Γ}. Because the electronic state is altered as $V_g^{\Gamma} \rightarrow V_e^{\Gamma} \rightarrow V_g^{\Gamma}$, by the pump–dump pulse sequences, this mechanism has an analogy to that of the flushing ratchet system. However, in the pump-dump laser-ignition method, quantum dynamics of wave packets of chiral motors are utilized for determining unidirectional motions as shown below.

The rotational direction is characteristic of the chirality of the molecular rotor. As shown in Fig. 6, the direction is determined by the gradient of $V_{\mathbf{e}}^{1}$ around its Franck–Condon region, and the gradient can be estimated using quantum chemistry calculations. We call this motion in the direction of rotation a regular rotation. If the molecule has a symmetric plane as in the case of an achiral molecule, the gradient should be zero. Therefore, an achiral molecule cannot produce unidirectional motion using a linearly polarized laser. As shown in Fig. 6, on the other hand, from symmetry considerations, the gradient has opposite signs between the (*R*)- and (*S*)- motors.

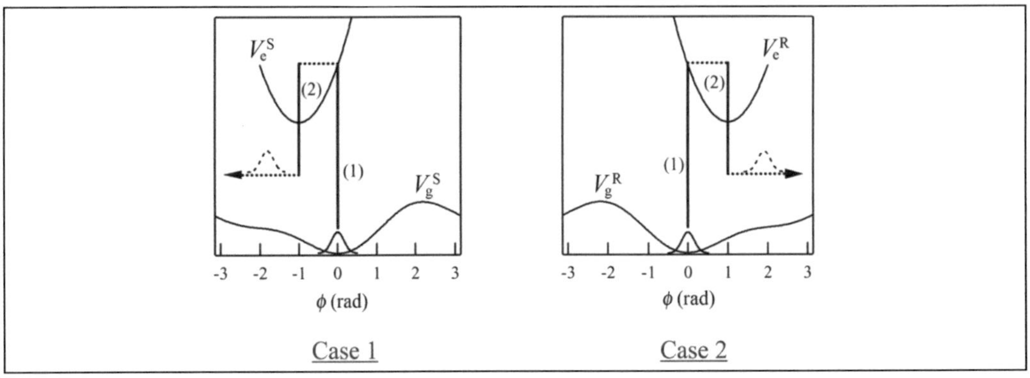

Fig. 6. Scheme for a pump-dump laser-ignition: an (*S*)-motor (left) and an (*R*)-motor (right).

The time evolution of the molecular motor is determined by the time-dependent Schrödinger equation within the two-electronic state (the electronic ground and excited states) model

$$i\hbar\frac{\partial}{\partial t}\begin{pmatrix}\psi_g(\phi;t)\\ \psi_e(\phi;t)\end{pmatrix}=\begin{pmatrix}\hat{T}+V_g(\phi) & -\boldsymbol{\mu}_{ge}(\phi)\cdot\boldsymbol{E}(t)\\ -\boldsymbol{\mu}_{eg}(\phi)\cdot\boldsymbol{E}(t) & \hat{T}+V_e(\phi)\end{pmatrix}\begin{pmatrix}\psi_g(\phi;t)\\ \psi_e(\phi;t)\end{pmatrix}, \tag{12}$$

where $\boldsymbol{\mu}_{ge}(\phi)$ ($=\boldsymbol{\mu}_{eg}(\phi)$) is the transition dipole moment between the two states, $V_n(\phi)$ is the potential energy in the electronic state n ($n = g$ and $n = e$) and $\hat{T}=-\frac{\hbar^2}{2I}\frac{\partial^2}{\partial\phi^2}$ with moment of inertia I was assumed to be common in both electronic states. $\boldsymbol{E}(t)$ is the electric field of vis/UV laser. $\psi_n(\phi; t)$ is the nuclear wave packet.

The electric field of the linearly polarized pump laser with central angular frequency ω_p and the linearly polarized dump laser with ω_d is expressed as

$$\boldsymbol{E}(t)=A_p f(t;t_p,t_p+T_p)\cos(\omega_p t)\boldsymbol{e}_Z+A_d f(t;t_d,t_d+T_d)\cos(\omega_d t)\boldsymbol{e}_Z \tag{13}$$

with the pulse amplitude A_j (j = p and d), the pulse duration T_j; the time delay t_j. $f(t)$ denotes the envelope function of the pulse,

$$f(t;t_1,t_2)=\begin{cases}\sin^2\pi\frac{t-t_1}{t_2-t_1} & (t_1\le t<t_2)\\ 0 & (t<t_1,\ t_2\le t)\end{cases}. \tag{14}$$

In order to estimate the magnitude of rotary motion of the motor, quantum-mechanical expectation value of the angular momentum in the electronic ground state $\ell_g(t)$, defined as

$$\ell_g(t)=\int_{-\pi}^{\pi}d\phi\,\psi_g^*(\phi;t)\left(-i\hbar\frac{\partial}{\partial\phi}\right)\psi_g(\phi;t), \tag{15}$$

was calculated.

Figure 7 shows the result of *ab initio* calculation of (*R*)-2-methylcyclopenta-2,4 dienecarbaldehyde in S_0 and S_1. The fully relaxed geometrical structure of the molecule was calculated at the MP2/6-31+G** level of theory at each point of the chosen reaction coordinate ϕ. The potential energy in the electronic ground-state S_0 and that in the first excited singlet state S_1 were calculated by means of multi-configuration methods (Ross, 1989).

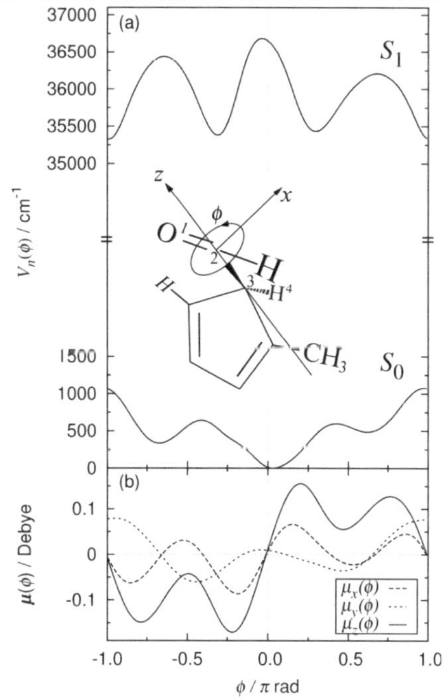

Fig. 7. (a) Potential energies in the ground and first excited singlet states of (R)-2-methyl-cyclopenta-2,4-dienecarbaldehyde. (b) Transition moments as a function of ϕ. Reproduced with permission from *J. Phys. Chem. B*, 108, 4916 (2004).

Figure 7a shows the PESs of the (*R*)-motor in S_0 and S_1. Because the S_1 PES slope in the Franck–Condon region is negative, a wave packet created by a pump pulse would run with a positive angular momentum. Note that the minimum energy position in the ground state is slightly shifted in a positive direction. This is reflected in the geometry of the molecule by a slight bend of the aldehyde group toward the methyl group by about 7 degrees.

We now apply the pump–dump laser-ignition method to drive a rotation of the -CHO group of (*R*)-2-methyl cyclopenta-2,4-dienecarbaldehyde. To analyze the rotational dynamics, we solved the time-dependent Schrödinger equation (12) by means of the split-operator method with the help of an FFT algorithm using 256 grid points for ϕ. Here, the motor–laser interaction, $-\mu_{ge}(\phi)\cdot E(t)$ was treated numerically using Pauli matrices (Choi, 1989).

The upper panels a, b, and c in Fig. 8 show the time evolution of the rotational wave packets created in S_0 by applying the pump–dump laser-ignition method. We omitted wave packets trapped in the S_0 potential well because they do not evolve after the dump pulse is turned off. Here, the parameters of pulses used were $A_p = A_d = 10^{10}$ V/m, $T_p = T_d$ =100 fs, t_p = 50 fs,

t_d = 180 fs, ω_p = 34,300 cm^{-1}, and ω_d = 32,800 cm^{-1}. The initial state was set on the lowest rotational level in the electronic ground state. It can be seen that in the early stage of the time evolution, the motor takes about 550 fs for one cycle, which corresponds to angular frequency ω of 1.1×10^{13} s^{-1}. From an analogy to classical mechanics, $\ell \approx \omega I$ and the expectation value of the angular momentum of the motor is estimated to be $\ell \approx 31\ \hbar$.

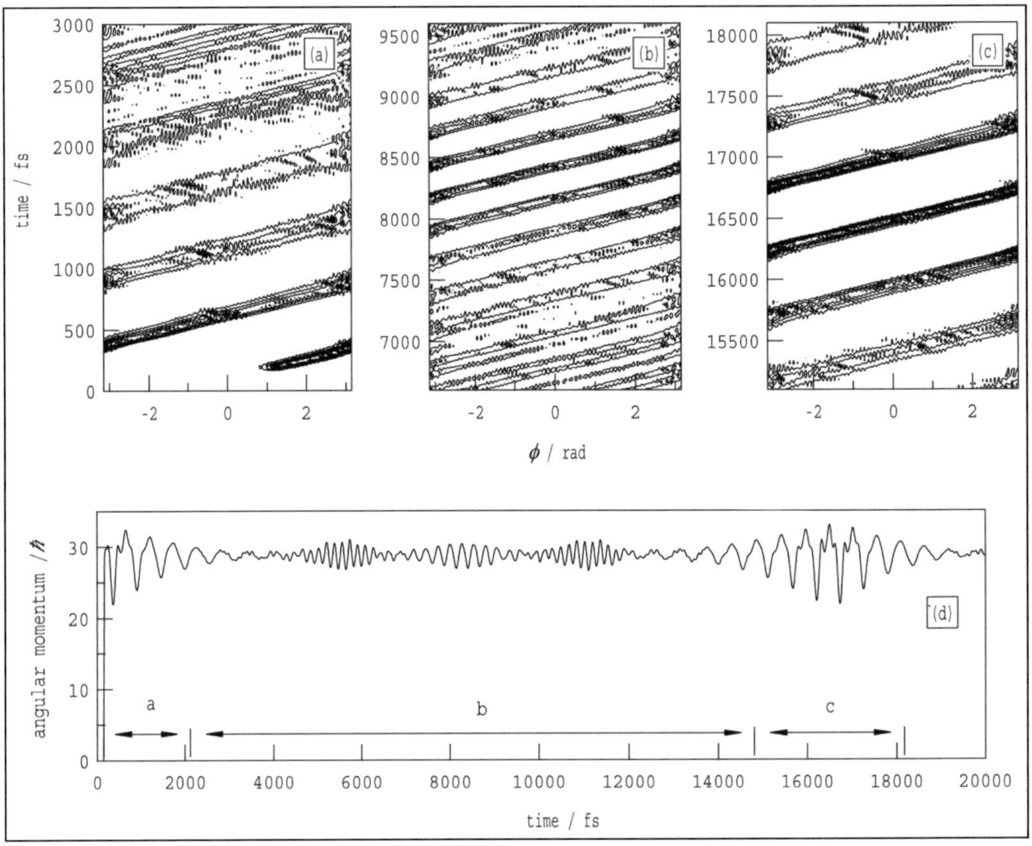

Fig. 8. Quantum dynamics behaviors of (R)-2-methyl-cyclopenta-2,4-dienecarbaldehyde, which were induced by a pump-dump laser-ignition method. Reproduced with permission from *J. Phys. Chem. B*, 108, 4916 (2004).

An interesting feature appears in panel b where the rotational wave packet is split into two wave packets. These wave packets converge to a single one again after about 1,500 fs, as shown in panel c. Panel d shows that the time-dependent expectation value of the angular momentum $\ell_g(t)$ can be divided into three regions, a, b and c, within rephrasing time of about 17,000 fs. These three regions are the same as those in the upper panels.

This feature can be explained by a free rotor model, where the eigenfunction is $1/\sqrt{2\pi}\exp(im\phi)$ with eigenvalue $\varepsilon_m = \hbar^2 m^2/2I$. The wave packet $\psi_g(\phi,t)$ can be expressed as

$$\psi_g(\phi;t) = \frac{1}{\sqrt{2\pi}} \sum_{m=m_a}^{m_b} C_m \exp(im\phi) \exp\left(-\frac{i}{\hbar}\varepsilon_m t\right). \tag{16}$$

Here, the coefficient C_m which represents the probability amplitude of the rotational quantum number m is determined at the time when the dump pulse is applied. Under the condition of a localized distribution of the angular momentum, the wave packet at half of the rephrasing time, $\pi I/\hbar$, can be expressed (Hoki et al., 2004) as

$$\psi_g(\phi;t=\pi I/\hbar)=\frac{1}{\sqrt{2}}\exp\left[-\frac{i\pi}{4}\right]\psi_g(\phi;t=0)+\frac{1}{\sqrt{2}}\exp\left[\frac{i\pi}{4}\right]\psi_g(\phi-\pi;t=0) \tag{17}$$

This expression indicates that the wave packet splits into two parts with the same magnitudes at half of the rephrasing time. The rephrasing time in the present motor system is roughly estimated to be $2\pi I/\hbar = 17{,}000$ fs, which is the same order as the numerically evaluated one, as shown in panel c. This explains why the splitting of the rotational wave packet into two parts takes place in panel b.

2.4 Method for detecting quantum dynamics of molecular motors in real time

In the preceding subsections, we adopted a quantum-mechanical expectation value of angular momentum to theoretically prove rotary motions of chiral molecular motors. However, it is not so simple to experimentally detect the expectation values of angular momenta of neutral molecular motors by using optical or magnetic methods. In this subsection, we briefly refer to observation of ultrafast motions of molecular motors in real time. There are several spectroscopic methods for observation of nuclear motions on reaction potential surfaces in the field of so-called femtosecond chemistry (Potter et al., 1992; Zewail & Bernstein, 1992). In this subsection, we adopt a time- and frequency-resolved ionization method. We show by using a time-frequency-resolved ionization simulation that the results obtained in Sec. 2.3 can be experimentally observed. The method consists of a femtosecond probe pulse of central frequency ω_3 for ionization of molecular motors in a nonstationary state in addition to pump and dump pulses. Ionization signals are plotted as a function of ω_3 after fixing a delay time t_3 between the dump and probe pulses.

Figure 9 shows calculated time-frequency-resolved spectra $I(\omega_3,t_3)$ of (R)-2-methyl-cyclopenta-2,4-dienecarbaldehyde (Hoki et al., 2004). The pulse duration was set to 100 fs, and the maximum value of the pulse amplitude was set to a small value (10^7 Vm^{-1}) in order to satisfy a one-photon ionization in a weak field limit. We can see an oscillatory behavior of the spectrum at ω_3 = 60,900 cm^{-1} corresponding to the ionization potential around ϕ = 0. Figure 9 shows wave packet motions of the molecular motor in the ground state. The time evolution of the intensity peaks reflects the energy of the ionization potential as a function of ϕ (Hoki et al., 2004). The time for one cycle oscillation was estimated to be 540 fs since two peaks of the oscillation are located at 380 and 920 fs.

The method for the time- and frequency-resolved spectrum described above can be applied to detect molecular motor dynamics not only for pump-dump laser-ignition method but also those for other methods described in this chapter.

3. Optimal control of unidirectional motions

There are two types of theoretical methods for designing optimal laser pulses. One is a local control method and the other is a global control method (Gordon & Fujimura, 2002). Here, "local" means that maximization of the target is carried out at each time. Therefore, the

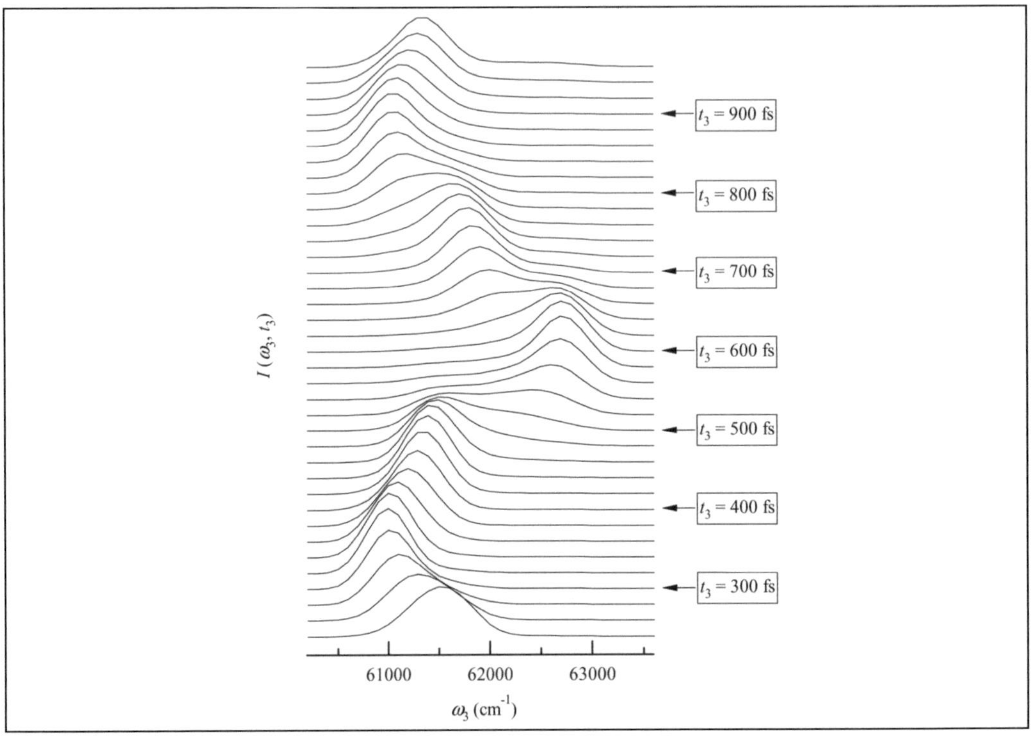

Fig. 9. Time-frequency-resolved spectra $I(\omega_3,t_3)$ of (*R*)-2-methyl-cyclopenta-2,4-dienecarbaldehyde. Reproduced with permission from *J. Phys. Chem. B*, 108, 4916 (2004).

algorithm is quite simple, and only one-sided propagation, forward or backward propagation, is needed to solve the time-dependent Schrödinger equation. A global control method is, on the other hand, general and both forward and backward propagations have to be carried out. However, it can be applied to any dynamical systems, though sometimes converging problems arise.

3.1 Local control of a molecular motor

In the local control method, electric field of laser pulses $\boldsymbol{E}(t)$ is given as

$$\boldsymbol{E}(t) = -2A\,\mathrm{Im}\langle\Psi(t)|\hat{W}\hat{\boldsymbol{\mu}}|\Psi(t)\rangle, \tag{18}$$

where A is a regulation parameter of the laser intensity, and $\hat{W}$ is the target operator. Substituting $|\Psi(t_0)\rangle = |0\rangle$ into Eq. (18), we obtain an electric field $\boldsymbol{E}(t_0)$ using Eq. (18) at the initial time t_0. Solving the time-dependent Schrödinger equation with the initial condition of $|\Psi(t_0)\rangle$, we obtain a wave function after an infinitesimally increased propagation time. With this form, the procedure described above is repeated until $\boldsymbol{E}(t_f)$ at the final time t_f is obtained. The locally optimized electric field $\boldsymbol{E}(t)$ guarantees a monotonic increase in the expectation value of the target operator $\hat{W}$ if it commutes with the molecular Hamiltonian, i.e., $\left[\hat{H}_0, \hat{W}\right] = 0$. This condition can be satisfied when the target operator has the form of $\hat{W} = \sum |n\rangle w_n \langle n|$, where $|n\rangle$ is an eigenstate of $\hat{H}_0$ and w_n is a waiting factor.

Consider a quantum control which yields a unidirectional rotational state $|T\rangle$ from the ground state $|0\rangle$. Unfortunately, we cannot set the target operator proportional to a projector $|T\rangle\langle T|$ because the projector does not commute with the Hamiltonian $\hat{H}_0$. To overcome this difficulty, we make use of time-reversal symmetry of the time-dependent Schrödinger equation. We design the locally optimal electric field **E**(*t*) by carrying out backward propagation starting from the target state $|T\rangle$ to yield the initial state $|0\rangle$ as much as possible. By doing so, we can set our target as the form of $\sum |n\rangle w_n \langle n|$, where the waiting factor w_n should satisfy condition $w_{n-1} > w_n$ to ensure sequential population transfer to the ground state $|0\rangle$. Once the optimized electric field is determined, the motor dynamics is evaluated by solving the time-dependent Schrödinger equation (18) in the forward propagation.

We now apply the local control procedure described above to (*R*)-2-chloro-5-methyl-cyclopenta-2,4-dienecarbaldehyde. We construct the target state by using two eigenstates as $|T\rangle = (|66\rangle + i|65\rangle)/\sqrt{2}$ for a counter-intuitive rotation toward the steep slope of the potential energy curve and $|T\rangle = (|66\rangle - i|65\rangle)/\sqrt{2}$ for an intuitive rotation toward the gentle slope of the potential energy curve. These two eigenstates are chosen because (i) the eigen energy of the 65th and 66th molecular states is higher than the potential barrier of 1,500 cm^{-1} and (ii) the energy difference is less than 0.001 cm^{-1}, so that these states practically degenerate in our observation time and correspond to the quantum number of $m = \pm 33$ of a free rotation system.

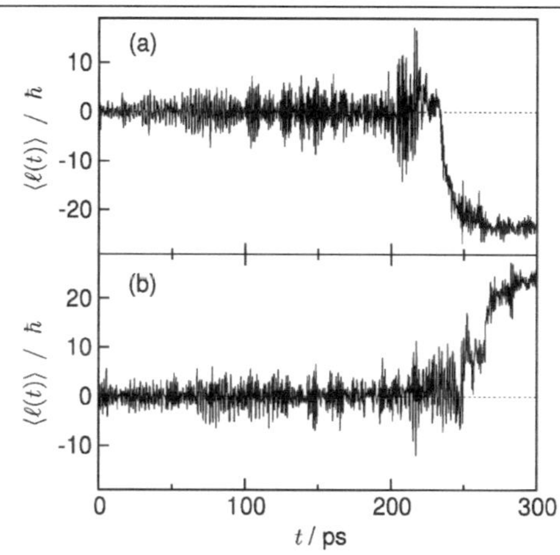

Fig. 10. (a) Instantaneous angular momentum of the (*R*)-motor driven by a local control method for the intuitive rotation and (b) for the counter-intuitive rotation. Reproduction with permission from *Phys. Chem. Chem. Phys.*, 7, 1900 (2005).

Figure 10 shows time-dependent behaviors of the instantaneous angular momentum $\ell(t)$ of the (*R*)-motor molecule defined in Eq. (15), where the direction of the linearly polarized electric field vector is set to the *x*-axis. This clearly shows that rotational directions of the motor are controlled well, i.e., at the final time of 300 ps, the instantaneous angular momentum becomes constant values of about $-23\hbar$ and $23\hbar$ in Fig. 10a and 10b, respectively. It should be noted that there exists a time difference in the initiation between the intuitive and counter-intuitive rotational directions. This is discussed by using the Fig. 11.

The figure in the left-hand side of Fig. 11 shows time- and frequency-resolved spectra of the locally optimized electric fields, $S(\omega,t)$, to rotate the motor in the intuitive and counter-intuitive directions shown in Fig. 10. A time- and frequency-resolved spectrum is defined as

$$S(\omega,t)=\left|\int d\tau E(\tau)g(t-\tau)\right|^2, \tag{19}$$

where $g(t)$ is a window function. We can see that the electric fields consist of four components (ε_1, ε_2, ε_3 and ε_4). The first two components, ε_1 and ε_2, simultaneously operate at the initial stage of motor initiation (0–160 ps). The third component, e_3, dominates in the low-frequency regime of the rotary motion whose potential is highly anharmonic. The third component bridges between the initial stage and the final stage of initiation (180–250 ps) at which unidirectional rotation starts. The fourth component, ε_4, accelerates the rotary motion. The frequency of this component is around 60 cm^{-1}, which is close to the frequency difference between two quasi-degenerate pairs n = 63, 64 and n = 65, 66.

Two features appear in the initial two components ε_1 and ε_2. One feature is that ε_1 consists of a central frequency of about 60 cm^{-1} and ε_2 consists of a central frequency of 120 cm^{-1}; that is, the latter is twice the former. This feature reflects optical transitions between eigenstates of a chiral molecule: the dipole moment of the chiral molecule is proportional to $\cos\phi$ and the transition moment between the kth and lth eigenstates involves both odd and even quantum transitions since the minimum of the asymmetric potential energy function is slightly shifted in the minus direction as can be seen in Fig. 3b because of the molecular chirality. The other feature is that the frequencies in both components are expressed by a negative chirp behavior. The negative chirp form of the electric fields originates from a gradual decrease in the frequency difference between two transitions with n = 0–16.

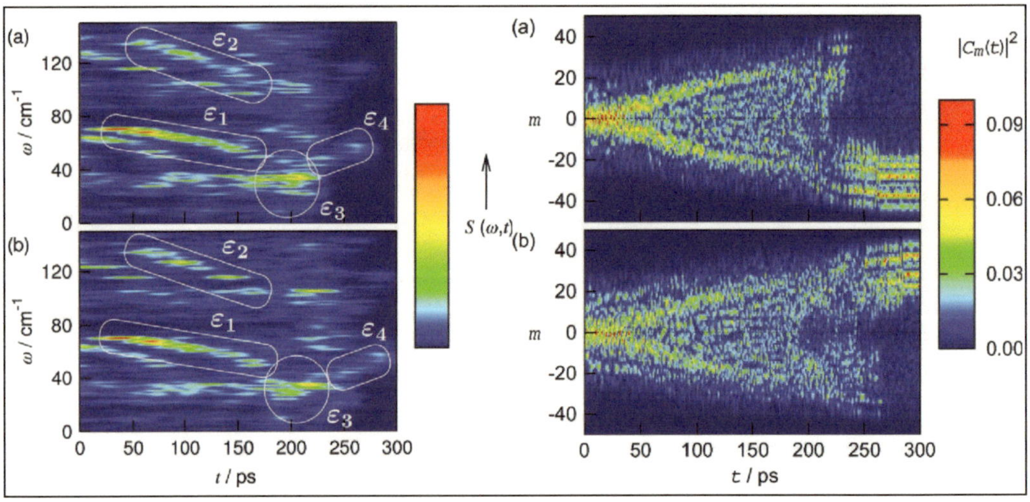

Fig. 11. (Left) Time- and frequency-resolved spectra of the locally optimized electric fields, $S(\omega,t)$, for (a) intuitive and (b) counter-intuitive directions. (Right) Time-dependent populations under the designed electric field $E_x(t)$ for (a) intuitive and (b) counter-intuitive directions. Reproduced with permission from *Phys. Chem. Chem. Phys.*, 7, 1900 (2005).

To clarify the mechanism of unidirectional rotation under the condition of irradiation of controlled laser pulses, we examined when unidirectional motion begins. In the right-hand side of Fig. 11, time-dependent populations under the designed electric field $E_x(t)$ are shown. We can see that a bunch of eigenstates $\{|n\rangle\}$ make a significant contribution to creation of a linear combination to compose the unidirectional rotational state. This is mainly due to the effect of the potential of the chiral motor. The intuitive rotation begins at about $t = 240$ ps, while the counter-intuitive rotation begins at about $t = 260$ ps. The difference of 20 ps between these two rotations corresponds to about 1.1 cm^{-1}. This frequency is close to the difference in frequency between the two eigenstates $n = 57$ and 58. This indicates that the direction of rotation is determined by the phase of a coherent superposition of rotational eigenstates created by locally optimized electric fields. Therefore, the time to change the direction of rotation can be selected by timing of designed pulses.

We note in Fig. 11 that timing of the ε_3 pulse component is different between intuitive and counter-intuitive directions. This is the essential factor for determining the rotational direction. Figure 10 clearly shows that creation of coherent states, i.e., rotational states in the intuitive direction or those in the counter-intuitive direction, depends on the electric fields estimated by using local control laser fields. Note that the intuitive rotation is controlled earlier than the counter-intuitive rotation by 20 ps. The earlier control of the rotary motion in the intuitive direction is related to the fact that the intuitive rotation is induced when a non-optimized laser field is applied.

3.2 Global control of a molecular motor via an excited state

In this subsection, we present results of application of a global optimization method to designing femtosecond pulses for unidirectional rotary motions of a chiral motor through an excited state.

A global control method tailors an electric field $\boldsymbol{E}(t)$ by solving an optimization problem of the objective functional

$$Y[\boldsymbol{E}(t)] = \langle \Psi(t_f)|\hat{W}|\Psi(t_f)\rangle - \int_0^{t_f} dt \frac{1}{\hbar A(t)}|\boldsymbol{E}(t)|^2, \tag{20}$$

Here, $\hat{W}$ is a target operator, t_f is control time and $A(t)$ is a regulation function for tuning laser intensity. Under the variation condition, $\delta Y/\delta \boldsymbol{E}(t) = 0$ for any t in the range $[0, t_f]$, an expression for the optimal field $\boldsymbol{E}$(t) is obtained as

$$\boldsymbol{E}(t) = -A(t)\mathrm{Im}\langle \Xi(t)|\hat{\boldsymbol{\mu}}|\Psi(t)\rangle, \tag{21}$$

where $|\Xi(t)\rangle$ is the time-dependent Lagrange multiplier that is a backward solution of the time-dependent Schrödinger equation with the final condition $|\Xi(t_f)\rangle = \hat{W}|\Psi(t_f)\rangle$. The optimal field is obtained by solving both the time-dependent Schrödinger equation with the initial condition $|\Psi(t=0)\rangle = |0\rangle$ and that with the final condition.

Again, we consider (*R*)-2-methyl-cyclopenta-2,4-dienecarbaaldehyde as a chiral molecular motor. The target operator for a unidirectional rotation with a positive angular momentum in the ground state, $\hat{W}_g^+$, is given in terms of the angular momentum basis set $|m\rangle$ as

$$\hat{W}_g^+ = \sum_{m>0} |m\rangle w_m \langle m|, \tag{22a}$$

where $w_m \geq 0$ is a weighting factor. In a similar way, the target operator for a unidirectional rotation with a negative angular momentum is given as

$$\hat{W}_g^- = \sum_{m<0} |m\rangle w_m \langle m| \,. \tag{22b}$$

Here, we note that the target operator for the regular rotation of the motor in the ground state is given as Eq. (22a) since the momentum created at the Franck-Condon region is positive as we discussed in the Sec. 2.3. The target operator for the reverse rotation is given as Eq. (22b).

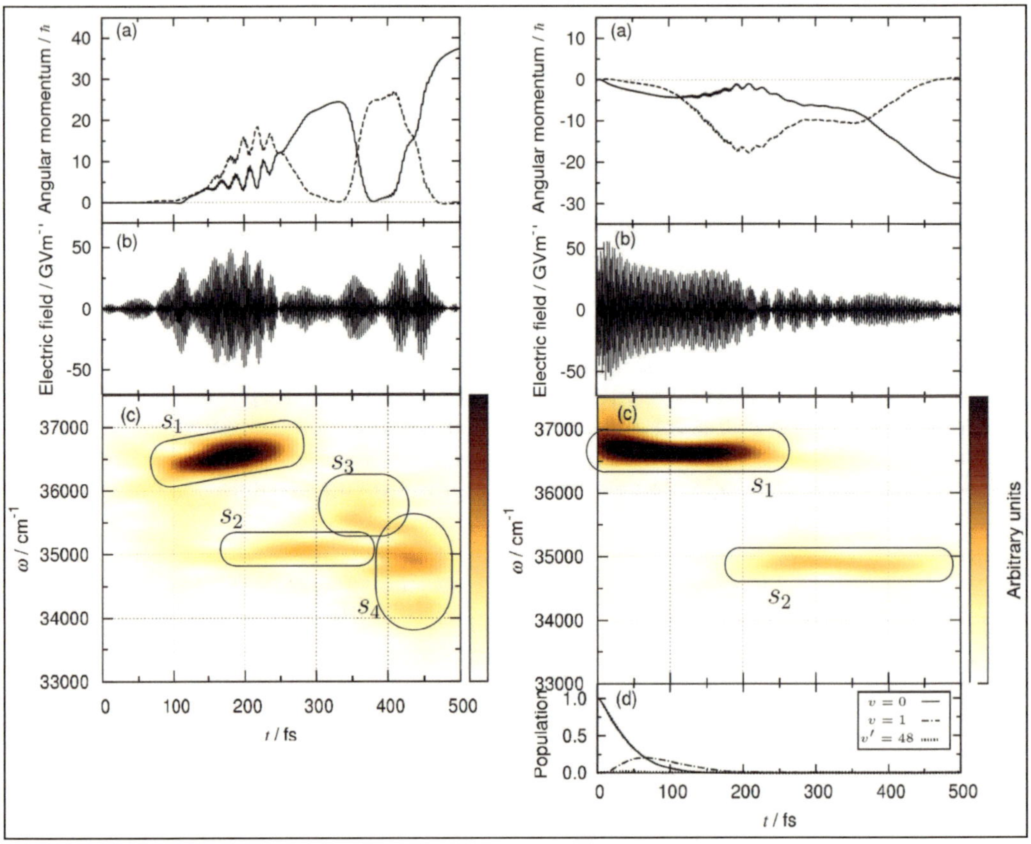

Fig. 12. (Left) Results for global control for driving (*R*)-2-methyl-cyclopenta-2,4-dienecarbaldehyde in the regular direction. (Right) Results for driving the motor in the reverse direction. Reproduced with permission from *Chem. Phys.*, 347, (2008).

Figure 12a (left-hand side) shows the instantaneous angular momentum $\ell_n(t)$ ($n = e$ or g) for the regular rotation which were obtained by using the optimal pulses in Fig. 12b. Here, the control time was set to $t_f = 500$ fs and a z-polarized linear electric field was used. The solid line denotes the instantaneous angular momentum in the ground state S_0, $\ell_g(t)$, and the dashed line denotes that in the first electronic excited state S_1, $\ell_e(t)$. At the final time, $\ell_g(t_f) \approx 37\hbar$ was obtained for the regular rotation. The angular momentum $\ell_e(t)$ increases faster than $\ell_g(t)$ in the early stage, $t < 250$ fs. This indicates that the direction of the regular rotation is determined in S_1 as explained in section 2.3.

Figure 12c shows the time- and frequency-resolved spectrum $S(\omega,t)$ for the optimal pulses in Fig. 12b. Here, the main components are labeled by s_1, s_2, s_3 and s_4. The first component for the regular rotation, s_1, with the central frequency of 36,400–36,800 cm^{-1} corresponds to the frequency range estimated from differences between $V_e(\phi)$ and $V_g(\phi)$. The feature of s_1 is a positive chirping that is the origin of the sharp increase in $\ell_e(t)$ between 100 and 250 fs in Fig. 12a. The second component, s_2, has a central frequency around 35,000 cm^{-1}, which corresponds to the energy difference between $V_e(\phi)$ and $V_g(\phi)$ at $\phi \approx 0.3\pi$ as well. We note in Fig. 12c that the pulse around 100–250 fs has a positive chirp character to control motion of the excited wave packet that has a wide distribution of torsional states in S_1 with torsional quantum numbers $v' = 32$–48. Except for the appearance of the chirping element in the optimal pulse, the pulses for optimal control of the rotation of the motor in the positive direction are basically the same as those obtained by using the pump–dump laser-ignition method described in section 2.3.

Results for optimal control of the motor in the negative direction are shown in the right-hand side of Fig. 12. At the final time, $\ell_g(t_f) \approx -24\hbar$ was obtained, the magnitude $|\ell_g(t)|$ being small compared with that for the regular rotation. The optimal electric field reaches the maximum amplitude of *ca* 50 GVm^{-1} in the early stage, which is about the same intensity as that for the regular rotation. Also, the large amplitude is maintained longer duration than that for the regular rotation. These results indicate the complexity in control for reverse rotation compared with the results for regular rotation.

Figure 12c (right-hand side) shows the time- and frequency-resolved spectrum of the optimal field for the reverse rotation. It consists of two main components in the optimal pulses, labeled s_1 and s_2. The pulse s_1 with central frequency of 36,700 cm^{-1} causes a coherent emission to torsional quantum states in S_0 in the early stage, which is a coherent Stokes pulse. The other component s_2 is a dump pulse and creates a rotational wave packet in S_0.

Figure 12d shows the time-dependent population of torsional eigenstates for the reverse rotation, $v = 0$, 1 in S_0 and $v' = 48$ in S_1. Here, $v' = 48$ is the dominant resonant state of the coherent Stokes scattering process. Figure 12d indicates that a linear combination of $v = 0$ and $v = 1$ is created around $t = 100$ fs. Note that a quarter of the oscillation period of the superposition state is 130 fs, so that the superposition state in S_0 could obtain sufficient displacement to change the Franck–Condon excitation as shown in Fig. 7.

It should be noted that a new mechanism for the reverse rotation was found by applying a global control method. The direction of the rotational motion is determined by the coherent Stokes scattering processes via excited states above energy levels over the threshold in S_1.

4. Laser-driven molecular machine

4.1 Model

In this section, we present results of a quantum dynamics simulation of a laser-driven molecular machine, which is an extension of the chiral molecular motors discussed in the preceding sections. Here, we adopt a real molecule, (*R*)-2-chloro-5-trifluoromethyl-cyclopenta-2,4-dienecarbaldehyde (cp-Cl-CF_3-CHO), as shown in Fig. 13. The molecule consists of three units: an aldehyde group (-CHO), a trifluoromethyl group (-CF_3) and a cyclopentadiene with a chlorine atom as a main body. The aldehyde group can be driven by laser pulses, but the trifluoromethyl group cannot be because it is optically inactive. In that sense, the molecule can be regarded as one of the smallest molecular machines: the motion of the aldehyde group as a motor and that of the trifluoromethyl group as a gear or a propeller. We note that there is no belt or chain which directly connects the motor and blades of the gear in contrast to a macroscopic fan. It would be interesting to see if the machine can work by irradiation of laser pulses and to determine how power is transmitted from the motor to the running propeller and what the transmission mechanism is if it works.

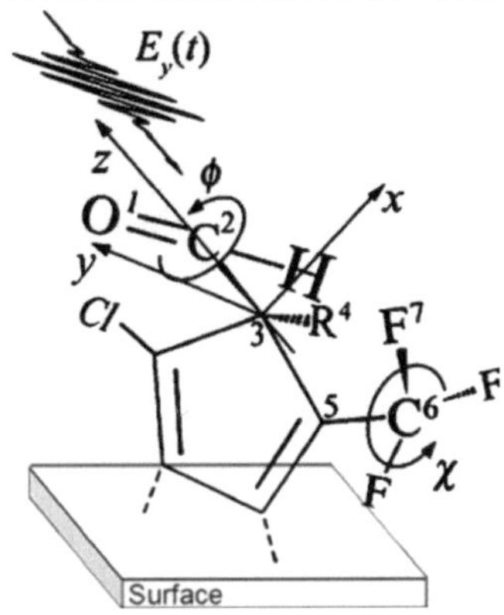

Fig. 13. (R)-2-chloro-5-trifluoromethyl-cyclopenta-2,4-dienecarbaldehyde attached at a surface as a molecular machine. The C^3 atom is a chiral center. The z-axis is defined to be along the C^3–C^2 bond. R^4 denotes an alkyl group. A linearly polarized laser pulse propagating along the y-axis $E_y(t)$ is applied. A torsional coordinate of the aldehyde group is denoted by ϕ and that of the trifluoromethyl group is labeled by χ. Reproduced with permission from *Phys. Chem. Chem. Phys.*, 11, 1662 (2009).

For the sake of simplicity, we treat the quantum dynamics simulation of the molecular machine in a two-dimensional model, in which one of the coordinates ϕ is regarded as that of the motor and another χ is regarded as a running propeller. The coordinate ϕ is defined as a dihedral angle of the O^1-C^2-C^3-R^4 group and χ is specified by a dihedral angle of the F^7-C^6-C^5-C^3 group as shown in Fig. 13. The z-axis is defined to be along the C^3-C^2 bond. The x-axis is defined to be on the C^2-C^3-R^4 plane. The cyclopentadiene group, which is the main body of the machine, was assumed to be fixed on a surface to reduce the role of entire molecular rotations. In the actual simulation, an alkyl group, -R^4, is replaced by -H for simplicity.

4.2 Results of quantum dynamics simulation

The two-dimensional potential energy surface of the molecular machine in the ground state, $V(\phi, \chi)$, was calculated with B3LYP / 6-31+G** (Becke, 1993) in the Gaussian 03 package of programs. All of the other structural parameters were optimized at every two dihedral angles. Three components of the dipole moment function, $\mu_x(\phi, \chi)$, $\mu_y(\phi, \chi)$ and $\mu_z(\phi, \chi)$, were calculated in the same way as that used for calculation of $V(\phi, \chi)$. Quantum chemical calculation shows strong ϕ dependence in $\mu_x(\phi, \chi)$ and $\mu_y(\phi, \chi)$, while χ dependence is fairly small. This indicates that the motion of ϕ is optically active but that of χ is not. The z component $\mu_z(\phi, \chi)$ was nearly constant so that the interaction term is negligible. Thus,

$\mu(\phi, \chi)$ can be expressed in the same analytical form as Eq. (4) with an amplitude μ = 2 Debye. Moments of inertia were assumed to be constant at the most stable molecular structure, $I_\phi = 2.8\times10^{-46}$ kg m^2 and $I_\chi = 1.5\times10^{-45}$ kg m^2. I_χ is about five-times heavier than I_ϕ. Figure 14 shows the results of quantum dynamical calculations of the light-driven molecular machine at a low temperature limit. Figure 14a shows the electric field of the pulse which is given as $\mathbf{E}(t) = f(t)\cos(\omega t)\mathbf{e}_y$ with envelope function $f(t)$ given by Eq. (11). Here, e_y is the unit vector along the y-axis as is defined in Fig. 13; frequency ω = 45 cm^{-1} was taken as a central frequency of a pulse; E_0 = 3.7 GVm^{-1} was taken as the amplitude of the envelope function $f(t)$ and t_p = 30 ps was taken as pulse length.

Figure 14b shows the instantaneous angular momenta, $L_\phi(t) = Tr[\hat{\ell}_\phi \hat{\rho}(t)]$ (in red) and $L_\chi(t) = Tr[\hat{\ell}_\chi \hat{\rho}(t)]$ (in blue), of the motor and propeller of the machine, respectively. We also defined "expectation values of rotational angles ϕ and χ", $\phi(t)$ and $\chi(t)$, as indexes of the rotations,

$$\phi(t) = \frac{1}{I_\phi}\int_0^t dt' L_\phi(t') \tag{23a}$$

and

$$\chi(t) = \frac{1}{I_\chi}\int_0^t dt' L_\chi(t') . \tag{23b}$$

They are shown in Fig. 14c in red and blue, respectively. We can clearly see correlated behaviors between the motor and propeller. We can also see how the rotational power is transmitted from the motor to the propeller. The molecule really acts as a single molecular machine.

The dynamic behaviors shown in Fig. 14 can be divided into three stages: early, transient and steady stages. In the early stage with the time range of 0 – 13 ps that ends just before the light pulse peak, the motor is subjected to a forced oscillation with large amplitudes in the torsional mode, which is induced by the light pulse, while the propeller just oscillate around the most stable structure with its small amplitudes. In other words, "idling" operates in this stage. This stage can be described by the one-dimensional model: as is the case with Sec. 2.2, it starts to rotate toward the gentle slope side of the asymmetric potential of the chiral molecule. In the transient stage where a bump is located in $\phi(t)$, the rotational direction of the motor is changed. Then $\chi(t)$ starts to increase, i.e., the propeller start to rotate. The rotational directions of the motor and propeller are opposite. This indicates that the aldehyde group and trifluoromethyl group play the role of a bevel gear at the molecular level, although they are not close to each other so as to have direct interactions as can be seen in macroscopic bevel gears. In the stationary stage after the pulse vanishes, the motor and propeller continue to rotate with a constant motion since there are no dephasing processes included.

Figure 14d shows the time-dependent expectation values of the following energies: the potential energy, $V(t) = Tr[\hat{V}\hat{\rho}(t)]$, the kinetic energies, $T_\phi(t) = Tr[\hat{T}_\phi \hat{\rho}(t)]$ and

$T_\chi(t) = Tr[\hat{T}_\chi \hat{\rho}(t)]$, and the sum of them, $H(t)=V(t)+T_\phi(t)+T_\chi(t)$. In the early stage, only the wave packet in the direction of ϕ is forced to oscillate by the pulse. This can be seen from Fig. 14d, in which both $V(t)$ and $T_\phi(t)$ begin to oscillate in a correlative way, while $T_\chi(t)$ does not change. In the next stage in which the motion of ϕ changes its direction, $T_\chi(t)$ begins to increase gradually. This is another proof that the motor and propeller are correlated and that the motion of propeller is induced not by laser pulse but by intramolecular interactions, i.e., non-linear interactions between two torsional modes, ϕ and χ.

Temperature effects on the dynamics of the molecular machine were also investigated (Yamaki et al., 2009).

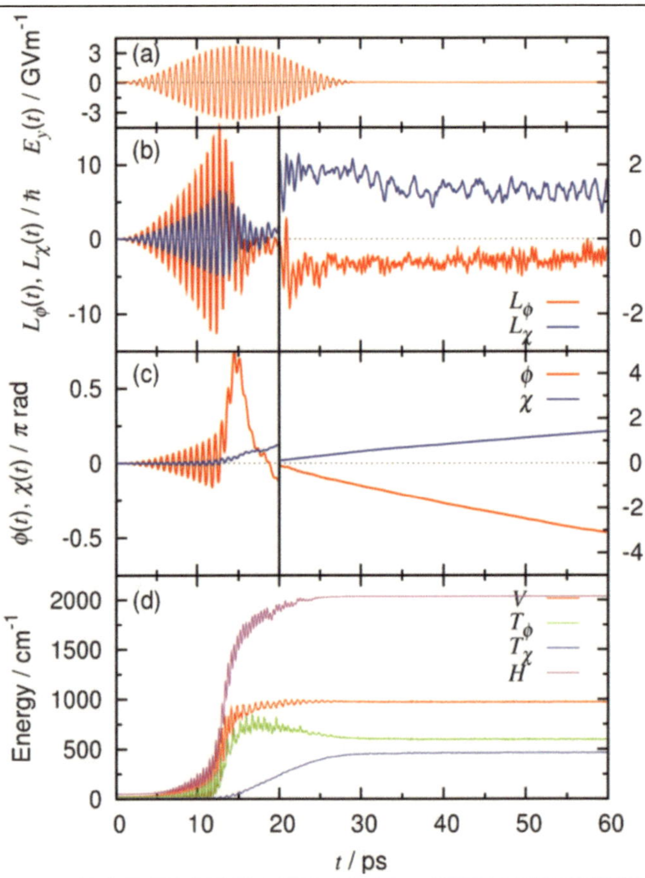

Fig. 14. (a) The y-component of the electric field of the pulse $E_y(t)$ used. (b) Quantum mechanical expectation values of angular momentum at T=0 K: that of the motor $L_\phi(t)$ (in red) and that of the propeller $L_\chi(t)$ (in blue). The scale of the vertical axis for t ≤ 20 ps is stretched compared with that for t ≥ 20 ps. (c) Rotational angle of the motor $\phi(t)$ (in red) and that of the propeller $\chi(t)$ (in blue). (d) Quantum mechanical expectation values of energies: potential energy $V(t)$ (in red), kinetic energy of f rotation, $T_\phi(t)$ (in green), and of χ, $T_\chi(t)$ (in blue), and the sum of them (in magenta). Reproduced with permission from *Phys. Chem. Chem. Phys.*, 11, 1662 (2009).

Finally, we briefly discuss the mechanism of formation of the bevel gear in the molecular machine. Quantum dynamics simulation shows that the rotational wave packet of the motor, which is created by a laser pulse, is transferred to that of the propeller. Such a correlated behavior can be quantum mechanically explained in terms of a rotational coherence transfer mechanism. We note that the correlated groups, the motor and propeller, are located at a distance of 2.3 Å. This is long compared with distance of 1.4 Å (1.5 Å) between carbon atoms of a double (single) bond. There may be two possible mechanisms: one originates from through-conjugation and the other from through-space interactions. It should be noted that the conjugation of the machine is restricted to its main body. Therefore, the through-space interaction mechanism is the most likely mechanism. Further detailed analysis is needed to confirm the transfer mechanism.

5. Summary and perspectives

Results of theoretical treatments on quantum dynamics and quantum control of laser-driven chiral molecular motors were presented. First, fundamental principles for unidirectional motions of chiral molecular motors driven by linearly polarized (nonhelical) laser pulses were described. Similarities and differences between the mechanism for driving directional motions in the case of Brownian motors for bio-motors and in the case of chiral molecular motors developed in our study were clarified. In bio-motors, the unidirectional motions are explained in terms of so-called nonequilibrium fluctuations of a Brownian motion with a saw-toothed ratchet potential, while chiral molecular motors, which are characterized by asymmetric potential similar to a saw-toothed ratchet potential, are driven in a unidirectional way by time-dependent periodic perturbations of linear polarized lasers with no angular momentum. Here, the magnitudes of the perturbations are large compared with those of interactions between molecular motors and heat bath modes, which makes the system different. Quantum dynamics simulations showed that the directional motion is determined by molecular chirality. This supports the mechanism for unidiredtional motions of chiral motors. We call the direction of the gentle slope of the asymmetric potential the intuitive direction for the unidirectional motion.

Secondly, after reviewing a quantum control theory for driving a molecular rotor with a designated unidirectional motion, we presented the results of quantum control of chiral molecular rotors. Pulse shapes for driving rotational motions in the intuitive direction or the counter-intuitive direction were found with the help of the quantum control theory. The mechanisms of the intuitive and counter-intuitive rotations were clarified by analyzing nuclear wave packet motions. We restricted ourselves to simple real molecules rather than complicated molecular systems to elucidate features of quantum control of molecular motors. We also presented an effective method for controlling unidirectional motions via an electronic excited state of chiral motors.

Thirdly, results of theoretical design of the smallest laser-driven molecular machine were presented. The smallest chiral molecular machine has an optically driven motor and a running propeller on its body. The mechanism of the transmission of driving forces from the motor to the propeller was clarified by using a quantum dynamical treatment.

In this chapter, the quantum control procedures were applied to small molecular motors with the rotary part consisting of a simple, optically active group connected to the body by a single bond. Molecular machines with nano-scale dimension have now been synthesized and wait for their operation by external forces. One of the next subjects is to demonstrate that these artificial machines can be driven by laser pulses. For example, laser pulses designed by quantum control procedures will be able to control their motions: acceleration or slowdown, forward or reverse motions and even turning directions. In principle, laser light can control coherent directed motions of assembled molecules as well. This can realize coherent collective precession of molecular rotors with chiral propellers (Kinbara & Aida, 2005; Tabe & Yokoyama, 2003). Similarly, it would be interesting to control a molecular motor in a cage, which is a model of molecular gyroscope (Bedard & Moore, 1995; Dominguez et al., 2002; Setaka et al., 2007). Another interesting subject is to apply control procedures described in this chapter to bio-systems with a micrometer dimension. For example, results of laser-induced rotational motions of both normal and malaria-infected red blood cells in various medium solutions have recently been reported (Bambardekar et al., 2010). The experiments were carried out by using linearly polarized laser pulses. It was found that the shape anisotropy of red blood cells induces rotations in optically trapped red blood cells. The rotational dynamics depends on the shape changes, which are realized by altering the experimental conditions such as osmolarity of the medium containing the cells. Differences in rotational motions between normal and malaria-infected red blood cells have been identified as well. Such a complicated rotational dynamics can be analyzed by using laser optimal control procedures, which can be used as a fast diagnostic method for malaria-infected red blood cells.

6. References

Abrahams, J. P.; Leslie, A. G. W.; Lutter, R. & Walker, J. E. (1994). Structure at 2.8 Å resolution of F_1-ATPase from bovine heart mitochondria, *Nature*, 370, (Aug 25 1994) 621-628, 0028-0836

Assion, A.; Baumert, T.; Bergt, M.; Brixner, T.; Kiefer, B.; Seyfried, V.; Strehle, M. & Gerber, G. (1998). Control of chemical reactions by feedback-optimized phase-shaped femtosecond laser pulses, *Science*, 282, 5390, (Oct 30 1998) 919-922, 0036-8075

Astumian, R. D. & Häanggi, P. (2002). Brownian motors, *Physics Today*, 55, 11, (Nov 2002) 33-39, 0031-9228

Balzani, V.; Credi, A. & Venturi, M. (2008). Molecular machines working on surfaces and at int erfaces, *Chem. Phys. Chem.* , 9, 2, (Feb 1 2008) 202-220, 1439-4235

Bambardekar, K.; Dharmadhikari, J.; Dharmadhikari, A. K.; Yamada, T.; Kato, T.; Kono, H.; Fujimura, Y.; Sharma, S. & Mathur, D., (2010). Shape anisotropy induces rotations in optically trapped red blood cells, *J. Biomed. Opt.*, 15. 4, (Jul/Aug, 2010)041504, 1083-3668

Bandrauk, A. D.; Fujimura, Y. & Gordon, R. J. (2002). *ACS Symposium Series 821, Laser Control and Manipulation of Molecules*, American Chemical Society, 0-8412-3786-7, Washington, DC

Becke, A. D. (1993). A new mixing of Hartree--Fock and local density--functional theories, *J. Chem. Phys.*, 98, 2, (1993) 1372--1377, 0021-9606

Bedard, T. C. & Moore, J. S. (1995). Design and synthesis of a molecular turnstile, *J.Am. Chem. Soc.*, 117, 43 (1995) 10662-10671, 0002-7863

Boyer, P. D. (1993). The binding change mechanism for ATP synthase – Some probabilities and possibilities, *Biochimica et Biophysica Acta*, 1140, 3, (Jan 8 1993) 215-250, 0005-2728

Choi, S. E. & Light, J. C. (1989). Use of the discrete variable representation in the quantum dynamics by a wave packet propagation: Predissociation of NaI($^1\Sigma_0^+$)$\rightarrow$NaI(0^+) $\rightarrow$Na(^{2}S)+I(^{2}P), *J. Chem. Phys.* 90, 5, (Mar 1 1989) 2593-2599, 0021-9606

Dominguez, Z.; Dang, H.; Strouse, M. J. & Garcia-Garibay, M. A., (2002). Molecular "Compasses" and "Gyroscopes". I. Expedient synthesis and solid state dynamics of an open rotor with a bis(triarylmethyl) frame *J. Am. Chem. Soc.*, 124, 11, (Mar 20 2002) 2398-2399, 0002-7863

Feringa, B. L.; van Delden, R. A.; Koumura, N. & Geertsema, E. M. (2000). Chiroptical molecular switches, *Chem. Rev.*, 100, 5, (May 2000) 1789-1816, 0009-2665

Frisch, M. J.; Trucks, G. W.; Schlegel, H. B.; Scuseria, G. E.; Robb, M. A.; Cheeseman, J. R.; Zakrzewski, V. G.; Montgomery, J. A.; Stratmann, R. E.; Burant, J. C.; Dapprich, S.; Millam, J. M.; Daniels, A. D.; Kudin; K. N.; Strain, M. C.; Farkas, O.; Tomasi, J.; Barone, V.; Cossi, M.; Cammi, R.; Mennucci, B.; Pomelli, C.; Adamo, C.; Clifford, S.; Ochterski, J.; Petersson, G. A.; Ayala, P. Y.; Cui, Q.; Morokuma, K.; Malick, D. K.; Rabuck, A. D.; Raghavachari, K.; Foresman, J. B.; Cioslowski, J.; Ortiz, J. V.; Stefanov, B. B.; Liu, G.; Liashenko, A.; Piskorz, P.; Komaromi, I.; Gomperts, R.; Martin, R. L.; Fox, D. J.; Keith, T.; Al-Laham, M. A.; Peng, C. Y.; Nanayakkara, A.; Gonzalez, C.; Challacombe, M.; Gill, P. M. W.; Johnson, B. G.; Chen, W.; Wong, M. W.; Andres, J. L.; Head-Gordon, M.; Replogle, E. S. & Pople, J. A. (1998). *Gaussian98 (RevisionA.7)*, Gaussian Inc., Pittsburgh, PA

Fujimura, Y.; González, L..; Kröner, D.; Manz, J.; Mehdaoui, I. & Schmidt, B. (2004). Quantum ignition of intramolecular rotation by means of IR+UV laser pulses, *Chem. Phys. Lett.*, 386, 4-6, (Mar 11 2004) 248-253, 0009-2614

Gordon, R. A. & Fujimura, Y. (2002). Coherent control of chemical reactions, In: *Encyclopedia of Physical Science and Technology*, Robert A. Meyers (Ed.), 207--231, Academic Press, 978-0-12-227410-7, San Diego

Gouliemakis, E.; Uiberacker, M.; Kienberger, R.; Baltuska, A.; Yakovlev, V.; Scrinzi, A.; Westerwalbesloh, T.; Kleineberg, U.; Heinzmann, U.; Drescher, M.; & Krausz, F., (2004). Direct measurement of light waves, *Science*, 305, (Aug 27 2004) 1267-1269, 0036-8075

Hoki, K.; Ohtsuki, Y. & Fujimura, Y. (2001). Locally designed pulse shaping for selective preparation of enantiomers from their racemate, *J. Chem. Phys.*, 114, 4,(Jan 22 2001) 1575-1581, 0021-9606

Hoki, K.; Yamaki, M.; Koseki, S. & Fujimura, Y. (2003) Molecular motors driven by laser pulses: Role of molecular chirality and photon helicity, *J. Chem. Phys.*, 118, 2, (Jan 8 2003) 497-504, 0021-9606

Hoki, K.; Yamaki, M. & Fujimura, Y. (2003). Chiral molecular motors driven by a nonhelical laser pulse, *Angew. Chem. Int. Ed.*, 42, 26, (Jul 7 2003) 2975-2978, 1433-7851

Hoki, K.; Yamaki, M.; Koseki, S. & Fujimura, Y. (2003). Mechanism of unidirectional motions of chiral molecular motors driven by linearly polarized pulses, *J. Chem. Phys.*, 119, 23, (Dec 15 2003) 12393-12398, 0021-9606

Hoki, K., Sato, M.; Yamaki, M.; Sahnoun, R; González, L.; Koseki, S. & Fujimura, Y. (2004). Chiral molecular motors ignited by femtosecond pump-dump laser pulses, *J. Phys. Chem. B*, 108, 15, (Apr 15 2004) 4916-4921, 1520-6106

Jülicher, F.; Ajdari, A. & Prost, J. (1997) Modeling molecular motors, *Rev. Mod. Phys.*, 69, 4, (Oct 1997) 1269-1281, 0034-6861

Kay, E. R.; Leigh, D. A. & Zerbetto, F. (2007). Synthetic molecular motors and mechanical machines, *Angew. Chem. Int. Ed.*, 46, 1-2, (2007) 72-191, 1433-7851

Kinbara, K. & Aida, T. (2005). Toward intelligent molecular machines: Directed motions of biological and artificial molecules and assemblies, *Chem. Rev.*, 105, 4, (Apr 2005) 1377-1400, 0009-2665

Kosloff, R.; Rice, S. A.; Gaspard, P.; Tersigni, S. & Tannor, D. J. (1989). Wavepacket dancing - Achieving chemical selectivity by shaping light-pulses, *Chem. Phys.*, 139, 1, (Dec 1 1989) 201-220, 0301-0104

Lindblad, G. (1976). Generators of quantum dynamical semigroups, *Commun. Math. Phys.*, 48, 2, (1976) 119--130, 0010-3616

Ohtsuki, Y.; Zhu, W. & Rabitz, H. (1999). Monotonically convergent algorithm for quantum optimal control with dissipation, *J. Chem. Phys.*, 110, 20, (May 22 1999) 9825-9832, 0021-9606

Potter, E. D.; Herek, J. L.; Pedersen, S.; Liu, Q. & Zewail, A. H. (1992). Femtosecond laser control of a chemical-reaction, *Nature*, 355, 6355, (Jan -2 1992) 66-68, 0028-0836

Reimann, P. (2002). Brownian motors: noisy transport far from equilibrium, *Physics Reports*, 361, 2-4, (Apr 2002) 57-265, 0370-1573

Rice, S. A. & Zhao, M. (2000). *Optimal Control of Molecular Dynamics*, John Wiley & Sons Inc, 978-0-471-35423-9, New York

Roncaglia, R. & Tsironis, G. P. (1998). Discrete quantum motors, *Phys. Rev. Lett.*, 81, 1, (Jul 6 1998) 10-13, 0031-9007

Ross, B. O. (1987). The complete active space self-consistent field method and its applications in electronic structure calculations, In: *Ab initio methods in quantum chemistry-II, volume 69 of advances in chemical physics, Lawly*, K. P. (ed.) 339-445, Wily, 0065-2385, New York

Setaka, W.; Ohmizu, S.; Kabuto, C. & Kira, M. (2007). A molecular gyroscope having phenylene rotator encased in three-spoke silicon-based stator, *Chem. Lett.*, 36, 8, (Aug 5 2007) 1076-1077, 0366-7022

Shapiro, M. & Brumer, P. (2000). Coherent control of atomic molecular, and electronic processes, In: *Advances in atomic molecular and optical physics Vol.42*, (2000), Ed. Bederson, B. & Walther, H., 287-345, Academic Press Inc., 9780120038428, San Diego

Shi, S. H. & Rabitz, H. (1990). Quantum-mechanical optimal-control of physical observables in microsystems, *J. Chem. Phys.*, 92 , 1, (Jan 1 1990) 364-376, 0021-9606

Sugawara, M. & Fujimura, Y. (1994). Control of quantum dynamics by a locally optimized laser field - application to ring puckering isomerization, *J. Chem. Phys.*, 100, 8, (Apr 15 1994), 5646-5655, 0021-9606

Tabe, Y.; Yokoyama, H. (2003). Coherent collective precession of molecular rotors with chiral propellers, *Nature Materials*, 2, 12, (Dec 2003) 806-809, 1476-1122

Yamaki, M; Hoki, K.; Ohtsuki, Y.; Kono, H. & Fujimura, Y. (2005). Quantum control of a chiral molecular motor driven by laser pulses, *J. Am. Chem. Soc.*, 127, 20, (May 25 2005) 7300-7301, 0002-7863

Yamaki, M; Hoki, K.; Ohtsuki, Y.; Kono, H. & Fujimura, Y. (2005). Quantum control of a unidirectional rotation of a chiral molecular motor, *Phys. Chem. Chem. Phys.* 7, 9, (May 7 2005) 1900-1904, 1463-9076

Yamaki, M; Hoki, K.; Kono, H. & Fujimura, Y. (2008). Quantum control of a chiral molecular motor driven by femtosecond laser pulses: Mechanisms of regular and reverse rotations, *Chem. Phys.*, 347, 1-3, (May 23 2008) 272-278, 0301-0104

Yamaki, M; Nakayama, S.; Hoki, K.; Kono, H. & Fujimura, Y. (2009). Quantum dynamics of light-driven chiral molecular motors, *Phys. Chem. Chem. Phys.* 11, 11, (2009) 1662-1678, 1463-9076

Zewail, A. & Bernstein, R. (1992). Real-Time Laser femtochemistry viewing the transition from reagents to products, In: *The Chemical Bond Structure and Dynamics*, Zawail, A., (Ed.) 223-279, Academic Press, Inc., 0-12-779620-7, San Diego, CA

Energy Approach to Atoms in a Laser Field and Quantum Dynamics with Laser Pulses of Different Shape

Alexander V. Glushkov[1,2,3], Ol'ga Yu. Khetselius[1],
Andrey A. Svinarenko[1,3] and George P. Prepelitsa[1]
[1]Odessa (OSENU) University, P.O.Box 24a, Odessa-9, 65009,
[2]Institute of Spectroscopy (ISAN), Russian Acad.Sci., Troitsk-Moscow, 142090,
[3]UK National Academy of Sciences, London SW1Y 5AG,
[1]Ukraine
[2]Russia
[3]United Kingdom

1. Introduction

In most branches of physics, a controlled manipulation of the considered system has proven to be extremely useful to study fundamental system properties, and to facilitate a broad range of applications. A prominent example for this is quantum optics or laser physics in general, for instance related to light-matter interactions on the level of single quantum objects (Letokhov, 1977, 1984; Delone & Kraynov, 1984, 1995, 1999; Allen & Eberly, 1987; Kleppner,et al, 1991; Fedorov, 1995; Scully & Zubairy, 1997; Friedberg, et al 2003; Popov, 2004; Ficek & Swain, 2005; Shahbaz et al, 2006; Burvenich et al, 2006; Müller et al, 2008; Glushkov et al, 2003, 2004, 2005, 2008, 2009) . Similar control is also possible at lower driving field frequencies, e.g., with NMR techniques in the microwave frequency region. Towards higher frequencies, in particular the development and deployment of high-intensity lasers have opened the doors to new fascinating areas of physics of light-matter interactions. Laser fields reach and succeed the Coulomb field strength experienced by the electrons due to the nucleus and thus give rise to a plethora of exciting phenomena. The above examples have in common that they focus on the interaction of the driving fields with the outer electron shell of the atoms. Now it is clear that direct laser-atom and nucleus interactions may indeed become of relevance in future experiments employing x-ray lasers, opening the field of high-intensity atomic and nuclear quantum optics. In particular, the coherence of the laser light expected from new sources such as TESLA XFEL is the essential feature which may allow to access extended coherence or interference phenomena reminiscent of atomic quantum optics. Such laser facilities, especially in conjunction with moderate acceleration of the target atoms and nuclei to match photon and transition frequency, may thus enable to achieve nuclear Rabi oscillations, photon echoes or more advanced quantum optical schemes in atoms, nuclei, molecules, clusters, bose-condensate etc .

The interaction of the atomic systems with the external alternating fields, in particular, laser fields has been the subject of intensive experimental and theoretical investigation (Holt et al, 1983; Delone & Kraynov, 1984, 1995, 1999; Ullrich et al, 1986; Allen & Eberly, 1987; Scully & Zubairy, 1997; Aumar-Winter, 1997; Becker & Faisal, 2002; Batani & Joachain, 2006; Glushkov, 2005, 2008; etc). The appearance of the powerful laser sources allowing to obtain the radiation field amplitude of the order of atomic field in the wide range of wavelengths results to the systematic investigations of the nonlinear interaction of radiation with atoms and molecules. Calculation of the deformation and shifts of the atomic emission and absorption lines in a strong laser field, definition of the k-photon emission and absorption probabilities and atomic levels shifts, study of laser emission quality effect on characteristics of atomic line, dynamical stabilization and field ionization etc are the most actual problems to be solved. Naturally, it is of the great interest for phenomenon of a multiphoton ionization. At present time, a progress is achieved in the description of the processes of interaction atoms with the harmonic emission field. But in the realistic laser field the according processes are in significant degree differ from ones in the harmonic field. The latest theoretical works claim a qualitative study of the phenomenon though in some simple cases it is possible a quite acceptable quantitative description. Among existed approaches it should be mentioned the Green function method (the imaginary part of the Green function pole for atomic quasienergetic state), the density - matrix formalism (the stochastic equation of motion for density - matrix operator and its correlation functions), a time-dependent density functional formalism, direct numerical solution of the Schrödinger (Dirac) equation, multi-body multi-photon approach etc. Decay probabilities of the hydrogen atom states in the super-strong laser field are calculated by the Green function method under condition that electron- proton interaction is very small regarding the atom-field interaction. Note that this approach is not easily generalized for multielectron atoms. Alternative approach is using the double-time Gell-Mann and Low formalism for the investigation of line-shape of a multi-ionized atom in the strong field of electromagnetic wave. The effects of the different laser line shape on the intensity and spectrum of resonance fluorescence from a two-level atom are intensively studied (Bjorkholm & Liao, 1975; Grance, 1981; Georges & Dixit, 1981; Zoller, 1982; Kelleher et al, 1985; Sauter et al, 1986; Glushkov-Ivanov, 1992, 1993; Friedberg et al, 2003; Glushkov et al, 2005, 2008, 2009 et al).

The laser model considered is that of an ideal single-mode laser operating high above threshold, with constant field amplitude and undergoing phase-frequency fluctuations analogous to Brownian motion. As a correlation time of the frequency fluctuations increases from zero to infinity, the laser line shape changes from Lorentzian to Gaussian in a continuous way. For intermediate and strong fields, the average intensity of fluorescence in the case of a resonant broadband Loretzian line shape is higher than that in the case of a Gaussian line shape with the same bandwidth and total power. This is in contrast to the weak- field case where the higher peak power of the Gaussian line shape makes it more effective than the Lorentzian line shape. In a case of a nonzero frequency correlation time (the non - Lorentzian line shape) an intensity of fluorescence undergoes the non-Markovian fluctuations . In relation to the spectrum of resonance fluorescence it is shown that as the line shape is varied from Lorentzian to Gaussian the following changes take place : in the case of off-resonance excitation, the asymmetry of the spectrum decreases; in a case of resonance excitation, the center peak to side-peak height ratio for the triplet structure increases. The predicted center - line dip, which develops in the spectrum in the case of broadband excitation when the Rabi frequency and the bandwidth are nearly equal, becomes increasingly deeper. In the modern experiment it has been found an anomalously strong nonlinear coupling of radiation with atoms which can not fully explained by the modern theoretical models. In any case the problem requires a consistent quantum electrodynamic approach.

Another important topic is a problem of governing and control of non-linear processes in a stochastic, multi-mode laser field (Grance, 1981; Lompre et al, 1981; Zoller, 1982; Glushkov & Ivanov, 1992). The principal aim of quantum coherent control is to steer a quantum system towards a desired final state through interaction with light while simultaneously inhibiting paths leading to undesirable outcomes. This type of quantum interference is inherent in non-linear multiphoton processes. Controlling mechanisms have been proposed and demonstrated for atomic, molecular and solid-state systems (Goldansky-Letokhov, 1974, Letokhov, 1977; Delone-Kraynov, 1984). Theoretical studies of the laser-atom non-linear interactions are often based on solving the time-dependent Schrödinger equation or using the time-independent Floquet formalism or special perturbation theories (Brändas & Floelich, 1974; Hehenberger et al, 1977; Silverstone et al, 1979; Delone-Kraynov, 1984; Glushkov-Ivanov, 1992, 1993, 2004; Popov, 2004;). It has been extended the non-Hermitian multi-state Floquet dynamics approach to treat one-electron atomic system to the case of general multi-electron ones. The result is a generalization of the R-matrix Floquet theory, developed by Burke *et al*, that allows for pulse shape effects whilst retaining the ab initio treatment of detailed electron correlation. The approach based on the eigenchannel R-matrix method and multichannel quantum-defect theory , introduced by Robicheaux and Gao to calculate two-photon processes in light alkaline-earth atoms has been implemented by Luc-Koenig et al, 1997 in j-j coupling introducing explicitly spin-orbit effects and employing both the length and velocity forms of the electric dipole transition operator. For example, the two-photon processes including above-threshold ionization in magnesium have been in details studied (Luc-Koenig et al, 1997). Nevertheless in many calculations there is a serious problem of the gauge invariance, connected with using non-optimized one-electron representation (in fact provided by not entire account for the multi-body interelectron correlations). The known example is non-coincidence of values for the length and velocity forms of the electric dipole transition operator (Grant, 2007; Glushkov & Ivanov, 1992).

In whole one can note that a problem of correct description of the non-linear atomic dynamics in a stochastic, multi-mode laser field is quite far from the final solution. It requires developing the consistent, advanced approaches to description of multi-photon dynamics and new schemes for sensing the stochasticity and photon-correlation effects. In this paper we present a new consistent method for studying the interaction of atom with the realistic laser field, based on the quantum electrodynamics (QED) and S-matrix adiabatic formalism Gell-Mann and Low. In relativistic case the Gell-Mann and Low formula expressed an energy shift δE through the QED scattering matrix including the interaction with as the laser field as the photon vacuum field (Ivanova et al, 1985; Ivanov-Letokhov, 1986; Glushkov-Ivanov, 1992, 1993; Glushkov et al, 1986, 2004, 2008, 2009). It is more natural to describe the interaction of atom with the realistic laser field by means of the radiation emission and absorption lines (Glushkov-Ivanov, 1986, 1992). Their position and shape fully determine the spectroscopy of atom in a laser field. The radiation atomic lines can be described by moments of different orders μ_n . The first moment are directly linked with the filed shift and width of the corresponding resonances. The main contribution into μ_n is given by the resonant range. The values μ_n can be expanded into perturbation theory (PT) series , though in resonant range the PT can't be used for the transition probabilities. The powerful Ivanov-Ivanova method (Ivanov-Ivanova, 1981;

Ivanova et al, 1985, 1988; Ivanov et al, 1988) is used for calculating the corresponding QED PT second order sums. As example we use the presented method for numerical calculation of the three-photon resonant, four-photon ionization profile of atomic hydrogen (1s-2p transition; wavelength =365 nm) and multi-photon resonance shift and width for transition 6S-6F in the atom of Cs (wavelength 1059nm) in a laser pulse of the Gaussian form. We consider also a quite exact approach to calculation of the characteristics of multi-photon ionization in atomic systems, which is based on the QED PT and use it for numerical calculating the above threshold ionization (ATI) characteristics for atom of magnesium in a intense laser field.

2. Structure of the multi-mode laser pulse

As it is well known, for a laser with more than one longitudinal mode, mode beating gives rise to intensity fluctuations within the laser pulse (eg. Kelleher et al, 1985). The beat frequencies for n modes range up to $nc/2L=B$, where L is the optical length of the laser oscillator. A detailed analysis of the mode structure of the typical dye laser shows that it has about 15 modes, separated by 1 GHz with a Gaussian amplitude distribution. Classically, the field can be written as follows:

$$E(t)=\varepsilon(t)e^{-i\omega t} + c.c., \quad (1)$$

where

$$\varepsilon(t)=\sum_i 0.5a_i(t)e^{-i(\Delta\omega_i t+\phi_i)} .$$

Each mode has amplitude a_i containing a gaussian time envelope, a frequency detuning $\Delta\omega_i$ from the central laser frequency and phase ϕ_i. As experimental study (Lompre et al, 1981; Kelleher et al, 1985;) of described laser pulse showed that there is no evidence of phase coherence in the temporal behavior of the laser pulse and thus it is usually assumed that the modes have random phases. Figure 1 shows the temporal variation of intensity for the multi-mode pulse of stochastic laser radiation with emission lines width b=0,1 см^{-1}, the coherence time -$3\cdot 10^{-10}$s.

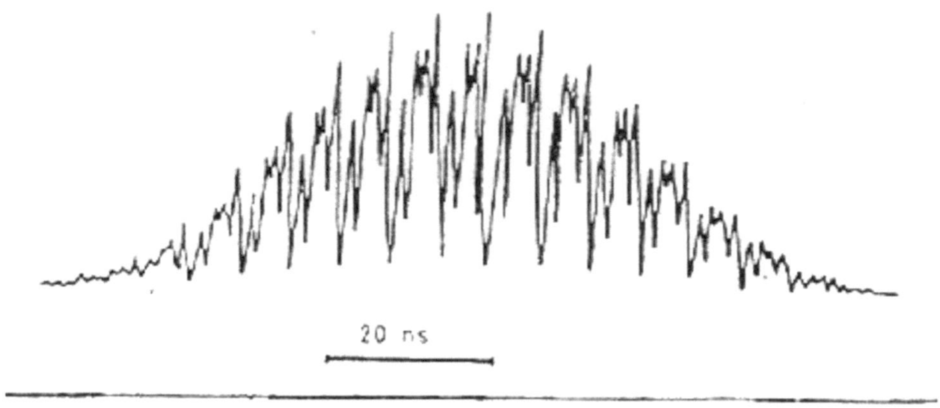

Fig. 1. The temporal variation of intensity for the multi-mode pulse of stochastic laser radiation with emission lines width b=0,1 см^{-1}, the coherence time -$3\cdot 10^{-10}$s.

Further to make sensing a stochastic structure of the multi-mode laser pulse one can consider an interaction "atomic system – stochastic multi-mode laser pulse". Below it will be shown that this interaction is influences by the specific chaotic, photon-correlation effects. New theoretical scheme for sensing stochasticity and photon-correlation features is based on the S-matrix energy approach (Glushkov & Ivanov, 1992, 1993) to calculating the multi-photon resonances spectra characteristics for atomic systems in a stochastic laser field.

3. S-matrix energy approach to atoms in a multi-mode laser field

Let us present the corresponding theoretical scheme. Following to (Glushkov & Ivanov, 1992, 1993; Glushkov et al, 2006, 2008, 2009), we describe the interaction of atom with the realistic laser field not by means the separated atomic levels shifts and by another set of characteristics, which are directly observed in the experiment. We are studying the radiation emission and absorption lines. Its position and shape fully determine the spectroscopy of atom in the field. It is natural to describe these lines by there moments of different orders μ_n. The moments μ_n are strongly dependent upon the laser pulse quality: intensity and the mode constitution. In particular, the k-photon absorption line center shift in the transition $\alpha \rightarrow p$ can not be obtained from the corresponding expressions for the "one"-photon absorption by the change $\omega_0 \rightarrow \omega_0/k$ and introduction of the multiplier $1/k$ (ω_0 - the central laser emission frequency). The difference arises already in the first non-appearing perturbation theory (PT) order and connects with the unusual behaviour of the dynamic polarizability of atom in the resonant range (Glushkov-Ivanov, 1986, 1992). Let us describe the interaction of atom with laser radiation by means the potential:

$$V(r,t)= V(r) \int d\omega f(\omega - \omega_0) \sum_{n=-\infty}^{\infty} \cos [\omega_0 t+ \omega_0 n\tau], \tag{2}$$

where n is the whole number. The potential V represents the infinite duration of laser pulses with known frequency τ. Here we consider the effects of interaction of the atom with the single pulse.. The representation $V(rt)$ as the infinite sequence of pulses is a formal moment connected with the application of the stationary PT formalism. The function $f(\omega)$ is a Fourier component of the laser pulse. The condition $\int d\omega f^2(\omega)=1$ normalizes potential $V(rt)$ on the definite energy in a laser pulse. Let us consider the pulse with Lorentzian shape (coherent 1-mode pulse): $f(\omega) = N/(\omega^2+\Delta^2)$, Gaussian shape (multi-mode chaotic laser pulse): $f(\omega) = N\exp[\ln 2(\omega^2/\Delta^2)]$, and soliton-like pulse of the following shape: $f(t) = N\, ch^{-1}[t/D]$. Further we will be interested by a cases of the Gaussian and soliton-like pulses. A case of the Lorentzian shape has been considered by Glushkov & Ivanov (1992). The further program resulted in the calculating an imaginary part of energy shift Im $E_\alpha(\omega_0)$ for any atomic level as the function of the laser pulse central frequency. An according function has the shape of the resonant curve. Each resonance is connected with the transition α-p, in which the definite number of photons are absorbed or radiated. Let us consider following situation: α-p transition with the absorption of k photons(α, p-discrete levels). For the resonance which corresponds to this transition, we calculate the following values:

$$\delta\omega(p\alpha \,|\, k) = \int' d\omega \,\mathrm{Im}\, E_\alpha(\omega) \,(\omega - \omega_{p\alpha} / k) \,/\, N, \tag{3}$$

$$\mu_m = \int' d\omega \,\mathrm{Im}\, E_\alpha(\omega) \,(\omega - \omega_{p\alpha} / k)^m \,/\, N, \tag{4}$$

where $\int' d\omega \, \mathrm{Im}\, E_\alpha$ is the normalizing multiplier; $\omega_{p\alpha}$ is position of the non-shifted line for atomic transition α-p, $\delta\omega(p\alpha|k)$ is the line shift under k-photon absorption and $\omega_{p\alpha} = \omega_{p\alpha} + k\cdot\delta\omega(p\alpha|k)$. The first moments μ_1, μ_2 and μ_3 determine the atomic line center shift, its dispersion and coefficient of the asymmetry. To calculate μ_m, we need to get an expansion of E_α to PT series: $E_\alpha = \sum E_\alpha^{(2k)}(\omega_0)$. To get this expansion, we use method, based on the Gell-Mann and Low adiabatic formula for δE_α (Ivanov et al, 1986, 1993; Ivanova et al, 1985, 1993). The representation of the S- matrix in the form of PT series induces the expansion for δE_α:

$$\delta E_\alpha(\omega_0) = \lim_{\gamma\to 0} \gamma \sum_{k_1 k_2 \ldots k_n} a(k_1, k_2, \ldots, k_n), \quad (5)$$

$$I_\gamma(k_1, k_2, \ldots, k_n) = \prod_{j=1} S_\gamma^{(kj)}, \quad (6)$$

$$S_\gamma^{(m)} = (-1)^m \int_{-\infty}^{0} d\, t_1 \ldots \int_{-\infty}^{t_m - 1} d\, t_m \langle \Phi_\alpha \mid V_1 V_2 \ldots V_m \mid \Phi_\alpha \rangle, \quad (7)$$

$$V_j = \exp(1H_0 t_j)\, V(rt_j) \exp(-1H_0 t_j) \exp(\gamma t_j). \quad (8)$$

Here H is the atomic hamiltonian, $a(k_1, k_2, \ldots, k_n)$ are the numerical coefficients. The structure of the matrix elements $S_\gamma^{(m)}$ is in details described (Glushkov & Ivanov, 1986, 1992, 1993). After sufficiently complicated one can get the expressions for the line moments. Let us present results for the Gaussian laser pulse:

$$\delta\omega(p\alpha \mid k) = \{\pi\Delta / (k+1)k\} \, [\, E(p, \omega_{p\alpha}/k) - E(\alpha, \omega_{p\alpha}/k)], \quad (9)$$

$$\mu_2 = \Delta^2/k$$

$$\mu_3 = \{4\pi\Delta^3 / [k(k+1)]\} \, [\, E(p, \omega_{p\alpha}/k) - E(\alpha, \omega_{p\alpha}/k)],$$

where

$$E(j, \omega_{p\alpha}/k) = 0{,}5 \sum_{p_i} V_{jpi} V_{pij} \left[\frac{1}{\omega_{jp_i} + \omega_{p\alpha}/k} + \frac{1}{\omega_{jp_i} - \omega_{p\alpha}/k} \right] \quad (10)$$

The summation in (10) is fulfilled on all states of atomic system. For the Lorentzian pulse the expressions were obtained by Glushkov & Ivanov (1986, 1992). In a case of the laser pulse with shape $ch^{-1}[t/D]$ it is necessary to carry out a direct numerical calculation (we did it) or use different approximations to simplify the expressions. Indeed, the last procedure may result in a great mistake.

Each term in equations (9) for $\delta\omega$ is formally similar to the known expression for off-resonant shift of atomic level (p or α) in the monochromatic emission field with frequency $\omega_{p\alpha}/k$. However, here these values have other physical essence. When $k \to \infty$ (an infinite little laser pulse central frequency) the formula for δE gives the correct expression for energy level shift in the stationary field.

The expressions (9),(10) for $\delta\omega$ and μ_n describe the main characteristics of the absorption line near resonant frequency $\omega_{p\alpha}/k$. One can see that these characteristics are determined not only by the radiation frequency, but also by the quantiness of the process. For example, the line shift is proportional $1/(k+I)$, but no – to value of $1/k$, as one can wait for. Under k=1 there is an additional non-standard term. It will be shown below that this approach allows getting the results in an excellent agreement with experiment. The details of the numerical procedure are given below and presented in refs. (Glushkov et al, 2004, 2005, 2006, 2008, 2009) too.

4. Ivanova-Ivanov approach to calculating the QED perturbation theory: second order sum

In this chapter we present the Ivanova-Ivanov approach to calculating sums of the second order of the QED perturbation theory (Ivanov & Ivanova, 1981; Ivanov et al, 1988, 1993; Ivanova et al, 1985, 1986; Glushkov et al, 2008, 2009). It will be used in calculation of the expressions (9), (10). In fact, speech is about determination of the matrix elements for operator of the interelectron interaction over an infinitive set of virtual states, including the states of the negative continuum. A sum on the principal quantum number is defined in quadratures of the Dirac function and auxiliary functions *x*, *x* (look below). All computational procedure results in solution of simple system of the ordinary differential equations with known boundary conditions under *r=0*. Exchange of the interelectron interaction operator $1/r_{12}$ on one-electron operator *V(r)* decreases a brevity of summation on the virtual states. In a one-particle representation the cited sums are expressed through sums of the one-electron matrix elements:

$$\sum_{n_1} \langle n\chi m|V|n_1\chi_1 m_1\rangle \langle n_1\chi_1 m_1|V|n\chi m\rangle / \left(\varepsilon_{n_1\chi_1 m_1} - \varepsilon\right), \tag{11}$$

where $\varepsilon = \varepsilon_{n\chi m} + \omega_{p\alpha}/k$ is the energy parameter. One-electron energies $\varepsilon_{n\chi m}$ include the rest energy $(aZ)^{-2}$. Let us note that here we use the Coulomb units (an energy in the Coulomb units [q.u.]: 1 q.u.=Z^2 a.u.e.[Z – a charge of a nucleus; a.u.e.= 1 atomic unit of energy).
Consider a scheme of calculating the sum (11). Fundamental solutions of one-electron Dirac equations with potential V_C =U(*r*) have the same asymptotics as and the Dirac equation with Coulomb potential under $r\to 0$ and $r\to\infty$. Let us consider a bi-spinor of the following form:

$$\Phi_{\chi_1 \overline{m_1}} = \sum_{n_1} \varphi_{n_1\chi_1 m_1} \langle n_1\chi_1 m_1|V|n\chi m\rangle / \left(\varepsilon_{n_1\chi_1 m_1} - \varepsilon\right) \tag{12}$$

The radial parts *F, G* of bi-spinor Φ satisfy to system of differential equations:

$$-F' / \alpha Z + (1+\chi_1) F/\alpha Z r + A_2 G = \Lambda_2 ,$$

$$G'/\alpha Z + (1-\chi_1) G/\alpha Z r + A_1 F = \Lambda_1 , \tag{13}$$

$$A_1(r) = U(r) + 1/(\alpha Z)^2 - \varepsilon$$

$$A_2(r) = U(r) - 1/(\alpha Z)^2 - \varepsilon. \tag{14}$$

The radial functions Λ_1, Λ_2 in a case of the dipole interaction are presented below. Solution of the system (13) can be represented as follows:

$$F(r) = \alpha Z\left[x(r)\tilde{f}(r) - \tilde{x}(r) f(r)\right]/2\gamma ,$$

$$G(r)=\alpha Z\left[x(r)\tilde{g}(r)-\tilde{x}(r)g(r)\right]/2\gamma,\ \gamma=\left[\chi^2-\alpha^2Z^2\right]^{1/2}, \tag{15}$$

A pair of functions f, g and $\tilde{f},\tilde{g}$ are two fundamental solutions of equations (13) without right parts; These functions satisfy to conditions: $f\sim r^{\gamma-1}$, f, $g\sim r^{-\gamma-1}$ under $r\to 0$. Here we introduce the following functions:

$$x=\alpha Z\int_0^r dr'r'^2\left[\Lambda_1(r')f(r')+\Lambda_2(r')g(r')\right]$$

$$\tilde{x}=\alpha Z\int_0^r dr'r'^2\left[\Lambda_1(r')\tilde{f}(r')+\Lambda_2(r')\tilde{g}(r')\right]+D. \tag{16}$$

Further let us define a constant D in expressions (16). Let us suppose that $\varepsilon<(\alpha Z)^{-2}$ (i.e. an energy lies below the boundary of ionization), but an energy does not coincide with any discrete eigen value of the Dirac equation. Then

$$D=-\alpha Z\int_0^\infty dr\,r^2\left(\Lambda_1\tilde{f}+\Lambda_2\tilde{g}\right). \tag{17}$$

Let an energy ε coincides with energy of some discrete level $n_0\chi_1m_1$. It is supposed that this state excludes from (11) and (13). Then a constant D can be found from condition:

$$\int_0^\infty dr\,r^2\left(Ff_{n_0\chi_1m_1}+Gg_{n_0\chi_1m_1}\right)=0. \tag{18}$$

Now let $\varepsilon>(\alpha Z)^{-2}$ (i.e. an energy lies above the boundary of ionization). Then a constant D can be found from the following condition:

$$\lim_{r\to\infty} r^2\int_r^{T+r} dr'\,r'^2\left(Ff_{\varepsilon\chi_1m_1}+Gg_{\varepsilon\chi_1m_1}\right)=0, \tag{19}$$

Here $\varepsilon\chi_1m_1$ is one-electron state of scattering with energy ε; T is a period of asymptotic oscillations of the functions f, g:

$$T=2\pi\left[\varepsilon^2-(\alpha Z)^{-2}\right]^{1/2}.$$

Let us give the corresponding expressions for functions Λ_1, Λ_2 in the most typical case of the dipole interaction of an atom with the laser field. The corresponding potential is as follows:

$$V(\mathbf{r})=(\mathbf{a},\alpha), \tag{20}$$

Here **a** is a vector of polarization of radiation; α is a vector of the Dirac matrices.
Let us remember that s usually the vectors $a_1=(1,i,0)$, $a_2=(1,-i,0)$ are corresponding to the circular polarization and the vector $a_3=(1, 00)$ is corresponding to linear one. Under definition

of the multi-photon resonance energies and widths there is a task of calculating the sums (11), where an index n_1 runs the whole spectrum of states or some state n_0 is excluded from the sum. In the first case the functions Λ_1 and Λ_2 are defined by the expressions:

$$\Lambda_1 = B\left(a\middle| j_1 l_1 m_1,\ j\tilde{l}m\right) g_{n\chi}/\alpha Z,$$

$$\Lambda_2 = B\left(a\middle| j_1 \tilde{l}_1 m_1,\ jlm\right) f_{n\chi}/\alpha Z \tag{21}$$

In the second case one can substitute the following functions to the right parts of (13):

$$\overline{\Lambda}_1 = \Lambda_1 - f_{n_0\chi_1 m_1} Y/\alpha Z,$$

$$\overline{\Lambda}_2 = \Lambda_2 - g_{n_0\chi_1 m_1} Y/\alpha Z,$$

$$Y = \int dr\, r^2 \left[f_{n_0\chi_1 m_1} g_{n\chi m} B\left(a\middle| j_1 l_1 m_1, j\tilde{l}m\right) - g_{n_0\chi_1 m_1} f_{n\chi m} B\left(a\middle| j_1\tilde{l}_1 m_1, jlm\right)\right] \tag{22}$$

Here the functions Λ_1 and Λ_2 defined by the expressions (21).
The angle functions are dependent upon a polarization vector and defined by the following formula:

$$B\left(a_1\middle| jlm,\ j'l'm'\right) = (-1)^{j'+l'-1/2}\, \delta_{u'}\delta_{m',m-1} b(-m, m'),$$

$$B\left(a_2\middle| jlm,\ j'l'm'\right) = (-1)^{j+l-1/2}\, \delta_{u'}\delta_{m',m+1} b(-m, m'),$$

$$B\left(a_3\middle| jlm,\ j'l'm'\right) = \delta_{u'}\delta_{mm'}\left[b(-m, -m) + (-1)^{j+j'} b(m, m')\right],$$

$$b(m, m') = \left[2\frac{\chi + 1/2 + m}{2\chi + 1} - \frac{\chi' + 1/2 + m'}{2\chi' + 1}\right]^{1/2}. \tag{23}$$

The final expression for the sum (11) can be written as follows:

$$\int dr\, r^2 \left[f_{n\chi_1} G B\left(a\middle| jlm, j_1\tilde{l}_1 m_1\right) + g_{n\chi} F \cdot B\left(a\middle| j\tilde{l}m_1, j_1 l_1 m_1\right)\right]. \tag{24}$$

Finally the computational procedure results in a solution of sufficiently simple system of the ordinary differential equations for above described functions and integral (24). In concrete numerical calculations the block "Super spinor" of the PC "Superatom" package (Ivanov-Ivanova, 1981; Ivanova et al, 1985, 1986, 2001; Glushkov et al, 2004, 2008, 2009) is used.

5. Energy QED approach to multiphoton resonances and above threshold ionization

In this section we consider a quite exact approach to calculation of the characteristics of multi-photon ionization in atomic systems, which is based on the QED perturbation theory

(Glushkov-Ivanov, 1992, 1993; Glushkov, 2006, 2008, 2009). Below we calculate numerically the above threshold ionization (ATI) cross-sections for atom of magnesium in a intense laser field. The two-photon excitation process will be described in the lowest QED PT order. This approach is valid away from any one-photon intermediate-sate resonance. We start from the two-photon amplitude for the transition from an initial state Ψ_0 with energy E_0 to a final state Ψ_f with energy E_f=E_0+2ω is:

$$T_{f0}^{(2)} = \lim_{\eta \to 0+} \int d\varepsilon < \Psi_f | D \cdot e | \varepsilon > (E_0 + \omega - \varepsilon + i\eta)^{-1} < \varepsilon | D \cdot e | \Psi_0 > \quad (25)$$

Here D is the electric dipole transition operator (in the length r form), e is the electric field polarization and ω is a laser frequency. It's self-understood that the integration in equation (25) is meant to include a discrete summation over bound states and integration over continuum states. Usually an explicit summation is avoided by using the Dalgarno-Lewis by means the setting (Luc-Koenig et al, 1997):

$$T_{f0}^{(2)} = C_f < \Psi_f || D \cdot e || \Lambda_p > \quad (26)$$

where <|| ||> is a reduced matrix element and C_f is an angular factor depending on the symmetry of the Ψ_f, Λ_p, Ψ_0 states. Λ_p, can be founded from solution of the following inhomogeneous equation (Luc-Koenig et al, 1997):

$$(E_0 + \omega - H) | \Lambda_p > = (D \cdot e) | \Psi_0 > \quad (27)$$

at energy $E_0+\omega$, satisfying outgoing-wave boundary condition in the open channels and decreasing exponentially in the closed channels. The total cross section (in cm^4W^{-1}) is defined as:

$$\sigma/I = \sum_J \sigma_J / I = 5{,}7466 \times 10^{-35} \cdot \omega_{au} \sum_J | T_{J,0}^{(2)} |^2 \quad (28)$$

where I (in W/cm^2) is a laser intensity. To describe two-photon processes there can be used different quantities: the generalized cross section $\sigma^{(2)}$, given in units of cm^{4}s, by

$$\sigma_{cm^4 s}^{(2)} = 4{,}3598 \times 10^{-18} \omega_{au} \sigma / I_{cm4/W} \quad (29)$$

and the generalized ionization rate $\Gamma^{(2)}/I^2$, (and probability of to-photon detachment) given in atomic units, by the following expression:

$$\sigma / I_{cm4/w} = 9{,}1462 \times 10^{-36} \omega_{au} \Gamma_{au}^{(2)} / I_{au}^2 \quad (30)$$

Described approach is realized as computer program block in the atomic numeric code "Super-atom" (Ivanov-Ivanova, 1981; Ivanova et al, 1985, 1986, 2001; Glushkov-Ivanov, 1992,1993; Glushkov et al, 2004, 2008, 2009), which includes a numeric solution of the Dirac equation and calculation of the matrix elements of the (17)-(18) type. The original moment is connected with using the consistent QED gauge invariant procedure for generating the atomic functions basis's (optimized basis's) (Glushkov & Ivanov, 1992). This approach allows getting results in an excellent agreement with experiment and they are more precise in comparison with similar data, obtained with using non-optimized basis's.

6. Some results and discussion

6.1 The multi-photon resonances spectra and above threshold ionization for atom of magnesium

Let us present the results of calculating the multi-photon resonances spectra characteristics for atom of magnesium in a laser field (tables 1,2). Note that in order to calculate spectral properties of atomic systems different methods are used: relativistic R-matrix method (R-метод; Robicheaux-Gao, 1993; Luc-Koenig E. etal, 1997), added by multi channel quantum defet method, K-matrix method (K-method; Mengali-Moccia,1996), different versions of the finite L^2 method (L^2 method) with account of polarization and screening effects (SE) (Moccia-Spizzo, 1989; Karapanagioti et al, 1996), Hartree-Fock configuration interaction method (CIHF), operator QED PT (Glushkov-Ivanov, 1992; Glushkov et al; 2004) etc.

Methods	E	Γ	σ/I
Luc-Koenig E. etal, 1997	Without	account	SE
Length form	68492	374	1,96 10-27
Velocity form	68492	376	2,10 10-27
Luc-Koenig E. etal, 1997	with	account	SE
Length form	68455	414	1,88 10-27
Velocity form	68456	412	1,98 10-27
Moccia and Spizzo (1989)	68320	377	2,8 10-27
Robicheaux and Gao (1993)	68600	376	2,4 10-27
Mengali and Moccia(1996)	68130	362	2,2 10-27
Karapanagioti et al (1996)	68470	375	2,2 10-27
Our calculation	68281	323	2,0 10^{-27}

Table 1. Characteristics for $3p^{2}\,{}^1S_0$ resonance of atom of the magnesium: E- energy, counted from ground state ($см^{-1}$), Γ- autoionization width ($см^{-1}$), σ/I- maximum value of generalized cross-section ($см^4W^{-1}$)

In table 1 we present results of calculating characteristics for $3p^{2}\,{}^1S_0$ resonance of Mg; E-energy, counted from ground state ($см^{-1}$), Γ-autoionization width ($см^{-1}$), σ/I- maximum value of generalized cross-section ($см^4W^{-1}$). R-matrix calculation with using length and velocity formula led to results, which differ on 5-15%, that is evidence of non-optimality of atomic basis's. This problem is absent in our approach and agreement between theory and experiment is very good.

Further let us consider process of the multi-photon ATI from the ground state of Mg. The laser radiation photons energies ω in the range of 0,28-0,30 a.u. are considered, so that the final autoionization state (AS) is lying in the interval between 123350 $см^{-1}$ and 131477$см^{-1}$. First photon provides the AS ionization, second photon can populate the Rydberg resonance's, owning to series *4snl,3dnl,4pnp* c J=0 and J=2.

In table 2 we present energies ($см^{-1}$; counted from the ground level of Mg $3s^2$) and widths ($см^{-1}$) of the AS (resonance's) *4snl,3dnl,*$4p^2$ ${}^1D_2$, calculated by the K-, R-matrix and our methods. In a case of ${}^1S_0$ resonance's one can get an excellent identification of these resonance's. Let us note that calculated spectrum of to photon ATI is in a good agreement with the R-matrix data and experiment. In a whole other resonances and ATI cross-sections demonstrate non-regular behaviour.

Studied system is corresponding to a status of quantum chaotic system with stochastization mechanism. It realizes through laser field induction of the overlapping (due to random

	R-method			QED approach			*K*-method	
$^1D_2$	*E*	*Γ*	$^1D_2$	*E*	*Γ*		*E*	*Γ*
4s3d	109900	2630	4s3d	109913	2645		110450	2600
$3d^2$	115350	2660	$3d^2$	115361	2672		115870	2100
4s4d	120494	251	4s4d	120503	259	(ds)	120700	170
3d5s	123150	1223	3d5s	123159	1235	(ds)	123400	2000
$4p^2$	124290	446	$4p^2$	124301	458		124430	500
3d4d	125232	400	3d4d	125245	430		125550	590
4s5d	126285	101	4s5d	126290	113	(ds)	126250	120
3d6s	127172	381	3d6s	127198	385	(ds)	127240	350
4s6d	127914	183	4s6d	127921	215		127870	1900
3d5d	128327	208	3d5d	128344	215			
4s7d	128862	18	4s7d	128874	24	(ds)	128800	30
3d5g	128768	4,5	3d5g	128773	5,2	3d5g	128900	2,2
3d7s	129248	222	3d7s	129257	235		129300	160
4s8d	129543	114	4s8d	129552	125	(ds)	129500	140
			3d6d	129844	115			
			4s9d	129975	64			
			4s10d	130244	5			
			3d8s	130407	114			
			4s11d	130488	118			
			4s12d	130655	28			
			3d7d	130763	52			
			4s13d	130778	36			
			4s14d	130894	14			
			4s15d	130965	7			

Table 2. Energies and widths (cm^{-1}) of the AS (resonance's) *4snl,3dnl,*$4p^2$ $^1D_2$ for Mg (see text)

interference and fluctuations) resonances in spectrum, their non-linear interaction, which lead to a global stochasticity in the atomic system and quantum chaos phenomenon. The quantum chaos is well known in physics of the hierarchy, atomic and molecular physics in external electromagnetic field. Earlier it has been found in simple atomic systems H, He, and also Ca. Analysis indicates on its existence in the Mg spectrum. Spectrum of resonance's can be divided on three intervals: 1). An interval, where states and resonances are clearly identified and not strongly perturbed; 2) quantum-chaotic one, where there is a complex of the overlapping and strongly interacting resonances; 3). Shifted one on energy, where behaviour of energy levels and resonances is similar to the first interval. The quantitative estimate shows that the resonances distribution in the second quantum-chaotic interval is satisfied to Wigner distribution as follows:

$$W(x)=x\exp(-\pi x^2/4). \quad (31)$$

At the same time, in the first interval the Poisson distribution is valid.

6.2 The three-photon resonant, four-photon ionization profile of atomic hydrogen

Below we present the results of calculating the multi-photon resonances spectra characteristics for atomic systems in a stochastic laser field and show the possibilities for

sensing a structure of the stochastic, multi-mode laser pulse and photon-correlation effects for atomic (and nano-optical) systems in this field (figure 2). We start from results of the numerical calculation for the three-photon resonant, four-photon ionization profile of atomic hydrogen (1s-2p transition; wavelength =365 nm).
In figure 2 we present the shift S (=δω) and width W of the resonance profile as the function of the mean laser intensity at the temporal and spatial center of the UV pulse: experimental data 3s, 3w (Kelleher et al, 1986; multi-mode Gauss laser pulse with bandwidth 0.25 cm^{-1}; full width at half of one), theoretical calculation results on the basis of the stochastic differential equations method 1s and 1w by Zoller (1982) and results of our calculation: 2s, 2w.
At first, one can see the excellent agreement between the theory and experiment. At second, a comparison of these results with analogous data for a Lorentzian laser pulse (Lompre et al, 1981; Glushkov & Ivanov, 1992) shows that the corresponding resonance shift obtained with the gaussian pulse is larger the shift, obtained with Lorentzian pulse at ~3 times. This is an evidence of the photon-correlation effects and stochasticity of the laser pulse.

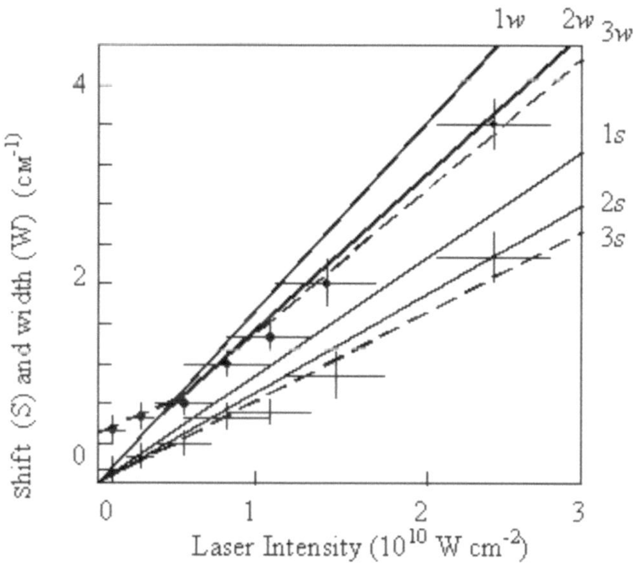

Fig. 2. Shift (S) and width (W) of resonant profile as laser intensity function: experiment - S_3, W_3 (Keller et al, 1981); theory of Zoller (1982)- S_1, W_1 and our results- S_2, W_2.

6.3 Calculation results of the multi-photon resonance width and shift for transition 6S-6F in the atom of Cs

Further let us consider the numerical calculation results for three-photon transition 6S-6F in the Cs atom (wavelength 1,059 μm; see figure 3). The detailed experimental study of the multi-photon processes in Cs atom has been carried out by Lompre et al (1981). Lompre et al experimentally studied a statistics of the laser radiation and there are measured the characteristics of the multi-photon ionization.

The lines shift is linear to respect to the laser intensity (laser intensity is increased from 1,4 to 5,7 10(7) W/cm^2) and is equal (a case of the gaussian multi-mode laser pulse): $\delta\omega(p\alpha \mid k)$ $=bI$ *with* $b=(5,6+-0,3)$ $cm^{-1}/GW{\cdot}cm^{-2}$ (*b* is expressed in terms of energy of the three-photon transition 6S-6F).
The corresponding shift obtained with coherent (one-mode) laser pulse is defined as follows: $\delta\omega 0(p\alpha \mid k) =aI$, $a=2$ $cm^{-1}/GW{\cdot}cm^{-2}$.Theoretical values, obtained with using no-optimized atomic basises, are as follows: i). for soliton-like laser pulse: $\delta\omega(p\alpha \mid k) =bI$, $b=6,7$ $cm^{-1}/GW{\cdot}cm^{-2}$; ii). for the gaussian multi-mode pulse (chaotic light): $\delta\omega(p\alpha \mid k) =bI$ *with* $b=5,8$ $cm^{-1}/GW{\cdot}cm^{-2}$; iii). for the coherent one-mode pulse: $\delta\omega 0(p\alpha \mid k)=aI$, $a=2,1$ $cm^{-1}/GW{\cdot}cm^{-2}$.
The analogous theoretical values, obtained in our calculation within described above S-matrix formalism, are the following:

i. the gaussian multi-mode pulse (chaotic light)

$$\delta\omega(p\alpha \mid k) =bI, \quad b=5,63\ cm^{-1}/GW{\cdot}cm^{-2};$$

ii. the coherent one-mode pulse:

$$\delta\omega 0(p\alpha \mid k)=aI, \quad a=2,02\ cm^{-1}/GW{\cdot}cm^{-2};$$

iii. the soliton-like laser pulse:

$$\delta\omega(p\alpha \mid k) =bI, \quad b=6,5\ cm^{-1}/GW{\cdot}cm^{-2} .$$

One can see that for the with multi-mode pulse, the radiation line shift is significantly larger (in ~ 3 times), then the corresponding shift, which is obtained for single-mode pulse. In fact the radiation line shift is enhanced by the photon-correlation effects. In figure 3 we present the results of calculation for the multi-photon resonance width for transition 6S-6F in the atom of Cs (wavelength 1059nm) in dependence upon the laser intensity.
We use the following denotations: S- for single-mode Lorentz laser pulse; M_1, M_3, M_4- for multi-mode Gauss laser pulse respectively with line band 0.03cm^{-1}, 0.08cm^{-1} and 0.15cm^{-1}; M_2, M_5- for multi-mode soliton-type with line band 0.03 cm^{-1} and 0.15cm^{-1}; ϕ-experimental data (Lompre et al, 1981). Lompre et al presented the experimental data for laser pulse of the Gaussian form with line band respectively 0.03cm^{-1}, 0.08cm^{-1}, 0.15cm^{-1}. In general there is a physically reasonable agreement between theory and high-qualitative experiment.
The detailed analysis shows that the shift and width of the multi-photon resonance line for interaction of atomic system with multimode laser pulse is greater than the corresponding resonance shift and width for a case of interaction between atom and single-mode laser pulse. This is entirely corresponding to the experimental data by Lompre et al. From physical point of view it is provided by action of the photon-correlation effects and influence of the multi-modity of the laser pulse (Lompre et al, 1981; Zoller, 1982; Kleppner,et al, 1991; Glushkov-Ivanov, 1992; Glushkov, 2004, 2005, 2008).

7. Modeling a population differences dynamics of the resonant levels in a rectangular form laser pulse: Optical bistability effect

Here we consider the following tasks (i) to simulate numerically a temporal dynamics of populations' differences at the resonant levels of atoms in a large-density medium in a nonrectangular form laser pulse and (ii) to determine possibilities that features of the effect of internal optical bistability at the adiabatically slow modification of effective filed intensity appear in the sought dynamics.

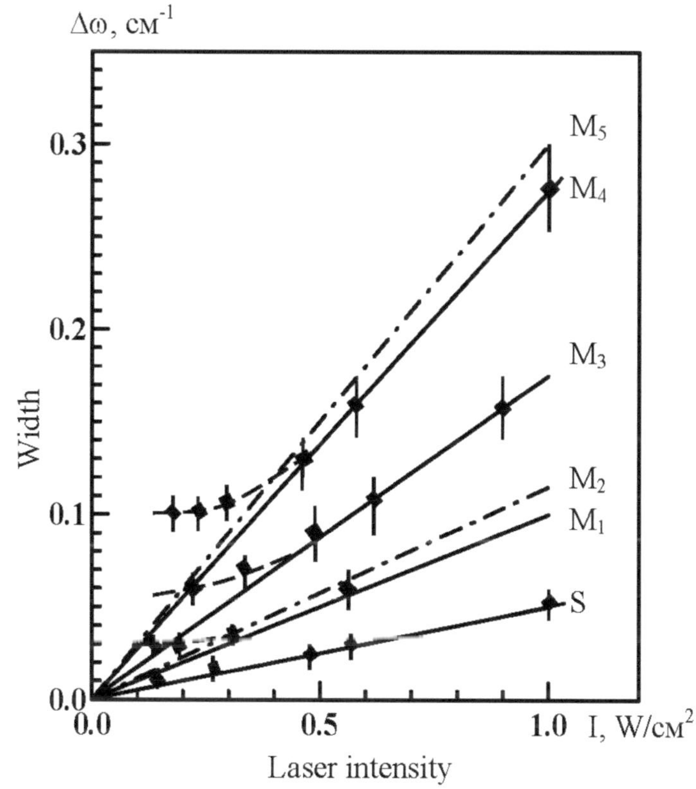

Fig. 3. The multi-photon resonance width for transition 6S-6F in the atom of Cs (wavelength 1059nm) in dependence upon the laser intensity I: theoretical data by Glushkov-Ivanov, 1992; Glushkov et al, 2008, 2009) S- for single-mode Lorentz laser pulse; M_1, M_3, M_4- for multi-mode Gauss laser pulse respectively with line band 0.03cm^{-1}, 0.08cm^{-1} and 0.15cm^{-1}; M_2, M_5- for multi-mode soliton-type with line band 0.03 cm^{-1} and 0.15cm^{-1}; ⍿-experiment (Grance, 1981; Lompre et al, 1981).

It is known that the dipole-dipole interaction of atoms in dense resonant mediums causes the internal optical bistability at the adiabatically slow modification of radiation intensity (Allen & Eberly, 1987; Scully & Zubairy, 1997; Afanas'ev & Voitikova, 2001; Ficek & Swain, 2005; Glushkov et al, 2008). The experimental discovery of bistable cooperative luminescence in some matters, in crystal of $Cs_3Y_2Br_9Yb^{3+}$ particularly, showed that an ensemble of resonant atoms with high density can manifest the effect of optical bistability in the field of strong laser emission.

The Z-shaped effect is actually caused by the first-type phase transfer. Most attractive potentialities of sought effect are associated with the development of new system for optical information processing as well as with the creation of optical digital and analog processors. The creation of optical computer with an optical radiation as the data carrier excludes the necessity in the multiple transformation of electric energy into optical one and vice-versa. This consequently leads to the energy saving and abrupt increase of computer speed. The progress in the stated areas is especially defined by the creation of optical elements for the computer facilities on basis of optical bistability phenomenon.

On basis of the modified Bloch equations, we simulate numerically a temporal dynamics of populations' differences at the resonant levels of atoms in the field of pulse with the nonrectangular $ch^{-1}t$ form. Furthermore, we compare our outcomes with the results (Afanas'ev & Voitikova, 2001; Glushkov et al, 2008), where there are considered the interaction between the ensemble of high-density atoms and the rectangularly- and sinusoidally-shaped pulses. The modified Bloch equations describe the interaction of resonance radiation with the ensemble of two-layer atoms taking into account the dipole-dipole interaction of atoms.

A fundamental aspect lies in the advanced possibility that features of the effect of internal optical bistability at the adiabatically slow modification of effective filed intensity for pulse of $ch^{-1}t$ form, in contrast to the pulses of rectangular form, appear in the temporal dynamics of populations' differences at the resonant levels of atoms.

The modified Bloch equations, which describes the interaction of resonance radiation with the ensemble of two-layer atoms subject to dipole-dipole interaction of atoms, are as follows:

$$\frac{dn}{d\tau} = \frac{i2\mu T_1}{\hbar}(E^*P - P^*E) + (1-n)$$

$$\frac{dP}{d\tau} = \frac{i2\mu T_1 n}{\hbar} - PT_1\frac{1-i(\delta+bn)}{T_2}, \tag{32}$$

where $n = N_1 - N_2$ are the populations' differences at the resonant levels, P is the amplitude of atom's resonance polarization, E is the amplitude of effective field, $b = 4\pi\mu^2 N_0 T_2/2h$ is the constant of dipole-dipole interaction, T_1 is the longitudinal relaxation time, $\delta = T_2(\omega - \omega_{21})$ is the offset of the frequency ω of effective field from the frequency of resonance transition ω_{21}, N_0 is the density of resonance atoms, μ is the dipole moment of transition, $\tau = t/T_1$. Analytical solution of the set (32) cannot be found in general case.

Therefore we carried out the numerical modeling using the program complex "Superatom" (Ivanov-Ivanova, 1981; Ivanova et al, 1985, 1986, 2001; Glushkov-Ivanov, 1992,1993; Glushkov et al, 2004, 2008, 2009). The temporal dynamics for the populations' differences at the resonant levels of atoms in a nonrectangular form pulse field:

$$E(\tau) = |E_0|^2\, ch^{-1}\frac{\pi\tau T_1}{T_2}. \tag{33}$$

was calculated.

In the numerical experiment τ varies within $0 \le \tau \le T_p/T_1$ and T_p is equal to $10T_1$. It is known (c.f. Afanas'ev & Voitikova, 2001) from general examination of set (32) that on the assumption of $b > 4$ and $b > |\delta|$ with $\delta < 0$ (the long-wavelength offset of incident light frequency is less than Lorenz frequency $\omega_L = b/T_2$) and if the intensity of light field has certain value ($I_0 = 4|E_0|^2\mu^2 T_1 T_2/h^2$) then there are three stationary states n_i (two from them with maximal and minimal value of n are at that stable). This can be considered as evidence and manifestation condition of the internal optical bistability effect in the system.

Figure 4 shows the results of our numerical modeling the temporal dynamics of populations' differences at the resonant levels of atoms for the nonrectangular form pulse (2).

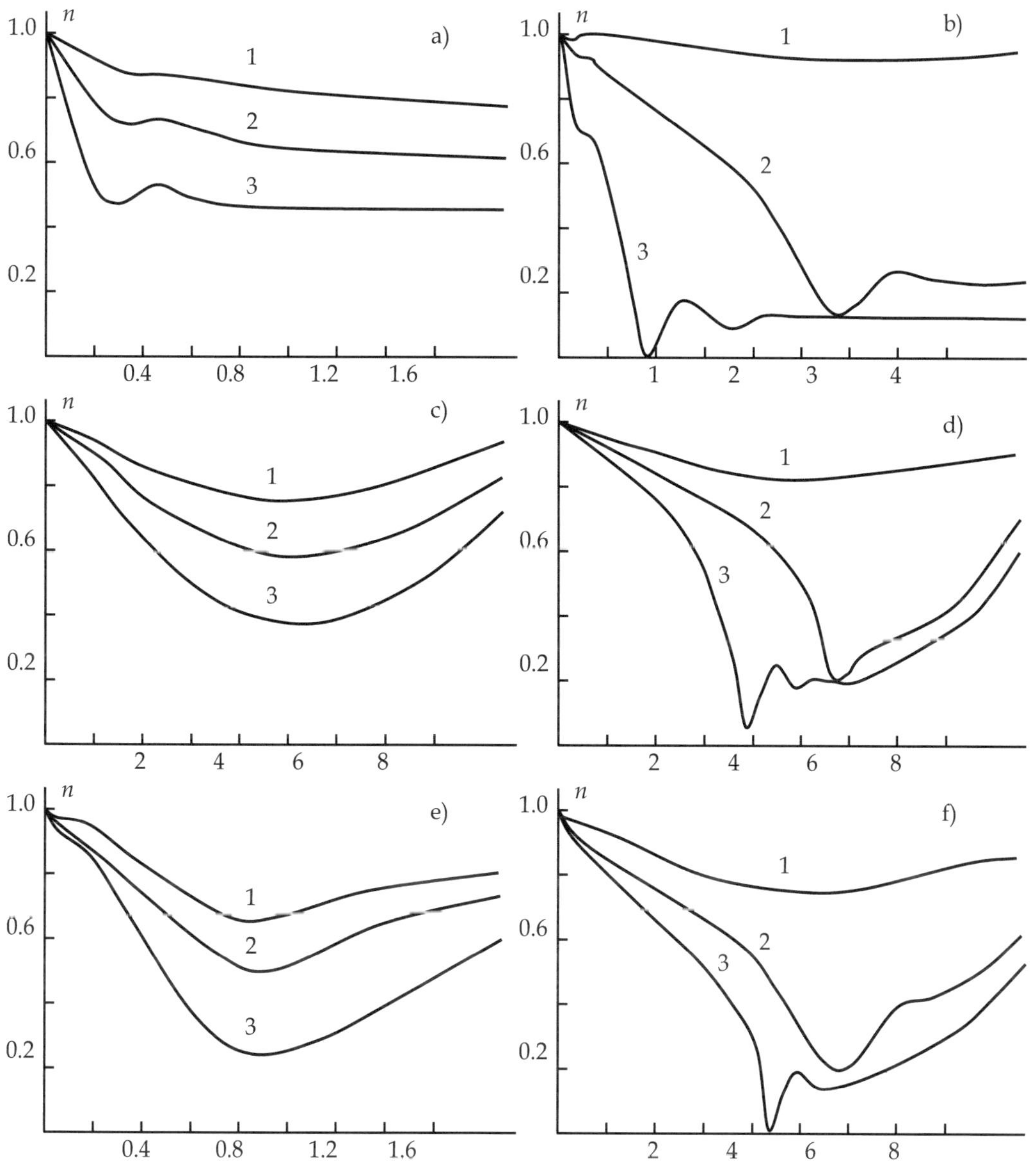

Fig. 4. Results of modeling temporal dynamics of populations' differences $n(\tau)$ at resonant levels of atoms for the pulses of rectangular (a, b) and sinusoidal (c, d) forms by method of by method of Allen & Eberly (1987) and Afanas'ev & Voitikova (2001), and for the pulse calculated by Eq. (32) with $\delta = 2$, $T_1 = 5T_2$; $b = 0$ (a, c, e); $b = 6.28$ (b, d, f); $I_0 = 2$ (1), 5 (2), and 10 (3)

For collation, Figure 4 also shows similar results but for rectangularly- and sinusoidally-shaped pulses. The increase of field intensity above certain value $I_0 = 2.5$ for selected parameters (shown in Fig. 4) leads to the abrupt increase of populations' differences. This fact represents the Z-shaped pattern of dependence $n(I)$ observed in the stationary mode. It is important to note that there is the significant difference between the model results for the pulses of various forms. For given values of rectangularly-shaped pulse intensity, which is equal to several values of T_1, the dependence $n(\tau)$ tends to stationary state with magnitude defined by zero values of right-hand terms in the set (32). For the sinusoidally-shaped pulse, the slow rise of intensity is typical, and the explicit hysteresis pattern for the dependence of populations' differences from the field intensity is obtained.

For the pulse calculated by Eq. (33), the sought effect looks more explicitly. This is especially important from the standpoint of using the sought effect to simulate efficient neural networks and their components. Substantial fact also is the implementation of hysteresis in the dependence of populations' differences from the field intensity if a threshold values for b and $\delta < 0$ have a place. This corresponds to the situation when the frequency of radiation ω is within the range, which is formed by the proper frequency ω_{21} and a frequency with the local-field correction:

$$\omega_{21}^{L} = \omega_{21} - 4\pi\mu^2 N_0/3h \tag{34}$$

Note that if above mentioned frequencies are almost equal or, e.g., a multimode electromagnetic field (chaotic light) is used, a stochastic resonance can be observed in the analyzed system.

8. Modeling Laser photoionization isotope separation technology and new principal scheme for γ -laser on quickly decayed nuclear isomers with autoionization sorting of highly excited atoms

To number of the very actual problem of modern nuclear technology, quantum and photoelectronics is related a search of the effective methods for isotopes and nuclear isomers separation and obtaining especially pure substances at atomic level (Letokhov, 1969, 1977; Letokhov et al, 1975-1977, 1979, 1980, 1985, 1986, 1992, 1990; Basov et al. 1969, 1977; Prokhorov, 1979, 2001; Janes et al, 1975; Solarz et al, 1976; Lisitsa, 1977; Singh et al, 1994; Duarte et al, 1990, 2002, 2003, 2010; Bokhan et al, 2006; etc). The basis for its successful realization is, at first, carrying out the optimal multi stepped photo-ionization schemes for different elements and, at second, availability of enough effective UV and visible range lasers with high average power (Letokhov, 1977, 1979, 1983; etc). The standard laser photo-ionization scheme may be realized with using processes of the two-step excitation and ionization of atoms by laser pulse. The scheme of selective ionization of atoms, based on the selective resonance excitation of atoms by laser radiation into states near ionization boundary and further photo-ionization of the excited states by additional laser radiation, has been at first proposed and realized by Letokhov et al (Letokhov, 1969, 1977). It represents a great interest for laser separation of isotopes and nuclear isomers. The known disadvantage of two-step laser photoionization scheme a great difference between cross-sections of resonant excitation σ_{exc} and photo-ionization σ_{ion} ($[\sigma_{exc}/\sigma_{ion}]>10^4\div10^8$). It requires using very intensive laser radiation for excited atom ionization. The same is arisen in a task of sorting the excited atoms and atoms with excited nuclei in problem of creation of γ -laser on quickly decayed nuclear isomers.

Originally, Goldansky and Letokhov (1974) have considered a possibility of creating a γ -laser, based on a recoiless transition between lower nuclear levels and shown that a γ -laser of this type in the 20-60 keV region is feasible. A feature of design is operation based on relatively short-lived isomer nuclear states with lifetime of 0,1 to 10 sec. These authors has estimated the minimal number of excited nuclei required for obtaining appreciable amplification and possibility of producing sufficient amounts of excited nuclei by irradiation of the target with a thermal neutron beam or by resonant γ -radiation. It is important that low-inertia laser selection of a relatively small friction of excited nuclei of a given composition from the target by the two-step method of selective laser photoionization of atoms with excited nuclei by the radiation from two lasers is principally possible. But, it is obvious that here there is a problem of significant disadvantage of the two-step selective ionization of atoms by laser radiation method. The situation is more simplified for autoionization resonance's in the atomic spectra, but detailed data about characteristics of these levels are often absent (Letokhov, 1977, 1983; Glushkov & Ivanov, 1986, 1992).

The key problems here are connected with difficulties of theoretical studying and calculating the autoionization resonance characteristics. Several new optimal schemes for the laser photo-ionization sensors of separating heavy isotopes and nuclear isomers are proposed (Letokhov, 1983; Glushkov et al, 2004, 2008). It is based on the selective laser excitation of the isotope atoms into excited Rydberg states and further autoionization and DC electric field ionization mechanisms. To carry out modelling the optimal scheme of the *U* and *Tm* isotopes (nuclei) sensing, the optimal laser action model and density matrices formalism were used. The similar schemes of laser photo ionization method are developed for control and cleaning the semiconductor substances (Glushkov et al, 2008). The optimal laser photo-ionization schemes for preparing the films of pure composition on example of creation of the 3-D hetero structural super lattices (layers of $Ga_{1-x}Al_xAs$ with width 10Å and GaAs of 60Å) have been proposed and new models of optimal realization of the first step excitation and further ionization of the Ga^+ ions in Rydberg states by electric field are calibrated. In this paper we give the further development of approach to construction for the optimal schemes of the laser photo-ionization isotope separation technology and to creation of new possible principal scheme of γ -laser on quickly decayed nuclear isomers with laser autoionization or electromagnetic field ionization sorting the excited atoms.

Let us remind that in a classic scheme the laser excitation of the isotopes and nuclear isomers separation is usually realized at several steps: atoms are resonantly excited by laser radiation and then it is realized photo ionization of excited atoms. In this case photo ionization process is characterized by relatively low cross section $\sigma_{ion}=10^{-17}$-10^{-18}cm^2 and one could use the powerful laser radiation on the ionization step. This is not acceptable from the energetics point of view (Letokhov, 1983; Buchanov, 2001; Stoll, 2001; Glushkov, 2005).

The alternative mechanism is a transition of atoms into Rydberg states and further ionization by electric field or electromagnetic pulse. As result, requirements to energetic of the ionized pulse are decreased at several orders. The main feature and innovation of the presented scheme is connected with using the DC electric field (laser pulse) autoionization on the last ionization step of the laser photoionization technology. There is a principal difference of the simple ionization by DC electric filed. The laser pulse ionization through the auto ionized states decay channel has the advantages (more high accuracy, the better energetics, universality) especially for heavy elements and isotopes, where the DC electric field ionization from the low excited states has not to be high effective. This idea is a key one in the realization of sorting the definite excited atoms with necessary excited nuclei of the A^+ kind, obtained by optimal method of selective photo-ionization of the A kind atoms at the first steps. The suitable objects for modelling laser photoionization separation technology are the isotopes of alkali element Cs, lanthanides and actinides.

We considered the isotopes of $_{55}Cs^{133}_{78}$ and $_{70}Yb^{171}_{101}$. For example, the resonant excitation of the Cs can be realized by means dye lasers with lamp pumping (two transitions wavelengths are: $6^2S_{1/2}\rightarrow 7\ ^2P_{3/2}$ 4555A and $6^2S_{1/2}\rightarrow 7\ ^2P_{1/2}$ 4593A). In table 3 there are listed the energy parameters for different states of the caesium, obtained in the different approximations (Derevianko & Porsev, 2003; Glushkov et al, 2008; Khetselius, 2009). It is useful to remind the corresponding hyperfine splitting energy (6 $^2S_{1/2}$, transition 4-3) of Cs: experimental data- $\Delta\nu(F,F')$= 9192,64MHz ; $\Delta E(F,F')$= 306,630· 10^{-3} cm^{-1} ; theortical data - $\Delta\nu(F,F')$= 9177,80MHz ; $\Delta E(F,F')$= 306,135· 10^{-3} cm^{-1} (Khetselius, 2009).

The next step is in the further excitation to the Rydberg S,P,D states with main quantum number n=31-37 (the optimal value n=35). Final step is the autoionization of the Rydberg excited atoms by a electromagnetic field pulse and output of the created ions. The scheme will be optimal if an atom is excited by laser radiation to state, which has the decay probability due to the autoionization bigger than the radiation decay probability. In figure 5 we present the numeric modeling results of the optimal form of laser pulse in the photoionization scheme with auto-or electric field ionization by solving the corresponding differential equations system (Glushkov et al, 2008).

State	ε^{RHF}	$\varepsilon^{RHF}+\delta\varepsilon^{RHF}$	ε^{QED}	ε^{Exp}
$6s_{1/2}$	0,12737	0,14257	0,14295	0,14310
$6p_{1/2}$	0,08562	0,09198	0,09213	0,09217
$6p_{3/2}$	0,08379	0,08951	0,08960	0,08964
$7s_{1/2}$	0,05519	0,05845	0,05862	0,05865
$7p_{1/2}$	0,04202	0,04385	0,04391	0,04393
$7p_{3/2}$	0,04137	0,04303	0,04309	0,04310

Table 3. Valent electron ionization energies (in atom. units) of the ^{133}Cs: ε^{RHF} –one-configuration Hartree-Fock data, relativistic Hartree-Fock (RHF); $\varepsilon^{RHF}+\delta\varepsilon^{RHF}$ – the same data, but with account for the correlation corrections (Derevianko & Porsev, 2005; ε^{QED} – QED perturbation theory data (Khetselius, 2009); ε^{Exp}- experimental data (see text)

The following definitions are used: δ+dashed line is corresponding to optimal form of laser pulse, curves 1 and 2 are corresponding to populations of the ground and excited states of Cs. The δ -pulse provides maximum possible level of excitation (the excitation degree is about ~0,25; in experiment (Letokhov, 1983) with rectangular pulse this degree was ~ 0,1). It is in great degree similar to analogous scheme with the DC electric field and stochastic collisional ionization mechanisms.

In fig.5 there is also presented the typical behaviour of the ground (curve 1) and highly excited (curve 2) states population. Let us remember data regarding the excitation and the ionization cross sections for studied system: the excitation cross section at the first step of the scheme is ~$10^{-11}cm^2$; the ionization cross-section from excited 7^2P_2 state: $\sigma_2=10^{-16}cm^2$, from ground state $\sigma_2=10^{-18}cm^2$ (Letokhov, 1983). One can see that the relation of these cross sections is 10^5 and 10^7 correspondingly. This fact provides the obvious non-efficiency of standard photoionization scheme.

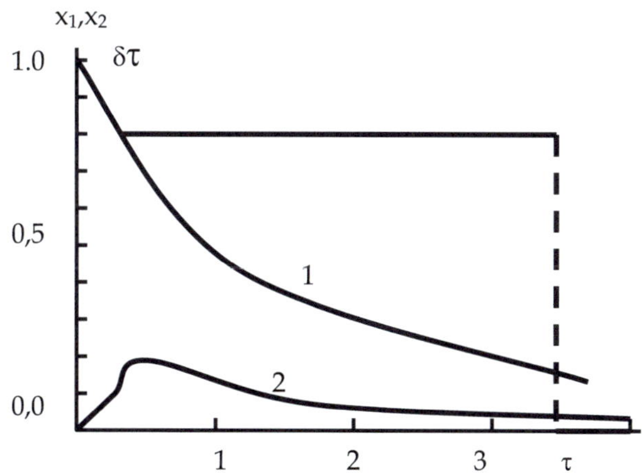

Fig. 5. Results of modelling Cs isotopes separation process by the laser photo-ionization method (δ+dashed - laser pulse optimal form; see text)

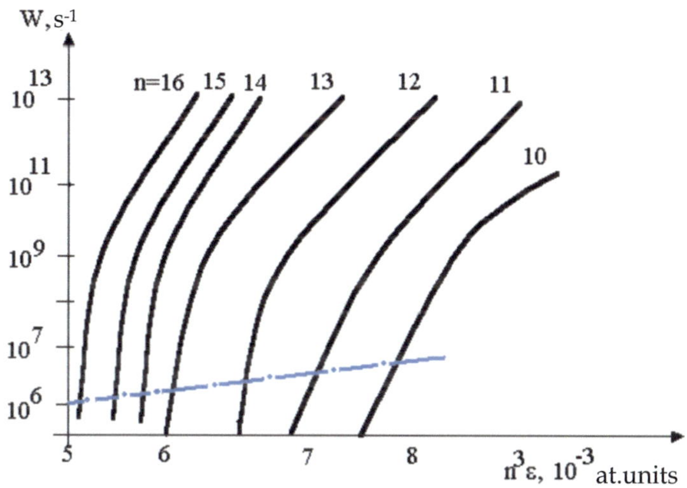

Fig. 6. Dependence of the velocity of ionization for high excited atoms of Cs upon the electric field strength for states with quantum numbers n=10-16, m=0,n_2=n-1.

In figure 6 we present the results of our calculating dependence of the ionization velocity for high excited atoms of Cs upon the electric field strength for states with quantum numbers n= 10-16,m=0,n2=n-1. The dashed line is corresponding to velocity of the radiative decay. The decay of Cs atoms and ions in the high-excited state demonstrates the properties of the H-like systems at the qualitative level. But, there is quite significant quantitative difference. We have found that the ionization velocity for states with n>14 is more than the radiative decay velocity in electric field with strength E less than 15 kV/cm. Our estimate for the Ga atom ionization cross section is $1{,}5 \cdot 10^{-13}$ cm^2 that is higher than the corresponding cross section of ionization process by laser pulse in the two- stepped photo ionization (Letokhov, 1977) scheme ($\sim 10^{-17} cm^2$).

Using δ-pulse provides a quick ionization, but the ionization yield will be less than 100% because of the sticking on intermediate levels. Experimentally obtained dependence of the critical ionization field strength E upon the effective quantum number n* is usually approximated by simple theoretical dependence $E_{cr}=(2n^*)^{-4}$.

Using the autoionization mechanisms at the final step for ionization of the Rydberg excited atoms provides more optimal scheme from energetic point of view. For example, for the $35^2S_{1/2}$ transition the corresponding cross section can reach the value ~10^{-13}cm^2. So, from energetic point of view, this type of ionization can be very perspective alternative to earlier proposed classical two-step and more complicated photoionization schemes (Letokhov, 1983). More suitable situation takes a place for the for Yb isotope separation.

It is very important that the proposed scheme can be easily implemented to the possible advanced scheme of the γ - laser on quickly decayed nuclear isomers with using laser photoionization sorting excited nuclei M^*_{k+1} with autoionization mechanism through the Rydberg states.

Figure 7 easily explains the principal moments of this scheme. It generalizes the known Goldansky-Letokhov (Goldansky & Letokhov, 1974) and other (Baldwin et al, 1981; Glushkov, 2005) schemes and has to be more efficient especially from energetics point of view. In this context it is worth to remind very impressive results of the last years, connected with

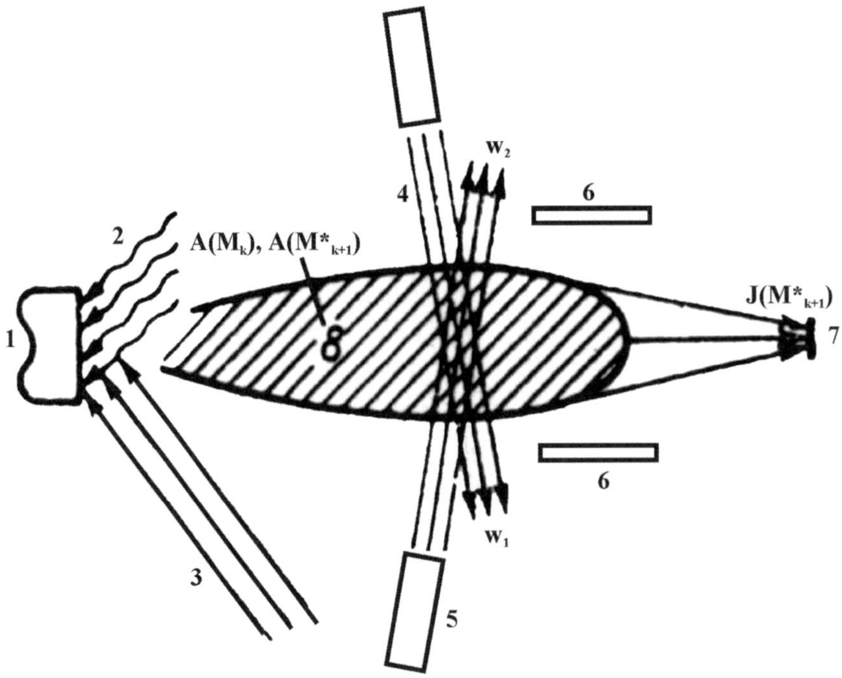

Fig. 7. Possible scheme γ - laser on quickly decayed nuclear isomers with using laser photoionization sorting excited nuclei M^*_{k+1} with electric field and auto- and electric field ionization mechanisms: 1 – target of atoms M_k; 2- flux of slow neutrons; 3 – laser ray for evaporation of target; 4 – laser ray for the first step excitation of atoms with excited nucleus $A(M^*_{k+1})$; 5 – laser ray for second-step excitation to highly excited atomic states and Rydberg autoionization by electromagnetic field; 6 – collector system; 7 - atoms with excited nucleus $A(M^*_{k+1})$; 8 – flux of evaporated atoms;

engineering atomic highly excited Rydberg states and correspondingly cooperative laser-gamma-muon-electron- nuclear states (transitions) with the laser (and raser) pulses. It is quite possible that cited new effects can be realized in the tasks considered here.
The laser photo ionization scheme with autoionization of the highly excited atoms (with optimal set of energetic and radiative parameters: pulse form, duration, energetic for laser and electric field pulses etc.) could provide significantly more high yield and effectiveness of the whole process of the isotope separation. It is especially worth for implementation to the possible principal scheme of γ -laser on quickly decayed nuclear isomers with autoionization sorting the excited atoms.

9. Conclusions

We presented a new consistent method for studying the interaction of atom with a realistic laser field, based on the quantum electrodynamics (QED) and S-matrix adiabatic formalism Gell-Mann and Low. In relativistic case the Gell-Mann and Low formula expressed an energy shift δE through QED scattering matrix including the interaction with as the laser field as the photon vacuum field. It is natural to describe the laser field-atom interaction by means of the radiation emission and absorption lines. Their position and shape fully determine the spectroscopy of atom in the field. The radiation atomic lines can be described by moments of different orders μ_n. The main contribution into μ_n is given by the resonant range. The values μ_n can be expanded into perturbation theory (PT) series. As example, the method has been used for numerical calculation of the three-photon resonant, four-photon ionization profile of atomic hydrogen (1s-2p transition; wavelength =365 nm) and multi-photon resonance width and shift for transition 6S-6F in the atom of Cs (wavelength 1059nm) in a laser pulse of the Gaussian and soliton-like forms.
The results of numeric calculation of population kinetics of resonant levels for atoms in the non-rectangular form laser pulse on the basis of the modified Bloch equations are presented. Cited equations describe an interaction between two-level atoms ensemble and resonant radiation with an account of the atomic dipole-dipole interaction. It has been found for a case of $ch^{-1}t$ laser pulse a strengthen possibility of manifestation of the internal optical bi-stability effect special features in the temporary dynamics of populations for the atomic resonant levels under adiabatic slow changing the acting field intensity in comparison with a case of the rectangular form pulses.
At last, we shortly presented an optimal scheme of the laser photo-ionization heavy isotopes (isomers) separation technology and the new possible principal scheme of γ -laser on quickly decayed nuclear isomers with autoionization sorting the highly excited heavy atoms.

10. Acknowledgements

The authors are very much thankful to Professors E.Ivanova, L.Ivanov, V.Letokhov, W. Kohn, E. Brandas, S.Wilson, I.Kaplan, J.Maruani, A.Theophilou for helpful discussion and comments. One of the authors (A.G.) acknowledges support of Russian Academy of Sciences (Troitsk-Moscow, Russia), Universities of Dresden and Frieburg (Germany), Universities of Geneva and Zurich (Switzerland) grants. One of authors (O.K.) acknowledges the support of the Abdus Salam International Centre for Theoretical Physics (Trieste, Italy). Authors would like to thank Mrs. J. Marusic for attention and help in preparing the manuscript too.

11. References

Afanas'ev, A. & Voitikova, M.V. Effect of the local field on transient processes in a dense ensemble of two-level atoms. *Opt. Spectr.*, Vol.90,N6 (2001) p.799-802.

Allen L.,& Eberly J. H. (1987) *Optical Resonance and Two-Level Atoms*. pp.1-256, Dover Publications, New York.

Aumar, F. & Winter, H. (1997) *Photonic, Electronic and Atomic Collisions,* (Eds), pp.1-700, World Scientific, Singapore

Baldwin, G.G., Salem, J.C., Goldansky, V.I., Approaches to development of gamma ray lasers. *Rev.Mod.Phys.*, Vol.53,N4 (1981) p.687-742.

Basov, N.G. & Letokhov, V.S., Optical frequency standards. *Sov.Phys.-Uspekhi,* Vol.11,N12 (1969) p.855-880.

Basov, N.G., Belenov, E., Isakov, V.A., Markin, E.A., New methods of isotope separation. *Sov.Phys.-Uspekhi,* Vol.121,N3 (1977) p.427-445.

Batani, D. & Joachain C.J. (2006*). Matter in super-intense laser fields*, pp.1-650, AIP Publ., New York.

Becker, A., Faisal, F.H.M. S-matrix analysis of coincident measurement of two-electron energy distribution for double ionization of He in an intense laser field. *Phys.Rev.Lett.*, Vol.89,N18 (2002) p.193003-1-193003-4.

Bjorkholm, J.E., Liao, P.F., AC Stark splitting of the two-photon spectra. *Opt. Communication,* Vol.21,N1 (1977) p.P.132-136.

Bokhan, P.A., Buchanov, V.V., Fateev, N.V., Kalugin, M.M., Kazaryan, M.A., Prokhorov, A.M., Zakrevskii, D.E. (2006) *Laser Isotope Separation in Atomic Vapor*. Wiley-VCH, Berlin.

Brändas, E. & Floelich, P., Pade approximants to Stark effect energies for excited states of hydrogen. Phys. Rev.A. Vol.16,N6 (1977) p.2207-2216.

Buchanov, V.V., Kazaryan, M., Kalugin, M., Prokhorov, A.M. (2001) Laser separation of Silicon Isotopes by the AVLIS Technology , In: *Atomic and Molecular Pulsed Lasers,* p.72, SO RAN, Tomsk.

Burvenich, T.J.,Evers. J., Keitel, C.H., Dynamic nuclear Stark shift in superintense laser fields. *Phys.Rev.C,* Vol.74,N6 (2006) p.044601 (12p.).

Delone, N.B. & Kraynov, V.P. (1984) Atom in a strong laser field, pp.1-280, Energoatomizdat, Moscow.

Delone, N.B. & Kraynov, V.P. Stabilization of atom in a field of laser radiation. Phys.Uspekhi., Vol.165,N11 (1995) p.1295-1321.

Delone, N.B. & Kraynov, V.P. Dynamical Stark shift of atomic levels. Phys.Uspekhi., Vol.169,N7 (1999) p.753-772.

Derevianko, A. & Porsev, S.G. Dressing lines and vertices in calculations of the matrix elements with the coupled-cluster method and determination of Cs atomic properties. *Phys.Rev. A.*, Vol.71, N8 (2005) p..032509-1-10

Duarte, F. J. & Hillman L.W. (eds.) (1990) *Dye Laser Principles* , ch.9, Academic, New York.

Duarte, F. J. , Olivares, I.E., Duarte, A.E., Saravia, E.A., Lithium isotope separation with tunable diode lasers, *Appl. Optics,* Vol.41 (2002) p.2973-2977.

Duarte, F. J. (2003) *Tunable Laser Optics,* pp.1-320, Elsevier Academic, New York.

Duarte, F. J., Tunable lasers for atomic vapor laser isotope separation: the Australian contribution. *Australian Physics,* Vol.47,N2 (2010) p.38-40.

Dunning, F.B., Mestayer, J.J., Reinhold, C.O., Yoshida, S., Burgdorfer, J., Engineering atomic Rydberg states with pulsed electric fields. *J. Phys. B: At. Mol. Opt. Phys.,* Vol.42 (2009) p.022001 (8p.).

Fedorov, M.V., Stabilization of atom in a strong laser field. *Physics Uspekhi.,* Vol.169,N1 (1999) p.66−71.

Ficek, Z. & Swain, S. (2005) *Quantum Interference and Coherence,* pp.1-400, Springer, Berlin.

Friedberg, R., Hartmann, S. R., Manassah, J. T., Frequency shifts in emission and absorption by resonant systems of two-level atoms. *Phys. Rep.*, Vol.37,N1 (2003) p.101-179.

Galagher, F.F., Hamprey, L.M., Field ionization of highly excited states of sodium. *Phys.Rev.A.*, Vol.16,N3 (1977) p.1098-1100.

Georges, A.T., Dixit, S.N., Laser line shape effects in the resonance fluorescence. *Phys.Rev.A.*, Vol.23,N5 (1981) p.2580-2593.

Gibbs, H. M. (1985) *Optical Bistability: Controlling Light with Light.* pp.1-471, Academic Press, New York.

Glushkov, A.V., Ivanov, L.N., Ivanova, E.P. (1986) Relativistic decay of excited atomic states. Generalized energy approach. In: *Autoionization Phenomena in Atoms*, pp.152-164, Moscow University Press, Moscow.

Glushkov, A.V., Ivanov, L.N., Letokhov, V.S. (1992) Nuclear quantum optics. *Preprint of Institute for Spectroscopy of the USSR Academy of Sciences* (ISAN), AS N4, 1-16, ISAN, Moscow-Troitsk.

Glushkov, A.V. Negative ions of inert gases. *JETP Lett.*, Vol.55, N2 (1992) p.97-100.

Glushkov, A.V. & Ivanov, L.N. Radiation Decay of Atomic States: atomic residue and gauge non-invariant contributions. *Phys. Lett .A.*, Vol.170,N1 (1992) p.33-37.

Glushkov, A.V. & Ivanov, L.N. (1992) Shift and deformation of radiation atomic lines in the laser emission field. Multiphoton processes. *Preprint of Institute for Spectroscopy of the USSR Academy of Sciences* (ISAN), AS N3, 1-12, ISAN, Moscow-Troitsk.

Glushkov, A.V. & Ivanov, L.N. (1992) DC Strong Field Stark effect: consistent quantum mechanical approach *Preprint of Institute for Spectroscopy of the USSR Academy of Sciences* (ISAN), AS N1, 1-16, ISAN, Moscow-Troitsk. Glushkov, A.V. & Ivanov, L.N. (1992) A broadening of the thulium atom autoionization resonances in a weak electric field *Preprint of Institute for Spectroscopy of the USSR Academy of Sciences* (ISAN), AS N3, 1-10, ISAN, Moscow-Troitsk.

Glushkov, A.V. & Ivanov, L.N. DC Strong-Field Stark-Effect: consistent quantum-mechanical approach. *J.Phys. B:At.Mol.Opt. Phys.*, Vol.26, N16 (1993) p.L379-396.

Glushkov, A.V. & Malinovskaya, S.V. (2003). Co-operative laser nuclear processes: border lines effects. In: *New projects and new lines of research in nuclear physics.* G.Fazio & F.Hanappe (Eds) pp.242-252, World Scientific, Singapore.

Glushkov, A.V., Ambrosov, S.V., Ignatenko, A.V., Korchevsky, D.A. DC Strong Field Stark Effect for Non-hydrogenic Atoms: New Consistent Quantum Mechanical Approach. *Int.Journ.Quant.Chem.* Vol.99,N5 (2004) p.936-939.

Glushkov, A.V.,Malinovskaya, S.V., Chernyakova, Yu.G., Svinarenko, A.A. Cooperative Laser-Electron-Nuclear Processes: QED Calculation of Electron Satellites Spectra for Multi-Charged Ion in Laser Field. *Int.Journ.Quant.Chem.* Vol.99,N5 (2004) p.889-893.

Glushkov, A.V. (2005) Atom in Electromagnetic Field, pp.1-450, KNT-Nauka, Moscow-Kiev.

Glushkov, A.V., Ambrosov, S.V., Loboda, A.V., Gurnitskaya, E.P., Prepelitsa, G.P. Consistent QED approach to calculation of electron-collision excitation cross-sections and strengths: Ne-like ions. *Int. Journ.Quant.Chem.* Vol.104, N4 (2005) p.562-569.

Glushkov, A.V., Malinovskaya, S.V., Loboda, A.V., Gurnitskaya, E.P., Korchevsky, D.A. Diagnostics of the collisionally pumped plasma and search of the optimal plasma parameters of x-ray lasing: Calculation of electron-collision strengths and rate coefficients for Ne-like plasma. *J.Phys.CS.*, Vol.178,N1 (2005) p.188-198.

Glushkov, A.V. (2008) Relativistic Quantum theory. Relativistic quantum mechanics of atomic systems, pp.1-900, Astroprint, Odessa.

Glushkov, A.V., Khetselius, O.Yu., Malinovskaya, S.V. Optics and spectroscopy of cooperative laser-electron nuclear processes in atomic and molecular systems - New trend in quantum optics. *Europ.Phys.Journ.*, Vol.T160,N1 (2008) p.195-204.

Glushkov, A.V., Khetselius, O.Yu., Loboda, A.V., Svinarenko, A.A. (2008) QED approach to atoms in a laser field: Multi-photon resonances and above threshold ionization. In: *Frontiers in Quantum Systems in Chemistry and Physics* Series: Progress in Theoretical Chemistry and Physics, Wilson, S.; Grout, P.J.; Maruani, J.; Delgado-Barrio, G.; Piecuch, P. (Eds.), Vol.18, pp.541-588, Springer, Berlin.

Glushkov, A.V., Khetselius, O.Yu., Malinovskaya, S.V. (2008) New laser-electron nuclear effects in the nuclear γ transition spectra in atomic and molecular systems In: *Frontiers in Quantum Systems in Chemistry and Physics* Series: Progress in Theoretical Chemistry and Physics, Wilson, S.; Grout, P.J.; Maruani, J.; Delgado-Barrio, G.; Piecuch, P. (Eds.), Vol.18, pp.523-540, Springer, Berlin.

Glushkov, A.V., Khetselius, O.Yu., Gurnitskaya, E.P., Loboda, A.V., Florko, T.A., Sukharev, D.E., Lovett, L., Gauge-invariant QED perturbation theory approach to calculating nuclear electric quadrupole moments, hyperfine structure constants for heavy atoms and ions In: *Frontiers in Quantum Systems in Chemistry and Physics* Series: Progress in Theoretical Chemistry and Physics, Wilson, S.; Grout, P.J.; Maruani, J.; Delgado-Barrio, G.; Piecuch, P. (Eds.), Vol.18, pp.501-522, Springer, Berlin.

Glushkov, A.V., Loboda, A.V., Gurnitskaya, E.P., Svinarenko, A.A., QED theory of radiation emission and absorption lines for atoms in a Strong Laser Field. *Physica Scripta.*, Vol.T.135,N1 (2009) p.014022 (6p.).

Glushkov, A.V., Khetselius, O.Yu., Lovett, L. (2009) Electron-β-Nuclear Spectroscopy of Atoms and Molecules and Chemical Environment Effect on the β-Decay parameters. In: *Advances in the Theory of Atomic and Molecular Systems Dynamics, Spectroscopy, Clusters, and Nanostructures.* Series: Progress in Theoretical Chemistry and Physics, Piecuch, P., Maruani, J., Delgado-Barrio, G., Wilson, S. (Eds), Vol.20, pp.125-172, Springer, Berlin.

Goldansky, V.I. & Letokhov, V.S. Effect of laser radiation on nuclear decay processes. *Sov. Phys. JETP.*, Vol.67,N6 (1974) p.513-516.

Grance, M., Multiphoton ionization near threshold: Effects of non-linearity. *J. Phys.B.: Atom. Mol.Opt. Phys.*, Vol.14,N22 (1981) p.4301-4306.

Hehenberger, M., McIntosh, H.V., Brändas, E. Weil's theory applied to the Stark effect in the hydrogen atom. *Phys.Rev.* Vol.10,N5 (1974) p.1494-1506.

Holt, C.R., Raymer, M.C., Reinhardt, W.P., Time dependence of the two-, three- and four photon ionization of atomic hydrogen. *Phys.Rev.A.*, Vol.27,N5 (1983) p.2971-2988.

Ivanov, L.N. & Ivanova E.P. Extrapolation of atomic ion energies by model potential method: Na-like spectra. *Atom.Data Nucl .Data Tabl.*, Vol.24,N2 (1979) p.95-121.

Ivanov, L.N. & Letokhov, V.S. Spectroscopy of autoionization resonances in heavy elements atoms. *Com.Mod.Phys.D.:At.Mol.Phys.*, Vol.4 (1985) p.169-184.

Ivanov, L.N., Ivanova, E.P., Knight, L. Energy Approach to consistent QED theory for calculation of electron-collision strengths. *Phys.Rev.A.*, Vol.48, N9 (1993) p.4365-4374.

Ivanova, E.P., Ivanov, L.N., Glushkov, A.V., Kramida, A.E., High-order corrections in Relativistic Perturbation Theory with model zero Approximation. *Physica Scripta.*, Vol.32,N4 (1985) p.512-524.

Ivanova, E.P. & Glushkov, A.V. Theoretical Study of Multicharged Ions Spectra of F-, Ne-Isoelectronic Sequences. *J. Quant. Spectr. Rad.Transfer.*, Vol.36, N2 (1986) p.127-145.

Ivanova, E.P., Ivanov, L.N., Aglitsky, E.V., Modern Trends in Spectroscopy of multi-charged Ions. *Physics Rep.*, Vol.166,N6 (1988) p.315-390.

Ivanova, E.P., Zinoviev, N.A. The possibility of X-ray lasers based on inner-shell transitions of Ne-like ions. *Phys. Lett.A.* ,Vol.274,N2 (2001) p.239-246.

Janes, G.S., Forsen, H.K., Levy, R.H. (1975) *The Exxon/Avco Program in Laser Enrichment: Uranium Enrichment*; Research Report, N422, p.-1-274, Avco Everett Res. Lab.

Janjusevic, M., Mittleman, M., Ionization of atomic hydrogen in an ultraintense laser field. *J.Phys.B.: Atom. Mol.Opt. Phys.*, Vol.21,N12 (1988) p.2279-2294.

Jones, R.R., Buckbaut, P.H., Ionization suppression of Stark states in intense laser filed. *Phys.Rev.Lett.*, Vol.67,N23 (1991) p.3215-3218.

Kelleher, D.E., Ligare, M., Brewer, L.R. Resonant four photon ionization of atomic hydrogen. *Phys.Rev.A.*, Vol.31,N4 (1985) p.2747-2751.

Khetselius, O.Yu., Relativistic Calculating the Hyperfine Structure Parameters for Heavy-Elements and Laser Detecting the Isotopes and Nuclear Reaction Products. *Physica Scripta.*, Vol.T.135,N1 (2008) p.014023 (8p.).

Khetselius, O.Yu., Relativistic perturbation theory of the hyperfine structure parameters for some heavy-element isotopes. *Int. Journ.Quant.Chem.* Vol.109,N14 (2009) p.3330-3335.

Kleppner D., Chun-Ho I., Welch G.R. (1990) In: *Irregular Atomic Systems and Quantum Chaos*, Gay J.C. (Ed.), pp.21-48 Kluwer, New York.

Liao, P.F., Bjorkholm, J.E., Direct observation of atomic levels shifts in two-photon absorption. *Phys.Rev.Lett.*, Vol. 34,N1 (1975) p.1-4.

Letokhov, V.S. (1969) On possibility of isotopes separation by methods of resonant photoionization of atoms and photodissociation of molecules by laser radiation. *Preprint of Physical Institute of the USSR Academy of Sciences* (FIAN), N1-69, p.1-72, FIAN, Moscow.

Letokhov, V.S. , Use of lasers to control selective chemical reactions, *Science*, Vol.180, N4085 (1973) p.451-458.

Letokhov, V.S. & Ivanov, L.N., Selective ionization of atoms in electric and laser fields. *Quantum Electronics* (USSR), Vol.2,N3 (1975) p.585-590.

Letokhov, V.S., Minogin, V.G., Pavlik, B.D., Cooling and trapping of atoms and molecules by resonant laser-field. *Optics Comm.* Vol.19 (1976) p.72-76.

Letokhov, V.S. (1977). *Laser Spectroscopy}*, pp.1-380, Academic Press, New York.

Letokhov, V.S. & Moore, C.B. (1977) *Chemical and Biochemical Applications of Lasers.* pp.1-670, New York, Acad.Press.

Letokhov, V.S., Laser isotope separation. *Nature*, N277, (22 February 1979) p.605-610.

Letokhov, V.S. (1979) Using lasers in atomic and nuclear physics. In: *Apllication of Lasers in Atomic, Molecular and Nuclear Physics*, Prokhorov A.M. & Letokhov V.S. (Eds.), pp.412-445, Nauka, Moscow.

Letokhov, V.S., Balykin, V.I., Mishin, V.I., Bekov, G.I., Laser detection of single atoms. *Sov.Phys.-Uspekhi,* Vol.132,N10 (1980) p.79-85.

Letokhov, V.S. (1983) *Nonlinear Selective Photo-processes in atoms and molecules*, pp.1-408, Nauka, Moscow.

Letokhov, V.S., Minogin V.G., Basises of laser spectroscopy. *Sov.Phys.-Uspekhi,*Vol.147,N9 (1985) p.79-85.

Letokhov, V.S., Laser induced processes in spectroscopy, isotope separation and photochemistry. *Sov.Phys.-Uspekhi* Vol.29 (1986) p.70-81.

Letokhov, V.S., Balykin V.I., The laser optics of neutral atomic beams. *Sov.Phys.-Uspekhi,*Vol.33 (1990) p.79-85.

Letokhov, V.S., Ivanov, L.N., Glushkov, A.V. (1992) Laser separation of heavy lanthanides and actinides isotopes: Autoionization resonances and decay in electric field.

Preprint of Institute for Spectroscopy of the USSR Academy of Sciences (ISAN), AS N5, p.1-18, ISAN, Moscow-Troitsk.

Letokhov, V.S., Principles of nonlinear optical spectroscopy. *Sov.Phys.-Uspekhi,*Vol.41 (1998) p.523-533.

Letokhov, V.S., Problems of nanooptics. *Sov.Phys.-Uspekhi,*Vol.42 (1999) p.241-282.

Letokhov, V.S. & Balykin, V.I., Information cooling of neutral atoms. *Phys. Rev. A.*, Vol.64 (2001) p.063410.

Lisitsa, V.S., New in the Stark and Zeemane effects for hydrogen atom. *Sov.Phys.-Uspekhi,* Vol.153,N3 (1977) p.379-422.

Lompre, L-A., Mainfray, G., Manus, C., Marinier, J.P., Laser Light statistics and band-width effects in the resonant multi-photon ionization of the caesium atoms at 1.059 μm. *J.Phys.B: At. Mol. Opt. Phys.*, Vol.14,N12 (1981) p.4307-4326.

Luc-Koenig, E., Lyras, A., Lecomte, J.-M., Aymar, M. The eigenchannel R-matrix study of two-photon processes including the above-threshold ionization in magnesium. *J.Phys. B: At. Mol. Opt. Phys.*, Vol.30, N12 (1997) p.5213-5232.

Maquet, A., Chu, S-I., Reinhardt, V.P., Stark ionization in dc and ac electric field: an L^2 complex coordinate approach. *Phys.Rev.A.*, Vol.26,N6 (1983) p.2246-2261.

Müller, C., Di Piazza, A., Shahbaz, A., Bürvenich, T.J., Evers, J., Hatsagortsyan, K.Z., Keitel, C.H. High-Energy, Nuclear, and QED Processes in Strong Laser Fields. *Laser Physics,* Vol.18,N 3 (2008) p. 175–184.

Popov, V.S., Tunnel and multiphoton ionization of atoms and ions in an intense laser field (Keldysh theory). Phys.Uspekhi., Vol.174,N9 (2004) p.921-951.

Prokhorov, A.M., Karlov, N.V., Krynetsky, B.B., Mishin, V.A., Selective photoionization of atoms and its applications for isotope separation and spectroscopy. *Sov.Phys.-Uspekhi,* Vol.137,N4 (1979) p.593-620.

Prokhorov, A.M., Buchanov, V.V., Kazaryan, M.A., Kalugin, M.M., (2001) Laser separation of silicon isotopes by the AVLIS technology. In: *Atomic and Molecular Pulsed Lasers.* p.72, So RAN Pub., Tomsk.

Sauter, Th., Neuhauser, W., Blatt, R., Toschek, P.E., Quantum jumps observed in the fluorescence of a single atom. *Opt.Commun.*, Vol.60,N5 (1986) p.287-292.

Scully, M.O. & Zubairy, M. S. (1997) *Quantum Optics,* pp.1-490, Cambridge University Press, Cambridge.

Shahbaz, A., Muller, C., Staudt, A., Burvenich, T.J., Keitel, C.H., Nuclear quantum optics with X-ray laser pulses. *Phys.Rev.Lett.*, Vol.98,N8 (2006) p.263901 (6p.).

Silverstone, H., Adams, B., Cizek, J., Otto, P. Stark effect in hydrogen: Dispersion relations, asymptotic formulas and calculation of the ionization rate via high-order perturbation theory. *Phys.Rev. Lett.* Vol.43,N20 (1979) p.1498-1501.

Singh, S., Dasgupta, K., Kumar, S., Manohar, K.G., Nair L.G., Chatterjee, U.K., High-power high-repetition-rate capper-vapor-pumped dye laser. *Opt. Eng.*, Vol.33 (1994) p.1894-1904.

Solarz, R.W., May, C.A., Carlson, L.R., Worden, E.F., Johnson, S.A., Paisner, J.A., Radziemski, L.J., Detection of Rydberg states in atomic uranium using time-resolved stepwise laser photoionization. *Phys.Rev.A.*, Vol.14, N3 (1976) p.1129-1136.

Stoll, W. (2001) Present Status of industrial Isotope separation by laser technology. In: *Atomic and Molecular Pulsed Lasers,* p.71, SO RAN, Tomsk.

Ullrich, C.A., Erhard, S., Gross, E.K.U., (1986) Density Functional Approach to Atoms in Strong Laser Pulses. In: *Superintense Laser Atoms Physics*. Ullrich, C.A. & Gross, E.K.U. (Eds.) pp.1-48, Kluwer, New York.

Zoller, P. Stark Shifts and resonant multi-photon ionization in multimode laser fields. *J.Phys.B: At.Mol.Opt.Phys.,* Vol.15,N8 (1982) p.2911-2933.

Part 2

Ultrashort Laser Pulse Emission and Applications

9

Second Harmonic Generation under Strong Influence of Dispersion and Cubic Nonlinearity Effects

Sergey Mironov[1], Vladimir Lozhkarev[1], Vladislav Ginzburg[1], Ivan Yakovlev[1], Grigory Luchinin[1], Efim Khazanov[1], Alexander Sergeev[1], and Gerard Mourou[2]

[1]Institute of Applied Physics, Nizhny Novgorod,
[2]Institut de la Lumière extrème (ILE), Palaiseau,
[1]Russia
[2]France

1. Introduction

High intensity contrast ratio (ICR) is required for majority of experiments with strong-field radiation. The demands to temporal intensity profile are determined by task of application of super strong radiation. As a rule, the experiments require ICR as higher as possible. SHG process of powerful femtosecond radiation is known to be an excellent approach to significant enhancement of a temporal intensity profile. The peak power of output radiation of modern laser complexes exceeds the petawatt level (Aoyama, Yamakawa et al. 2003; Lozhkarev, Freidman et al. 2007; Liang, Leng et al. 2007 ; Yanovsky, Chvykov et al. 2008) and power density in the radiation is in a range of 10TW/cm^2. At such high intensities cubic polarization of frequency doubling media is needed to be taken into consideration. The polarization leads to nonlinear phase accumulation of interacted waves (Razumikhina, Telegin et al. 1984; Choe, Banerjee et al. 1991; Chien, Korn et al. 1995; Ditmire, Rubenchik et al. 1996; Mironov, Lozhkarev et al. 2009) and small-scale self-focusing (SSSF) appearance (Bespalov and Talanov 1966; Rozanov and Smirnov 1980; Lowdermilk and Milam 1981; Kochetkova, Martyanov et al. 2009; Poteomkin, Martyanov et al. 2009). The present research is devoted to demonstration of a possibility of SHG process implementation for ICR enhancement.

In the chapter, we present theoretical and experimental results of SHG of output radiation of petawatt level femtosecond laser (Lozhkarev, Freidman et al. 2007). The influence of cubic polarization and dispersion effects are taken into account in the theoretical model. The coincidence of experimental results and the observed theoretical model are thoroughly discussed in section 2. Numerical simulation of SHG process has made it possible to demonstrate the possibility of ICR enhancement. We suggested a technique of additional pulse compression after SHG process. Results of the theoretical explorations are presented in section 3.

The model of linear stage of plane wave instability in media with quadratic and cubic nonlinearity is developed. The model is the generalization of the classical theory of beam filamentation (Bespalov and Talanov 1966). Linear equations and methodic of their implementation for estimations of critical level of spatial noise in powerful laser beams are presented in section 4. We propose an original approach to small-scale self-focusing (SSSF) suppression in experiments with super intense radiation. The created theoretical model of the phenomena clearly explains the suppression, and its coincidence with an obtained experimental data is discussed in section 5. We present results of mathematical modeling of influence of surface dusts to generation of spatial noise and its further amplification during the SHG of super intense radiation.

2. Theoretical aspects and experimental results of SHG of super strong femtosecond radiation

SHG process of super intense femtosecond radiation in the frame of the second order of dispersion theory is described by the system of differential equations:

$$\begin{aligned}
&\frac{\partial A_1}{\partial z}+\frac{1}{u_1}\frac{\partial A_1}{\partial t}-\frac{ik_2^{(1)}}{2}\frac{\partial^2 A_1}{\partial t^2}=-i\beta\cdot A_1A_2^{*}e^{-i\Delta kz}-i\gamma_{11}\left|A_1\right|^2\cdot A_1-i\gamma_{12}\left|A_2\right|^2\cdot A_1\\
&\frac{\partial A_2}{\partial z}+\frac{1}{u_2}\frac{\partial A_2}{\partial t}-\frac{ik_2^{(2)}}{2}\frac{\partial^2 A_1}{\partial t^2}=-i\beta\cdot A_1^{2}e^{i\Delta kz}-i\gamma_{21}\left|A_1\right|^2\cdot A_2-i\gamma_{22}\left|A_2\right|^2\cdot A_2,
\end{aligned} \tag{1}$$

where A_1 and A_2 are the complex amplitudes of the fundamental and the second harmonic fields, z is the longitudinal coordinate of wave propagation, and $\Delta k=k(\omega_2)-2k(\omega_1)$ is the linear wave vector phase mismatch, u_1 and u_2 group velocities of fundamental and second harmonic pulses, γ_{ij} $(i,j=1,2)$ are the coefficients of third-order nonlinear wave coupling. The inertia of cubic nonlinearity is not taken into consideration in the equations, because in our work we observe optical pulses with durations more than 5-10fs (Akhmanov, Vysloukh et al. 1988).
The cubic polarization of frequency doubling media leads to nonlinear phase (B-integral) accumulation of interacted fundamental and second harmonic waves. As a consequence of it, the waves acquire additional phase mismatch and the energy conversion efficiency decreases. We demonstrated in the frame of plane wave approximation (Mironov, Lozhkarev et al. 2009), that the detuning direction of fundamental beam propagation from phase matching angle on the amount (for oo-e type):

$$\Delta\theta=\frac{\Delta n}{n_1^{3}\left(n_1^{-2}-n_o^{-2}\right)}\sqrt{\frac{n_1^{-2}-n_o^{-2}}{n_e^{-2}-n_1^{-2}}}, \tag{2}$$

significantly increases energy efficiency significantly. Here $\Delta n=\lambda A_{10}^2\cdot\left(2\gamma_{11}+2\gamma_{12}-\gamma_{21}-\gamma_{22}\right)/\left(8\pi\right)$, n_1 is the refractive index of an ordinary fundamental wave, n_o, n_e are the main values of refractive indexes for the second harmonic, A_{10} - amplitude of entrance to nonlinear element electric field. The detuning angle linearly depends on fundamental intensity, and for I=5 TW/cm^2 and 910nm central wavelength, the detuning angle in a KDP crystal is $\Delta\theta=-0.5^{\circ}$.
The other important feature of SHG of femtosecond radiation is dispersion effects. As far as, refractive index of frequency doubling media depends on wave length, group velocities of fundamental and second harmonic waves are different. The fact is a cause of reducing area

of interaction of pulses. The phenomenon is characterized by the length of group velocities mismatch:

$$L_{12} = T \cdot \left| \frac{1}{u_1} - \frac{1}{u_2} \right|^{-1} = T \cdot P(\lambda),$$

here T – is fundamental pulse duration, $P(\lambda)$ – dispersion parameter. Also, dispersion effects lead to pulse broadening. The spatial scale of the effect (the length of group velocity dispersion) can be found:

$$L_i = \frac{T^2}{4\ln(2)|D_i|}.$$

Here $D_i = \left.\frac{d^2k}{d\omega^2}\right|_{\omega_i}$; $i=1$ for fundamental and $i=2$ for second harmonic. In order to decrease influence of the effects it is necessary to satisfy the conditions to the crystal length – L:

$$L_{nl} < L < L_{12}, L_1, L_2$$

here L_{nl}–$(\beta \cdot A_{10})^{-1}$ – typical scale of energy conversion to the second harmonic, β – quadratic nonlinearity. Dispersion effects significantly depend on properties of frequency doubling media (at fundamental and second harmonic wave) and the fundamental pulse duration.

Experimental implementation of the SHG of femtosecond petawatt pulses requires a large aperture (more than 10cm) ultra-thin (hundreds of microns) nonlinear elements with wideband phase-matching at 800nm (Ti: sapphire lasers) and 910nm (optical parametrical amplification lasers). At present, KDP and DKDP crystals are the only crystals satisfying these requirements. Further analysis will be done for the crystals. We summarize characteristic lengths in Table.1.

λ_1 nm	T=20fs			T=50fs			T/L_{12}, fs/mm	L_{nl}, mm (5TW/cm²)	Deuteration, level D
	L_1, mm	L_2,mm	L_{12},mm	L_1,mm	L_2, mm	L_{12}, mm			
800	5.3	1.36	0.26	33	8.52	0.65	77	0.15	0
910	12.7	1.65	0.55	79.4	10.4	1.38	36	0.18	0
800	4.06	1.34	0.21	25.38	8.38	0.52	97	0.15	1
910	6.02	1.62	0.33	30.76	10.14	0.82	61	0.18	1

Table 1. Characteristic lengths for KDP (D=0) and DKD crystals.

As can be seen from table 1, the effect of group velocities mismatch is more significant for SHG in comparison with the pulses broadening. The length of group velocity mismatch for 910nm fundamental wave length is two times more the one for 800nm. The difference is significant for super intense femtosecond radiation, because the mutual influence of cubic polarization and dispersion effects starts to be crucial. The efficiency of energy conversion vs KDP crystal length is presented in fig.1. The results were obtained in the assumption, that angle of beam propagation is optimal in accordance with formula (2).

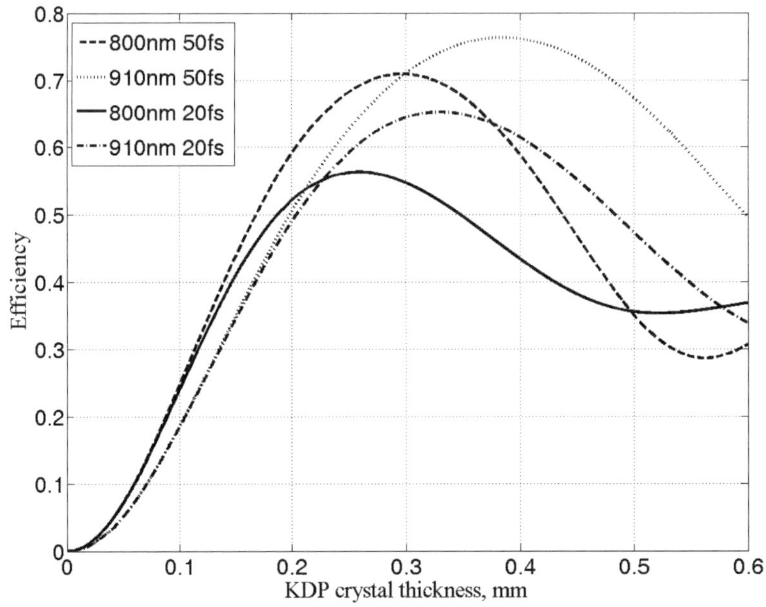

Fig. 1. The energy conversion efficiency vs KDP thickness, fundamental intensity $5TW/cm^2$

As can be seen from fig.1 optimal KDP crystal for effective SHG of 910nm, 50fs and $I=5TW/cm^2$ fundamental radiation is about 0.4mm, but the one for 800nm and the same parameters is about 0.3mm. The shorter pulses at the intensity level require thinner crystals. For instance, 20fs and 910nm the crystal length should be about 0.35mm, but for 800nm it is about 0.25mm. The right choice of nonlinear element thickness gives opportunity to obtain more than 50% efficiency of energy conversion even for 20fs, $5TW/cm^2$ fundamental radiation in KDP crystal.

The length of group velocity mismatch can be varied by means of changing parameters of frequency doubling element. Deuteration factor - D in DKDP crystals can be chosen during their growth stage. Refractive index of the crystal depends on the deuteration factor, and hence group velocities of fundamental and second harmonic pulses can be varied. In fig.2 we present dependence dispersion parameter $P(\lambda)$ from fundamental wave length at different level of deuteration factor in DKDP crystal. We used DKDP properties from the work(Lozhkarev, Freidman et al. 2005).

According to fig.2, dispersion parameter $P(\lambda)$ (calculated for DKDP crystal) is equal to zero only for one fundamental wave length. In this case group velocities of fundamental and second harmonic pulses are equal. Evidently, the situation is optimal for frequency doubling process. For the deuteration factor D=0 (KDP crystal) optimal central wave length of fundamental pulse is 1033nm. The increasing of deuteration factor leads to optimal wave length varying. The D range 0 - 1 corresponds to diapason of optimal wave lengths 1033–1210 nm. So, the level variation of the deuteration can be used for managing of dispersion properties of frequency doubling nonlinear element. The dependence of refractive index for KDP and DKDP crystals from temperature can not be used for managing of dispersion properties. At the present, the crystals are optimal for SHG of super powerful ultra short laser pulses. As far as, created from the crystals nonlinear elements can be done with large aperture (about 10 cm) and half millimeter thickness or less. The properties are crucial for the application.

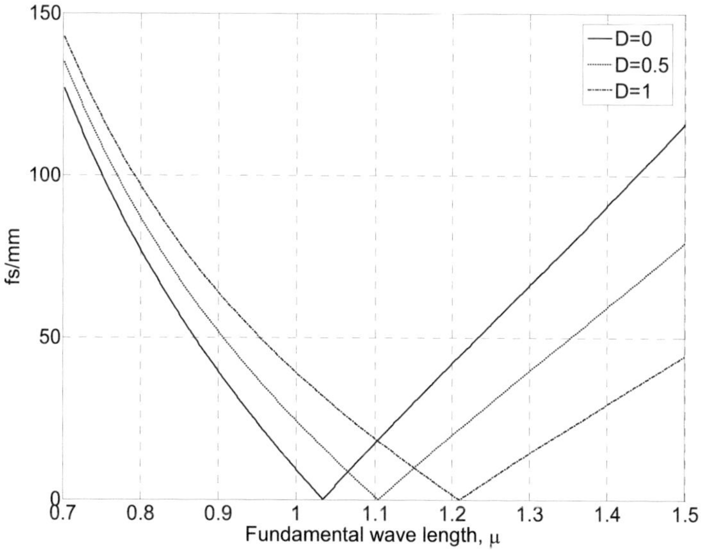

Fig. 2. Dispersion parameter P(λ) versus fundamental wave length at deuteration level D=0,0.5 and 1.

In order to verify theoretical model of SHG process under strong influence of cubic polarization effects we implemented modeling experiments. In the experiments we used output radiation of front end system of petawatt level femtosecond OPCPA laser (Lozhkarev, Freidman et al. 2007) as a fundamental beam. The parameters of the radiation incident on the SHG crystal (0.6 or 1.0mm thickness) were the following: beam diameter 4.3mm, pulse duration 65fs, energy range 1÷18 mJ, and central wavelength 910 nm. It is necessary to point out that the beam quality was not good, as a result at 18mJ energy the average over cross-section intensity was 2TW/cm^2 and the peak intensity was 5TW/cm^2. All measurements were done in vacuum, because for the range of fundamental intensities nonlinear beam self-action in air is important, even at several centimeters of propagation distance.

We have measured the energy efficiency of SHG in 0.6mm-thick KDP crystal at different detuning external angles Δθ, see fig.3. The main goal was to experimentally verify the fact that for efficient SHG different intensities require different optimal angles of beam propagation.

As can be seen from fig.3, a perfect phase matching (i.e. Δθ=0mrad) is optimal for SHG efficiency at low (2÷4mJ) and medium (10mJ) input energies. But at high energies (18mJ) SHG is more efficient at the optimal detuning angle Δθ=- 3.1mrad, because the phase induced by third-order nonlinearity and linear phase mismatching compensate each other. A negative value of optimal detuning angle is also clearly seen from comparison of data for Δθ=-6.2mrad and Δθ=6.2mrad: SHG efficiency is almost the same at 2-4mJ and is noticeably different at 16-18mJ. The experimental results are in a good agreement with formula (2), which gives for 18mJ energy (average intensity 2TW/cm^2) Δθ=-3.5mrad.

Relatively low SHG efficiency and large spread of the experimental data are explained by poor quality of both the beam and the ultra thin KDP crystal. In a 1mm-thick KDP crystal we reached 41% of SHG energy efficiency at such high intensities.

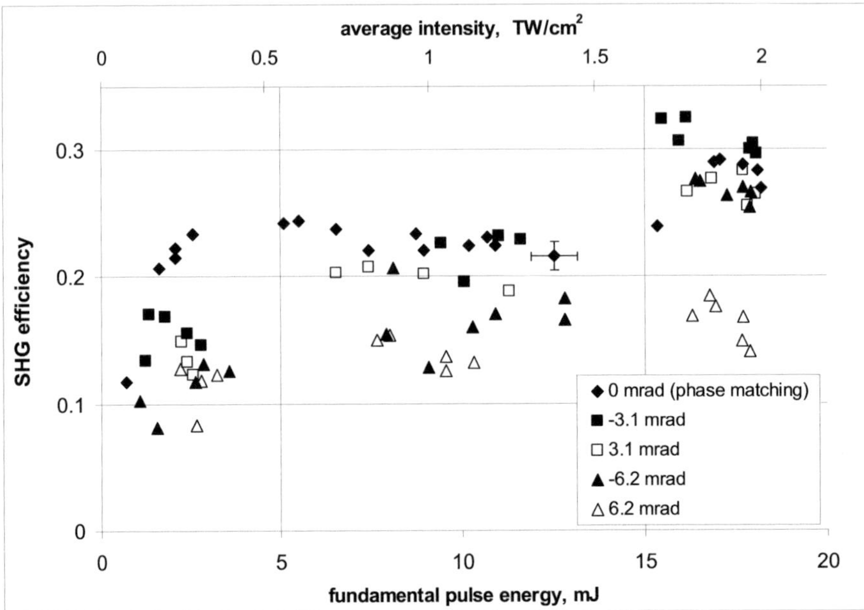

Fig. 3. The energy conversion efficiency *versus* input fundamental pulse energy at different detuning (external angles) from linear phase matching direction.

3. Pulse shortening and ICR enhancement

Cubic polarization leads to fundamental and second harmonic pulses spectrum modification and widening. The phenomenon depends on fundamental pulse intensity, cubic nonlinearity, fundamental central wave length and nonlinear element thickness. The output second harmonic pulse is not Fourier transform limited. In the case of optimal nonlinear element thickness, duration of second harmonic radiation approximately equals to the one of input pulse. Additional correction of spectrum phase of output second harmonic pulse makes it possible to significantly reduce pulse duration. The simplest way is the second order phase correction:

$$A_{2comp}(t) = F^{-1}\left[e^{iS\omega^2} F\left[A_2(z=L,t)\right]\right],$$

here F, F^{-1} are the direct and inverse Fourier transforms, $A_2(z=L,t)$, $A_{2\mathrm{comp}}(t)$ are the electric fields of second harmonic radiation before and after phase correction, and S is the coefficient of quadratic spectral phase correction. The electric field $A_2(z=L,t)$ is obtained by the numerical solution of (1). The results are presented in fig.4 for optimal detuning angle $\Delta\theta$ (2) and for 5TW/cm^2 fundamental Gaussian pulse (20fs FWHM). The coefficient S was chosen to minimize the pulse duration. Maximums of all the three pulses were shifted to zero time for clarity.

In accordance with fig.4, SHG process increases temporal ICR on pulse wings. Additional spectrum phase correction allows significantly reduce pulse duration. For instances, for fundamental wavelength Gaussian pulse with duration 20fs (FWHM) and intensity 5TW/cm^2, the second harmonic pulse may be compressed to 12fs (800nm, 0.2mm-thick

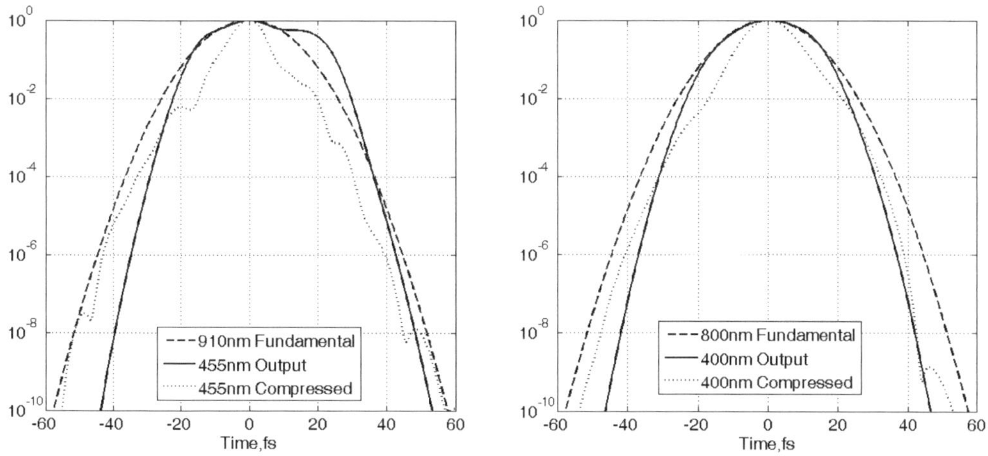

Fig. 4. Shapes of incident and second harmonic pulses before and after spectral phase correction. (a) fundamental wavelength 910 nm, KDP thickness 0.4 mm; (b) fundamental wavelength 800 nm, KDP thickness 0.2 mm.

KDP) and to 9fs (910nm, 0.4mm-thick KDP). The 50fs input pulse may be compressed even more efficiently in a 0.4mm-thick KDP: to 18fs at 800nm and to 16fs at 910nm.
The spectrum phase correction decreases ICR of second harmonic pulse in comparison with uncompressed, but it is higher than fundamental. As a result, SHG together with additional spectrum phase correction is suitable for peak intensity increasing and improvement of temporal intensity profile of super strong laser radiation.

4. Plane wave instability in mediums with quadratic and cubic nonlinearities

The other negative manifestation of cubic polarization is small-scale self focusing (SSSF). The process is the main cause of laser beam filamentation and nonlinear element destructions. The theoretical aspects of the phenomenon in media with cubic polarization are observed in literature (Bespalov and Talanov 1966; Rozanov and Smirnov 1980; Lowdermilk and Milam 1981; Poteomkin, Martyanov et al. 2009). The main goal of the section is to develop theoretical approach to describe the process in media with quadratic and cubic nonlinearity. The model (Ginzburg, Lozhkarev et al. 2010) is necessary for estimations of critical level of spatial noise in super strong laser beams.
Let's assign three fundamental plane waves on the input surface of frequency doubling nonlinear element (waves 1, 3, 4 see fig 4). Angles α_1 and α_2 determine directions of harmonic disturbances of laser beam in non critical and critical planes of frequency doubling nonlinear element. Let's consider, that waves 3 and 4 has equal by amount, but anti directional transverse wave vectors. The wave 1 is significantly more powerful, than waves 3 and 4, and it runs in the optimal direction for energy conversion (in accordance with formula 2). Second harmonic wave (wave 2), which is generated by the wave number 1, runs at the same direction too.
Weak fundamental waves (3, 4) interact with strong waves (1, 2) and generate harmonic disturbances of second harmonic radiation (waves 5 and 6) see fig 4. The transversal wave vectors of fundamental and second harmonic waves should be equal. The requirement is a

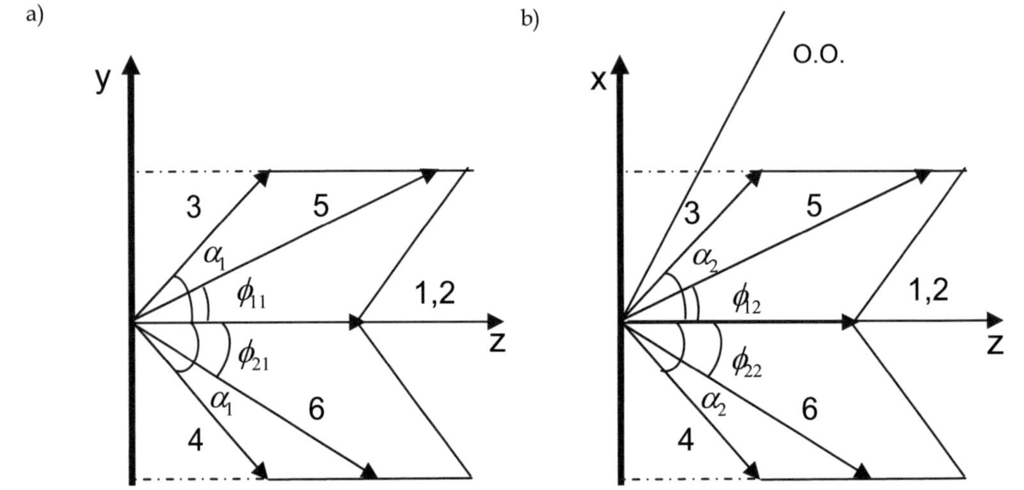

Fig. 4. The scheme of runs strong fundamental (wave 1) and second harmonic (wave 2) waves and their harmonic disturbances (waves 3,4 and 5,6 correspondingly) in non critical (a) and critical plane of frequency doubling nonlinear element.

consequence of Maxwell equations. The directions of second harmonic beam disturbances in non critical (ϕ_{11}, ϕ_{21}) and critical (ϕ_{12}, ϕ_{22}) planes are determined by boundary conditions (3):

$$\begin{array}{ll} k_1 \sin(\alpha_2) = k_5 \sin(\phi_{12}) & k_1 \sin(\alpha_2) = k_6 \sin(\phi_{22}) \\ k_1 \cos(\alpha_2)\sin(\alpha_1) = k_5 \cos(\phi_{12})\sin(\phi_{11}) & k_1 \cos(\alpha_2)\sin(\alpha_1) = k_6 \cos(\phi_{22})\sin(\phi_{21}) \end{array} \quad (3)$$

So, the beam disturbances 3, 4, 5 and 6 have equal transversal wave vectors and their amplitudes satisfy the demands:

$$|\varepsilon_i| << |\varepsilon_1|, |\varepsilon_2| \quad (i = 3..6) \quad (4)$$

Assume, that on the boundary of the frequency doubling nonlinear element (z=0) the amplitudes of strong waves are the following:

$$\varepsilon_1(z=0)=\varepsilon_{10} \text{ and } \varepsilon_2(z=0)=0 \quad (5)$$

Here ε_{10} – the electric field of the fundamental string wave (1). The conditions for harmonic disturbance amplitudes:

$$\varepsilon_3(z=0) = \varepsilon_4(z=0) = \varepsilon_{30} \cdot e^{i\varphi}, \varepsilon_{5,6}(z=0) = 0, \quad (6)$$

here ε_{30} and φ the initial electric strength of waves 3,4 and their phase on the entrance surface of nonlinear element. Assume, that amplification of harmonic disturbances is not crucial for strong wave interaction, i.e. amplitudes of wave 1 and 2 satisfy the system of differential equations:

$$\begin{aligned}
\frac{d\varepsilon_1}{dz} &= -i\cdot\beta\cdot\varepsilon_2\cdot\varepsilon_1^*\cdot e^{-i\Delta k\cdot z} - i\cdot\gamma_{11}\cdot\left|\varepsilon_1\right|^2\cdot\varepsilon_1 - i\cdot\gamma_{12}\cdot\left|\varepsilon_2\right|^2\cdot\varepsilon_1, \\
\frac{d\varepsilon_2}{dz} &= -i\cdot\beta\cdot\varepsilon_1^2\cdot e^{i\Delta k\cdot z} - i\cdot\gamma_{21}\cdot\left|\varepsilon_1\right|^2\cdot\varepsilon_2 - i\cdot\gamma_{22}\cdot\left|\varepsilon_2\right|^2\cdot\varepsilon_2,
\end{aligned} \tag{7}$$

Now, observe mathematical description of plane wave instability in media with quadratic and cubic nonlinearity. Implementation of standard linearization procedure to quasi-optical equations, which describe dynamic of each frequency component, and grouping items with equal transverse wave vectors, gives opportunity to obtain equations for amplitudes of harmonic disturbances:

$$\begin{aligned}
\frac{d\varepsilon_3}{dz} &= \frac{-i}{\cos(\alpha_1)\cos(\alpha_2)}\left[\beta\left[E_1^*E_5 + E_4^*E_2\right] + \gamma_{11}\left[E_1^2E_4^* + 2\left|E_1\right|^2 E_3\right] + \gamma_{12}\left[\left|E_2\right|^2 E_3 + E_1E_2E_6^* + E_1E_5E_2^*\right]\right]e^{ik_1\cos(\alpha_1)\cos(\alpha_2)z} \\
\frac{d\varepsilon_4}{dz} &= \frac{-i}{\cos(\alpha_1)\cos(\alpha_2)}\left[\beta\left[E_1^*E_6 + E_3^*E_2\right] + \gamma_{11}\left[E_1^2E_3^* + 2\left|E_1\right|^2 E_4\right] + \gamma_{12}\left[\left|E_2\right|^2 E_4 + E_1E_2E_5^* + E_1E_6E_2^*\right]\right]e^{ik_1\cos(\alpha_1)\cos(\alpha_2)z} \\
\frac{d\varepsilon_5}{dz} &= \frac{-i}{\cos(\phi_{11})\cos(\phi_{12})}\left[2\beta E_3E_1 + \gamma_{21}\left[\left|E_1\right|^2 E_5 + E_1E_2E_4^* + E_1^*E_2E_3\right] + \gamma_{22}\left[\left|E_2\right|^2 E_6^* + 2\left|E_2\right|^2 E_5\right]\right]e^{ik_5\cos(\phi_{11})\cos(\phi_{12})z} \\
\frac{d\varepsilon_6}{dz} &= \frac{-i}{\cos(\phi_{21})\cos(\phi_{22})}\left[2\beta E_4E_1 + \gamma_{21}\left[\left|E_1\right|^2 E_6 + E_1E_2E_3^* + E_1^*E_2E_4\right] + \gamma_{22}\left[\left|E_2\right|^2 E_5^* + 2\left|E_2\right|^2 E_6\right]\right]e^{ik_6\cos(\phi_{21})\cos(\phi_{22})z}
\end{aligned} \tag{8}$$

Here $E_i = \varepsilon_i e^{-ik_{zi}z}$, k_1– magnitude of fundamental wave vector, k_5, k_6 – wave vectors of second harmonic wave disturbances. In the frame of the model, the disturbances in the second harmonic beam appear from interaction between fundamental disturbances and strong fundamental wave (1). The second harmonic beam modulation is amplified over cubic polarization.

The amplification of harmonic disturbances depends on entrance fundamental intensity, quadratic and cubic nonlinearity, linear wave vectors mismatch, initial phase φ on the entrance surface of frequency doubling media. The gain factors of fundamental and second harmonic disturbances can be determined *(i=3..6)*:

$$G_i(z,\alpha_1,\alpha_2,\varphi) = \frac{\left|\varepsilon_i(z,\alpha_1,\alpha_2,\varphi)\right|^2}{\left|\varepsilon_{30}\right|^2}.$$

As a rule, initial phase φ is a random parameter, and hence the averaged gain factors are mostly interested in theoretical investigations:

$$G_{avi}(z,\alpha_1,\alpha_2) = \frac{1}{2\pi}\int_0^{2\pi} G_i(z,\alpha_1,\alpha_2,\varphi)d\varphi$$

The averaged gain factors $G_{av1,2}$ versus harmonic disturbances propagation directions (α_1, α_2) are presented in fig.4 for 0.5mm KDP crystal, 1.5 TW/cm² fundamental intensity and assumption of optimal direction of beam propagation (in accordance with formula 2).

In accordance with fig 5, maximum of averaged by initial phase gain factors of fundamental and second harmonic disturbances, which were calculated for I=4.5 TW/cm² and 0.5mm KDP thickness, are the following G_{av1}=14, G_{av2}=270. Angular detuning in critical plane imposes restrictions on amplification of harmonic disturbances. As a result of it, angular

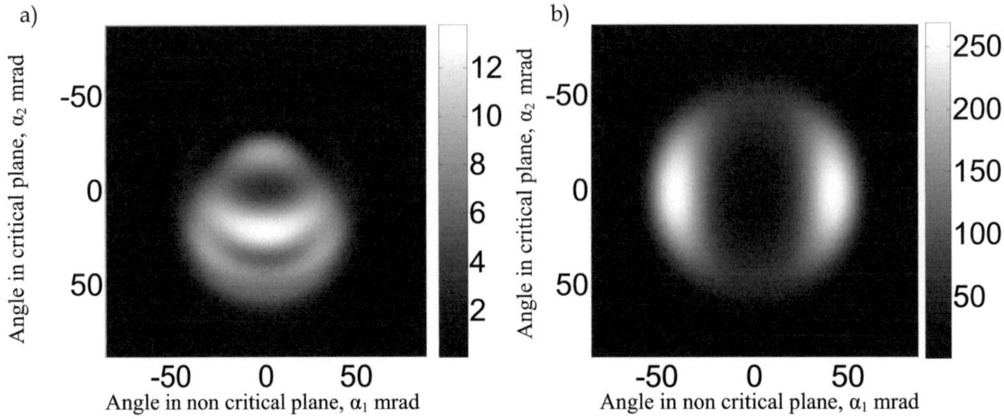

Fig. 5. The averaged gain factors of harmonic disturbances versus angles (α_1, α_2) a) G_{av} fundamental and b) G_{av} second harmonic. The diagrams were obtained by numerical solution of systems differential equations (7) and (8) with boundary conditions 5 and 6, KDP thicknesses 0.5mm and fundamental intensity I=4.5 TW/cm^2.

diagram in fig. 5 are symmetrical in non critical plane, i.e. $G_{avi}(z,-\alpha_1,\alpha_2)=G_{avi}(z,\alpha_1,\alpha_2)$ and non symmetrical in critical plane $G_{avi}(z,\alpha_1,-\alpha_2)\neq G_{avi}(z,\alpha_1,\alpha_2)$. Maximum amplification of harmonic disturbances of second harmonic wave (2) takes place at angles α_1=42 mrad and α_2=0 from direction of strong waves (1 and 2) propagation.

Note that for media with only cubic nonlinearity the gain factor can be found analytically (Rozanov and Smirnov 1980):

$$G_{th}=\frac{1}{4}\left(2\cosh^2(B_{11}x)+\left(\frac{2B_{11}x}{\kappa_1^2-4B_{11}}\right)^2\sinh^2(B_{11}x)+\left(\frac{\kappa_1^2-4B_{11}}{2B_{11}x}\right)^2\sinh^2(B_{11}x)\right),$$

here $B_{11}=\gamma_{11}A_{10}^2L$ is B-integral, $x^2=\frac{\kappa_1^2}{B_{11}}-\frac{\kappa_1^4}{4B_{11}{}^2}$, $\kappa_1=k_{1\perp}\sqrt{L/k_1}$ - normalized transversal wave vector.

The other important parameter, which characterizes SSSF, is integral by spatial spectrum gain factor of harmonic disturbances. Let's determine it for the task by the following way:

$$G_{\text{int}_j}=\frac{1}{\pi\alpha_{cr}^2}\iint_{\Omega}G_{av_j}d\alpha_{j1}d\alpha_{j2}$$

here j=1,2 indexes of gain factors of noise amplification of fundamental and second harmonic wave, α_{cr} – angle in non critical plane, which corresponds to decreasing the gain factor of second harmonic disturbances in e times from the maximum value; Ω – is the circle of radius α_{cr} . Despite the complicated determination of the integral gain factors, they are physically enough, because it takes into account anisotropic topology of the gain structure (see fig.5). Calculated by the way integral gain factors for fundamental radiation I=4.5 TW/cm^2 and 0.5 mm KDP thickness are equal to G_{int1}=5, G_{int2}=107. Integral gain factors versus B-integral are presented in fig 6.

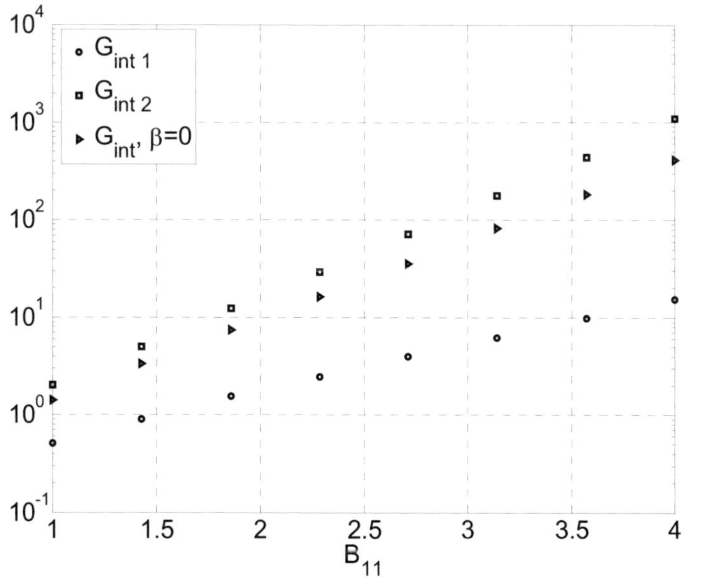

Fig. 6. Integral gain factors $G_{int1,2}$ fundamental and second harmonic disturbances versus B-integral

As can be seen from fig 6, the harmonic disturbances of fundamental and second harmonic waves are amplified significantly during second harmonic generation process. The fact can be a cause of a nonlinear element corruption.

Let's found critical level of spatial noise in the fundamental beam on the entrance surface of nonlinear element. Peak amount and mean square deviation intensity I_{rms} from average in beam profile are connected with relative noise power P_n/P by the following empiric formulas (Rozanov and Smirnov 1980):

$$\begin{aligned} I_{peak}/I_{av} &= \left(1+5\sqrt{P_n/P}\right)^2 \\ I_{rms}/I_{av} &= \left(1+\sqrt{P_n/P}\right)^2 - 1 \end{aligned} \tag{9}$$

In accordance with (Kumar, Harsha et al. 2007), KDP crystal can stand under intensity about I_{peak}=18.5 TW/cm², 100fs pulse duration and central wave length λ=795 nm. Let's assume that the peak intensity is the threshold level for crystal destruction. In this case $K_{th} = I_{peak}/I_{av} = 4.1$ for average intensity I_{av}=4.5 TW/cm². Noise power on the output surface of frequency doubling nonlinear element can be found like $P_{nout} = G \cdot P_n$. Critical level of noise power $K_n = P_n/P$ of the fundamental entrance beam (by means of 9) is the following:

$$K_n = \frac{1}{G}\left(\frac{1}{5}\left(\sqrt{K_{th}} - 1\right)\right)^2.$$

For gain factor G=107 the amount is $K_n = 4 \cdot 10^{-4}$ and $I_{rms}/I_{av} = 4 \cdot 10^{-2}$. The influence of SSSF effects can be diminished by means of beam quality improving and self filtering implementation, see next section.

5. Small-scale self-focusing suppression

The presented estimations of noise power critical level are overstated. The question is that, the observed theoretical model assumes the presence of high amplifying harmonic disturbances in nonlinear element and in the area of strong field. According to the observed theoretical model directions, where noise components start to be more intensive, depends on many factors such as fundamental intensity and central wave length, quadratic and cubic nonlinearity, direction of strong wave propagation with respect to optical axe, and thickness of frequency doubling nonlinear element. For fundamental intensity 4.5TW/cm^2 and 0.5mm KDP crystal the optimal angle for amplification of harmonic disturbance is 42 mrad (fig. 5). Such high angles make it possible to use free space propagation to cut off dangerous spatial components from strong light beam and there is no necessity to use spatial filters. The main sources of harmonic disturbances are mirror surfaces. Hence, the distance between the last mirror and nonlinear element is important parameter for spatial spectrum clipping (see fig 7).

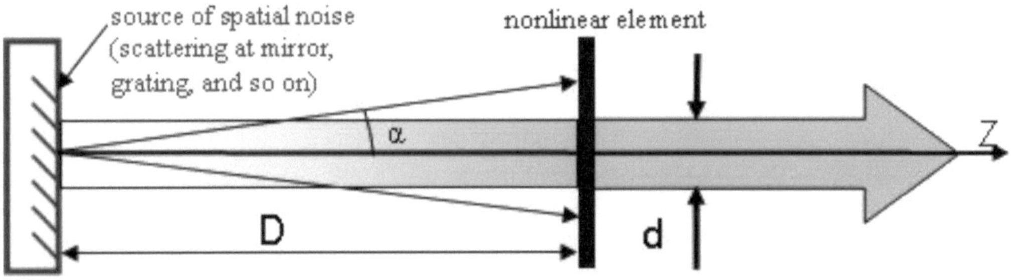

Fig. 7. The idea of small-scale self-focusing suppression. d is beam diameter, D is free-space propagation distance, α is angle of propagation of noise wave.

Differential equations 7 and 8 make it possible to find angle of optimal spatial noise increasing – α_{max}. The safety distance is $D \geq \frac{d \cdot \sqrt{1-n^2 \cdot \sin^2(\alpha_{max})}}{2 \cdot n \cdot \sin(\alpha_{max})}$, there d is the entrance beam diameter, n – refractive index. This idea of small-scale self-focusing suppression by beam self-filtering due to free propagation before SHG may be used in any high intensity laser. The key parameters of the task are B-integral and angle of view d/D.

The power of noise on the entrance surface can be calculated by the following expression:

$$P_o = \iint_{\Omega} P_{\alpha o} d\alpha_1 d\alpha_2 = \pi \alpha^2 P_{\alpha o}$$

here P_{ao} – angular power density, a_1 и a_2 angles in non critical and critical plane, Ω – area in angular parameter space. On the entrance surface angular power density is homogeneous and the Ω is the circle with radius a and center in the origin of coordinates. On the output surface the noise power is the following:

$$P_{out} = \iint_{\Omega} P_{\alpha o} G(\alpha_1, \alpha_2) d\alpha_1 d\alpha_2 = P_{\alpha o} \cdot \pi \cdot \alpha^2 \cdot G_{\text{int}}$$

Output noise power is proportional to $\alpha^2 G_{int}$. Maximum angle of integration is determined by the optical scheme geometry and refraction angle of frequency doubling nonlinear element.

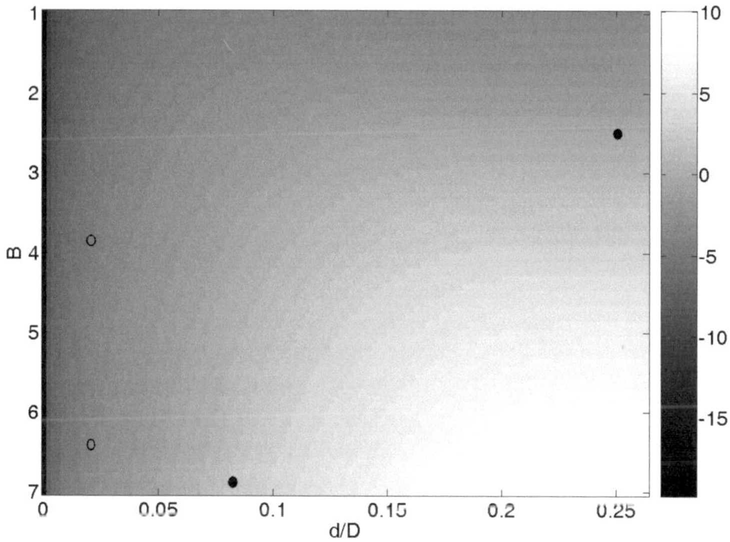

Fig. 8. Logarithm of normalized power of noise *ln(F)* versus B-integral and angle of view d/D at intensity 5TW/cm^2. Open circles show experiments without filamentaion and fill circles show experiment with filamentation.

The dependence of normalized power of the spatial noise $F=P_{out}(B,d/D)/ P_{out}(B=2.5, d/D=1)$ at the output of nonlinear element is presented in fig. 8. At a large angle of view d/D (no self-filtering) all noise waves reach the nonlinear element and noise power at its output is the highest. In this case filamentation appears at B≈2.5 (F≈1), the fact was observed in our experiments in accordance with other papers (Bespalov and Talanov 1966; Bunkenberg, Boles et al. 1981; Speck 1981; Vlasov, Kryzhanovskiĭ et al. 1982). When d/D reduces to about 0.2 the noise power decreases as well. It drops sharply when d/D<0.1. As on can see from fig.3, experimental points with 0.6mm and 1mm thick crystal (d/D=0.02) are in a safety region: F<<1 even though B=3.8 and 6.4 correspondingly. On the other hand for experimental points where filamentaion was observed (d/D=0.08) the F parameter is above unity, see fig.8.

Note, that in nanosecond lasers the typical intensity is 1GW/cm^2, hence, self-filtering takes place at the angle of view d/D<0.003, because the angle is proportional to the square root of laser beam intensity. Such a small value limits practical use of self-filtering for 1GW/cm^2 intensity laser pulses. The critical angle for medium with cubic nonlinearity only can be found in accordance with the following formula:

$$\alpha_{cr} = \sqrt{\frac{2\gamma I}{n}}$$

Here γ- cubic nonlinearity coefficient [cm^2·watt^{-1}], I – intensity [cm^{-2}·watt] and n - refractive index.

6. Surface dust influence to SHG process

Surface dusts are conductive to harmonic disturbance generation. The observed small-scale self-focusing suppression scheme can not be used for clipping spatial spectrum of noises, which are generated on surface. Generated on nonlinear element surface harmonic components have the same initial phase, which is equal to zero. The dynamic of plane wave instability can be described by system of differential equations 7 and 8. In this case boundary conditions are the following:

$$\varepsilon_1(z=0)=\varepsilon_{10}\,,\varepsilon_2(z=0)=0,\;\; \varepsilon_{3,4}(z=0)=\varepsilon_{10}\,,\; \varepsilon_{5,6}(z=0)=0$$

Gain factors versus of angles of harmonic disturbance propagation are presented in fig.9.

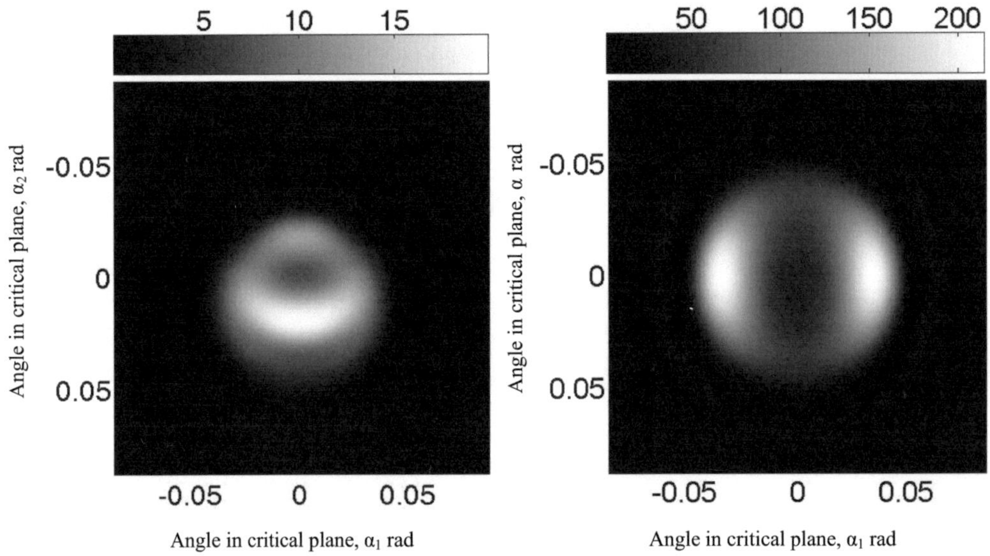

Fig. 9. Gain factors of harmonic disturbances a) fundamental and b) second harmonic waves. Calculated for 0.5mm KDP crystal and intensity 4.5TW/cm^2, angle of strong wave's propagation is optimal according to (2).

In accordance with fig.9 gain factors of harmonic disturbances of fundamental and second harmonic waves, which appears from scattering strong waves on surface dusts have topological structure the same to the averaged by initial phase gain factors. In this case preferable direction for noise amplification lies in non critical plane too. For fundamental intensity 4.5TW/cm^2 and 0.5mm KDP crystal the maximum gain factor of fundamental disturbances is 20, as for second harmonic noise it is 216 by optimal angle ±35mrad in non critical plane.

Thus, gain factors of harmonic disturbances of second harmonic wave, which is generated on surface dusts, are big sufficiently. In case of use of self-filtering technique described in Section 5, the input surface of SHG crystal is the main source of the noises. It is needed to be taken into consideration during experimental investigations.

7. Conclusion

We presented experimental conformation of theoretical model of SHG of super strong femtosecond radiation under strong influence of dispersion and cubic polarization effects. Despite the poor quality of the fundamental beam, the energy conversion efficiency of 35% (41%) was achieved in 0.6 (1.0) mm-thick KDP crystal.

SHG process is well for a increasing of temporal intensity contrast ratio significantly. The suggested additional spectrum phase correction of output second harmonic radiation is a way to reduce pulse duration more than 3 times for 50fs, 5TW/cm^2 and 0.4mm KDP crystal. We demonstrated significant difference between 910nm and 800nm fundamental waves for SHG process and established the privilege of the first one.

The developed theoretical model of plane wave instability in mediums with quadratic and cubic nonlinearity makes it possible to estimate critical level of spatial noises power in super strong femtosecond laser beams for safety realization of SHG process. The suggested scheme of small-scale self-focusing suppression is theoretically explained and experimentally verified: no manifestation of self-focusing at B-integral above 6. The obtained experimental results are in a good agreement with the model. In the conclusion it is necessary to point out two additional sources of spatial noises as apply to SHG process. The first one is surface inhomogeneous and the second one is internal refractive index disturbances. The methodic of minimization of the factors is the improvement of nonlinear element quality: surface and homogeneity of refractive index.

8. References

Akhmanov, S. A., V. A. Vysloukh, et al. (1988). Optika femtosekundnykh lazernykh impul'sov. Moscow, Nauka.

Aoyama, M., K. Yamakawa, et al. (2003). "0.85-PW, 33-fs Ti:sapphire laser." Optics Letters 28(17): 1594-1596.

Bespalov, V. I. and V. I. Talanov (1966). "Filamentary structure of light beams in nonlinear liquids." JETP Letters 3: 307-310.

Bunkenberg, J., J. Boles, et al. (1981). "The omega high-power phosphate-glass system: design and performance." IEEE Journal of Quantum Electronics QE-17(9): 1620-1628.

Chien, C. Y., G. Korn, et al. (1995). "Highly efficient second-harmonic generation of ultraintense Nd:glass laser pulses." Optics Letters 20(4): 353–355.

Choe, W., P. P. Banerjee, et al. (1991). "Second-harmonic generation in an optical medium with second- and third-order nonlinear susceptibilities." J. Opt. Soc. Am. B 8(5): 1013–1022.

Ditmire, T., A. M. Rubenchik, et al. (1996). "The effect of the cubic onlinearity on the frequency doubling of high power laser pulses." J. Opt. Soc. Am. B 13(14): 649–655.

Ginzburg, V. N., V. V. Lozhkarev, et al. (2010). "Influence of small-scale self-focusing on second harmonic generation in an intense laser field." Quantum Electronics 40(6): 503-508.

Kochetkova, M. S., M. A. Martyanov, et al. (2009). "Experimental observation of the small-scale self-focusing of a beam in the nondestructive regime." Quantum Electronics 39(10): 923-927.

Kumar, R. S., S. S. Harsha, et al. (2007). "Broadband supercontinuum generation in a single potassium di-hydrogen phosphate (KDP) crystal achieved in tandem with sum frequency generation." Appl. Phys. B 86: 615-621.

Liang, X., Y. Leng, et al. (2007). "Parasitic lasing suppression in high gain femtosecond petawatt Ti:sapphire amplifier." Optics Express 15(23): 15335-15341.

Lowdermilk, W. H. and D. Milam (1981). "Laser-induced surface and coating damage." IEEE Journal of Quantum Electronics QE-17(9): 1888-1902.

Lozhkarev, V. V., G. I. Freidman, et al. (2007). "Compact 0.56 petawatt laser system based on optical parametric chirped pulse amplification in KD*P crystals." Laser Physics Letters 4(6): 421-427.

Lozhkarev, V. V., G. I. Freidman, et al. (2005). "Study of broadband optical parametric chirped pulse amplification in DKDP crystal pumped by the second harmonic of a Nd:YLF laser." Laser Physics 15(9): 1319-1333.

Mironov, S. Y., V. V. Lozhkarev, et al. (2009). "High-efficiency second-harmonic generation of superintense ultrashort laser pulses." Applied Optics 48(11): 2051-2057.

Poteomkin, A. K., M. A. Martyanov, et al. (2009). "Compact 300 J/ 300 GW frequency doubled neodimium glass laser. Part I: Limiting power by self-focusing." IEEE Journal of Quantum Electronics 45(4): 336-344.

Razumikhina, T. B., L. S. Telegin, et al. (1984). "Three-frequency interactions of high-intensity light waves in media with quadratic and cubic nonlinearities." Sov. J. Quantum Electron 14.

Rozanov, N. N. and V. A. Smirnov (1980). "Small-scale self-focusing of laser radiation in amplifier systems." Soviet Journal of Quantum Electronics 10(2): 232-237.

Speck, D. R. (1981). "The shiva laser-fusion facility." IEEE Journal of Quantum Electronics QE-17(9): 1599-1619.

Vlasov, S. N., V. I. Kryzhanovskiĭ, et al. (1982). "Use of circularly polarized optical beams to suppress selffocusing instability in a nonlinear cubic medium with repeaters " Soviet Journal of Quantum Electronics 12(1): 7-10.

Yanovsky, V., V. Chvykov, et al. (2008). "Ultra-high intensity- 300-TW laser at 0.1 Hz repetition rate." Optics Express 16(3): 2109-2114.

10

Temporal Stretching of Laser Pulses

Rajeev Khare and Paritosh K. Shukla
Raja Ramanna Centre for Advanced Technology
Department of Atomic Energy
India

1. Introduction

Many lasers operate predominantly or only as pulsed sources and the pulse duration is determined by the duration of effective pumping, intra-resonator energy extraction rates, etc. However, in certain applications of the pulsed lasers, it is necessary to extend the duration of the laser pulses without reducing its pulse energy. The duration of laser pulses is increased by using laser pulse stretchers, which stretch the pulses temporally. An ideal laser pulse stretcher increases the duration of the laser pulse without introducing losses so the peak power of the laser is reduced without reducing its average power. The temporal stretching of laser pulses is vital for many applications of pulsed lasers. The temporal stretching of oscillator pulses of high beam quality is required in oscillator–amplifier systems for achieving high output power. The temporal stretching of pump laser pulses leads to an increase in energy conversion efficiency and tuning range along with a decrease in linewidth of the tunable dye lasers. While launching the laser pulses of high energy into optical fibers, the temporal stretching of laser pulses is done to reduce the peak power, without reducing the pulse energy, to save the input faces of optical fibers from damage. Thus the temporal stretching of laser pulses also increases the upper limit of transmission of pulse energy in the optical fibers. The optical pulse stretchers are used for removing laser-induced plasma spark generation in spontaneous Raman-scattering spectroscopy by reducing the peak power. In the guide star experiments, the pulse stretching of laser pulses is required to avoid saturation effects. The temporal pulse stretching is required in material processing because a stretched pulse is more efficient for heating the material. The temporal stretching of laser pulses is needed in optical microlithography to avoid degradation of semiconductor materials as well as optics. In the medical application of lasers, the pulse stretching of laser pulses is done to reduce the high peak intensities, which generally damage the tissues. The temporal stretching of laser pulses is done in holographic interferometry for removing the boiling effect, which is detrimental to the quality of the photographs. The ultrashort pulses are temporally stretched in the chirped pulse amplification (CPA) to avoid the nonlinear effects that lead to catastrophic damage. There are many more important applications, where temporal stretching of laser pulses is done. According to the type of the laser and the requirements of the application, various types of laser pulse stretchers, both passive and active, are developed for temporal stretching of laser pulses.

2. Techniques of temporal stretching of laser pulses

Based on the physics and technology, the techniques of temporal stretching of laser pulses can be broadly classified into four categories; (a) optical pulse stretching, (b) pulse stretching by intracavity nonlinear materials (c) electronic pulse stretching and (d) pulse stretching by dispersion.

2.1 Optical pulse stretching

The technique of optical pulse stretching uses a configuration of optical elements that split the amplitude of the incident laser pulse and introduce optical delays among them and also recombine the temporally delayed portions of the pulse to provide a temporally stretched laser pulse. These optical pulse stretchers are generally passive and configured external to the laser system for temporal stretching of nanosecond long laser pulses. Several types of optical components can be configured in different ways in optical pulse stretchers to obtain temporal pulse stretching. A typical optical pulse stretcher is an optical ring cavity formed by several 100% reflecting mirrors and at least one partially reflecting beam-splitter.

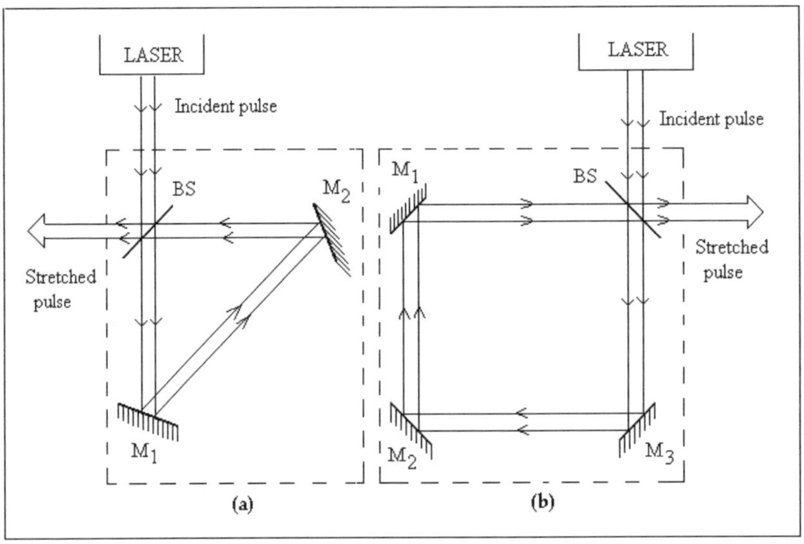

Fig. 1. Schematics of optical pulse stretchers of ring cavities with (a) three mirrors and (b) four mirrors. M_1, M_2 and M_3: 100% reflecting plane mirrors; BS: beam-splitter.

The optical ring cavity of a pulse stretcher, shown in Fig. 1(a), is a triangle ring cavity consisting of a beam-splitter, *BS*, and two 100% reflecting plane mirrors, M_1 & M_2. The optical ring cavity, shown in Fig. 1(b), is a square ring cavity consisting of a beam splitter, *BS*, and three 100% reflecting plane mirrors, M_1, M_2 & M_3. The beam-splitter and the 100% reflecting mirrors of the optical ring cavity can also be arranged in other shapes, like rectangle, etc. The laser pulse, which is to be stretched, is incident on the beam-splitter, *BS*. If the absorption losses of the beam-splitter are zero, then it splits the amplitude of the incident laser pulse into two parts such that one part is reflected while the remaining transmitted part is stored in the ring cavity to traverse an additional path. The stored part is then released in each roundtrip by the beam-splitter, *BS*, where it gets spatially superimposed on the reflected part as well as on the earlier released parts to form a single pulse of longer duration. Thus the optical ring cavity divides an initially large amplitude laser pulse into many smaller amplitude pulses and recombines them after introducing optical delays among them. A spatial overlapping between the successive parts of the laser pulse at the beam-splitter is essential in an optical pulse stretcher. The optical ring cavity, shown in Fig. 1(a), is a right-angled triangle ring cavity because this configuration permits the optimum superposition of the input and output pulses upon exiting the cavity by use of a 45° incidence beam-splitter. The beam-splitter reflectivity, the optical cavity delay time, the laser beam pointing stability and the laser beam divergence are the key parameters, which affect the performance of a optical pulse stretcher. The optical ring cavity traps and stores a portion of the circulating laser pulse, subsequently releasing the stored pulse over a longer period of time as determined by the optical delay time and the intracavity leakage rate. The optical delay time, τ, is the roundtrip propagation time of the ring cavity and is given by

$$\tau = L/c \tag{1}$$

where c is the speed (i.e., group velocity) of light in air and L is the length of the ring cavity. Suppose the reflectivity of the *BS*, whose absorption losses are zero, is R% and p(t) is the instantaneous power of the pulse, which is incident on the pulse stretcher, then *BS* divides the pulse such that reflected part, Rp(t), goes out of the pulse stretcher and transmitted part (1–R)p(t), circulates inside the ring cavity. After 1st roundtrip, which is completed in time τ, *BS* transmits a part of pulse, $(1-R)^2p(t-\tau)$, out of the cavity, which gets superimposed on the earlier reflected part. Now the reflected part of the pulse, $R(1-R)p(t-\tau)$, circulates inside the ring cavity. After 2nd roundtrip, which is completed in time 2τ, *BS* transmits another part of amplitude, $(1-R)^2Rp(t-2\tau)$, out of the pulse stretcher and reflects the remaining part, $R^2(1-R)p(t-2\tau)$, to circulate inside the ring cavity. The process repeats and the ring cavity divides an initially large amplitude laser pulse into many smaller amplitude pulses such that the successive pulses exiting from the pulse stretcher are temporally shifted with respect to the initial pulse by $\tau, 2\tau, \ldots, n\tau$, where n is an integer representing the number of round trips in the cavity. The integrated instantaneous power of the pulse, obtained as the output of the optical pulse stretcher is given by

$$\begin{aligned} P(t) &= Rp(t) + (1-R)^2p(t-\tau) + R(1-R)^2p(t-2\tau) + R^2(1-R)^2p(t-3\tau) + R^3(1-R)^2p(t-4\tau) \\ &\quad + R^4(1-R)^2p(t-5\tau) + \cdots + R^{n-1}(1-R)^2p(t-n\tau) + \cdots \\ &= Rp(t) + \sum_{n=1}^{\infty} R^{n-1}(1-R)^2p(t-n\tau) \end{aligned} \tag{2}$$

In Eq. (2), the 1st term is initial reflection of *BS*, the 2nd term is transmission of *BS* after 1st roundtrip, the 3rd term is transmission of *BS* after 2nd roundtrip, and so on. There are infinite numbers of terms in the expression, however only first few terms are effective because the pulse amplitude, oscillating in the ring cavity, becomes negligibly small after few roundtrips. Several configurations of passive optical pulse stretchers are developed. A passive optical pulse stretcher of square ring cavity configuration was used in a copper vapor laser MOPA system, where it stretched the oscillator pulse of duration of about 34 ns to about 50 ns (Amit *et al.*, 1987). A similar optical pulse stretcher of rectangular ring cavity was set up to stretch the pulses of a copper vapor laser from 60 ns to 72 ns at base (Singh *et al.*, 1995).

The technique of optical pulse stretching is suitable for temporal stretching of pulses of durations of few tens of nanoseconds. A very compact optical ring cavity would be required for temporal stretching of pulses of very short durations. A Fabry–Perot interferometer, which consists of two closely spaced parallel mirrors of high reflectivity, is a compact optical cavity and can also be used for temporal stretching of subnanosecond pulses. The degree of pulse stretching depends on the spacing between the two mirrors and their reflectivity. A Fabry-Perot interferometer with mirrors of reflectivity 98% was used for temporal stretching of 290 ps pulses of a mode-locked Nd:YAG to two to five times by changing the mirror spacing (Martin, 1977). If the pulse duration of the pulse, to be stretched by the optical pulse stretcher, were smaller than the roundtrip time of the ring cavity then a series of separated pulses would be obtained rather than a single stretched pulse. An optical pulse stretcher, consisting of three 100% reflecting plane mirrors and a beam-splitter of reflectivity 40% in a rectangular ring cavity of length 3.3 m (roundtrip time = 11 ns), applied to a Q-switched Nd:YAG laser to stretch its 7 ns pulses, generated a series of separate pulses instead of a single stretched pulse (Beyrau *et al.*, 2004). If a series of separate pulses are not desired, then a laser pulse stretcher with multiple optical ring cavities can be used to obtain a single stretched pulse. In this arrangement, several basic ring cavities are arranged in tandem such that the output of the first ring cavity is used as the input to the second ring cavity and so on. This configuration of multiple ring cavities not only increases the stretching factor but also recombines all the pulses in such a way that a single stretched pulse with longer temporal width is obtained. The roundtrip time of the shortest ring cavity, the number of ring cavities and the ratio of roundtrip times of different ring cavities of the pulse stretcher are chosen in such a way that the a stretched pulse is single and smooth. It is shown that a pulse stretcher, which consists of multiple ring cavities such that the roundtrip times of subsequent cavities decrease in geometric progression and the roundtrip time of the shortest cavity equals to the pulse width (FWHM) of the original pulse, is the best to recombine multiple pulses into a smooth single output pulse (Kojima *et al.*, 2002). The advantage of pulse stretcher with multiple ring cavities is that each additional cavity further stretches the original pulse, however it also increases the optical propagation distance. An optical pulse stretcher, consisting of three triangular ring cavities with roundtrip times of 35.2 ns, 16.7 ns and 9.03 ns (in the ratio of ~ 4:2:1) temporally stretched a 8.4 ns laser pulse of a Q-switched Nd:YAG laser to a 150 ns long (FWHM = 75 ns) single pulse with a peak power reduction of 0.10× and 83% efficiency (Kojima *et al.*, 2002). The configuration of multiple ring cavities in the pulse stretcher is such that the removal of the beam-splitter from the path of laser beam makes that ring cavity ineffective. Therefore the beam-splitters of the ring cavities of the pulse stretcher can be mounted on linear translation stages to have different combinations of the ring cavities to operate to provide stretched pulses of different lengths without disturbing the whole optical layout. An optical pulse stretcher, consisting of two ring cavities of 12 and 6 m long, not only stretched the 24 ns pulses of the 193 nm ArF excimer laser but also allowed a fast switching between different pulse lengths (24, 60, 63, and 122 ns) by operating the motorized translation stages on which the beam splitters were mounted (Burkert *et al.*, 2010).

In the optical pulse stretchers, discussed so far, plane mirrors are used and consequently the size of that part of the beam, which traverses longer path length, becomes larger at the beam-splitter due to finite beam divergence of the laser beam. The gradual growth of beam diameter in each roundtrip results in imperfect spatial overlapping between different parts of the laser beam at the beam-splitter. This also causes requirement of large diameter optics

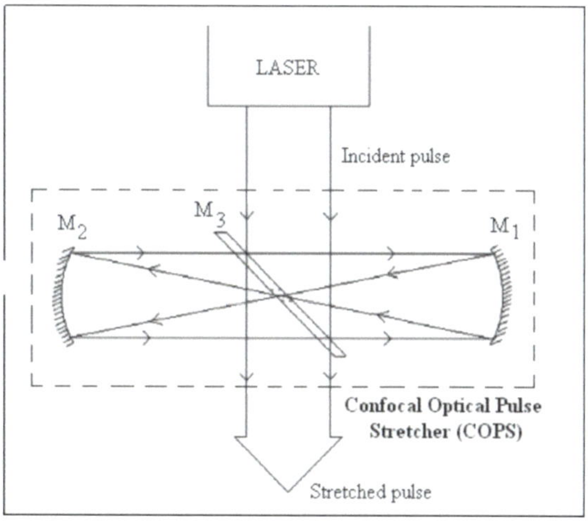

Fig. 2. Schematics of the confocal optical pulse stretcher (COPS). M_1, M_2: 100% reflecting concave mirrors of equal focal lengths; M_3: partially reflecting plane mirror with a hole at the center (scraper beam-splitter).

because the size of the laser beam exiting from the pulse stretcher becomes significantly larger than the original size. The large size of the temporally stretched beam requires an additional telescopic optics for reducing the size of the beam to inject it in the amplifier in typical oscillator–amplifier experiments. To overcome these problems a confocal optical pulse stretcher (COPS) was designed where the increase of the beam size due to beam divergence was compensated by the confocal nature of the pulse stretcher (*Khare et al.*, 2009). Irrespective of the beam divergence of the laser beam, the sizes of the successive parts of the beams remain same at the beam-splitter in the COPS, which ensures perfect spatial overlapping of laser pulses and also removes the requirement of additional telescopic optics. The optical configuration of the COPS is shown in Fig. 2. Two 100% reflecting concave mirrors M_1 and M_2, of equal focal lengths, f, are used to form confocal cavity of COPS. A partially reflecting plane mirror with a hole at the centre (scraper beam-splitter), M_3, is used as a beam-splitter in the COPS. The central hole of M_3 is made at an angle of 45° with respect to normal to the surface and M_3 is configured at the common focal plane at an angle of 45° with respect to the axis of the confocal cavity of the COPS. The COPS is configured such that the laser beam falls on M_3 at angle of 45° with respect to the normal to the surface. Thus the axis of COPS becomes perpendicular to the direction of propagation of the laser beam. The beam-splitter, M_3 splits the incident laser beam in two parts such that the transmitted part goes out of the COPS while the reflected part traverses towards M_1. The concave mirror, M_1, reflects and focuses the beam. The focused beam passes fully through the hole of M_3 without obstruction because M_3 is configured at an angle of 45° to the axis of COPS at the focal plane and the hole is also made at an angle of 45° with respect to normal to the surface of M_3. The hole removes the problems of mirror damage by the focused beam. After passing through the hole of M_3, the beam diverges and falls on M_2, which reflects and collimates the beam. The scraper beam-splitter, M_3, reflects a part of the beam out of COPS, which gets combined with the earlier transmitted beam. The part of the laser beam, which was transmitted by M_3, falls on the mirror M_1 again and the process repeats. The optical delay time, τ, is the round-trip propagation time of COPS and is given by

$$\tau = 4f/c \tag{3}$$

where c is the speed (i.e., group velocity) of light in air. The intracavity leakage rate of COPS depends on reflectivity of M_3 only, if absorption losses of M_3 are zero. Suppose the reflectivity of M_3, whose absorption losses are zero, is R% and p(t) is the instantaneous power of the pulse, which is incident on COPS, then M_3 divides the pulse such that transmitted part (1–R)p(t), goes out of COPS and reflected part, Rp(t), circulates inside COPS. The process repeats and the COPS divides an initially large amplitude laser pulse into many smaller amplitude pulses such that the successive pulses exiting from the COPS are temporally shifted with respect to the initial pulse by τ, 2τ,. . ., nτ, where n is an integer representing the number of round trips in the cavity. The integrated instantaneous power of the pulse, obtained as the output of the COPS is given by

$$\begin{aligned} P(t) &= (1-R)p(t) + R^2p(t-\tau) + R^2(1-R)p(t-2\tau) + R^2(1-R)^2p(t-3\tau) + R^2(1-R)^3p(t-4\tau) \\ &\quad + R^2(1-R)^4p(t-5\tau) + \cdots + R^2(1-R)^{n-1}p(t-n\tau) + \cdots \\ &= (1-R)p(t) + \sum_{n=1}^{\infty} R^2(1-R)^{n-1}p(t-n\tau) \end{aligned} \tag{4}$$

In Eq. (4), the 1st term is initial transmission of M_3, the 2nd term is reflection of M_3 after 1st roundtrip, the 3rd term is reflection of M_3 after 2nd roundtrip, and so on. The temporal stretching of pulses by COPS of fixed length is determined by the reflectivity of M_3. The reflectivity of M_3 is chosen such that more number of terms of Eq. (4) are effective such that energy content of the successive pulses, exiting from COPS, are significant to contribute effectively to have maximum temporal stretching. A simple reasoning indicates that the temporal stretching of the pulse would be a maximum for that value of R of M_3, which splits the pulses such that the energy content of the later split pulses are more than the earlier spilt pulses. In case of COPS, the condition that the 2nd split pulse has more energy than the 1st split pulse is given by

$$R^2 > (1-R) \Rightarrow R^2 + R - 1 > 0 \Rightarrow R > 0.62 \tag{5}$$

The condition that the 3rd split pulse has more energy than that of the 2nd split pulse is given by

$$R^2(1-R) > R^2 \tag{6}$$

However, Eq. (6) cannot be satisfied for any value of R lying in the range $0 < R < 1$. In fact, it is evident from Eq. (3) that the nth split pulse cannot have more energy the (n–1)th pulse for all values of n except for n = 1. The value of R, which provides maximum energy in the 3rd split pulse, is given by

$$\frac{\partial}{\partial R}\left[R^2(1-R)\right] = 0 \Rightarrow R = 0.67 \tag{7}$$

The value of R, which provides maximum energy in the 4th split pulse, is given by

$$\frac{\partial}{\partial R}\left[R^2(1-R)^2\right] = 0 \Rightarrow R = 0.5 \tag{8}$$

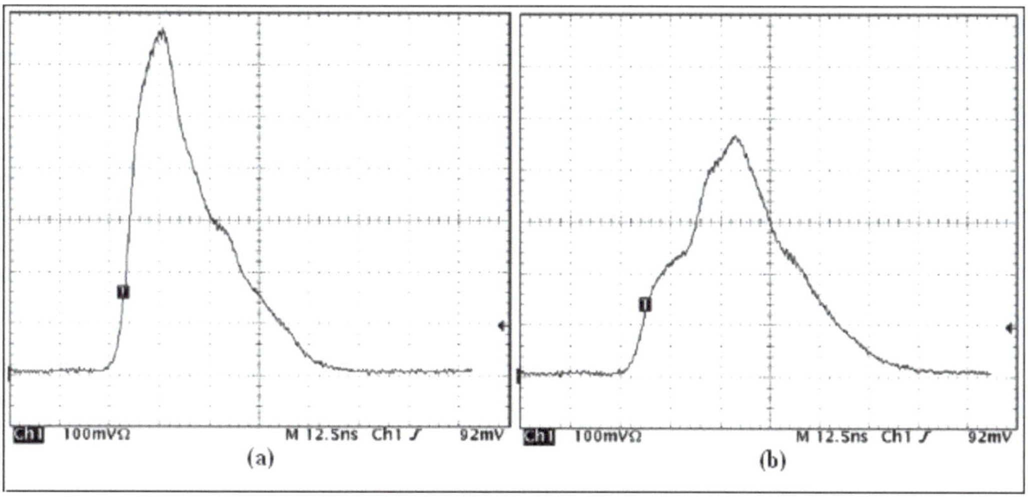

Fig. 3. Incident pulse (a) and stretched pulse (b) of the copper vapor laser by the COPS.

The value of R, given by Eq. (8) clashes with earlier requirements, given by Eq. (5). Therefore a value of R = 0.5 would not allow the 2nd split pulse to have more energy than 1st split pulse and the 3rd split pulse would have only 12.5% of incident pulse energy. Furthermore in this case, the energy of the 4th split pulse would be only 0.0625% of incident pulse energy, which would not contribute significantly in the process of pulse stretching. Consequently the value of R = 0.5 would not result in optimum stretching of the incident pulse. The value of R = 0.67 of M_3 would provide maximum stretching of incident pulse by COPS. Fig. 3 shows the performance of a COPS, consisting of two 100% reflecting concave mirrors of focal lengths 1.0 m each and a scraper beam-splitter of 70% reflectivity, which stretched a 40 ns pulse of a copper vapor laser to 55 ns, without loss of pulse energy (Khare *et al.*, 2009).

The ABCD matrix of the COPS for each roundtrip is given by

$$\begin{pmatrix} 1 & f \\ 0 & 1 \end{pmatrix}\begin{pmatrix} -1 & 2f \\ 0 & -1 \end{pmatrix}\begin{pmatrix} 1 & f \\ 0 & 1 \end{pmatrix} = \begin{pmatrix} -1 & 0 \\ 0 & -1 \end{pmatrix} = -\begin{pmatrix} 1 & 0 \\ 0 & 1 \end{pmatrix} \qquad (9)$$

The RHS of Eq. (1) is a unit matrix with a negative sign. Thus the size of the beam remains unchanged in COPS in each round trip but the spatial profile is inverted. A two fold 1:1 imaging at the beam splitter of the pulse stretcher can be implemented by using 4f confocal arrangement of four concave mirrors of equal focal lengths, aligned in the off axis configuration and a partially reflecting plane mirror as the beam splitter. This configuration ensures complete true-sided overlap of the spatial profile of the delayed pulses with the original pulse at the beam splitter. A 4f confocal pulse stretcher, consisting of four concave mirrors of focal lengths 1.5 m and a beam splitter of reflectivity 60%, stretched the 24 ns pulse of an ArF excimer laser to 60 ns (Burkert *et al.*, 2010).

Most of the optical pulse stretchers use beam-splitters and mirrors in an optical ring-cavity to obtain a temporally stretched laser pulse. The optical fibers, which are very convenient tool for transmission of laser beams, can also be used in place of beam-splitters and mirrors to design a passive optical pulse stretcher. The optical fibers of different lengths are bundled together at both ends. In this configuration of optical fibers, a single laser pulse, injected at one end of the fiber bundle would be split spatially into many parts among the various fibers and each fraction would traverse a different length of optical fiber to reach at the other end. At the other end of the bundle, the optical fibers are grouped together and the parts of the laser pulse are recombined spatially into a single beam. Therefore a laser pulse, which is injected at one end of bundle of optical fibers, would emerge from the other end as a temporally stretched pulse. This optical pulse stretcher is flexible and its efficiency is limited primarily by the core-to-cladding ratio of the fibers and packing density of the bundle. Two optical pulse stretchers consisting of a bundle of 80 optical fibers of core diameter of 200 μm and outside diameter of 230 μm, with lengths varying from 0.5 to 4.0 m and 0.5 to 8.0 m at an increment of 0.5 m, provided a temporal stretching of 15 ns and 30 ns respectively with 40% transmission efficiency (Hanna & Mitchell, 1993).

Prisms, instead of mirrors, can also be configured to introduce an optical delay in a passive optical pulse stretcher. When a laser beam falls on a prism, a part of the beam is reflected and the remaining part is transmitted through it. The prisms can be arranged in such a way that the transmitted parts of laser beam traverse a closed path and then join the earlier reflected as well as transmitted parts. This configuration of prism generates partial pulses from the original laser pulse and introduces optical delays between them to provide a temporally stretched pulse. The reflectivity of the prism surfaces, the angle of incidence of the laser beam on the prism, polarization selectivity of the glass/air, the physical dimensions of the prisms and their separation are the key parameters, which determine the stretching factor. Here the dispersion characteristics of the prism, which play an important role in dispersive pulse stretchers, are not used. The 100 ps pulses of a mode locked Nd:YAG laser were stretched to 200 ps by a passive optical pulse stretcher using a pair of prisms, which formed a unidirectional light loop based on total internal reflections and polarization selectivity of the glass/air interface (Tóth, 1995).

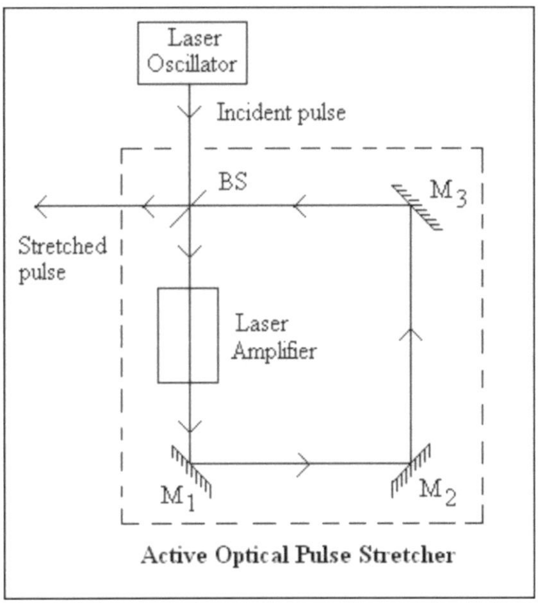

Fig. 4. Schematics of the active optical pulse stretcher. M_1, M_2, M_3: 100% reflecting plane mirrors; BS: Beam-splitter.

Generally the optical pulse stretchers operate in the passive mode, however the optical pulse stretchers can be configured in active mode also by providing gain inside the ring cavity. The optical configuration of an active pulse stretcher that incorporates a laser amplifier is shown in Fig. 4. Thus the ring cavity of the active pulse stretcher divides the initial large amplitude laser pulse from the laser oscillator into many smaller amplitude laser pulses and releases them over a longer period of time in such a way that each part is amplified by the laser amplifier. The amplified pulses, which are released over a longer period of time, are recombined at the beam-splitter. The active pulse-stretchers are more effective in comparison to passive pulse stretchers in temporal stretching of incident pulses due to inherent amplification of each part during successive roundtrips in the ring cavity. An active optical pulse stretcher of cavity length of 6 m, applied to copper vapor laser MOPA system, temporally stretched the 18 ns (FWHM) oscillator pulse of 510.6 nm to 80 ns (FWHM), while a similar passive optical pulse-stretcher under identical conditions provided temporal stretching to 50 ns only (*Kundu et al.*, 1995).

2.2 Pulse stretching by intracavity nonlinear materials

Temporal pulse stretching by intracavity nonlinear materials are used in Q-switched lasers for incresaing the duration of laser pulses. The Q-switched lasers emit pulses of durations of several tens of nanoseconds. In these nonlinear materials, the absorption of light increases more than linearly with increasing intensity of light. The absorption as a function of laser intensity of a typical nonlinear absorber is shown in Fig. 5. If such a nonlinear material is inserted in the cavity of a Q-switched laser, the intracavity losses increase nonlinearly with building up of photon flux, which prevents rapid depletion of inverted gain population. The existence of gain for a prolonged period in a Q-switched laser leads to temporal pulse stretching of the laser pulses with reduced peak powers. The optical process, which gives rise to nonlinear absorption in the material depends upon the material used and the intensity of the laser within the cavity. The nonlinear mechanism increases the nonlinear losses with increasing intracavity power and thus limits the power and lengthens the pulse duration, whereas the output energy remains constant under ideal conditions. There are several nonlinear optical mechanisms, such as two-photon absorption, excited-state absorption, harmonic generation, parametric oscillation, polarization ellipse rotation, stimulated scattering processes including Rayleigh, Raman and Brillouin, etc., which introduce nonlinear absorption in the materials.

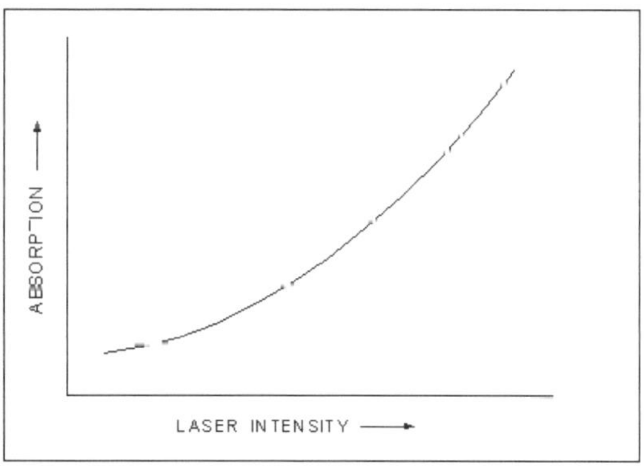

Fig. 5. The absorption characteristics of a nonlinear absorber.

The two-photon absorption (TPA) is a nonlinear optical process in which two photons are absorbed simultaneously, such that the energy of the photons adds up to the energy of the excited atom or molecule. The rate of TPA is proportional to the square of the radiation intensity. The pulse duration of a Q-switched laser has been increased by insertion of a two-photon absorbing semiconductors in the laser cavity. An important advantage of the two-photon nonlinearity in semiconductors is its localized nature and instantaneous response, which not only lengthens a pulse but also smoothens out its spatial and temporal inhomogeneities. A theoretical and experimental investigation confirmed that the pulse stretching of a Q-switched ruby laser by insertion of a CdS crystal in the cavity was due to the two-photon absorption in CdS (Hordvik, 1970). The pulse durations were increased by one to two orders of magnitude by two-photon absorption in the nonlinear optical crystals of GaAs and CdS in the laser cavities of Q-switched Ruby and Nd-glass lasers respectively (Schwartz *et al.*, 1967). The pulse stretching depends on the nature of the nonlinear losses and on the rate of Q-switching. The insertion of nonlinearly absorbing semiconductors CdP_2 and ZnP_2 into the resonators of ruby and neodymium lasers increased the pulse durations by different amounts. The 20 ns pulses of ruby laser were increased to 360 ns and 290 ns by CdP_2 and ZnP_2 respectivly while the 25 ns pulses of neodymium laser were increased to 190 ns and 150 ns repectively (Lisitsa *et al.*, 1974). The pulse duration of 30 ns of Q-swiched ruby and neodymium lasers were stretched to continuously variable duration up to 350 ns and 400 ns respectively by insersion of two-photon absorbing semiconductor plates, cut from single crystals of CdS and CdSe, at Brewster angle in the respective cavities (Arsen'ev *et al.*, 1972). The use of readily available semi-insulating form of GaAs as a two photon absorber in a Q-switched Nd:YAG laser stretched the pulses from 30 ns to 150 ns while high-purity n-type GaAs stretched the pulses up to 1.5 μs (Walker & Alcock, 1976). The solution of Rhodamine 6G was used as a two-photon absorber in the cavity of a Q-switched Nd:glass laser and a pulse stretching of about 8 times (FWHM ~ 350 ns) of 45 ns pulse was obtained (Bergamasco *et al.*, 1993). A crystal of ZnSe was used as a two-photon absorber in a Q-switched alexandrite laser (λ = 750 nm) to achieve temporal stretching of pulse-widths (FWHM) up to 1.7 μs of 5 ns pulses (Rambo *et al.*, 1999). The excited state absorption (ESA) is a nonlinear optical process in which the absorption cross-section of an excited state is more than that of the ground state and the rate of ESA increases nonlinearly with the radiation intensity. Insertion of a single crystal of GaAs at the Brewster angle in the cavity of a Q-switched Nd:glass laser produced temporal pulse stretching with greater spatial homogeneity due to ESA (Schwarz *et al.*, 1967).

The second harmonic generation (SHG) is a nonlinear optical process, in which a radiation interacts with a nonlinear material to generate a radiation with twice the frequency of the incident radiation. The nonlinear crystal inside the cavity of the Q-switched lasers acts as a variable output coupler due to the nonlinear power conversion from the fundamental wave to the second harmonic. The second harmonic generation by the intracavity nonlinear crystal introduces nonlinear power dependent internal loss, which leads to temporal pulse stretching of the laser pulses. In applications where the desired output is the second harmonic of the pumping laser, this approach is particularly advantageous because the pulse stretching is achieved at no expense in output energy (Murray & Harris, 1970). A lithium iodate ($LiIO_3$) crystal was placed inside the cavity of a Q-swiched Nd:YAG laser, operating at 0.946 μm, to provide both output coupling and an easily adjustable, nonlinear loss mechanism for stretching pulses in the range of 200 ns to 1 μs at 0.473 μm at constant energy (Young *et al.*, 1971). A temporal pulse stretching in the range of 100 ns to 2 μs was obtained in a Q-switched Yb:YAB laser by self frequency doubling in the Yb:YAB crystal itself (Dekker *et al.*, 2005). The temporal pulse stretching to durations (FWHM) 650 ns and 3.2 μs was achieved without significant loss of pulse energy in a Q-switched Nd:YLF laser by using the effect of overcoupled intracavity second harmonic generation with intracavity KTP crystals of lengths 5 and 10 mm respectively while the pulses were stretched to 600 ns and 1.5 μs by the intracavity LBO crystals of lengths 10 and 20 mm respectively in the same set-up (Kracht & Brinkmann 2004).

The polarization ellipse of a laser beam rotates as the laser beam propagates through a nonlinear medium (Maker & Terhune, 1965). The rotation arises from the tensor nature of the third-order nonlinear susceptibility and the magnitude of the rotation increases with increasing laser intensity. The induced optical birefringence and ellipse rotation is used for intensity-dependent intracavity loss in Q-switched lasers to slow down the release of optical energy from the laser cavity. The field-induced loss limits the cavity intensity and consequently the excess energy stored in the inverted population appears in the form of a longer pulse. To use the rotation of polarization ellipse as an inducible loss in the laser cavity, a nonlinear device, consisting of an optical nonlinear medium through which intensity-dependent ellipse rotation can occur, a polarizer followed by a retardation plate to elliptically polarize the beam before it enters the nonlinear medium, and an analyzer after the nonlinear medium to observe the ellipse rotation in the nonlinear medium, is used. These elements are arranged in such a way that the cavity loss is a minimum without ellipse rotation. As ellipse rotation increases with the cavity field intensity, transmission through the analyzer decreases, leading to an increase in the cavity loss. A nematic liquid crystal is a transparent liquid that causes the polarization of light waves to change as the waves pass through the liquid. The extent of the change in polarization depends on the intensity of the applied electric field. Pulse stretching was demonstrated in a Q-switched ruby laser by using a nematic liquid crystal EBBA (p-ethoxy-benzylidene-p-butylaniline) as the nonlinear optical ellipse rotation medium in the cavity to introduce intensity dependent cavity loss due to rotation of polarization ellipse (Murphy & Chang, 1977). Stretching of pulses of 70 ns duration by a factor of 4 has been observed by placing liquid crystal MBBA (n-p-methoxybenzylidene-p-butylaniline) in the cavity of a Q-switched ruby laser along with other polarizing elements for self-induced ellipse rotation (Hsu & Shen, 1982).

A light scattering process is said to be stimulated if the fluctuations in the optical properties that cause the light scattering are induced by the presence of the light field. Stimulated light scattering is much more efficient than spontaneous light scattering. The stimulated scattering causes a nonlinear loss at the laser frequency since part of the laser energy is transferred to a different frequency. Stimulated Rayleigh-Wing scattering (SRWS) is the stimulated scattering process resulting from the tendency of anisotropic molecules to become aligned along the electric field direction of an applied optical wave. SRWS leads to the generation of a wave shifted in frequency by approximately the inverse of the orientational relaxation time of the anisotropic molecules (Miller *et al.*, 1990). The 40 ns pulses of a Q-switched ruby laser were temporally stretched to 150-400 ns by introduction of a benzene cell, which acted as a Rayleigh-wing scatterer, into the laser cavity (Callen *et. al.*, 1969). The stimulated Brillouin scattering (SBS) is a third-order nonlinear effect, which arises from the interaction of light with propagating density waves or acoustic phonons. Intracavity stimulated Brillouin scattering (SBS) in hexafluorethane (C_2F_6) was used for temporal stretching of 60 ns Q-switched pulses from a phase-conjugated Nd:YAG laser to 300 ns (Seidel & Phillipps, 1993).

2.3 Electronic pulse stretching

The technique of electronic pulse stretching is based on electronic feedback circuits, which control the electro-optical devices of *Q*-switched lasers in such a way that the laser pulses are temporally stretched. This is an active pulse stretching technique, where the electronic feedback loop modifies the switching of the electro-optic shutter to control the temporal losses in the Q-switched lasers. In several Q-switched lasers, an electronic circuit is used to switch voltages to the electro-optical device (Pockels or Kerr cell), placed inside the laser resonator, to produce controllable birefringence. At the beginning of the Q-switching process, the optical resonator is closed (high losses) by applying a high voltage to the electro-optical device for population inversion to reach to a level far above the threshold for normal lasing action. Then the voltage applied to the electro-optical device is removed to restore the high Q of the cavity and consequently the stored energy is suddenly released in the form of a very short pulse of high energy. This short pulse of high energy can be temporally stretched by applying a negative feedback to the electro-optical shutter, used for Q-switching, in such a way that its reopening does not remain rapid and the stored energy is released for a longer duration. This is the basic principal of operation of the electronic pulse stretchers. The block diagram of a simple feedback controlled electro-optical circuit for electronic pulse stretching of a Q-switched laser is shown in Fig. 6. The Q-switched laser consists of a gain medium, *G*, polarizing optics, *PO*, and the electro-optical device, *EO*. The mirrors M_1 (100%) and M_2 (partially transmitting) form the optical cavity of the laser. The gain medium is optically pumped by a flashlamp. A beam-splitter, *BS*, is inserted in the optical cavity of the Q-switched laser to direct a small part (5% to 10%) of the intracavity photon flux into a photodiode, *PD*. *R* is the load resistor. Negative feedback on the circulating power is applied through the action of the photodiode derived voltage on the electro-optical device, *EO*. In this way the stored energy in the gain medium, *G*, is released at a controlled rate.

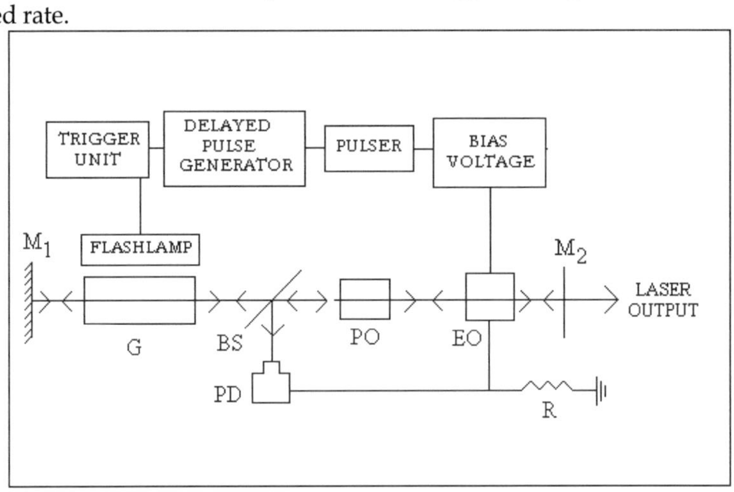

Fig. 6. Block diagram of an electronic pulse stretcher. G: gain medium of the Q-switched laser; PO: polarizing optics; EO: electro-optical device; M_1, M_2: resonator mirrors; BS: beam-splitter; PD: photodiode; R: resistor.

Several designs of electronic circuits with sophisticated feedback loops have been designed to regulate the voltages on the electro-optical devices for temporal stretching of pulses of the Q-switched lasers. A feedback circuit connected to the Kerr cell of a Q-switched ruby laser was used to temporally stretch pulses of 5 ns duration to 5 μsec (Thomas & Price, 1966). The most promising of this kind of active pulse stretching techniques use a compound feedback circuit with negative and positive feedbacks to have stretched pulses of desired shapes. This allows the laser oscillation to build up, and a sustained laser action is ensured by the balance between the negative and positive feedbacks. These techniques are quite sophisticated and circuit parameters are highly optimized because the net high voltage applied to the Pockels or Kerr cell must follow precisely a prescribed waveform. A 40 ns pulse of a Q-switched ruby laser was stretched to a flat-top pulse of 240 ns by using a compound control circuit having negative and auxiliary positive feedback loops having a photodiode, a KD*P (potassium dihydrogen phosphate) Pockels cell, a Krytron switch, and the optimized values of inductance and shunt capacitance (Lovberg *et al.*, 1975). An electronic feedback circuit with controllable parameters of resistance across the photodiode, Pockels cell bias voltage, Pockels-cell open time, and delay time of the delay pulse generator resulted in stretching the 200 ns pulses a Q-switched Nd: glass laser to 50 μs pulses of smooth near-rectangular shape and otherwise stretchable up to 400 μs (Panarella & Brdley, 1975). The pulses of a Q-switched Nd:YAG laser with an intracavity polarizer, a quarterwave plate, and a Pockels cell driven by a high voltage planar triode were temporally stretched from 30 ns to 650 ns with flatness within ±7% for 505 ns by a dual feedback circuit, which provided sufficient feedback to put the cavity above threshold, but keeping the cavity *Q* low enough to generate a pulse with a moderate growth rate (Schmid, 1980). A compound feedback circuit and a clamping circuit, comprising of a vacuum photodiode and thyristors as switching elements, were used to control the voltage on a Pockels cell of a flashlamp pumped Q-switched ruby laser oscillator for stretching its 30 ns pulses to flat pulses of durations up to 100 μs by varying the oscillator voltage, Pockels cell voltage, clamping circuit voltage and the delay between firing of the flashlamp and opening of the Pockels cell (Harigel *et al.*, 1986). A closed loop controller circuit with feedback and feedforward portions was designed to apply voltages of appropriate waveforms to the opposite sides of the Pockels cell of a Q-switched ruby laser to stretch its pulses of 30 ns (FWHM) duration to fairly flat pulses with adjustable duration between 200 ns to 16 μs (Verluyten *et al.*, 1990). A pulse stretching circuit, consisting of a pin photodiode, a transimpedance amplifier, a derivative element, a push–pull amplifier, and field effect transistors in the feedback loops, controlled the net voltage on the Pockels cell of a Q-switched ruby laser to stretch its pulses to durations from 200 ns to 1 μs (Pflüger *et al.*, 1996).

2.4 Pulse stretching by dispersion

In case of ultrashort pulses, the technique of pulse stretching by dispersion is used. A dispersive pulse stretcher is suitable for temporal stretching of ultrashort pulses because the spectral bandwidth of the ultrashort pulses is very large. The dispersive pulse stretcher is a configuration of dispersive optical elements that introduce wavelength dependent optical delays to introduce frequency chirp for temporal stretching of ultrashort laser pulses. The term frequency chirp means temporal arrangement of the frequency components of the laser pulse.

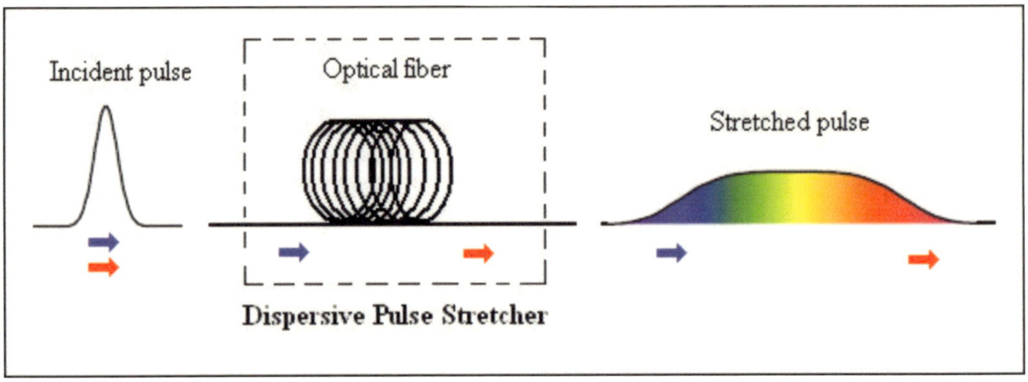

Fig. 7. Schematics of a dispersive pulse stretcher using an optical fiber.

Generally optical fibers or diffraction gratings are used for frequency chirping. In a medium with positive dispersion, the higher frequency components of the pulse travel slower than the lower frequency components and the pulse becomes positively-chirped or up-chirped, increasing in frequency with time. In a medium with negative dispersion, the higher frequency components travel faster than the lower ones and the pulse becomes negatively chirped or down-chirped, decreasing in frequency with time.

A dispersive pulse stretcher using an optical fiber is shown in Fig. 7. In this configuration, the long-wavelength (red) components of the pulse propogate faster than its short-wavelength (blue) components and the pulse is temporally stretched. The effect of pulse stretching in a dispersive pulse stretcher is generation of a 'temporal rainbow' with longer pulse duration and lower peak power. A 1.3 km long single mode silica fiber of 9 μm core was used to temporally stretch 55 ps pulses of Nd:YLF laser to 300 ps across a 3.5 nm bandwidth for chirped pulse amplification (Maine *et al.*, 1988). A multimode optical fiber with length of 50 m and core of 200 μm diameter was used to temporally stretch laser pulse of 160 fs duration of a mode-locked Ti:sapphire laser to nearly 300 ps for combined laser-SR (synchrotron radiation) spectroscopy (Azuma *et al.*, 2001).

Diffraction gratings provide large stretching factors and by using diffraction gratings based dispersive pulse stretchers, ultrashort optical pulses are stretched to more than 1000 times to their original duration. The dispersive pulse stretcher, shown in Fig. 8, is based on parallel configuration of a pair of gratings, G_1 and G_2. The ultrashort pulse, incident on the first grating, G_1, is separated into its component wavelengths. The beams of different wavelengths propagate along different directions in the space between the two gratings. After diffraction by the second grating, G_2, the beams of different wavelengths become parallel but they have traversed different distances. The pulses of longer wavelengths (red) have traversed longer distance in comparison to pulses of shorter wavelengths (blue). Thus the pulses of higher frequency arrive earlier than lower frequency pulses in this configuration of parallel gratings.

The dispersive pulse stretcher, shown in Fig. 9, consists of two reflection gratings, two achromatic lenses and a retro-mirror pair. In this arrangement the two gratings, G_1 and G_2, are aligned in anti-parallel configuration. Between the gratings, the two lenses, L_1 and L_2, of equal focal lengths, f, are placed at a distance $2f$ from each other to form a telescope of magnification 1. The telescopic arrangement is used to ensure that the dispersing radiation returns parallel after passing the two dispersive gratings. The telescope reverses the sign

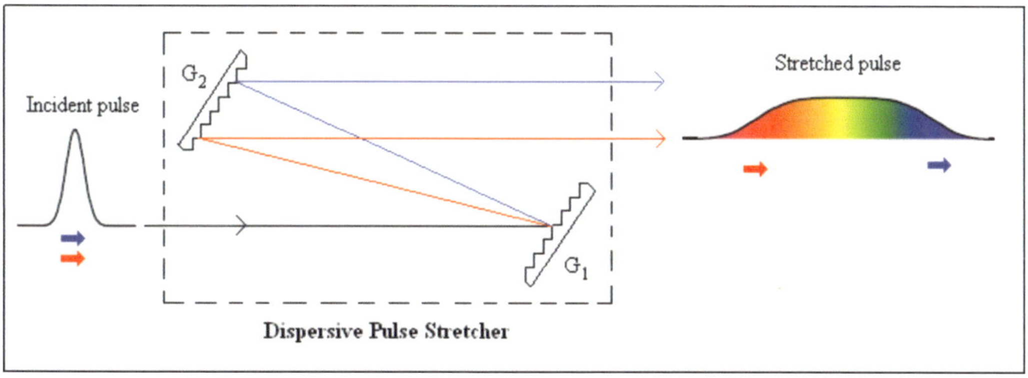

Fig. 8. Schematics of a dispersive pulse stretcher with a pair of parallel gratings. G_1, G_2: gratings.

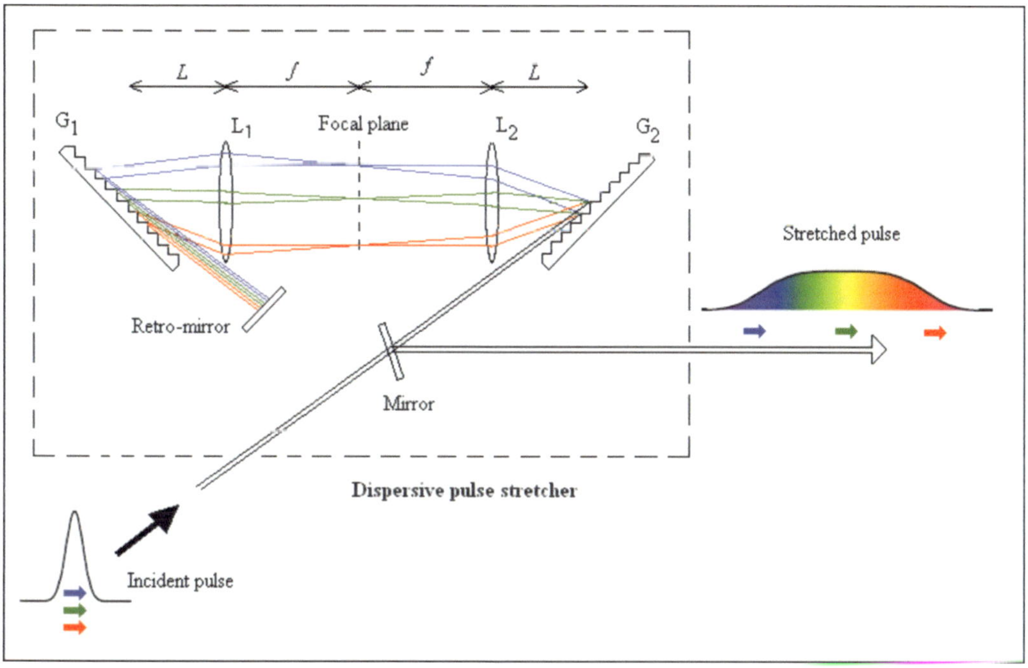

Fig. 9. Schematics of a dispersive pulse stretcher with a pair of anti-parallel gratings. L_1, L_2: lenses of focal lengths f; G_1, G_2: gratings.

of the angles so that the beams will meet the second grating at the same angle as they leave the first one. This arrangement provides positive dispersion if the lenses are placed at a distance L from the gratings such that $L < f$ (Martinez, 1987). In this configuration of gratings, the pulses of longer wavelengths (red) traverse shorter distances in comparison to pulses of shorter wavelengths (blue). The retro-mirror is used for retro-reflecting the dispersed beam back in the optical arrangement for removing the lateral walk and also to retain the beam shape. The double pass arrangement increases the pulse stretching by two-fold. This type of dispersive pulse stretcher was used to temporally stretch laser pulses of duration 250 fs (FWHM) from a Nd:flouro phosphate glass oscillator to 240 ps (FWHM) by placing a pair of identical holographic gratings (groove density = 1740 l/mm) symmetrically outside of two identical achromatic lenses (focal length = 500 mm) at a distance of 250 mm and using a retro-reflector to obtain a linear chirp of 44 ps/nm at central wavelength of 1054 nm (Sharma *et al.*, 2005).

3. Significance of temporal stretching of laser pulses

The temporal stretching of laser pulses plays a significant role in many applications of pulsed lasers. The necessity of temporal pulse stretching in certain applications of pulsed lasers along with use of different laser pulse stretchers is discussed.

Temporal pulse-streching of the laser beam from master oscillator is required in master oscillator power amplifier (MOPA) systems, which are used to obtain high power with good beam quality and good efficiency. In the MOPA system, the beam quality (linewidth, divergence, coherence, etc.) of the laser beam is controlled by the master oscillator while the amplifier is used to increase the power of the beam. Generally the pulses of the oscillator are temporally shorter than the amplifier pulses and this leads to incomplete overlap of oscillator pulses with the duration for which gain is available in the amplifier. Consequently the MOPA system suffers from poor extraction efficiency along with generation of amplified spontaneous emission (ASE), which degrades the beam quality of the amplified output. Thus a laser pulse stretcher is generally configured in a MOPA system for achieving maximum extraction efficiency and elimination of amplifier ASE. A passive optical pulse stretcher was used in the MOPA system, which increased the output from 18 W to 25 W along with elimination of ASE because of increase of duration of oscillator pulses from 34 ns to 50 ns (Amit *et al.* 1987). A passive optical pulse stretcher was configured in a MOPA chain, consisting of an unstable oscillator and two amplifiers, to obtain an output of 50 W due to stretching of pulses of the copper vapor laser to 72 ns (Kumar *et al.* 1995).

In dye lasers pumped by pulsed lasers, the temporal stretching of pump laser pulses increases the energy conversion efficiency and tuning range of the dye laser along with a decrease in its linewidth. The linewidth of the dye laser output depends mainly on the passive linewidth of the resonator and the number of roundtrips taking place in the resonator (Khare *et al.*, 1995). The number of round-trips depends upon the pulse duration of the pump laser. A passive optical pulse stretcher was configured in a copper vapor laser pumped dye laser system to stretch the pulses of copper vapor laser from 60 ns to 72 ns, which resulted in increasing the pulse duration of dye laser from 20 ns to 28 ns and this led to increasing the energy conversion efficiency and spectral tuning range of the dye laser from 8% to 10%, and 37 nm to 47 nm respectively along with decreasing the linewidth from 0.08 cm^{-1} to 0.06 cm^{-1} (Singh *et al.*, 1995). The temporal stretching of the pump laser by a passive optical pulse stretcher in a copper vapor laser pumped dye amplifier, not only increased the extraction efficiency of the amplifier but also smoothened the pulse shapes due to increase in tolerance on synchronization between pump and signal pulses (Singh *et al.*, 1994).

The temporal stretching of the pump laser pulses is done in the experiments of pulsed amplification of a continuous-wave (cw) laser, which combines the power of the pulsed laser with the spectral qualities of the cw laser. The temporal stretching of the duty cycle of the pump laser pulses reduces the linewidth of the amplified pulses. In the pulsed amplification of cw He–Ne laser by a frequency doubled Nd:YAG laser, a reduction in the linewidth of the amplified pulses is obtained by temporal stretching of duration of pump laser pulses (Black & Valentini, 1994). The pulse stretching of the pump laser pulses from 8 ns to 12 ns in the pulsed amplification of cw ring dye laser by a frequency doubled Nd:YAG laser, reduced the bandwidth of the amplified pulses to 130 MHz from 180 MHz (Lee & Hahn, 1996).

The optical fibers are very convenient and useful tool for transmission and delivery of high power laser beams. The key problem in fiber-optic transmission is that the high peak-power beams cannot be coupled into fiber optics because of the damage of the end faces. (Seidel & Phillipps, 1993). The light intensities of *Q*-switched pulses focused to a diameter of 600 μm reach to 10^{11} W/cm^2 and are beyond the damage threshold of optically polished glass and fused silica (Lowdermilk & Milam, 1981). The material of the optical fiber can be prevented from intensity dependant damages by temporal stretching of laser pulses. The temporal stretching of laser pulses of a *Q*-switched ruby laser by an electronic pulse stretcher increased the upper limit of transmission of pulse energy by a factor of 4 and pulses of 300 mJ were transmitted in quartz fibers with a 600 μm core diameter without fiber damage (Pflüger *et al.*, 1996). Another disadvantage of high peak-power systems is the reduced applicability for material cutting because high peak power produces plasma on the surface, which absorbs most of the incident energy. A temporally stretched pulse containing the same output energy becomes more efficient for heating the material (Seidel & Phillipps, 1993). Thus the use of stretched pulses permits delivery of full pulse energy at reduced peak power, which not only avoids the damage of the optical fiber but also controls the laser-material interactions for improvement.

Excimer lasers provide nanosecond pulses in the ultraviolet spectral range with high pulse energies. Excimer lasers are widely used for optical microlithography however its high peak power degrades the optical material of the projection lens and adversely affects its imaging property. The high peak intensities lead to breakthrough in gases and damages in optical materials such as lenses and beam splitters. A cumulative damage mechanism is known in fused silica in addition to the well-known single pulse damage threshold (*Burkert et al.*, 2007). Sufficiently high pulse energies far below the single pulse damage threshold can lead to a densification of the glass, which is called compaction (*Liberman et al.*, 1999). Such density variations in a lens cause beam aberrations. Excimer lasers in combination with pulse stretchers provide pulses with lower peak intensity at the same pulse energy. Gas breakthrough and damages in optical materials can be reduced or at least retarded by pulse stretching. The temporal stretching of lasers pulses reduces the peak power and increases the lifetime of the scanner lens (Pätzel & Stamm, 2005).

The Spontaneous Raman-scattering (SRS) spectroscopy with pulsed lasers provides a quantitative measurement of major chemical species concentrations and temperatures in turbulent and chemically reacting flow processes such as combustion. To collect high-quality SRS data in gas phase flows, high-energy pulsed lasers are required to compensate for the weak signal levels generated by the Raman effect. However, high-energy Q-switched lasers often suffer from laser-induced plasma spark generation at the focused probe volumes. The strong optical emission from the plasma spark overwhelms the weak Raman-scattering signal, making spatially resolved measurements with high-energy Q-switched lasers difficult. The SRS signal is linearly proportional to the total energy of the laser pulse and not the intensity. Thus the SRS signals are measured by reducing the peak-power of the Q-switched laser in such a way that the total pulse energy is maintained. Hence temporal stretching of laser pulses was done in spontaneous Raman-scattering spectroscopy for removing laser-induced plasma spark generation by reducing peak power (Kojima & Nguyen, 2002). Temporal stretching of laser pulses is done by configuring a passive optical pulse stretcher in the coherent anti-Stokes Raman spectroscopy to increase signal-to-noise ratio and to improve precision of the determined temperatures and concentrations while avoiding window damage and laser-induced breakdown (Beyrau *et al.*, 2004).

One of the important applications of lasers is its use in medical sciences. In ophthalmology, the pulses of a Q-switched laser operating in the green spectral region is temporally stretched to obtain microsecond long pulses, which are most suited for to selectively target dysfunctional retinal pigment epithelium (RPE) while preserving the adjacent tissues, especially the photoreceptors and the neuronal retina (Kracht & Brinkmann 2004). The high peak intensities of shorter pulses of Q-switched lasers generally damage the tissues and thus the pulses were temporally stretched by using intracavity pulse stretcher with two-photon absorber (Rambo *et al.*, 1999). The pulse durations are stretched and optimized for lithotripsy (kidney stone fragmentation) and other medical applications.

Pulsed lasers are used in the studies of holographic interferometry. Pulse stretching of Q-switched ruby laser was required to reduce instantaneous power at constant energy to avoid the boiling of the chamber liquid, which adversely affected the quality of the photographs, taken to record high-energy neutrino interactions in bubble chamber holography (Harigel *et al.*, 1986). A remarkable asset of stretched pulses is the considerably increased coherence length compared with the Q-switched operation (Harigel *et al.*, 1986). The lasers are used in creating artificial guide stars for adaptive optics for ground-based astronomy by permitting correction of the optical aberrations due to the atmosphere. Pulse stretching of dye laser pulses was required in the sodium guide star experiment for correcting the adaptive optics system of large astronomical telescopes, which were used to study the effects of atmospheric turbulence. The optical pulse-stretcher stretched the 40 ns (FWHM) pulses of dye laser by a factor of 15 to lower the laser flux and avoid saturation effects in the sodium guide star experiment (Friedman *et al.*, 1992).

The chirped-pulse amplification (CPA) is the most important technique to amplify ultrashort pulses (*Salin et al.*, 1992). In the amplifiers, the peak intensities associated with the ultrashort pulses becomes so high that excessive nonlinear pulse distortions and optical damage take place. Dispersive pulse stretchers are used for temporal stretching of the ultrashort pulses in CPA, where the stretched pulses are amplified to moderate peak power levels to prevent damage of the amplification medium and then the amplified stretched pulses are compressed to its Fourier-transform limit to obtain high peak power pulses.

4. Conclusion

The physics and technology of a wide variety of techniques, developed for temporal stretching of laser pulses, is described. The significance of temporal stretching of laser pulses and their applications in different fields are discussed.

5. References

Amit, M.; Lavi, S.; Erez, G. & Miron, E. (1987). Temporal and spatial properties of an oscillator-amplifier copper vapor laser, *Optics Communications*, Vol. 62, No. 2, pp. 110-114.

Arsen'ev, V. V.; Dneprovskii, V. S. & Klyshku, D. N. (1972). Control of laser pulse duration by nonlinear absorption in semiconductors, *Soviet Journal of Quantum Electronics*, Vol. 2, No. 1, pp. 23-26.

Azuma, J.; Asaka, S.; Tsujibayashi, T.; Itoh, M.; Watanabe, M.; Arimoto, O.; Nakanishi, S.; Itoh, H. & Kamada, M. (2001). Optical detection system using time structure of UVSOR for combined laser–SR experiments. *Nuclear Instruments and Methods in Physics Research A*, Vol. 467–468, pp. 1455–1457

Beyrau, F.; Weikl, M. C.; Seeger, T. & Leipertz, A. (2004) Application of an optical pulse stretcher to coherent anti-Stokes Raman Spectroscopy. *Optics Letters*, Vol. 29, No. 20, pp. 2381-2383

Bergamasco, G.; Calveli, P.; Cecchetti, W. & Polloni, R. (1993). Passive pulse stretching in a Nd: glass laser. *Optics and Quantum Electronics*, Vol. 25, pp. 271-273

Black, J. F. & Valentini, J. J. (1994). Compact, high-gain pulsed dye amplifier for weak cw sources. *Applied Optics*, Vol. 33, No. 18, pp. 3861-3864

Burkert, A.; Triebel, W.; Natura, U. & Martin, R. (2007). Microchannel formation in fused silica during ArF excimer laser irradiation. *Physics and Chemistry of Glasses – European Journal of Glass Science and Technologt Part B*, Vol. 48, No. 3, pp. 107-112

Burkert, A.; Bergmann, J.; Triebel, W. & Natura, U. (2010). Pulse stretcher with variable pulse length for excimer laser applications. *Review of Scientific Instruments*. Vol. 81, pp. 033104-1 - 033104-5.

Callen, W. R.; Pantell, R. H. & Warszawski, J. (1969). Pulse stretching of *Q*-switched lasers. *Opto-electronics*, Vol. 1, pp. 123– 127.

Dekker, P.; Dawes, J. M. & Piper, J. A. (2005). 2.27-W *Q*-switched self-doubling Yb:YAB laser with controllable pulse length. *Journal of Optical Society of America B*, Vol. 22, No. 2 pp. 378-384

Friedman, H. W.; Horton, J. A.; Kuklo T. J. & Wong, N. J.(1992). Design of pulse stretching cell for a sodium guide star optical system. *Conference Proceedings of Lasers and Electro-Optics Society Annual Meeting (OPTICON '92)*, Boston

Hanna, D. & Mitchell, J. W. (1993). A laser pulse stretcher made from optical fibres. *Nuclear Instruments and Methods in Physics Research A*, Vol. 324, pp. 14-18

Harigel, G.; Baltay, C.; Bregman, M.; Hibbs, M.; Schaffer, A; Bjelkhagen, H; Hawkins, J.; Williams, W.; Nailor, P.; Michaels, R.; & Akbari, H. (1986). Pulse stretching in a *Q*-switched ruby laser for bubble chamber holography. *Applied Optics*, Vol. 25, No. 22, pp. 4102–4110

Hordvik, A.; (1970). Pulse stretching utilizing two-photon-induced light absorption, *IEEE Journal of Quantum Electronics*, Vol. QE-6, No. 4, pp. 199 - 203.

Hsu, R. & Shen, Y. R. (1982). Effects of self-induced ellipse rotation and self-focusing in a Q-switched laser cavity. *Applied Optics*, Vol. 21, No. 14, pp. 2608-2615

Khare, R.; Daultabad, S. R.; Vora, H. S. & Bhatnagar, R. (1995). A tunable three-wavelength copper-vapor laser pumped dye laser with collinear output. Optics Communications, Vol. 114, pp. 275-279

Khare, R.; Shukla, P. K.; Mishra, G. K.; Mukherjee, C.; Talwar, S.; Dubey, V. K.; Saxena, P. & Mittal, J. K (2009). A novel confocal optical pulse stretcher for laser pulses. *Optics Communications*, Vol. 282, p.p. 3850-3853.

Kojima, J. & Nguyen, Q-V. (2002). Laser pulse stretching with multiple optical ring cavities. *Applied Optics*, Vol. 41, No. 30, pp. 6360-6370

Kracht, D. & Brinkmann, R. (2004). Green Q-switched microsecond laser pulses by overcoupled intracavity second harmonic generation. *Optics Communications*, Vol. 231, pp. 319–324

Kumar, S.; Kundu, S.; Singh, S.; Monga, J. C.; Dasgupta, K.; Manohar, K. G.; Natesh R. & Nair, L. G. (1995). Two stage amplification of CVL unstable oscillator pulses. *Proceedings of National Laser Symposium*, I.R.D.E.-Dehradun, India, pp. 23-24

Kundu, S.; Kumar, S.; Dasgupta, K.; Singh, S.; Manohar, K. G.; Natesh R. & Nair, L. G. (1995). Pulse stretching by regenerative amplification in a copper vapour laser, *Proceedings of National Laser Symposium*, I.R.D.E.-Dehradun, India, pp.27-28

Lee, E. S. & Hahn, J. W. (1996). Four-pass amplifier for the pulsed amplification of a narrow-bandwidth continuous-wave dye laser. *Optics Letters*, Vol. 21, No. 22, pp. 1836-1838

Liberman, V; Rothschild, M.; Sedlacek, J. H. C.; Uttaro, R. S.; Grenville, A.; Bates, A. K. & Van Peski, C. (1999). Excimer-laser-induced degradation of fused silica and calcium fluoride for 193-nm lithographic applications. *Optics Letters*, Vol. 24, No. 1, pp. 58-60

Lisitsa, M. P.; Mozol P. E. & Fekeshgazi, I. V. (1974). Laser pulse lengthening with the aid of nonlinearly absorbing semiconductors CdP_2 and ZnP_2, *Soviet Journal of Quantum Electronics*, Vol. 4, No. 1, pp. 110-111

Lovberg, R. V.; Wooding, E. R. & Yeoman, M. L. (1975). Pulse stretching and shape control by compound feedback in a Q-switched ruby laser. *IEEE Journal of Quantum Electronics*, Vol. QE-11, No. 1, pp. 17–21

Lowdermilk, W. & Milam, D. (1981). Laser-induced surface and coating damage. *IEEE Journal of Quantum Electronics*. Vol. QE-17, No. 9, pp. 1888–1903

Maine, P.; Strickland, D.; Bado, P.; Pessot, M. & Mourou G. (1988). Generation of ultrahigh peak power pulses by chirped pulse amplification. *IEEE Journal of Quantum Electronics*, Vol. 24, No. 2, pp. 398-403

Maker, P. D. & Terhune, R. W. (1965). Study of optical effects due to an induced polarization third order in the electric field strength. *Physical Review*, Vol. 137, No. 3A, pp. A801-A818

Martin, W. E.; (1977). Pulse stretching and spectroscopy of subnanosecond optical pulses using a Fabry-Perot interferometer. *Optics Communications*, Vol. 12, No. 1, pp. 8-12

Martinez, O. E. (1987). 3000 times grating compressor with positive group velocity dispersion: Application to fiber compensation in 1.3-1.6 μm region. *IEEE Journal of Quantum Electronics*, Vol. QE-23, No. 1, pp. 59-64

Miller, E. J.; Malcuit, M. S. & Boyd, R. W. (1990). Simultaneous wave-front and polarization conjugation of picosecond optical pulses by stimulated Rayleigh-wing scattering, *Optics Letters*, Vol. 15, No. 21, pp. 1188-1190.

Murphy, D. V.; & R. K. Chang, (1977). Pulse stretching of Q-switched laser emission by intracavity nonlinear ellipse rotation. *Optics Communications*, Vol. 23, No. 2, pp. 268-272

Murray, J. E. & Harris, S. E. (1970). Pulse lengthening via overcoupled internal second Harmonic generation. *Journal of Applied Physics*, Vol. 41, No. 3, pp. 609–613

Panarella, E. & Brdley, L. L. T. (1975). Contolled timewise redistibution of laser energy. *IEEE Journal of Quantum Electronics*, Vol. QE-11, No. 5, pp. 181 - 185.

Pätzel, R.; & Stamm, U. (2005). Excimer lasers in microlithography, *Excimer Laser Technology*, Basting, D.; & Marowsky, G. (Eds.) pp. 89-103, Springer, ISBN-10 3-540-20056-8, Berlin

Pflüger, S.; Sellhorst, M.; Sturm, V. & Noll R. (1996). Fiber-optic transmission of stretched pulses from a *Q*-switched ruby laser. *Applied Optics*, Vol. 35, No. 25, pp. 5165-5169

Rambo, P.; Schwarz, J.; Bernstein, A.; Giuggioli, L.& Diels, J.-C. (1999). Pulse stretching in an alexandrite laser. *Proceeding of Conference on Lasers and Elecro-optics (CLEO-99)*, Baltimore, Marryland, pp. 400-401.

Salin, F.; Squier, J. & Mourou, G. (1992). Large temporal stretching of ultrashort pulses. *Applied Optics*, Vol. 31, No. 9, pp. 1225-1228

Schwarz, J.; Naiman, C. S. & Chang, R. K. (1967). Effects of internal induced absorption on laser emission. *Appllied Physics Letters*, Vol. 11, No. 7, pp. 242-244

Schmid, W. (1980). Pulse stretching in a Q-switched Nd: YAG laser. *IEEE Journal of Quantum Electronics*, Vol. 16, No. 7, pp. 790-794

Seidel, S. & Phillipps, G. (1993). Pulse lengthening by intracavity stimulated Brillouin scattering in a Q-switched, phase-conjugated Nd:YAG laser oscillator. *Applied Optics*, Vol. 32, No. 36, pp. 7408-7417

Sharma, A. K.; Raghuramaiah, M.; Mishra, K. K.; Naik, P. A.; Kumbhare, S. R. & Gupta, P. D. (2005). Characteristics of a stable, injection Q-switched Nd:phosphate glass regenerative amplifier for a chirped pulse amplification based Table Top Terawatt laser system. *Optics Communications*, Vol. 252, pp. 369-380

Singh, S.; Dasgupta, K.; Kumar, S.; Manohar, K. G.; Nair, L. G. & Chatterjee, U. K. (1994). High-power high-repetition-rate copper-vapor-pumped dye laser. Optical Engineering, Vol. 33, No. 6, pp. 1894-1904

Singh, S.; Kumar, S. & Nair, L. G. (1995). A narrowband dye oscillator pumped by a temporally stretched CVL pulse. *Proceedings of National Laser Symposium*, I. R. D. E.-Dehradun, India, pp. 25-26

Thomas, C. H. & Price, E. V. (1966). Feedback Control of a Q-Switched Ruby Laser, *IEEE Journal of Quantum Electronics*, Vol. QE-2, No. 9, pp. 617-623

Tóth, C; (1995). Simple Optical Pulse Lengthening Setup for the Subnanosecond Range. *Applied Optics*, Vol. 34, Issue 34, pp. 8070-8071.

Verluyten, L.; Willocq, S.; Harigel G. G.; Hawkins J. K.; Smart W. M.; Williams, W. A.; Peters M. W. & Akbari, H. (1990). Laser pulse stretching via enhanced closed loop control with slow Q-switching. *Nuclear Instruments and Methods in Physics Research A*, Vol. 292, pp. 313-318

Walker, A. C. & Alcock, A. J. (1976). Picosecond resolution, real-time, linear detection system for 10-μm radiation. *Review of Scientific Instruments*, Vol. 47, No.8, pp. 915-920

Young, J. F.; Murray, J. E.; Miles, R. B. & Harris, S. E. (1971). *Q* switched laser with controllable pulse length, *Applied Physics. Letters*. Vol. 18, No. 4, pp. 129–130

11

Ultrafast Laser Pulse Synchronization

Heping Zeng, Ming Yan and Wenxue Li
State Key Laboratory of Precision Spectroscopy
East China Normal University
China

1. Introduction

Progress in ultrafast science and technology has pushed the temporal resolution to sub-femtosecond and attosecond region (Kienberger el at., 2002; Fleischer & Moiseyev, 2006; Corkum & Krausz, 2007; Goulielmakis el at., 2007) and technology exploration in ultrashort pulse generation has made it possible to equip femtosecond lasers that are only a few cycles in duration (Xu el at., 1996; Shirakawa el at., 1999; Rullière, 2005; Mauritsson el at., 2006; Sansone el at., 2006). Coherent control experiments now use precisely shaped optical pulses with sub-femtosecond timing jitters to steer coherent evolution governed by the interaction of coherent light with matter towards quantum paths for desired dynamic processes (Kitzler el at., 2002;Tong el at., 2003; Bandrauk el at., 2005). Precise control and synchronization of ultrashort pulses are essential for many other applications in ultrafast laser spectroscopy (Drescher el at., 2002; Hannaford, 2005; Cavalieri el at., 2007), Fourier synthesis of light pulses or optical waveforms (Jones el at., 2000; Shelton el at., 2001; Supradeepa el at., 2008), multi-color pump-probe investigations (Schoenlein el at., 2000; Son el at., 2002; Manzoni el at., 2006; Gu el at., 2009), coherent anti-Stokes Raman scattering microscope (Jones el at., 2002; Potma el at., 2002), difference frequency mixing for mid-infrared generation (de Barros el at., 1995; Kaindl el at., 2000) and terahertz source generation (Kitaeva, 2008; a. Sell el at., 2008; b. Sell el at., 2008), and so forth. Recently, significant breakthrough has been made in ultrafast laser technology and precision frequency metrology (Hentschel, el at., 2001;Udem el at., 2002; Schibli el at., 2004) with the innovation concept of simultaneous control in both time and frequency domain developed in femtosecond combs (Shelton el at., 2001; Ma el at., 2004). This technique is versatile in controlling ultrashort mode-locked lasers by translating precision control over mode frequencies into ultra-stable control in the time domain, which may eventually facilitate accurate synchronization of multicolor mode-locked lasers desired for coherent optical synthesis to extend the pulsed laser spectrum (Jones el at., 2000; Shelton et al., 2001; Bartels el at., 2004).

Rapidly growing interest has been focused in recent years on synchronizing independently mode-locked multi-color lasers with improved synchronization accuracy and robustness, sufficiently low time jitters, as well as large tolerance to disturbance (Kawamura el at., 2000; Rudd el at., 2005; Baudelet el at., 2006). In general, pulse synchronization requires independent lasers to be mode-locked at exactly the same repetition rate. As the repetition rate (f_r) depends linearly on the effective cavity length (L_{eff}), $f_r=c/L_{eff}$ (c, the speed of light in vacuum), a feed-back locking system is indispensible for laser synchronization to compensate for the unavoidable cavity variations. In accordance with the feedback methods, laser synchronization scheme can be classified into active, passive and hybrid synchronization. In the active synchronization scheme, the feedback is typically provided by electronic phase-locking-loop devices (Schibli el at., 2003), which offer the advantage of long-term stability and independence on the oscillating wavelengths in laser operation. However, such a scheme suffers from limited bandwidth of electronic circuits and the time jitter among synchronized lasers is critically dependent on the control precision of the electronic circuits. Passive synchronization is typically achieved by using all-optical methods on the basis of nonlinear optical effects within the concerned laser cavities (Bartels el at., 2003; Yoshitomi el at., 2006). Although the passive technique is limited by the walk-off length of the synchronous lasers, it provides relatively simple laser configurations and in most cases, results in small timing jitters, making it attractive and popular in laser synchronization experiments. Besides active and passive synchronization schemes, the active-passive hybrid synchronization technique combines the advantages of the both schemes (Yoshitomi el at., 2005). The walk-off length of the synchronous lasers is reduced by using active feedbacks with electronic phase-locking-loop devices, while small time jitters among synchronized lasers are guaranteed by nonlinear optical couplings within the synchronizing laser cavities. Correspondingly, the hybrid synchronization scheme becomes more complicated than the other two schemes.

The stabilization of the synchronization between separate lasers depends critically on cavity length mismatch. With large cavity length mismatch tolerance, a synchronized laser system can be expected to work stably in tough environments. In principle, cavity length variations (ΔL) and offset frequency drifts (Δf_r) between different lasers can be effectively balanced by modulating the intracavity dispersion and nonlinearity. For instance, the cavity length mismatch can be effectively compensated by cross phase modulation (XPM) effect. By this means, synchronous mode-locking can be achieved in coupled-cavity lasers by sharing the same Kerr-type nonlinear medium (Wei el at., 2002), or by injecting the master laser pulses into the slave laser in the master-slave configuration for independently mode-locked lasers (Yoshitomi el at., 2006). Until now, a record cavity length mismatch about 140 μm was realized in master-slave fs fiber lasers (a. Rusu el at., 2004). However, it should mention that in the coupled-cavity lasers, a large length variation of the coupled cavity may destabilize the mode-locking and pulse-train synchronization. Once the cavity lengths are largely mismatched in comparison with the walk-off length, XPM-based passive schemes may fail in enforcing synchronous mode-locking. Since XPM is limited within the walk-off length between the interacting pulses, high peak powers are required to induce sufficient optical nonlinearity for a tight synchronization. Furthermore, in the master-slave laser injection configuration, the injection power of the master laser is seriously limited by the light coupling equipments like the fiber coupler, the wavelength-division multiplexing (WDM), Lens and mirrors, and so on. Therefore, the relatively low-intensity master pulse can only induce limited wavelength shifts and round-trip group delay changes for the slave laser catching up with the master laser pulses. To achieve a large cavity mismatch tolerance, a hybrid configuration with the master-slave and the coupled-cavity schemes is recommended. The hybrid passive synchronization scheme entails XPM an additional function in the slave laser to start mode-locking rather than merely to shift the central wavelength and modulate the intracavity group delay, from which influences of the master laser injection can be greatly enhanced. As a consequence, XPM mode-locking in the hybrid master-slave configuration exhibits advantageous features inherited from both coupled-cavity and master-slave schemes, i.e., remote control of slave lasers with independently mode-locked master lasers, sensitive influence of the master injection on XPM mode-locking.

It is of broad interest to synchronize pulses of variable pulse durations from fs to ns. Nanosecond laser pulses are typically obtained by Q-switching techniques and synchronized by external electronic triggering with timing jitters in the ns range. It is still a technical challenge to construct coherent ns-duration pulse arrays with an ultrahigh resolution in the time domain due to the lack in efficient techniques for precise temporal and spectral control of ns pulses. All the synchronization schemes developed so far are inapplicable for coherent phase-locking of ns optical pulses. On the other hand, high-energy ns lasers precisely synchronized with the same triggering fs laser could support the development of phase-locked high-energy ns laser arrays for promising applications in particle acceleration and fast ignition (Kodama el at., 2001), and ultrashort laser synchronization with x-rays or electron beams from synchrotrons (Schoenlein el at., 2000). Furthermore, fs laser synchronization with a high-energy ns laser could facilitate lasers of different pulse durations to interact with the target under a precise control in the time domain (Kodama el at., 2001; Vozzi el at., 2009), which is also quite useful for the development of robust and stable optical parametric chirped pulse amplifiers (Butkus el at., 2004; Rudd el at., 2005; Zhang el at., 2006). It has been recently demonstrated that square ns mode-locking could be established in a fiber laser of a long fiber cavity (Potma el at., 2002), and that the square ns pulses could be scaled to hundred watts with negligible temporal and spectral distortion by using large-mode-area fiber amplifiers (b. Yan el at., 2009). Interestingly, square ns mode-locking could be passively synchronized with an ultrashort laser in the master-slave laser configuration (Li el at., 2009). Since the long fiber cavity indispensable for square ns mode-locking typically produces a large cavity length variation, sufficiently strong master injection pulses are required to trigger a synchronous square ns mode-locking of the slave laser (Vozzi el at., 2009). A large cavity mismatch tolerance is especially required for the precise control over mode frequencies in a long fiber cavity that allows synchronized square mode-locking in the ns region.

In this chapter, we review our recent experimental development in synchronizing multicolor mode-locked fiber lasers of different pulse durations. A robust technique based on cross-absorption modulation (XAM) was demonstrated efficiently for ultrashort fiber laser synchronization in the master-slave configuration. For fiber lasers mode-locked through nonlinear polarization rotation (NPR) in the fiber laser cavity, XAM induced an observable change of the NPR in the slave fiber laser cavity due to the nonlinear couplings between the slave laser pulses and the externally injected fs master laser pulses, resulting in a trigger for square ns mode-locking in slave fiber laser that generated ns output pulses of square temporal profiles with the pulse duration dependent linearly on the master injection power. Square ns mode-locking synchronized with a fs Ti:sapphire master laser was achieved in both Yb- and Er-doped slave fiber lasers by using additional Er-doped fibers as the saturable absorber. The synchronization of the square ns pulses with the fs master injection was implemented with the rising edges of the ns pulses tightly synchronous to the injected fs pulses, with remarkably low time jitters in the range of hundreds femtoseconds. Furthermore, such an XAM-based synchronization could be maintained with the record cavity-length mismatches exceeding 8.2 and 3.6 cm for the Yb- and Er-doped slave fiber lasers, respectively. Interestingly, the XAM-based synchronization could be used to synchronize mode-locked lasers at a relatively low repetition rate of sub-MHz (a. Yan el at., 2009; b. Yan el at., 2009), while the square ns mode-locked laser pulses could be controlled to output tunable flat-top pulses with the pulse rising edges tightly synchronized with the triggering fs pulses, and high-power amplification could be realized by using large-mode-area photonic crystal fiber amplifiers for the synchronized pulses. The synchronized square

ns pulses could be easily controlled by varying the master injection power, confirming that the square ns mode-locking worked with observable intra-cavity power clamping and that the clamped power in the slave fiber laser could be varied by XAM-induced changes of the optical nonlinearity of the Er-doped fiber. Note that the square ns pulses were controlled by this synchronous mode-locking method with an ultrahigh temporal resolution at least four orders of magnitude better than the corresponding pulse duration. As a result, the synchronously triggered square ns mode-locking provides a robust method not only for the precise control of the square ns pulses in the time domain, but also for the convenient experimental realization of synchronized pulses variable from fs to ns in pulse duration.
This review chapter is organized as follows. After this brief introduction, we present the basic theory for passive synchronization in Section 2. Section 3 mainly compares XPM- and XAM-based synchronization schemes applicable for ultrafast laser synchronization and square nanosecond mode-locked laser synchronization. The amplification of synchronous high-energy laser pulses is discussed at the end of this section. We conclude the review in section 4 by emphasizing that the XAM-based synchronization technique is a promising supplement to the XPM-based passive synchronization.

2. Basic theory for XPM- and XAM-based synchronization

Nonlinear effects in a Kerr medium or in fiber competing with light dispersion within the laser cavity provide an effective feed-back effect for laser pulse going against cavity variation and staying at a same repetition rate (or round-trip frequency) with another laser. Thus, to analyze passive synchronization mechanism, we need start with the nonlinear effects like cross-phase modulation and cross-absorption modulation. An established XPM theory has been reported for several decades, while very few information about XAM is reported. In this Section, we concentrate on the detailed discussion of the XPM-based passive laser synchronization, and introduce some basic understanding about the XAM-based synchronization under investigation.

2.1 Cross phase modulation

To analyze the supporting role of cross-phase modulation in passive synchronization, we assume that two lasers (laser-1 and laser-2) share the same nonlinear medium with a given length of L inside their cavities. The nonlinear refractive index for this medium is denoted as n_2. Before being synchronized, the two lasers can be independently mode-locked to produce ultrashort pulse trains at different center wavelengths. The round-trip frequencies for laser-1 and laser-2 are f_1 and f_2, respectively. Therefore, the two laser pulses can meet each other in the nonlinear medium with a frequency of $\Delta f=|f_1-f_2|$. Once the two pulses co-propagate inside the shared medium, there exists a possibility for the two lasers to run at the same round-trip frequency as the nonlinear couplings may produce offset frequency drifts so that different lasers can be effectively matched. Assume that $E_1\ (z,t)$ and $E_2\ (z,t)$ represent the envelope of the electric field of laser-1 and -2, respectively. Since the two pulses have different wavelengths as well as different group velocities (v_{G1} and v_{G2}), the two pulses will co-propagate in the medium within a walk-off length of $L_{off}=\frac{T_1}{d}$, where $d=\frac{(v_{c1}-v_{c2})}{v_{c1}v_{c12}}$ and T_1 is the pulse duration of Laser-1 at the starting position of interaction region. If $L > L_{off}$, the pulse interaction inside the nonlinear medium can be theoretically expressed in terms of the nonlinear propagation equations as (a. Agrawal, 1989; Zhu el at., 2005):

$$\frac{\partial}{\partial z}E_1(z,t)+d\frac{\partial}{\partial t}E_1(z,t)+i\frac{\beta_1}{2}\frac{\partial^2}{\partial t^2}E_1(z,t)+\alpha_1 E_1(z,t)$$
$$=i\gamma_1\left[|E_1(z,t)|^2 E_1(z,t)+2|E_2(z,t)^2 E_1(z,t)\right], \tag{1}$$

$$\frac{\partial}{\partial z}E_2(z,t)+d\frac{\partial}{\partial t}E_2(z,t)+i\frac{\beta_1}{2}\frac{\partial^2}{\partial t^2}E_2(z,t)+\alpha_2 E_2(z,t)$$
$$=i\gamma_2\left[|E_2(z,t)|^2 E_2(z,t)+2|E_1(z,t)^2 E_2(z,t)\right], \tag{2}$$

where z and t denote the laser pulse propagation distance and time. Subscripts 1 and 2 denote the laser-1 and -2. a is the linear absorption coefficients, β is the intra-cavity group velocity dispersion (GVD), and $\gamma=2\pi n_2/(\lambda A_{eff})$ is the nonlinear coefficient, where A_{eff} is the effective beam cross section and λ is the center wavelength. In these equations, the contribution of GVD is described by the third term on the left-hand side. As to the right-hand side, the first term is due to the self-phase modulation (SPM) effect and the second term for the XPM effect. By analytically resolving these two equations, one can finally get soliton-like solutions (Baldeck el at., 1988, Agrawal el at., 1989).

For the sake of simplicity, we merely consider the effects of GVD and XPM while omitting the influence from the self-phase modulation. As a result, the maximum phase shift $\Delta\varphi_{21}$ of laser-1 induced by laser-2 can be expressed as (Lisak el at., 1990):

$$\Delta\varphi_{21}=\frac{\gamma_1 P_2 L_{off}}{\pi T_1}, \tag{3}$$

and correspondingly, the laser-1 induced maximum phase shift $\Delta\varphi_{12}$ to laser-2 is:

$$\Delta\varphi_{12}=\frac{\gamma_2 P_1 L_{off}}{\pi T_2}, \tag{4}$$

where P is the peak power of laser pulse.

In order to understand the feed-back mechanism of XPM effect in passive synchronization, we need consider the XPM-induced phase shift in frequency domain. Supposing the laser-1 and -2 having instantaneous optical frequencies of ω_1 and ω_2, respectively, the XPM-induced frequency change can be simply described as (Wei el at., 2002):

$$\Delta\omega_1 \propto -n_2\frac{\partial I_2}{\partial t}, \tag{5}$$

$$\Delta\omega_2 \propto -n_2\frac{\partial I_1}{\partial t}, \tag{6}$$

where I is the intra-cavity laser intensity. In general, the two lasers have a quite similar round-trip frequency, but not exactly equal. The relative fluctuation Δf for the two lasers may come from any small perturbations during the laser operation, such as small cavity-length drifts, thermal and temperature fluctuation, random polarization change, pump instability and so on. As affected by the perturbations, the initially co-propagated laser-1 may go behind or ahead of the laser-2, as shown in Fig. 2-1.

In the case that the laser 1 goes behind the laser 2, the rising-edge of laser-1 will cross with the falling-edge of laser-2. Because the slope of intensity I_2 is negative in its falling-edge for laser-1, the spectrum of laser-1 will be blue-shifted ($\Delta\omega_1>0$) while laser-2 will be red-shifted ($\Delta\omega_2>0$) due to the positive intensity slope of I_1, according to relation (5) and (6). Thus, in a medium with negative group dispersions for the two lasers, the blue-shifted light will go

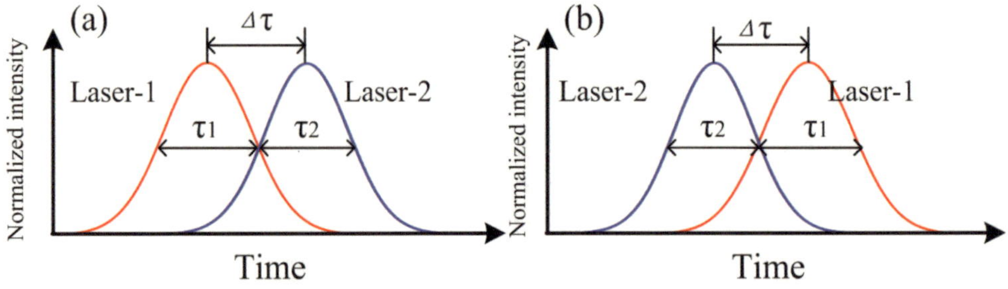

Fig. 2-1. The propagation schematic diagram of Laser-1 (a) behind and (b) ahead Laser -2.

faster than the red-shifted one, and therefore the delayed pulse (laser-1) can catch up with the advancing pulse (laser-2). In the case that laser-1 is ahead of laser-2, the XPM will lead laser-1 to be red-shifted and laser-2 to be blue-shifted in spectrum, which may eventually make both pulses be maximally overlapped in the time domain. Once the pulses are maximally overlapped, the spectral shifts for both pulses will be reduced to zero, since the laser intensity profile is supposed to be temporally symmetric. In this situation, the two lasers operate at a synchronization state with the same round-trip frequency.

For evaluating a passive synchronization system, there are two crucial parameters: 1) the ability for the system compensating for environmental perturbations which is estimated by the value of cavity mismatch tolerance, and 2) the precision for the synchronization which is defined as timing jitter between the two laser pulses (Shelton el at., 2002). Besides, the synchronization stability is also an important factor to decide how long lasers are operated at synchronization state. Recently, experimental results have shown that the jitter for two separate fs pulses has been reduced into attosecond region and the mismatch tolerance is extended to several centimeters. The details for experimentally measuring these two parameters will be presented in Section 3-2.

As an automatic feed-back effect, XPM is considered to be an effective method to passively realize robust laser synchronization for last decades. However, XPM is not the only mechanism employed in synchronization scheme. Recently, a novel effect based on cross absorption modulation has been demonstrated quite useful and robust to synchronize two independently mode-locked lasers of ultra-long fiber cavity, which supports large cavity length mismatch tolerance. The next Section 2.2 will focus on the details of the XAM effect and its application background.

2.2 Cross absorption modulation

Cross absorption modulation was first reported in the Reference (a. Yan el at., 2009) for synchronizing a ytterbium-doped and a erbium-doped fiber lasers to an fs Ti:sapphire laser at a repetition rate of nearly 240 kHz. It was well demonstrated that the cavity length mismatch could be compensated by XPM in coupled-cavity lasers sharing the same Kerr-type nonlinear medium, or independently mode-locked lasers in the configuration with master injection into the slave laser. However, XPM is limited within the walk-off length between the interacting pulses for purpose of synchronizing a fiber laser with an 800-m-long cavity (corresponding to a repetition rate of 240 kHz). This produces the technical challenge to passively synchronize ns-duration fiber lasers with an ultra-short laser. The long fiber cavity indispensable for square ns mode-locking produces a little difficulty to achieve a stable and robust passive synchronization. As the fiber cavity-length is much longer than the walk-off length, one cannot rely on XPM-based passive schemes.

Cross absorption modulation is a kind of modulation on slave light imposed by master light through a co-propagating medium which periodically absorbs the master light and then switches from its ground state to an excited state with a change of its refractive index or nonlinear coefficient. XAM is normally ignored in non-resonant interaction media as the nonlinear absorption coefficient is typically two-order smaller than the nonlinear coupling coefficient. Nevertheless, it may be comparable to or even larger than XPM in the near-resonant media.

In the resonant medium, the modulation can be enhanced by propagating pulse stimulating the medium from its ground state to an excited one. The enhanced XAM adjusts the group velocity of the co-propagating slave pulses through changing the nonlinear refractive index of the resonant medium to match the repetition rate of the slave laser with that of the master laser. Thus, the slave pulse polarization can be rotated owing to the changed nonlinear refractive index. For a mode-locked fiber laser, since the intra-cavity polarization state is changed, the temporal or spectral characteristics of the slave pulse must be also changed with the master laser. However, the instinct mechanism of XAM is under investigation and thus a systematic theory about XAM-based synchronization is still lack. In section 3.3 the novel XAM-synchronization will be described from the aspect of its experimental realization.

3. Passive synchronization techniques

As mentioned in the introduction part of this review chapter, active synchronization with an electronic feedback device suffers from the limitation of the timing jitters of detectors, filters, mixers and piezo transducers and so on (Spence el at., 1993; Crooker el at., 1996). Despite a record timing jitter of 300 as was achieved between Cr:Forsterite and Ti:Sapphire lasers by the active synchronization (Vozzi el at., 2009), the complexity of such an electronic system makes the technique unpopular. Unlike the electronically supported synchronization, passive technique permits an all-optical method to obtain synchronous laser pulses without the limitations from the complicated electronic feedback scheme. By the passive synchronization scheme (Yoshitomi el at., 2006), timing jitter of 3.7 fs was reported for synchronizing an Er-doped fiber laser to a mode-locked Cr:Forsterite. To date, a record timing jitter as low as 100 as has been achieved by using an active-passive hybrid synchronization scheme (Yoshitomi el at., 2005). Thus, the passive synchronization technique is concerned as an alternative or a co-operator for the active one. Most of these reported passive synchronization systems relay on XPM to modulate the intra-cavity dispersion and nonlinearity for matching the cavity lengths and offset frequency drifts of different lasers. Recently, the XAM-based technique has also been reported to be able to passively synchronize two lasers at a relatively low repetition rate of sub-MHz (a. Yan el at.,

2009). In this section, we will focus in what follows on various experimental implementations of these passive techniques. Section 3.1 aims at the precise XPM-synchronization for ultra-fast lasers at high repetition rate. Section 3.2 discusses an application of the XPM technique in the synchronization between ps and ns lasers. A mode-locked ns pulse generation technique will be also introduced in this section. After preliminary discussion on a few examples of different synchronization configurations, we present experimental measurements of the mismatch tolerance and the RMS timing jitter. Section 3.3 is concentrated on the XAM-based synchronization and its experimental results, while Section 3.4 concerns the synchronous pulse amplification and its impacts on the synchronization precision.

3.1 Accurate synchronization among ultra-fast lasers

During the last decades, the XPM effect has been employed as the major method to passively synchronize individually operated lasers. Such an XPM-technique can be realized (1) in a shared laser cavity or a shared nonlinear medium, or (2) in a master-slave injection configuration in which the two laser pulses interacted with each in a segment of single mode fiber inside the slave laser cavity. Usually case (1) appears in solid laser system. Since light intensity can be hardly improved to a large extent in free space, the required XPM is mainly provided by the large nonlinear coefficient of the shard medium. In this case, the two lasers play the equal role in synchronization and no distinction between master and slave laser. In this kind of synchronization scheme, the two lasers are cross-mode-locked at the same repetition rate (or round-trip frequency) to produce dual-wavelength laser light. As a distinct contrast with the case (1), the two lasers in case (2) show obviously distinguished roles as a master and slave laser. Generally, the slave laser is made by a fiber laser for a large intensity in constrained space and it oscillates dependently on the master laser, while the master works at a relatively independent state. The difference for the two cases is that case (1) is more sensitive to the environmental fluctuation while the construction of case (2) is much simple. In this Section, we will first introduce a fraction spectrum amplification technique to generate synchronous ultra-fast pulses. This technique is not widely used as XPM technique, but it can easily produce dual-wavelength pulses with ultra-low timing jitter in a certain situation. We will then give some particular examples to show how to experimentally obtain synchronous pulses with XPM technique in case (1) and case (2), respectively.

3.1.1 Synchronous pulses from fraction spectrum amplification

In many cases, there exists a situation that a mode-locked laser oscillator operates with non-continuous spectrum. In this case, the output of the oscillator can be spectrally separated into two (or more) parts: the main part (master source) is operating at λ_1 with spectral width of $\Delta\lambda_1$, and a small fraction (slave source) simultaneously works at λ_2 ($\lambda_2 \neq \lambda_1$) with spectral width of $\Delta\lambda_2$. Note that the two parts are spectrally separated but temporally overlapped. By spectrally detaching the two parts and amplifying the slave fraction, one can easily obtain two synchronous pulses centering at two different wavelengths. This method for obtaining synchronous pulses at various wavelengths is dubbed as fraction spectrum amplification (FSA). Since the two lasers come from the same cavity in FSA, the timing jitter between the two synchronous pulses can easily controlled to quite small values, even as small as sub-fs. As a typical example for FSA synchronous pulses generation, we introduce here a synchronous FSA of a few-cycle Ti:Sapphire fs laser with the experimental setup as schematically illustrated in Fig. 3-1 (a. Li el at., 2009).

FSA is a useful way to generate synchronous pulse trains, but its realization requires a special laser source. The source used in this example is a commercialized Ti:sapphire laser oscillator (Rainbow). The specialty for this laser is that it deliveries fs pulse trains at a center wavelength of 800 nm with the spectral width of ~100 nm and at a near-IR fraction center of 1040 nm [shown in Fig. 3-2 (a)]. The 1040-nm light of 150 μW is much weaker than the 800-nm light of 200 mW. Considering the large difference in average power of the two parts, FSA becomes a favorable choice to easily realize synchronous pulses in this Ti:sapphire laser system. As the master source, the 800-nm light is temporally compressed into 10-fs region. And the 1040-nm fraction covering a spectra range from 980 nm to 1070 nm is employed as the slave source.

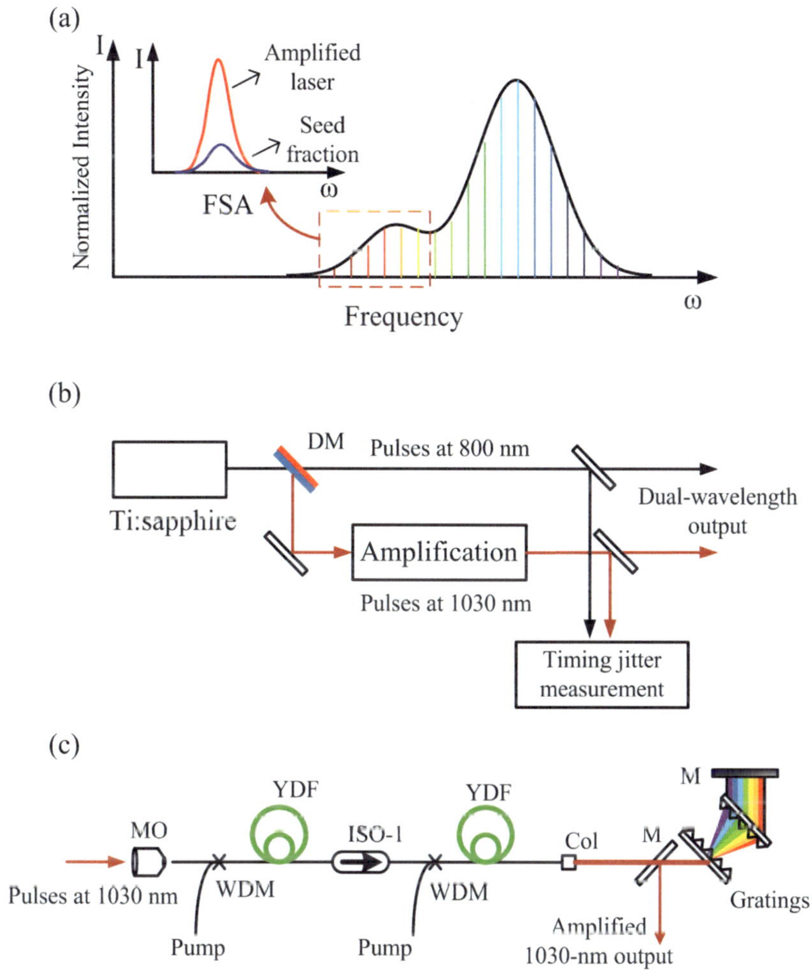

Fig. 3-1. (a) Spectral schematic of FSA, (b) experimental schematic of FSA and (c) experiment setup of the amplification section. MO, micro-objective; ISO, optical isolator; WDM, wavelength-division multiplexing (980/1064 nm); YDF, ytterbium-doped fiber; DM, dual-wavelength mirror.

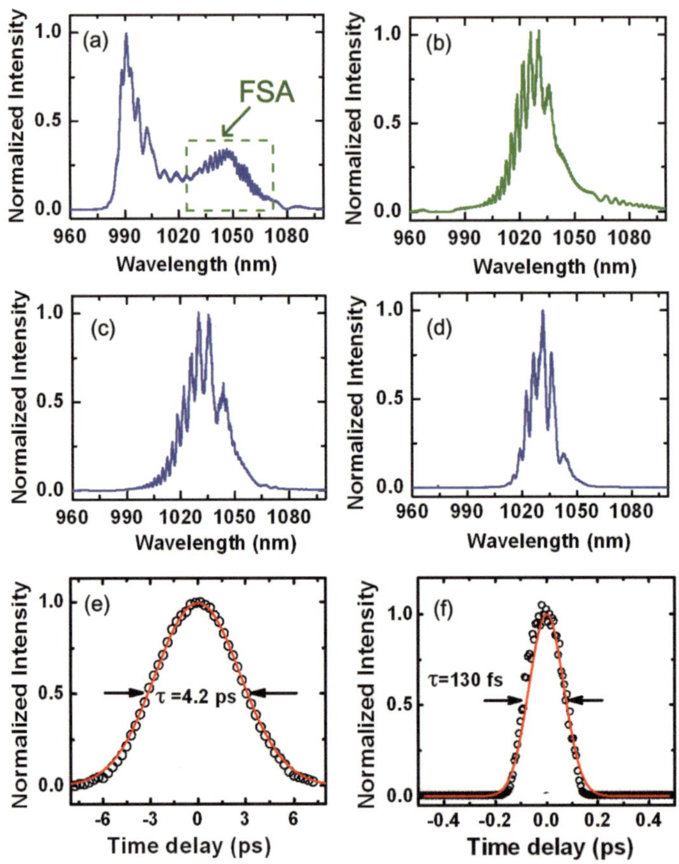

Fig. 3-2. The fraction spectrum of the few-cycle Ti:sapphire laser pulse (a), the gain spectrum of the Yb-doped fiber (b), the spectrum of the first-stage (c) and the second-stage (d) Yb-doped fiber amplifiers, the pulse duration of the amplified laser at 1030 nm before (e) and after (f) the grating compression.

The selection of amplifiers for the FSA is dependent on many factors such as the wavelength of the seed pulse, the setup complexity, the incident pulse power, gain bandwidth, and so on. By considering the near-IR spectral property and the weak power of the 1040-nm fraction source, ytterbium-doped fiber amplifier is recommended as an advantageous selection for amplifying such weak seed light pulses. As shown in Fig. 3-1 (c), a two-stage fiber amplification system is utilized for the FSA. It is difficult to directly amplify a weak signal into high power by one stage amplifier due to the large amplified spontaneous emission (ASE) in the small-signal amplification. The 1040-nm pulse trains are selected from the Ti:Sapphire laser by a dichroic mirror, and are first amplified to the average power of 2.5 mW by a first-stage fiber amplifier. Due to the limited gain bandwidth of the Yb-doped fiber [shown in Fig. 3-2 (b)], the spectrum of the amplified pulses is narrowed to 22 nm after the first stage [shown in Fig. 3-2 (c)]. In the second stage amplifier, the output spectrum is further narrowed to a full-width at half-maximum (FWHM) of 13.8 nm, and the power is amplified up to 140 mW under diode pump of 200 mW.

In order to obtain fs pulses at 1033 nm, a diffraction-grating compressor based on transmission gratings with a grating period of 1250 lines/mm is used to externally compress the amplified laser pulses. For the compressor working at its maximum diffraction efficiency at 1064 nm, the grating pair is placed at 41.7± Littrow angle. Finally, the FWHM duration of the amplified pulse is compressed to 130 fs, which is 32 times smaller than the uncompressed amplified pulse of 4.2 ps [shown in Fig. 3-2 (e) and (f)].

As mentioned above, since the synchronous pulse trains obtained by the FSA are generated from the same oscillator, the cavity-variation induced synchronization instability can be effectively avoided. In the experiment, the timing jitter is measured as low as 0.55 fs.

3.1.2 Master-slave injection configuration for laser synchronization

Master-slave configuration is most wildly used for synchronizing two individual lasers. In this configuration, the master pulses are injected into the salve laser cavity. And the master co-propagates and interacts with the slave pulse inside the slave cavity. The operation of the slave laser is dependent on the master pulse injection due to XPM effect induced by the master laser. Due to the intensity-dependent XPM, master-slave configuration is more favorable to be applied into a fiber laser. Since the small diameter of single mode fiber providing higher light intensities in the fiber core, XPM effect will be largely enhanced inside the fiber cavity to support a robust timing synchronization. As a typical example for the master-salve configuration, Figure 3-3 presents an experiment on synchronizing an Er-doped fiber laser to an Yb-doped laser source (Li el at., 2009).

With the master-slave configuration, the 1030-nm laser light (generated by using fraction spectrum amplifier) is synchronized to 1560-nm pulse train at a repetition rate of ~80 MHz, as schematically illustrated in Fig. 3-3 (a). When the three lasers operated at free-running mode, the longitudinal frequencies of the lasers varied, as shown in Fig. 3-3 (b). However, when the lasers were synchronized, they would oscillate at a same repetition rate or round trip frequency $f=f_1=f_2=f_3$. In this case, the three laser beams could be treat as one beam with a combined spectral distribution. The experimental setup is shown in Fig. 3-3 (c). The 1030-nm pulses are chosen as the master source. As an independent laser, the Er-doped fiber laser can be mode-locked by carefully aligning the quarter- and half-wave plates inside a unidirectional ring cavity to change the nonlinear polarization evolution. One of the collimators is mounted on a translation stage inside the slave laser cavity so that the cavity length can be slightly changed with its repetition rate to match the corresponding master repetition rate. It should be mentioned that the repetition-rate match is a quite important part to realize the passive synchronization. The repetition rate of the Er-doped fiber laser is designed to be 40 MHz, a half of that of master laser. The output pulse is centered at 1560 nm with the pulse width of ~290 fs [Fig. 3-4 (a) and (b)]. As a slave laser under the case of the master pulses injection, the Er-doped fiber laser can be locked at the same repetition rate of the master laser, which is the second harmonic of its own fundamental repetition rate. The radio frequency of the slave laser before and after being synchronized is given in Fig. 3-4 (c) and (d).

Because the two synchronous lasers come from two spatially-separated oscillators in the master-slave configuration, the relative variation of the two cavities limits the synchronization precision to a large extent. Therefore, the relative jitter of the two synchronized lasers is larger than that of the FSA. The integrated timing jitter in the Fig. 3-3 (c) setup is nearly 8.5 fs, which is 15 times larger than that in the FSA experiment.

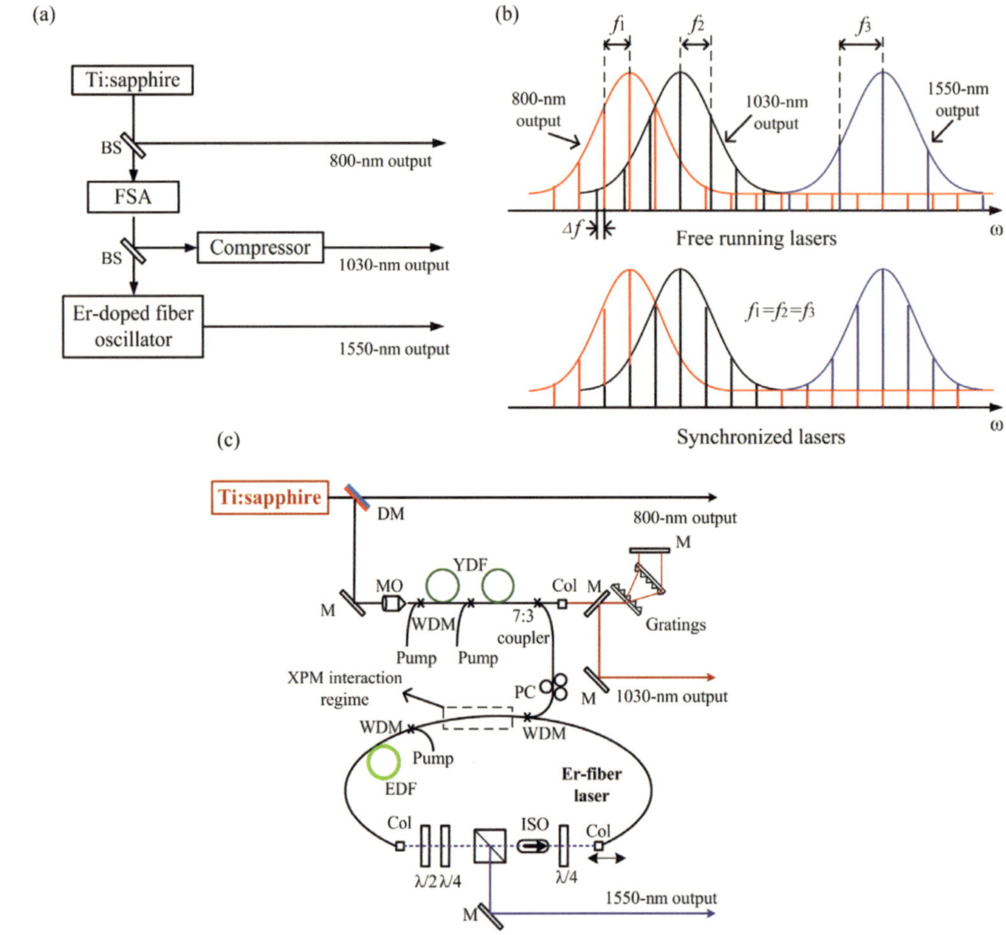

Fig. 3-3. Experimental structure (a), spectral schematic of the synchronized three-color lasers (b) with the experiment setup (c). DM: dichroic mirror (HT at 800 nm and HR at 1030 nm), MO: micro-objective, WDM: wavelength division multiplexing, YDF: ytterbium-doped fiber, EDF: erbium-doped fiber, ISO1: fiber isolator, ISO2: free space isolator, PC: fiber polarization controller, COL: fiber collimator, λ/2 and λ/4: half-wave plate and quarter-wave plate, PBS: polarization beam splitter.

3.1.3 Synchronization achieved by using Kerr nonlinear medium

For fiber lasers, master-slave configuration is considered to be a simple but efficient scheme because single-mode fiber has a very small core diameter to restrict light in a narrow area resulting high light intensity. However, in free space, it is difficult to keep light in a small area for a distance as long as the walk-off length. Thus, to realize a passive synchronization between two solid lasers, Kerr-type nonlinear medium is required to support a strong XPM effect for two synchronous pulses interacting with each other (Apolonski el at., 1993; de Barros & Becker, 1993; Fuerst el at., 1996; Telle el at., 1999; Jones el at., 2000; Apolonski el at., 2000; b Rusu el at., 2004). This is because the Kerr medium can provide large nonlinearity compensating for the disadvantage of low light intensity.

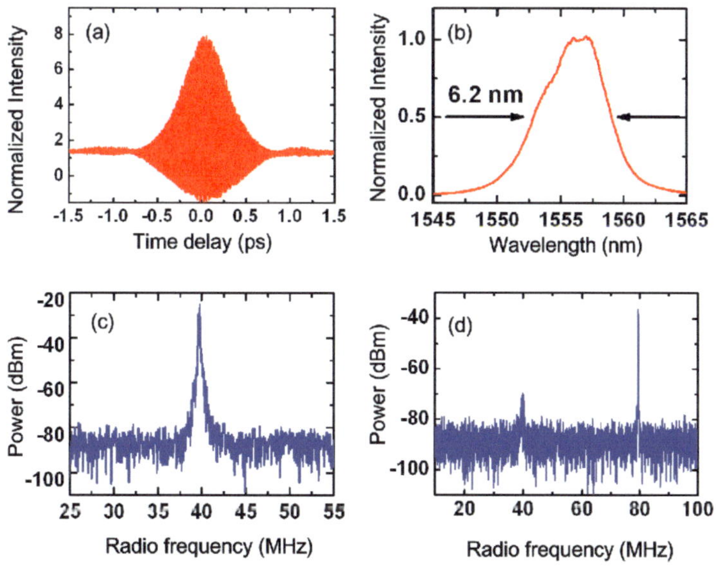

Fig. 3-4. Pulse duration of the synchronized mode-locked EDFL pulses (a) with the corresponding spectrum (b) and the radio frequency spectrum of the mode-locked EDFL pulses before (c) and after (d) synchronization.

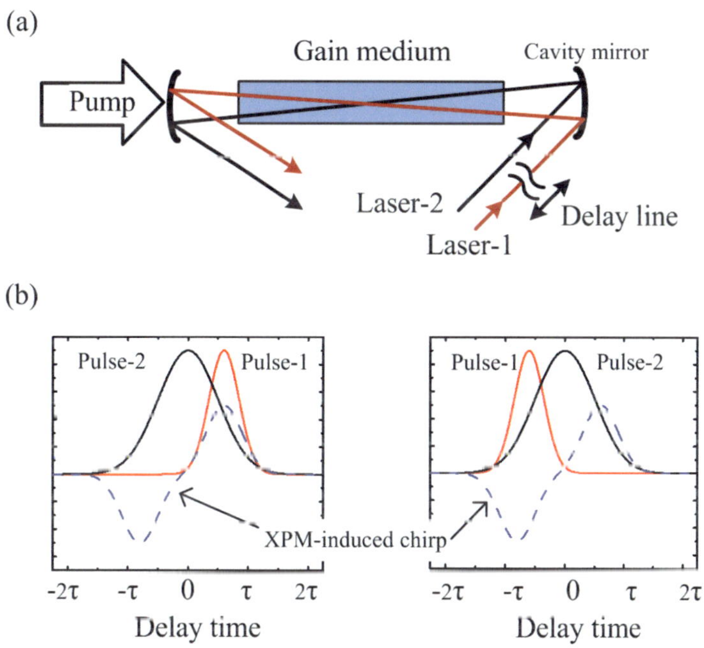

Fig. 3-5. The experimental scheme for synchronization by sharing a same gain medium (a) and the XPM induced frequency chirp for Laser 2 (b). The red line is for laser-1 and black one is for laser-2.

Cavity-shared laser synchronization can be realized by using a setup as shown in Fig. 3-5 (a), where the two lasers share the same gain medium of Ti:sapphire crystal. Due to the broad gain spectrum of the Ti:sapphire crystal, the two lasers can oscillate at different wavelengths. Since the gain medium exhibits Kerr nonlinearity, the two lasers can interact with each other in such a high-nonlinearity medium resulting in a XPM-synchronization as discussed in Section 2.1. In this case, the two lasers are cross-mode-locked at matched cavity-lengths. Once Pulse-1 goes before (behind) Pulse-2 caused by the cavity variation, it will interact with the falling (rising) edge of Pulse-2, as shown in Fig. 3-5 (b). As a result, Pulse-1 obtains a positive (negative) frequency chirp from Pulse-2, which compensates for the cavity variation when Pulse-1 propagating in normal dispersion cavity.

3.2 Mode-locked nanosecond pulse generation and synchronization

Synchronization of ns pulse trains and even precisely phase-locked ns laser arrays are required in many high-energy physics experiments, such as in the development of high-energy laser pulses for particle acceleration, and laser synchronization with x-rays or electron beams from synchrotrons (Schoenlein el at., 1996; Baum & Zewail, 2007). Conventionally, the synchronous ns laser pulses can be obtained by Q-switching technique and active synchronization scheme with a complicated electronic feedback system. In this section, we will introduce a simpler scheme to passively synchronize a mode-locked ns laser with a ps laser by using XPM and peak-power clamping effects.

3.2.1 Peak-power clamping effect

Recently, it is found that Erbium-doped fiber laser with a long cavity can generate ns square mode-locked pulses by the peak-power clamping effect. In order to discuss this effect, a simplified Er-fiber laser scheme is illustrated in Fig. 3-6 (a) with a cavity-length of L.

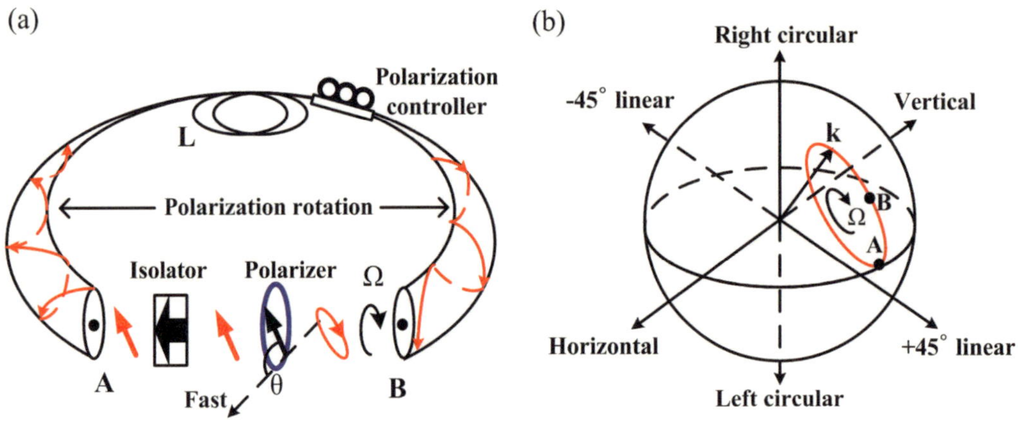

Fig. 3-6. The simplified Er-doped fiber laser cavity (a) and a corresponding poincaré sphere diagram (b).

After passing through the polarization dependent isolator (ISO), the round-trip transmission of the laser pulse can be expressed by (Matsas el at., 1992; b. Li el at., 2009):

$$T_p = \cos^2(\Omega) - \frac{1}{2}\sin(2\theta)\sin[2(\theta-\Omega)]\times[1-\cos(2\pi L / L_{b0})] \tag{7}$$

where Ω is the rotation angle induced by both polarization controllers and fiber intrinsic linear birefringence and θ the azimuth angle of the polarization-dependent isolator with respect to the fast axis. The beat length L_b is power-dependent in a case of light power being high enough to bring in a nonlinear effect. Assuming that the pulse is linearly polarized at θ =45°, the beat length will be (b. Li el at., 2009):

$$\frac{L_b}{L_{b0}} = (\frac{3}{8} + \frac{5}{8}\sqrt{1+p^2})^{-1/2} \tag{8}$$

where L_{b0} is the linear beat length of the birefringent element (in this case, the total power-independent birefringence of the cavity) and the normalized power P is defined as $P = 2n_2I/3\Delta n$, where I is the light intensity, n_2 is the nonlinear refractive index and Δn is the refractive difference between the two birefringent axes. According to Eq. (8), the requirement for the lowest normalized power that maximizes the round-trip transmission in Eq. (7) can be deduced as:

$$(\frac{3}{8} + \frac{5}{8}\sqrt{1+p^2} - \frac{2}{2+L_{b0}/L} \tag{9}$$

$$P_{sw} = \sqrt{\frac{1}{25}(2(2+L_{b0}/L)^2 - 3)^2 - 1} \tag{10}$$

The switching power decreases with the fiber length according to Eq. (10). Note that the peak power will be clamped to maintain the condition when the pulse power is sufficient to meet the maximum round-trip transmission. In this situation, without the injection of the master laser, the slave laser could be self mode-locked with the square ns mode-locking mode by adjusting the polarization.

In the situation where a master pulse is injected into the slave laser, the injected pulse induces a nonlinear phase shift of the slave laser between two orthogonal polarization modes as (Agrawal, 2001)

$$\Delta\varphi_{XPM} = \frac{4\pi n_2 |E_p|^2 L_{eff}}{3\lambda} \tag{11}$$

where L_{eff} is the effective interaction length of XPM coupling between the master and slave lasers, E_p the electric field of the injection master laser in the slave laser cavity, and λ the central wavelength of the slave laser. As the XPM-induced nonlinear phase shift is merely related to the power of the master pulse, it just behaves as an equivalent linear polarization rotating element Ω in Eq. (7). As a result, the master laser injection functions as an optical trigger to synchronize the slave laser, while the power for the slave reaching its first maximum round-trip transmission is still maintained. To specifically describe the process of ns pulse generation, an experiment of synchronizing an ns Er-doped fiber laser (slave laser) to a ps Yb-doped fiber laser (master laser) is discussed (b. Li el at., 2009).

In this experiment, a master-slave configuration is employed as shown in Fig. 3-7. The master [Fig. 3-7 (a)] is a passively mode-locked Yb-fiber laser with a repetition rate of 1.91 MHz. The initial pulse width of the master laser is 47 ps centering at 1053 nm. Before being injected into the slave laser, the master laser is at first amplified to 150 mW by an Yb-doped fiber amplifier. As a slave laser [Fig. 3-7 (b)], an Er-doped fiber laser can be mode-locked at its fundamental repetition rate of 956 kHz, which is half of the master laser's repetition rate, when it is pumped by a 450 mW fiber-pigtailed diode laser at 980 nm. In order to increase the peak-power clamping effect for generating ns pulses, a 200-meter-long single-mode fiber is installed in the slave cavity.

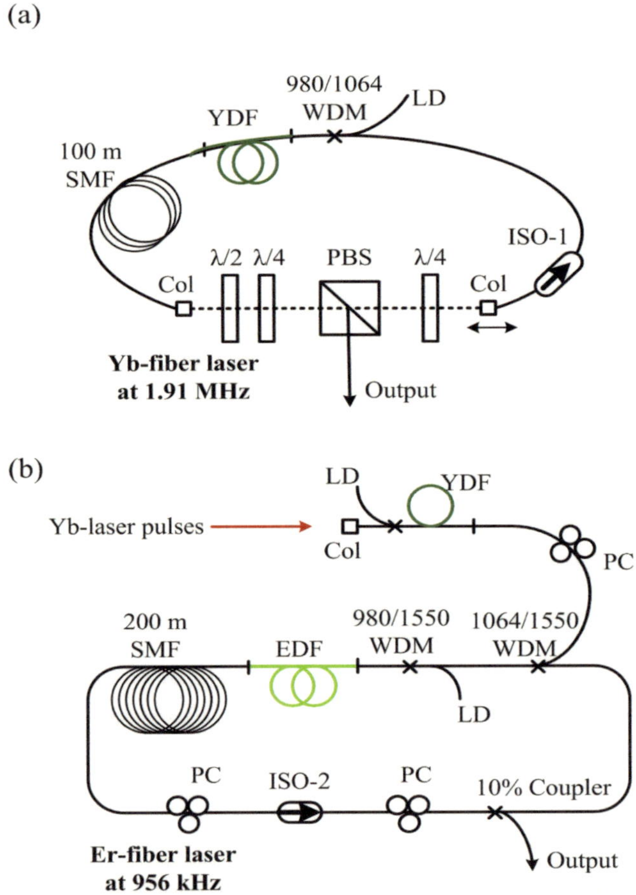

Fig. 3-7. Experimental setup of Yb-doped master fiber laser (a) and Er-doped slave fiber laser (b). YDF: Yb-doped fiber; YDFA: Yb-doped fiber amplifier; EDF: Er-doped fiber; Col: collimator; ISO1 & ISO2: isolators; WDM: wavelength-division multiplexing; PBS: polarization beam splitter; PC: polarization controllers.

Here, we intentionally make the two lasers working at different repetition rates, because the ns fiber laser needs long cavity for offering large peak-power clamping effect while the ps laser requires short one for inducing less dispersion. When the synchronization is achieved between the two lasers, the repetition rate of the slave laser will jump to its second harmonic mode equaling to that of the master laser, while its pulse duration will be reduced to half of the original value according to the peak-power clamping effect, as illustrated in Fig. 3-8. In this experiment, the effects induced by the peak-power clamping in ns-ps synchronization can be summarized as follows.

Firstly, the pump power threshold for the ns pulse generation is inversely proportional to the laser repetition rate. In the experiment, the pump power threshold for the free-running square mode-locking is 200 mW. With the master injection, the Er-laser is synchronized to run at its second harmonic repetition rate due to the XPM-induced nonlinear polarization rotation, and the threshold for the mode-locking is decreased to 50 mW. Meanwhile, with the maximum pump power of 450 mW, the free-running pulse duration is 11 ns as shown in Fig. 3-9 (a), the corresponding spectra is shown in Fig. 3-9 (b). While, the synchronous square ns pulses exhibits a duration of 5.5 ns [Fig. 3-9 (c) with corresponding spectrum at

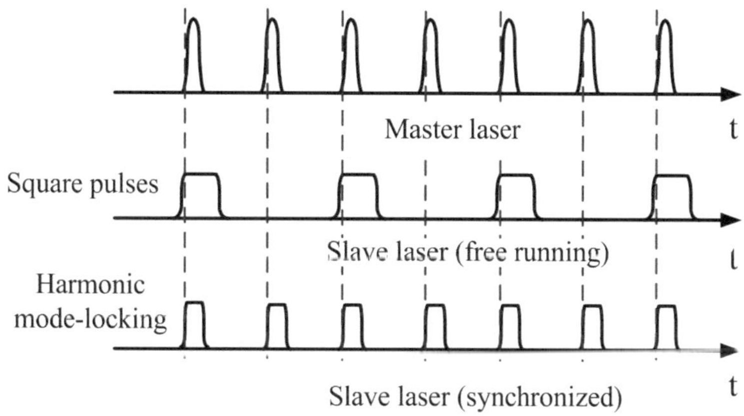

Fig. 3-8. Schematic diagram of laser pulse trains for the ps-ns synchronization experiment.

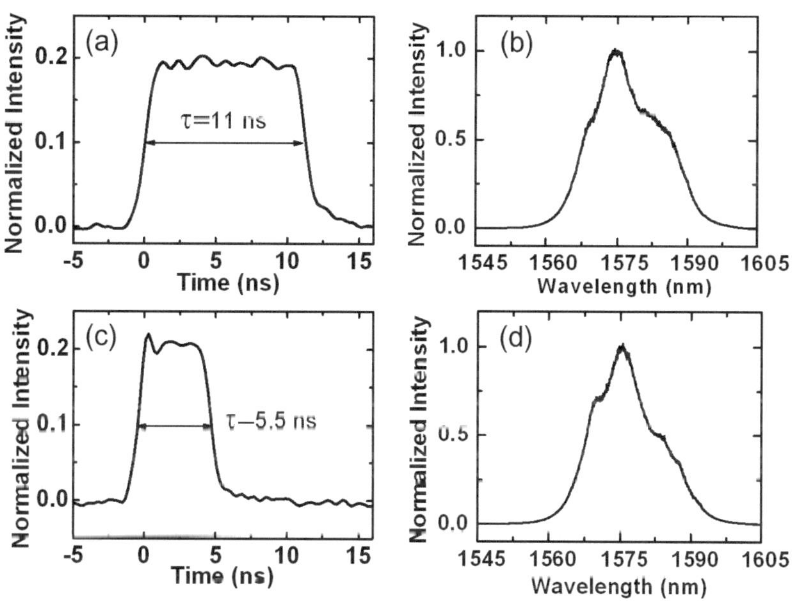

Fig. 3-9. Free-running square mode-locking waveform of the slave Er-laser (a) and its corresponding spectrum (b); Synchronized mode-locking waveform of the slave Er-laser (c) and its corresponding spectrum (d).

Fig. 3-9 (d)] at the pump power of 450 mW, exactly half that of the free-running pulses as an indicative of peak-power clamping at the same level due to the induced switch from the fundamental to the second harmonic mode-locking.

Secondly, the pulse shapes of the synchronously mode-locked Er-laser critically depend on the pump power. As shown in Fig. 3-10, when the pump power is below the threshold value, the output pulse is not a square one because the energy stored in the pulse is not sufficient to sustain the square wave. In this case, the peak power is increased with the pump power. When the pump power reaches the threshold, square mode-locked pulses are generated with an obvious rising of the pulse tail part (Fig. 3-10). As the pump power increases further, the pulse is stretched linearly with the pulse energy due to the peak-power clamping effect.

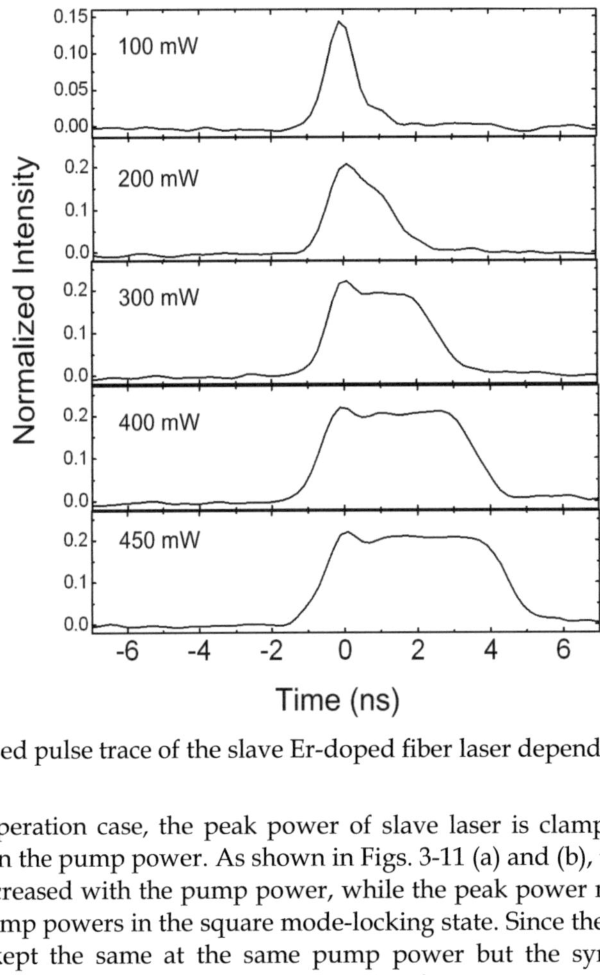

Fig. 3-10. Synchronized pulse trace of the slave Er-doped fiber laser dependent on the pump power.

Thirdly, in the ns operation case, the peak power of slave laser is clamped in a constant value independent on the pump power. As shown in Figs. 3-11 (a) and (b), the output power is almost linearly increased with the pump power, while the peak power maintains around 3.3 W at different pump powers in the square mode-locking state. Since the energy stored in the laser cavity is kept the same at the same pump power but the synchronized pulse repetition rate is doubled, the pulse energy in the free-running state is halved in synchronized mode-locking state and the peak power of each pulse is clamped at the same value. As a result, the pulse duration of synchronous Er-laser is half of that in the free-running state as shown in Fig. 3-9.

Fourthly, the pulse duration of the synchronous slave laser does not change with the injection master laser power. Therefore, in the whole experiment, the peak-power clamping effect is the main reason for square-shaped pulse and the injected master pulse just acts as a trigger to synchronize the square ns mode-locking. As a direct consequence of the peak-power clamping effect, much longer pulse duration can be achieved with higher pump powers.

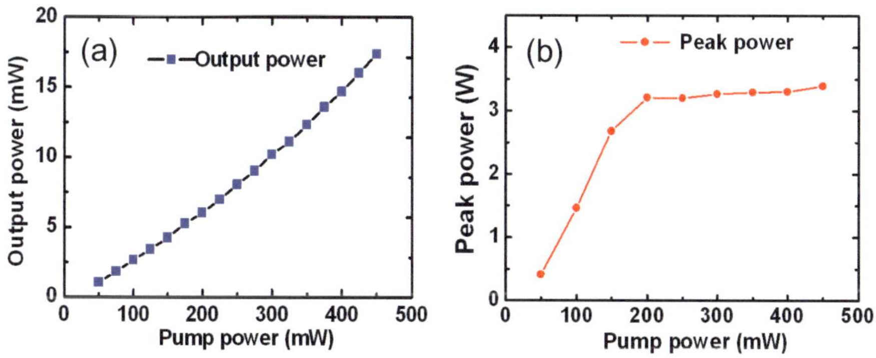

Fig. 3-11. Output power (a) and the peak power (b) of the slave laser as a function of the pump power.

3.2.2 Mismatch tolerance in synchronization

As mentioned before, mismatch tolerance is a crucial index for judging a passive synchronization system. Larger tolerance value means a stronger capability for system standing against variation of cavity length induced by instability of environments. Usually, when a passive synchronization is achieved, the slave laser will run at the same repetition rate with the master laser. In synchronization region, the repetition rate of slave laser is independent with its cavity-length. However, once the cavity-length exceeds the tolerance range, the slave laser will operate in a free-running mode, which means the two lasers are unsynchronized. To measure the mismatch tolerance value, at least one of the two lasers should have a length tunable cavity. The following example will be used to expatiate upon the measurement of mismatch tolerance (Li & Gu el at., 2009).

In the ps-ns synchronization experiment, part of the output pulses from the master and the slave are detected independently in order to monitor the synchronization of the two lasers. The oscilloscope is triggered by the master pulse trains. Only when the synchronization is achieved, the slave pulse trains can be clearly displayed on the oscilloscope. In order to tune the master cavity length, a translation stage with precision of 10 nm is placed inside the master cavity, as shown in Fig. 3-12 (a) (the red dashed box). When the master cavity length mismatch is changed from the −1.3 to 1.3 mm around zero position, the repetition rate of the slave laser keeps the same as the master laser [Fig. 3-12 (b)]. However, beyond the ±1.3 mm range, the slave laser will jump back to its fundamental repetition rate of 956 kHz. In this case, the synchronization is ceased. This maximum mismatch range of 2.6 mm is a quite large value for this system against environmental vibrations in comparison with previous XPM-synchronization experiments (Wei el at., 2002; Yoshitomi el at., 2006).

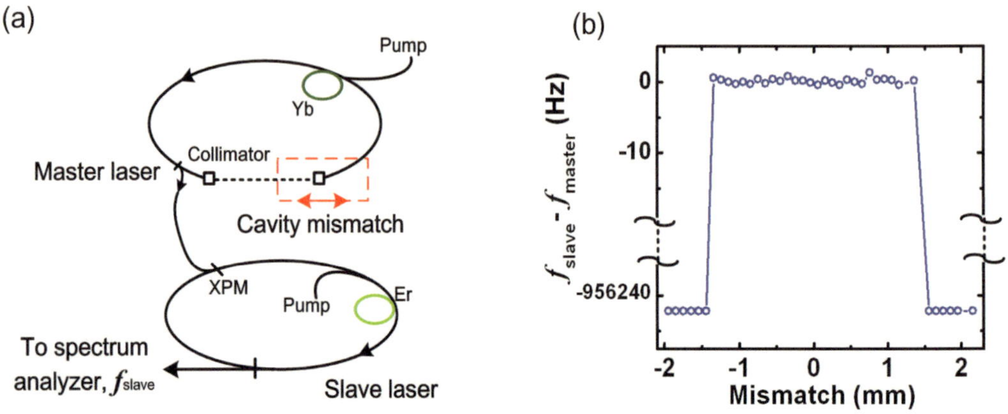

Fig. 3-12. (a) Schematic setup of cavity mismatch measurement and (b) cavity mismatch for the master and slave fiber laser: f_{master} and f_{slave} are the repetition rates of the master and slave lasers, respectively. In the synchronization region, the repetition rate of the slave laser equals to that of the master laser. While beyond that region, the slave laser will jump back to its fundamental repetition rate.

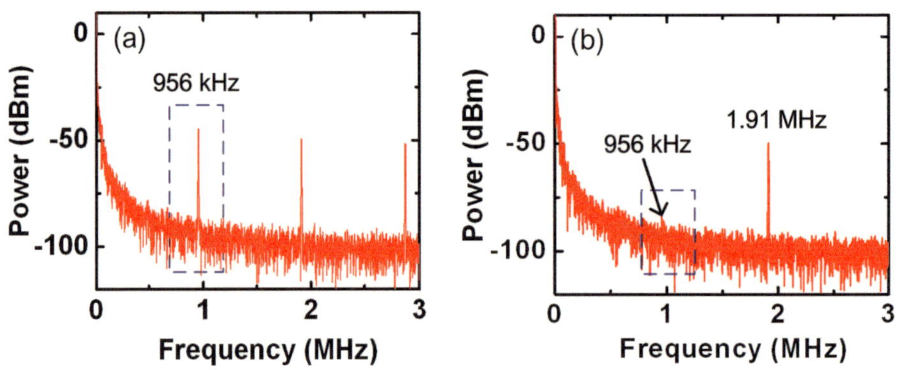

Fig. 3-13. (a) RF spectrum of the free-running repetition rate and (b) synchronous repetition rate.

To further monitor the synchronization state, the output signal of the slave Er-fiber laser is inputted into a spectrum analyzer with a resolution of 1 Hz as the master cavity length is tuned. In the free-running regime, the slave laser oscillates at its fundamental repetition rate of ~956 kHz [Fig. 3-13 (a)]. While in the synchronization regime, the fundamental repetition rate of the slave laser is fully restrained (at least 50 dB) and only its harmonic signal can be seen as shown in Fig. 3-13 (b).

3.2.3 The measurement of timing jitter

Usually, jitter is a concept in the fields like electronics and telecommunications to evaluate the time variation of a periodic signal in relation to a reference clock source. As in frequency domain, the concept of jitter is represented as "phase noise" caused by temporal instabilities. The concept of jitter or timing jitter has been extended to explain the relative time fluctuation between two synchronous optical pulses. Until now, many methods have been reported to measure the timing jitter for synchronous lasers. As a widely-used method for timing jitter measurement, optical cross correlation technique is employed to indirectly measure the ultra-fast timing jitter (Paschotta, 2004; Foreman el at., 2007; Chen el at., 2006). The theory and practical details of such kind of jitter measurement is discussed with an example based on the FSA synchronization experiment (Li el at., 2009). The experimental setup is shown in Fig. 3-14 (a). In general, the cross-correlation based timing jitter measurement is designed to linearly connect the time fluctuation, $\Delta\tau$, on the order of femtosecond between the two synchronous pulses with the intensity fluctuation, Δi, of the sum frequency signal, as shown in Fig. 3-14 (b). This is because the time fluctuation in fs region is usually too fast to be directly measured by lots of electronic equipments. However, with cross-correlation technique, the fast time variation can be reflected by the measurable intensity changes of the sum frequency signal and then be recorded by a spectrum analyzer for calculating the exact value of the time fluctuation (timing jitter).

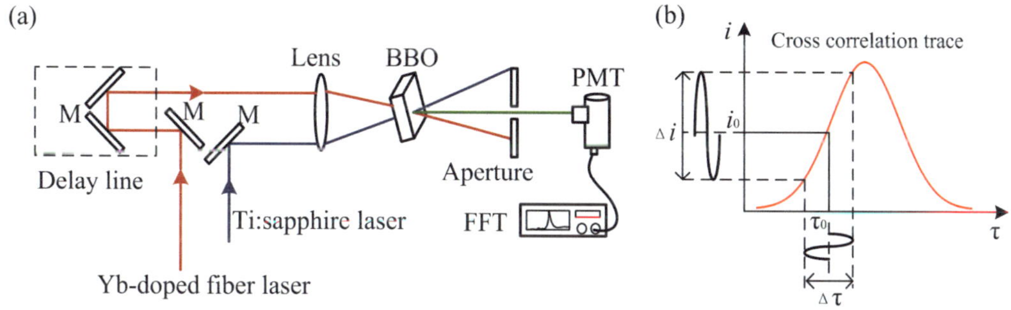

Fig. 3-14. Experiment layout (a) and schematic diagram of the timing jitter measurement with optical cross correlation. M: mirror; BBO: β-barium borate crystal; FFT: fast Fourier transformed spectrum analyzer; PMT: photomultiplier tube.

In the measurement, the Yb-doped fiber laser beam (1030 nm) is crossed with a part of the rest of Ti: sapphire laser beam (800 nm) in a 0.5-mm β-barium borate (BBO) crystal to generate the sum frequency signal at ~452 nm. The Ti: sapphire laser beam passes through a time-delay line. When the delay is scanned, the SFG is detected by a photomultiplier tube (PMT, 10-KHz bandwidth) and the cross-correlation trace between the two fs pulses is recorded and shown in Fig. 3-15 (a).

In order to obtain the jitter power spectral density and its integrated RMS timing jitter, the time delay of the two pulses is positioned at the half-maximum of the cross-correlation signal where the signal can change linearly with the delay time. The Fourier-transformed spectrum of the fluctuation of the correlation intensity is recorded by an FFT spectrum analyzer (SRS, SR760), as shown in Fig. 3-15 (c).

The noise level is normalized against carrier power and bandwidth resolution and expressed in units of $dBc/Hz^{1/2}$. The contribution of timing jitter comes mainly from the band within 1~10 kHz. The noise sidebands are related to amplitude noise and timing jitter by (Chen el at., 1996; Wilcox el at., 2006):

$$S_n(f) = S_E(f) + (2\pi n f_0) S_{JE}(f) + (2\pi n f_0)^2 S_J(f)\,, \tag{12}$$

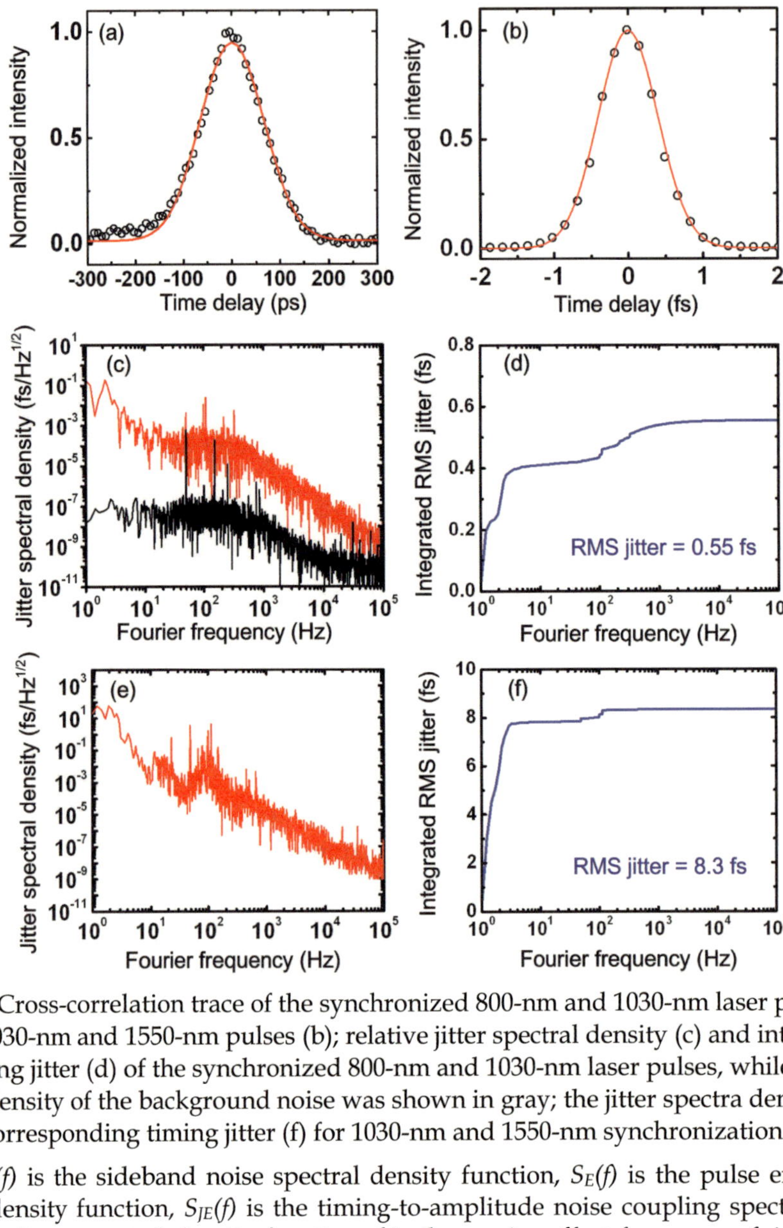

Fig. 3-15. Cross-correlation trace of the synchronized 800-nm and 1030-nm laser pulses (a) and for 1030-nm and 1550-nm pulses (b); relative jitter spectral density (c) and integrated RMS timing jitter (d) of the synchronized 800-nm and 1030-nm laser pulses, while the jitter spectral density of the background noise was shown in gray; the jitter spectra density (e) and the corresponding timing jitter (f) for 1030-nm and 1550-nm synchronization.

where $S_n(f)$ is the sideband noise spectral density function, $S_E(f)$ is the pulse energy noise spectral density function, $S_{JE}(f)$ is the timing-to-amplitude noise coupling spectral density, $S_J(f)$ is the jitter spectral density function, f is the carrier offset frequency, f_0 is the cavity repetition rate, and n is the harmonic number. The experimentally measured sidebands are integrated over the measurement bandwidth, yielding the quantity:

$$\int_{f_1}^{f_2} S_n(f)df = \frac{\sigma_E^2}{2} + (2\pi n f_0)\frac{C_{JE}}{2} + (2\pi n f_0)^2 \frac{\sigma_J^2}{2}, \tag{13}$$

where f_1 and f_2 are the start value and end value for integration, respectively, σ^2_n is the total noise power, σ_E is the RMS normalized pulse energy fluctuation, σ_J is the total timing jitter and C_{JE} is the cross-correlation term between pulse timing jitter and normalized pulse energy fluctuation. The factor of 2 is for the single sideband noise. For the sake of simplicity, we merely consider the contribution from the term of timing jitter at the fundamental mode of $n=0$, and then transform the spectral density to the time-related one by (Jiang el at., 2002; Haus & Mecozzi, 1993; Hönninger el at., 1999; Eliyahu el at., 1997):

$$T(f) = \frac{S(f)}{2\pi f_0} \text{ (s/Hz}^{1/2}\text{)}. \tag{14}$$

Thus, the RMS timing jitter σ_{RMS} can be calculated by:

$$\sigma_{rms} = \sqrt{\int_{f_1}^{f_2} [T(f)]^2 df} \text{ (s)}. \tag{15}$$

By this way, the timing jitter between the 800-nm pulses and the 1030-nm pulses is calculated to be 0.55 fs, as shown in Fig. 3-15 (d), according to the integration from 1 Hz to 100 kHz. Meanwhile, with the same method, the cross-correlation trace and timing jitter are also measured for the 1030-nm pulses and 1550-nm pulses obtained in section 3.1.2 as shown in Figs. 3-15 (b) and (e), respectively. The jitter between the 1030-nm and 1550-nm lasers is nearly 8.3 fs [Figs. 3-15 (f)]. Actually, for achieving a three-color laser source, one can easily use fraction spectra amplification technique to obtain synchronous pulse trains at 800 nm and 1030 nm, and then synchronize the 1030-nm pulses to 1550-nm laser by utilizing the master-slave configuration for fiber lasers.

3.3 XAM-based synchronization scheme

Up to now, all the passive synchronization schemes discussed above are applicable to ultrashort mode-locked lasers at repetition rate higher than MHz, but inapplicable to mode-locked ns-duration lasers generated at sub-MHz repetition rate. The importance of such synchronous ns pulses has been discussed in the introduction part of this review chapter. Because the sub-MHz ns pulses are conventionally obtained by Q-switching technique, the corresponding synchronization employs active methods. However, the combination of Q-switching and active synchronization has to share same electronic triggers with a quite large timing jitter as limited by the electronic circuits. Even the ns pulses can be achieved in a fiber laser with a long ring cavity, the XPM in the fiber is still limited by the walk-off length and high peak power required to induce sufficient nonlinearity for a tight synchronization. Efficient all-optical techniques are desired to precisely synchronize sub-MHz laser pulse trains.

Recently, a synchronization scheme of XAM was found to permit synchronization at low repetition rates with large tolerable cavity-length mismatches (a. Yan el at., 2009). Compared to XPM-based synchronization, XAM relaxes the restrictions on peak powers of interacting pulses. In the XAM-based synchronization scheme, the master-slave modulation effect is largely enhanced by using a resonant absorption medium in the slave cavity. Thus, synchronization can be achieved in a 800-m-long fiber laser with a repetition rate of ~250 kHz. We will present here an experimental example of XAM-based synchronization for individual lasers at low repetition rates. The XAM triggers synchronous square ns mode-locking sensitive to the injected master laser power, which is unique for the XAM-based synchronization and differs from the XPM-based synchronization as discussed in the previous sections.

3.3.1 Synchronization between Sub-MHz femtosecond and nanosecond lasers

In general, XAM is very weak in non-resonant media, while it becomes comparable to or even larger than XPM in near-resonant media. As an example, we present XAM-based synchronization for low-repetition-rate lasers in the master-slave configuration as schematically shown in Fig. 3-14. A tight synchronization between a 250-kHz Ti:sapphire laser (TiS) and an ns Yb-doped fiber laser is achieved by using enhanced-XAM in an Er-doped fiber inside the Yb-doped fiber laser cavity. In the Er-fiber, the TiS pulse at 800 nm behaves as pump as well as controlling pulse (optical trigger) causing a transition of Er^{3+} from energy level $4I_{15/2}$ to $4I_{9/2}$. In the absence of the trigger pulse, Er^{3+} ions will drop back to and then stay at the level $4I_{15/2}$ through spontaneous emission. The trigger-induced transitions between the energy levels introduce a periodical modulation for the refractive index in the Er-fiber, resulting in the robust synchronization between the trigger pulse and the slaved pulse.

How does the XAM induce a feedback mechanism in the synchronization system? In fact, as the resonant media at 800 nm, the Er- fiber absorbs the TiS pulses with a corresponding refractive index change, resulting in a XAM-induced nonlinear polarization rotation (NPR) for the co-propagating Yb-fiber laser pulses. For the slave pulse, pulse spectral shift accompanying with XAM-induced NPR compensates for the fiber cavity variations to keep the robust synchronization. The details of the theory for XAM applying in synchronization are still under investigation.

As schematically shown in Fig. 3-16 (a), a mode-locked TiS laser (RegA 9000 from coherent Inc.) operates as a master laser to deliver a 250-KHz, 70-fs pulse train. Its spectrum width is ~60 nm with a center wavelength of 800 nm. The salve laser is an Yb-doped fiber oscillator pumped by a 976-nm, 300-mW diode laser. The center wavelength of the salve is at 1041 nm with pulse power of ~3-mW. To oscillate at the same repetition rate with the master, the fiber laser stretched its cavity with ~800 m of single-mode fiber. Besides, the fiber cavity includes 1.5-m-long Yb-doped fiber as the gain medium and 1-m-long Er-doped fiber (unsaturated absorption at 800 nm, ~3.0 dB/m; dopant concentration, 5.4×10^{24} m^{-3}) for inducing enhanced XAM. Without synchronization, the fiber laser pulse duration can be tuned from sub-ns to ~10 ns roughly by adjusting two sets of polarization controllers and accurately by rotating a half-wave plate placed between the two collimators.

In order to induce XAM into the fiber laser, a portion of ~200 mW of the TiS laser is used and nearly 20 mW is injected into a 0.5-m-lomg single-mode fiber with a microscope objective (×40). The injected pulse is further coupled into the fiber laser cavity through 800/1064-nm wavelength-division multiplexer (WDM) immediately followed by the 1-m-long Er-doped fiber. Through the 800/1064-nm WDM only 10-mW mater power is left. After the injection, the synchronous pulses can be obtained by carefully adjusting the polarization controllers inside the fiber cavity. Finally, timing jitter of ~0.6 fs is achieved and the cavity mismatch tolerance is extended to 8.2 cm, which is the largest value ever recorded in passive experiments, due to the long cavity and absorption effect. During cavity mismatch measurement, it is also found that the ns Yb-doped fiber laser can keep the synchronization state by shifting the center wavelength from 1038 nm to 1043 nm.

Usually, there are three keys important for the XAM synchronization. First, for a fiber laser with hundreds-meters-long fiber cavity, the fiber laser pulse must be highly chirped due to the dispersion effect inside the cavity. Considering the normal dispersion for the near-infrared laser light, the slave pulse will propagate with red head and blue tail. Secondly, the master laser at 800 nm is able to excite Er3+ from energy level $I_{15/2}$ to level $I_{9/2}$ [shown in

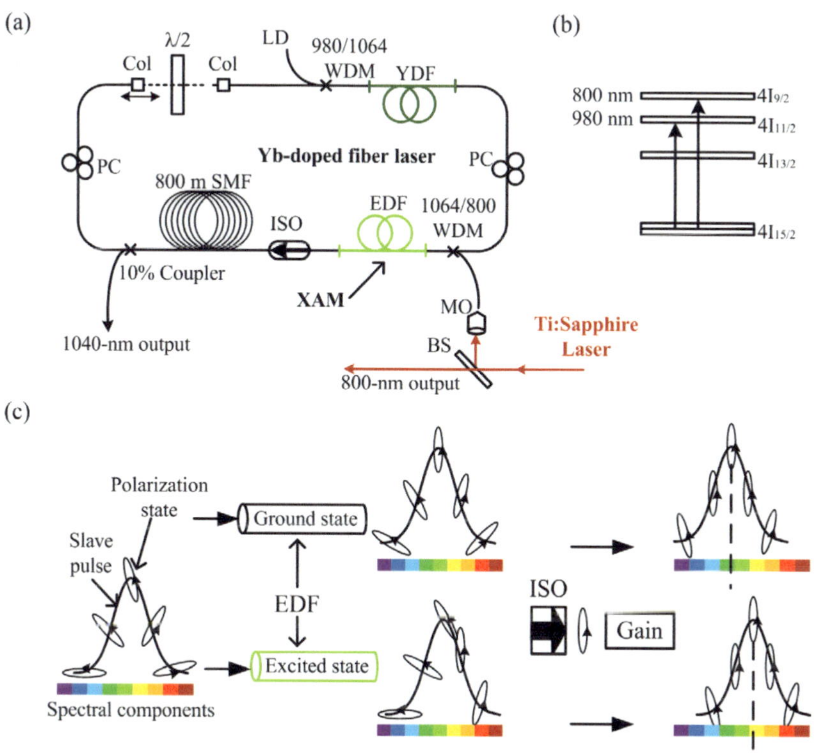

Fig. 3-16. Experimental setup of XAM-based synchronization (a), energy structure of Er-doped fiber (b) and the principle schematic of the XAM synchronization. BS, beam splitter; Col, collimator; OC, output coupler; MO, micro-objective; WDM, wavelength division multiplexer; PC, polarization controller; ISO, Isolator; λ/2, half-wave plate. In (c), when the slave pulse passes through an excited Er-doped fiber, its polarization states will be changed comparing to going through the ground-state Er-fiber. And under the cooperation of the polarization-dependent isolator and the gain medium, the spectral center of the polarization-rotated slave pulse is shifted. Note that this spectral shift is crucial for compensating for relative variation of the repetition rates in the XAM synchronization.

Fig. 3-16 (b)], which changes the birefringence index of the Er-doped fiber. As a result, the polarization state of the co-propagating slave pulse will be changed as illustrated in Fig. 3-16 (c). Thirdly, the polarization-dependent isolator inside the fiber cavity only permits certainly polarized light pass through with a minimum loss. The light with other polarization states will be isolated. The certain polarization state can be achieved by adjusting the polarization controllers inside the fiber cavity, when the salve laser is synchronized. Since the fiber laser cavity is sensitive to environmental fluctuations, the slave pulse will go ahead or behind of the master pulse in the Er-doped fiber. If the slave pulse go ahead of the master pulse, the blue tail of the slave pulse which overlaps with the master pulse will be affected by XAM effect with a polarization rotation so that the blue part could maximally transmit the isolator and be amplified by the Yb-doped gain fiber while the red part is isolated. As a result, the center wavelength of the slave laser is blue shifted, which means in the fiber cavity of normal dispersion the slave pulse is slowed down to match the master pulse in the time domain. While, if the slave pulse falls behind of the master pulse, a spectral red-shift will be induced to the salve laser for catching up with the master pulse. In the XAM experiment, total spectral shifts of ~3 nm were found to balance with the cavity-mismatch of ~8.2 cm.

3.3.2 Sensitivity of slaved laser to injection power of master laser

In the XAM-based synchronization experiment, the pulse duration of the slave fiber laser can be affected by the injection pulse power of master laser, which is different with the case in the XPM-based synchronization experiment (Section 3.2.1). In the XPM experiment, the master injection influence very little to the ps-ns synchronization realized by peak-power clamping effect, while the injection power plays an important role in the XAM experiment to affect the slave pulse. Thus, to study this sensitivity is helpful for people understanding the mechanism of XAM applied in synchronization experiment.

As we know, in a fiber laser mode-locked by nonlinear polarization rotation (Fermann el at., 1997), the pulse duration is related with the intra-cavity pulse polarization. In the XAM-triggered synchronous square ns mode-locking experiment, when the Yb-doped fiber laser oscillates in the ns regime, the polarization-related pulse duration and its changes can be easily observed on the oscilloscope with a fast photo-detector.

Interestingly, the fiber laser pulse duration recorded on the oscilloscope is highly sensitive to the coupled power (as shown in Fig. 3-15). This sensitivity indicates that the injected master laser behaves as a power-dependent polarization controller by changing the refractive index of the resonant medium through the XAM process. The polarization at different injection power is estimated by comparing its pulse duration with the case of rotating the half-wave plate at a fixed injection power of 7.4 mW. And then the rotation angle is fixed at 326° of the half-wave plate while the master injection power is changed. As a result, it is found that the injected power of 1.8, 3.4, 6.3 and 7.4 mW correspond to the polarization angle of 339°, 335°, 330° and 326°, respectively. These angles are directly read from the half-wave plate. With these angles, we can simply estimated the XAM induced refractive index change $\Delta n(\lambda)$ at wavelength of λ by (Fekete el at., 2009):

$$\Delta n(\lambda) = \frac{\lambda}{\pi L_{Er}} \Delta\theta \tag{16}$$

where θ is the polarization angle, $\Delta\theta$ means the changes of polarization angle which is twice of the change of rotation angle of half-wavelength plate and L_{Er} is the length of the 1-m Er-doped fiber. Therefore, the corresponding change of the refractive index is measured to be nearly -7.5×10^{-8} as the injection power is changed from 1.8 mW to 7.4 mW. Note that 1.8 mW is the lowest injection power for starting the ns synchronization.

Until now, we merely introduce the basic concept of XAM and its phenomenon in the synchronization experiment. The intrinsic mechanism and full theory for the XAM-based synchronization require further exploration. With XAM, we successfully realize the ns laser source synchronized with an fs laser. By amplifying this ns source, we can obtain high-energy synchronous laser sources at sub-MHz repetition rates.

3.3.3 Generation of high-power synchronous laser source

In previous sections, we discuss the realization of different synchronization schemes. Considering many physics researches relaying on the high-power synchronous light source, we focus here on the amplification of the laser light.

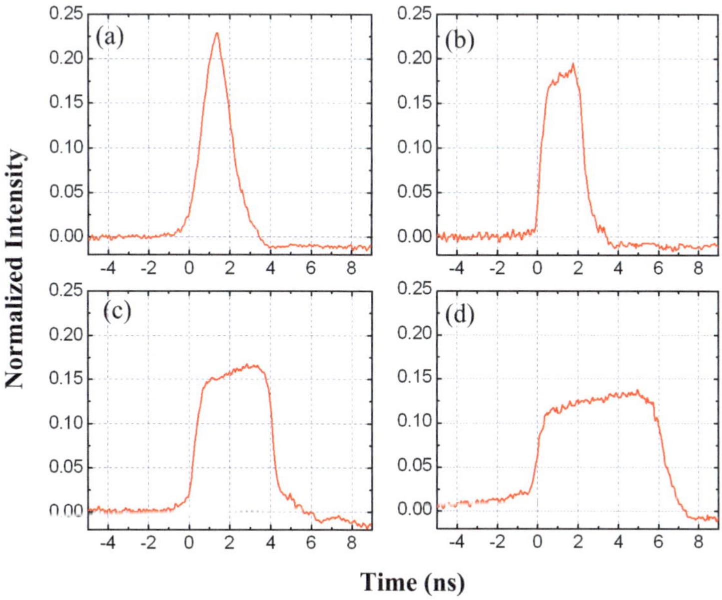

Fig. 3-17. The XAM-induced slave laser sensitivity to the master injection powers of (a) 1.9 mW, (b) 4.4 mW, (c) 6.1 mW and (d) 8.3 mW. The pulses are detected by a 3-GHz photo-detector and recorded by a oscilloscope with bandwidth of 6 GHz.

Recent progress in fiber lasers opens up a new way for high-power laser oscillators and amplifiers. The advent of double-clad fiber technology benefits high-power lasers and amplifiers from kHz to GHz repetition rate (Hao el at., 2007; Papadopoulos el at., 2007). The special fabrication of double-clad fiber not only provides an effective way to transfer the energy from diode lasers into fiber core where the signal pulses propagate, but also ensures diffraction-limited beam quality (a. Limpert el at., 2003). The long signal-pump laser interaction distance can afford a high optical-to-optical efficiency, while the large surface-to-volume ratio results in excellent heat dissipation (b. Limpert el at., 2003). Moreover, large-mode-area (LMA) photonics-crystal double-clad fiber upgrades the threshold of nonlinear effects such as stimulated Raman and Brillouin scattering, while its inner core provides a single-mode operation during amplification (c. Limpert el at., 2003). Therefore, it is a natural idea to combine the temporal synchronization of multi-color lasers with the LMA double-clad fiber amplification technology to attain high-power synchronous lasers.
However, to design a well-performed amplification system, many factors should be taken into account. First, the amplified pulses will suffer from the spectral and temporal distortion, nonlinear phase shifts and nonlinear polarization evolution during the amplification. Especially, some undesirable optical nonlinearity may occur with serious nonlinear phase shifts, since the amplifying laser is tightly guided in the inner cores of the double-clad fibers. Secondly, amplified spontaneous emission (ASE) noises may degrade the synchronization accuracy by inducing detrimental influence into the timing jitter. Thus, in further amplification, special care should be taken to make the seed light having sufficient pulse energy dominating over ASE noises. For this reason, multi-stage amplifiers in cascade can be employed to reduce ASE.

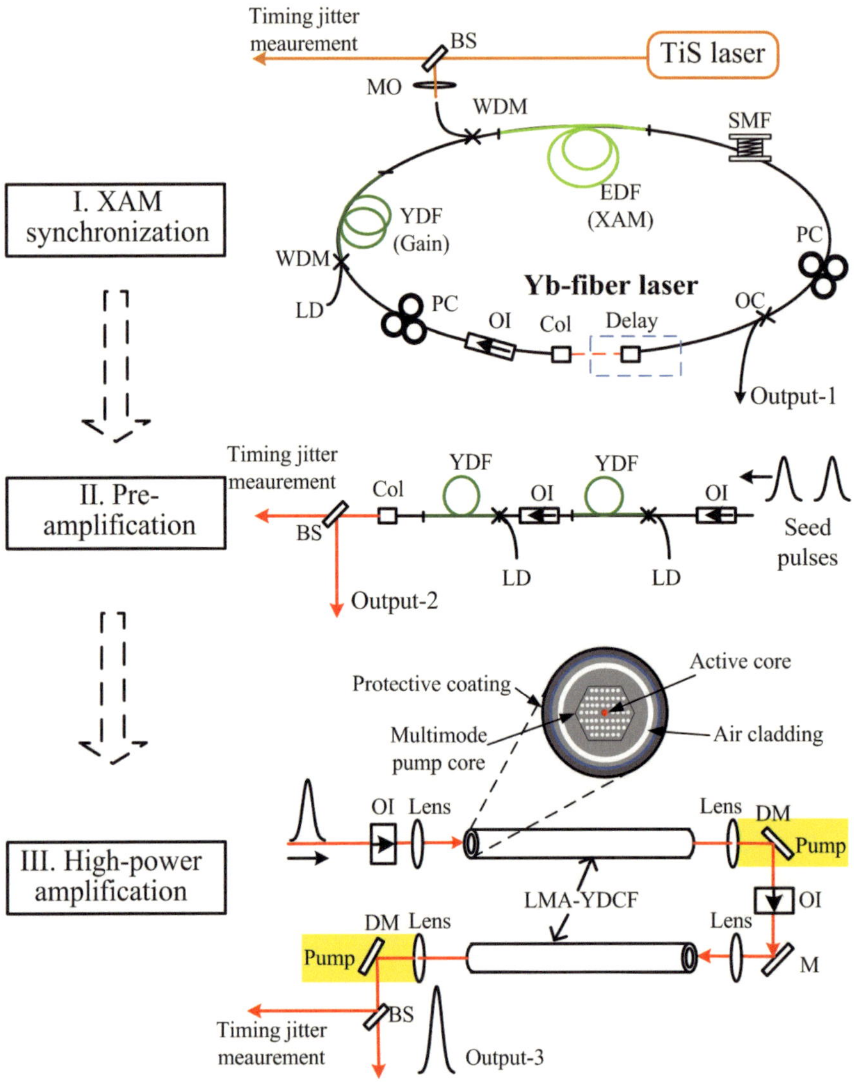

Fig. 3-18. Experimental setup of high-power synchronous ns laser source. BS: beam splitter; MO: micro-objective (×40); WDM: wavelength division multiplexing (980/1041 nm); YDFA: ytterbium-doped fiber amplifier; EDF: erbium-doped fiber; YDF: ytterbium-doped fiber; SMF: single mode fiber; Col: collimator at 1041 nm; OC: 10% output coupler; PC: polarization controller; OI: optical isolator; LD: laser diode ; LMA-YDCF: large mode area Ytterbium-doped crystal fiber; PMT: photomultiplier, FFT: fast Fourier transformed spectrum analyzer.

Previously, based on the XAM-scheme, we achieved a synchronous ns pulse train at repetition rate of 250 kHz. In the following example, the obtained ns pulse is amplified to 131 W (0.55 *mJ per pulse*) by a four-stage amplification system. Meanwhile, the influence of the amplification system on the seed light properties and timing jitter will be studied (b. Yan el at., 2009).

The experimental setup is shown in Fig. 3-18. The ns fiber laser is first synchronized to the TiS laser, and then injected into a multi-stage amplification system which includes a two-stage Yb-doped fiber pre-amplifier and a two-stage LMA-YDCF power amplifier in cascade. In the pre-amplifier, ytterbium-doped single-mode fiber (YbDF350, OFS) of 0.6 m and 1.5 m are used for the first and second stages, respectively, both are pumped by diode lasers at 976 nm. After the pre-amplifier, the fiber laser is amplified to 180 mW. For the first-stage power amplifier, a 1.5-m-long LMA Yb-doped double-clad photonic-crystal fiber is used as a gain medium. Its pump absorption is 10 dB/m at 976 nm with an active core diameter of 40 μm (NA=0.03) and an inner cladding diameter of 200 μm (NA=0.55). The second-stage power amplifier employs 0.85 m-long Yb-doped rod-type photonic-crystal fibers, of which the pump absorption is 30 dB/m at 976 nm, with an active core diameter of 70 μm (NA=0.02) and an inner cladding diameter of 200 μm (NA=0.6). In order to suppress parasitic lasing, the LMA-YDCF ends are sealed to protect the capillaries from environmental influences and polished at an angle of 8°. For the high-power amplifiers, 75% of pump energy is coupled into the inner clad and 60% of the seed power is coupled into the fiber core. With the first-stage power amplifier, the fiber laser pulses are amplified to 4 W, and then are finally boosted to 131 W by the second-stage power amplifier. The slope efficiency of the last amplifier is measured to be 49.5%.

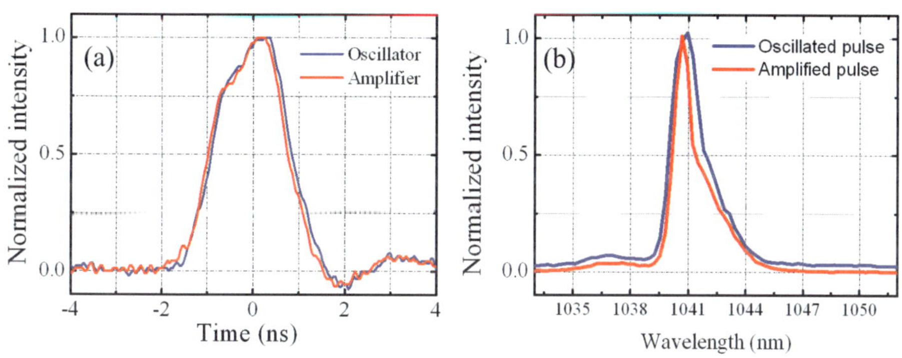

Fig. 3-19. The measured temporal profiles of the synchronized seed fiber laser and the amplified pulses (a) and their corresponding spectra (b).

The corresponding single pulse energy is about 0.55 mJ. This is to the best of our knowledge the highest single-pulse energy for passively synchronized ns lasers. In addition, since the ns light is narrow-band operated, the amplification system brings no obvious spectral variation and temporal distortion into the ns seed pulse, as depicted in Fig. 3-19.

In order to investigate the impacts on the synchronization precision from the power amplification system, the timing jitters between the fs laser pulses and the ns fiber laser pulses before and after the amplification system are measured by virtue of optical cross correlation technique discussed in Section 3.2.3. The Fourier-transformed spectrum of the fluctuation of the correlation intensity is shown by the blue line in Fig. 3-20 (a) with a RMS jitter of about 0.8 ps. The timing jitter between the Ti: sapphire laser beam and the ns fiber laser beam after the last high-power amplifier approximated 13 ps, as shown in Fig. 3-20 (b). Note that at least two main reasons are responsible for the increase of RMS jitter. Firstly, the two-stage LMA amplifier with a complicated water chilling system is individually placed at an optical table spatially apart from the pre-amplifier and the synchronization system. It is quite possible for the spatial disjunction increasing environmental disturbances (air variation, temperature fluctuation, etc.) to the amplified pulses. Secondly, the pump power fluctuations of the whole four-stage amplifier contribute the RMS jitter augment. It should also mention that the spike in Fig. 3-18 (a), the red line, is mainly caused by the fluctuation of fiber-based amplification system at a frequency of several Hz. Fortunately, compared with the pulse duration as large as several ns, the 13-ps RMS jitter is small enough not to threaten the applications of the high-energy synchronized ns pulses.

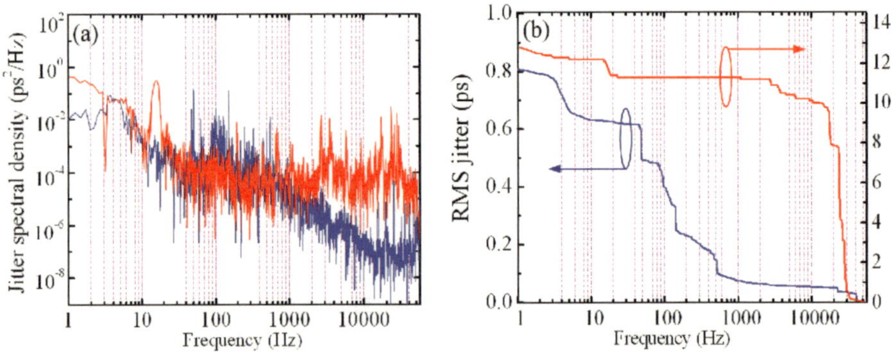

Fig. 3-20. Power spectral density of the jitter (a) for Ti: sapphire laser and ns fiber laser before (blue line) and after (red line) high-power amplifiers and their corresponding integrated RMS jitter from 50 KHz to 1 Hz (b). The narrow peaks in the (a) were due to the 50-Hz electro-disturbance.

4. Conclusion

In this review chapter, passive synchronization between independently mode-locked lasers is introduced with nonlinear effects of XPM and XAM. The mechanism of laser synchronization is mainly due to the reliability of its feed-back system. In the XPM-synchronization, the feed-back system is provided by XPM in coupled-cavity lasers sharing the same Kerr-type nonlinear medium, or independently mode-locked lasers in the configuration with master injection into the slave laser, to compensate the cavity length mismatch. In the XAM-synchronization, the feed-back is realized in a resonant medium for the master laser modulating the medium at different electronic states and indirectly changing the co-propagating slave laser. As the most popular synchronization method, XPM scheme is widely used for ultra-fast lasers with timing jitter as small as several fs, or even hundreds of as. Besides, since it is found that fiber laser with long ring cavity can directly produce mode-locked ns laser pulses, the combination between the long-cavity fiber laser and passive synchronization methods can be used to generate synchronous ns laser pulses which have a lot of applications in physical researches.

5. References

a. Agrawal, G. P. *Nonlinear Fiber Optics* (Academic, New York, 1989), pp. 199-206.
b. Agrawal, G. P.; Badeck P. L. & Alfano, R. R. (1989). *Phys. Rev. B* 40, 5060-5072.
Agrawal, G. P. *Applications of Nonlinear Fiber Optics* (Academic, New York, 2001).
Apolonski, A.; Poppe, A.; Tempea, G.; Spielmann, C.; Udem, T.; Holzwarth, R.; Hänsch, T. W. & Krausz, F. (2000). *Phys. Rev. Lett.* 85, 740-743.
Baldeck, P. L.; Alfano, R. R. & Agrawal, G. P. (1988). *Appl. Phys. Lett.* 52, 1939-1941.
Bandrauk, A.D.; Chelkowski, S. & Goudreau S. (2005). *Journal of Modern Optics* 52 (2), 411-428.
Bartels, A.; Diddams, S.A.; Ramond, T.M. & Hollberg, L. (2003). *Opt. Lett.* 28, 663-665.
Bartels, A.; Newbury, N. R.; I. Thomann,; Hollberg, L. & Diddams, S. A. (2004), *Opt. Lett.* 29, 403-405.
Baudelet, M.; Guyon, L.; Yu, J.; Wolf, J.-P.; Amodeo, T.; Frejafon, E. & Laloi, P. (2006). *Appl. Phys. Lett.* 88, 063901.
Baum, P. & Zewail, A. H. (2007). *PNAS* 104, 18409-18414.
Butkus, R.; Danielius, R.; Dubietis, A.; Piskarskas, A. & Stabinis, A. (2004). *Appl. Phys. B* 79, 693-700.
Cavalieri, A. L.; Müller, N.; Uphues, Th.; Yakovlev, V. S.; Baltuka, A.; Horvath, B.; Schmidt, B.; Blümel, L.; Holzwarth, R.; Hendel, S.; Drescher, M.; Kleineberg, U.; Echenique, P. M.; Kienberger, R.; Krausz, F. & Heinzmann, U. (2007), *Nature* 449, 1029-1032.
Chen, L.P.; Wang, Y. & Liu, J.M. (1996). *IEEE J. Quantum Electron.* 32, 1817.
Chen, Y.; Jiang, J. & Jones, D. J. (2006). *Opt. Express* 14, 12134-12144.
Corkum, P. B. & Krausz, F. (2007). *Nature Physics* 3, 381 - 387.
Crooker, S.A.; Betz, F.D.; Levy, J. & Awschalom, D.D. (1996). *Rev. Sci. Instrum.* 67, 2068.
De Barros, M. R. X. & Becker, P. C. (1993). *Opt. Lett.* 18, 631-633.
De Barros, M. R. X.; Miranda, R. S.; Jedju, T. M. & Becker, P. C. (1995), *Opt. Lett.* 20, 480.
Drescher, M.; Hentschel, M.; Kienberger, R.; Uiberacker, M.; Yakovlev, V.; Scrinzi, A.; Westerwalbesloh, Th.; Kleineberg, U.; Heinzmann, U. & Krausz, F. (2002). *Nature* 419, 803-807.
Eliyahu, D.; Salvatore, R.A. & Yariv, A. (1997). *J. Opt. Soc. Am. B* 14, 167.
Evans, J. M.; Spence, D. E.; Burns, D. & Sibbett, W. (1993). *Opt. Lett.* 18, 1074-1076.
Fekete, J.; Cserteg,A.& Szipocs, R. (2009). *Laser Phys. Lett.* 6, 49-53.
Fermann, M.E.; Galvanauskas, A.; Suchal G. & Harter, D. (1997). *Appl. Phys. B* 65, 259-275.
Fleischer, A. & Moiseyev, N. (2006). *Phys. Rev. A* 74 (5), 053806.
Foreman, S.M.; Holman, K.W.; Hudson, D. D.; Jones, D. J. & Ye, J. (2007). *Rev. Sci. Instrum.* 78, 021101.
Fuerst, C.; Leitenstorfer, A. & Laubereau, A. (1996). *IEEE J. Sel. Top. Quantum Electron.* 2, 473-479.
Goulielmakis, E.; Yakovlev, V. S. Cavalieri, A. L.; Uiberacker, M.; Pervak, V., Apolonski, A.; Kienberger, R.; Kleineberg U.; & Krausz, F. (2007). *Science* 317, 769-775.
Gu, X.; Li, Pan, Y. H.; Wu, E. & Zeng, H. (2010). *IEEE J. Sel. Topics Quantum Electron.*, in press.
Hannaford, P. *Femtosecond Laser Spectroscopy* (Springer, New York, 2005) ISBN 0-387-23293-1.
Hao, Q.; Li, W. & Zeng, H. (2007) *Opt. Express* 15, 16754-16759.

Haus, H.A. & Mecozzi, A. (1993). *IEEE J. Quant. Elect.* QE-29, 983.

Hentschel, M.; Kienberger, R.; Spielmann, Ch.; Reider, G. A.; Milosevic, N.; Brabec, T.; Corkum, P.; Heinzmann, U.; Drescher M. & Krausz F. (2001). *Nature* 414, 509-513.

Hönninger, C.; Paschotta, R.; Morier-Genoud, F.; Moser, M. & Keller, U. (1999). *J. Opt. Soc. Am. B* 16, 46.

Jiang, L.A.; Wong S.T. & Grein, M.E. (2002). *IEEE J. Quantum Electron.* 38, 1047-1052.

Jones, D. J.; Diddams, S. A.; Ranka, J. K.; Stentz, A.; Windeler, R. S.; Hall, J. L. & Cundiff, S. T. (2000). *Science* 288, 635–639.

Jones, D. J.; Potma, E. O.; Cheng, J.; Burfeindt, B.; Pang, Y.; Ye, J. & Xie, X. S. (2002), *Rev. Sci. Instrum.* 73, 2843.

Kaindl, R. A.; Wurm, M.; Reimann, K.; Hamm, P.; Weiner, A. M. & Woerner, M. (2000). *J. Opt. Soc. Am. B* 17, 2086–2094.

Kawamura, K.; Ogawa, T.; Sarukura, N.; Hirano, M. & Hosono, H. (2000). *Appl. Phys. B* 71, 119–121.

Kienberger, R.; Hentschel, M.; Uiberacker, M.; Spielmann, Ch.; Kitzler, M.; Scrinzi, A.; Wieland, M.; Westerwalbesloh, Th.; Kleineberg, U.; Heinzmann, U.; Drescher, M. & Krausz, F. (2002). *Science* 297, 1144-1148.

Kitaeva, G. K. (2008). *Laser Phys. Lett.* 5, 559–576.

Kitzler, M.; Milosevic, N.; Scrinzi, A.; Krausz, F. & Brabec, T. (2002). *Phys. Rev. Lett.* 88 (17), 173904.

Kodama, R.; Norreys, P.A.; Mima, K.; Dangor, A.E.; Evans, R.G.; Fujita, H.; Kitagawa, Y.; Krushelnick, K.; Miyakoshi, T.; Miyanaga, N.; Norimatsu, T.; Rose, S.J.; Shozaki, T.; Shigemori, K.; Sunahara, A.; Tampo, M.; Tanaka, K.A.; Toyama, Y.; Yamanaka, T. & Zepf, M. (2001). *Nature* 412, 798–802.

a. Li, Y.; Li, W.; Hao, Q.; Yan, M.; Zhou, H. & Zeng, H. (2009). *Laser Phys. Lett.* DOI 10.1002, 1–5.

b. Li, Y.; Gu, X.; Yan, M.; Wu, E & Zeng, H. (2009). *Opt. Express* 17, 4526–4531.

a. Limpert, J.; Liem, A. Zellmer,; H. & Tünnemann, A. (2003). *Electron. Lett.* 39, 645-647.

b. Limpert, J.; Schreiber, T.; Liem, A.; Nolte, S.; Zellmer, H.; Peschel, T.; Guyenot, V. & Tünnermann, A. (2003). *Opt. Express* 11, 2982-2990.

c. Limpert, J.; Schreiber, T.; Nolte, S.; Zellmer, H.; Tünnermann, A.; Iliew, R.; Lederer, F.; Broeng, J.; Vienne, G.; Petersson,A. & Jakobsen, C. (2003). *Opt. Express* 11, 818-823.

Lisak, M.; Hook, A. & Anderson, X. (1990). *J. Opt. Soc. Am. B* 7, 810–814.

López-Martens, R.; Varjú, K.; Johnsson, P.; Mauritsson, J.; Mairesse, Y.; Salières, P; Gaarde, M. B.; Schafer, K. J.; Persson, A.; Svanberg, S.; Wahlström, C. & L'Huillier, A. (2005), *Phys. Rev. Lett.* 94, 033001.

Ma, L.-S.; Bi, Z.; Bartels, A.; Robertsson, L.; Zucco, M.; Windeler, R. S.; Wilpers, G.; Oates, C.; Hollberg, L. & Diddams S. A. (2004). *Science* 303, 1843-1845.

Manzoni, C.; Polli, D. & Cerullo, G. (2006), *Rev. Sci. Instrum.* 77, 023103.

Matsas, V. J.; Newson, T. P. & Zervas, M. N. (1992). *Opt. Commun.* 92, 61-66.

Mauritsson, J.; Johnsson, P.; Gustafsson, E.; L'Huillier, A.; Schafer,K. J. & Gaarde, M. B. (2006). *Phys. Rev. Lett.* 97, 013001.

Papadopoulos, D. N.; Zaouter, Y.; Hanna, M.; Druon, F.; Mottay, E.; Cormier, E. & Georges, P. (2007). *Opt. Lett.* 32, 2520-2522.

Paschotta, R. (2004). *Appl. Phys. B* 79, 163–173.

Potma, E.O.; Jones, D.J.; Cheng, J.-X.; Xie, X.S. & Ye, J. (2002). *Opt. Lett.* 27, 1168–1170.

Rudd, J.V.; Law, R.J.; Luk, T.S.; & Cameron, S.M. (2005). *Opt. Lett.* 30, 1974–1976.

Rullière, C. *Femtosecond laser pulses: principles and experiments, 2nd ed.* (Springer, New York, 2005) ISBN 978-0-387-01769-3.

a. Rusu, M.; Herda, R. & Okhotnikov, O. G. (2004). Optics Express 12(20), 4719-4724.

b. Rusu, M.; Herda R. and Okhotnikov, O. G. (2004) *Opt. Lett.* 29, 2246-2248.

Sansone, G.; Benedetti, E.; Calegari, F.; Vozzi, C.; Avaldi, L.; Flammini, R.; Poletto, L.; Villoresi, P.; Altucci, C.; Velotta, R.; Stagira1, S.; De Silvestri, S. & Nisoli, M. (2006). *Science* 314, 443-446.

Schoenlein, R. W.; Chattopadhyay, S.; Chong, H. H. W.; Glover, T. E.; Heimann, P. A.; Shank, C. V.; Zholents, A. A. & Zolotorev, M. S. (2000) *Science* 287, 2237-2240.

Shelton, R. K.; Ma, L.-S.; Kapteyn, H. C.; Murnane, M. M.; Hall, J. L. & Ye, J. (2001) *Science* 293, 1286-1289.

Schibli, T.R.; Kim, J.; Kuzucu, O.; Gopinath, J.T.; Tandon, S.N.; Petrich, G.S.; Kolodziejski, L.A.; Fujimoto, J.G.; Ippen, E.P. & Kaertner, F.X. (2003). *Opt. Lett.* 28, 947–949.

Schibli, T.R.; Minoshima, K.; Hong, F.-L.; Inaba, H.; Onae, A.; Matsumoto, H.; Hartl, I. & Fermann, M. E. (2004). *Opt. Lett.* 29, 2467-2469.

Schoenlein, R. W.; Leemans, W. P.; Chin, A. H.; Volfbeyn, P.; Glover, T. E.; Balling, P.; Zolotorev, M.; Kim, K.-J.; Chattopadhyay, S. & Shank, C. V. (1996). *Science* 274, 236–238.

a. Sell, A.; Leitenstorfer, A. & Huber, R. (2008). *Opt. Lett.* 33, 2767–2769.

b. Sell, A.; Scheu, R.; Leitenstorfer, A.; & Huber, R. (2008). *Appl. Phys. Lett.* 93, 251107.

Shelton, R.K.; Ma, L.-S.; Kapteyn, H.C.; Murnane, M.M.; Hall, J.L.; & Ye, J. (2001). *Science* 293, 1286–1289.

Shelton, R.K.; Foreman, S.M.; Ma, L.-S.; Hall, J.L.; Kapteyn, H.C.; Murnane, M.M.; Notcutt, M. & Ye, J. (2002). *Opt. Lett.* 27,312–314.

Shirakawa, A.; Sakane, I.; Takasaka, M. & Kobayashi, T. (1999). *Appl. Phys. Lett.* 74, 2268.

Son, D. H.; Kambhampati, P.; Kee, T. W. & Barbara, P. F. (2002), *J. Phys. Chem. A* 106, 4591–4597.

Spence, D.E.; Sleat, W.E.; Evans, J.M.; Sibbett, W. & Kafka, J.D. (1993). *Opt. Commun.* 101, 286.

Supradeepa, V. R.; Huang, C.; Leaird, D. E. & Weiner, A. M. (2008), *Opt. Express* 16, 11878-11887.

Telle, H. R.; Steinmeyer, G.; Dunlop, A. E.; Stenger, J.; Sutter, D. H. & Keller, U. (1999). *Appl. Phys. B* 69, 327–332.

Tong, X. M.; Zhao, Z. X. & Lin, C. D. (2003). *Phys. Rev. Lett.* 91, 233203.

Udem, Th.; Holzwarth, R. & Hänsch, T. W. (2002), *Nature* 416, 233-237.

Vozzi, C.; Calegari, F.; Ferrari, F.; Lucchini, M.; De Silvestri, S.; Svelto, O.; Sansone, G.; Stagira, S. & Nisoli, M. (2009). *Laser Phys. Lett.* 6, 259–267.

Wei, Z.; Kaboyashi Y. & Torizuka, K. (2002). *Appl. Phys. B* 74, S171–S176.

Wilcox, K.G.; Foreman, H.D.; Roberts J.S. & Tropper, A.C. (2006). *IEEE Electronic Lett.* 42, 1-3.

Xu, L.; Spielmann, C.; Krausz, F. & Szipöcs R. (1996). *Opt. Lett.* 21, 1259-1261.

a. Yan, M.; Li, W.; Hao, Q.; Li, Y. & Zeng H. (2009). *Opt. Lett.* 34, 2018–2020.

b. Yan, M.; Li, W.; Hao, Q.; Li, Y.; Yang, K.; Zhou H. & Zeng, H. (2009). *Opt. Lett.* 34, 3331–3333.

Yoshitomi, D.; Kobayashi, Y.; Takada, H.; Kakehata, M. & Torizuka, K. (2005). *Opt. Lett.* 30, 1408–1410.

Yoshitomi, D.; Kobayashi, Y.; Kakehata, M.; Takada, H.; Torizuka, K.; Onuma, T.; Yokoi, H.; Sekiguchi,T. & Nakamura, S. (2006). *Opt. Lett.* 31, 3243–3245.

Zhang, X.M.; Fan, D.Y.; Zeng, X.M.; Wei, X.F.; Huang, X.J.; Wang, X.; Zhu, Q.H. & Qian, L.J. (2006). *Opt. Lett.* 31, 646–648.

Zhu, C.; Wang, Y.; He, J.; Wang, S. & Hou, X. (2005). *J. Opt. Soc. Am. B* 22, 1221-1227.

12

Carrier-Envelope Phase Stabilization of Grating Based High-Power Ultrafast Laser

Shouyuan Chen, Yi Wu, Kun Zhao and Zenghu Chang
Kansas State University, Manhattan, KS 66503
USA

1. Introduction

At a fixed point of the space, the electric field of a Fourier transform-limited laser pulse can be written as:

$$E(t) = A(t)\exp(i\omega_0 t + \varphi_{CE}), \tag{1}$$

where $A(t)$ is the pulse amplitude, which is peaked at $t = 0$, and ω_0 is the laser pulse central frequency. Carrier-envelope (CE) phase, φ_{CE}, denotes the offset between the peak of the pulse envelope and the nearest peak of the carrier-wave electric field as shown in Fig. 1.

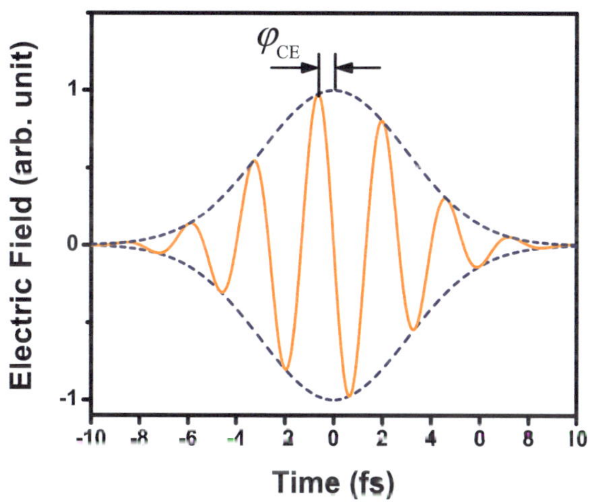

Fig. 1. A few-cycle laser pulse showing the carrier-envelope offset, φ_{CE}.

When the width of $A(t)$ approaches few or single-cycle duration, CE phase starts to play an important role in the laser-matter interactions becuase the electric-field amplitude changes rapidly within half of the cycle depending on the different CE phase. This fast variation of the electric field within the pulse envelope is the origin of CE phase effects in a variety of high-field processes, such as above-threshold ionization, high-order harmonic generation and molecular dissociation (Paulus et al., 2001; Haworth et al., 2007; Kling et al., 2006). The importance of CE phase are even identified in terahertz emission spectroscopy with few-

cycle pulses (Kreb et al., 2006). Advances in CE phase control also make it possible to controlling such processes as injected photocurrents in semiconductors (Fortier et al., 2004) and in sub-single- cycle pulse trains generated with Raman sidebands (Chen et al., 2008).
One of the important applications of the CE phase stabilization is the isolated attosecond generation (Hentschel et al., 2001; Sansone et al., 2006; Mashiko et al., 2008; Feng et al., 2009). For generating attosecond pulses using double optical gating (DOG) or general double optical gating (GDOG), the effective electric field inside the gate can be expressed as

$$E(t) = g(t)\cos(\omega_0 t + \varphi_{CE}), \tag{2}$$

where $g(t)$ is the gating function with a duration of a fraction of a laser cycle and the center of the gate occurs at $t = 0$. In these cases, the opening time of the gate for single attosecond pulse extraction is of the order of half to one cycle (Sansone et al., 2006; Mashiko et al., 2008). Thus, it is crucial to stabilize and control the CE phase. In Section 3 of this paper, the effects of CE phase on polarization gating and double optical gating will be explored in detail.

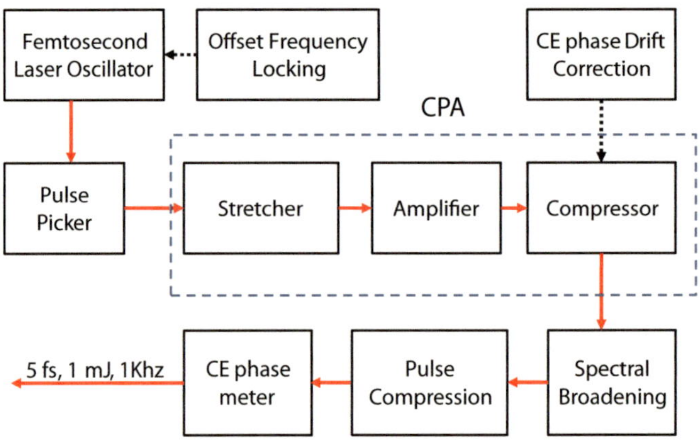

Fig. 2. A schematic diagram of a CE phase stable laser system producing few-cycle pulses.

A typical laser system for generating few-cycle CE phase-stabilized pulses is shown in Fig. 2. First, the CE phase evolution in the oscillator is stabilized and the pulses with the same CE phase are selected and sent to the chirped pulse amplification (CPA) system, where the laser pulses are temporally stretched, amplified, and recompressed (Strickland & Mourou (1985)). The CE phase drift introduced by the amplifier stage is also corrected. The amplified laser pulses are then spectrally broadened through a nonlinear process and compressed to only a few cycles in duration. Finally, the CE phase of the few-cycle pulses is measured and stabilized.
The CE phase stabilization of the laser pulse generated from an amplifier with dispersive material based stretcher and compressor had been successfully demonstrated before (Baltuška et al., 2003). However, the low damage threshold of the dispersive material limits the laser pulse energy to about 1 mJ, which ultimately limits the flux of the attosecond pulse generated from such laser system. It is of much current interest to increase the attosecond pulse flux for many attosecond experiments and extend the attosecond physics to nonlinear regime (Schultze et al., 2010; Tzallas et al., 2003). On the other hand, the grating based amplifier has been successfully demonstrated to generate PW class ultrafast pulse (Kiriyama et al., 2010), which can be used for high flux attosecond pulse generation, and thus it is

important to study the CE phase evolution in such laser system and stabilize the CE phase of the laser pulse of the grating based amplifier.

2. CE phase stabilization by controlling the grating separation

2.1 CE phase shift caused by grating shift in stretcher and compressor

For the grating based stretcher and compressor, it has been shown that the pointing stability of the laser beam on the gratings may introduce CE phase noise to the amplified femtosecond pulses (Kakehata et al., 2002; Thomann et al., 2004). It was soon shown that the grating separation could also cause the CE phase drift of amplified laser pulses (Chang, 2006).

To understand the CE phase drift caused by the stretcher and compressor, we start with the analysis of the laser pulse phase variation after it propagate through a stretcher. A typical setup of the stretcher is shown in Figure 3. The analysis is also valid for grating compressors.

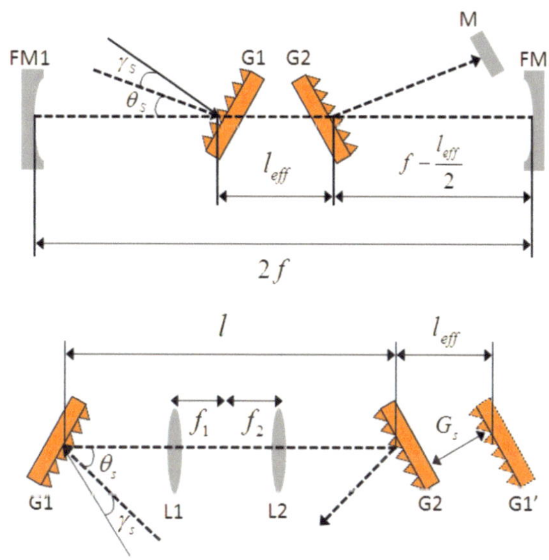

Fig. 3. Grating stretcher. The top figure uses mirrors to form the telescope and the bottom figure uses lenses to form the telescope. In both figures: *G*1 and *G*2 are the gratings, *FM*1 and *FM*2 are the focusing mirrors, γ_s: the incidence angle on the first grating, θ_s: the angle between the diffracted beam and incident beam, l_{eff}: the effective grating separation, *G*1: the image of *G*1 in the bottom figure (Moon et al., 2010).

If we assume the pulses to be transform limited at the stretcher, in time domain, the pulses can be expresses as Eq. 1. In spectral domain, the electric filed is

$$E(\omega) = E_0(\omega)\exp[i(\varphi_{CE} + \phi(\omega))], \quad (3)$$

where $|E_0(\omega)|$ is the power spectrum and $\phi(\omega)$ is the spectral phase, which is also equal to zero. After the pulse propagates through the double-pass grating compressor, the CE phase shift was:

$$\Delta\varphi_{CE} = \omega_0\tau(\omega_0) - \varphi'(\omega_0) = 4\pi\left(\frac{G_s}{d_s}\tan[\gamma_s - \theta_s(\omega)]\right), \quad (4)$$

Where $\omega_0\tau(\omega_0)$ is the group delay, $\phi'(\omega_0)$ is the phase delay, G_s is the effective perpendicular distance between gratings, d_s is the grating constant, γ_s is the angle of incidence, and $\theta_s(\omega)$ is the diffraction angle. From the geometry,

$$G_s = -l_{eff}\cos(\gamma_s - \theta_s), \tag{5}$$

where l_{eff} is the effective linear distance between the gratings, the CE phase shift becomes:

$$\Delta\varphi_{CE} = -4\pi\frac{\Delta l_{eff}}{d_s}\sin(\gamma_s - \theta_s). \tag{6}$$

Considering the incident angle is close to the Littrow angle, and the grating constant is of the order of wavelength (grating Constance d = $1200g/mm$ is commonly used), the CE phase approximates:

$$\frac{\Delta\varphi_C E}{\Delta l_{eff}} = 2\pi\frac{\Delta l_{eff}}{d_s^2} \approx \frac{2\pi}{\lambda}. \tag{7}$$

Thus, Eq. 7 shows that a change in the grating separation comparable to a wavelength will yield a CE phase change of 2π. Eqs. 5 and 7 also realized how the CE phase can be controlled and stabilized by changing the grating separation. λ

2.2 CE phase stabilization of multi-pass amplifier

The experimental setup used in determining the effect of the grating separation on the CE phase is carried on Kansas Light Source (KLS) laser system, which contains a multi-pass amplifier, as shown in Fig. 4. One of the mirrors in the stretcher was placed in a piezoelectronic transducer (PZT) controlled mount. CE-phase stable pulses from the oscillator were sent to the CPA system. An f-to-2f interferometer (Kakehata et al., 2002), spectrometer, and computer measured the spectral interferometry signal from which the CE phase was extracted. In the f-to-2f, the laser was focused into a sapphire plate for spectral broadening and then the infrared components were frequency doubled by a BBO crystal. Finally, a polarizer was used to select a common polarization and the beam was sent to a spectrometer. The resulting interferogram was analyzed by a computer to retrieve the CE-phase drift. For each data point, 50 laser shots were integrated. In the experiment, a 60-V sinusoidal voltage was applied to the PZT and observed. As a comparison, a DC voltage was applied to the PZT. The results are shown in Fig. 5 (Li, Moon & Chang, 2006). It was determined from the measurement that a 1 μm change in grating separation introduced a 3.7 ± 1.2S rad phase shift, which is consistent with the calculation of Eq. 7.
The grating separation in the stretcher was then used as a feedback control to stabilize the CE-phase drift of the amplified pulses. This is shown in Fig. 6, where the CE-phase error was kept to 160 mrad rms over 800 s. 50 laser shots were integrated for each data point. This was the standard experimental procedure for all CE-phase measurements in the KLS laboratory. The bottom plot shows the PZT movement during the same period. The concept of controlling the stabilized CE phase is illustrated in Fig. 7 (Li et al., 2006). In Fig. 7, the grating separation was precisely controlled to scan the phase over a range of 2π. In the experiment, the setpoint for locking the phase was changed from $1.1\pi$to 0.9πin steps of 0.2π.

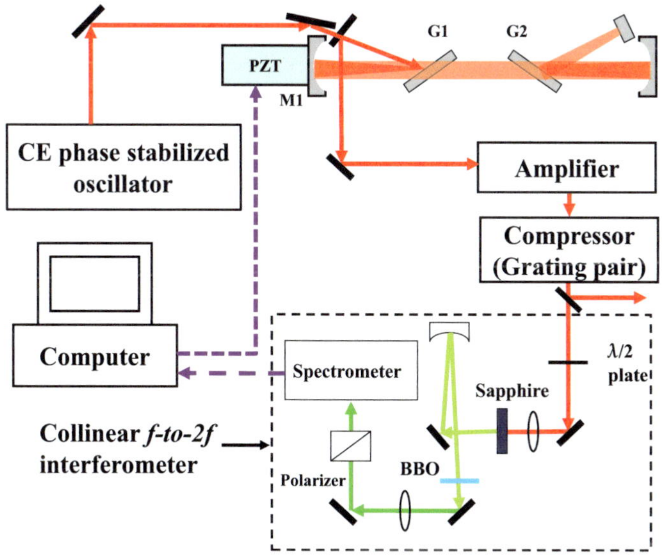

Fig. 4. Kansas Light Source laser system for testing the effects of the grating separation of the stretcher on the CE-phase stability. G1 and G2 are the gratings. M1 is one of the telescope mirrors driven by a piezoelectronic transducer (PZT) (Moon et al., 2010).

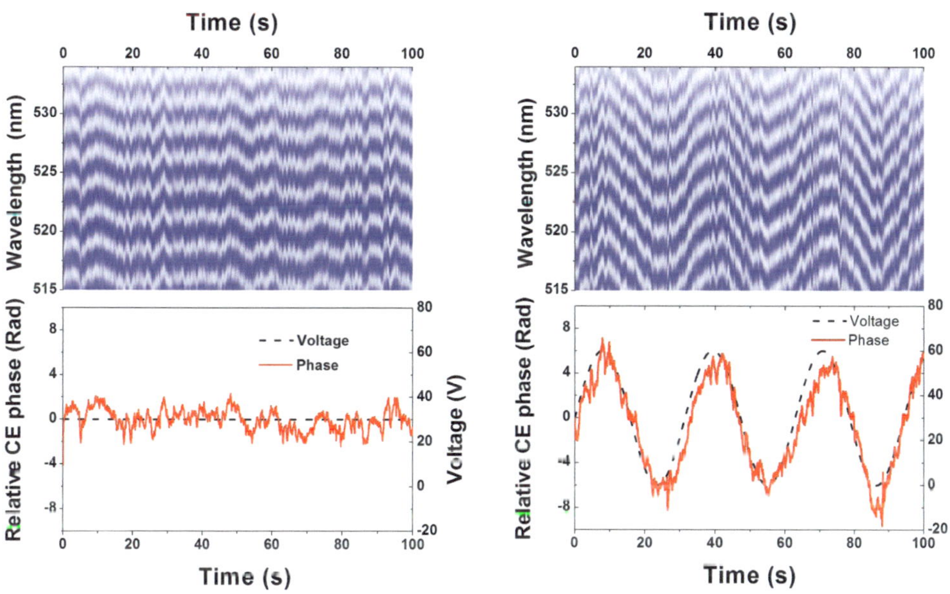

Fig. 5. Dependence of the CE phase of the amplified pulses on the grating separation. (a) Fringe pattern and (b) corresponding relative CE phase obtained with a 30-V DC voltage applied to the PZT. (c) Fringe pattern and (d) corresponding relative CE phase obtained with a 60-V sinusoidal voltage applied to the PZT, which caused the PZT to move 3.6 μm (Li, Moon & Chang, 2006).

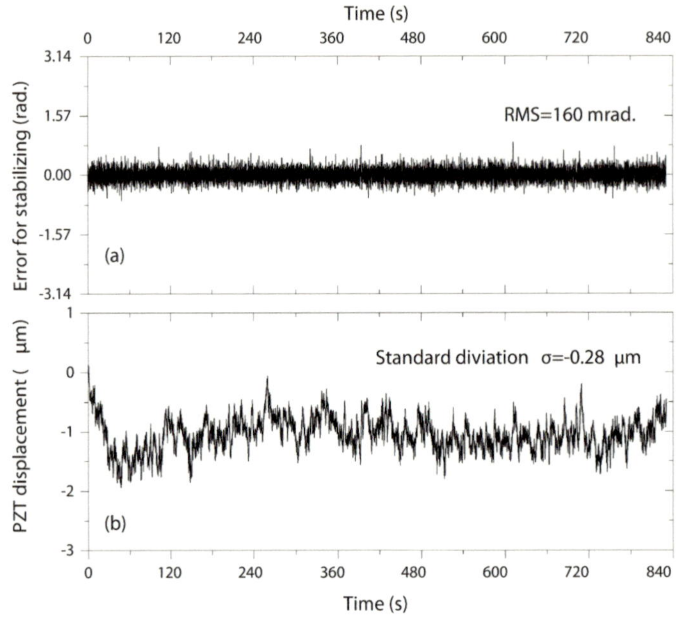

Fig. 6. (a) The error signal for the slow feedback stabilization, (b) the displacement of PZT when the set-point was shifted (Li et al., 2006).

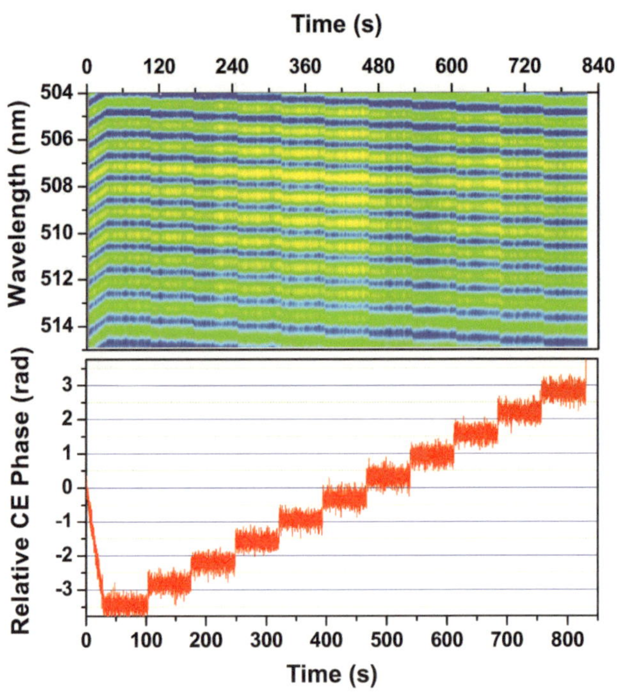

Fig. 7. Precisely controlling the CE phase in amplified pulses. Top: the temporal evolution of interference fringes measured with a collinear f-2f interferometer. Bottom: the swept CE phase (Li et al., 2006).

The relative CE phase was kept at each setpoint for 1 minute and then moved to the next setpoint. During the process, the phase error was smoothly locked to an average of 161 mrad rms, which showed how well the gratings could stabilize the CE phase and change the CE phase simultaneously. Note that in the algorithm used, the first detected fringe pattern was phase zero. Therefore, the CE phase was changed relative to the first detected phase, which had an unknown value.

Similarly, the grating separation in a compressor can also be used to control and stabilize the CE phase. The size of the optic to be used as a control mechanism should be considered when choosing to use either the stretcher or compressor. A large optic, such as a large mirror in the stretcher, would be harder to move using a PZT than a smaller optic. Also, a larger bandwidth of CE-phase noise could be suppressed with a smaller optic. The effects of controlling the grating separation in a compressor arrangement were investigated (Li et al., 2008). The experimental setup used to investigate the compressor control is shown in Fig. 8. Controlling the compressor grating separation was found to stabilize the CE phase to 230 mrad rms over 270 s, which was nearly the same as the stretcher grating performance. Fig. 9 shows the performance of the compressor grating separation control. The top plot shows the difference between the situation where the feedback control was turned on and when it was inactive. The bottom plot shows the fast Fourier transform of the phase drift. The plot shows how the feedback control corrects CE-phase error under 4 Hz. This is expected since the drift of CE phase-stable pulses through an amplifier is slow, since the oscillator stabilization corrects the majority of the fast drift. However, as in (Moon et al., 2006), locking the path-length difference in the oscillator CE-phase stabilization interferometer will reduce the fast noise (>3 Hz) by over 40%, which can improve the overall CE-phase-locking quality of the system.

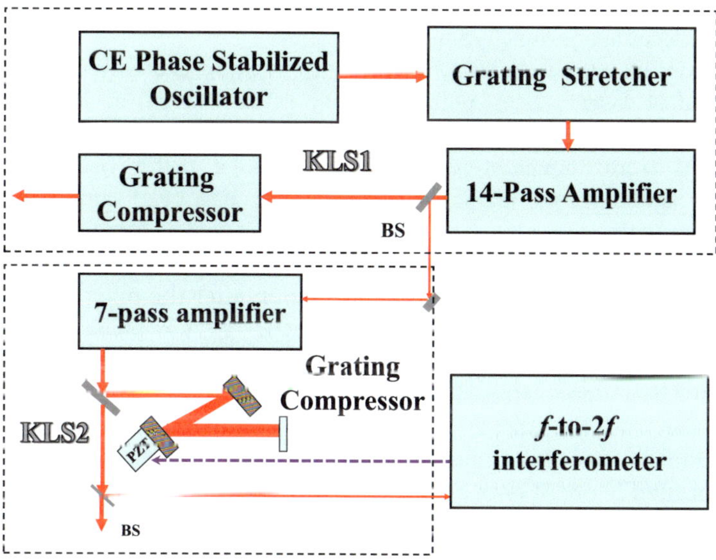

Fig. 8. Experimental setup for controlling the CE phase of the amplified laser pulses using the compressor grating (Li et al., 2008). BS: beamsplitter.

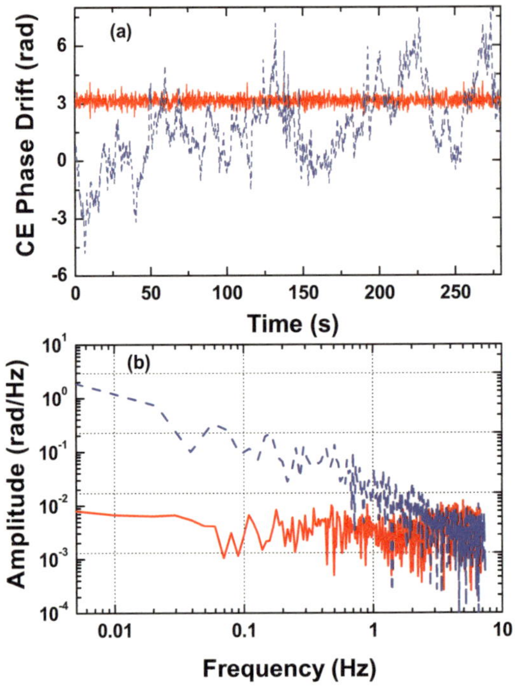

Fig. 9. (a) The evolution of the freely drifting (dotted line) and stabilized (solid line) CE phase. (b) The fast Fourier transform of the CE-phase drift under the free-running (dotted line) and stabilized conditions (solid line) (Li et al., 2008).

2.3 CE stabilization of regenerative amplifier

Besides the multi-pass amplifier, regenerative amplification is another attractive choice for generating ultrafast laser pulses with multi-mJ pulse energy at high repetition rates (Vaillancourt et al., 1990; Barty et al., 1996). Compared with multi-pass amplifiers, the laser pulses generated from regenerative amplification have better beam profile, pointing stability, power stability and extraction efficiency. It has been commonly used as the preamplifier for high energy femtosecond laser systems. However, the CE phase stabilization has been demonstrated previously only on multi-pass lasers (Baltuška et al., 2003; Li et al., 2008). Although the CE phase evolution after the regenerative amplifier has been explored to some extent (Kakehata et al., 2004), the CE phase drift caused by the amplification stage has not been corrected. Furthermore, no high-field CE phase dependent experiments have been demonstrated with a regenerative amplifier. We demonstrate the CE phase stabilization and control of the Manhattan Attosecond Radiation Source (MARS) regenerative amplification laser system and unambiguously confirm the stability by measurement of the CE phase dependence of the high order harmonic spectrum generated by double optical gating (Chen et al., 2009; Mashiko et al., 2008).
The MARS laser system consists of a commercially available Ti:Sapphire Coherent Legend Elite Duo™ chirped pulse amplifier (CPA) operating at 1 KHz, seeded by a Femtolasers Rainbow™ oscillator. The oscillator operates at 78 MHz repetition rate with a 200mWoutput power. The carrier-envelope offset frequency (f_{CEO}) of the oscillator is stabilized using the

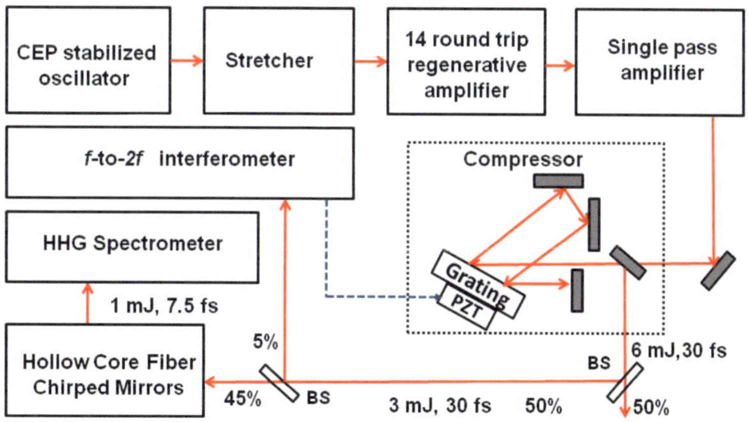

Fig. 10. Layout of the MARS laser system Chen et al. (2009). BS: beam splitter.

monolithic carrier-envelope phase-stabilization scheme (Fuji et al., 2005). With the assistance of the temperature feedback control, f_{CEO} of this oscillator can be locked about 12 hours on a daily basis (Yun et al., 2009). As shown in Figure 10, the amplifier consists of a grating based stretcher, two amplification stages and a grating based compressor.

The stretcher stretches the pulse to about 160 ps. The first amplification stage is a 14 round trip regenerative amplifier, which amplifies the laser pulse to 4 mJ. The second stage is a single pass amplifier to boost the laser pulse energy to 8 mJ. The Ti:Sapphire crystals in both amplification stages are thermoelectricity cooled to -12 °C. The regenerative amplifier and single pass amplification stages are each pumped by 50% of a 45WCoherent Evolution HE™ pump laser. After compression, the final output pulse energy is 6 mJ with a central wavelength of 800 nm and spectral bandwidth of 37 nm, which supports a fourier transform limited pulse duration of 28 fs. The pulse duration as measured with FROG is 30 fs. The CE phase drift of the amplified pulse is measured by sending partial of the output beam to the f-to-2f interferometer (Kakehata et al., 2001). The feedback signal is then sent to a piezoelectric transducer (PZT) stage on the compressor grating to stabilize CE phase slow drift by controlling the grating separations (Li et al., 2008; Chang, 2006). Because of the high output power of the MARS laser system, the laser beam is typically split by a 50/50 beam splitter to support two experiments simultaneously. During the CE phase stabilization process, it was found that CE phase after this amplifier system is very sensitive to mechanical vibration and acoustic noise. One reason for this sensitivity is that In Regen cavity, all the laser passes follow the exact same beam path. Any perturbation of air in the cavity will be amplified by the pass number. It is also the major difference between the Regen and Multi-pass amplifier. In the later configuration, the laser beam follow different pathes, and thus has less sensitivity than Regen on the acoustic noise. However, by moving the pump laser further away from the amplifier, padding the laser cover with sound absorption materials and improving the stability of the optical mounts in the stretcher and compressor, the CE phase stability was achieved.

The RMS error of the locked CE phase was measured to be 90 mrad over a period of 4.5 hours with 50 ms spectrometer integration time as shown in Fig. 11. The locking duration is sufficient for many low count rate experiments, such as attosecond streaking and laser ion beam interaction experiments, which can require several hours of CE phase stabilization (Wang et al., 2009; McKenna et al., 2008). To control the CE phase in the experiment, the grating separation of the compressor was stabilized at preset values. As shown in Fig. 12, the CE phase was swept from $-\pi$ to π for many cycles, which can be used to observe the CE phase effects in high harmonic and attosecond pulse generation experiments (Li et al., 2006). Some regions in the Fig. 11a and Fig. 12a show transient reductions of the fringe visibility, which are likely due to acoustic noise and vibration in the laboratory which could not be avoided over such a long locking period. However, such small disturbances did not affect the quality of the CE phase control.

To unambiguously confirm the CE phase stabilization of MARS laser system, the dependence of the high order harmonic spectrum generated in Argon gas from DOG was measured using an XUV transmission grating spectrometer (Shan & Chang, 2001). To generate the short pulse (<10 fs) required by DOG, half of the laser pulse with an energy of 3 mJ was sent to a 1 m long hollow-core fiber with an inner diameter of 400 m and filled with

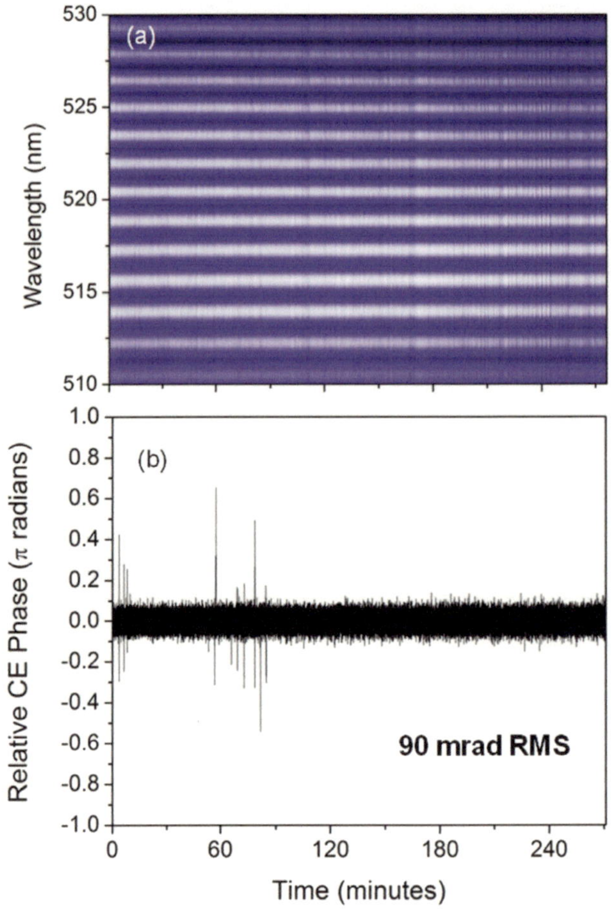

Fig. 11. Long term CE phase stabilization. (a) f-to-2f interference fringes. (b) Retrieved CE phase from the fringes, which has 90 mrad RMS averaged over 50 laser shots Chen et al. (2009).

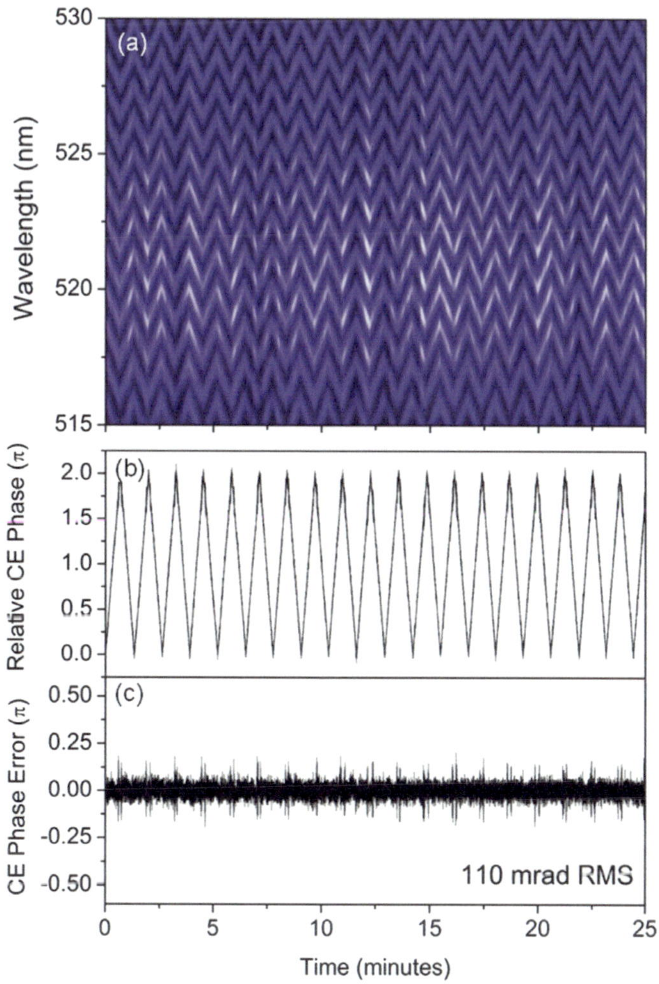

Fig. 12. Periodic CE Phase sweep from $-\pi$ to π. (a) f-to-2f interferometer fringes, (b) The scanning phase value retrieved from the fringes, (c) CE Phase error of retrieved phase relative to the preset scanning phase. The periodic increases in error are due to the PZT response at the turning points of control curve Chen et al. (2009).

Neon gas at a pressure of 2 atm. After self-phase modulation, the laser pulse spectrum was broadened from 600 nm to 950 nm. The Frequency-Resolved-Optical-Gating measurement shows the pulse duration is 7.5 fs after it was compressed by the chirped mirrors, which is short enough for generating the single isolated attosecond pulse (Gilbertson et al., 2008). The final output power of the short pulse is 1 mJ after the chirped mirrors. The short pulse was then focused by a 300 mm spherical mirror into a 1.5 mm gas cell filled with Argon at 40 torr to generate the high order harmonics. When the CE phase was scanned linearly from 0 to 8π, the harmonic spectra varied with a 2π periodicity as shown in Fig. 13, which is consistent with the 2π periodicity of the electric field generated by DOG (Chang, 2007). The total time to perform the CE phase scan was 30 minutes.

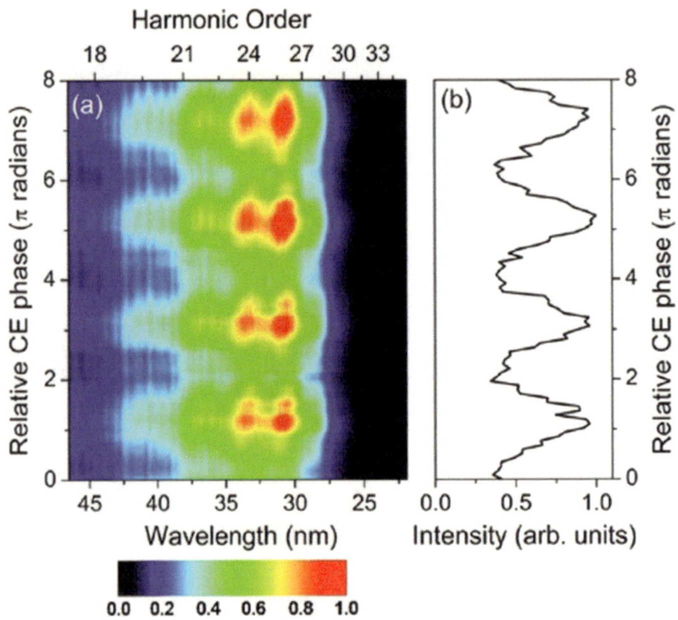

Fig. 13. (a) DOG harmonic spectra taken with the CE phase scanned from 0π to 8π. (b) Line out of the normalized integrated spectrum. The integration range is from 48 nm to 20 nm. The 2π periodicity is consistent with the asymmetric electric field of DOG Chen et al. (2009).

3. Isolated attosecond pulse generation with CE phase stabilized high-power laser

As mentioned in the previous section, one of the important applications of CEP stabilized laser is to generate isolated attosecond pulses.

Attosecond pulse generation is usually interpreted in the semi-classical re-collision model (three-step model) (Corkum, 1993; Corkum & Chang, 2008). Briefly, as a strong near infrared (NIR) laser pulse strikes an atom, a free electron wave packet is produced by ionization. Once freed, the wave packet moves away from the atom. However, when the oscillating laser electric field reverses direction, half of the packet is driven back towards the parent ion. This return electron can recombine with the parent ion, emitting an extreme ultraviolet (XUV) photon, which is the origin of attosecond XUV pulses. In general, a multi-cycle laser will produce an attosecond XUV pulse every half of a laser cycle. The result is a train of attosecond pulses. It is obvious that the CEP is critical in the isolated attosecond pulse generation with a gating technique. It is preferred that the CEP of the NIR laser pulse is stabilized so that the center of the gate always overlaps with a single attosecond XUV pulse in the pulse train. If the CEP is not optimized, the pulse energy of the single attosecond pulse would be reduced or, in the worst scenario, multiple attosecond pulses will be generated instead of an isolated attosecond pulse.

To study the relation between the CE phase and attosecond pulse generation, the isolated attosecond pulse generation and characterization experiments were performed in the KLS lab (Feng et al., 2009; Gilbertson et al., 2010).

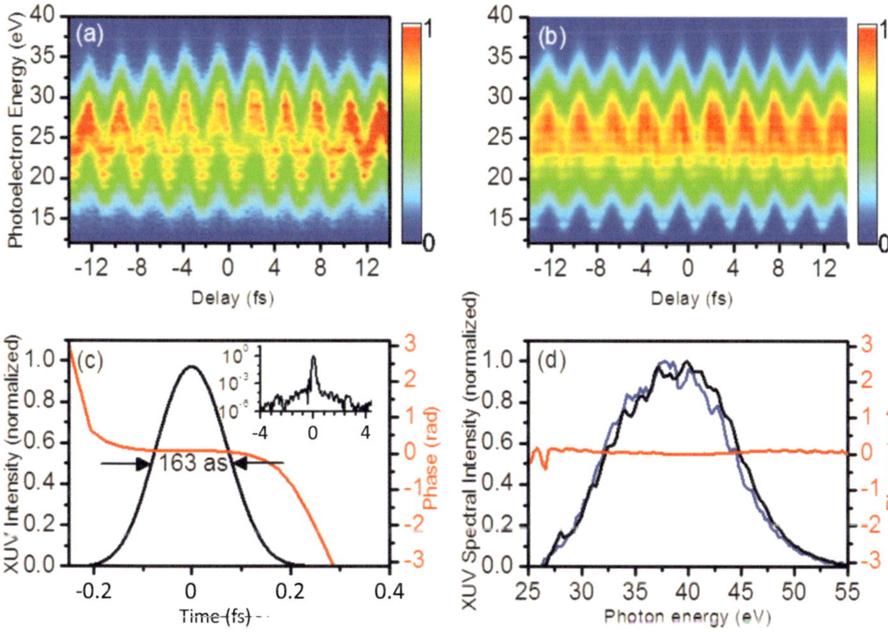

Fig. 14. The experimentally obtained (a) and retrieved (b) spectrograms of isolated attosecond pulses streaked by multicycle laser pulses. The temporal profile (solid line) and phase (dotted line) are shown in (c). The inset figure shows the same temporal profile but over an extended range. The pre- and post-pulses located at ±2600 as are less than 0.1% of the main pulse. Panel (d) shows the experimental (dashed line) and retrieved (solid line) XUV-only spectrum. The dashed-dotted line shows the spectral phase and indicates that the pulse is nearly transform limited (Gilbertson et al., 2010).

Figure 14 shows the results of the temporal characterization of isolated attosecond pulses produced by GDOG technique using a streak camera setup (Feng et al., 2009; Gilbertson et al., 2010) and the frequency resolved optical gating for the complete reconstruction of attosecond bursts (FROG-CRAB) method (Mairesse & Quéré, 2005). Figures 14(a) and (b) show the experimental and reconstructed streaked spectrograms, respectively. Figure 14(c) shows the temporal profile of the pulse (solid line) and the temporal phase (dotted line). The full width at half maximum (FWHM) of the pulse is about 163 as. The inset figure shows the temporal profile over an extended range, which indicates the contributions from pre- and post pulses are less than 0.1% of the main peak. This shows that the pulse is indeed an isolated attosecond pulse. Figure 14(d) shows a comparison between the experimental XUV-only spectrum (dashed line) and the retrieved spectrum (solid line) from the retrieved temporal profile and phase shown in Fig. 14(c). This marginal check indicated the reconstructed results can be trusted and the pulse is nearly transform-limited.

The gate width of the GDOG in the above experiment was set equal to one optical cycle, or roughly 2.5 fs. This is the upper limit for generating isolated attosecond pulses with a proper CE phase. The gate width can be further reduced so that it is much less than one optical cycle. Figures 15(a) and (b) show the electric field of the driving laser with two values of the CE phase within the gate. The color gradient indicates the ellipticity of the generating laser pulse with white being the most linear. Here, the gate width was chosen to

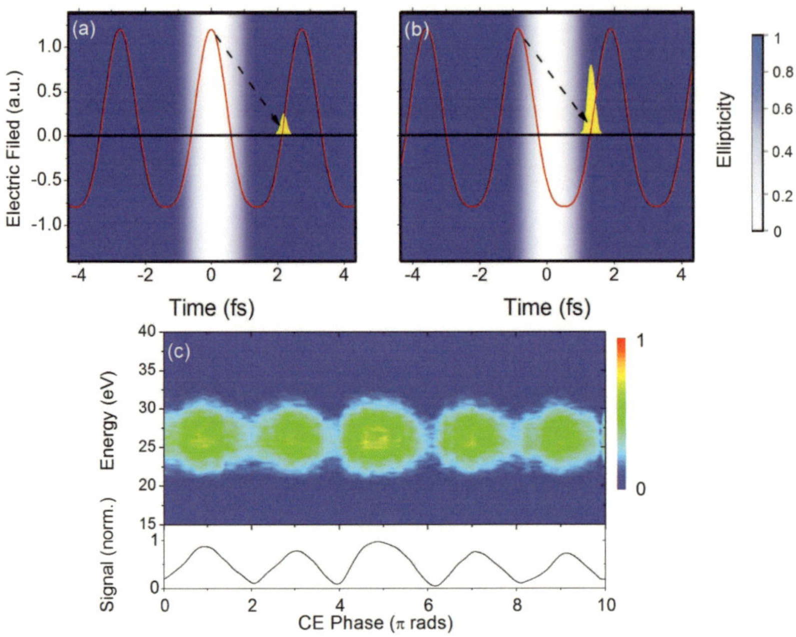

Fig. 15. The effect of a narrow gate width (~1 fs) on the generated attosecond pulse. In (a), the CE phase of the NIR laser forces the freed electron recombines in a field of high ellipticity, severely limiting its recombination probability. In (b), the CE phase is more favorable for highflux attosecond pulse emission since the electron experiences a linear field for its full lifetime. In the figures, the color gradient represents the ellipticity of the field with blue being the most elliptical and white the most linear. The experimental evidence for this effect is shown in (c). The upper figure shows the energy spectrum as a function of the CE phase of the NIR laser while the lower plot shows the total signal integrated along the energy axis. The 2π periodic structure is the effects of the two-color gating in GDOG (Gilbertson et al., 2010).

be ~1 fs (about half of a laser cycle) and is where the attosecond pulse is produced. In Fig. 15(a), the freed electron is born during a strongly linearly polarized portion but recombines to emit an XUV photon in a field that is increasingly elliptical. This reduces the recombination probability so that the attosecond XUV photon flux would be low. In Fig. 15(b), the electron spends all of its excursion time away from the parent ion in a mainly linearly polarized field so that the attosecond photon flux would be maximized. In both cases, since the gate width is much smaller than the spacing between two adjacent attosecond pulses in the pulse train, it is not possible to generate two attosecond pulses per laser shot. The CE phase only affects the flux of the isolated pulses.

Figure 15(c) shows the experimental evidence for this effect. For this portion of the experiment, a 9 fs laser pulse was produced by the 2 mJ, 25 fs NIR pulse from the CEP-locked amplifier passing through a Ne filled hollow-core fiber and a chirp-mirror compressor. The laser power fluctuates less than 1%. This beam then passed through the GDOG optics consisting of a 530 μm quartz plate, a 0.5 mm Brewster window, a 440 μm quartz plate and a 141 μm BBO, and was focused by an f=375 mm spherical mirror into a 1.4 mm long Ar gas target. The gate width for these parameters was calculated to be about 1.4 fs.

The upper figure in Fig. 15(c) shows the energy spectrum of the photoelectrons liberated by an attosecond XUV pulse as a function of the CE phase of the input NIR laser. The CE phase was continuously shifted from 0 to 2π. Typically, the CE phase stability is better than 250 mrads after the hollow-core fiber (Mashiko et al., 2007). Two features of the spectrogram are obvious. First, the spectrum is a continuum for all CE phase values, which satisfies the necessary condition for generating isolated attosecond pulses. Second, the intensity of the spectrum strongly depends on the CE phase, which is expected for such a narrow gate width. The lower figure shows the total counts (integrated over the energy spectrum) as a function of the CE phase. The modulation depth is an indication of the width of the linear polarization gate. For narrower gate widths, the modulation depth would become even stronger while for wider gate widths, the modulation would become shallower and eventually the energy spectrum would exhibit modulations indicative of multiple pulses within the gate (Sola et al., 2006).

The attosecond XUV pulses generated under different CEP values are also characterized by the attosecond streak camera. A streaked spectrogram similar to the one shown in Fig. 14 was obtained when the CE phase is unlocked. The carrier of the laser field is not smeared out since the attosecond pulse is automatically locked to the driving laser oscillation in time. The temporal profile and phase as reconstructed by FROG-CRAB are also similar to the ones in Fig. 14. The pulse duration was found to be about 182 as.

Then, streaked spectrograms for four different values of the CE phase of the input laser were taken, as Figure 16 shows. The CE phase was locked to a 200 mrad RMS. The differences in count rates are attributed to the different values of the CE phase and hence the different fluxes of the attosecond XUV photons. Figure 17(a) shows the XUV spectrum at each

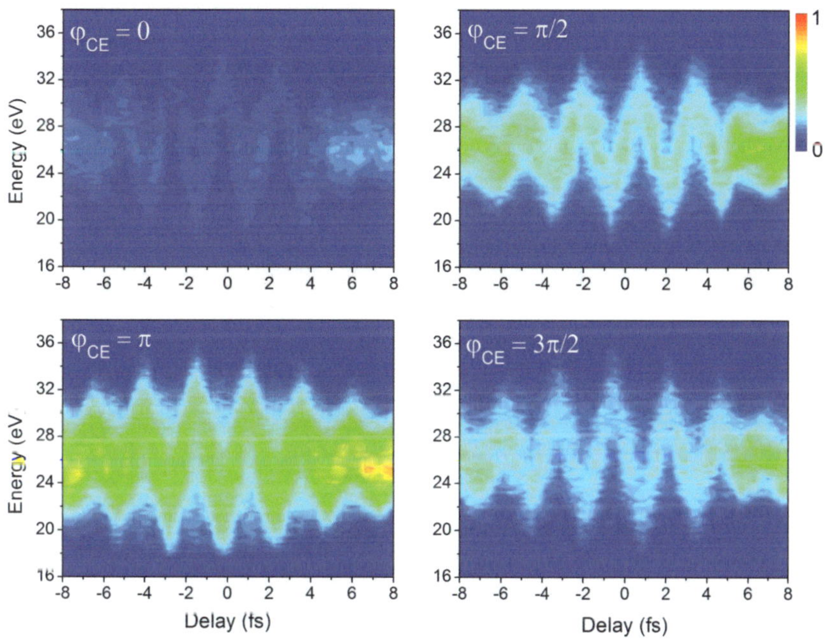

Fig. 16. Streaked photoelectron spectrograms for four different values of the CE phase, ~0 rad, ~ $\pi/2$ rad, ~ π rad, and ~ $3\pi/2$ rad. The images are normalized to the peak counts of the ~ π rad spectrogram Gilbertson et al., (2010).

value of the CE phase. The temporal profiles and phases for the spectrograms in Fig. 16 were reconstructed with FROG-CRAB (Mairesse & Quéré, 2005) and all the pulse durations are about 180 as. Finally, each streaked spectrogram was Fourier filtered to extract the oscillating NIR field. Figure 17(b) shows the results, where the CE phase of the 9 fs laser pulse can be easily seen.

To improve the utility of this result, attosecond pulses were produced using 25 fs NIR pulses directly from the chirped pulse amplifier. Figure 18 shows streaked spectrograms for two different values of the CE phase. Again, the count rate is different between the two cases in agreement with the attosecond pulse dependence on the CE phase. Reconstructions with FROG-CRAB show both have nearly identical durations of 190 as and phase shapes. The signal ratio between the two cases is not as extreme as the short pulse case. This can

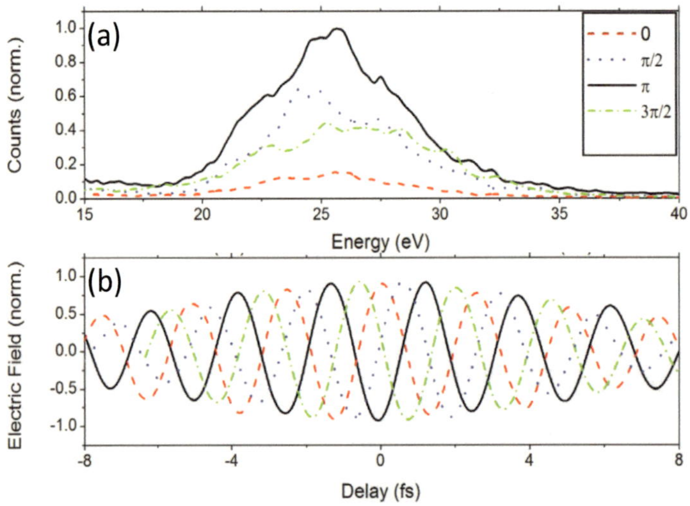

Fig. 17. Panel (a) shows the photoelectron energy spectrum for each of the streaked spectrograms in Fig. 16. Panel (b) shows the extracted NIR laser electric fields corresponding to each of the spectrograms in Fig. 16 Gilbertson et al., (2010).

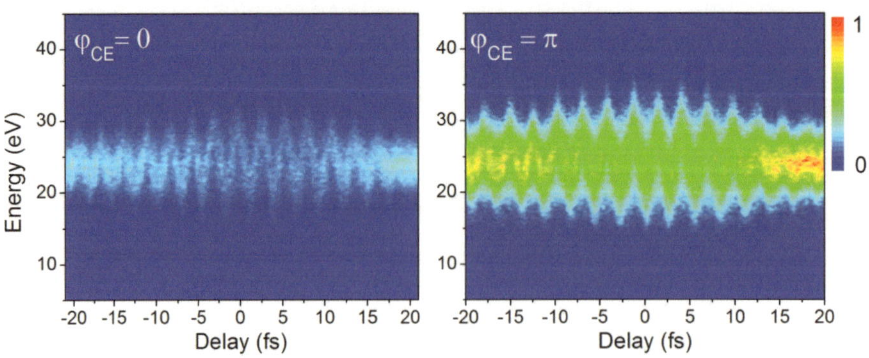

Fig. 18. Streaked spectrograms of attosecond pulses produced directly from an amplifier with an approximately π CEP shift between them Gilbertson et al., (2010).

possibly be explained by the gate width being slightly wider than the short pulse case. This is in excellent agreement with the CE phase unlocked reconstruction of 190 as.
These results show that the CEP locking plays a key role in single attosecond XUV pulse generation with a gating technique, DOG or GDOG (Feng et al., 2009; Gilbertson, Wu et al., 2010; Gilbertson, Khan et al., 2010). Although the single attosecond pulses produced under different CEP have almost identical pulse duration and phase profile, the photoelectron count rate or the flux of the XUV photos in the isolated attosecond pulses varies significantly as the CEP changes. As we extend the HHG spectrum to higher energy range to generate even shorter XUV pulses, 25 as, for example, which is about one atomic unit of time (Mashiko et al., 2009), the efficiencies of both XUV photon emission in attosecond generation and photoelectron emission in the streaking experiment drop significantly. Therefore, it would become even more important to lock the CEP at its optimum value to maximize the photon/photoelectron counts for the generation and characterization of 25 as XUV pulses, as well as for attosecond nonlinear experiments and any other attosecond experiments which require high photon flux.

4. Conclusion

In summary, the CE phase of the multi-pass and regenerative amplifier was both stabilized by changing the grating separation in stretcher or compressor. The grating-based CPA and CE-phase control methods increased the energy of the CE phase stabilized laser pulse to the multi-mJ level and the CE phase could be precisely controlled. The CE phase stabilization and control of these laser system are unambiguously confirmed by experimental observation of the 2π periodicity of the high order harmonic spectrum generated by double optical gating. Therefore, CE-phase stable and controllable high-energy pulses are now a viable technology for studying ultrafast science. We have also demonstrated that the almost identical attosecond pulses can be generated at different CE phase values given the sufficient narrow gate width. However, the photon flux drops significantly if the CEP is tuned away from its optimum value for attosecond XUV pulse generation. This is true for both 9 fs and 23 fs lasers, where the 23 fs NIR pulses were produced directly from a CPA amplifier. These studies pave the way for the realization of high-power CE phase stabilized lasers and high-flux single-isolated attosecond pulse generation, which are critical steps toward the study of nonlinear physics and pump probe experiments with single attosecond pulses.
Challenges do lie ahead for CE-phase-stabilization technology. For example, adaptive pulse shaping is a method where the phase of the laser pulse can be manipulated. If this method is combined with CE-phase stabilization and control, it could allow for the generation of ultra-short pulses with precise control of the absolute phase. Also, no group has actively stabilized and controlled the CE phase of even higher power laser system, such as TW class laser. This is also one of the major challenges future CE phase research. Thus, there is room to improve in the area of CE-phase stabilization and control of Ti:sapphire laser amplifiers.

5. References

Baltuška, A., Udem, T., Uiberacker, M., Hentschel, M., Goulielmakis, E., Gohle, C., Holzwarth, R., Yakovlev, V. S., Scrinzi, A., Hänsch, T. W. & Krausz, F. (2003). Attosecond control of electronic processes by intense light fields, *Nature* 421(6923): 611–615.
Barty, C. P. J., Guo, T., Blanc, C. L., Raksi, F., Rose-Petruck, C., Squier, J.,Wilson, K. R., Yakovlev, V. V. & Yamakawa, K. (1996). Generation of 18-fs, multiterawatt pulses

by regenerative pulse shaping and chirped-pulse amplification, *Opt. Lett.* 21(9): 668–670. URL: http://ol.osa.org/abstract.cfm?URI=ol-21-9-668

Chang, Z. (2006). Carrier-envelope phase shift caused by grating-based stretchers and compressors, *Appl. Opt.* 45(32): 8350–8353. URL: http://ao.osa.org/abstract.cfm?URI=ao-45-32-8350

Chang, Z. (2007). Controlling attosecond pulse generation with a double optical gating, *Phys. Rev. A* 76(5): 051403.

Chen, S., Chini, M., Wang, H., Yun, C., Mashiko, H., Wu, Y. & Chang, Z. (2009). Carrierenvelope phase stabilization and control of 1 khz, 6 mj, 30 fs laser pulses from a ti:sapphire regenerative amplifier, *Appl. Opt.* 48(30): 5692–5695. URL: http://ao.osa.org/abstract.cfm?URI=ao-48-30-5692

Chen, W.-J., Hsieh, Z.-M., Huang, S., Su, H.-Y., Lai, C.-J., Tang, T.-T., Lin, C.-H., Lee, C.-K., Pan, R.-P., Pan, C.-L. & A. H. Kung, Phys. Rev. Lett. 100, . . (2008). Carrier-envelope phase-controlled quantum interference of injected photocurrents in semiconductors, *Phys. Rev. Lett.* 100: 163906.

Corkum, P. B. (1993). Plasma perspective on strong field multiphoton ionization, *Phys. Rev. Lett.* 71: 1993.

Corkum, P. B. & Chang, Z. (2008). The attosecond revolution, *Opt. Photonics News* 19: 24.

Feng, X., Gilbertson, S., Mashiko, H., Wang, H., Khan, S. D., Chini, M., Wu, Y., Zhao, K. & Chang, Z. (2009). Generation of isolated attosecond pulses with 20 to 28 femtosecond lasers, *Phys. Rev. Lett.* 103: 183901.

Fortier, T. M., Roos, P. A., Jones, D. J., Cundiff, S. T., Bhat, R. D. R. & Sipe, J. E. (2004). Carrier-envelope phase-controlled quantum interference of injected photocurrents in semiconductors, *Phys. Rev. Lett.* 92: 147403.

Fuji, T., Rauschenberger, J., Apolonski, A., Yakovlev, V. S., Tempea, G., Udem, T., Gohle, C., Hänsch, T. W., Lehnert, W., Scherer, M. & Krausz, F. (2005). Monolithic carrierenvelope phase-stabilization scheme, *Opt. Lett.* 30(3): 332–334. URL: http://ol.osa.org/abstract.cfm?URI=ol-30-3-332

Gilbertson, S., Khan, S. D., Wu, Y., Chini, M. & Chang, Z. (2010). Isolated attosecond pulse generation without the need to stabilize the carrier-envelope phase of driving lasers, (*Phys. Rev. Lett.*) 105, 093902.

Gilbertson, S., Mashiko, H., Li, C., Moon, E. & Chang, Z. (2008). Effects of laser pulse duration on extreme ultraviolet spectra from double optical gating, *Applied Physics Letters* 93(11): 111105. URL: http://link.aip.org/link/?APL/93/111105/1

Gilbertson, S., Wu, Y., Khan, S. D., Chini, M., Zhao, K., Feng, X. & Chang, Z. (2010). Isolated attosecond pulse generation using multicycle pulses directly from a laser amplifier, *Phys. Rev. A* 81(4): 043810.

Haworth, C. A., Chipperfield, L. E., Robinson, J. S., Knight, P. L., Marangos, J. P. & Tisch, J. W. G. (2007). Half-cycle cutoffs in harmonic spectra and robust carrier-envelope phase retrieval, *Nature Phys.* 3: 52–57.

Hentschel, M., Kienberger, R., Spielmann, C., Reider, G. A., Milosevic, N., Brabec, T., Corkum, P., Heinzmann, U., Drescher, M. & Krausz, F. (2001). Attosecond metrology, *Nature* 414: 509.

Kakehata, M., Fujihira, Y., Takada, H., Kobayashi, Y., Torizuka, K., Homma, T. & Takahashi, H. (2002). Measurements of carrier-envelope phase changes of 100-hz amplified laser pulses, *Applied Physics B: Lasers and Optics* 74(0): s43–s50. URL: http://dx.doi.org/10.1007/s00340-002-0880-1

Kakehata, M., Takada, H., Kobayashi, Y., Torizuka, K., Fujihira, Y., Homma, T. & Takahashi, H. (2001). Single-shot measurement of carrier-envelope phase changes by spectral interferometry, *Opt. Lett.* 26(18): 1436–1438.
URL: http://ol.osa.org/abstract.cfm?URI=ol-26-18-1436

Kakehata, M., Takada, H., Kobayashi, Y., Torizuka, K., Takamiya, H., Nishijima, K., Homma, T., Takahashi, H., Okubo, K., Nakamura, S. & Koyamada, Y. (2004). Carrier-envelopephase stabilized chirped-pulse amplification system scalable to higher pulse energies, *Opt. Express* 12(10): 2070–2080.
URL: http://www.opticsexpress.org/abstract.cfm?URI=oe-12-10-2070

Kiriyama, H., Mori, M., Nakai, Y., Shimomura, T., Sasao, H., Tanoue, M., Kanazawa, S., Wakai, D., Sasao, F., Okada, H., Daito, I., Suzuki, M., Kondo, S., Kondo, K., Sugiyama, A., Bolton, P. R., Yokoyama, A., Daido, H., Kawanishi, S., Kimura, T. & Tajima, T. (2010). High temporal and spatial quality petawatt-class Ti:sapphire chirped-pulse amplification laser system, *OPTICS LETTERS* 35(10): 1497–1499.

Kling, M. F., Siedschlag, C., Verhoef, A. J., Khan, J. I., Schultze, M., Uphues, T., Ni, Y., Uiberacker, M., Drescher, M., Krausz, F. & Vrakking, M. J. J. (2006). Control of Electron Localization in Molecular Dissociation, *Science* 312(5771): 246–248.
URL: http://www.sciencemag.org/cgi/content/abstract/312/5771/246

Kreb, M., Löffler, T., Thomson, M. D., Dörner, R., Gimpel, H., Zrost, K., Ergler, T., Mashammer, R., Morgner, U., Ullrich, J. & Roskos, H. G. (2006). Determination of the carrierenvelope phase of few cycle laser pulses with terahertz-emission spectroscopy, *Nature Phys.* 2: 327–331.

Li, C., Mashiko, H., Wang, H., Moon, E., Gilbertson, S. & Chang, Z. (2008). Carrier-envelope phase stabilization by controlling compressor grating separation, *Applied Physics Letters* 92(19): 191114. URL: http://link.aip.org/link/?APL/92/191114/1

Li, C., Moon, E. & Chang, Z. (2006). Carrier-envelope phase shift caused by variation of grating separation, *Opt. Lett.* 31(21): 3113–3115.
URL: http://ol.osa.org/abstract.cfm?URI=ol-31-21-3113

Li, C., Moon, E., Mashiko, H., Nakamura, C. M., Ranitovic, P., Maharjan, C. M., Cocke, C. L., Chang, Z. & Paulus, G. G. (2006). Precision control of carrier-envelope phase in grating based chirped pulse amplifiers, *Opt. Express* 14(23): 11468–11476.
URL: http://www.opticsexpress.org/abstract.cfm?URI=oe-14-23-11468

Mairesse, Y. & Quéré, F. (2005). Frequency-resolved optical gating for complete reconstruction of attosecond bursts, *Phys. Rev. A* 71: 011401(R).

Mashiko, H., Gilbertson, S., Chini, M., Feng, X., Yun, C., Wanga, H., Khan, S. D., Chen, S. & Chang, Z. (2009). Extreme ultraviolet supercontinua supporting pulse durations of less than one atomic unit of time, *Opt. Lett.* 34: 3337.

Mashiko, H., Gilbertson, S., Li, C., Khan, S. D., Shakya, M. M., Moon, E. & Chang, Z. (2008). Double optical gating of high-order harmonic generation with carrierenvelope phase stabilized lasers, *Phys. Rev. Lett.* 100: 103906.

Mashiko, H., Nakamura, C. M., Li, C., Moon, E., Wang, H., Tackett, J. & Changa, Z. (2007). Carrier-envelope phase stabilized 5.6 fs, 1.2 mj pulses, *Appl. Phys. Lett.* 90: 161114.

McKenna, J., Sayler, A. M., Anis, F., Gaire, B., Johnson, N. G., Parke, E., Hua, J. J., Mashiko, H., Nakamura, C. M., Moon, E., Chang, Z., Carnes, K. D., Esry, B. D. & Ben-Itzhak, I. (2008). Enhancing high-order above-threshold dissociation of *h*2+ beams with few-cycle laser pulses, *Phys. Rev. Lett.* 100(13): 133001.

Moon, E., Li, C., Duan, Z., Tackett, J., Corwin, K. L., Washburn, B. R. & Chang, Z. (2006). Reduction of fast carrier-envelope phase jitter in femtosecond laser amplifiers, *Opt. Express* 14(21): 9758–9763.
URL: http://www.opticsexpress.org/abstract.cfm?URI=oe-14-21-9758

Moon, E., Wang, H., Gilbertson, S., Mashiko, H., Chini, M. & Chang, Z. (2010). Advances in carrier-envelope phase stabilization of grating-based chirped-pulse amplifiers, *Laser & Photonics Reviews* 4(1): 160–177.

Paulus, G. G., Grasbon, F., Walther, H., Villoresi, P., Nisoli, M., Stagira, S., Priori, E. & Silvestri, S. D. (2001). Absolute-phase phenomena in photoionization with few-cycle laser pulses, *Nature* 414: 182–184.

Sansone, G., Benedetti, E., Calegari, F., Vozzi, C., Avaldi, L., Flammini, R., Poletto, L., Villoresi, P., Altucci, C., Velotta, R., Stagira, S., Silvestri, S. D. & Nisoli, M. (2006). Isolated single-cycle attosecond pulses, *Science* 314: 443.

Schultze, M., Fiess, M., Karpowicz, N., Gagnon, J., Korbman, M., Hofstetter, M., Neppl, S., Cavalieri, A. L., Komninos, Y., Mercouris, T., Nicolaides, C. A., Pazourek, R., Nagele, S., Feist, J., Burgdorfer, J., Azzeer, A. M., Ernstorfer, R., Kienberger, R., Kleineberg, U., Goulielmakis, E., Krausz, F. & Yakovlev, V. S. (2010). Delay in Photoemission, *Science* 328(5986): 1658–1662.
URL: http://www.sciencemag.org/cgi/content/abstract/328/5986/1658

Shan, B. & Chang, Z. (2001). Dramatic extension of the high-order harmonic cutoff by using a long-wavelength driving field, *Phys. Rev. A* 65(1): 011804.

Sola, I. J., Mével, E., Elouga, L., Constant, E., Strelkov, V., Poletto, L., Villoresi, P., Benedetti, E., Caumes, J.-P., Stagira, S., Vozzi, C., Sansone, G. & Nisoli, M. (2006). Controlling attosecond electron dynamics by phase-stabilized polarization gating, *Nat. Phys.* 2: 319.

Strickland, D. & Mourou, G. (1985). Compression of amplified chirped optical pulses, *Optics Communications* 55(6): 447 - 449.
URL: http://www.sciencedirect.com/science/article/B6TVF-46JGW3M-1BX/2/ef583949f08c6e26c541eeff02c32fa1

Thomann, I., Gagnon, E., Jones, R., Sandhu, A., Lytle, A., Anderson, R., Ye, J., Murnane, M. & Kapteyn, H. (2004). Investigation of a grating-based stretcher/compressor for carrier-envelope phase stabilized fs pulses, *Opt. Express* 12(15): 3493–3499.
URL: http://www.opticsexpress.org/abstract.cfm?URI=oe-12-15-3493

Tzallas, P., Charalambidis, D., Papadogiannis, N. A., Witte, K. & Tsakiris, G. D. (2003). Direct observation of attosecond light bunching, *Nature* 426(6964): 267–271.
URL: http://dx.doi.org/10.1038/nature02091

Vaillancourt, G., Norris, T. B., Coe, J. S., Bado, P. & Mourou, G. A. (1990). Operation of a 1-khz pulse-pumped ti:sapphire regenerative amplifier, *Opt. Lett.* 15(6): 317–319.
URL: http://ol.osa.org/abstract.cfm?URI=ol-15-6-317

Wang, H., Chini, M., Khan, S. D., Chen, S., Gilbertson, S., Feng, X., Mashiko, H. & Chang, Z. (2009). Practical issues of retrieving isolated attosecond pulses, *Journal of Physics B: Atomic, Molecular and Optical Physics* 42(13): 134007.
URL: http://stacks.iop.org/0953-4075/42/i=13/a=134007

Yun, C., Chen, S., Wang, H., Chini, M. & Chang, Z. (2009). Temperature feedback control for long-term carrier-envelope phase locking, *Appl. Opt.* 48(27): 5127–5130.
URL: http://ao.osa.org/abstract.cfm?URI=ao-48-27-5127

13

The Generation and Characterisation of Ultrashort Mid-Infrared Pulses

J. Biegert[1,2], P.K.Bates[1] and O.Chalus[1]
[1]*ICFO – Institut de Ciencies Fotoniques*
[2]*ICREA – Institució Catalana de Recerca i Estudis Avançats*
Spain

1. Introduction

Over the past decade, ultrashort pulsed light sources have become an indispensable tool both in the laboratory and over a wider range of applications in the medical, industrial and telecommunication sectors. The availability of energetic sub-100 fs pulses, combined with the stability and usability of solid-state laser amplifiers, has opened up entire new fields such as femtochemistry, laser micro-drilling and knife-less laser eye surgery. However, while current ultrashort pulse sources are highly developed, their central wavelengths almost exclusively lie in the near-infrared spectral range below 1000 nm.

Specifically, coherent pulses of mid-infrared (mid-IR) radiation, i.e. at wavelengths longer than 3 microns, are intensely sought for a range of applications in the life sciences, spectroscopy and environmental sensing but have not readily been available due to various technical challenges. These challenges are related not only to detecting and handling mid-IR radiation but also to the scarcity of mid-IR sources. Much effort has been invested in developing appropriate sources and technology to enable reliable production of such sources, but, even 50 years after the invention of the laser, a large portion of the mid-IR spectrum remains inaccessible, especially if one is interested in ultrashort pulsed sources. It is just within the last years that optical technology has made a major step forward; recent advances in fiber technologies are becoming available and reliable nonlinear media are now accessible. However, the current generation of mid-IR sources is not yet nearly as advanced as those in the near-IR. The various approaches and techniques often cover very narrow spectral ranges, come with very low output power, or are unable to provide short pulses of radiation. The last point in particular is common to the majority of mid-IR sources commonly used to date. Another drawback is that these very specific sources are typically designed as a specialists tool for a particular application. Very few systems have been designed to offer a robust, all-round performance in a flexible, upgradeable format. Thus, mid-IR sources often lack flexibility, and, with each source optimised for a very narrow set of applications (or perhaps even just one application), mid-IR source development has fractured into different specialist areas, resulting in a lack of coherence across the field, and ultimately thwarting the advancement of mid-IR science and technology.

In this chapter we will restrict ourselves to sources of ultrashort pulses in the mid-IR spectral range. We will begin by motivating the development of sources as well as some technical limitations, mention some available sources as well as describe our new platform for ultrashort pulses and describe why it promises to even surpass the performance of the current state-of-the art NIR systems.

The range of applications of such a mid-IR source is immense, particularly in bio-medical and biological research. Figure 1 shows a compilation of information relative to these fields, absorption curves of tissue and water as well as absorption bands of molecules which constitute the building blocks of life.

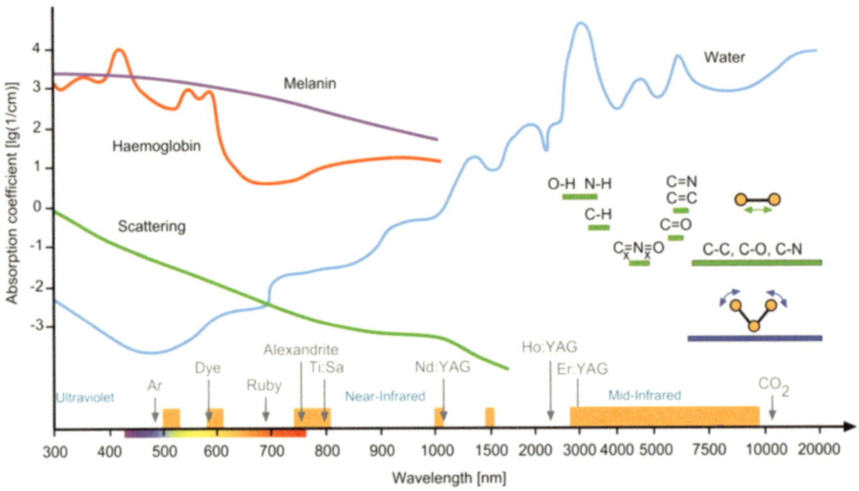

Fig. 1. Shown are common (long pulse) laser sources and the wavelength range accessible in the mid-IR by the system described in this chapter. Overlaid are the the absorption curves and scattering in tissue together with major absorption bands of marker molecules and compounds of interest; adapted from Peng et al. (2008)

Figure 1 clearly demonstrates that, while visible wavelengths are more suitable for imaging applications, due to a longer penetration depth, mid-IR wavelengths hold a clear advantage in terms of selectivity and rapid (and localized) absorption. In fact, femtosecond lasers can function as a pair of nano-scissors in sub-cellular surgery and have potential applications in a single organelle or chromosome dissection, inactivation of specific genomic regions on individual chromosomes and highly localised gene and molecular transfer. The major advantage of pulsed laser nano-surgery is the well-controlled and non-invasive capability of severing sub-cellular structures with high accuracy in time and three-dimensional space.

Spectroscopy of cellular compounds or volatile components in human breath will have its highest sensitivity and selectivity in the mid-IR since those wavelengths cover most of the molecular absorption bands and since each molecule or compound has its specific fingerprint. By closely monitoring the spectral shifts or changes in line strength, it will become possible to see how those compounds behave in their environment. Ordinary human breath is teeming with bio-molecules that can reveal the presence or absence of certain diseases or metabolic processes. To date, researchers have identified over 1000 different compounds contained in human breath that have both endogenous and exogenous origins, and provide information about physiological processes occurring in the body, as well as environment-related ingestion or absorption of contaminants. Just as bad breath can indicate dental problems, the identification and measurement of molecules in exhaled breath can provide a window into the metabolic state of the human body. While the presence and concentration of many of these molecules are currently not well understood, many biomarker molecules have been correlated to specific diseases and metabolic processes. Such correlations provide the potential for non-invasive methods of health screening for a wide variety of medical conditions, including detecting the presence of cancer, monitoring respiratory diseases, assessing liver and kidney function, and determining exposure to toxins. For example: excess methylamine may signal liver and kidney disease; ammonia may be a sign of renal failure; elevated acetone levels can indicate diabetes; and nitric oxide levels may be used to diagnose asthma. More sensitive, earlier detection of disease is obviously highly desirable in all cases, but in many conditions this can spell the difference between life and death.

While many of the above-mentioned applications can be covered by continuous wave or long-pulse sources, some applications will significantly benefit from ultrashort pulsed sources. This can be due either to the fact that shorter pulses usually go hand in hand with high achievable intensities, as required for nanosurgery applications, or to the pulse's broad spectral bandwidth, which allows easy detection over many absorption bands instead of arduous scanning. In particular, high intensities and well controlled electrical fields are the basic requirement for investigations in fundamental strong field physics, when the laser electric field strength approaches that of the atomic binding energy in the matter. Almost all such investigations are extremely sensitive to the electric field structure of the laser pulse, require high repetition rates due to the low probability of the processes under investigation, and are particularly sensitive to the wavelength of the driving laser.

Many strong field physics experiments involve the measurement of photoionised electrons, which makes mid-IR pulses very interesting, since they allow for a much clearer discrimination between tunnelling and multi-photon ionisation, whereas current ultrashort NIR laser sources operate in a mix of multi-photon and tunnelling ionisation regimes. The lower photon energy of mid-IR pulses can be used to create strong field experiments that clearly involve tunnelling ionisation only, allowing investigation of fundamental atomic processes with unprecedented clarity.

Another growing area of interest is the production of attosecond (10^{-18}s) pulses from ultrashort intense femtosecond lasers. Attosecond pulses with a carrier frequency corresponding to extreme ultraviolet wavelength can be produced from short-pulse laser systems, using high order harmonic generation (HHG) as coherent up-shifting mechanism from the near-IR drive laser (Mcpherson et al. (1987); Ferray et al. (1988)). The availability of few-cycle mid-IR light pulses for this purpose should yield shorter attosecond pulses due to a square of wavelength dependence of the shortest wavelength reachable via HHG (Sheehy et al. (1999); Gordon and Kaertner (2005)). Recent experiments have confirmed this scaling of the harmonic cutoff with drive wavelength, while showing that predicted losses in harmonic yield (Tate et al. (2007)) can be compensated by taking advantage of more favourable HHG phase matching at longer wavelengths (Popmintchev et al. (2008)). Based on their results we expect a 3 μm source to generate harmonic spectra extending to a photon energy well above 1 keV.

A unique feature of the source we present here is its ability to operate at extremely high repetition rates. Higher repetition rates help to improve signal to noise ratio for most experiments, but they are also essential for some in strong field physics; for instance, particle coincidence experiments with reaction microscopes (COLTRIMS) (Moshammer et al. (1996)) permit the investigation of atomic and molecular processes with unprecedented scrutiny, but are limited mainly by the stability of current lasers due to the low cross sections of the processes under investigation; the measurement time is, in practice, nearly always longer than the time over which the best lasers can deliver constant performance. Using a 100 kHz repetition rate, experiments taking six days with a 1 kHz system can be completed in 90 min, greatly reducing the demands on the laser system stability.

Many of the demands of strong field physics are extremely challenging for any laser system. In particular, the generation of single attosecond pulses requires driving pulse durations of only a few cycles of the underlying electric field, and a stable carrier-to-envelope phase (CEP). The CEP is the offset between the peak of the pulse intensity envelope and the peak of the underlying electric field, as shown in Fig. 2. For a pulse whose duration is many cycles of the electric field, this parameter is relatively unimportant, but for a few-cycle pulse such as the one shown in Fig. 2, the structure of the electric field can vary strongly with the CEP value, and can adversely affect an experiment. CEP stability is therefore necessary to maintain the electric field shape between successive laser pulses, and is a considerable technical challenge for even the most advanced systems.

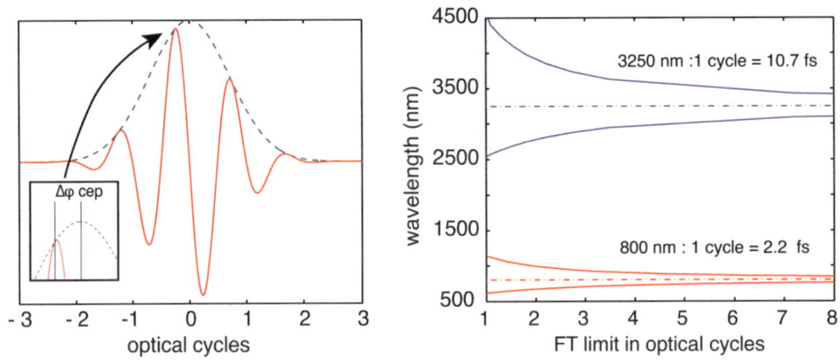

Fig. 2. The left panel shows the intensity envelope (blue, dashed) and electric field structure (red, solid) of a few-cycle pulse, with the definition of the carrier envelope phase shown in the inset. The CEP must be controlled to ensure the electric field structure is constant from pulse to pulse. The immense bandwidth needed to support such a few-cycle pulse is shown in the right panel, for pulses centered at 800 nm (red) and 3250 nm (blue), plotted against pulse duration in optical cycles.

Current high-energy feedback stabilised systems are capable of CEP locked operation for several hours at most. Total CEP stability of the laser source is essential for many experiments, both in attoscience measurements which typically involve large scans of pump-pulse delay times, and even more so for photoionisation experiments. For example, the measurement of double ionisation demands CEP stability over about 12 hours with a 1 kHz repetition rate; an unrealistic requirement from current electronically stabilised systems. As we will see in the following sections, moving central wavelengths to the mid-IR allows us to use a method of passive CEP stabilisation that has been proven to operate with no slow drifts over >240 hours. Finally, generating ultrashort pulses in the mid-IR necessarily requires the generation of large bandwidths, with few-cycle mid-IR pulse spectra covering hundreds of nanometers - the bandwidths of pulses at 800 and 3250 nm are shown in Fig. 2, normalised to the pulse duration in optical cycles. These spectra have the ability to cover simultaneously many vibrational transitions in important molecules, and this combined with the intrinsic potential CEP stability opens a wide range applications (Thorpe and Ye (2008)). Generating and amplifying such a bandwidth requires careful management of dispersion throughout the laser system, in a wavelength range where many materials have anomalous dispersion, poorly characterized dispersion curves or limited transmission bandwidth. Control of the bandwidth and spectral phase is essential for few-cycle pulse generation, as is an accurate method of pulse characterisation.

2. Few-cycle mid-IR pulse generation

The development of any ultrashort pulsed source in the mid-IR should not only match but ideally surpass the abilities of current NIR sources. State of the art pulse durations at centre wavelengths in the visible to NIR currently lie in the few-cycle range at repetition rates up to several kHz. Sources are nearly exclusively based on Ti:Sapphire chirped pulse amplification (CPA) systems, combined with spectral broadening via gas-filled hollow fibres (Nisoli et al. (1998)) or filamentation (Hauri et al. (2004)) and compression in the visible to near-IR (NIR) with chirped mirrors (Schenkel et al. (2003)) to routinely generate pulses with durations of a few cycles of the electric field. These systems are intrinsically limited to the NIR by their reliance on Ti:Sapphire CPA, and as such cannot be directly reproduced in the mid-IR. The way to access ultrashort pulses in the mid-IR proceeds nearly exclusively via three wave mixing in nonlinear crystals and specifically parametric amplification to overcome the limited gain bandwidth of Ti:Sapphire or other solid state gain media.

We would like here to distinguish between optical parametric amplification (OPA) and optical parametric chirped pulse amplification (OPCPA) approaches, even though, strictly speaking, all sources of ultrashort pulses are OPCPA due to the near impossibility of avoiding chirp. The distinction is made therefore by labelling OPA as an approach without intended and pre-defined chirp in contrast to OPCPA where the seed pulse to be amplified has to acquire a pre-defined chirp. This distinction is nevertheless important, as OPA based approaches are limited in energy due to the high peak powers of the pump pulse used in the process.

Probably the most established method to access ultrashort mid-IR pulses is via non-collinear OPA of some white light continuum or frequency shifted output, from Ti:Sapphire (Wilhelm et al. (1997)) or, more recently, Yb-based fiber CPA systems (Schriever et al. (2008)). For the broad spectra of hollow fibre broadened Ti:Sapphire lasers DFG can also be used (Vozzi et al. (2006)), followed by amplification in an OPA using the Ti:Sapphire system as a pump. Different implementations of these various approaches have generated few-cycle pulses at 1.2 - 3 μm (Vozzi et al. (2006); Cirmi et al. (2008); Zhang et al. (2009)) and recently this has been extended to longer wavelengths, delivering 25 fs duration pulses at ~ 3 μm, with pulse energies of 2 μJ (Brida et al. (2008)). The latter system amplifies a white-light continuum with a Ti:Sapphire pump laser to generate an amplified signal at 1.3 μm. In a second stage OPA this signal is amplified further, and the idler from the interaction is extracted, which has 2 μJ infrared energy at 3 μm. The CEP stability of this system has yet to be proven, and scaling to higher energies is limited by the un-chirped nature of the OPA interaction, however it is an interesting source of low energy mid-IR pulses. The low energy output from such frequency converted systems is very applicable to ultrafast spectroscopy, where it has found many uses (Nibbering and Elsaesser (2004)).

Mixing of amplified ultrashort pulses from a Ti:Sapphire laser with longer pulses at around 1 μm wavelength has been shown to produce ultrashort pulses in the mid-IR (Sheehy et al. (1999); Rotermund et al. (1999)), but is limited to roughly the duration of the driving laser pulse, and suffers from low efficiency. This technique has been a workhorse of ultrafast mid-IR spectroscopy, but is unlikely to be scalable to higher energy or few-cycle pulse durations, and does not provide CEP stable pulses.

A more exotic, but elegant, approach to few-cycle mid-IR pulse generation (Fuji et al. (2006)), uses a four wave mixing process generated inside a filament in air. The interaction of the 800 nm fundamental of a Ti:Sa system and its second harmonic results in an 13 fs 1.3 cycle pulse with 1.5 μJ energy at a wavelength of 3.4 μm and extremely broad bandwidth. However, this system has poor efficiency, requiring 1.8 mJ of fundamental to generate just over 1 μJ. Moreover, the repetition rate is low at 1 kHz, and the stability of such a system is unclear. Furthermore the scalability is inherently limited to μJ energies due to the intensity clamping in the filament, which limits the pump energy to ~1 mJ for ~ 30 fs pulses.

2.1 OPCPA in the mid-IR

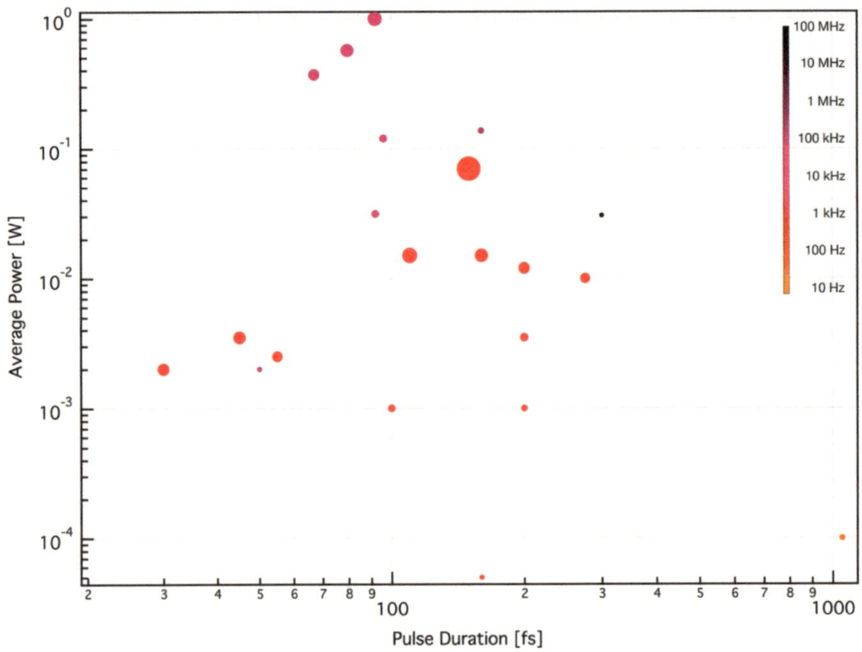

Fig. 3. **Mid-IR ultrashort pulse sources.** A summary of the ultrashort pulses available in the mid-IR. The colour scale represents repetition rate, while the size of each circle corresponds to the energy per pulse. The system described in this chapter lies in the top left quadrant of the picture.

An alternative approach to frequency-shifting of NIR laser systems is direct amplification of ultrashort mid-IR pulses using optical parametric chirped pulse amplification (OPCPA). OPCPA involves the amplification of broad-bandwidth chirped seed pulses using a narrowband, typically pico-to-nanosecond pump laser. This approach allows amplification across a huge range of central wavelengths in the NIR and mid-IR with ultra-broad gain bandwidths that make possible the direct amplification of few-cycle pulses. Indeed, NIR OPCPA sources have demonstrated that they can directly produce amplified few-cycle pulses as short as 5.5 fs (Adachi et al. (2008)). OPCPA systems have already been reported at 2.1 μm (Fuji et al. (2006)) and in the mid-IR at 3.2 μm (Chalus et al. (2009)).

Unlike the gain-storage media used in traditional CPA amplifiers, no energy is deposited in OPCPA, meaning that the possible pulse repetition rates are limited only by available pump laser technologies. In contrast to OPA based schemes, OPCPA uses a long pump and chirped seed pulse, allowing the energy of the systems to be scaled up even to joule level energies (Chekhlov et al. (2006); Lozhkarev et al. (2006)). The previous two references are NIR OPCPA systems pumped by the second harmonic of the pump laser, and so moving to the mid-IR where pumping with the fundamental is possible should already increase the output energies. OPCPA is the only technology that currently offers the possibility of scaling up ultrashort mid- IR pulse energies to the Joule level.

2.2 OPCPA pump laser selection

OPCPA fundamentally is a nonlinear three-wave mixing process, and as such requires adequate (high) pump intensities to generate gain in reasonable crystal lengths. We can identify three main regimes for such pump sources: femtosecond, picosecond and nanosecond. Thus, a significant challenge is the selection of an appropriate pump laser for the OPCPA process.

Femtosecond systems with significant pump pulse energy typically employ CPA. The advantage of such an approach is that an OPCPA could serve as a back-end to simply extend the CPA's wavelength regime. The significant drawback is a highly complex system which inherits any problem that the CPA system might already have. Additional issues that one might have to address are the synchronisation between pump and seed over long path lengths as well as short pump pulses which could be beneficial in terms of achieving high pulse contrasts but as well limiting achievable efficiency.

Nanosecond durations are easily available from well developed pump sources, especially Q-switched Nd:YAG sources. Such lasers are very simple and reliable but their nanosecond duration requires very large seed stretch factors to be efficient in OPCPA. Especially the recompression to few-cycle pulse duration is far from trivial and could come with penalties in achievable contrast. Injection seeding or some form of synchronization of such Q-switched lasers is required due to their large pulse to pulse jitter. They typically also require longer crystal lengths to achieve significant gain, which can limit the bandwidth. However, nanosecond systems can produce energies orders of magnitude greater than femtosecond systems for a similar price.

Picosecond systems present a good compromise in terms of readily achieving pump pulse intensities for OPCPA whilst requiring moderate seed stretch factors and avoiding the complexity of CPA based pumps. Master-oscillator power-amplifier (MOPA) pump lasers with picosecond duration are commercially available in a wide range of configurations, with excellent performance characteristics and at repetition rates up to a few hundred kHz. These higher repetition rates can increase signal to noise ratios in experiments, and reduce data collection times, but only if the laser stability is not degraded by the increased repetition rate. As we have mentioned before, OPCPA as a technique is virtually repetition rate insensitive as no energy is deposited in the crystals, however, the pump laser's stability has a direct influence on the stability of the OPCPA, such that this parameter becomes extremely important. For example, during investigation of the change of absorption in a material as in (Gertsvolf et al. (2008)) fluctuations over a few percent already limited the measurement. As of today, OPCPA sources have achieved stabilities from 1.5% and higher (Tavella et al. (2010); Ishii et al. (2005)) while solid state lasers perform on a better level.

It should be noted that recent developments in high repetition rate and high energy fibre laser systems offer an interesting option for pumping OPCPA systems. These systems typically offer only a few hundred microjoules of energy but operate at repetition rates of a few hundred kHz (Roser et al. (2005)) or even MHz repetition rates (Boullet et al. (2009)). These systems allow the use of small stretch factors in the OPCPA chain, and their low energy means that the short pulse duration does not lead to unreasonable requirements for large crystal apertures. They do offer the possibility of extremely high average powers, and more importantly near alignment-free OPCPA systems.

3. Experimental implementation of a mid-IR OPCPA source

In the remainder of this chapter we will describe an implementation of the sort of mid-IR ultrashort pulsed source motivated by the applications described in the introduction. The source has been designed to provide an extremely stable, high repetition rate pulsed source, with stable CEP, and capability for few cycle durations. In this implementation we have not explored the high energy capability of mid-IR OPCPA, but the possibility of upgrading the source is there, simply through the addition of extra amplifier stages. The source is compact, stable, easy to operate and we believe this approach leads to a source that can fulfil the key criteria needed across a wide range of applications in biology, spectroscopy, and strong field physics.

3.1 Generation of a CEP stable mid-IR seed pulse

There are currently no available broadband oscillators operating in the mid-IR, and thus the seed pulse for our system must be generated from a shorter wavelength oscillator and a nonlinear process. This in fact is very advantageous for a ultrashort long wavelength system, as it allows us to use a combination of standard, well-developed oscillator technology, and also to passively stabilise the CEP via difference frequency generation (DFG).

By mixing pulses with central wavelengths of 1050 nm and 1550 nm in an appropriate nonlinear crystal, a pulse can be generated via DFG at 3200 nm central wavelength. This pulse is the idler of the three wave mixing interaction, and the phase of the pump, signal and idler pulses (ϕ_p, ϕ_s & ϕ_i) in the interaction can be expressed as the following

$$\phi_s = \phi_s(0) - \frac{\Delta kz}{2} + \frac{\Delta k\gamma_s^2}{2}\int\frac{dz}{f+\gamma_s^2} \tag{1}$$

$$\phi_p = \phi_p(0) - \frac{\Delta k}{2}\int\frac{f dz}{1-f} \tag{2}$$

$$\phi_i = \phi_p(0) - \phi_s(0) - \pi/2 - \frac{\Delta kz}{2} \tag{3}$$

Where $\Delta k = k_s + k_i - k_p$ is the wave-vector mismatch, $\phi_x(0)$ is the input phase of the pulses, f is the fractional depletion of the pump energy, γ is a gain coefficient dependent on the crystal parameters, and z is the crystal length. In the case of perfect phasematching $\Delta k = 0$, the expression becomes simply

$$\phi_s = \phi_s(0) \qquad \phi_p = \phi_p(0) \qquad \phi_i = \phi_p(0) - \phi_s(0) - \pi/2 \tag{4}$$

If the pump and seed pulses in the interaction originate from the same laser oscillator, they will have the same, although rapidly changing, CEP value. In the DFG interaction, the difference between $\phi_p(0)-\phi_s(0)$ is therefore constant, and the idler phase is passively stabilised to a constant value. This principle has been successfully demonstrated experimentally (Baltuska et al. (2002)), and allows locking of the CEP phase to a fixed value with much less complexity than active-feedback systems commonly used in e.g. Ti:Sapphire oscillator systems. The effect of imperfect phasematching is to couple the output CEP to fluctuations in the pump laser intensity via the f parameter, but for a stable pump laser and a correctly aligned OPA this does not affect the CEP stability in a drastic way (Renault et al. (2007)).

In our experimental realisation, the seed for our OPCPA system is derived from a two-color fibre laser system (Toptica Photonics) which delivers amplified and phase-coherent ultrashort pulses at 1050 nm (48 fs, 16 mW) and 1550 nm (75 fs, 180 mW). The use of fibre laser ensures excellent timing stability between the two arms, alignment free and hands-off operation over long operation times. To generate the required ultrabroad mid-IR seed pulse, we use a difference-frequency generation (DFG) stage: DFG between the frequency shifted pulses from the fibre system allows generation of a seed pulse in the mid-IR spectral region which in our case stretches from 3000-4000 nm at the $1/e^2$ level. This configuration should passively stabilise the CEP of the generated idler pulses as described above, and measurements using the same fibre oscillator have shown timing jitter between the two arms to be less than 21 as over 200 hours, corresponding to a CEP drift of less than 90 mrad over this time, without complicated locking electronics or feedback loops (Adler et al. (2007)). There is also no need for octave spanning oscillators nor seed bandwidths and as a consequence complexity is reduced significantly.

DFG is achieved with a simple, 2 mm long, periodically poled lithium niobate crystal (PPLN) which yields a sub-picosecond duration mid-IR pulse with a spectrum covering 400 nm of bandwidth at the FWHM level with a power of about 1.5mW at 100MHz, corresponding to a transform limited pulse duration of 33 fs (Fig. 4). The PPLN crystal is poled in a fan-out geometry to allow fine-tuning of the phasematching bandwidth; the spatial variation of the fan-out poling is however chosen to vary slowly enough in order to avoid noticeable spatial chirp across the generated mid-IR beam. In order to preserve the CEP of the optically stabilized seed pulse, care must be taken with the system design. The entire OPCPA is enclosed in an air-tight insulated box, with a beam height of just 63 mm chosen to minimise mechanical vibrations of the mounts. The optics are mounted on 25 mm diameter stainless steel pedestals for optimum stability. The consideration of the CEP

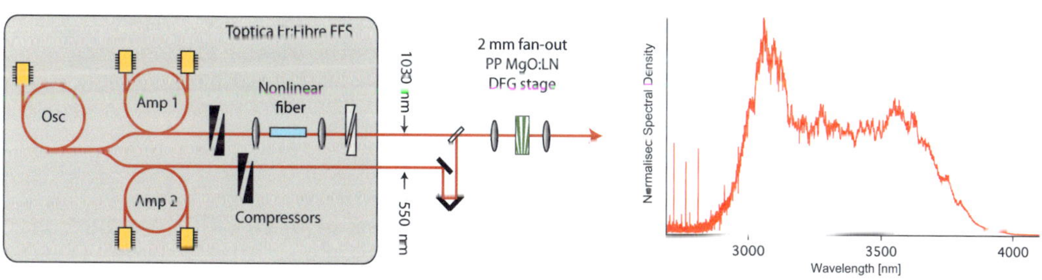

Fig. 4. **Mid-IR seed generation.** The two colour output from a commercial fiber MOPA system (FFS, Toptica Photonics) generates, via DFG, self-CEP stable, 3.2 μm radiation.

stability also defined our choice of stretcher system, as reports in the literature have identified the sensitivity of CEP stabilised systems to mechanical drifts in grating based stretcher or compressor systems. As such we prefer to use bulk stretching in a block of sapphire to avoid sensitivity to these drifts. Simulations of the DFG output show that the mid-IR pulse is already negatively chirped (Chalus et al. (2008)) to approximately 200 fs duration, and that a 5 cm long block of undoped Sapphire is sufficient to further negatively stretch the pulse to 6 ps compared to the pulse duration of 9 ps.

The stretched pulse duration must be a significant fraction of the pump pulse duration for good energy extraction, however, in high gain OPCPA stages the temporally varying intensity of a gaussian pump pulse can cause reduced gain for the edges of the chirped seed spectrum. A balance needs to be found between the pulse stretching and the effect on the bandwidth (Moses et al. (2009)). By modelling the relationship between gain and stretched pulse duration in our system, we have found that a combination of different stretched pulse durations in our amplifiers is the best configuration for our system. Because we are using a bulk stretcher, we can easily split the stretching into three separate stages, allowing the use of a short 1 ps stretched pulse in the first amplifier to optimise bandwidth and the cost of only a small reduction in energy extraction, while stepping up the pulse duration to 4 ps in the second and 6 ps in the third amplifier, where the low gain has less effect on the spectral width, and good temporal overlap allows efficient energy extraction. The stretched mid-IR seed pulse is difficult to characterise temporally, and we estimate the ratio between the pump duration and seed duration by changing their timing overlap in the first OPCPA stage and monitoring the spectral shift and idler energy.

The pump laser used for the OPCPA is a picosecond high-average-power pump laser from Lumera Laser GmbH. It operates at 1064 nm with 100 kHz at 40W output power and with pulse duration of 8 ps. Its spatial mode is close to $M^2 \approx 1.2$ and it has stability better than from the most advanced Ti:Sa CPA systems; power fluctuations of <0.4% pulse-to-pulse and <0.1% RMS over 15 hrs are *routinely* observed. The fiber oscillator is used as master oscillator and the pump laser's oscillator is slaved to it to better than 350 fs rms over 6 hours via an electronic synchronization unit (Menlo Systems). As will become evident for the results we present for this system, the timing jitter between the pump and seed pulses does not prevent the generation of extremely stable mid-IR pulses. (Chalus et al. (2009))

For optimum stability, mid-IR OPCPA systems should make use of optical synchronization of pump and seed pulses, such as that used in (Teisset et al. (2005); Fuji et al. (2006)). While electronic stabilisation systems have worked well here and in other OPCPA systems (Witte et al. (2005)), passive optical stabilisation offers a simpler, more robust way to cleanly synchronise the pump and signal pulses without drift for many hours. It is particularly easy in the mid-IR, where the long wavelength of the seed pulse means that the frequency shifting needed to seed the pump laser is towards higher frequency, which is usually easier to achieve and more efficient than shifting to lower frequencies.

3.2 The OPCPA amplification chain

The OPCPA amplifier chain in our system (Fig. 5) consists of three OPCPA stages, each configured for a different gain level. The choice of three stages allows us to optimise bandwidth, stability and energy extraction: by using the first stage to generate high gain with little depletion of the pump, the second stage to slightly deplete the pump and the final stage run well into pump depletion, we can take advantage of the fact that a strongly depleted OPCPA amplifier shows approximately linear coupling between pump intensity fluctuations and amplified seed fluctuations (Ross et al. (2007)). Using a nearly-depleted and depleted stage in series ensures that over a reasonable range of pump laser fluctuations we can stay in the optimum range of depletion to maximise the seed stability. The use of a three stage OPCPA system also allows us to increase the amplified bandwidth by tuning the first and second crystals to amplify slightly different parts of the spectrum. This does reduce the total output of the system somewhat, but allows us to increase bandwidth by nearly 100 nm.

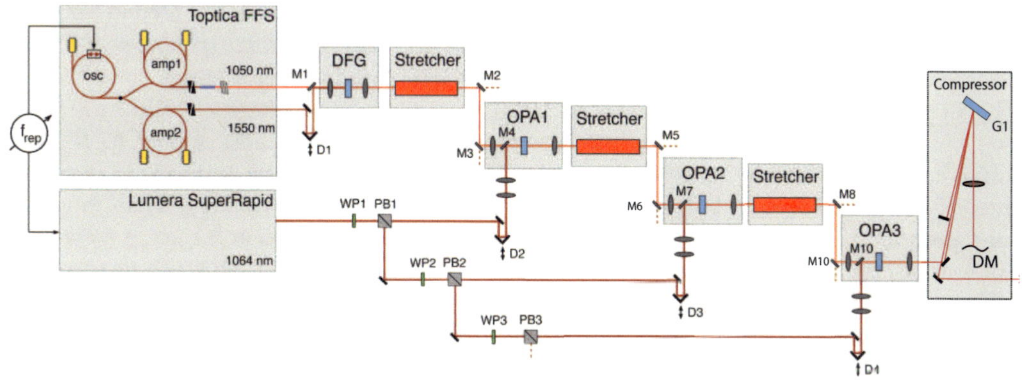

Fig. 5. **Mid-IR OPCPA source layout.** The two colour output from a commercial fiber MOPA system (FFS, Toptica Photonics) generates, via DFG, self-CEP stable, 3.2 μm radiation. These pulses are then stretched and amplified by a triple stage OPCPA pumped by a Nd:YVO_4 laser (SuperRapid, Lumera Laser) and finally compressed by a Martinez-type compressor. The compressor includes a linear deformable mirror with which we fine tune dispersion.

The operating parameters for the first stage are crucial since the highest gain is achieved here. We refrain from operating at maximum possible gain in order to achieve a good balance between amplification and parametric fluorescence background. The seed-pump spatial ratio is set to ~ 3 : 2. The first stage is a 2 mm fan-out periodically poled MgO:LN crystal pumped at an intensity of 60 GWcm^{-2}, giving an energy gain of 8 × 10^3, close to the small signal gain value calculated at 1.1×10^4. The pump power of 2.1W at 100 kHz results in an amplified idler at 3.2 μm centre wavelength, (note that our seed is the idler wave) with a bandwidth of 200 nm FWHM and approximately 80 nJ energy. (Fig. 6). The extraction efficiency of <1% into the idler is low due to the small pump-signal overlap in time and low

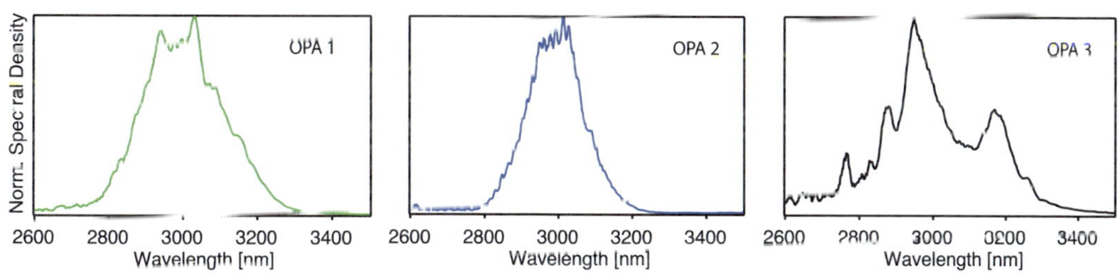

Fig. 6. **The amplified mid-IR spectra after each OPA stage in the system**. The first OPA is biased slightly towards shorter wavelengths, while the amplification in the second and third OPA broadens the spectrum and shifts it to longer wavelengths

depletion, however operating such a high gain stage with a longer chirped pulse of 6 ps led to a reduction in the amplified bandwidth of 50 nm. In any case, the three-stage design does not require high efficiency in this first stage.

The second OPCPA stage uses a crystal identical to the first, pumped with 5.1W average pump power to give a pump intensity of 57 $GWcm^{-2}$. The seed pulse is propagated from the first to the second stage collimated, using dielectric mirrors and filters to reject the residual pump light and amplified signal from the first OPA at 1550 nm. It passes through the second 1 cm sapphire block to stretch the duration to 4 ps, allowing for more efficient energy extraction without affecting the bandwidth. The energy gain in the second amplifier is 40, leading to an amplified idler energy of 1.2 μJ, and and amplified spectrum of 250 nm bandwidth FWHM centered at slightly longer wavelength than the first stage.

The amplified seed pulse from the second OPCPA stages is again filtered by dielectric filters, and passes through the third sapphire block to give a pulse duration of 6 ps as the final amplifier seed. The final OPCPA stage is again an identical crystal pumped at the same intensity of ~ 55GW/cm^2, but uses 20Wof average power to generate over 1Wof amplified idler power. This corresponds to 10 μJ per pulse in the idler centered at 3.1 μm, a conversion efficiency of 5% which however compares well to other OPCPA systems when the quantum efficiency of the process is taken into account. The combination of an idler at 3.1 μm, a signal at 1.5 μm and a pump at 1 μm means that the signal:idler photon ratio should be approximately 3:1. The conversion efficiency of 5% therefore implies a total of 5+10=15% conversion into the signal and idler from the pump, and is not unreasonable for a broadband OPCPA system pumped by a gaussian spatial and temporal profile pump. These values correspond to the optimum energy configuration for the OPCPA system, but the bandwidth can be increased as mentioned above, by altering the tuning of the second and third stage. By optimising for a maximum bandwidth of 350 nm FWHM, the power drops to 550 mW or 5.5 μJ.

Parametric super-fluorescence has been observed to be a problem from previously reported OPCPA systems (Tavella et al. (2005)) and it is vitally important to minimise it in any OPCPA setup. To test the parametric superfluorescence of our system, we employed two different techniques. Simply blocking the seed, we measured the power output of the full OPCPA chain with all pumps set to their operational values. Alternatively, we could use the electronic synchronization to delay the seed relative to the pump by 20 ps-5 ns, so that no amplification was observed, but any long timescale radiation that might seed a parasitic process was still preserved. In both cases, no measurable fluorescence was observed in the reflectivity bandwidth of our dielectric mirrors, which runs from 2.8-3.6 μm. In fact, even when running with gold-coated mirrors, no measurable or visible fluorescence could be seen from the amplifiers. The dynamic range of this measurement is approximately four orders of magnitude. We attribute this to a careful choice of low small signal gain in the amplifiers, careful filtering between the amplifiers, and the cleanliness of our pump laser in space and time. The low parametric fluorescence is a particularly important feature of our system, as previous ultrashort OPCPA systems have shown decreased pulse contrast on a long timescale due to excessive parametric superfluorescence in the system (Tavella et al. (2005))

3.3 Pulse compression

The amplified idler pulse from the system is negatively chirped, and as such requires a positive dispersion compression system. As mentioned before, a non-grating based or bulk compressor is preferable to ensure the best CEP stability, but at our mid-IR wavelength this

type of compressor is difficult to implement. Transmissive materials with appropriate dispersion in the mid-IR such as Silicon or Germanium exhibit high refractive index and high absorption (on the order 3200 cm^{-1}), making their use energetically costly even if Fresnel losses are removed with proper coatings. For power-scaling of these systems, bulk compressors can become extremely costly, and show thermal effects due to their high absorption. In contrast, a relatively simple solution is to use a Martinez-type grating based system, similar to the stretchers typically employed in the visible to near-IR range.
Gold gratings with 200 l/mm are both cheap and readily available, and are appropriate for compression in this system, while due to the slow change in refractive index across our bandwidth in the mid-IR, CaF2 lenses can be used to create the imaging optics. Such a system is easily scaleable to higher power levels. As mentioned before, grating based systems are not ideal for CEP stabilised systems, however the sensitivity scales inversely with wavelength and is further reduced for large line spaced gratings, meaning that for our system should exhibit less CEP variation due to the compressor than NIR Ti:Sa systems. Many of these NIR grating compressor systems have been successfully CEP stabilised (Kakehata et al. (2004)) and yield performance similar to that of prism-based compressor systems.
Controlling the dispersion across the full 400 nm window of our pulse bandwidth is highly challenging, and we have chosen to use a programmable dispersive system to enable fine control of the spectral phase. To achieve this, we have placed a 1D deformable mirror with a silver-coated membrane (OKO Technology) in the Fourier plane of the 4-f compressor setup. It operates with 19 actuators and provides a maximum displacement of 9 μm which corresponds to a delay of about 60 fs or a phase of 12π our wavelength range. Each actuators's range is discretized in 4096 steps of about 2 nm addressed individually. The mirror is computer controlled and its optimum configuration was obtained by the use of a genetic algorithm to converge to the shortest pulse.
The final compressor setup uses two 200 line/mm gold coated gratings and a M=-1 telescope created by two 2″ diameter 250 mm focal length CaF2 lenses, with the deformable mirror used to fold the compressor in the Fourier plane between the lenses. The compressor is designed to support 600 nm spectral bandwidth and exhibits a measured transmission efficiency of 70% for our current 350 nm bandwidth FWHM. Second order phase is solely adjusted through its grating separation with the deformable mirror operating passively without any deformation; minimum achievable pulse durations were measured to 85 fs.
To optimise the pulse duration, we measure the SHG conversion efficiency of the compressed pulse and compare it to that of the uncompressed one. This gives an extremely rapid feedback on the compressed pulse intensity, while ensuring the measurement is independent of any pulse intensity variations. The algorithm considered a population of 60 individuals, crossover and mutation were included and we took care that no excessively steep gradients were applied to neighboring actuators across the membrane. In our case the spectrum was spread over approximately 8 actuators only and convergence was achieved after about 180 generations.
Just measuring the SHG intensity not adequate for characterising the resulting pulse duration, and we therefore use our FROG measurement device, which will be described in the following section, to fully characterise the temporal electric field of the pulse. Figure 7(a) shows a FROG measurement of the shortest pulses obtained with the above-mentioned procedure and (b) and (c) the retrieved temporal profile and spectrum. The shortest measured pulse duration is 67 fs with the transform limit supporting 57 fs, while the compressed energy at the output of the system for the shortest pulse was 3.8 μJ. We believe the discrepancy can be assigned to a

number of factors: firstly, the use of SHG conversion efficiency as the feedback to the genetic algorithm limits the discrimination between different pulse durations and may not provide an adequate guide to finally remove all spectral phase errors. An alternative is to first optimize the SHG, then use feedback via the measured pulse duration of the FROG. This provides a very strong criteria for the genetic algorithm, but is significantly slower. Additionally, the dispersed spectrum covers only 8 actuators in the current system, and changing the lenses of the compressor would allow this to be extended to cover more actuators, hence a more accurate addressing of the spectral phase.

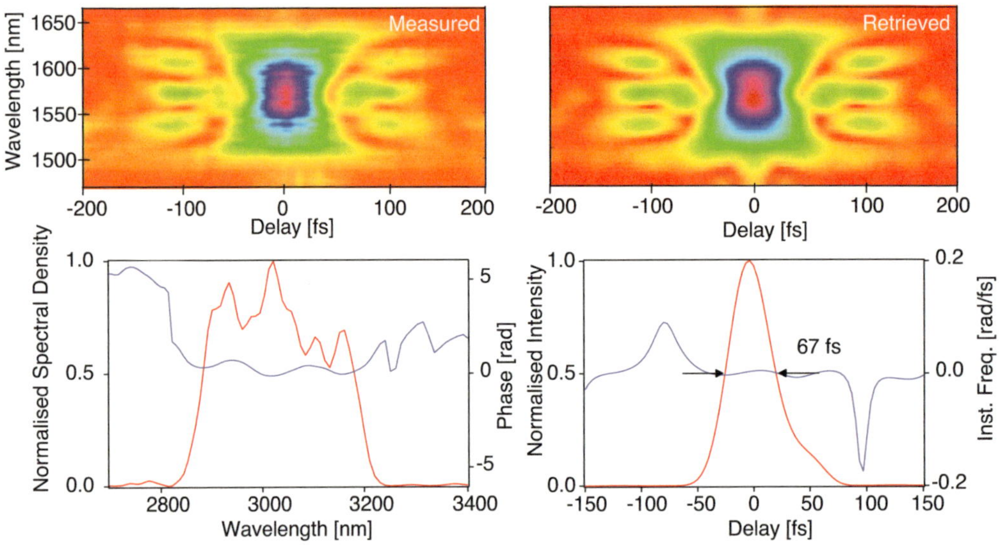

Fig. 7. Measured and retrieved SHG FROG traces of the compressed pulse (FROG Error =0.41%). Left: the measured FROG and retrieved spectrum and spectral phase. Right: the spectra after each nonlinear stage.

4. Mid-IR pulse characterisation

To use few-cycle pulses in any experiment, full and accurate characterisation of the pulse is essential, and measurement of e.g. just the temporal intensity of the pulse is not adequate. Techniques such as auto-or cross correlation can provide limited information, and for clear understanding and control of the pulses we require a measurement that resolves the temporal electric field structure of the pulse, such as frequency resolved optical gating (FROG) (Trebino (2000)) and spectral phase interferometry for the direct reconstruction of electric fields (SPIDER) (Iaconis and Walmsley (1998)). Below we describe a SHG-FROG characterization device for mid-infrared pulses, optimised for measurement of true few-cycle (< 20 fs) mid-IR pulses. It has a working bandwidth of 1000 nm, can resolve femtosecond timescale structures over a temporal range of 100 ps and does not suffer from time reversal ambiguity as is seen with other SHG-FROG systems. The detector allows us to have high spectral resolution and the combined system enables measurement of few-cycle to picosecond durations without reconfiguration.

Characterisation of pulses at centre wavelengths up to 2 μm is relatively straightforward using standard techniques but at longer wavelengths, e.g. in the mid-IR, characterisation becomes increasingly arduous. Array detectors are either unavailable, expensive, or come at very low resolution and require cryogenic cooling. Optics have to be carefully chosen to provide accurate response to the immense bandwidths; note that a 2 cycle pulse at 650 nm is associated with a $1/e^2$ bandwidth of 251 nm whereas it is 1212 nm at 3200 nm. While dispersion is low at 3 μm wavelengths, it contributes significantly to few-cycle pulses and has to be taken account of in the system design.

To date there have been several successful implementations of techniques to measure the full temporal electric field of ultrashort mid-IR pulses. The XFROG technique has characterized 13 fs mid-IR pulses (Fuji and Suzuki (2007)), however it requires a well synchronized ultrashort, precisely characterised laser pulse to use as a gate. Free-space electro-optic sampling (Grischkowsky et al. (1990)), also requires a synchronised shorter sampling pulse, which would have a duration of sub-10 fs if we used this technique at ~ 3 μm. For frequency converted Ti:Sa systems these pulses are often available, but with OPCPA the pulses would have to be generated through a complex frequency shifting process, and we may prefer a self-referencing technique such as polarisation-gating (PG-) or second-harmonic generation (SHG-) FROG, and SPIDER (Naganuma et al. (1989); Trebino (2000); Iaconis and Walmsley (1998)). SPIDER is suited for rapid acquisition (Kornelis et al. (2003)) whereas FROG has a simpler experimental arrangement and a wider temporal range. PG-FROG is unsuitable due to the finite response time of the nonlinear medium (Delong et al. (1995)) which limits its application for few-cycle pulses. SHG-FROG is suitable for those pulses (Akturk et al. (2004)); it can also cover a wide range of wavelengths and time durations, and is sensitive enough to be used with low energy pulses.

Temporal characterisation of a mid-IR FEL has been demonstrated via SHG-FROG (Richman et al. (1997)) for a 2 ps pulse with a small spectral width of ~40 nm, and a 25 fs OPA at 3.2 μm (Brida et al. (2008)). We present here a system whose range extends past both these pulse durations, while also operating over a wide range of central wavelengths.

The setup (Fig. 8) consists of only a Michelson interferometer and measures just 30×30 cm, for maximum stability and easy transportation. We use a pellicle beam splitter (Thorlabs) with negligible dispersion and a constant splitting ratio over the bandwidth, and a retroreflector mounted on a high resolution scanning stage (New Scale Technologies). All reflective optics are gold-coated to preserve the broad bandwidth. The stage can scan 15 mm with a resolution of 20 nm, corresponding to a time delay of 100 ps with 0.12 fs resolution. The corner cube gives a fixed lateral offset for a non-collinear, background free FROG signal

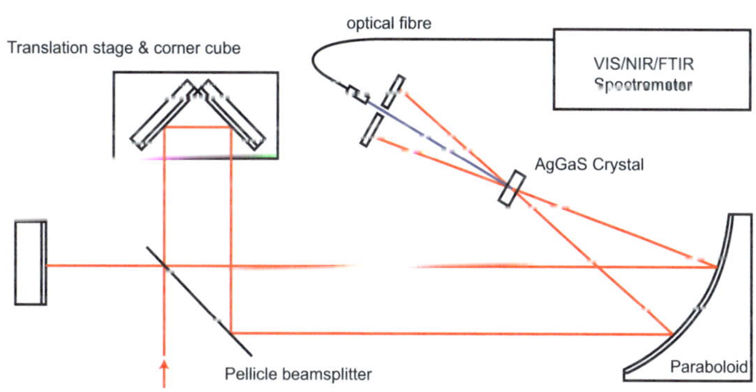

Fig. 8. **Layout of the SHG-FROG system** for mid-IR ultrashort pulse characterisation (see text).

while preventing any unwanted beam offset during long range scans. The two parallel output beams are focused with a 50mm focal length, off-axis parabolic mirror into the nonlinear crystal. The resulting non-collinear SHG is selected by an iris and coupled to a spectrometer via an optical fiber. This setup gives us the flexibility to measure broad bandwidths over a wide spectral range, with large high resolution time scans allowing accurate detection of pre- or post-pulses at picosecond delays, while simultaneously detecting a well-resolved femtosecond main pulse. Note that, even though this setup lends itself to single-shot measurements - at least up to 3500 nm fundamental wavelength - with InGaAs cameras, we refrain from doing so since the temporal dynamic range would be severely hampered by the SHG-FROG beam imaging geometry.

Crystal	Transparency Range (nm)	θ (deg)	d_{eff} (pm/V)	$\Delta\theta$ (mrad)	$\Delta\lambda$ (nm)
Ag_3AsS_3	600-13000	19.4	23.4	103.5	1100
$AgGaGeS_4$	700-13000	57.3	17.21	328	3400
$AgGaSe_2$	710-18000	76.5	35.1	1230	865
$AgGaS_2$	500-13000	35.7	9.11	312.5	2350
Ag_3SbS_3	700-14000	39.5	14.1	112	424
$Li{:}NbO_3$	330-5500	57.0	3.68	230.5	565
KTA (Type II e-axis)	330-4000	64.4	2.6	270	635

Table 1. Characteristics of SHG crystals at 3200 nm with 200 μm thickness. Data calculated with SNLO (Smith (2009)) θ is the phasematching angle, $\Delta\theta$ is the angular acceptance and $\Delta\lambda$ is the acceptance bandwidth following Smith (2009).

Table 1 shows a compilation of nonlinear crystals with a transparency range covering the fundamental as well as SH wavelengths, a suitably high d_{eff} for efficient conversion, and preferably Type I phase-matching to simplify the experimental layout. A pulse duration of 15 fs, or 1.5 cycles at 3 μm, requires a gaussian spectrum of ~1 μm FWHM, so we need to have substantially more than this to avoid strong spectral reshaping during SHG. A thin crystal is therefore warranted to support such immense bandwidths, but with the trade-off of reduced conversion efficiency. For a spot size of 35 μm, a five-cycle pulse with as low energy as 1 nJ results in an intensity of 1 GW/cm^2; this is sufficient to generate an adequate SH signal. We also need to accommodate a non-collinear angle of 100 mrad and an angular content due to focussing of 20 mrad for each of the generating beams.
KTA will phase-match in type II, and is limited by its transparency range and bandwidth, while neither the Ag_3AsS_3, $AgGaSe_2$, Ag_3SbS_3, nor the $Li{:}NbO_3$ have acceptance bandwidths that support few-cycle pulses. $AgGaGeS_4$ and $AgGaS_2$ (Silver Thiogallate) support a few-cycle pulse bandwidth, have adequate angular acceptance for our focussing geometry and a broad transparency range. The higher d_{eff} and broader bandwidth of the $AgGaGeS_4$ appears to make it a better candidate, however this crystal is not readily available, and so we resort to using $AgGaS_2$.
By exchanging the fibre-coupled spectrometer and crystal, we can use our FROG for 0.4-5 μm central wavelengths, with a Si-detector based spectrometer (Ocean Optics HR 4000) for SHG signals from near-IR radiation from 200-1100 nm, an InGaAs spectrometer (Ocean optics NIR 256) for SHG from 1000-2500 nm and a Fourier Transform Infra Red spectrometer (FTIR) (Oriel MIR8025) for 0.9-10 μm. The first two spectrometers provide quick response times, while the FTIR can measure over very large spectral ranges with extremely high resolution of up to 0.5 cm^{-1}. The spectrometers and motorised stage are

controlled through a home-built LabView program, which also reconstructs the pulse duration from the SHG-FROG trace using the PCGPA algorithm on a grid with sizes of up to 2048 x 2048 points. Such a large grid is necessary not only to measure and to retrieve with high enough resolution and to prevent aliasing but also to converge with a meaningfully small FROG error.

The first test measurement (Fig. 9) shows a richly featured mid-IR pulse from our OPCPA system. The center wavelength is at 3200 nm, the pulse energy is 1.2 μJ and its spectrum covers about 600 nm. We scanned the delay stage in steps of 6 fs, over a range of 3 ps, corresponding to 500 points of delay with a step size of 1 μm. High spectral resolution was ensured by the FTIR spectrometer, which recorded and summed 20 spectra at each delay point, sampled at a resolution of 1 cm^{-1}. To remove the SHG-FROG time ambiguity, we insert an 8 mm Silicon plate in the beam path and observe the sign of the phase change.

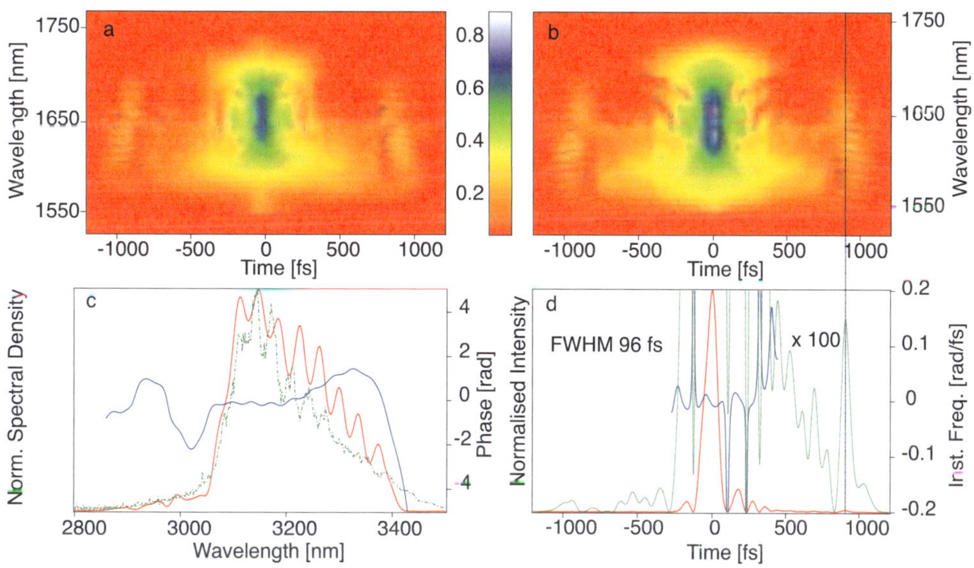

Fig. 9. **(a)**Measured FROG trace at 3.2 μm- colour shows the square root of intensity **(b)** Retrieved FROG trace. **(c)** Retrieved (solid) & measured (dotted) spectrum and phase. **(d)** Retrieved temporal pulse - also shown magnified ×100 (dotted) to highlight the well-resolved post pulse at 900 fs

Figure 9 a) shows the measured FROG trace with many complex femtosecond features across a >2 ps timescale. The long scan range and excellent signal to noise ratio ensures the detection of the post pulses at ~1 ps, while the high resolution FTIR captures the corresponding spectral fringes. The low intensity features of the trace are both detected and retrieved accurately as shown in Fig. 9 b). In contrast, a measurement with the low-resolution InGaAs spectrometer captures neither the post pulse nor the fringing. Reconstruction with high spectral and temporal resolution requires the use of a large grid size (2048x2048), and we achieve excellent agreement, with FROG error 0.0025. The retrieved pulses have a duration of 96 fs, corresponding to 9.0 cycles at 3.2 μm, and these pulses cover a 45 fs, 4.2 cycle transform limited bandwidth (Fig. 9 c) and d).

As a further test we measured a ~100 fs duration and 3.2 μJ energy pulse, centered at 3250 nm, before and after passing through a 1 cm thick sapphire plate. The delay stage was scanned in 400 steps of 6 fs for the directly measured pulse, and in 500 steps of 13 fs for the one with added dispersion; while integrating over 6 spectra for each delay point. Figure 10 a) & b) show the measured FROG traces, Fig. 10 c) & d) show the retrieved temporal profiles and instantaneous frequency, while Fig. 10 e) & f) show the reconstructed spectrum and spectral phase. The plate adds negative dispersion, and increases the retrieved pulse duration from 110 fs to 153 fs. We have calculated the expected phase for the pulse propagated through the sapphire, by adding the calculated phase of the sapphire to the retrieved phase of the undispersed pulse. This is plotted as a dotted line in Fig. 10 f), and we

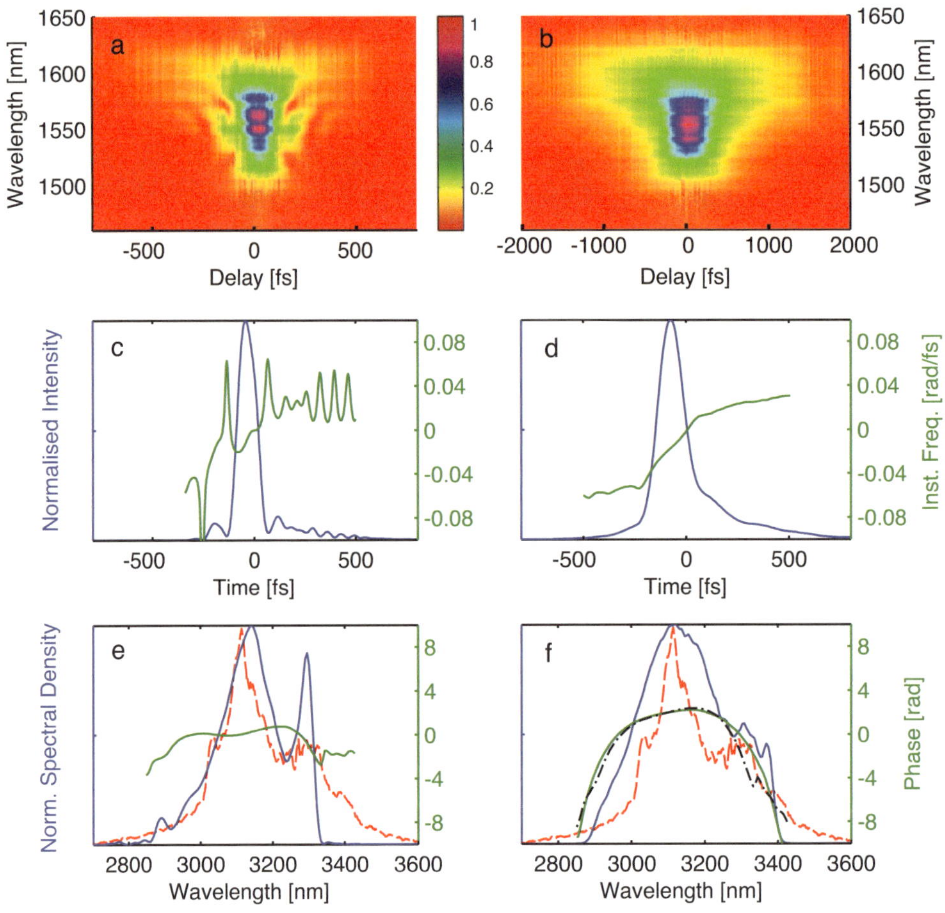

Fig. 10. **(a)** FROG trace of undispersed pulse - colour shows the square root of intensity. **(b)** FROG trace after propagation through 1 cm sapphire (note different timescale). **(c)** Retrieved intensity and instantaneous frequency (right scale) from **a**. **(d)** Retrieved intensity and instantaneous frequency from **b**. **(e)** Retrieved (solid) & measured (dashed) spectrum (left scale) and spectral phase (right scale) of unchirped pulse. **(f)** Retrieved & measured spectrum and spectral phase after 1 cm sapphire. The dash-dot line shows the calculation of the expected spectral phase

can see that there is good agreement - the overall shape is accurately retrieved and the deviation of the spectral phase is less than a radian from 2850-3270 nm. We note that calculation of the phase in this region is difficult due to lack of reliable values for material refractive indices, and further the difference in the temporal domain from the calculated and measured phases corresponds to just over half an optical cycle, or 5.4 fs.

4.1 Stability

We monitored the power of the compressed pulse over half an hour by sending the full beam onto a pyroelectric detector from OPHIR (3A head, acquisition rate 3Hz) and measured a power stability of 0.75% RMS over 30 minutes - see also Fig. 11. The stability is excellent and - to our knowledge - signifies the lowest jitter observed from any OPCPA system, which exhibit from 1.5% RMS fluctuations and higher (Tavella et al. (2010); Ishii et al. (2005)). We note that our measurement represents a low-pass filtered result due to the inadequate response times of the detector, especially for the 100 kHz repetition rate. Such an analysis is analog to most measurements quoted in the literature but higher peak to peak fluctuations might exist which we plan to investigate in a future measurement. The inset in Fig. 11 shows a knife edge measurement of the compressed pulse and the high spatial quality, which is sufficient for focussing close to the diffraction limit.

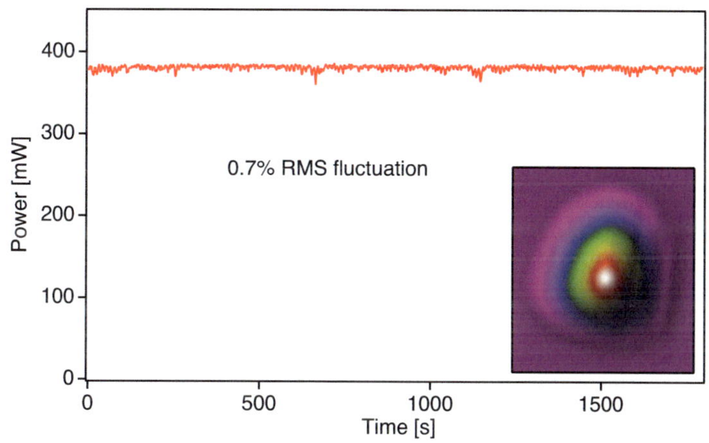

Fig. 11. **Stability of the system and profile.** The stability of the OPCPA system after compression is excellent with fluctuation under 0.75% over 30 minutes. Insert is the beam profile after compression

5. Conclusion

Sources of ultrashort pulses in the mid-IR are in high demand, as can be seen from the many examples we have presented here, from a range of very different fields: life sciences to strong field physics. OPCPA is a technique which frees many operating parameters from the restrictions of the current CPA based systems. We have, as example, presented a source which is capable of satisfying the needs of a wide variety of fields: some require high repetition rate, others few cycle pulses, yet others CEP stability, while strong field physics ideally needs all of these features combined. This system currently signifies the highest energy shortest pulse OPCPA system in the mid-IR with compressed pulse energy of 3.8 μJ at 100 kHz, a pulse duration of 67 fs (6.3 cycles) and a central wavelength at 3.2 μm. This system is in principle CEP stable, and offers an important proof of principle, showing that OPCPA can be used to operate at extremely high average powers in the mid-IR.

We have also described a broad-bandwidth, high resolution SHG-FROG device, for few-cycle pulse measurement in the mid-IR. The system is capable of measuring pulses over a wide wavelength range spanning from 800 nm to 5 μm and for large temporal ranges up to 100 ps. This system has been used to measure a complex sub-10-cycle (sub-100-fs) mid-IR pulse and also to retrieve the change introduced in a ~100 fs mid-IR pulse by the dispersion of a sapphire plate. The retrieved and experimental FROG traces show rich detail and excellent agreement. The ability to characterise few-cycle mid-IR laser pulses is a pre-requisite not only for optimization of such laser systems, but also for the application of ultrashort mid-IR pulses to experiments in the described wide range of fields.

Ultrafast pulsed light sources in the mid-IR have a very bright future, due to growing interest in this area combined with a growing range of applications. We have presented here a prototype for the next generation of ultrashort mid-IR sources. Our novel source, with its combination of passively stabilised CEP and broad bandwidth OPCPA, provides an ideal template for the development of high energy, high repetition rate mid-IR systems. The use of OPCPA should in the near future allow the production of few-cycle pulse durations without the need for nonlinear broadening, while the high rep rate and stability already are ideal for ultrafast spectroscopy. OPCPA is a scaleable technique, and the pulses from this system could easily be increased to mJ level at lower repetition rates, by including additional OPCPA stages pumped by high energy, commercially available pump lasers. The availability of such sources will open up a new regime of strong field and spectroscopic experiments driven by mid-IR pulses.

6. References

Qian Peng, Asta Juzeniene, Jiyao Chen, Lars O Svaasand, Trond Warloe, Karl-Erik Giercksky, and Johan Moan. Lasers in medicine. *Reports on Progress in Physics*, 71(5):056701, 2008.

A. Mcpherson, G. Gibson,H. Jara, U. Johann, T. S. Luk, I. Mcintyre, K. Boyer, and C. K. Rhodes. Studies of multiphoton production of vacuum-ultraviolet radiation in the rare gases. *J. Opt. Soc. Am. B*, 4:595, 1987.

M. Ferray, A. L'Huillier, X. F. Li, L. A. Lompr, G. Mainfray, and C. Manus. Multiple-harmonic conversion of 1064 nm radiation in rare gases. *J. Phys. B*, 21:L31–L35, 1988.

B Sheehy, J.d.d.Martin, L. F. Dimauro, P Agostini, K. J. Schafer, M Gaarde, and K. C. Kulander. High harmonic generation at long wavelengths. *Phys. Rev. Lett.*, 83:5270, 1999.

A. Gordon and F.X. Kaertner. Scaling of kev hhg photon yield with drive wavelength. *Opt. Exp.*, 13(8):2941–2947, 2005.

J. Tate, T. Auguste, H. G. Muller, P. Salieres, P. Agostini, and L. F. DiMauro. Scaling of wave-packet dynamics in an intense midinfrared field. *Physical Review Letters*, 98(1):013901, January 2007.

Tenio Popmintchev, Ming-Chang Chen, Oren Cohen, Michael E. Grisham, Jorge J. Rocca, Margaret M. Murnane, and Henry C. Kapteyn. Extended phase matching of high harmonics driven by mid-infrared light. *Opt. Lett.*, 33(18):2128–2130, 2008.

R. Moshammer, M Unverzagt, W Schmitt, J Ullrich, and H Schmidt-Bocking. A 4 pi recoil-ion electron momentum analyzer: a high-resolution "microscope" for the investigation of the dynamics of atomic, molecular and nuclear reactions. *Nuc. Instrum. Meth. Phys. Res. Sec. B*, 108(4):425–445, 1996.

M. J. Thorpe and J. Ye. Cavity-enhanced direct frequency comb spectroscopy. *Applied Physics B-Lasers And Optics*, 91(3-4):397–414, June 2008.

M Nisoli, S Stagira, S De Silvestri, O Svelto, S Sartania, Z Cheng, G Tempea, C Spielmann, and F Krausz. Toward a terawatt-scale sub-10-fs laser technology. *IEEE J. Sel. Top. Quant. Electron.*, 4(2):414–420, 1998.

C.P. Hauri, W. Kornelis, F.w. Helbing, a. Heinrich, a. Couairon, a. Mysyrowicz, J. Biegert, and U. Keller. Generation of intense, carrier-envelope phase-locked few-cycle laser pulses through filamentation. *Appl. Phys. B*, 79(6):673–677, 2004.

B Schenkel, J Biegert, U Keller, C Vozzi, M Nisoli, G Sansone, S Stagira, S De Silvestri, and OSvelto. Generation of 3.8-fs pulses froma daptive compression of a cascaded hollow fiber supercontinuum. *OPTICS LETTERS*, 28(20):1987–1989, OCT 15 2003.

T. Wilhelm, J. Piel, and E. Riedle. Sub-20-fs pulses tunable across the visible from a blue-pumped single-pass noncollinear parametric converter. *Opt. Lett.*, 22(19):1494–1496, 1997.

Christian Schriever, Stefan Lochbrunner, Patrizia Krok, and Eberhard Riedle. Tunable pulses from below 300 to 970 nm with durations down to 14 fs based on a 2 mhz ytterbiumdoped fiber system. *Opt. Lett.*, 33(2):192–194, 2008.

C. Vozzi, G. Cirmi, C. Manzoni, E. Benedetti, F. Calegari, G. Sansone, S. Stagira, O. Svelto, S. De Silvestri, M. Nisoli, and G. Cerullo. High-energy, few-optical-cycle pulses at 1.5 mu m with passive carrier-envelope phase stabilization. *Optics Express*, 14(21): 10109–10116, October 2006.

G. Cirmi, C.Manzoni, D. Brida, S. De Silvestri, and G. Cerullo. Carrier-envelope phase stable, few-optical-cycle pulses tunable from visible to near ir. *Journal Of The Optical Society Of America B-Optical Physics*, 25(7):B62–B69, July 2008.

Chunmei Zhang, Pengfei Wei, Yansui Huang, Yuxin Leng, Yinghui Zheng, Zhinan Zeng, Ruxin Li, and Zhizhan Xu. Tunable phase-stabilized infrared optical parametric amplifier for high-order harmonic generation. *Opt. Lett.*, 34(18):2730–2732, 2009.

D. Brida, M. Marangoni, C. Manzoni, S. De Silvestri, and G. Cerullo. Two-optical-cycle pulses in the mid-infrared from an optical parametric amplifier. *Optics Letters*, 33(24):2901– 2903, December 2008.

Erik T. J. Nibbering and Thomas Elsaesser. Ultrafast vibrational dynamics of hydrogen bonds in the condensed phase. *Chemical Reviews*, 104(4):1887–1914, April 2004.

F. Rotermund, V. Petrov, F. Noack, M. Wittmann, and G. Korn. Laser-diode-seeded operation of a femtosecond optical parametric amplifier with mgo:linbo3 and generation of 5- cycle pulses near 3 μm. *J. Opt. Soc. Am. B*, 16(9):1539–1545, 1999.

T. Fuji, N. Ishii, C. Y. Teisset, X. Gu, T. Metzger, A. Baltuska, N. Forget, D. Kaplan, A. Galvanauskas, and F. Krausz. Parametric amplification of few-cycle carrier-envelope phase-stable pulses at 2.1 mu m. *Optics Letters*, 31(8):1103–1105, April 2006.

S. Adachi, N. Ishii, T. Kanai, A. Kosuge, J. Itatani, Y. Kobayashi, D. Yoshitomi, K. Torizuka, and S. Watanabe. 5-fs, multi-mj, cep-locked parametric chirped-pulse amplifier pumped by a 450-nm source at 1 khz. *Optics Express*, 16(19):14341–14352, September 2008.

O. Chalus, P. K. Bates, M. Smolarski, and J. Biegert. Mid-ir short-pulse opcpa with micro-joule energy at 100 khz. *Optics Express*, 17(5):3587–3594, March 2009.

O. V. Chekhlov, J. L. Collier, I.N. Ross, P. K. Bates, M. Notley, C. Hernandez-Gomez, W. Shaikh, C. N. Danson, D. Neely, P. Matousek, and S. Hancock. 35 j broadband femtosecond optical parametric chirped pulse amplification system. *Optics Letters*, 31(24):3665- 3667, December 2006.

V. V. Lozhkarev, G. I. Freidman, V. N. Ginzburg, E. V. Katin, E. A. Khazanov, A. V. Kirsanov, G. A. Luchinin, A. N. Mal'shakov, M. A. Martyanov, O. V. Palashov, A. K. Poteomkin, A. M. Sergeev, A. A. Shaykin, I. V. Yakovlev, S. G. Garanin, S. A. Sukharev, N. N. Rukavishnikov, A. V. Charukhchev, R. R. Gerke, and V. E. Yashin. 200 tw 45 fs laser based on optical parametric chirped pulse amplification. *Optics Express*, 14(1):446- 454, January 2006.

M. Gertsvolf, H. Jean-Ruel, P. P. Rajeev, D. D. Klug, D. M. Rayner, and P. B. Corkum. Orientation-dependent multiphoton ionization in wide band gap crystals. *Phys. Rev. Lett.*, 101(24):243001, Dec 2008.

F. Tavella, A. Willner, J. Rothhardt, S.Hädrich, E. Seise, S. D¨usterer, T. Tschentscher, H. Schlarb, J. Feldhaus, J. Limpert, A. Tünnermann, and J. Rossbach. Fiber-amplifier pumped high average power few-cycle pulse non-collinearopcpa. *Opt. Express*, 18(5):4689- 4694, 2010.

N. Ishii, L. Turi, V. S. Yakovlev, T. Fuji, F. Krausz, A. Baltuska, R. Butkus, G. Veitas, V. Smilgevicius, R. Danielius, and A. Piskarskas. Multimillijoule chirped parametric amplification of few-cycle pulses. *Optics Letters*, 30(5):567–569, March 2005.

F. Roser, J. Rothhard, B. Ortac, A. Liem, O. Schmidt, T. Schreiber, J. Limpert, and A. Tunnermann. 131 w 220 fs fiber laser systems. *Optics Letters*, 30(20):2754–2756, October 2005.

Johan Boullet, Yoann Zaouter, Jens Limpert, Stéphane Petit, Yann Mairesse, Baptiste Fabre, Julien Higuet, Eric Mével, Eric Constant, and Eric Cormier. High-order harmonic generation at a megahertz-level repetition rate directly driven by an ytterbium-doped- fiber chirped-pulse amplification system. *Opt. Lett.*, 34(9):1489–1491, 2009.

A Baltuska, T Fuji, and T Kobayashi. Controlling the carrier-envelope phase of ultrashort light pulses with optical parametric amplifiers. *Phys. Rev. Lett.*, 88(13):133901, 2002.

A. Renault, D. Z. Kandula, S. Witte, A. L. Wolf, R. T. Zinkstok, W. Hogervorst, and K. S. E. Eikema. Phase stability of terawatt-class ultrabroadband parametric amplification. *Optics Letters*, 32(16):2363–2365, August 2007.

F. Adler, A. Sell, F. Sotier, R. Huber, and A. Leitenstorfer. Attosecond relative timing jitter and 13 fs tunable pulses froma two-branch er: fiber laser. *Optics Letters*, 32(24):3504–3506, December 2007.

O. Chalus, P. K. Bates, and J. Biegert. Design and simulation of few-cycle optical parametric chirped pulse amplification at mid-ir wavelengths. *Optics Express*, 16(26):21297–21304, December 2008.

Jeffrey Moses, Cristian Manzoni, Shu-Wei Huang, Giulio Cerullo, and Franz X. Kaertner. Temporal optimization ofultrabroadband high-energy opcpa. *Opt. Express*, 17(7):5540- 5555, 2009.

C. Y. Teisset, N. Ishii, T. Fuji, T. Metzger, S. Kohler, R. Holzwarth, A. Baltuska, A.M. Zheltikov, and F. Krausz. Soliton-based pump-seed synchronization for few-cycle opcpa. *Optics Express*, 13(17):6550–6557, AUG 22 2005.

S. Witte, R. T. Zinkstok, W. Hogervorst, and K. S. E. Eikema. Generation of few-cycle terawatt light pulses using optical parametric chirped pulse amplification. *Optics Express*, 13 (13):4903–4908, Jun 27 2005.

I. N. Ross, G. H. C. New, and P. K. Bates. Contrast limitation due to pump noise in an optical parametric chirped pulse amplification system. *Optics Communications*, 273(2):510- 514, May 2007.

F. Tavella, K. Schmid, N. Ishii, A. Marcinkeviius, L. Veisz, and F. Krausz. High-dynamic range pulse-contrast measurements of a broadband optical parametric chirped-pulse amplifier. *Applied Physics B: Lasers and Optics*, 81(6):753–756, October 2005.

M. Kakehata, H. Takada, Y. Kobayashi, K. Torizuka, H. Takamiya, K. Nishijima, T. Homma, H. Takahashi, K. Okubo, S. Nakamura, and Y. Koyamada. Carrier-envelope-phase stabilized chirped-pulse amplification system scalable to higher pulse energies. *Optics Express*, 12(10):2070–2080, May 2004.

R. Trebino. *Frequency-Resolved Optical Gating*. Kluwer Academic Publishers, 2000. C. Iaconis and I. A. Walmsley. Spectral phase interferometry for direct electric-field reconstruction of ultrashort optical pulses. *Optics Letters*, 23(10):792–794, May 1998.

T. Fuji and T. Suzuki. Generation of sub-two-cycle mid-infrared pulses by four-wave mixing through filamentation in air. *Optics Letters*, 32(22):3330–3332, November 2007.

D. GRISCHKOWSKY, S. KEIDING, M. VANEXTER, and C. FATTINGER. Far-infrared time-domain spectroscopy with terahertz beams of dielectrics and semiconductors. *Journal Of The Optical Society Of America B Optical Physics*, 7(10):2006–2015, October 1990.

K NAGANUMA, K. MOGI, and H. YAMADA. General-method for ultrashort light-pulse chirp measurement. *Ieee Journal Of Quantum Electronics*, 25(6):1225–1233, June 1989.

W. Kornelis, J. Biegert, J. W. G. Tisch, M. Nisoli, G. Sansone, C. Vozzi, S. De Silvestri, and U. Keller. Single-shot kilohertz characterization of ultrashort pulses by spectral phase interferometry for direct electric-field reconstruction. *Optics Letters*, 28(4):281–283, February 2003.

K. W. DELONG, C. L. LADERA, R. TREBINO, B. KOHLER, and K. R. WILSON. Ultrashortpulse measurement using noninstantaneous nonlinearities - raman effects in frequency-resolved optical gating. *Optics Letters*, 20(5):486–488, March 1995.

Selcuk Akturk, Mark Kimmel, and Rick Trebino. Extremely simple device for measuring 1.5 micron ultrashort laser pulses. *Opt. Express*, 12(19):4483–4489, 2004.

B. A. Richman, M. A. Krumbugel, and R. Trebino. Temporal characterization of mid-ir free-electron-laser pulses by frequency-resolved optical gating. *Optics Letters*, 22(10):721- 723, May 1997.

A. Smith. Snlo version 5.0, nonlinear optics code. available from A.V. Smith, AS-Photonics, Albuquerque, N.M., 2009.

14

Contrast Improvement of Relativistic Few-Cycle Light Pulses

László Veisz
Max-Planck-Institut für Quantenoptik
Germany

1. Introduction

Laser-plasma interaction has been a widespread and popular field of research since the birth of the laser. It involves numerous interesting phenomena such as inertial confinement fusion; generation and acceleration of electron, positron, neutron, proton and ion particle beams; source of electro-magnetic radiation in the IR, visible, UV and x-ray range, just to mention some important applications. The continuously increasing power and the development of chirped pulse amplification lead to electric fields in which oscillating free electrons reach almost the speed of light. This so called relativistic optics regime starts at intensities about 10^{18} W/cm^2 at 1 μm wavelength. The interaction of relativistically intense laser pulses with solid state -possibly liquid- target material is always accompanied by the creation of inhomogeneous plasmas, as a given amount of extended preplasma is formed in front of the target before the main laser pulse arrives. This preplasma is produced by undesired laser light impinging onto the target before the short main pulse. Its density decreases rapidly farther from the target due to the hydrodynamic expansion of the hot plasma during the time till the interaction pulse comes. The extension of the preplasma significantly influences the interaction and leads to completely different types of processes depending on its value. Typical lasers produce an extended preplasma before the main pulse, because they amplify pre- and postpulses and pedestals from various origins. Therefore a certain types of processes involving interaction of intense laser pulse with extended low density plasmas dominate these investigations. On the other hand, practically preplasma-free environment is required by various relativistic laser-plasma experiments involving ultra-intense laser and high-density plasma interaction like surface high harmonic generation Monot et al. (2004); Thaury et al. (2007), laser-driven proton and ion acceleration Hegelich et al. (2006) or laser interaction with few nanometer thick diamond-like-carbon foils, so called nanofoils. This preplasma formation is influenced by the contrast, which characterizes the laser intensity before and after the main pulse relative to this pulse. Therefore precise characterization and improvement of the contrast are essential to successfully conduct these types of experiments.

Laser pulses with a duration of only a few optical cycles at moderate intensities opened up the new era of attosecond physics Brabec & Krausz (2000); Krausz & Ivanov (2009) via the controlled reproducible generation of weak XUV pulses with a duration of hundred(s) of attoseconds precisely synchronized to the laser pulse. These pulses allow the investigation of electron motion in atoms, molecules and solid states in an XUV pump and visible / near infrared probe configuration. Relativistic intensity few-cycle sources hold the promise to generate XUV pulses with unprecedented energy and thus form the basis of the novel research

field of XUV pump XUV probe investigations with single attosecond bursts.
This chapter concentrates on the contrast enhancement of few-cycle light sources. The main topics discussed are the characterization of laser contrast, implementation and application of contrast enhancement techniques. It should be noted already here that these methods are perfectly suited for longer -from many optical cycle long up to picosecond- laser pulses as well. Two successful and efficient techniques, the plasma mirror and the optical parametric (chirped pulse) amplification, were applied for the generation of few-cycle pulses with extreme contrast. The first section starts with the introduction of the most important physical quantities and overview of the various techniques followed by a description of the above mentioned two methods and also the cross-polarized wave generation (XPW). This section closes with the description of contrast measurement techniques. The results in the next section present realization, characterization and some practical tips to contrast enhancement. Discussion, future work and conclusions are at the end of this chapter.

2. Overview and methods

In the first part of this section the most important parameters are summarized and quantified relevant to the characterization of the high-dynamic range contrast of ultrahigh-intensity laser systems and the contrast requirements by experiments are discussed. Thereafter the most general contrast enhancement techniques are shortly introduced with their advantages and shortcomings.

2.1 Overview of the important physical phenomena

The interaction of relativistic intensity - $I \geq 10^{18}$ W/cm^2 - laser pulses with solid or liquid targets results in the ionization of the target material much before the main intense pulse, changing the plasma properties and the interaction itself as mentioned in the introduction. The generated plasma is heated rapidly to hot temperatures and starts to expand creating typically an exponential plasma density ramp in front of the target as shown in Fig. 1. A fully ionized solid state target has a maximal electron plasma density (n_{max}) of about 10^{24} cm^{-3}.

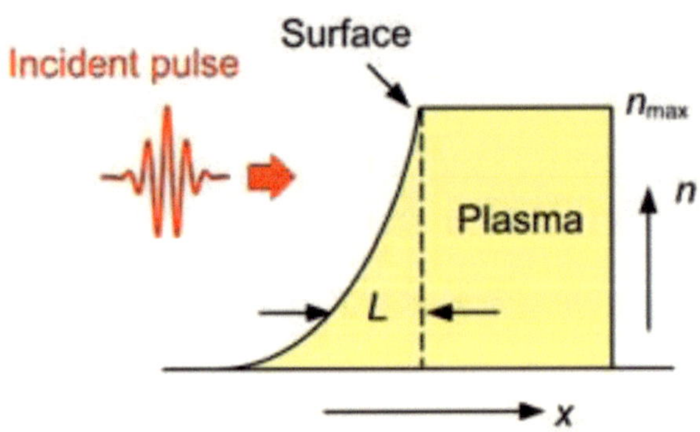

Fig. 1. Typical plasma density (n) from laser irradiated solid state or liquid targets as a function of the perpendicular coordinate to the surface (x). As the maximal electron density (n_{max}) corresponding to the solid density is much higher than the critical density, the light is reflected and interacts with the preplasma characterized by its extension (L), the so-called electron density scale length.

The critical density -where the plasma frequency reaches the laser frequency- is a very specific electron density:

$$n_c = \omega_0^2 m_e \varepsilon_0 / e^2; \qquad n_c(cm^{-3}) = 1.1 \times 10^{21} \lambda^{-2}(\mu m) \tag{1}$$

where λ, $\omega_0 = 2\pi c/\lambda$ are the laser wavelength and frequency, c is the speed of light in vacuum, ε_0 is the vacuum permittivity and $2m_e$, e are the electron mass and charge, respectively. The critical density has a value of 1.7×10^{21} cm^{-3} at $\lambda = 800$ nm laser wavelength indicating that the plasma near to a typical target gets a few 100 times "overdense" which means that the plasma density correspondingly exceeds the critical density. The light propagating in an inhomogeneous plasma upward the density gradient is reflected and partially absorbed at the critical density depending on the preplasma extension and cannot penetrate deeper to even higher densities Kruer (1988). The preplasma extension has an important role and it is characterized by the so-called *electron plasma density scale length*:

$$L = \left| n_e \left(\frac{dn_e}{dx} \right)^{-1} \right|_{x_0} \tag{2}$$

which is generally taken at the critical density ($x_0 = x_{cr}$). It is usual to normalize this value to the laser wavelength as this is the real physical measure of the plasma. The laser electric field is evanescent in the overdense plasma and its penetration, i.e. how high densities will the laser light reach, is characterized by this normalized quantity. Typically a normalized scale length of 0.1 is a good value allowing efficient implementation of surface high harmonic generation or ion acceleration, around 1 it is an average value suppressing / prohibiting the previous experiments and significantly higher than 1 is poor value involving mainly laser interaction with underdense plasmas like in gas targets and generation of electrons and bremsstrahlung. The normalized density scale length depends on the target material and the temporal structure and amount of laser light impinging onto the target before the main pulse. The high-dynamic-range temporal structure of laser light is described by the contrast, which is defined by the ratio of the laser main pulse intensity to the intensity at a given time instant. The reciprocal value is also used as contrast for example in the autocorrelation measurement. Fig. 2. depicts a typical temporal structure of ultrahigh intensity laser pulses. Sources that generate these pulses are based on the principle of chirped pulse amplification (CPA) Strickland & Mourou (1985); Yanovsky et al. (2008) or optical parametric chirped pulse amplification (OPCPA) systems Herrmann et al. (2009); Lozhkarev et al. (2007); Hernandez-Gomez et al. (2010). The temporal structure contains pedestals -long background pulses of different duration-, pre- and postpulses -short weak pulses- and the "foot" of the main laser pulse. Nanosecond pedestals with a contrast of 10^{-5} to 10^{-8} -or 5 to 8 orders of magnitude- can for example originate from amplified spontaneous emission (ASE) in the CPA laser amplifier; while the pre- and postpulses with typically 2 to 6 orders of magnitude contrast come from birefringence of optical elements, double internal reflections in optics or amplification of a previous oscillator / regenerative amplifier pulses and have temporal distance of a few 100 fs to many ns to the main pulse. A typical source of the foot of the pulse with 3 to 6 orders of magnitude contrast is the imperfect compression, but the noisy spectrum itself from amplification might also be a source.
The damage threshold is approximately the start of irreversible processes and plasma formation. It depends on the target material and laser pulse duration and varies from 10^{10} W/cm^2 for ns pulses to 10^{13} W/cm^2 for fs pulses in dielectrics Stuart et al. (1995; 1996); Tien et al. (1999) and it can reach even lower values for metals and liquids.

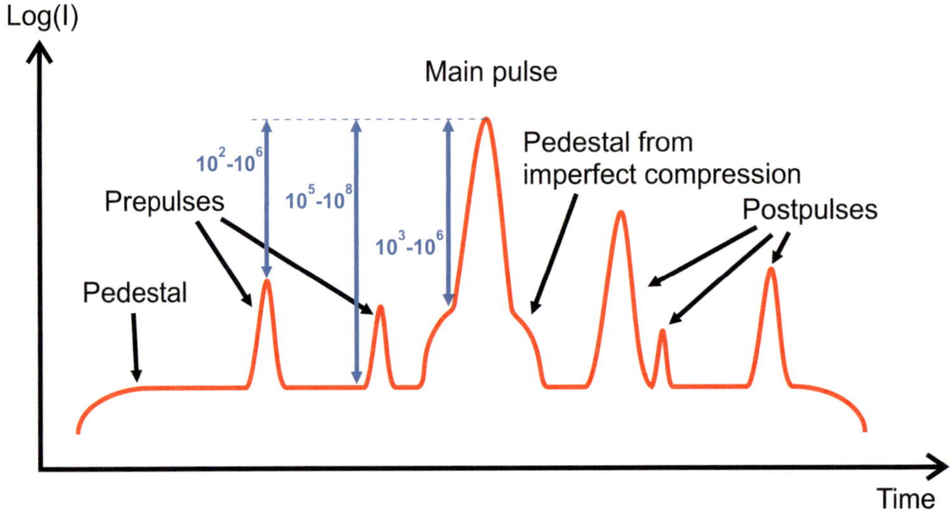

Fig. 2. Typical temporal structure of a laser pulse from a chirped-pulse amplification (CPA) or optical parametric chirped pulse amplification (OPCPA) system. Aside from the main pulse, it generally has pre- and postpulses. Pedestals usually arise from amplified spontaneous emission (ASE) and imperfect compression. Contrast of the pulse is defined by the ratio between the intensities of the main pulse and prepulses/pedestal. Typical values of the contrast are indicated for prepulses and pedestals.

Correspondingly, $10^{-8} - 10^{-12}$ contrast is required for preplasma sensitive experiments with presently reachable peak intensities of $10^{18} - 10^{22}$ W/cm^2 Yanovsky et al. (2008). The current demand of typically 8 - 12 orders of magnitude contrast on laser systems is already very high. Future multi-PW systems have even more enormous expectations of 12 - 14 orders of magnitude contrast and in the same time an ultra-broad bandwidth requirement supporting even few-cycle pulse duration. Nowadays widespread chirped pulse amplification (CPA) lasers and newly appearing optical parametric chirped pulse amplification (OPCPA) systems without extra contrast improvement technology are not capable to fulfill these requirements.

2.2 Summary of the relevant technologies

As the required contrast is higher than the typically available one various *contrast enhancement techniques* have been developed. Here we shortly summarize the most known methods that can be divided into three categories. The first category contains various approaches that work with only a limited input energy -on the order of 1 mJ- and have relative low conversion efficiency -generally 10%-, but enhance the contrast by a higher value -from 3 to 9 orders of magnitude maximal improvement. These techniques are not applicable to the final laser output due to their limitations therefore they are applied in a double chirped pulse amplification scheme Kalashnikov et al. (2005), where after the first amplifiers the pulses are compressed cleaned and later stretched, amplified and compressed a second time. The following approaches belong to this category:

- non-linear Sagnac interferometer Renault et al. (2005). It applies a Sagnac interferometer -containing a beam splitter and two mirrors- a filter and a Kerr nonlinear medium. The efficiency is about 10-20 %, contrast enhancement is 4 order of magnitude (OOM), the quality of output profile is decreased.

- saturable absorber Itatani et al. (1998); Kiriyama et al. (2010), is applied after some preamplifier stages at the μJ energy level in compressed pulses with $\sim$ 20% efficiency. The contrast is bettered by only 2 OOM.
- elliptical polarization rotation in air Homoelle et al. (2002); Jullien et al. (2004); Kalashnikov et al. (2004) is based on the nonlinear induced birefringence. The efficiency is 25-50 % depending on the configuration -single vs. multi pass- and on the chirp of the input pulse and the contrast enhancement is 3-4 OOM. The stability of the output pulses is reduced.
- cross-polarized wave generation Jullien et al. (2005); Chvykov et al. (2006). This technique is discussed in more detail in the next section.
- self-diffraction process in a Kerr nonlinear medium Liu & Kobayashi (2010) was suggested as this third order nonlinearity generates the cleaned beam propagating in a new different direction than the input beams. It has about 10 % efficiency and an improvement of 4.5 OOM corresponding to cubic temporal intensity dependence has been demonstrated with a potential of even better values.

The second category of contrast enhancement techniques have no input energy limitation, have higher efficiency -in the range of 35-80%- and consequently are applicable at the end of the laser systems. Therefore they improve the contrast on target, i.e. prepulses and pedestals even from compression is improved. The drawback of these approaches is that their efficiency influences directly the final energy of the system and it cannot be regained by further amplification. One of the oldest idea to improve the contrast is the frequency doubling of the pulses Marcinkevičius et al. (2004); Yuan et al. (2010). Theoretically the contrast is squared applying frequency doubling in typically KDP crystals, but the energy conversion efficiency is limited (35%-80%), the quality of the beam profile and wavefront is decreased. The effort to increase the efficiency requires thick crystals that limit the bandwidth and so the lowest pulse duration to above approx. 100 fs and reduces the contrast improvement around the main pulse. Plasma mirrors in a single / double configuration are another example of this category having 70 / 50% efficiency, 2-3 orders of magnitude improvement and better spatial beam profile.
The third category includes the optical parametric amplification (OPA) and optical parametric chirped pulse amplification (OPCPA) based methods. They are connected to the amplification and are not dedicated only to the contrast and so are not always classified in the two previous categories. The following examples are based on OP(CP)A:

- OP(CP)A preamplifier stage is used for idler generation Shah et al. (2009); Lozhkarev et al. (2006). At the beginning of a ultrahigh-intensity system a non-collinear OPA / OPCPA stage is installed. The idler is used for further amplification from this stage, which is not generated just when the pump and the seed are also present. Very high contrast improvement factors are realized, but the angular chirp must be (pre-)compensated and so it is not practical for few-cycle pulses.
- non-saturated OP(CP)A preamplifier stage substituting complex laser amplifiers in hybrid OPCPA CPA laser systems Gaul et al. (2010); Kiriyama et al. (2008). This technique applying typically a nanosecond pump laser is generally not improving the contrast just preserving a good input value and substitutes a large laser preamplifiers that reduce the contrast.
- OPCPA system with short pump pulse duration (from 1 ps up to $\sim$100 ps) Herrmann et al. (2009); Gu et al. (2009); Major et al. (2009); Dorrer et al. (2007). This technique is discussed in more detail in the next section.

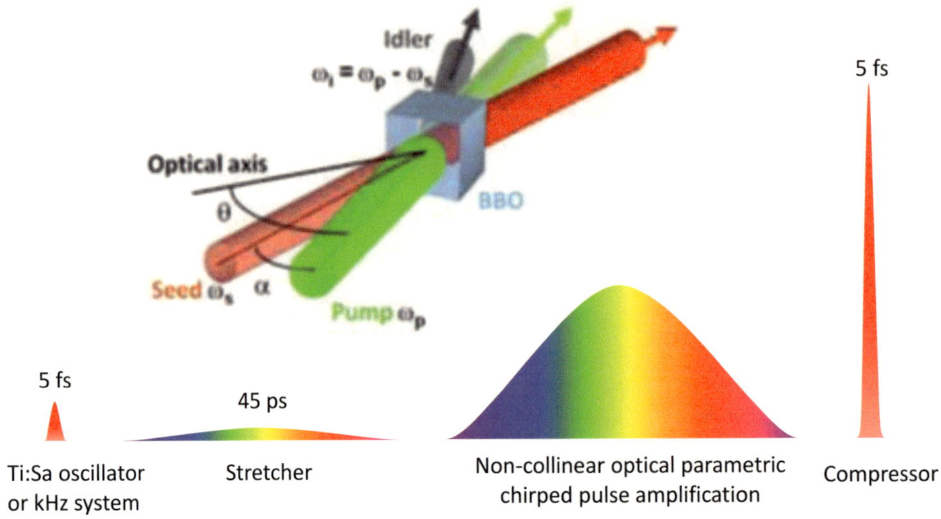

Fig. 3. Schematics of an optical parametric chirped pulse amplification system. The short and weak pulses from a Ti:sapphire oscillator or multipass amplifier are stretched in time, amplified via noncollinear OPCPA with "short" pump pulses and compressed in a suitable stretcher. Inset: basic setup of an OP(CP)A stage with the weak input seed, strong and short input pump and the generated amplified signal and idler.

A lot of strategies were developed to improve the contrast as listed above, but their applicability to few-cycle light sources was not or rarely investigated. After this summary of the contrast enhancement techniques two of them applied -and one planned to be applied- in our few-cycle systems are discussed in more detail.

2.2.1 Optical parametric amplification and optical parametric chirped pulse amplification

The goal of this section is to introduce the basic principles of optical parametric amplification (OPA) and optical parametric chirped pulse amplification (OPCPA) schemes and show their advantage in contrast enhancement. It is far from a complete description of these approaches. For more complete reviews see Refs. Cerullo & De Silvestri (2003); Dubietis et al. (2006). OPA is a second order nonlinear process in which from an input beam called pump, energy is transferred to another input beam called signal and as a byproduct a third output beam called idler at the difference frequency of the pump and the signal frequency is generated as shown in Fig. 3 inset. This parametric amplification process of the signal is taking place in nonlinear optical crystals such as BBO, LBO, KDP, etc. and is characterized by phase matching -originating from momentum conservation- between the three involved waves which determines the spectral bandwidth and decreases with increasing crystal thickness and an exponential amplification of the signal wave. The nonsaturated gain is proportional to the crystal thickness and the root of the pump intensity -the amplification is an exponential function of the gain. OPCPA Dubietis et al. (1992) is a technique, where a short signal pulse, generated for example by a Titanium:sapphire oscillator or a kHz amplifier, is stretched temporally to match the pump pulse duration. Thereafter a -typically- noncollinear OPA stage with the chirped pulses -termed OPCPA stage- amplifies the pulses that are compressed at the end of the system as shown in Fig. 3.

The OPCPA has various advantages over conventional laser amplifiers:

- Very broad gain bandwidth reaching ¿300 nm, which is not available from conventional amplification by lasers. This supports pulses with few optical cycle duration.
- Huge single pass gain reaching a value of 10^6. However typically much lower gain values are used to preserve the good contrast.
- Negligible thermal load in the amplifier crystals in contrary to laser amplifiers. No cooling is needed, which is one of the biggest challenges in lasers.
- Good contrast achievable if the pump pulse is relative short (1 ps to $\sim$100 ps) as the gain exists only during the short pump pulse. Any kind of signal -as stretched seed or undesired background- is correspondingly only amplified in the temporal window of the pump.

The following challenges are connected to OPCPA:

- Stretching and compression of huge spectral bandwidth -relevant mainly for few-cycle pulses- with a precise spectral phase control and high compressor throughput. These expectations limit the stretching ratio.
- The pump laser easily becomes a complex system when its duration is much shorter than a few nanosecond duration of commercially available sources. The short pump duration is required due to the limited stretching and the good required contrast of signal.
- Synchronization of pump and seed is getting difficult with decreasing pump pulse duration, requiring special techniques as optical synchronization Teisset et al. (2005) or even active delay stabilization.
- Amplification of the optical parametric fluorescence (OPF) also called superfluorescence Kleinman (1968), which generates an incoherent background like ASE in lasers and increases with the gain and pump intensity.
- Carrier envelope phase stabilization Baltuska et al. (2003) of the few-cycle pulses turns into an inevitable though challenging task.

As discussed above the finite pump pulse duration corresponds to a finite temporal gain window. Outside of this window the undesired background like ASE, OPF or prepulses is not amplified and so the contrast gets better outside by a factor of the parametric gain. Presently OPCPA with short pump pulses -in the ps range- is mainly used for the generation of few-cycle pulses Herrmann et al. (2009); Gu et al. (2009); Major et al. (2009), that are unachievable with other methods. The good contrast is a extra benefit from OPCPA and can be well utilized only in ultra-high intensity few-cycle systems.

2.2.2 Plasma mirror

Intense focused laser pulse impinging onto a transparent target for the laser wavelength -typically glass- start to generate plasma when the intensity exceeds the damage threshold of the material. The target is ionized and an expanding plasma is generated on the surface with an expansion velocity about the plasma sound speed (0.1 nm/fs at 1 keV plasma temperature). A high density plasma layer is formed and as soon as the electron density in the plasma exceeds the critical density for the incident wavelength as defined by Eq. 1 its reflectivity for the incident pulse switches to a high value. The remaining part of the laser pulse and everything afterward are reflected. The low intensity prepulses and the pedestal are transmitted through the transparent substrate before the plasma formation. This way, a low reflectivity is used for the prepulses and the pedestal, while a several orders of magnitude higher reflectivity value is applied for the main pulse. This fast plasma shutter is well suited for suppression of unwanted light before the main pulse. Consequently the contrast of the pulse is increased by the ratio of the plasma reflectivity to cold or Fresnel surface reflectivity of the material. The contrast improvement is typically 2 to 3 orders of magnitude with AR coated targets and *s* incident polarization or in a geometry with an incidence angle close to Brewster's angle and *p*-polarization. If the plasma scale length -see Eq. 2- exceeds the laser wavelength the plasma starts to absorb and distort the phasefront of the reflected pulse leading to lower reflectivity and the loss of beamed specular reflection Hörlein et al. (2008). The principle of the plasma mirror is illustrated in Fig. 4.

The plasma mirror Kapteyn et al. (1991) is used to improve the laser pulse after amplification and compression and provides higher throughput without limitation on the input energy Gibbon (2007). Since it is used after the whole laser system, the plasma mirror can be implemented without any modification to the system itself. Further advantages are wide bandwidth acceptance as will be discussed later Nomura et al. (2007), and spatial filtering effect if the plasma mirror is in the vicinity of the laser focus Gold (1994); Doumy et al. (2004a); Hörlein et al. (2008), but no smoother beam profile or even degradation was reported using the target in the near-field Dromey et al. (2004); Hörlein et al. (2008). Several investigations in different geometries Backus et al. (1993); Ziener et al. (2002); Doumy et al. (2004a) as normal, 45° and Brewster's angle of incidence were conducted to study the reflectivity of the plasma mirror yielding 50-80% overall -time- and space-integrated- energy reflectivity and a measured contrast enhancement of 50-100 for *s*-polarization and antireflection coated targets Dromey et al. (2004); Monot et al. (2004) and 25-50% energy throughput and 50-400 enhancement for *p*-polarization and Brewster's angle Backus et al. (1993); Nomura et al. (2007). The temporally resolved reflectivity during the plasma mirror is formed was measured to be 300-1000 fs determined with 100-500 fs laser pulses Bor et al. (1995); von der Linde et al. (1997); Grimes et al. (1999). Some studies pursued the application possibility of the plasma mirror: improving the repetition rate by using a liquid jet as the target Backus et al. (1993) and

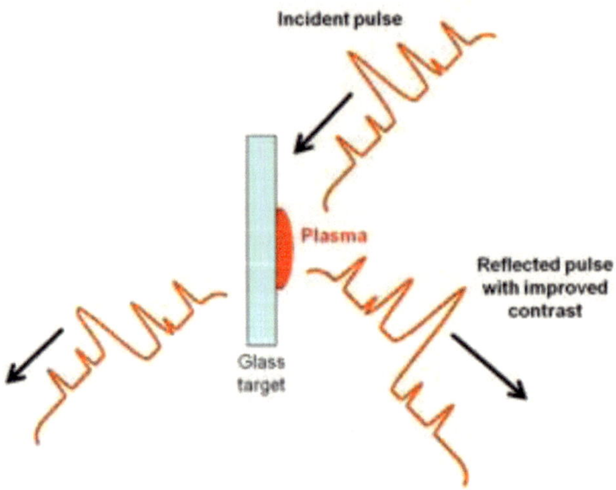

Fig. 4. Working principle of the plasma mirror. The incident low intensity prepulses and pedestal are transmitted through the transparent glass target, while the foot of the high intensity main pulse generates a plasma, which reflects the main pulse.

cascading two plasma mirrors with an overall reflectivity of 31-50% to improve the contrast by $10^4 - 5 \times 10^4$ to reach a required level in the experiments Wittmann et al. (2006); Lévy et al. (2007); Thaury et al. (2007); Doumy et al. (2004b). All previous studies used pulses with 25 fs of duration or longer and only our investigations Nomura et al. (2007) and others shown later applied sub-10-fs pulses. On the other hand, intense few-cycle pulses with a sufficiently high contrast would open up a new prospect for many applications as intense single attosecond pulse generation Tsakiris et al. (2006). Therefore it has great significance to study the possibility to obtain high-contrast few-cycle pulses using a plasma mirror.

2.2.3 Cross-polarized wave generation

Light propagating in nonlinear optical crystals experiences the partial conversion into light with perpendicular polarization. This additional component is called the cross-polarized wave (XPW) Minkovski et al. (2004; 2002). There are two different processes leading to XPW generation: the nonlinear polarization rotation -an elliptic polarization state remains elliptic with the same ellipticity just the main elliptical axis is rotated- and the induced ellipticity -the ellipticity changes, but the main elliptical axis stays the same. XPW generation is a third order nonlinear effect originating in practice from the dominant real part of $\chi^{(3)}$. The XPW efficiency is proportional to the product of $\chi^{(3)}_{xxxx}$ and the anisotropy of the $\chi^{(3)}$ tensor Minkovski et al. (2004). It has perfect and simultaneous phase- and group-velocity matching due to the same frequencies of input and output beams and propagation along the optical axis, which results in high efficiencies. Typically BaF_2 or LiF is used in the experiments since it has moderate $\chi^{(3)}_{xxxx}$ and high anisotropy leading to high-efficiency XPW generation ($\geq 10\%$) without significant self-phase modulation, which depends only on $\chi^{(3)}_{xxxx}$. The XPW process was applied to femtosecond pulse cleaning as the temporal third order nonlinearity suppresses low intensity light surrounding the main laser pulse. Typical schematics of the XPW setup is shown in Fig. 5. The polarization of the beam input with an energy from a few μJ to a few mJ is cleaned by a polarizer and it is focused to reach the required $3 - 7 \times 10^{12}$ W/cm^2 intensity in the BaF_2 crystal, which is typically not in the focus. Here the orthogonally polarized component is generated with 10% efficiency if the angle β between the laser polarization and the x axis of BaF_2 is optimized, which for [001] or z-cut crystals weakly depends on the intensity for high intensities. Subsequently the beam is collimated and send through an analyzer to remove the original polarization. The contrast after the filter neglecting saturation Jullien et al. (2006b):

$$C_{out} = C_{in}^3 + C_{in} KR/\eta_{eff}, \tag{3}$$

where $C_{in/out}$ is the contrast at the input/output of the contrast filter ($C_{in} = 10^{-2} - 10^{-8}$), R is the polarizer extinction ratio ($R = 10^{-2} - 10^{-5}$), η_{eff} is the internal energy efficiency ($\eta_{eff} = 0.1 - 0.2$) and $K = \eta_{eff}/\eta_{peak} \sim 0.2$ is an integration constant connecting the effective efficiency and the peak efficiency (η_{peak}) and originating from temporal and spatial profiles. This equation indicates that the output contrast is proportional to the third power of the input contrast, but the improvement is limited by the polarizer extinction ratio. Therefore high quality polarizers with low extinction ratios and good input contrast provides a better enhancement. This might be slightly influenced by saturation very near to the pulse peak.
The XPW leads to 3-5 OOM enhancement and 10-11 OOM laser contrast Jullien et al. (2005); Chvykov et al. (2006). A double crystal scheme was also applied to increase the efficiency to 20-30% due to the nonlinear self focusing that increases the intensity in the second crystal, the different corresponding Gouy phase shift between fundamental and XPW providing an

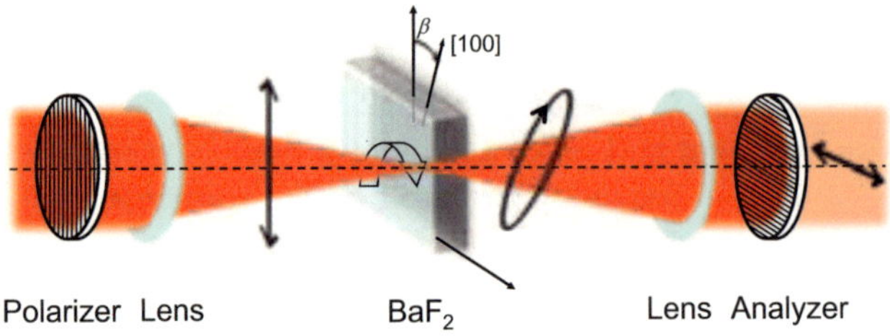

Fig. 5. Schematics of cross-polarized wave generation

optimal phase difference at the second crystal and the possibility of independent optimization of β Chvykov et al. (2006); Jullien et al. (2006a;b). BaF_2 with holographic cut orientation [011] further increases the efficiency. 11.4% and 28% were demonstrated in single and double crystal scheme as the coupling coefficient is slightly higher in this case Canova et al. (2008a). Further advantages of the holographic cut is that β is not intensity dependent allowing better phase matching at high intensities. XPW in BaF_2 is suitable for a broad wavelength range from UV to near-IR Canova et al. (2008b); Cotel et al. (2006); Jullien et al. (2006a). A significant smoothing and a $\sqrt{3}$ broadening of the spectrum is generated by the XPW as it is a third order temporal nonlinearity, which was observed experimentally in the case of optimal compression Jullien et al. (2007); Canova et al. (2008c). An even a larger broadening and pulse shortening of a factor of 2.2 was measured with a spatially super-Gaussian beam from a Ti:sapphire laser having 23% -even up to 28%- internal efficiency as a consequence of an interplay between cross- and self-phase modulation of the XPW and fundamental waves and the strong saturation Jullien et al. (2008). XPW with few-cycle pulses was also demonstrated Jullien et al. (2009; 2010), it shows spectral intensity and phase smoothing and preserves the carrier envelope phase Osvay et al. (2009). Up to now only a limited (2 OOM) contrast improvement of XPW with few-cycle pulses was experimentally supported Jullien et al. (2010). Reaching high efficiency needs $\sim$mm crystal thickness which changes significantly the pulse duration of sub-10-fs pulses during propagation in the crystal due to dispersion. Therefore it is not clear whether the XPW technique is applicable to few-cycle pulses and a higher contrast improvement accessible.

2.2.4 Characterization of contrast

Various measurement techniques of laser contrast are discussed in this session. The difficulties in measuring the contrast are the required high dynamic range of higher than 8 OOM and the good temporal resolution approaching the pulse duration of the main pulse. A normal photo diode for example has a dynamic range of 3-4 OOM and a temporal resolution of about 100 ps. None of these properties is suitable for a detailed contrast determination. Principally a simple second harmonic autocorrelation measurement routinely applied for pulse duration measurement delivers already information about the foot of the pulse with 3-4 OOM dynamics Roskos et al. (1987); Antonetti et al. (1997) and under certain conditions this measurement limit can be extended to 7-9 OOM for example using Lock in detection Braun et al. (1995); Curley et al. (1995). The time ambiguity is certainly present in these investigations using the second harmonic and so the leading and trailing edges are not distinguishable. To this end autocorrelation based on the surface-enhanced third harmonic signal with Lock in detection was used with a 1 kHz system providing a dynamics of 10^5 Hentschel et al. (1999). Still the required measurement dynamics is not reached and typical ultrahigh intensity lasers

have low repetition rate (~10 Hz) prohibiting the use of Lock in detection. Cross correlation based on third harmonic generation (THG) in two subsequent nonlinear crystals provides both high dynamic > 10 OOM and free from time ambiguity Luan et al. (1993); Antonetti et al. (1997); Aoyama et al. (2000); Tavella et al. (2005). Even a single shot version of this cross-correlator was realized for low repetition rate high energy laser systems Dorrer et al. (2008); Ginzburg et al. (2008). Nowadays THG cross-correlation is the most popular method to characterize contrast. An alternative way is the optical parametric amplifier correlator (OPAC) Divall & Ross (2004); Witte et al. (2006), which is based an optical parametric amplification of the fundamental in a short temporal window defined by the frequency doubled pump. The detection limit is 11 OOM with a theoretical value of 15 OOM. Recently specular reflectivity of overdense plasmas applied to estimate the contrast Pirozhkov et al. (2009) giving a measure of the preplasma generated by the general preceding background. An extended preformed plasma leads to beam breakup and increased absorption so a sufficiently good contrast gives a high reflectivity even at ultra-relativistic intensities.
We applied a THG cross-correlator, the upgraded version of that in Ref. Tavella et al. (2005), capable to measure 10-11 OOM to determine the contrast improvement separately by the implemented techniques.

3. Results and discussion

In this chapter various efforts to improve the contrast on two different few-cycle light sources will be discussed. The first system is a Titanium:sapphire laser with 1 kHz repetition rate Verhoef et al. (2006) and the second is an OPCPA system, called Light Wave Synthesizer 20 Herrmann et al. (2009). A plasma mirror was realized and characterized with the first system described in chapter 3.1, while short pump OPCPA was "implemented" in LWS-20 and XPW and plasma mirror are planned to be implemented in the near future to obtain a unique contrast as discussed in chapter 4.

3.1 Plasma mirror with a kHz Titanium:sapphire laser

A plasma mirror was implemented in a few-cycle laser system and characterized in detail Nomura et al. (2007); Nomura (2008). The reflectivity and the focusability were determined in *s*- and *p*-polarization and the time resolved contrast improvement was also measured. The source was a broadband 1 kHz Ti:sapphire laser system based on chirped pulse amplification with three multi-pass amplifier stages and a hollow-fiber compressor Verhoef et al. (2006). The system typically delivered pulses with 550 μJ energy, a spectrum extending from 550 to 900 nm with a central wavelength of 730 nm and 7 fs duration at 1 kHz repetition rate as shown in Fig. 6. The output beam was guided through a vacuum beamline to the target chamber. The energy on the target was 350-400 μJ.
The experimental setup is shown in Fig. 7. Either *p*- or *s*-polarization of the incident beam could be set by changing the alignment of a periscope before entering into the target chamber. The beam with 50 mm diameter was focused onto a 120 mm diameter BK7 glass target with an f_{eff} = 150 mm, 90o silver off-axis paraboloid mirror (F/3) leading to a focus full width at half maximum (FWHM) diameter of 7-8 μm. Three motorized stages allowed to rotate the target and translate it parallel to the surface and parallel to the incident beam (defined as z-direction). At 1 kHz repetition rate a target lasted approximately for an hour. The reflected beam from the target was refocused with a thin achromatic lens and sent to a detector outside the vacuum chamber. We measured the reflected energy using a power meter as detector; the spatial peak reflectivity by imaging the beam profile around the focus of the incident and

the reflected beam with a microscope objective onto a charge-coupled device (CCD) camera; and the temporal structure with high dynamics of the incident and also of the reflected pulses using a third-order correlator.

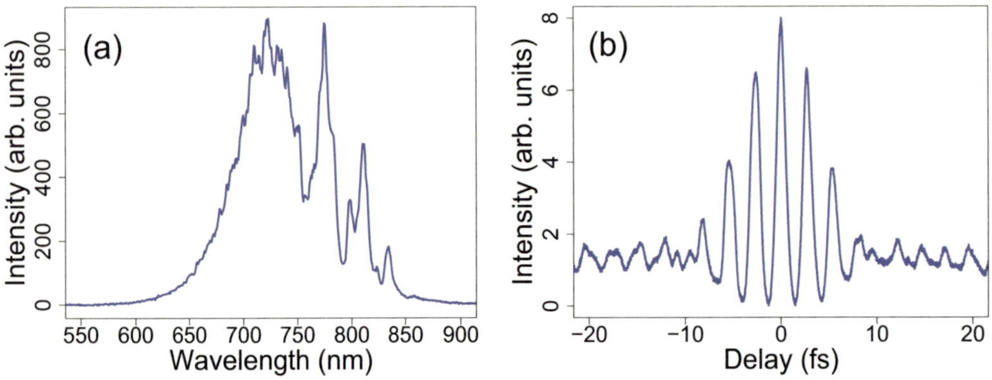

Fig. 6. Typical spectrum (a) and interferometric second-order autocorrelation (b) of the Ti:sapphire laser pulses used in the first plasma mirror experiment. The pulse duration is about 7 fs.

The plasma mirror efficiency was characterized by the energy throughput, i.e. the spatially integrated or average reflectivity, and the peak reflectivity. We calculate the peak reflectivity as the ratio of the peak fluences, which are obtained from the measured beam profiles on the target and energies. As we will see, this gives the same as the ratio of the peak intensities, which is the definition of the reflectivity. The energy measured with the power meter was averaged over some thousand shots. The incident fluence was changed by either moving

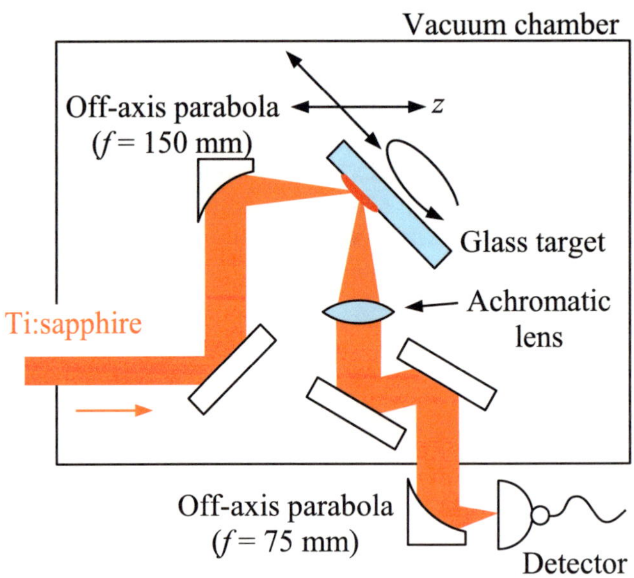

Fig. 7. Experimental setup

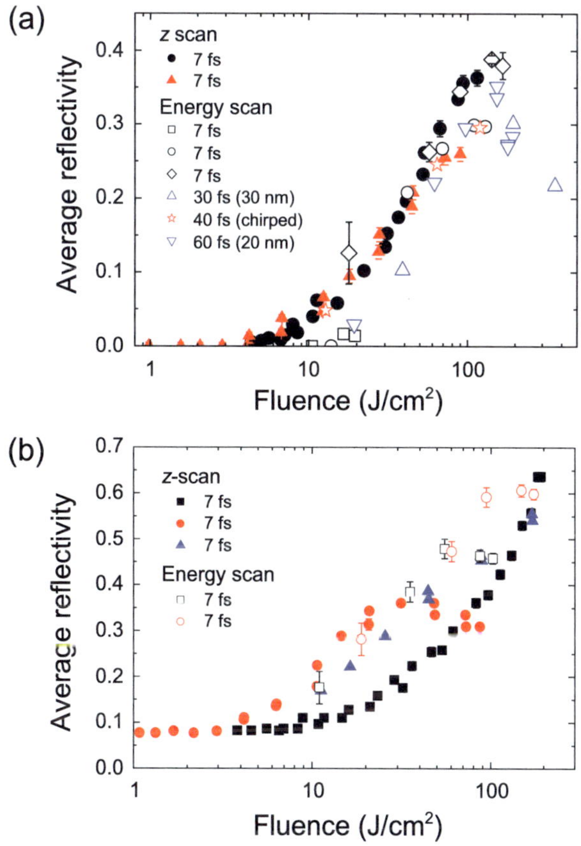

Fig. 8. Average reflectivity of the plasma mirror for (a) p-polarization and (b) s-polarization as a function of the average incident fluence. Different symbols represent different sets of measurements containing also runs with elongated pulses due to chirp or clipped spectrum. For p-polarization, the highest and lowest reflectivity measured are $\sim 40\%$ and $\sim 0.5\%$, respectively, therefore a contrast improvement of two orders of magnitude is expected.

the target out of focus (z-scan) or decreasing the energy of the incident pulse (energy scan). Different sets of measurements are shown with different symbols in Fig. 8. The measurements were well reproducible and gave the same results for z-scan and for energy scan. We also measured the average reflectivity with longer pulse durations, which was achieved by either chirping the pulse or clipping the spectrum. Therefore, we plotted the reflectivity as a function of the incident fluence in Figs. 8, 9.

Fig. 8 (a) shows the average reflectivity for p-polarization as a function of the average incident fluence, which is determined with respect to the spatial full width at half maximum (FWHM) area of the focused beam. The highest average reflectivity reached up to $\sim 40\%$ between 100 and 150 J/cm^2, whereas the lowest reflectivity was as low as $\sim 0.5\%$ because the 45° incidence angle was close to Brewster's angle ($\sim 56°$). From these values, a contrast improvement of two orders of magnitude is expected. The pulse duration was increased up to 60 fs, i.e., a factor of 9, but no significant change was observed in the behavior of the reflectivity versus fluence dependence. The average reflectivity measured for s-polarization is plotted in

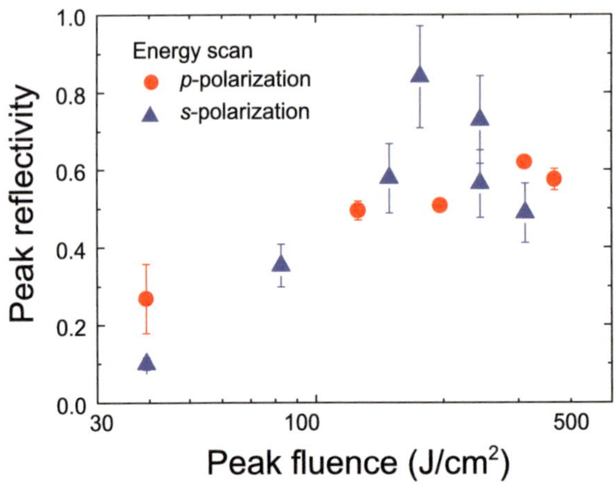

Fig. 9. Spatial peak reflectivity of the plasma mirror for *p*- and *s*-polarization plotted against the spatial peak incident fluence.

Fig. 8 (b). The highest reflectivity reached up to $\sim$ 65% and might be even higher for higher fluence on target unavailable in this experiment. In spite of the higher average reflectivity, the expected contrast improvement is only one order of magnitude due to the relatively high Fresnel reflectivity at *s*-polarization, which is $\sim$ 8% at 45° angle of incidence for our target material. The results plotted in Fig. 8 (b) had larger fluctuations than those in Fig. 8 (a) due to the different laser conditions. Reducing the reflectivity with antireflection (AR) coated targets can boost the contrast improvement up to factor of 300 and have maximal throughput. Using *p*-polarized light allows us to use cheaper uncoated glass targets at the cost of decreased throughput ($\sim$ 40%). The contrast improvement factors are in the same order for *s*-polarized light with AR-coated targets and for *p*-polarized light with ordinary targets, at 45° incidence angle. Using Brewster's angle increases the improvement factor for *p*-polarization even more, although the alignment is more sensitive.

The spatial peak reflectivity for *p*- and *s*-polarized pulses is depicted in Fig. 9 as a function of the peak fluence. The maximum value was above 60% for *p* and above 80% for *s* polarization. The spectra of the incident and reflected pulses were also measured, but they were almost identical and no significant blue shift was observed.

It is important for applications of the plasma mirror that the reflected light is still focusable and the wavefront and beam profile are not degraded. To investigate the spatial characteristics of the reflected beam, we collimated it with an achromatic lens (f = 150 mm) and refocused with an f = 75 mm off-axis parabola. The image of the refocused spot was magnified with a microscope objective and captured by a CCD beam profiler. The target was moved in the focal (z) direction and the imaging system was adjusted for each measurement. The measured spot diameters are plotted in Fig. 10 (a). The horizontal lines indicate the spot diameter without activating the plasma mirror, i.e., with low input energy. The different focus diameters for *s*- and *p*-polarizations are due to different alignments of the beamline. A horizontal and a vertical lineout of the refocused beam profile are plotted for *s*-polarization with (solid) and without (dashed) plasma mirror in Fig. 10 (b) when the target was in the focus (z = 0). We observed two effects on the reflected beam: cleaner smoothed near-field beam profile and smaller refocused spot. Both changes can be explained by the fluence-dependent reflectivity of the plasma mirror. The plasma mirror reflects more efficiently at the central part of the beam, while the reflection at the surrounding area is relatively suppressed, which acts as

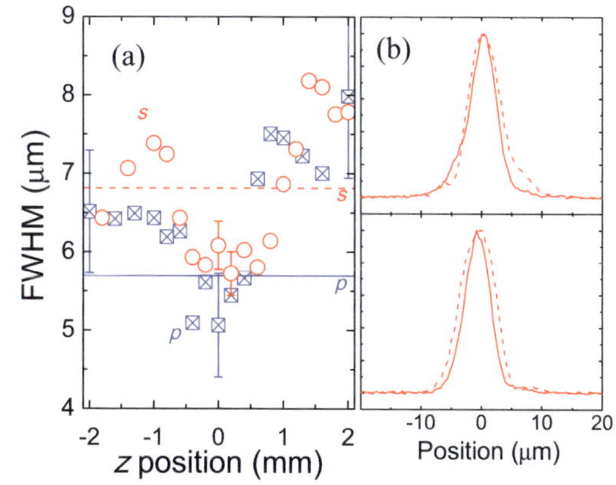

Fig. 10. (a) Refocused spot size (FWHM) as a function of the plasma-mirror position in the focal (z) direction. The polarization of the incident beam was p (blue square) or s (red circle). Horizontal lines indicate the reference spot size without activating the plasma mirror for p (solid) and s (dashed) polarization. (b) Horizontal and vertical lineouts of the refocused beam profile with the target in the focus ($z = 0$) for s-polarization with (solid) and without (dashed) plasma mirror.

a spatial filter resulting in a cleaner beam profile Moncur (1977). At the same time, this fluence-dependent reflectivity makes the peak narrower, which results in a smaller spot size on the plasma mirror and consequently a smaller refocused spot size.

The most important property of a plasma mirror is the contrast enhancement factor that is estimated based on cold and hot plasma reflectivity in general, but it is rarely verified experimentally. We present a complete high-dynamic-range third-order correlation of the reflected pulses, which allows us to obtain the time-resolved reflectivity and contrast enhancement of the plasma mirror. The polarization of the beam incident to the target was set to p to realize a better contrast improvement. The fluence on the plasma mirror was estimated to be ~ 60 J/cm^2 corresponding to about 30% average reflectivity. The reflected beam was recollimated and sent into a home-made third-order correlator Tavella et al. (2005). Fig. 11 shows the measured third-order correlation of the reflected pulse together with that of the incident pulse. The negative delay represents the leading edge of the pulse as before. Although the measured contrast was limited by the low energy of about 50 μJ sent into the correlator, the expected contrast improvement of two orders of magnitude at the pulse front is striking, for example, around -2 or at -8.5 ps. The peak appearing at -1.5 ps is an artefact from a post pulse, which appears due to the nature of correlation measurements. Also a pulse steepening effect is evident on the rising edge. On the other hand, no effect is observed on the falling edge of the pulse. Since 100 μm thick crystals were used in the correlator to gain a stronger signal, the third-order correlation does not reflect the short pulse duration.

Fig. 12 depicts the time-resolved reflectivity of the plasma mirror for p-polarization obtained by dividing the correlation of the reflected pulse by that of the incident pulse. We normalized the curve by setting the average reflectivity between 0 and 4 ps to the expected peak reflectivity of 50%.

A steep rise in the reflectivity is clearly seen at -500 fs. This steep rise indicates that the plasma is generated 400-500 fs before the main pulse. Therefore, the plasma mirror is efficiently

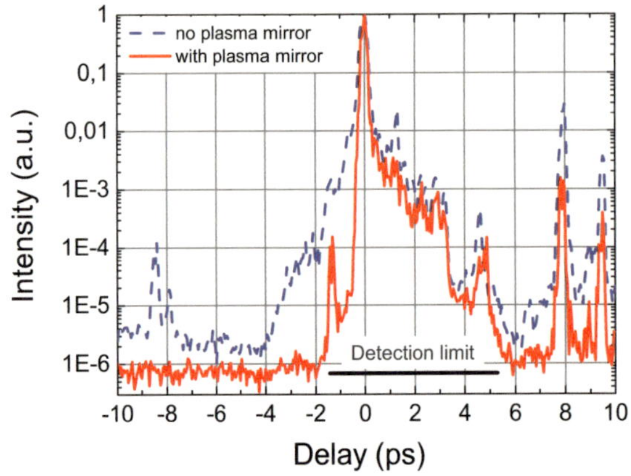

Fig. 11. Measured contrast without (black) and with (red) the plasma mirror using p-polarization. Although the measured contrast was limited by the low input energy ($\sim 50\mu$J), contrast improvement of two orders of magnitude is seen in the leading edge, for example, around -2 ps.

generated with the pedestal of our sub-10-fs pulses, similarly to the previous experiments with longer pulses. It is apparent that the reflectivity is constant during the pulse, hence the way we attained the peak reflectivity using the fluences is correct.A decrease in the reflectivity is also visible ~ 6 ps after the main pulse.

Hydrodynamic simulation of the preformed plasma expansion with a simulation code MEDUSA Christiansen et al. (1974) was performed to further understand the physical process. The input pulse used for the simulation was a 7 fs Gaussian pulse sitting on a 1.7 ps Gaussian

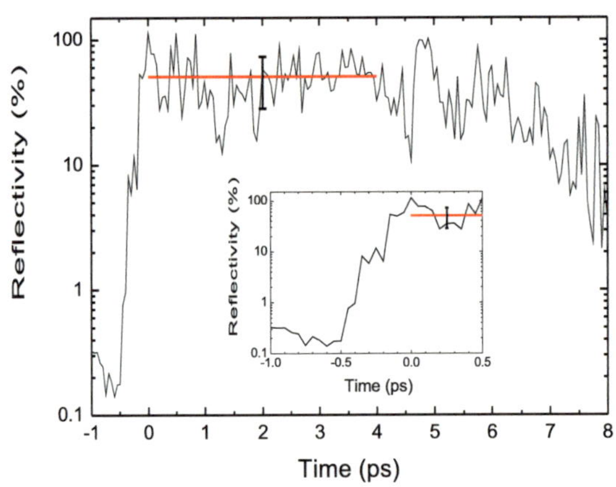

Fig. 12. Time-resolved peak reflectivity of the plasma mirror calculated from the correlations in Fig. 11. The horizontal red line is the average value of the peak reflectivity between 0 and 4 ps and the error bar corresponds to the standard deviation. Inset: the fast increase of the reflectivity at the leading edge.

pedestal with 2×10^{-4} contrast and a 7 fs Gaussian prepulse 8.5 ps before the main peak with 10^{-4} contrast as shown in Fig. 13. These parameters were determined by fitting the third-order correlation trace measured without the plasma mirror. The result of the simulation is also shown in Fig. 13. The simulation shows that the scale length of the plasma (Eq. 2) is $\sim 0.03\lambda$ at the critical density (Eq. 1) when the peak of the pulse arrives. If the scale length is too large, a plasma mirror acts similarly to a chirped mirror because different wavelengths are reflected at different depths in the plasma surface, owing to the different critical densities. With this scale length, however, this chirping effect is negligible and the pulse duration stays the same after the plasma mirror. The simulation also shows that the scale length exceeds 0.1λ around +2 ps after the main peak. Above this scale length, the process of resonant absorption starts Gibbon & Bell (1992), and reaches its maximum efficiency around $L = 0.3\lambda$ Kruer (1988). The simulation shows that this scale length is reached around +4 ps, which explains the decrease of the reflectivity around 6 ps.

In spite of the detailed measurements the preservation of the few-cycle pulse duration by the plasma mirror was just indirectly supported. In chapter 4 this important property will also be further discussed.

3.2 Contrast improvement of an OPCPA system

The second few-cycle light source, in which we applied contrast enhancement is the 8-fs, 16-TW OPCPA system, Light Wave Synthesizer 20 (LWS-20) Herrmann et al. (2009). This chapter describes the results from the short pump pulse OPCPA. Later in the next chapter we will discuss the potential if XPW and plasma mirror are also implemented. LWS-20 is the first optical parametric chirped pulse amplifier (OPCPA) system with few-cycle pulse duration and ~20 TW peak power. OPCPA generally offers a unique alternative to conventional lasers with much broader amplification bandwidth and correspondingly much shorter pulses reaching the sub-10-fs range, much higher gain, and low thermal load as analyzed before. In our OPCPA system as shown in Fig. 14 pulses from an ultra-broadband oscillator (Rainbow, Femtolasers), producing ~5.5 fs pulses with 80 MHz repetition rate, are split for optical synchronization. One part is wavelength shifted to 1064 nm to seed a commercial pump

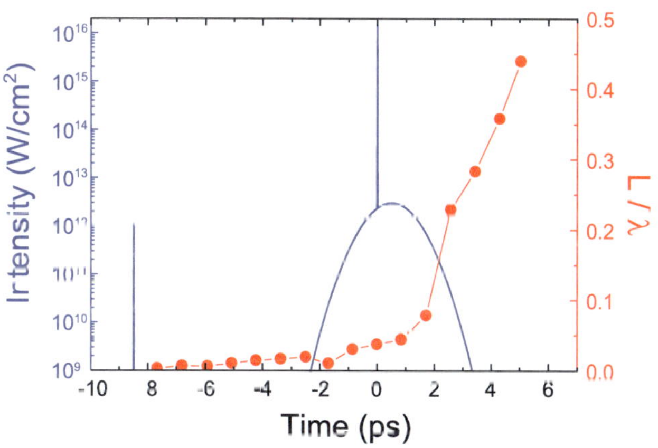

Fig. 13. Evolution of plasma scale length calculated with MEDUSA. Temporal profile of the input pulse (blue curve) estimated from the measurement. Evolution of the plasma scale length (red circles). It stays almost unchanged as the main pulse arrives and starts to increase after most of the pedestal has passed.

laser (EKSPLA) producing up to 1 J, 75 ps, 10 Hz pulses at 532 nm. The main part of the oscillator energy is amplified in a Femtopower Compact Pro 1 kHz Ti:sapphire CPA laser, which tightens the bandwidth and produces 25 fs long pulses after compression in the prism compressor.

These pulses with 750-800 μJ energy are sent into a neon filled tapered-hollow-core fiber to broaden the spectrum to seed the amplifier stages. After an optional XPW stage for contrast enhancement the pulses are stretched to 45 ps -group delay between blue and red spectral components- with a specially designed negative dispersion grism stretcher. An acousto optic programmable dispersive filter (Dazzler, Fastlight) serves the purpose of optimizing and fine tuning the spectral phase. The slightly compressed pulses -to about 30 ps after the Dazzler- are amplified in two non-collinear optical parametric chirped pulse amplifier stages based on type I BBO nonlinear optical crystals. The first stage is pumped by 15 mJ and amplifies the few-μJ seed pulses to about 1 mJ and the second stage is pumped with an energy of up to 800 mJ and delivers up to 170 mJ. The supported wavelength range of the OPA is from 700 nm up to 1050 nm, but due to practical limitations in the Dazzler, only spectral components up to about 980 nm can be used for compression, which corresponds to a Fourier limited pulse duration of 8 fs. The pulses are compressed in a high transmission compressor containing bulk glasses of 160 mm SF57 and 100 mm quartz and by four chirped mirrors to approx. 8 fs. After the compressor a pulse energy of up to 130 mJ is reached with 10 Hz repetition rate. A Shack-Hartmann wavefront sensor (Imagine Optic) and an adaptive mirror in a closed loop configuration are used to optimize the wavefront and so the focusing properties of the laser to reach $\gg 10^{18}$ W/cm^2 relativistic intensity on target. The system is ideally suited for electron acceleration in the non-linear laser wakefield acceleration regime with high efficiency and

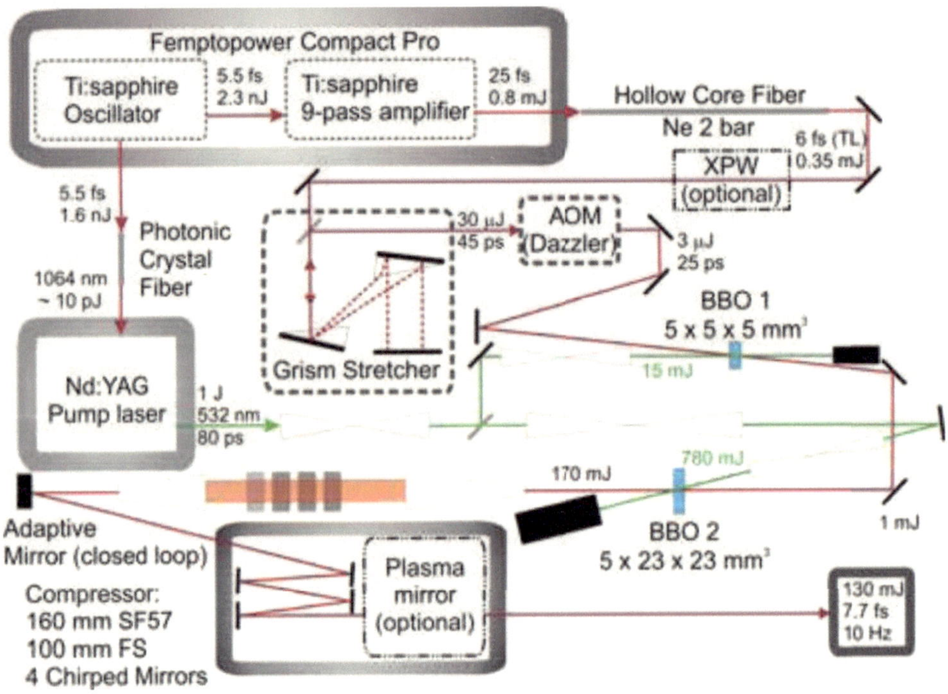

Fig. 14. Setup of the Light Wave Synthesizer 20 (LWS-20) OPCPA system.

stability to generate monoenergetic electrons Schmid et al. (2009) as well as for high harmonic generation towards a single attosecond pulse generation on plasma surfaces Heissler et al. (2010) and gas jets. Carrier envelope phase (CEP) measurements are also envisaged for CEP stabilization that will be necessary to generate single attosecond bursts.
As discussed before the contrast is improved in a short pulse (75 ps in our case) OPCPA system outside the pump duration. In LWS-20 the input contrast from the kHz front end is between 7-8 orders of magnitude (OOM) and it is conserved in approx. ±40 ps temporal window and many orders of magnitude better outside this window as shown in Fig. 15 blue dashed line. There is a 5 ps pedestal originating from stretching and compression. This is suppressed to 10^{-8} in the best case without other contrast enhancement as will be discussed later. The background from -5 ps up to -20 ps is the ASE from the front end amplified in the OPCPA stages. After the main pulse a longer continuously decreasing pedestal coming from the hollow-core fiber follows. The expected contrast enhancement > 40 ps before the pulse peak is 10^5 as the amplification increases the energy from about 1 μJ to on the order of 100 mJ. Although the third order correlator is capable of measuring 10 OOM it is still not enough to correctly determine the improvement in the contrast outside the pump temporal extension. Therefore we misaligned the front end -attenuated the multipass seed- to reduce the ASE contrast to deliver 5-6 OOM contrast. This reduced contrast is preserved in the OPCPA chain (at -6 ps 10^{-5}), but suppressed before the pump (at -45 ps 10^{-10}) as indicated by the red curve in Fig. 15. As a conclusion the OPCPA with short pump pulses improves the contrast corresponding to the gain coefficient by 5 OOM.

4. Conclusion and future work

In conclusion, the contrast improvement of sub-10-fs pulses by using a plasma mirror and OPCPA are demonstrated. The reflected pulses from the plasma mirror were cleaned both spatially and temporally. The spatial peak reflectivity reached ≥ 80% (≥ 60%) and the energy

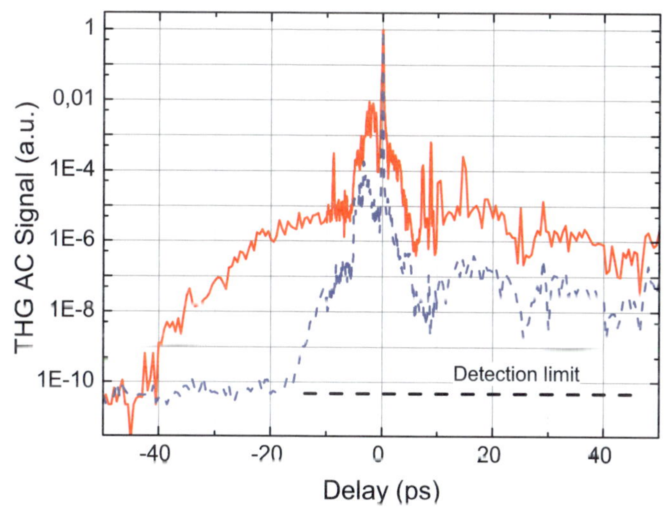

Fig. 15. Contrast of the LWS-20 OPCPA system (blue dashed line) and contrast with misaligned frond end to visualize the 10^5 enhancement between -6 and -45 ps due to OPCPA (red solid line).

throughput had a value of $\sim 65\%$ ($\sim 40\%$) for *s*- (*p*-) polarization at 45° angle of incidence. Using AR coated targets and *s*-polarization an average reflectivity of 70-80% is expected. The first measurement of the complete high-dynamic-range correlation revealed the temporal structure of the pulses reflected from the plasma mirror. The time-resolved reflectivity of the plasma mirror was determined with the help of these results, showing the contrast improvement of two orders of magnitude and the pulse steepening at the leading edge. This enhancement can be further increased to min. 2.5 orders of magnitude with AR coated targets. Improving the contrast with the plasma mirror will lead to better performances in experiments such as high-order harmonic generation on plasma surfaces or ion acceleration. The plasma mirror reflectivity is found to be independent on the chirp of the the incident pulses, which allows to optimize the pulse duration on a second target. The pulse spectrum was practically the same before and after the plasma mirror. Therefore the fact whether the plasma mirror pertains the short duration is not significant. On the other hand, the final size of the plasma mirror target will impose a limit on the number of laser shots in one experimental run. The use of the plasma mirror should be determined by weighing the benefits gained by the contrast improvements against the drawback of the limited number of shots. In the case of a moderate energy system (~ 100 mJ) many hours operation with 10 Hz repetition rate is principally possible.

The OPCPA technique with short pump pulses has among others also a big advantage in background suppression. Using moderate saturation a contrast improvement corresponding to the gain is achievable outside of the pump pulse duration. In our OPCPA system, LWS-20, an enhancement of 10^5 is realized with 80 ps pump pulses. Using even shorter pump lasers (~ 1 ps) this window is significantly reduced, but other difficulties as pump seed synchronization or non-linear effects in air and other optical components may arise. Hybrid laser systems utilize this advantage and the final high-energy laser amplification, which is presently a challenge for the short pulse pump laser. Comparing the plasma mirror to the OPCPA technique both of them have advantages and draw backs. The OPCPA amplifies already with an improved contrast, but only outside the pump window is the contrast better while the plasma mirror enhances the contrast also directly before the pulse peak, steepens the rising edge and removes background generated after the front end very near to the main pulse. The XPW technique is robust and has a large improvement, but enhances the input contrast into the amplifier and removes background just from the front end and cannot affect reasons for worse contrast that are generated later. The decision which of the methods is best suited in a given system is not easy to answer and can depend from case to case.

To further improve the contrast for experiments with LWS-20 a cross-polarized wave (XPW) generation cleaner stage (see Chapter) and a plasma mirror are planned to be implemented. The structure of LWS-20 is ideally suited to implement XPW after the hollow-core fiber and before the grism stretcher. This structure makes it practically to a double-CPA system Kalashnikov et al. (2005) with an OPCPA instead of CPA as the second amplification part. The expected contrast improvement using Eq. 3 and a Glan-Laser polarizer with an extinction ration of better than $2 - 5 \times 10^{-4}$ is up to 10^{-4}. The plasma mirror with AR coated targets having 0.2% reflectivity and an estimated plasma mirror reflectivity of 60% is expected to enhance a contrast by about 3×10^{-3} and also steepen the rising edge if the pulses. After the implementation of XPW about 10^{-17} and the implementation of the plasma mirror about 10^{-19} contrast is expected 45 ps before the pulse peak. These values and the good contrast also before this delay makes the LWS-20 system an ideal candidate as a front end of future multi-Petawatt to Exawatt lasers.

5. Acknowledgments

The author gratefully acknowledge the work on the laser system in Vienna of A. J. Verhoef, J. Seres, E. Seres, G. Tempea and the work done on the plasma mirror by J. Nomura, K. Schmid, T. Wittmann and J. Wild. Furthermore the work on LWS-20 or its predecessors by D. Herrmann, R. Tautz, F. Tavella, A. Marcinkevičius, V. Pervak, N. Ishii, A. Baltuška is acknowledged as well as the users who contributed to the system significantly as A. Buck, J. M. Mikhailova, K. Schmid, C. M. S. Sears, Y. Nomura. Furthermore grateful thanks are due to G. Tsakiris. Extra thanks to Prof. F. Krausz for his support. A. Buck, J. M. Mikhailova, T. Wittmann are acknowledged for reading and correcting the manuscript.

6. References

Antonetti, A., Blasco, F., Chambaret, J. P., Cheriaux, G., Darpentigny, G., Le Blanc, C., Rousseau, P., Ranc, S., Rey, G. & Salin, F. (1997). A laser system producing 5×10^{19} w/cm^2 at 10 hz, *Appl. Phys. B* 65: 197–204.

Aoyama, M., Sagisaka, A., Matsuoka, S., Akahane, Y., Nakano, F. & Yamakawa, K. (2000). Contrast and phase characterization of a high-peak-power 20-fs laser pulse, *Appl. Phys. B* 70: S149–S153.

Backus, S., Kapteyn, H. C., Murnane, M. M., Gold, D. M., Nathel, H. & White, W. (1993). Prepulse suppression for high-energy ultrashort pulses using self-induced plasma shuttering from a fluid target, *Opt. Lett.* 18(2): 134.

Baltuska, A., Udem, T., Uiberacker, M., Hentschel, M., Goulielmakis, E., Gohle, C., Holzwarth, R., Yakovlev, V. S., Scrinzi, A., Hansch, T. W. & Krausz, F. (2003). Attosecond control of electronic processes by intense light fields, *Nature* 421(7075): 611–615.

Bor, Z., Racz, B., Szabo, G., Xenakis, D., Kalpouzos, C. & Fotakis, C. (1995). Femtosecond transient reflection from polymer surfaces during femtosecond uv photoablation, *Appl. Phys. A* 60: 365–368.

Brabec, T. & Krausz, F. (2000). Intense few-cycle laser fields: Frontiers of nonlinear optics, *Rev. Mod. Phys.* 72(2): 545 591.

Braun, A., Rudd, J. V., Cheng, H., Mourou, G., Kopf, D., Jung, I. D., Weingarten, K. J. & Keller, U. (1995). Characterization of short-pulse oscillators by means of a high-dynamic-range autocorrelation measurement, *Opt. Lett.* 20(18): 1889–1891.

Canova, L., Albert, O., Forget, N., Mercier, B., Kourtev, S., Minkovski, N., Saltiel, S. & LopezMartens, R. (2008c). Influence of spectral phase on cross-polarized wave generation with short femtosecond pulses, *Appl. Phys. B* 93: 443–453.

Canova, L., Kourtev, S., Minkovski, N., Jullien, A., Lopez-Martens, R., Albert, O. & Saltiel, S. M. (2008a). Efficient generation of cross polarized femtosecond pulses in cubic crystals with holographic cut orientation, *Appl. Phys. Lett.* 92(23): 231102.

Canova, L., Kourtev, S., Minkovski, N., Lopez-Martens, R., Albert, O. & Saltiel, S. M. (2008b). Cross-polarized wave generation in the uv region, *Opt. Lett.* 33(20): 2299–2301.

Cerullo, G. & De Silvestri, S. (2003). Ultrafast optical parametric amplifiers, *Rev. Sci. Instrum.* 74(1): 1–18.

Christiansen, J. P., Ashby, D. E. T. F. & Roberts, K. V. (1974). MEDUSA a one-dimensional laser fusion code, *Comput. Phys. Commun.* 7(5): 271–287.

Chvykov, V., Rousseau, P., Reed, S., Kalinchenko, G. & Yanovsky, V. (2006). Generation of 10^{11} contrast 50 TW laser pulses, *Opt. Lett.* 31(10): 1456–1458.

Cotel, A., Jullien, A., Forget, N., Albert, O., Chriaux, G. & Le Blanc, C. (2006). Nonlinear temporal pulse cleaning of a 1-m optical parametric chirped-pulse amplification system, *Appl. Phys. B* 83: 7–10.

Curley, P. F., Darpentigny, G., Cheriaux, G., Chambaret, J.-P. & Antonetti, A. (1995). High dynamic range autocorrelation studies of a femtosecond ti:sapphire oscillator and its relevance to the optimisation of chirped pulse amplification systems, *Opt. Commun.* 120(1-2): 71 – 77.

Divall, E. J. & Ross, I. N. (2004). High dynamic range contrast measurements by use of an optical parametricamplifier correlator, *Opt. Lett.* 29(19): 2273–2275.

Dorrer, C., Begishev, I. A., Okishev, A. V. & Zuegel, J. D. (2007). High-contrast optical-parametric amplifier as a front end of high-power laser systems, *Opt. Lett.* 32(15): 2143–2145.

Dorrer, C., Bromage, J. & Zuegel, J. D. (2008). High-dynamic-range single-shot cross-correlator based on an optical pulse replicator, *Opt. Express* 16(18): 13534–13544.

Doumy, G., Dobosz, S., D'Oliveira, P., Monot, P., Perdrix, M., Quéré, F., Réau, F., Martin, P., Audebert, P., Gauthier, J. C. & Geindre, J. P. (2004b). High order harmonic generation by non-linear reflection of a pedestal-free intense laser pulse on a plasma, *Appl. Phys. B* 78: 901–904.

Doumy, G., Quéré, F., Gobert, O., Perdrix, M., Martin, P., Audebert, P., Gauthier, J. C., Geindre, J.-P. & Wittmann, T. (2004a). Complete characterization of a plasma mirror for the production of high-contrast ultraintense laser pulses, *Phys. Rev. E* 69(2): 026402.

Dromey, B., Kar, S., Zepf, M. & Foster, P. (2004). The plasma mirror—a subpicosecond optical switch for ultrahigh power lasers, *Rev. Sci. Instrum.* 75: 645.

Dubietis, A., Butkus, R. & Piskarskas, A. P. (2006). Trends in chirped pulse optical parametric amplification, *IEEE J. Sel. Topics Quantum Electron.* 12(2): 163–172.

Dubietis, A., Jonusauskas, G. & Piskarskas, A. (1992). Powerful femtosecond pulse generation by chirped and stretched pulse parametric amplification in bbo crystal, *Opt. Commun.* 88(4-6): 437–440.

Gaul, E. W., Martinez, M., Blakeney, J., Jochmann, A., Ringuette, M., Hammond, D., Borger, T., Escamilla, R., Douglas, S., Henderson, W., Dyer, G., Erlandson, A., Cross, R., Caird, J., Ebbers, C. & Ditmire, T. (2010). Demonstration of a 1.1 petawatt laser based on a hybrid optical parametric chirped pulse amplification/mixed nd:glass amplifier, *Appl. Opt.* 49(9): 1676–1681.

Gibbon, P. (2007). Plasma physics: Cleaner petawatts with plasma optics, *Nature Phys.* 3: 369 – 370.

Gibbon, P. & Bell, A. R. (1992). Collisionless absorption in sharp-edged plasmas, *Phys. Rev. Lett.* 68(10): 1535–1538.

Ginzburg, V. N., Didenko, N. V., Konyashchenko, A. V., Lozhkarev, V. V., Luchinin, G. A., Lutsenko, A. P., Mironov, S. Y., Khazanov, E. A. & Yakovlev, I. V. (2008). Third-order correlator for measuring the time profile of petawatt laser pulses, *Quantum Electron.* 38(11): 1027–1032.

Gold, D. M. (1994). Direct measurement of prepulse suppression by use of a plasma shutter, *Opt. Lett.* 19(23): 2006–2008.

Grimes, M. K., Rundquist, A. R., Lee, Y.-S. & Downer, M. C. (1999). Experimental identification of "vacuum heating" at femtosecond-laser-irradiated metal surfaces, *Phys. Rev. Lett.* 82(20): 4010–4013.

Gu, X., Marcus, G., Deng, Y., Metzger, T., Teisset, C., Ishii, N., Fuji, T., Baltuška, A., Butkus, R., Pervak, V., Ishizuki, H., Taira, T., Kobayashi, T., Kienberger, R. & Krausz, R. (2009).

Generation of carrier-envelope-phase-stable 2-cycle 740-μj pulses at 2.1-μm carrier wavelength, *Opt. Express* 17(1): 62–69.

Hegelich, B. M., Albright, B. J., Cobble, J., Flippo, K., Letzring, S., Paffett, M., Ruhl, H., Schreiber, J., Schulze, R. K. & Fernández, J. C. (2006). Laser acceleration of quasi-monoenergetic MeV ion beams, *Nature* 439(7075): 441–4.

Heissler, P., Hörlein, R., Stafe, M., Mikhailova, J. M., Nomura, Y., Herrmann, D., Tautz, R., Rykovanov, S. G., Földes, I. B., Varjú, K., Tavella, F., Marcinkevičius, A., Krausz, F., Veisz, L. & Tsakiris, G. D. (2010). Towards single attosecond pulses using harmonic emission from solid density plasmas, *Appl. Phys. B* p. accepted.

Hentschel, M., Uemura, S., Cheng, Z., Sartania, S., Tempea, G., Spielmann, Ch. & Krausz, F. (1999). High-dynamic-range pulse-front steepening of amplified femtosecond pulses by third-order dispersion, *Appl. Phys. B* 68(1): 145–148.

Hernandez-Gomez, C., Blake, S. P., Chekhlov, O., Clarke, R. J., Dunne, A. M., Galimberti, M., Hancock, S., Heathcote, R., Holligan, P., Lyachev, A., Matousek, P., Musgrave, I. O., Neely, D., Norreys, P. A., Ross, I., Tang, Y., Winstone, T. B., Wyborn, B. E. & Collier, J. (2010). The vulcan 10 pw project, *J. Phys.: Conf. Ser.* 244(3): 032006.

Herrmann, D., Veisz, L., Tautz, R., Tavella, F., Schmid, K., Pervak, V. & Krausz, F. (2009). Generation of sub-three-cycle, 16 tw light pulses by using noncollinear optical parametric chirped-pulse amplification, *Opt. Lett.* 34(16): 2459–2461.

Homoelle, D., Gaeta, A. L., Yanovsky, V. & Mourou, G. (2002). Pulse contrast enhancement of high-energy pulses by use of a gas-filled hollow waveguide, *Opt. Lett.* 27(18): 1646–1648.

Hörlein, R., Dromey, B., Adams, D., Nomura, Y., Kar, S., Markey, K., Foster, P., Neely, D., Krausz, F., Tsakiris, G. D. & Zepf, M. (2008). High contrast plasma mirror: spatial filtering and second harmonic generation at 10^{19} w/cm^{-2}, *New J. Phys.* 10(8): 083002.

Itatani, J., Faure, J., Nantel, M., Mourou, G. & Watanabe, S. (1998). Suppression of the amplified spontaneous emission in chirped-pulse-amplification lasers by clean high-energy seed-pulse injection, *Opt. Commun.* 148: 70–74.

Jullien, A., Albert, O., Burgy, F., Hamoniaux, G., Rousseau, J.-P., Chambaret, J.-P., Augé-Rochereau, F., Chériaux, G., Etchepare, J., Minkovski, N. & Saltiel, S. M. (2005). 10^{-10} temporal contrast for femtosecond ultraintense lasers by cross-polarized wave generation, *Opt. Lett.* 30(8): 920–922.

Jullien, A., Albert, O., Chériaux, G., Etchepare, J., Kourtev, S., Minkovski, N. & Saltiel, S. M. (2006a). A two crystal arrangement to fight efficiency saturation in cross-polarized wave generation, *Opt. Express* 14(7): 2760–2769.

Jullien, A., Augé-Rochereau, F., Chériaux, G., Chambaret, J.-P., d'Oliveira, P., Auguste, T. & Falcoz, F. (2004). High-efficiency, simple setup for pulse cleaning at the millijoule level by nonlinear induced birefringence, *Opt. Lett.* 29(18): 2184–2186.

Jullien, A., Canova, L., Albert, O., Boschetto, D., Antonucci, L., Cha, Y.-H., Rousseau, J., Chaudet, P., Chriaux, G., Etchepare, J., Kourtev, S., Minkovski, N. & Saltiel, S. (2007). Spectral broadening and pulse duration reduction during cross-polarized wave generation: influence of the quadratic spectral phase, *Appl. Phys. B* 87: 595–601.

Jullien, A., Chen, X., Ricci, A., Rousseau, J.-P., Lopez-Martens, R., Ramirez, L., Papadopoulos, D., Pellegrina, A., Druon, F. & Georges, P. (2010). High-fidelity front-end for high-power, high temporal quality few-cycle lasers, *Appl. Phys. B* pp. 1–6.

Jullien, A., Durfee, C., Trisorio, A., Canova, L., Rousseau, J.-P., Mercier, B., Antonucci, L., Chriaux, G., Albert, O. & Lopez-Martens, R. (2009). Nonlinear spectral cleaning of few-cycle pulses via cross-polarized wave (xpw) generation, *Appl. Phys. B* 96: 293–299.

Jullien, A., Kourtev, S., Albert, O., Chriaux, G., Etchepare, J., Minkovski, N. & Saltiel, S. (2006b). Highly efficient temporal cleaner for femtosecond pulses based on cross-polarized wave generation in a dual crystal scheme, *Appl. Phys. B* 84: 409–414.

Jullien, A., Rousseau, J.-P., Mercier, B., Antonucci, L., Albert, O., Chériaux, G., Kourtev, S., Minkovski, N. & Saltiel, S. M. (2008). Highly efficient nonlinear filter for femtosecond pulse contrast enhancement and pulse shortening, *Opt. Lett.* 33(20): 2353–2355.

Kalashnikov, M. P., Risse, E., Schönnagel, H., Husakou, A., Herrmann, J. & Sandner, W. (2004). Characterization of a nonlinear filter for the front-end of a high contrast double-CPA Ti:sapphire laser, *Opt. Expr.* 12(21): 5088–5097.

Kalashnikov, M. P., Risse, E., Schönnagel, H. & Sandner, W. (2005). Double chirped-pulse-amplification laser: a way to clean pulses temporally, *Opt. Lett.* 30(8): 923–925.

Kapteyn, H. C., Murnane, M. M., Szoke, A. & Falcone, R. W. (1991). Prepulse energy suppression for high-energy ultrashort pulses using self-induced plasma shuttering, *Opt. Lett.* 16(7): 490.

Kiriyama, H., Michiaki, M., Nakai, Y., Shimomura, T., Sasao, H., Tanaka, M., Ochi, Y., Tanoue, M., Okada, H., Kondo, S., Kanazawa, S., Sagisaka, A., Daito, I., Wakai, D., Sasao, F., Suzuki, M., Kotakai, H., Kondo, K., Sugiyama, A., Bulanov, S., Bolton, P. R., Daido, H., Kawanishi, S., Collier, J. L., Hernandez-Gomez, C., Hooker, C. J., Ertel, K., Kimura, T. & Tajima, T. (2010). High-spatiotemporal-quality petawatt-class laser system, *Appl. Opt.* 49(11): 2105–2115.

Kiriyama, H., Michiaki, M., Nakai, Y., Shimomura, T., Tanoue, M., Akutsu, A., Kondo, S., Kanazawa, S., Okada, H., Motomura, T., Daido, H., Kimura, T. & Tajima, T. (2008). High-contrast, high-intensity laser pulse generation using a nonlinear preamplifier in a ti:sapphire laser system, *Opt. Lett.* 33(7): 645–647.

Kleinman, D. A. (1968). Theory of optical parametric noise, *Phys. Rev.* 174(3): 1027–1041.

Krausz, F. & Ivanov, M. (2009). Attosecond physics, *Rev. Mod. Phys.* 81(1): 163–234.

Kruer, W. L. (1988). *The Physics of Laser Plasma Interactions*, Addison-Wesley, Redwood City, CA.

Lévy, A., Ceccotti, T., D'Oliveira, P., Réau, F., Perdrix, M., Quéré, F., Monot, P., Bougeard, M., Lagadec, H., Martin, P., Geindre, J.-P. & Audebert, P. (2007). Double plasma mirror for ultrahigh temporal contrast ultraintense laser pulses, *Opt. Lett.* 32(3): 310–312.

Liu, J. & Kobayashi, T. (2010). Temporal contrast improvement of femtosecond pulses by a self-diffraction process in a kerr bulk medium. Frontiers in Optics 2010, Rochester, NY, USA, October 24-28 2010.

Lozhkarev, V. V., Freidman, G. I., Ginzburg, V. N., Katin, E. V., Khazanov, E. A., Kirsanov, A. V., Luchinin, G. A., Mal'shakov, A. N., Martyanov, M. A., Palashov, O. V., Poteomkin, A. K., Sergeev, A. M., Shaykin, A. A. & Yakovlev, I. V. (2006). Compact 0.56 petawatt laser system based on optical parametric chirped pulse amplification in kd*p crystals, *Opt. Expr.* 14(1): 446–454.

Lozhkarev, V. V., Freidman, G. I., Ginzburg, V. N., Katin, E. V., Khazanov, E. A., Kirsanov, A. V., Luchinin, G. A., Mal'shakov, A. N., Martyanov, M. A., Palashov, O. V., Poteomkin, A. K., Sergeev, A. M., Shaykin, A. A. & Yakovlev, I. V. (2007). Compact 0.56 petawatt laser system based on optical parametric chirped pulse amplification in kd*p crystals, *Laser Phys. Lett.* 4(6): 421–427.

Luan, S., Hutchinson, M. H. R., Smith, R. A. & Zhou, F. (1993). High dynamic range third-order correlation measurement of picosecond laser pulse shapes, *Meas. Sci. Technol.* 4(12): 1426–1429.

Major, Z., Trushin, S. A., Ahmad, I., Siebold, M., Wand, C., Klingebiel, S., Wa, T., Fülöp, T., Henig, A., Kruber, S., Weingartner, R., Popp, A., Osterhoff, J., Hörlein, R., Hein, J., Pervak, V., Apolonski, A., Krausz, F. & Karsch, S. (2009). Basic concepts and current status of the petawatt field synthesizer – a new approach to ultrahigh field generation, *Rev. Laser Eng.* 37(6): 431–436.

Marcinkevičius, A., Tommasini, R., Tsakiris, G., Witte, K. J., Gaizauskas, E. & Teubner, U. (2004). Frequency doubling of multi-terawatt femtosecond pulses, *Appl. Phys. B* 79(5): 547–554.

Minkovski, N., Petrov, G. I., Saltiel, S. M., Albert, O. & Etchepare, J. (2004). Nonlinear polarization rotation and orthogonal polarization generation experienced in a single-beam configuration, *J. Opt. Soc. Am. B* 21(9): 1659–1664.

Minkovski, N., Saltiel, S. M., Petrov, G. I., Albert, O. & Etchepare, J. (2002). Polarization rotation induced by cascaded third-order processes, *Opt. Lett.* 27(22): 2025–2027.

Moncur, N. K. (1977). Plasma spatial filter, *Appl. Opt.* 16: 1449–1451.

Monot, P., Doumy, G., Dobosz, S., Perdrix, M., D'Oliveira, P., Quéré, F., Réau, F., Martin, P., Audebert, P., Gauthier, J.-C. & Geindre, J.-P. (2004). High-order harmonic generation by nonlinear reflection of an intensehigh-contrast laser pulse on a plasma, *Opt. Lett.* 29(8): 893–895.

Nomura, Y. (2008). *Temporal characterization of harmonic radiation generated by intense laser-plasma interaction*, PhD thesis, Ludwig-Maximilians-Universität, München, Germany.

Nomura, Y., Veisz, L., Schmid, K., Wittmann, T., Wild, J. & Krausz, F. (2007). Time-resolved reflectivity measurements on a plasma mirror with few-cycle laser pulses, *New J. Phys.* 9(1): 9.

Osvay, K., Canova, L., Durfee, C., Kovács, A. P., Börzsönyi, A., Albert, O. & Lopez-Martens, R. (2009). Preservation of the carrier envelope phase during cross-polarized wave generation, *Opt. Express* 17(25): 22358–22365.

Pirozhkov, A. S., Choi, I. W., Sung, J. H., Lee, S. K., Yu, T. J., Jeong, T. M., Kim, I. J., Hafz, N., Kim, C. M., Pae, K. H., Noh, Y.-C., Ko, D.-K., Lee, J., Robinson, A. P. L., Foster, P., Hawkes, S., Streeter, M., Spindloe, C., McKenna, P., Carroll, D. C., Wahlström, C.-G., Zepf, M., Adams, D., Dromey, B., Markey, K., Kar, S., Li, Y. T., Xu, M. H., Nagatomo, H., Mori, M., Yogo, A., Kiriyama, H., Ogura, K., Sagisaka, A., Orimo, S., Nishiuchi, M., Sugiyama, H., Esirkepov, T. Z., Okada, H., Kondo, S., Kanazawa, S., Nakai, Y., Akutsu, A., Motomura, T., Tanoue, M., Shimomura, T., Ikegami, M., Daito, I., Kando, M., Kameshima, T., Bolton, P., Bulanov, S. V., Daido, H. & Neely, D. (2009). Diagnostic of laser contrast using target reflectivity, *Appl. Phys. Lett.* 94(24): 241102.

Renault, A., Augé-Rochereau, F., Planchon, T., D'Oliveira, P., Auguste, T., Chériaux, G. & Chambaret, J.-P. (2005). ASE contrast improvement with a non-linear filtering Sagnac interferometer, *Opt. Commun.* 248(4–6): 535–541.

Roskos, H., Seilmeier, A., Kaiser, W. & Harvey, J. D. (1987). Efficient high-power optical pulse compression with logarithmic wing analysis, *Opt. Commun.* 61(1): 81–86.

Schmid, K., Veisz, L., Tavella, F., Benavides, S., Tautz, R., Herrmann, D., Buck, A., Hidding, B., Marcinkevicius, A., Schramm, U., Geissler, M., Meyer-ter Vehn, J., Habs, D. & Krausz, F. (2009). Few-cycle laser-driven electron acceleration, *Phys. Rev. Lett.* 102(12): 124801.

Shah, R. C., Johnson, R. P., Shimada, T., Flippo, K. A., Fernandez, J. C. & Hegelich, B. M.

(2009). High-temporal contrast using low-gain optical parametric amplification, *Opt. Lett.* 34(15): 2273–2275.

Strickland, D. & Mourou, G. (1985). Compression of amplified chirped optical pulses, *Opt. Commun.* 56(3): 219–221.

Stuart, B. C., Feit, M. D., Herman, S., Rubenchik, A. M., Shore, B. W. & Perry, M. D. (1996). Nanosecond-to-femtosecond laser-induced breakdown in dielectrics, *Phys. Rev. B* 53(4): 1749–1761.

Stuart, B. C., Feit, M. D., Rubenchik, A. M., Shore, B. W. & Perry, M. D. (1995). Laser-induced damage in dielectrics with nanosecond to subpicosecond pulses, *Phys. Rev. Lett.* 74(12): 2248–2251.

Tavella, F., Schmid, K., Ishii, N., Marcinkevičius, A., Veisz, L. & Krausz, F. (2005). High-dynamic range pulse-contrast measurements of a broadband optical parametric chirped-pulse amplifier, *Appl. Phys. B* 81(6): 753–756.

Teisset, C., Ishii, N., Fuji, T., Metzger, T., Köhler, S., Holzwarth, R., Baltuška, A., Zheltikov, A. & Krausz, F. (2005). Soliton-based pump-seed synchronization for few-cycle opcpa, *Opt. Express* 13(17): 6550–6557.

Thaury, C., Quere, F., Geindre, J.-P., Levy, A., Ceccotti, T., Monot, P., Bougeard, M., Reau, F., d/'Oliveira, P., Audebert, P., Marjoribanks, R. & Martin, P. (2007). Plasma mirrors for ultrahigh-intensity optics, *Nature Phys.* 3: 424–429.

Tien, A.-C., Backus, S., Kapteyn, H., Murnane, M. & Mourou, G. (1999). Short-pulse laser damage in transparent materials as a function of pulse duration, *Phys. Rev. Lett.* 82(19): 3883–3886.

Tsakiris, G. D., Eidmann, K., Meyer-ter Vehn, J. & Krausz, F. (2006). Route to intense single attosecond pulses, *New J. Phys.* 8(1): 19.

Verhoef, A.-J., Seres, J., Schmid, K., Nomura, Y., Tempea, G., Veisz, L. & Krausz, F. (2006). Compression of the pulses of a Ti:sapphire laser system to 5 femtoseconds at 0.2 terawatt level, *Appl. Phys. B* 82(4): 513–517.

von der Linde, D., Sokolowski-Tinten, K. & Bialkowski, J. (1997). Laser-solid interaction in the femtosecond time regime, *Appl. Surf. Sci.* 109-110: 1 – 10.

Witte, S., Zinkstok, R. T., Wolf, A. L., Hogervorst, W., Ubachs, W. & Eikema, K. S. E. (2006). A source of 2 terawatt, 2.7 cycle laser pulses based on noncollinear optical parametric chirped pulse amplification, *Opt. Express* 14(18): 8168–8177.

Wittmann, T., Geindre, J. P., Audebert, P., Marjoribanks, R., Rousseau, J. P., Burgy, F., Douillet, D., Lefrou, T., Ta Phuoc, K. & Chambaret, J. P. (2006). Towards ultrahigh-contrast ultraintense laser pulses—complete characterization of a double plasma-mirror pulse cleaner, *Rev. Sci. Instrum.* 77: 083109.

Yanovsky, V., Chvykov, V., Kalinchenko, G., Rousseau, P., Planchon, T., Matsuoka, T., Maksimchuk, A., Nees, J., Cheriaux, G., Mourou, G. & Krushelnick, K. (2008). Ultra-high intensity- 300-tw laser at 0.1 hz repetition rate, *Opt. Express* 16(3): 2109–2114.

Yuan, P., Xie, G., Zhang, D., Zhong, H. & Qian, L. (2010). High-contrast near-ir short pulses generated by a mid-ir optical parametric chirped-pulse amplifier with frequency doubling, *Opt. Lett.* 35(11): 1878.

Ziener, Ch., Foster, P. S., Divall, E. J., Hooker, C. J., Hutchinson, M. H. R., Langley, A. J. & Neely, D. (2002). Specular reflectivity of plasma mirrors as a function of intensity, pulse duration, and angle of incidence, *J. Appl. Phys.* 93: 768.

15

Modeling the Interaction of a Single-Cycle Laser Pulse With a Bound Electron Without Ionization

Ufuk Parali and Dennis R. Alexander
*Department of Electrical Engineering,
University of Nebraska Lincoln
Lincoln, NE 68588,
USA*

1. Introduction

Ultrashort light pulse research has led to the creation of laser systems generating pulses only a few cycles in duration (Couairon et al., 2006). Now that these ultrashort few-cycle EM pulses exist experimentally, the need for mathematical models to describe these short pulse interactions with matter becomes very important (Porras, 1999). Questions arise on what is the meaning of the index of refraction of a material during a single cycle pulse interaction. There is a growing need to model and to understand the interaction of single ultrashort pulses or a train of ultrashort pulses with matter below the point where strong field effects dominate. This need is driven by the advances made in femtosecond (fs) and attosecond (as) laser technologies. Applications of these ultra short pulses range from free space communications, 3D depth profiling in biological samples, optical communication, high resolution/precision atomic and molecular scale imaging, high speed electronics and optoelectronics in terahertz (THz) regime, behavior of electrons in quantum structures, relativistic physics, high-energy physics, astrophysics to medical applications. Furthermore, ultrafast few cycle lasers are expected to be a promising solution to probe the fastest events in atomic, molecular, biochemical, and solid state systems due to their unique property of being the shortest controlled bursts of energy ever developed (Corkum, 2007; Zewail, 2000; Niikura, 2002; Itatani et al., 2004; Krauss et al., 2009; Couairon et al., 2006; Yan et al., 1985; Steinmeyer et al., 1999; Akimoto, 1996).

Basic physics of the pulse-matter interaction depends strongly on the ratio of the pulse duration and the characteristic response time of the medium (as well as on the pulse intensity and energy). This ratio is the key term in the polarization response of the medium from a classical point of view. The goal of this book chapter is to provide insight in the linear polarization response of dispersive materials to ultrashort single cycle pulses. This book chapter is concerned with the case where the electric field strength is low and thus would not produce ionization. Since the energy is below the ionization threshold of the medium, there is not any plasma effect during the interaction of the applied field with the material Understanding the linear polarization response is extremely crucial in order to formulate a realistic field integral. This realistic field integral will provide a more realistic propagation model of optical pulses through dispersive media (Joseph et al. ,1991; Dvorak & Dudley, 1995; Kozlov & Sazabov, 1997; Wilkelmsson, et al., 1995, Kinsler, 2003; Eloy &Wilhelmsson,

1997; Pietrzyk et al., 2008; Macke & Segard, 2003; Zou & Lou, 2007; Xiao &Oughstun 1999; Hovhannisyan, 2003. The interaction of an ultra short pulse with matter involves the interaction of the incident electric field with the electrons of the material. In this book chapter, classical approaches to this problem are modified in two separate cases for solving the interaction of a single-cycle ultrashort laser pulse with a bound electron without ionization. In this book chapter , interaction of an ultrashort single-cycle pulse (USCP) with a bound electron without ionization is compared for two different assumptions on the movement of the electron and the applied field. For a more realistic mathematical description of USCPs, Hermitian polynomials and combination of Laguerre functions are used for two different single cycle excitation cases. These single cycle pulse models are used as driving functions for the classical approach to model the interaction of a bound electron with an applied electric field. A new novel time-domain technique was developed for modifying the classical Lorentz damped oscillator model in order to make it compatible with USCP excitation (Parali & Alexander, 2010). This modification turned the Lorentz oscillator model equation into a Hill-like function with non-periodic time varying damping and spring coefficients. In section two of this book chapter, we extend earlier work (Parali & Alexander, 2010) by introducing a convolution of the applied electric field with the time dependent position of the electron. This latter model provides a continuous updating of the applied electric field convoluted with the time dependent position of the electrons motion. The two models vary in the complexity of the assumptions being applied to the computations. For the sake of completeness, we have chosen to include both pieces of work in this book chapter.

1.2 Mathematical model

In order to make an original contribution for the analysis of the interaction of an ultrashort single-cycle pulse (USCP) with a bound electron without ionization, first it is necessary to find a realistic model for a USCP. Such pulses have a rather different structure from conventional modulated quasi-monochromatic signals with a rectangular or Gaussian envelope (Shvartsburg, 1998; Wang et al., 1997; Shvartsburg, 1996; Shvartsburg, 1999). Due to the following main reasons associated with USCPs, combination of Laguerre functions and Hermitian polynomials (Mexican Hat) are used in this study for modeling applied EM field:

i. Arbitrary transient steepness: The rising and the falling times of the signal can be essentially unequal.
ii. Varying zero spacing: The distances between zero-crossing points may be essentially unequal.
iii. Both the waveform envelope and its first spatial and temporal derivatives are continuous.
iv. Arbitrary envelope asymmetry: USCP waveforms can be classified conventionally for two groups.
    1. The sharply defined zero-crossing point at the pulse leading edge as initial point (combination of Laguerre functions).
    2. The sharply defined narrow maximum against a background of comparatively long tails (Hermitian polynomials - Mexican Hat) (Shvartsburg, 1998; Wang et al., 1997; Shvartsburg, 1996; Shvartsburg, 1999).

Although delta function or the Heaviside step function are widely used, they assume zero signal duration and zero relaxation time. These assumptions are not suitable for modeling the waveform of a USCP. There are some other more realistic models, such as modulated Gaussian or rectangular transients, but these models assume equally spaced zeros which is not suitable for a USCP, neither (Shvartsburg, 1998; Wang et al., 1997; Shvartsburg, 1996; Shvartsburg, 1999).

The combination of Laguerre functions for defining the spatiotemporal profile of a USCP is defined as $E_m(t)=B\left(L_m(t)-L_{m+2}(t)\right)$ where $L_m(x)=\left(\exp(x/2)/m!\right)\frac{d^m}{dx^m}\left[\exp(-x)x^m\right]$ is a single Laguerre function with order m and $x=\left(t-zc^{-1}\right)/t_0$. Here, c is the velocity of light in vacuum, z is the propagation direction and t_0 is the time scale of the pulse. In this study, the combination of 2nd and 4th order Laguerre functions are used to define a single USCP:

$$E_2(\alpha)=\exp\left(-(7.5\alpha)^2\right)\left[-\frac{1}{24}\alpha^4+\frac{15}{24}\alpha^3-\frac{5}{2}\alpha^2+2\alpha\right], \tag{1}$$

where the phase term is defined as $\alpha=\left(t-\phi-zc^{-1}\right)/t_0$ in which ϕ is the initial phase [Fig. 1(a)].

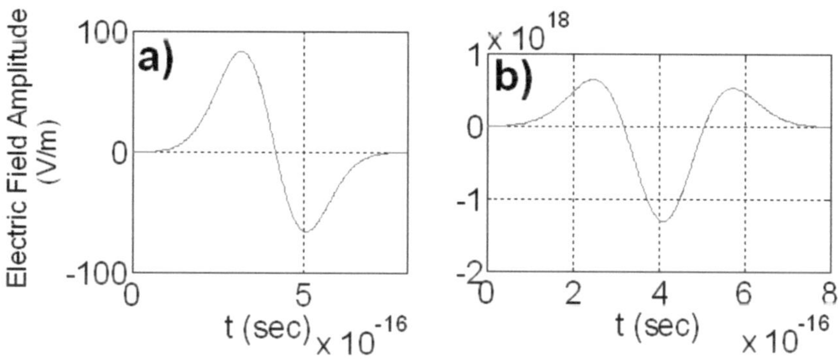

Fig. 1. (a) Applied Laguerre USCP with pulse duration τ_p=8x10^{-16}. (b) 1st derivative of the LaguerreUSCP.

Fig. 1(b) shows the first derivative of the applied field and it is clearly seen that the analytical expression $E(\alpha)$ in Eq. 1 satisfies the conditions of arbitrary transient steepness and arbitrary envelope asymmetry. From Fig. 1(a), it is also clearly seen that it satisfies the condition of varying zero spacing for a USCP. In addition to these, time profile of the Laguerre USCP is almost fulfilling the integral property:

$$\int_0^{\infty} E_2(\alpha)\,d\alpha=0. \tag{2}$$

For the Hermitian (Mexican Hat) USCP [Fig. 2(a)], the following definition is used:

$$E(\alpha)=\left(1-\alpha^2\right)\exp\left(-\alpha^2/2\right). \tag{3}$$

Fig. 2(b) illustrates that the Hermitian pulse satisfies the above concerns.

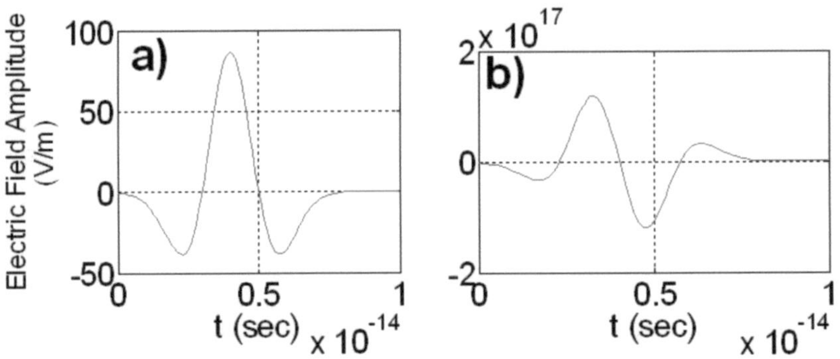

Fig. 2. (a) Applied Hermitian USCP with pulse duration $\tau_p=8x10^{-15}$, (b) 1st derivative of the Hermitian USCP.

In addition to the question how to formulate ultrashort single cycle transients, it is also natural to ask how these pulses propagate in optical medium. In this study, USCP means the smallest possible single cycle piece (unity source) of a wave packet. It is the part of an actual carrier field and does not contain any other carrier fields in itself. For a USCP, it is difficult to introduce the concept of an envelope and it is not possible to define a group velocity. For such short pulses the distinction between carrier oscillations and slowly varying envelope (SVE), which have two different temporal scales that are peculiar to quasi-monochromatic pulses, becomes diffuse or meaningless (Xiao & Oughstrun, 1999; Rothenberg, 1992; Humagai et al., 2003; Crisp, 1970). Jumping from many cycle optical waves to single cycle optical pulses in dealing with light-matter interaction, the mathematical treatments should be revised. The traditional analysis of pulsed EM phenomena is questionable (Shvartsburg, 1998; Wang et al., 1997; Shvartsburg, 1996; Shvartsburg, 1999). If the applied field is a USCP, the shortest possible field as explained above, then it is impossible to separate the applied source into pieces to find the effect of each part (or piece) by superposing as being suggested in the models explained in many fundamental textbooks (Scaife, 1989).

In order to understand the USCP-medium interaction phenomenon, we must acquire certain special features such as operating directly with Maxwell equations beyond the scope of Fourier representations [(Shvartsburg, 1998; Wang et al., 1997; Shvartsburg, 1996; Shvartsburg, 1999). Since the situations occur where the time scale of the pulse is equal or shorter than the relaxation time of the medium, material has no time to establish its response parameters during the essential part of the pulse continuance (Gutman, 1998; Gutman 1999; Daniel 1967; Shvartsburg 2005; Shvartsburg 2002). These parameters, which govern the polarization response of the media, change their values during the pulse continuance (Gutman, 1998; Gutman, 1999). Thus, solutions of Maxwell equations with time-dependent coefficients are required for the analysis of the wave dynamics (Shvartsburg, 2005; Shvartsburg, 2002).

In our study, we consider an approach such that under a single USCP excitation, the change in the relative position of a bound electron to its parent atom without ionization will change the amplitude of the dipole in the atom and so forth the instantaneous polarization. As a result of this fluctuation in the polarization, the index of refraction will change in the duration of the single USCP excitation during which the propagation dynamics of the same applied USCP and the other USCPs coming after the first one will be evaluated. So physically, we consider a case where the medium is including the source. This is a common situation especially in optical communication. In addition to this, we can associate this approach to some diagnostic techniques in ultrafast optics such as pump-probe experiments where both pump and probe pulses propagate and evaluate the time varying physical parameters of the medium. But before diving into Maxwell equations, we have to figure out how the polarization response of the medium must be handled for the interaction of a USCP EM field with a bound electron. Understanding the polarization response of the material under the excitation of a USCP EM field is one of the most important, not clearly answered yet, core question of today and near future ultrafast laser engineering.

Polarization is a very crucial physical term, especially for optical communication, since it defines the change in the index of refraction in the material due to the applied field (Gutman, 1998; Gutman, 1999; Daniel, 1967; Cole & Cole, 1941; Djurisic & Li, 1998). In terms of permittivity, we can write index of refraction (for a nonmagnetic material) as:

$$n = \left(1 + P_{pol}(t) / \varepsilon_o E(t)\right)^{1/2}, \tag{4}$$

where ε_o is the permittivity of free space, $E(t)$ is the applied electric field, and $P_{pol}(t)$ is the electronic polarization. The polarization response of the medium gives the change in the index of refraction. This change or this polarization response affects the temporal and spatial evaluation of the propagating pulse (Couairon et al., 2006; Steinmeyer et al., 1999; Blanc et al., 1993; Agrawal & Olsson 1998; Schaffer 2001).

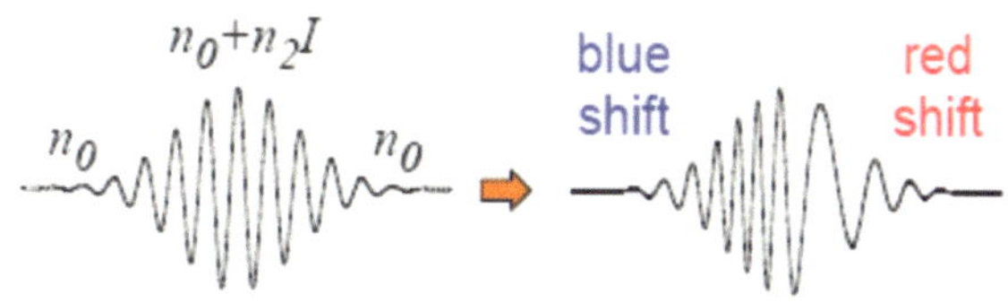

Fig. 3. Schematic representation of self modulation (pulse chirping). Although we are interested in the low intensity applied fields for linear polarization in this study, temporal dependence of the intensity profile of the applied field can still cause a temporal dependence in the refractive index (Schaffer, 2001).

The starting point of all these dynamics is the inhomogeneous wave equation:

$$\frac{\partial^2 E(z,t)}{\partial z^2} - \frac{1}{c_0^2}\frac{\partial^2 E(z,t)}{\partial t^2} = \mu_o \frac{\partial^2 P_{pol}}{\partial t^2}, \tag{5}$$

where the polarization is the source term of the governing differential equation. In order to find the polarization, we must find the oscillation field (displacement) of the bound electrons. According to the Lorentz damped forced oscillator model:

$$m_e \frac{d^2 x(t)}{dt^2} + m_e \gamma_o \frac{dx(t)}{dt} + k_o x(t) = eE(t), \tag{6}$$

$x(t)$ is the time dependent displacement or the oscillation field of a bound electron with respect to the applied field $E(t)$, γ_o is the damping constant, k_o is the spring constant of the material and m_e is the mass of electron.

For USCP excitation, unlike the long pulse excitation fields, the response (oscillation) of the electron must be handled in a different manner. Since, both due to the mass of inertia of the electron and the shortness of the USCP compared to the relaxation time of the medium, the electron will not sense the applied field exactly at the leading edge point of the pulse. The response of the electron to the applied field will increase gradually. During this time period, the electron will not follow the oscillation profile of the applied electric field. So, the oscillation field of the electron will not only have a difference in the phase but also will have a different time profile (time-dependency) with the applied field. In regular cases, if the applied field is in the form of e^{jwt} time-dependency, then we assume that the oscillation of the electron will be in the same time-dependency form. In the literature, Lorentz oscillator model is directly used in e^{jwt} time-dependency (Oughstun & Sherman 1989). But for a USCP excitation, not only the time-dependency e^{jwt} is not valid, but also the oscillation field will have a different waveform than the applied field waveform (time-dependency). This means that, the $x(t)$ term in Eq. (6), that is the oscillation motion field of the electron, will have a modified form of time-dependency with respect to the applied USCP. In order to define the modified function $x(t)$, we developed a new time domain technique that we call "Modifier Function Approach". In this approach, we define the oscillation field of the electron as the multiplication of the applied USCP with the modifier function:

$$x(t) = x_o(t)E(t), \tag{7}$$

where $x_o(t)$ is the modifier function. It has a unit of (meter)²/volt which is equivalent to Coulomb*meter/Newton. So physically, modifier function defines dipole moment per unit force. Plugging Eq. (7) into Eq. (6), we obtain

$$m_e \frac{d^2\left(x_o(t)E(t)\right)}{dt^2} + m_e\gamma_o \frac{d\left(x_o(t)E(t)\right)}{dt} + k_o\left(x_o(t)E(t)\right) = eE(t). \tag{8}$$

Performing the necessary calculations in Eq. (8), we obtain Eq. (9) and(10):

$$\begin{aligned} &m_e \frac{d^2x_o(t)}{dt^2}E(t) + 2m_e \frac{dx_o(t)}{dt}\frac{dE(t)}{dt} + m_e x_o \frac{d^2E(t)}{dt^2} + m_e\gamma_o \frac{dx_o(t)}{dt}E(t) + \\ &m_e\gamma_o x_o(t)\frac{dE(t)}{dt} + k_o x_o(t)E(t) = eE(t), \end{aligned} \tag{9}$$

$$m_e E(t)\frac{d^2x_o(t)}{dt^2} + m_e\left(2\frac{dE(t)}{dt} + \gamma_o E(t)\right)\frac{dx_o(t)}{dt} + m_e\left(\frac{d^2E(t)}{dt^2} + \gamma_o\frac{dE(t)}{dt} + \frac{k_o}{m_e}E(t)\right)x_o(t) = eE(t)$$

$$\frac{d^2x_o(t)}{dt^2} + \left(\frac{2}{E(t)}\frac{dE(t)}{dt} + \gamma_o\right)\frac{dx_o(t)}{dt} + \left(\frac{1}{E(t)}\frac{d^2E(t)}{dt^2} + \frac{\gamma_o}{E(t)}\frac{dE(t)}{dt} + \frac{k_o}{m_e}\right)x_o(t) = \frac{e}{m_e}. \tag{10}$$

We can briefly write Eq. (10) as:

$$\frac{d^2x_o(t)}{dt^2} + P(t)\frac{dx_o(t)}{dt} + Q(t)x_o(t) = \frac{e}{m_e}, \tag{11}$$

where

$$P(t) = \frac{2}{E(t)} \frac{dE(t)}{dt} + \gamma_o, \tag{12}$$

$$Q(t) = \frac{1}{E(t)} \frac{d^2E(t)}{dt^2} + \frac{\gamma_o}{E(t)} \frac{dE(t)}{dt} + \frac{k_o}{m_e}. \tag{13}$$

It is seen from Eq. (11) that this equation has a similar form to that of a Hill type equation where for a regular Hill equation, $P(t)$ and $Q(t)$ terms are periodic and the right side is zero. A linear equation of this type occurs often when a system exhibiting periodic motion is perturbed in some way (Hand & Finch, 2008). This type of equation was first derived by G.W. Hill to describe the effect of perturbations on the orbit of the Moon, and it occurs in many other places in physics, including the quantum motion of electrons in a periodic potential of a crystal (Hand & Finch, 2008) . The band theory of solids is based on a similar equation, as is the theory of propagating electromagnetic waves in a periodic structure (Hand & Finch, 2008). Other applications include parametric amplifiers. Although $P(t)$ and $Q(t)$ terms are periodic in a Hill equation, in our case they are not. So, in our model, Eq. (11) is a Hill-like equation which has a dc source on its right side and a time-dependent damping coefficient (see Eq. 12) and a time-dependent spring coefficient (see Eq. 13) in terms of a damped forced oscillator model. The objective of Eq. (11) is to find the modifier function which can be then used to define the oscillation field (polarization response) of the material. Due to the time-dependent damping and spring coefficients, the modifier function is totally coupled with the time dependency or time profile of the applied field.

Eq. (6) could also have been solved directly in the temporal domain, in which case we would have lost the analogy with the Hill-like equation. But the appropriateness of using the more complicated approach with the modifier function has solid physical reasons. In the case of a USCP excitation, the polarization response of the material is not unique all through the pulse continuance. Due to the shortness of the duration of the applied USCP comparing to the relaxation time of the bound electron, the interaction dynamics and the ability of the material to sense and follow the applied USCP field during its continuance will be completely different than the conventional matter-field interaction approach. In Eq. (6), physical parameters (damping and spring coefficients) are constant. However, the interaction dynamics will not be constant during the USCP excitation. So, in order to penetrate the effect of the applied field into the oscillator model via these physical parameters to have a better understanding of the oscillation response of the material under USCP excitation, we must find the definition of these physical parameters in terms of the applied field and the physical constants of the system (material). Eq. (12) and Eq. (13) are these definitions. They are being used in Eq (11) to find the modifier function which has been embedded into Eq. (6). The physical dimension of the modifier function is a dipole moment per unit force. It frames the time dependency and the phase delay of the oscillation field of the bound electron under USCP excitation.

1.3 Numerical results and discussions for first case assumptions

In Fig. 4, different interaction characteristics of Laguerre and Hermitian pulses are shown for a fixed, relatively low value of damping constant ($\delta_o = 1x10^{14}$ Hz). Due to the definition:

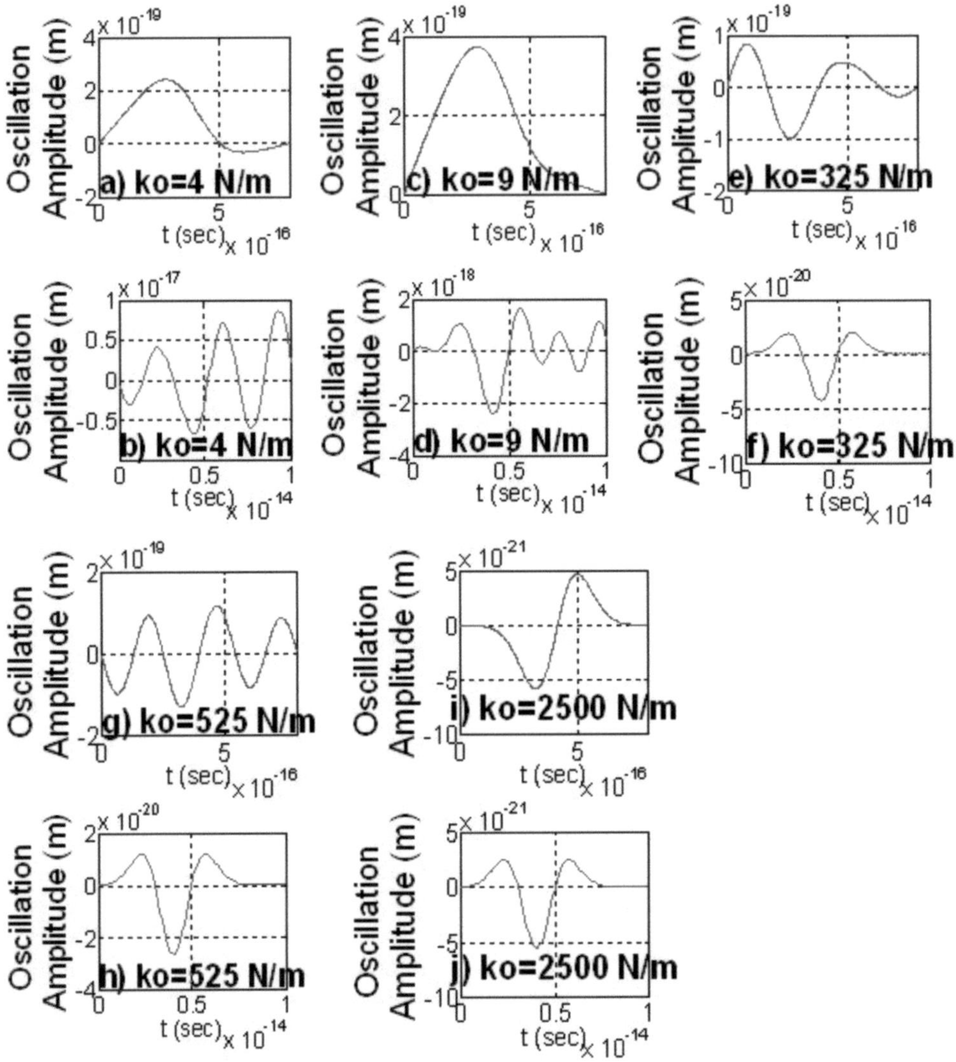

Fig. 4. Bounded electron motion under Laguerre USCP excitation ((a), (c), (e), (g), (i)) and Hermitian USCP excitation ((b), (d), (f), (h), (j)) for various values of spring constant (k_o) with a fixed damping constant ($\delta_o = 1x10^{14}$ Hz).

$w_0 = \sqrt{k_o/m_e}$, (m_e is the mass of electron, k_o is the spring constant for bound electron), the free oscillation frequency of material is in UV range for spring constant values of 4 N/m, 9 N/m, 325 N/m, 525 N/m [Figs. 4(a), 4(b), 4(c), 4(d), 4(e), 4(f), 4(g), 4(h)], 650 N/m [Fig. 8(b)] and 750 N/m [Figs. 5(b), 8(c)]. For spring constant values of 1500 N/m [Fig. 5(c)], 2500 N/m [Figs. 4(i), 4(j)] and 7500 N/m [Fig. 8(d)], the free oscillation frequency is in X-ray range. As it is clearly seen in Fig. 4, the Hermitian interaction has a more tendency to oscillation than the Laguerre interaction for relatively low values of spring constant [see Figs. 4(a), 4(b), 4(c), 4(d)]. As the spring constant is increased, Laguerre interaction gains a more oscillatory profile [see Figs. 4(e), 4(g)] while the oscillation due to the Hermitian pulse interaction stabilizes and its time profile settles down into the inverted phase time profile of the excitation pulse (inverted Mexican Hat) [see Figs. 4(f), 4(h), 4(j)]. Here, the amplitude of oscillation or the amplitude of trembling-like motion of the electron is in the range of 10^{-20} m - 10^{-21} m which is in the scale of electron radius length. Finally, as the spring constant is increased to relatively higher values, the Laguerre interaction settles down into the inverted phase time profile of the excitation pulse, too (inverted Laguerre pulse) [see Fig. 4(i)]. Fig. 4 shows a very clear distinction between the interaction characteristics of Laguerre and Hermitian USCPs until the spring constant is 2500 N/m (after this value, we obtain only the inverted phase time profile of the excitation source for the oscillation). The oscillation characteristics of bound electron under different single USCP sources originates from modifier function approach. The Hill-like equation, which is the result of the modification on the classic Lorentz damped oscillator model with the modifier function approach, causes the time varying physical parameters to come into play during the interaction process. Since these physical parameters (time varying damping and spring coefficients) are absolutely source dependent, they behave differently in the pulse duration of each different USCP source. As a result of this, we see different oscillation profiles for a bound electron under a single Laguerre and Hermitian USCP excitations.

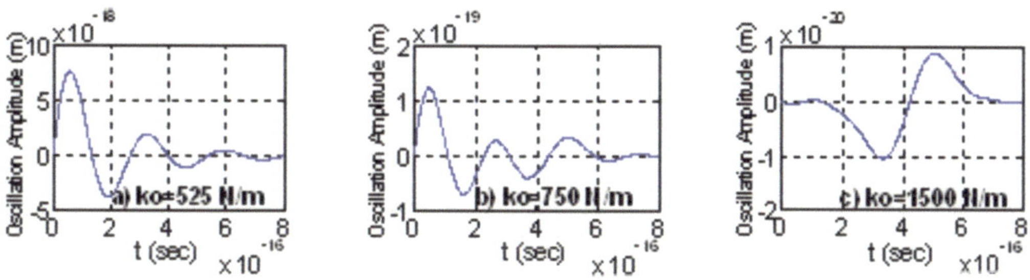

Fig. 5. Laguerre pulse excitation oscillations for damping constant: $\delta_o = 1x10^{16}$ Hz.

In Fig. 5, response of a bound electron is shown for a Laguerre pulse excitation for varying values of spring constant with a fixed, relatively higher damping constant value ($1x10^{16}$) than the previous case (Fig.4). An interesting feature here in Fig. 5(a) and Fig. 4(g) is that although they are at the same spring constant value, they show different oscillation characteristics. Due to a higher dampimg coefficient in Fig. 5(a), while the oscillation attenuates quicker at the second half cycle of the Laguerre USCP than in Fig. 4(g), it hits to a higher peak at the first half cycle of the excitation pulse than in Fig. 4(g). So, for a reasonable value of spring constant, while relatively higher damping coefficient makes the first half cycle of the Laguerre USCP more efficient in the means of interaction, it makes the second half cycle less efficient. In order to compare oscillation results more detailly between Figs. 5(a) and 4(g), it is necessary to look at their physical parameter solutions such as time varying damping and time varying spring coefficients. As it is explained above, these time varying parameters come into play due to the nature of "Modifier Function Approach". In Fig. 6, time varying damping coefficient, time varying spring coefficient and the modifier function solutions of

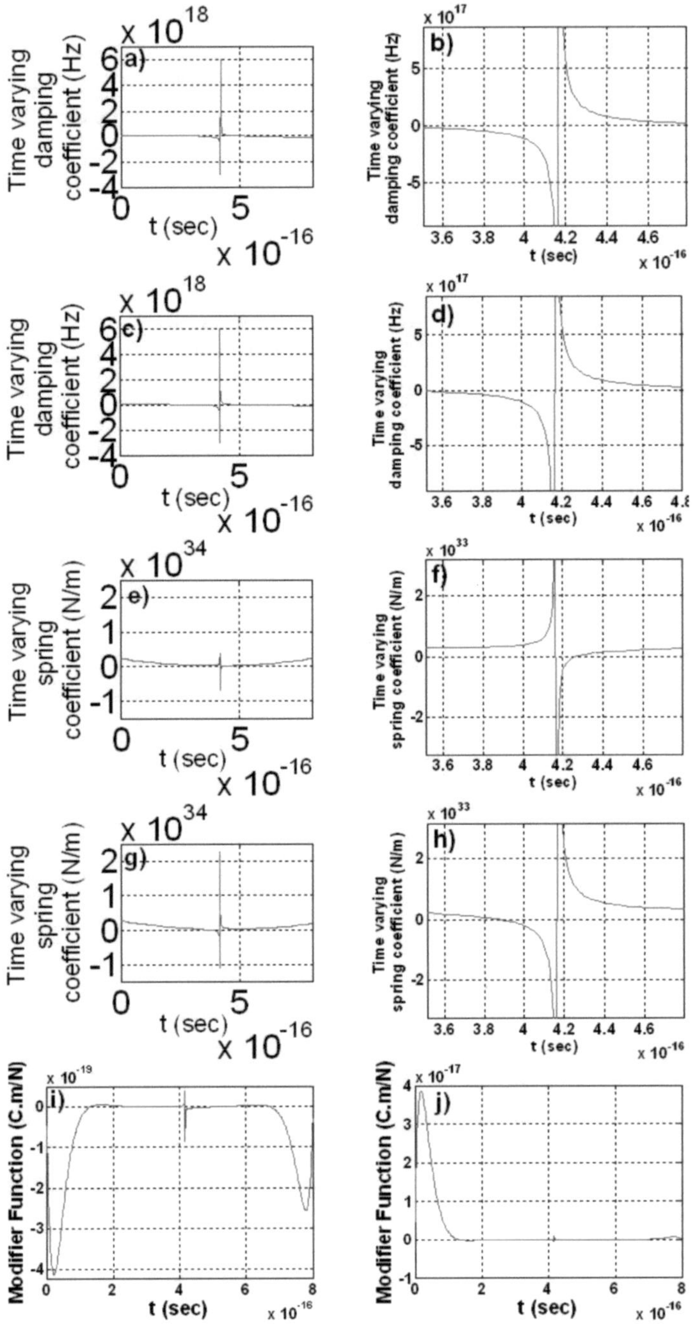

Fig. 6. Laguerre pulse excitation physical parameter solutions for spring constant $k_o = 525$ N/m. (a), (b), (e), (f) and (i) are the solutions of Fig. 4(g) (damping constant $\delta_o = 1x10^{14}$ Hz). (b) and (f) are the magnified views of (a) and (e) respectively. (c), (d), (g), (h) and (j) are the solutions of Fig. 5(a) (damping constant $\delta_o = 1x10^{16}$ Hz). (d) and (h) are the magnified views of (c) and (g) respectively.

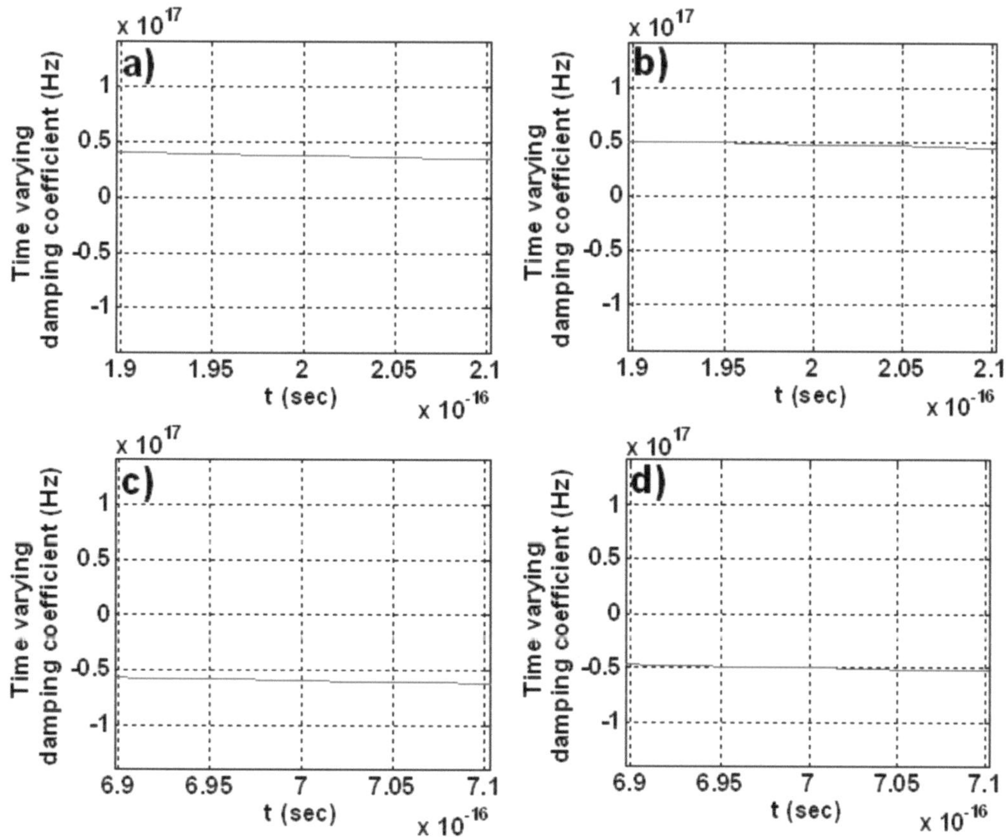

Fig. 7. (a) – (b): Magnified views of left wings of Figs. 6(a)-6(c). (c) – (d): Magnified views of right wings of Figs. 6(a)-6(c).

Figs. 4(g) and 5(a) are shown respectively for two different damping constant values with a fixed spring constant at 525 N/m. In Figs. 6(a) and 6(c), a sudden jump is seen in the time varying damping coefficient profiles at the time point where the excitation pulse changes its polarization direction. Although they look identical, the magnified views [see Figs. 7(a), 7(b), 7(c), 7(d)] of the left and right wings of the damping coefficient show the difference between two different damping constant cases. Here, the left wing corresponds to the first half cycle, right wing corresponds to the second half cycle of the Laguerre excitation pulse. Comparing the amount of the change on the y-axis with the time duration on the x-axis between Figs. 7(a) - 7(b), and 7(c) - 7(d), it is easy to see the reasonable amount of difference to affect the solution of modifier function [see Figs. 6(i), 6(j)]. For time varying spring coefficients [see Figs. 6(e), 6(g)], a significant difference is seen in the time profile although the spring constant values are the same for both cases. The jump in Fig. 6(g) hits a higher peak than the jump in Fig. 6(e). This can be a reasonable explanation for a relatively low oscillation tendency in the second half cycle of Fig. 5(a) than the Fig. 4(g). It can be said that, due to the dissipation of higher energy, this jump causes a lower oscillation profile for the bound electron during its interaction with the second half cycle of the Laguerre pulse in Fig. 5(a) than in Fig. 4(g). In Fig. 5(c), as the spring constant is increased to a relatively higher values, same as in Fig. 4(i), the oscillation profile settles down into the inverted time phase profile of the excitation pulse. Different from Fig. 4(i), the oscillation settles down at a relativley lower spring constant value. So, it can be said that, for a higher damping constant, a lower spring constant is enough to stabilize the oscillation profile in time domain.

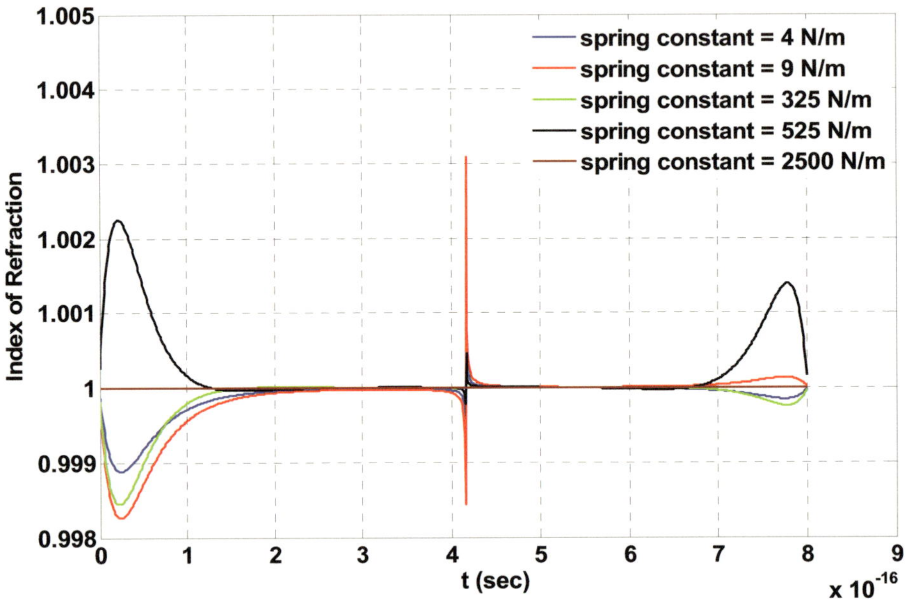

Fig. 8. Time dependent index of refraction during the interaction of a single Laguerre USCP with a bound electron without ionization for different spring constant values with a fixed damping constant value ($\delta_o = 1x10^{14}$ Hz). It is obtained from Eq. (4) where $P_{pol}(t) = -Nex(t)$. Here $N = 6.02x10^{23}$ and $-e$ is the electron charge.

Fig. 8 shows the perturbation effect of an applied single Laguerre USCP on the index of refraction during its continuance for varying spring constants with a fixed damping constant value. As it is clearly seen in Fig. 8, for all spring constant values except the relatively higher case (2500 N/m), there are three regions where the perturbation effects are dominant. These are the trailing and leading regions of the pulse and the time region where the applied electric field changes its polarization sign. The change in the index of refraction around the trailing and leading edges is not as sharp as the change at the point where the polarization sign of the field changes. To see this sudden effect more clearly, the zoomed view of this region is shown in Fig. 9.

The same type of perturbation behavior seen in Fig. 8, is seen in the interaction of a single Hermitian USCP with a bound electron, too (see Fig. 10). Both of these figures have the same damping constant value. The only difference in the time dependent perturbation of index of refraction between these two cases is that since there are two points where the Hermitain USCP field changes its polarization sigh, we have sudden changes in the perturbation of index of refraction twice around these point. The zoomed view of these regions shows the sudden effects more clearly in Fig 11.

In Fig. 12, we see a similar type of change in the time dependent index of refraction for damping constant $\delta_o = 1x10^{16}$ Hz.

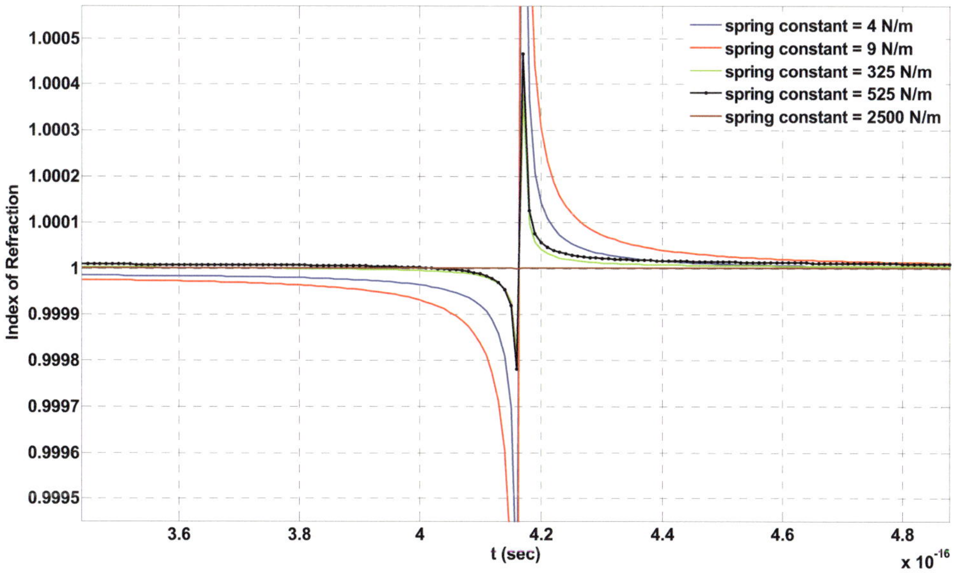

Fig. 9. The jump in the time dependent index of refraction where the electric field changes its polarization sign.

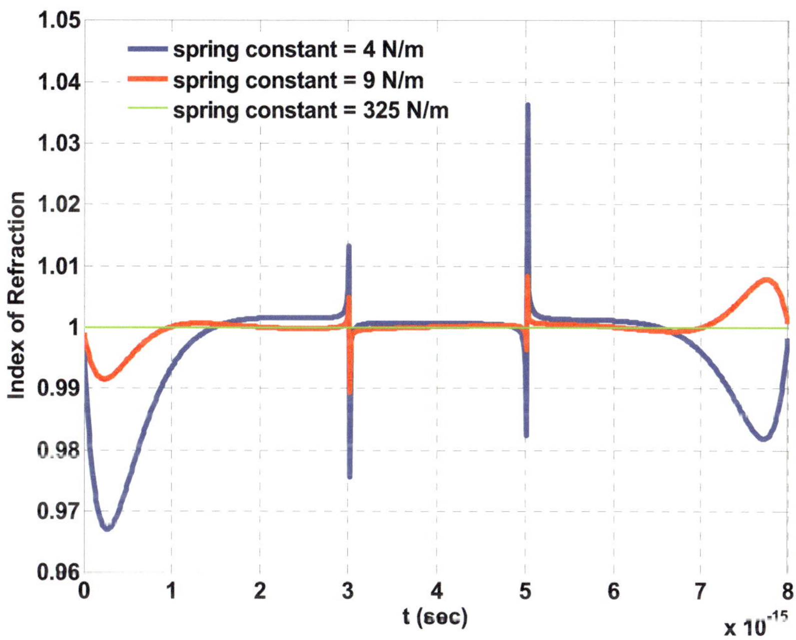

Fig. 10. Time dependent index of refraction during the interaction of a single Hermitian USCP with a bound electron without ionization for different spring constant values with a fixed damping constant value ($\delta_o = 1x10^{14}$ Hz).

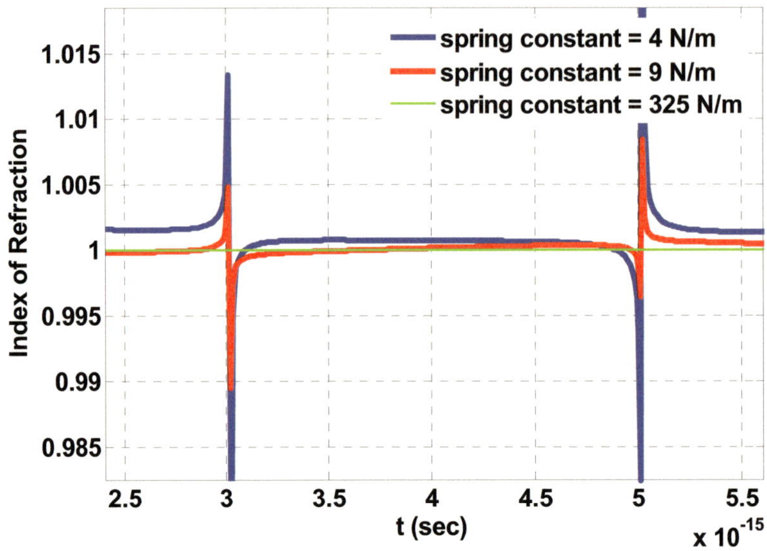

Fig. 11. The jump in the time dependent index of refraction where the electric field changes its polarization sign for single Hermitian USCP interaction.

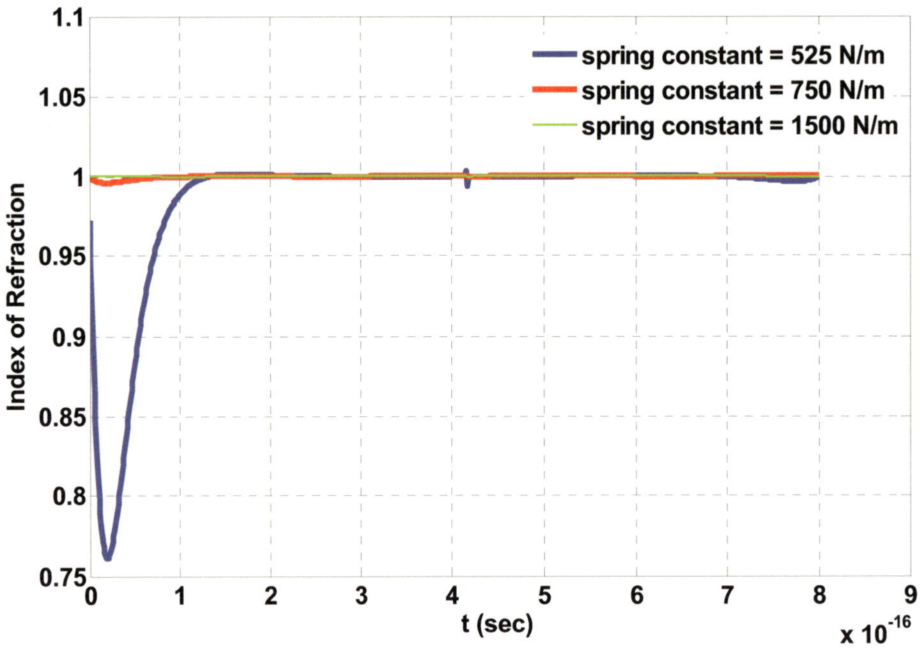

Fig. 12. Time dependent index of refraction during the interaction of a single Laguerre USCP with a bound electron without ionization for different spring constant values with a fixed damping constant value ($\delta_o = 1x10^{16}$ Hz).

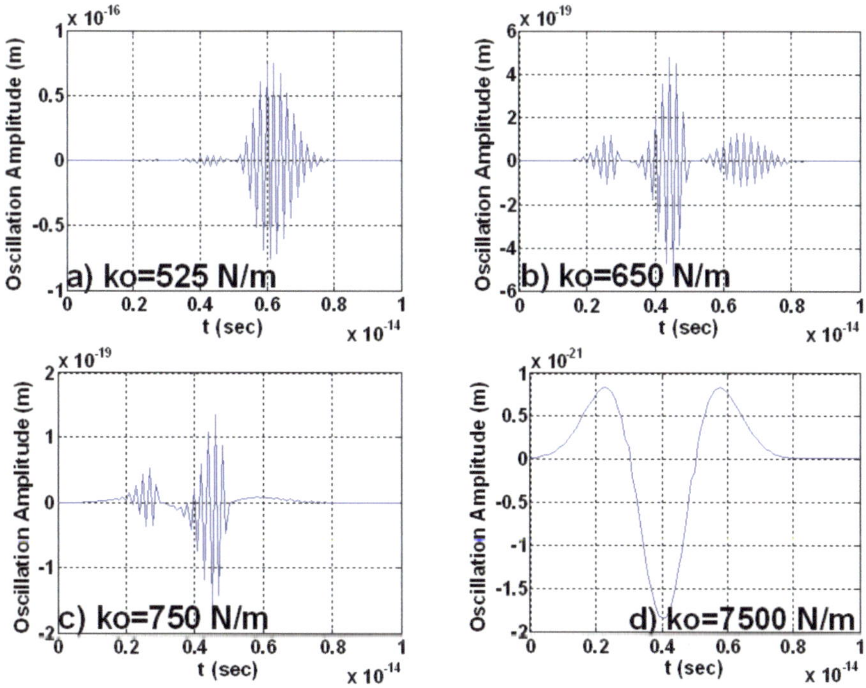

Fig. 13. Hermitian pulse excitation oscillations for damping constant: $\delta_o = 1x10^{17}$ Hz

For a damping constant value of 1x10^{17} (Fig. 13), very different oscillation behaviors are seen than the previous cases (Fig. 4) of Hermitian pulse excitation. The most prominent feature in Figs. 13(a), 13(b) and 13(c) is the high frequency oscillation profile with a phase delay with respect to the excitation pulse. In Fig. 13, spring constant is increased gradually from 13(a) to 13(c) while keeping the damping value constant. For a relatively low value of spring constant in Fig. 13(a), the main lobe and the trailing tail of the excitation pulse have almost no effect on the oscillation of the electron. The bound electron starts sensing the leading tail of the Hermitian excitation after a phase delay of 5 fs. In Fig. 14, the modifier function solutions for the Hermitian pulse excitation for Fig. 13 are shown. As it is clearly seen in Fig. 14(a), modifier function suppresses the interaction effect of main lobe and the trailing tail of Hermitian function. As a result of this, the bound electron starts sensing the excitation pulse with a phase delay [Fig. 13(a)] associated with the modifier function. Same behaviour of the modifier function is seen in Figs. 14(b) and 14(c), too. As a result of this, approximately 2 fs phase delay occurs in Figs. 13(b) and 13(c). In Fig. 14(d), the type of modifier function is seen that gives a completely phase inverted time profile of the excitation pulse for the oscillation of the bound electron. In Fig. 13(d), the stabilized oscillation profile is seen as a result of this modifier function.

In Fig. 15, as in the Fig. 13, there is a high oscillation frequency behaviour in the perturbation effect of the single Hermitian USCP on the index of refraction. Especially, the magnitude of the perturbation effect is more significant around the main lobe and the trailing edge regions than the leading edge region of the applied field. The effect of the Hermitian USCP on the index of refraction decreases as the spring constant increases for the given fixed damping constant value.

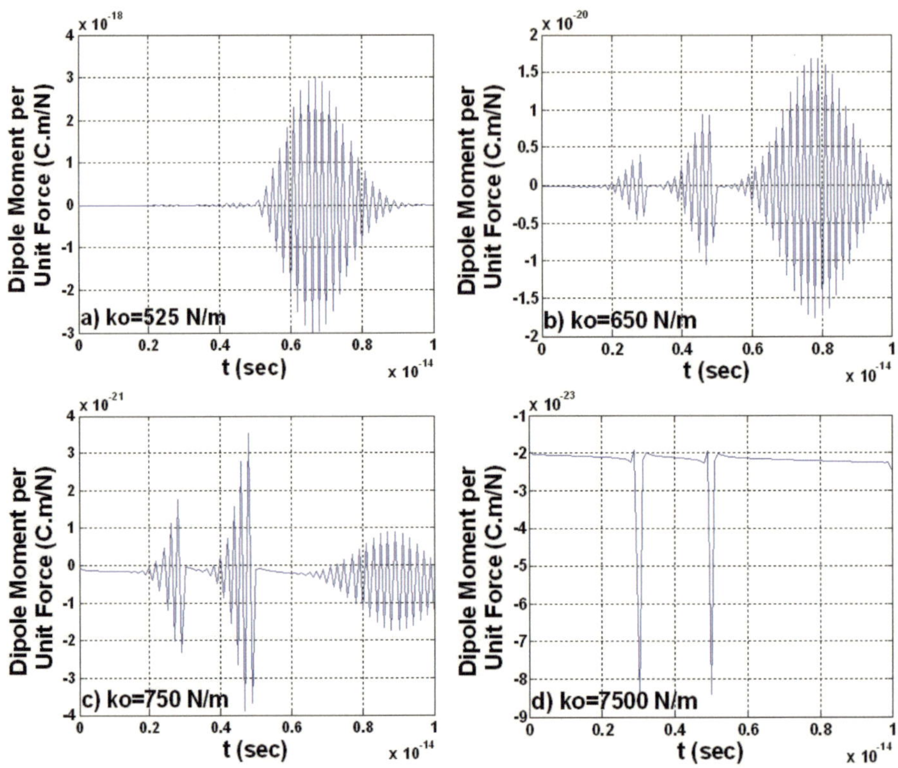

Fig. 14. Hermitian pulse excitation modifier functions for damping constant: $\delta_o = 1x10^{17}$ Hz.

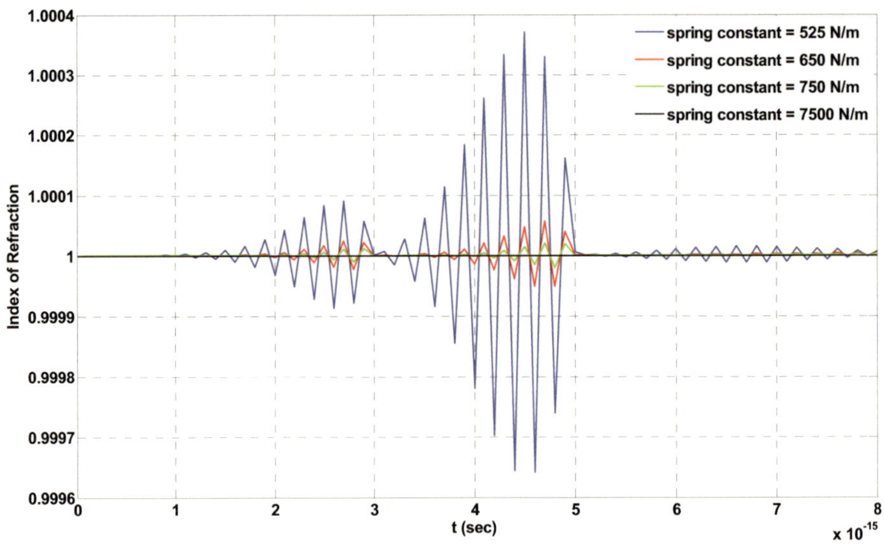

Fig. 15. Time dependent index of refraction during the interaction of a single Hermitian USCP with a bound electron without ionization for different spring constant values with a fixed damping constant value ($\delta_o = 1x10^{17}$ Hz).

2.1 Mathematical model for convolutional modifier function approach

In section 1.2, we explained why the oscillation field of the bound electron under single USCP exposure must be defined in terms of the multiplication of the applied USCP with a modifier function. In a more realistic approximation, we need to include a constant updating between the electron motion and the time dependent applied field. This is the major difference between approaches used in sections 1.2 and 2.1. Suppose that we are applying two different USCPs ranging in different spectral content on to the same type of material at different points. If we assume that the majority of the spectral content of one of these USCPs is relatively closer to the natural oscillation frequency of the bound electron of the material than the spectral content of the other USCP (see Fig. 16), then it will not be realistic to consider exactly the same type of time domain USCP interaction mechanism (modifier function approach that has been explained in section 1.2) for both of these two different USCPs.

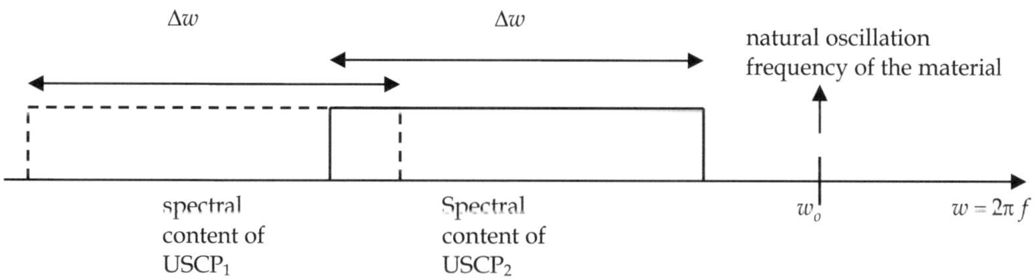

Fig. 16. Spectral content of two different USCPs with the same pulse duration. They are being applied to different points on a material which has a natural oscillation frequency of w_o.

As it is seen in Fig. 16, we note that since the majority of the spectral content of $USCP_2$ is closer to w_o than the majority of the spectral content of $USCP_1$, in the context of interaction efficiency, the interaction of $USCP_2$ will be relatively more intense than the interaction of $USCP_1$ for the given spectral content and for the given natural oscillation frequency. Given the formulation provided in section 1.2, we are just directly masking (multiplying) the modifier function (that we found from Eq. 11) on to the time domain profile of $USCP_1$ to find the oscillation field of the bound electron during the continuance of this pulse. If we follow the same procedure to calculate the oscillation field of the bound electron under $USCP_2$ excitation, this will cause us to miss the cumulative tendency due to the memory effect of the oscillation field of the bound electron in time domain due to the interaction with single $USCP_2$ compared to the interaction with single $USCP_1$. In order to take into consideration the cumulativeness effect under $USCP_2$ excitation, instead of defining oscillation as in Eq. 7, we need to define the time dependent electron motion with a convolution operation: since a convolution can be considered as an operation that shows the effect of current and past inputs to the current output of a system:

$$x(t) = x_o(t) * E(t). \tag{14}$$

If we plug Eq. (14) into Eq. (6), we obtain

$$\frac{d^2}{dt^2}\left(x_o(t) * E(t)\right) + \gamma_o \frac{d}{dt}\left(x_o(t) * E(t)\right) + \frac{k_o}{m_e}\left(x_o(t) * E(t)\right) = \frac{e}{m_e} E(t). \tag{15}$$

Eq. 15 allows us to obtain the oscillation field after the pulse (wake-field) due to the nature of convolution operation in Eq. 14. The modifier function is a hidden function that must be evaluated first to find the oscillation field caused by the USCP excitation where the source duration is much shorter than the relaxation dynamics of the material. Due to the nature of convolution operation in Eq. 14, although the USCP actually vanishes at $t=\tau$ (where τ is the pulse duration), the modifier function will still exist after the end of the pulse and our technique evaluates the oscillation field after the pulse duration due to the memory effect of the convolution operation.

In order to find the modifier function in Eq. 15, different mathematical solution techniques can be used. For the work in this book chapter, let us use Eq. 14 in the following form:

$$f(t)=x_o(t)+x_o(t)*E(\tau), \tag{16}$$

$$f(t)=x_o(t)+\int_0^t x_o(t-\tau)E(\tau)d\tau, \tag{17}$$

which is called Volterra Integral Equation (VIE) of the second kind where the source function $f(t)$ and the kernel function $E(\tau)$ are given and $x_o(t)$ is the unknown function. There are many existing state of the art numerical techniques for solving the VIE in the literature (e.g., Tang et al., 2008). Future work will be undertaken to solve this equation following numerical techniques developed specifically for the approximate solution of VIE. In future work it is anticipated that publications will compare numerical solutions to our simpler mathematical solution approach. However, in this book Chapter we will follow a simpler mathematical procedure in order to obtain physical understanding and insight of differences between convolutional modifier function approach (section 2.1) and the modifier function approach used in section 1.2. Let's define the convolution integral in Eq. 17 as:

$$\int_0^t x_o(t-\tau)E(\tau)d\tau=f(t)-x_o(t), \tag{18}$$

where $f(t)$ is going to be a reasonable trial function that will be defined for finding the modifier function in Eq. 15. By plugging the definition in Eq. 18 into Eq. 15, we obtain:

$$\frac{d^2}{dt^2}x_o(t)+\gamma_o\frac{d}{dt}x_o(t)+\frac{k_o}{m_e}x_o(t)=F(t), \tag{19}$$

where

$$F(t)=\frac{d^2}{dt^2}f(t)+\gamma_o\frac{d}{dt}f(t)+\frac{k_o}{m_e}f(t)-\frac{e}{m_e}E(t). \tag{20}$$

While in Eq. 11 in section 1.2 we are calculating the modifier function for time dependent damping and spring coefficients, in Eq. 19 we calculate the modifier function for constant damping and spring coefficients with a time dependent source term modified by the trial function $f(t)$. This approach allows us to incorporate the cumulative tendency of the oscillation field and memory effect originating from the spectral content of the USCP and to have constant damping and spring coefficients during the pulse continuance.

2.2 Numerical results for second case

For our numerical calculations, we used the following forms as two trial functions simultaneously for the Laguerre USCP excitation case:

$$f_1(t) = f_{o1}\left(\exp\left(-\beta^2\right) - Sin\left(\beta^a\right)\right), \tag{21}$$

$$f_2(t) = f_{o2}\left(\exp\left(-\beta^2\right) + Sin\left(\beta^a\right)\right), \tag{22}$$

where a is ranging from 1 to n and $\beta = {t - zc^{-1}}/{to}$. So, at the end of the calculations, the total oscillation field has been evaluated as:

$$x(t) = \frac{1}{2n}\sum_{i=1}^{n}\left[x_{o1i}(t) * E(t) + x_{o2i}(t) * E(t)\right], \tag{23}$$

where $x_{o1i}(t)$ is calculated for $f_1(t)_{a=i}$ and $x_{o2i}(t)$ is calculated for $f_2(t)_{a=i}$ from Eq. 19.
For the Hermitian USCP excitation case, we used the following form as the trial function in the numerical calculations:

$$f_3(t) = f_{o3}\left(1 - \beta^{-a}\right), \tag{24}$$

and the total oscillation field has been evaluated as:

$$x(t) = \frac{1}{n}\sum_{i=1}^{n}\left[x_{o3i}(t) * E(t)\right], \tag{25}$$

where $x_{o3i}(t)$ is calculated $f_3(t)_{a=i}$ from Eq. 19.
The values of the amplitude constants f_{o1}, f_{o2}, and f_{o3} are dependent on the trial function and the number of trial functions that are chosen for the solution of the modifier function.
In Fig. 17, we see some important results of the convolutional modifier function approach on the oscillation field of the bound electron under Laguerre and Hermitian USCP excitation and both have close spectral content to the natural oscillation frequency of the material. Although there is not much difference in the oscillation frequency compared to the Fig. 4 in section 1.3, there is a significant difference in the oscillation amplitude where the convolutional modifier function approach has higher amplitudes. In addition to this (different than Fig. 4), in Fig. 17 we see some phase delay in the oscillation field with respect to the applied USCP for both Laguerre and Hermitian excitations (see Figs. 17(a), 17(b), 17(c), 17(e), 17(g), 17(h), 17(i) and 17(j)). Another significant result shown in Fig. 17, due to the nature of the convolution operation, we can see the oscillation in the wake-field after the continuance of the USCP.
For Fig 18, we have a higher oscillation amplitude and almost the same oscillation frequency as compared to Fig. 5. Also in Figs. 18(a), 18(b) and 18(c) there is a phase delay which is not seen in Figs. 5(a) and 5(b). It is observed comparing Fig. 17 to Fig. 18 there is a significant difference in the wake field oscillations which are attenuated much quicker in Fig. 18 after the end of the pulse continuance.
In Figs. 19 and 20, we plot the real and imaginary part of the perturbation effect of an applied single Laguerre USCP for the convolutional modifier function approach. The common behavior that we note in Figs. 8, 19 and 20 is that there is a sudden jump for real

and imaginary parts of the index of refraction at the point where the USCP field changes its polarization sign. In comparison (see Fig. 8) the real part of the perturbation effect of the applied USCP vanishes at some regions of the Laguerre USCP as illustrated in Fig 19. As can be seen in Fig. 20, at the regions where the real part vanishes, the imaginary part of the perturbation effect on the index of refraction comes into play.

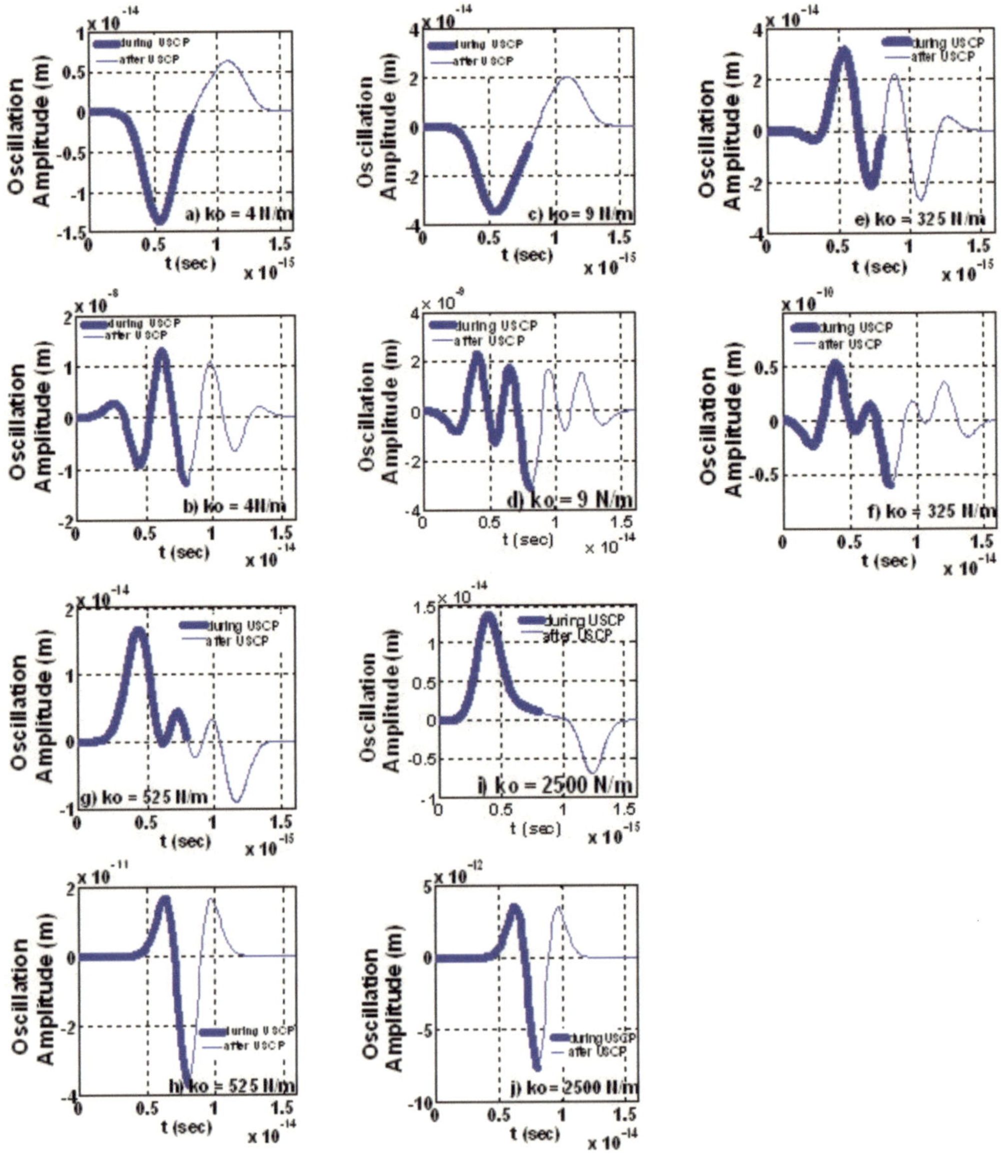

Fig. 17. Bounded electron motion for the convolutional modifier function approach under Laguerre USCP excitation ((a), (c), (e), (g), (i)) and Hermitian USCP excitation ((b), (d), (f), (h), (j)) for various values of spring constant (k_o) with a fixed damping constant ($\delta_o = 1x10^{14}$ Hz).

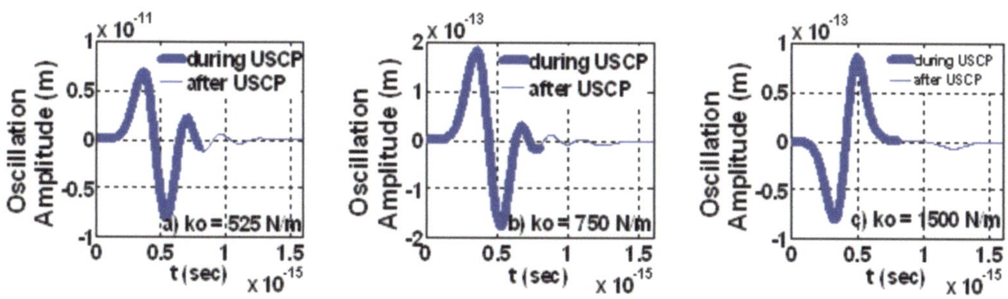

Fig. 18. Bounded electron motion for the convolutional modifier function approach under Laguerre USCP excitation for various values of spring constant (k_o) with a fixed damping constant ($\delta_o = 1x10^{16}$ Hz).

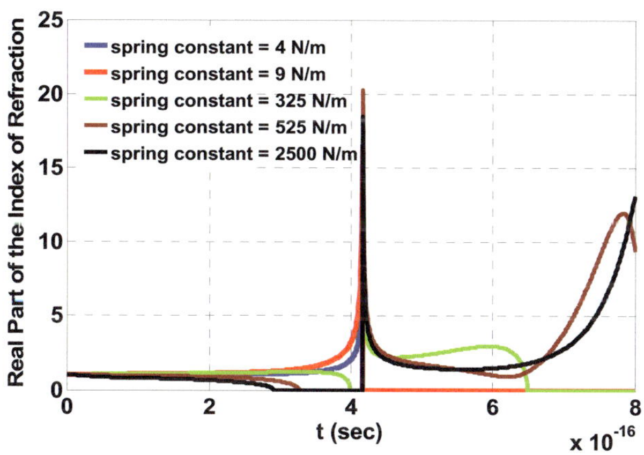

Fig. 19. Real part of the time dependent index of refraction during the interaction of a single Laguerrc USCP with a bound electron without ionization for different spring constant values with a fixed damping constant ($\delta_o = 1x10^{14}$ Hz)[see Eq. 4].

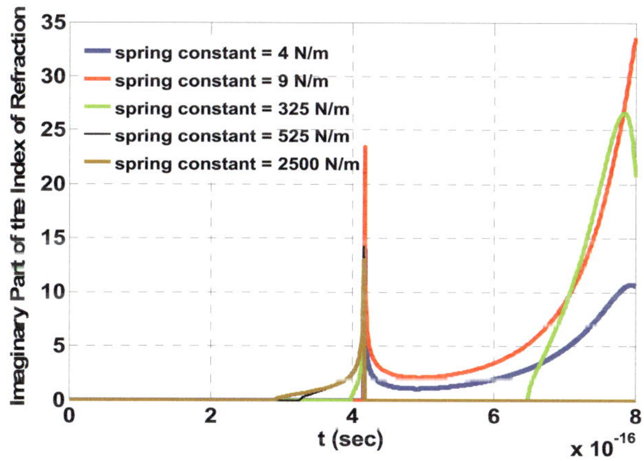

Fig. 20. Imaginary part of the time dependent index of refraction during the interaction of a single Laguerre USCP with a bound electron without ionization for different spring constant values with a fixed damping constant ($\delta_o = 1x10^{14}$ Hz)[see Eq. 4].

3. Conclusion

The results of this work indicate that if the applied field is a USCP, then it is not possible to separate the field into pieces to find the polarization effect of each part of the applied field on a bound electron since the USCP can not be further broken down into separate pieces of the applied field. The traditional Fourier method of multiplying the Delta function response with the applied field and integrating (superposing) this product in time can only be used for SVE approximation which is not realistic for single cycle pulses of unity femtosecond and attosecond applied fields. In a USCP case, the Lorentz oscillator model must be modified in order to find the polarization effect of a single USCP. Since a USCP is extremely broadband, it is not realistic to use a center frequency in the calculations as is done in the Fourier series expansion approach. Results in this work are presented on the transient response of the system during the USCP duration without switching to frequency domain. In order to accomplish this mathematically, we developed a new technique we label as the "Modifier Function Approach". The modifier function is embedded in the classic Lorentz damped oscillator model and by this way, we upgrade the oscillator model so that it is compatible with the USCP on its right side as the driving force. Results of this work also provide a new modified version of the Lorentz oscillator model for ultrafast optics. The results also indicate that the time response of the two models used to represent the USCP can alter the time dependent polarization of the material as it interacts with a single cycle pulse.

As a second model, we chose to provide a convolution of the applied field and the movement of the electron for a further refinement of the classical Lorentz damped oscillator model. The convolution approach allows one to incorporate previous motion of the electron with the interacting applied field. Results are compared for the motion of the electron for each case and the observed change in the index of refraction as a function of time for two different cases. As expected the index of refraction is not a constant in the ultra short time time domain under the assumptions applied in these studies. The motion of the electron is also highly dependent on the type of input single cycle pulse applied (Laguerre or Hermitian).

In future work, we plan on providing chirp to the pulse and performing the necessary calculations to show the motion of the electron and the effects on the index of refraction as a function of time.

5. References

Agrawal, G. P. Olsson, N. A. (1989). Self-phase modulation and spectral broadening of optical pulses in semiconductor laser amplifiers, *IEEE Journal of Quantum Electronics*, Vol. 25., No. 11., (November 1989).

Akimoto, K. (1996). Properties and Applications of ultrashort electromagnetic mono- and sub- cycle waves. *Journal of the Physical Society of Japan*, Vol. 65., No. 7., (2020-2032).

Blanc, S. P. Sauerbrey, R. Rae, S. C. Burnett, K. (1993). Spectral blue shifting of a femtosecond laser pulse propagating through a high-pressure gas, *Journal of Optical Society of America*, Vol. 10, No. 10, (October 1993).

Cole, K. S. Cole, R. H. (1941). Dispersion and absorption in dielectrics, *Journal of Chemical Physics*, Vol. 9., (341-351), (April 1941).

Corkum P. B. (200). Attosecond science *Nature Physics* 3, (381 – 387).

a. Couairon, A. Biegert, J. Hauri, C. P. Kornelis, W. Helbing, F. W. Keller, U. Mysyrowicz, A. (2006). Self-compression of ultrashort laser pulses down to one optical cycle by filamentation, *Journal of Modern Optics*, Vol. 53., No. 1-2., (January 2006).

b. Couairon, A. Biegert, J. Hauri, C. P. Kornelis, W. Helbing, F. W. Keller, U. Mysyrowicz, A. (2006). Self-compression of ultra-short laser pulses down to one optical cycle by filamentation, *Journal of Modern Optics*, Vol. 53., Issue 1&2., (January 2006), (75 - 85).

Crisp, M. D. (1970). Propagation of small-area pulses of coherent light through a resonant medium. *Physical Review A*, Vol. 1., No. 6., (June 1970).

Daniel, V. V. (1967). Dielectric Relaxation. *Academic Press, New York.*

Djurisic, A. B. Li, E. H. (1998). Modeling the index of refraction of insulating solids with a modified Lorentz oscillator model. *Applied Optics*, Vol. 37., No. 22., (August 1998).

Dvorak, S. L. Dudley, D. G. (1995). Propagation of ultrawideband electromagnetic pulses through dispersive media. *IEEE Transaction of Electromagnetic Compatibility*, Vol. 37., No. 2., May 1995.

Eloy, J. F. Moriamez, F. (1992). Spectral analysis of EM ultrashort pulses at coherence limit. Modelling. *SPIE Intense Microwave and Particle Beams III*, Vol. 1629.

Eloy, J. F. Wilhelmsson, H. (1997). Response of a bounded plasma to ultrashort pulse excitation. *Physica Scripta*, Vol. 55., (475-477).

Gutman, A. L. (1998). Electrodynamics of short pulses for pulse durations comparable to relaxation times of a medium. *Doklady Physics*, Vol. 43., No. 6, 1998., (343-345).

Gutman, A. L. (1999). Passage of short pulse throughout oscillating circuit with dielectric in condenser. *Ultra-Wideband, Short-Pulse Electromagnetics 4, Kluwer Academic / Plenum Publishers, New York.*

Hand, L. N. Finch, J. D. (2008). Analytical Mechanics. *Cambridge University Press*, 7th edition, Cambridge.

Hovhannisyan, D. (2003). Propagation of a femtosecond laser pulse of a few optical oscillations in a uniaxial crystal. *Microwave and Optical Technology Letters*, Vol. 36., No. 4., (February 2003).

Itatani, J. Levesque, J. Zeidler, D. Niikura, H. Pépin, H. Kieffer, J. C. Corkum, P. B. Villeneuve, D. M. (2004). Tomographic imaging of molecular orbitals. *Nature* 432, (867–871).

Joseph, R. M. Hagness, S. C. Taflove, A. (1991). Direct time integral of Maxwell's equations in linear dispersive media with absorption for scattering and propagation of femtosecond electromagnetic pulses. *Optics Letters*, Vol. 16., No. 18., (September 1991).

Kinsler, P. New, G. H. C. (2003). Few-cycle pulse propagation. *Physical Review* A 67, 023813

Kozlov S. A., Sazanov S. V. (1997). Nonlinear propagation of optical pulses of a few oscillations duration in dielectric media. *JETP* 84 (2), (February 1997).

Krauss, G. Lohss, S. Hanke, T. Sell, A. Eggert, S. Huber, R. Leitenstorfer, A. (2009). Synthesis of a single cycle of light with compact erbium-doped fibre technology. *Nature Photonics* 4., (33-36).

Kumagai, H. Cho, S. H. Ishikawa, K. Midorikawa, K. Fujimoto, M. Aoshima, S. Tsuchiya, Y. (2003). Observation of the comples propagation of a femtosecond laser pulse in a dispersive transparent bulk material, *Journal of Optical Society of America*, Vol. 20., No. 3., (March 2003).

Macke, B. Segard, B. (2003). Propagation of light pulses at a negative group velocity, *European Physical Journal D*, Vol. 23., (125-141).

Niikura H. (2002). Sub-laser-cycle electron pulses for probing molecular dynamics. *Nature* 417, (917–922).

Oughstun, K. E. Sherman, G. C. (1989). Uniform asymptotic description of electromagnetic pulse propagation in a linear dispersive medium with absorption (the Lorentz medium), *Journal of Optical Society of America A*, Vol. 6., No. 9., (September 1989), (1394-1420).

Parali, Ufuk Alexander,Dennis R. (2010). Interaction of a single-cycle laser pulse with a bound electron without ionization. *Optics Express*. Vol. 18., No. 14., (July 2010).

Pietrzyk, M. Kanattsikov, I. Bandelow, U. (2008). On the propagation of vector ultrashort pulses, *Journal of Nonlinear Mathematical Physics*, Vol. 15., No. 2., (162-170).

Porras M. A. (1999). Nonsinusoidal few-cycle pulsed light beams in free space, *Journal of Optical Society of America B*, Vol. 16., No. 9., (September 1999).

Rothenberg, J. E. (1992). Space-time focusing: Breakdown of the slowly varying envelope approximation in the self-focusing of femtosecond pulses. *Optics Letters*, Vol. 17., No. 19., (October 1992)

Scaife, B.K.P. (1989). Principles of Dielectrics, Oxford University Press, Oxford.

Schaffer, C. B. (2001). Interaction of femtosecond laser pulses with transparent materials, Ph.D. Thesis. Harvard University.

Shvartsburg, A. B. (1996). Time-Domain Optics of Ultrashort Waveforms. *Clarendon Press*, Oxford

Shvartsburg, A. B. (1998). Single-cycle waveforms and non-periodic waves in dispersive media (exactly solvable models). *Physics – Uspekhi*, Vol. 41., No. 1., (77-94).

Shvartsburg, A. B. (1999). Impulse Time-Domain Electromagnetic of Continuos Media. *Birkhauser Verlag*, Boston.

Shvartsburg, A. B. (2005). Optics of nonstationary media, *Physics – Uspekhi*, Vol. 48., No. 8., (797-823)

Shvartsburg, A. B. Petite, G. (2002). Progress in Optics, Vol. 44 (Ed. E Wolf), p. 143., Elsevier Sci

Steinmeyer, G. Sutter, D. H. Gallmann, L. Matuschek, N. Keller, U. (1999). Frontiers in Ultrashort Pulse Generation: Pushing the Limits in Linear and Nonlinear Optics. *Science*, Vol. 286., (November 1999).

Tang, T.; Xu, X. & Cheng, J. (2008). On spectral methods for Volterra integral equations and the convergence analysis, Journal of Computational Mathematics, Vol.26, No. 6., (825-837).

Wang, Z. Zhang, Z. Xu, Z. Lin, Q. (1997). Space-time profiles of an ultrashort pulsed Gaussian beam, *IEEE Journal of Quantum Electronics*, Vol. 33., No. 4., (April 1997).

Wilkelmsson, H. Trombert, J. H. Eloy, J. F. (1995). Dispersive and dissipative medium response to an ultrashort pulse: A green's function approach. *Physica Scripta*, Vol. 52., (102-107).

Xiao, H. Oughstun, K. E. (1999). Failure of the group velocity description for ultrawideband pulse propagation in a casually dispersive, absorptive dielectric, *Journal of Optical Society of America B*, Vol. 16., No. 10., (October 1999).

Yan, Y. Gamble, E. B. Jr. Nelson, K. A. (1985). Impulsive stimulated scattering: General importance in femtosecond laser pulse interactions with matter, and spectroscopic applications. *J. Chem. Phys.* Vol. 83., No. 11., (December 1985).

Zewail A. (2000). Femtochemistry: atomic-scale dynamics of the chemical bond, *Journal of Physical Chemistry A, Vol.* 104, (5660–5694).

Zou, Q. Lu, B. (2007). Propagation properties of ultrashort pulsed beams with constant waist width in free space. *Optics and Laser Technology*, Vol. 39, (619-625).

16

Ultrashort, Strongly Focused Laser Pulses in Free Space

Alexandre April
Centre d'optique, photonique et laser, Université Laval
Québec, Canada

1. Introduction

Technological advances in ultrafast optics now allow the generation of laser pulses whose duration is as short as a few optical cycles of the electric field; furthermore, these pulses can be focused to a spot size comparable to the wavelength. These strongly focused, ultrashort laser pulses have found applications, for instance, in high-resolution microscopy, particle trapping and electron acceleration. In order to characterize the spatiotemporal behavior of such ultrashort, tightly focused pulses, one needs the expressions of their electromagnetic fields.

Ultrafast nonparaxial pulsed beams must be modeled as exact solutions to Maxwell's equations. Many studies on the propagation of a pulsed beam are based on a scalar paraxial theory, which provides an accurate description of the pulsed beam propagation when the beam divergence angle is small and the beam spot size is much larger than the wavelength for each spectral component. However, the analysis of tightly focused laser beams requires expressions of optical beams that extend beyond the paraxial approximation. Moreover, the vector nature of light cannot be neglected to properly describe tightly focused beams. Also, the appropriate spectrum amplitude must be employed in order to model ultrashort pulses.

Many authors have proposed expressions for the electromagnetic fields of laser pulsed beams, but most of these models are incomplete. For example, Wang and co-workers presented scalar paraxial pulsed Gaussian beams that have a Gaussian spectrum (Wang *et al.*, 1997), but their expressions are not suitable to describe ultrashort pulses, as reported by Porras (Porras, 1998). Caron and Potvliege suggested forms of spectra, which are appropriate to characterize pulses of very small duration, but the expressions for their vectorial nonparaxial ultrashort pulses are written in terms of numerically calculated angular spectra (Caron & Potvliege, 1999). Lin *et al.* presented closed-form expressions for subcycle pulsed focused vector beams that are exact solutions to Maxwell's equations obtained in the context of the so called complex-source point method, but they used an unsuitable Gaussian spectrum (Lin *et al.*, 2006). Recently, an der Brügge and Pukhov have provided solutions for ultrashort focused electromagnetic pulses found with a more appropriate spectral amplitude, but the expressions hold true only in the paraxial regime (an der Brügge & Pukhov, 2009).

The aim of this chapter is to provide a simple and complete strategy to correctly model strongly focused, ultrashort laser pulses. Three main tools are employed to find the expressions for the fields of such pulsed beam. First, the Hertz potential method is used in order to efficiently obtain the spatiotemporal expressions for the electromagnetic fields that rigorously satisfy Maxwell's four equations. Then, the complex source/sink model is exploited to determine an exact solution to the Helmholtz equation that describes a physically realizable nonparaxial beam that generalizes the standard Gaussian beam. Finally, the so-called Poisson-like spectrum is employed to characterize ultrashort pulses whose duration could be as short as one optical cycle. The combination of these three main ingredients leads to closed-form expressions that accurately describe the electromagnetic fields of laser pulsed beams in free space.

This chapter is divided as follows. In Section 2, the traditional theories used to characterize laser pulsed beams are briefly exposed. In Section 3, the Hertz potential method, the complex-source/sink model, and the Poisson-like spectrum are introduced. In Section 4, the method presented in this chapter is applied to selected types of laser pulses. Finally, in Section 5, one of these special case is investigated in detail to shed light on features related to the propagation of tightly focused, ultrashort pulsed beams.

2. The traditional theories of pulsed beams

Well-established theories for laser pulsed beams are available, but many of them remain accurate only in some specific regimes. A number of authors have treated the propagation of ultrashort, nonparaxial laser pulses with a scalar analysis, although the vector nature of light cannot be ignored for strongly focused beams (Porras, 1998; Saari, 2001; Lu *et al.*, 2003). Some authors have given solutions for ultrashort pulsed beams within the paraxial approximation, whose validity may be questioned for pulses with spectral distributions extending to very low frequencies (Feng & Winful, 2000; an der Brügge & Pukhov, 2009). Others have presented solutions for ultrashort nonparaxial electromagnetic pulsed beams having a Gaussian spectrum, which is not suitable to describe such pulses (Wang *et al.*, 1997; Lin *et al.*, 2006). In fact, the scalar treatment, the paraxial approximation and the Gaussian spectrum are not adequate to model ultrashort, tightly focused pulsed beams. In this section, the shortcomings encountered with these traditional approaches are explored.

2.1 The scalar wave function

To theoretically describe the spatiotemporal behavior of ultrashort, nonparaxial pulses, one needs expressions of their electromagnetic fields that are exact solutions of the wave equation. The electric field $\mathbf{E}(\mathbf{r},t)$ and the magnetic field $\mathbf{H}(\mathbf{r},t)$ of a laser pulse must satisfy Maxwell's equations. In differential form, these fundamental equations in free space are given in Table 1.

Faraday's law	Ampère-Maxwell law	Gauss's law for **E**	Gauss's law for **H**
$\nabla\times\mathbf{E}=-\mu_0\dfrac{\partial\mathbf{H}}{\partial t}$	$\nabla\times\mathbf{H}=\varepsilon_0\dfrac{\partial\mathbf{E}}{\partial t}$	$\nabla\bullet\mathbf{E}=0$	$\nabla\bullet\mathbf{H}=0$

Table 1. Maxwell's equations in free space.

Here, μ_0 and ε_0 are the permeability and the permittivity of free space, respectively. The principle of duality applies in free space: the substitutions $\mathbf{E}\rightarrow\eta_0\mathbf{H}$ and $\mathbf{H}\rightarrow-\mathbf{E}/\eta_0$, where $\eta_0=(\mu_0/\varepsilon_0)^{1/2}$ is the intrinsic impedance of free space, leave Maxwell's four equations unchanged. From Maxwell's equations, one can obtain the wave equations in free space for the electric and the magnetic fields:

$$\nabla^2 \mathbf{E} - \frac{1}{c^2}\frac{\partial^2 \mathbf{E}}{\partial t^2} = \mathbf{0}, \tag{1a}$$

$$\nabla^2 \mathbf{H} - \frac{1}{c^2}\frac{\partial^2 \mathbf{H}}{\partial t^2} = \mathbf{0}, \tag{1b}$$

where $c = (\mu_0 \varepsilon_0)^{1/2}$ is the speed of light in free space. Thus, each Cartesian component of the electric and the magnetic fields must satisfy the scalar wave equation.

The electromagnetic fields can be analyzed in the frequency domain by taking the Fourier transform of Eqs. (1a) and (1b): the temporal derivatives $\partial/\partial t$ are then converted to $j\omega$, where $\omega = kc$ is the angular frequency of the spectral component and k is its wave number. The Fourier transforms of the electric and the magnetic fields, denoted by $\tilde{\mathbf{E}}$ and $\tilde{\mathbf{H}}$ respectively, must satisfy the vector Helmholtz equations $\nabla^2\tilde{\mathbf{E}} + k^2\tilde{\mathbf{E}} = \mathbf{0}$ and $\nabla^2\tilde{\mathbf{H}} + k^2\tilde{\mathbf{H}} = \mathbf{0}$.

It is often assumed that a laser beam is a transverse electromagnetic (TEM) beam, that is, the electric and the magnetic fields are always transverse to the propagation axis, which is the z-axis in this chapter. However, the only true TEM waves in free space are infinitely extended fields. For example, consider a x-polarized beam for which the y-component E_y of its electric field is zero; the x-component E_x of its electric field satisfies the scalar wave equation $\nabla^2 E_x - c^{-2}\,\partial^2 E_x/\partial t^2 = 0$, from which a solution for E_x may be found. One can estimate the longitudinal electric field component of this x-polarized optical beam by applying Gauss's law for **E** to such a beam, giving an expression for the z-component E_z of the beam:

$$E_z = -\int \frac{\partial E_x}{\partial x} \mathrm{d}z\,. \tag{2}$$

Since an optical beam has a finite spatial extent in the plane transverse to the direction of propagation, the component E_x must depend on the transverse coordinate x and, therefore, E_z must be different from zero. Thus, even if it only exhibits a small beam divergence angle, an optical beam always has a field component that is polarized in the direction of the propagation axis. The same argument applies to the magnetic field. In some cases, the strength of the longitudinal component of the fields of a tightly focused laser beam can even exceed the strength of its transverse components. As a result, in order to accurately characterize laser beams or pulses, a vectorial description of their electromagnetic fields is needed and will be discussed in Section 3.1.

2.2 The paraxial approximation

In many applications in optics, the light beam propagates along a certain direction (here, along the z-axis) and spreads out slowly in the transverse direction. When the beam divergence angle is small, the beam is said to be paraxial. Specifically, the electric field of a paraxial beam in the frequency domain is a plane wave $\exp(-jkz)$ of wavelength $\lambda = 2\pi/k$ modulated by a complex envelope that is assumed to be approximately constant within a neighborhood of size λ. The phasor of the x-component of a paraxial beam is therefore written as $\tilde{E}_x = \tilde{A}\exp(-jkz)$, where $\tilde{A}$ is the complex envelope that is a slowly varying function of position. The complex enveloppe must satisfy the paraxial Helmholtz equation (Siegman, 1986):

$$\frac{\partial^2 \tilde{A}}{\partial x^2} + \frac{\partial^2 \tilde{A}}{\partial y^2} - 2jk\frac{\partial \tilde{A}}{\partial z} = 0, \tag{3}$$

provided that the condition $\left|\partial^2\tilde{A}/\partial z^2\right| << \left|2k\partial\tilde{A}/\partial z\right|$ is verified. This condition is called the slowly varying envelope approximation or simply the paraxial approximation. When it applies, the use of this approximation considerably simplifies the analysis of optical beams in many applications.

To model a laser beam, the Gaussian beam is often used. The phasor of the paraxial Gaussian beam, whose envelope is a solution to the paraxial Helmholtz equation, is (Siegman, 1986)

$$\tilde{u}(\mathbf{r},\omega) = F(\omega)\frac{jz_R}{\tilde{q}(z)}\exp\left[-jk\left(\frac{r^2}{2\tilde{q}(z)}+z\right)\right], \tag{4}$$

where $F(\omega)$ is an arbitrary function of the frequency ω only, r and z are the radial and the longitudinal coordinates, respectively, $\tilde{q}(z) = z + jz_R$ is the complex radius of curvature, $z_R = \frac{1}{2}kw_0^2$ is the Rayleigh range, and w_0 is the waist spot size of the beam. The beam divergence angle is given by $\delta \equiv \arctan\left(w_o/z_R\right)$.

The envelope of the Gaussian beam is one solution of the paraxial Helmholtz equation among the infinite number of solutions of this differential equation. Well-known solutions are the envelopes of the higher-order Gaussian modes which include standard and elegant Hermite–Gaussian or Laguerre–Gaussian beams. The elegant beams were introduced by Siegman and they differ from the standard beams because the former contain polynomials with a complex argument, whereas in the latter the argument is real (Siegman, 1986). Physically, the standard beams constitute the natural modes of a stable laser resonator with mirrors having uniform reflectivity, while the elegant beams describe modes generated by a laser resonant cavity that includes soft Gaussian apertures. Both modes form an eigenfunction basis to the paraxial Helmholtz equation. While the Hermite–Gaussian modes are adequate to describe optical beams with rectangular geometry, the Laguerre–Gaussian modes are more appropriate to describe beams with cylindrical symmetry. The phasor of the paraxial elegant Laguerre–Gaussian beam is (April, 2008a)

$$\tilde{u}_{p,m}^{e}(\mathbf{r},\omega) = F(\omega)\left(\frac{jz_R}{\tilde{q}(z)}\right)^{p+m+1}\frac{r^m}{w_o^m}L_p^m\left(\frac{jkr^2}{2\tilde{q}(z)}\right)\exp\left[-jk\left(\frac{r^2}{2\tilde{q}(z)}+z\right)\right]\cos(m\phi), \tag{5}$$

where p = 0,1,2,... is the radial mode number, m = 0,1,2,... is the angular mode number, $L_p^m(\cdot)$ is the associated Laguerre polynomial, and ϕ is the azimuthal angle. Superscript "e" in solutions $\tilde{u}_{p,m}^{e}$ stands for even modes, with the even function $\cos(m\phi)$ for the azimuthal dependence. Odd modes $\tilde{u}_{p,m}^{o}$ are obtained by replacing $\cos(m\phi)$ in Eq. (5) by $\sin(m\phi)$. If $p = m = 0$, then Eq. (5) reduces to Eq. (4), i.e. $\tilde{u}_{0,0}^{e}(\mathbf{r},\omega) = \tilde{u}(\mathbf{r},\omega)$. Both Eqs. (4) and (5) are accurate if the paraxial approximation holds, i.e. when the waist spot size w_0 is not too small with respect to the wavelength λ or more precisely when $w_0 >> \lambda/(\pi\sqrt{2})$.

If the envelope is not a slowly varying function of position, the paraxial approximation does not apply. In fact, when the waist spot size of an optical beam is smaller than the wavelength, the beam is said to be nonparaxial. Moreover, some spectral components of an ultrashort pulsed beam can be considered paraxial while others in the same pulse are nonparaxial. In brief, to accurately describe ultrashort strongly focused pulses, the nonparaxial effects have to be taken into account; thus, exact solutions to the wave equation for their electromagnetic fields are required and will be provided in Section 3.2.

2.3 The Gaussian spectrum

In many cases, it is convenient to use a Gaussian spectrum to model a physical laser pulse. However, for an ultrashort pulsed beam, which has a very broad spectrum, the Gaussian spectrum is no longer appropriate, because the spectral content cannot physically extend in negative frequencies (Caron & Potvliege, 1999). In fact, while it accurately describes the beamlike behavior near the optical axis, the amplitude distribution becomes boundless for large values of the transverse coordinate.

In order to briefly investigate this shortcoming, consider a paraxial Gaussian pulse that has the following Gaussian spectrum:

$$F(\omega) = T\sqrt{\pi}\exp\left[-\tfrac{1}{4}T^2\left(\omega-\omega_o\right)^2 + j\phi_o\right], \tag{6}$$

where T is the duration of the pulse, ω_o is the frequency of the carrier wave, and ϕ_o is a constant phase. The analytic signal $u(\mathbf{r},t)$ of the Gaussian pulse in the temporal domain is obtained by taking the inverse Fourier transform of the function $\tilde{u}(\mathbf{r},\omega)$ given by Eq. (4), i.e.

$$u(\mathbf{r},t) = \frac{1}{2\pi}\int_{-\infty}^{\infty}\tilde{u}(\mathbf{r},\omega)\exp(j\omega t)\mathrm{d}\omega\,. \tag{7}$$

When Eqs. (4) and (6) are substituted in Eq. (7), an integral over ω remains to be solved; the dependence on ω in $\tilde{u}(\mathbf{r},\omega)$ comes from $F(\omega)$ and k (because $k=\omega/c$).

We now consider a so-called isodiffracting pulse (Wang *et al.*, 1997; Caron & Potvliege, 1999; Feng & Winful, 2000). For this type of pulse, all the frequency components have the same Rayleigh range z_R. It may be argued that a mode-locked laser produces isodiffracting pulses, because the Rayleigh range of the generated optical beam is determined by the geometry of the laser cavity only and is thus independent of the frequency ω. In fact, many authors have pointed out that isodiffracting pulses are natural spatiotemporal modes of a curved mirror laser cavity. For isodiffracting pulses, the complex radius of curvature $\tilde{q}(z)$ is frequency independent and, thus, the inverse Fourier transform of Eq. (7) can be easily carried out:

$$u(r,z,t) = \frac{jz_R}{\tilde{q}(z)}\exp\left[j(\omega_o t+\phi_o) - \frac{1}{T^2}\left(t-\frac{z}{c}-\frac{r^2}{2c\tilde{q}(z)}\right)^2 - jk_o\left(z+\frac{r^2}{2\tilde{q}(z)}\right)\right]. \tag{8}$$

Here, $k_o=\omega_o/c$ is the wave number of the carrier wave and c is the speed of light in free space. The physical pulsed beam is the real part of Eq. (8). This equation shows that there is spatiotemporal coupling, i.e. there exists a coupling among the beam parameters in space and time. In fact, the spatial coordinates are involved in the temporal shape of the pulse, whereas the duration of the pulse is involved in the spatial distribution of the pulsed beam.

The pulsed beam modeled by Eq. (8) is not a well-behaved solution: the amplitude profile is boundless for large values of the transverse coordinate r. As a consequence, the energy carried by the beam is infinite. To show this drawback explicitly, consider Eq. (8) when the pulse is in the beam waist ($z = t = 0$):

$$u(r,0,0) = \exp\left[\frac{r^4}{(2cTz_R)^2} - \frac{k_o r^2}{2z_R} + j\phi_o\right]. \tag{9}$$

This amplitude profile grows as $\exp(r^4)$ for large values of r - more precisely, for $r >> T(2\omega_0 z_R c)^{1/2}$. According to this condition, the amplitude growth is not encountered if the pulse is a pulsed plane wave (for which $z_R \rightarrow \infty$) or if the Gaussian spectrum is narrow enough (i.e. if $\omega_0 T >> 1$). The reason of this unphysical growth is a consequence of the existence of negative frequencies in the spectral content of the pulsed beam. In fact, the Gaussian spectrum $F(\omega)$ does not vanish for $\omega < 0$, and the amplitudes of the spectral components with negative frequency, however small they are, grow exponentially for sufficently large values of the transverse coordinate. It must be concluded that a Gaussian spectrum is not suitable to characterize arbitrarily short laser pulses. A suitable spectrum will be introduced in Section 3.3.

3. The three tools to model nonparaxial, ultrashort laser pulses

To find adequate expressions that correctly characterize the fields of ultrashort nonparaxial electromagnetic pulses in all regimes, three main tools are used. First, to obtain all the electromagnetic fields components that satisfy Maxwell's equations exactly, the Hertz potential method is employed. Second, to solve the Helmholtz equation rigorously, the complex-source/sink method is exploited. Third, to model ultrafast pulses whose duration could be as short as one optical cycle of the electric field, a Poisson-like spectrum is used.

3.1 The Hertz potential method

As mentioned in Section 2, when the beam divergence angle becomes sufficently large, not only the paraxial approximation does not hold, but a scalar treatment is no longer adequate. To accurately desbribe a strongly focused beam, the phasors of its electromagnetic fields must be exact solutions to Maxwell's equations. Many authors have proposed expressions for the electric field of an optical beam that is a rigorous solution to Maxwell's equations. Richards and Wolf developed an integral representation of the electric field of a tightly focused beam (Richards & Wolf, 1959); nevertheless, the integrals have to be solved numerically in general. Another method, developed by Lax, Louisell and McKnight, consists in adding corrections to the phasor of the paraxial beam (Lax *et al.*, 1975). The resulting phasor is therefore expressed as a troncated power series; the larger the number of terms is, the more accurate is the expression. According to the methods of Richards and Wolf as well as of Lax *et al.*, the vector wave equation is solved for the electric field. This approach is rather complicated since the electric field of an optical beam generally has three nonzero components. The Hertz potential method allows to solve Maxwell's equations in a more efficient way.

The physical fields that have to be determined, for a given laser pulse, are the electric field $\mathbf{E}(\mathbf{r},t)$ and the magnetic field $\mathbf{H}(\mathbf{r},t)$. However, it is often useful to introduce the vector magnetic potential $\mathbf{A}(\mathbf{r},t)$ and the electric potential $V(\mathbf{r},t)$, which are defined by $\mathbf{H} \equiv (1/\mu_0)\nabla \times \mathbf{A}$ and $\mathbf{E} \equiv -\nabla V - \partial\mathbf{A}/\partial t$. Because a vector is entirely defined only if its divergence and its curl are specified, the divergence of the vector magnetic potential must be defined and it is usually determined with the Lorenz condition for potentials: $c^2\nabla \bullet \mathbf{A} + \partial V/\partial t = 0$, where c is the speed of light in vacuum. With the Lorenz condition, the vector magnetic potential and the electric potential both satisfy the wave equation in free space. The vector magnetic potential and the electric potential are introduced in order to simplify the computation of electromagnetic fields; often, potentials are easily computed and then electromagnetic fields are directly deduced from the definitions of these potentials.

The Hertz potential method is a powerful tool that can be used to determine the spatiotemporal expressions for the electromagnetic fields of pulsed beams. The electric and magnetic Hertz potentials, $\boldsymbol{\Pi}_e$ and $\boldsymbol{\Pi}_m$ respectively, are defined in terms of the electromagnetic potentials: $V \equiv -\nabla \bullet \boldsymbol{\Pi}_e$ and $\mathbf{A} \equiv (1/c^2)\partial\boldsymbol{\Pi}_e/\partial t + \mu_0 \nabla \times \boldsymbol{\Pi}_m$. This particular choice for the Hertz potentials is such that the Lorenz condition is identically verified, since the divergence operator and the temporal derivative commute and since the divergence of a curl vanishes. The Hertz potentials may be seen as "super-potentials" in the sense that they are vector potentials from which other potentials can be obtained. Similarly to the electromagnetic fields and the electromagnetic potentials, the Hertz potentials are also chosen to satisfy the wave equation in free space:

$$\nabla^2 \boldsymbol{\Pi}_e - \frac{1}{c^2}\frac{\partial^2 \boldsymbol{\Pi}_e}{\partial t^2} = \mathbf{0}, \tag{10a}$$

$$\nabla^2 \boldsymbol{\Pi}_m - \frac{1}{c^2}\frac{\partial^2 \boldsymbol{\Pi}_m}{\partial t^2} = \mathbf{0}. \tag{10b}$$

If Eqs. (10a) and (10b) are satisfied, then the electric field $\mathbf{E}(\mathbf{r},t)$ and the magnetic field $\mathbf{H}(\mathbf{r},t)$ may be obtained with the following relationships (Sheppard, 2000):

$$\mathbf{E} = \nabla \times \nabla \times \boldsymbol{\Pi}_e - \mu_0 \frac{\partial}{\partial t} \nabla \times \boldsymbol{\Pi}_m \tag{11a}$$

$$\mathbf{H} = \nabla \times \nabla \times \boldsymbol{\Pi}_m + \varepsilon_0 \frac{\partial}{\partial t} \nabla \times \boldsymbol{\Pi}_e \tag{11b}$$

It can be verified by straightforward calculations that Eqs. (11a) and (11b) satisfy identically Maxwell's four equations in free space if $\boldsymbol{\Pi}_e$ and $\boldsymbol{\Pi}_m$ are both solutions of the vector wave equation. Thus, if the Hertz potentials are known for a given pulsed beam, then the electromagnetic fields are deduced by applying Eqs. (11a) and (11b).

The Hertz potential method consists in assuming that the Hertz potentials are linearly polarized. Accordingly, the nonzero Cartesian component of $\boldsymbol{\Pi}_e$ and $\boldsymbol{\Pi}_m$ obeys a scalar wave equation whereas the electromagnetic fields still have to obey a vector wave equation, since all the components of the fields are nonzero in general. It can be seen, therefore, that working with the Hertz potentials has the advantage of simplifying the determination of a solution to the wave equation. In fact, dealing directly with the electric and the magnetic fields, it is necessary to solve a vector wave equation for the fields instead of a scalar wave equation for the Hertz potentials.

The appropriate choice of the nonzero components of the Hertz potentials depends on the state of polarisation of the given pulsed beam. Among others, three states of polarization can be easily generated with the Hertz potential method: transverse magnetic (TM), transverse electric (TE), and linearly polarized (LP) beams. TM beams may be obtained with an electric Hertz potential oriented along the propagation axis ($\boldsymbol{\Pi}_e = \hat{\mathbf{a}}_z \Psi(\mathbf{r},t)$ and $\boldsymbol{\Pi}_m = \mathbf{0}$, where $\Psi(\mathbf{r},t)$ is a scalar function with $\mathrm{V \cdot m}$ units), whereas TE beams may be obtained with a magnetic Hertz potential oriented along the z-axis ($\boldsymbol{\Pi}_e = \mathbf{0}$ and $\boldsymbol{\Pi}_m = \hat{\mathbf{a}}_z \eta_0^{-1} \Psi(\mathbf{r},t)$, where η_0 is the intrinsic impedance of free space). Also, a linearly polarized beam can be produced by a combination of an electric dipole and a magnetic dipole, oriented along the x- and the y-axes, respectively, or in other words by setting $\boldsymbol{\Pi}_e = \hat{\mathbf{a}}_x \Psi(\mathbf{r},t)$ and $\boldsymbol{\Pi}_m = \hat{\mathbf{a}}_y \eta_0^{-1} \Psi(\mathbf{r},t)$.

Hence, we are looking for a rigorous solution to the scalar wave equation for the nonzero Cartesian component $\Psi(\mathbf{r},t)$ of the Hertz potentials.

3.2 The complex source/sink model

The methods from Richards and Wolf and from Lax *et al.* previously mentioned give solutions to the wave equation for the electric field of an optical beam. The former leads to an integral representation of the electric field while the latter gives an infinite-series expansion of the field. However, integral representations or series expansion of an optical beam become computationally onerous or increasingly inaccurate as the beam divergence angle grows. Hence, closed-form solutions for the electromagnetic fields of a nonparaxial beam would be interesting to avoid such an inconvenience. First, the so-called complex point-source method is introduced; second, the confocal parameter that characterizes the beam divergence is clearly defined; third, the complex source/sink model which does not exhibit the shortcomings of the complex point-source method is explained; finally, the nonparaxial higher-order beams are presented.

The complex point-source model

The complex point-source method is a simple approach to obtain a rigorous solution to the Helmholtz equation, expressed in a simple closed form, that describes a nonparaxial beam. Deschamps has been the first to introduce the complex source-point method, which consists in assuming that the beam is generated by a source located at an imaginary distance along the propagation axis (Deschamp, 1971). Mathematically, it means that the longitudinal coordinate of the phasor of the wave is replaced by a complex quantity whose imaginary part is closely related to the beam divergence angle. The complex source-point method turns out to be a useful technique to convert a spherical wave into a nonparaxial Gaussian beam.
Couture and Bélanger have shown that, in the context of the perturbative method of Lax *et al.*, the sum of all the corrections to the paraxial Gaussian beam transforms the Gaussian beam into the complex source-point spherical wave (Couture & Bélanger, 1981). Thus, the complex source-point method is equivalent to the approach of Lax *et al.*, provided that the boundary condition is such that the corrections are zero along the optical axis of the beam. The complex source-point method allows to analytically write the phasor of a nonparaxial optical beam in a closed form, without having to deal explicitly with a series expansion.
It is well known that the phasor of the paraxial Gaussian beam [Eq. (4)] can be formally obtained if it is assumed that it consists in a paraxial spherical wave (a parabolic wave) emitted by a point source positioned at the imaginary distance $-jz_R$ along the propagation axis, where z_R is the Rayleigh range of the beam (Siegman, 1986). With a similar approach applied to the phasor of a nonparaxial spherical wave, one can obtain the phasor of a nonparaxial Gaussian beam.
The components of the Hertz potentials are now considered in the spectral domain. The Fourier transform of the function $\Psi(\mathbf{r},t)$, denoted by $\tilde{\Psi}(\mathbf{r},\omega)$, must then satisfy the scalar Helmholtz equation $\nabla^2\tilde{\Psi}+k^2\tilde{\Psi}=0$, where $k=\omega/c$ is the wave number of the spectral component of angular frequency ω (Fig. 1). The phasor of the spherical wave is a rigorous solution to the scalar Helmholtz equation; it is expressed as $\exp(-jkR)/R$, where $R=[x^2+y^2+(z-z_s)^2]^{1/2}$ is the spherical radius of curvature of the wave and z_s is the axial location of the point source. We now convert z_s into a pure imaginary number, i.e. $z_s=-ja$, where a is real constant called the confocal parameter (Sheppard & Saghafi, 1999a). The

complex spherical radius is therefore $\tilde{R} = [r^2 + (z + ja)^2]^{1/2}$, where $r = (x^2 + y^2)^{1/2}$ is the transverse coordinate. The phasor of the nonparaxial Gaussian beam is then

$$\tilde{\Psi}^{+}(\mathbf{r},\omega) = \exp(-ka)\exp(-jk\tilde{R})/\tilde{R}, \tag{12}$$

where exp(-*ka*) is a standard normalization constant, which ensures the continuity of the solution between the paraxial and the nonparaxial regimes. The superscript "+" recalls that the beam is diverging from the origin. The phasor $\tilde{\Psi}^{-}(\mathbf{r},\omega) = \exp(-ka)\exp(+jk\tilde{R})/\tilde{R}$ represents a beam that is converging toward the origin.

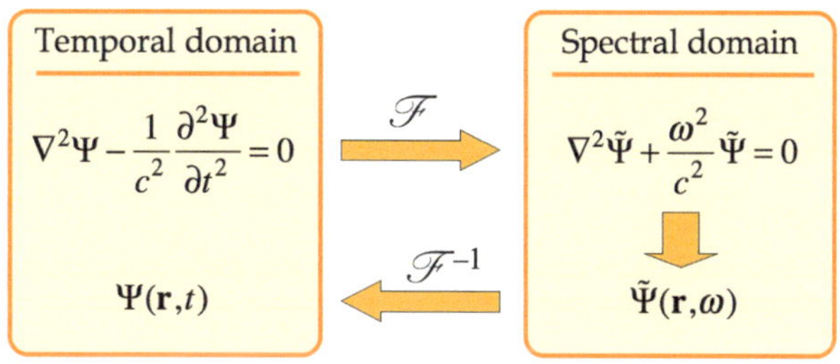

Fig. 1. By a Fourier transformation, the wave equation for the Hertz potential is converted into the Helmholtz equation, which is solved in the spectral domain; thanks to an inverse Fourier transformation, the phasor obtained is then converted into an exact spatiotemporal solution to the wave equation for the component of the Hertz potential.

The confocal parameter

The phasor of the nonparaxial Gaussian beam obtained with the help of the complex source-point method depends on the parameter *a*, which is the confocal parameter of the oblate spheroidal coordinates (Landesman & Barrett, 1988). In fact, it turns out that the oblate spheroidal coordinates (ξ,η,ϕ) are the ones in which it is natural to express the phasor of the nonparaxial Gaussian beam. Consider a system of mutually orthogonal, confocal ellipses and hyperbolas in the sense that the ellipses and the hyperbolas share the same foci and intersect at right angles. The distance between the origin and each focus is *a*. The surfaces of the oblate spheroidal coordinate system are formed by rotating the system of confocal ellipses and hyperbolas about the minor axis of the ellipse (Fig. 2). The rotation axis is *z* and the resulting focus is a ring of radius *a* in the *x*-*y* plane.

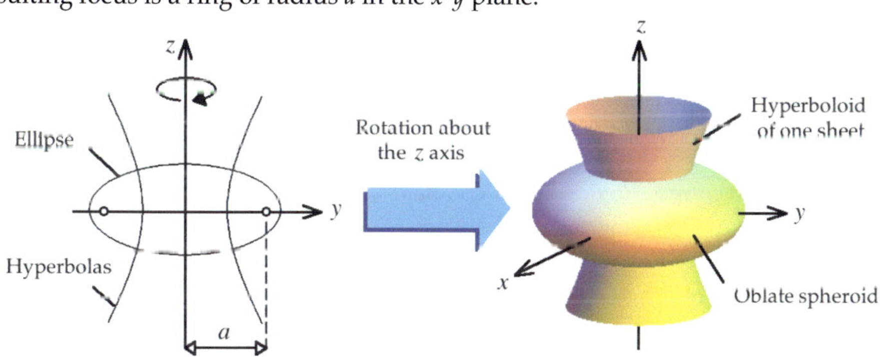

Fig. 2. The surfaces of the oblate spheroidal coordinate system are formed by rotating a system of confocal ellipses and hyperbolas about the *z*-axis.

The Cartesian coordinates (x,y,z) and the oblate spheroidal coordinates (ξ,η,ϕ) are related by the parametric equations $x=a[(1+\xi^2)(1-\eta^2)]^{1/2}\cos\phi$, $y=a[(1+\xi^2)(1-\eta^2)]^{1/2}\sin\phi$, and $z=a\xi\eta$, where $\xi\geq 0$, $-1\leq\eta\leq 1$, and $0\leq\phi\leq 2\pi$, and where a is the confocal parameter. The real and imaginary parts of the complex spherical radius $\tilde{R}$ can be easily expressed in terms of the oblate spheroidal coordinates. Substituting the parametric equations relating the Cartesian coordinates and the oblate spheroidal coordinates in $\tilde{R}=[x^2+y^2+(z+ja)^2]^{1/2}$ yields $\tilde{R}=a(\xi+j\eta)$. The real part of $\tilde{R}$ is $a\xi$ while its imaginary part is $a\eta$. It is therefore convenient to express the coordinates ξ and η in terms of the Cartesian coordinates [Berardi, 2004]:

$$\xi=\frac{1}{\sqrt{2}a}\left\{\left(R^2-a^2\right)+\left[\left(R^2-a^2\right)^2+4a^2z^2\right]^{1/2}\right\}^{1/2}, \tag{13a}$$

$$\eta=\sqrt{2}z\left\{\left(R^2-a^2\right)+\left[\left(R^2-a^2\right)^2+4a^2z^2\right]^{1/2}\right\}^{-1/2}, \tag{13b}$$

where $R^2=x^2+y^2+z^2$. The parameter a, which characterizes the divergence of the beam, is related to the Rayleigh range of the beam. The distance between the origin and the foci of the hyperbolas is a whereas the angle between the z-axis and the asymptotes is δ (Fig. 3).

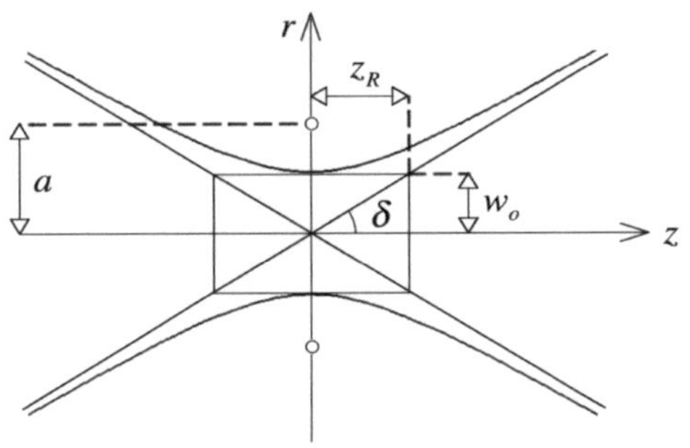

Fig. 3. The waist spot size, the Rayleigh range and the divergence angle of the nonparaxial beam are characteristics of the hyperboloid that defines the oblate spheroidal coordinates.

The waist spot size w_o is defined as the length of the semi-major axis of the hyperbola while the Rayleigh range is defined by the length of the semi-minor axis (Fig. 3). From the geometry of the hyperbola, the Rayleigh range is given by $z_R\equiv\frac{1}{2}kw_o^2=(a^2-w_o^2)^{1/2}$ (Rodríguez-Morales & Chávez-Cerda, 2004). The last equality provides a quadratic equation in w_o^2. Solving this equation for w_o^2 results in $w_o^2=(2/k^2)\{[1+(ka)^2]^{1/2}-1\}$. Consequently, for a given wave number $k=2\pi/\lambda$, the confocal parameter can be written in terms of the waist spot size w_o, the Rayleigh range $z_R=\frac{1}{2}kw_o^2$, or the beam divergence angle $\delta\equiv\arctan(w_o/z_R)$. All the relationships between these parameters are listed in Table 2.
Large values of ka refer to the paraxial regime and small values of ka correspond to the nonparaxial regime. The phasor of the Gaussian beam tends to the uniform plane wave if ka is very large, whereas it becomes the phasor of the spherical wave if $ka=0$. The threshold between the paraxial and the nonparaxial regimes cannot be clearly defined. Nonetheless, it is usually accepted that the beam divergence angle of a paraxial beam must

not exceed 30°. Therefore, as a rule of thumb, the paraxial approximation can be used as far as ka is greater than 7 (Rodrígez-Morales & Chávez-Cerda, 2004). Note that, in the limit $ka >> 1$, the confocal parameter a tends to the Rayleigh range z_R.

Waist spot size w_o	$w_o = \left(\frac{2z_R}{k}\right)^{1/2}$	$w_o = \frac{2}{k\tan\delta}$	$w_o = \frac{\sqrt{2}}{k}\left\{\left[1+(ka)^2\right]^{1/2} - 1\right\}^{1/2}$
Rayleigh range z_R	$z_R = \frac{1}{2}kw_o^2$	$z_R = \frac{2}{k\tan^2\delta}$	$z_R = \frac{\left[1+(ka)^2\right]^{1/2} - 1}{k}$
Beam divergence angle δ	$\tan\delta = \frac{2}{kw_o}$	$\tan\delta = \left(\frac{2}{kz_R}\right)^{1/2}$	$\cos\delta = \frac{\left[1+(ka)^2\right]^{1/2} - 1}{ka}$
Confocal parameter a	$a = \frac{2}{k\sin\delta\tan\delta}$	$a = z_R\left(1+\frac{2}{kz_R}\right)^{1/2}$	$a = w_o\left[1+(\frac{1}{2}kw_o)^2\right]^{1/2}$

Table 2. Relationships between the waist spot size, the Rayleigh range, the beam divergence angle, and the confocal parameter.

The complex source/sink model

The complex source/sink model is a simple approach to find an exact solution to the Helmholtz equation that describes a physically realizable nonparaxial beam (Ulanowski & Ludlow, 2000). In fact, it has been pointed out that the phasor of the nonparaxial Gaussian beam as defined by Eq. (12) has two shortcomings: an axial discontinuity and a circular singularity of radius a occur in the plane of the beam waist (Fig 4a). On the one hand, the axial discontinuity is due to the choice for the branch of the square root in the complex spherical radius $\tilde{R}$; the height of the discontinuity in the amplitude distribution on the z-axis is $\exp(-2ka)$ and it becomes significant when $ka > 1$. On the other hand, the circular singularity can be explained because Eq. (12) tends to infinity when its denominator vanishes; this happens if $r = a$ in the plane $z = 0$. Nevertheless, such a singularity does not

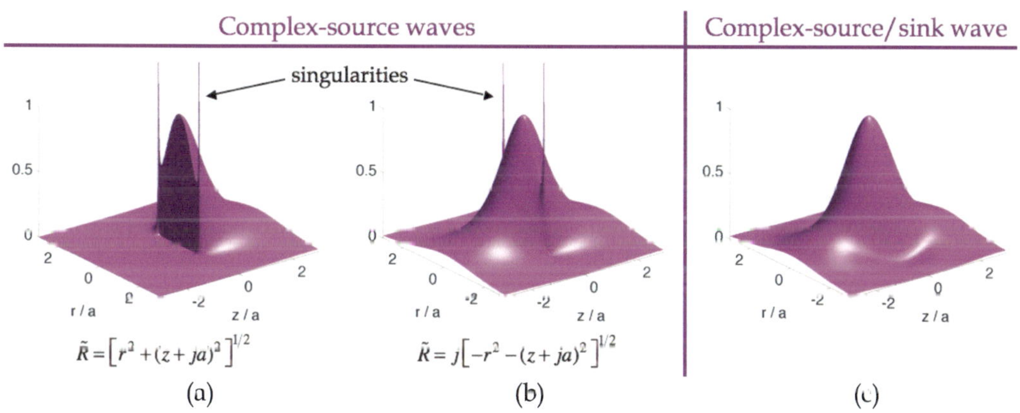

Fig. 4. The square modulus of the phasor of the nonparaxial Gaussian beam near the plane of the beam waist, with ka = 2, (a) has both an axial discontinuity and a singularity with the choice I for the value of the complex spherical radius, (b) exhibits only the singularity with the choice II, and (c) is well-behaved in the context of the complex source/sink model.

have a significant impact on the behavior of the beam if a is large enough with respect to the wavelength λ, i.e. in the paraxial regime ($ka >> 1$).
In the phasor of the nonparaxial Gaussian beam defined by Eq. (12), the choice for the value of the complex spherical radius $\tilde{R}$ is relevant. Two choices can be specified:

[Choice I] $$\tilde{R} = \left[r^2 + (z + ja)^2\right]^{1/2}, \tag{14a}$$

[Choice II] $$\tilde{R} = j\left[-r^2 - (z + ja)^2\right]^{1/2}. \tag{14b}$$

The value of $\tilde{R}$ for which the real part $a\xi$ is positive corresponds to the choice I, whereas the one for which the imaginary part $a\eta$ is positive is associated to the choice II. In particular, on the optical axis (r = 0), Eq. (14a) reduces to $\tilde{R} = z + ja$ if $z > 0$ and $\tilde{R} = -z - ja$ if $z < 0$ while Eq. (14b) reduces to $\tilde{R} = z + ja$ for all z, considering a as a nonzero positive real number. The choice I in Eq. (12) leads to a wave that radiates outward from the plane of the beam waist (z = 0), while the choice II in Eq. (12) gives a beam traveling from negative z to positive z like a purely traveling beam. Thus, although these two representations are both solutions of the Helmholtz equation, they describe different complex-source waves. The use of the choice II in Eq. (12) removes the axial discontinuity in the phasor of the nonparaxial beam (Fig. 4b). Nonetheless, neither choice of the branch for the square root $\tilde{R}$ in Eq. (12) removes the nonphysical singularity of radius a in the plane of the beam waist. Both complex-source waves in the Figs. 4a and 4b have a singularity at $r = a$ when $z = 0$, where the square modulus of the phasor tends to infinity. It may be argued that this drawback originates from the description of the field as due to a source (even though it is located at an imaginary coordinate), which is inherently contradictory, since the field is physically source-free in the spatial region under consideration.
Sheppard and Saghafi as well as Ulanowski and Ludlow have shown that the superposition of two counter-propagating beams can remove both the axial discontinuity and the circular singularity (Sheppard & Saghafi, 1998; Ulanowski & Ludlow, 2000). In fact, a singularity-free nonparaxial Gaussian beam is proportional to the superposition $\tilde{\Psi}^-(\mathbf{r},\omega) - \tilde{\Psi}^+(\mathbf{r},\omega)$. An exact solution to the Helmholtz equation that generalizes the Gaussian beam and that is valid in all space in the nonparaxial regime is therefore

$$\tilde{\Psi}(\mathbf{r},\omega) = \tilde{\Psi}_0 F(\omega)\frac{\exp(-ka)}{2j}\left[\frac{\exp(jk\tilde{R})}{\tilde{R}} - \frac{\exp(-jk\tilde{R})}{\tilde{R}}\right] = \tilde{\Psi}_0 F(\omega)\exp(-ka)\frac{\sin(k\tilde{R})}{\tilde{R}} \tag{15}$$

where $\tilde{\Psi}_0$ is a constant amplitude and $F(\omega)$ is an arbitrary well-behaved function that represents the spectral amplitude of the pulse. According to the superposition principle, the phasor of Eq. (15) is a rigorous solution of the source-free Helmholtz equation. Eq. (15) is a singularity-free phasor, because it is finite at $\tilde{R} = 0$, and thus describes a physically realizable optical beam (Fig. 4c). Eq. (15) is the same whichever choice is taken for the complex spherical radius $\tilde{R}$. Thus, no great care is needed concerning the specific choice between Eqs. (14a) or (14b) in calculating the phasor of the nonparaxial Gaussian beam of the Eq. (15). Also, it can be shown that the nonparaxial Gaussian beam reduces to the phasor of the paraxial Gaussian beam in the paraxial limit, i.e. Eq. (15) reduces to Eq. (4) when $ka >> 1$ (April, 2008a).

From another point of view, it is seen that the removal of the nonphysical singularity has been accomplished by combining a sink to the source; this leads to what is called the complex source/sink method. In fact, Eq. (15) may be viewed as the superposition of an outgoing beam, produced by the source located at $z_s = -ja$, and an incoming beam, absorbed by the sink placed at the same position. This optical wave, consisting in a superposition of two counter-propagating beams, results in a standing-wave component near the $z = 0$ plane. The complex-source/sink wave provides a rigorous solution to the wave equation in free-space over all space. In summary, the complex source/sink model, as opposed to the complex point-source model, yields an expression for the phasor of a physically realizable (singularity-free) beam.

Due to the partially standing-wave nature of such a solution, producing a complex-source/sink beam requires a focusing element that subtends a solid angle greater than 2π, such as a 4Pi microscope or the parabolic mirror of large extent schematically illustrated in Fig. 5. Qualitatively, it can be seen that incident rays on the parabolic mirror for which $r < r_0 = 2f$ contribute to the propagating beam, whereas rays for which $r > r_0$ contribute to the counter-propagating one, where f is the focal length of the parabolic mirror (Fig. 5). If the beam is focused by a focusing element that subtends a solid angle less than 2π, then the counter-propagating component of the beam cannot be produced physically. Nevertheless, in that case, the complex-source/sink solution can then be regarded as a rigorous solution to an approximate model.

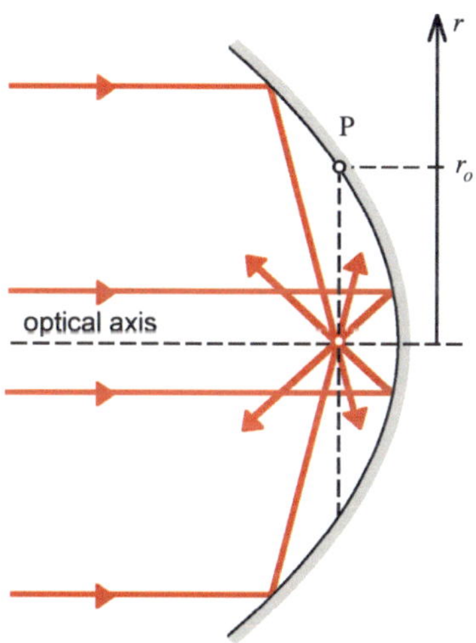

Fig. 5. Focusing a collimated beam with a parabolic mirror of large extent leads to propagating as well as counter-propagating contributions in the electromagnetic fields in the focal region of the mirror.

Higher-order nonparaxial Gaussian beams

The phasor of the fundamental nonparaxial Gaussian beam constitutes one among the infinite number of exact solutions to the Helmholtz equation. In fact, some of these additional solutions are the higher-order nonparaxial Gaussian modes. Shin and Felsen used the approach of Deschamps with a complex multipole source to produce higher-order beams that reduce to the elegant Hermite–Gaussian modes in the paraxial limit (Shin & Felsen, 1977). Phasors of nonparaxial beams, that may describe elegant higher-order beams in cylindrical coordinates, were proposed by Couture and Bélanger. These phasors are expressed in terms of associated Legendre functions and spherical Hankel functions of complex arguments (Couture & Bélanger, 1981). Landesman and Barrett obtained the same solutions using the oblate spheroidal coordinates (Landesman & Barrett, 1988). Such phasors can be viewed as being generated with the complex point-source method, leading to nonphysically realizable beams.

Singularity-free phasors of higher-order nonparaxial beams can be generated with the help of the complex source/sink method. The higher-order nonparaxial beams proposed by Ulanowski and Ludlow are in turn expressed in terms of associated Legendre functions and spherical Bessel functions of the first kind (Ulasnowski & Ludlow, 2000):

$$\tilde{\psi}^{e}_{n,m}(r,\phi,z) = \exp(-ka)\, j_n(k\tilde{R}) P^m_n(\cos\tilde{\theta})\cos(m\phi)\,, \tag{16}$$

where $\cos\tilde{\theta} \equiv (z + ja)/\tilde{R}$, $j_n(k\tilde{R})$ is the spherical Bessel function of the first kind of order n, and $P^m_n(\cos\tilde{\theta})$ is the associated Legendre function. Eq. (16) is an exact solution to the Helmholtz equation. Spherical Bessel functions in Eq. (16) are preferred to spherical Hankel functions suggested by Couture and Bélanger to ensure the absence of singularity in the plane of the beam waist ($z = 0$) at $r = a$. The explicit form of the fundamental mode ($n = m = 0$) in Eq. (16) is $\tilde{\psi}^{e}_{0,0} = \exp(-ka)\, j_0(k\tilde{R}) = \exp(-ka)\sin(k\tilde{R})/(k\tilde{R})$ and it is proportional to the singularity-free nonparaxial Gaussian beam [Eq. (15)], as expected.

While Shin and Felsen have shown that the complex point-source method leads to the elegant nonparaxial Hermite–Gaussian modes, Seshadri employed the method to find the differential and the integral representations of the nonparaxial cylindrically symmetric elegant Laguerre–Gaussian modes (Seshadri, 2002). Two years later, Bandres and Gutiérrez-Vega presented the same analysis without the restriction on the cylindrical symmetry, providing the complete differential and integral representations of the nonparaxial elegant Laguerre–Gaussian modes, denoted by $\tilde{U}^{\sigma}_{p,m}(\mathbf{r},\omega)$ where $\sigma = \{e,o\}$ is the parity (Bandres & Gutiérrez-Vega, 2004). However, these expressions exhibit the axial discontinuity as well as the circular singularity in the plane of the beam waist. Closed-form expressions for the phasor $\tilde{U}^{\sigma}_{p,m}(\mathbf{r},\omega)$ of the singularity-free nonparaxial elegant Laguerre–Gaussian beam can be written as a finite sum (April, 2008a):

$$\tilde{U}^{\sigma}_{p,m}(\mathbf{r},\omega) = F(\omega) 2^{p+2} \left(\frac{ka}{2}\right)^{p+1+m/2} \sum_{s=0}^{p} \binom{p+m}{s+m} \frac{(4s+2m+1)(2s-1)!!}{(2p+2s+2m+1)!!} \tilde{\psi}^{\sigma}_{2s+m,m}(r,\phi,z)\,. \tag{17}$$

This particular linear combination of functions $\tilde{\psi}^{\sigma}_{n,m}(r,\phi,z)$, as given by Eq. (17), has the property to reduce to the elegant Laguerre–Gaussian beam defined by Eq. (5) in the paraxial limit. Actually, another function denoted by $\tilde{V}^{\sigma}_{p,m}(\mathbf{r},\omega)$ has the same property. Whereas $\tilde{U}^{\sigma}_{p,m}(\mathbf{r},\omega)$ is written as a linear combination of functions $\tilde{\psi}^{\sigma}_{n,m}(r,\phi,z)$ for which $n - m$ is even, $\tilde{V}^{\sigma}_{p,m}(\mathbf{r},\omega)$ is a linear combination of functions $\tilde{\psi}^{\sigma}_{n,m}(r,\phi,z)$ for which $n - m$ is odd:

$$\tilde{V}^{\sigma}_{p,m}(\mathbf{r},\omega) = -jF(\omega) 2^{p+2} \left(\frac{ka}{2}\right)^{p+1+m/2} \sum_{s=0}^{p} \binom{p+m}{s+m} \frac{(4s+2m+3)(2s+1)!!}{(2p+2s+2m+3)!!} \tilde{\psi}^{\sigma}_{2s+m+1,m}(r,\phi,z)\,. \tag{18}$$

It is clear that $\tilde{U}^{\sigma}_{p,m}(\mathbf{r},\omega)$ and $\tilde{V}^{\sigma}_{p,m}(\mathbf{r},\omega)$ in Eqs. (17) and (18) are also exact solutions of the Helmholtz equation, because they are written as a linear combination of functions $\tilde{\psi}^{\sigma}_{n,m}(r,\phi,z)$, which are themselves rigorous solutions of the Helmholtz equation. In the paraxial limit (*ka* >> 1), both solutions reduce to the phasor of the paraxial elegant Laguerre–Gaussian beam (April, 2008a):

$$\lim_{ka>>1} \tilde{U}^{\sigma}_{p,m}(\mathbf{r},\omega) = \lim_{ka>>1} \tilde{V}^{\sigma}_{p,m}(\mathbf{r},\omega) = \tilde{u}^{\sigma}_{p,m}(\mathbf{r},\omega). \tag{19}$$

Both solutions $\tilde{U}^{\sigma}_{p,m}(\mathbf{r},\omega)$ and $\tilde{V}^{\sigma}_{p,m}(\mathbf{r},\omega)$ form a complete eigenfunction basis for the Helmholtz equation. These solutions are expressed as a simple linear combination of spherical Bessel functions and associated Legendre functions of complex arguments.

3.3 The Poisson-like spectrum

As mentioned in Section 2.3, the Gaussian spectrum is not appropriate to characterize arbitrarily short pulses, because it contains spectral components of appreciable amplitude with negative frequencies when the spectrum is broad enough. A suitable spectrum $F(\omega)$ whose spectral content does not extend in the negative frequencies must be chosen to adequately describe an ultrashort pulse. Here, we choose the Poisson-like spectral amplitude (also called the power spectrum), defined by (Caron & Potvliege, 1999; Feng & Winful, 2000)

$$F(\omega) = 2\pi \exp(j\phi_0)\left(\frac{s}{\omega_0}\right)^{s+1} \frac{\omega^s \exp(-s\omega/\omega_0)}{\Gamma(s+1)}\theta(\omega), \tag{20}$$

where s is a real positive parameter, ϕ_0 is the absolute phase of the pulse, ω_0 is the frequency for which the spectral amplitude is maximum, $\Gamma(\cdot)$ is the gamma function, and $\theta(\omega)$ is the unit step function which ensures that the pulse does not exhibit negative frequencies. This will make the time-domain complex fields analytic functions that are well behaved for all time and all points in space. Since $F(0)=0$, it follows that the pulse does not have a dc component. The parameter s controls the shape and the width of the spectrum (Fig. 6a). Chirped pulses may be modeled by taking ω_0 as a complex number.

Spectra of the form expressed in Eq. (20) are often observed in terahertz experiments and may also be used to describe femtosecond laser pulses. Moreover, spectra described by Eq. (20) lead to closed-form expressions for the electromagnetic fields of isodiffracting pulsed beams in terms of elementary functions. The inverse Fourier transform of Eq. (20) gives

$$f(t) \equiv \frac{1}{2\pi}\int_{-\infty}^{\infty} F(\omega)\exp(j\omega t)\mathrm{d}\omega = \exp(j\phi_0)\left(1-\frac{j\omega_0 t}{s}\right)^{-(s+1)} \tag{21}$$

The real part of $f(t)$ provides the temporal shape of the pulse. A pulse for which s is close to unity is a single-cycle pulse (Fig. 6b).

It can be shown that the spectral amplitude as well as the temporal shape of the pulse reduce to Gaussian functions in the limit of a narrow spectrum, i.e. when s is very large (Caron & Potvliege, 1999):

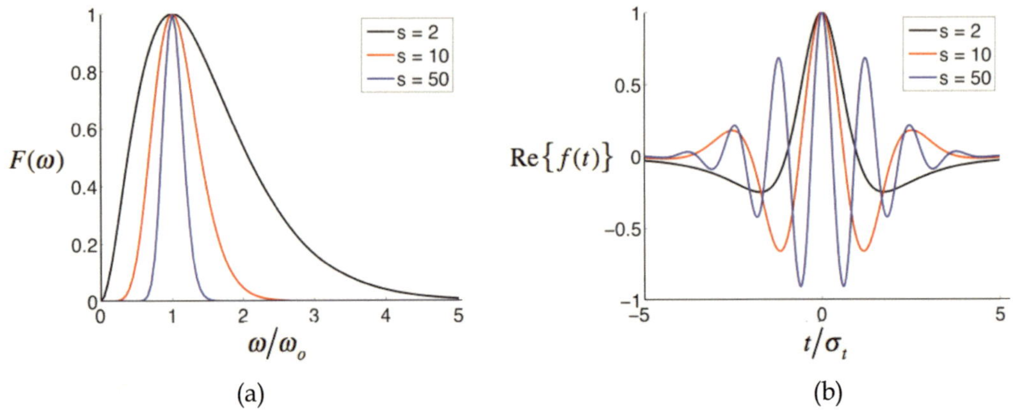

Fig. 6. (a) Spectral and (b) temporal shape of the pulse for different values of parameter s, with a zero absolute phase ϕ_o .

$$F(\omega) \sim \sqrt{\pi} T \exp\left[-\tfrac{1}{4}T^2(\omega-\omega_o)^2 + j\phi_o\right]\theta(\omega) \tag{22a}$$

and

$$f(t) \sim \exp\left[-\frac{t^2}{T^2} + j(\omega_o t + \phi_o)\right], \tag{22b}$$

if $s >> 1$, where $T \equiv (2s)^{1/2}/\omega_o$ is the duration of the Gaussian pulse. These results are coherent with the fact that the Fourier transform of a Gaussian function is also a Gaussian function. Thus, the Poisson-like spectrum is an interesting alternative to describe a pulse whose spectrum reduces to a Gaussian function if its duration is sufficiently long.

Hereafter, the parameter σ_t, defined as the root mean square (RMS) width of the temporal distribution $|f(t)|^2$, will be used as the expression of the pulse duration. Also, the RMS width σ_ω of the spectral intensity $|F(\omega)|^2$ will be employed to evaluate the width of the spectrum. Their expressions are explicitly

$$\sigma_t = \frac{1}{\omega_o}\frac{s}{(2s-1)^{1/2}}, \tag{23a}$$

$$\sigma_\omega = \frac{\omega_o}{2}\frac{(2s+1)^{1/2}}{s}. \tag{23b}$$

According to Eqs. (23a) and (23b), the inequality $\sigma_t\sigma_\omega \geq 1/2$ is verified as it must be for every pair of Fourier transforms; furthermore, $\sigma_t\sigma_\omega = 1/2$ in the limit $s \to \infty$, in which case the distributions become Gaussian functions.

4. Characterization of some laser pulses

Combining the Hertz potential method, the complex-source/sink model and the Poisson-like spectrum, it is possible to obtain the expressions for electromagnetic fields that are rigorous solutions to Maxwell's equations. In principle, any kind of pulsed beam can be generated with this approach. In particular, transverse magnetic, transverse electric, and linearly polarized pulses will be analyzed. The isodiffracting pulse will then be discussed as an important special case.

4.1 Previous works on laser pulses

Many authors have studied models of pulsed beams, which are unfortunately incomplete or intricate. For instance, Kiolkowski and Judkins explored the simple case of a paraxial Gaussian pulsed beam, where they have considered that the waist spot size w_0 is the same for all the spectral components of the pulse (Kiolkowski & Judkins, 1992). As a consequence, they were not able to get a general analytical expression for their Gaussian pulsed beam applicable for all z; only results valid in the limiting cases of the far field ($z >> z_R$) and the near field ($z << z_R$) have been found. Wang *et al.* found in turn the expression of a paraxial Gaussian pulsed beam that satisfies the so-called isodiffracting condition, according to which the Rayleigh range z_R is the same for all spectral components (Wang *et al.*, 1997). Thus, they obtained a general and remarkably simple closed-form expression for the Gaussian pulsed beam, only valid for paraxial beams and long enough pulses. A year later, Porras found the expression for an ultrafast Gaussian pulsed beam (Porras, 1998), shedding light on the spatiotemporal couplings that occur when a pulse propagates in free space (the diffractional effects depend on time and the temporal shape of the pulse depends on the spatial distribution of the pulse).

In the context of the complex point-source method, Heyman and Felsen determined the expression of a nonparaxial pulse, using the isodiffracting condition that, according to them, means that the confocal parameter a is frequency independent (Heyman & Felsen, 2001). It should be noted that, in the paraxial limit where $a \approx z_R$, this definition of the isodiffracting condition is equivalent to the one from Wang *et al.* With the help of the complex-source/sink model, Saari provided an expression for a scalar singularity-free nonparaxial pulse, written in terms of the oblate spheroidal coordinates (Saari, 2001).

The previously mentioned authors did not take into account the vector nature of their pulses. But Lu *et al.* presented the expressions for the electromagnetic fields of a nonparaxial pulsed beam; these expressions were obtained with a perturbative method and the results are expressed as a truncated series written in terms of convolutions (Lu *et al.*, 2003). In principle, the method of Lax *et al.* is used in order to add spatial corrections to the paraxial beam. In 2006, Varin *et al.* succeeded in generalizing this method to the spatiotemporal corrections to the paraxial vector pulsed beam (Varin *et al.*, 2006); using their approach, they have obtained the expression of a special nonparaxial ultrashort pulsed beam, expressed as an infinite series.

4.2 Some special pulses

The complex source/sink method yields a rigorous solution $\tilde{\Psi}(\mathbf{r},\omega)$ to the Helmholtz equation in the frequency domain, which represents a nonzero component of a Hertz potential. To describe a well-behaved pulse, the spectrum of $\tilde{\Psi}(\mathbf{r},\omega)$ is chosen so that it does not contain spectral components of negative frequency. The scalar field $\tilde{\Psi}(\mathbf{r},\omega)$ can be proportional to the nonparaxial Gaussian beam defined by Eq. (15). Otherwise, the nonparaxial elegant Laguerre–Gaussian beam $\tilde{U}^{\sigma}_{p,m}(\mathbf{r},\omega)$ may be used to describe higher-order beams. Once the solution for $\tilde{\Psi}(\mathbf{r},\omega)$ is known, the spatiotemporal component of the Hertz potential $\Psi(\mathbf{r},t)$ is recovered by taking its inverse Fourier transform (Fig. 1):

$$\Psi(\mathbf{r},t)=\frac{1}{2\pi}\int_{-\infty}^{\infty}\tilde{\Psi}(\mathbf{r},\omega)\exp(j\omega t)\mathrm{d}\omega\,. \tag{24}$$

For a given state of polarization, both electric and magnetic Hertz potentials can be constructed. Three states of polarization can easily be generated with the Hertz potential method: the transverse magnetic (TM), the transverse electric (TE), and the linearly polarized (LP) pulsed beams (Table 3).

TM pulsed beam	TE pulsed beam	LP pulsed beam in the x-direction
$\Pi_e=\hat{\mathbf{a}}_z\Psi(\mathbf{r},t)$	$\Pi_e=\mathbf{0}$	$\Pi_e=\hat{\mathbf{a}}_x\Psi(\mathbf{r},t)$
$\Pi_m=\mathbf{0}$	$\Pi_m=\hat{\mathbf{a}}_z\eta_0^{-1}\Psi(\mathbf{r},t)$	$\Pi_m=\hat{\mathbf{a}}_y\eta_0^{-1}\Psi(\mathbf{r},t)$

Table 3. Electric and magnetic Hertz potentials for TM, TE and LP pulsed beams.

Then, the expressions for the electromagnetic fields can be determined with the help of Eqs. (11a) and (11b). Since the spectrum of $\tilde{\Psi}(\mathbf{r},\omega)$ in Eq. (24) is one-sided, the resulting fields $\mathbf{E}(\mathbf{r},t)$ and $\mathbf{H}(\mathbf{r},t)$ are expressed in their complex analytical signal representations. The physical fields are simply obtained from the real part of $\mathbf{E}(\mathbf{r},t)$ and $\mathbf{H}(\mathbf{r},t)$. The lowest-order member of each family of pulsed beams listed in Table 3 is analyzed briefly in the following paragraphs.

The TM_{01} pulse

The TM_{01} beam is the lowest-order radially polarized beam. Therefore, the fundamental Gaussian beam ($p = m = 0$) must be chosen as the rigorous solution of the Helmholtz equation for the axial component of the electric Hertz potential. Since it is cylindrically symmetric, the function $\Psi(\mathbf{r},t)$ does not depend on the azimuthal angle ϕ, and Eqs. (11a) and (11b) can be simplified to give the nonzero cylindrical components of the electric and magnetic fields in terms of the electric Hertz potential (Table 4).

Electric field components		Magnetic field component
$E_r=\dfrac{\partial^2\Psi}{\partial z\partial r}$	$E_z=\dfrac{\partial^2\Psi}{\partial z^2}-\dfrac{1}{c^2}\dfrac{\partial^2\Psi}{\partial t^2}$	$H_\phi=-\varepsilon_0\dfrac{\partial^2\Psi}{\partial t\partial r}$

Table 4. The nonzero electromagnetic components of a TM_{01} pulsed beam.

A TM_{01} pulse is said to be radially polarized since the azimuthal component of the electric field is zero, i.e. $E_\phi=0$. Furthermore, the pulse is transverse magnetic (TM), because the magnetic field of the optical pulse does not have a longitudinal component, as opposed to its electric field. In the paraxial regime, the electric energy density (defined by $W\equiv\frac{1}{2}\varepsilon_0|\mathbf{E}|^2$) on the axis is small compared to its maximum value, giving to the beam a "doughnut" shape; in the nonparaxial regime, the longitudinal component of the electric field dominates the radial component, so that the electric energy density is maximum at the center of the beam. The longitudinal component of the electric field of a strongly focused TM_{01} pulse can be exploited for electron acceleration (Varin *et al.*, 2005).

The TE_{01} pulse

The expressions for the electromagnetic components of a TE_{01} pulsed beam may be easily derived from those of a TM_{01} pulsed beam by means of the previously mentioned duality transformation $\mathbf{E} \to \eta_0 \mathbf{H}$ and $\mathbf{H} \to -\mathbf{E}/\eta_0$ (Table 5).

Electric field component	Magnetic field components	
$E_\phi = \frac{1}{c}\frac{\partial^2 \Psi}{\partial t \partial r}$	$H_r = \frac{1}{\eta_0}\frac{\partial^2 \Psi}{\partial z \partial r}$	$H_z = \frac{1}{\eta_0}\left(\frac{\partial^2 \Psi}{\partial z^2} - \frac{1}{c^2}\frac{\partial^2 \Psi}{\partial t^2}\right)$

Table 5. The nonzero electromagnetic components of a TE_{01} pulsed beam.

A TE_{01} pulse is a special case of the family of the azimuthally polarized pulses; in fact, the radial component of the electric field is zero. Also, the pulse is said to be transverse electric (TE), since the electric field of the pulse does not have a longitudinal component while the longitudinal component of its magnetic field is nonzero. This beam is characterized by a transverse electric energy density profile of doughnut shape, in both the paraxial and nonparaxial regimes. Hence, an azimuthally polarized pulsed beam always has a zero intensity at the center of its transverse intensity distribution. As a result, a TE_{01} pulsed beam may have interesting applications in stimulated emission depletion (STED) microscopy (Deng *et al.*, 2007).

The LP_{01} pulse

A linearly polarized beam is produced by a combination of crossed electric and magnetic dipoles located at an imaginary distance, which is called a LP beam. (Sheppard & Saghafi, 1999a). Because the LP_{01} beam is the lowest-order member of the family of the linearly polarized beams, the nonparaxial Gaussian beam defined by Eq. (15) is chosen as the exact solution of the Helmholtz equation for the nonzero Cartesian components of the Hertz potentials. Applying Eqs. (11a) and (11b), the six electromagnetic components of a LP_{01} pulsed beam in the x-direction can be computed (Table 6).

Electric field components	Magnetic field components
$E_x = \frac{\partial^2 \Psi}{\partial x^2} - \frac{1}{c^2}\frac{\partial^2 \Psi}{\partial t^2} + \frac{1}{c}\frac{\partial^2 \Psi}{\partial t \partial z}$	$H_x = \frac{1}{\eta_0}\frac{\partial^2 \Psi}{\partial x \partial y}$
$E_y = \frac{\partial^2 \Psi}{\partial y \partial x}$	$H_y = \frac{1}{\eta_0}\left(\frac{\partial^2 \Psi}{\partial y^2} - \frac{1}{c^2}\frac{\partial^2 \Psi}{\partial t^2} + \frac{1}{c}\frac{\partial^2 \Psi}{\partial t \partial z}\right)$
$E_z = \frac{\partial^2 \Psi}{\partial z \partial x} - \frac{1}{c}\frac{\partial^2 \Psi}{\partial t \partial x}$	$H_z = \frac{1}{\eta_0}\left(\frac{\partial^2 \Psi}{\partial z \partial y} - \frac{1}{c}\frac{\partial^2 \Psi}{\partial t \partial y}\right)$

Table 6. The electromagnetic components of a LP_{01} pulsed beam.

In the paraxial regime, the x-component of the electric field and the y-component of the magnetic field of the LP_{01} pulsed beam dominate the other field components, and the energy density profile has a Gaussian shape (it is sometimes called a TEM_{00} pulsed beam). In the nonparaxial regime, the power transferred from the transverse components of the electric field of the LP_{01} pulsed beam to its longitudinal component increases as the value of ka

decreases; the longitudinal component of the electric field is not cylindrically symmetric and, as a consequence, the focal spot becomes asymmetrically deformed and elongated in the direction of the polarization (here, in the x-direction).

4.3 Isodiffracting pulses

For isodiffracting pulses, all the frequency components have the same wavefront radius of curvature (Melamed & Felsen, 1998). For each traveling beam $\exp(\pm jk\tilde{R})/\tilde{R}$ that constitutes the expression of Eq. (15), the wavefront is an oblate spheroid with a radius of curvature whose length is given by the real part $a\xi$ of the complex spherical radius $\tilde{R}$. It is seen from Eq. (13a) that ξ depends only on the spatial coordinates and on the parameter a. Therefore, the confocal parameter a of an isodiffracting pulse must be frequency independent (Heyman & Felsen, 2001). Note that, in the paraxial limit, it means that all the spectral components of an isodiffracting pulse have the same Rayleigh range z_R, as it is well known.
Closed-form expressions for the electromagnetic fields of an isodiffracting strongly focused, ultrafast laser pulse can be obtained. With the isodiffracting condition, the inverse Fourier transform of Eq. (15) can easily be carried out. If $f(t)$ is the inverse Fourier transform of $F(\omega)$, then substituting Eq. (15) in Eq. (24) and performing the integration yields

$$\Psi(\mathbf{r},t) = \frac{\Psi_o}{\tilde{R}}\left[f(\tilde{t}_+) - f(\tilde{t}_-) \right], \tag{25}$$

where $\Psi_o \equiv \tilde{\Psi}_o/2j$ is a constant amplitude, $\tilde{t}_\pm \equiv t \pm \tilde{R}/c + ja/c$, and c is the speed of light in vacuum. The time $t = 0$ corresponds to the instant for which the pulse is in the plane of the beam waist, which is located at $z = 0$. The introduction of the complex temporal variables $\tilde{t}_\pm$ in Eq. (25) shows that in general there is a spatiotemporal coupling. Moreover, Eq. (25) contains all the information about the spatial and temporal behaviors of the pulsed beam.
An understanding of the properties of isodiffracting pulses is particularly relevant for studying the spatiotemporal behavior of mode-locked laser pulses. As Eq. (25) shows, isodiffracting pulses have the advantage of being easily analyzed with simple closed-form expressions. However, if one has to characterize a nonisodiffracting pulsed beam (whose frequency components do not have all the same confocal parameter), the method presented in this chapter is still applicable, but a closed-form analytical solution may be not obtained; in general, numerical integrations have to be performed.

5. The isodiffracting TM_{01} pulsed beam

To explore the method proposed in this chapter, let us consider the TM_{01} mode-locked laser pulse. The TM_{01} beam in particular is analyzed in detail because of its practical importance comparatively to other TM, TE or LP beams. Quabis, Dorn and co-workers have shown that smaller spot sizes can be achieved with a radially polarized beam instead of a linearly polarized beam (Quabis *et al.*, 2000). Because of their remarkable focusing properties, TM_{01} beams are of considerable interest, for example, in high-resolution microscopy. Moreover, when it is strongly focused, the electric field of the TM_{01} pulsed beam has a significant longitudinal component that can be exploited in particle trapping and electron acceleration (Varin *et al.*, 2005). First, some techniques to generate TM_{01} beams are briefly discussed; second, the expressions of the electromagnetic fields of a TM_{01} pulsed beam are presented; then, the focusing properties of an isodiffraction TM_{01} pulse are briefly explored.

5.1 Generation of TM_{01} beams

The TM_{01} beam can be viewed as a coherent combination of two orthogonally polarized elegant Laguerre-Gaussian modes of order (0,1); the first mode is horizontally polarized and has a $\cos\phi$ angular dependence, where ϕ is the azimuthal angle, and the second mode is vertically polarized with a $\sin\phi$ azimuthal dependence (Fig. 7). The result of this superposition is radially polarized and its electromagnetic fields has transverse components that are proportional to elegant Laguerre–Gaussian modes of order (0,1).

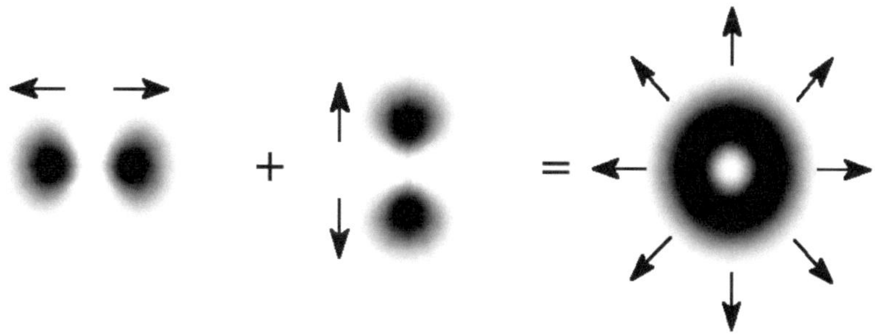

Fig. 7. A TM_{01} laser beam can be seen as a coherent superposition of two orthogonally polarized Laguerre-Gaussian modes of order (0,1). The arrows represent the spatial distribution of the instantaneous electric vector field.

Many approaches have been demonstrated to produce TM_{01} beams in laboratory. To name a few, a TM_{01} beam can be generated interferometrically, outside the resonator, with a Mach–Zehnder interferometer which allows the coherent superposition of two orthogonally polarized Laguerre–Gaussian beams of order (0,1) of different parity with the same beam waist (Tidwell *et al.*, 1990). Radially polarized beams may be generated directly from a laser by inserting in the laser cavity axially-symmetric optical elements with suitable polarization selectivity; such elements include a conical reflector used as a resonator mirror, a conical Brewster window and a birefringent *c*-cut laser crystal (Kawauchi *et al.*, 2008). Other techniques to produce pseudo-radially polarized beams involve, for instance, a polarization converter consisting in four half-wave plates, one in each quadrant (Dorn *et al.*, 2003).

5.2 Electromagnetic fields of a nonparaxial TM_{01} pulsed beam

The analytical expressions for the electromagnetic fields can be obtained by computing the derivatives presented in Table 4 with Eq. (25):

$$E_r(\mathbf{r},t)=\frac{\Psi_o\sin(2\tilde{\theta})}{2\tilde{R}}\left[\frac{3}{\tilde{R}^2}\left(f(\tilde{t}_+)-f(\tilde{t}_-)\right)-\frac{3}{c\tilde{R}}\left(\frac{\partial f(\tilde{t}_+)}{\partial t}+\frac{\partial f(\tilde{t}_-)}{\partial t}\right)+\frac{1}{c^2}\left(\frac{\partial^2 f(\tilde{t}_+)}{\partial t^2}-\frac{\partial^2 f(\tilde{t}_-)}{\partial t^2}\right)\right] \quad (26)$$

$$E_z(\mathbf{r},t)=\frac{\Psi_o}{\tilde{R}}\left\{\frac{(3\cos^2\tilde{\theta}-1)}{\tilde{R}}\left[\frac{f(\tilde{t}_+)-f(\tilde{t}_-)}{\tilde{R}}-\frac{1}{c}\left(\frac{\partial f(\tilde{t}_+)}{\partial t}+\frac{\partial f(\tilde{t}_-)}{\partial t}\right)\right]-\frac{\sin^2\tilde{\theta}}{c^2}\left(\frac{\partial^2 f(\tilde{t}_+)}{\partial t^2}-\frac{\partial^2 f(\tilde{t}_-)}{\partial t^2}\right)\right\} \quad (27)$$

$$H_\phi(\mathbf{r},t)=\frac{\varepsilon_0\Psi_o\sin\tilde{\theta}}{\tilde{R}}\left[\frac{1}{\tilde{R}}\left(\frac{\partial f(\tilde{t}_+)}{\partial t}-\frac{\partial f(\tilde{t}_-)}{\partial t}\right)-\frac{1}{c}\left(\frac{\partial^2 f(\tilde{t}_+)}{\partial t^2}+\frac{\partial^2 f(\tilde{t}_-)}{\partial t^2}\right)\right], \quad (28)$$

where the complex angle $\tilde{\theta}$ is still defined such as $\cos\tilde{\theta} \equiv (z + ja)/\tilde{R}$. The resulting pulsed beam is isodiffracting in the sense that all the frequency components have the same confocal parameter a. It should be pointed out that the use of Eq. (20) as the spectral amplitude of the pulse is consistent with the net-force condition $\int_{-\infty}^{\infty} \mathbf{E}(\mathbf{r},t)dt = \mathbf{0}$ that all laser pulses must satisfy (Milosevic *et al.*, 2006). Also, Eqs. (26)–(28) are consistent with the self-induced blueshift reported by Lin *et al.*, which implies that, for a ultrashort pulsed beam, the instantaneous frequency is highest at the center of the pulse and always higher than the carrier frequency ω_o (Lin *et al.*, 2006).

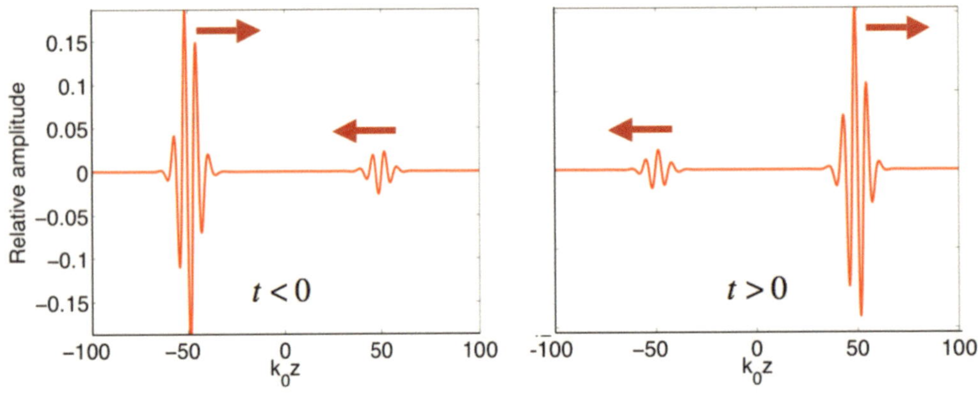

Fig. 8. The numerical calculations of the on-axis (r = 0) propagation of the pulse described by the real part of Eq. (27), with $k_o a = \omega_o a/c = 2$, illustrate that the optical wave consists in the superposition of two counter-propagating pulses of different amplitudes.

It is apparent from Eqs. (26)–(28) that the electric field of the pulse is made of two contributions: the first one is due to terms involving $f(t)$ and its derivatives evaluated at $\tilde{t}_+$, and the second one to terms evaluated at $\tilde{t}_-$. In the case of the choice I for the value of $\tilde{R}$ [Eq. (14a)], the terms evaluated at $\tilde{t}_+$ describe two counter-propagating pulses of different amplitudes converging toward the plane of the beam waist (z = 0) for $t < 0$ and they vanish identically for $t > 0$. On the other hand, terms involving $\tilde{t}_-$ are zero for $t < 0$ whereas they represent the two counter-propagating pulses diverging from the plane of the beam waist for $t > 0$. In the case of the choice II for the value of $\tilde{R}$ [Eq. (14b)], the terms evaluated at $\tilde{t}_+$ describe a pulse traveling in the $+z$ direction for all t, while terms involving $\tilde{t}_-$ represent a pulse of weaker amplitude traveling in the $-z$ direction. In both case, the addition of these two contributions leads to Eqs. (26)–(28), which describe a strong purely traveling pulse superimposed to a counter-propagating traveling pulse of weaker amplitude for all t (Fig. 8). Such a field distribution certainly occurs when the nonparaxial pulse is generated by a 4π focusing system such as a parabolic mirror of large extent (see Fig. 5). Nonetheless, it can be verified that the amplitude of the counter-propagating pulse is very small compared to the amplitude of the propagating pulse even when the beam divergence angle is as large as 50°. Therefore, the amplitude of the counter-propagating pulse is negligible when the divergence angle of the pulsed beam is reasonably small.

The analytic signal (the complex-valued) electric field of the tightly focused, ultrafast TM_{01} pulse is obtained by using Eq. (21) in Eqs. (26)–(28). The expression of the physical electric field is found by taking the real part of the resulting analytic signal. Some profiles of electric energy density, defined by $W(\mathbf{r},t)=\frac{1}{2}\varepsilon_0|\mathbf{E}(\mathbf{r},t)|^2$, of a TM_{01} pulse with different propagation parameters are shown in Fig. 9.

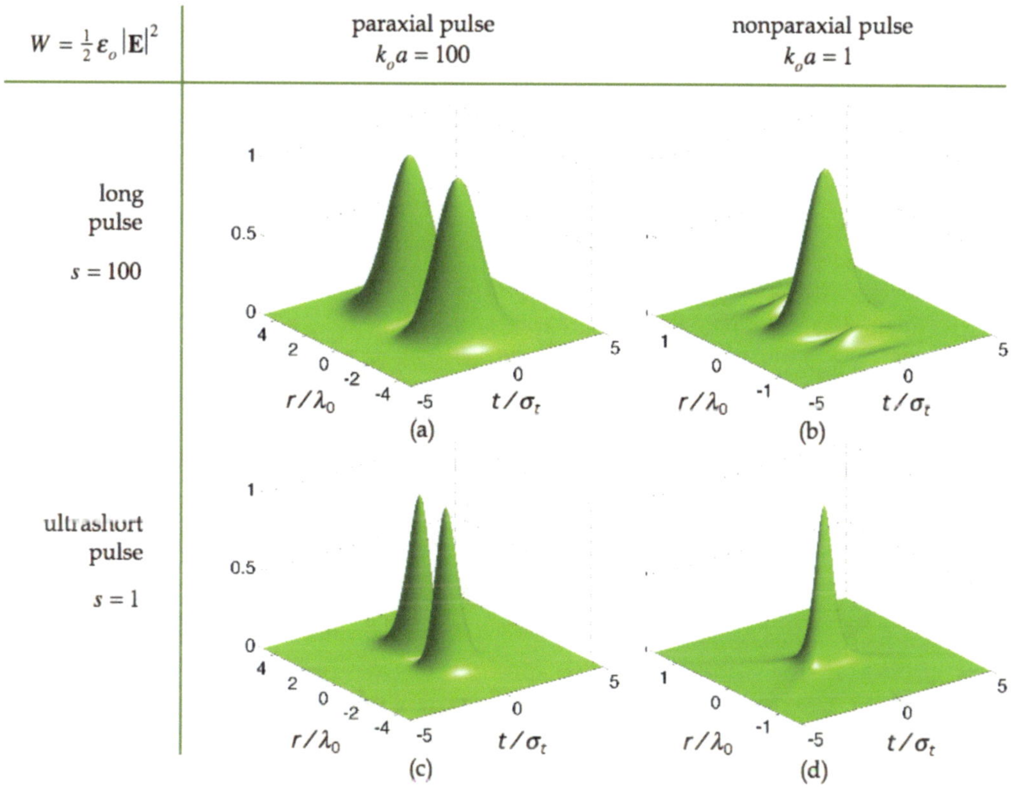

Fig. 9. The spatiotemporal shape of the electric energy density of a nonparaxial TM_{01} pulse is strongly dependent on the type of propagation regime.

It is instructive to compare the theoretical results to experimental data in order to validate the approach presented in this chapter. The intensity profile of a strongly focused TM_{01} pulsed beam has been obtained experimentally. The intensity distribution calculated numerically is in agreement with the experimental data (Fig. 10). For this experiment, a Ti:sapphire laser source emitting 140-fs pulses at 800 nm was used; this corresponds approximately to $\omega_0 = 2{,}4\times10^{15}$ rad/s and to $s=2\times10^5$. To produce the TM_{01} laser beam, a polarization converter was employed to convert a linearly polarized Gaussian beam into a radially polarized beam. Then, the pulsed beam was focused by a water immersion objective having a 1.2 numerical aperture, which corresponds to $k_0 a = 1$.

Very long pulses can be treated as continuous wave (CW) beams of frequency ω, for which the spectrum is infinitesimally narrow centered on the frequency ω. The spectrum of such a beam can be modeled by a Dirac delta centered on the frequency ω; the temporal shape of the beam reduces to $f(t)=\exp(j\omega t)$. Substituting this function in Eqs. (26)–(28) and computing the expressions of the electromagnetic fields of the nonparaxial TM_{01} beam, one can show that the results may be written as (Sheppard & Saghafi, 1999b; April, 2008b):

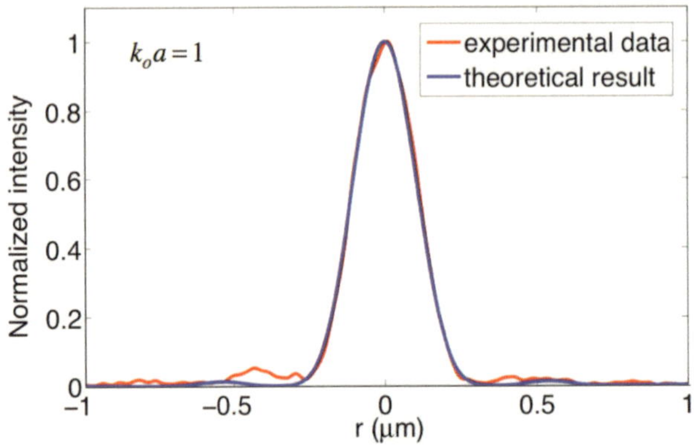

Fig. 10. The theoretical electric energy density profile in the beam waist is in agreement with the experimental data obtained with a Ti:Sapphire laser pulse for which $k_o a = 1$, $\omega_o = 2,4 \times 10^{15}$ rad/s and $s = 2 \times 10^5$. (Figure courtesy Harold Dehez, COPL, Université Laval)

$$E_r(\mathbf{r},t) = \tfrac{3}{4} E_o \exp(-ka + j\omega t) j_2(k\tilde{R}) \sin(2\tilde{\theta}), \tag{29}$$

$$E_z(\mathbf{r},t) = E_o \exp(-ka + j\omega t)\left[j_0(k\tilde{R}) + j_2(k\tilde{R}) P_2(\cos\tilde{\theta})\right], \tag{30}$$

$$H_\phi(\mathbf{r},t) = j\tfrac{3}{2} H_o \exp(-ka + j\omega t) j_1(k\tilde{R}) \sin\tilde{\theta}, \tag{31}$$

where $E_o = \frac{4}{3} j \Psi_o k^3$ is a constant amplitude, $H_o = E_o/\eta_0$, $k = \omega/c$ is the wave number of the beam, $j_n(kR)$ is the spherical Bessel function of the first kind of order n, and $P_2(\cos\tilde{\theta}) = \frac{1}{4}[1 + 3\cos(2\tilde{\theta})]$ is the Legendre polynomial of degree 2.

5.3 Focusing properties of isodiffracting TM_{01} pulses

The longitudinal component of the electric field of a TM_{01} beam is rotationally symmetric and exhibits an intensity distribution with a narrow central peak. When the TM_{01} beam is tightly focused, the contribution of the longitudinal component of its electric field dominates the contribution of the transverse components. Therefore, the electric energy density distribution in the plane of the beam waist is a sharp peak (Fig. 10). Moreover, it has been verified that the width of the electric density energy distribution of a TM_{01} beam is smaller than the one that could be obtained with a strongly focused LP_{01} beam (Quabis *et al.*, 2000).

The propagation of ultrashort pulses can differ significantly from that of the quasi-monochromatic light. In fact, the isodiffracting TM_{01} pulse exhibits interesting focusing properties: the width of its focal point becomes remarkably small as the duration of the pulse shortens (Fig. 11).

The full-width at half-maximum (FWHM) of the electric energy density of the nonparaxial TM_{01} pulse is $w_{\text{FWHM}} = 0,401\lambda_o$ for the long pulse in Fig. 11a, where λ_o is the carrier wavelength, whereas it is $w_{\text{FWHM}} = 0,145\lambda_o$ for the single-cycle pulse in Fig. 11c. Thus, according to this model, it appears that the radial extent of the pulse in the beam waist decreases as the pulse becomes shorter. Isodiffracting, tightly focused, ultrafast TM_{01} pulses turn out to have interesting focusing properties that may be exploited for superresolution and may have potential applications in high-resolution microscopy.

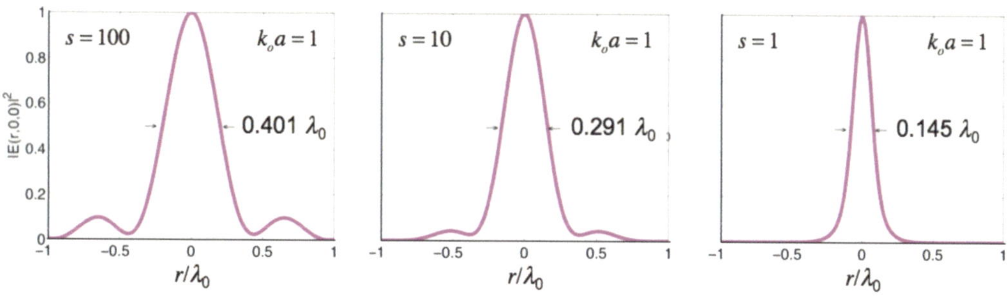

Fig. 11. As the pulse duration decreases (i.e. as s decreases), the width of the electric energy density profile of a tightly focused TM_{01} pulse with $k_o a = 1$ (i.e. a beam divergence angle of approximately 66°) in the plane of the beam waist ($z = 0$) decreases.

6. Conclusion

To conclude, we have presented a procedure to find the expressions for the electromagnetic fields of ultrashort, nonparaxial electromagnetic pulses that are rigorous solutions of the wave equation. The three main ingredients to build such solutions are (1) the Hertz potential method used to efficiently find an exact solution to Maxwell's four equations, (2) the complex-source/sink model employed to determine exact solutions to the Helmholtz equation that can be seen as the nonparaxial generalizations of the fundamental Gaussian beam and the higher-order Gaussian modes, and (3) the Poisson-like spectral amplitude of the form $\omega^s \exp(-s\omega/\omega_0)$ that allows the description of ultrafast pulses whose duration could be as short as one optical cycle. The analytical expressions for the electromagnetic fields found with this approach may be employed to investigate the spatiotemporal behavior of tightly focused, ultrafast laser pulses in free space.

As a continuation of this work, the treatment of nonisodiffracting pulses could be explored as well as the detailed study of higher-order pulsed beams. Finally, it would be interesting to extend the work presented in this chapter to the propagation of strongly focused, ultrashort pulses in a dispersive and/or nonlinear medium.

7. Acknowledgements

This work was supported by grants from Natural Sciences and Engineering Research Council of Canada (NSERC), the Fonds de recherche sur la nature et les technologies (FQRNT), Québec, the Canadian Institute for Photonic Innovations (ICIP/CIPI), and the Centre d'optique, photonique et lasers (COPL), Québec. The author thanks Michel Piché for helpful discussions and Harold Dehez for his contribution for the experimental data.

8. References

An der Brügge, D. & Pukhov, A. (2009). Ultrashort focused electromagnetic pulses. *Physical Review E*, Vol. 79, (January 2009) pp. 016603-1–016603-5, ISSN 1539-3755.

April, A. (2008a). Nonparaxial elegant Laguerre–Gaussian beams. *Optics Letters,* Vol. 33, No. 12, (June 2008) pp. 1392–1394, ISSN 0146-9592.

April, A. (2008b). Nonparaxial TM and TE beams in free space. *Optics Letters,* Vol. 33, No. 14, (July 2008) pp. 1563–1565, ISSN 0146-9592.

Bandres, M. A. & Gutiérrez-Vega, J. C. (2004). Higher-order complex source for elegant Laguerre–Gaussian waves. *Optics Letters,* Vol. 29, No. 19, (October 2004) pp. 2213–2215, ISSN 0146-9592.

Berardi, C.; Erricolo, D. & Uslenghi, P. L. E. (2004). Exact Dipole Radiation for an Oblate Spheroidal Cavity Filled With Isorefractive Material and Aperture-Coupled to a Half Space. *IEEE Transactions on Antennas and Propagation,* Vol. 52, No. 9, (September 2004) pp. 2205–2213, ISSN 0018-926X.

Caron, C. F. R. & Potvliege, R. M. (1999). Free-space propagation of ultrashort pulses: space-time couplings in Gaussian pulse beams. *Journal of Modern Optics,* Vol. 46, No. 13, (1999) pp. 1881–1891, ISSN 0950-0340.

Couture, M. & Bélanger, P.-A. (1981). From Gaussian beam to complex-source-point spherical wave. *Physical Review A,* Vol. 24, No. 1, (July 1981) pp. 355–359.

Deng, S. ; Liu, L. ; Li, R. & Xu, Z. (2007). STED microscopy with the azimuthally-polarized depletion beam, *Proceedings of SPIE,* Vol. 6826, pp. 682621-1-682621-7, Beijing, China, November 2007, SPIE.

Deschamps, G. A. (1971). Gaussian beam as a bundle of complex rays. *Electronics Letters,* Vol. 7, No. 23, (November 1971) pp. 684–685.

Dorn, R.; Quabis, S. & Leuchs, G. (2003). Sharper Focus for a Radially Polarized Light Beam. *Physical Review Letters,* Vol. 91, No. 23, (December 2003) pp. 233901-1–233901-4, ISSN 0031-9007.

Feng, S. & Winful, H. G. (2000). Spatiotemporal structure of isodiffracting ultrashort electromagnetic pulses. *Physical Review E,* Vol. 61, No. 1, (January 2000) pp. 862–873, ISSN 1063-651X.

Heyman, E. & Felsen L. B. (2001). Gaussian beam and pulsed-beam dynamics: complex-source and complex-spectrum formulations within and beyond paraxial asymptotics. *Journal of the Optical Society of America A,* Vol. 18, No. 7, (July 2001) pp. 1588–1611, ISSN 1084-7529.

Kawauchi, H; Kozawa, Y. & Sato, S. (2008). Generation of radially polarized Ti:sapphire laser beam using a c-cut crystal. *Optics Letters,* Vol. 33, No. 17, (September 2008) pp. 1984–1986, ISSN 0146-9592.

Landesman, B. T. & Barrett, H. H. (1988). Gaussian amplitude functions that are exact solutions to the scalar Helmholtz equation. *Journal of the Optical Society of America A,* Vol. 5, No. 10, (October 1988) pp. 1610–1619, ISSN 0740-3232.

Lax, M.; Louisell, W. H. & McKnight, W. B. (1975). From Maxwell to paraxial wave optics. *Physical Review A,* Vol. 11, No. 4, (April 1975) pp. 1365–1370.

Lin, Q.; Zheng, J. & Becker, W. (2006). Subcycle pulsed focused vector beams. *Physical Review Letters,* Vol. 97, (December 2006) pp. 253902-1–253902-4, ISSN 0031-9007.

Lu, G.; Guo, H.; Deng, D.; Yi, F. & Kong, H. J. (2003). Nonparaxially corrected solution for isodiffracting sub-cycle pulsed-beam propagation in free space. *Journal of the Korean Physical Society,* Vol. 42, No. 5, (May 2003) pp. 627–630, ISSN 0374-4884.

Melamed, T. & Felsen, L. B. (1998). Pulsed-beam propagation in lossless dispersive media. I. Theory. *Journal of the Optical Society of America A,* Vol. 15, No. 5, (May 1998) pp. 1268–1276, ISSN 0740-3232.

Milosevic, D. B.; Paulus, G. G.; Bauer, D. & Becker, W. (2006). Above-threshold ionization by few-cycle pulses. *Journal of Physics B: Atomic, Molecular and Optical Physics,* Vol. 39, No. 5, (July 2006) pp. R203–R262, ISSN 0953-4075.

Porras, M. A. (1998). Ultrashort pulsed Gaussian light beams. *Physical Review E,* Vol. 58, No. 1, (July 1998) pp. 1086–1093, ISSN 1063-651X.

Quabis, S.; Dorn, R.; Eberler, M.; Glöckl, O. & Leuchs G. (2000). Focusing light to a tighter spot. *Optics Communications,* Vol. 179, (May 2000) pp. 1–7, ISSN 0030-4018.

Richards, B. & Wolf, E. (1959). Electromagnetic diffraction in optical systems. II. Structure of the image field in an aplanatic system. *Proceedings of the Royal Society of London. Series A, Mathematical and Physical Sciences,* Vol. 253, No. 1274, (December 1959) pp. 358–379.

Rodríguez-Morales, G. & Chávez-Cerda, S. (2004). Exact nonparaxial beams of the scalar Helmholtz equation. *Optics Letters,* Vol. 29, No. 5, (March 2004) pp. 430–432, ISSN 0146-9592.

Saari, P. (2001). Evolution of subcycle pulses in nonparaxial Gaussian beams. *Optics Express,* Vol. 8, No. 11, (May 2001) pp. 590–598, ISSN 1094-4087.

Seshadri, S. R. (2002). Virtual source for a Laguerre–Gauss beam. *Optics Letters,* Vol. 27, No. 21, (November 2002) pp. 1872–1874, ISSN 0146-9592.

Sheppard, C. J. R. & Saghafi, S. (1998). Beam modes beyond the paraxial approximation: A scalar treatment. *Physical Review A,* Vol. 57, No. 4, (April 1998) pp. 2971–2979, ISSN 1050-2947.

Sheppard, C. J. R. & Saghafi, S. (1999a). Electromagnetic Gaussian beams beyond the paraxial approximation. *Journal of the Optical Society of America A,* Vol. 16, No. 6, (June 1999) pp. 1381–1386, ISSN 0740-3232.

Sheppard, C. J. R. & Saghafi, S. (1999b). Transverse-electric and transverse-magnetic beam modes beyond the paraxial approximation. *Optics Letters,* Vol. 24, No. 22, (November 1999) pp. 1543–1545, ISSN 0146-9592.

Sheppard, C. J. R. (2000). Polarization of almost-plane waves. *Journal of the Optical Society of America A,* Vol. 17, No. 2, (February 2000) pp. 335–341, ISSN 0740-3232.

Shin, S. Y.; Felsen, L. B. (1977). Gaussian beam modes by multipoles with complex source points. *Journal of the Optical Society of America,* Vol. 67, No. 5, (May 1977) pp. 699–700.

Siegman, A. E. (1986). *Lasers,* University Science Books, ISBN 0-935702-11-3, Sausalito, CA

Tidwell, S. C.; Ford, D. H. & Kimura, W. D. (1990). Generating radially polarized beams interferometrically. *Applied Optics,* Vol. 29, No. 15, (May 1990) pp. 2234–2239, ISSN 0003-6935.

Ulanowski, Z. & Ludlow, I. K. (2000). Scalar field of nonparaxial Gaussian beams. *Optics Letters,* Vol. 25, No. 24, (December 2000) pp. 1792–1794, ISSN 0146-9592.

Varin, C.; Piché, M. & Porras, M. A. (2005). Acceleration of electrons from rest to GeV energies by ultrashort transverse magnetic laser pulses in free space. *Physical Review E,* Vol. 71, (February 2005) pp. 026603-1–026603-10, ISSN 1539-3755.

Varin, C.; Piché, M. & Porras, M. A. (2006). Analytical calculation of the longitudinal electric field resulting from the tight focusing of an ultrafast transverse-magnetic laser beam. *Journal of the Optical Society of America A,* Vol. 23, No. 8, (August 2006) pp. 2027–2038, ISSN 1084-7529.

Wang, Z.; Zhang, Z.; Xu, Z. & Lin, Q. (1997). Space–time profiles of an ultrashort pulsed Gaussian beam. *IEEE Journal of Quantum Electronics,* Vol. 33, No. 4, (April 1997) pp. 566–573, ISSN 0018-9197.

Ziolkowski, R. W. & Judkins, J. B. (1992). Propagation characteristics of ultrawide-bandwidth pulsed Gaussian beams. *Journal of the Optical Society of America A,* Vol. 9, No. 11, (November 1992) pp. 2021–2030, ISSN 0740-3232.

17

Interaction of Short Laser Pulses with Gases and Ionized Gases

Stephan Wieneke[1], Stephan Brückner[1] and Wolfgang Viöl[2]
[1] *Laser-Laboratorium Göttingen, Hans-Adolf-Krebs-Weg 1, 37077 Göttingen*
[2] *University of Applied Sciences and Arts, Von-Ossietzky-Str. 99, 37085 Göttingen*
Germany

1. Introduction

In this chapter, the interaction mechanisms between short laser pulses and gases, respectively pre-ionized gases are described in detail. Ordinarily, gases are transparent for electro-magnetic radiation. This is not valid any more when the intensity of radiation goes beyond a certain value. This barrier depends on the pressure, the gas type, the pulse length, the pulse repetition frequency, the beam diameter and the laser wavelength used. In order to understand the interaction mechanisms, knowledge about the propagation of electromagnetic radiation as well as knowledge about the absorption processes (for example: inverse bremsstrahlung) is necessary. Further, the mathematic models for the absorption processes and the following effects (as, for example, heating) of short laser pulses in gases and pre-ionized gases are explained. The appearing effects such as breakdown intensity or heating play an important role in the generation of new radiation sources. With specific laser pulses, individual radiation sources can be created (such as UV, EUV, soft X-ray).

2. Fundamentals of ionized gas generation

An ionized gas is called plasma to lay the focus on its consistence of electrified particles. This characteristic of plasma particles defines their major difference to a neutral gas. The most established way to generate a plasma is the electric gas discharge, for example in a spark or a flash of lighting. The ignition of an electric field occurs when a discharge between two electrodes starts to generate enough charge carriers to sustain itself. Due to the usage of an high performance laser, the so called laser induced plasmas can develop. For this, an intense laser pulse is focused on an existing target. The target is ionized this way; it is converted to plasma. To gain a laser induced plasma, solid targets, liquid targets or gas targets can be used. Because of their low vapor pressure, solid targets are particularly suitable for the use in vacuum. However it has to be ensured that a new area of the target will be hit with the laser pulse, for the material in the focal plane of the laser is destroyed during the plasma generation. The emission would be strongly decreasing after some pulses, because the laser would "drill" a hole in the material whereby no material would be left in the focus area (1). When the laser pulse meets the target, a shock wave is generated. This shock wave centrifuges particles out of the surface. Furthermore, particles are released because of thermal evaporation. A smaller amount of debris is generated when a liquid target or a gas target are used as medium. With liquid targets, particle densities similar to those in solid targets, can be reached (4; 5). The liquid target is injected to the vacuum with the help of a nozzle assembly. The beam diameter here typically amounts to 10 - 30 μm, in which the liquid target is blasted with a pressure between 15 and 60 bar into the vacuum. Compared to solid targets, the debris appearing here during the plasma generation is smaller - about several orders of magnitude. In general it can be retained that a less debris is generated when smaller beam diameter is used. Ideally, the beam diameter has the size of the laser pulse focus diameter. Because the liquid target is made to follow up constantly through the nozzle assembly, it has to be removed from the vacuum system to perpetuate the vacuum conditions. Gas targets can be added intermittently through a valve. The laser pulses and gas shocks are synchronized so that the pulses are focused to a gas cloud. The disadvantage here lies in the fact that the density of gas has a lower value than the density of all-solids, so that an efficient absorption of laser beams is not possible.

2.1 Propagation of electromagnetic waves in a plasma

To understand the interaction between a plasma and laser beams, knowledge about the propagation electromagnetic waves in a plasma plays an important role. Here, a two liquid model can be used as a starting point for a precise analysis. Only the equation for the electrons are observed here, for the ions in a high frequency light source can be considered as immotile. For transversal verbs of small intensity, the current term is linearly simplified to $J = -e_0 n_e \bar{v}_e$ in the Maxwell-equations. All non-linear terms in order velocity moments can be neglected as well. The movement of the electrons in a plasma can be writen as (6):

$$m_e \frac{\partial \bar{v}_e}{\partial t} = -e_0 \vec{E}. \tag{1}$$

Using this equation, the following wave equation for electromagnetic waves in plasma arises from the Maxwell-equations:

$$\left(\frac{\partial^2}{\partial t^2} + \omega_p^2 - c_0^2 \nabla^2 \right) \vec{E} = 0. \tag{2}$$

The plasma frequency and the Debye length are important parameters of plasma. The plasma frequency describes the collective electron motion and equates to the natural frequency of electrons in a plasma. For a plane electromagnetic wave like $\vec{E} = E_0 e^{i(kx - \omega t)}$ the dispersion relation below is following up:

$$\omega^2 = \omega_p^2 + k^2 c_0^2. \tag{3}$$

It varies from the dispersion relation in vacuum in the summand ω_p^2. The most important interaction mechanism between plasma and laser field in of mild to moderate intensities (I $< 10^{15}$ W/cm^2 for 1 μm laser wave length) is the so called shock absorption. Again, the two liquids model offers a basis for a theoretical analysis if a shock term with the shock frequency between ions and electrons is included to the equation of motion for electrons: ν_{ei}.

$$m_e \frac{\partial \bar{v}_e}{\partial t} = -e_0 \vec{E} - m \nu_{ei} \bar{v}_e. \tag{4}$$

Thereby, a modified dispersion relation arises (for $\nu_{ei} << \omega$):

$$\omega^2 = \omega_p^2 + k^2 c_0^2 - i \omega_p^2 \frac{\nu_{ei}}{\omega^2}. \tag{5}$$

The real part matches with the solution without shocks. The negative imaginary part takes into consideration that the electromagnetic wave is damped during its dispersion in the low-density plasma. In the process, the wave induces electron oscillations whose energy leads to an increase of the average electron energy because of the shocks of electrons with ions. Because of that, the electron temperature in the plasma is raised. Furthermore, higher ionization levels can be reached with the ions through shocks. This mechanism has a major importance for the generation of hot plasmas and is called *collisional absorption* or *inverse bremsstrahlung* according to the contrary process of bremsstrahlung, where kinetic electron energy is converted to radiation energy. In Figure 1 this process is shown schematically. A laser pulse meets a plasma with an electron density gradient. While the light wave passes the low-density plasma, it is damped and heats the plasma with the inverse bremsstrahlung. The wave is just able to spread up to the critical density in the plasma, there it is reflected. The area of the reflection point is the most effective are for an energy transfer from the radiation field to the plasma with the help of collisional absorption (7).

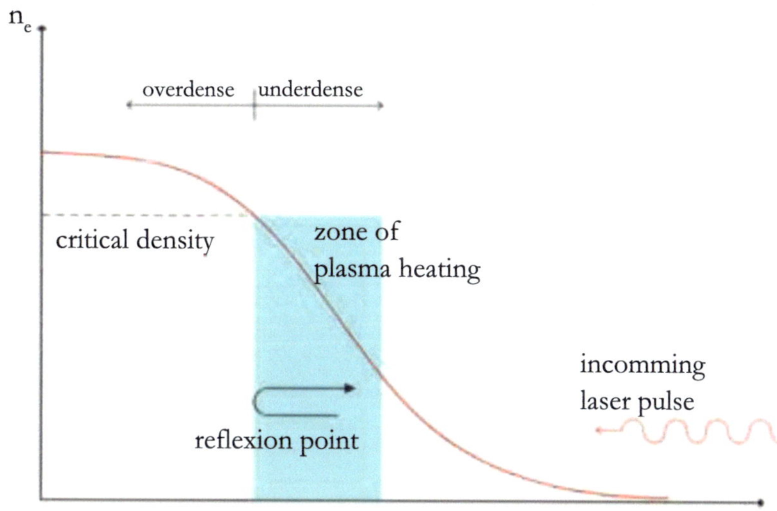

Fig. 1. One-dimensional illustration of the collisional absorption effect.

2.2 Collisional absorption - inverse bremsstrahlung

The absorption of laser radiation in a plasma, and thus the heating-up of the plasma, is affected by the inverse bremsstrahlung in the regime considered here. In this process a free electron accepts a radiation quantum of the laser beam. For reasons of the impulse conservation this is only possible in interaction with a neutral atom or an ion. The atoms or ions only absorb a small amount of the radiation energy and are primarily excited and heated indirectly by collisions with the accelerated electrons. While primarily neutral atoms exist at the beginning of the ignition, the electron-atom inverse bremsstrahlung dominates at that time. The according absorption coefficient α_{ea} is given as

$$\alpha_{ea} = \frac{1}{4\pi\epsilon_0} \cdot \frac{e^2 \nu_{ea} n_e \lambda_L^2}{\pi m_e c^3}. \tag{6}$$

Here n_e is the electron density, λ_L the laser wavelength and ν_{ea} is the impact frequency for impulse-exchanging collisions between electrons and neutral atoms. In an already ignited discharge, however, the electron-ion inverse bremsstrahlung plays the decisive role for the heating-up of the plasma. The absorption coefficient α_{ei} for electron-ion inverse bremsstrahlung applies to

$$\alpha_{ei} = \frac{n_e^2 e^6}{6\sqrt{3}n\epsilon_0^3 c\hbar\omega_L^3 m_e^2} \cdot \sqrt{\frac{m_e}{2\pi k_B T_e}} \cdot \left(1 - e^{-\frac{\hbar\omega_L}{k_B T_e}}\right) \bar{g}, \tag{7}$$

where ω_L is the laser frequency and n is the real part of the refractive index. The averaged Gaunt-factor $\bar{g}$, as adjustment to relativistic and quantum mechanical effects, applies to:
In this paper, the Nd:YAG laser is considered. In that case, a more precise expression for the Gaunt-factors can be written as:

$$\bar{g}_{Nd:YAG} = 6.2 \times 10^{-3} \left(\frac{T_e}{K}\right)^{0.42} + 0.86. \tag{8}$$

Figure 2 shows the linear absorption coefficient as a function of the plasma temperature for different electron densities n_e. These graphs clarify that, in order to achieve an effective absorption of laser radiation, the plasma requires a high electron density and a low electron temperature. Janulewicz *et al.* found that the absorption medium demands a concave electron density distribution with the maximum on the axis, against what the temperature distribution is to behave in reverse (2). This distribution is fulfilled in z-pinch plasmas and is experimentally determined in (3).

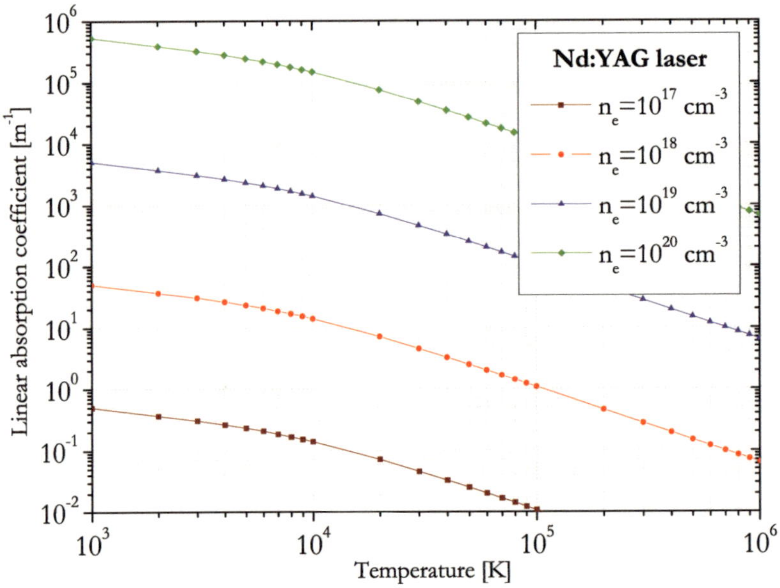

Fig. 2. Linear absorption coefficient as a function of temperature for varoius electron densities and for a wavelength of 1.06 μm.

2.3 Ignition behavior of laser based discharges

Ordinarily, gases are transparent for electromagnetic radiation. This does not apply any more when the radiation intensity exceeds a certain value, as it can happen in an high-energy laser. Just as it would happen in an ordinary discharge, the gas is ionized and the transmitted laser beam is damped. The minimal intensity needed to initiate a gas breakthrough can be found between $10^7 - 10^{11}$ W/cm^2. This threshold depends on the pressure, the gas type, the pulse length, the pulse repetition frequency, the beam diameter and the used wavelength. The ionization can be effected by one or several of the following processes. If the field intensity is very high, the Columbo barrier is deformed and Tunnel ionization occurs. In the Tunnel ionization, electrons can be freed up the atom during one vibration of the electric field. In the multi photon ionization, an electron in the ground state is able to absorb several laser photons; it can be stimulated even beyond the ionization energy. This process is almost independent from the gas pressure. However, neither the Tunnel ionization nor the multi photon ionization plays a role when an Nd:YAG laser with its moderate field intensity and the small photon energy about 1.06 μm is used. The ionization is effected because of a cascade process; therefore, free electrons are needed. These electrons can be generated with multi photon ionization, cosmic radiation, a residual ionization accrued because of the high repetition rate of the laser pulses, or an electric discharge ignited before (9). Due to elastics shocks with neutral atoms, the electrons can absorb energy in the laser field and set free more electrons when they exceed the ionization energy with inelastic shocks with atoms or ions. These new electrons are integrated into the process as well. Referring to the cascade theory, the ignition intensity needed has a minimum at a certain pressure; it increases with the pressure. For the ignition intensity I_z in the focus of a laser needed to ignite a plasma in a gas, the following relation is given (8):

$$I_z \geq \frac{\epsilon_0 m_e c_0 E_I}{e_0^2 \ln 2} \cdot \frac{\omega_L^2 + \nu_{ea}^2}{\nu_{ea}} \cdot \left\{ \frac{1}{\tau_L} \cdot \ln\left(\frac{n_{eb}}{n_{e0}}\right) + \frac{D}{\Lambda_d^2} + h_a \nu_{ea} + \nu_{ve} + \nu_{vi} \right\}. \tag{9}$$

According to equation (9), the ignition threshold is proportional to the ionization energy E_I of the gas. The terms inside of the brackets do take various loss terms into consideration. They are described in the following passage (10): If the first term is dominant here because of a small laser pulse duration, it is called *time-determined* ignition. The ignition threshold is reciprocally proportional to the laser pulse duration τ_L and therefore defined through the whole energy flow. Here, n_{eb} is the electron density where the electron-ion bremsstrahlung is much more effective than the electron-atom bremsstrahlung. In the case under consideration, the ratio n_{eb}/n_{e0} accounts for 10^{13} (8). According to this model, a significant decrease of the ignition intensity needed because of the higher electron density n_{e0} at the beginning of the laser pulse can merely be expected in the case of a time-determined ignition. If the second term prevails (as, for example, in small pressures or focus volume) the ignition is called a *diffusion-determined* ignition. For the diffusion coefficient D it holds (8):

$$D = \frac{\langle v^2 \rangle}{3\nu_{ea}}. \tag{10}$$

Where the average quadratic speed $\langle v^2 \rangle$ of the electrons shall be such that the average electron energy corresponds with half of the ionization energy of gas:

$$\left\langle v^2 \right\rangle = \frac{2\langle E_e \rangle}{m_e} = \frac{E_I}{m_e}. \tag{11}$$

In case of a cylinder arrangement, the following equation is used for the diffusion length Λ_d (15):

$$\frac{1}{\Lambda_d^2} = \left(\frac{2{,}405}{R}\right)^2 + \left(\frac{\pi}{L}\right)^2. \tag{12}$$

R is taken as radius for the cylinder assumed here, L is its length. The equation (12) can be transcribed to a laser generated cylinder volume:

$$\frac{1}{\Lambda_d^2} = \left(\frac{2.405}{w_f}\right)^2 + \left(\frac{\pi}{2z_R}\right)^2 = \left(\frac{2.405}{w_f}\right)^2 + \left(\frac{\lambda_L}{2w_f^2}\right)^2. \tag{13}$$

Here, R is replaced by the radius of the laser beam waist w_f. The length L is replaced by the doubled Rayleigh lenght z_R in the discharge volume. If the ignition is diffusion-determined the ignition intensity scales with $1/w_f^2$. In the third term, charge carrier losses because of attachment are described. Due to the low xenon attachment coefficient h_a, this term can be neglected here. Losses based on elastic materials are described with the help of the fourth term. Here, $\langle E_e \rangle = E_I/2$ is assumed:

$$\nu_{ve} = \frac{2m_e \langle E_e \rangle \ln 2}{m_a E_I} \cdot \nu_{ea}. \tag{14}$$

The last term stands for the energy loss due to inelastic materials. It applies:

$$\nu_{vi} = \left(\tilde{\alpha} + \frac{\tilde{\beta}}{\Lambda_d^2}\right) \cdot \nu_{ea}. \tag{15}$$

This term is important for molecular gases, for they have a huge amount of excited states. For xenon, the material frequency $\nu_{ea} = 2 \cdot 10^8\ s^{-1} \cdot p$ can be presumed. Fig. 3 shows the behavior of the ignition intensity for xenon dependent from the pressure, for a Nd:YAG laser with a wavelength of $\lambda = 1.06\ \mu m$. It is calculated with the help of equation (9). The experimentally achieved values according to (11), for a laser pulse duration about $\tau_L = 35$ ns as well as the experimentally achieved values following the work of (12) for a laser pulse duration of $\tau_L =$ 10 ps were also simulated with this model. As a whole, an ignition intensity of $10^{16}\ W/cm^2$ is calculated for xenon with a pressure of 13 Pa, an assumed beam focus of $2w_f = 50\ \mu m$ and a laser pulse duration of $\tau_L = 9$ ns. These experimental determined values accord with the simulated values very well. However we found out in the experiment that the gas ignition was not possible below a pressure og about 750 Pa.

3. Modelling of the interaction of laser puleses and neutral gases

3.1 Influence of the target density

To identify the influence of the target density on the achievable electron temperature, a laser pulse is coupled to a target at different target densities. The achievable maximum electron temperature is determinded numerically. As input parameter, a fit function of the experimentally determined laser pulse profile is used. With a repetition rate about 10 Hz, the laser system generates short NIR laser pulses with a maximum pulse energy about 750 mJ. The pulse width is ca. 9 ns in the FWHM. A lense with the focal width of $f = 150$ mm is used for the focusing of the laser pulse; it generates a focus diameter about 50 μm. The simulation results are shown in fig. 4. Here it is clearly to see that relevant electron temperatures cannot be generated with a laser pulse energy about 750 mJ in the relevant low pressure area about 10 Pa. This was already shown in the experiments in section 2.3. An operating pressure >1 kPa with a high initial particle density necessary to generate an adequate ionization and thereby electron temperature in a laser induced plasma. Because of that, liquid targets or solid targets are mostly used for laser induced plasmas. Another possibility to gain an efficient ionization with low operating pressure is the use of a stronger laser pulse. For an operating pressure of about 10 Pa, a laser pulse with the pulse energy of > 10 J is needed.

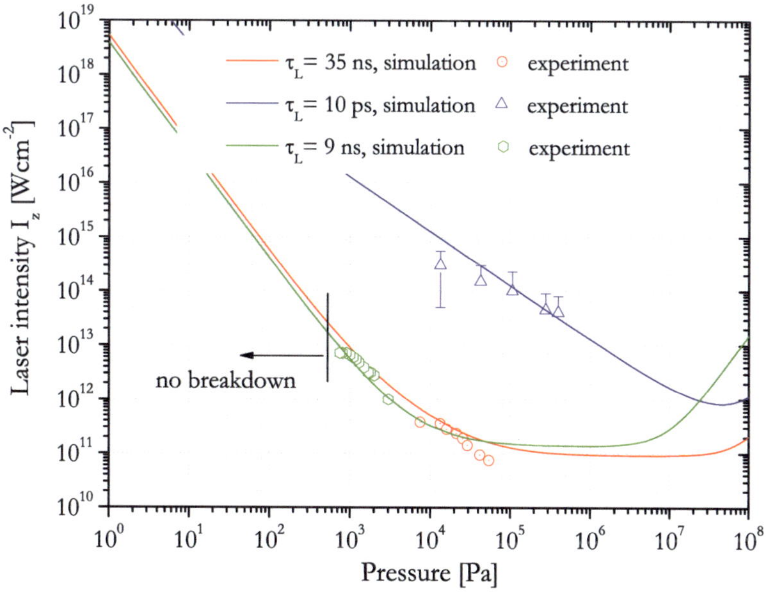

Fig. 3. Dependence of the breakdown intensity in addition to the pressure for different pulse duration times and a wavelength of 1.06 μm.

3.2 Influence of the laser pulse duration

The influence of the laser pulse duration on an efficient heating will be investigated in the following passage. As input parameter for the simulations a Gauss profile of the laser pulse is taken. Its width is varied between 0.5 - 30 ns. These simulations are carried out for different power densities (from 0.5 TW/cm^2 to 9.2 TW/cm^2). Here, the laser is coupled to a Xenon target at a gas density of 6 mg/cm^3. The fact that higher electron temperatures can initially be reached with increasing pulse rates (while the intensity stays the same) indicates a collisional absorption (inverse bremsstrahlung) as main mechanism for the plasma heating. The reason for this fact can be found through an observation of the electron density at the maximum electron temperature (red circles in figure 5).The most efficient plasma heating happens with a constant electron density at about $7{\cdot}10^{19}$cm^{-}3. This electron density is to find about one order of magnitude below the critical electron density of the laser wave length used. Above this electron density, the laser beam does not heat the plasma efficiently any more. With a growing laser intensity, this electron density of $7{\cdot}10^{19}$ cm^{-3} can be reached more quickly because of a higher target ionization. Supposedly, the optimal pulse duration postpones from a falling pulse intensity to shorter pulse durations.

3.3 Energy distribution and conversion efficiency

The time dependant behavior of the single shares of energy (kinetic energy, radiation energy, intern energy) of the plasma is important for an efficient change of the laser pulse to radiation energy of the plasma. A high amount of kinetic energy in the plasma leads to a high contamination rate of the components (debris), a high radiation component leads to a high conversion efficiency. For a better impact assessment, the different shares of energy for initial electron densities ne0 are calculated in a time-resolved manner. In fig 6 one can see the temporal relation of the energy distribution (radiation energy, kinetic energy and intern

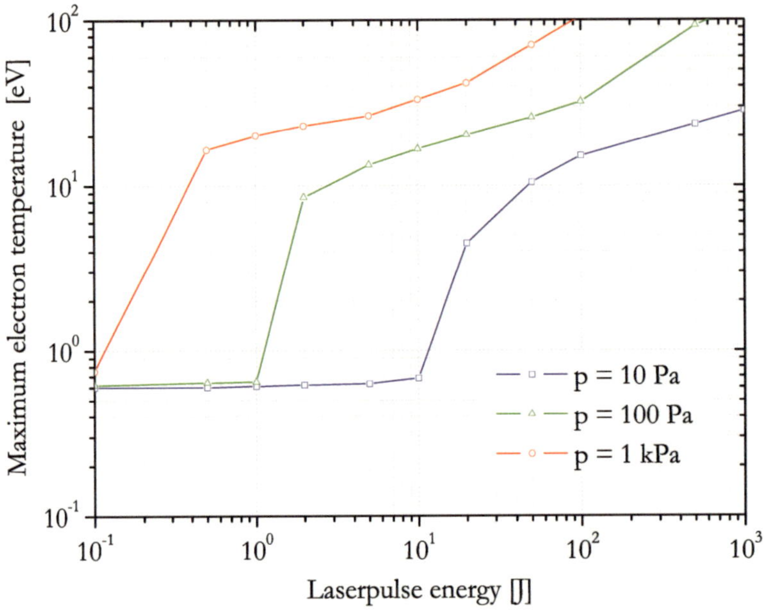

Fig. 4. Maximal reachable electron temperature in the plasma, dependent from the laser pulse energy for different pressures (target density).

energy in the plasma) for a xenon plasma heated with an Nd:YAG laser, E=1 J and 9 ns pulse duration, with an initial electron density of n_{e0}=10^{18} cm^{-3}. It is obvious that about 8 % of the laser pulse energy absorb by the plasma is emitted after 30 ns. In comparison, only 8 % are turned to kinetic energy and 4 % are turned to intern energy of the plasma (thermic + ionization-/ excitation energy). A huge part of the leftover intern energy is turned to kinetic energy, which means that the plasma is cooling down and expands. The same situation in shown in fig 7 for an initial electron density of 10^{19} cm^{-3}. Here, the attention is attracted to the fact that only 60 % are changed to radiation energy; therefore the kinetic energy increases to 17 % and the intern energy increases to 23 %. When the initial electron density is further increased, the radiation production decreases and the kinetic energy rises up until it reaches the critical density. However, if the initial electron density is reduced, the radiation production increases to a maximum, but the laser radiation is not completely coupled to the plasma. For a better description of this discovery, an absolute conversion efficiency is determined.

4. Modelling of the interaction of laser puleses and pre-ionized gases

In order to be able to provide information concerning the propagation of electromagnetic waves in a gas which is already ionized (pinch-plasma) and the heating connected to this, a modeling is done with Comsol Multiphysics. Here, the laser pulse is coupled to the face surface of a plasma assumed already generated. For the laser pulse, a Gauss profile is assumed. Fig. 8 shows a construction drawing and the model area. The laser pulse is focused to the electrically generated pinch-plasma through a lense. To facilitate the modeling, it is executed in the 2D axially symmetric mode.

In order to meet statements on the temperature field of a laser-intensified plasma, the knowledge of the spatial and temporal intensity $I(r,z,t)$ of the short laser pulse is necessary.

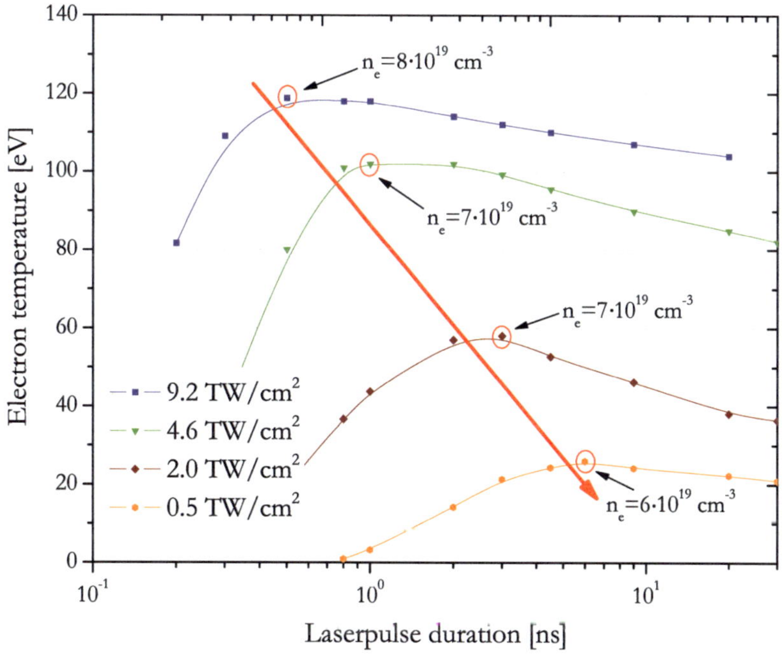

Fig. 5. Maximum reachable electron temperature of a laser induced xenon plasma (target density 6 mg/cm^3), dependent from the laser pulse duration for various laser intensities.

In vacuum, this distribution is given by

$$I(r,z,t) = \frac{I_0}{1+\left(\frac{z}{z_0}\right)^2} \exp\left(\frac{-2r^2}{w_f^2}\left(1+\left(\frac{z}{z_0}\right)^2\right)\right) \cdot \exp\left(\frac{-4\ln(2)(t-\frac{z}{c_0})^2}{\tau_p^2}\right), \tag{16}$$

where I_0 is applied to:

$$I_0 = \frac{4\sqrt{\ln(2)}J_0}{\sqrt{\pi}\tau_p \pi w_f^2}. \tag{17}$$

The definition for the radius of the beam waist is

$$w_f = \frac{f\lambda_L}{w_0\pi}. \tag{18}$$

Here f is the focal length of the optical system, w_0 is the $1/e^2$-beam radius at the lens, τ_p is the laser pulse duration, z_0 is the Rayleigh-length, J_0 is the energy of the laser pulse and z and r are the propagation directions, respectively. The propagation of an electromagnetic wave in a plasma is accompanied by its absorption in accordance with the Beer-Lambert law:

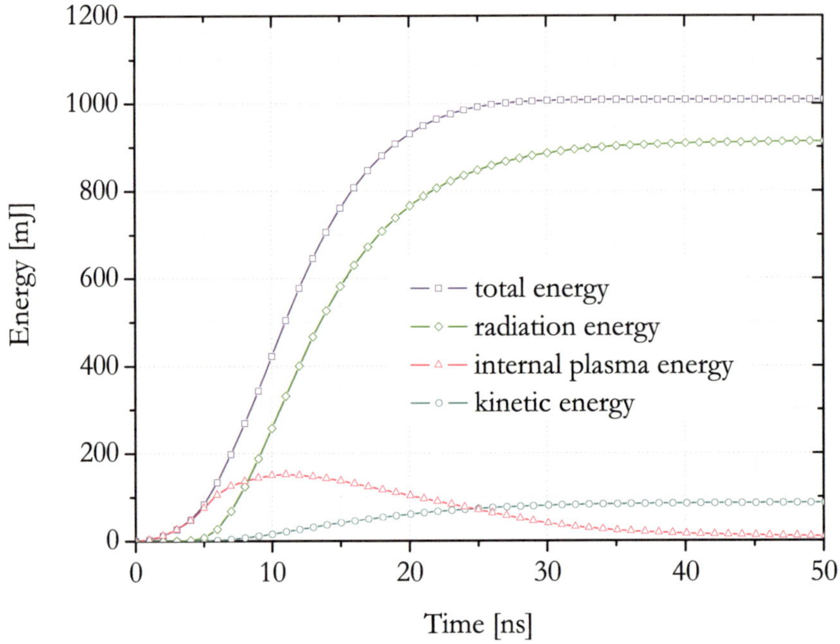

Fig. 6. Time dependency of the energy distribution of a xenon plasma, heated with a Nd-YAG laser pulse, with an initial electron density of n_{e_0}=10^{18} cm^{-3}.

$$\ln\left(\frac{I_L(r,z,t)}{I(r,z,t)}\right) = -\int_0^z \alpha_{ei} dz, \tag{19}$$

where α_{ei} is the absorption coefficient and z is the length. The main mechanism of the absorption of probe radiation in the absence of absorption bands and lines is, as already mentioned in 2.2, the inverse bremsstrahlung. During the quick heating-up of the electrons, due to the laser pulse, no expansion work is achieved. The amount of heat dQ_L is supplied to the electrons per time unit. Standarized to the volume, with the absorption coefficient of the electrons for inverse bremsstrahlung α_{ei}, the heat source strength P_L results:

$$P_L(r,z,t) = \frac{1}{V}\frac{dQ_L}{dt} = \alpha_{ei} I_L(r,z,t). \tag{20}$$

The electron temperature changes agreeable to

$$\frac{dT_e}{dt} = \frac{2}{3n_e k_B}\alpha_{ei} I_L(r,z,t). \tag{21}$$

For the absorption coefficient α_{ei}, equation (7) is used. Because of the proportionality of the absorption coefficient to n_e^2, under certain circumstances, a possible rest ionisation from the previous laser pulse may play an important role. A high repetition rate and the effect of the magnetic field on the remaining ionisation can have a positive influence on the absorption of the plasma in this issue.

For the thermal conduction out of the central area of the discharge the continuity equation obtains:

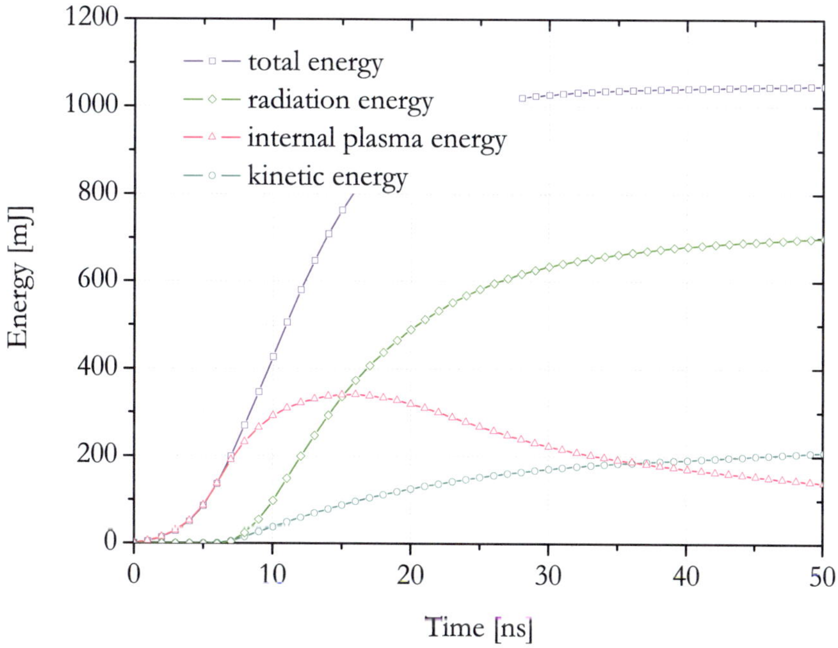

Fig. 7. Time dependency of the energy distribution of a xenon plasma, heated with a Nd-YAG laser pulse, with an initial electron density of n_{e_0}=10^{19} cm^{-3}.

$$\frac{dq}{dt} + \nabla \vec{j}_w = 0. \tag{22}$$

The heat flux obtains:

$$\vec{j}_w = -\kappa \nabla T_e, \tag{23}$$

where κ is the thermal conductivity. The temporal change of the electron temperature due to thermal conduction obtains:

$$\frac{dT_e}{dt} = \frac{2}{3n_e k_B} \nabla(\kappa \nabla T_e). \tag{24}$$

For the thermal conductivity, we use the expression

$$\kappa(T) = 4 \cdot 10^4 \left(9 - 8 \langle Z \rangle^{-\frac{1}{3}}\right) \left(\frac{Tk_B}{e_0}\right)^{\frac{5}{2}} \frac{k_B}{e_0 \langle Z \rangle \ln \Lambda}. \tag{25}$$

Here e_0 ist the elementary charge of an electron. The term $\ln \Lambda$ describes the so-called Coulomb logarithm, which applies to

$$\ln \Lambda = \ln \left[12\pi \left(\frac{\epsilon_0}{e_0}\right)^{3/2} n_e^{-1/2} T_e^{3/2} \right]. \tag{26}$$

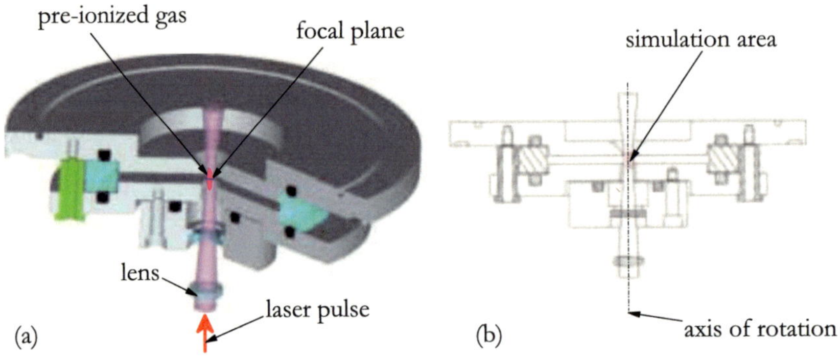

Fig. 8. (a) 3D scheme of the experimental setup. The laser pulse is focused onto the pre-ionized gas. (b) 2D scheme of the simulation setup. The red box shows the area modeled by the simulations.

4.1 Influence of the electron density

The coupling of laser beams and with that the heating of the plasma is strongly dependent from the absorption coefficient for inverse bremsstrahlung, i.e. from the electron density. The influence of the initial electron density on the reachable temperatures, respectively the propagation of laser radiation in the plasma is quantitatively simulated with the help of Comsol Multiphysics. Here, the laser pulse is coupled to the cross sectional area of the plasma cylinder with the help of a focusing optic. The focus diameter amounts 50 μm here. In this simulation, it is possible to give a constant, homogenous electron density to the plasma due to the smaller diameter of the laser beam in comparison with the pinch diameter. As an input parameter for the simulation, the initial electron density n_e is used. Fig. 9 shows the temporal and spacial temperature development in plasma at an initial electron density about 10^{17} cm^{-3}. Here one can see that the value chosen for the electron density is too small. The laser beam pervades the plasma almost unhampered (without absorption); only a small heating of the plasma occurs (about maximal 10 eV). With an increase of the initial electron density the absorption coefficient increases as well, so the laser beam can heat the plasma more efficiently. The maximum heating with the laser pulse is reached at an initial electron density of n_e = $7{\cdot}10^{19}$ cm^{-3}. Fig. 10 shows that a laser beam can propagate exactly to its focal level, to the area of the highest intensity. Here, a local heating of the electrons up to 100 eV occurs. Another increase of the electron density, only an untimely absorption of the laser pulse would occur. This situation is shown in fig. 11 for an initial electron density of $n_e = 7 \cdot 10^{20}$ cm^{-3}. Here one can see that, due to the high electron density, the laser radiation is absorbed strongly from the dense plasma and cannot spread completely. The maximum temperature reached is generated far before the area with the highest power density (focal area).

Fig. 12 shows a quantitative diagram of the dependency of the electron temperature reached with different electron densities and laser pulses. It is obvious that the optimal heating can be reached with a plasma electron density of $^1/_{10}$ of the cut-off density of the particular laser wave length. This behavior was already found in section 3. If there is a higher electron density in the plasma, the laser beam cannot enter the plasma in an optimal way and thereby not heat it efficiently (16). As a result it can be said that knowledge concerning the occurring electron density in the plasma (development of free electrons thorough electric stimulation and the generation of free electrons with the laser pulse) is of major importance to gain an efficient plasma heating with a laser beam.

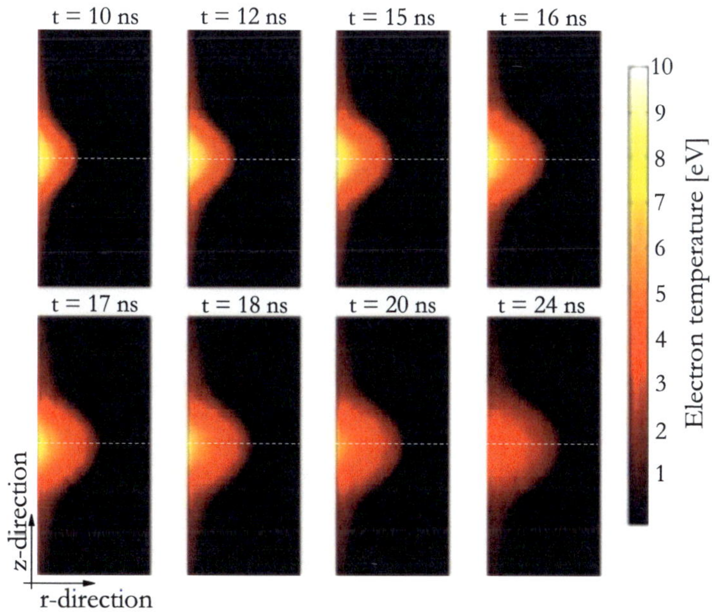

Fig. 9. Time development of the temperature distribution of a laser heated plasma with an initial electron density of n_e= 10^{17} cm^{-3}. The interrupted line symbolizes the focal area. Here, a maximum electron temperature of about 10 eV is reached.

4.2 Influence of the distance of time between laser pulse and pinch-plasma

The distance of time between the generation of a pinch-plasma and the laser pulse plays a key role for an efficient combination of the two methods. Here, the basic differences for the result can be generated here. If on one hand the laser pulse is brought to the plasma exactly at the time of the pinch moment, the plasma is experiencing a further heating. On the other hand, the plasma can emit a double pulse generated in the extreme ultraviolett spectrum (EUV) when the laser pulse is timely staggered with the pinch moment. These guesses are to be examined with the help of simulations. For HELIOS-CR has only limited possibilities to simulate laser and pinch-plasma combination, some simplifications and assumptions are to be made. The basic idea is it to describe the pinch plasma as a pre-pulse with a defined energy. This pre-pulse is used o generate the plasma. The main pulse (laser pulse), timely staggered, is used to heat the plasma. The plasma expands isotherm during the radiation process and adiabatic without radiation, which means that it is possible to influence the density profile and the temperature of the plasma with a timing of the pre-pulse and the main pulse and phase them to the highest conversion efficiency for the desired wavelength. The pinch-plasma is described as a pre-pulse with a duration about 10 ns (time duration of the pinch) and an energy of 600 mJ as Gau intensity profile. The main pulse follows after a variable temporal shift with a pulse duration of 9 ns and an energy of 750 mJ. Figure 13 shows the temporal development of electron density temperature and density for three different, timely staggered laser pulses. In the figures 13a and 13d, the laser pulse follows the pinch moment with about 100 ns. Here, the plasma is not heated to a point above the temperature generated by the pinch-plasma (maximal 19 eV) because the plasma did already cool down after about 20-30 ns. However the laser pulse causes another plasma heating up to 16 eV. This way, two sequenced

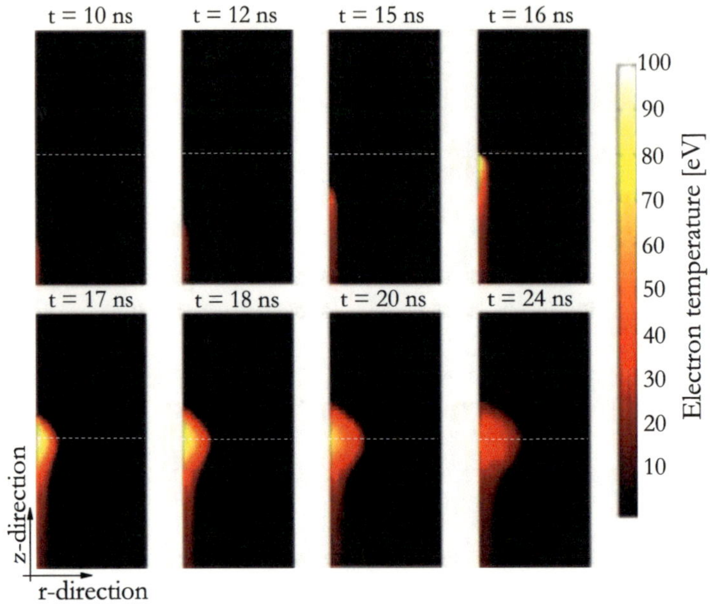

Fig. 10. Time development of the temperature distribution of a laser heated plasma with an initial electron density of n_e=7 · 10^{19} cm^{-3}. The interrupted line symbolizes the focal area. Here, a maximum electron temperature of about 100 eV is reached.

radiation pulses in the extreme ultraviolet spectral range can be generated. In the figures 13c and 13f, the laser pulse is coupled to the plasma simultaneous to the pinch moment. This generates a further plasma heating. The maximum temperature that can be reached is about 38 eV for this special case. These results do point out the importance of the time difference Δt between the pinch moment and the laser pulse coupling. On the one hand, a further plasma heating is possible, and on the other hand, two two sequenced radiation pulses in the extreme ultraviolet spectral range can be generated.

5. Experimental investigation

For all experimental investigation methods, an active mode locked Nd:YAG laser with two additional amplifier stages is used. It generates pulses with a half-width about 9 ns at maximal 0.8 J pulse energy. The laser runs at a maximum repetition frequency of 10 Hz, or it operates with single pulses. For the experiments, the pinch plasma has a voltage about 7 kV. The total capacity is about 46 nF at a total inductivity of 9.2 nH. The total energy of the hollow cathode triggered Z-pinch adds up to 1.1 J. Fig 14 shows a scheme of the experimental setup for a synchronization of the laser pulse and the hollow cathode triggered Z-pinch discharge. Due to the unsteady ignitions of the Z-pinch discharges, some measures for a synchronization of the laser pulse with the hollow cathode triggered Z-pinch discharges are necessary. The additional laser pulse heating of the plasma needs a fast reproducible laser triggering with a close relation to the pinch moment. To hit the plasma in a compacted state (durability about 10 ns) with the laser, a sufficiently strong and jitter free trigger signal is necessary about 100 ns before the main discharge occurs. For the avalanche breakdown of the hollow cathode triggered Z-pinch has a huge jitter of 50 μs, the laser pulse timing cannot be carried

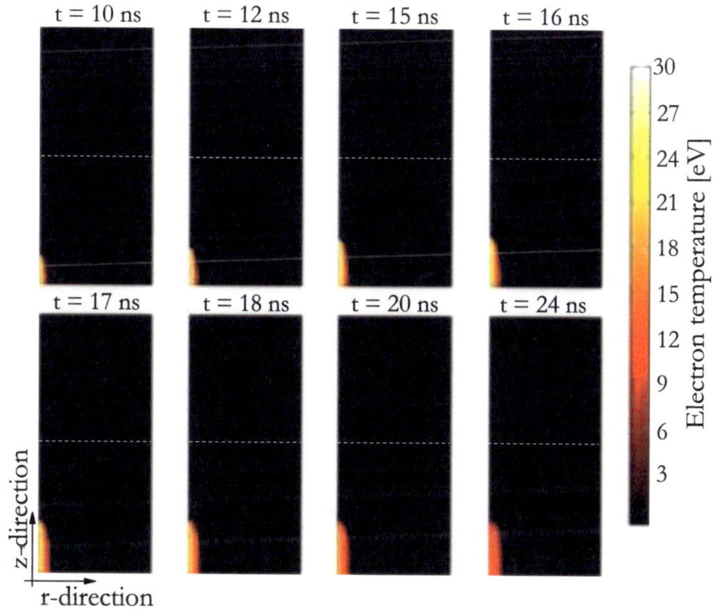

Fig. 11. Time development of the temperature distribution of a laser heated plasma with an initial electron density of n_e=10^{20} cm^{-3}. The interrupted line symbolizes the focal area. Here, a maximum electron temperature of about 30 eV is reached.

out with the 4-Channel Delay-Generator. The trigger event has to emerge from discharge course to be hit. The trigger signal cannot be generated by the control elements of the hollow cathode triggered Z-pinch because of the unavoidable delay caused by the 4-Channel Delay Generator, some meters of coaxial cable and the laser electronics itself. Through, one of the hollow cathode discharge characteristics modeled (13) and experimentally approved (14) is the fact that it emits an intense electron beam shortly before the avalanche breakdown occurs. Because of that, a Faraday cup is used in the experiment to collect the discharges. It is added close to the anode bore. The electric potential of the Faraday cup becomes negative because of the appearance of the intense electron beam. When a high-impedance resistor (ordinarily 10^9 - 10^{11} Ω) is used, a measurable voltage increases. This voltage gives a sufficient signal about 100-200 ns [1] before the main discharge occurs, with a jitter about maximal 5 ns.

5.1 Laser-induced re-heating of pre-ionized gases

Figure 15 shows the experimentally determined xenon spectra with and without laser pulse heating. The laser pulse is coupled to the plasma about 90 ns after the avalanche breakdown. The spectra show that the spectrum intensity duplicates, but no new lines are generated with the laser pulse heating. The reason for this is that the pinch-plasma did already cool down because of the expansion. It seems that the laser pulse can only effect another plasma heating. According to this, two hot plasmas with a time delay ($\Delta t \approx$ 90 ns) and almost similar electron temperature and radiation power emerge. Due to the exposure time of the CCD camera (t=20 ms), two radiation events are integrated to the extreme ultraviolet spectral range; with that a higher intensity can be reached. This behavior of the timely delayed laser plasma

[1]dependent from the gas used: N_2, O_2, Xe, Ar, CO_2

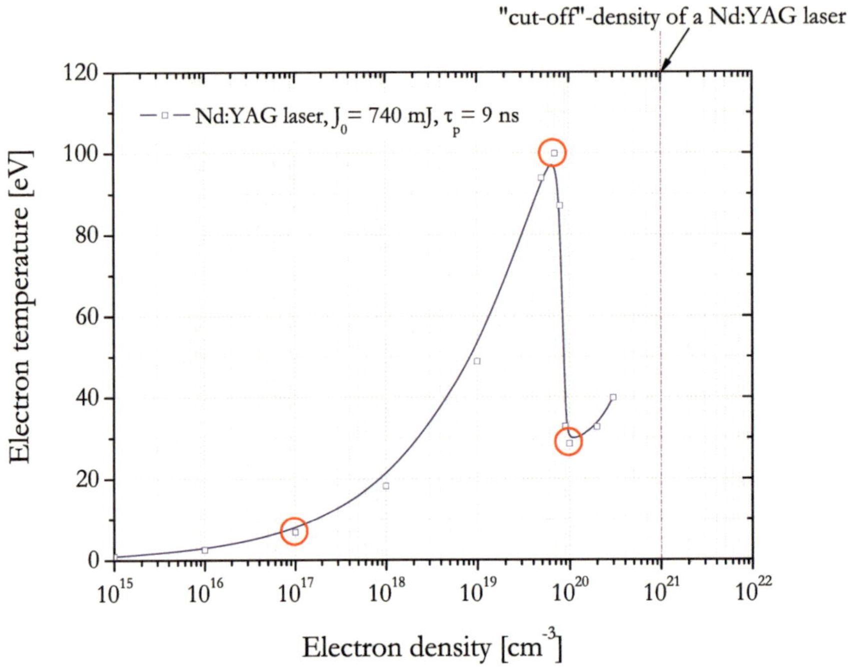

Fig. 12. Development of the electron temperature in plasma, dependent from the electron density during the plasma heating with the Nd-YAG laser pulse. The red circles show the values simulated in Fig. 9 - 11.

coupling was already simulated in section 4.2. The timely delayed EUV radiation pulses can be used as starting points for new application fields, such as, for example, the "pump and probe methods". Figure 15 shows the experimentally determined extreme ultraviolet spectra of a xenon z-pinch plasma combined with a short laser pulse. The time difference between the z-pinch plasma (in pinched state) and the incident laser pulse is about 100 ns. As a consquence the laser does not hit the plasma at peak compression but rather the residuals of the discharge. As shown the intensity over the whole spectral range increases at about a factor of two compared to the sole z-pinch plasma, whereas the sole laser without a discharge has no effect (green line). The spectra are taken from a single shot each. A comparison of the spectral line intensities of each spectra, as shown in (16), leads to an estimated electron temperature of $T_e \approx 70$ eV in both cases.

5.2 Laser-induced additional heating of pre-ionized gases

Figure 16 shows an experimentally determined spectrum order. Here, the pinch-plasma is run at a repetition rate of 1 Hz and one spectrum is taken, respectively. After 20 pulses, a laser pulse is additionally coupled to the plasma (20 pulses). The timely delay between avalanche breakdown and laser pulse here indeed only amounts to about 10-20 ns. The figure shows that the additional plasma heating occurs as desired. Furthermore one can see that not every laser pulse causes an additional plasma heating because of timely instabilities of the laser pulses. The newly generated spectral lines do mostly derive from helium-like nitrogen-ions. The strongest lines in the 1s3d and 1s4d change-over at 17.3865 nm and 13.0286 nm. A comparison of the intensity ratio of the lines from different ionization stage with simulated spectra gives an electron temperature about 57 eV in the plasma. When this temperature is reached, the intensity of the emitting line in the water window at 2.786 nm is about 50 times higher than the strongest line at 16.255 nm. This shows that it was possible to generate an efficient emitter in the area of the water window.

6. References

[1] Rolf, F. (2005). Entwicklung eines Rastermikroskopes für den Einsatz an Laborquellen im EUV Spektralbereich, *Phd Thesis, Bayerische Julius-Maximilians-Universität Würzburg*, 2005

[2] Janulewicz, K. A. (2004). Review of state-of-the-art and output characteristics of table-top soft x-ray lasers, *X-Ray Spectrom.*, Vol. 33, No. 4, 2004, 262-266

[3] Brückner, S. and Wieneke, S. and Viöl, W. (2008). Theoretical and experimental investigations of the suitability of low-current z-pinch plasma as an absorption medium for laser radiation, *Contrib. Plasma Phys.*, Vol. 48, No. 8, 2008, 577-585

[4] Rymell, L. and Hertz, H. M. (1993). Droplet target for low-debris laser-plasma soft X-ray generation, *Opt. Commun.*, Vol. 103, No. 105, 1993, 110

[5] Richardson, M. and Torres, D. and Depriest, C. and Jin, F. and Shimkaveg, G. (1998). Mass-limited, debris-free laser-plasma EUV source, *Opt. Commun.*, Vol. 145, No 109, 1998, 112

[6] Attwood, D. T. (2000), *Soft X-rays and Extreme Ultraviolet Radiation - Principles and Applications*, Cambridge University Press

[7] Vogt, U. (2002). Röntgenemission aus laserinduzierten Plasmen: Einfluss von Laserintensität und Pulsdauer bei verschiedenen Targetsystemen, *Phd Thesis, Fakultät Mathematik und Naturwissenschaften der Technischen Universität Berlin*, 2002

[8] Chan, C. H. (1973). Significant loss mechanismus in gas breakdown at 10.6 μm, *J. Appl. Phys.*, Vol. 44, No. 3, 1973, 1179-1188

[9] Viöl, W. (1988). Hochleistungs-CO_2-Laserpulse hoher Repetitionsfrequenz zur Erzeugung optischer Entladungen, *Phd Thesis, Mathematisch-Naturwissenschaftliche Fakultät der Universität Düsseldorf*, 1988

[10] Burger, M. (2003). Spektroskopische Untersuchung und Modellierung eines lasererzeugten Heliumplasmas im starken Magnetfeld, *Phd Thesis, Matematisch-Naturwissenschaftliche Fakultät der Heinrich-Heine-Universität Düsseldorf*, 2003

[11] Bergqvist, T. and Kleman, B. (1966). Breakdown in gases by 1.06 μm laser radiation, *Ark. Fys.*, Vol. 31, No. 2, 1966, 177-189

[12] R. K. Avery (1984). Interpretation of picosecond laser-induced breakdown in argon and xenon, *J. Phys. D: Appl. Phys.*, Vol. 17, 1984, 1657-1663

[13] Boeuf, J.P. and Pitchford, L.C. (1991). Pseudospark discharges via computer simulation, *IEEE Trans. Plasma Sci.*, Vol. 19, No. 2, 1991, 286-296

[14] Benker, W. and Christiansen, J. and Frank, K. and Gundel, H. and Hartmann, W. and Redel, T. and Stetter, M. (1989). Generation of intense pulsed electron beams by the pseudospark discharge. *IEEE Trans. Plasma Sci.*, Vol. 17, No. 5, 1989, 754-757

[15] Raizer, Y. P. (1997). Breakdown of Gases in Fields of Various Frequency Ranges, In: *Gas Discharge Physics*, Springer-Verlag, Berlin

[16] Wieneke, S. and Brückner, S. and Viöl, W. (2008). Simulating the heating of z-pinch plasmas with short laser pulses, *Journal of Plasma Physics*, Vol. 74, No. 3, 2008, 361-369

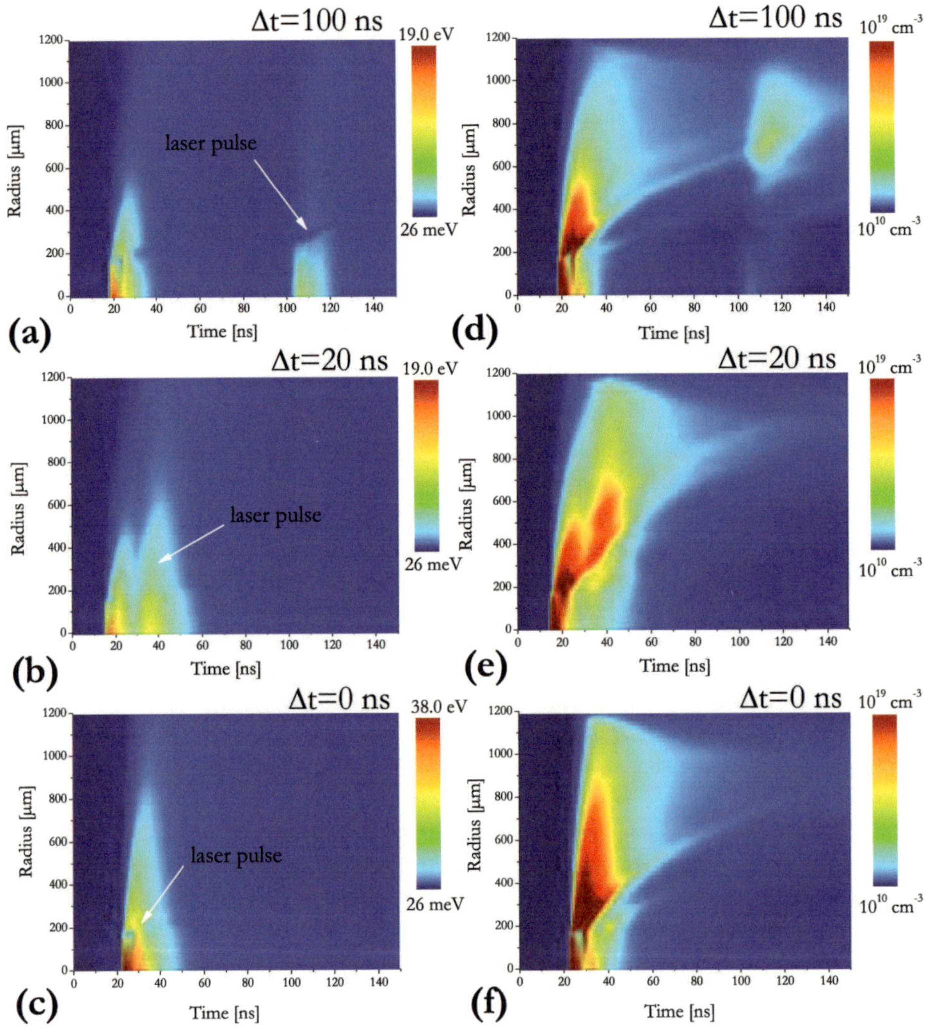

Fig. 13. Simulated time dependence of a laser heated xenon pinch plasma for different delay times of the laser pulse for the electron temperature (a-c) and electron density (d-f).

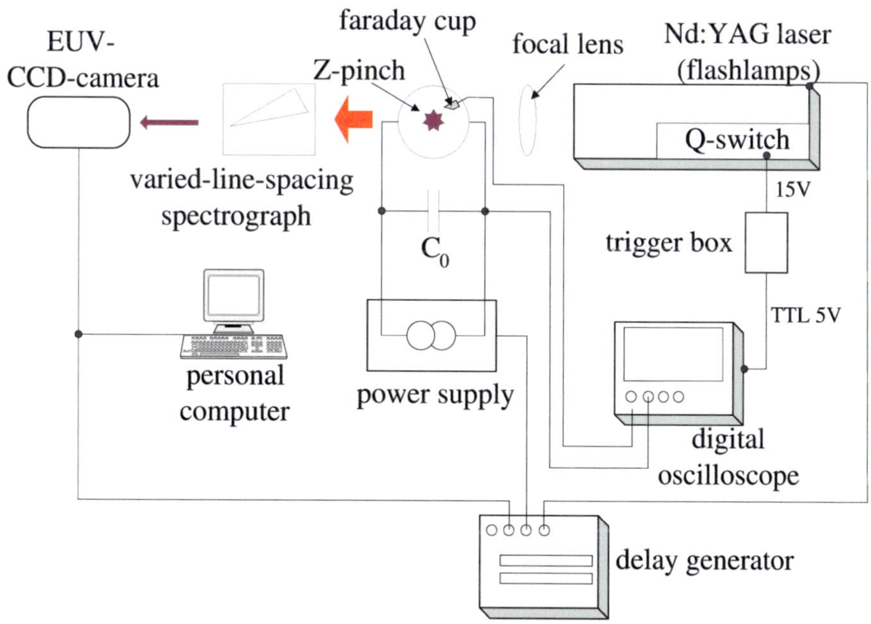

Fig. 14. Experimental setup for a synchronization of the Z-pinch discharge and the laser pulse

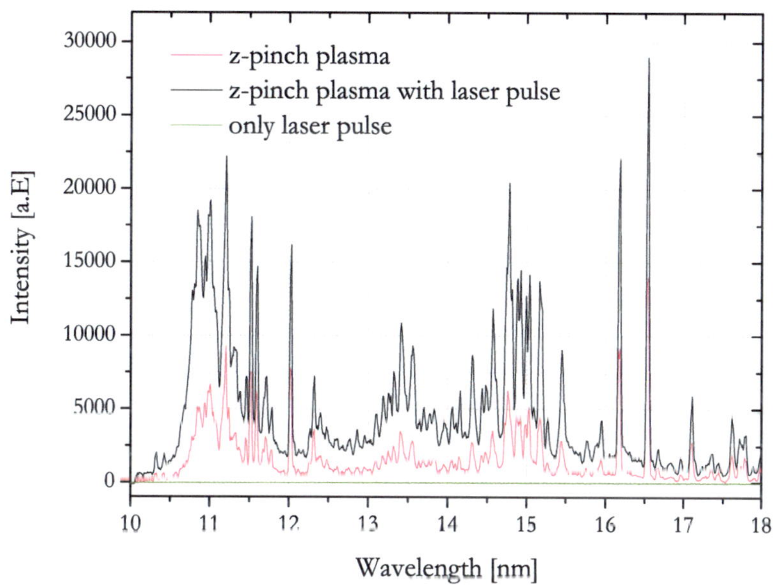

Fig. 15. Experimentally determained EUV spectra of a xenon z-pinch plasma (black) and a combined laser pulse reheated z-pinch plasma (red). In comparison the sole laser pulse does not create any EUV radiation, because the charge carrier density is to low for the breakdown of the discharge (green).

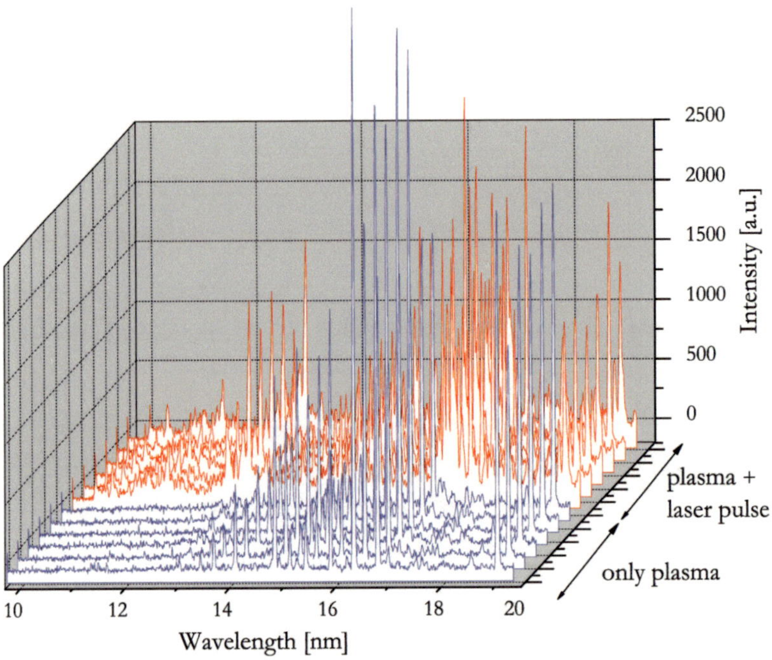

Fig. 16. Experimentally determined emission spectra of a nitrogen plasma with and without laser pulse heating. The spectra are continuously gathered at a repetition rate about 1 Hz (20 pulses are taken with and without laser pulse, respectively). The charging voltage is about 7 kV at a gas pressure about 6 Pa.

18

Characterisation and Manipulation of Proton Beams Accelerated by Ultra-Short and High-Contrast Laser Pulses

Sargis Ter-Avetisyan[1], Mathias Schnürer[2] and Peter V Nickles[2,3]

[1]*School of Mathematics and Physics, The Queen's University of Belfast,*
[2]*Nonlinear optics and short pulse spectroscopy, Max Born Institute, Berlin,*
[3]*Gwangju Institute of Science and Technology, WCU Department,*
[1]*UK*
[2]*Germany*
[3]*Rep. of Korea*

1. Introduction

Laser-driven acceleration of particle beams is a burgeoning field of research, based on the possibility of creating ultra-large electric fields in plasma, largely exceeding current limits of conventional acceleration technology. Research in this area has led to very significant progression regarding the acceleration of electrons and ions. Laser-driven ion sources have unique properties: high brightness (about 10^{13} protons/ions per shot), high current (in kilo-Ampere range), ultra-low emittance, and short pulse duration (less than 1 ps), opening prospects for a broad range of applications. Recent advances in the laser technology have led to even further enhancements in the provision of extremely short and high-intensity pulses, which can further improve the accelerated ion beam specifications particularly regarding maximum ion energy and ion flux which are demanding most of the potential applications. These developments have stimulated an emergence of advanced diagnostics for measuring complex plasma effects.

The absorption mechanisms of laser radiation at the target are the basic processes, which indeed defines the whole ion acceleration scenario. For laser intensities, where the classical normalized momentum of electrons quivering in the laser electric field: $a = 8.53\times10^{-10}$ (μm) $I^{1/2}$ (W/cm^2) > 1, the electrons become relativistic and the effect of the laser magnetic field is no longer negligible. The perpendicular component of the Lorentz force $e\mathbf{v}\times\mathbf{B}$ couples with the electric force to drive the electrons in the laser propagation direction (Wilks et al., 1992, Lefebvre & Bonnaud, 1997, Kruer & Estabrook, 1985) in contrast to inverse bremsstrahlung and resonance absorption, which causes the quiver motion of the electrons in the laser field. The ponderomotive force drives the electrons with a step- or plateau-like density profile and has a strong directionality along the laser propagation direction. Electron temperatures about 1 MeV have been measured (Malka & Miquel, 1996). The laser energy transfer to the hot electrons can also be out carried by fast plasma waves through the nonlinear ponderomotive force (Tajima & Dawson, 1979) and by laser field itself (Pukhov et al., 1999).

Whereas the ions from the target front will be accelerated normal to the target front surface in the ambipolar expansion of the plasma, the hot electron component created directly by the laser pulse in the plasma plume will propagate through the target. It has typically a divergence between 5°-50°, density of the order of the critical density (10^{20} - 10^{21} cm^{-3}) and a temperature of the order of the laser ponderomotive potential.

The free motion of this hot electron beam through the target requires a return current that locally compensates the flow of the hot electrons (Passoni et al., 2004). It will be provided by the target material (for metallic target - conduction electrons, insulators - the background free electron population created by field and thermal ionization). Since the density of the background electron population in both cases is of the order of the solid density, much bigger than the fast electron density, the required velocity for current neutralization is small and their temperature is much lower than that of the hot electrons (Tikhonchuk, 2002). Finally the physical parameters and dynamics of these two electron populations will define the electrostatic sheath field which is created at the vacuum - solid interface and accelerates ions to high energies.

The ions are created and accelerated either at the target rear surface (Snavely et al., 2000, Mackinnon et al., 2001, Hegelich et al., 2002) through the self-consistent electrostatic accelerating field generated by fast electrons escaping in vacuum (so-called target normal sheath acceleration—TNSA mechanism) or at the target front surface, illuminated by the laser (Clark et al., 2000a, Maksimchuk et al., 2000, Clark et al., 2000b). Particle-in-cell (PIC) simulations suggest a variety of mechanisms that may be responsible for acceleration at the front surface: formation of multiple collisionless electrostatic shocks (Denavit, 1992, Silva et al., 2004, Wei et al., 2004); a solitary wave produced by shock-wave decay in a plasma slab (Zhidkov et al., 2002); or a mechanism wherein the ponderomotive pressure of the short laser pulse displaces the background electrons, and the ions are accelerated by the electrostatic field of the propagating double layer (Shorokhov & Pukhov, 2004). However, all these mechanisms are relying on ion acceleration in the electrostatic field created due to charge displacement driven by the laser field.

These scenarios are not mutually exclusive. Their relative contributions depend strongly upon the particular target and laser parameters and can contribute to the generation of electrons and, in turn, to ion acceleration mechanisms. Particle-in-cell (PIC) simulations by Wilks et al., (2001), and Pukhov, (2001), and observations by Zepf et al., (2003), and Karsch et al., (2003), show that ions can be produced at the target front and the rear sides simultaneously, even if the generation processes are quite different. Here the laser pulse contrast has a profound effect on accelerated ions and their cut-off energy (Mackinnon et al., 2002, Kaluza et al., 2004, Lindau et al., 2005).

The acceleration is most effective on light ions (specifically protons), which are usually present on target surfaces in the form of contaminants like hydrocarbons and water, or can be present among the constituents of the solid target (e.g. as in plastic targets). The heaviest ion population of the target provides a positive charge, which offers much more inertia and makes the charge separation responsible for the huge accelerating field. Part of this heavy population can also be effectively accelerated, on a longer time scale, if the protons are not enough to acquire most of the energy contained in the electric field, or if protons are removed before the arrival of the laser pulse.

The complex, non-linear nature of laser-driven plasma dynamics and ion acceleration phenomena requires the development of an innovative diagnostic complex allowing simultaneous measurements of different plasma parameters (using visible- and XUV-light, X-rays, Gamma-rays, ion and electrons) with high temporal, spectral and spatial resolution together with laser pulse parameters. This requirement is especially important because of both: shot-to-shot fluctuations of laser pulse parameters and inherent shot-to-shot variations in the local target parameters, which can derogate the whole plasma dynamics.

In laser-matter interaction studies one of the important research tasks is to investigate the energy distributions of emerging charged particles, in which the different laser energy absorption mechanisms are hidden and the energy redistribution among the plasma components become apparent. As a basic diagnostic the Thomson spectrometer has been widely and successfully used for analyzing the energy spectra of laser accelerated charged particles. In Fig.1 a typical experimental setup of laser plasma interaction experiments with Thomson parabola spectrometers and, as an example, spectral traces recorded with absolute calibrated micro-channel-plate (MCP) detector (Ter-Avetisyan, et al., 2005) are shown.

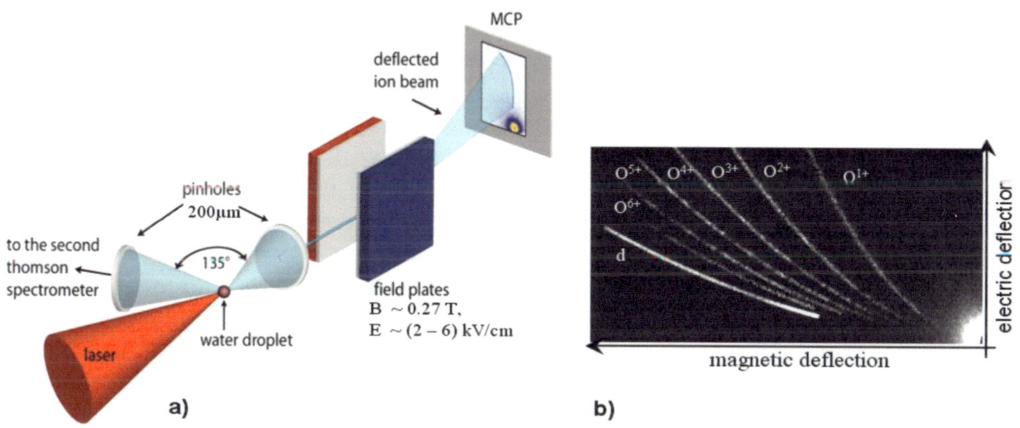

Fig. 1. a) A typical experimental setup of laser plasma interaction experiments with Thomson parabola spectrometers and, as an example, b) measured accelerated ion spectra from heavy water droplet irradiated at 35 fs, ~1 J, ~10^{19} W/cm^2 intensity is shown.

Several modifications of the basic spectrometer design allow a comprehensive and precise analysis of ion acceleration. In particular, (i) the simultaneous measurement of ion and electron spectra in the same direction allow to understand the measured particularities in the ion spectra resulting from the evolution of a two-electron component (hot and cold) plasma (Ter-Avetisyan et al., 2004a). (ii) The complexity of the temporal (Ter-Avetisyan, et al., 2005) and spatial characteristics (Schreiber et al., 2006) of laser driven ion source could be demonstrated by precise measurement of the proton/ion trajectories, and its applicability for proton deflectometry (Ter-Avetisyan et al., 2008, Sokollik et al., 2008), and (iii) proton source tomography revealed detailed properties of the laser driven ion source (Ter-Avetisyan et al., 2009a). Latest is crucial in view of planning proton beam steering systems.

The comprehensive set of on-line diagnostics, which are complementary each other and can be used simultaneously, is a very powerful tool for laser plasma interaction studies. In each experiment only various diagnostics would allow unambiguous demonstration of ion acceleration processes in their whole complexity, providing a set of data also for theoretical interpretation.

2. Specific features of accelerated ion and electron dynamic

2.1 Dips in ion emission spectrum and electron dynamics

Several physical mechanisms have been considered for the appearance of high-energy electrons, and have been proposed as a way to understand the generation of ions with kinetic energies of several tens of MeV during the short laser-pulse interaction with dense plasmas, (Malka & Miquel, 1996, Clark et al., 2000b, Forslund & Brackbill, 1982, Fews et al., 1994, Beg et al., 1997, Gitomer et al., 1986). A theory has been developed (Wickens et al., 1978, Wickens &.Allen, 1979) for the free expansion of the laser plasma with hot (T_h) and cold (T_c) electron temperature components, as a way to treat these non-equilibrium effects and to describe the ion energy distribution. It was shown that the energy fraction carried by fast ions depends on the temperature and concentration of the electrons in the plasma. This leads to an ion-emission velocity spectrum whose most notable feature is a pronounced dip in the distribution (Kishimoto et al., 1983). The slopes of the upper ($-(M/ZkT_c)^{1/2}$) and lower ($-(M/ZkT_h)^{1/2}$) asymptotes in the ion velocity spectrum makes possible a determination of the effective absolute hot- and cold-electron temperatures (Wickens &.Allen, 1981). (Here, M is ion mass, Z is charge state, kT is cold- or hot-electron energy).

The dip in the velocity distribution corresponds to an internal electrostatic sheath appearing due to hot- and cold-electron isothermal expansion, where ions are strongly accelerated in a small region. This dip develops in a region of self–similar flow where the ions experience rapid acceleration due to an abrupt increase in the electric field. This increase occurs at the location in the expanding plasma where most of the cold electrons are reflected, corresponding to a step in the ion charge density (Wickens &.Allen, 1981). The depth of the dip as a function of the peak field is a sensitive function of the hot-to-cold electron temperature ratio T_h/T_c in the ion spectra, while the position in the spectrum depends on the hot-to-cold electron density ratio n_h/n_c.

In a very short pulse (35 fs) and high intensity laser plasma several groups of electrons with different temperatures can be generated (Zhidkov et al., 2001), which could then cause multiple dips in the ion energy spectra. No hint of this has been observed so far, and therefore detailed measurements of the ion and electron spectra are required to find correlations between these processes.

Here we report on precise measurements of the spectral density distribution of the ion emission from plasmas created by 35 fs laser pulses at intensities of $(0.8 - 1.2)\times10^{19}$ W/cm^2. A pulse from a multi-TW Ti: Sapphire laser (Kalachnikov et al., 2002) with a maximum energy of 750 mJ, was focused with an f/2.5 off-axis parabolic mirror onto water droplets of ~ 20 μm diameter (Hemberg et al., 2000). Measurements were made of ion spatial and energy distributions, X-ray spectral properties, the electron spectrum emitted, and the correlation of maximum electron energies with the cut-off energies of proton and deuteron spectra.

The ion energies have been measured with a Thomson parabola spectrometer, into which the ions enter through a 200 μm aperture. The ions are detected by a 40 mm MCP coupled to a phosphor screen. The signal from the phosphor screen is imaged with a charge-coupled device (CCD- camera). An ^{241}Am α -particle source (energy 5.4 MeV) with known emittance is used to calibrate the setup (Ter-Avetisyan, et al., 2005). This sensitive, calibrated detection technique allows the measurement of an ion spectrum from a single laser shot in absolute terms.

A typical camera picture taken with a single laser shot, showing ion traces from a heavy water droplet, is depicted in Fig.2 and the deduced deuteron spectra is shown in the inset.

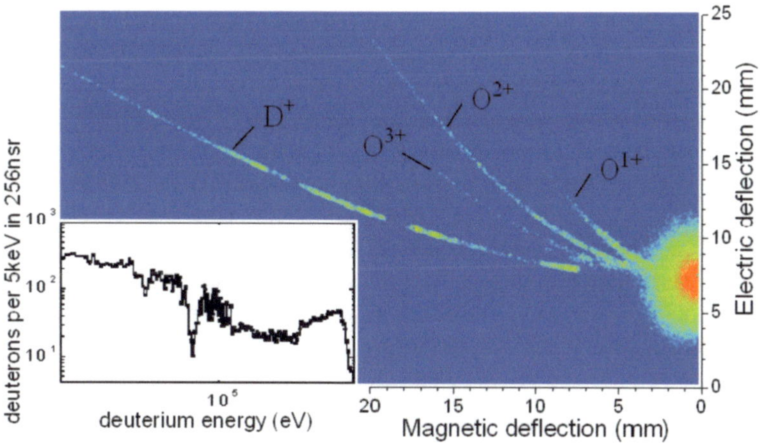

Fig. 2. Color image from the MCP-phosphorous screen of an emitted ion spectra (deuterons and oxygen ions, right blob – "zero" point: radiation impact along spectrometer axis) from a heavy water droplet taken from a single laser shot in backward (135^0 to laser propagation) direction with the Thomson spectrometer. In inset the deuteron spectrum is plotted.

The most interesting features in the picture are the clearly visible dips along the deuteron trace. As mentioned before, the same feature has been observed for proton emission from water droplets, and has been observed in backward (135° to the laser axis) as well as in forward (laser propagation direction) emission. The occurrence of the spectral dips was reproducible in the experiment, although the exact position, depth, and fine structure varied from shot to shot due to small variations in the laser parameters and beam alignment in our setup. This is, to our knowledge, the first observation of these dips in ultra-short (sub 50 fs) high intensity laser-plasma interaction experiments.

The generated hot electron spectrum was measured with a GAFchromic film (HD-810) in a direction transverse to the laser axis using a 0.27 T magnet spectrometer. The film is sensitive to electrons above 10 keV due to its layer characteristic (Busch, et al., 2003). The measured time-integrated hot electron spectrum (Fig.3a) shows a maximum at an energy of about 0.7 MeV with a tail expanding to 2 or 3 MeV.

From the measured electron spectra, one can deduce a hot electron component with a temperature of (0.63 ± 0.03) MeV. This fits well with the energy the electrons can acquire from the ponderomotive force F_p of the laser pulse (Kruer & Estabrook, 1985): $F_p = -dU_p / dz$, $U_p(eV) = 9.33 \times 10^{-14} I(W / cm^2)\lambda^2(\mu m)$, where U_p is the ponderomotive energy. This gives a potential energy of 0.6 MeV for our laser intensity, proving a rather efficient laser energy transfer to the electrons. Because up to 20% of the laser energy can be absorbed in energetic electrons (Wilks et al., 1992), a significant number of electrons with energies of several hundreds of keV is produced. The electron impact ionization cross section at the energies 400-500 keV is about 10^{-19} cm^{-2} (http://physics.nist.gov/cgi-bin/Ionization/table.pl?ionization=H2O), therefore these highly energetic electrons can cross the target without being significantly slowed. A space-charge field accumulated in the droplet captures the hot electrons with energies below about 200 keV (low energy cut off in Fig.3a). The potential due to these electrons and the estimated electron density is sufficient to create electrostatic acceleration fields of the order of 1 MV/μm (Busch, et al., 2003) which, in turn, can accelerate ions to MeV energies. Electrostatic or magnetic fields around the

target will influence the directionality of ion emission. However, for our spherical target, we expect a relatively isotropic ion emission distribution, which can be shown by measurements (Busch, et al., 2003, Karsch et al., 2003).
A strong correlation between the maximum ejected electron energies and the deuteron cutoff energies could be directly established (Fig.3b) with the help of a re-designed Thomson spectrometer allowing to measure in one laser shot the emitted ion and electron spectrum in a same direction. For that the second MCP detector was added to that side of the spectrometer where the electrons are deflected. In Fig.3b, a variation in the maximum energies of the ejected electrons by only a factor of 1.2 changes the deuteron cutoff energy by a factor of about 5. This extreme sensitivity emphasizes the predominant role of the energy transfer to the electrons for the ion acceleration.

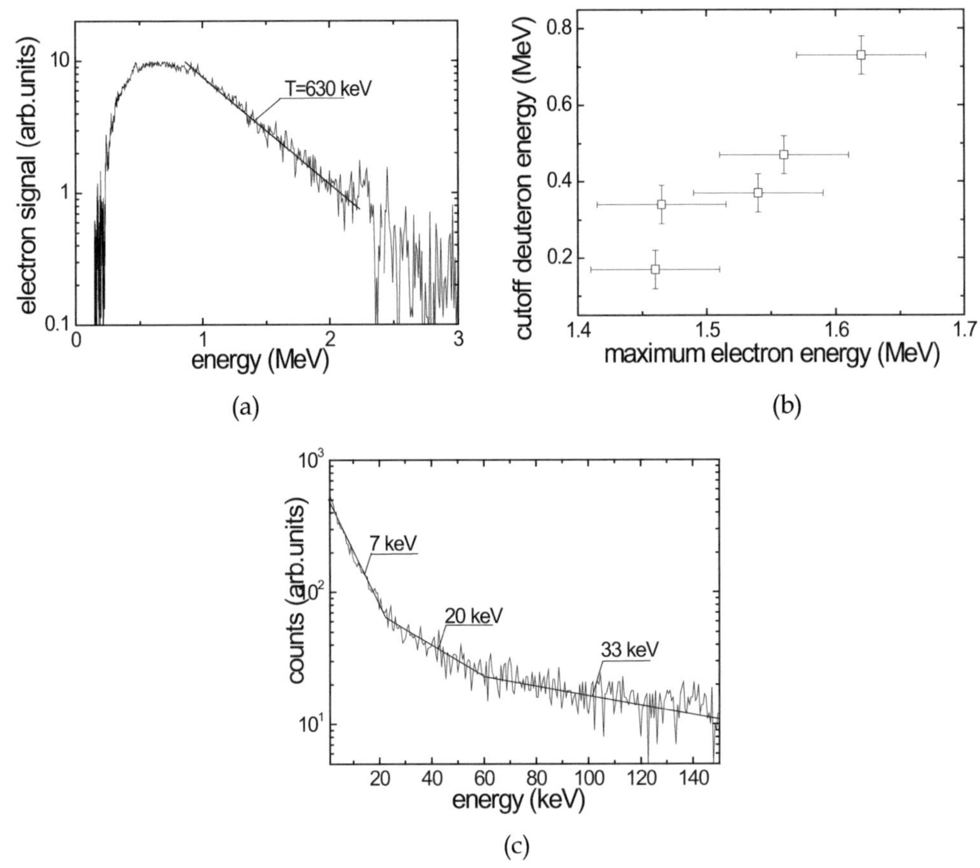

Fig. 3. Electrons, ions and x-rays emitted of heavy water micro droplet. a) Hot electron energy distribution (superposition of ~50 000 shots). Slope fits to an exponential decay with a temperature T= 630 keV parameter. b) Correlated maximum energies of emitted electrons and deuterons in same laser shot. c) X-ray emission spectra from heavy water droplet.

The space charge field built up by the trapped hot electrons in the target is responsible for ion acceleration properties (Fig.3a), therefore, these electrons have been studied further in details. According to a simple bremsstrahlung model (Griem, 1964), hard X-ray emission

from the plasma is defined by the electron density distribution inside the target. For highly energetic electrons crossing the droplet, the target is "thin", so their bremsstrahlung is weaker and the spectral intensity is constant (Blochin, 1957).
A calibrated X-ray CCD-camera operating in a single-photon detection mode was used for energy-dispersive X-ray measurement (Ter-Avetisyan, et al., 2003). In the experiments, the CCD-camera was mounted at an angle of 90° to the laser propagation direction and at a distance of 100 *cm* from the plasma source. A 200 *nm Zr* filter and a beam aperture was used to block the optical light and the scattered X-rays. The resolution of our spectral diagnostics is about 0.5 *keV*. A typical X-ray spectrum is shown in Fig.3c. The slope of the distribution shows the existence of multi-temperature components, and can be fitted assuming three different effective temperatures of about (7±0.3) *keV*, (20±4) *keV*, and (33±12) *keV*.

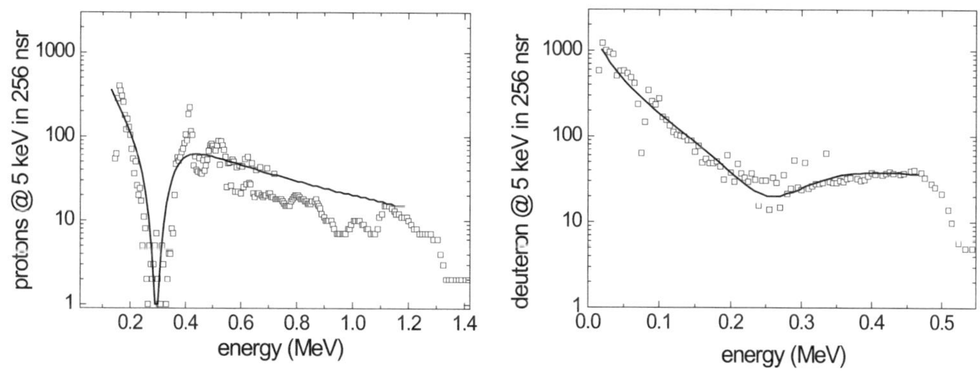

(a) (b)

Fig. 4. Single shot proton a) and deuteron b) spectrum with typical dips along the trace: squares - experiment, line - simulation (parameter cf. text)

Recently, a fluid model based on a *single electron temperature* approximation was applied successfully for high intensity laser-driven ion acceleration (Mora, 2003). Accurate results could be obtained for the structure of the ion front, the ion energy spectrum, and the cutoff ion energy. In the present letter, on the other hand we explain the dips in the emitted ion spectrum (Fig.2) by relying on the fact that we have an electron spectrum characterized by *several electron temperatures*. This is the precondition for an application of the fluid model (Wickens et al., 1978, Wickens &.Allen, 1979, Kishimoto et al., 1983). Figures 4a and 4b compare experimental proton and deuteron energy distributions with calculations based on the theory of Wilkens et al. (1978). A reasonable fit for the depth and position of the dip in the proton spectrum (Fig.4a) is obtained when the hot-to-cold electron temperature ratio T_h / T_c is assumed to be about 9.8, and the hot-to-cold electron density ratio n_h/n_c is about 1/100. Individual electron temperatures T_c = 7.5 keV and T_h = 74 keV compare quite well to the range of temperatures derived from the X-ray emission. Here, T_h is lower by a factor of about 2, but this can be due to the restricted linearity range (< 50 keV) in the X-ray measurement. Also, if one takes into account that bulk ion energy scales with the hot electron temperature as E_{ion}= 4.5 T_h (Wickens &.Allen, 1979), a mean ion energy E_{ion}=330 keV would be derived, in remarkably good agreement with the ion temperature inferred from the ion slope (Fig.4a). This is somewhat different from the original model where the ion energy is predicted to be similar to the hot cf. cold electron energies. A sharp proton cutoff energy occurs at about 1.3 MeV. Notice that the analytical velocity distribution derived in (Wilkens et al. 1978) breaks down if the temperature ratio T_h/T_c exceeds 9.9. In order to overcome this problem, a more complex electron velocity distribution was included in (Kishimoto et al., 1983) [33].

For a heavy water droplet, presented in Fig.4b where the cutoff energy occurs at about 0.55 *MeV*, the model with a hot-to-cold electron temperature ratio T_h/T_c =7.7 and an electron density ratio n_h/n_c=1/25 fits the measurement quite well. The individual temperatures are T_c =11 *keV* and T_h = 85 *keV*. The model shows how small changes in T_e and n_e have large effects on spectral slopes.

From the structure of the emitted ion spectrum such important parameters as hot- and cold-electron temperatures and their density ratio can be determined. Due to the short laser pulse the hot electron population shows a multi-temperature behaviour. This, in turn, can cause multi-dips in the ion spectrum. It is worth noting that the results demonstrated here could open a way to tailor the ion spectra (Nishihara at al., 2001) from short pulse laser-driven plasmas by choosing proper electron distributions appropriate to particular applications.

2.2 Laser pulse contrast and electron dynamic

Two different laser energy absorption mechanisms at the front side of a laser-irradiated foil have been found to occur, such that two distinct relativistic electron beams with different properties are produced. One beam arises from the ponderomotively driven electrons propagating in the laser propagation direction, and the other is the result of electrons driven by resonance absorption normal to the target surface. These properties become evident at the rear surface of the target, where they give rise to two spatially separated sources of ions with distinguishable characteristics when ultra-short (40 fs) high-intensity laser pulses irradiate a foil at 45° incidence. The two, clearly distinguishable branches of electron trajectories have been measured by Čerenkov diagnostics. This correlates with proton emission from two separated sources, which have been resolved with a high-spatial-resolution Thomson spectrometer (Schreiber et al., 2006). The crucial parameters of the experiment are the laser pulse intensity and the contrast ratio.

In the experiments, a 40 fs pulse from a multi-TW Ti: Sapphire laser was focused onto a thin (6 μm) aluminum foil target at 45° with an intensity of about $2x10^{19}$ W/cm^2. The temporal contrast of the laser pulse was characterized by a scanning third-order cross correlator with a dynamic range of 10^{10}, having a resolution of 150 fs and a scanning range of ± 200 ps. The pulse shape several ns before the main pulse was controlled by a fast photodiode with temporal resolution of about 300 ps. In typical operating conditions, the amplified spontaneous emission (ASE) pedestal of the laser pulse, several picoseconds before the pulse peak, was at a level of $(0.8 - 5)\times10^{-7}$ relative to the peak intensity. This is termed "low" contrast. The ASE pedestal could be reduced by driving the Ti: Sapphire laser amplifiers in specific delayed pump modes. This led to a reduction of pulse energy by up to 550 mJ, but it permitted an improvement of the ASE level down to $(1 - 3)\times10^{-8}$, which is referred to as "high" contrast. In both cases, no pre-pulses were observed.

The Čerenkov method, applied as an electron diagnostic, uses the partial conversion of the electron bunch energy into a flux of photons in a medium where the electron velocity is higher than the light velocity (Stein et al., 2004). A 50 μm tesa foil (polypropylene with chemical composition $(- C_3H_6 -)_n$, index of refraction - 1.49) used as a Čerenkov radiator was attached with very thin Acrylat glue to the target rear, and imaged with an objective (f = 8 cm) and a magnification of about 15 to a gated charge-coupled device (CCD) camera. It is assumed that a 50 μm path length in this low Z substance has a negligible effect on the energy and main direction of the electrons. Propagating through the foil, the electron beam undergoes filamentation (Manclossi et al., 2006), but nonetheless with unaltered beam directions emerging from the target. To block any light from the laser pulse and its harmonics, a 6 mm Schott BG18 filter and a 3 mm VG8 filter were inserted into the beam path. These filters yield a spectral window centered at 510 nm with a half width of 70 nm. A weak and uniformly distributed background signal from transition radiation is omitted as a result. For the medium and the geometry used, the detected Čerenkov light is associated with electrons in the 180 – 240 keV energy interval.

Measurements of the ion energies were carried out with a Thomson spectrometer (Ter-Avetisyan et al., 2005). In order to achieve high spatial resolution, we set up a Thomson spectrometer in a 1:12 imaging mode (Schreiber et al., 2006), providing spatial resolution of about 20 μm. This estimation takes into account the laminarity of the proton beam, and is measured from the spreading of a spectral trace which is a consequence of the divergence of the proton beamlet intercepted by a screen with a 30 μm pinhole.

The Čerenkov medium behind the target made it impossible to use both diagnostics simultaneously for the single laser shot. Therefore, correlations between the data sets are statistically relevant. Due to a 10 Hz repetition rate of the Ti:Sapphire laser, and using on-line detection systems for both diagnostics, the statistical uncertainties could be minimized.

Figure 5 shows spatially resolved CCD pictures of Čerenkov radiation behind the 6 μm Al target at 45° laser incidence (a - c), and it shows the evolution of the Čerenkov signal as a function of target thickness (d). The arrows inserted in the pictures show the target irradiation and Čerenkov emission geometry. In the pictures, one of the two components of the hot electron current produced is identified as a consequence of hot electrons driven along the target normal due to resonance absorption, and the other component along the laser propagation axis is due to the ponderomotive force. Whereas in Fig.5a both electron components make a comparable contribution to the resulting Čerenkov emission, resonantly and ponderomotively driven electrons are dominant in Figs.5b and 5c, respectively. The appearances of the emission patterns are statistical, and correlated with the statistics of the laser pulse contrast shot-to-shot fluctuation data.

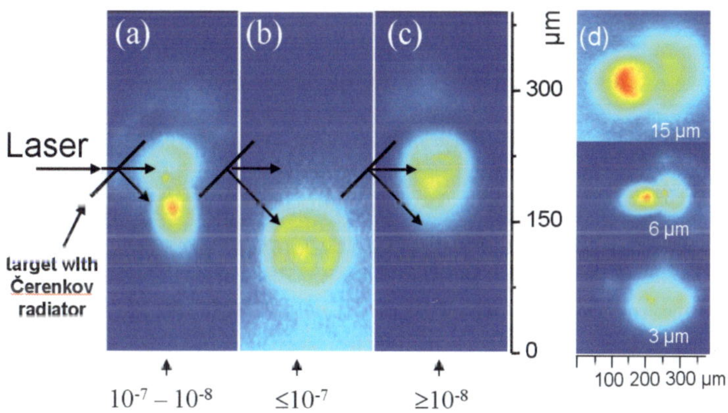

Fig. 5. CCD pictures of Čerenkov radiation (arrows show emission direction) at 45° laser incidence resulting from electrons propagating: (a) in the laser direction and normal to the target; or (b) perpendicular to the laser direction only; and (c) in the laser direction only. (d) Čerenkov light distributions as a function of Al-target thickness.

The evolution of the Čerenkov signal as a result of target thickness is shown in Fig.5d. With a 3 μm target the two components are merged together, while at 6 μm they are clearly separated and at the target thickness is 15 μm, further broadening of the electron beams propagating inside the target occurs, and blurs the Čerenkov emission. In the experiments reported in Ref. (Stein et al., 2004), these two groups of separately moving hot electrons at the target rear side also were observed. Here we investigate the dynamics of the hot electron populations, and the consequences for the ion acceleration processes at "low" and "high" laser pulse contrasts.

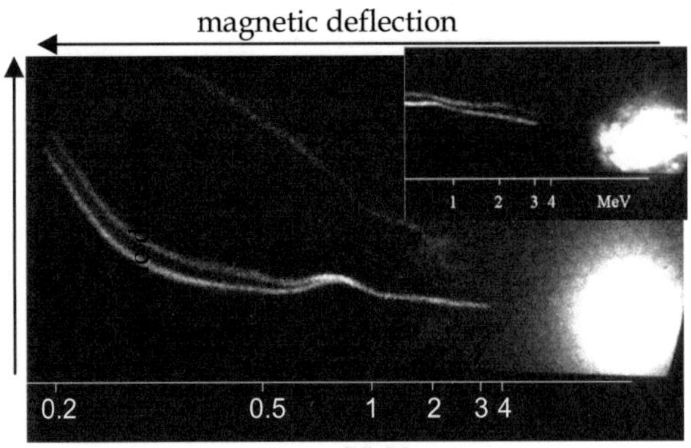

Fig. 6. A CCD picture of a proton spectrum at 45° laser incidence on a 6 μm Al target, measured in the direction of the target normal. The inset shows proton spectra from a 3 μm target thickness at similar irradiation conditions.

The resonantly and ponderomotively driven electron populations at the target rear side create two spatially separated electrostatic fields, from where the ions/protons can be accelerated. Therefore, one can expect ion emission from two spatially separated sources if the laser pulse contrast is between 10^{-8}–10^{-7}, where both laser energy absorption mechanisms are equivalent. Indeed, the CCD picture of the proton spectrum in Fig.6, measured in the direction of the target normal, shows two parallel proton parabolas, thus confirming the existence of the two sources of proton emission. The lower trace results from the source created by electrons from resonance absorption, and the upper one from the source created by ponderomotively driven electrons. The appearance of the second source (or second proton trace) is in good agreement with the detection of two electron populations by Čerenkov diagnostics (Fig.5a). This feature is never observed at normal laser incidence on the target. Figure 6 also shows that the parabolas are not perfect. This coincides with the previous findings (Schreiber et al., 2006) that the source position could be variable. Note that the two parallel traces in Fig.6 change in a similar manner, while the proton spectrum of a 3 μm thick target (inset of Fig.6) shows an independent character, which indicates a complex source dynamics.

In order to see the differences in the electron acceleration processes and the implications for the following ion acceleration, we performed two-dimensional (2D) particle-in-cell (PIC) simulations of laser irradiation onto targets with different scales of pre-plasmas. The targets are 2 μm thick and have densities of 40 n_c. The pre-plasma is assumed to have an exponentially decreasing profile with two scale lengths. For the overdense region where the density is higher than 4 n_c, the scale length is 0.1 μm. For the lower density region resulting from an expanding corona plasma, the scale lengths are 0.3 μm for the small pre-plasma case and 3.5 μm for the large pre-plasma case. The initial electron temperature is 1 keV, and ions are kept immobile. The targets are irradiated by p-polarized laser pulses from the left boundary with wavelengths of 1μm. The laser pulses ramp up in five laser cycles, and sustain a peak intensity of 2×10^{19} W/cm^2 that is uniform in the y-direction. The pulse duration is 40 fs. The boundary conditions are absorbing in the x-direction, and periodic in the y-direction.

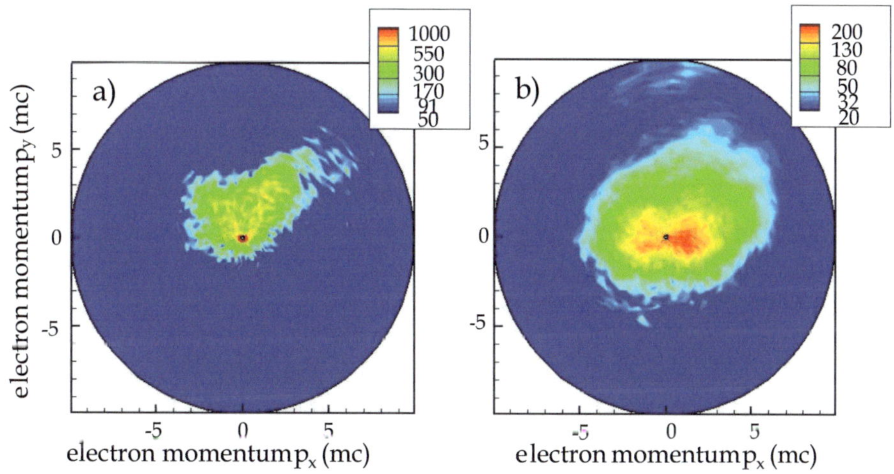

Fig. 7. Electron momentum distribution observed in PIC simulation at the target rear surface for a) "small" L = 0.3 μm scale length pre-plasma case, and for b) L = 3.5 μm "large" scale length pre-plasma case. Laser pulses irradiate the target from the left boundary.

The differences resulting from the dominant acceleration processes are seen from the angular distribution of accelerated electrons. In Fig.7, the angular electron momentum distributions observed in the PIC simulation at the target rear surface are shown for the a) "small" L=0.3 μm scale-length pre-plasma case, and for the b) L=3.5 μm "large" scale-length pre-plasma case. In the small pre-plasma case, ponderomotive acceleration and vacuum heating play dominant roles in energy absorption by electrons. Due to vacuum heating, laser energy is absorbed by low energy bulk electrons that are mainly accelerated in the target-normal direction. High energy electrons generated by ponderomotive acceleration propagate in the laser irradiation direction, and are reflected back toward the target at the specular angle, as can be seen clearly in the figure. For the large pre-plasma case, high energy electrons are mainly accelerated by resonance absorption in the target normal direction and are reflected back also in the normal direction.

These high energy electrons induce a strong magnetic field around the critical surface. In the small pre-plasma case, a magnetic field with uniform intensity is induced along the critical surface. This surface magnetic field prevents low energy electrons from propagating into the target, and makes them flow along the surface (Nakamura et al., 2004). In the large pre-plasma case, a magnetic field is formed via Weibel instability (Weibel, 1959), which is in the target-normal direction. The surface field tends to make electrons flow along the surface, which leads to a uniform structure of the interaction surface. The target-normal magnetic field acts to collimate electron jets and return currents that are periodic along the surface. As a result, the critical density surface is rather smooth in the small pre-plasma case and modulated in the large pre-plasma case.

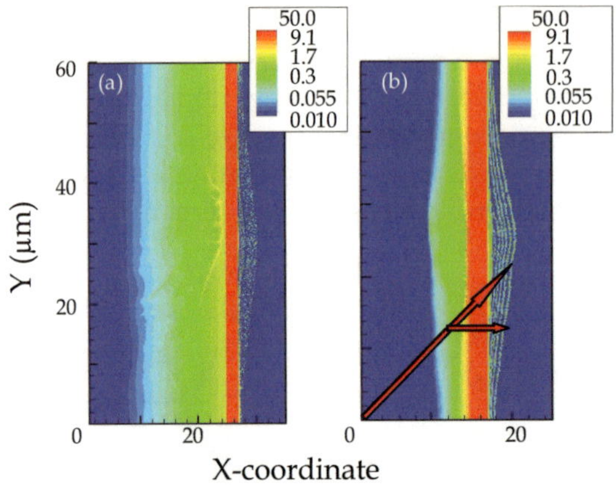

Fig. 8. Ion density distribution for a) large (L = 3.5 μm) and b) small (L = 0.3 μm) pre-plasma cases. Laser irradiation is from the left boundary, where the spot size is 15 μm, and the spot centre corresponds to y = 8.5 μm at the left boundary. Two arrows in b) indicate the trace of the laser spot centre and the target-normal projection of the spot centre.

These different electron acceleration processes lead to differences in the lateral distributions of the rear side sheath field and proton acceleration as follows. The target size is 2 μm in thickness and 50 μm in width. The targets consist of 50% protons and 50% C^{4+}. Laser conditions are the same as above except that the laser pulses have a spot size of 15 μm with a super-Gaussian distribution. The boundary conditions are absorbing in the x- and y-directions. Ion density profiles at 150 fs after the pulse has ended are shown in Fig.8.

For the large pre-plasma case, the ion distribution is symmetric with respect to the laser axis passing through the laser focal point at the critical surface. Since a large part of the hot electrons propagate in the target-normal direction, the sheath field structure is similar to the normal incidence case. For the small pre-plasma case, however, the proton distribution is non-uniform and asymmetric. As is seen in previous simulations, very energetic electrons are accelerated in the laser propagation direction, and lower energy electrons are generated in the target-normal direction via vacuum heating. These two components result in two acceleration fronts of protons, which are clearly seen in Fig.8b and detected in the experiment. The distance between the two acceleration points depends on target thickness and size of pre-plasma. In PIC simulation, the target is thinner and the density is lower than in the experiments. Thus the separation distance is considered to be larger than is visible in PIC simulation. Also, in Fig.8 we show the angular distributions for pre-plasma scales of 0.3 and 3.5 μm in order to see clear differences, but similar results are seen at much smaller differences also. The important thing is that the dominant acceleration process changes when the scale length becomes smaller than the wavelength.

A PIC simulation of the experimental parameters shows that the dominant electron acceleration process depends on the profile of the pre-plasma. For plasmas where the scale length of the pre-plasma is larger than the laser wavelength, a plasma wave is resonantly excited at the critical density surface, which results in the generation of fast electrons propagating in the target-normal direction (Estabrook & Kruer, 1978). If the pre-plasma scale length is smaller than the laser wavelength, a well-collimated electron beam is accelerated by ponderomotive forces in the propagation direction of the laser (Wilks et al., 1992, Kruer & Estabrook, 1985). Correspondingly, protons are accelerated in a target-normal direction from two spatially separated sources. The laser intensity and the contrast determine the laser absorption process and thus they control the entire acceleration scenario. PIC simulations confirm these findings.

2.3 Energy resolved spatial characteristics of laser accelerated protons

As we have seen, interactions of relativistic-intensity laser pulses with targets with and without pre-plasmas are significantly different from the perspective of the dominant electron acceleration processes and the energy coupling to plasmas.

To clarify the proton acceleration process that is uniquely observed for small pre-pulses, oblique irradiation, and in thin targets we incorporate particle diagnostics which provide a microscopic measurement of the proton acceleration with PIC simulations. We found that the laser-accelerated proton source drifts on the rear surface of the target in the direction of projection of the laser wave vector onto the surface. The drift of the proton source is explained by the lateral movement of the sheath field. This moving proton source leads to the observation of a limited range of proton energies at any given point along the target surface.

The proton source was investigated with an experimental setup and geometry similar to those in section 1.2. The ion emission spectrum was measured with an absolutely calibrated Thomson parabola spectrometer (Ter-Avetisyan et al., 2005) set up in magnifying imaging mode (Schreiber et al., 2006). The spectrum is imaged through a 30 μm pinhole, located at a distance of 5 cm from the source, on a detector screen 75 cm behind the pinhole, at 45° laser incidence on a 13 μm aluminum target; see Fig.9.

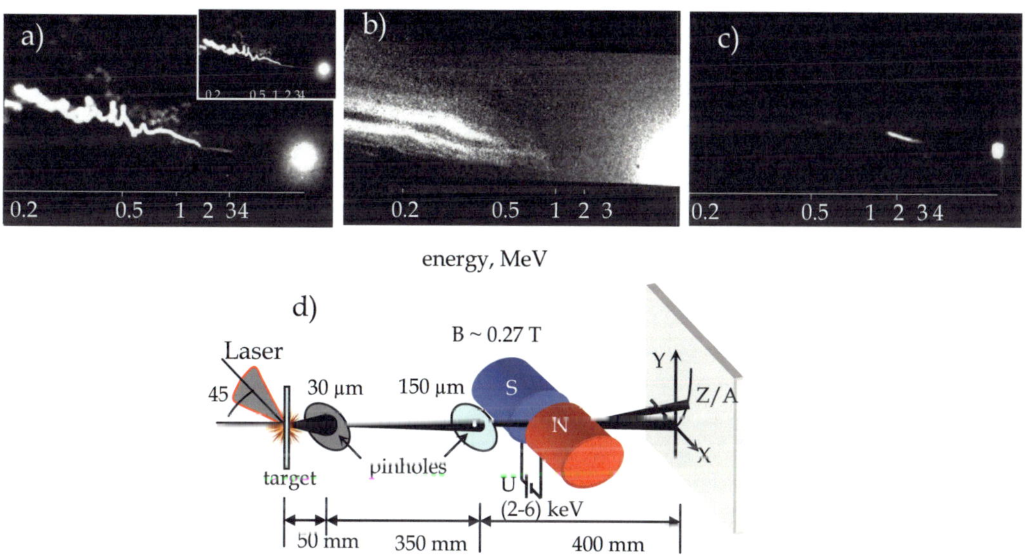

Fig. 9. a), b), c) CCD picture of the proton spectrum at 45^0 laser incidence on a target measured in the direction of target normal with a magnified imaging ratio of 1:15. The ideal parabola trace is the solid line in the figure a) inset. b) Proton spectrum from a Mylar target. c) Proton spectrum from Al target when proton beam passes through two pinholes. d) Schematic figure of the geometry of two pinhole measurement.

The striking features of this figure are that the whole low-energy part of the spectrum is shifted from the ideal parabolic trace calculated according to our geometry, whereas the high-energy part obeys the calculated parabolic trace (solid line in Fig.9a inset) assuming that the ions are emitted from a fixed ion source, and finally that the trace of the low-energy part shows strong oscillations.

The imaging factors of the setup include the geometry and the emitted proton beam characteristics, such as partial beam divergence. In Fig.9a it is found that the trace spread decreases with increasing proton energy. The trace spread and the partial beam divergence calculated for the present geometry (distance from the source to the detector) decrease from 0.13 to 0.11 mrad when the proton energy increases from 0.2 to 3 MeV. Therefore, in the present geometry, the image covers only the diameter of an 18 μm area on the target with a magnification factor of about 15. The observed phenomena are attributed to the continuous movement of the proton emission source up to about 20 μm. When electron beam filamentation occurs during electron propagation inside the target, as it is the case for insulators (Manclossi et al., 2006, Fuchs et al., 2003), several proton emission points and, correspondingly, multiple parabolic traces would be measured. The proton spectra from 13 and 20 μm Mylar targets taken at a relatively low laser pulse contrast ($\sim 10^{-7}$) indicate many separate proton emission points on the target. In Fig.9b is given, as an example, a proton spectrum from a 20 μm Mylar target. Here the energy scale is calculated assuming that the source point is in the middle of the right blob in Fig.9b.

The continuous proton spectrum in Fig.9a suggests that one proton source point is continuously moving on the surface. In order to give additional proof of this assumption, a second pinhole with a diameter of 150 μm was installed at 35 cm behind the first pinhole (see Fig.9d), in order to detect only the protons emitted from an area of 20 μm diameter with straight trajectories along this axis. Particles that are emitted from the other parts of the source and even have passed the first pinhole will be blocked by the second one. A continuous decrease of the size of the second pinhole led to truncation of the spectrum and finally only the high-energy part of the spectrum remains, which lies on the expected ideal parabolic trace as shown in Fig.9c, indicating that they are accelerated from the indicated point source. Here, another interesting point is that the source position from which the high energy protons are emitted does not emit the low-energy ones in the same direction. It has to be mentioned that these phenomena are observed only at 45° laser irradiation case.

2D PIC simulations were carried out to see the temporal evolution of the sheath field on the rear surface. The simulation conditions are as follows. The targets are 13 μm in thickness and 70 μm in width with density of 40 n_c, where n_c denotes the critical density for the laser wavelength of 0.8 μm, and consist of electrons and Al^{3+} ions. The proton source is a contaminant layer, which is modelled as a buried proton layer at the rear surface with thickness of 0.1 μm and density of $1n_c$, i.e., the 0.1 μm layer from the rear surface is composed of 39 n_c of Al^{3+} and $1n_c$ of protons, with 40 n_c electrons. The electrons initial temperature is set at 1 keV, and the ions are initially cold. The system size is 50 μm in the x direction and 70 μm in the y direction and the cell number is 2245 × 3369. 12 particles are used in one mesh for each species and the total number of particles is about 8×10^7. The density profile of the pre-plasma is assumed to be exponential with two scale lengths. For the overdense region where the density is higher than 3 n_c the scale length is 0.1 μm. For the lower-density region, lower than 3 n_c, the scale length is 0.5 or 3.5 μm. A p-polarized laser pulse whose spot diameter is 10 μm irradiates the target from the left boundary with the incident angle of 45°. The laser pulse ramps up in five laser cycles and sustains its peak intensity of 2×10^{19} W/cm^2 for a duration of 40 fs. The laser axis irradiates the target surface at y ≅ 17 μm. On the boundaries, particles are reflected back to the system by reducing their energy to the thermal one, and the electromagnetic fields are damped at the boundaries.

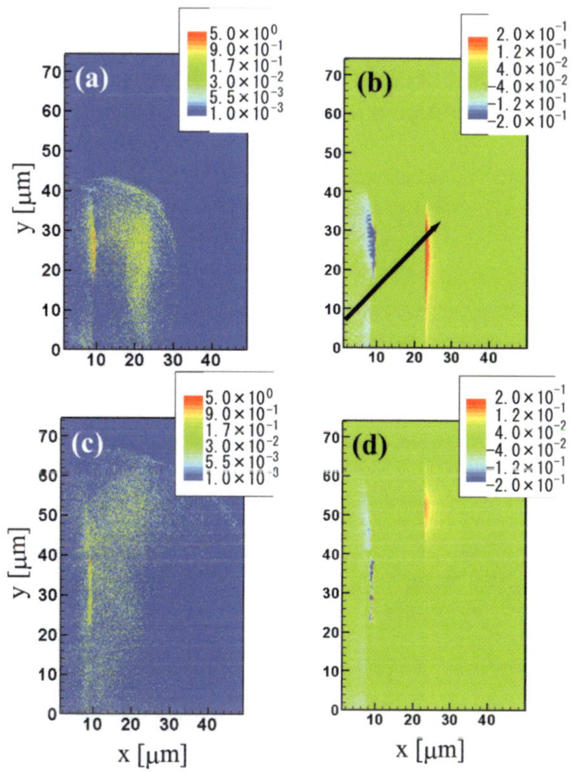

Fig. 10. Temporal evolution of the electron energy density and the longitudinal electric field. The electron energy density is normalized by $n_c m_c^2$, and observed at a) 140 fs, and c) 220 fs. The electric field is normalized by laser electric field, and observed at b) 140 fs, and d) 220 fs. The target width is 13 μm and is located at 10 μm < x < 23 μm. The black line in b) indicates the laser axis.

The temporal evolution of the longitudinal electric field E_x and electron energy density distribution for the small pre-plasma case (0.5 μm length scale in the lower-density region) are shown in Fig.10. The region where electron energy density is high moves upward; this direction of movement is explained as follows. The high-energy electrons are accelerated in the laser propagation direction by ponderomotive acceleration for the small-pre-plasma case. When they exit from the rear surface, very energetic electrons escape from the target, but a large number of the electrons are dragged back and reinjected into the target toward the front side. As a result, electrons move upward, being trapped along the target and therefore the sheath field also moves in the same direction, as is seen in Fig.10, which results in the movement of the proton acceleration point.

There are two components of the sheath fields showing different time evolutions. The sheath field induced by electrons accelerated in the target-normal direction by vacuum heating is symmetric around the target-normal position and expands radially, quickly reducing its magnitude. Another component is generated by electrons accelerated by ponderomotive acceleration, with greater magnitude than the previous one, since the ponderomotively accelerated electrons have much higher energy than target-normal electrons as is shown in Fig.11 and also in Ref. (Sheng et al., 2000).

At the early time of irradiation, a rather symmetrically expanded sheath field is induced. Shortly thereafter, a stronger sheath field localized at the ponderomotive position (y ~ 29 μm) is induced, while the symmetric field is weakened by diffusion (Fig.10d). The sheath field is strongest at the ponderomotive position and has a size comparable to the laser spot size.

The temporal evolutions of the maximum of the sheath field intensities at different positions are compared in Fig.12a, where the ponderomotive position (y ~ 29 μm), and 11 and 22 μm upward to the position (i.e., y = 41 and 52 μm) are chosen. As is seen from the figure, the intensity of the sheath field decreases as it propagates upward with velocity roughly close to the speed of light. The energy spectra of accelerated protons observed at those positions with 2 μm width are shown in Fig.12b. In the figure, the time delay of the proton acceleration due to the sheath movement is taken into account. It is seen that protons with the highest energies ($2.3 \le E \le 3.0$ MeV) are accelerated at the ponderomotive position by the strong sheath field, and protons accelerated in the upper position by the weakened field have lower energies.

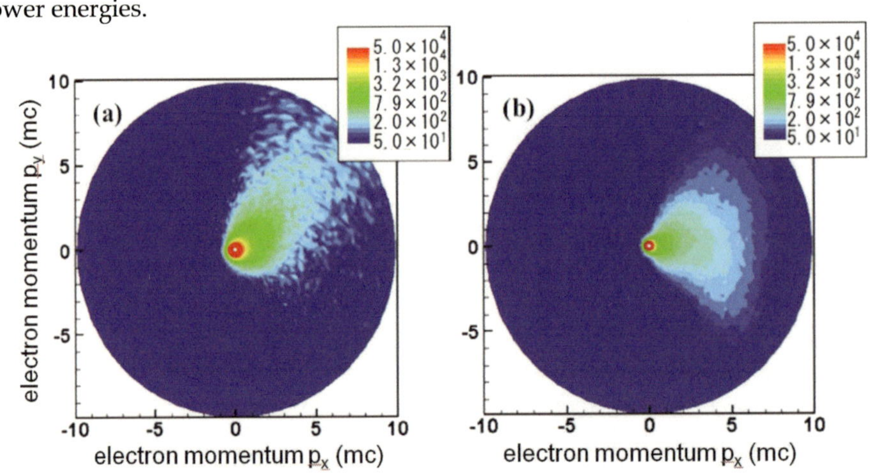

Fig. 11. Angular distribution of laser-accelerated electrons for scale length of preplasma a) 0.5 and b) 3.5 μm.

Since the lateral movement of the region of high electron energy density is due to the lateral movement of ponderomotively accelerated electrons, the scale length of the pre-plasma is crucial for the drift since the dominant acceleration process depends on the scale length. The angular distributions of electron momentum inside the targets are compared for two different scale lengths, i.e., 0.5 and 3.5 μm, shown in Figs.11a and 11b. In the small-pre-plasma case, the high-energy electrons are well collimated and accelerated along the laser propagation direction by the ponderomotive acceleration. On the other hand, in the large-pre-plasma case, the electrons are dominantly accelerated toward the target-normal direction by the resonance absorption. In this case, the electrons propagating away from the rear surface return to the target in every direction like a fountain, which results in the

symmetric expansion of the sheath field. As a result, in the large pre-plasma case, directional lateral drift of the sheath field and proton source movement are not observed.
The target thickness is also an important parameter. The lateral movement of the sheath field becomes less effective as the target thickness increases, since the electron energy density at the target surface decreases due to the expansion of the high-energy electrons into the target. Therefore, the target thickness is chosen to be comparable to or less than the laser pulse length for effective lateral drift of the sheath field. Also, the laser intensity is required to be relativistic so that the ponderomotive acceleration becomes dominant, generating a directional electron beam toward the laser propagation direction. A number of experiments on laser-proton acceleration have been carried out with similar parameters in terms of the laser intensity, irradiation angle, pulse duration, and target material and thickness (Hegelich et al., 2002, Fuchs et al., 2006, Cowan et al., 2004). But lateral drift of the sheath field has not been observed. The critical difference is the level of ASE. In the above cited experiments, the ASE level is mentioned as ~ 4×10^{12} W/cm^2, and in our experiments the ASE is reduced to ~1×10^{11} W/cm^2 by use of a specifically delayed pump mode with a comparable duration.

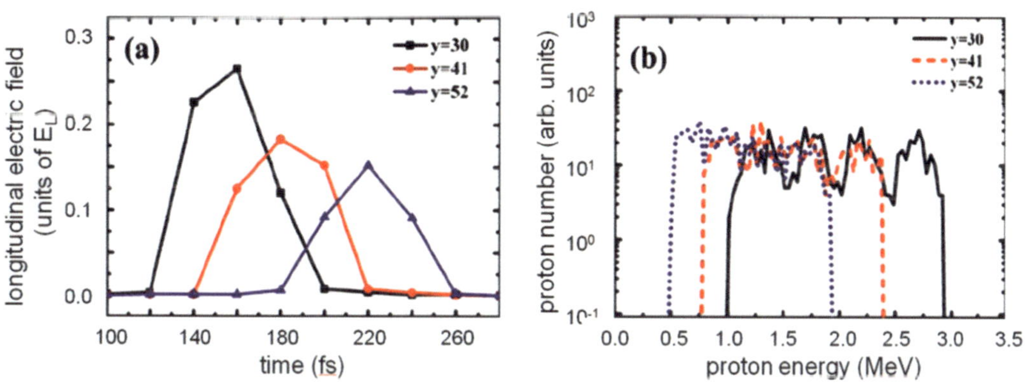

Fig. 12. a) Temporal evolutions of maxima of sheath field intensity at different positions y = 30, 41, and 52 μm. b) Proton energy spectra observed at three positions chosen the same as in a). The observation timings are delayed for the upper position such that t = 400, 440, and 480 fs at y = 30, 41, and 52 μm, respectively, since there is a time delay before the sheath field reaches and starts to accelerate protons.

Thus the amount of energy of the ASE is lowered by roughly 1/40 in our experiments, which leads to the estimated scale length being lowered by 1/6. In the numerical simulations by Fuchs et al., (2006), which show good agreement with the experimental results, the scale length is set at 3 μm, which is consistent with the scale length of 0.5 μm in our experimental conditions. Finally, we want to mention the oscillation in the parabolic trace for the low-energy part. This might be attributable to the complex electron transport coupled to self-induced electric and magnetic fields on the rear surface. In the above simulations, a magnetic field of ~30 MG and electric field of ~1 TV/m are observed, induced by the lateral transport of high-energy electrons with an estimated current of roughly tens of kiloamperes. This surface current would be disturbed by a Weibel-like instability (Weibel, 1959), whose wave vector lies dominantly in the z direction, which is not taken into account in a two-dimensional (2D) simulation. Its understanding and consideration of 3D effects are left for future work.

In conclusion, we have shown a scenario of proton acceleration by a fast-moving rear surface sheath field. This is observed as a directional shift of a single parabolic line, indicating that the proton emission point is drifting along the surface, which is proved by a two-pinhole measurement. This drift is confirmed by 2D PIC simulations which clarify the mechanism and conditions for the drift. The drift is only observed when intense laser pulses obliquely irradiate targets with small pre-plasmas. In this case, ponderomotively accelerated electrons propagate along the target and induce a commoving sheath field. This drift of the acceleration position leads to a selection of the protons by their energies as a function of the observation point, which might shed light on the control of laser accelerated proton energy.

3. Tomography of an ultrafast laser driven proton source

3.1 Correlation of spectral, spatial and angular characteristics of proton source

Even if the measurements discussed in previous sections show a clear feature of the source, however, those are measurements of the ion emission from a very small area as compared to the whole source extension, which is about a few hundred μm. The complete scenario of the ion acceleration phenomena in its complexity could be given only if the source characteristics as a whole or at least a more extended part of it are investigated. Such understanding is especially important for modelling and ion beam tracing.

In order to image a broader area of the source the principle of tomography was employed. Tomography is an imaging method. The object is imaged section by section (or sectioning). In this work we apply sectional imaging of the proton source by using an array of pinholes in combination with a Thomson spectrometer (Ter-Avetisyan et al., 2009a). A technique which provides sectional images of an object is usually referred to as a tomography, and, in principle, the superposition of tomographic cross sections coupled to a reconstruction algorithm allows the visualization of the internal structure of an object.

Because the laser accelerated proton beam is laminar (Borghesi, et al., 2004) and irregularities at small scale do not disrupt the general beam characteristic it is sufficient to substitute the entrance pinhole of the spectrometer with a multi-pinhole array. It is installed perpendicular to the dispersion direction: parallel to magnetic and electric fields. This way the proton beam is divided into small beam-lets and the obtained spectral images combine the simultaneous record of many independent spectrometers which have the same dispersion constant (Fig.13). Because the source has a finite size and the emission is laminar the spectrometers record the ion spectra corresponding to different emission zones of the source according to their emission angle. This detection principle allows the reconstruction of the source, which we call here proton source tomography. It gives an energy resolved topographic picture of the source.

In the experiment 40 fs pulses from a multi 10 TW Ti: Sapphire laser system were focused down to a 10 μm perfectly round spot onto a 5 μm Ti target. The maximum laser intensity was about 2×10^{19} W/cm^2. The ASE pedestal of the laser pulse, several picoseconds before the pulse peak, was at a level of 10^{-8} relative to the peak intensity.

The proton emission spectrum was measured with a Thomson parabola spectrometer calibrated in absolute terms (see more details in Ter-Avetisyan et al., 2005). The applied magnification (Schreiber et al., 2006), provides additionally high spatial resolution. The spectra were obtained from 21 beam-lets produced with a row of 21 pinholes having 30 μm in diameter each and a period of d = 530 μm (Fig.13).

This pinhole mask is positioned parallel to target surface at a distance of a = 85 mm, parallel to the magnetic and electric fields of the Thomson spectrometer, and lie in the plane of laser

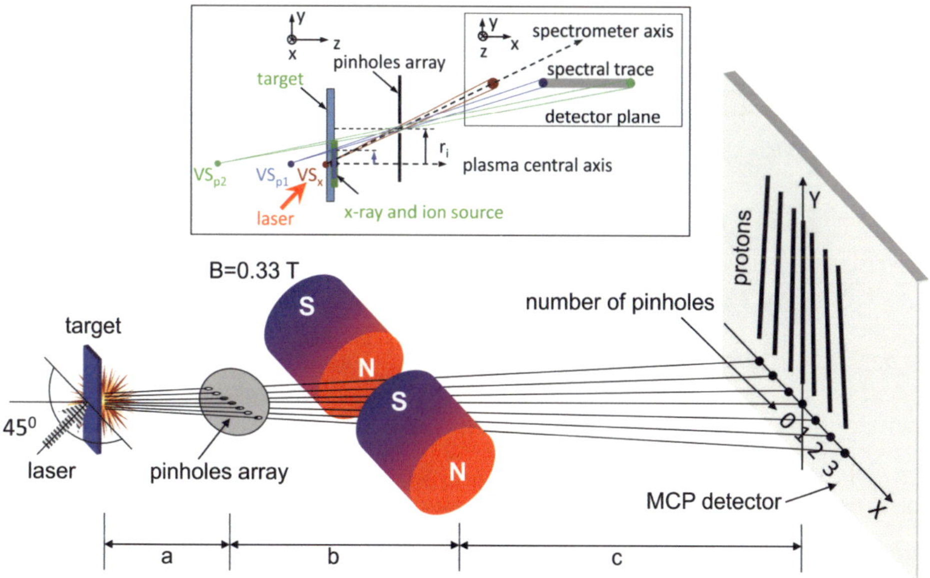

Fig. 13. Experimental set up. Figure inset shows imaging geometry: ions propagate along Z-axis, energy dispersion is along Y-axis, and analyzed emission coordinate is along X-axis. XY detector plane is flipped in YZ cartoon plane. Shift of pinholes (r_i) and all distances are known from the geometry (cf. text). VS_{p1}, VS_{p2}, and VS_x, are the virtual source points for protons with different energies ($E_{p1} > E_{p2}$) and x-rays, respectively.

beam polarization (p-polarized). A magnetic field of about 0.33 T between 4 cm separated poles was applied in the spectrometer. The spectrum was imaged on the MCP detector coupled to the phosphor screen located at a distance of b = 120 mm, c = 275 mm from the pinholes (Fig.14a). All precautions mentioned in (Ter-Avetisyan et al., 2004b) have been taken for the measurements. Additionally, a careful scan of the magnetic field distribution in all three dimensions between the magnetic poles and outside was carried out. The derived instrument function was conformed with an additional test experiment to exclude any artifact on the spectral traces caused by field inhomogeneities.

At this geometry $\alpha = 2.3^o$ is the largest possible angle of protons entering the spectrometer. The restriction comes mainly due to the finite size of MCP detector. The magnetic field disperses the protons acting on their v_z velocity component perpendicular to magnetic field which is $v_z = \vec{v} \times \cos\alpha \cong v$. Therefore a possible difference in dispersion for protons passing through different pinholes is negligible and well below our resolution limit. It was not necessary to apply the electric field in order to separate the ions because of the ultra-short and high temporal contrast of the laser pulse, a pure proton beam was produced (Schnürer et al., 2007, Nickles et al., 2007).

A typical spectral image is shown in Fig.14a. As mentioned, this picture combines in principle the record of many independent spectrometers (in our case - 21) located under a different angle to the source and having the same dispersion constant and the bright spots in the right side of the figure are formed by energetic photons passing through pinholes and representing the axis of the spectrometers. The ϕ^o in the right side of Fig.14a are the tilt angles of the measured spectra from the spectrometers axes. The spectrum along the "central axis" has a 0° tilt. The spectrometers axes can be exactly defined with the measured

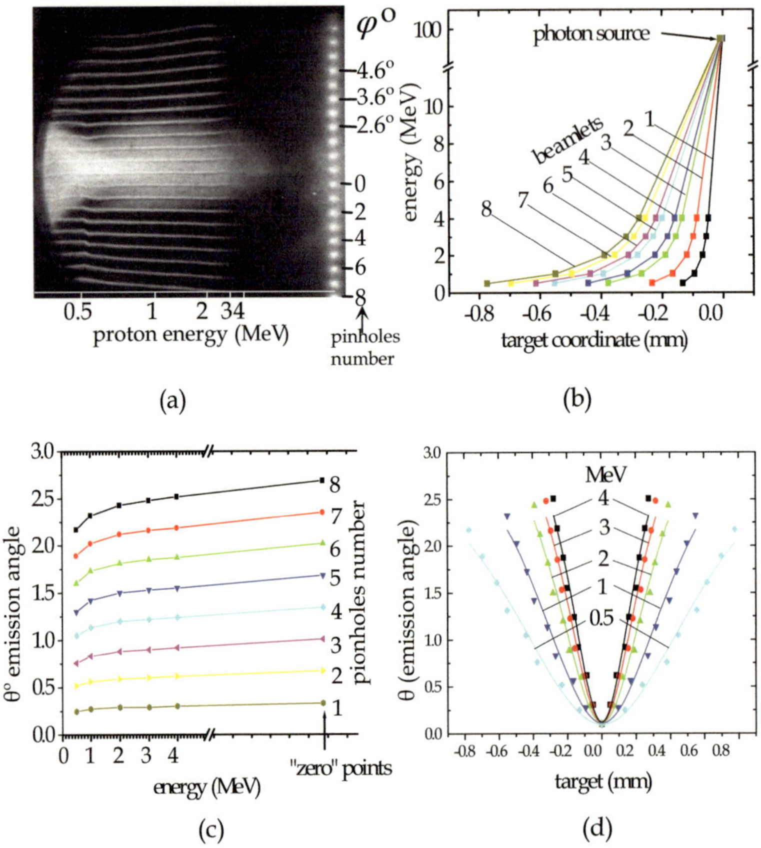

Fig. 14. (a) The imaged spectra through a row of pinholes are shown. Because the figure is symmetric to ϕ° = 0 line, on one half of the figure the dependence on the tilted angle ϕ° from axes is given (right upper side) and on the other half the numbering of the traces (right low side); (b) Energy of the emitted protons as function of target coordinate in the beamlets from 1st up to 8th is depicted; (c) Proton emission angle in each beam-let depending on proton energies; (d) emission angle of protons with 0.5, 1, 2, 3, 4 MeV energies dependent on emission target coordinate.

energetic photons which are registered over a whole angle of ± 4.6° from a source with an extension of about 20 μm. This means that initially the laser drives the hot electrons through the target with a divergence angle of about 45°. Having the axis of each pinhole-spectrometer and the angle of departure of the proton trace from this axis, the coordinate of the proton emission at the target surface and the emission angle can be calculated.

The striking features of Fig.14a are: (i) A clear spectral trace from each pinhole is observed. There are no crossing lines. This confirms that the beams of different energies are strongly laminar. (ii) The spectral traces are tilted from their axes. The "central" trace lies exactly on its spectrometer axis. Symmetrically, on both sides of the "central" trace all spectral traces have a tilt which increases with the distance to the central point with the angle ϕ°. Because of the good symmetry in the following figures we will show the spectra located only on one side from "central axes" in order to do not overload them.

The analysis of the experimental data and the reconstruction of proton trajectories are straightforward. With the position of the pinholes along the linear array and the measured coordinates of the "zero points" both the target coordinate and the angle of the emitted protons could be calculated. This was done for different energies contributing to the spectral trace and by applying a simple geometrical relation. However, the scan of the magnetic field showed that the magnetic field lines are slightly curved in x,y and x,z planes and the existence of B_y and B_z components tilt the spectral traces. The protons will be deflected in ± x direction due to both $(v_z \times B_y)$ and $(v_y \times B_z)$, where the v_z and v_y velocity component appears inside the magnet due to the dispersion. This was verified in a test experiment where a single proton beam was injected at ten different positions inside the magnet. The resulting instrument function was determined and used to unfold the measured raw data.

In Fig.14b the emitted proton energy is depicted as a function of the target coordinate derived from the 1st up to the 8th beam-let. It can be seen, that at the same target coordinate protons are emitted with different energies, while protons with the same energy arise from different target coordinates. As higher the energy of protons as smaller the region from where the emission occurs.

The proton emission coordinate as a function of the emitted proton energies (Fig.14b) of the ith beamlet can be fitted with an exponential function as:

$$x_{E_i} = m_i = n_i \times \exp(-E_i / 0.93) \tag{1}$$

where m_i and n_i are implicit functions of the target coordinates in relation to the pinhole i ; m_i is the target coordinate from where the high energy protons have passed through the pinholes and therefore it is defined by the geometry of the pinhole array. It can be found that m_i is a parabolic function of the period (d) of the pinhole array: $m_i(mm) = -0.018 - 0.038d - 0.003d^2$, where 0.018 mm is the target coordinate from where the high energy protons have passed through the pinhole number zero.

Similarly, n_i is also defined by the pinhole array but it shows how strong the target emission coordinate changes with the energy of the protons which are passing through a same pinhole. The n_i is also a parabolic function of the pinhole array period: $n_i(mm) = -0.06 - 0.11d - 0.01d^2$. Therefore, the relation between these two parameters is: $m_i = 0.36 \times n_i$.

The proton trajectory in each beamlet depends on proton energy as shown in Fig.14c. The related emission angle of the protons is continuously changing with its kinetic energy. At the same angle only protons with different energies are emitted. Additionally, protons with low energy are emitted at a smaller angle compared to protons with high energy. The emission along the symmetry axis of the source is unique; the protons emitted under 0° to the target normal are bearing any energy according to the spectral distribution.

The dependence of the proton emission angle on energy (Fig.14c) in different beamlets can be fitted with an exponential function as:

$$\theta_{n,E_i} = \alpha_n - \beta_n \times \exp(-E_i / 0,93) \tag{2}$$

where α_n is the emission angle of measured highest energy protons passing the nth pinhole. The α_n is a linear function from the pinhole array period: $\alpha_n = -0,02 + 0.65d$, showing a linear increase of the emission angle of high energy protons with the target coordinate and 0.02° is the opening angle of the beam which is passing through the pinhole number 0

(Fig.13). The parameter β_n shows how strong the emission angle of protons changes if they are passing through the same pinhole with different energies. The β_n is the parabolic function of the pinhole array period (d) and can be expressed as $\beta_n = -0.04 - 0.07d + 0.005d^2$. It means that going with increasing distance to the centre and from one pinhole to the next one the angle increases in respect to a similar change of the energy.
The emission angles of protons with the same energy but with different emission coordinates in relation to the 16 beamlets which are symmetrically located on both sides of the "central axes" are shown in Fig.14d. For each proton energy the data can be fitted with the Lorentz function and it can be found that the width of the source (Δx) is an exponential function of the proton energies: $\Delta x \sim E^{-0.5}$.
With these relations one can reconstruct the proton trajectories as they are emerging from the target. One can see in Fig.15a, where as an example 0.5 and 4 MeV proton trajectories are shown, that protons with a same energy are accelerated as an expanding laminar beam. The emission angle increases with the target coordinate (or the distance from the source centre) and higher energies belong to a large emission angle if the same target region is analyzed. Additionally, the measured high energy protons are emitted from a smaller target region than the lower energetic once.

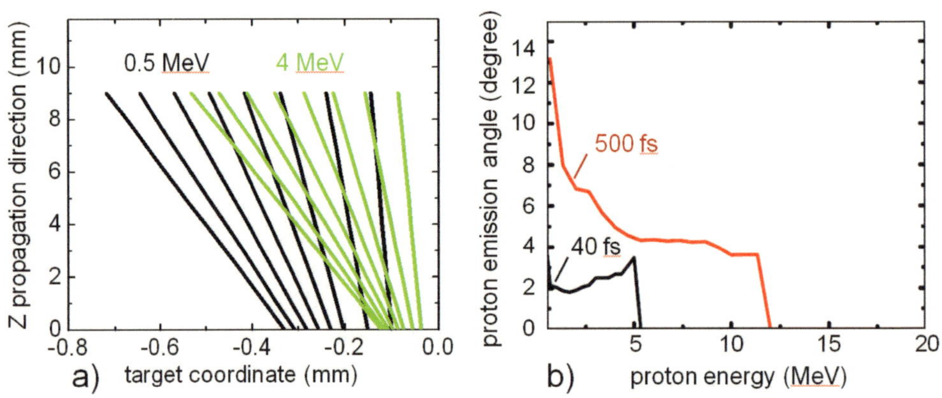

Fig. 15. a) Reconstructed proton trajectories as they are emerging from the target. As an example are shown 0.5 and 4 MeV proton trajectories. b) Simulated angular spread of proton emission depending on proton energy for 40 fs and 500 fs laser pulses.

The discussion above makes clear, that each beam-let is formed by an unique combination of energy, target coordinate and emission angle of the protons. The low energy protons are emitted with a smaller angle than the high energetic ones (Fig.15a). At first glance this observation seems to be in contradiction with a statement in (Cowan et al., 2004) that the angular envelope of the protons with increasing proton energy is decreasing. There are two things which make a decisive comparison to other experiments and its conclusions difficult. First, there are no data known to us from the literature with a similar resolution in order to make a direct comparison concerning the emission of the central part of the beam as we performed. Second, due to the present limitation of the observed range of angles we did not look to the maximum extension of the source as a function of energy. The latter one is in other works mainly attributed to the source size. If one calculates the ratio between the emission coordinates (source size) and the emission angle as a function of the ion kinetic energy one finds a qualitative similar functional dependence of our data and those from

other work. Nevertheless a final conclusion for different experimental parameters realized in different work should be drawn from data with similar resolution. However, the discussed features unambiguously follow directly from the experimental findings. Also recent experiments showed (Roth et al., 2005) that the highest energy protons are emitted in the central, high density portion of the sheath distribution, which agrees with the present results (Fig.14b). Nevertheless, it is likely that both the used long pulse (350 - 850 ps) and high laser energy (20 - 30 J) (Cowan et al., 2004), in contrast to our ultra-short pulse (40 fs) with "low" energy (0.7 J), and more importantly the high temporal laser pulse contrast (10^{-7} - 10^{-8}) (Nickles et al., 2007) are the decisive parameters for the proton source formation and emission characteristics of the accelerated particles.
In order to look particularly to proton acceleration scenarios in case of long and high energy laser pulses in comparison to short and low energy laser pulses 2D particle-in-cell (PIC) simulations were performed. The simulation box had a size of 150 μm × 150 μm. The 10 μm thick and 150 μm long target was located at the coordinate 12.0 < x(μm) < 22.0. The pre-plasma was set 0.5 μm in front of the target and composed of Al^{3+} cold ions and 500 eV electrons with a density 20 n_c. The n_c denotes the critical density. A proton contamination layer was buried within a distance of 0.5 μm from the rear surface. The laser with a wavelength of 0.8 μm and focused to 10 μm was irradiating the target under 0°. The laser pulse durations were set as 40 fs and 500 fs and had Gaussian profiles both in space and time. The electromagnetic field was damped at the boundary of the simulation box, and the particles were reflected with thermal energy.
The simulation results of the proton energy dependence on the emission angle are shown in Fig.15b. The full simulation time was 900 fs for the short, and 1400 fs for the long laser pulse. The difference is apparent: For the 500 fs laser pulse irradiation the angular spread of the accelerated protons is gradually increasing with decreasing proton energies, as it was observed in (Cowan et al., 2004). This is in contrast to the case of the 40 fs laser pulse irradiation where the higher energetic protons have the largest angular spread, and the emission angle decreases with the proton energy. In the long pulse case, the sheath field accelerates protons for roughly 500 fs, where the acceleration is described by hydrodynamic plasma evolution (Mora, 2003). In short pulse case the maximum sheath field is sustained only about 40 fs while protons are about to be accelerated. When the proton acceleration starts, the sheath field is still expanding and the accelerating front has a large curvature, which results in larger angular spread for higher energetic protons. The simulated energies and the angular spreads are quite comparable to that presented here and to the experimental results in (Cowan et al., 2004).
Additionally, results of simulation for the angular behaviour of protons emitted from different target coordinates are similar to those shown in Fig.15b for both 40 fs and 500 fs laser irradiation cases. Therefore, one can conclude, that the observed features in the present experiments is an unique property of the ultrashort and high contrast laser pulse acceleration scenario.
On base of the found interconnections between the target coordinate, emission angle and energy of emitted protons, one can constitute the following rules:

- The proton emission is directed to the target normal and it is symmetric to the laser axis.
- Protons with different energies are emitted such that: as higher the proton energy as smaller the source size but as bigger the emission angle in relation to the source size

- Protons with the same energy are emitted from different target coordinates such that: as far away from the centre as larger the emission angle
- Protons which are emitted from the same target coordinate but with different energies obey to the following: as higher the energy as larger the emission angle.

3.2 Correlation of spectral, spatial and angular characteristics of proton source

The measurements discussed in section I.4 were made around the central symmetry axis of the source. Due to the limited size of the detector only protons with 0.5 to 4 MeV energies emitted within a 2-3 degree angle were analysed. In order to acquire an understanding about the whole source behaviour, broader proton energy ranges have to be covered. Additionally, it is of interest and necessary to explore to what extent the measured proton source properties near the central symmetry axis are applicable to the whole and much more extended proton source. Particularly in view of planning proton beam steering systems it is necessary to know up to which point the measured energy dependence of the angular emission characteristic is preserved.

Here, the investigation of the source emission characteristics was extended at larger angles from the central symmetry axis and in a rather large range of proton kinetic energies from 0.06 MeV up to 4 MeV.

The measurement of the proton emission spectrum was carried out similar to section 3.1 Being restricted by the size of the detector a good compromise between the requirements of high magnification and of an extended spectral range was found by locating the pinhole array at a distance of about a = 30 mm from the target surface and the detector screen at b = 195, c = 255 mm from the pinholes (Fig.13). The inset of Fig.13 shows the imaging geometry which allowed a complete characterisation of the source in the whole spectral range.

In our previous section 3.1 the pinhole array was aligned in such a way that the central symmetry axis of the plasma crossed the central pinhole. In the measurements reported here we applied an imaging geometry with higher magnification and, distinctively, the pinhole array was shifted along the Y axis (i.e. parallel to the target surface and away from the ion beam central axis) (see inset of Fig.13). This shift, of about 2 mm, gave a corresponding shift of about 40 mm for the positions of the "zero" points on the detector. The observation angle extended up to 4.77 degrees relative to central symmetry axis of the plasma. In fact, with this geometry we imaged a source area centred at about 70 μm from the source centre.

The respective spectral traces cover the high energy part of the spectra (Fig.16a) in the same way as in section 3.1 but extend the source imaging region to cover energies down to 0.2 MeV protons. Fig.16a looks very similar to Fig.14a in section 3.1 where a lower magnification was applied. The data analysis confirmed that the observed spatial and angular characteristics of the emitted protons are exactly the same as in the case of axis-symmetric detection (section 3.1) and can be treated in a similar manner. The tilts of the spectra from their axes are still symmetric relative to the zero degree trace (i.e. the trace due to the pinhole on-axis with the interaction point). For the 45° incidence on target as employed in the experiment the focus is elliptically shaped and has two main symmetry axes and therefore one should expect the source area to show also this elliptical symmetry. We have measured along the short main axis of the elliptically shaped source area. As a result also the tilts of the trajectories for the corresponding energies from the main axis are symmetric relative to the "central" trace. The data analysis was carried out by treating the signal observed as the result of 21 independent spectrometers as defined by the pinholes.

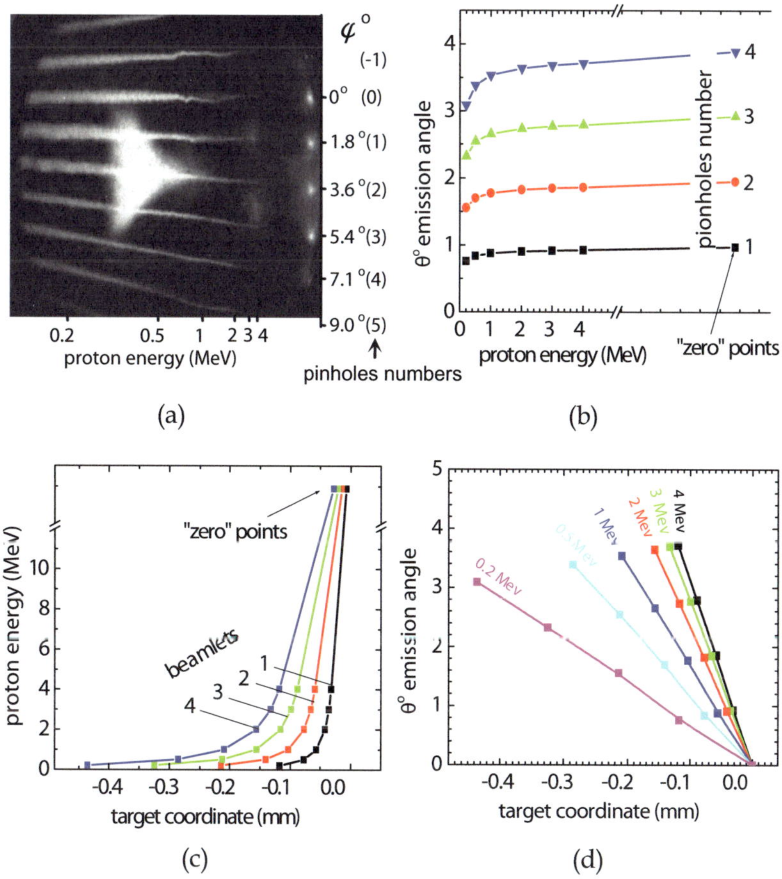

Fig. 16. a) Spectral traces of a row of pinhole images: On the right side of the figure the tilt angle φ° of the spectral traces from the central symmetry axes corresponding to φ° = 0 line (target normal) and respective numbers of beam-lets (in brackets) are given. b) Proton emission angle in each beam-let depending on proton energies in beamlets from 1st up to 4th is depicted. c) Emitted proton energy as function of target coordinates in beamlets from 1st up to 4th. d) Emission angle of protons with 0.2, 0.5, 1, 2, 3, and 4 MeV energies dependent on target emission coordinate is exhibited.

Each spectrometer axis was defined by the coordinate of the "zero" point and the coordinate of the corresponding pinhole (Fig.13). If protons within the observed energy range are all emitted from the same source point the projection of their spectrum on the X-axis of the detector (Fig.13) should be a point, i.e. the line describing the spectrum should be along the Y axis. The measurement showed instead that the lines described by the spectra in the detector plane are tilted with respect to the Y axis i.e. the projection on the X axis is different for different energies (Fig.16a). From the X coordinate of the projection for a particular energy the corresponding pinhole coordinate and assuming straight propagation of the particles, the coordinate on the target surface from where the corresponding proton is emitted (inset of Fig.13) can be calculated.

The "zero" points coupled with the position of the corresponding pinhole, are used as a reference to define the axes of the spectrometers. When using MCP detectors, the "zero" points can be due to X-rays (Ter-Avetisyan et al., 2009b) and neutral particles (Busch, et al., 2003, McKenna et al., 2007) which are undeflected by the magnetic field in the spectrometers. The data analysis shows that the trend observed in section 3.1 of the proton emission angle decreasing with decreasing proton energy is preserved in these measurements (Fig.16b). Additionally, higher energy protons are emitted from a smaller target region (Fig.16c) with higher divergence (Fig.16d) while the proton source for lower energies becomes larger and has a lower divergence. These trends confirm the assumption that "each source point behaves similarly" which allows one to characterise the general source behaviour independently from the specific area of the source under observation. Therefore a similar approach for the calculation of the proton emission trajectories (section 3.1) has been applied.

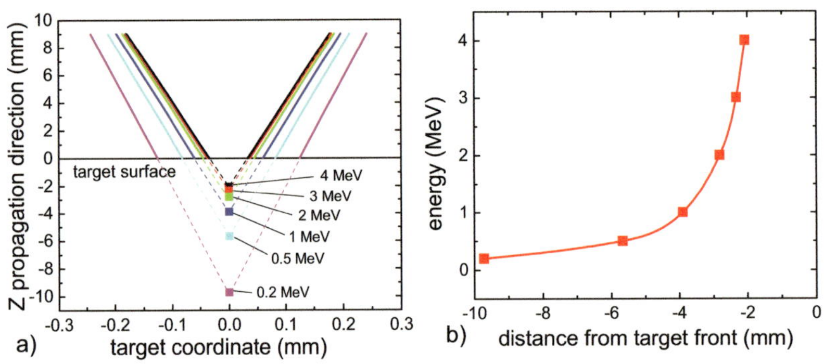

Fig. 17. Reconstructed proton trajectories as they are emerging from the target. As an example a) proton trajectory passing through 1 and -1 pinholes are shown. The virtual source position in front of the target is deduced by linear interpolation of the proton trajectory lines (dashed lines). b) The virtual source position in front of the target for different proton energies is shown.

The trajectories of protons with different energies, which pass for example through the pinholes +1 and -1 are deduced from the spectral traces and shown in Fig.17a. One can adopt a virtual source concept to describe the behaviour of a laminar ion source, as discussed in (Borghesi, et al., 2004). If one assumes that the propagation of the proton beam after the acceleration process is ballistic and there is no ion interaction within the beam, the proton trajectories are straight lines (inset of Fig.13) and one can trace back these trajectories to a virtual source (VS_p) in front of the target as shown in Fig.17a (dash lines).

The virtual source position in front of the target changes for different proton energies (Fig.17b), which means that the protons with different energies are emitted with a different curvature of the accelerating sheath field. In the inset of Fig.13 the virtual source points for protons with different energies, where $E_{p1} > E_{p2}$, are shown. This hints at a change of the sheath field curvature during the acceleration process.

As a next step measurements with a larger displacement of the observation axis were realized (i.e. the pinhole array was shifted sideways so that the central pinhole was further away from the interaction axis). This set of measurements was aimed to explore the emission characteristics far from the source centre. Furthermore, the measurement was extended to protons with even lower energy. A displacement of the observation axis of about 1 mm from the central source axis was set, which corresponds (for the central pinhole of the array) to an observation angle of about 4.88 degree relative to the central symmetry axis of the plasma. The measured proton spectra are shown in Fig.18a.

Due to the limited size of the detector the "zero points" are not visible in this case. As the "zero point" coordinates are needed for the trajectory reconstruction now the deflecting electric field of the Thomson spectrometer was switched on in separate shots to perform an energy calibration by using deflection of protons in an applied electric field. From the parabolic traces the corresponding "zero points" were extrapolated and these extrapolated coordinates have been used to analyze the measurements for the respective pinhole and spectrometer setting. The proton emission angle depends on proton energies as shown in Fig.18b. The data show clearly (Fig.18b) the continuous decrease of the proton emission angles relative to target normal with decreasing proton energies. This tendency holds down to proton energies of about 0.14 MeV which are emitted along the target normal; therefore one can say that they are "collimated". The proton beam at energies lower than 0.14 MeV is

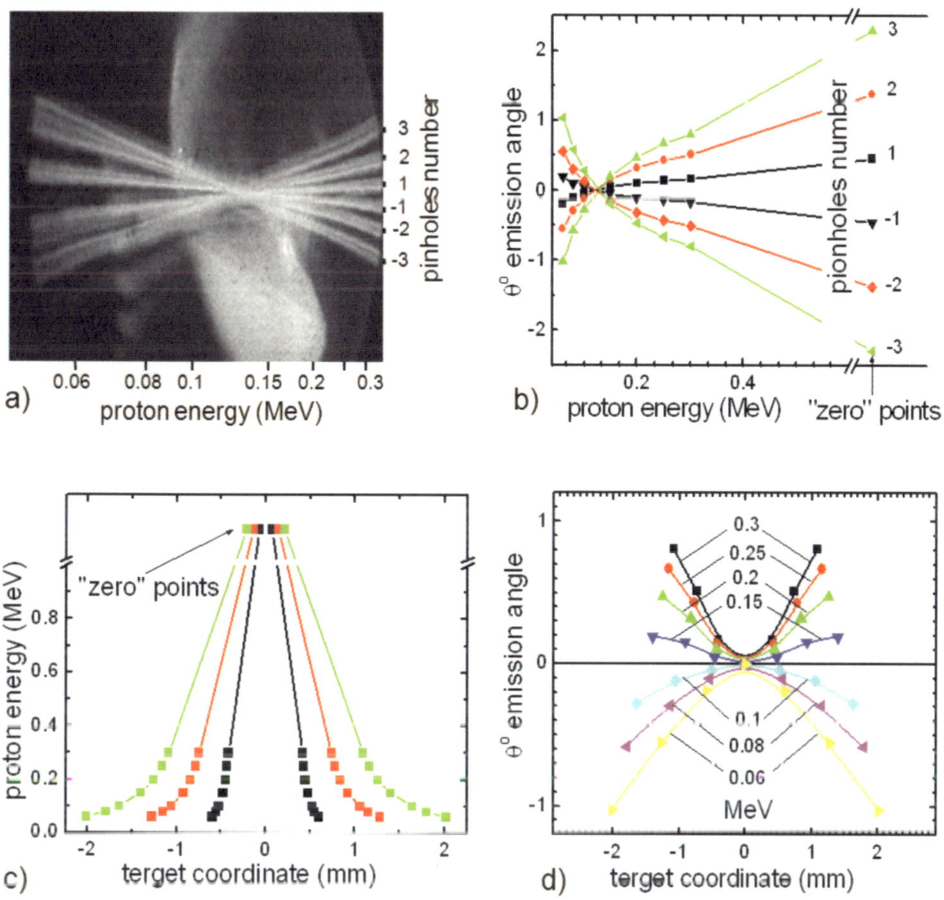

Fig. 18. a) The imaged spectra through a row of pinholes are shown. On the right side the numbers of beam-lets are given. b) Proton emission angle in each beam-let depending on proton energies in beam-lets is depicted. c) Emitted proton energy as function of target coordinate and (d) emission angle of protons with 0.06, 0.08, 0.1, 0.15, 0.2, 0.25 and 0.3 *MeV* energies dependent on emission coordinate is exhibited. $\theta^0 = 0$ is the angle of target normal.

convergent and in the detector plane one can see the crossing of trajectories at about 0.12 MeV (Fig.18b). For such low proton energies the source size becomes as large as a few mm (Fig.18c). Concerning this largely extended source area we thus found that the beam becomes converging (Fig.18d) as much as the proton energy decreases. On basis of this measurement geometry the proton trajectories are calculated and traced back to the target surface (cf. Fig.19a).

To our knowledge this is the first observation showing that a proton beam driven by an ultra-short laser pulse consists of a divergent, a "collimated" and a converging component, with the divergence varying continuously as a function of energy.

The structure of the transverse electric field obtained in 2D PIC simulations performed at the conditions as in section 3.1 gave some qualitative indication of possible reasons for the observed divergence-to-convergence behaviour of the proton beam. The simulation starts at t= 0 and the laser peak reaches the target surface at t = 120 fs. The structure of the transverse electric field is plotted in Fig.19b. The bell-shaped red lines (4 are visible) represent iso-density contour plots of protons in the accelerated bunch.

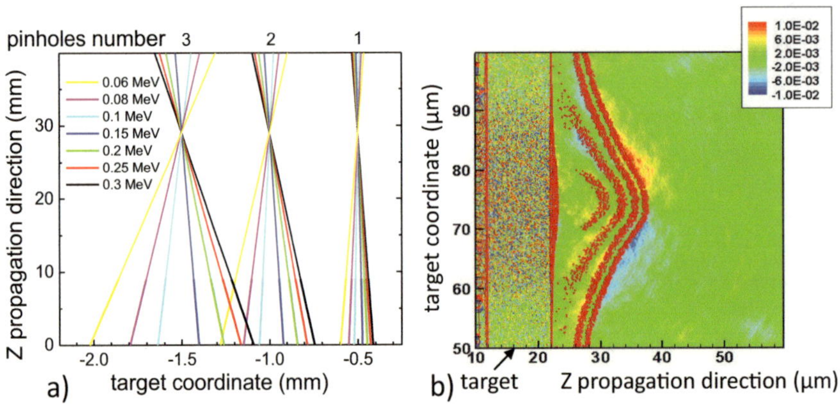

Fig. 19. a)Reconstructed proton trajectories as they are emerging from the target with the energies 0.06, 0.08, 0.1, 0.15, 0.2, 0.25 and 0.3 MeV and passing through the 1st, 2nd, and 3rd pinholes are shown. b) 2D PIC simulation results: transverse component of sheath field at t= 500 fs (the parameters of the simulation are explained in the text).

The protons propagate from the left to the right. At the ion front we observe transverse field components which are opposite with respect to the y = 75 symmetry line: the field is directed up for y > 75 (red colour code) and down for y < 75 (colour code blue). The transverse field is most intense at the acceleration front which causes radial deflection and angular spreading of the higher energy protons. Behind the ion front we observe weaker field strength and finally a change in the field polarity (at x < 25): For the region y > 75 we observe now light blue coded areas (i.e. field directed down) and for y < 75 yellow ones (field directed up), respectively. Thus the slower protons, which are behind the fast ion front, experience a transverse field component which can in principle result in converging trajectories, as observed in the experiment for rather low energetic protons.

The situation might be different in the case of ions accelerated by long (ps) laser pulses, where the longitudinal momentum gained by the ions is higher and transverse fields may be less effective in spreading the high energy ions.

Summarizing the emission properties of a proton beam accelerated by a high contrast ultra-short laser pulse have to be understood as follows. A target source point emits protons within a broad range of energies. Taken two energies E_1 and E_2 such that $E_2 < E_1$, the relation of the proton emission angles is $\theta_{E_1} < \theta_{E_1}$. Any point located at a distance d from the target rear surface and located within the total beam cone will be crossed by the trajectories of ions with different energies which are stemming from different target-source coordinates. This means, only a certain target area (S) is contributing to the ion spectrum which is measured at some point in the beam cone and at a certain distance from the target rear surface. As a result, the spectrum of the protons in that point can be controlled by changing the proton source area. It might be that our results apply also to the results of Ref. Schwoerer et al., (2006), where a proton beam with a small energy spread was detected when the proton source area was restricted.

The data also indicates that the proton beam is emitted from the target so that for high energies it is divergent and becomes convergent for low energies (below 0.2 MeV). 2D PIC simulations provide indications on how the effect of field components transverse to the beam propagation axis could lead to these features. It will be interesting to test in future experiments how this behavior will scale if the cut-off energies in the ion beam are significantly enhanced using a laser driver providing much higher energy and intensity.

Additionally comparative investigations with longer driver pulses may clarify whether (as it is likely) the divergence behavior measured here is an exclusive feature of short pulse interactions.

4. Conclusion

In this article recent studies of phenomena in relativistic proton acceleration have been reviewed. Experimental results have shown unique properties of protons accelerated by ultrashort laser pulses, which opens prospects for a broad range of applications (Borghesi et al., 2006).

Significant advances in laser technology are likely to expect for the next years, leading to further enhancement of both the intensity of the extremely short pulses and the average power. These developments will surely stimulate the elaboration of new ideas and more advanced diagnostic developments for measuring the effects not even thought of today. All of this will open up new research areas, pushing current activities to new frontiers, and generate further excitement in the field of laser-matter physics.

5. References

Allen, M., Sentoku, Y., Audebert, P., Blazevic, A., Cowan, T., et al., (2003). Proton spectra from ultraintense laser-plasma interaction with thin foils: Experiments, theory, and simulation. Phys Plasmas 10, 3283-3289.

Beg, F.N., Bell, A.R., Dangor, A.E., Danson, C.N., Fews, A.P., et al., (1997). A study of picosecond laser-solid interactions up to 10^{19} W cm^{-2}. Phys Plasmas 4, 447-457.

Blochin, M.A., (1957). Physik der Röntgenstrahlen. Akademischer Verlag, Berlin, pp. 70.

Borghesi, M., Fuchs, J., Bulanov, S.V., Mackinnon, A.J., Patel, P., & Roth, M. (2006). Fast ion generation by high intensity laser-irradiation of foils and applications. Fus. Sci Tech 49, 412-439.

Borghesi, M., Mackinnon, A.J., Campbell, D.H., Hicks, D.G., et al., (2004). Multi-MeV proton source investigations in ultraintense laser-foil interactions. Phys Rev Lett 92, 55003.

Busch, S., Schnurer, M., Kalashnikov, M., Schonnagel, H., Stiel, H., et al., (2003). Ion acceleration with ultrafast lasers. Appl Phys Lett 82, 3354-3356.

Clark, E.L., Krushelnick, K., Davies, J.R., Zepf, M., Tatarakis, et al., (2000a). Measurements of energetic proton transport through magnetized plasma from intense laser interactions with solids. Phys Rev Lett 84, 670-673.

Clark, E.L., Krushelnick, K., Zepf, M., Beg, F.N., Tatarakis, M., et al., (2000b). Energetic heavy ion and proton generation from ultraintense laser-plasma interactions with solids. Phys Rev Lett 85, 1654-1657.

Cowan, T.E., Fuchs, J., Ruhl, H., Kemp, A., Audebert, P., et al., (2004). Ultralow emittance, multi-MeV proton beams from a laser virtual-cathode plasma accelerator. Phys Rev Lett 92, 204801.

Denavit, J., (1992). Absorption of high-intensity subpicosecond lasers on solid density targets. Phys Rev Lett 69, 3052-3055.

Estabrook, K., & Kruer, W.L., (1978). Properties of resonantly heated electron distributions. Phys Rev Lett 40, 42-45.

Fews, A.P., Norreys, P.A., Beg, F.N., Bell, A.R., Dangor, A.E., et al., (1994). Plasma Ion emission from high-intensity picosecond laser-pulse interactions with solid targets. Phys Rev Lett 73, 1801-1804.

Forslund, D.W., & Brackbill, J.U., (1982). Magnetic-field-induced surface transport on laser-irradiated foils. Phys Rev Lett 48, 1614-1617.

Fuchs, J., Cowan, T.E., Audebert, P., Ruhl, H., Gremillet, L., et al., (2003). Spatial uniformity of laser-accelerated ultrahigh-current MeV electron propagation in metals and insulators. Phys Rev Lett 91, 255002.

Gitomer, S.J., Jones, R.D., Begay, F., Ehler, A.W., Kephart, J.F., & Kristal, R., (1986). Fast ions and hot-electrons in the laser-plasma interaction. Phys Fluids 29, 2679-2688.

Griem, H.R., (1964). Plasma Spectroscopy. McGraw-Hill, New York, pp. 113.

Hegelich, M., Karsch, S., Pretzler, G., Habs, D., Witte, et al., (2002). MeV ion jets from short-pulse-laser interaction with thin foils. Phys Rev Lett 89, 085002.

Hemberg, O., Hansson, B.A.M., Berglund, M., & Hertz, H.M., 2000. Stability of droplet-target laser-plasma soft x-ray sources. Journal of Applied Physics 88, 5421-5425.

Kalachnikov, M.P., Karpov, V., Schonnagel, H., & Sandner, W., (2002). 100-terawatt titanium-sapphire laser system. Laser Physics 12, 368-374.

Kaluza, M., Schreiber, J., Santala, M.I.K., Tsakiris, G.D., et al., (2004). Influence of the laser prepulse on proton acceleration in thin-foil experiments. Phys Rev Lett 93, 045003.

Karsch, S., Dusterer, S., Schwoerer, H., Ewald, F., Habs, D., et al., (2003). High-intensity laser induced ion acceleration from heavy-water droplets. Phys Rev Lett 91, 015001.

Kishimoto, Y., Mima, K., Watanabe, T., & Nishikawa, K., (1983). Analysis of fast-ion velocity distributions in laser plasmas with a truncated maxwellian velocity distribution of hot-electrons. Physics of Fluids 26, 2308-2315.

Kruer, W.L., & Estabrook, K., (1985). Jxb Heating by very intense laser-light. Physics of Fluids 28, 430-432.

Lefebvre, E., & Bonnaud, G., (1997). Nonlinear electron heating in ultrahigh-intensity-laser plasma interaction. Phys Rev E 55, 1011-1014.

Lindau, F., Lundh, O., Persson, A., McKenna, P., Osvay, K., et al., (2005). Laser-accelerated protons with energy-dependent beam direction. Phys Rev Lett 95, 175002.

Mackinnon, A.J., Sentoku, Y., Patel, P.K., Price, D.W., Hatchett, S., et al., (2002). Enhancement of proton acceleration by hot-electron recirculation in thin foils irradiated by ultraintense laser pulses. Phys Rev Lett 88, 215006.

Mackinnon, A.J., Borghesi, M., Hatchett, S., Key, M.H., Patel, P.K., et al., (2001). Effect of plasma scale length on multi-MeV proton production by intense laser pulses. Phys Rev Lett 86, 1769-1772.

Maksimchuk, A., Gu, S., Flippo, K., Umstadter, D., & Bychenkov, V.Y., (2000). Forward ion acceleration in thin films driven by a high-intensity laser. Phys Rev Lett 84, 4108-4111.

Malka, G., & Miquel, J.L., (1996). Experimental confirmation of ponderomotive-force electrons produced by an ultrarelativistic laser pulse on a solid target. Phys Rev Lett 77, 75-78.

Manclossi, M., Santos, J.J., Batani, D., Faure, J., Debayle, A., et al., (2006). Study of ultra-intense laser-produced fast-electron propagation and filamentation in insulator and metal foil targets by optical emission diagnostics. Phys Rev Lett 96, 125002.

McKenna, P., Carroll, D.C., Clarke, R.J., Evans, R.G., Ledingham, K.W.D., et al., (2007). Lateral electron transport in high-intensity laser-irradiated foils diagnosed by ion emission. Phys Rev Lett 98, 145001.

Mora, P., (2003). Plasma expansion into a vacuum. Phys Rev Lett 90, 185002.

Nakamura, T., Kato, S., Nagatomo, H., & Mima, K., (2004). Surface-magnetic-field and fast-electron current-layer formation by ultraintense laser irradiation. Phys Rev Lett 93, 265002.

Nickles, P.V., Ter-Avetisyan, S., Schnuerer, M., Sokollik, T., Sandner, W., et al., (2007). Review of ultrafast ion acceleration experiments in laser plasma at Max Born Institute. Laser Part Beams 25, 347-363.

Nishihara, K., Amitani,H., Murakami,M., Bulanov, S.V. & Esirkepov, T.Zh., (2001). High energy ions generated by laser driven Coulomb explosion of cluster. Nucl Instrum and Meth A 464, 5.

Passoni, M., Tikhonchuk, V.T., Lontano, M., & Bychenkov, V.Y., (2004). Charge separation effects in solid targets and ion acceleration with a two-temperature electron distribution. Phys Rev E 69, 026411.

Pukhov, A., Sheng, Z.M., & Meyer-ter-Vehn, J., (1999). Particle acceleration in relativistic laser channels. Phys Plasmas 6, 2847-2854.

Pukhov, A., (2001). Three-dimensional simulations of ion acceleration from a foil irradiated by a short-pulse laser. Phys Rev Lett 86, 3562-3565.

Roth, M., Brambrink, E., Audebert, P. et al., (2005). The generation of high-quality, intense ion beams by ultra-intense lasers. Plasma Phys Control Fusion 47, B841-B850.

Schnurer, M., Ter-Avetisyan, S., Nickles, P.V., & Andreev, A.A., (2007). Influence of target system on the charge state, number, and spectral shape of ion beams accelerated by femtosecond high-intensity laser pulses. Phys Plasmas 14, 033101.

Schreiber, J., Ter-Avetisyan, S., Risse, E., Kalachnikov, M.P., Nickles, P.V.,et al., (2006). Pointing of laser accelerated proton beams. Phys Plasmas 13, 033111.

Schwoerer, H., Pfotenhauer, S., Jackel, O., Amthor, K.U., Liesfeld, B., et al., (2006). Laser-plasma acceleration of quasi-monoenergetic protons from microstructured targets. Nature 439, 445-448.

Shorokhov, O., & Pukhov, A., (2004). Ion acceleration in overdense plasma by short laser pulse. Laser Part Beams 22, 175-181.

Silva, L.O., Marti, M., Davies, J.R., Fonseca, R.A., Ren, C., Tsung, F.S., & Mori, W.B., (2004). Proton shock acceleration in laser-plasma interactions. Phys Rev Lett 92, 015002.

Snavely, R.A., Key, M.H., Hatchett, S.P., Cowan, T.E., et al., (2000). Intense high- energy proton beams from petawatt-laser irradiation of solids. Phys Rev Lett 85, 2945-2948.

Sokollik, T., Schnurer, M., Ter-Avetisyan, S., et al., (2008). Transient electric fields in laser plasmas observed by proton streak deflectometry. Appl Phys Lett 92, 091503.

Stein, J., Fill, E., Habs, D., Pretzler, G., & Witte, K., (2004). Hot electron diagnostics using X-rays and Čerenkov radiation. Laser Part Beams 22, 315-321.

Tajima, T., & Dawson, J.M., (1979). Laser Electron-Accelerator. Phys Rev Lett 43, 267-270.

Ter-Avetisyan, S., Schnurer, M., Stiel, H., & Nickles, P.V., (2003). A high-density sub-micron liquid spray for laser driven radiation sources. J Phys D: Appl Phys 36, 2421-2426.

Ter-Avetisyan, S., Schnurer, M., Busch, S., Risse, E., et al., (2004a). Spectral dips in ion emission emerging from ultrashort laser-driven plasmas. Phys Rev Lett 93, 155006.

Ter-Avetisyan, S., Schnürer, M., Busch, S., & Nickles, P.V., (2004b). Negative ions from liquid microdroplets irratiated with ultrashort and intense laser pulses. J Phys B: At Mol Opt Phys 37, 3633 - 3640.

Ter-Avetisyan, S., Schnurer, M., & Nickles, P.V., (2005). Time resolved corpuscular diagnostics of plasmas produced with high-intensity femtosecond laser pulses.J Phys D: Appl Phys 38, 863-867.

Ter-Avetisyan, S., Schnurer, M., Nickles, P.V., Sokollik, T., Risse, E., et al., (2008). The Thomson deflectometer: A novel use of the Thomson spectrometer as a transient field and plasma diagnostic. Rev Sci Instrum 79, 033303.

Ter-Avetisyan, S., Schnurer, M., Nickles, P.V., Sandner, W., Nakamura, T., & Mima, K., (2009a). Correlation of spectral, spatial, and angular characteristics of an ultrashort laser driven proton source. Phys Plasmas 16, 043108.

Ter-Avetisyan, S., Ramakrishna, B., Doria, D., Sarri, G., Zepf, M., et al., (2009b). Complementary ion and extreme ultra-violet spectrometer for laser-plasma diagnosis. Rev Sci Instrum 80, 103302.

Tikhonchuk, V.T., (2002). Interaction of a beam of fast electrons with solids. Phys Plasmas 9, 1416-1421.

Wei, M.S., Mangles, S.P.D., Najmudin, Z., Walton, B., Gopal, A., et al., (2004). Ion acceleration by collisionless shocks in high-intensity-laser-underdense-plasmainteraction. Phys Rev Lett 93, 155003.

Weibel, E.S., (1959). Spontaneously growing transverse waves in a plasma due to an anisotropic velocity distribution. Phys Rev Lett 2, 83-84.

Wickens, L.M., Allen, J.E., & Rumsby, P.T., (1978). Ion emission from laser-produced plasmas with 2 electron temperatures. Phys Rev Lett 41, 243-246.

Wickens, L.M., & Allen, J.E., (1979). Free expansion of a plasma with 2 electron temperatures. J Plasma Phys 22, 167-185.

Wickens, L.M., & Allen, J.E., (1981). Ion emission from laser-produced, multi-ion species, 2-electron temperature plasmas. Phys Fluids 24, 1894-1899.

Wilks, S.C., Kruer, W.L., Tabak, M., & Langdon, A.B., (1992). Absorption of ultra-intense laser-pulses. Phys Rev Lett 69, 1383-1386.

Wilks, S.C., Langdon, A.B., Cowan, T.E., Roth, M., Singh, M., et al., (2001). Energetic proton generation in ultra-intense laser-solid interactions. Phys Plasmas 8, 542-549.

Zepf, M., Clark, E.L., Beg, F.N., Clarke, R.J., Dangor, A.E., et al., (2003). Proton acceleration from high-intensity laser interactions with thin foil targets. Phys Rev Lett 90, 064802

Zhidkov, A.G., Sasaki, A., Fukumoto, I., Tajima, T., Auguste, T., et al., (2001). Pulse duration effect on the distribution of energetic particles produced by intense femtosecond laser pulses irradiating solids. Phys Plasmas 8, 3718-3723.

Zhidkov, A., Uesaka, M., Sasaki, A., & Daido, H., (2002). Ion acceleration in a solitary wave by an intense picosecond laser pulse. Phys Rev Lett 89, 215002.

19

Picosecond Laser Pulse Distortion by Propagation through a Turbulent Atmosphere

Josef Blazej, Ivan Prochazka and Lukas Kral
Czech Technical University in Prague
Czech Republic

1. Introduction

We have been investigating the influence of atmospheric turbulence on the propagation of a picosecond laser pulse. The figure of merit of presented results is the time of propagation, its absolute delay and jitter. Phase wavefront deformation or beam profile changes were not studied. The correlation of the atmospheric turbulence with the propagation delay fluctuation was measured. The research was motivated by the needs of highly precise laser ranging of ground, air, and space objects; and highly precise and accurate time transfer ground-to-space and ground-to-ground by means of picosecond optical laser pulse.

Firstly for comparison, lets briefly summarize the effects of a turbulent atmosphere to continuous laser beam. The total effect of atmospheric turbulences on a continuous laser beam propagation is a highly complex subject. Atmospheric turbulences can be defined as random spatial variations in the refraction index of the atmosphere resulting in a distortion of the spatial phase fronts of the propagating signal. Spatial phase front distortion induces the variable path of light energy and thus all effects described later on. Variations of the refraction index are caused by the turbulent motion of the atmosphere due to the variations in temperature and gradients in the water vapour. Following (Degnan, 1993), the optically turbulent atmosphere produces three effects on low power laser beams: 1) beam wander, 2) beam spread and 3) scintillations. Severe optical turbulence can result in a total beam break-up. Beam wander refers to the random translation of the spatial centroid of the beam and is generally caused by the larger turbulent eddies through which the beam passes. In astronomical community it is usually referred as seeing. Beam spread is a short term growth in the effective divergence of the beam produced by smaller eddies in the beam path. The two effects are often discussed together in terms of a "long term" and "short term" beam spread. The "long term" beam spread includes the effects of beam wander, whereas the "short term" beam spread does not. For more details, see (Degnan, 1993). Maximum turbulence occurs at mid-day in the desert (low moisture) under clear weather conditions. For the usual laser wavelength of 532 nm one can expect 2.4-4.6 cm for the coherence length at zenith angles of 0° and 70° respectively. At the tripled Nd:YAG wavelength (355 nm) the corresponding values are 3.1 and 1.6 cm (Degnan, 1993). Turbulence induced beam spreading will only have a significant impact on beam divergence (and hence signal level) if the coherence length is on the order of, or smaller than, the original effective beam waist radius. Since a typical 150 µrad beam implies an effective waist radius of 2.26 mm, the effect of beam spread on signal level for such systems is relatively small, i.e. a few percent.

Atmospheric turbulence produces a fluctuation in the received intensity at a point detector. During satellite laser ranging aperture averaging, which occurs at both the target retro-reflectors and at the ground receiving telescope, tends to reduce the magnitude of the fluctuations. Thus the round trip propagation geometry must be considered when evaluating theoretical scintillation levels. The effect of scintillation is significant under conditions of strong turbulence.

In contrast with above mentioned, we have been investigating the influence of atmospheric turbulence on the propagation of a picosecond laser pulse. In this case, the fluctuation should be not described as a coherence length, but typically as a time jitter of absolute delay of laser pulse propagated trough the atmosphere. The research was motivated by the needs of highly precise laser ranging of ground, air and space objects. The ground targets laser ranging with picosecond single shot resolution revealed the fact, that the resulting precision in influenced, among others, by the atmospheric index of refraction fluctuations. The influence of the atmospheric refraction index fluctuations on the star image is known for a long time, it is called seeing (Bass, 1992). It has been studied for more than a century. It represents a serious limitation in the astronomical images acquisition. The angular resolution of large astronomical telescope is limited by the seeing, its influence is much larger in comparison to a diffraction limit. Recently, numerous techniques exist for seeing compensation by means of adaptive optics (Roddier, 1998) active and nowadays also passive.

The interesting point of view is the comparison of propagation delay between microwave and optical region. Due to the refractive index and its variations within the troposphere, the microwave signal is also naturally delayed as the optical laser pulse propagated. Typically, the total delay of the radio signal is divided into "hydrostatic" and "wet" components. The hydrostatic delay is caused by the refractivity of the dry gases in the troposphere and by the nondipole component of the water vapour refractivity. The main part, or about 90 % of the total delay, is caused by the hydrostatic delay and can be very accurately predicted for most of the ranging applications using surface pressure data. The dipole component of the water vapour refractivity is responsible for the wet delay and amounts to about 10% of the total delay. This corresponds to 5-40 cm (above 1 ns) for the very humid conditions. The mapping function is used to transform the zenith troposphere delay to the slant direction. In recent years, the so-called Niell Mapping Function served as a standard for processing microwave measurements. It was built on one year of radiosonde profiles primarily from the northern hemisphere (Niell, 1996). Compared to the microwave technique, the main advantages of the SLR measurements are the insensitivity to the first and higher order ionospheric propagation effects, and the relatively high accuracy with which water vapour distribution can be modelled. Ions are too heavy and sluggish to respond to optical frequencies in the 300 to 900 THz band. Laser wavelengths in the visible and ultraviolet bands are typically far from strong absorption feature in the water vapour spectrum. Signal delay due to the water vapour in atmosphere is significantly different in the optical versus the microwave band. The ratio is about 67, meaning that the typical "wet component" in the zenith direction of about 5-40 cm (above 1 ns) for the microwave band (GPS) corresponds to the delay of about 0.1-0.6 cm (2 ps) for optical band. Since the effect is relatively small, about 80 % of the delay can be modelled by means of surface pressure, temperature and humidity measured on the station. Recently GNSS-based measurements offered new and promising possibilities, the global IGS network and dense regional GNSS networks developed all around the world provide high temporal information on the integrated atmospheric water vapour.

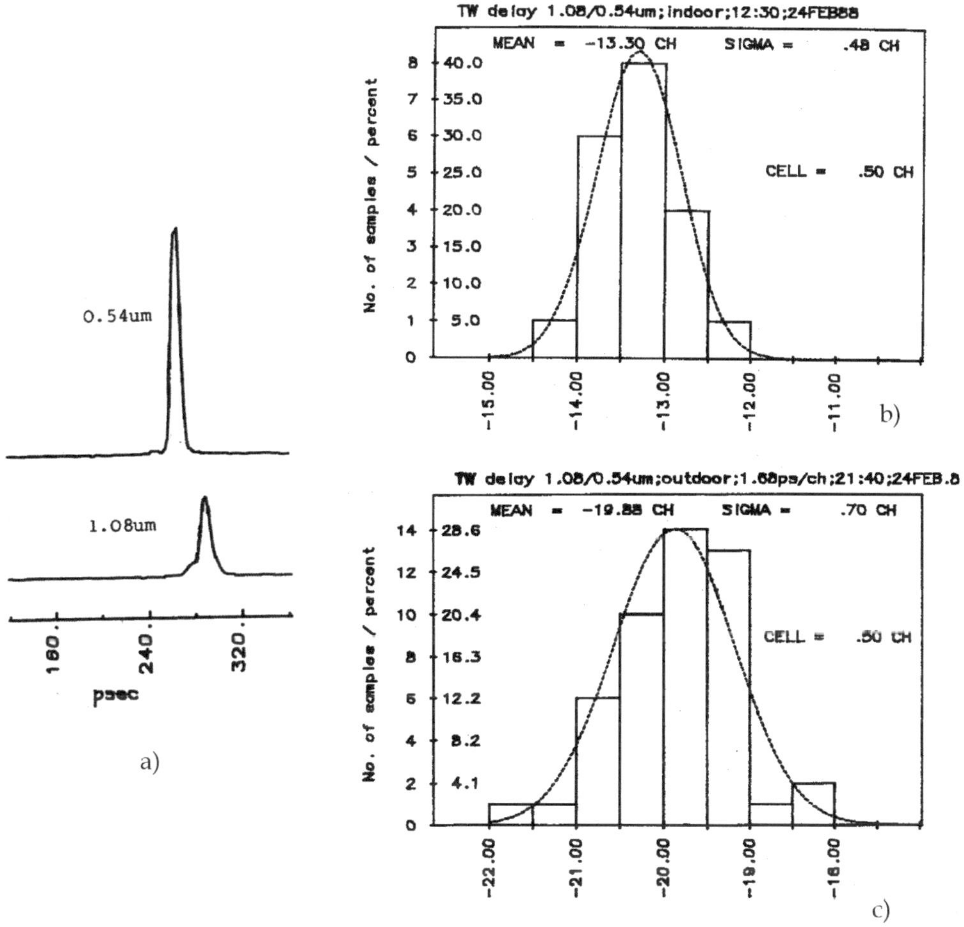

Fig. 1. Results of two wavelength ranging experiment (Hamal et al., 1988); pulse temporal profiles recorded by the linear streak camera (a) and delay histograms from indoor (b) and outdoor (c) ranging.

In contrast to astronomical imaging trough turbulent atmosphere, the picosecond pulse propagation and its distortion in a time domain has been studied just recently, once the picosecond lasers, detection and timing techniques became available. The effect has been observed for the first time (Hamal et al., 1988), when laser pulses 10 picoseconds long at the wavelength of 1079 nm and 539 nm were propagated different atmospheric path length two way, see figure 1. The pulses were transmitted simultaneously using passively mode locked Nd:YAP laser, part of the energy was converted to the second harmonic, pulses were propagated to the ground target formed by corner cube retroreflector at distances ranging from 1 to 200 meters. The returned optical signal was analysed using a linear streak camera. The streak camera together with image processing enabled to monitor simultaneously the returned signal beam direction fluctuations and fluctuations of the time interval between the two wavelength pulses. The timing resolution of the technique was high – typically 0.5 picosecond. The experiments showed the dependence of the pulse propagation delay fluctuation on both propagation distance and atmospheric fluctuation conditions. The

propagation delay fluctuations caused by the turbulent atmosphere were in the range of 0 to 1.5 ps for the propagation length 1 to 200 meter two way.
The experiment described above provided encouraging results, however, the technique (Hamal et al., 1988) was not suitable for routine measurements over longer baselines.

2. Theoretical models

The atmospheric turbulence - mixing of air of different temperatures, which causes random and rapidly changing fluctuations of air refractive index and hence unpredictable fluctuations from standard models of atmospheric range correction. We tried to estimate the atmospheric contribution to the ranging jitter using

1. an existing numerical modeling code (physical optics approach)
2. an analytical model developed by C. S. Gardner (geometric optics approach).

We used the commercial version of the General Laser Analysis and Design (GLAD) code (AOR, 2004). GLAD is an extensive program for modelling of diffractive propagation of light through various media and optical devices. The light is considered to be monochromatic and coherent (or partially coherent). The electromagnetic field in GLAD is described by its two-dimensional transversal distribution. Two arrays of complex numbers (one for each polarization state) represent the intensity and phase at each point in x and y axis. The propagation is done by the angular spectrum method. That means the field distribution is decomposed into a summation of plane waves, these plane waves are propagated individually and then resumed into resulting distribution. A user specifies a starting distribution at first and then applies aberrations, apertures, etc., and finally performs diffractive propagation of the distribution to some distance. At the end, the resulting distribution can be analysed. Using GLAD, we developed a model of atmospheric light propagation according to recommendations in GLAD Theoretical Description (AOR, 2004). It consists of alternating steps of random aberration and diffractive propagation applied to the initial plane wave.
After many attempts with different input parameters this model gives always pathlength RMS only several micrometers, i.e. negligible. What is even more surprising, the computed pathlength RMS does not significantly increase with L_0, as was expected from theory, although the wavefront size was always selected large enough ($10 \times L_0$) to model even the lowest-frequency aberrations. Therefore we have found this model not well describing the satellite laser ranging signal delay although the far field intensity profile has been modeled correctly. The origin of the problem has not been identified. The GLAD atmospheric model and its results correspond well to the "adaptive optics problem"; the corrections applied in adaptive optics are of the order of micrometers, just the values predicted by the model.
It is interesting discrepancy between wavefront shift necessary to correct the beam position and absolute propagation delay even of corrected laser beam.
In ref. (Gardner, 1976) derived analytical formulae that allow us to predict the turbulence--induced random pathlength fluctuations, directly for the case of satellite laser ranging, or generally for propagation delay. He also computed some concrete results and predicted that the RMS path deviations could reach millimeters, and at some extreme situations even several centimeters. However, Gardner used a very rough model of C_n^2 height dependence, which resulted in larger values of C_n^2 than are recently observed. We evaluated the Gardner's formulae using the recent model of C_n^2 height profile. For ground-to-space paths, we have selected the Hufnagel-Valley (Bass, 1992) model. This approach is predicting

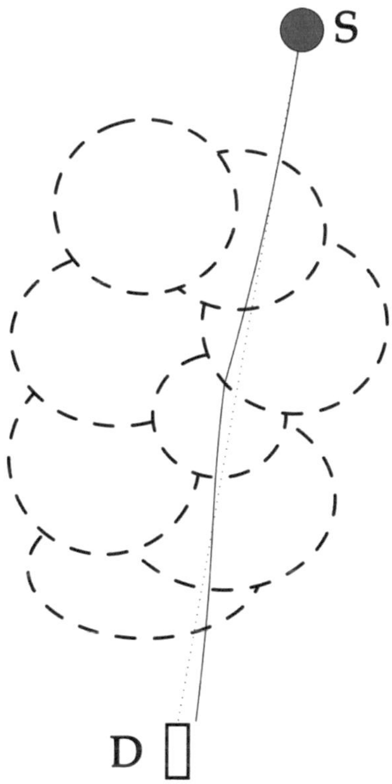

Fig. 2. The ideal (dotted) and real (solid) path of laser beam from source S (retro-reflector, start, artificial star) to detector.

realistic values of the atmospheric seeing induced range fluctuation of the order of millimeters.

It allows us to predict the turbulence-induced random fluctuations of optical path length, i.e. the turbulence-induced ranging jitter:

$$\sigma_{turb}^2 = 26.3 \cdot C_n^2(\xi = 0) \cdot L_0^{5/3} \cdot L_e \tag{1}$$

(eq. 20 in the Gardner's article, using the Greenwood-Tarazano spectral model of turbulence). σ_{turb} is the turbulence-induced ranging jitter, $C_n^2(\xi = 0)$ is turbulence strength at the beginning of the beam path (ξ is the distance from the observatory measured along the beam propagation path), L_0 is the turbulence outer scale (must be estimated) and L_e is effective pathlength given by

$$L_e = \frac{1}{C_n^2(\xi = 0)} \int_0^L C_n^2(\xi) d\xi \tag{2}$$

That means if we want to predict the turbulence-induced ranging jitter on a given path, we have to know integral of the turbulence strength C_n^2 along the path, and the outer scale L_0. The integral can be determined from measurement of astronomical seeing (FWHM of long exposure stellar image profile). To derive the relation between seeing and turbulence-induced ranging jitter, we used the two following relations

$$FWHM = \frac{\lambda}{r_0} \tag{3}$$

$$r_0 = 2.1 \cdot \left[1.46 \cdot k^2 \int_0^L C_n^2(\xi) \left(\frac{L-\xi}{L} \right)^{5/3} d\xi \right]^{-3/5} \tag{4}$$

where *FWHM* represents the value of seeing, r_0 is Fried's parameter, λ is wavelength of the seeing measurement, k is optical wavenumber equal to $2\pi/\lambda$, and L is one-way target distance. Using these relations, we were able to derive a relation allowing us to predict the turbulence-induced ranging jitter from the seeing measurement:

$$\sigma_{turb} = 1.28 \cdot L_0^{5/6} \cdot \lambda^{1/6} \cdot FWHM^{5/6} \tag{5}$$

for a slant path to space, and

$$\sigma_{turb} = 2.11 \cdot L_0^{5/6} \cdot \lambda^{1/6} \cdot FWHM^{5/6} \tag{6}$$

for a horizontal path. In the case of slant path to space, a star located at the same elevation as the ranging target can be used to measure the seeing FWHM. In the case of horizontal path, a ground-based point light source can be used, located in the same direction and the same distance as the ranging target (otherwise a correction for different distances must be applied).

3. Experimental setup

The experimental part was carried out at the Satellite Laser Ranging (SLR) station in Graz, Austria. The site is located 400 meters above the sea level. The laser ranging system consists of Nd:YAP diode-pumped laser with second harmonic generation (wavelength 532 nm, pulse width 8 ps), 10 cm transmitter telescope and 50 cm receiver telescope. The echo signal is detected by C-SPAD (Kirchner et al., 1997) (single photon avalanche detector with time walk compensation) and the time intervals are measured using event timer ET (Kirchner, Koidl, 2000). The laser operates at 2 kHz repetition rate, giving us sufficient sampling rate for the atmospheric influence investigation. The single shot precision of the whole system is 1 mm RMS (tested by ground target ranging). Such high repetition rate and ranging precision were necessary for the investigation of the turbulence influence, since the expected turbulence-induced jitter was of the order of one millimeter (maximum) and the fluctuation frequencies were expected up to 1 kHz.

We used three different types of laser ranging ground-based cube-corner retroreflector, a mobile retroreflector mounted on an airplane, and Earth orbiting satellites equipped by corner cube retroreflectors, see figure 3. In parallel, the atmospheric seeing was measured for a horizontal path of 4.3 km and a star in elevation close the satellite path. The standard Differential Image Motion Monitor (DIMM) technique (Beaumont, 1997) was employed.

The ground-based target was a cube-corner retroreflector mounted on a mast located 4.3 kilometers from the observatory. The laser beam path was horizontal and led over a hilly terrain covered with forests and meadows, with average height above the surface about 50 meters.

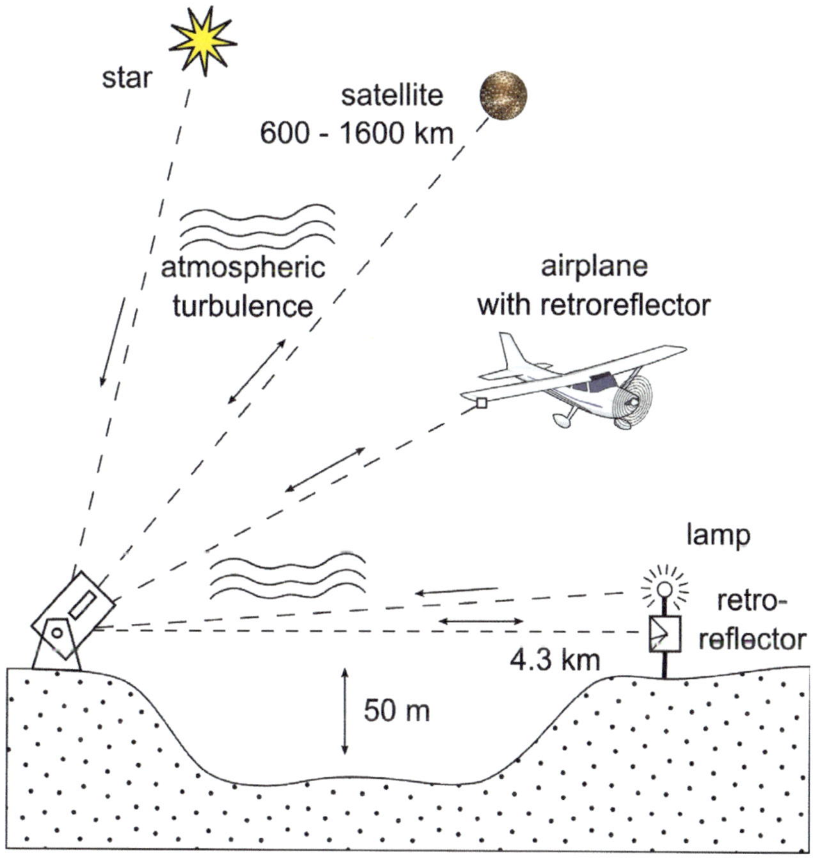

Fig. 3. Laser ranging to different targets and simultaneous seeing measurement to monitor atmospheric condition.

In the case of satellite ranging, we selected two satellites with low signature (not spreading the laser pulse in time) and high return energy, which leads to the best achievable ranging precision: ERS-2 and Envisat. We analyzed selected segments of their passes corresponding to different elevation above the horizon. Unfortunately, the laser ranging to an airplane based retro reflector did not provide sufficiently high quality data due to difficulties of optical tracking of such a target.

The typical measurement series consisted of hundred thousand of range measurements, normally distributed around the mean value. Since not every returns came from the retro (noise, prepulses), the typical sampling rate was around 1 kHz. This means the dataset covered about 100 seconds in time. However, the jitter of the measured range was sum of the instrumental jitter (stop detector, electronics etc.) and the turbulence-induced jitter:

$$\sigma^2 = \sigma_{inst}^2 + \sigma_{turb}^2 \quad (7)$$

Thus we had to extract the pure turbulence contribution σ^2_{turb} from the overall jitter σ (sigma denotes standard deviation). We took advantage of the knowledge that the instrumental jitter is completely random from shot to shot (behaves as white noise), whereas the atmospheric fluctuations are typically correlated over several neighboring shots (their time spectrum is spread from 0 to some maximum frequency fmax, lower than the sampling frequency of 1 kHz).

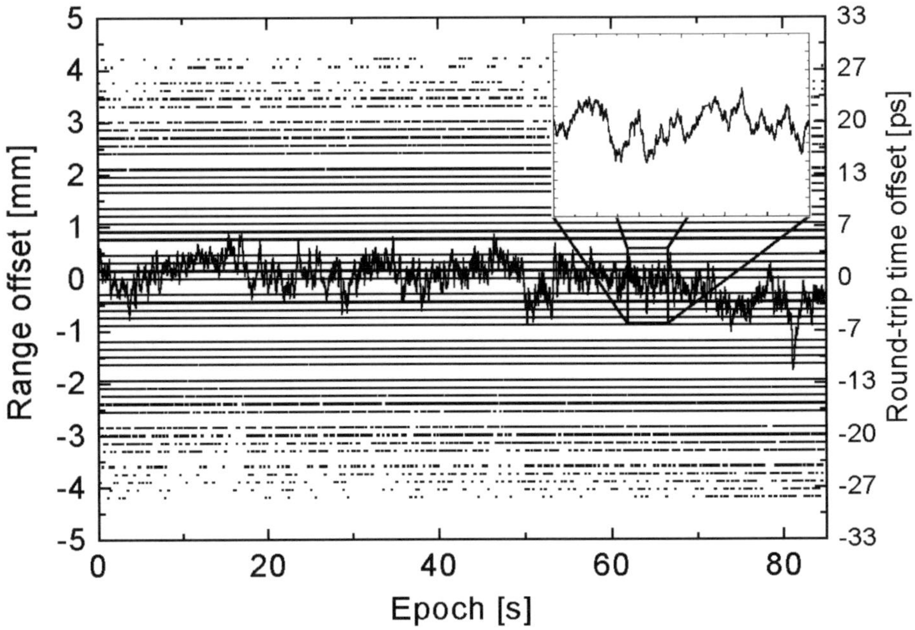

Fig. 4. Example of 4.3 km distant ground target ranging data (points) and 200-point moving average (line). Note the relatively fast turbulent fluctuations, and the long-term trend, later removed by polynomial fitting. The "pixelation" of orig. data is caused by a rounding of non-integer resolution of event timer – 1.2 ps.

If we compute averages from every N_a points of the dataset (normal points analogy from satellite laser ranging), the instrumental jitter will decrease root-square-of-N_a times, whereas the jitter of turbulence-induced fluctuations will remain the same, if the averaging interval will not be too wide. This is a similar situation to a sine wave combined with random noise – if the averaging interval will be shorter than approximately ¼ of the sine period, the sine wave will not be influenced by the averaging, whereas the random noise will be lowered. Now we can write an equation for the jitter σ_{avg} after the averaging:

$$\sigma_{avg}^2 = \left(\frac{\sigma_{inst}}{\sqrt{N_a}} \right)^2 + \sigma_{turb}^2 \tag{8}$$

now we have two equations (1) and (2) for two unknown variables σ_{turb} and σ_{inst}. Hence, the result will be

$$\sigma_{turb} = \sqrt{\frac{N_a \sigma_{turb}^2 - \sigma^2}{N_a - 1}} \tag{9}$$

This is the way, how to find out the pure turbulence contribution σ_{turb} to the overall ranging jitter σ.

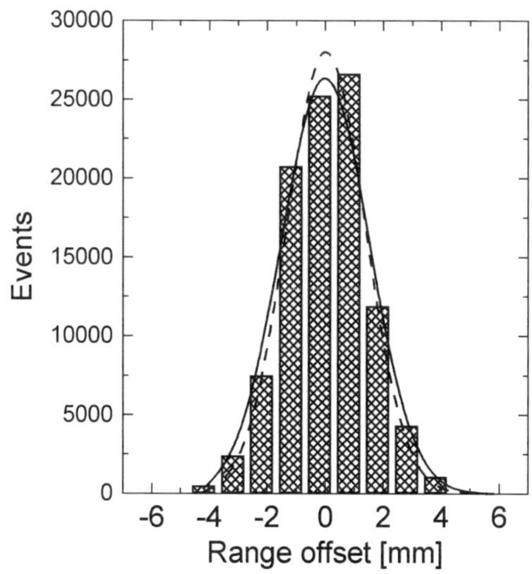

Fig. 5. Histogram from the dataset plotted at left figure 3. The solid line is gaussian fit for 3σ data editing criterion (RMS 1.4 mm), the dashed line is fit for 2.2σ criterion (RMS 1.2 mm).

However, there still remains the task to find out the maximum possible averaging interval Na. From above, the time interval corresponding to N_a must be shorter than approximately ¼ of the period of the fastest atmospheric fluctuation. Considering the spectral distribution of atmospheric fluctuations (Kral et al., 2006) the value $N_a = 3$ was used. The long-term trends in ranging data, caused by slow temperature and pressure changes during the measurement, see figure 4, were removed by polynomial fitting and computing of the residuals before further analysis, see figure 5. For the time spectrum see figure 6. Data was measured at SLR station Graz on May 10, 2004.

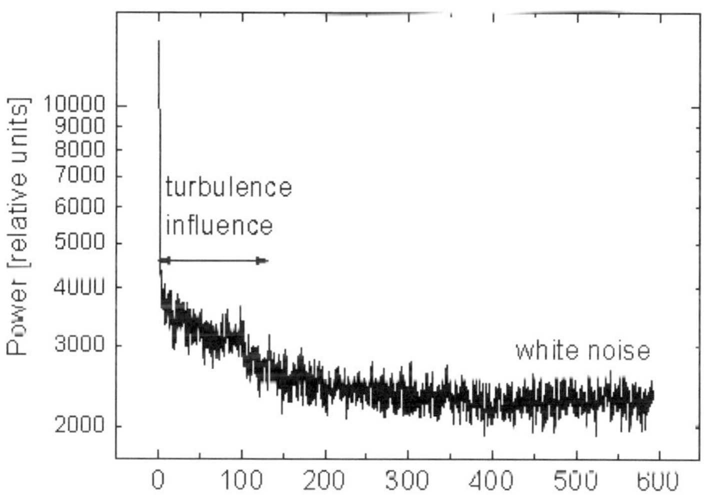

Fig. 6. Typical time spectrum of the fluctuations of measured range of the ground target. The turbulence significantly contributes at lower frequencies up to approx. 130 Hz. The sampling rate was 1.2 kHz. The same data like in figure 4.

4. Results

The computed values of the atmospheric turbulence contribution to the laser ranging fluctuation are summarized on figure 7 and 8. The figure 7 corresponds to the horizontal beam propagation, the figure 8 corresponds to the slant path to space for elevation range between 15 and 65 degrees. The measured values - filled squares - are plotted over the theoretical curves computed for different values of the outer scale parameter L_0.

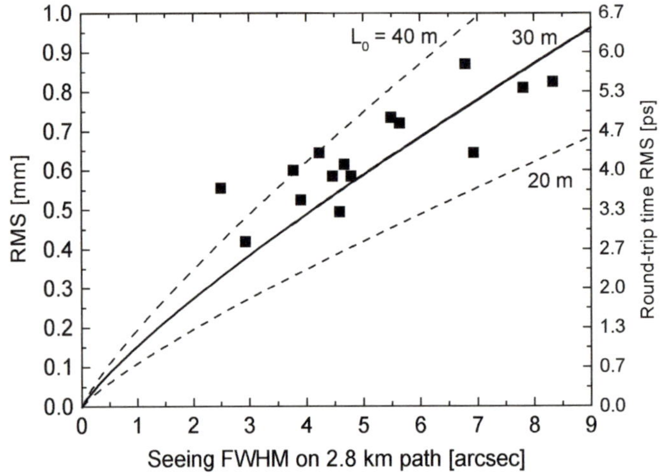

Fig. 7: The turbulence-induced ranging jitter as a function of turbulence strength (measured by the seeing). The graph was constructed from measurements of the 4.3 km distant ground target (horizontal path), taken under various meteorological conditions.

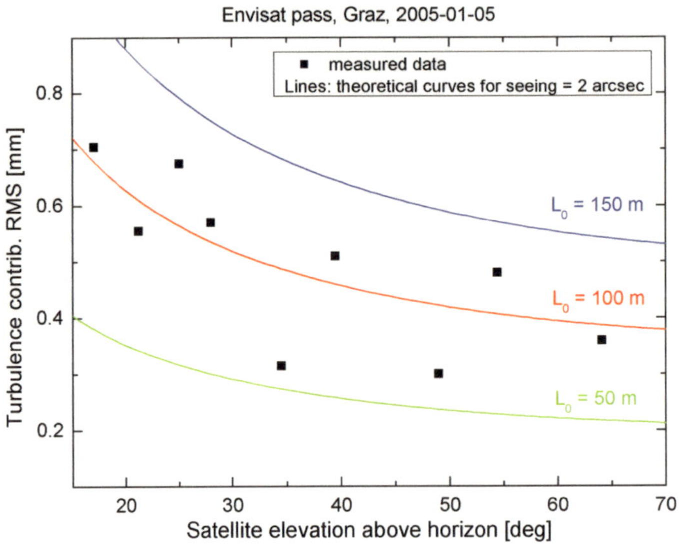

Fig. 8. The turbulence-induced ranging jitter as a function of satellite elevation. The graph was constructed from satellite measurements by slant path to space.

From these two figures one can conclude, that the values of 30 meters and 100 meters fit best the measured values for the horizontal and slant path to space respectively. This is a first experimental determination of the outer scale parameter. The outer scale L_0 is key to measure, and still not well understood. By measurement of seeing parameter together with determination of the laser ranging jitter from satellite laser ranging data, the outer scale L_0 can be determined. However, to carry out such a measurement, the high repetition rate laser ranging system (2 kHz rate is a minimum) with (sub) millimeter single shot instrumental ranging precision is required. These are quite challenging system requirements.

5. Future outlook

As it was described in the previous chapters, the instrumental precision of the laser ranging system is a key to the atmospheric turbulence influence on the laser pulse propagation studies. Recently, new technologies are emerging and becoming available, which will improve the instrumental resolution of the laser ranging chain, namely new timing systems and improved echo signal detectors.

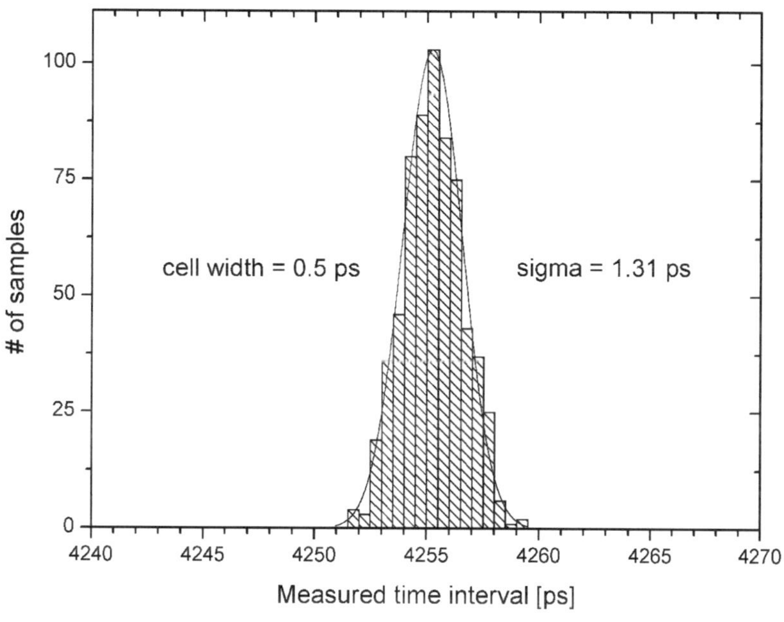

Fig. 9. N-PET timing device temporal resolution, two channel cable delay test.

The new sub-picosecond resolution event timing (N-PET) system has been developed by our group (Panek & Prochazka, 2007). It provides the single shot timing resolution of 920 femtoseconds per channel, see figure 9, and excellent timing linearity and temporal stability, see figure 10, of the order of hundreds of femtoseconds.

This novel timing system has been tested at the laser ranging facility in Graz and provided better instrumental resolution of the system along with ranging data distribution closed to the normal one.

The second key contributor to the instrumental resolution limitation is the echo signal detector. The avalanche photodiode based detector operating in the single and multi-photon counting regime is routinely used (Prochazka et al., 2004). Recent achievements in the detector chip signal processing (Blazej & Prochazka, 2008) will enable to lower the error correlated with the signal strength fluctuation and hence further improve the instrumental resolution namely for ranging to space targets.

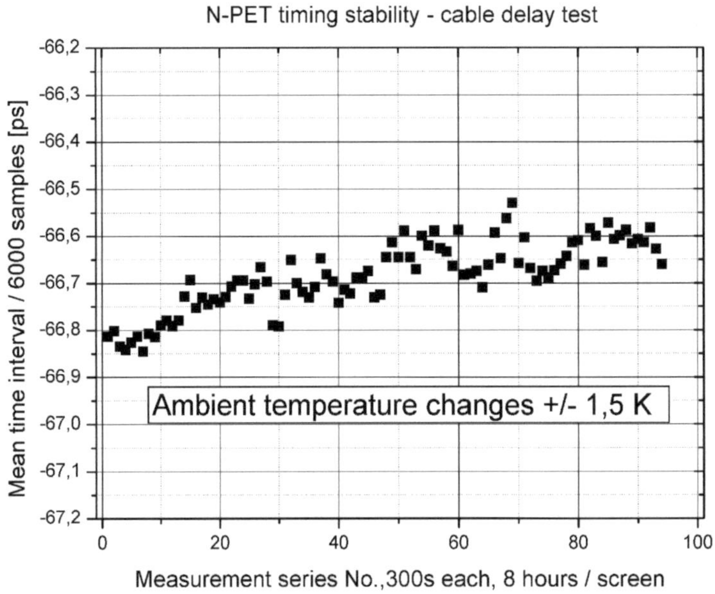

Fig. 10. N-PET timing device temporal stability.

6. Conclusion

We are presenting the results of the studies related to propagation of ultrashort optical pulse through the turbulent atmosphere. Three independent types of path configurations have been studied: horizontal path, slant path at elevation 10 – 80 degrees to a flying target and slant path from ground to space. The correlation of the atmospheric turbulence with the propagation delay fluctuation was measured. The appropriate theoretical model was found and matched to the experimental results. The entirely different approach in comparison to adaptive optics was developed to describe the effect. The experiments described enabled us for the first time to determine the outer scale parameter L_0 on the basis of direct measurement. The recent achievements in the field of pulsed lasers, fast optical detectors and timing systems enable us to resolve the effects of propagation differences monitoring on the level of units of picosecond propagation time. Additionally, new techniques of optical receivers signal processing give a way to distinguish the atmospheric fluctuations contribution from the energy dependent detection delay effects.

7. Acknowledgement

Authors would like to express their thanks to Georg Kirchner and Franz Koidl, Graz SLR station. This research has been supported by the Research framework of Czech Ministry of Education № MSM6840770015.

8. References

Applied Optics Research (AOR), (2004). *GLAD Users Guide*, ver. 4.5, Applied Optics Research, Washington.

Applied Optics Research (AOR), (2004). *GLAD Theoretical Description*, ver. 4.5, Applied Optics Research, Washington.

Bass, M. (1992). *Handbook of Optics*, McGraw-Hill Professional, ISBN 978-0070477407, vol. 1, Chapter 44.

Beaumont, H. et al. (1997). Image quality and seeing measurements for long horizontal overwater propagation. *Pure Applied Optics*, Vol. 6, pp. (15–30), ISSN 0963-9659.

Blazej, J., Prochazka, I., (2008). Photon number resolving detector in picosecond laser ranging and time transfer in space, *Technical Digest «Modern problems of laser physics»*, p.183., Novosibirsk, Russia, August 24 - 30 2008, SB RAS, Novosibirsk, Russia.

Degnan, J. J. (1993). Millimeter Accuracy Satellite Laser Ranging: A Review, *Contributions of Space Geodynamics: Technology*, D. E. Smith and D. L. Turcotte (Eds.), AGU Geodynamics Series, Vol. 25, pp. (133-162). [online] <http://ilrs.gsfc.nasa.gov/reports/degnan/index.html>

Gardner, C. S. (1976). Effects of random path fluctuations on the accuracy of laser ranging systems, *Applied Optics*, Vol. 15, No. 10, p. 2539, ISSN 0003-6935.

Hamal, K., Prochazka, I., Schelev, M., Lozovoi, V., Postovalov, V. (1988). Femtosecond Two wavelength Laser Ranging to a Ground Target, Proceedings of SPIE Vol. 1032, pp. 453-456, ISBN 9780819400673, Xian, People's Republic of China, August 28 - September 2 1988, SPIE, Bellingham, WA, USA.

Kirchner, G., Koidl, F., Blazej, J., Hamal, K., Prochazka, I. (1997). Time-walk-compensated SPAD: multiple-photon versus single-photon operation, *Proceedings of SPIE Vol. 3218*, p. 106. ISBN 9780819426505, London, UK, September 24 - 26 1997, SPIE, Bellingham, WA, USA.

Kirchner, G., Koidl, F. (2000). Graz Event Timing system E. T., *Proceedings of the 12th International Workshop on Laser Ranging*, Matera, Italy, November 13 - 17, 2000 [online] <http://geodaf.mt.asi.it/html_old/news/iwlr/Kirchner_et_Koidl_ET.pdf>, ILRS, USA.

Kral, L., et al., (2006). Random fluctuations of optical signal path delay in the atmosphere, *Proceedings of SPIE 6364*, p. 0M, ISBN: 9780819464590, Stockholm, Sweden, 11 September 2006, SPIE, Bellingham, WA, USA.

Niell, A. E. (1996). Global mapping functions for the atmosphere delay at radio wavelengths, *Journal of Geophysical Research*, Vol. 101, No. B2, pp. (3227-3246), ISSN 0148-0227.

Panek, P., Prochazka, I. (2007). Time interval measurement device based on surface acoustic wave filter excitation, providing 1 ps precision and stability, *Review of Scientific Instruments*, Vol. 78, No. 9, pp. (78-81), ISSN 0034-6748.

Prochazka, I., Hamal, K., Sopko, B. (2004). Recent Achievements in Single Photon Detectors and Their Applications, *Journal of Modern Optics*, Vol. 51, No. 9, pp. (1298-1313), ISSN 0950-0340.

Roddier, F. (1998), Curvature sensing and compensation: a new concept in adaptive optics, *Applied Optics*, Vol. 27, No. 7, p. 1223, ISSN 0003-6935.

20

Comparison between Finite-Difference Time-Domain Method and Experimental Results for Femtosecond Laser Pulse Propagation

Shinki Nakamura
Ibaraki University
Japan

1. Introduction

There has recently been significant interest in the generation of single-cycle optical pulses by optical pulse compression of ultrabroad-band light produced in fibers. There have been some experiments reported on ultrabroad-band pulse generation using a silica fiber (Nakamura et al, 2002a), (Karasawa et al, 2000) and an Ar-gas filled hollow fiber (Karasawa et al, 2001) , and the optical pulse compression by nonlinear chirp compensation (Nakamura et al, 2002a), (Karasawa et al, 2001). For these experiments on generating few-optical-cycle pulses, characterizing the spectral phase of ultrabroad-band pulses analytically as well as experimentally is highly important.

Conventionally, the slowly varying-envelope approximation (SVEA) in the beam propagation method (BPM) has been used to describe the propagation of an optical pulse in a fiber (Agrawal, 1995). However, if the pulse duration approaches the optical cycle regime (<10 fs), this approximation becomes invalid (Agrawal, 1995). It is necessary to use the finite-difference time-domain (FDTD) method (Joseph & Taflove, 1997), (Kalosha & Herrmann, 2000) without SVEA (Agrawal, 1995). Previous reports by Goorjian (Goorjian et al., 1992), (Joseph et al., 1993), Joseph (Joseph & Taflove, 1997), (Goorjian et al., 1992), (Joseph et al., 1993), Taflove (Joseph & Taflove, 1997), (Goorjian et al., 1992), (Joseph et al., 1993), (Taflove & Hagness., 2000) and Hagness (Goorjian et al., 1992), (Taflove & Hagness., 2000) (JGTH) proposed an excellent FDTD algorithm considering a combination of linear dispersion with one resonant frequency and nonlinear terms with a Raman response function.

We performed an experiment of chirped 12 fs optical pulse propagation as described in Section 3. In order to compare FDTD calculation results with the experimentally measured ultrabroad-band spectra of such an ultrashort laser pulse, we extend the JGTH algorithm to that considering all of the exact Sellmeier fitting values for ultrabroad-band spectra. Because of the broad spectrum of pulses propagating in a fiber, it becomes much more important to take the accurate linear dispersion into account. It is well known that at least two resonant frequencies are required for the linear dispersion to fit accurately to the refractive index data. A recent report by Kalosha and Herrmann considers the linear dispersion with two resonant frequencies and the nonlinear terms without the Raman effect (Kalosha & Herrmann, 2000). For the single-cycle pulse generation experiment, we must use at least the shortest pulse of 3.4 fs (Yamane et al., 2003) or sub-5 fs (Karasawa et al, 2001), (Cheng et al., 1998) or the commercially available 12 fs pulses. Such a time regime is comparable to the

Raman characteristic time of 5 fs (Agrawal, 1995) in a silica fiber. Therefore, it is very important to consider not only the accurate linear dispersion of silica but also the Raman effect in a silica fiber in the few-optical-cycles regime. In addition, because of the high repetition rate and pulse intensity stability, in particular, ultrabroad-band supercontinuum light generation and few-optical-cycles pulse generation by nonlinear pulse propagation in photonic crystal fibers (Ranka et al., 2000) and tapered fibers (Birks et al., 2000), which are both made of silica, have attracted much attention. In this work, we have extended the FDTD method with nonlinear polarization P_{NL} involving the Raman response function (JGTH-algorithm) to 12 fs ultrabroadband pulse propagation in a silica fiber with the consideration of linear polarization P_L, including all exact Sellmeier-fitting values of silica with three resonant frequencies, in order to compare the calculation results with our experimental results (Nakamura et al, 2002a), (Karasawa et al, 2000). We have already compared the extended FDTD method (Nakamura et al., 2002b) with BPM by applying the split-step Fourier (SSF) method which is the solution of a generalized nonlinear Schrödinger equation (GNLSE) with SVEA(Agrawal, 1995), and the extended FDTD agreed better with the experimental results than with BPM. However, we have not shown the details of the calculation algorithm and temporal characteristics, and did not consider a chirp of the initial incident pulse to a fiber. In this chapter, we show the details of the the extended FDTD calculation algorithm (Nakamura et al., 2002c), (Nakamura et al., 2003) temporal characteristics of the pulse, and consider a chirp of the initial incident pulse to a fiber, and finally, we demonstrate the group delay compensation which generates the compressed pulse. Since 2004, the extended FDFD is called as the auxiliary differential Equation (ADE)-FDTD (Fujii et al., 2004). Additionally, we compared between the extended FDTD calculation and experimental result in dual wavelengths pulses propagation in a fiber. Finally, we investigated the slowly varying envelope approximation breakdown by comparing between BPM and the extended FDTD numerical results.

2. Extended FDTD algorithm

For simplicity, the electric and magnetic fields are expressed by Ey and Hx and one-dimensional propagation along the z direction is considered. The optical fiber is assumed to be isotropic and nonmagnetic. If a linear configuration is assumed, Maxwell's equations are as follows:

$$\begin{aligned} \frac{\partial H_x}{\partial t} &= \frac{1}{\mu_0}\frac{\partial E_y}{\partial z}, \\ \frac{\partial D_y}{\partial t} &= \frac{\partial H_x}{\partial z}, \end{aligned} \tag{1}$$

where μ_0 is the permeability in a vacuum and Dy is the dielectric flux density. By means of Yee's central difference method, Eq. (1) can be expressed by the following, in which the time and spatial steps are shifted by 1/2 step:

$$\begin{aligned} H_x\Big|_{i+1/2}^{n+1/2} &= H_x\Big|_{i+1/2}^{n-1/2} + \frac{\Delta t}{\mu_0 \Delta z}\left(E_y\Big|_{i+1}^{n} - E_y\Big|_{i}^{n} \right), \\ D_y\Big|_{i}^{n+1} &= D_y\Big|_{i}^{n} + \frac{\Delta t}{\Delta z}\left(H_x\Big|_{i+1/2}^{n+1/2} - H_x\Big|_{i-1/2}^{n+1/2} \right), \end{aligned} \tag{2}$$

where n is the time step number and i is the spatial step number. They are $t = n\,\Delta t$ and $z = i\Delta z$, respectively.
As the third step, let us use

$$D(\omega) = \varepsilon_0 \varepsilon_r(\omega) E(\omega) \tag{3}$$

to derive a new $E_y\big|_i^{n+1}$. Then, the iteration algorithm of the FDTD is complete. The authors have introduced the linear polarization P_L and the nonlinear polarization P_{NL}, corresponding to the Sellmeier fitting equation into the third step above, at the same time. First, the total polarization is expressed as

$$\begin{aligned} P &= \varepsilon_0\left(\chi^{(1)}E + \chi^{(2)}E^2 + \chi^{(3)}E^3 + \cdots\right) \\ &= P_L + P_{NL}. \end{aligned} \tag{4}$$

For simplicity, only the linear polarization is considered at first. The linear polarization is expressed as

$$P_L = \sum_{i=1}^{3} P_i, \tag{5}$$

$$P_i = \varepsilon_0 \int_0^t \chi_i^{(1)}(t-\tau) E(\tau)\, d\tau, \tag{6}$$

where $\chi_i^{(1)}$ has a Lorenz form with respect to frequency and is described as follows (Karasawa et al, 2001), (Agrawal, 1995):

$$\chi_i^{(1)}(\omega) = \frac{G_i \omega_i^2 (\varepsilon_s - 1)}{\omega_i^2 + 2j\omega\delta_i^2 - \omega^2}, \ (i = 1, 2, 3). \tag{7}$$

Here, j is the imaginary unit, ε_s is the permittivity in an electrostatic field, ω_i is the resonant angular frequency, δ_i is the attenuation constant with respect to resonant absorption, and $\sum_{i=1}^{3} G_i = 1$. If the above is inversely Fourier-transformed by assuming that ω_i is constant, the following differential equation is obtained:

$$\left[\chi_i^{(1)}\right]'' + 2\delta_i\left[\chi_i^{(1)}\right]' + \omega_i^2 \chi_i^{(1)} = 0 \tag{8}$$

where $\chi_i^{(1)}(t=0) = 0$ and $\left[\chi_i^{(1)}\right]'(t=0) = G_i\omega_i^2(\varepsilon_s - 1)$. If the optical fiber is sufficiently short that the attenuation term δ_i due to absorption can be neglected, then $\delta_i = 0$, so that Eq. (7) can be described as follows:

$$\chi_i^{(1)}(\omega) = \frac{G_i(\varepsilon_s - 1)\cdot \omega_i^2}{\omega_i^2 - \omega^2}, \ (i = 1, 2, 3). \tag{9}$$

Then, $\varepsilon_r(\omega)=1+\sum_{i=1}^{3}\chi_i^{(1)}(\omega)$ and the Sellmeier equation (Nakamura et al, 2002) is

$$\varepsilon_r(\omega)=n(\omega)^2=1+\sum_{i=1}^{3}\frac{b_i\cdot\omega_i^2}{\omega_i^2-\omega^2}. \tag{10}$$

Therefore,

$$\chi_i^{(1)}(\omega)=\frac{G_i(\varepsilon_s-1)\cdot\omega_i^2}{\omega_i^2-\omega^2}=\frac{b_i\cdot\omega_i^2}{\omega_i^2-\omega^2} \tag{11}$$

where $b_i=G_i(\varepsilon_s-1)$. Also, since $\sum_{i=1}^{3}b_i=\sum_{i=1}^{3}G_i(\varepsilon_s-1)=(\varepsilon_s-1)\sum_{i=1}^{3}G_i=(\varepsilon_s-1)$, we have

$$\varepsilon_s=1+\sum_{i=1}^{3}b_i\ . \tag{12}$$

The derivation above is summarized as follows:

$$\begin{aligned}D(\omega)&=\varepsilon_0E(\omega)+P(\omega)\\&=\varepsilon_0\left(1+\sum_{i=1}^{3}\frac{b_i\cdot\omega_i^2}{\omega_i^2-\omega^2}\right)E(\omega)\\&=\varepsilon_0E(\omega)+\varepsilon_0\left(\sum_{i=1}^{3}\frac{b_i\cdot\omega_i^2}{\omega_i^2-\omega^2}\right)E(\omega).\end{aligned} \tag{13}$$

If the nonlinear polarization is taken into account by returning to Eq. (4), then Eq. (13) can be treated as follows:

$$E=\frac{1}{\varepsilon_0}\left(D-\sum_{i=1}^{3}P_i-P_{NL}\right) \tag{14}$$

where

$$P_{NL}=\varepsilon_0\chi^{(3)}E\int_{-\infty}^{\infty}\left[\alpha\delta(t-\tau)+(1-\alpha)g_R(t-\tau)\right]\left[E(\tau)\right]^2d\tau\,. \tag{15}$$

Here, $\delta(t-\tau)$ is the delta function and α is the ratio of the intensities of the Kerr effect and the Raman effect. Also, $g_R(t)=\left[\left(\tau_1^2+\tau_2^2\right)/\tau_1\tau_2^2\right]e^{-t/\tau_2}\sin(t/\tau_1)$. When Eqs. (11) and (12) are taken into consideration in Eqs. (14) and (15), we obtain

$$P_i''+\omega_i^2P_i=\omega_i^2b_i\left(D-\sum_{i=1}^{3}P_i-P_{NL}\right). \tag{16}$$

Further, if the nonlinear polarization P_{NL} in Eq. (15) is described as

$P_{NL}=\chi^{(3)}E(t)\left[\varepsilon_0\alpha E(t)^2+(1-\alpha)G\right]$, then

$$G=\varepsilon_0\int_{-\infty}^{\infty}g_R(t-\tau)\left[E(\tau)\right]^2d\tau. \tag{17}$$

Also, if Eq. (16) is replaced by G, we obtain

$$P_i''+\omega_i^2P_i=\omega_i^2b_i\left(D-\sum_{i=1}^{3}P_i-\chi^{(3)}(1-\alpha)E\cdot G-\varepsilon_0\chi^{(3)}\alpha E^3\right) \tag{18}$$

leading to the form

$$\begin{aligned}
&P_1''+\omega_1^2(1+b_1)P_1+\omega_1^2b_1P_2+\omega_1^2b_1P_3\\
&+\omega_1^2b_1\chi^{(3)}\cdot\left[(1-\alpha)EG+\varepsilon_0\alpha E^3\right]=\omega_1^2b_1D\\
&P_2''+\omega_2^2(1+b_2)P_2+\omega_2^2b_2P_1+\omega_2^2b_2P_3\\
&+\omega_2^2b_2\chi^{(3)}\cdot\left[(1-\alpha)EG+\varepsilon_0\alpha E^3\right]=\omega_2^2b_2D\\
&P_3''+\omega_3^2(1+b_3)P_3+\omega_3^2b_3P_1+\omega_3^2b_3P_2\\
&+\omega_3^2b_3\chi^{(3)}\cdot\left[(1-\alpha)EG+\varepsilon_0\alpha E^3\right]=\omega_3^2b_3D.
\end{aligned} \tag{19}$$

Further, applying a central difference scheme centered at time step n, these equations yield

$$\begin{aligned}
&a_1P_1|^{n+1}+c_1P_2|^{n+1}+c_1P_3|^{n+1}\\
&\quad=c_1\left(D|^{n+1}+D|^{n-1}\right)-c_1\left(P_2|^{n-1}+P_3|^{n-1}\right)\\
&\qquad+4P_1|^n+g_1P_1|^{n-1}-\omega_1^2b_1\chi^{(3)}\cdot\left[(1-\alpha)E|^nG|^n+\varepsilon_0\alpha\left(E|^n\right)^3\right]\\
&c_2P_1|^{n+1}+a_2P_2|^{n+1}+c_2P_3|^{n+1}\\
&\quad=c_2\left(D|^{n+1}+D|^{n-1}\right)-c_2\left(P_1|^{n-1}+P_3|^{n-1}\right)\\
&\qquad+4P_2|^n+g_2P_2|^{n-1}-\omega_2^2b_2\chi^{(3)}\cdot\left[(1-\alpha)E|^nG|^n+\varepsilon_0\alpha\left(E|^n\right)^3\right]\\
&c_3P_1|^{n+1}+c_3P_2|^{n+1}+a_3P_3|^{n+1}\\
&\quad=c_3\left(D|^{n+1}+D|^{n-1}\right)-c_3\left(P_1|^{n-1}+P_2|^{n-1}\right)\\
&\qquad+4P_3|^n+g_3P_3|^{n-1}-\omega_3^2b_3\chi^{(3)}\cdot\left[(1-\alpha)E|^nG|^n+\varepsilon_0\alpha\left(E|^n\right)^3\right],
\end{aligned} \tag{20}$$

Where

$$a_i = 2 + \omega_i^2 \Delta t^2 (1 + b_i)$$

$$c_i = \omega_i^2 \Delta t^2 b_i \tag{21}$$

$$g_i = -2 - \omega_i^2 \Delta t^2 (1 + b_i) .$$

Here, $E|^n$ and $G|^n$ have already been updated in a previous time step. From the above, the unknown $P_1|^{n+1}$, $P_2|^{n+1}$ and $P_3|^{n+1}$ can be derived. Here, by using

$$\begin{aligned} F = P_L &= \sum_{i=1}^{3} P_i \\ &= \sum_{i=1}^{3} \varepsilon_0 \int_0^t \chi_i^{(1)} (t - \tau) E(\tau) d\tau, \end{aligned} \tag{22}$$

the linear polarization P_L is replaced by $P_L = F$. Then,

$$F|^{n+1} = P_L|^{n+1} = P_1|^{n+1} + P_2|^{n+1} + P_3|^{n+1} \tag{23}$$

is explicitly obtained. Hence, an unknown linear polarization can be found. The reason why only one resonant frequency can be taken into account for the linear polarization in the JGTH algorithm (Joseph & Taflove, 1997), (Goorjian et al., 1992), (Joseph et al., 1993) (Taflove & Hagness., 2000) is as follows. Since we take G as $G|^n$ in applying the central difference to Eq. (19) to obtain Eq. (20), we can obtain the linear polarization explicitly. However $(G|^{n+1} + G|^{n-1})/2$ was used in the previous JGTH algorithm to improve the finite-difference accuracy. In this case, the unknown $G|^{n+1}$ is necessary, and therefore the successive solution of the nonlinear polarization after the linear polarization is derived is not possible algorithmically. Therefore, in the JGTH algorithm, the linear polarization is restricted to one so that symmetry is provided to the differential equations for linear polarization and nonlinear polarization. This is followed by the solution of the simultaneous finite difference equations to obtain two unknowns simultaneously. Since only the known $G|^n$ is used by our method in the finite difference equation for the linear polarization, it is possible to derive the nonlinear polarization after the linear polarization is found. The accuracy of the finite difference method is sufficient in this procedure. As described in detail below, the use of only one resonant frequency for the linear polarization when the results are compared with experimental results is a serious problem due to its more significant approximation of dispersion, as follows.

Let us now describe the approximation of the dispersion in the conventional FDTD method. The equation used in Refs. (Joseph & Taflove, 1997)-(Taflove & Hagness., 2000), in a series of conventional FDTD methods, called the JGTH algorithm, is the case of i = 1 in Eq. (7) and $\varepsilon_r(\omega) = n(\omega)^2 = \varepsilon_\infty + b_1\omega_1/(\omega_1^2\text{-}\omega^2)$ in Eq. (10). Here, ε_∞ is the relative permittivity when the frequency is infinite. It is not possible in principle to approximate the refractive index data of fused silica in the entire frequency range in this equation. However, it is possible within a limited frequency range. For example, if an approximation is introduced such that the error of the refractive index is 0.1% at 250-517 THz, then the parameters can be determined as ε_∞ =

0.9560, b_1 = 1.130 and ω_1 = 1.1639 × 10^{16} rad/s. Using these values, the values of β_1, β_2 and β_3 are found to be 4.901 × 10^{-9} s/m, 7.287 × 10^{-26} s^2/m and 3.469 × 10^{-41} s^3/m, respectively. $\beta_n(\mathrm{n}=1,2,3,\cdots)$ is the n-th order derivative of propagation constant β_0 at an incident center angular frequency ω_0 defined as $\beta_n = \left(d^n\beta/d\omega^n\right)_{\omega=\omega_0}$ (Agrawal, 1995) . In contrast, the Sellmeier equation with three resonant frequencies used in the extended FDTD method by the authors accurately approximates the refractive index in the entire frequency range where measurement is possible (Malitson., 1961). If this equation is used, β_1 = 4.894 ×10^{-9} s/m, β_2 = 3.616 × 10^{-26} s^2/m and β_3 = 2.750 × 10^{-41} s^3/m. Since the conventional method and the extended FDTD method approximately agree with regard to β_1, the results for the group velocity v_g = $(\beta_1)^{-1}$ can be relied upon to a certain extent in the conventional method (Joseph & Taflove, 1997), (Taflove & Hagness., 2000). This implies that the conventional method can be used for the calculation of the propagation of a pulse of nanosecond order, for which the spectral width is sufficiently narrow and the effect of the group velocity dispersion is negligible. However, since the group velocity dispersion β_2 = $\partial(v_g)^{-1}/\partial\omega$ differs from the value obtained from the refractive index data by a factor of 2, the comparison of the conventional method with the propagation experiment of a pulse of sub-nanosecond order or shorter is not possible. Therefore, in the conventional FDTD method of the JGTH algorithm (Joseph & Taflove, 1997) (Goorjian et al., 1992), (Taflove & Hagness., 2000), calculation of the propagation of femtosecond pulses can be carried out only for a hypothetical medium after the dispersion value is set to an easily calculated value and $\chi^{(3)}$ is set to an unrealistic value that is 10^{20} times the usual value. The comparison between simple JGTH algorithm with parameters of ε_∞ = 0.9560, b_1 = 1.130 and ω_1 = 1.1639 × 10^{16} rad/s and the extended FDTD method is shown as temporal and spectral characteristics in section IV.
For the above reason, the nonlinear polarization is derived next from the linear polarization $F|^{n+1}$ derived from Eq. (23). The definition of the nonlinear polarization has already been presented in Eq. (15). When Eq. (15) is Fourier-transformed, we obtain

$$\tilde{P}_{NL}(\omega) = \varepsilon_0\chi_0^{(3)}\tilde{E}(\omega)\times\left\{\alpha\tilde{E}^2(\omega)+(1-\alpha)\tilde{g}_R\tilde{E}^2(\omega)\right\} \tag{24}$$

$$\tilde{g}_R(\omega) = \frac{\bar{\omega}_0^2}{\bar{\omega}_0^2 - j\omega\bar{\delta} - \omega^2}, \tag{25}$$

where $\bar{\delta} = 1/\tau_2$ and $\bar{\omega}_0^2 = (1/\tau_1)^2 + (1/\tau_2)^2$. Next, if Eq. (17), the definition of G, is Fourier-transformed, we obtain

$$\tilde{G}(\omega) = \varepsilon_0\tilde{g}_R(\omega)\tilde{E}^2(\omega). \tag{26}$$

When this is used, Eq. (25) becomes

$$\left(\bar{\omega}_0^2 - j\omega\bar{\delta} - \omega^2\right)\tilde{G}(\omega) = \varepsilon_0\bar{\omega}_0^2\tilde{E}^2(\omega). \tag{27}$$

Hence, the equation to be solved is expressed as follows by taking the inverse Fourier transform of Eq. (27):

$$\frac{1}{\bar{\omega}_0^2}\frac{\partial^2 G}{\partial t^2}+\frac{\bar{\delta}}{\bar{\omega}_0^2}\frac{\partial G}{\partial t}+G-\varepsilon_0 E^2=0. \tag{28}$$

From the relationship between the dielectric flux density and the electric field,

$$\begin{aligned}\tilde{D}&=\varepsilon_0\tilde{E}+\tilde{P}\\ &=\varepsilon_0\tilde{E}+\tilde{P}_L+\tilde{P}_{NL}\\ &=\varepsilon_0\tilde{E}+\varepsilon_0\chi^{(1)}\tilde{E}+\varepsilon_0\chi^{(3)}\tilde{E}\left[\alpha\tilde{E}^2+(1-\alpha)\tilde{g}_R\tilde{E}^2\right]\\ &=\varepsilon_0\left(1+\alpha\chi^{(3)}\tilde{E}^2\right)\tilde{E}+\tilde{F}+(1-\alpha)\chi^{(3)}\tilde{E}\tilde{G}.\end{aligned} \tag{29}$$

Eq. (28) can be modified to the following:

$$\begin{aligned}&\frac{1}{\bar{\omega}_0^2}\frac{\partial^2 G}{\partial t^2}+\frac{\bar{\delta}}{\bar{\omega}_0^2}\frac{\partial G}{\partial t}\\ &\quad+\left[1+\frac{(1-\alpha)\chi^{(3)}\left(E_y\right)^2}{1+\alpha\chi^{(3)}\left(E_y\right)^2}\right]\cdot G\\ &\quad+\left[\frac{E_y}{1+\alpha\chi^{(3)}\left(E_y\right)^2}\right]\cdot F\\ &\quad=\left[\frac{E_y}{1+\alpha\chi^{(3)}\left(E_y\right)^2}\right]\cdot D_y.\end{aligned} \tag{30}$$

Further, applying a central difference scheme centered at time step n, these equations yield

$$\begin{aligned}&\left(\frac{1}{\bar{\omega}_0^2\Delta t^2}\right)\cdot\left(G\big|_i^{n+1}-2G\big|_i^{n}+G\big|_i^{n-1}\right)+\left(\frac{\bar{\delta}}{\bar{\omega}_0^2\Delta t}\right)\cdot\left(G\big|_i^{n+1}-G\big|_i^{n-1}\right)\\ &\quad+\left[1+\frac{(1-\alpha)\chi^{(3)}\left(E_y\big|_i^n\right)^2}{1+\alpha\chi^{(3)}\left(E_y\big|_i^n\right)^2}\right]\cdot\frac{\left(G\big|_i^{n+1}+G\big|_i^{n-1}\right)}{2}\\ &\quad+\left[\frac{E_y\big|_i^n}{1+\alpha\chi^{(3)}\left(E_y\big|_i^n\right)^2}\right]\cdot\frac{\left(F\big|_i^{n+1}+F\big|_i^{n-1}\right)}{2}\\ &\quad=\left[\frac{E_y\big|_i^n}{1+\alpha\chi^{(3)}\left(E_y\big|_i^n\right)^2}\right]\cdot\frac{\left(D_y\big|_i^{n+1}+D_y\big|_i^{n-1}\right)}{2}.\end{aligned} \tag{31}$$

When $F|^{n+1}$ derived from Eq. (23) is substituted into Eq. (31) together with the known $F|^{n-1}$, $G|^{n}$, $G|^{n-1}$ and $E|^{n}$, the unknown $G|^{n+1}$ can be derived. Here, an explicit form of updating G is presented as

$$G|_i^{n+1} = \left[d+e+f\right]^{-1} \times \left\{ 2d \cdot G|_i^{n} + \left(e - d + \frac{f}{2}\right) \cdot G|_i^{n-1} - g \cdot \frac{\left(F|_i^{n+1} + F|_i^{n-1}\right)}{2} + g \cdot \frac{\left(D_y|_i^{n+1} + D_y|_i^{n-1}\right)}{2} \right\}. \quad (32)$$

where

$$d = \left(\frac{1}{\overline{\omega}_0^2 \Delta t^2}\right)$$
$$e = \left(\frac{\overline{\delta}}{\overline{\omega}_0^2 \Delta t}\right)$$
$$f = \left[1 + \frac{(1-\alpha)\chi^{(3)}\left(E_y|_i^{n}\right)^2}{1+\alpha\chi^{(3)}\left(E_y|_i^{n}\right)^2}\right] \quad (33)$$
$$g = \left[\frac{E_y|_i^{n}}{1+\alpha\chi^{(3)}\left(E_y|_i^{n}\right)^2}\right].$$

When the obtained $G|^{n+1}$ and $F|^{n+1}$ are substituted into the following, the unknown $E|^{n+1}$ can be derived and the iteration algorithm for the FDTD is terminated. Hence,

$$E_y^{p+1}\Big|^{n+1} = \frac{D_y|^{n+1} - F|^{n+1} - (1-\alpha)\chi_0^{(3)} E_y^p\big|^{n+1} G|^{n+1}}{\varepsilon_0\left[1+\alpha\chi_0^{(3)}\left(E_y^p\right)^2\right]}. \quad (34)$$
$$p = 0,1,2,\cdots$$

Note that the initial condition is $E_y^0\big|^{n+1} = E_y|^{n}$. Here, Eq. (34) is derived from Eq. (29).

3. Femtosecond single pulse propagation

3.1 Experimental Setup

Figure 1 shows the setup used for the authors' experiments (Nakamura et al, 2002a), (Karasawa et al, 2000). The 12 fs and 10 nJ pulses (sech^2) at the center wavelength of 800 nm were generated from a mode-locked Ti:sapphire laser (Femtolaser GmbH, Femtosource M-

1). The incident pulse width was measured by a fringe-resolved autocorrelator (FRAC). The 12 fs pulses were coupled into a 2.5 mm silica fiber by a 36x reflective objective (Ealing). The advantage of this kind of reflective objective is that no additional group-delay dispersion (GDD) or third-order dispersion (TOD) is introduced to the pulses. However, the 12 fs pulses are slightly chirped because the pulse width of the full width at half maximum (FWHM) resulting from the Fourier transformation of the fiber input spectrum is 8.5 fs.

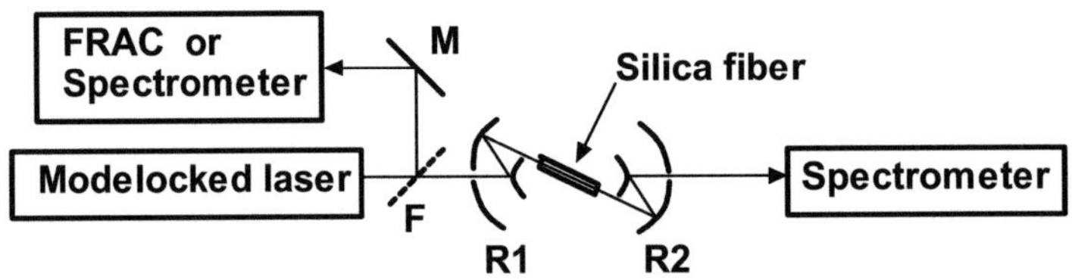

Fig. 1. Layout of the experimental setup (Nakamura et al., 2002a). M, silvercoated plane mirrors; F, flipper beam splitter, R1, R2, reflective objectives (36x); FRAC, fringe resolved autocorrelator.

The input pulse is obtained by calculating the inverse Fourier transform of the spectrum of the input pulse. However, the transform-limited pulse width (8.5 fs) is smaller than the experimentally measured pulse width (12 fs) obtained by a fringe-resolved autocorrelator. To use the pulse with the correct spectrum and the pulse width in the calculation, the temporal duration is adjusted by adding the quadratic spectral phase in the form $\varphi(\omega) = c(\omega - \omega_0)^2$ to the experimental spectrum, to render the inverse Fourier transform pulse width the same as that of the measured one. This gave the values of c = -9.043 fs^2/rad and $\partial^2\varphi(\omega) / \partial\omega^2 |_{\omega=\omega_0}$ = -113.6 fs^2. The peak power of an input pulse was 175 kW. A single-mode fused silica fiber (Newport F-SPV) with a core diameter of 2.64 μm was used. The output from the 2.5 mm fiber was collimated by another reflective objective and measured by a spectrometer. The obtained input and output spectra of the fiber are shown in Fig. 2.
The dashed line in Figure 2 shows the input spectrum and the solid line shows the output spectrum. The dashed line and short-dashed line in Fig. 3 show the assumption of an input spectrum whose form is a sech^2 shape, using this as field excitations in the extended FDTD calculation (dashed line) and the simpler JGTH-FDTD calculation (short-dashed line). There is no difference between the spectra of the extended FDTD and JGTH-FDTD. The solid line in Fig. 3 shows the experimentally obtained input pulse spectrum. Note that the intensity at

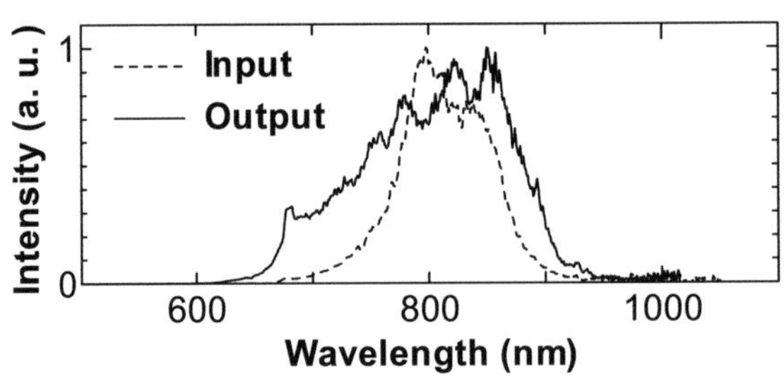

Fig. 2. Input and output spectra measured in the 12 fs laser propagation experiment in a silica fiber (Karasawa et al, 2000). Dashed line shows the input spectrum and solid line shows the output spectrum.

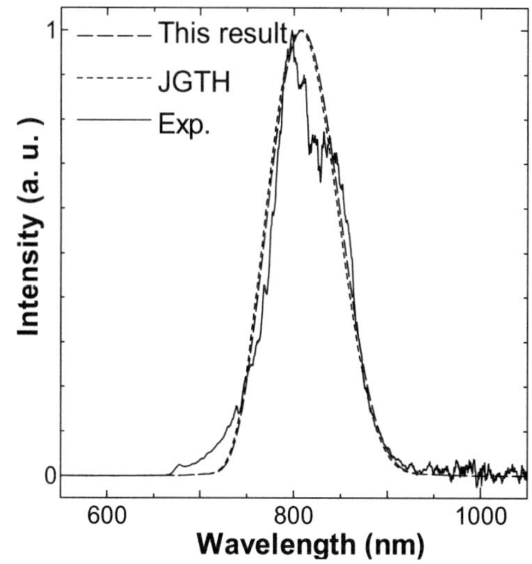

Fig. 3. Spectra of 12 fs incident laser pulse into a 2.5-mm silica fiber, under the assumption of having slight chirp, using this as field excitations in the extended FDTD (dashed line) (Nakamura et al., 2005a) and the simpler JGTH-FDTD (short-dashed line) (Nakamura et al., 2005a) , and our previously reported experimental result (Karasawa et al, 2000) (solid line).

the 650-750 nm of the assumed spectrum is zero for the solid line, however, the intensity in the wavelength region is represented by a dashed line, and disagreement could be expected at 650-750 nm in the calculated results. The inverse Fourier transformation of the assumed input spectrum (dashed line in Fig. 3) indicates a pulse width of 8.5 fs (FWHM), which is the same as that of the experimental spectrum (solid line in Fig. 3). In addition, the output pulse was characterized by our second-harmonic generation frequency-resolved optical gating (SHG-FROG) apparatus set up in place of the spectrometer. Figures obtained by the SHG-FROG are shown in section 3.2-A.

3.2 Numerical results

A. *Comparison between Experimental and Numerical Results*

In our calculation for a fused silica fiber, the parameters in Eq. (10) are set as b_1 = 0.6961663, b_2 = 0.4079426, b_3 = 0.8974794, λ_1 = 0.0684043 μm, λ_2 = 0.1162414 μm, and λ_3 = 9.896161 μm (Agrawal, 1995) , where $\lambda_i = 2\pi c/\omega_i$ and c is the velocity of light in vacuum. We used the value of the nonlinear refractive coefficient $n_2 = 2.48\times10^{-20}$ m^2/W from Ref. (Taylor, 1996), and the third-order susceptibility $\chi^{(3)}$ was found to be $\chi^{(3)} = 1.85\times10^{-22}$ m^2/V^2 at 800 nm, as given by $\chi^{(3)} = (4/3)\varepsilon_0 c n(\omega_0)^2 n_2$, where ω_0 is the center angular frequency of the optical pulse. The parameters α, τ_1, and τ_2 in Eq. (15) are set to be α = 0.7, τ_1 = 12.2 fs and τ_2 = 32 fs (Agrawal, 1995) . A single time step of the finite difference is set as $\Delta t = 4.4475215\times10^{-17}$ s at the wavelength of 800 nm. The time step of Δt at 800 nm is defined as the optical cycle at 800 nm of 2.6666667 fs divided by 60. In the extended FDTD calculation, we set all parameters to be the same as in our experiment (Nakamura et al, 2002a), (Karasawa et al, 2000).

We compare the FDTD calculation with the experimental result for the pulse peak power of 175 kW in order to generate an ultrabroad spectrum which can finally be compressed to 7.1 fs (Nakamura et al, 2002) or even shorter. The total fiber length of L = 2.5 mm corresponds

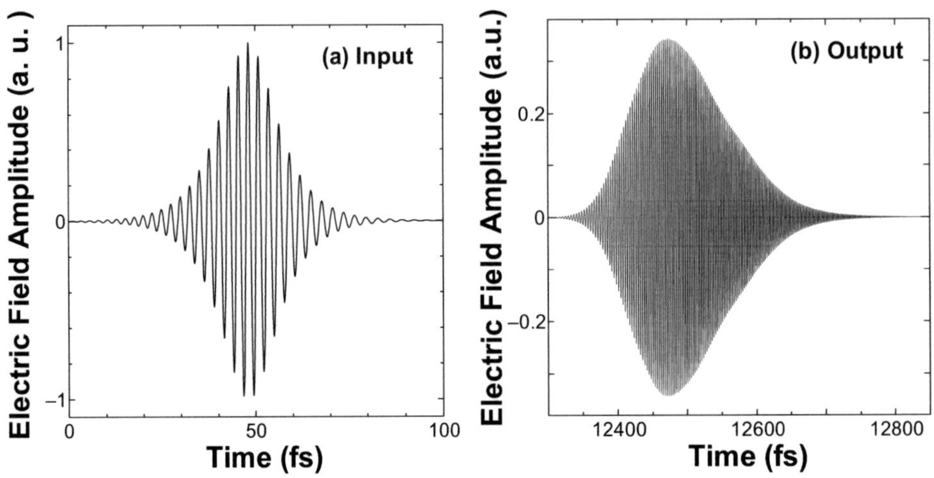

Fig. 4. Input (a) and output (b) temporal electric fields of a 2.5 mm silica fiber calculated by the extended FDTD Maxwell equation method considering all orders of dispersions and the Raman response [α = 0.7 in Eq. (15)]. (Nakamura et al., 2005a)

to 136,500 spatial steps, which means that L = 136,500×Δz, where Δz is a unit spatial step in the z direction. We need 293,000 time steps to measure the electric field until the pulse tail passes completely. The peak power of an input pulse is set to be 175 kW (soliton number N = 2.09). The effective core area A_{eff} is set to be 5.47 μm^2. The iteration number in Eq. (34) is three. An iteration number of two is sufficient, as shown in Refs. (Joseph & Taflove, 1997) (Goorjian et al., 1992), (Taflove & Hagness., 2000). Figure 4 shows the input (a) and output (b) temporal electric fields of the fiber. Since the chirp is very small ($\partial^2\varphi(\omega)/\partial\omega^2 |_{\omega=\omega_0}$ = -113.6 fs^2), it is seen that the pulse has no chirp in Fig.4(a), while Fig. 4(b) seems to have large chirp because of the linear and nonlinear effects in fiber propagation. Figure 5 shows the temporal intensity profile of the pulse after propagation of the fiber, where the solid line indicates measurement by FROG in the previous experiment, and the dashed line and the short-dashed line indicate the results calculated by the extended FDTD Maxwell equation method and the simpler JGTH-FDTD method, respectively.

The solid line shows a ringing behavior but generally agrees with the calculated result except for the ringing. The envelope of the ringing appears to be the dashed line. The feature of a steep increase (from -100 fs to 0 fs) and a slow decrease (from 0 fs to 200 fs) of the dashed line is also seen in the solid line. The ringing of the experimental result occurs because of the interference between core-mode light and cladding-mode light. Because our simulation is performed by 1D calculation, the cladding mode is not considered. Hence, there is no ringing in the numerical result. The experimentally measured pulse width (FHWM) is 92 fs. Numerically, the pulse width of FWHM is obtained as 108 fs. Therefore, the pulse widths are in good agreement with each other. However, the pulse width (FWHM) obtained by the simpler JGTH-FDTD method is 146 fs, which is 1.5 times the experimental result. The pulse width obtained by JGTH-FDTD does not agree with the experimental result. This disagreement is caused by the approximation of dispersion in JGTH-FDTD as described in section 3.2. Hence, the extended FDTD is superior to the simpler JGTH-FDTD method.

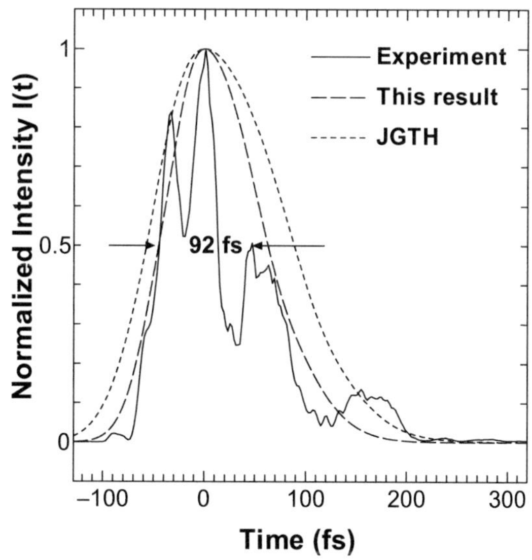

Fig. 5. Temporal intensity profile obtained experimentally (Nakamura et al., 2002) (solid line) and numerically by the extended FDTD (dashed line) and the simpler JGTH-FDTD (short-dashed line) as a function of time for 12 fs laser transmission through a 2.5-mm silica fiber. (Nakamura et al., 2005a)

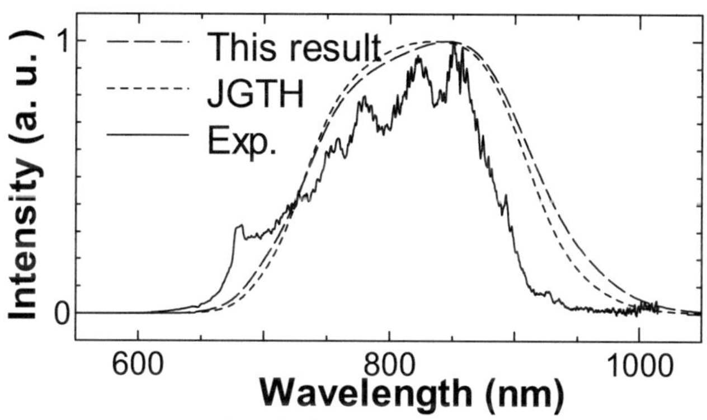

Fig. 6. Spectra after propagation of a slightly chirped 12 fs laser pulse through a 2.5-mm silica fiber, calculated by the extended FDTD (dashed line) (Nakamura et al., 2005a) and the simpler JGTH-FDTD (short-dashed line) (Nakamura et al., 2005a), and our previously reported experimental result (Karasawa et al., 2000) (solid line).

Figure 6 shows the results calculated by the extended FDTD Maxwell equation method (dashed line) and the simpler JGTH-FDTD method (short-dashed line), and our previously reported experimental result (solid line, experiment [Exp.]) (Karasawa et al, 2000). The spectra obtained by the extended FDTD and the simpler JGTH-FDTD show similar results. The trapezium form of the FDTD results is roughly similar to that of the experimental result. Several peaks in the experimental result appear due to interference between core-mode light and cladding-mode light. In BPM solving GNLSE by SSF with SVEA, the second derivative of the electric field with respect to z, $\partial^2 E_y / \partial z^2$, is neglected, which corresponds to neglecting the backward propagation wave. On the other hand, our extended FDTD Maxwell equation method accurately includes the delayed Raman response and all orders of the dispersion in silica using Sellmeier's equation, and does not use SVEA. The shortest wavelengths (intensity of 1%) of the spectra obtained by the extended FDTD, the simpler JGTH-FDTD, and the experimental result (Exp.) are 653 nm, 660 nm and 626 nm, respectively. The longest wavelengths of the extended FDTD, the simpler JGTH-FDTD, and the experimental result (Exp.) are 1046 nm, 1027 nm, and 972 nm (maximum measured wavelength of the spectrometer), respectively. The spectral bandwidths $\Delta\lambda$ (1% of peak intensity) of the extended FDTD, the simpler JGTH-FDTD, and the experimental result (Exp.) are 393 nm, 367 nm, and 346 nm, respectively. Thus, the spectrum bandwidth of the experimental result (Exp.) is narrower than those of the extended FDTD and the simpler JGTH-FDTD. Note that the intensity at 650-750 nm of the assumed spectrum is lower than the experimental result, and also that the intensity at 650-750 nm of the assumed spectrum of the input pulse is zero; however, the intensity of the region of the wavelengths is not zero in the experimental result of the input pulse. This affects the weak intensity at 650-750 nm in the fiber output spectrum obtained by the extended FDTD and the simpler JGTH-FDTD. In considering wavelengths longer than 800 nm in Fig. 6, the intensity of the experimental data is lower than those of the extended FDTD and the simpler JGTH-FDTD results. We assume that the sensitivity of the spectrometer detector is low at wavelengths longer than 800 nm because the detector is made of silicon. We believe our extended FDTD result is correct. We suggest that it is better to use an IR detector made of a material such as InGaAs for the spectrometer.

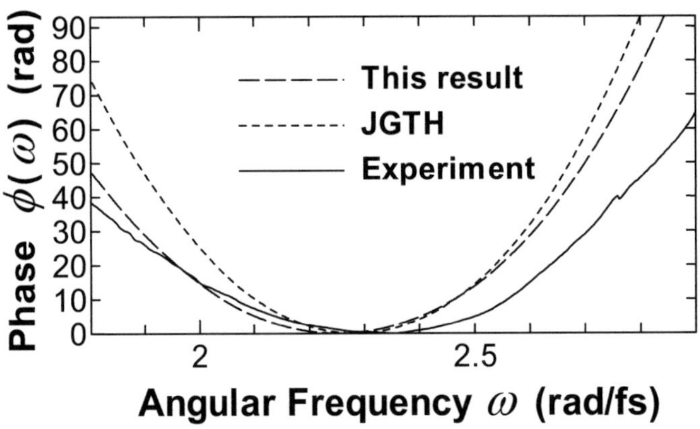

Fig. 7. Spectral phase as a function of angular frequency after a chirped 12 fs pulse propagation through a 2.5-mm silica fiber previously obtained experimentally (solid line)(Nakamura et al., 2002) and numerically by the extended FDTD (dashed line) (Nakamura et al., 2005a) and the simpler JGTH-FDTD (short-dashed line) (Nakamura et al., 2005a)

Figure 7 shows the spectral phase obtained experimentally by the FROG (solid line) and numerically by the extended FDTD (dashed line) and the simpler JGTH-FDTD (short-dashed line) as a function of angular frequency for 12 fs laser transmission through a 2.5-mm silica fiber. From our previous experimental result corresponding to the solid line, we determine the group-delay dispersion (GDD) and the third-order dispersion (TOD) values to be 342 fs^2 and 187 fs^3, respectively. From our calculation corresponding to the dashed line, we determine the GDD and TOD values to be 518 fs^2 and 333 fs^3, respectively. The calculated GDD and TOD values do not agree with the experimental results. We have not yet clarified the reason for this disagreement. We must consider assuming the $sech^2$ temporal shape of the input pulse, but not the complex waveform obtained by the inverse Fourier transform of the experimentally observed spectrum. However, the calculated result by the extended FDTD (dashed line) from the angular frequency of 1.8 rad/fs to 2.4 rad/fs (lower frequency) appears to be in good agreement with the experimental result. On the other hand, the simpler JGTH-FDTD (short-dashed line) shows a large difference in not only the higher angular frequency but also the lower angular frequency. The phase values at the lowest angular frequency of the extended FDTD, the simpler JGTH-FDTD, and the experimental result are 47.4 rad, 74.2 rad, and 38.9 rad, respectively. In this comparison, the phase value of the simpler JGTH-FDTD is twice that of the experimental result. This marked difference between the simpler JGTH-FDTD and the experimental result is caused by the dipersion problem of the simpler JGTH-FDTD, which is described in section II and above in this section. The spectral phase is very important for single-cycle pulse generation via phase compensation of the nonlinear-chirped supercontinuum generated in a silica fiber. Hence, the extended FDTD is superior to the simpler JGTH-FDTD in predicting the spectral phase for the phase compensation.

B. *Compensation of group delay by using SLM*

We assume the model in which the output beam from the fiber is collimated by a spherical mirror and is directed to a 4-f system by the SLM. The 4-f system consisted of two spherical mirrors and two reflective gratings (Nakamura et al, 2002a), (Karasawa et al, 2001), (Yamane et al., 2003). The optical path lengths from the gratings to the spherical mirrors with a focal length f and those from the spherical mirrors to the SLM are all set to be f. The liquid crystal SLM (Citizen Company) (Karasawa et al, 2001) consisted of 648 97–µm-wide pixels, with a 5 µm gap between adjacent pixels.

The SLM is used as a phase modulator, which can impose a phase variation of 2π on light whose wavelength is less than 1400 nm with a resolution of 192 (Karasawa et al, 2001). The phase $\phi_{SLM}(x)$ from the SLM at position x is given by a polynomial of the form

$$\phi_{SLM}(x) = \frac{\alpha}{2}\left[\omega(x) - \omega_0\right]^2 + \frac{\beta}{6}\left[\omega(x) - \omega_0\right]^3 + \frac{\gamma}{24}\left[\omega(x) - \omega_0\right]^4, \tag{35}$$

where ω_0 is the center angular frequency for a Taylor expansion, $\omega(x)$ is the angular frequency at position x and is given by $\omega(x) = 2\pi c/\{d \sin[\tan^{-1}(x/f) + \sin^{-1}(\lambda_c/d)]\}$, c is the speed of light, and λ_c = 800 nm is the wavelength at the central position of the SLM, where x – 0. λ_c does not have to be equal to the center wavelength for the Taylor expansion λ_0, where $\lambda_0 = 2\pi c/\omega_0$. α, β, and γ are GDD, TOD, and the fourth-order dispersion (FOD), respectively, at ω_0. The group delay $t_{d,SLM}(x)$ imposed by the SLM is

$$t_{d,SLM}(x) = \frac{\partial \varphi_{SLM}(\omega)}{\partial \omega}$$

$$= \alpha\left[\omega(x) - \omega_0\right] + \frac{\beta}{2}\left[\omega(x) - \omega_0\right]^2 + \frac{\gamma}{6}\left[\omega(x) - \omega_0\right]^3. \tag{36}$$

Parameters α, β, and γ were initially estimated from the total group delay $t_d(\omega)$ of a fiber (including the self-phase modulation). $t_d(\omega)$ is fitted to the form of Eq. (36), and α, β, and γ are obtained. Then the negative values of these fitted parameters were initially imposed by the SLM to satisfy $t_{d,SLM}(\omega) + t_d(\omega)$ = constant (zero in our case) in the entire frequency range. Theoretically, this condition should produce the shortest pulses. Use of the Taylor expansion in Eq. (35) is found to be the most suitable way to perform fine adjustment of the phase. Parameters α, β, and γ are fine-tuned by a computer to yield the shortest pulse, with the constant group delay dispersion optimized as a function of wavelength. Figure 8 shows the calculated group delay after the pulse propagation of a fiber (solid curves) and the compensated group delay (dashed curve) with the negative value of the optimized group delay applied by the SLM. The group delay after fiber propagation is added by applying its negative value to the SLM, and then the total of these values is shown as a dashed line for which the shortest pulse (a nearly transform-limited pulse) is obtained for the condition where λ_0 is set to 800 nm. The temporal intensity profile of this shortest pulse is shown in Fig. 9. The pulse width (FWHM) is estimated to be 6.9 fs. This pulse width agrees with the compressed pulse of 7.1 fs (Nakamura et al, 2002a) previously obtained experimentally by using a combination of prism compression and a 256 pixel SLM. This finding indicates that the combination of this extended FDTD and group delay compensation simulation is robust for estimating the shortest pulse characteristics after the propagation of a fiber, and useful for constructing a fiber-compression system for obtaining a single-cycle pulse.

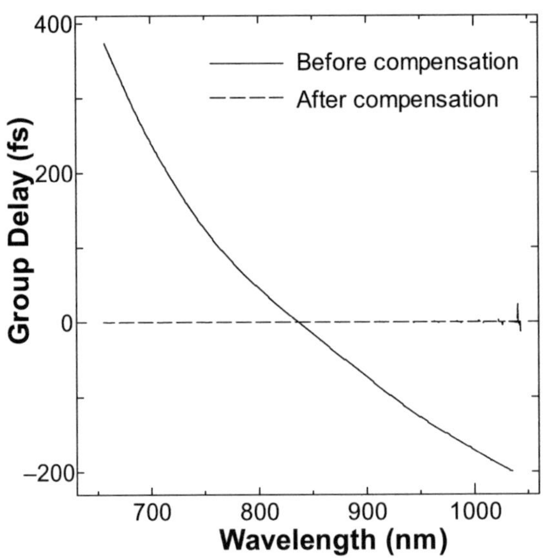

Fig. 8. Calculated group delay after pulse propagation of a fiber (solid curves) and the compensated group delay (dashed curve) with the negative of the optimized group delay applied by the SLM. (Nakamura et al., 2005a)

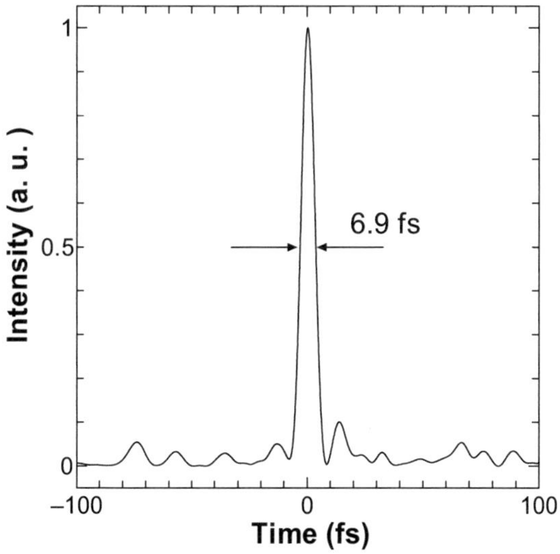

Fig. 9. Temporal time profile compensated optimally by SLM after propagation of a fiber. The pulse width (FWHM) is 6.9 fs. (Nakamura et al., 2005a)

3.3 Summary of single pulse propagation

We have extended the nonlinear FDTD method with a JGTH algorithm to one with exact Sellmeier's fitting values in order to compare the experimental and calculated results of nonlinear femtosecond ultrabroadband-pulse propagation in a silica fiber, and have analyzed the temporal and spectral characteristics of the propagated pulse. This extended method is robust against the breakdown of the SVEA. The spectrum, temporal intensity profile, and pulse width after phase compensation results as calculated by the extended FDTD method agree well with those obtained in our previous experiment. These findings indicate that the combination of this extended FDTD method and the group delay compensation simulation is robust for estimating the spectrum, the temporal pulse profile after propagation through a fiber and the shortest pulse obtained by compensation, and is useful for estimating the shortest pulse when the fiber-compression system is designed to obtain a single-cycle pulse (Yamane et al., 2003).

When the extended FDTD method is established as the ultrashort optical pulse propagation analysis technique, its application to the characteristic analysis of an ultrahigh-speed optical switch, the propagation characteristic analysis of the optical pulse in a photonic crystal, and nonlinear propagation analysis in a photonic crystal fiber or a taper fiber are expected.

4. Dual wavelengths femtosecond pulses propagation

4.1 Induced-phase modulation

Recently, there has been significant interest in the generation of single-cycle optical pulses by optical pulse compression of ultrabroadband light produced in fibers. The spectra-broadening by the induced-phase modulation (IPM) is one of these methods. There have been some reports of experiments on ultrabroadband-pulse generation using IPM in a silica fiber (Xu et al., 1999), (Karasawa et al., 2001b), (Shibata et al., 2002), (Yamashita et al., 2004) and an Ar-gas-filled hollow fiber, (Karasawa et al., 2000b) and also on optical pulse compression by nonlinear chirp compensation. (Yamashita et al., 2004) For these experiments on generating few-optical-cycle pulses, characterizing the spectral phase of ultrabroadband pulses analytically as well as experimentally is very important.

In IPM calculation, details of the phase characteristics at the connecting point of the two pulses are not clearly understood because the beam propagation method (BPM) has been used to describe the propagation of an optical pulse in a fiber. (Agrawal, 1995) In the BPM, each phase curve is obtained independently as a function of frequency for each center frequency of the two pulses. (Agrawal, 1995) The two phase curves are connected at the same phase value, as reported in refs. (Yamashita et al., 1998) and (Yamashita et al., 1996). Recently, we extended (Nakamura et al., 2002b) the finite-difference time domain (FDTD) method with nonlinear polarization P_{NL} involving the Raman response function to 12 fs ultrabroadband-pulse propagation in a silica fiber with the consideration of linear polarization P_L, including all exact Sellmeier-fitting values of silica with *three resonant frequencies,* in order to compare the calculated results with our experimental results. (Nakamura et al., 2002a), (Karasawa et al., 2000) This extended FDTD method (Nakamura et al., 2002b) is more accurate than previous FDTD methods (Joseph & Taflove, 1997), (Kalosha & Herrmann), (Goorjian et al., 1992), (Joseph et al., 1993), (Taflove & Hagness, 2000). Recently, a similarly accurate algorithm, which assumes that the linear and nonlinear dispersions are modeled by three and one Lorentzian resonances, respectively, was suggested by Goojian and Cundiff (Goojian & Cundiff, 2004). On the point of phase observation, the extended FDTD is superior to BPM (Yamashita et al., 1998) and (Yamashita et al., 1996) because the extended FDTD method includes no assumption for the two-pulse case, which implies that it can be used to calculate two different pulses simultaneously. We have recently numerically obtained the phase when two different pulses were *simultaneously* injected into a fiber and IPM occurred, and we clarified how the two pulse phases maintain their connection with each other (Nakamura et al., 2004a). We did not observe the four-wave mixing but only SPM and IPM with the simultaneous injection of two-color pulses into a fiber (Nakamura et al., 2004a). In this study, we observed the anti-Stokes light of the degenerate four-wave mixing (DFWM) in the IPM spectral phase obtained by the extended FDTD method with an initial delay of -80 fs. The chirped first and second pulses broadening with dispersion interact with each other, and the nonlinear interaction length is prolonged with this initial delay. Then the phase matching condition for DFWM at the end of the fiber is satisfied with an initial delay of -80 fs, which corresponds to the situation of two pulses meeting at the end of the fiber via a different group velocity of the two pulses. This is the first observation of the DFWM with IPM by FDTD. Furthermore, we compared the extended FDTD method (Nakamura et al., 2002b), (Nakamura et al., 2004b), (Nakamura et al., 2005) with *experimental results* (Xu et al., 1999) for nonlinear propagation with the IPM effect of two-color sub-picosecond laser pulses in a silica fiber that have *different initial delays* between the two pulses. To the best of our knowledge, this is the first comparison between FDTD calculation and *experimental results* for nonlinear propagation with the IPM effect of two-color 120 fs laser pulses in a silica fiber. The extended FDTD method clarifies how the two pulse phases maintain their connection with each other with *three initial time delays.*

4.2 Experimental setup

A schematic of our experimental setup (Xu et al., 1999) is shown in Fig. 10. The first pulse was generated from a Ti:sapphire regenerative amplifier at a 1 kHz repetition rate.

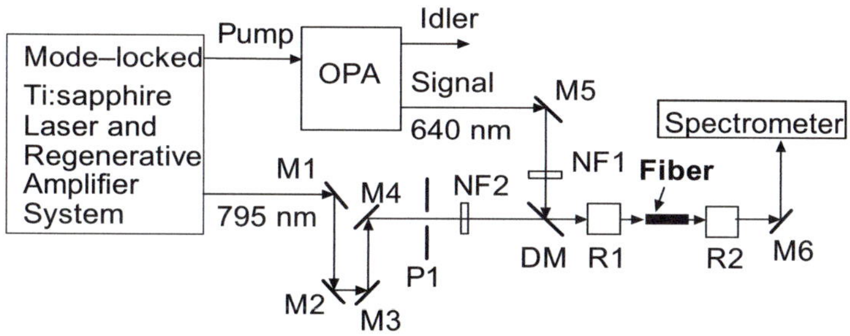

Fig. 10. Layout of the experimental setup.(Xu et al., 1999) M, silver-coated plane mirrors; R1 and R2, reflective objectives (36x); DM, dichroic mirror; NF1 and NF2, valuable optical attenuators; OPA, optical parametric amplifier.

The pulse duration was about 120 fs full width at half maximum (FWHM) with a bandwidth of 12 nm FWHM centered at 795 nm, and the output energy was 750 μJ, henceforth called the fundamental pulse. The main output energy from the regenerative amplifier was used to generate a continuum wave and pump an optical parametric amplifier (OPA) for the second pulse generation. The tuning range in the OPA was from 480 to 740 nm with an output energy of over 10 μJ at 1 kHz. In this experiment, we set the wavelength of the second pulse at 640 nm with a 120 fs FWHM in the pulse duration and a 12 nm FWHM in the spectral width, henceforth called the signal pulse. Since the continuum wave was generated by the SPM of the fundamental pulse under the single-filament condition and was selectively amplified using its second harmonic pulse by the optical parametric process with the phase matching condition, the phase of the signal wave should have a relationship to the phase of the fundamental wave. This condition is critical for generating a stable combined spectrum with IPM in a single-mode glass fiber, which will be confirmed later. The calculated walk-off length (Agrawal, 1995) L_w between the signal pulse and the fundamental pulse was about 2.7 mm. The reason the second pulse wavelength was selected to be 640 nm lies in the fact that the walk-off length L_w of these two pulses in a fused silica fiber (L_w = 2.7 mm) is around the length that we are able to cut easily, about 3 mm, and the IPM effect between the two pulses can be observed clearly at this length. The two pulses were combined by a dichroic mirror, DM, which has a high transmission for the fundamental pulse and a high reflection for the signal pulse. The optical path of the fundamental pulse was varied by an optical delay stage formed by M2 and M3 so that the delay of the two pulses in the fiber could be adjusted. N1 and N2 are variable neutral-density filters and the energies of the two pulses can be controlled independently. The combined two pulses, which have the same linear polarizations, were coupled into a 3 mm single-mode polarization-preserving fused-silica fiber (2.7 μm core diameter, 550 ± 50 nm cutoff wavelength) by a $36\times$ reflective objective, R1. The output from the fiber was collimated by the same type of reflective objective, R2, to a spectrometer that monitored the output spectrum.

4.3 Numerical results

In our calculation for a fused silica fiber, the parameters in eq. (10) are set as b_1 = 0.6961663, b_2 = 0.4079426, b_3 = 0.8974794, λ_1 = 0.0684043 μm, λ_2 = 0.1162414 μm, and λ_3 = 9.896161 μm (Agrawal, 1995), where $\lambda_i = 2\pi c/\omega_i$ and c is the velocity of light in vacuum. We used the value of the nonlinear refractive coefficient $n_2^I = 2.48 \times 10^{-20}$ m^2/W from ref. (Taylor et al., 1996), and the third-order susceptibility $\chi^{(3)}$ was found to be $\chi^{(3)} = 1.85 \times 10^{-22}$ m^2/V^2 at 800 nm, given by $\chi^{(3)} = (4/3)\varepsilon_0 cn(\omega_0)^2 n_2^I$, where ω_0 is the center angular frequency of the optical pulse. The parameters α, τ_1, and τ_2 in eq. (15) are set to be $\alpha = 0.7$, $\tau_1 = 12.2$ fs and $\tau_2 = 32$ fs. (Agrawal, 1995) A single time step of the finite difference is set as $\Delta t = 7.116034310 \times 10^{-17}$ s. The time step of Δt is defined as the single optical cycle divided by 60. In the extended FDTD calculation for two-pulse propagation in a silica fiber, we set all other parameters to be the same as those in the previous IPM experiment. (Xu et al., 1999) Both pulses had the same pulse width of 120 fs and the same energy of 20 nJ. The fiber was 3-mm long with a core diameter of 2.7 μm. The fiber length of 3 mm nearly equals the walk-off length of 2.7 mm. The effective core area A_{eff} was 5.47 μm^2. Two laser pulses at the center wavelengths of 795 and 640 nm (the fundamental and signal pulse) propagated in a 3 mm silica fiber with the delay of the signal pulse for the fundamental pulse, T_d = 0, -40, -80 fs. The delay of –40 fs corresponds to the situation in which the two pulses are launched simultaneously into the same fiber with an optimum initial delay between the two pulses that meet at the center of the fiber, to be discussed later, because of the IPM effect of the two pulses, the two spectra meet each other. The delay of –80 fs corresponds to the situation which two pulses meet at the end of the fiber. Since every delay case is performed close to the walk-off length, the nonlinear interaction length is sufficient.

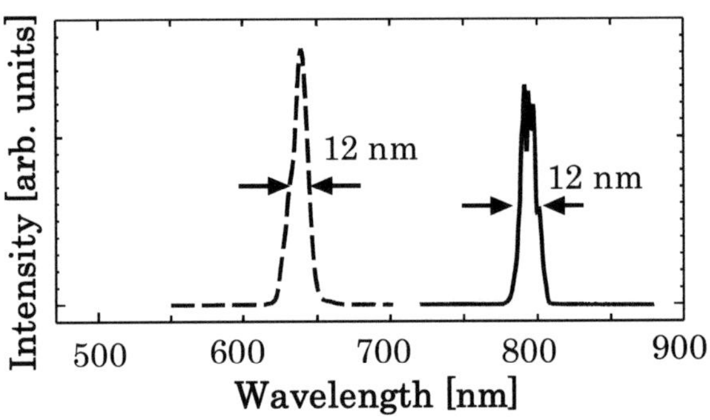

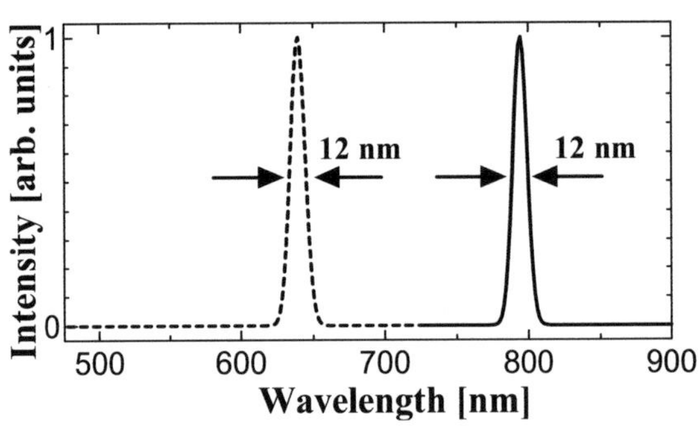

Fig. 11. Spectra obtained (a) experimentally (Xu et al., 1999) and (b) numerically (Nakamura et al., 2005b) for two pulses at center wavelengths of 795 nm (solid line) and 640 nm (dashed line) before propagation in a 3 mm silica fiber.

Figure 11 shows (a) the experimentally measured and (b) the numerically assumed spectra of the input fundamental (solid line) and signal (dashed line) pulses. The FWHM spectral width of $\Delta\lambda$ is 12 nm for both fundamental and signal pulses. The Fourier transform of a Gaussian spectrum with the FWHM spectral width of $\Delta\lambda$ =12 nm for the fundamental and signal pulses leads to a pulse width of 80.0 and 50.0 fs, respectively, which does not agree with the input pulse width of 120 fs. This fact indicates that the input pulses are chirped, which accords with the second derivation of the spectral phase of $\partial^2\varphi(\omega)/\partial\omega^2|_{\omega=\omega_0} = 4651$ and 2382 fs^2 for the fundamental and signal pulses, respectively. Fig. 11(b) is assumed by the Fourier transformation of the pulse width of 120 fs with the chirp values. Due to the group velocity difference between the two optical pulses propagating in the fiber, they travel at different speeds and after some propagation distance they are separated. We measured the IPM-induced spectra with different initial time delays between the two pulses.

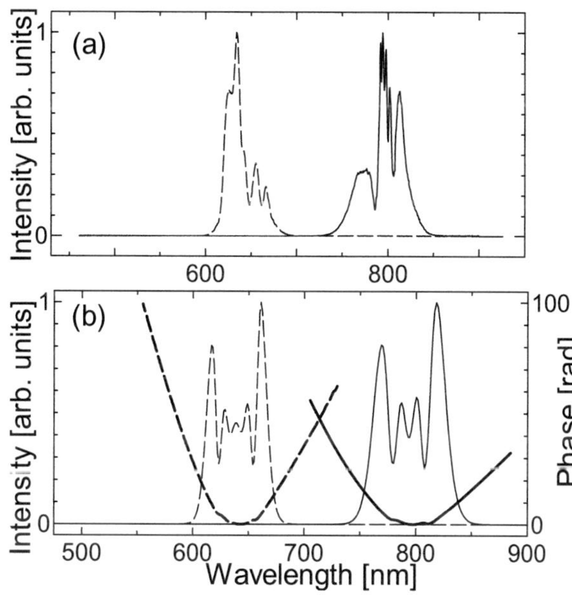

Fig. 12. (a) Experimentally (Xu et al., 1999) and (b) numerically (Nakamura et al., 2005b) obtained spectra induced by SPM in 3 mm single-mode fused-silica fiber. The thin solid line represents the spectrum induced by the SPM of the pulse at the center wavelength of 795 nm and the thin dashed line represents the spectrum induced by the SPM of the pulse at the center wavelength of 640 nm at the pulse duration of 120 fs and the energy of 20 nJ in the 3 mm fiber for each pulse. The thick lines in (b) are spectral phases obtained numerically as a function of wavelength for the 3 mm fiber-propagated pulses at center wavelengths of 795 nm (thick solid line) and 640 nm (thick dashed line).

Fig. 12 depicts the spectra of the two pulses with SPM alone at the energy of 20 nJ in the fiber for each pulse, which are obtained (a) experimentally and (b) numerically. The solid line represents the SPM-induced spectrum of the fundamental pulse and the dashed line represents the SPM-induced spectrum of the signal pulse. From Figs. 12(a) and 12(b) we observe the separated spectra for the two pulses with only the SPM effect at 20 nJ energy levels. Here, we define the wavelengths in which the spectral intensity becomes 1/100 of the maximum intensity as λ_{min} and λ_{max} for the shortest and longest wavelength, respectively. Then, we define the spectral width as $\Delta\lambda_{1/100} = \lambda_{max} - \lambda_{min}$. The spectral widths $\Delta\lambda_{1/100}$ of the dashed lines in Figs. 12(a) and 12(b) are 93.8 and 83.7 nm, respectively. The spectral widths $\Delta\lambda_{1/100}$ of the solid lines in Figs. 12(a) and 12(b) are 117.5 and 110.9 nm, respectively. The spectral widths in Fig. 12 caused by only SPM are smaller than those caused by SPM and IPM of Figs. 13 and 14, this will be discussed later. The difference in the spectral width between the experimental and numerical results is less than 10 nm (<10%), which indicates that the difference is small. Blue shift due to self-steepening is not observed but breakdown of the spectral symmetry is observed. The intensity of the spectrum is higher for wavelengths longer than the center wavelengths in Fig. 12(b). In particular, the solid line (fundamental pulse) in Fig. 12(b) shows asymmetrical behavior and three- or four-peak behavior, which are similar to those in the solid line in Fig. 12(a). The extended FDTD result and the experimental result in Fig. 12 are similar from the point of view of spectral widths. The thick lines in Fig. 12(b) show the numerically obtained spectral phase wherein the dashed line shows the phase of the fundamental pulse and the solid line shows that of the signal pulse. For both the fundamental and signal pulse, the phase curve is a symmetric parabolic curve. The spectral phases of the two pulses cross around the wavelength of 713 nm, where the phase value is 48.0 rad.

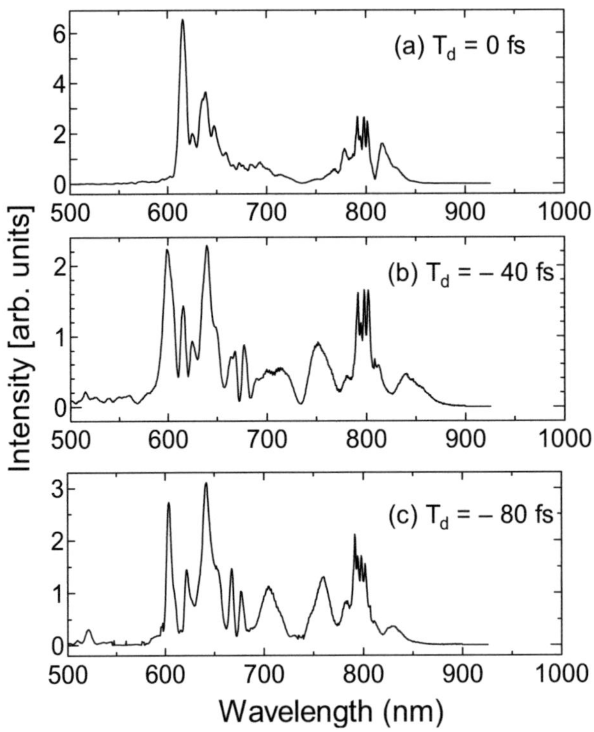

Fig. 13. Experimentally measured spectra (Xu et al., 1999) induced by IPM for two pulses at center wavelengths of 795 and 640 nm with different initial time delays after propagation in 3 mm silica fiber. (a) Spectrum when two pulses meet at entrance of fiber, (b) spectrum when two pulses meet at center of fiber and (c) spectrum when two pulses meet at end of fiber.

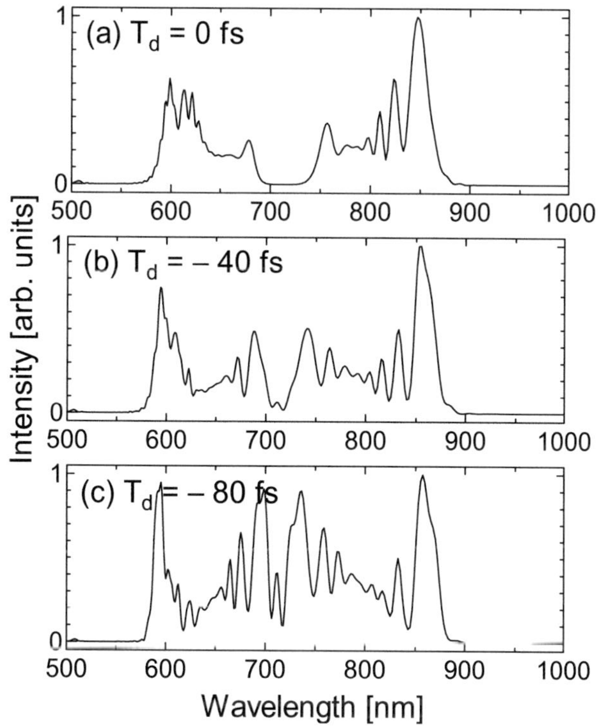

Fig. 14. Numerically obtained spectra induced by IPM for two pulses at center wavelengths of 795 and 640 nm with different initial time delays after propagation in 3 mm silica fiber. (a) Spectrum when two pulses meet at entrance of fiber, (b) spectrum when two pulses meet at center of fiber and (c) spectrum when two pulses meet at end of fiber. (Nakamura et al., 2005b)

Figures 13 and 14 show the IPM-induced spectra obtained by the previous experiment (Xu et al., 1999) and the extended FDTD calculation with different initial time delays between the two pulses, these figures indicate spectra after propagation of 3 mm fiber with the initial delays T_d of (a) 0, (b) –40, and (c) –80 fs, respectively. The pulse energies after fiber propagation of the two pulses are the same value, 40 nJ. The initial time delay between the two pulses, (a) T_d = 0 fs, represents that the two pulses coincide at the fiber entrance, and the delays, (b) T_d = -40 fs and (c) T_d = -80 fs, represent that the two pulses meet at the center and the end of the fiber, respectively. Figures 13 and 14 indicate that the spectra are all broader than that of Fig. 12, because not only SPM but also IPM occurs. The spectral bandwidth of Fig. 13(a) is $\Delta\lambda_{1/100}=\lambda_{max}-\lambda_{min}$= 851.1 - 568.9 = 282.2 nm. The spectral bandwidth of Fig. 14(a) is $\Delta\lambda_{1/100}=\lambda_{max}-\lambda_{min}$= 883.8 - 571.5 = 312.3 nm. Quantitatively, there is a difference between the experiment [Fig. 13(a)] and the calculation [Fig. 14(a)]. However, a larger gap between the two spectra is observed in both cases. This shows that the calculation qualitatively agrees well with the previous experimental results. In our case, when the two pulses meet at the entrance of the fiber T_d = 0 fs, the leading edge of the signal pulse interacts with the trailing edge of the fundamental pulse because the fundamental pulse travels faster than the signal pulse in the fused silica fiber. As a result, both the signal pulse and the fundamental pulse have IPM-induced chirp. This leads to a larger modulation of the signal pulse spectrum at shorter wavelengths and a larger spectral modulation of the fundamental pulse at longer wavelengths. Hence, the modulation of the signal pulse at longer wavelengths and of the fundamental pulse at shorter wavelengths is depressed. Then, the larger gap between the fundamental and signal spectra is formed. In contrast, when the two pulses meet at the end of the fiber T_d = -80 fs, the trailing edge (longer wavelengths) of the signal pulse mainly interacts with the leading edge (shorter wavelengths) of the fundamental pulse. As a result, the IPM-induced spectrum of the signal pulse shifts towards its longer wavelengths and the IPM-induced fundamental pulse spectrum shifts towards its shorter wavelengths and a larger overlap of the two pulse spectra is observed in both Figs. 13(c) and 14(c). This shows that the calculation qualitatively agrees well with the previous experimental results. The spectral bandwidth of Fig. 13(c) is $\Delta\lambda_{1/100}=\lambda_{max}-\lambda_{min}$= 861.8 - 582.9 = 278.9 nm. The spectral bandwidth of Fig. 14(c) is $\Delta\lambda_{1/100}=\lambda_{max}-\lambda_{min}$= 895.9 - 579.1 = 316.8 nm. It is found that when the two pulses meet at the center of the fiber, the fundamental pulse passes through the signal pulse in a symmetric manner. As a result, the spectra are broadened by the IPM effect to shorter and longer wavelengths simultaneously for both pulses and the broadest combined spectrum is generated in both Figs. 13(b) and 14(b). This shows that the calculation qualitatively agrees well with the previous experimental results. The spectral bandwidth of Fig. 13(b) is $\Delta\lambda_{1/100}=\lambda_{max}-\lambda_{min}$= 887.3 - 567.3 = 320.0 nm. The spectral bandwidth of Fig. 14(b) is $\Delta\lambda_{1/100}=\lambda_{max}-\lambda_{min}$= 896.6 - 577.6 = 319.0 nm. Figures 13(b) and 14(b) generally have the same value of $\Delta\lambda_{1/100}$. This case is referred to as the optimum initial time delay to produce the broadest spectrum. In Figs. 13 and 14, it is the same feature that the most broadened spectra are Fig. 13(b) and Fig. 14(b) with an initial delay of T_d = -40 fs. Consequently, the extended FDTD results and our previous experimental results have the same behavior of spectral bandwidth, which indicates the same initial delay dependence. That is, the extended FDTD calculation qualitatively agrees well with the previous experimental results although it does not agree quantitatively. This quantitative disagreement might be due to the low sensitivity of the detector of the spectrometer at wavelengths longer than 800 nm because the detector is made of Si. We suggest that InGeAs should be used for the detector, which is highly sensitive in the detection of the infrared region light.

Figure 15 shows the spectral phase obtained numerically as a function of wavelength for mixed two pulses at center wavelengths of 795 and 640 nm after simultaneous copropagation with an initial delay of (a) 0, (b) -40, and (c) -80 fs in a 3 mm silica fiber. Around the center wavelength (795 nm) of the fundamental pulse, the phase curve is parabolic and its symmetry axis is (a) 774.14, (b) 749.45, and (c) 775.58 nm. Around the center wavelength (640 nm) of the signal pulse, the phase curve is parabolic and its symmetry axis is (a) 673.99, (b) 676.90, and (c) 687.78 nm. The symmetry axis of every parabolic curve in Fig. 15 shifts toward the wavelength of 720 nm, which is the center wavelength between 795 and 640 nm. This feature is seen when the IPM effect occurs, so that this feature is not found only in the SPM case of Fig. 12. In addition, it is observed in Fig. 15 that the spectral phases of the two pulses at the wavelength λ of (a) 711.7, (b) 713.17, and (c) 713.17 nm are continuously connected at a phase value of $\phi(\lambda)$ = (a)17.76, (b) 10.46, and (c) 17.94 rad without any discrete point. Furthermore, regarding wavelengths shorter than 640 nm, one more parabolic curve is found between 500 and 570 nm. This curve is found to be that of the anti-Stokes light (theoretically 535.58 nm) generated by the induced DFWM where the pump light is 640 nm and the Stokes light is 795 nm. The chirped first and second pulses broadening by dispersion interact each other, and the nonlinear interaction length is prolonged with an initial delay of -80 fs. The DFWM is robust for chirping. (Cundiff et al., 1999), (Geraghty et al., 1998)

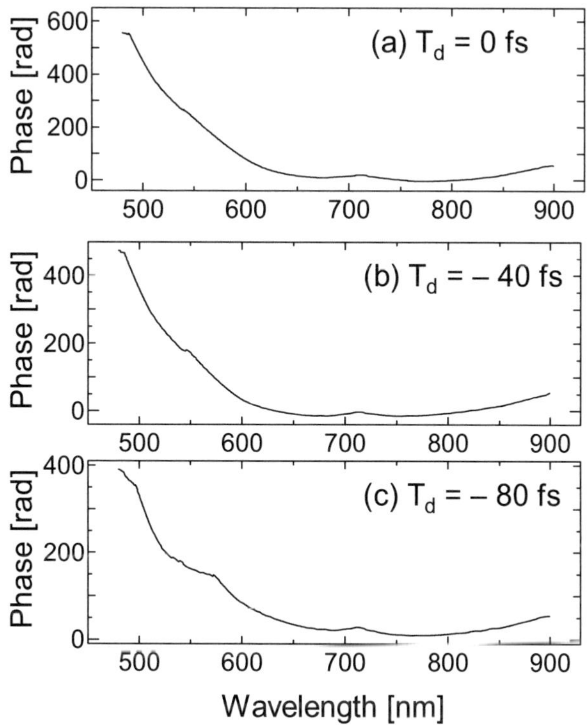

Fig. 15. Spectral phase obtained numerically as function of wavelength for mixed two pulses at center wavelengths of 795 and 640 nm after simultaneous copropagation with an initial delay of (a) 0, (b) -40, and (c) -80 fs in 3 mm silica fiber. (Nakamura et al., 2005b)

When the two pulses meet at the end of the fiber T_d = 80 fs, the trailing edge of the signal pulse mainly interacts with the leading edge of the fundamental pulse and the chirped pulses gradually become overlapping at the places where the pulses are temporally broadened; therefore, the interaction length is long. As a result, the four-wave mixing effect is enhanced and a larger parabolic phase curve of the anti-Stokes light in the spectral phase is observed. Then the phase matching condition for DFWM at the end of the fiber is satisfied with an initial delay of -80 fs, which corresponds to the situation in which two pulses meet at the end of the fiber via a different group velocity of the two pulses. Looking again at Fig. 15, the anti-Stokes light appears faintly in the phase with an initial delay of -40 as well, but no anti-Stokes light appears with an initial delay of 0 fs. The anti-Stokes light of DFWM appears when long chirped pulses finally overlap at the end of the fiber, the IPM effect is strong when short pulses overlap at the beginning of the fiber, and IPM and DFWM cross-interact and spectral broadening is maximum when the pulses overlap at the center of the fiber. Such precise information on the spectral phase of the pulse propagated in a fiber is very important for the wavelength conversion with spectral broadening or the generation of a monocycle pulse using a spatial light modulator (SLM) that can compensate the spectral phase of a superbroadened continuum. The shortest pulse attainable by phase correction of this ultrabroad spectrum is obtained by the Fourier transform of the spectrum of Fig. 14 and assuming a constant spectral phase. This yields a transform-limited pulse. Figure 16 shows the temporal profiles of the transform-limited pulses obtained numerically for the mixed two-pulse spectra shown in Fig. 14 at the center wavelengths of 795 and 640 nm after simultaneous copropagation with an initial delay of (a) 0, (b) -40, and (c) -80 fs in a 3 mm silica fiber. The FWHM pulse widths are (a) 4.1, (b) 4.0, and (c) 4.7 fs. The pulse width under an initial delay of –40 fs is 4.0 fs, which is the same value, 4 fs (Xu et al., 1999), obtained by Fourier transform of the experimental result of Fig. 15(b) reported by Xu et al.

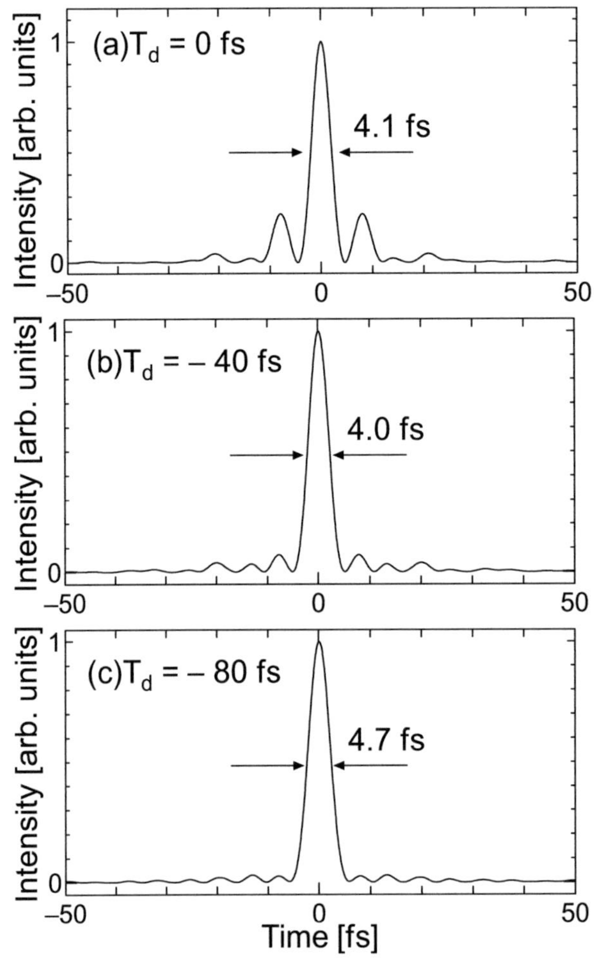

Fig. 16. Temporal profiles of transform-limited pulses obtained numerically for mixed two-pulse spectra shown in Fig. 6 at center wavelengths of 795 and 640 nm after simultaneous copropagation with initial delay of (a) 0, (b) -40, and (c) -80 fs in a 3 mm silica fiber. (Nakamura et al., 2005b)

4.4 Summary of dual wavelengths femtosecond pulses propagation

The ultrabroad spectrum generation based on the IPM of two optical pulses copropagating in a single-mode fused-silica fiber has been numerically demonstrated, which is compared with our previous experimental results. To the best of our knowledge, this is the first comparison between the extended FDTD calculation and the experiment. In the extended FDTD, it was found that the spectral phases of the two pulses around the wavelength of 710 nm are continuously connected at a phase value of $\phi(\lambda)$ = 10-20 rad without any discrete point. This fact was unknown when BPM was used previously. In the extended FDTD simulation, two 120 fs pulses with wavelengths at 640 and 795 nm were coupled into a 3 mm single-mode fiber. At a pulse energy of 20 nJ coupled into the fiber for both pulses, an ultrabroad coherent spectrum induced by IPM which covered the range from 567 to 897 nm was obtained and was compared with our previous experiment. This result generally agrees well with the previous experimental results at the optimum initial delay of -40 fs, which corresponds to the situation in which two pulses meet at the center of the fiber, where the spectral width is $\Delta\lambda_{1/100}$ = 320 nm. The Fourier transform of the spectrum can yield a 4.0 fs transform-limited pulse that is the same pulse width of the Fourier-transform-limited pulse of the previous experimental results. It opens a path for generating a single-cycle high-energy optical pulse in the near future. It is necessary to use not only fundamental (795 nm) and signal (640 nm) pulses, but also the idler pulse (1067 nm) of the OPA in order to obtain a single-cycle (2.66 fs at the center wavelength of 795 nm), which means three-color pulse propagation in a fiber.

Finally, we obtained the spectral phase after fiber propagation with the calculation. The extended FDTD method clarified how the two pulse phases maintain their connection with each other because it includes no assumption for the two-pulse case, which implies that the method can be used to calculate two different pulses simultaneously with three delays. We reconfirmed that the spectral phases of two different pulses are connected continuously in any case of *three initial delays*. We found in the IPM spectral phase that the DFWM occurs with an initial delay of -80 fs, which corresponds to the situation in which two pulses meet at the end of the fiber. To the best of our knowledge, this is the first simultaneous observation of DFWM and IPM by FDTD simulation.

5. Slowly varying envelope approximation breakdown in fiber propagation

5.1 Slowly varying envelope approximation breakdown

There has recently been significant interest in the generation of single-cycle optical pulses by optical pulse compression of ultrabroad-band light produced in fibers. There have been some experiments reported on ultrabroad-band-pulse generation using a silica fiber (Nakamura et al., 2002a), (Karasawa et al., 2000) and an Ar-gas-filled hollow fiber (Karasawa et al., 2001), and optical pulse compression by nonlinear chirp compensation (Nakamura et al., 2002a), (Karasawa et al., 2001) . For these experiments on generating few-optical-cycle pulses, characterizing the spectral phase of ultrabroad-band pulses analytically as well as experimentally is highly important.

Conventionally, the slowly varying envelope approximation (SVEA) in the beam propagation method (BPM) has been used to describe the propagation of an optical pulse in a fiber. (Agrawal, 1995) However, if the pulse duration approaches that of the optical cycle regime (<10 fs), this approximation becomes invalid. (Agrawal, 1995) It is necessary to use the finite-difference time-domain (FDTD) method (Joseph & Taflove, 1997), (Kalosha & Herrmann, 2000) without SVEA. (Agrawal, 1995) In previous reports, Goorjian and coworkers (Goorjian et al., 1992), (Joseph et al., 1993)), Joseph and coworkers (Joseph & Taflove, 1997) (Goorjian et al., 1992), (Joseph et al., 1993)), Taflove and coworkers (Joseph & Taflove, 1997) (Goorjian et al., 1992)-(Taflove & Hagness, 2000) and Hagness and coworkers (Goorjian et al., 1992), (Taflove & Hagness, 2000), (JGTH) proposed an excellent FDTD algorithm considering a combination of linear dispersion with one resonant frequency and nonlinear terms with a Raman response function.

We performed an experiment of 12 fs optical pulse propagation, as described in §4. In order to compare FDTD calculation results with the experimentally measured ultrabroad-band spectra of such an ultrashort laser pulse, we extend the JGTH algorithm to that considering all of the exact Sellmeier fitting values for ultrabroad-band spectra. Because of the broad spectrum of pulses propagating in the fiber, it becomes much more important to take an accurate linear dispersion into account. It is well known that at least two resonant frequencies are required for the linear dispersion to accurately fit the refractive index data. In a recent report, Kalosha and Herrmann considered the linear dispersion with two resonant frequencies and the nonlinear terms without the Raman effect.(Kalosha & Herrmann, 2000). For the single-cycle pulse-generation experiment, we must use at least the shortest pulse of 3.4 fs (Yamane et al., 2003) or sub-5 fs (Karasawa et al., 2001), (Cheng et al., 1998) or 7.1 fs (Nakamura et al., 2002a) or the commercially available 12 fs pulses. Such a time regime is comparable to the Raman characteristic time of 5 fs (Agrawal, 1995) in a silica fiber. Therefore, it is very important to consider not only an accurate linear dispersion of silica but also the Raman effect in a silica fiber in the few-optical-cycles regime. In addition, because of the high repetition rate and pulse intensity stability in particular, ultrabroad-band supercontinuum light generation and few-optical-cycles pulse generation by nonlinear pulse propagation in photonic crystal fibers (Ranka et al., 2000) and tapered fibers (Birks et al., 2000), which are both made of silica, have attracted much attention. We have extended the FDTD method, (Nakamura et al., 2002b) with nonlinear polarization P_{NL} involving the Raman response function (JGTH-algorithm) to 12 fs *ultrabroad-band*-pulse propagation in a silica fiber with the consideration of linear polarization P_L, including *all exact Sellmeier-fitting values* of silica with *three resonant frequencies*, in order to compare the calculation results with our experimental results(Nakamura et al., 2002a), (Karasawa et al., 2000b). In this section, we describe the details of the calculation algorithm of the extended FDTD method (Nakamura et al., 2002b) and we also compare the extended FDTD method (Nakamura et al., 2002b) with BPM by applying the split-step Fourier (SSF) method, which is the solution of a modified generalized nonlinear Schrödinger equation (MGNLSE) (Sone et al., 2002), with SVEA, precisely considering the same Raman response function as that of the extended FDTD method, and up to the fifth-order dispersion. Then, in the calculations, the pulse width is gradually shortened from 12 fs to 7 fs to 4 fs. Moreover, the soliton number N is established as 1 or 2. To the best of our knowledge, this is the first observation of the breakdown of SVEA as a function of the laser pulse width and soliton number by comparison between the extended FDTD (Nakamura et al., 2002b) and BPM calculations for the nonlinear propagation of a very short (< 12 fs) laser pulse in a silica fiber.

5.2 Beam propagation method

Conventionally, BPM for solving the generalized nonlinear Schrödinger equation (GNLSE) including SVEA has been used to describe the ultrashort-laser-pulse propagation in an optical fiber. The GNLSE is expressed as

$$\begin{aligned}\frac{\partial A}{\partial z} &= -\frac{j}{2!}\beta_2\frac{\partial^2 A}{\partial T^2}+\frac{1}{3!}\beta_3\frac{\partial^3 A}{\partial T^3}\\ &-\frac{\alpha_a}{2}A+jN^2\left[|A|^2 A+\frac{j}{\omega_0}\frac{\partial}{\partial T}\left(|A|^2 A\right)-T_R A\frac{\partial |A|^2}{\partial T}\right],\end{aligned} \tag{37}$$

$$N^2 = \gamma P_0 T_0^{\,2} / |\beta_2| \,,\; \gamma = n_2 \omega_0 / (c A_{\text{eff}}) \,, \tag{38}$$

where A is the slowly varying amplitude of the pulse envelope of an electric field, which is normalized by N, α_a is the absorption coefficient, j is an imaginary unit, T_0 is the incident pulse width which is expressed using the full width at half maximum (FWHM) pulse width t_p as $T_0 = t_p / (2\sqrt{\ln 2})$ for a Gaussian pulse and $T_0 = t_p / \{2\ln(1+\sqrt{2})\}$ for a sech^2 pulse, c is the velocity of light in vacuum, ω_0 is the center angular frequency of the incident pulse, n_2 is the nonlinear refractive index (m^2/W), P_0 is the incident pulse peak power, A_{eff} is the effective core area, T_R is the Raman time constant and T_R =5 fs. (Agrawal, 1995) β_n $(n = 1, 2, 3, \cdots)$ is the n-th-order derivative of the propagation constant β_0 at an incident center angular frequency ω_0 is as follows:

$$\beta_n = \left(\frac{d^n \beta}{d\omega^n} \right)_{\omega=\omega_0} . \tag{39}$$

T is defined as

$$T = t - \frac{z}{v_g} - t - \beta_1 z \,, \tag{40}$$

where v_g is the group velocity of the optical pulse, t is time and z is the distance from the incident edge of a fiber. In our previous letter (Nakamura et al., 2002b), we also used this GNLSE for comparison with the extended FDTD method. However, a linear approximation of the actual Raman gain curve using the Raman time constant of T_R = 5 fs (Agrawal, 1995), which is related to the slope of the Raman gain, and a dispersion approximation up to only the third-order dispersion terms are included in GNLSE. In this section, we use a more precise version of the nonlinear Schrödinger equation, MGNLSE (Sone et al., 2002), (Gross & Manassah, 1992), that includes a Raman response function which is also found in the extended FDTD and up to fifth-order dispersion as follows:

$$\begin{aligned} \frac{\partial A}{\partial z} &= -\frac{j}{2!}\beta_2 \frac{\partial^2 A}{\partial T^2} + \frac{1}{3!}\beta_3 \frac{\partial^3 A}{\partial T^3} + \frac{j}{4!}\beta_4 \frac{\partial^4 A}{\partial T^4} - \frac{1}{5!}\beta_5 \frac{\partial^5 A}{\partial T^5} \\ &- \frac{\alpha_a}{2} A + jN^2 \left[|A|^2 A + \frac{j}{\omega_0}\frac{\partial}{\partial T}\left(|A|^2 A \right) + \Im^{-1}\left[\chi_R \Im\left[|A|^2 \right] \right] A \right] , \end{aligned} \tag{41}$$

where the symbols $\Im$ and $\Im^{-1}$ are the operators of Fourier transformation and inverse Fourier transformation, and χ_R is the generalized Raman-scattering susceptibility. The generalized Raman-scattering susceptibility (Gross & Manassah, 1992) can be approximated in the harmonic oscillator model for the molecular vibrations as

$$\chi_R(\Omega) = \frac{\chi_0(\Omega_R \Gamma_R)}{\Omega_R^{\,2} - \Omega - j\Gamma_R \Omega_R} , \tag{42}$$

where $\Omega = (\omega - \omega_0) T_0$ is the angular frequency normalized to the pulse width, and Ω_R (=13.2 THz$\times$ $2\pi T_0$) (François, 1991) is the Raman shift normalized to the pulse width, i.e., the molecular vibrational frequency multiplied by T_0, and Γ_R (=15.44 THz × $2\pi T_0$) (François, 1991) is the normalized phenomenological linewidth.

5.3 Experiment

Figure 1 shows the setup used for the our experiments. (Nakamura et al., 2002a), (Karasawa, et al., 2000). The 12 fs and 10 nJ pulses (sech^2) at the center wavelength of 800 nm were generated from a mode-locked Ti:sapphire laser (Femtolaser GmbH, Femtosource M-1). The incident pulse width was measured by a fringe-resolved autocorrelator (FRAC). The 12 fs pulses were coupled into a 2.5 mm silica fiber by a 36x reflective objective (Ealing). The advantage of this kind of reflective objective is that no additional group-delay dispersion (GDD) or third-order dispersion (TOD) is introduced to the pulses. The peak power of the input pulse was 175 kW. A single-mode fused silica fiber (Newport F-SPV) with a core diameter of 2.64 μm was used. The output from the 2.5 mm fiber was collimated by another reflective objective and measured by a spectrometer (Ocean Optics, S-2000). The obtained input and output spectra of the fiber are shown in Fig. 2.

5.4 Numerical results

A. *Comparison between experimental and numerical results*

In our calculations for a fused silica fiber, the parameters in eq. (10) are set as b_1 = 0.6961663, b_2 = 0.4079426, b_3 = 0.8974794, λ_1 = 0.0684043 μm, λ_2 = 0.1162414 μm, and λ_3 = 9.896161 μm (Agrawal, 1995), where $\lambda_i = 2\pi c/\omega_i$ and c is the velocity of light in vacuum. We use the value of the nonlinear refractive coefficient $n_2 = 2.48\times10^{-20}$ m^2/W from ref. (Taylor et al., 1996), and the third-order susceptibility $\chi^{(3)}$ is found to be $\chi^{(3)} = 1.85\times10^{-22}$ m^2/V^2 at 800 nm, as given by $\chi^{(3)} = (4/3)\varepsilon_0 c n(\omega_0)^2 n_2$, where ω_0 is the center angular frequency of the optical pulse. The parameters α, τ_1, and τ_2 in Eq. (15) are set to be $\alpha = 0.7$, τ_1 = 12.2 fs and τ_2 = 32 fs (Agrawal, 1995). A single time step of the finite difference is set as $\Delta t = 4.4475215\times10^{-17}$ s at the wavelength of 800 nm. The time step of Δt at 800 nm is defined as the optical cycle at 800 nm of 2.6666667 fs divided by 60. The time step for the wavelength of 1550 nm is defined by the same rule as described above.

In the extended FDTD calculation, we set all parameters to be the same as those in our experiment (Nakamura et al., 2002a), (Karasawa et al., 2000). We compare the results of the extended FDTD calculation and the SSF calculation with the experimental result for the pulse peak power of 175 kW in order to generate an ultrabroad spectrum which can finally be compressed to 7.1 fs (Nakamura et al., 2002a) or shorter. The total fiber length of L = 2.5 mm corresponds to 136,500 spatial steps, which means that $L = 136{,}500\times\Delta z$, where Δz is a unit spatial step in the z direction. We need 293,000 time steps to measure the electric field up to the complete passage of the pulse tail. The peak power of the input pulse is set to be 175 kW (soliton number N = 2.09). The initial temporal pulse form is assumed to be Fourier-transformed sech^2 because the mode-locked pulse generally has the sech^2 shape. Hence the input spectrum is naturally assumed to have a sech^2 shape. The initial pulse width is 12 fs (FWHM). The effective core area A_{eff} is set to be 5.47 μm^2.

Figure 17(a) shows the results calculated by the extended FDTD Maxwell equation method (A), the solution of MGNLSE obtained using the SSF method with SVEA (B) (up to the fifth-order dispersion terms with the Raman term using the Raman response function), and our previously reported experimental result (Karasawa et al., 2000) (C). It is seen that with

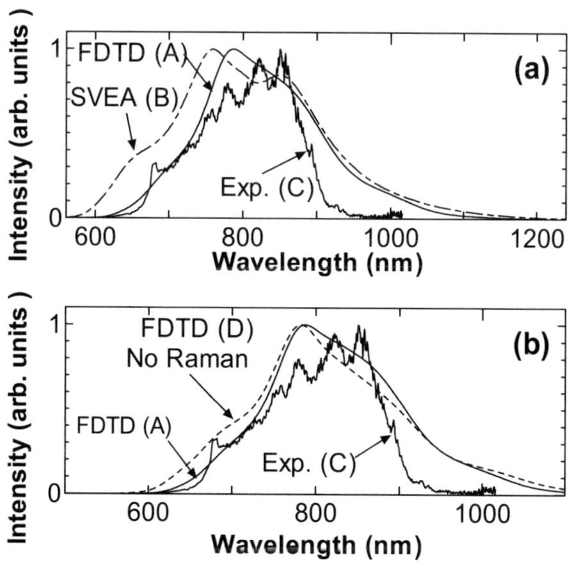

Fig. 17. (a) Spectra of 12 fs laser transmission through a 2.5 mm silica fiber calculated using (A) the extended FDTD Maxwell equation considering all orders of dispersions and the Raman response [α = 0.7 in eq. (15)] and (B) the solution of MGNLSE obtained using the SSF method with SVEA (considering up to 5th-order dispersion terms and the Raman term using the Raman response function), and (C) our previously reported experimental result (Karasawa et al., 2000), where the incident laser intensity corresponds to the soliton number of 2.09. (b) Spectra calculated by the FDTD Maxwell equation method (D) without the Raman response [α = 1 in eq. (15)] and (A) with the Raman response. (A) and (C) are the same as those in (a). (Nakamura et al., 2004)

SVEA (B), the spectral intensity at short wavelengths and at long wavelengths is much higher and slightly higher than those for FDTD (A) and the experimental result (C), respectively. The shortest wavelengths (intensity of 1%) of the spectra of FDTD (A), SVBA (B) and the experimental result (C) are 600 nm, 560 nm and 600 nm, respectively. The longest wavelengths of FDTD (A), SVBA (B) and the experimental result (C) are 1160 nm, 1240 nm and 1016 nm (maximum measurable wavelength of spectrometer), respectively. The FWHM spectral bandwidths $\Delta\lambda$ of FDTD (A), SVBA (B) and the experimental result (C) are 172 nm, 214 nm and 136 nm, respectively. Thus the spectral bandwidth of the experimental result (C) is narrow and that of FDTD (A) is closer to the experimental result (C) than is that of SVEA (B). This indicates that the extended FDTD directly solving Maxwell equation is superior to BPM in which MGNLSE (Sone et al., 2002) is solved by SSF with SVEA. The Raman response function (Gross et al., 1992), which is also included in the extended FDTD, and up to the fifth-order dispersion terms are accurately included in BPM for solving MGNLSE (Sone et al., 2002) by SSF with SVEA (B). However, in BPM for solving MGNLSE by SSF with SVEA, the second derivative of the electric field with respect to z, $\partial^2 E_y / \partial z^2$, is neglected, which corresponds to neglecting the backward-propagating wave. On the other hand, our extended FDTD Maxwell equation method (A) accurately includes the delayed Raman response and all orders of dispersion in silica using Sellmeier's equation, and does not require SVEA. Thus, the difference between (A) and (B) is considered to be due to the higher order (more than 6th order) dispersion effect, or the backward-propagating wave. For wavelengths longer than 800 nm in Fig. 17(a), the extended FDTD result and the BPM result are similar but the intensity of the experimental data is lower than those of both calculated results. We assume the sensitivity of the spectrometer detector to be low at wavelengths longer than 800 nm because the detector is made of silicon. We believe our extended FDTD result is accurate. We suggest that it is better to use an IR detector made of a material such as InGaAs for the spectrometer.

Next, in order to clarify the importance of the Raman response, we performed a calculation using (D) the FDTD Maxwell equation method without the Raman response [$\alpha = 1$ in eq. (15)], as shown in Fig. 17(b), where (A) and (C) show the same data as those in Fig. 1(a). In Fig. 17(b), the spectrum for case (A) is closer to the experimental result (C) than that of the case of FDTD in which the Raman effect is not considered (D). It is evident that by including the Raman term (A), the spectral intensity at a shorter wavelength is lower, and the agreement between the experimental and calculated results becomes better than that in the case of (D). For example, the spectral intensity at 700 nm in (D) is 48% higher than that in (A), which is almost the same as that in the experimental result (C). On the other hand, at a longer wavelength, for example, 850 nm, the spectral intensity of (A) is 15% higher than that of (D). This feature of (A) shows a tendency analogous to that of (C) because there is a higher peak at 850 nm than at the center wavelength of 800 nm in (C). These tendencies of the spectral characteristics indicate that it is important to include the Raman term.

B. *Observations of breakdown of SVEA*

We calculate the time profiles and spectra in silica fibers with lengths of 0 to $2L_D$, where L_D is the dispersion length, with Gaussian input pulses of 12 fs, 7 fs and 4 fs duration (FWHM) at the center wavelengths of 800 nm and 1550 nm which simulate the normal group-velocity dispersion (GVD) and anomalous GVD, respectively. The dispersion lengths L_D at 800 nm and 1550 nm are 1.4364 mm and 1.8586 mm for the 12 fs pulse, respectively. Power regions in this calculation are selected to be sufficiently small to enable comparison with results based on the previous theory (Karasawa et al., 200b), (Agrawal, 1995), which means that we normalize the peak power of the input pulse by a soliton parameter of $N = 1$.

Figures 18(a) and 18(b) show the time profiles of the output pulses from the silica fiber with incident pulse width of 12 fs and fiber lengths of 0 to $2L_D$, where L_D is the dispersion length (Agrawal, 1995), calculated by the extended FDTD and BPM with SVEA, respectively. The dispersion length L_D for 12 fs pulses at 800 nm is 1.4364 mm. The pulse widths simply broaden with propagation in the normal dispersion regime, as shown in Figs. 18(a) and 18(b), because both GVD and self-phase modulation (SPM) (Agrawal, 1995) produce up-chirp in normal dispersion. Figures 18(c) and 18(d) show the Fourier-transformed spectra of Figs. 18(a) and 18(b), where ν and ν_0 are the frequency and the center frequency of the pulse, respectively. The symmetric spectral broadening due to the SPM effect is seen in the normal dispersion regime, as shown in Figs. 18(c) and 18(d).

Figure 19 show the case of the pulse width of 7 fs under the same definitions of (a) to (d) as in Fig. 18, where the dispersion length L_D for 7 fs pulses at 800 nm is 0.48939 mm. This figure indicates that pulse width broadening by positive GVD and spectral broadening by the SPM effect are greater than those in Fig. 18, and there is no difference between the results calculated using FDTD [Figs. 19(a) and 19(c)] and BPM [Figs. 19(b) and 19(d)]. This means that there is no breakdown of SVEA. Figure 20 shows the case of the pulse width of 4 fs under the same definitions of (a) to (d) as in Figs. 18 and 19, where the dispersion length L_D for 4 fs pulses at 800 nm is 0.15980 mm. This figure also indicates that pulse width broadening and spectral broadening effects are greater than those in Figs. 18 and 19, and there is again no difference between the results calculated using FDTD [Figs. 20(a) and 20(c)] and BPM [Figs. 20(b) and 20(d)]. However, focusing on the case of $z = 2L_D$ in Fig. 20(a) calculated by the extended FDTD, we see that the main peak is slightly doubled or slightly dipped. On the other hand, only one peak exists in the results of BPM in Fig. 20(b). In this case, a very small difference is found but it need not be taken into account. That is, there is no breakdown of SVEA even in the case of a pulse width of 4 fs. It is very useful to know that we can use BPM even for extremely short pulses of 4 fs without breakdown of SVEA in the case of a soliton number of 1 in the normal GVD region.

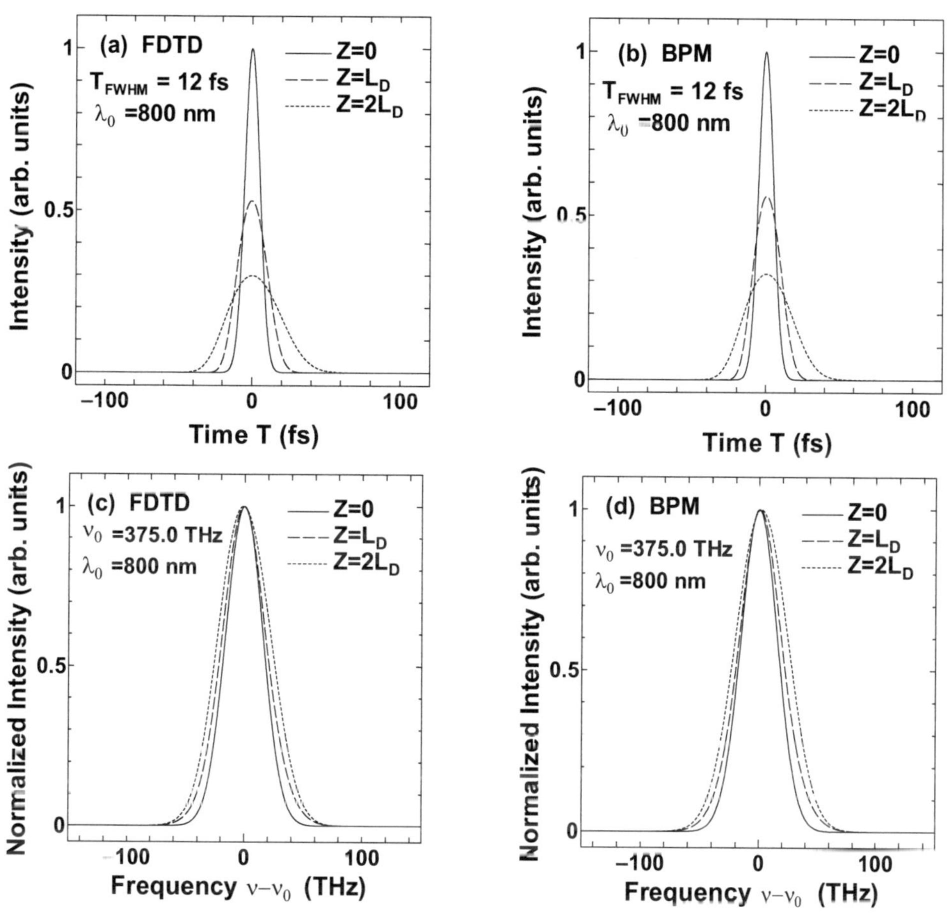

Fig. 18. (a), (b) Temporal profiles and (c), (d) spectra numerically obtained by (a), (c) the extended FDTD method and (b), (d) BPM for 12 fs, 800 nm laser propagation through a silica fiber of up to twice the dispersion length L_D, where the laser intensity corresponds to the soliton number of 1. (Nakamura et al., 2004)

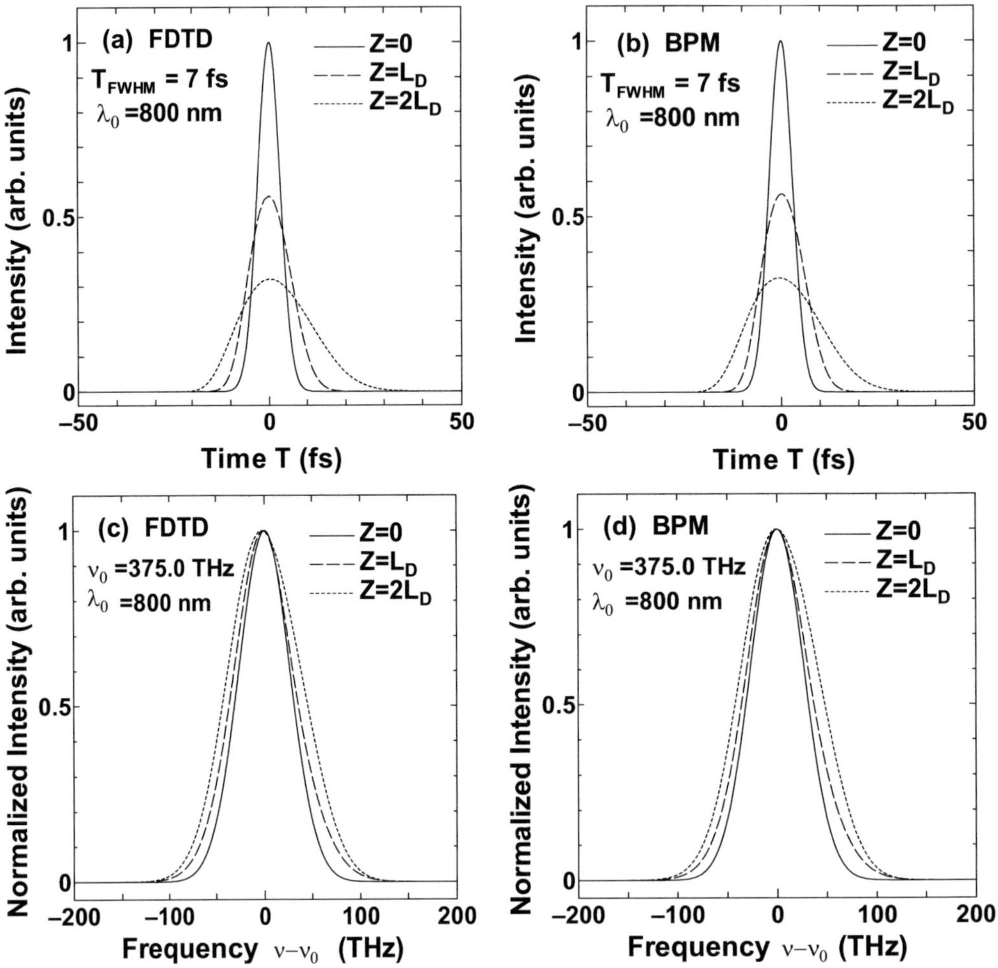

Fig. 19. (a), (b) Temporal profiles and (c), (d) spectra numerically obtained by (a), (c) the extended FDTD method and (b), (d) BPM for 7 fs, 800 nm laser propagation through a silica fiber of up to twice the dispersion length L_D, where the laser intensity corresponds to the soliton number of 1. (Nakamura et al., 2004)

On the other hand, in the anomalous dispersion regime, it is seen in Figs. 21(a) and 21(b) that the pulse widths do not vary with pulse evaluation because the soliton effects (Agrawal, 1995) occur as a result of interplay between SPM and anomalous GVD. The dispersion length L_D for 12 fs pulses at 1550 nm is 1.8586 mm. However, the details of Figs. 21(a) and 21(b) show that the pulse peaks are delayed and appear shifted toward the trailing side with propagation. This is due to self-steepening. (Agrawal, 1995) There are three small peaks following the main peak. These are formed by third-order or higher dispersions. In Figs. 21(a) and 21(b), the second largest peaks slightly differ from each other, and this difference also exists in Figs. 22(a) and 22(b). Figures 21(c) and 21(d) show the Fourier-transformed spectra of Figs. 21(a) and 21(b), where ν and ν_0 are the frequency and the center frequency of the pulse, respectively. There is no obvious change of spectral broadening, as shown in Figs. 21(c) and 21(d), because the fundamental soliton is generated, as already described.

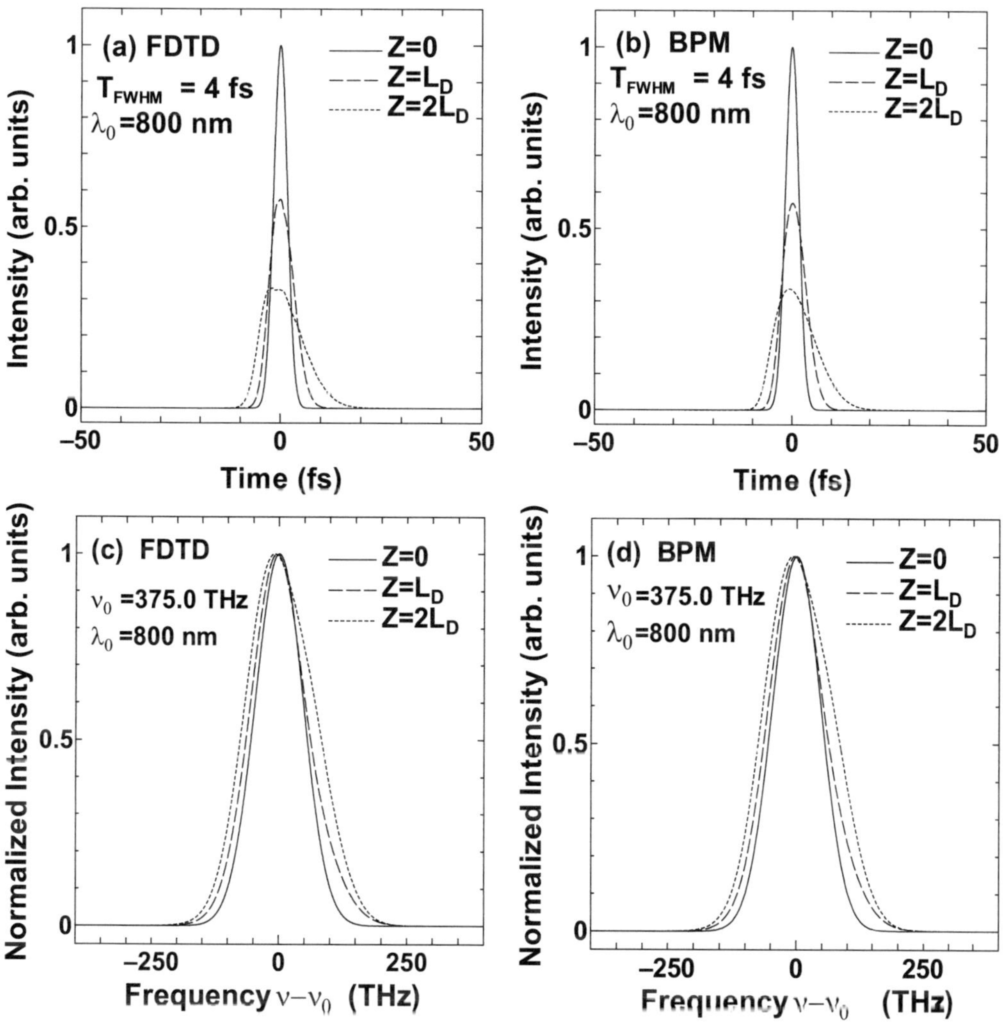

Fig. 20. (a), (b) Temporal profiles and (c), (d) spectra numerically obtained by (a), (c) the extended FDTD method and (b), (d) BPM for 4 fs, 800 nm laser propagation through a silica fiber of up to twice the dispersion length L_D, where the laser intensity corresponds to the soliton number of 1. (Nakamura et al., 2004)

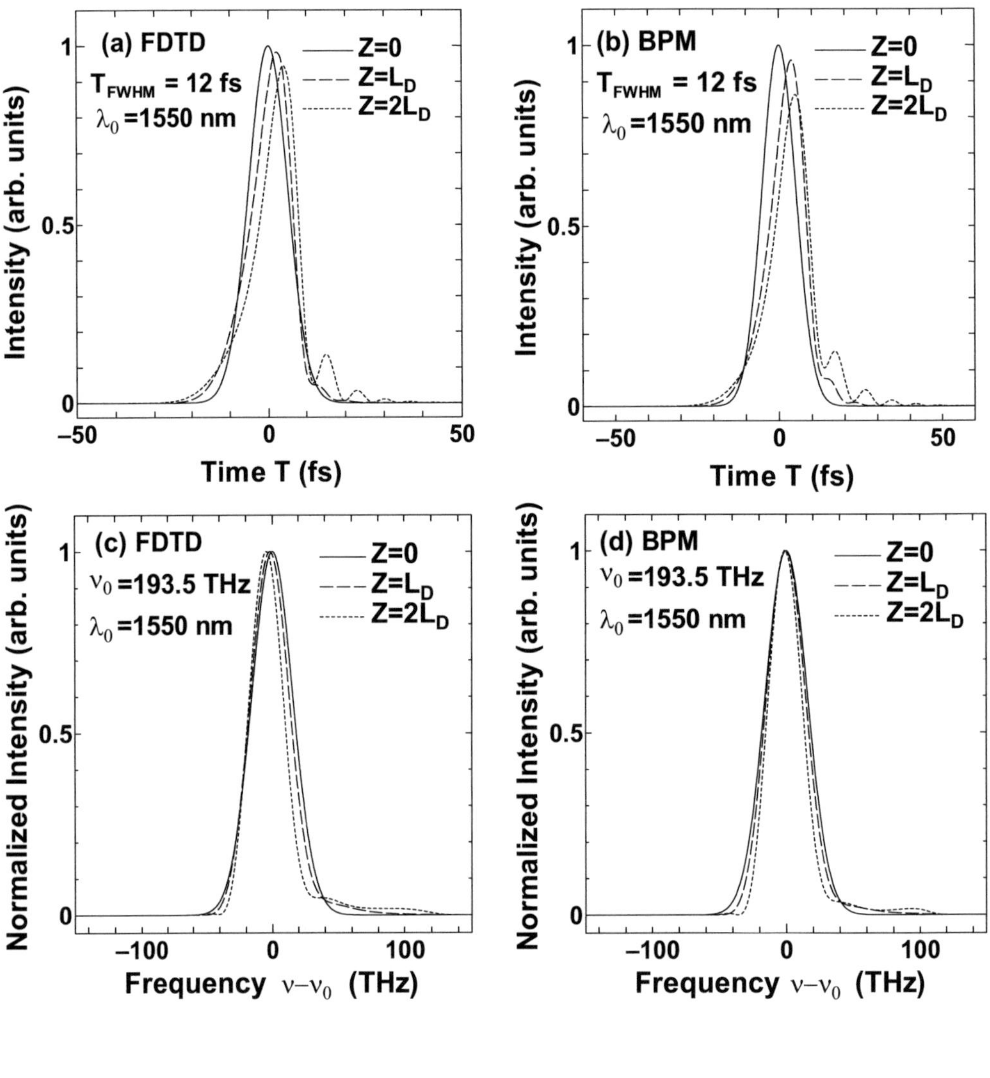

Fig. 21. (a), (b) Temporal profiles and (c), (d) spectra numerically obtained by (a), (c) the extended FDTD method and (b), (d) BPM for 12 fs, 1550 nm laser propagation through a silica fiber of up to twice the dispersion length L_D, where the laser intensity corresponds to the soliton number of 1. (Nakamura et al., 2004)

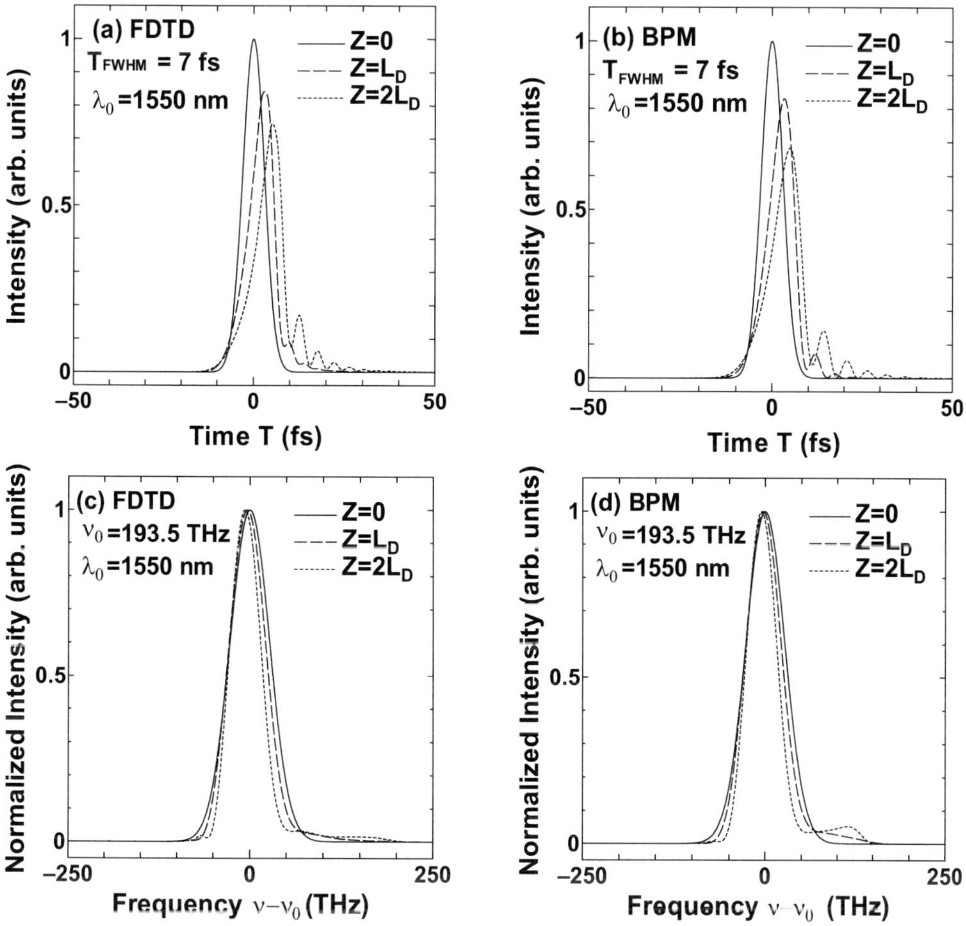

Fig. 22. (a), (b) Temporal profiles and (c), (d) spectra numerically obtained by (a), (c) the extended FDTD method and (b), (d) BPM for 7 fs, 1550 nm laser propagation through a silica fiber of up to twice the dispersion length L_D, where the laser intensity corresponds to the soliton number of 1. (Nakamura et al., 2004)

However, it is seen that there is a small spectral continuum component from $\nu - \nu_0$ = 20 to 120 THz in Figs. 22(c) and 22(d). This feature is due to blue shift (Agrawal, 1995) by self-steepening. This effect becomes more obvious in the case of the pulse widths of 7 fs [Figs. 22(c) and 22(d)] and 4 fs [Figs.23(c) and 23(d)], where the dispersion lengths L_D for 7 fs and 4 fs pulses at 1550 nm are 0.63324 mm and 0.20677 mm, respectively. We also note that the spectra at z = 2L_D in Figs. 21(c) and 21(d) shift slightly to a lower frequency due to the Raman effect. (Agrawal, 1995) The Raman shift in Fig. 21(c) is larger than that in Fig. 21(d). This Raman effect also become more obvious in the case of the pulse widths of 7 fs [Figs. 22(c) and 22(d)] and 4 fs [Figs.23(c) and 23(d)], and the Raman shift difference between (c) and (d) become smaller with decreasing pulse width. In terms of the Raman effect, FDTD and BPM result in analogous behavior. On the other hand, the behavior of blue shift differs between Fig. 23(c) and Fig. 23(d). The second largest slow perceptible peak exists from ν–ν_0 = 100 to 200 THz in Fig. 23(d), but the frequency region is a small continuum in Fig. 23(c).

From the total viewpoint of the anomalous region, the breakdown of SVEA is tolerant in 12 fs, however, in the time region shorter than 7 fs, the breakdown cannot be ignored.
Through these calculations, we successfully demonstrated the dispersion and SPM effect (Agrawal, 1995), soliton propagation at 1550 nm, self-steepening, and the Raman effect. Furthermore it was found that BPM and the extended FDTD results are almost the same in the normal GVD region, and SVEA does not break down in the cases of incident pulse widths of 12 fs, 7 fs, and 4 fs and soliton number of $N = 1$ in the normal dispersion frequency region, as well as in the case of the incident pulse width of 12 fs and soliton number of $N = 1$ in the anomalous dispersion frequency region.

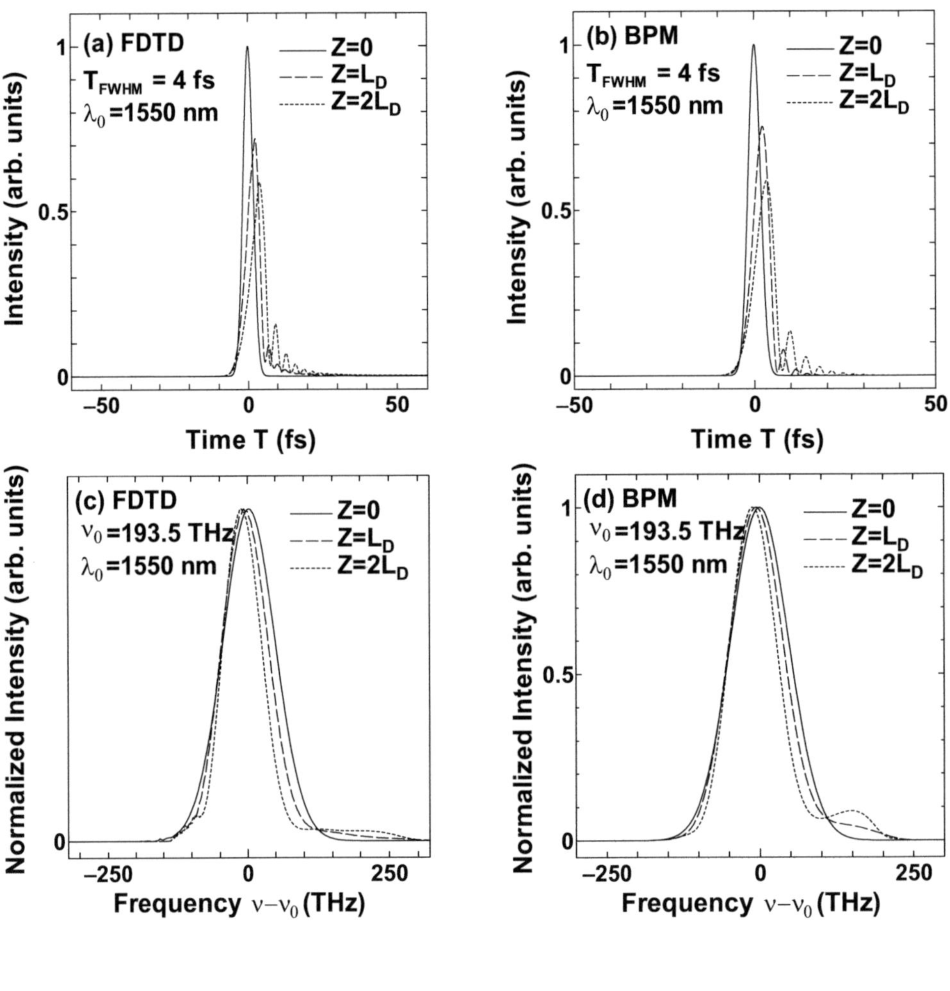

Fig. 23. (a), (b) Temporal profiles and (c), (d) spectra numerically obtained by (a), (c) the extended FDTD method and (b), (d) BPM for 4 fs, 1550 nm laser propagation through a silica fiber of up to twice the dispersion length L_D, where the laser intensity corresponds to the soliton number of 1. (Nakamura et al., 2004)

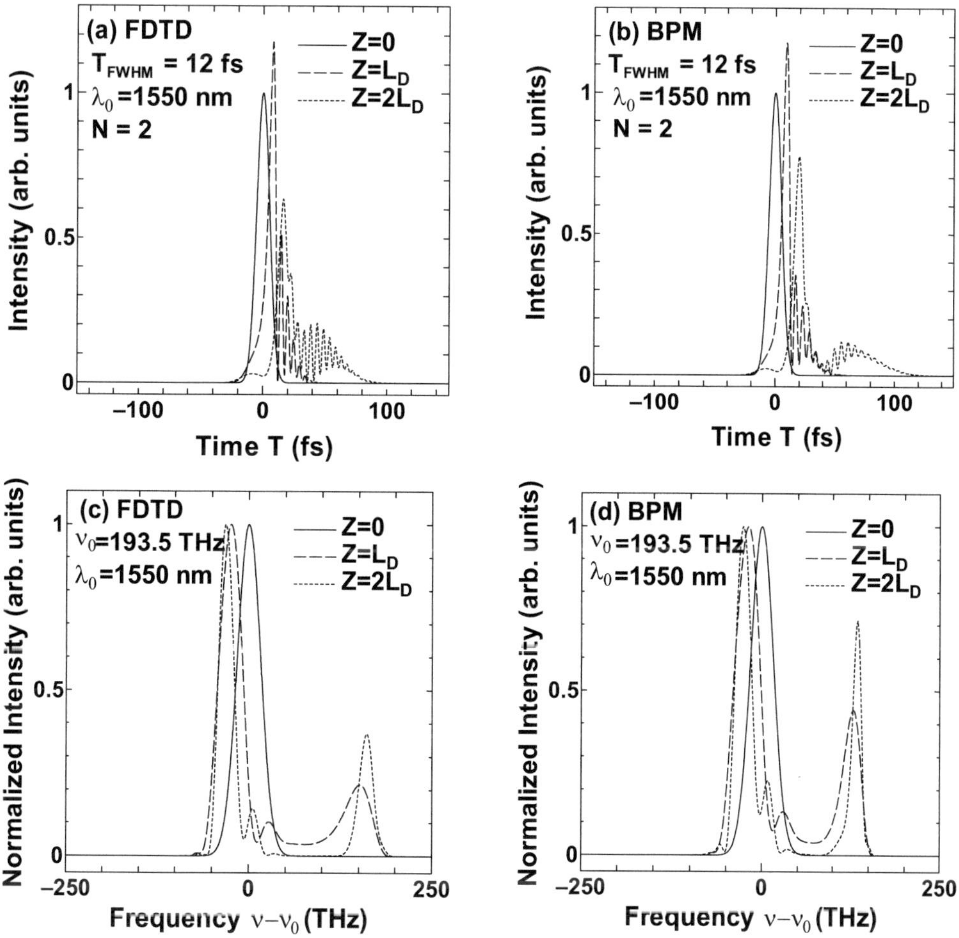

Fig. 24. (a), (b) Temporal profiles and (c), (d) spectra numerically obtained by (a), (c) the extended FDTD method and (b), (d) BPM for 12 fs, 1550 nm laser propagation through a silica fiber of up to twice the dispersion length L_D, where the laser intensity corresponds to the soliton number of 2. (Nakamura et al., 2004)

Although we obtained extended FDTD results which could be explained by the previous theory (Agrawal, 1995) and could verify the validity of our calculation method, the physical behavior of very short pulses will be different when the peak power is four times larger than that in the case of $N = 1$, which makes the breakdown of SVEA possible even for long pulses. We have already shown the breakdown of SVEA in a comparison of experimental and numerical results for the positive dispersion area with $N = 2.09$ for 12 fs pulse propagation in the fiber. Thus, we perform calculations for $N = 2$ and the anomalous dispersion area.
Figure 24 shows (a), (b) temporal profiles and (c), (d) spectra numerically obtained by (a), (c) the extended FDTD method and (b), (d) BPM for 12 fs, 1550 nm laser propagation through a silica fiber of up to $z = 2L_D$, where the laser intensity corresponds to the soliton number N of 2. This figure shows that the breakdown of SVEA occurs in BPM for even a long pulse width of 12 fs. For the spectra shown in Figs.24(c) and 24(d), the height of the second largest peak

in Fig. 24(d) is twice that in Fig. 10(c). In the temporal profiles of Figs. 24(a) and 24(b), the main peaks have the same form, however numerous small-peak characteristics from 20 fs to 80 fs are different. The small peaks in Fig. 24(a) are larger and more obvious than those in Fig. 24(b). These results imply that SVEA of MGNLSE in BPM, as assumed in Figs. 24(b) and 24(d), breaks down in the case of $N = 2$ in the anomalous GVD region for 12 fs pulse propagation.

5.5 Summary of SVEA breakdown

We have extended the nonlinear FDTD method with a JGTH algorithm to one with exact Sellmeier fitting values in order to compare the experimental and calculated results of nonlinear femtosecond ultrabroad-band-pulse propagation in a silica fiber, and have analyzed the spectral characteristics of the propagated pulse. This extended method is robust against the breakdown of the SVEA. The spectrum obtained in our previous experiment agrees better with the spectrum calculated by the extended FDTD method than with that calculated by the solution of MGNLSE obtained using the SSF method with SVEA. Consequently, we confirmed that the various nonlinear physical phenomena in an optical fiber can be described by FDTD methods even in a nearly single-cycle time regime, and that SVEA in BPM is invalid in the case of pulse widths less than 4 fs with the soliton number of 1 at 800 nm, which is the normal dispersion regime, and in the case of pulse widths less than 7 fs with the soliton number of 1 at 1550 nm, which is the anomalous dispersion regime. In the case of the soliton number of 2, the breakdown of SVEA occurs for 12 fs pulses in both the normal and anomalous dispersion frequency regimes. To the best of our knowledge, this is the first observation of the breakdown of SVEA as a function of laser pulse width and soliton number by comparison of the extended FDTD and BPM calculations for the nonlinear propagation of an ultrashort (< 12 fs) laser pulse in a silica fiber. If the extended FDTD method can be established as an ultrashort-optical-pulse propagation analysis technique, its application to analyses of the characteristics of an ultrahigh-speed optical switch, the propagation characteristic of optical pulses in a photonic crystal, and the nonlinear propagation in a photonic crystal fiber or a tapered fiber can be expected.

6. Conclusion

Conventionally, the BPM for solving the GNLSE including the SVEA has been used to describe the ultrashort laser pulse propagation in an optical fiber. However, if the pulse duration approaches the optical cycle regime (<10 fs), this approximation becomes invalid. Then, it becomes necessary to use the FDTD method for solving the Maxwell equation with the least approximation. In order to both experimentally and numerically investigate nonlinear femtosecond ultrabroad-band-pulse propagation in a silica fiber, we have extended the FDTD calculation of Maxwell's equations with nonlinear terms to that including all exact Sellmeier-fitting values. We have compared the results of this extended FDTD method with experimental results for the nonlinear propagation of a very short (12-fs) chirped laser pulse in a silica fiber. We obtain the fiber output pulse compressed to 7 fs by the simulation of group delay compensation under the assumption of using an SLM. To our knowledge, this is the first comparison between FDTD calculation and experimental results for nonlinear propagation of a very short (12-fs) chirped pulse in a silica fiber.

Next, experimental results of nonlinear propagation with the IPM effect of two-color 120 fs laser pulses with three different initial delays in a silica fiber were analyzed with FDTD calculations, including SPM, IPM, and DFWM in the fused-silica fiber. Initial delay dependence by the FDTD qualitatively agrees with the previous experimental results. An ultrabroad spectrum induced by IPM mostly agrees with the previous experimental results that had the optimum initial delay. A Fourier transform of this spectrum yields a 4 fs optical pulse, which agrees with the previous experimental results. The spectral phases of two different pulses with three different initial delays are connected continuously. The IPM spectral phase obtained by the extended FDTD shows that the DFWM occurs. To the best of our knowledge, this is the first observation of DFWM with IPM by the FDTD simulation.
Finally, We compared the results of this extended FDTD method with the solution of the BPM that includes the Raman response function, which is the same as in the extended FDTD method, up to fifth-order dispersion with the SVEA, as well as with the experimental results for nonlinear propagation of a 12 fs laser pulse in a silica fiber. Furthermore, in only the calculation, pulse width was gradually shortened from 12 fs to 7 fs to 4 fs to observe the breakdown of the SVEA in detail. Moreover, the soliton number N was established as 1 or 2. To the best of our knowledge, this is the first observation of the breakdown of the SVEA by comparison between the results of the extended FDTD and the BPM calculations for the nonlinear propagation of an ultrashort (< 12 fs) laser pulse in a silica fiber.

7. References

Agrawal, G. A. (1995). *Nonlinear Fiber Optics* 2nd ed., Academic Press, ISBN 978-0123743022, San Diego CA, Chaps. 1 and 2.

Birks, T. A.; Wadsworth, W. J. & Russell, P. St. J. (2000). Supercontinuum generation in tapered fibers. *Opt. Lett.*, Vol. 25, No. 19, (Oct., 2000) 1415-1417, ISSN 0146-9592.

Cheng, Z.; Tempea,; Brabec, G. T.; Ferencz, K.; Spielman, C. & Krausz, F. (1998). Generation of Intense Diffraction-Limited White Light and 4-fs Pulses. in *Ultrafast Phenomena XI*, Elsaesser, T.; Fujimoto, J. G.; Wiersma D. A. & Zinth, W., 8-10, Springer, ISBN 978-3540654308, Berlin.

Cundiff, S. T.; Collings, B. C.; Boivin, L.; Bergman, M. C.; Knox W. H. & Evangelides, Jr. S. G. (1999). Propagation of highly chirped pulses in fiber-optic communications systems. *IEEE J. Lightwave Technol.*, Vol. 17, No. 5, (May, 1999) 811–816, ISSN 0733-8724.

François, P. L. (1991). Nonlinear propagation of ultrashort pulses in optical fibers: total field formulation in the frequency domain. *J. Opt. Soc. Am. B*, Vol. 8, No. 2 (Feb. 1991) 276-293, ISSN 0740-3224.

Fujii, M.; Tahara, M.; Sakagami, I; Freude, W. & Russer, P. (2004). High-Order FDTD and Auxiliary Defferntial Equation Formulation of Optical Pulse Propagation in 2-D Kerr and Raman Nonlinear Dispersive Media. *IEEE Journal of Quantum Electronics*, Vol. 40, No. 2, (Feb., 2004) 175-182, ISSN 0018-9197.

Geraghty, D. F.; Lee, R. B.; Verdiell, M.; Ziari, M.; Mathur, A & Vahala, K. J. (1998). Wavelength conversion by four-wave mixing in semiconductor optical amplifiers. *IEEE Photonics Technol. Lett.*, Vol. 10, No. 1, (Jan., 1998) 69-71, ISSN 1041-1135.

Goorjian, P. M.; Taflove, A.; Joseph, R. M. & Hagness, S. C. (1992). Computational modeling of femtosecond optical solitons from Maxwell's equations. *IEEE J. Quantum. Electron.*, Vol. 28, No. 10, (Oct. 1992) 2416-2422, ISSN 0018-9197.

Goorjian, P. M.& Cundiff, S. T.(2004). Nonlinear effects on the carrier-envelope phase. *Opt. Lett.*, Vol. 29, No. 12, (June, 2004) 1363-1365, ISSN 0146-9592.

Gross, B. & Manassah J. T. (1992). Supercontinuum in the anomalous group-velocity dispersion region *J. Opt. Soc. Am. B*, Vol. 9, No. 10, (Oct., 1992) 1813-1818, ISSN 0740-3224.

Joseph, R. M.; Goorjian, P. M. & Taflove, A. (1993). Direct time integration of Maxwell's equations in 2-D dielectric waveguides for propagation and scattering of femtosecond electromagnetic solitons. *Optics Letters*, Vol. 18, No. 7, (Apr., 1993), 491-493, ISSN 0146-9592.

Joseph, R. M. & Taflove, A. (1997). FDTD Maxwell's equations models for nonlinear electrodynamics and optics. *IEEE Trans. Antenas Propagat.*, Vol. 45, No. 3, (Mar., 1997), 364-374, ISSN 0018-926X.

Kalosha, V. P. & Herrmann, J. (2000). Self-phase modulation and compression of few-optical-cycle pulses. *Phys. Rev. A*, Vol. 62, No. 1, (June, 2000) R11804.1-R11804.4, ISSN 1050-2947.

Karasawa, N.; Nakamura, S.; Morita, R.; Shigekawa, H. & Yamashita, M. (2000) Comparison between theory and experiment of nonlinear propagation for 4.5-cycle optical pulses in a fused-silica fiber. *Nonlinear Opt.*, Vol. 24, (Dec. 2000). 133-138, ISSN 1058-7268.

Karasawa, N.; Morita, R.; Shigekawa, H. & Yamashita M. (2000b). Generation of intense ultrabroadband optical pulses by induced phase modulation in an argon-filled single-mode hollow waveguide. *Opt. Lett.*, Vol.25, No. 3, (Feb., 2000) 183-185, ISSN 0146-9592.

Karasawa, N.; Li, L.; Suguro, A.; Shigekawa, H.; Morita, R. & Yamashita, M. (2001) Optical pulse compression to 5.0 fs using only a spatial light modulator. *J. Opt. Soc. Am. B*, Vol. 18, No. 11, (Nov., 2001) 1742-1746, ISSN 0740-3224.

Karasawa, N.; Nakamura, S.; Nakagawa, N.; Shibata, M.; Morita, R.; Shigekawa, H.& Yamashita, M. (2001b). Comparison between theory and experiment of nonlinear propagation for a-few-cycle and ultrabroadband optical pulses in a fused-silica fiber *IEEE J. Quantum Electron*, Vol. 37, No.3, (Nov. 2001) 398-404, ISSN 0018-9197.

Malitson, I. H. (1965). Interspecimen comparison of the refractive index of fused silica. *J. Opt. Soc. Am.*, Vol. 55, No. 10, (Oct. 1965), 1205-1208, ISSN 0740-3224.

Nakamura, S.; Li, L.; Karasawa, N.; Morita, R.; Shigekawa, H. & Yamashita, M. (2002a). Measurements of Third-Order Dispersion Effects for Generation of High-Repetition-Rate, Sub-Three-Cycle Transform-Limited Pulses from a Glass Fiber. *Jpn. J. Appl. Phys.*, Vol. 41, No. 3A, (Mar. 2002) 1369-1373, ISSN 0021-4922.

Nakamura, S.; Koyamada, Y.; Yoshida, N.; Karasawa, N.; Sone, H.; Ohtani, M.; Mizuta, Y.; Morita, R.; Shigekawa, H. & Yamashita M. (2002b). Finite-Difference Time-Domain Calculation with All Parameters of Sellmeier's Fitting Equation for 12-fs Laser Pulse Propagation in a Silica Fiber. *IEEE Photon. Technol. Lett.*, Vol. 14, No.4, (Apr. 2002) 480-482, ISSN 1041-1135.

Nakamura, S.; Takasawa, N.; Saeki, T. & Koyamada, Y. (2004a). Numerical Analysis of Induced Phase Modulation between Two Ultrashort Pulses in a Silica Fiber by Using the Extended Finite-Difference Time-Domain Method. *Opt. Rev.*, Vol. 11, No. 2, (Mar./Apr., 2004) 68-71, ISSN 1340-6000.

Nakamura, S; Saeki, T & Koyamada, Y. (2004b). Observation of Slowly Varying Envelope Approximation Breakdown by Comparison between the Extended Finite-Difference Time-Domain Method and the Beam Propagation Method for Ultrashort-Laser-Pulse Propagation in a Silica Fiber. *Japanese Journal of Applied Physics*, Vol. 43, No. 10A, (Oct., 2004) 7015-7025, ISSN 0021-4922.

Nakamura, S; Takasawa, N. & Koyamada, Y. (2005a). Comparison between Finite-Difference Time-Domain Calculation with All Parameters of Sellmeier's Fitting Equation and Experimental Results for Slightly Chirped 12 fs Laser Pulse Propagation in a Silica Fiber. *IEEE Journal of Lightwave Technology*, Vol. 23, No. 2, (Feb., 2005) 855-863, ISSN 0733-8724.

Nakamura, S; Takasawa, N. & Koyamada, Y.; Sone, H.; Xu, L.; Morita, R. & Yamashita, M. (2005b). Extended Finite Difference Time Domain Analysis of Induced Phase Modulation and Four-Wave Mixing between Two-Color Femtosecond Laser Pulses in a Silica Fiber with Different Initial Delays. *Japanese Journal of Applied Physics*, Vol. 44, No. 10, (Oct., 2005) 7453-7459, ISSN 0021-4922.

Ranka, J. K.; Windeler, R. S. & Stentz, A. J. (2000). Visible continuum generation in air-silica microstructure optical fibers with anomalous dispersion at 800 nm. *Opt. Lett.*, Vol. 25, No. 1, (Jan., 2000), 25-27, ISSN 0146-9592.

Shibata, M.; Hirasawa, M.; Nakagawa, N.; Morita, R.; Suguro, A.; Shigekawa, H. & Yamashita, M. (2002). Experimental demonstration of phase-dispersion compensation for ultra-broadband femtosecond optical pulses generated by induced-phase modulation. *Appl. Phys. B*, Vol. 74, No. 1, (July, 2002) s291-s294, ISSN 0946-2171.

Sone, H.; Arai, T.; Imai, M. & Imai Y. (2002). Modal Birefringence Dependent Supercontinuum Due to Cross-Phase Modulation in a Dispersion-Flattened/Decreasing Fiber. *Optical Review*, Vol. 9, No. 3, (May/Jun., 2002) 89-93, ISSN 1340-6000.

Taflove, A. & Hagness, S. C. (2000). *Computational Electrodynamics: The Finite-Difference Time-Domain Method*, 2nd ed., Artech House, ISBN 978-1580538329, Norwood MA, Chapter 9, Section 9.6 (pp. 398-401).

Taylor, A. J.; Rodriguez, G. R. & Clement, T. S. (1996). Determination of n_2 by direct measurement of the optical phase. *Opt. Lett.*, Vol. 21, No. 22, (Nov., 1996), 1812-1814 ISSN 0146-9592.

Xu, L; Karasawa, N.; Nakagawa, N.; Morita, R.; Shigekawa, H. & Yamashita, M. (1999). Experimental generation of an ultra-broad spectrum based on induced-phase modulation in a single-mode glass fiber. *Opt. Commun*, Vol. 162, No. 4-6, (Apr., 1999) 256-260, ISSN 0946-2171.

Yamane, K.; Zhang, Z.; Oka, K.; Morita, R.; Yamashita, M. & Suguro, A. (2003). Optical pulse compression to 3.4 fs in the monocycle region by feedback phase compensation. *Opt. Lett.*, Vol. 28, no. 22, (Nov. 2003) 2258 2260, ISSN 0146-9592.

Yamashita, M; Hirasawa, M.; Nakagawa, N.; Yamamoto, K.; Oka, K.; Morita, R. & Suguro A. (2004). Pulse compression of white-light continuum generated by induced phase modulation in a conventional glass fiber. *J. Opt. Soc. Am. B*, Vol. 21, No. 2, (Feb., 2004) 458-462, ISSN 0740-3224.

Yamashita, M.; Sone, H.; Morita, R.; & Shigekawa, H. (1998). Generation of monocycle-like optical pulses using induced-phase modulation between two-color femtosecond pulses with carrier phase locking. *IEEE J. Quantum Electron*, Vol. 34, No. 11, (Nov., 1998) 2145-2149, ISSN 0018-9197.

Yamashita, M.; Sone, H. & Morita, R. (1996). Proposal for Generation of a Coherent Pulse Ultra-Broadened from Near-Infrared to Near-Ultraviolet and Its Monocyclization. *Jpn. J. Appl. Phys.*, Vol. 35, (Aug., 1996) L1194-L1197, ISSN 0021-4922.

21

Non Perturbative Time-Dependent Density Functional Theory, TDDFT: Study of Ionization and Harmonic Generation in Linear Di-(N_2) and Tri-(CO_2, OCS, CS_2) Atomic Molecules with Ultrashort Intense Laser Pulses-Orientational Effects

Emmanuel Penka Fowe and André Dieter Bandrauk*
Laboratoire de Chimie Théorique, Faculté des Sciences, Université de Sherbrooke, Sherbrooke, Québec J1k 2R1
Canada

1. Introduction

High-order harmonic generation, HOHG, resulting from the interaction of intense ultrashort laser pulses with atoms has been extensively studied in recent years offering potential application as a source of coherent ultrashort radiation in the extreme ultraviolet and soft X-ray regions.[1-3] For molecules, molecular high order harmonic generation, MHOHG[3] offers the possibility of synthesizing attosecond (1 attosecond=10^{-18} second) pulses for creating and controlling coherent electron wavepackets, CEWP.[4] An intuitive and efficient theoretical picture based on a classical three step electron recombination trajectory[5-6] model in the laser field has helped to elucidate this process. According to this model, an electron tunnels out from the atom or molecule and may recombine with the parent[5,7] or neighbouring[8] ion emitting a high energy photon, after undergoing laser–driven motion in the continuum. If the tunnelling electron under the laser field does not return to the parent ion position, it is completely ionized and this is referred to as molecular ionization. The electron can also recombine with a neighbouring ion, an example of laser induced electron transfer, LIET.[9] Theoretical and experimental work on HOHG has been mostly devoted to atoms. The study of MHOHG in molecules is at the early stages. In contrast to atoms, for molecules, the returning wave packet in the semiclassical picture of HOHG encounters a core comprising two or more nuclei, which are presumed to behave as pointlike source potentials leading to interference in the MHOHG spectrum. Additionally, it has been suggested that the electron recollision cross-sections responsible for MHOHG relates to the projection of the valence orbitals with respect to the direction of the propagation of the recolliding electron, making possible a new molecular orbital tomography[10]. It is therefore, natural to ask if there is any orientation

* Canada Research Chair – Computational Cheistry and molecular photonics.
Andre.Dieter.Bandrauk@USherbrooke.ca

dependence of the closely related process of MHOHG spectra. Also, due to the shapes and the symmetries of different molecular orbitals, we are interested to know how molecular orbitals can come into play into the MHOHG and molecular ionization yield processes when different laser intensities and molecular angle versus laser polarisation are taken into account.
While theoretical approaches based on the accurate 3D time dependent Schrödinger equation, TDSE, are limited to the one electron sytem[9, 11-12] and 2D models for two electrons system[13] because of limitation of memory and computational time, the major difficulties in the theoretical study of molecular systems in a laser field resides on the multi-center nature of the molecule and also the treatment of the multi electron-electron interactions. Here, we focus on time-dependent density functional theory, TDDFT, methods beyond linear response theory[14] as a tool for studying the nonperturbative response of molecules to intense ultrashort laser pulses.[15] The time independent Density Functional Theory, DFT, has become an ubiquitous method[16] of solving ground state electronic problems in atoms and molecules. Extension to the nonperturbative regime through time-dependent methods has emerged by the existence of a rigorous theorem relating the exact time dependent density to external time-dependent potentials[17] and has been extended to linear response via a linear perturbative version of TDDFT.[18] We focus here on a non-linear non perturbative TDDFT.[15]
For the visual understanding of electron motion in our molecules during the time propagation under strong laser pulses, we will use the time dependent approach of the electron localization function, TDELF.[19-22] Conceptually, the physical idea of the electron localization function, ELF, is based on the fact that a highly localized electron repels other electrons with same spin very strongly due to the Pauli exclusion. The ELF uses the probability to find a second spin-like electron in the vicinity of the first one as a measure for electronic localization. Numerical values of the probability are conveniently mapped on the interval]0,1] facilitating analysis and interpretation. So, in areas where two electrons of the same spin have a high probability to be found (in reference to the homogeneous gas) the function should tend to 0 by construction. It follows that, in areas where the probability of finding two electrons of opposite spin is high (in reference to the homogeneous gas) the function tends to 1. Consequently, a region of the space with a high value of the ELF corresponds to a region of chemical bond or lone pair of the Lewis theory. Thus, TDELF will allow us to make a clear separation between the HOMO of our molecules and the rest of the valence molecular orbitals[23], and therefore will enable us to visualise their behaviour in the presence of a laser pulse, such as an example laser induced electron transfer, LIET, which has recently been shown to be controllable through the pulse shape and phase.[9, 24]
In the present work, non perturbative TDDFT methods[15] as opposed to linear response TDDFT[18] are used to study the orientational dependence of ionization and molecular high order harmonic generation, MHOHG, in tri-atoms CO_2, OCS, CS_2 and the di-atomic N_2 molecules as a function of laser intensity, $I_0 \geq 3.54 \times 10^{14}$ W/cm^2 =10^{-2} au (1au; Io=3.54×10^{16} W/cm^2, E_o=5×10^9 V/cm) for few cycle 800 nm laser pulses. This work is organized as follow: First we briefly recall the TDDFT formalism, followed by the procedure for calculating ionization rates and the time dependent electron localization function, TDELF. The next sections are devoted to results and discussions. The paper is ended by some concluding remarks for future research.

2. Methods: Nonperturbative TDDFT Kohn-Sham formalism

The TDDFT method provides the most detailed and feasible *ab-initio* approach for tackling many-body problems in external time dependent field. Density functional theory (DFT) as first introduced by Hohenberg and Kohn[25], and extended by Kohn and Sham[26] (KS) is based on the existence of an exact mapping between one-particle density and external potentials. This leads to the density of the interacting system being obtained from the density of an auxiliary system of non-interacting particles moving in an effective local single-particle potential. A time dependent generalisation of DFT was provided by Runge and Gross[17], showing that there is a one-to-one correspondence between the external (time dependent) potential, $\upsilon_{ext}(r,t)$, and the time dependent one-electron density,$n(r,t)$, for many-body systems evolving from a fixed initial state. The time dependent electronic density is written as:[27]

$$n(r,t)=\sum_{\sigma=\uparrow,\downarrow} n_\sigma(r,t)=\sum_{\sigma=\uparrow,\downarrow}\sum_{i}^{N_\sigma}\psi_{i\sigma}(r,t)|^2, \tag{1}$$

where, N_σ is the number of the occupied KS orbital in the spin state σ and $\psi_{i\sigma}(r,t)$ is the orbital obtained through the time dependent KS equations (in a.u.):

$$i\frac{\partial}{\partial t}\psi_{i\sigma}(r,t)=\left[-\frac{1}{2}\nabla^2+\upsilon_{eff}(r,t)\right]\psi_{i\sigma}(r,t) \tag{2}$$

where

$$\upsilon_{eff}(r,t)=\upsilon_{ext}(r,t)+\upsilon_h(r,t)+\upsilon_{xc,\sigma}(r,t) \tag{3-a}$$

The first term is the external potential, due to the interaction of the electron with an external laser field and the nuclei, while the second term accounts for the classical Hartree electrostatic interaction between electrons, and the third term is the exchange correlation (*xc*) potential includes all non–trivial many body effects, and has an extremely complex (and essentially unknown) functional dependence on the density. When an intense electric with a low frequency field is applied to a molecule, electrons tend to be displaced out by ionization; so, the higher ionization potential, IP, or barriers, the more this electronic displacement will be hindered. Therefore, the choice of an approach which accurately reproduces the experimental IP is useful for our analysis. Local density approximation (LDA) has been widely used in strong fields[15] due to its simplicity and applicability to various systems with relatively lower computational cost. However, LDA is constructed from the homogenous gas system and suffers from a wrong asymptotic behaviour originating from the incomplete cancellation of the self-interactions, SI. For these reasons, we have used the Van Leeuwen and Baerends[28] potential, LB94, which introduces a gradient correction to the LDA, exchange correlation so as to reproduce the Coulomb asymptotic behaviour of the potential.

$$v_x^{\sigma LB94}(\mathrm{r})=v_x^{\sigma LDA}(\mathrm{r})-\beta n_\sigma^{1/3}(\mathrm{r})\frac{x_\sigma^2}{1+3\beta x_\sigma\ln(x_\sigma+\sqrt{(x_\sigma^2+1)})} \tag{3-b}$$

where,

$$x_\sigma=\frac{|\nabla n_\sigma(\mathrm{r})|}{n_\sigma(\mathrm{r})^{4/3}};\ \beta=0.05, \tag{3-c}$$

A convenience of the LB94 potential lies in its explicit dependency on the local density and its gradient. This potential yields excellent estimates of the IP for atoms and small molecules, especially in the framework of the linear response TDDFT excitation energy.[29] This model potential significantly corrects the higher moments of the density, notably for diffuse outer orbitals involved in excitations and in our case in ionization. The LB94 also includes partial and considerable self interaction correction, SIC, resulting from an interaction of an electron with itself, of importance to electronic transport.[30] Such corrections are known to produce spurious oscillations in photoionization calculations,[31] but little is known about the corrections for strong fields in the nonperturbative regime. Our results show that the LB94 potential offers accurate IP's for all orbitals; an essential starting point for accurate ionization. Furthermore, at high intensities, excited states have very short lifetime thus, reducing considerably non adiabatic effects. The neglect of nonadiabatic effects, which are a major difficulty of linear[32] TDDFT are implicit in the strong field approximation, SFA[6], a current simple nonpertubative model.

3. Computational methods

a. Ionization -KS orbitals

In the present work, the TDDFT KS equation, Eq.2, was discretized in space using finite-difference (FD) grid techniques. For numerical efficiency we represent the electron-ion interaction by the norm conserving non-local Troullier–Martins pseudopotential[33], which allows to explicitly treat many fewer electrons, and to avoid the fast oscillations produced by the core electrons. The pseudopotentials are generated such that the pseudo-wavefunctions imitate the all-electron atomic wavefunctions. The bond length has been set to its experimental[1, 34-36] value, for OCS bonds length are CS=0.156 nm, CO =0.1157 nm, CS= 0.155 nm for CS_2, CO=0.1162 nm for CO_2 and NN=0.1041nm for N2. The molecule is placed in a large three-dimensional (3D) rectangular grid cell (atomic units au, are used) of dimension $|z_{max}|$ = 90 au (1a.u. = 0.0529nm) and $|x_{max}| = |y_{max}|$ =25 au if the laser is oriented parallel to the molecule. The uniform FD grid spacing, $\Delta z = \Delta x = \Delta y = 0.28$ au was used and the convergence of the calculations was checked against results which use a smaller grid spacing. The size of the grid is determined by the maximum radius $a = E/m\omega^2$ of an electron in a time dependent electric field of maximum amplitude E. Thus at $E = 0.1$ a.u. corresponding to an intensity I= 3.50×10^{14} W/cm^2 (1a.u: Io= 3.50×10^{16} W/cm^2, Eo= 5×10^9 V/cm), $a = 30$ au at $\lambda = 800$ nm, $\omega = 0.057$ au. Thus since a is the size of the recolliding electron trajectory, z_{max} = 90 a.u. >> a ensures that all recolliding electrons responsible for MHOHG are collected in the numerical procedure. The laser intensity is related to the field strength by $I = \frac{1}{2}\varepsilon_0 c E^2$, ε_0 is the permittivity of free space and c is the speed of the light. The angle, θ, between the linear polarized laser and the main molecular axis varies from 0 to 90°. The total pulse area condition $\int_{-\infty}^{+\infty} E(t)dt = 0$ is verified to ensure no spurious static effects for pulses satisfying Maxwell's equations.[4, 37] The external potential created by an intense laser field is assumed to arise from an oscillating electric field with a cosine envelope. When the laser is linearly polarized parallel to the molecular axis z, $\upsilon_{ext}(r,t)$ is given by

$$\upsilon_{ext}(r,t) = zE\cos^2\left(\frac{\pi t}{T}\right)\sin(\omega t) \quad (0 \prec t \prec T) \tag{4}$$

T determines the laser pulse duration. The time-dependent equations, Eq.(2) are solved using the Crank-Nicholson scheme[34, 36] with Δt=0.02 a.u (1a.u=24x10^{-18}s) as the time step. All calculations have been done using eight optical cycles pulse duration on a multiprocessors parallel machine. To absorb the ionized charge density and avoid reflections at the boundary of the box due to ionization during the propagation, we employ a mask function. By using this technique, the KS wave-function is multiplied at each time-step at grid boundaries by a function *M(z)* which is 1 in the inner simulation region and gradually goes to 0 at the boundaries. The masking function in one dimension is defined as:

$$M(z)=\begin{cases}\sin^{\frac{1}{4}}\left(\dfrac{z_{\max}-|z|}{2a}\pi\right), (z_{\max}-|z|\prec a), \\ 1 \qquad (otherwise).\end{cases} \quad (5)$$

$|z_{max}|$=90 au is the extension of the box in the z direction, a=10 au is the width of the mask function. It is assumed that the electrons in the $z_{\max}\succ|z|$ domain are completely ionized and cannot return to the nuclei. A different approach to remove outgoing flux density consists in adding an imaginary potential[38] which causes an exponential decay of the wavefunction or in using the exterior complex scaling coordinate transformation as absorbing boundary. All methods produce identical results for sufficiently large grids. The time-dependent ionization probability, $P_{i,\sigma}(t)$, of an individual spin-orbital is calculated as:[15]

$$P_{i,\sigma}(t)=1-N_{i,\sigma}(t), \quad (6)$$

where $N_{i,\sigma}(t)=\langle\psi_{i,\sigma}(r,t)|\psi_{i,\sigma}(r,t)\rangle$ is the time dependent population (survival probability) of the *i,σ*-th spin-population.
The number of electrons inside the grid of volume *V* is given as:

$$N_{bound}(t)=\int_V d^3r\, n(r,t) \quad (7)$$

and the normalised remaining electron population as:

$$\beta(t)=\frac{\int_V d^3r\, n(r,t)}{\int_V d^3r\, n(r,0)}=e^{-\Gamma t}, \quad (8)$$

Γ is the total ionization rate and $n(r,0)=2\sum_i^{N_\sigma}|\psi_{i\sigma}(r,0)|^2$ is the total number of electrons in the initial state (without external field). The total ionization yield probability, $\gamma(t_f)$, computed at the final time, t_f, of the propagation is then obtained from the following difference:

$$\gamma(t_f)=1-\beta(t_f), \quad (9)$$

b. MHOHG

MHOHG is a highly nonlinear nonperturbative effect resulting from two essential steps : (i) ionization at the peaks of a slowly varying (low frequency) electromagnetic pulse *E(t)* of maximum amplitude E (Eq.4), followed by (ii) laser induced recollision after 2/3 of a cycle with maximum energy of harmonics of order N_m: $N_m\omega = IP + 3.17\,U_p$, where IP is the ionization potential, N_m is the order of the harmonics and $U_p = I/4m\omega^2$ is the ponderomotive energy.[5, 8] Calculations of MHOHG with ultrashort (few cycles) intense, $I \geq 10^{14}$ W/cm^2 requires special attention to grid sizes and gauges as mentioned in recent work.[39] In the present work, the MHOHG spectrum is calculated during the dynamical propagation as follows.[15] The power spectrum of the dipole acceleration $\ddot{d}_z(t)$ or equivalently a time-dependent Hellman-Feyman theorem in a given direction yields the predicted MHOHG spectra:[7, 39]

$$S_z(\omega) \propto |a(\omega)|^2 = \left|\int\left(\frac{d^2}{dt^2}d_z(t)\right)e^{i\omega t}dt\right|^2 = \left|\int\left\langle\psi(t)\left|\frac{\partial H}{\partial t}\right|\psi(t)\right\rangle e^{i\omega t}dt\right|^2 . \tag{10}$$

Recent work has shown that the most accurate method for generating the MHOHG spectra when dealing with molecular systems.[4, 12] is to calculate the dipole acceleration, $\ddot{d}_z(t)$, from the exact time dependent function according to Ehrenfest's theorem.[12]

c. Time-Dependent Electron Localization Function (TDELF)

The ELF is a method for the mapping of electron pair probability in multi-electron systems. It was first introduced by Becke and Edgecombe[22] as a descriptor of electronic localization, using arguments based on the conditional same spin pair probability function. Here, we briefly recall the definitions of the extension of the ELF for time dependent processes.[20-21] The time-dependent ELF or TDELF, $\eta(r,t)$, is given as:

$$\eta(r,t) = \frac{1}{1+\left(\dfrac{C(r,t)}{C_h(r,t)}\right)^2} \tag{11}$$

where

$$C(r,t) = \sum_\sigma \left\{\sum_{i=1}^{N_\sigma} |\nabla\psi_{i\sigma}(r,t)|^2 - \frac{1}{4}\frac{[\nabla n_\sigma(r,t)]^2}{n_\sigma(r,t)} - \frac{j_\sigma^2(r,t)}{n_\sigma(r,t)}\right\}, \tag{12}$$

$$C_h(r,t) = \frac{3}{5}(6\pi^2)^{2/3} n_\sigma(r,t)^{5/3} \tag{13}$$

j_σ^2 is the squared modulus of the current density and is the time-dependent correction to the original ELF.[22] $C(r,t)$ measures the probability of finding one electron in the near vicinity of a reference like-spin electron at position r and time t. Thus, the Pauli repulsion between two like-spin electrons, described by the smallness of $C(r,t)$, is taken as a measure of the electron localization. $C_h(r,t)$ is the corresponding factor found for a uniform electron gas, i.e., a homogeneous electron gas whose density is the same as that of the real system. The TDELF will be used for the time-resolved observation of the molecular ionization, and can thus provide a visual understanding of the electron *density* transfer dynamics (most notably lone pair and bond electrons) under strong laser pulses.

4. Electronic analysis and ionization

One of the relevant information regarding the ground state geometry of these molecules is that they are linear. The electronic structure from the DFT-KS calculation is depicted in Fig.1, where the three highest occupied KS MO's HOMO, HOMO-1, HOMO-2, with their phases and energies are presented. For the tri-atomic molecules, one finds that the highest occupied molecular orbital, HOMO, is an anti-bonding type, $1\pi_g$, formed of π lobes lying perpendicular to the inter-nuclear axis and located on the sulphur or oxygen atoms for respectively CS_2 and CO2, while the anti-bonding orbitals forming the HOMO, 2π, of the OCS is located on the sulphur and the oxygen atom. The HOMO of the diatomic N_2 molecule is a bonding σ_g formed mostly by the overlap between the $2p_z$ orbitals of the nitrogen atoms. The computed values of the IP from various *ab-initio* theory levels are presented in Table 1.

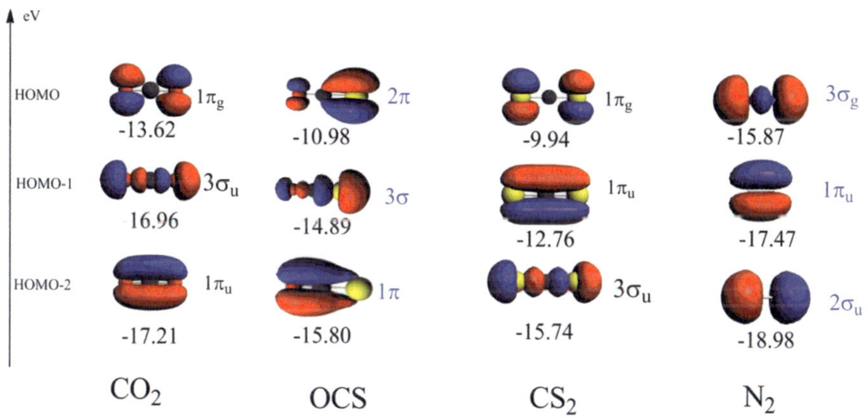

Fig. 1. (Color online) DFT/LB94 images of the molecular orbitals of CS_2, OCS, CS_2 and N_2. Only the three highest relevant occupied molecular orbitals, HOMO, HOMO-1 and HOMO-2 are shown. Atoms colors are: Carbon (black), Sufur (yellow) and oxygen (red)

Molecule	Orbital	HF	LDA	B3LYP	LB94	Expt
OCS	HOMO	11.49	7.68	8.40	10.98	11.20
	HOMO-1	17.38	11.80	12.97	14.89	16.04
	HOMO-2	18.03	11.94	13.08	15.80	15.08
CS_2	HOMO	10.16	6.91	7.51	9.94	10.09
	HOMO-1	14.42	9.77	10.64	12.76	12.69
	HOMO-2	15.92	10.72	11.78	15.74	14.47
CO2	HOMO	14.86	9.25	10.24	13.62	13.8
	HOMO-1	20.23	12.84	14.33	16.96	18.1
	HOMO-2	19.52	13.01	14.46	17.21	17.6
N_2	HOMO	17.28	10.89	12.35	15.87	15.58
	HOMO-1	16.82	12.84	12.87	17.47	17.20
	HOMO-2	21.24	13.01	15.35	18.98	18.73

Table1. Comparison of most relevant orbital absolute binding energies computed as $-\varepsilon_{KS\text{-}MO}$ and experimental ionization potentials (in eV).

A survey of Table 1 reveals that for the tri-atomic molecules, the IP calculated with the HF approach are higher (around 7%) than the experimental data. This deviation is most related to the absence of the correlation in the HF approach. Using DFT, the serious underestimation of the ionization potential, IP at the LDA level is clearly noticeable. Computed values are approximately 43% lower than the experimental ones. The reason of this large discrepancy lies as we mentioned in the previous theoretical section 2), is the wrong asymptotic behaviour of the LDA exchange-correlation potential which originates from the SI, as opposed to the HF approach (SI free eliminated by the exact exchange). Results using partially corrected SI through the hybrid functional B3LYP moderately improve the calculated value of the IP, the relative error drops from 43% with the LDA to 25%. For the N_2 molecule, the HOMO and HOMO-1 are nearly degenerate in energy, as consequence, the HOMO and HOMO-1 order is inverted in the HF results and we pointed out that CASSCF do not remove this error in N_2.[55] Again, the LDA and B3LYP functional underestimate by at least 20% the experimental value of the IP. These observations reflect the increasing correlation effect when molecular orbitals are nearly degenerate. Regarding the results with the LB94 potential, one sees that it performs much better, the global agreement with the experiment being satisfied within 3% of error. A convenience of the LB94 potential lies in its explicit dependence on the local density and a gradient correction to the LDA exchange potential so as to reproduce the correct asymptotic behaviour. We note that for the OCS molecule the HOMO-1 and HOMO-2 eigenvalues are nearly degenerate with an inverse molecular orbital order with respect to the experimental data, whereas the symmetric CS_2 (as in N_2) respects the experimental order with negative orbital energies nearly equal to the corresponding IP's, thus suggesting that these orbitals should be close to Dyson orbitals.[40]

a. Tri-atomic molecules (CO_2, CS_2, OCS,)

To gain a better understanding of the molecular ionization, we have investigated the time evolution of some relevant KS orbital using Eq.6. Three angular orientations are considered $\theta=0^o$, 45°, 90°. Results using three laser intensities range from lower 3.50×10^{14}, to higher 1.40×10^{15} and highest 2.99×10^{15} W/cm^2 are displayed in Fig.2a, Fig.2b and Fig.2c for respectively the CO_2, CS_2 and the OCS molecule. Evidence of a strong dependence of the KS remaining orbital population, $N_{i,\sigma}(t)$, with the laser intensity is observed. It emerges as expected that in general the KS MO ionization is most important for CS_2 due to its lower IP as compared to OCS. The smaller is the ionization potential, IP, of the electronic shell, the easier it can be ionized. When $\theta=0^o$, for symmetry reason the two components of each 2π or $1\pi_g$ orbital have the same behaviour as they are degenerate. For the lower laser intensity Io= 3.50×10^{14} W/cm^2, one sees that the HOMO, shows as expected the dominant response to the laser field. It is followed respectively by the inner shell orbitals $3\sigma/3\sigma_u$ (HOMO-1 for OCS and CO_2 and HOMO-2 for CS_2). As the laser intensity increases, it emerges that the HOMO does not remain the most affected by the external laser field. Instead, our TDDFT/LB94 calculation shows that the ionization of the inner $3\sigma/3\sigma_u$ molecular orbital dominates so that it exceeds that of the HOMO and thus presents the dominant response to the field. This behaviour of the HOMO under high laser intensity is related to their symmetry which has a nodal plane containing the molecular axis. This gives rise to a low ionization yield when aligned parallel to the laser electric field, while the $3\sigma_u/3\sigma$ has the right symmetry (MO shape) to interact with the laser field as its density is maximum parallel to the field. One notes also that the ionization of the inner bonding $1\pi_u/1\pi$

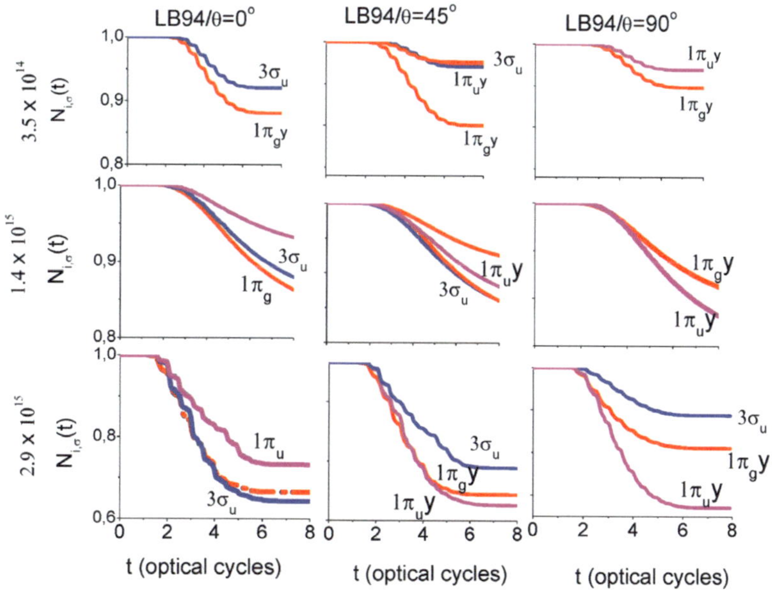

Fig. 2a. (Color online) TDDFT/LB94 orbital population, $N_{i,\sigma}(t)$, of CO_2 at different laser intensities for selective fixed angle, θ, between the molecular axis and the laser polarization direction. Only the relevant KS orbitals which possess an important response to the laser field are shown with their label.

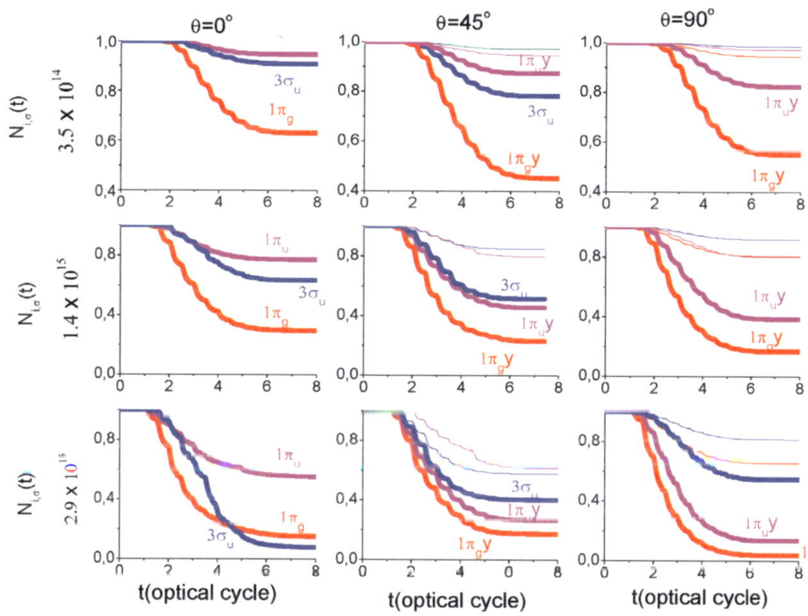

Fig. 2b. (Color online) TDDFT/LB94 orbital population, $N_{i,\sigma}(t)$, of CS_2 at different laser intensities for selective fixed angle, θ, between the molecular axis and the laser polarization direction. Only the relevant KS orbitals which possess an important response to the laser field are shown with their label.

(HOMO-2 for OCS and CO_2 and HOMO-1 for CS_2) is increasing with the laser intensity. At angles θ=45° and θ=90°, the symmetry behaviour of the different components of the π orbitals is broken (degeneracy is removed) and we have renamed them as follows: components of the $1\pi_g$ of CS_2 and CO_2 are $1\pi_g x$, $1\pi_g y$ and those of the 1π of the OCS are $1\pi x$, $1\pi y$. When θ=45°, independently of the laser intensity applied on these molecules, evidence of the ionization enhancement (compare to θ=0° case) of the HOMO is clearly visible. The high ionization of the HOMO is followed by that of the inner $1\pi_u/1\pi$ and $3\sigma_u/3\sigma$ MO orbital. The enhanced ionization of the HOMO is the result of the combined effects of its lower IP and its alignment along the laser polarization axis at θ=45°. The total orbital ionization yield follows the trend CS_2 >OCS > CO_2 inversely proportional to their IP. Finally, when θ=90°, once again, one finds that the KS MO's of CS_2 ionize more than the others. One finds also

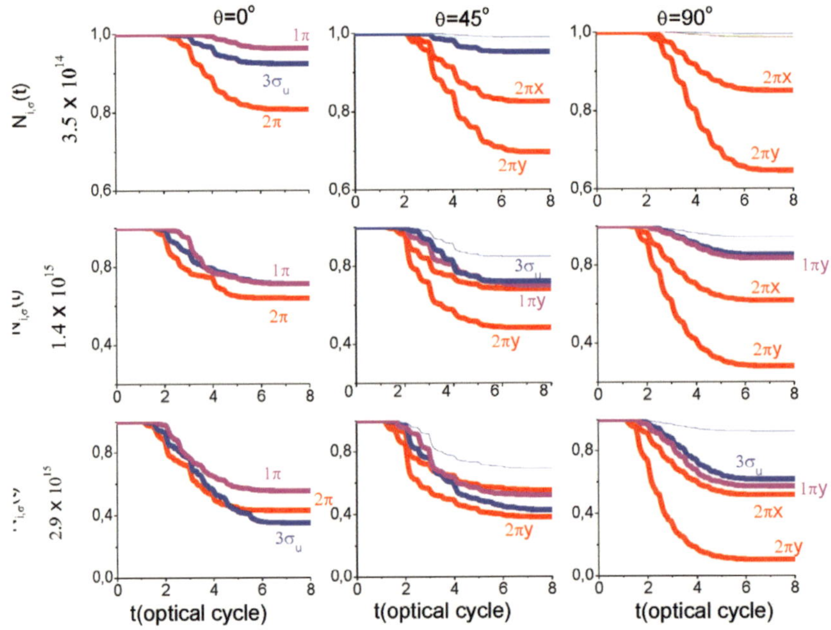

Fig. 2c. (Color online) TDDFT/LB94 orbital population, $N_{i,\sigma}(t)$, of OCS at different laser intensities for selective fixed angle, θ, between the molecular axis and the laser polarization direction. Only the relevant KS orbitals which possess an important response to the laser field are shown with their label.

that independently of the laser intensity, the HOMO remains the most affected by the field. For higher laser peak intensities, 1.40×10^{15} and 3.0×10^{15} W/cm^2, the ionization of the $3\sigma_u/3\sigma$ HOMO-2 of CS_2 or HOMO-1 of OCS and CS_2 is so important that, it tends to exceed that of the HOMO. More remarkably, for CS_2 and CO_2, our TDDFT calculations predict that KS orbital ionization is lower for θ=90° than for θ=0° as the laser intensity is increasing mainly due to the contribution of the bonding $1\pi_u/1\pi$ orbital. Although this orbital has a larger IP than the HOMO $1\pi_g/2\pi$, its density dominates at 90° whereas the HOMO, $1\pi_g$, has a node (zero density) at 90° in the center of the molecule. We conclude that orbitals which have maximum density in the direction of laser polarization, such orbitals will influence or dominate in ionization and in the MHOHG processes. However, for the OCS molecule, one sees that as the laser is increasing, the two components of the HOMO ($2\pi y$, $2\pi x$) present the dominant response to the laser field. This is because in comparison with the CS_2 and CO_2 MOs', the HOMO and the HOMO-2 of the OCS have almost the same shape (density), as consequence, its HOMO ionizes faster as the laser intensity is increasing due to its lower IP. In conclusion, we find that the KS HOMO ionization increases from θ=0° and reaches a maximum when the molecule is aligned at an angle around 45° with respect to the direction of the laser polarization, and then decreases at larger angles towards 90°. To see how the molecular orientation as well as each KS orbital contributes to the total molecular ionization, we display in Table II the total molecular ionization yield (see Eq.9) as a function of the laser intensity and the angle molecular orientation, θ.

	I=3.5x10^14 W/cm^2			I=1.4x10^15 W/cm^2			I=2.99x10^15 W/cm^2		
	0°	45°	90°	0°	45°	90°	0°	45°	90°
CS_2	13.3	12.9	9.1	34.1	32.3	25.3	58.8	54.5	39.3
CO_2	2.0	2.2	1.2	9.0	9.2	7.5	24	23	21.2
OCS	7.3	7.3	6.8	24.8	22.12	19.3	41.0	37.9	30.5
N_2	3		1.6	16		15.6	27		27.5

Table 2. Computed total ionization yield (%), $\gamma(t_f)$, for selected values of θ and laser intensities. Values are given in percentage

In general the ionization yield follows the trend CS_2>OCS>CO_2 due to their IP difference. The ionization is significantly enhanced at higher laser intensity in the case θ=0° than the case of θ=90°. This is attributed to the KS MO geometry of the linear molecule. In fact, at θ=0° the molecular ionization yield is high due to the fact that the two components of the HOMO behave in the same manner (same ionization rate) and due to the additive contribution played by the inner $3\sigma_u/3\sigma$ MO whose densities are aligned along the laser polarization axis whereas for θ=90°, only one component of the HOMO as well as HOMO-2 is very active in the presence of the laser field due to degeneracy removal.

b. N_2 molecule

We add for comparison the two center N_2 molecule for which the LB94 potential MO IP's are in excellent agreement with experiment (Table 1). Fig.2d illustrates ionization calculations done at I=3.50×10^14, 1.40×10^15, and 3.0×10^15 W/cm² laser peak intensity for θ=0°, 45° and 90°. At 3.50×10^14 W/cm² and θ=0°, the bonding HOMO, $3\sigma_g$, shows as expected the dominant response to the field followed by the inner, $2\sigma_u$, HOMO-2, anti-bonding, while the ionization of the $1\pi_u$ HOMO-1 is not relevant.

At angles θ=45° and θ=90°, the symmetry behaviour of the different components of the $1\pi_u$ orbital is broken (degeneracy is removed) and we have renamed them as $1\pi_u x$, $1\pi_u y$. For θ=45°, again the HOMO, $3\sigma_g$, is again the most affected by the laser field followed by the inner $2\sigma_u$, HOMO-2 which is closely also followed by the $1\pi_u y$ component of the HOMO-1. At θ=90°, while the HOMO, $3\sigma_g$, is still the most influenced by the electric field, it is now followed by the $1\pi_u y$ component of the HOMO-1 and the contribution of the others are irrelevant. At I=1.40×10^15 W/cm², for θ=0°, both the HOMO, $3\sigma_g$, and $2\sigma_u$, HOMO-2 present the dominant response to the field during the first four optical cycle and finally the ionization of the $3\sigma_g$ exceeds that of the inner $2\sigma_u$ during the rest of the propagation. At the highest, I=3.0×10^15 W/cm² laser peak intensity, the ionization of both the $3\sigma_g$, and $2\sigma_u$, MOs

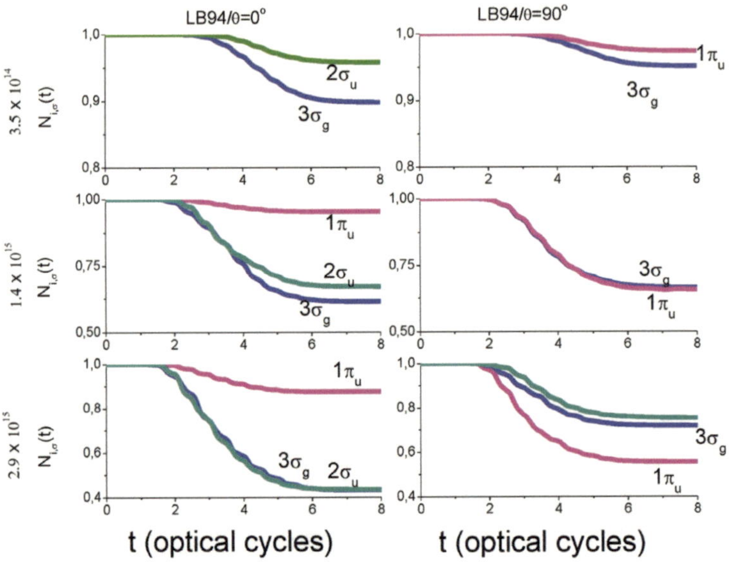

Fig. 2d. (Color online) TDDFT/LB94 orbital population, $N_{i,\sigma}(t)$, of N_2 at different laser intensities for selective fixed angle, θ, between the molecular axis and the laser polarization direction. Only the relevant KS orbitals which possess an important response to the laser field are shown with their label.

ionization remain nearly the same despite the 3eV energy gap between them. It turns out that when aligned parallel to the laser field, the $3\sigma_g$, and $2\sigma_u$, whose electron density is maximal along the laser polarization are the most affected by the field. For the lower I= 3.50×10^{14} W/cm^2, the ionization of the HOMO, $3\sigma_g$, is most important due to its lower IP, however, at high laser intensities I= 1.40×10^{15} and $3{,}0\times10^{15}$ W/cm^2, the $2\sigma_u$ HOMO-2 presents the dominant response to the field due to its antibonding nature (ionized fast) and the symmetry coupling with the HOMO $3\sigma_g$. At θ=90° and higher laser intensities, I=1.40×10^{15} and 2.99×10^{15} W/cm^2, the $1\pi_u y$ component of the HOMO-1 show the dominant response to the field because its electron density is maximal along the laser polarization axis, similar to the CO_2 results.

4. TDELF analysis for tri-atomic molecules

The ground state ELF images are shown in Fig.3a for η= 0.85 (i) corresponding to highly localized pair electrons such as lone pairs and the two-dimensional representation of the ELF isosurfaces in the plane containing the three atoms of the molecule (ii). At η= 0.85 (i), the image exhibits only the lone pair (noted a) of the oxygen atom while the CO bond localization is absent. This is because the lone pair electron on the oxygen is the most localized in the CO_2 molecule. By representing the ELF in the plane containing the three atoms (ii), one can more easily identify the CO chemical bond region (note b) Fig.3b exhibits for comparison in a two-dimensional representation the ELF of CO_2, OCS and CS_2 molecule. The highly localized bond region (CO, SO, CS) and the lone pair region are clearly identifiable. From the electronic structure analysis (Fig.1), it is evident that the HOMO, $1\pi g/2\pi$ is essentially located in a diffuse lone pair region (a) while the HOMO-1 and the HOMO-2 are essentially located in localized bonding region (b). In Fig.3b, the lobe representing the lone pair on the oxygen (one lobe on the main axis of the three atoms)

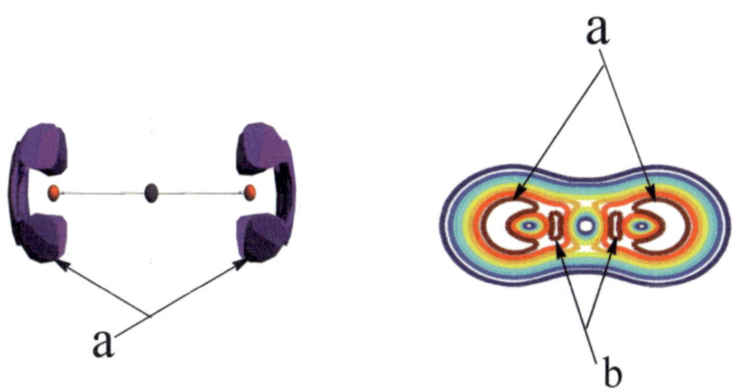

(i) a: Lone pair on oxygen (ii)
b: CO bond

Fig. 3a. DFT/LB94 representation of the ELF ground state of CO_2 molecule

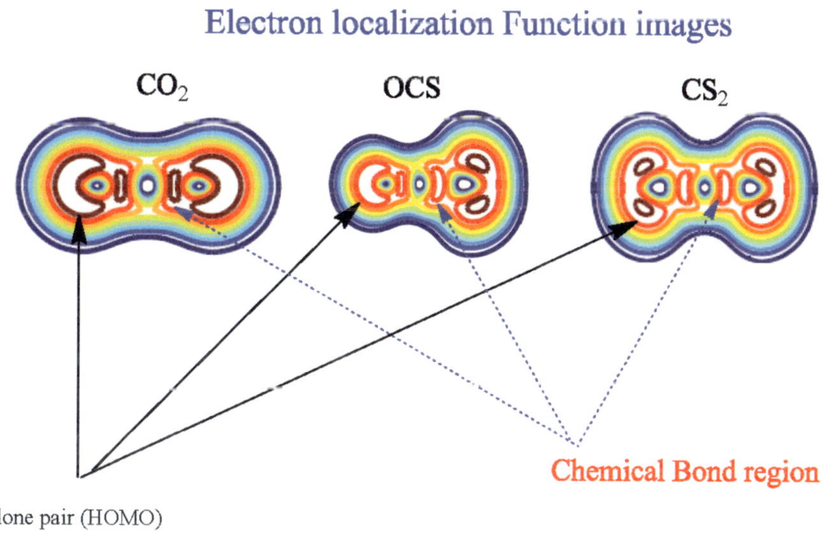

Fig. 3b. DFT/LB94 representation of the ELF ground state of the OCS, CS_2 and CO_2 molecule

and the sulphur (two lobes symmetrically distributed along the main axis of the three atoms) and the laser time evolution are shown in detail. Such clear separation between HOMO and relevant inner orbitals enables us to get a deep insight on the main factors (shape and IP) of the KS ionization process as described in the KS ionization section-4. The TDELF images evolve in real time and space for $\theta=0°$ and $\theta=90°$ at 3.50×10^{14} W/cm^2 laser peak intensity respectively for the molecules to follow.

a. CO_2 and CS_2

When $\theta=0°$, at the initial time step i.e. 0 optical cycle (o.c.), the ELF picture is that of the ground state. As the time propagation evolves, the first major perturbation occurs on the lone pair region, i.e. the oxygen for CO_2 or the sulphur for CS_2 after 1.75 optical cycles.

The electric field being negative LIET is thus expected to occur from left to right. The major perturbation consists of a lengthening of the lone pair region followed by the appearance of a hole, reflecting the depletion/ionization of the electron density. The ionization is found as expected higher in CS_2 (large hole) than in CO_2. The highest perturbation in the CS_2 is due to its lower IP. At 2.25 optical cycles, (the electric field is positive), a perturbation as exactly described below arises now on the other atom of the oxygen or sulphur. The same asymmetric perturbation behaviour is observed respectively at 3.75 and 4.25 optical cycles with the perturbation getting larger. The asymmetric distribution of the electron density on the two oxygens (for CO_2) or sulphur (CS_2) atoms shows that the two atoms do not experience simultaneously or instantaneously the same laser field strength, and that, depending on the sign of the electric field, the perturbation affects successively each of the oxygen/sulfur atom (the picture shows that at 1.75 o.c, for negative electric field the atom at

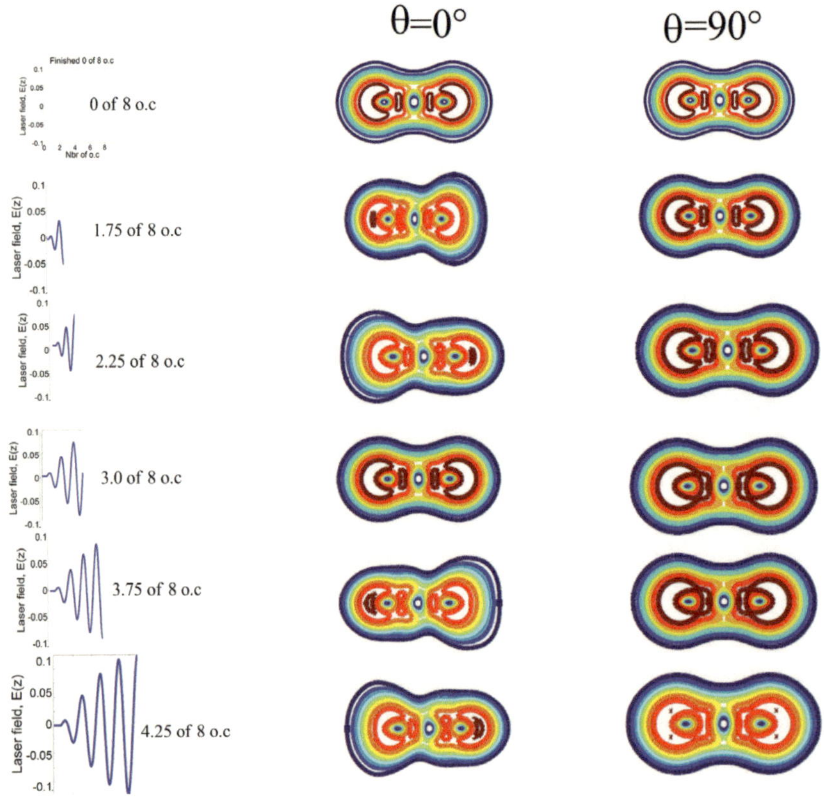

Fig. 3c. Snapshot of the isosurface of the TDELF images for Io = 3.5 x 10^{14} W/cm^2 as the CO_2 molecule evolves in the laser field. Calculations are done for with $\theta=0^o$ and $\theta=90^o$.

previous KS ionization analysis, the picture shows that the major part of the perturbation is located on the region corresponding to the lone pair region ($1\pi_g$). As the time propagation evolves the intensity of the perturbation on the lone pair region increases as well, reaching a maximum for the extrema (minimum i.e 3.75 o.c (E<0) or maximum i.e 4.25 o.c (E>0)) of the external electric field and seeming to disappear around zero values of the external field (e.g at 3.0 o.c). The weak deformation of the CO bond domain reflects the weak contribution of the inner orbital to the molecular ionization process in agreement with the previous KS ionization finding. For the perpendicular configuration $\theta=90^o$, the first noticeable perturbation occurs after 1.75 o.c and corresponds to the lengthening or increase in diffuseness of the lone pair region on the oxygen atoms of CO_2 or the sulphur atoms for CS_2. The perturbation is symmetrically distributed on the system; this effect reflects the fact that the two oxygen or sulphur atoms feel simultaneously the same effect of the laser field. As the time propagation evolves (3.75 o.c and 4.25 o.c), one also sees that the dominant response comes from the lone pair region. Most notably, one observes that during the propagation, the electronic perturbation is mainly characterized by an interaction between the outer lone pair and the inner bonding CO region. This may be regarded as the manifestation of the enhancement of the ionization of the bonding $1\pi_u$ molecular orbital localized on the CO bond region. Evidence of the laser field intensity influence is noticeable, the stronger is the intensity peak value, and higher is the perturbation. Once again, one

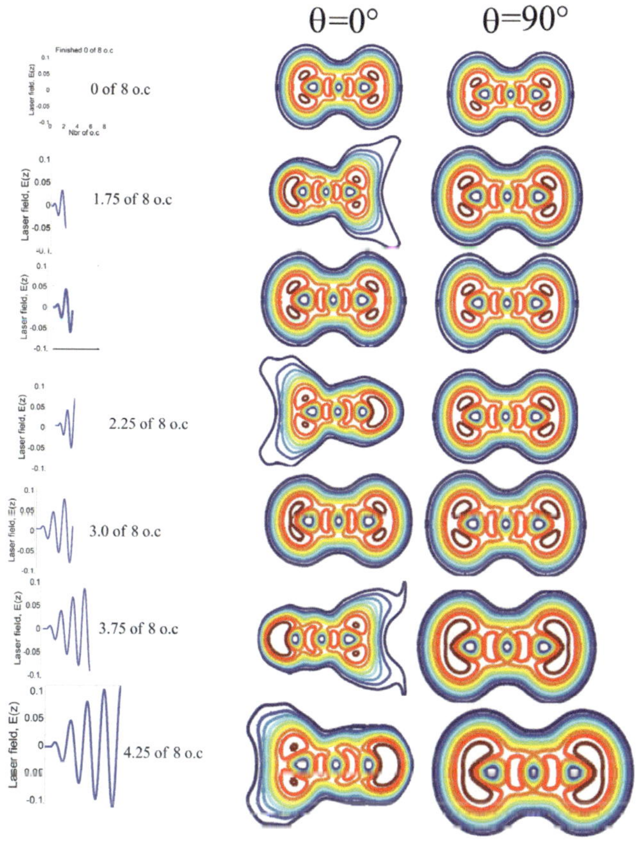

Fig. 3d. Snapshot of the isosurface of the TDELF images for Io = 3.5 x 10^{14} W/cm^2 as the CS_2 molecule evolves in the laser field. Calculations are done for with θ=0° and θ=90°.

observes that the ionization is found as expected higher on CS_2 than CO_2, and the perturbation from the bonding region is also higher. This suggests that for the case of CS_2, both the HOMO and the inner orbitals are significantly affected at I=3.5×10^{14} W/cm^2. We next illustrate results at higher I= 3.0×10^{15} W/cm^2 laser peak intensity in Fig.3e for CO_2.
For θ=0°, one finds that the first major perturbation starts closer to the beginning of the time propagation and it is mostly localized on one of the lone pair regions of the molecule.

Thereafter, the asymmetric perturbation appears instantaneously on the CO bond and the lone pair domain (1.75o.c), its effect increasing with the duration of the propagation. Therefore, it is difficult to identify which part of the system presents the dominant response to the field; this reflects the contribution of both the HOMO KS orbital (due to its lower IP) and the inner $3\sigma_u$ MO (due to its shape or its electron density which is maximal along the polarization of the electric field). These results corroborate our finding in the previous section concerning the KS ionization. When θ=90°, at the beginning of the time propagation, the electron depletion occurs on the lone pair region and is characterised by diffuseness of this region. The image shows that at 1.75 optical cycles for which E<0, the interaction between the lone pair density and that of the CO bond is visible and an amount of electron depletion is noticeable in the lone pair region. The symmetric distribution of the electron

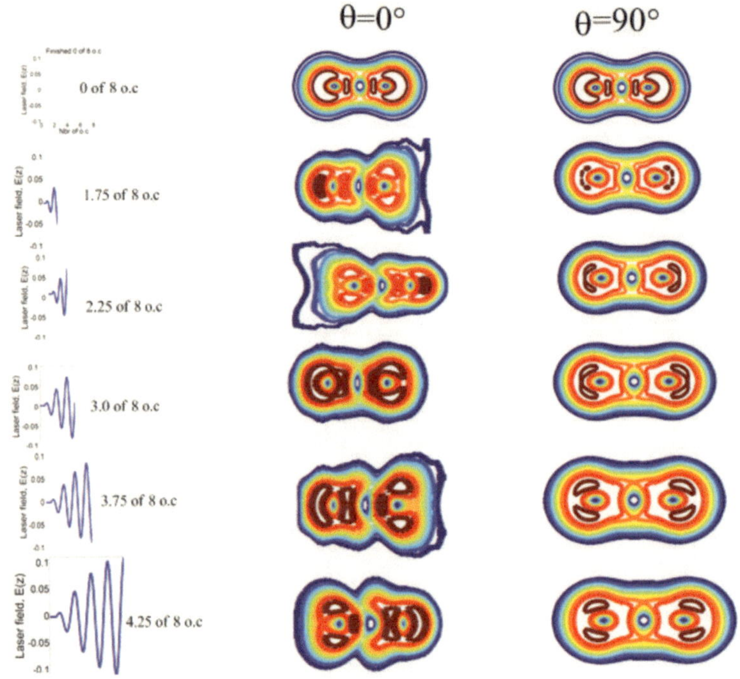

Fig. 3e. Snapshot of the isosurface of the TDELF images for Io = 2.5×10^{15} W/cm² as the CS_2 molecule evolves in the laser field. Calculations are done for with θ=0° and θ=90°.

density perturbation on the two oxygen atoms is clearly evident. As the time propagation evolves, the bi-partition of the perturbation domain of the lone pair orbital of each oxygen atom is observable and tends to enhance the overlap between the lone pair and the CO bonding region. This can be seen as the strong perturbation (ionization) of the inner $1\pi_u$ KS orbital. These observations are consistent with our results in Fig.2a and Fig.2b which show that the IP and the shape of the KS orbital are the main factors which define its ionization. At lower laser peak intensity, the IP is the main factor which governed the ionization process while at higher laser peak intensity; the ionization of the KS orbital is influenced by its shape.

b. OCS

The most interesting feature in the study of the non symmetric OCS molecule with the TDELF function is the fact that the HOMO is located on two different atoms; the oxygen and

the sulphur. As found in the previous section, the two atoms ionize differently due to their IP difference. The oxygen is easily recognized with one lobe while the sulphur bears two. For $\theta=0^o$, at 1.75 (E<0) optical cycles, the perturbation of the electron density appears on the lone pair position of the oxygen atom, while its homolog lone pair on the sulphur atom seems less sensitive to the negative electric field. The ELF after 2.25 (E>0) optical cycle shows the perturbation appears now on the lone pair region of the sulphur atom while the corresponding lone pair region on the oxygen atom remains nearly unchanged. As the propagation is evolving, the intensity of the perturbation on the lone pair region is increasing as well; reaching a maximum at the electric field maxima or minimal. For the orientation $\theta=90^o$, an overview of the TDELF image at 1.75 and 2.25 optical cycle shows an amount of electron density leaving the system. While the two atoms are feeling exactly the same laser intensity at the same time, the ionized electron density is much higher from the sulphur moiety than the oxygen one, due to the electronegativity difference of the sulphur and the oxygen atom. As expected, the perturbation on the different lone pair region is getting higher as the laser intensity is increasing. Fig.3f shows that the OCS behaves in the presence of the laser as superposition of the moieties, one from CS_2 and the other from CO_2, giving rise to a non-symmetric perturbation on the system when the laser is oriented parallel or perpendicular to the laser field.
In conclusion, we find that at the lower $I=3.50\times10^{14}$ W/cm^2 (10^{-2}a.u.) laser peak intensity, the lone pair region (HOMO) of linear molecules is the most affected by the field, in good agreement with our finding in the KS-MO ionization analysis in section-4. We show next that harmonics emitted by the molecule at that particular laser peak intensity is also mostly due to the HOMO of each molecule, whereas at higher intensities, inner shell ionization must be considered.[41]

6. Molecular High Order Harmonic Generation-MHOHG

In Fig.4a, Fig.4b, we present the MHOHG power spectra obtained by, Eq.11, emitted respectively by CS_2 and CO_2 at $\theta=0°$, 45° and 90°. Calculations are done at $I=3.50\times10^{14}$ W/cm^2 laser intensity and the ionization yield is given in Table II to see the connection between the molecular ionization and MHOHG spectrum signal. One finds in general that the overall shape of the power spectra resemble each other. For all orientations, features of these spectra strongly resemble those of harmonic spectra from atoms, i.e., a sharp decrease of the first few harmonics from bound orbital electrons, followed by a “plateau” and ending with the ‘cut-off’ due to recollision of an ionized electron that usually determines the highest harmonic order achievable[5, 8-9, 42]and is given by the classical law, $N_m\omega=IP+3.17U_p$. The computed cut-off number, N_m, which is in agreement with the classical law is found approximately at the 55th harmonic for CS_2 and 57th for CO_2; the difference lies in their IP. The figures show that the harmonic peaks in the plateau region are in general smaller in the case of $\theta= 0°$ (parallel polarization) than other orientations. The difference can be as high as one order of magnitude, which is the case for the harmonics between the 35th and 45th for CO_2. The most interesting feature of these spectra appears when one has a close look at the plateau region. There is a broad region with suppressed harmonic emission rate (indicated by the arrow on the figures). This has been interpreted in terms of two center interference between the contributions from two atomic centers.[11-12, 43-44] In fact, a CO_2 or CS_2 molecule can be regarded from their HOMO as an elongated diatomic molecule where the point emitters are located on the two sulphur nuclei for CS_2 or the two oxygen nuclei for

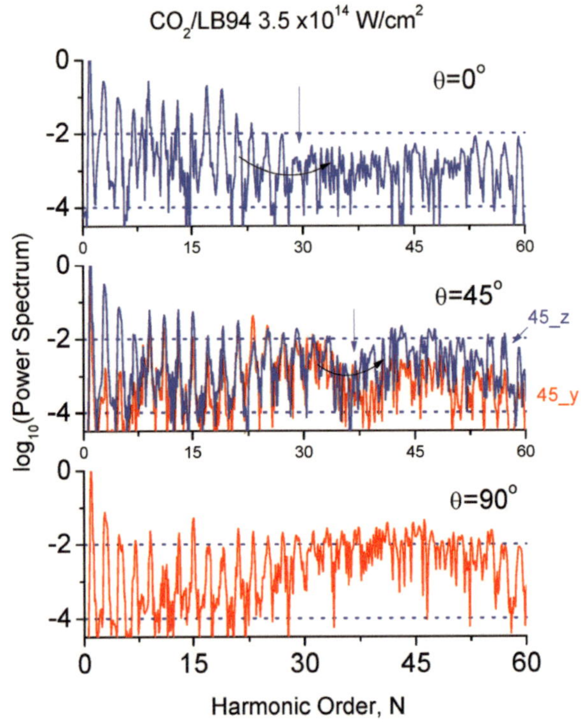

Fig. 4a. (Color online) TDDFT/LB94 harmonic spectra of CO_2 for some angles, between the molecular axis and the laser polarization direction. Laser pulse duration, eight optical cycles, peak intensity I= 3.50×10^{14} W/cm^2, and wavelength, 800nm.

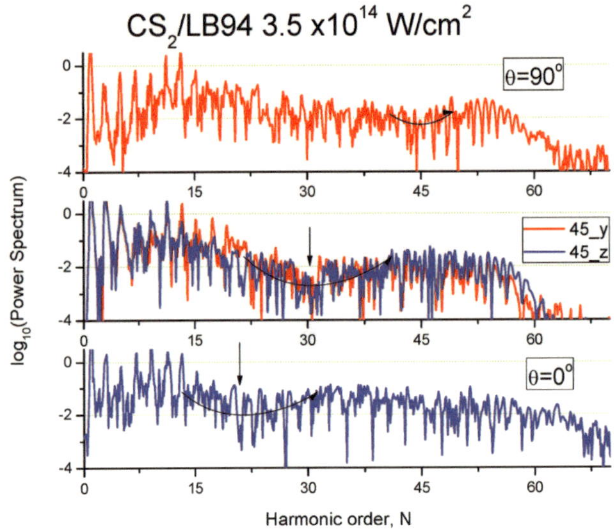

Fig. 4b. (Color online) TDDFT/LB94 harmonic spectra of CS_2 for some angles, between the molecular axis and the laser polarization direction. Laser pulse duration, eight optical cycles, peak intensity I= 3.50×10^{14} W/cm^2, and wavelength, 800nm.

CO_2. The suppression comes from destructive interference between the contributions from the two atomic centers discussed below.

a. CO_2

The total molecular ionization yield (Table II) is respectively 2.0, 2.3 and 1.2 % when θ=0°o, 45°, 90°. From Fig.4a, one sees that when θ=0°, the spectrum shows the presence of a low signal in agreement with the experimental observation[45-47], but with slight different laser peak intensities centered around the 29th harmonic. When the molecular axis is 45° with respect to the laser polarisation, two plots are done, showing the projection of the time dependent dipole acceleration along the z and y axis. The MHOHG spectrum is split into contributions from the z-axis (noted by 45_z on the figure) and the corresponding perpendicular x-axis (noted by 45_x). One sees that the minimum (the destructive interference between the contributions from the two atomic centers) is shifted to the harmonic order N around 35 for both. Furthermore the MHOHG signal increases significantly compared to what is obtained at θ=0° due to the strong HOMO $1\pi_g$ orbital ionization. Evidence of the elliptical polarization of the MHOHG spectra is visible due to the different signal amplitude from the two component of the harmonic spectrum as shown in the Fig.4a. The interference minimum of the z component of the harmonic spectrum is located near the harmonic order N=31 while it shifts to N= 37 for the x component. It also appears that the harmonic spectrum signal from the z component is higher than that of the corresponding x component, due to the different contribution of the inner $3\sigma_u$ and $1\pi_u$ molecular orbitals. When θ=90°, no obvious minimum is observable, in comparison to the case θ=0°. Instead, one observes an enhancement of the signal of harmonics above N=25. This is mainly the result of the constructive interference from the two oxygen atoms. In fact, they have the same potential and are feeling exactly at the same time, the same electric field.

b. CS_2

For CS_2, the total molecular ionization yield (Table II) is respectively 13.3, 12.9 and 9.1% when θ=0°, 45 °, 90°. The harmonic spectrum at $3.5\text{x}10^{14}$ W/cm^2 is given in Fig.4b. A close view reveals that the destructive interference (weak MHOHG signal spectrum amplitude) takes place near N=25 when θ=0°, it moves to harmonic order 31 when θ=45°. Evidence of the elliptical polarization of the MHOHG spectra is again visible due to the different signal amplitude from the z and x component of the harmonic spectrum near the cut-off. The data shows for harmonics order N<31, the signal from the x component is higher than that of the z component, while the opposite trend is found for N near the cut-off. When θ=90°, a shallow minimum is observed near the cut-off, in contrary to the case θ=0°, so the evidence of the constructive interference is not clearly observable on the spectrum while the two sulphur atom have the same potential and are experiencing the same laser intensity. This comes from the lower IP of the KS MOs of CS_2, which gives rise to high ionization involving inner MOs at I=3.50×10^{14} W/cm^2 laser peak intensity as shown in table II. To check this, we have further investigated the influence of the laser intensity or the inner orbitals on the interference effect, by repeating the calculation on CS_2 at the low laser intensity Io=1.72×10^{14} W/cm^2 for θ=90° and 0° in Fig.4c. The computed total ionization yield, Eq.9 is found around 6.5% for the parallel case and less than 4.4% for the perpendicular one. We assume that for this lower laser intensity, the contribution of the inner orbital is unimportant as shown on the orbital population ionization analysis in Fig.4c.

The corresponding MHOHG spectrum shows a cut-off around the 35th harmonic (Fig.4d). As in Fig.4a, the harmonic signals in the plateau region are higher for the perpendicular orientation than the parallel one. For θ=0°, a pronounced minimum is visible around the 27th

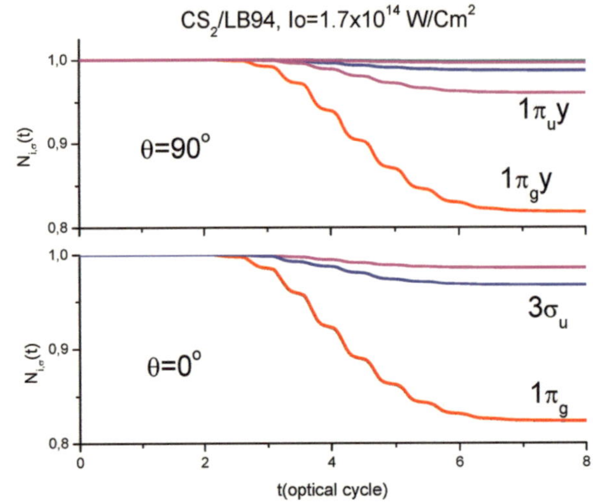

Fig. 4c. (Color online) TDDFT/LB94 orbital population, $N_{i,\sigma}(t)$, of CS_2 for selective fixed angle, θ, between the molecular axis and the laser polarization direction and I= 1.72×10^{14} W/cm^2. Only the relevant KS orbitals which possess an important response to the laser field are shown with their label.

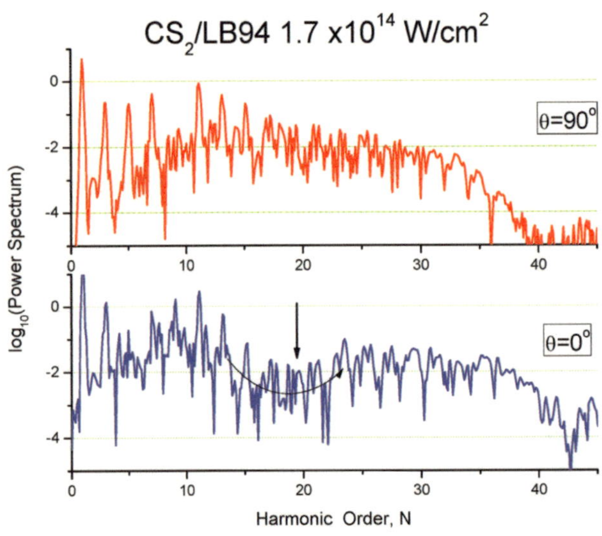

Fig.4d. (Color online) TDDFT/LB94 harmonic spectra of CO_2 for some angles, between the molecular axis and the laser polarization direction. Laser pulse duration, eight optical cycles, peak intensity I= 1.72×10^{14} W/cm^2, and wavelength, 800nm.

harmonic while when θ=90° the harmonic spectrum signal is significantly higher as expected resulting from a constructive interference as found in the CO_2 and the H_2^+ case.[48]

c. OCS

Considering the nonsymmetric OCS case, one finds that the amount of ionization, computed using Eq.9 is nearly the same at the end of the eight cycle laser pulse (around 7%) independent of the laser molecule angle θ. However, the difference between theses orientations on the harmonic spectra (Fig.4e) is remarkable.

For the parallel orientation, θ=0°, the spectra shows in the plateau region a striking pronounced minimum or destructive interference as for the CS_2 and CO_2 case around N= 39. For θ=90°, unlike what we find for the case of CS_2, a minimum is found around N=25. This interference minimum is rather broad and as deep as is found for θ=0°. Although the oxygen and the sulfur atom experience exactly the same external electric field, the minimum observed is due to the electronegativity and the IP difference between the two atoms. To explain these results, we invoke the semiclassical picture of MHOHG. When the driving laser field is polarized perpendicular to the molecular axis, if the two nuclei are identical (CS_2 and CO_2), they experience the same tunnelling ionization and the returning electron wavepacket can interact simultaneously with all the ionic cores. Then, both nuclei at the same time emit radiation of the same intensity, frequency and phase[12, 49], thus leading to enhanced harmonic yield. However, if the two nuclei are different (different atomic electronegativity or atomic ionization potential, IP as for the OCS case), although they experience the same laser intensity, both nuclei emit different radiation of different intensity, giving rise to the presence of both minima and maxima in the spectrum. In contrast, the electron wavepacket if they are driven by a laser field parallel to the molecular axis can miss the molecular core. In fact the nuclei screen each other from the returning electron wavepacket. Thus the interference of harmonics emitted from each nucleus occurs over a wide spread of harmonic order thus giving rise to a smaller efficiency for HHG in the case of parallel orientation.

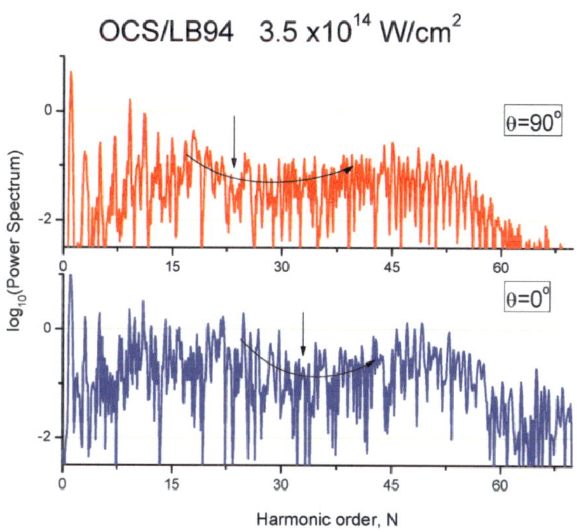

Fig. 4e. (Color online) TDDFT/LB94 harmonic spectra of OCS for some angles, between the molecular axis and the laser polarization direction. Laser pulse duration, eight optical cycles, peak intensity I= 3.50×10[14] W/cm², and wavelength, 800nm.

d. N_2

In Fig.4f we present the MHOHG power spectra obtained by, Eq.11, emitted respectively by N_2 at θ=0°and 90°. Calculations are done at I=3.50×10[14] W/cm² laser intensity and the ionization yield is respectively 3% at θ=0° and 5% at θ=90°. The 'cut-off' due to recollision of an ionized electron that usually determines the highest harmonic order achievable, N_m, is found approximately at the 55th harmonic. When the molecule is aligned parallel to the laser polarization (θ=0°), the spectrum displays two shallow minima, the first at N=25 in agreement with the experimental measurements[45-47] at I=2.3 x 10^{14} W/cm^2 and the second one near the cut-off at N=43. This latter is absent in the experimental data most probably due to the different laser intensity used or due to the fact that experimentally the noise level is much higher near the cut-off. The difference with the corresponding harmonic spectrum amplitude for θ=90° is remarkable. An interesting future is the complete absence of a minimum interference and the harmonic signal is strongest near the cut-off region of the spectrum. This reflects the constructive interference in the MHOHG recombination step that results in a maximum in the high harmonic signal originating from the HOMO-1 of symmetry π_u (although it is bound by an additional 1.6 eV) due to its shape and also its large recombination dipole as shown in Fig.1 and Fig.3d.

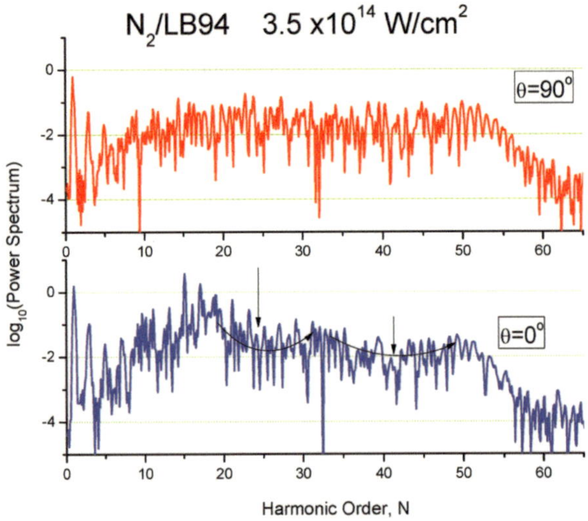

Fig. 4f. ((Color online) TDDFT/LB94 harmonic spectra of N_2 for some angles, between the molecular axis and the laser polarization direction. Laser pulse duration, eight optical cycles, peak intensity I= 3.50×10^{14} W/cm^2, and wavelength, 800nm.

e. Comparative MHOHG

In figures 4a-f, we have illustrated MHOHG spectra for CO_2, CS_2, OCS and N_2, i.e. 3-center and 2-center molecules. The TDELF images, Fig 3a-f show that OCS behaves like a superposition of two 2-center moieties, CO and CS. The dominant features of the MHOHG spectra are intensity minima at certain laser-molecule angles, θ. Thus in Fig.4a, CO_2 exhibits a minimum around N=37 at θ=45°. CS_2 z and y harmonic components have minima around N=30, Fig.4c and similarly for OCS Fig.4d. For comparison the 2-center N_2 shows a minimum around N=25 and 41 at θ=0° which disappears for perpendicular ionization and recombination at θ=90°, Fig.4f. CO_2 (Fig.4a), CS_2 (Fig 4b) and N_2 (Fig.4f) show no significant minima in their MHOHG spectrum at θ=90° whereas OCS (Fig.4d) manifests at θ=90° a broad minimum centered around N=30. The symmetric molecules N_2, CO_2, CS_2 can be treated as 2-center ionization and recombination systems with emitter interference patterns for the multiphoton ionization and photon recombination.[4, 12] Thus for molecular bonding orbitals sums of atomic orbital, σ_g and π_u, both ionization and recombination amplitude will have 2-center interference modulation proportional to cos($\mathbf{p}_e$•$\mathbf{R}$/2) whereas antibonding antisymmetric combinations σ_u and π_g produce a $\sin(\mathbf{p}_e \bullet \mathbf{R}/2)$ interference[4] for electron momentum $\mathbf{p}_e$ and internuclear distance R. Thus for perpendicular laser-molecule orientation, $\theta=90^o$, $\cos(\mathbf{p}_e \bullet \mathbf{R}/2)=1$ and $\sin(\mathbf{p}_e \bullet \mathbf{R}/2)=0$ for electrons ionized or recombining along the laser polarization. CO_2, OCS and CS_2 ionization probabilities are in fact minimal at $\theta=90^o$, whereas for N_2 at higher intensities these are equal (Table 2). This result is confirmed by the HOMO antibonding combination of atomic orbitals of the triatomics π_g as compared to the bonding $3\sigma_g$ HOMO of N_2. Similarly, lower higher IP bonding π_u orbitals in the triatomics have dominant ionization at $\theta=90^o$ and higher intensities where $\cos(\mathbf{p}_e \bullet \mathbf{R}/2) = 1$, Fig.2. In the case of harmonics, one observes in general little structure or interference at $\theta=90^o$ for symmetric molecule due to both $\cos(\mathbf{p}_e \bullet \mathbf{R}/2)$ and $\sin(\mathbf{p}_e \bullet \mathbf{R}/2)$ modulating factors. Clear MHOG intensity minima are found for CO_2 around N=37 at $\theta=45^o$ Fig.4a, for CS_2 and N=30 and $\theta=45^o$ Fig.4b, and N_2 at N=25 for $\theta=0^o$, Fig.4f. Harmonics at orders N=25 are produced by electrons with momentum $p_e=1.69$, since $p_e^2/2=N\omega$, with $\omega=0.057$ au (800nm). The corresponding electron wavelength is $\lambda_e= 2\pi/p_e=0.197$ nm, or equivalently 2R for N_2 (R=0.104 nm). Since recombination occurs to the bonding $3\sigma_g$, HOMO, $\cos(\mathbf{p}_e \bullet \mathbf{R}/2) =\cos(\pi R/\lambda_e)=\cos(\pi/2)=0$, thus explaining the minimum. The same exercise at $\theta=45^o$ for CO_2 gives $\lambda_e=R(CO_2)=0.2$ nm and $\sin(\mathbf{p}_e \bullet \mathbf{R}/2)= \sin(\pi R/\lambda_e)=0$. The same result is found for CS2, since in both cases recombination occurs in an antibonding π_g HOMO. In these three symmetric molecular cases, two center interferences, $\cos(\mathbf{p}_e \bullet \mathbf{R}/2)$ for bonding and $\sin(\mathbf{p}_e \bullet \mathbf{R}/2)$ for antibonding HOMO's regulate MHOHG spectral intensities. The nonsymmetric OCS MHOHG spectra present anomalous intensities as seen in Fig.4d. A clear minimum appears at order N=35 at $\theta=0^o$ and an unexpected minimum around N=30 at $\theta=90^o$. The charge asymmetry in this molecule is responsible for more complex recombination and MHOHG spectra.

7. Conclusion

The nonlinear nonperturbative TDDFT calculations presented in this paper for symmetric CO_2, CS_2 and non-symmetric OCS tri-atomics and the di-atomic N_2 address the TDELF analysis and the MHOHG process occurring from MO ionization at high laser intensities where linear TDDFT is not applicable. As major results, we find that at equilibrium distance R and at intensities $I_0 > 3.5\times10^{14}$ W/cm^2, lower inner highest occupied molecular orbitals contribute significantly to ionization and to the MHOHG process. This is due to the symmetry of these orbitals. Even though such lower inner shell orbitals have higher ionization potentials and MHOHG processes occur when orbital densities are maximal with laser polarization direction. Our simulations also reveal that the direction of the laser polarization, θ, with respect to the molecular axis of the linear molecule can have a significant effect on MHOHG. At some angles, ionization probabilities of different orbitals cross in magnitude.[50] So, while maxima (constructive) and minima (destructive) due to intramolecular interference are found in the dependence of the harmonic intensities on the orientation of the asymmetric OCS molecule, only maxima interference are found for symmetric CS_2 and CO_2 molecules when the laser polarization is perpendicular to the molecular axis. The relative position of the minima interference increases to high harmonic order, N, when the laser-molecule angle, θ, increases. This is explained in section (6-e) by a simple general formula based on recombination of a continuum electron with a HOMO.[4] These findings are confirmed with the time dependent electron localization function, TDELF, representation through the analysis in term of density perturbations appearing on the TDELF images of each molecule.[23] For θ < 90° and at lower laser intensity (Io =10^{14} W/cm^2), one sees that the HOMO is the most affected by the laser field and a large asymmetry density is found, i.e., we clearly see that during each half cycle, the perturbation occurs alternatively from one nucleus to another (favouring minimal interference) while for θ=90°, both nuclei simultaneously feel the same perturbation from the laser field (favouring maximal interference).[51] The present TDDFT simulations show also that the local LB94 potential gives reliable ionization energies[52], IP, of individual orbitals, thus suggesting that the KS orbitals are close to Dyson orbitals[40] which are the exact many body orbitals for ionization. The relation between these Dyson and the LB94 KS orbitals requires further exploration. Furthermore, the MHOHG spectra show strong dependence on the laser intensity and the laser-molecular angle θ. This suggests that MHOHG can be a potentially powerful way of studying molecular structure such as orbital tomography.[10] This is made possible by observing and analyzing the shape of the MHOHG spectrum in the plateau region whereas electron recombination is predicted to occur near zero electric field.[49] We also show that elliptical polarization of the MHOHG spectra are influenced by inner shell ionization. In general, for intensities above I=3.5x10^{14} W/cm^2, inner shell orbitals, i.e. lower highest occupied molecular orbitals HOMO-1 and HOMO-2 with larger ionization potentials, IP, than the highest occupied orbital, HOMO, can contribute considerably to total ionization. The main reason is that these inner-valence orbitals have fewer nodes than the HOMO and therefore for certain laser polarization, the density of these lower (but higher IP) orbitals at the laser-molecule angle θ is much larger than the HOMO. At angle θ=45°, the degeneracy of π orbitals is removed by the laser, resulting in different MHOHG polarization components. The phase dependence of these different component harmonics and their relation to electron dynamics has not yet been explored.[49] As a general rule, we note that the ionization is generally preferred as the molecule is aligned along the major axis of the electron distribution in the active molecular orbital. It should be emphasized that while the local LB94 potential yields good molecular orbital ionization potentials IP, further study of MHOHG including the nucleus separation distance effects such as recently reported[53] for H_2 and using more accurate exchange potentials based on long range-short range corrected model[54] offer scope for future accurate characterization of MHOHG processes. The effects of the atomic position (bond length) on the molecular ionization and MHOHG signal are left for future study in order to treat new nonlinear phenomena in molecules such as Charge Resonance Enhanced Ionization, CREI.[51]

8. References

[1] M. I. Al-Joboury, et al., *J. Chem. Soc.*, 6350 (1965).
[2] M. Hentschel, et al., *Nature* 414, 509 (2001).
[3] M. Drescher, et al., *Nature* 419, 509 (2002).
[4] A. D. Bandrauk, et al., in *Progress in Ultrafast Intense Laser Science*, edited by K. Yamanouchi, et al. (Springer, Tokyo, 2006), Vol. III.
[5] P. B. Corkum, *Phys. Rev. Lett.* 71, 1994 (1993).
[6] M. Lewenstein, et al., *Phys. Rev. A* 49, 2117 (1994).
[7] S. Chelkowski, et al., *J. Phys. B: At. Mol. Opt. Phys.* 39, S409 (2006).
[8] A. D. Bandrauk, et al., *Phys. Rev. A* 56, R2357 (1997).

[9] A. D. Bandrauk, et al., *Phys. Rev. Lett.* 98, 0131001 (2007).
[10] J. Itatani, et al., *Nature (London)* 432, 867 (2004).
[11] G. L. Kamta and A. D. Bandrauk, *Phys. Rev. A* 70, 011404 (2004).
[12] G. L. Kamta and A. D. Bandrauk, Phys. Rev. A 71, 053407 (2005).
[13] A. D. Bandrauk and H. Z. Lu, *Phys. Rev. A* 72, 023408 (2005).
[14] R. Bauernschmitt and R. Ahlrichs, *Chem. Phys. Lett.* 256 (454) (1996).
[15] C. A. Ullrich and A. D. Bandrauk, in *Time-Dependent Density Functional Theory*, edited by M. A. L. Marques, et al. (Springer, Berlin, 2006), pp. 357-374.
[16] W. Kohn, *Rev. Mod. Phys.* 71, 1253 (1999).
[17] E. Runge and E. K. U. Gross, *Phys. Rev. Lett.* 52, 997 (1984).
[18] M. E. Casida, et al., in *Time-Dependent Density Functional Theory*, edited by M. A. L. Marques, et al. (Springer, Berlin, 2006), pp. 243-257.
[19] A. Savin, et al., *Angewandte Chemie-International Edition in English*, 30, 409 (1991).
[20] M. Erdmann, et al., *J. Chem. Phys.* 121, 9666 (2004).
[21] T. Burnus, et al., *Phys. Rev. A* 71, 010501 (2005).
[22] A. D. Becke and K. E. Edgecombe, *J. Chem. Phys.* 92, 5397 (1990).
[23] E. F. Penka and A. D. Bandrauk, *Can. J. Chem* (in press) (2010).
[24] X.-B. Bian and A. D. Bandrauk, *Phys. Rev. Lett.* (to appear 2010).
[25] P. Hohenberg and W. Kohn, *Phys. Rev.* 136, B864 (1964).
[26] W. Kohn and L. J. Sham, *Phys. Rev.* 140, A1133 (1965).
[27] H. S. Nguyen, et al., *Phys. Rev. A.* 69, 063415 (2004).
[28] R. V. Leeuwen and E. J. Baerends, *Phys. Rev. A* 49, 2421 (1994).
[29] A. Banerjee and M. K. Harbola, *Phys. Rev. A* 60, 3599 (1999).
[30] C. Toher, et al., *Phys. Rev. Lett.* 95, 146402 (2005).
[31] M. E. Madjet, et al., J. Phys. B 34, L345 (2001).
[32] H. O. Wijewardane and C. A. Ullrich, *Phys. Rev. Lett.* 100, 056404 (2008).
[33] N. Trouillier and J. L. Martins, *Phys. Rev. B* 43, 1993 (1991).
[34] C. R. Brundle and D. W. Turner, *Int. J. Mass Spectrom. Ion Phys.* 1, 285 (1968).
[35] D. W. Turner and D. P. May, *J. Chem. Phys.* 46, 1156 (1967).
[36] C. R. Brundle and D. W. Turner, *Int. J. Mass Spectrom. Ion Phys.* 2, 195 (1969).
[37] T. Brabec and F. Krausz, Rev. Mod. Phys. 72, 545 (2000).
[38] F. He, et al., *Phys. Rev. A* 75, 053407 (2007).
[39] A. D. Bandrauk, et al., *Phys. Rev. A* 79, 023403 (2009).
[40] J. V. Ortiz, *Int. J. Quantum Chem.* 100, 1131 (2004).
[41] A. Talebpour, et al., *Chem. Phys. Lett.* 313, 789 (1999).
[42] A. D. Bandrauk, et al., *J. Mod. Opt.* 52, 411 (2005).
[43] M. Lein, et al., *Phys. Rev. A* 66, 023805 (2002).
[44] M. Lein, et al., Phys. Rev. A 67, 23819 (2003).
[45] Y. Mairesse, et al., *N. J. Phys.* 10, 025015 (2008).
[46] T. Kanai, et al., *Nature (London)* 435, 470 (2005).
[47] O. Smirnova, et al., *Nature* 460, 972 (2009).
[48] G. L. Kamta and A. D. Bandrauk, *Phys. Rev. A* 80, 041403 (2009).
[49] A. D. Bandrauk and L.-J. Yuan, *Phys. Rev. A* 81, 063412 (2010).
[50] E. F. Penka and A. D. Bandrauk, *Phys. Rev. A* 81, 023411 (2010).

[51] T. Zuo, et al., *Chem. Phys. Lett.* 259, 313 (1996).
[52] S.-K. Son and S.-I. Chu, *Chem. Phys.* 91, 366 (2009).
[53] A. D. Bandrauk, et al., *Phys. Rev. Lett.* 101, 153901 (2008).
[54] J. Toulouse, et al., *Phys. Rev. A* 70, 062505 (2004).
[55] R de Vivie-Riedle, private communication

22

Femtosecond Fabrication of Waveguides in Ion-Doped Laser Crystals

Andrey Okhrimchuk
Fiber Optics Research Center of RAS
Russian Federation

1. Introduction

Permanent refractive index change (PRICE) in dielectrics by means of the femtosecond laser pulses is a novel enabling technology in photonics. A wide range of photonic structures manufactured using this method has been demonstrated in glasses and crystals since the first observation of the underling phenomenon (Davis, 1996). While numerous waveguides, waveguide lasers and amplifiers, couplers and Bragg gratings were fabricated on the basis of this phenomenon, PRICE theory is far from accomplishment yet, although some basic principles are beyond any doubt. To date it is obvious that to understand femtosecond modification of a transparent dielectric the process should be separated in two stages. The first stage consists of the non-linear absorption of a femtosecond pulse and electron plasma generation. The second stage consists of energy transfer from the electron plasma to ions and structural changes in a dielectric. The first stage seems to be very analogous both in glasses and crystals, as it deals with electronic excitation and only material parameters required for its description are energy gap width and coefficients of multiphonon absorption (MPA).

The second stage still rises many questions and should be considered as an incomplete chapter in PRICE theory. To date it is not clear whether PRICE proceeds in the same manner both in crystals and glasses. It is generally understood for glasses and associated with melting and densification (Glezer, 1997; Streltsov & Borelli, 2001). As a rule it gives positive refractive index change in the exposed region, and its magnitude can be as high as 10^{-2} (Allsop, 2010). Thus a straightforward way for waveguide inscription is open for glasses.

Contrary to glasses PRICE is rather more complicated and intrigued in crystals. As a rule refractive index change is negative in the exposed region. For example, a widely accepted point of view relays on the assumption that a crystal undergoes amorphisation in the exposed region, and this causes stresses and positive refractive index change in the surrounding area (Gorelik, 2003; Apostolopoulos, 2004). Thus a waveguide is usually created in the area adjacent to tracks written by femtosecond beam. Since in this case waveguiding is due to an indirect effect accompanying femtosecond modification of crystal lattice, magnitude of refractive index change in the waveguide is not so high as in glass waveguides and does not exceed $1*10^{-3}$ (Nejadmalayeri, 2005; Torchia, 2008; Siebenmorgen, 2009; Silva, 2010; Bookey, 2007; Burghoff, 2007). This value is not enough to build compact waveguide lasers with diode pumping. Meantime in many cases crystals are more a attractive media for femtosecond fabrication of compact waveguide lasers in comparison to glasses, because they have better thermo-conductivity, high optical damage threshold and allow high doping level of rare-earth and transition metal ions without significant degradation of spectroscopic characteristics. Finally negative refractive index change could be altered to an advantage, as it allows writing a cladding, while a waveguide core is composed of unperturbed crystal region. In such architecture degradation of spectroscopic parameter is excluded and scattering loss is basically lower.

This paper is devoted to investigation of the processes relating to femtosecond writing of waveguides in laser crystals doped by rare-earth and transition metal ions, as well as to design and fabrication of crystalline waveguide lasers. The author proposes his own point of view on processes taking place in crystal, and methods for inscription of low loss waveguides in ion-doped YAG crystals. An example of a waveguide laser will be presented.

2. Waveguide fabrication technique and architecture

Numerous femtosecond lasers operating in NIR (Ti: sapphire or Yb-based lasers) or VIS (harmonic of Ti:sapphire laser), and even in Mid-IR (OPO, 2400 nm) are used for waveguide fabrication in transparent dielectrics. They are either oscillators providing pulses with high rep rate from range of 5 MHz to 25 MHz, or regenerative amplifiers providing lower rep rate nearly less than 5 MHz. A sample is mounted on high-precision 3-D translation stage, and laser beam is tightly focused in the sample volume at a depth of 20-500 μm under polished surface of the sample. Microscopic lenses with numerical aperture NA= 0.4 - 1.4 are usually used for focusing, providing beam waist as low as 0.5 - 2 μm. Focusing depth is first of all constrained by aberrations which a converging beam suffered at air-dielectric interface (they could be reduced by implementing immersion liquid). Another constrain is the lens working distance. While the sample is moving in the direction perpendicular to the laser beam, beam waist produces a permanent stable track of modified material with altered refractive index, or permanent refractive index change (PRICE). Velocity of the translation in dependency of laser rep rate may be in the range of 0.01 - 50 mm/s.

Under moderate pulse energy, suitable for waveguide fabrication, PRICE is positive in fused silica and most of glasses (Davis, 1996; Miura, 1997; Homoelle, 1999; Schaffer, 2001; Streltsov, 2001). Thus even a single track recorded in glass is a waveguide core, while surrounding area is a cladding (Fig.1a). Transverse cross section of such waveguide is generally elliptical. Ellipticity is high for low rep rate inscription. Special beam shaping is applied in order to make waveguide with a circular cross section at low rep rate laser systems (Osellame, 2003; Moh, 2005; Marshall, 2006). High rep rate lasers provide circular waveguide cross section even without beam shaping due to strong cumulative heating effect (Eaton, 2005). However it holds true for glasses with low thermal conductivity (borosilicate, boroaluminosilicate, etc.), and does not work for fused silica.

Situation is more complicated in crystals. Positive refractive index change in the exposed crystal area was observed only under certain restricted conditions. Refractive index change is positive in $LiNbO_3$ crystal for extraordinary axis under low pulse energy (Burghoff, 2006; Burghoff, 2007). Second harmonic generation in a waveguide written in this regime in z-cut PPLN crystal was demonstrated (Lee, 2006). A track has a pronounced elliptical cross section with width of about 1 μm and was elongated in direction of femtosecond beam propagation by 10 μm and more. Because of this a multiscan technique was applied in order to fabricate a more symmetrical waveguide core. That is, 6 tracks separated by 0.7 μm in direction perpendicular to a track and beam propagation were written (Fig.1b). A near-circular single mode waveguide of analogous type was fabricated in z-cut $LiNbO_3$ single crystal (Bookey, 2007). In this work the modified region of the crystal is consisted of two parts: rectangular damaged region and a waveguide of 8.9 μm x 21 μm cross section underneath the damaged region (Fig.1c).

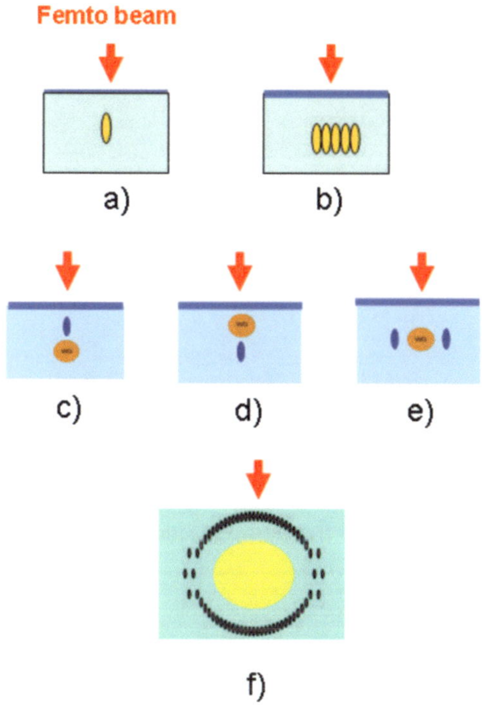

Fig. 1. Architectures of femtosecond written waveguides. Direct writing of a core: a) a single track, b) a multiple track core. Stress induced waveguides: c) and d) a waveguide is below and above the track, e) waveguide is in between two tracks. f) A micro-structured depressed cladding waveguide.

Sign of PRICE induced by 11-MHz rep rate Ti-Sapphire oscillator in YAG:Cr^{4+} crystal depends upon speed of crystal translation and pulse energy (Okhrimchuk, 2008). In Fig.2 the waveguide written in the regime of positive PRICE with a high aperture focusing lens (NA=0.8) is presented. The waveguide core of a rectangular 35x55 μm cross-section consistes of 30 parallel tracks separated in horizontal and vertical transverse dimensions by 3.5 and 20 μm correspondingly. Refractive index profile of the waveguide recorded by QPm technique (http://www.iatia.com.au) under observation in the direction of femtosecond laser beam propagation is shown in Fig.3.

The 1 kHz Ti:Sapphire laser always produces a negative PRICE in YAG single crystals (Okhrimchuk, 2005-a, 2005-b; Siebenmorgen, 2009). It was reported that in the ceramic YAG doped with 2 at.% of Nd^{3+} ions PRICE is negative too (Torchia, 2008). However there is refractive index increase in the area closely adjacent to a track (Dubov, 2004; Siebenmorgen, 2009) that is due to elasto-optical effect. The waveguide is produced due to this phenomenon in these works. An advanced waveguide architecture consists of two parallel tracks separated by 15 - 25 μm inscribed with a 1 kHz rep rate lasers (Fig. 1e). Although PRICE is negative inside tracks, mechanical stress induced between them provides positive PRICE. This way a single mode waveguide with almost circle mode profile was manufactured both in single

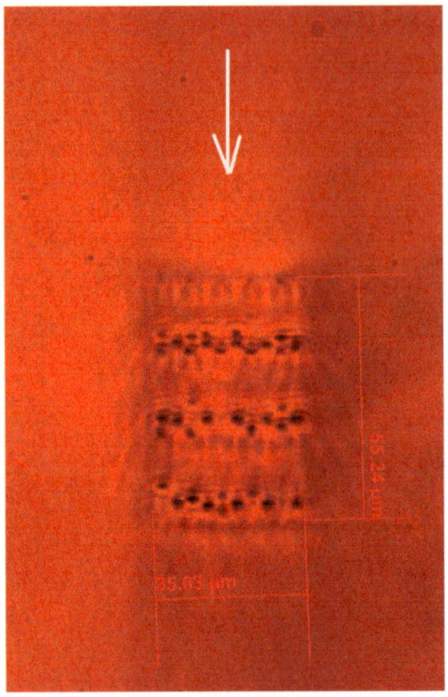

Fig. 2. End view of a waveguide written by Ti-Sapphire oscillator operating at 11-MHz rep rate. The arrow indicates direction of the femtosecond beam.

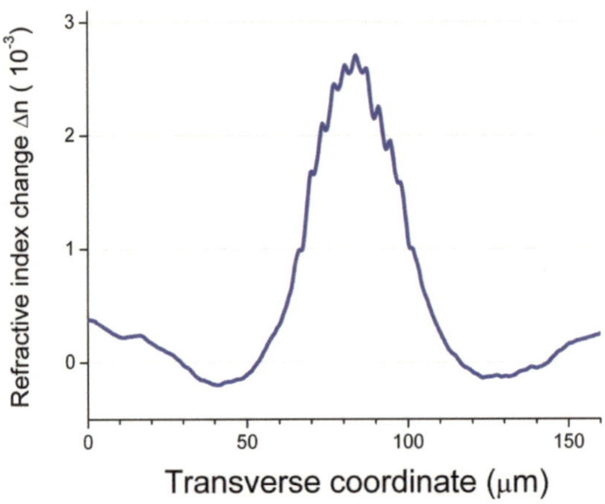

Fig. 3. Refractive index profile of a waveguide presented in Fig.2.

crystal and in ceramic YAG (Torchia, 2008; Siebenmorgen, 2009). In this design waveguide core is the stressed area which is not exposed by the femtosecond beam. An identical scheme was implemented earlier in x-cut $LiNbO_3$ single crystal (Burghoff, 2006), and z-cut PPLN (Thomas, 2007) and then in YVO_4:Nd single crystal (Silva, 2010).

Refractive index alteration was found in $Al_2O_3:Ti^{3+}$ single crystal under irradiation with 1 kHz rep rate Ti:sapphire laser. PRICE is negative inside a track analogously to YAG crystal, and a waveguide arises in the area below or above the track (Apostolopoulos, 2004) (Fig.1 c,d). No refractive index change was observed in $Al_2O_3:Ti^{3+}$ under irradiation with a high rep rate laser. Waveguide fabrication in crystalline silicon was made by irradiation at wavelength of 2400 nm with OPO pumped by chirped-pulse-amplified mode-locked Ti:sapphire laser (Nejadmalayeri,

2005). Plate of crystalline silicon was coved by oxide layer of 20 μm thickness. Femtosecond beam was focused inside silicon through this layer. An inscribed track produces mechanical stress, which causes index increase in the restricted area located between the track and silica-silicon interface. This way a waveguide with strong birefringence was fabricated.
An original approach based on negative PRICE was proposed by the author (Okhrimchuk, 2005-a; Okhrimchuk, 2009). A single mode and multimode waveguides were fabricated with a depressed cladding in YAG:Nd (1 at.%), YAG:Cr^{4+}, diffusion bonded YAG:Nd/YAG:Cr^{4+} and YAG:Yb(15 at.%) crystals. A femtosecond laser beam from a Ti-sapphire laser system operating at 800 nm with 1 kHz repetition rate (Spitfire, Spectra Physics) was used for waveguides fabrication. The laser produced pulses with 115 fs pulse width (FWHM). A

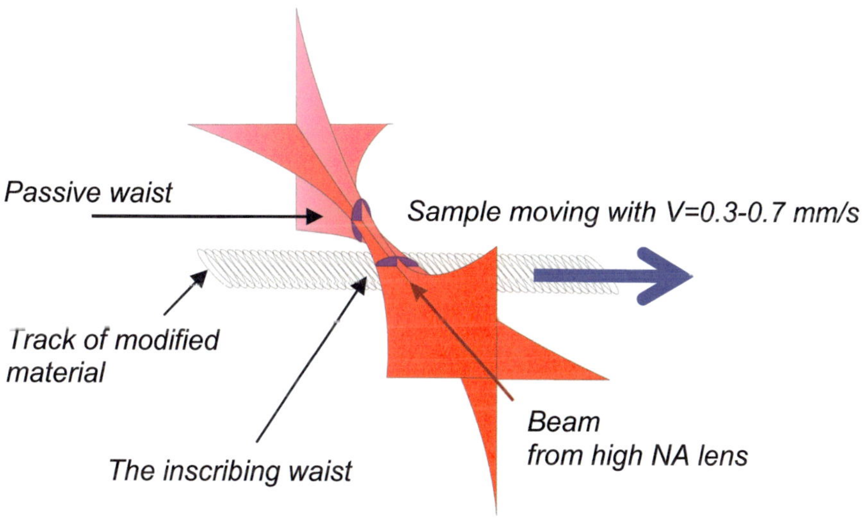

Fig. 3. Focusing of femtosecond beam with an elliptical cross section.

crystal sample was placed on a high precision 3-D Aerotech translation stage. Laser beam was focused under the crystal surface with Carl Zeiss (NA=0.65) or Mitutoyo lens (NA = 0.55). The crystal was translated at a constant speed of 0.3 - 0.7 mm/s to inscribe smooth tracks with lowered refractive index. A femtosecond laser beam with elliptic cross-section was used for the inscription. It proves to produce much more homogeneous tracks in YAG crystals compared to a standard beam of circular cross-section (Okhrimchuk, 2009). Such an approach leads to a dramatic reduction of scattering loss. To achieve this, a plane-concave cylindrical lens was placed before the high aperture lens to produce a waist of elliptical cross-section. Normally, two elliptical waists with mutually orthogonal orientations of principal axes were formed with ratio of major to minor diameters of nearly 10 (Fig.3). Typically, more than a half of input pulse energy was absorbed in the first waist nearest to the focusing lens. Therefore, for the moderate input energies used in the experiments, the inscription threshold was not reached in the second waist. Thus, tracks were inscribed by the first waist having a principal axis parallel to an inscribing track.
An elemental structural unit (a track) was produced by straight moving of the crystal with constant velocity so that a femtosecond beam waist crosses crystal from one edge to another. Cross section of a track is close to an oval with large to small axes ratio being larger then 10. In dependency on crystal doping it has refractive index lowered by (1 - 5)×10^{-3} in comparison with unperturbed value in the bulk of crystal, thus tracks all together formed a

waveguide cladding, while a core corresponded to a non-perturbed region of the crystal (Fig.1-f). Author would like to stress that a waveguide is formed due to lowed refractive index in the cladding. Unlike to alternative design with one or two tracks (Fig.1 c,d,e), author considers that contribution of mechanical stress in the core is negligible for this method of inscription and waveguide architecture.

Basing on the above approach a rectangular shaped, 10 mm long waveguide with core dimensions of 100 μm x13 μm was fabricated in $YAG:Nd^{3+}$ (0.8 at.%) as shown in Fig. 5. (Okhrimchuk, 2005-a). Propagation loss for the fundamental mode was estimated to be as low as 0.2 dB/cm at wavelength of 1064 nm.

Single mode waveguide with mode field size comparable with standard Yb-doped fiber was fabricated in $YAG:Cr^{4+}$ crystal (Okhrimchuk, 2009-b). End view of this waveguide is shown in Fig.6. It works as a waveguide saturable absorber insuring Q-switch operation of an Yb-fiber laser.

Multimode waveguides efficiently guiding light of multimode high power laser diodes were inscribed in diffusion bonded $YAG:Nd/YAG:Cr^{4+}$ and YAG:Yb(15 at.%) crystals. (Fig.7 & 8). The tracks were written perpendicular the $YAG:Nd/YAG:Cr^{4+}$ interface plane, thus waveguide crossed the bonded crystal from its Nd-part to Cr-part. Total number of tracks creating cladding equals 66 for both crystals. Numerical mode analysis of the waveguide shown in fig.7 was performed with a commercial code (www.comsol.com). The electromagnetic eigenvalue problem was solved by the finite-element method. It was found

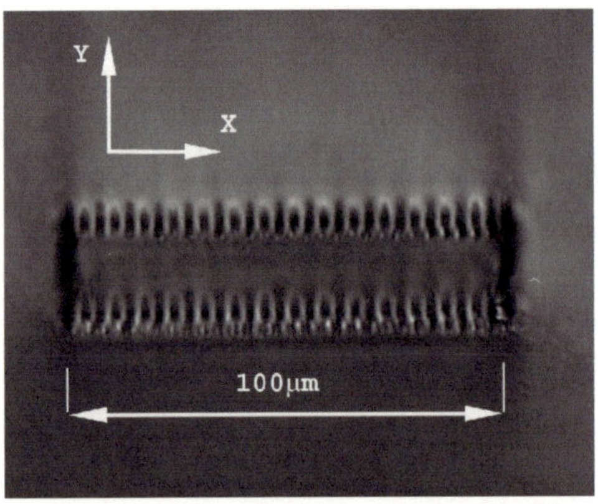

Fig. 5. End view of a depressed cladding waveguide inscribed in $YAG:Nd^{3+}$.

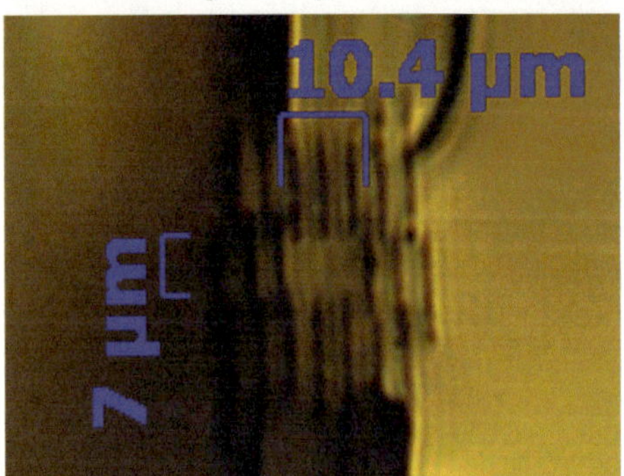

Fig. 6. End view of a depressed cladding waveguide inscribed in $YAG:Cr^{4+}$

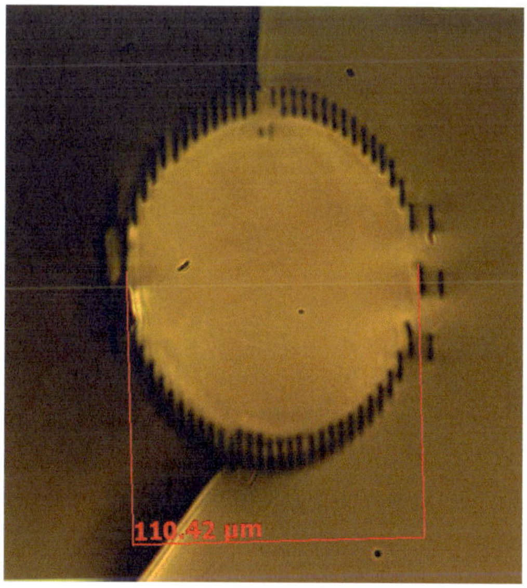

Fig. 7. End view of a depressed cladding multimode waveguide inscribed in diffusion bonded YAG:Nd/YAG:Cr^{4+} crystal. The refractive index is lowered by nearly 3×10^{-3} in each track compared to the unperturbed value in the bulk of crystal.

That the structure maintains 24 leakage modes at wavelength of 1064 nm. The propagation loss for two fundamental modes with mutually orthogonal polarisations was derived from the imaginary part of the propagation constant, and was determined to be less then $4\cdot10^{-3}$ cm^{-1}. It was experimentally found that both waveguides are strong enough to keep pump emission in the waveguide core (Fig.9 & 22). Finally an efficient Q-switch operation was obtained in YAG:Nd/YAG:Cr^{4+} crystal (see chapter 6).

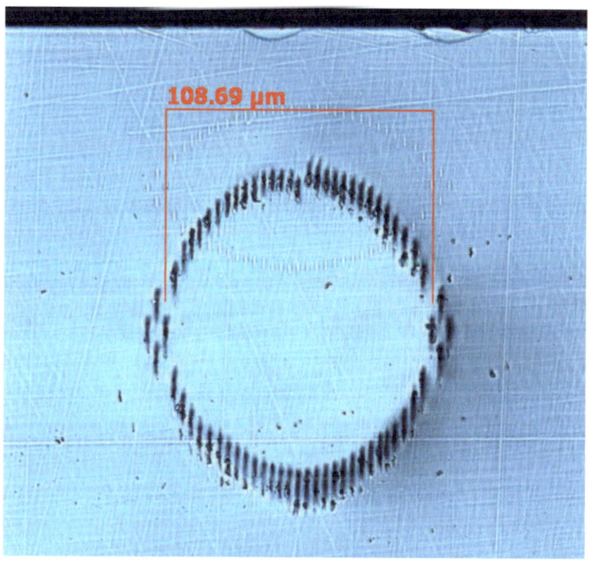

Fig. 8. End view of a depressed cladding waveguide inscribed in YAG:Yb^{3+}

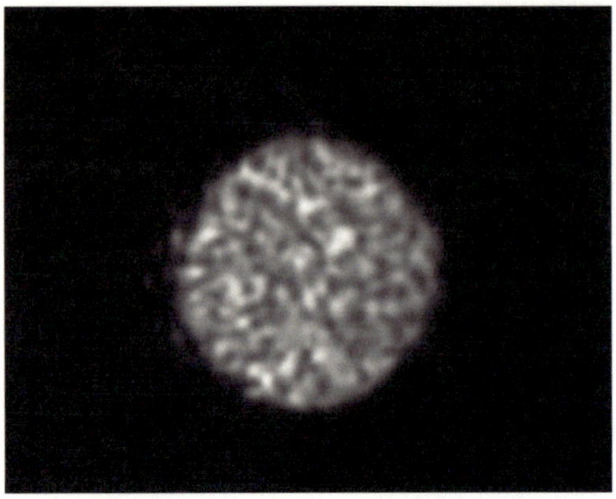

Fig. 9. Near field image of pump light at the output of the waveguide in YAG:Yb shown in Fig.8.

Propagation attenuation for pump emission in the waveguide inscribed in YAG:Yb crystal was measured in dependency upon numerical aperture (NA) of the input pump light at 965 nm. Experimental setup for measurement is shown in Fig.10. The total pump attenuation in the waveguide was found to be 5 cm^{-1} for pumping light with NA≤0.09, while absorption coefficient due Yb^{3+} ions is as high as 3.5 cm^{-1}. Thus ratio of bulk absorption coefficient of the crystal to the total waveguide propagation attenuation coefficient is as high as 0.7 for input light cone corresponding to NA=0.09. Thus pump should be efficiently utilized in the waveguide.

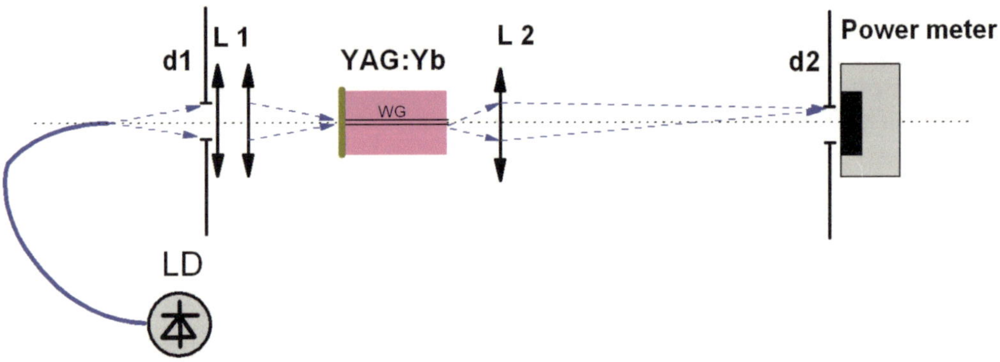

Fig. 10. Experimental setup for pump light attenuation measurements.

3. Grating inscription in YAG crystals

Two techniques of Bragg gratings inscription were tested both at fundamental wavelength of the femtosecond laser system (800 nm), and at the second harmonic (400 nm). First technique is a direct "point-by-point" (P-by-P), and the second is based on beam intensity modulation by Quartz acoustic modulator (AOM) (2008, Marshall). Differential interference contrast microscopy (DIC) was applied for inscribed structures inspection.

It was reasonably expected that a minimal inscribable grating pitch would be of the order or even slightly lower than a wavelength of a femtosecond laser due to sufficiently non-linear character of light-matter interaction. This consideration was supported by grating inscription in fused silica, where smallest grating with pitch of 260 nm was obtained by P-by-P technique with beam wavelength of 800 nm (Dubov, 2006;). However it was found in the course of grating inscription experiments that grating pitch in YAG crystals can not be inscribed smaller than 800 nm even with inscribing beam wavelength of 400 nm. When author tried to write grating with pitch size smaller than 800 nm, chaotic overcoming of grating marks was obtained. Microscopic pictures of gratings with minimal inscribable pitches for different inscription regimes are shown in Fig. 11 for inscription wavelength of 800 nm and in Fig.12, 13 for the wavelength of 400 nm.

The expected pitch size should be calculated by formula:

$$\Delta = \frac{V}{f}, \tag{1}$$

where V is sample moving velocity, and f is whether laser repetition rate, or AOM modulation frequency in dependency of technique applied. The inscribed grating pitch size obeys formula (1) when P-by-P technique was applied under moving velocity V=0.8 mm/s or higher and pitch size of 800 nm or higher is obtained (Fig.11 a,b), and no gratings were inscribed when V<0.8 mm/s. Analogous behaviour was observed for AOM modulation technique. For example, when we tried to inscribe pitch of Δ=410 nm a grating of nearly double size than predicted by formula (1) was really obtained (Fig.11- c).

It was a surprise that the smallest inscribable pitch size was found to be around 900 nm for femtosecond beam wavelength of 400 nm and P-by-P technique, and this pitch size is even larger, than for beam wavelength of 800 nm. It is interesting that a refractive index structuring was found when crystal was very slowly moving so that according formula (1) Δ<100 nm (Fig.12). Moreover nearly the same period of refractive index modulation of 900 nm was obtained. These facts coerce to suppose that some self-organising processes take place in this case.

Attempts were made to inscribe gratings with pitches sizes of Δ= 410, 670 and 760 nm under AOM modulation at wavelength of 400 nm. Instead gratings with nearly the same pitch sizes of 1100 - 1200 nm were really obtained (Fig.13).

Obtained experimental results open possibility for 1-st order Bragg grating writing for wavelength of 2.9 μm (2nΔ), which are promising for fabrication of waveguide Er:YAG lasers.

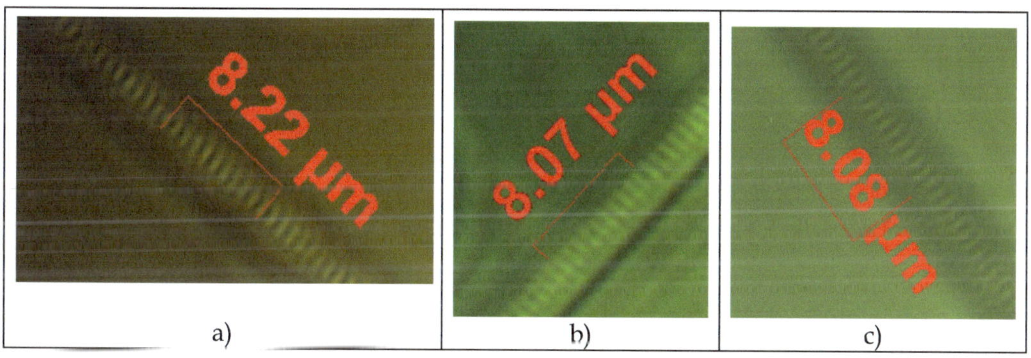

Fig. 11. Gratings inscribed in YAG:Cr^{4+} crystal at wavelength of 800 nm with point by point technique (a,b) and AOM modulation (c):

(a) beam of a circular cross section; V=0.80 mm/s; f=1 kHz, E_{in}=0.2 μJ.

(b)beam of an elliptical cross-section; V=0.81 mm/s; f=1 kHz, E_{in}=1.1 μJ.

(c) beam of an elliptical cross-section; V=0.0094 mm/s; f=23 Hz, E_{in}=1.8 μJ

Fig. 12. Self-organised gratings inscribed in YAG:Cr^{4+} crystal at wavelength of 400 nm with slow moving (V<0.015 mm/c).

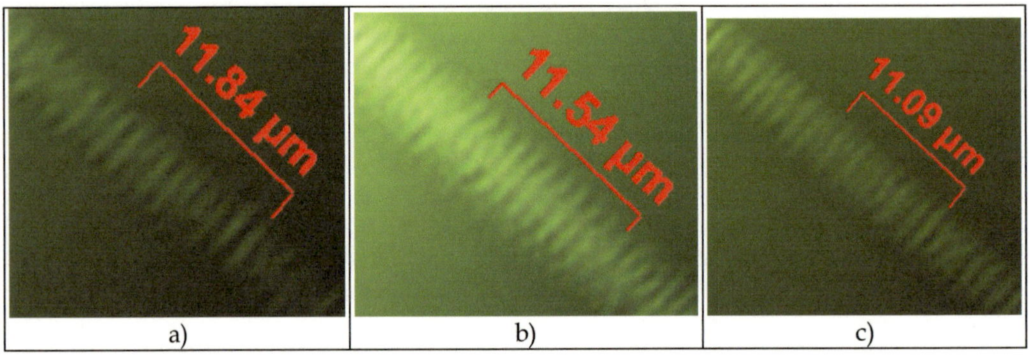

Fig. 13. Gratings inscribed in YAG:Cr^{4+} crystal at wavelength of 400 nm by AOM modulation technique with beam of an elliptical cross-section; E_{in}=0.56 μJ:
(a) f=23 Hz, V=0.0094 mm/s;
(b)f=23 Hz, V=0.0154 mm/s
(c) f=23 Hz, V=0.0174 mm/s.

4. Non-linear interaction of a femtosecond laser beam with dielectrics.

There are no doubt that understanding of processes concerning interaction of femtosecond pulses with a crystal lattice plays a central role in successful advancement of the technology. The first stage of this interaction comprising non-linear absorption of femtosecond pulses is rather clear and described in details in literature. This description is common for crystals and glasses, and even liquids (for wideband dielectrics).

It is a widely accepted that the process of energy absorption dealing with material modification by pulses shorter than 1 ps is determined by i) multi-photon absorption and ii) resistive heating of the generated plasma. These processes can be mathematically described by a popular simplified model for the evolution of the scalar electric field envelope $A(x,y,z,t)$ in paraxial approximation and electron density $\rho(x,y,z,t)$. Although the paraxial approximation has certain restrictions in exact quantitative description of femtosecond beam propagation dealing with practically used conditions of beam focusing (a high NA lens is generally used), it helps to understand basic character and peculiarities of the process, and to get a qualitative visual at least. It comprises nonlinear partial differential equations that account for the major propagation effects such as diffraction, Kerr nonlinearity, group velocity dispersion, multi-photon absorption, impact ionization and absorption and defocusing by the generated electron-hole plasma (Feng, 1997, Berge, 2007; Turitsyn, 2007):

$$i\frac{\partial A}{\partial z}+\frac{1}{2k}\left(\frac{\partial^2}{\partial x^2}+\frac{\partial^2}{\partial y^2}\right)A-\frac{k''}{2}\frac{\partial^2 A}{\partial t^2}+k_0 n_2|A|^2 A=-\frac{i\sigma_{bs}}{2}(1+i\omega\tau_e)\rho A-\frac{i\beta_K}{2}|A|^{2K-2}A \quad (2a)$$

$$\frac{\partial\rho}{\partial t}=\frac{1}{n^2}\frac{\sigma_{bs}}{E_g}\rho|A|^2+\frac{\beta_K}{K\hbar\omega}|A|^{2K} \quad (2b)$$

The last three terms in the left-hand side of Eq.(2a) describe effects of beam diffraction, group velocity dispersion (GVD), and Kerr nonlinearity, respectively. The latter term alone is responsible for a catastrophic self-focusing which is limited by the effects described by terms on the right-hand side of Eq.(2a), namely plasma absorption and defocusing, and multi-photon absorption. In Eq.(2a) the laser beam propagation along the z axis is assumed and this equation is essentially a reduced paraxial approximation of the wave equation for the complex electric field envelope $\boldsymbol{A}$ with a carrier frequency $\boldsymbol{\omega}$ in the moving frame of coordinates. $k= nk_0= n\omega/c$ is the propagation vector, n is the material linear refraction index, k'' is the GVD parameter, n_2 is the nonlinear refractive index describing Kerr effect, σ_{bs} is the cross section for inverse Bremsstrahlung, τ_e is the electron-phonon collision time in conduction band, and the quantity β_K describes the K-photon absorption ($K>1$). Equation (2b) implements the simplest Drude model for electron-hole plasma in dielectrics and describes the evolution of the electron density ρ. The first term on the right-hand of eq. (2b) side is responsible for the avalanche impact ionization and the second term - for the ionization resulting from multi-photon absorption (MPA). E_g is the energy gap between valence and conduction bands.

Self-focusing dealing with Kerr non-linearity plays an essential role in the inscription process. This role is obviously destructive for smart modification of crystal, and apparently it can play a positive role in some cases for inscription of structures in glasses of sub-wavelength sizes (Dubov, 2006). Self-focusing leads to filamentation and fragmentation of a laser beam and this is a one course of scattering loss in waveguides written in this regime. Authors of (Turitsyn, 2007) discussed regimes of inscription with a diminishing role of self-focusing. As a result of theoretical analysis they offered to use a non-Gaussian ring-shaped beam in order to increase self-focusing threshold, while PRICE threshold remains unchanged.

In the work (Okhrimchuk, 2009-a) beam of elliptical cross-section is suggested as a tool diminishing the self-focusing. A negative cylindrical lens with large focal length was arranged in the femtosecond beam path before it passes to the focusing lens. It is considered that breaking of axial symmetry of the beam postpones self-focusing to higher input energies, which partly takes place without the cylindrical lens. Self-focusing easily deforms axially-symmetric beam profile in such a way that the central (part) core of the beam suffers from spatial collapse caused by non-linear focusing. The collapsed part could be effectively absorbed, thus PRICE threshold is reached in this small central part, while the rest of the beam exhibits loss-less propagation. When a cylindrical lens is inserted, PRICE threshold is reached in wider area of beam cross-section, and the beam experiences a more uniform absorption, which decreases a transmittance, as it was observed in experiment (Table.1). This result suggests that breaking of axial symmetry of the femtosecond beam is a promising technological method to prevent destructive self-focusing at small energies, and thus it may lead to a better control over the energy deposition and a waveguides inscription process. The track made with the cylindrical lens is smooth, without any observable defects. Such tracks form a depressed cladding waveguide as it was demonstrated (Okhrimchuk, 2005-a, 2009-b). Understanding of relative contribution of the MPA the MPA and the plasma absorption and the plasma absorption in the process of energy deposition leading to PRICE is ultimately required for optimization of the inscription process. Transmittance of laser crystal and glasses in dependency on input p*u*lse energy was measured with this purpose (Okhrimchuk, 2009-a). Focusing conditions in these experiments were identical to conditions of waveguide inscription. Ti:sapphire oscillator-amplifier with repetition rate was of 1 kHz was used in these experiments. PRICE threshold was measured in the same conditions too (Table.1).

Material	MPA order K (λ, nm)	MPA coefficient β_K,	w/o cyl. lens E_{in} (nJ)			Transmittance at the threshold	
			Inscription thresholds	with cyl. lens E_{in} (nJ).	Intensity threshold, (10^{13} W/cm^2) (with & w/o cyl. Lens)	w/o cyl. lens.	with cyl. lens
YAG	**5** (800)	$3.1 \cdot 10^{-51}$ cm^7/W^4	61	610	**4.8**	0.64	0.49
YAG:Nd^{3+}			55	580	**4.5**	0.66	0.54
YAG:Cr^{4+}	(800)		46	1200	**4...9**		
	(400)		50	500			
$RbPb_2Cl_5$:Dy^{3+}				90	**0.66**		
ZnSe:Cr^{2+}			10^4		**850**		
ZnSe (polycr.)	**2**		n/r	n/r		-	-
ZnS:Cr^{2+} (polycr.)	**2**		250*	n/r up to 7000			
ZnS:Cr^{1+} (polycr.)	**2**		1250*				
Fused silica	**5**	$1.2 \cdot 10^{-51}$ cm^7/W^4	55	460	**3.6**	0.83	
	6	$2.7 \cdot 10^{-65}$ cm^9/W^5					
BK7	**3**	$2.7 \cdot 10^{-24}$ cm^3/W^2	55	550	**4.4**	0.71	0.55
	4	$6.9 \cdot 10^{-38}$ cm^5/W^3					
	5	$1.9 \cdot 10^{-51}$ cm^7/W^4					

Table 1. Result of measurement of the non-linear transmittance in dependency upon input pulse energy and the inscription thresholds. n/r - threshold is not reached for the investigated input energy range.

It is instructive to present the results of the transmittance experiments in terms of the absorbed pulse energy versus input energy in double logarithmic coordinates as shown in Fig. 14 for the beam of circular cross section and in Fig15 for the beam of elliptical cross section. Very similar dependences for the transmittance and values of threshold input energes were obtained for undoped YAG, YAG:Nd^{3+}, YAG:Cr^{4+} crystals, fused silica and BK7 glass. Strongly difference dependence was found for ZnSe crystals, and inscription threshold is not reached at all in range of input energies investigated.

Different behaviour is seen for the absorption in YAG crystals for low and high energy regions. It is natural to assume that different mechanisms are responsible for absorption in these regions. It is well established that for 100 fs pulse width and moderate electric field strength the non-linear absorption in wide band insulators is initiated by MPA process (Berge, 2007; Turitsyn, 2007). Following this it is reasonable to consider MPA process dominates in the low energy region, while the plasma absorption prevails for higher energies. Confirmative results were obtained while investigating electronic excitations in SiO_2, Al_2O_3 and MgO crystal by measurement of induced phase shift of a delayed probe beam (Quere, 2001). They found a transition from MPA absorption regime to free carrier absorption at the input pulse intensity nearly corresponding to the dependence curve bend in Fig. 14 &15. As phase shift saturates authors of the last paper consider that there is no more electronic density rise for input intensities higher than 10^{14} W/cm^{2,} and the electronic avalanche process is an ineffective in comparison with MPA. This is completely in agreement with a tendency of the absorption curve to be linear for high input pulse energies (Fig.14 &15).

So as MPA is dominant in non-linear absorption up to inscription threshold , and it is a main channel of supplying a conduction band by electrons, we can estimate free electron concentration at the threshold and above as:

$$\rho = \frac{\Delta E_{abs}}{E_g V_m}, \tag{3}$$

E_g is the energy gap, where $\boldsymbol{\Delta E_{abs}}$ is energy absorbed due to MPA at the threshold, $\boldsymbol{V_m}$ is volume occupied by free electrons. It is reasonably to consider that region occupied by free electrons is converted to the modified region (region of PRICE), and these regions nearly coincide with each other. Sizes of PRICE region were directly measured with a microscope: $\boldsymbol{V_m}$= 1μm x10μm x 12μm = 1.2*10^{-10} cm^2 for YAG:Nd(0.5%) crystal modified in the setup with the cylindrical lens. $\boldsymbol{\Delta E_{abs}}$= 0.04 μJ as could be defined from fig.15. Finally we obtain for free electron density $\boldsymbol{\rho}$= 3*10^{20} cm^{-3}.

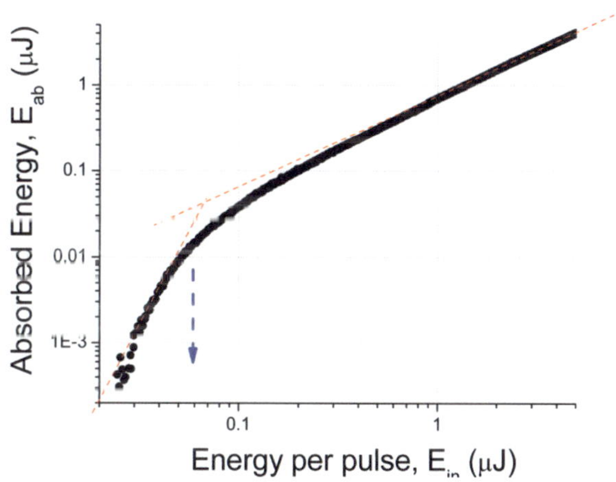

Fig. 14. Dependence of pulse absorbed energy in undoped YAG crystal in dependency upon input pulse energy. The arrow indicates PRICE threshold. Inscribing beam has cross section of circular symmetry.

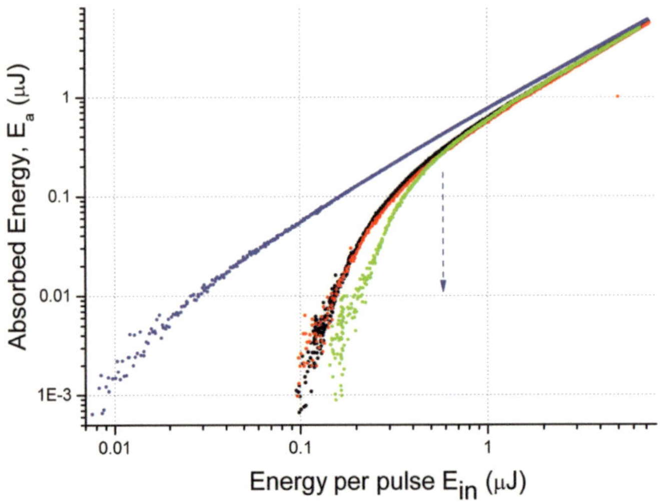

Fig. 15. Dependence of pulse absorbed energy in undoped YAG and YAG:Nd(0.5%) crystals (black and red points correspondingly) ZnSe crystal (blue points) and BK7 glass (green points) in dependency upon input pulse energy. The arrow indicates PRICE threshold for YAG and BK7. Inscribing beam has cross section of an elliptical symmetry.

Under approximations of Drude electron plasma density can be related with high frequency relative permittivity by following formula (Ashcroft & Mermin, 1976; Martin, 1997):

$$\varepsilon(\omega) = \varepsilon_c - \frac{4\pi\tau_e^{\,2}e^2\rho}{m_{eff}(1+\omega^2\tau_e^{\,2})} + i\frac{4\pi\tau_e e^2\rho}{m_{eff}\,\omega(1+\omega^2\tau_e^{\,2})}, \tag{4}$$

where ε_c is permittivity of unexcited crystal, m_{eff} is effective electron mass. Absorption coefficient of free electrons a is defined by imaginary part of permittivity:

$$\alpha = \frac{4\pi\tau_e e^2\rho}{m_{eff}nc(1+\omega^2\tau_e^{\,2})}, \tag{5}$$

where c is speed of light. The experimental transmittance data gives estimation of free electron absorption coefficient a as high as 1000 cm^{-1}. Now we can estimate with formula (5) free electron time τ_e. Accepting the effective mass equalled to an electron mass m_e we find a reasonable value τ_e= 2 fs. Thus it confirms correctness of our previous estimations of free electron density ρ and a determinative role of MPA in generation of free electrons in conduction band. Parameters concerned inscription in YAG are summarised in table 2.

Free electron density ρ	$3*10^{20}$ cm^{-3}
Free electron time τ_e	2 fs
Effective electron mass, m_{eff}	m_e
Plasma absorption coefficient, a	1000 cm^{-1}
MPA coefficient β_5,	$3.1 \cdot 10^{-51}$ cm^7/W^4

Table 2. Material parameters concerned femtosecond inscription in YAG crystals.

Experimental transmittance data in the low energy region can be described analytically under conditions of smallness of the loss terms in the right hand side of Eq. (2a) and sub-critical propagation regime when the peak pulse power is lower than the critical value for self-focusing (Okhrimchuk, 2009-a). These assumptions justify self-similar quasi-linear approximation for pulse propagation (Turitsyn, 2007). Self-similar Gaussian shape of the pulse is a good approximation. Under this simplification the following analytical formula for transmittance ***T(E$_{in}$)*** corresponding to pure MPA absorption can be derived:

$$
\begin{aligned}
&T(E_{in}) = \left(1 + (K-1)a(K)E^{K-1}\right)^{-\frac{1}{K-1}} \\
&\alpha(K) = \beta_K \cdot \left(\frac{2}{\pi}\right)^{3/2\cdot(K-1)} \cdot \frac{\pi \cdot \mu(K) \cdot n}{K^{3/2} \cdot \lambda \cdot \tau_p^{K-1}} \cdot \frac{1}{\omega_0{}^{2\cdot(K-2)}}, \\
&K = \{2,3,4,5,6\}, \quad \mu(K) = \pi \cdot \left\{1, \frac{1}{2}, \frac{3}{8}, \frac{5}{16}, \frac{35}{128}\right\}
\end{aligned}
\tag{6}
$$

where λ is a central wavelength, $2\tau_p$ is laser pulse width at $1/e^2$. Applicability of the formula (6) is not limited to low energy region. In fact, it describes MPA for wider range of transmittance, if the beam remains its Gaussian shape. The restriction of formula (6) arises from plasma related absorption and beam defocusing. The dependences for transmittance measured for un-doped YAG crystal, fused silica and BK7 glass were fitted with formula (6) (Okhrimchuk, 2009-a). The only fitting parameters were the order of MPA ***K***, coefficient of MPA β_K. Results of fitting are summarised in Table 1.

5. Nature of refractive index change in YAG crystals.

Dependence of PRICE upon input pulse energy was investigated at wavelength of 800 nm at the experimental setups designed for waveguide inscription (see chapter 2). Two different setups were used: the first is based on a high rep rate oscillator (11MHz), and the second setup comprises a low rep rate oscillator – regenerative amplifier system (1 kHz).

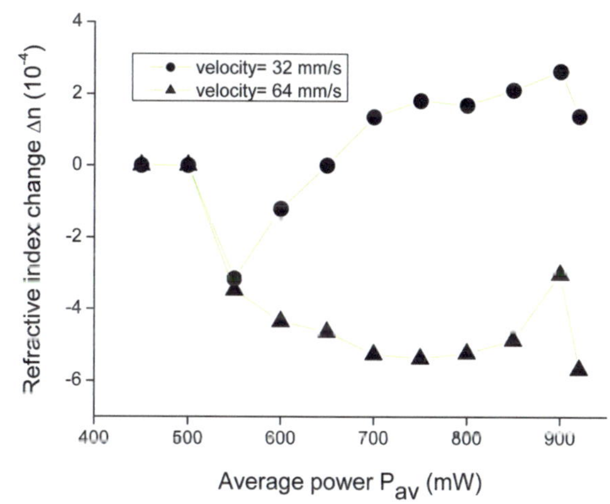

Fig. 16. Dependencies of PRICE and track height in YAG:Cr^{4+} crystal upon average input power. 11 MHz rep rate femtosecond laser.

Sign of PRICE induced by 11-MHz rep rate Ti-Sapphire oscillator in YAG:Cr^{4+} crystal depends upon speed of crystal translation and pulse energy (Okhrimchuk, 2008). For example at velocity V=32 mm/s it is positive for pulse energy exceeding 50 nJ, and is negative for pulse energies ranging from PRICE threshold to 50 nJ (Fig. 16).

Setup with 1 kHz laser includes Mitutoyo lens and a negative cylindrical lens with focal lens of 34 cm. The beam was focused at the depth of 60 μm. Translation speed was of 0.5 mm/s. Series of tracks written under different input pulse energies was inscribed. End view of a typical track is shown in Fig.17. Refractive index profiles of tracks were measured my QPm method at wavelength of 605 nm (http://www.iatia.com.au). PRICE is always negative for 1 kHz rep rate laser system. Typical index profile is shown in Fig17. Both PRICE magnitude and height of the track depend upon the input pulse energy, corresponding dependences for YAG:Nd(0.3 at.%) crystal are shown in Fig.18.

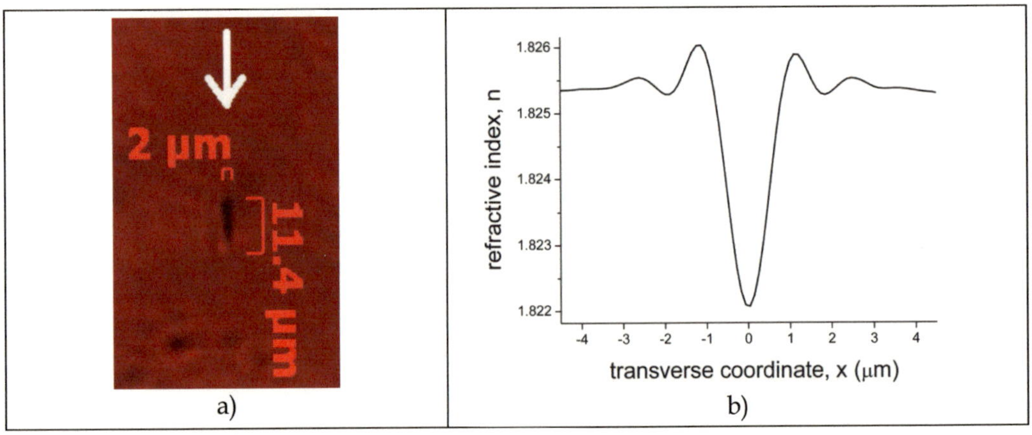

Fig. 17. a) Microscopic end view of a track inscribed in YAG:Nd(0.3%) crystal at a depth of 60 μm. b) Refractive index profile of the same track. The arrow indicates the direction of the femtosecond beam and of observation when index profile was measured.

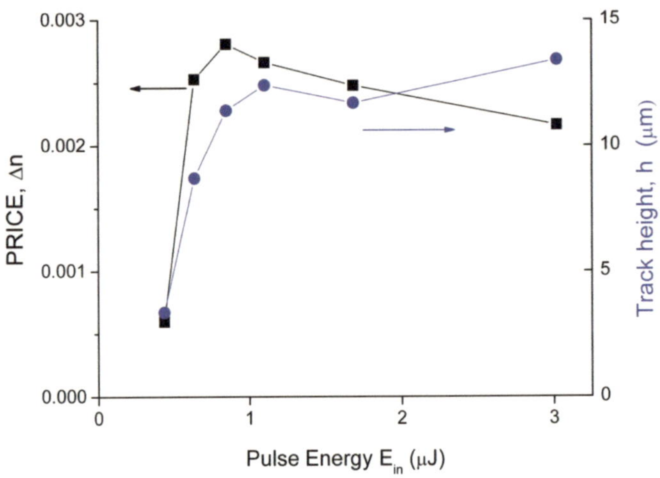

Fig. 18. Dependencies of PRICE and track height in YAG:Nd(0.3%) crystal upon input pulse energy.

Similar dependencies were found for YAG:Nd and YAG:Cr crystals. Dependence of refractive index change upon pulse energy has strong threshold character, and it saturates when the pulse energy exceeds the threshold by nearly factor of two. The saturation level depends upon doping level. Fig. 19 shows a dependence of the index change amplitude on the total concentration of dopant ions the crystal, measured for fixed input energy of 1 μJ. At this level of exposure refractive index change in a normally undoped YAG crystal strongly varied from sample to sample, and probably is determined by real purity of the crystal. Fig.19 includes data for one of the undoped YAG crystal investigated only, which probably has highest quality.

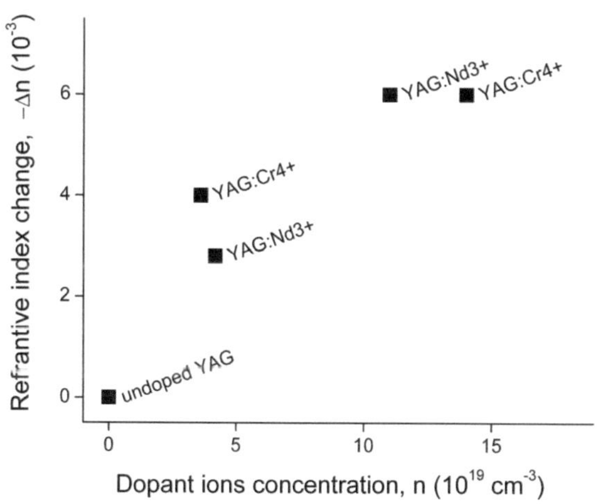

Fig. 19. Dependence of PRICE upon total dopant concentration in YAG crystal under input pulse energy of inscription of 1 μJ.

Reflectance spectrum of the modified region of the YAG:Nd crystal was investigated (Okhrimchuk, 2005-b). A rectangular area of 2000x2000 μm inside a crystal volume was processed by a focused femtosecond laser beam by writing a number of parallel tracks, inscribed with a spacing of 3 μm and laying in one plane. The beam was focused by microscope objective with NA=0.65 at a depth of 200 μm under the crystal surface and had intensity of 0.5 PW/cm^2 in focus. The thickness (height) of the modified volume was approximately 10 μm. Then unexposed layer of material between the exposed volume and the surface was removed by polishing. As a result, the processed volume of material could be directly accessed and characterised.

Reflectance spectra corresponding to the vibration modes of the crystal were recorded by using a standard FTIR spectrophotometer (Bruker, model Vector-22), customised for the measurements of reflected signals. The reflection spectra, measured with a resolution of 2 cm^{-1} are presented in Fig.20.

Firstly, the general profile of the spectrum corresponding to the exposed material is very similar to that of the unprocessed material, indicating that the garnet crystal lattice remains intact after ultrafast laser exposure. At the same time the spectrum lines of the processed material exhibit a shift to higher frequency which non-uniformly varies with frequency and shows a tendency to increase at higher frequencies. The maximal frequency shift equals 5 cm^{-1} approximately. Therefore, one can state that processing by an ultrafast laser causes a positive shift in vibration frequencies of YAG:Nd lattice. The above measurement of positive shift in vibration frequencies is in line with the negative change of refractive index, observed

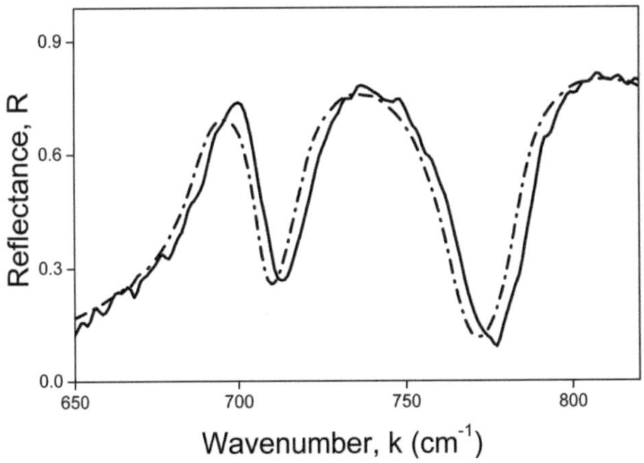

Fig. 20. Reflectance spectra of exposed (solid line) and unexposed (dashed-dotted line) of the YAG:Nd crystal.

earlier in phase delay measurements. Indeed, according to the Kramers-Kronig analysis, dielectric susceptibility $\varepsilon = \varepsilon' + i \cdot \varepsilon''$ in the region of transparency could be estimated as (Landay & Lifshits, 1982):

$$\varepsilon(\omega) = a - \frac{b}{\omega^2}$$
$$b = \frac{2}{\pi} \int_0^{\omega_1} x\varepsilon^{//}(x) \qquad (7)$$

where a - constant, ω_1 - cut-off frequency of vibration band. Results of the spectral measurements indicate that integral b increases in the processed material as a result of the general positive frequency shift in the vibration spectrum. For the refractive index change dn, one can obtain from (7):

$$dn = d\sqrt{\varepsilon} = -db \cdot \frac{1}{2\omega\sqrt{a\omega^2 - b}} < 0 \qquad (8)$$

Unchanged state of the reflectance spectrum indicates that notwithstanding a widely spreading hypnosis of melting the mechanism of YAG crystal modification under low repetition rate exposition differs from those taking place in glasses and crystalline quartz (Gorelik, 2003). The following simple estimation of crystal heating under exposition by one femtosecond pulse supports this statement. Temperature rise due to absorbed energy can be estimated under assumption that there are no channels of heat dissipation outside of electron plasma region:

$$\Delta T = \frac{\Delta E_{abs}}{c_p \rho_{YAG} V_m} \qquad (9)$$

Accepting c_p= 0.84 J/K (T=900 K), ρ_{YAG}=4.5 g/cm^3, V_m= 1.2*10^{-10} cm^3 (see chapter 4), we obtain that ΔT= 1300 K. Such temperature rise is not enough for YAG melting (T_m= 2220 K).

Moreover, if one takes into account energy loss due to irradiation, an estimation of temperature change even decreases. Definitely that temperature rise plays a higher role in the case of the high repetition rate inscription (Fig. 16.). However a luminescence spectrum of Cr^{4+} ions recorded in the regime of guiding of pump and luminescence emission does not differ from those recorded for unperturbed crystal area (Okhrimchuk, 2008). This fact excludes any amorphisation of YAG crystal for the high repetition rate inscription too.
Another mechanisms excluding melting should be found to explain PRICE in YAG crystal. It is evident from the results presented in Fig. 19 that reaction of YAG crystals to the femtosecond laser irradiation strongly depends upon concentration of defects in the crystal. This can be explained by the assumption that the point defects existing in the crystal define the possibility of modification of YAG crystal lattices without optical damage, resulting in the well-defined refractive index change. Such behavior arises from phenomenon that dopants present in a YAG crystal generate several types of lattice defects, and its concentration is rather higher, than in an undoped crystal. These defects interact with crystal lattice and each other and form a system with week binding. The flexible defect system of a doped crystal readily reacts to moderate external perturbation such as a high-intensity optical pulse, unlike regular ions of a crystal. Thus modifications without crystal lattice destroy can take place for some restricted range of pulse energies. There are no room for similar changes in ideal crystals and its modification probably occurs in a form of crystal lattice destroy only. The defect system has remarkable contribution to dielectric susceptibility of the crystal. Modification of the defect system is accompanied by changing of dielectric susceptibility of a crystal which manifests itself, in particular, as a change of the refractive index.
Recent investigation of some femto exposed complex glasses by means of electron probe microanalyzer (EPMA) (Kanehira, 2008; Liu, 2009) and electron dispersive x-ray spectroscopy (Liu, 2008) have revealed migration of positive ions out of an exposed region, while oxygen ions concentrate inside this region. Author of these papers explain the effect by temperature diffusion. Although temperature rise should have strong effect on ion mobility, let us to suppose another explanation. Diffusion expansion of electron plasma can be a driving force for ions migration. Indeed, average electron velocity in the conduction band can be calculated accordingly (Ashcroft & Mermin, 1976):

$$V_e = \frac{e|A|}{m_{eff}\omega} \tag{10}$$

Under pulse energy of 1 μJ, we get $V_e \approx 1.5 * 10^8$ cm/s. Average diffusion length of a free electron before its trapping by a hole or a lattice defect is calculated by the formula:

$$\langle l_e \rangle = V_e \sqrt{2\tau_e t_{tr}} \tag{11}$$

where t_{tr} is the electron trapping time. The last parameter was measured for Al_2O_3 and SiO_2 crystals (Guizard, 1995). It was found to be about 100 ps in Al_2O_3. So as chemical bonds in Al_2O_3 and in YAG crystals are similar we can accept this value for estimation of the average diffusion length in YAG. Taking into account τ_e=2 fs (table 2) we obtain $<l_e>$= 1 μm. Thus increase of electron plasma cloud size is comparable with its initial size. This could explain the reason of impossibility of inscription grating pitch smaller that 800 nm (chapter 3). The expanding electron plasma could drag some weekly bonded positive ions outside the exposed area. In this case even inscription with a shorter wavelength does not reduce a minimal inscribable pitch, because minimal pitch size not defined by size of the exposed area, but by size of the electron plasma expanding for a long time after laser pulse is over. Accordingly to this assumption another situation should take place in SiO_2, as electron trapping time is found to be rather shorter in this material (150 fs). Thus electrons is trapped very quickly and there are no plasma expansion. This is in line with successful inscription of a very small pitch in fused silica (Dubov, 2006).

The model of electron plasma dragging ions explains positive effect of implementation of the cylindrical lens (Okhrimchuk, 2009-a). When the lens is used the electron plasma cloud has a shape of a flattened ellipsoid, and a normal to its surface is primary directed perpendicular to direction of an inscribed track. Thus each pulse drags ions predominantly outside area of modification. This is not the case for beam of circular cross section, when remarkable part of ions are pushed in the direction of the track, and each following pulse redistributes ions arranged by previous one. This could provoke instabilities in the track writing process. On the contrary pulses push ions in the same direction, when the cylindrical lens is applied. In other words, beam of an elliptical cross-section works as an "optical plug".

6. Microchip waveguide laser operating in Q-switch mode

In this chapter an example of successful application of the femtosecond inscription technology is presented. A waveguide was fabricated with a depressed cladding in a diffusion bonded YAG:Nd^{3+}/YAG:Cr^{4+} crystal under a crystal surface at 110 µm depth (for an optical axis of the waveguide). The YAG:Nd^{3+} and YAG:Cr^{4+} crystals were diffusion bonded at high temperature close to the melting point. Concentration of Nd^{3+} ions was 1 at.%. The calcium co-doping was used to switch valence of chromium to 4+. An absorption coefficient in the YAG:Cr^{4+} crystal at 1064 nm was of 4 cm^{-1}. The experimental setup for waveguide inscription is described in chapter 2. Cross section of a track is close to an oval with large to small axes ratio being more than 10. The refractive index was lowered by nearly 3×10^{-3} compared to the unperturbed value in the bulk of crystal. As a result, the array of tracks formed a waveguide cladding, while a core was corresponded to a non-perturbed region of the crystal.

Crystal surfaces with the waveguide ends have been polished, and dielectric coating was applied in order to complete a laser cavity. The neodymium end was provided with a diachronic mirror reflecting 100% at 1064 nm and transmitting 98% at 808 nm. An output coupler (OC) with the reflection coefficient of 65% at 1064 nm was made at the chromium end. Small signal transmittance of YAG:Cr^{4+} crystal was of 57 %. The total waveguide length was of 5.4 mm; length of the neodymium part was of 4 mm. Microscopic image of the waveguide end view is shown in Fig 7.

The scheme of laser setup is shown in Fig. 21. In this design a pump power is delivered from a laser diode (Lumics) through the fiber of NA=0.15 and core diameter of 105 µm. Pump fiber core diameter is only slightly lower than the waveguide laser core diameter (110 µm). It was found the waveguide is strong enough to keep pump emission at the waveguide core. Near field image of OC end is taken for pump power below Q-switch oscillation threshold (Fig.22), thus it corresponds to pump emission only. Diameter of the pump emission spot exactly matches to diameter of the waveguide core. No sufficient leakage of pump is observed (small leakage is noticed on the right side that corresponds to discontinuity in the waveguide cladding). Oscillation was investigated in a pulsed pump mode with repetition rate ranging from 1 Hz to 10 kHz and pump pulse duration of 100 - 250 μs. Q-switch pulse energy and duration were found to be independent upon repetition rate. At repetition rate of 1 kHz and pump pulse duration of 250 μs output pulse energy is as high as 10 μJ, and threshold pump energy is 200 μJ (pump power is 0.85 W). It is important to note that nearly 3-times larger pump energy is required to produce the same output energy in traditional bulk microchip laser without waveguide but with the same other parameters (the OC and YAG:Cr^{4+} transmittances, the cavity length). Time profile of the Q-switch pulse posses a single pick shape without any additional shoulder at the pulse tail for investigated repetition rates (Fig 23). Pulse duration was found to be 1 ns (FWHM) for entire investigated repetition rate range.

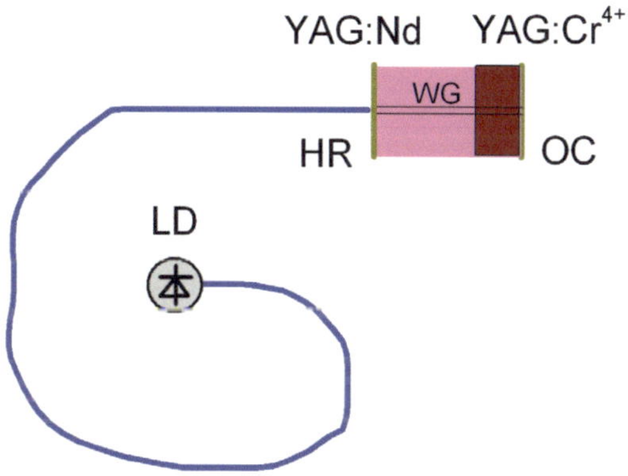

Fig. 21. Scheme of laser setup.

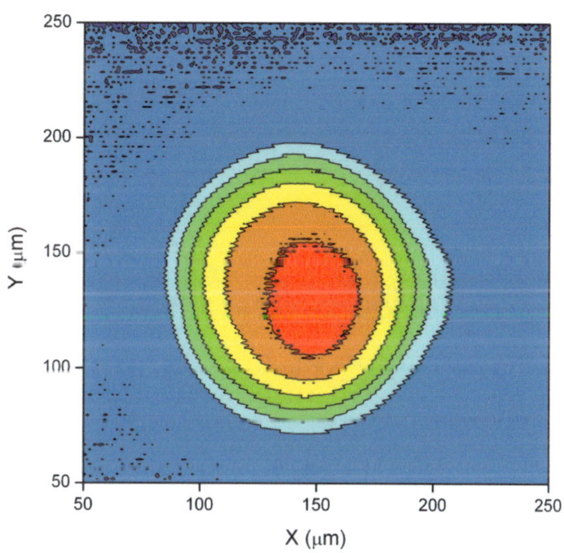

Fig. 22. Near field image of pump emission taken at OC location.

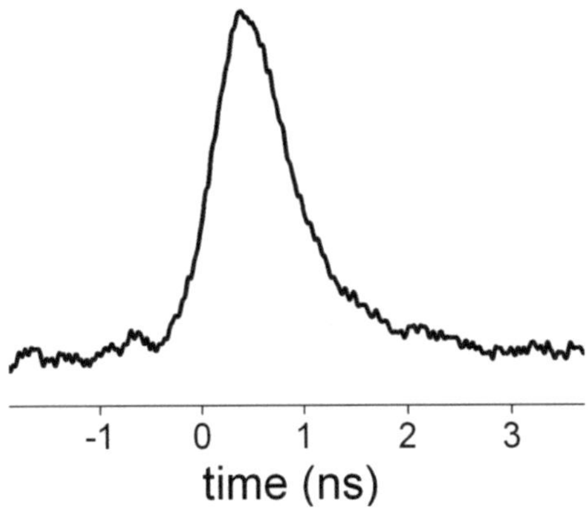

Fig. 23. Oscilloscope trace of a Q-switch pulse.

7. Conclusion

Micro-modification by means of femtosecond laser pulses is an effective tool for fabrication of low threshold and efficient waveguide lasers in YAG crystals. A beam with elliptical cross section is very effective tool for femtosecond inscription of depressed cladding waveguides in a low repetition rate regime. The technique is flexible for waveguide inscription with cross section of a predetermined shape. Both MPA and electron plasma absorptions are responsible for energy deposition and refractive index modification in YAG crystals. The presence of doping ions facilitates smart modification of crystal structure without optical damage and, as a result, refractive index change is not accompanied by disintegration of a crystal lattice during femtosecond laser inscription.

8. References

Allsop, T.; Dubov, M.; Mezentsev, V and Bennion, I. (2010), Inscription and characterization of waveguides written into borosilicate glass by a high-repetition-rate femtosecond laser at 800 nm. *Applied Optics*, Vol.49 (1938-1950).

Apostolopoulos, V.; Laversenne, L.; Colomb, T.; Depeursinge, C.; Salathé, R. P.; and Pollnau, M. (2004). Femtosecond-irradiation-induced refractive-index changes and channel waveguiding in bulk Ti^{3+}:Sapphire. *Appl. Phys. Lett.*, Vol. 85 (1122-1124).

Ashcroft, N.W. & Mermin, N.D. (1976). *Solid State Physics*, Holt, Rinehart, and Winston, New York.

Berg´e, L.; Skupin, S.; Nuter, R., Kasparian, J. and Wolf, J-P. (2007). Ultrashort filaments of light in weakly ionized, optically transparent media. *Rep. Prog. Phys.* Vol. 70 (1633–1713).

Bookey, H.T.; Thomson, R.R.; Psaila, N.D.; Kar, A.K.; Chiodo, N.; Osellame, R. and Cerullo, G. (2007). Femtosecond Laser Inscription of Low Insertion Loss Waveguides in Z-Cut Lithium Niobate. *IEEE Photonics Technology Letters*, Vol. 19 (892-894).

Burghoff, J.; Grebing, C.; Nolte, S. & Tunnermann, A. (2006). Efficient frequency doubling in femtosecond laser-written waveguides in lithium niobate. *Appl. Phys. Lett.*, Vol. 89 (081108).

Burghoff, J.; Hartung, H.; Nolte, S. & Tunnermann,A. (2007). Structural properties of femtosecond laser-induced modifications in $LiNbO_3$. *Appl. Phys. A*, Vol. 86 (165–170).

Davis, K.M.; Miura, K.; N. Sugimoto, N.; and Hirao, K. (1996), Writing waveguides in glass with a femtosecond laser. *Optics Letters*, Vol. 21 (1729-1731).

Dubov, M.V.; Krushchev, I.; Bennion, I.; Okhrimchuk, A. G. and Shestakov, A. V. (2004) Waveguide inscription in YAG:Cr^{4+} crystals by femtosecond laser irradiation, *in Conference on Lasers and Electro-Optics/International Quantum Electronics Conference and Photonic Applications Systems Technologies, Technical Digest (CD),* (Optical Society of America), paper CWA49.

Dubov, M.; Mezentsev, V. and Bennion, I. Femtosecond inscription of the first order bragg gratings in pure fused silica, (2006) in 2006 NSTI Nanotechnology Conference and Trade Show - NSTI Nanotech 2006 Technical Proceedings, Vol.3 (229–232), Nano Science and Technology Institute, (Boston, MA, United States of America).

Eaton, S.M.; Zhang, H.; Herman, P.R.; Yoshino, F.; Shah, L.; Bovatsek, J. & Arai, A. Y. (2005). Heat accumulation effects in femtosecond laser written waveguides with variable repetition rate. *Optics Express*, Vol. 13 (4708- 4716).

Feng, Q.; Moloney, J. V.; Newell, A. C.; Wright, E. M.; Cook, K.; Kennedy, P. K.; Hammer, D. X.; Rockwell, B. A. and Thompson, C. R. (1997). Theory and Simulation on the Threshold of Water Breakdown Induced by Focused Ultrashort Laser Pulses. *IEEE Journal of Quantum Electronics*, Vol. 33 (127-137).

Glezer, E.N. and Mazur, E. (1997). Ultrafast-laser driven micro-explosions in transparent materials. *Appl. Phys. Lett.*, Vol. 71 (882-884).

Gorelik, T.;Will, M.; Nolte,S.; Tuennermann,A. & Glatzel, U. (2003). Transmission electron microscopy studies of femtosecond laser induced modifications in quartz. *Appl. Phys. A*, Vol. 76 (309-311).

Guizard, S.; Martin, P. Daguzan, Ph; Petite; G.; Audeber, P.; Geindre, J. P.; Dos Santo, A. and Antonnett, A. (1995). Contrasted Behaviour of an Electron Gas in MgO, Al_2O_3 and SiO_2. *Europhys. Lett.*, Vol. 29 (401-406).

Homoelle, D.; Wielandy, S.; Gaeta, A.L.; Borrelli, N. F. and Smith, C. (1999). Infrared photosensitivity in silica glasses exposed to femtosecond laser pulses. *Optics Letters*, Vol. 24 (1311-1313).

Kanehira, S.;_ Miura, K. and Hirao, K. (2008). Ion exchange in glass using femtosecond laser irradiation. *Appl. Phys. Lett.*, Vol. 93 (023112).

Landay, L.D. & Lifshits, E.M. (1982*) Theoretical Physics*, Vol.8, Nauka, Moscow (in Russian).

Lee, Y. L.;_ Yu, N. E.; Jung, C.; Yu, B.-A.; Sohn, I.-B.; Choi, S.-C.; Noh, Y.-C.; Ko, D.-K.; Yang, W.-S.; Lee, H.-M.; Kim, W.-K. and Lee, H.-Y. (2006). Second-harmonic generation in periodically poled lithium niobate waveguides fabricated by femtosecond laser pulses. *Appl. Phys. Lett.*, Vol. 89 (171103).

Liu, Y.; Zhu, B.; Wang, L.; Qiu, J.; Dai, Y. and Ma, H. (2008). Femtosecond laser induced coordination transformation and migration of ions in sodium borate glasses. *Appl. Phys. Lett.*, Vol. 92 (121113).

Liu, Y.; Shimizu, M.; Zhu, B.; Dai, Y.; Qian, B.; Qiu, J.; Shimotsuma, Y.; Miura, K. and Hirao K. (2009). Micromodification of element distribution in glass using femtosecond laser irradiation. *Optics Letters*, Vol. 34 (136-138).

Martin, P.; Guizard, S.; Daguzan, Ph.; Petite; G.; D'Oliveira, P.; Meynadier, P. and Perdrix, M. (1997-I) Subpicosecond study of carrier trapping dynamics in wide-band-gap crystals. *Phys. Rev. B*, Vol.55 (5799- 5809).

Miura, K.; Qiu, J.; Inouye, H.; Mitsuyu,T. & Hirao, K. (1997). Photowritten optical waveguides in various glasses with ultrashort pulse laser. *Appl. Phys. Lett.*, Vol. 71 (3329-3331).

Marshall, G.D.; Ams, M. and Withford, M.J. (2006). Direct laser written waveguide–Bragg gratings in bulk fused silica. *Optics Letters*, Vol. 31 (2690-2691).

Marshall, G.D.; Dekker, P.; Ams, M.; Piper, J.A. and Withford, M.J. (2008). Directly written monolithic waveguide laser incorporating a distributed feedback waveguide-Bragg grating. *Optics Letters*, Vol. 33 (956-958).

Moh, K.J.; Tan, Y.Y.; Yuan, X.-C.; Low, D.K.Y. & Li, Z.L. (2005). Influence of diffraction by a rectangular aperture on the aspect ratio of femtosecond direct-write waveguides. *Optics Express*, Vol. 13 (7288- 7297).

Nejadmalayeri, A.H.; Herman, P.R.; Burghoff, J.; Will, M.; Nolte, S. and Tünnermann, A. (2005). Inscription of optical waveguides in crystalline silicon by mid-infrared femtosecond laser pulses. *Optics Letters*, Vol. 30 (964-966).

Okhrimchuk, A.G.; Shestakov, A. V.; Khrushchev, I. and Mitchell, J. (2005 -a). Depressed cladding, buried waveguide laser formed in a YAG:Nd^{3+} crystal by femtosecond laser writing. *Optics Letters*, Vol. 30 (2248-2250).

Okhrimchuk, A.G.; Shestakov, A. V.; Khrushchev, I. & Bennion, I. (2005 -b). Waveguide Laser Inscribed in Neodimium YAG Crystal by Femtosecond Writing. TOPS V.98 (836-841), OSA, Ed. I. Sorokina and C. Denman, P.

Okhrimchuk, A.; Dubov, M.; Mezentsev, V. and Bennion, I. (2008). Permanent Refractive Index Change Caused by Femtosecond Laser Beam in YAG Crystals. *Proceedings of 3-d EPS-QEOD Europhoton conference,* TUp.15, Paris, France, August-September 2008.

Okhrimchuk, A.G.; Mezentsev, V.K.; Schmitz, H.; Dubov, M. and Bennion, I. (2009-a). Cascaded nonlinear absorption of femtosecond laser pulses in dielectrics", *Laser Physics*, Vol.19, (1415-1422).

Okhrimchuk, A.G.; Mezentsev, V.K.; Dvoyrin, V.V.; Kurkov, A.S.; Sholokhov, E.M.; Turitsyn, S.K.; Shestakov, A.V. and Bennion, I. (2009-b). Waveguide-saturable absorber fabricated by femtosecond pulses in YAG:Cr^{4+} crystal for Q-switched operation of Yb-fiber laser. *Optics Letters*, Vol. 34 (3881-3883).

Osellame, R.; Taccheo, S.; Marangoni, M.; Ramponi, R.; Laporta, P.; Polli, D.; De Silvestri, S. and Cerullo, G. (2003). Femtosecond writing of active optical waveguides with astigmatically shaped beams. *J. Opt. Soc. Am. B*, Vol.20 (1559-1567).

Schaffer, C.B.; Brodeur, A.; García, J.F. and Mazur, E. (2001). Micromachining bulk glass by use of femtosecond laser pulses with nanojoule energy. *Optics Letters*, Vol. 26 (93-95).

Siebenmorgen, J.; Petermann, K.; Huber, G.; Rademaker, K., Nolte, S. & Tünnermann, A. (2009). Femtosecond laser written stress-induced Nd:$Y_3Al_5O_{12}$ (Nd:YAG) channel waveguide laser. *Appl. Phys. B*, Vol. 97 (251–255).

Silva, W. F.; Jacinto, C.; Benayas, A.; Vazquez de Aldana, J. R.; Torchia, G. A.; Chen, F.; Tan, Y. and Jaque, D. (2010). Femtosecond-laser-written, stress-induced Nd:YVO_4 waveguides preserving fluorescence and Raman gain. *Optics Letters*, Vol. 35 (916-918).

Streltsov, A.M. and Borrelli, N.F. (2001). Fabrication and analysis of a directional coupler written in glass by nanojoule femtosecond laser pulses. *Optics Letters*, Vol. 26 (42-43).

Thomas, J.;_ Heinrich, M.; Burghoff, J.; Nolte, S.;. Ancona, A & Tünnermann, A. (2007). Femtosecond laser-written quasi-phase-matched waveguides in lithium niobate. *Appl. Phys. Lett.*, Vol. 91 (151108).

Torchia, G. A.;_ Rodenas, A.; Benayas, A.; Cantelar, E.; Roso, L. and Jaque, D. (2008). Highly efficient laser action in femtosecond-written Nd:yttrium aluminium garnet ceramic waveguides. *Appl. Phys. Lett.*, Vol. 92 (111103).

Turitsyn, S.K.; Mezentsev, V.K.; Dubov, M.; Rubenchik, A. M., Fedoruk, M. P. and Podivilov, E.V. (2007). Sub-critical regime of femtosecond inscription. *Optics Express*, Vol. 15 (14750-14764).

Qurerre, F.; Guizard, S. and Martin, Ph. (2001). Time-resolved study of laser-induced breakdown in dielectrics. *Europhys. Lett.*, Vol. 56 (138-144).

Heat Absorption, Transport and Phase Transformation in Noble Metals Excited by Femtosecond Laser Pulses

Wai-Lun Chan[1] and Robert S. Averback[2]
[1]*University of Texas at Austin, Texas 78712,*
[2]*University of Illinois at Urbana-Champaign, Illinois 61801, USA*

1. Introduction

With the rapid development of short pulse lasers, femtosecond (fs) lasers have been used in technologies such as materials modification (Vorobyev & Guo, 2008), micromachining (Liu et al., 1997; Gattass & Mazur, 2008), and surface patterning (Miyaji & Miyazaki, 2008). However, details on energy absorption, heat dissipation, and phase transformation during the initial excitation of these intense pulses are poorly understood, despite their importance in determining the final microstructure and morphology of the laser-processed materials. Scientifically, intense laser pulses drive materials into highly non-equilibrium states, which provide us new opportunities in studying the fundamental properties of these states. By carefully controlling the relaxation of the excited materials, it is also possible to create novel metastable structures.

Before we can achieve these goals, we need a detailed knowledge on how materials respond to these intense pulses. This is a non-trivial task since the optical, thermodynamics and transport properties can change drastically upon intense laser irradiation. In this chapter, we summarize our recent works on Ag to illustrate some of the complexities involved in both heat transport and absorption at laser fluencies near the melting and ablation threshold. We observe significant heat confinement and absorption enhancement at high laser fluences. These differences derive mainly from the excitation of d-band electrons at high electron temperatures, which significantly changes the electronic properties, although many of the details remain poorly understood. From a more applied consideration, we show that by using fs-pulses, we are able to confine melting and ablation to the very top layers of the materials, which has important implications for pulsed laser deposition of ultra-thin layers and micromachining. Furthermore, we demonstrate that by utilizing the ultrafast quenching of the surface layer after laser excitations, we can create a pure metallic liquid at a temperature as low as 0.6 T_m, where T_m is the melting temperature of the metal. This degree of undercooling has not been achieved by other experimental techniques in pure metals and the transformation kinetics of this far-from-equilibrium state is not well-understood. Indeed, our results show that classical solidification theories cannot explain the experimental results and new atomistic mechanisms are needed in order to explain the measured kinetics.

2. Non-linear heating by a single laser pulse

Before we discuss the heat transport and phase transformation in noble metals irradiated by fs-lasers, we first need to consider how heat is absorbed by the materials during a single pulse irradiation. This is not a simple problem because, for a high intensity pulse with a finite width, the electrons excited by the early part of the pulse can alter significantly the reflectance during the latter part of the pulse. This integral effect makes the total absorbed energy depends not only on the transient reflectivity of the materials, but also on the pulse duration and the laser fluence. There are numerous time-resolved measurements on reflectivity changes in metals induced by fs-laser (see e.g. Norris et al., 2003; Eesley, 1983; Tas & Maris, 1994), but most of these studies are performed at relatively low excitations. Absorption measurements have been done at very high fluencies for studying the dynamics of electrons in the plasma state (Milchberg et al., 1988; Cerchez et al., 2008) and the energy that resides in the sample after laser ablation (Vorobyev & Guo, 2004). However, at a more moderate excitation, where processes such as melting and desorption occur, the change in absorption can still be significant, but they are often neglected. In the following, we summarize our absorption measurements in noble metals and show that excitation of d-band electrons significantly changes the optical properties of the materials, and therefore, the total energy absorbed. Knowing the actual amount of energy that is absorbed by the materials is important, for example, in determining how much materials can be melted by a single laser pulse.

2.1 Heat absorption measured by calorimetry

Instead of measuring the transient optical properties of the metals, we choose to measure the energy absorbed by the samples directly, using calorimetry method. A laser pulse with a wavelength of 800 nm and FWHM pulse duration of 140 fs was used to excite the sample. The same laser system was used for experiments described in the remainder of this chapter. The samples were 400 nm-thick metal thin films sputter deposited on a Si substrate. We thermally isolated the sample from the surrounding by holding the sample with a very fine (0.001") thermocouple wire. This allows us to measure the temperature change of the sample induced by the laser pulses before the heat is dissipated to the surrounding. Figure 1(a) shows a schematic of the setup. The absorption coefficient $A(f)$, defined as the absorbed energy divided by the incident energy, can be determined analytically by measuring the temperature change as a function of time. The detailed analysis can be found in (Chan et al., 2009a). Here, we discuss only our main results.

Figure 1(b) shows the $A(f)$ as a function of incident energy (solid square) for Ag. For comparison, the peak fluence f of the pulses is shown on the top axis. At low fluences, the measured $A(f)$ agrees well with the optical absorption calculated using the dielectric constant of Ag, which is ≈ 0.014. However, at higher incident energies, $A(f)$ deviates significantly from this theoretical absorption. Furthermore, since the measured $A(f)$ represents the *average* absorption coefficient over the entire Gaussian beam. Spatially, at the center of the laser beam, the fluence is much higher. Therefore, the actual absorption coefficient $a(f)$ at the center of the pulse can be much more sensitive to the change in the fluence as compared to $A(f)$. Mathematically, the actual $a(f)$ can be related to the measured $A(f)$ by:

$$A(f)f2\pi\sigma^2 = \int_0^\infty a\left[f\exp\left(-\frac{r^2}{2\sigma^2}\right)\right]f\exp\left(-\frac{r^2}{2\sigma^2}\right)2\pi r dr\,, \quad (1)$$

where r is the radial distance from the center of the pulse. For a Gaussian beam profile, $a(f)$ can be solved analytically (Chan et al., 2009b) and the calculated $a(f)$ is shown as the solid line in Fig. 1(b). We note that near the melting threshold (~ 0.4 J cm^{-2}), the actual absorbed energy can be 4 times as high as the absorption predicted by the optical constants of Ag.

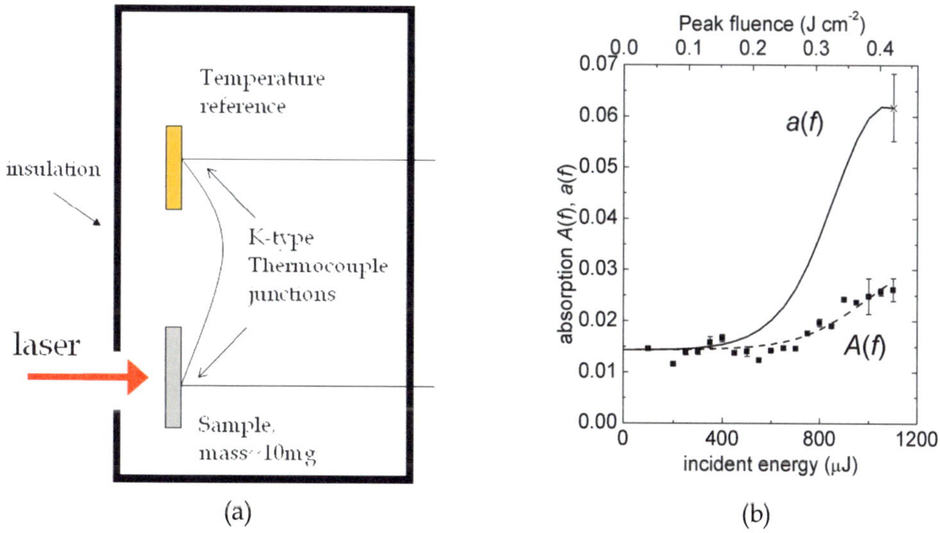

Fig. 1. (a) Schematics of the calorimetry setup. (b) The absorption coefficient as a function of laser incident energy in Ag. (Figure reprinted from Chan et al., 2009a)

2.2 Physical origin of the increase in absorption

In noble metals, the optical properties in the visible range can be described by the Drude model of free electrons (Pines, 1999). In this model, the reflectivity depends on the electron scattering rate, which decreases with an increase in the scattering rate. The abrupt increase in the optical absorption in Fig. 1(b) can be explained by a sudden increase in the electron scattering rate. Using the band structure of Ag (Fuster et al., 1990) and assuming the electrons maintain a Fermi-Dirac distribution, it can be shown that the fluence where the absorption begins to increase corresponds to the excitation of the d-band electrons. Furthermore, we note that the conducting s-electrons are scattered far more strongly by d-band holes compared to other s-electrons. Indeed, the s-d electron scattering rate is ≈ 2 orders of magnitudes higher than the s-s electron scattering rate (Mott, 1936; Baber, 1937). As the d-holes are created, therefore, the electron-scattering time shortens significantly. This results in an abrupt increase in the energy absorption, as observed in our calorimetry experiment.

According to the above picture, there should not be a comparable increase in the absorption in noble metals that already have a significant number of d-band holes at room temperature. To create such a sample, for instance, we can dope Ag with transition metals to create "doped"d-band holes (Coles & Taylor, 1962). We conduct absorption experiments on dilute Ag alloys doped with a few atomic percent of Pt. The results are shown in Fig. 2. The absorption coefficient $a(f)$ is determined from the experimentally measured $A(f)$ using Eq. (1). To provide a direct comparsion of the laser-induced d-band holes and the doped d-band holes, we plot $a(f)$ as a function of absorbed fluence. The population of the excited d-band holes (for Ag) can be estimated for a given absorbed fluence by assuming the electrons maintain a Fermi-Dirac distribution. They are shown as verticle lines in Fig. 2.

At low Pt concentrations, the laser-induced d-band holes seem to be equivalent to the d-band holes produced by alloying. As shown by the dotted line, the absorption at low fluences of the 2.7 % alloy is similar to the absorption of Ag at an absorbed fluence that can create ≈ 3% d-band holes. This particularly good agreement is probably somewhat fortuitous, considering that the laser-induced d-band holes are not physically the same as doped holes even though their effects can be similar. In addition, the calculated number of laser-induced d-band holes is the number of holes after the entire pulse is absorbed; the number of holes within the pulse duration will probably be less. Nevertheless, Fig. 2 suggests that the nature of the two kinds of holes is similar. For the fluence dependence, similar to Ag, the 2.7% alloy shows an increase in absorption at higher fluences. For the 8% alloy, we do not see a measurable increase in absorption. This simply illustrates that for a sample that already has a significant number of d-band holes, the laser fluence has little effect on increasing the overall optical absorption.

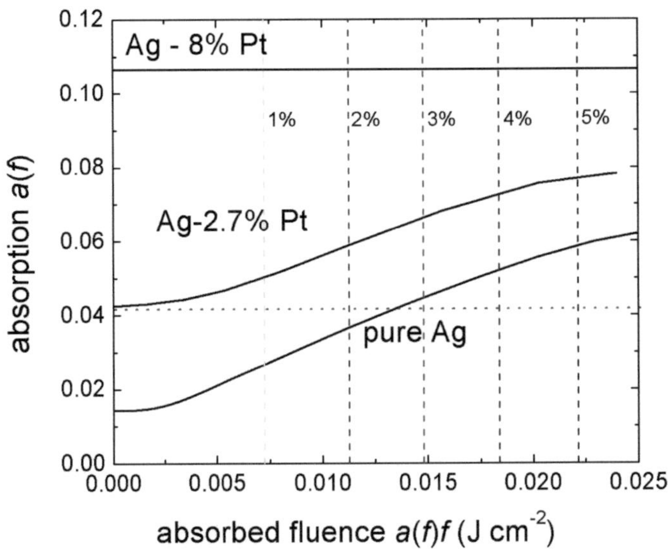

Fig. 2. Optical absorption as a function of absorbed fluence for diluted Ag alloys. The vertical lines represent the calculated population of laser-induced d-band holes. (Figure reprinted from Chan et al., 2009a)

From the above measurements, we can see that the optical absorption can vary significantly with the laser fluence. This is because electron bands under the Fermi-level, which normally remain filled at low excitations, can be excited at higher laser fluences. This significantly changes the optical response of the materials. We have qualitatively explained this absorption change by d-band excitation in Ag. Unfortunately, formulating a more quantitative model is challenging, even for simple materials such as noble metals. This is because our understanding on electron properties at these high excitations is limited.

3. Ultrafast melting in silver

The strong excitation of electrons not only changes the optical properties of the materials, but also the transport properties. The transport properties of these excited electrons determine how heat is dissipated from the initially excited region. This in turn controls how much materials is heated up and, in some cases, transformed to a different phase in the regions within and beyond the excitation depth of the laser. Technologically, fs-lasers are believed to be better tools for laser-machining and materials modification compared to ps- or ns-lasers. This is because the faster heating of fs lasers results in a better heat confinement near the surface region. In the other words, it creates less damage to materials beyond the optical absorption depth of the laser. Understanding the heat deposition profile after fs-laser irradiation is critical to these applications. However, the heat dissipation mechanisms are difficult to measure and still poorly understood.

In this section, we present our recent measurements on the melting dynamics of Ag. We measure the melt-depth as a function of time. The melting dynamics cannot be explained by the well-known transport properties of noble metals. More specifically, the melt-region is much shallower than what is predicted by the hot electron transport properties of Ag. Additional heat confining mechanisms are needed to explain the observed dynamics. As we have mentioned, the d-band electrons, even though they are ≈ 3 eV below the Fermi level in Ag, can be excited at high fluences. They affect the electron transport properties and the observed melting dynamics.

3.1 Measurements on the propagation of the melt-front into a superheated solid

To measure the depth of melt as a function of time after laser excitation, we performed time-resolved optical third harmonic (TH) generation measurements. Optical TH generation is extinct in an isotropic phase (e.g. liquid) but usually not in a crystalline phase (Butcher & Cotter, 1990). This gives us large measurement contrast between the two phases. The experimental details and the calibration methods can be found in (Chan et al., 2008). Here, we will briefly summarize the experimental method.

All the measurements were done on single crystal Ag (001) grown on MgO substrates. Optical pump-probe experiment was done to determine the melt-depth as a function of time. A pump pulse was first used to excite the solid and a time-delayed probe beam with a beam diameter 10 times smaller than the pump pulse was used to determine the melt-depth at the center of the pump pulse. The intensity of TH generation from the sample surface, which induced by the probe pulse, was measured using a sensitive CCD camera. Ultra-fast melting usually creates a continuous liquid layer on top of the solid phase. Since the liquid phase does not generate TH light, the depth of the melt $d(t)$ can be related to the normalized intensity I through the exponential relationship:

$$I(t) = \exp\left[-d(t)\left(\frac{3}{d_I} + \frac{1}{d_{III}}\right)\right] \quad (2)$$

where d_I and d_{III} are the extinction depths of the fundamental and TH light. The 3 in the exponential term arises from the cubic dependence of the TH signal on the probe laser intensity. Equation (2) neglects the difference in the dielectric constant between the liquid and solid phase. This, however, is a good approximation for metals such as Ag, for which the two phases have a very similar dielectric constants (Miller, 1969). The combined extinction depth $(3/d_I+1/d_{III})^{-1}$ for high laser fluences is not known *apriori*, but we measured it separately in our control experiment with multilayer films with known thicknesses.

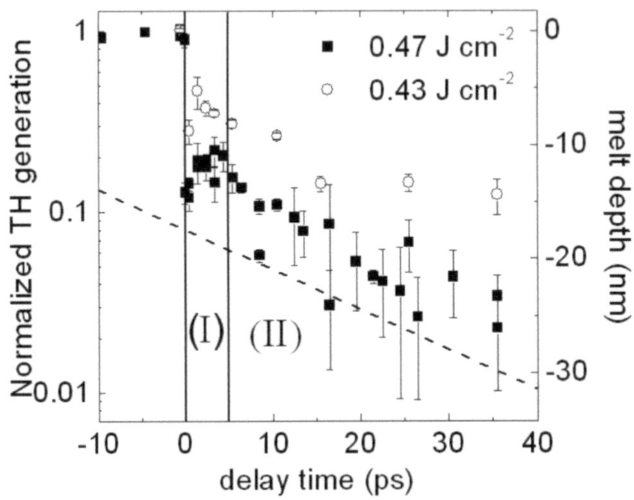

Fig. 3. Normalized TH generation as a function of delay time for a 200 nm Ag film. The axis on the right shows the corresponding melt depth. The melting can be divided into two stages: (I) initial melting due to transfer of energy from electron to phonon; (II) propagation of the melting front into the superheat solid. The dotted line represents a melting speed of 350 m s^{-1}. (Figure reprinted from Chan et al., 2008)

Figure 3 plots the melt-depth as a function of time at two different pump fluences. The data are plotted on a semi-log scale. According to Eq. (2), the slopes of the curves represent the velocities of the melt-crystal interface. The calculated melt-depth is shown on the axis on the right. The melting can be roughly divided into two stages. In the first stage, rapid melting occurs within first 5 ps. This corresponds to the electron-phonon coupling time in Ag (Ernstorfer et al., 2007; Lin et al., 2008b); i.e. time needed for the excited electrons to transfer the energy to the cold lattice. Following the initial fast melting, the melt-front propagates steadily into the bulk at a velocity ≈ 350 m s^{-1} (the slope for this velocity is shown as the dashed line in Fig. 3) for another 25 ps. The combined maximum melt-depth for both stages is ≈ 25-30 nm, which occurs at t ≈ 30 ps. Then, the melt resolidifed with much slower dynamics (the solidifcation kinetics will be addressed in Sec. 4).

A more surprising result is the melt-depth as a function of laser fluences for experiments performed on Ag films with different thicknesses. The results are shown in Fig. 4. We note that non-equilibrium electrons excited by the laser can carry heat away from the optical extinction depth before they couple to the phonons. This measurement characterizes the dynamics of these 'hot' electrons. Similar ideas have been used in (Suarez et al., 1995; Bonn et al. 2000) to study the fast electron transport at lower fluences. For the results represented in Fig. 4, we fix the delay time at 25 ps (onset before resolidification occurs) and measure the TH intensity as a function of pump fluence. The film does not melt until a threshold is reached. The amount of melt then increases approximately linearly with respect to the additional fluence. The threshold for melting and the amount of melt remains nearly independent of film thickness, until the thickness falls below 100 nm. Since heat transport across the Ag-MgO interface is negligible in 25 ps, the melting threshold should start to decrease when the film thickness falls below the depth of heat deposition. Therefore, the result indicates that at t = 25 ps, most of the heat is confined within the top 50 - 100 nm of the Ag.

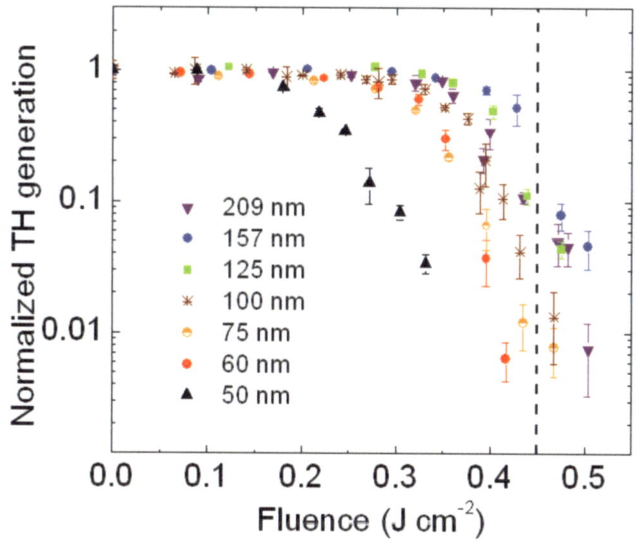

Fig. 4. TH generation as a function of pump fluence for different film thicknesses. The delay time is fixed at 25 ps. (Figure reprinted from Chan et al., 2008)

Experiments performed at lower fluencies (where d-band electrons are not excited) have indicated that hot electrons can carry heat a few times deeper than what we observe here via the so-call ballistic transport (Suarez et al., 1995; Bonn et al. 2000). Therefore, when the fluence is high enough to induce melting, the electron transport must change significantly to produce the heat confinment observed in Fig. 4.

3.2 Heat transport mechanisms

In order to understand what aspects of the heat transport can create the spatial confinement following, we use a two temperature model (TTM) to model the thickness dependences observed in Fig. 4. TTM (Anisimov et al., 1974) is a standard technique to model the evolution of electron and phonon temperatures. The details of our TTM can be found in (Chan et al., 2008) and will not be discussed here. Note that the model is not aimed at being rigorous for the first few ps after the excitations since such a treatment on the electron transport with d-band excitation that includes the consideration of non-equilibrium electrons, is beyond the scope of this chapter. The model described here is used to estimate the amount of heat confinement induced by excited d-band holes and to explore possible heat confining mechanisms.

It has been reported in other studies that the excitation of d-band electrons changes the electron heat capacity and electron-phonon coupling constant by an order of magnitude at high electron temperatures (Lin et al., 2008b). First, we include these effects into our TTM, but do not account for the effects of d-band excitation on electron transports. This result is shown as solid-squares in Fig. 5, which plots the maximum melt depth as a function of film thickness at a given laser fluence. The experimental results are shown as solid-circles for comparison. We see that this model highly underestimates the melt-depth in the thick films. It also predicts that the melt-depth decreases significantly with an increase in film thickness. This contradicts to the experiment in which the melt-depth does not change significantly for

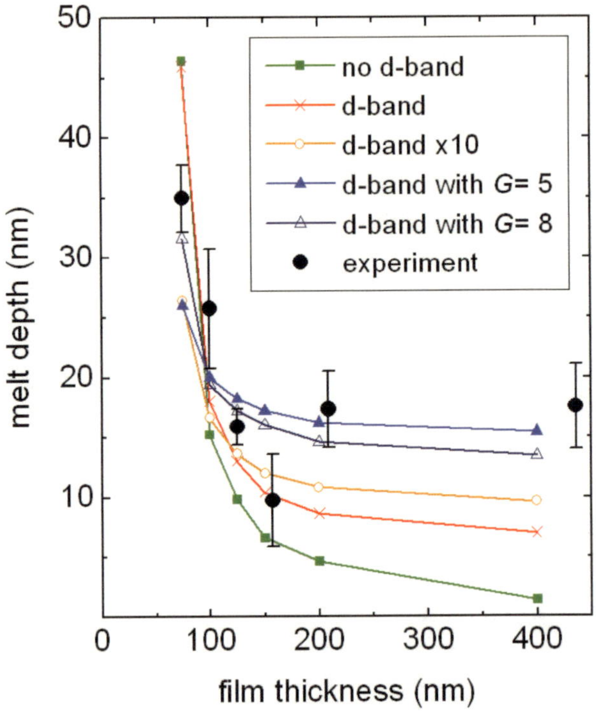

Fig. 5. Melt-depth as a function of film thickness at delay time equals to 25 ps. The experimental points (black-circles) are determined from Fig. 4, at fluence = 0.45 J cm^{-2} (the vertical dashed line in Fig. 4). For the TTM calculation, the absorbed fluence is fixed at 0.025 J cm^{-2}. The modeling parameters used are: green-line with squares – no enhancement in e-e scattering rate from d-band holes [C = 0 in Eq. (3)]; red-line with crosses – enhancement in e-e scattering modeled by Eq. (3), with C = 150; orange-line with open circles – C = 1500; blue-lines with triangles – C = 150, with the addition of an effective interface impedance G (see the text). Only G = 5 and 8 GW m^{-2} K^{-1} are shown. (Figure reprinted from Chan et al., 2008)

film thickness larger than 100 nm. This discrepancy is caused by neglecting the d-band excitation on thermal transport, which causes significant heat confinement. The overestimated heat conductivity allows heat to dissipate quickly from the surface region to the deeper region of the film, which reduces the overall melt-depth.

In Sec. 2, we have mentioned that the presence of d-band holes can increase the electron scattering rate. To account for the effect of d-band holes on electron scattering, we add the contribution of s-d scattering to the total scattering rate by the following phenomenological relationship:

$$A_{ee} = A_{ee,0}\left[1 + CN_h(T_{el})\right]. \tag{3}$$

In this equation, $A_{ee,0}$ is the e-e scattering coefficient in the low temperature limit, which for Ag is 0.13 m MW^{-1} (MacDonald & Geldart, 1980). N_h is the number of d-band holes per atom, which is determined by integrating the Fermi distribution. The constant C represents the ratio of the s-d electron scattering rate to the s-s electron scattering rate; we roughly estimate that C = 150, based on A_{ee} being about 2 orders of magnitudes larger in Ni than in

Ag (Mott, 1936; Baber, 1937). The TTM result with the inclusion of this reduced electron conductivity is shown as crosses in Fig. 5. Although the new results agree much better with the experiments, it still underestimates the melt-depth at larger film thicknesses. Increasing the constant C by another order of magnitude (open-circle) does not significantly change the result. An additional heat confinement mechanism is needed to explain the experimental data.

We explore possible additional mechanisms that can create the heat confinement necessary to explain the experimental results using the TTM; we do this by adding a thermal interface impedance. This is rationalized by recognition of the sharp interface that is created by the d-band excitations in the surface region. Since a large temperature gradient exists over a distance that is small compared to the electron mean free path, λ, of Ag [e.g. $\lambda \approx 50$ nm at RT (Kittle, 2005)], electron conductance on the 'cold' side of the interface can be overestimated in the above TTM. In other words, the excited d-band electrons should have influence on κ_{el} in the un-excited region, extending to a distance on the order of λ. A similar argument for non-local heat transport has been employed previously in (Mahan & Claro, 1988) to describe the reduction in phonon conductivity when the curvature in the temperature depth profile is large. We place the interface at a depth of 20 nm and assume its conductance varies inversely with the number of d-band holes. Since the number of d-band holes increases approximately linearly with electron temperature T_e above some critical temperature (≈ 4000 K), we take the interface conductance as $10000 \times G/(T_s\text{-}4000)$, where T_s is the surface temperature and G is a fitting constant with the unit of interface conductance. In the first 3-4 ps, T_s falls from around 15,000 K to a temperature below 4,000 K. Therefore, the interface is only active during stage I, shown in Fig. 3. We vary G from 2 to 12 GW m^{-2} K^{-1}. The melt-depth as a function of film thickness is shown in Fig. 5 (blue-lines with triangles). The prediction of the model agrees well with the experiment for $G \approx 5 - 8$ GW m^{-2} K^{-1}. For comparison, we note that the conductance of Cu-Al interface is 4 GW m^{-2} K^{-1} (Gundrum et al., 2005), and a conductance of 8 GW m^{-2} K^{-1} effectively reduces electron conductivity in a 50 nm-thick region in Ag by half. Therefore, the scattering of the conducting electrons by the d-band holes explains the heat confinement observed in experiments.

3.3 Melting dynamics and its implications

After the cooling down of the excited d-band holes (the end of stage I), the surface region is highly superheated. The melt-front continues to travel into the superheated solid but at the same time, heat is lost to deeper regions and the melting eventually stops. These dynamics result in the stage II melting as depicted in Fig. 3. To understand the melting dynamics in this stage, we plot the isotherms in the sample in which the phonon temperature T_p equals to 1234 K (the melting temperature T_m of Ag) and 1620 K as a function of time in Fig. 6. For regions in the sample for which $T_p > 1620$ K, we assume that they melt immediately because there is no energy barrier to melting. This treatment is consistent to the observations in MD simulations, which show that metals melt homogenously for $T_p > 1.2\ T_m$ (Ivanov et al., 2007; Delogu 2006). The actual position of the melt-front is therefore within the two isotherms. Note that the constant temperature contours can retract to the surface as the sample cools down, but the melting-front should only move forward and eventually stop when the temperature of the front is less than T_m. For reference, we include the experimentally measured velocity (≈ 350 m s^{-1}) as the dotted line in Fig. 6. The melting stops as the temperature of the front falls below T_m, which occurs at $t \approx 20$-25 ps. This is consistent with the experiment in which the melting continues until $t \approx 30$ ps.

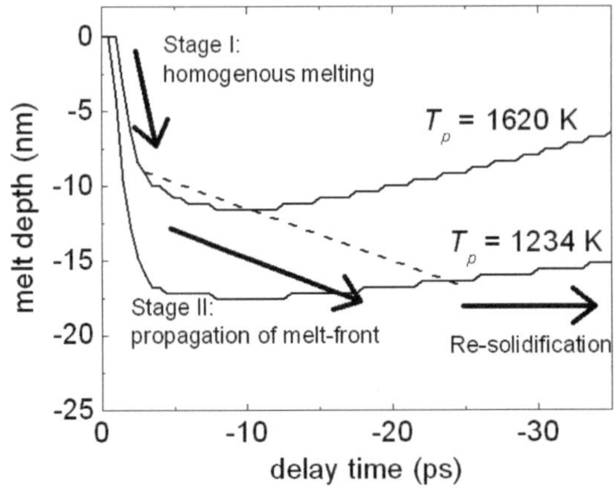

Fig. 6. Constant temperature contours at 1234 K and 1620 K calculated by TTM. In stage I melting, the solid melt homogenously along the $T = 1620$ K contour. In stage II, the melt-front propagates within these two contours. (Figure reprinted from Chan et al., 2008)

One of the interesting observations worth mention here is that during the stage II, when heat is transported away from a thin surface region (< 30 nm), a strong decoupling occurs between the phonon and electron system. This produces a situation characterized by 'hot' phonons and 'cold' electrons at the surface region, in contrast to 'hot' electrons and 'cold' phonons observed in stage I. To rationalize this interesting phenomenon, we note that heat is mainly carried away by the electrons. The heat from the hot phonons at the surface is first transferred to the electrons. Then, the heat is transported to the deeper regions by electrons. Hence, the rate of heat removal can be limited by the slow electron-phonon coupling of Ag. The decoupling can readily be observed from the results of the TTM. For example, in a 200 nm thick film, electron temperature T_e at the surface is just ≈ 100 K higher than T_e at back for $t > 5$ ps, regardless of the larger temperature difference (up to 1000K) in T_p. While different transport processes are included implicitly in the TTM, we can illustrate the phenomenon more clearly by calculating the effective conductance for heat carried away from a hot region of thickness h through electron G_{el} and phonon G_{ph}. The effective heat conductance for phonons, by dimensional analysis, is equal to κ_p /h (κ_p is phonon heat conductance). A thinner surface hot layer (i.e., smaller h) implies a steeper temperature gradient, which increases the effective heat conductance.

The conductance for electrons is more complicated. Heat must first be transferred from phonons in the hot surface layer to electrons, and it is carried away by electrons through diffusion. The transfer of heat back into the phonon system in the cold region is not the rate limiting process, assuming the cold reservoir is always much thicker than the hot region. The first step depends on the e-p coupling constant g and is proportional to the thickness of the hot layer. The conductance for this step is $G_{el\text{-}ep} = gh$. The conductance for the second step (electron diffusion) is $G_{el\text{-}d} = \kappa_{el}/h$. The conductance of electrons is the combination of the two conductances in series, which is given by

$$G_{el} = \frac{1}{\left(G_{el-ep}^{\ -1} + G_{el-d}^{\ -1}\right)} = \frac{1}{1/gh + h/\kappa_{el}}. \tag{4}$$

Normally $G_{el-ep} >> G_{el-d}$, which gives $G_{el} \approx G_{el-d}$. However, for small h and g, G_{el-ep} can be small enough such that it will dominate the heat transport. In our case, taking h = 25 nm yields $G_{el-d} \approx$ 16 GW m^{-2} K^{-1}, $G_{el-ep} \approx$ 0.9 GWm^{-2}K^{-1}. It is thus clear that G_{el} is limited by e-p coupling and its value, calculated by Eq. (4) is 0.85 GW m^{-2} K^{-1}. In addition, this small value of G_{el} is the same order of magnitude as the phonon conductance G_{ph}. Using κ_p in our model, we find G_{ph} = 0.27 GW m^{-2} K^{-1}. Although G_{ph} is still smaller than G_{el}, it is no longer negligible as most studies have assumed. The phonon conduction, moreover, becomes increasingly more important as h falls below 25 nm.

The above behaviors should be general to other noble metals (Cu and Au) that have electronic structures similar to Ag. To summarize, the above analysis on Ag allows us to estimate the amount of materials that can be melted by fs-laser pulses before ablation occurs. This is important to applications such as micromachining where a precise control on the laser damaging depth is needed. For the group of noble metals discussed, the excitation of d-band electrons in stage I limits the depth of the initial heat deposition to approximately the optical absorption depth of the material. Subsequently, transport of heat by electrons from the excited region in stage II is limited by the weak e-p coupling. Although this limitation lengthens the melt lifetime, it is ineffective in increasing the total melt-depth, since most of the heat removed from the surface layer is evenly redistributed over the remainder of the film, i.e., a large temperature drops in the hot region is compensated by a small temperature rise in the cold region. Increasing the laser fluence does not increase the melt-depth appreciably since the extra energy results in ablation before the heat can spread into the bulk. For example, we have found that in our MD simulations, a Cu lattice becomes unstable at $T_p \approx$ 4000 K (i.e. ablation will occur). Since the surface phonon temperature in our TTM calculation already reaches 2000 K and the heat confinement increases non-linearly as the laser fluence increases, we estimate that the maximum melt thickness is not larger than 30-40 nm before ablation becomes significant.

For comparison, we note that the above scenario can be much different if the effects of d-band excitation on thermal conductivity are not taken into account. The predicted melt-depth can be 3-5 times larger before the onset of ablation if the transport properties of noble metals at lower fluencies are used to model the melting dynamics.

4. Solidification of deeply-quenched melts

In the last part of this chapter, we will discuss the use of fs-laser pulses to study the ultrafast solidification dynamics of undercooled liquid Ag. This serves as an example in which we can use fs-lasers to produce a highly non-equilibrium phase. Furthermore, the time-resolved relaxation dynamics of the undercooled liquid can be studied quantitatively. Our experimental results do not agree with classical solidification theories (Chalmers, 1964), but are consistent with recent results from MD simulation. The MD simulation shows that a defect mechanism can describe the solidification behavior in a highly undercooled melt (Ashkenazy & Averback, 2007).

Quenching a pure metal into its glassy state has been a challenge to materials scientists over the last few decades (Turnbull & Cech, 1950). Two common ways to achieve this is either by removing the heterogeneous nucleation sites or by quenching the metals fast enough such that solidification does not have enough time to take place. Using traditional techniques, a pure metal can at most be quenched to $\approx$ 0.8 T_m because of its extremely fast solidification kinetics. Ultrafast lasers provide a new way to achieve this goal because it can confine the

melt in a very few surface region (10s of nm) while keeping the remainders of the sample cold. As we will see below, quenching rates as fast as 5×10^{12} K s^{-1} can be achieved.
There have been some earlier attempts to use ps or ns lasers to undercool liquids. However, in the ps-laser studies (MacDonald et al., 1989; Agranat et al., 1999), only resolidification time can be measured quantitatively. Important parameters such as surface temperature and solidification velocity remain unknown. In the ns-laser studies, the pulses are too long and a thick layer of materials can be heated up within the pulse duration. Hence, no significant undercooling can be achieved (Tsao et al., 1986; Smith & Aziz, 1994). In our current experiment with fs-lasers, we are able to quench the liquid with large undercooling and measure the solidification velocity quantitatively using the optical TH generation described above. The undercooling temperature is modeled by TTM with a high accuracy. As a result, we can measure the solidification velocity as a function of temperature down to $\approx 0.6\ T_m$.

4.1 Ultrafast quenching and solidification of undercooled liquid

Single crystals Ag grown on MgO substrates were used in the experiment. The details of the experiments can be found in (Chan et al., 2009b). A schematics diagram for the experiment is shown in Fig. 7. A thin layer of Ag is melted by the fs-laser pulse. The optical TH generation technique discussed in Sec. 3.1 was used to measure the position of the crystal-melt interface as a function of time. The rate of resolidification depends on the undercooling of the liquid Ag. The degree of undercooling during solidification can be readily controlled by simply changing the thickness of the thin films, which will be discussed below.

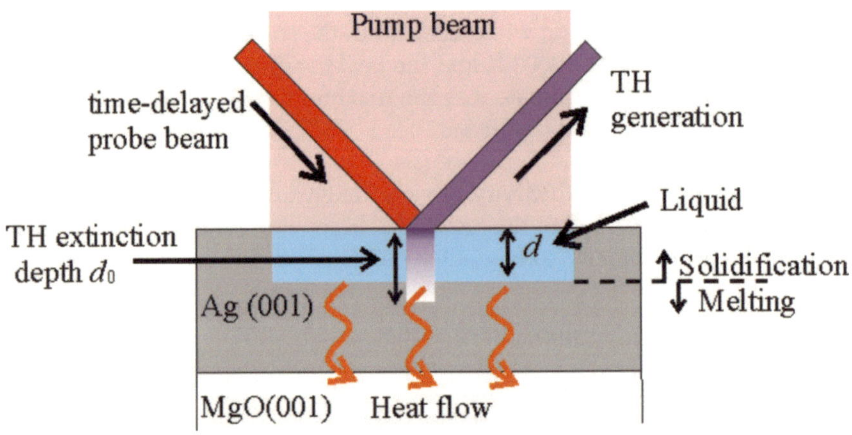

Fig. 7. A schematic of the experiment setup. The pump beam, which is ≈ 10 times larger in size than the probe beam, is used to melt the Ag. Optical TH generation is used to measure the thickness d of the liquid layer. The cooling rate is controlled by varying the thickness of the Ag layer. (Figure reprinted from Chan et al., 2009b)

Figure 8 shows the results for three Ag films with different film thicknesses. After the initial melting, the TH intensity recovers steadily for $t > 50$ ps, which represents resolidification of the liquid phase. The slope of the solid line represents the average interfacial velocity v_{ave}. The solidification process is completed by $t \approx 200$ - 300 ps. The signal does not fully recover at t ≈ 1 ns, but it does so, however, before $t \approx 1$ s. We attribute the degradation in signal at t = 1 ns to the production of quenched-in defects (primarily vacancies) during solidification, such defects have been observed in MD simulations (Lin et al., 2008a).

Note that the solidification velocity varies with the film thicknesses. The conductance of heat through the thin Ag film is much faster than through the Ag-MgO interface and the MgO substrate. During the solidification, therefore, the heat spreads rapidly across the entire Ag film, but only a small amount of heat can transport across the Ag-MgO interface. Larger undercoolings (or low temperatures) are thus achieved in thicker films.

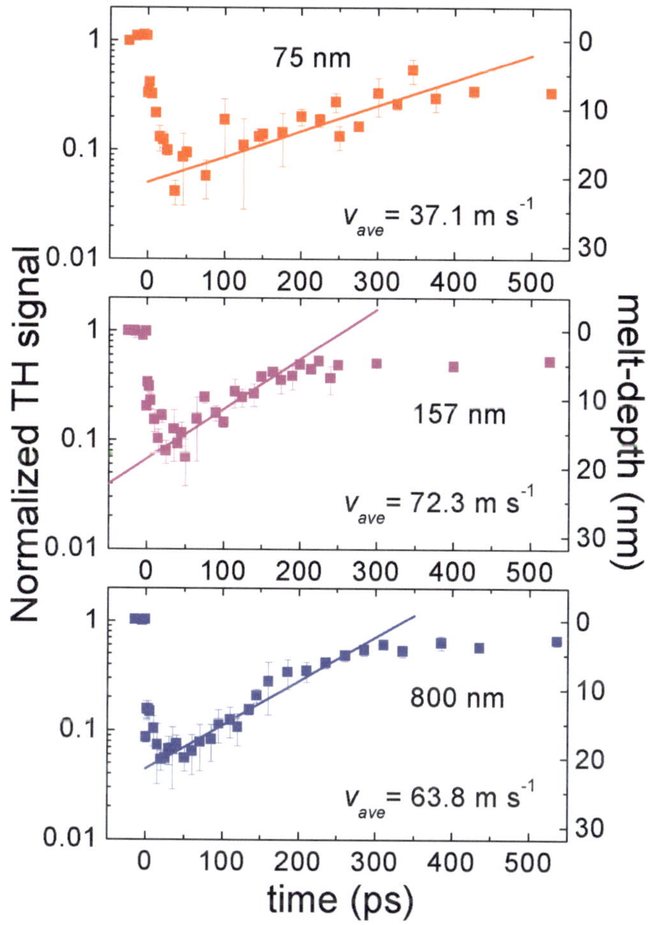

Fig. 8. The TH signal as a function of time measured for samples with three different thicknesses of the Ag layer. The converted melt-depth is shown on the axis on the right. The average resolidification velocity is indicated by the solid-lines. (Figure reprinted from Chan et al., 2009b)

We determine the temperature of the crystal-melt interface using TTM. For the solidification process, since the electronic system has already restored the Fermi-Dirac distribution, the TTM model is aimed at determining the interface temperatures with high accuracy. The details of the model can be found in (Chan et al., 2009b). Here, we note that the parameter in the model that has the strongest effect on the calculated temperatures is the total energy initially absorbed by the samples. Instead of modeling this parameter, we have measured it directly using the calorimetry setup discussed in Sec. 2. Figure 9 shows the interface temperature as a function of time for different film thicknesses. By combining the average temperature determined from the TTM and the v_{ave} found in experiment, we can plot the solidification velocity as a function of temperature, which is shown in Fig. 10.

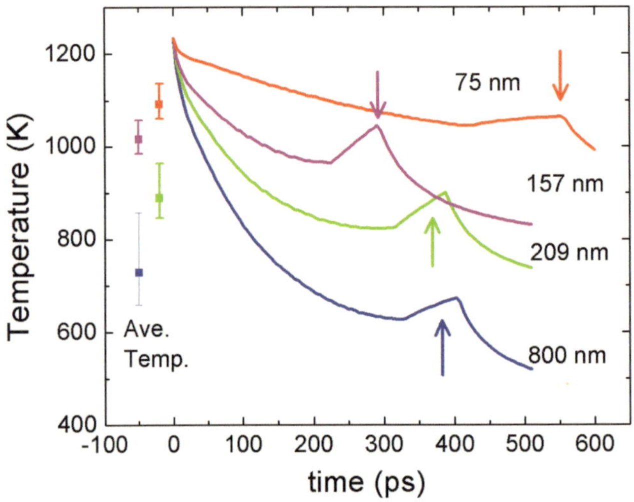

Fig. 9. The interface temperature as a function of time. The arrows indicate the end of the solidification. The average temperatures over the whole period of solidification are indicated by the solid-symbols on the left. The error bars above (below) the symbols represent the mean deviations from the average temperature during the period with temperatures above (below) the average temperature. (Figure reprinted from Chan et al., 2009b)

The solidification velocity is also obtained as a function of temperature using MD simulation (Ashkenazy & Averback 2007; 2010); these data are shown in Fig. 9 as circles. The agreement between experiment and simulation is quite good; note that there are no adjustable parameters. The velocity increases approximately linearly from T_m to 0.85 T_m, and then it becomes insensitive to temperature with further decrease in temperature. The long plateau observed in Fig. 10 explains why the experimental solidification velocity remains nearly constant as a function of time even though Fig. 9 shows that the crystal-melt interface temperature can vary by ≈ 200 – 300 K during solidification.

4.2 Kinetics mechanisms for solidification

Although continuum models for solidification have been developed for decades, none of these models have been experimentally verified in a pure metal at deep undercooling. This is mainly due to the difficulty in quenching a pure metallic liquid far below its melting point. The classical model assumes that the solidification rate in pure is controlled by collision-limited kinetics (MacDonald et al., 1989; Coriell & Turnbull, 1983; Broughton et al., 1982). Furthermore, it is often accepted that there is no energy barrier for an atom to move across the liquid-solid interface. Under these assumptions, the velocity can be written as

$$v(T) = C\sqrt{\frac{3kT}{m}}\left[1 - \exp\left(-\Delta\mu/kT\right)\right], \tag{5}$$

where C is a geometric factor on the order of 1, m is the atomic mass and $\Delta\mu$ is the free energy difference between the solid and liquid phase. This relation is shown as the solid line in Fig. 9. Because of the weak $(T)^{1/2}$ dependence, the velocity continues to increase until $T < 0.3\ T_m$; this clearly disagrees with the experimental and simulation data shown in Fig. 10.

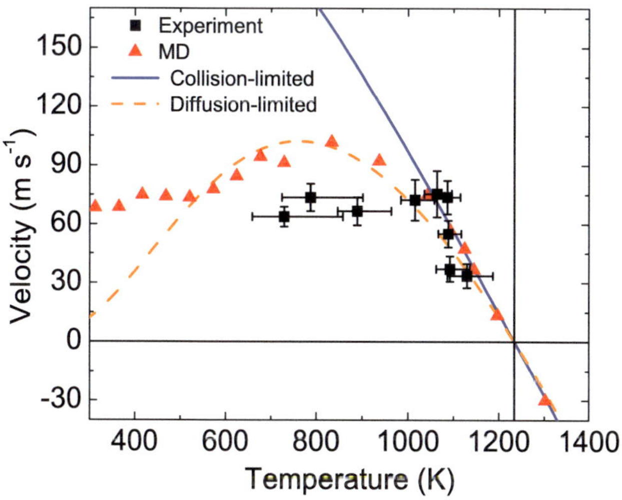

Fig. 10. Solidification velocities verse the temperatures. The experimental data (squares) show reasonable agreement to the MD simulations (triangles). However, it clearly deviates from the collision-limited model (the blue solid line). If we assume the motion of the atom across the liquid-solid interface is thermally activated (with an activation barrier = 0.12 eV), the predicted velocity is shown as the orange line (dashed-line). (Figure reprinted from Chan et al., 2009b)

If a barrier exists in the energy landscape for an atom to move from the liquid to the solid, one can replace the square root term in Eq. (5) by an exponential term $A\exp(-E/kT)$ (Frenkel, 1946), where E represents the barrier height. By setting E = 0.12eV and A = 1300 m s^{-1}, indicated by the orange (dashed) line in Fig. 10, we see that above 600 K, the velocity agrees well with the MD simulation and the experimental data. The existence of an activation barrier, therefore, can explain why the solidification velocity reaches its maximum at a relatively high temperature. Such a barrier, however, indicates that the velocity should diminish at lower temperatures, which disagrees with the MD simulations. We note that in Ag the glass transition temperature $T_g \approx$ 600K. The discrepancy between the continuum models and the MD data perhaps suggests that the solidification mechanisms for the liquid and glass states may be very different.

Our recent MD simulation (Chan et al., 2010) shows that atoms at the interface transform into the crystalline phase cooperatively instead of individually as assumed in the classical models. In our simulations, the transformation is often induced by 1 or 2 atoms making exceptional long jumps, the nearest neighbors surrounding these atoms then transform into the crystalline phase cooperatively. Interestingly, the magnitude and directionality of these long jumps is very similar to the motion of an interstitial defect in the crystalline phase. These observations agree with the model proposed earlier by Ashkenazy and Averback (Ashkenazy & Averback, 2007), in which the solidification kinetics is controlled by interstitial-like motions at the crystal-melt interface. To prove this model experimentally, we need to quench the liquid below its glass transition temperature. Currently, we are not able to achieve these deep undercoolings, but with more carefully designed experimental systems, more tunable laser systems, or advance characterization techniques such as time-resolved diffraction, this goal appears within reach.

5. Conclusion

In this chapter, we have presented a comprehensive study on the heat absorption, transport, and phase transformation kinetics in Ag irradiated by fs-lasers. Although the current study is focused on Ag, but similar behaviors are expected to be observed in Cu and Au as well. We have shown that a lot of complexities on the optical and transport properties can arise at fluences close to the melting and ablation threshold. These complexities come from the excitation of electron bands that are below the Fermi level. Although noble metals are among the most-studied materials, at these high excitations, many fundamental issues such as the relaxation of non-equilibrium hot electrons, the thermal transport under extremely high temperature gradients, and the dynamics of superheated solid are still not well-resolved. We can expect similar complex situations can be found in other transition metals as well.

We also demonstrated that by the using of fs-lasers, we can induce ultrafast quenching in Ag and measure the transformation kinetics of the undercooled liquid quantitatively. This only serves as one of the examples in which we can use fs-laser to drive materials into structurally unstable phases. With the rapid development of the laser technologies and time-resolved characterization techniques, we can study phases that are inaccessible before. This not only improves our understanding of materials under extreme environments, but also provides us new ways to create metastable materials that have novel structural and electronic properties.

6. Acknowledgement

We gratefully acknowledge the supports by the U.S. Department of Energy - NNSA under Grant No. DE-FG52-06NA26153 and the U.S. Department of Energy-BES under Grants No. DE-FG02-05ER46217.

7. References

Agranat, M. B.; Ashitkov, S. I.; Fortov, V. E.; Kirillin, A. V.; Kostanovskii, A. V.; Anisimov, S. I. & Kondratenko, P. S. (1999). Use of optical anisotropy for study of ultrafast phase transformations at solid surfaces. *Appl. Phys. A*, 69, 6, 637-640 , ISSN: 0947-8396

Anisimov, S. I.; Kapeliovich, B. L. & Perel'man, T. L. (1974). Electron emission from metal surfaces exposed to ultrashort laser pulses. *Sov. Phys. JETP*, 39, 375-377, ISSN: 1063-7761

Ashkenazy, Y. & Averback, R. S. (2010). Kinetic stages in the crystallization of deeply undercooled body-centered-cubic and face-centered cubic metals. *Acta Materialia*, 58, 2, 524-530, ISSN: 1359-6454

Ashkenazy, Y. & Averback, R. S. (2007). Atomic mechanisms controlling crystallization behavior in metals at deep undercoolings. *EPL*, 79, 2, 26005, ISSN: 0295-5075

Baber, W. G. (1937). The Contribution to the Electrical Resistance of Metals from Collisions between Electrons. *Proc. Roy. Soc. A* , 158, 383-396, ISSN 1364-5021

Bonn, M.; Denzler, D. N.; Funk, S. ; Wolf, M. ; Svante Wellershoff, S. & Hohlfeld, J. (2000). Ultrafast electron dynamics at metal surfaces: Competition between electron-phonon coupling and hot-electron transport. *Phys. Rev. B*, 61, 2, 1101-1105, ISSN 1098-0121

Broughton, J. Q.; Gilmer, G. H. & Jackson, K. A. (1982). Crystallization Rates of a Lennard-Jones Liquid. *Phys. Rev. Lett.*, 49, 20, 1496-1500, ISSN 0031-9007

Butcher, P. N. & Cotter D. (1990). *The Elements of Nonlinear Optics*, Cambridge University Press, ISBN: 0-521-34183-3, Cambridge

Cerchez, M.; Jung, R.; Osterholz, J.; Toncian, T.; Willi, O.; Mulser, R. & Ruhl, H. (2008). Absorption of Ultrashort Laser Pulses in Strongly Overdense Targets. *Phys. Rev. Lett.*, 100, 24, 245001, ISSN 0031-9007

Chalmers, B. (1964). *Principles of solidification*, John Wiley & Son, ISBN: 0471143251, New York

Chan, W. –L.; Averback, R. S. & Ashkenazy, Y. (2010). Anisotropic atomic motion at undercooled crystal/melt interfaces. *Phy. Rev. B*, 82, 2, 020201(R), ISSN 1098-0121

Chan, W. L.; Averback, R. S. & Cahill, D. G. (2009a). Nonlinear energy absorption of femtosecond laser pulses in noble metals. *Appl. Phys. A*, 97, 2, 287-294, ISSN: 0947-8396

Chan, W. -L.; Averback, R. S.; Cahill, D. G. & Ashkenazy, Y. (2009b). Solidification Velocities in Deeply Undercooled Silver. *Phys. Rev. Lett.*, 102, 9, 095701, ISSN 0031-9007

Chan, W.-L.; Averback, R. S.; Cahill, D. G. & Lagoutchev A. (2008). Dynamics of femtosecond laser-induced melting of silver. *Phys. Rev. B*, 78, 21, 214107, ISSN 1098-0121

Coles, B. R. & Taylor, J. C. (1962). The Electrical Resistivities of the Palladium-Silver Alloys. *Proc. Roy. Soc. A*, 267, 139-145, ISSN 1364-5021

Coriell, S. R. & Turnbull, D. (1982). Relative roles of heat transport and interface rearrangement rates in the rapid growth of crystals in undercooled melts. *Acta. Metall.*, 30, 12, 2135-2139, (1982), ISSN: 0001-6160

Delogu, F. (2006). Homogeneous melting of metals with different crystalline structure. *J. Phys.: Condens. Matter*, 18, 24, 5639 5654, ISSN: 0953 8984

Eesley, G. L. (1983). Observation of Nonequilibrium Electron Heating in Copper. *Phys. Rev. Lett.*, 51, 23, 2140-2143, ISSN 0031-9007

Ernstorfer, R.; Harb, M.; Dartigalongue, T.; Hebeisen, C. T.; Jordan, R. E.; Zhu, L. & Miller, R. J. D. (2007). Femtosecond Electron Diffraction Study on the Melting Dynamics of Gold, In: *Springer Series in Chemical Physics Vol. 88: Ultrafast Phenomena XV*, Corkum, P.; Dwayne Miller, R. J.; Jones, D. M. & Weiner, A. M. (Ed.), 755-757, Springer, ISBN: 978-3-540-68779-5, Berlin

Fuster, G.; Tyler, J. M.; Brener, N. E.; Callaway, J. & Bagayoko, D. (1990). Electronic structure and related properties of silver. *Phys. Rev. B*, 42, 12, 7322-7329, ISSN 1098-0121

Frenkel, J (1946) *Kinetic Theory of Liquids*, Oxford University Press, Oxford

Gattass, R. R. & Mazur E. (2008). Femtosecond laser micromachining in transparent materials. *Nature photonics*, 2, 4, 219-225, ISSN 1749-4885

Gundrum, B. C.; Cahill, D. G. & Averback, R. S. (2005). Thermal conductance of metal-metal interfaces. *Phys. Rev. B*, 72, 24, 245426, ISSN 1098-0121

Ivanov, D. S. & Zhigilei, L. V. (2007). Kinetic Limit of Heterogeneous Melting in Metals. *Phys. Rev. Lett.*, 98, 19, 195701, ISSN 0031-9007

Kittle, C. (2005). *Introduction to Solid States Physics – 8th ed.*, John Wiley & Son, ISBN: 0-471-41526-X, New York

Lin, Z.; Johnson, R. A. & Zhigilei, L. V. (2008a). Computational study of the generation of crystal defects in a bcc metal target irradiated by short laser pulses. *Phys. Rev. B*, 77, 21, 214108, ISSN 1098-0121

Lin, Z.; Zhigilei, L. V. & Celli, V. (2008b). Electron-phonon coupling and electron heat capacity of metals under conditions of strong electron-phonon nonequilibrium. *Phys. Rev. B*, 77, 7, 075133, ISSN 1098-0121

Liu, X.; Du, D. & Mourou, G. (1997). Laser ablation and micromachining with ultrashort laser pulses. *IEEE J. Quantum Electronics*, 33, 10, 1706-1716, ISSN 0018-9197

MacDonald, A. H. & Geldart, D. J. W. (1980). Electron-electron scattering and the thermal resistivity of simple metals. *J. Phys. F: Metal Phys.*, 10, 4, 677-692, ISSN: 0305-4608

MacDonald, C. A.; Malvezzi, A. M. & Spaepen, F. (1988). Picosecond time - resolved measurements of crystallization in noble metals. *J. Appl. Phys.*, 65, 1, 129-136, ISSN 0021-8979

Mahan, G. D. & Claro, F. (1988). Nonlocal theory of thermal conductivity. *Phys. Rev. B*, 38, 3, 1963-1969, ISSN 1098-0121

Milchberg, H. M.; Freeman, R. R.; Davey, S. C. & More, R. M. (1988). Resistivity of a Simple Metal from Room Temperature to 10^6 K. *Phys. Rev. Lett.*, 61, 20, 2364-2367, ISSN 0031-9007

Miller, J. C. (1969). Optical properties of liquid metals at high temperatures. *Phil. Mag.*, 20, 168, 1115-1132, ISSN: 1478-6435

Miyaji, G. & Miyazaki K. (2008). Origin of periodicity in nanostructuring on thin film surfaces ablated with femtosecond laser pulses. *Optics express*, 16, 20, 16265-16271, ISSN 1094-4087

Mott, N. F. (1936). The Electrical Conductivity of Transition Metals. *Proc. Roy. Soc. A*, 153, 699-717, ISSN 1364-5021

Norris, P. M.; Caffrey, A. P.; Stevens, R. J.; Klopf, J. M.; McLeskey Jr., J. T. & Smith, A. N. (2003). Femtosecond pump–probe nondestructive examination of materials (invited). *Rev. Sci. Instrum.*, 74, 400-406, ISSN 0034-6748

Perry, M. D.; Stuart, B. C.; Banks, P. S.; Feit, M. D.; Yanovsky, V. & Rubenchik, A. M. (1999). Ultrashort-pulse laser machining of dielectric materials. *J. Appl. Phys.*, 85, 9, 6803-6810, ISSN 0021-8979

Pines, D. (1999). *Elementary excitations in solids*, Perseus Books, ISBN: 0738201154, Massachusetts

Smith, P. M. & Aziz, M. J. (1994). Solute trapping in aluminum alloys. *Acta. Metall. Mater.*, 42, 10, 3515-3525, ISSN: 0956-7151

Suarez, C.; Bron, W. E. & Juhasz, T. (1995). Dynamics and Transport of Electronic Carriers in Thin Gold Films. *Phys. Rev. Lett.*, 75, 24, 4536-4539, ISSN 0031-9007

Tas, G. & Maris, H. J. (1994). Electron diffusion in metals studied by picosecond ultrasonics. *Phys. Rev. B*, 49, 21, 15046-15054, ISSN 1098-0121

Tsao, J. Y.; Aziz, M. J.; Thompson, M. O. & Peercy, P. S. (1986). Asymmetric Melting and Freezing Kinetics in Silicon. *Phys. Rev. Lett.*, 56, 25, 2712-2715, ISSN 0031-9007

Turnbull, D. & Cech, R. E. (1950). Microscopic Observation of the Solidification of Small Metal Droplets. *J. Appl. Phys.*, 21, 8, 804-810, ISSN 0021-8979

Vorobyev, A. Y. & Guo, C. (2004). Direct observation of enhanced residual thermal energy coupling to solids in femtosecond laser ablation. *Appl. Phys. Lett.*, 86, 1, 011916, ISSN 0003-6951

Vorobyev, A. Y. & Guo, C. (2008). Colorizing metals with femtosecond laser pulses. *Appl. Phys. Lett.*, 92, 4, 041914, ISSN 0003-6951

24

Probing Ultrafast Dynamics of Polarization Clusters in $BaTiO_3$ by Pulsed Soft X-Ray Laser Speckle Technique

Kai Ji[1] and Keiichiro Nasu[2]
[1]*Institute for Theoretical Physics and Heidelberg Center for Quantum Dynamics*
University of Heidelberg
Philosophenweg 19, 69120 Heidelberg
Germany
[2]*Photon Factory, Institute of Materials Structure Science*
High Energy Accelerator Research Organization
Graduate University for Advanced Studies
Oho 1-1, Tsukuba, Ibaraki 305-0801
Japan

1. Introduction

The technical development of ultrashort laser pulses covering infrared to extreme ultraviolet has opened a door to study the photo-induced dynamic physical and chemical processes. By using an optical excitation at a particular wavelength, exotic states can be trigged coherently in atoms, molecules, clusters as well as complex condensed systems. The evolution of excited states then can be imaged in the real-time domain by subsequent single or trains of pulse. This laser pump-probe technique features an unprecedented spatial and temporal resolution, thus not only allows us a fundamental insight into the microscopic ultrafast dynamics, but also brings about a potential of selective controlling on the microstructures at atomic scale (Krausz & Ivanov, 2009).

In this chapter, a recent advance of soft x-ray laser speckle pump-probe measurement on barium titanate ($BaTiO_3$) is reviewed, with primary concerns on the theoretical description of the photo-matter interaction and photo-induced relaxation dynamics in the crystal. Here, the observed time-dependent speckle pattern is theoretically investigated as a correlated optical response to the pump and probe pulses. The scattering probability is calculated based on a model with coupled soft x-ray photon and ferroelectric phonon mode of $BaTiO_3$. It is found that the speckle variation is related with the relaxation dynamics of ferroelectric clusters created by the pump pulse. Additionally, a critical slowing down of cluster relaxation arises on decreasing temperature towards the paraelectric-ferroelectric transition temperature. Relation between the critical slowing down, local dipole fluctuation and crystal structure are revealed by a quantum Monte Carlo simulation.

This chapter is organized as follows. In Section 1, the properties ferroelectric material and experimental techniques are introduced. The theoretical model and methods are elaborated in Section 2. In Section 3, the numerical results on speckle correlation, relaxation dynamics of polarization cluster and critical slowing down are discussed in details. In Section 4, a summary with conclusion is presented finally.

1.1 General properties of $BaTiO_3$

As a prototype of the ferroelectric perovskite compounds, $BaTiO_3$ undergoes a transition from paraelectric cubic to ferroelectric tetragonal phase at Curie temperature T_c=395 K. As schematically shown in Fig. 1, above T_c, the geometric centers of the Ti^{4+}, Ba^{2+} and O^{2-} ions coincide, giving rise to a non-polar lattice. Below T_c, the unit cell is elongated along the c axis with a ratio c/a~1.01. The Ti^{4+} and Ba^{2+} ions are displaced from their geometric centers with respect to the O^{2-} ions, to give a net polarity to the lattice. The formation of spontaneous polarization by the displacement of ions is along one of the [001] directions in the original cubic structure. Thus, below T_c, there are two kinds of ferroelectric domain developed with mutually perpendicular polarization (Yin et al., 1006). The paraelectric-ferroelectric phase transition and domain induced static surface corrugation have been well resolved by the means of atomic force microscopy (Hamazaki et al., 1995), scanning probe microscopy (Pang et al., 1998), neutron scattering (Yamada et al., 1969), and polarizing optical microscopy (Mulvmill et al., 1996).

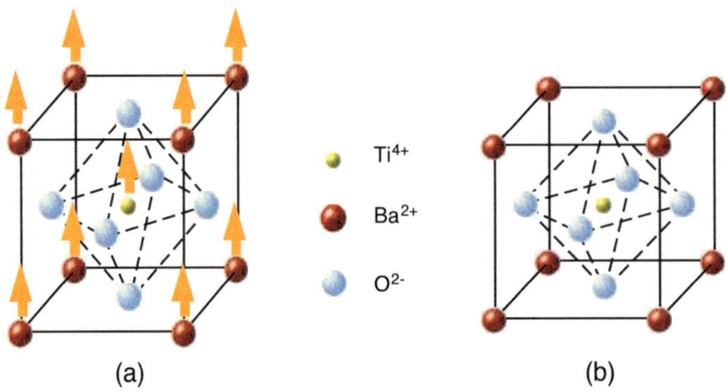

Fig. 1. The crystal structure of $BaTiO_3$ (a) below the Curie temperature the structure is tetragonal with Ba^{2+} and Ti^{4+} ions displaced relative to O^{2-} ions; (b) above the Curie temperature the cell is cubic.

Ferroelectric materials have a variety of functional capabilities in the electronic devices, which include non-volatile memory, high permittivity capacitor, actuator and insulator for field-effect transistor (Polla et al., 1998). Application of ferroelectric materials has attracted a great deal of attention in recent years to enhance the performance of implementation. This requires a comprehensive knowledge concerning the behaviours of ferroelectrics at the nanoscale, such as the roles of strain, deploarization fields, domain configurations and motions. Besides, there is also an increased demand to understand the mechanism of paraelectric-ferroelectric phase transition. In the case of $BaTiO_3$, it is generally considered that the transition is a classic displacive soft-mode type, which is driven by the anharmonic lattice dynamics (Harada et al., 1971; Migoni et al., 1976). However, recent studies have also suggested that there might be an order-disorder instability which coexists with the displacive transition (Zalar et al., 2003; Völkel & Müller, 2007), making this issue still controversial to date.

The nature of the phase transition is believed to be unveiled in the precursor phenomena. For $BaTiO_3$, this is the emergence of dipole fluctuations with regional uniform alignments, i.e. the polarization clusters (Takagi & Ishidate, 2000). While, it is still not clear how these polarization fluctuations condense into long range ferroelectric correlations as temperature decreases. Therefore, it is of great significance to directly observe the creation and evolution of polarization cluster in the vicinity of T_c.

1.2 Experimental methodology

Since the aforementioned conventional time-average-based measurements cannot be applied to detect the ultrafast transient status of dipole clusters, the diffraction speckle pattern of $BaTiO_3$ crystal captured by the picosecond soft x-ray laser has turned out to be an efficient way.

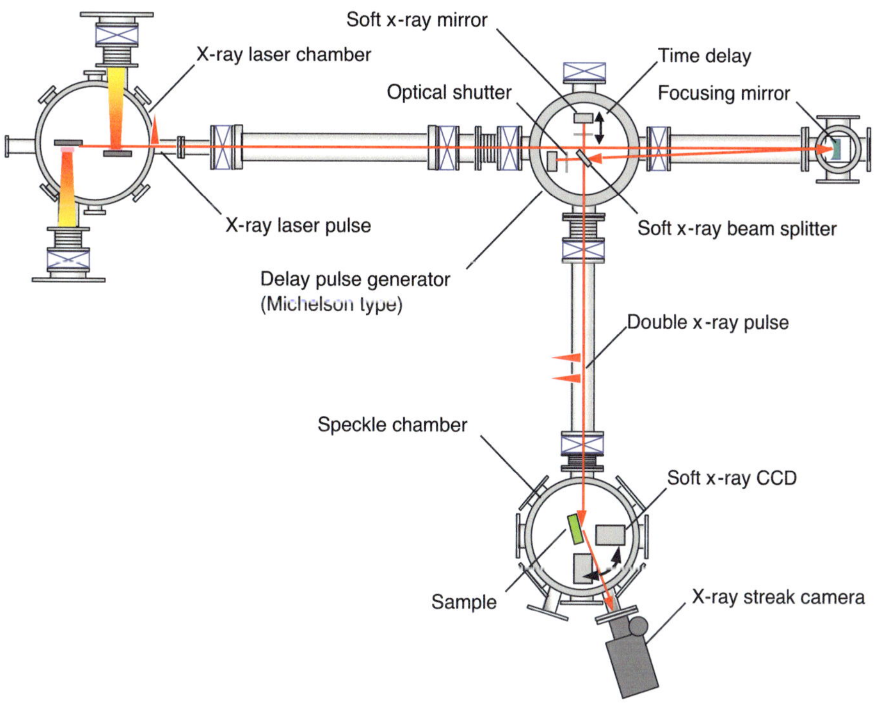

Fig. 2. Schematic diagram of soft x-ray speckle pump-probe spectroscopy system using a Michelson type delay pulse generator and a soft x-ray streak camera. All the optical components are set in a vacuum (~10^{-4} Pa). The delay time of the second x-ray pulse from the first one can be manipulated by changing the delay path length between the beam splitter and x-ray mirror.

Speckle is the random granular pattern produced when a coherent light is scattered off a rough surface. It carries information of the specimen surface, for the intensity and contrast of the speckle image vary with the roughness of surface being illuminated (May, 1977). Numerous approaches have been devised to identify surface profiles by either the speckle contrast or the speckle correlation method (Goodman, 2007). Recent application of pulsed soft x-ray laser has improved the temporal and spatial resolution to a scale of picosecond and nanometer. By this means, dynamics of surface polarization clusters of $BaTiO_3$ across the Curie temperature has been observed (Tai et al., 2002; Tai et al., 2004), which paves a new way to study the paraelectric-ferroelectric phase transition.

Very recently, Namikawa (Namikawa et al., 2009) study the polarization clusters in $BaTiO_3$ at above T_c by the plasma-based x-ray laser speckle measurement in combination with the technique of pump probe spectroscopy. In this experiment, as shown in Fig. 2, two consecutive soft x-ray laser pulses with wavelength of 160 Å and an adjustable time difference are generated coherently by a Michelson type beam splitter (Kishimoto et al., 2010). After the photo excitation by the pump pulse, ferroelectric clusters of nano scale are created in the paraelectric $BaTiO_3$ and tends to be smeared out gradually on the way back to the equilibrium paraelectric state. This relaxation of cluster thus can be reflected in the variation of speckle intensity of the probe pulse as a function of its delay time from the first pulse. It has been observed that the intensity of speckle pattern decays as the delay time increases. Moreover, the decay rate also decreases upon approaching T_c, indicating a critical slowing down of the cluster relaxation time. Hence, by measuring the correlation between two soft x-ray laser pulses, the real time relaxation dynamics of polarization clusters in $BaTiO_3$ is clearly represented. In comparison with other time-resolved spectroscopic study on $BaTiO_3$, for example the photon correlation spectroscopy with visible laser beam (Yan et al., 2008), Namikawa's experiment employs pulsed soft x-ray laser as the light source. For this sake, the size of photo-created ferroelectric cluster is reduced to a few nanometers, and the cluster relaxation time is at a scale of picosecond. This measurement, thus, uncovers new critical properties of the ultrafast relaxation dynamics of polarization clusters.

2. Theoretical model and methods

In this and next sections, we examine the newly reported novel behaviours of ferroelectric cluster observed by Namikawa from a theoretical point of view, aiming to provide a basis for understanding the critical nature of $BaTiO_3$. Theoretically, the dynamics of a system can be adequately described by the linear response theory, i.e., to express the dynamic quantities in terms of time correlation functions of the corresponding dynamic operators. In general, the path integral quantum Monte Carlo method is computationally feasible to handle the quantum many body problems, for it allows the system to be treated without making any approximation. However, simulation on real time dynamics with Monte Carlo method is still an open problem in computational physics because of the formidable numerical cost of path summation which grows exponentially with the propagation time. The common approach to circumvent this problem is to perform imaginary time path integration followed by analytic continuation, and to compute the real time dynamic quantities using Fourier transformation. In the present study, the real time correlation functions and real time dependence of speckle pattern are investigated following this scheme (Ji et al., 2009).

2.1 Model Hamiltonian

Theoretical interpretations for structural phase transition and domain wall dynamics have be well established in the framework of Krumhansl-Schrieffer model (also known as φ^4 model) (Krumhansl & Schrieffer, 1975; Aubry, 1975; Schneider & Stoll, 1978; Savkin et al., 2002). In this model, the particles are subject to anharmonic on-site potentials and harmonic inter-site couplings. The on-site potential is represented as a polynomial form of the order parameter such as polarization, displacement, or elasticity, which displays a substantial change around T_c. Since the φ^4 model is only limited to second-order transitions, in the present work we invoke a modified Krumhansl-Schrieffer model (also called φ^6 model) (Morris & Gooding, 1990; Khare & Saxena, 2008) to study the first-order ferroelectric phase transition of $BaTiO_3$. In this scenario, the Hamiltonian of $BaTiO_3$ crystal ($\equiv H_f$) is written as (here we let $\hbar$=1),

$$H_f = -\frac{\omega_0}{2}\sum_l \frac{\partial^2}{\partial^2 Q_l^2} + U_0 + U_i \,, \tag{1}$$

$$U_0 = \frac{\omega_0}{2}\sum_l \left(Q_l^2 - c_4 Q_l^4 + \frac{c_6}{3} Q_l^6 \right), \tag{2}$$

$$U_i = -\frac{\omega_0 d_2}{2}\sum_{<l,l'>} Q_l Q_{l'} \,, \tag{3}$$

where, U_0 and U_i are the on-site potential and inter-site correlation, respectively. Q_l is the coordinate operator for the electric dipole moment due to a shift of titanium ions against oxygen ions, i.e., the T_{1u} transverse optical phonon mode. ω_0 is the dipole oscillatory frequency, l labels the site, and $<l,l'>$ in Eq. (3) enumerates the nearest neighboring pairs.
In order to describe the optical response of $BaTiO_3$ due to x ray scattering, we design a theoretical model to incorporate the radiation field and a weak interplay between radiation and crystal. The total Hamiltonian reads,

$$H = H_p + H_f + H_{pf} \,, \tag{4}$$

$$H_p = \sum_k \Omega_k a_k^+ a_k \,, \; \Omega_k = c|k| \,. \tag{5}$$

Eq. (5) represents the polarized light field, where a_k^+ (a_k) is the creation (annihilation) operator of a photon with a wave number k and an energy Ω_k, and c is the light velocity in vacuum. In Namikawa's experiment, the wave length of x-ray is 160 Å, thus the photon energy is about 80 eV. Denoting the odd parity of T_{1u} mode, the photon-phonon scattering is of a bi-linear Raman type,

$$H_{pf} = \frac{V}{N}\sum_{q,q',k} a^+_{k+\frac{q}{2}} a_{k-\frac{q}{2}} Q_{q'-\frac{q}{2}} Q_{-q'-\frac{q}{2}} \,, \tag{6}$$

where V is the photon-phonon coupling strength, Q_q ($\equiv N^{-1/2}\Sigma_l e^{-iql} Q_l$) the Fourier component of Q_l with a wave number q. Without losing generalitivity, here we use a simple cubic lattice, and the total number of lattice site is N.

2.2 Optical response to pump and probe photons

Since there are two photons involved in the scattering, the photon-phonon scattering probability can be written as,

$$P(t) = \sum_{k_1,k_1'} \left\langle\left\langle a_{k_0}(0) a_{k_1}^+(\Delta) a_{k_0}(t) a_{k_1'}^+(\Delta+t) a_{k_1'}(\Delta+t) a_{k_0}^+(t) a_{k_1}(\Delta) a_{k_0}^+(0) \right\rangle\right\rangle, \tag{7}$$

where

$$\langle\langle\cdots\rangle\rangle = \mathrm{Tr}\left(e^{-\beta H}\cdots\right) / \mathrm{Tr}\left(e^{-\beta H}\right) \tag{8}$$

means the expectation, $\beta(\equiv 1/k_B T)$ is the inverse temperature, and the time dependent operators are defined in the Heisenberg representation,

$$O(t) = e^{itH} O e^{-itH}, \tag{9}$$

Here, t denotes the time difference between two incident laser pulses as manifested in Fig. 3, and k_0 the wave number of incoming photon. After a small time interval Δ, the photon is scattered out of the crystal. k_1 and k'_0 are the wave numbers of the first and second outgoing photons, respectively.

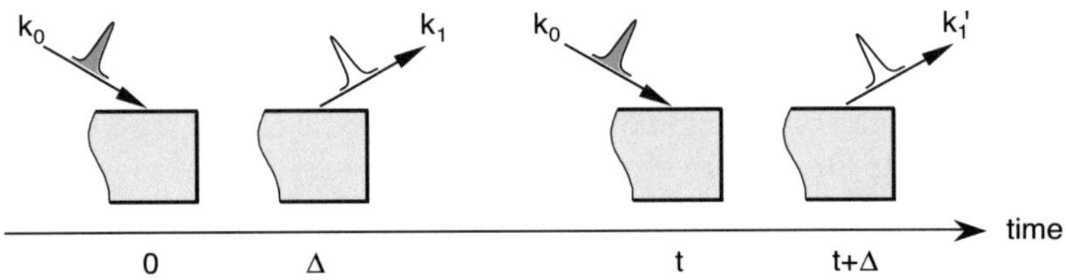

Fig. 3. Pulse sequence in an x-ray laser speckle experiment. The pump and probe pulses of k_0 creates and detects ferroelectric clusters in the sample of paraelectric $BaTiO_3$, respectively, and generate new x-ray fields in the direction k_1 and k'_0 after a short time interval Δ.

Treating H_{pf} as a perturbation, we separate Hamiltonian of Eq. (4) as,

$$H = H_0 + H_{pf}, \tag{10}$$

where

$$H_0 = H_p + H_f, \tag{11}$$

is treated as the unperturbed Hamiltonian. By expanding the time evolution operator in Eq. (9) with respect to H_{pf},

$$e^{-itH} \to e^{-itH_0}\left[1 - i\int_0^t dt_1 \hat{H}_{pf}(t_1) + \cdots\right], \tag{12}$$

we find that the lowest order terms which directly depend on t are of fourth order,

$$P(t) \to \int_0^\Delta dt_1 \int_0^\Delta dt_2 \int_0^\Delta dt'_1 \int_0^\Delta dt'_2 \sum_{k_1,k'_1} \Big\langle\Big\langle a_{k_0} \hat{H}_{pf}(t'_1) e^{i\Delta H_0} a^+_{k_1} e^{i(t-\Delta)H_f} a_{k_0} \hat{H}_{pf}(t'_2) e^{i\Delta H_0} a^+_{k'_1}$$

$$\times a_{k'_1} e^{-i\Delta H_0} \hat{H}_{pf}(t_2) a^+_{k_0} e^{-i(t-\Delta)H_f} a_{k_1} e^{-i\Delta H_0} \hat{H}_{pf}(t_1) a^+_{k_0} \Big\rangle\Big\rangle, \tag{13}$$

where the operators with carets are defined in the interaction representation,

$$\hat{O}(t) = e^{itH_0} O e^{-itH_0}. \tag{14}$$

Fig. 4 represents a diagram analysis for this phonon-coupled scattering process, where photons (phonons) are depicted by the wavy (dashed) lines, and the upper (lower) horizontal time lines are corresponding to the bra (ket) vectors (Nasu, 1994). Diagram (a) illustrates the changes of wave number and energy of photons due to the emitted or absorbed phonons. This is noting but the Stokes and anti-Stokes Raman scattering. Whereas, diagrams (b)-(f) are corresponding to the exchange, side band, rapid damping, rapid exchange and mutual damping effects, respectively.

Obviously, diagram (c) brings no time dependence, while diagrams (d), (e) and (f) only contributes a rapid reduction to the time correlation of two laser pulses because of the duality in phonon interchange. In this sense, the time dependence is primarily determined by the diagrams (a) and (b). Thus, the scattering probability turns out to be,

$$P(t) = \int_0^\Delta dt_1 \int_0^\Delta dt_2 \int_0^\Delta dt_1' \int_0^\Delta dt_2' \frac{2V^4}{N^4} \sum_{q,q'} \Big\langle\Big\langle a_{k_0} e^{it_1' H_p} a_{k_0}^+ a_{k_0-q} e^{-i(t_1'-\Delta)H_p} a_{k_0-q}^+$$

$$\times a_{k_0} e^{it_2' H_p} a_{k_0}^+ a_{k_0+q} e^{-i(t_2'-\Delta)H_p} a_{k_0+q}^+ a_{k_0+q} e^{i(t_2-\Delta)H_p} a_{k_0+q}^+ a_{k_0} e^{-it_2 H_p} a_{k_0}^+$$

$$\times a_{k_0-q} e^{i(t_1-\Delta)H_p} a_{k_0-q}^+ a_{k_0} e^{-it_1 H_p} a_{k_0}^+ \Big\rangle\Big\rangle \Big\langle\Big\langle \hat{Q}_{q'}(t_1') \hat{Q}_{q-q'}(t_1') \hat{Q}_{-q+q'}(t+t_2')$$

$$\times \hat{Q}_{-q'}(t+t_2') \hat{Q}_{q'}(t+t_2') \hat{Q}_{q-q'}(t+t_2) \hat{Q}_{-q+q'}(t_1) \hat{Q}_{-q'}(t_1) \Big\rangle\Big\rangle, \qquad (15)$$

where the photons and phonons are decoupled, and it becomes evident that the origin of the *t*-dependence is nothing but the phonon (dipole) correlation.

Since the photonic part in Eq. (15) is actually time-independent, and in the case of forward x-ray scattering we have $|k_0| \approx |k_1| \approx |k_1'|$, the normalized probability can be simplified as,

$$\frac{P(t)}{P(0)} = \frac{\sum_{q,q'} \left| \left\langle\left\langle Q_q^2 \right\rangle\right\rangle G_{q+q'}(t) \right|^2}{\sum_{q,q'} \left| \left\langle\left\langle Q_q^2 Q_{q+q'}^2 \right\rangle\right\rangle \right|^2}, \qquad (16)$$

where

$$G_q(t) = -i2\left\langle\left\langle T\hat{Q}_q(t)\hat{Q}_{-q}(0)\right\rangle\right\rangle, \qquad (17)$$

is the real time Green's function of phonon, and *T* the time ordering operator. In deriving Eq. (16), we have also made use of the fact that the light propagation time in the crystal is rather short. The Fourier component of Green's function,

$$G_q(\omega) = \int_{-\infty}^{\infty} dt\ G_q(t)\ e^{-i\omega t}, \qquad (18)$$

is related to the phonon spectral function [$\equiv A_q(\omega)$] through (Doniach & Sondheimer, 1998),

$$G_q(\omega) = \int_{-\infty}^{\infty} \frac{d\omega'}{2\pi} \frac{A_q(\omega')}{1-\exp(-\beta\omega')} \left[\frac{1}{\omega-\omega'+i0^+} - \frac{\exp(-\beta\omega')}{\omega-\omega'-i0^+} \right], \qquad (19)$$

The phonon spectral function describes the response of lattice to the external perturbation, yielding profound information about dynamic properties of the crystal under investigation.

Once we get the spectral function, the scattering probability and correlation function can be readily derived.

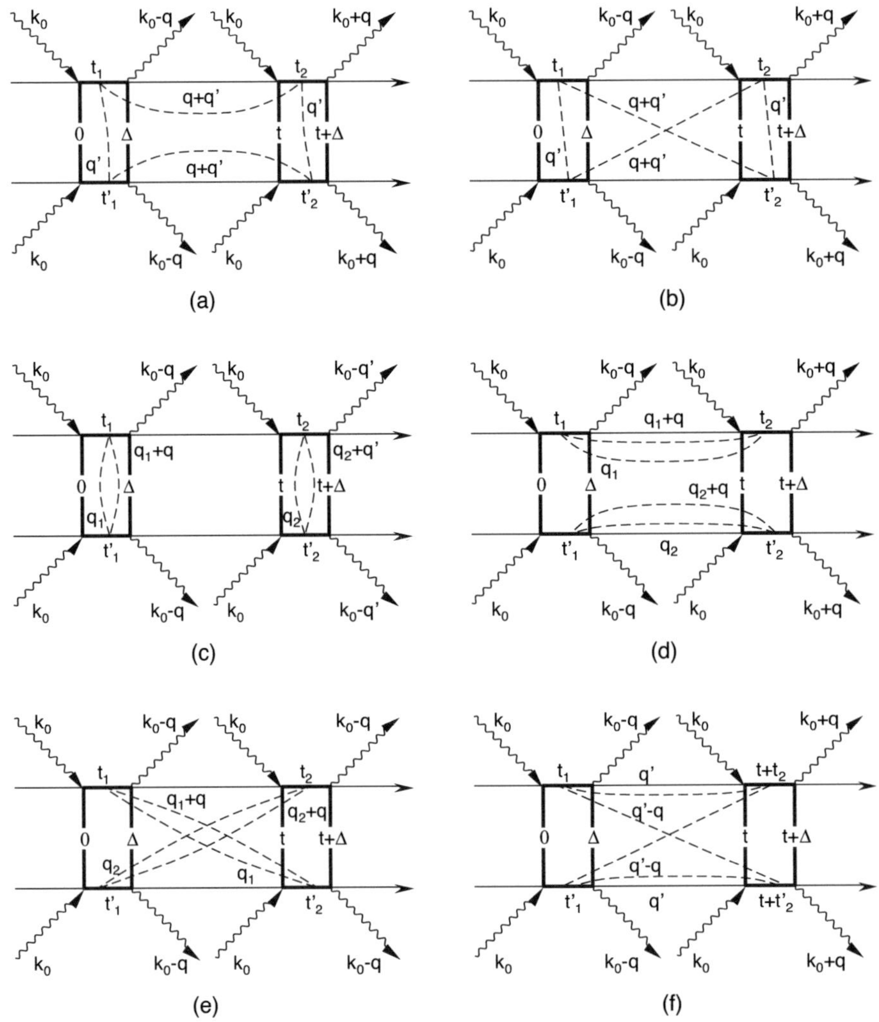

Fig. 4. Double-sided Feynman diagrams for scattering process of photon with electric dipole moment (phonon). The photons and phonons are denoted by the wavy and dashed lines, respectively. In each diagram, the upper and lower horizontal time lines represent the bra and ket vectors, respectively.

2.3 Dynamics of crystal

A mathematically tractable approach to spectral function $A_q(\omega)$ is to introduce an imaginary time phonon Green's function, for it can be evaluated more easily than its real time counterpart. In the real space, the imaginary time Green's function is defined as,

$$G_{ll'}(\tau) \equiv -2\left\langle\left\langle T\hat{Q}_l(\tau)\hat{Q}_{l'}(0)\right\rangle\right\rangle, \tag{20}$$

where $\tau(\equiv it)$ is the argument for imaginary time (unless otherwise noted, we use Roman t for real time and Greek τ for imaginary time). The imaginary time dependence of an operator in the interaction representation is given by

$$\hat{O}(\tau) = e^{i\tau H_0} O e^{-i\tau H_0} . \tag{21}$$

Under the weak coupling approximation, and by using the Suzuki-Trotter identity, the Green's function can be rewritten into a path integral form (here we assume τ>0) (Ji et al., 2004),

$$G_{ll''}(\tau) = \int Dx\ e^{-\beta[\Phi_f(x)-\Phi_f]}\left[-2x_l(\tau)x_{l''}(0)\right]. \tag{22}$$

where x_l is the eigenvalue of Q_l,

$$Q_l\left|x_l\right\rangle = x_l\left|x_l\right\rangle , \tag{23}$$

$\Phi_f(x)$ is the path-dependent phonon free energy,

$$e^{-\beta\,\Phi_f(x)} = e^{-\int_0^\beta d\tau\ \Omega_f[x(\tau)]} , \tag{24}$$

with

$$\Omega_f = -\sum_l\left[\frac{1}{2\omega_0}\left(\frac{\partial x_l}{\partial \tau}\right)^2 + \frac{1}{2}\omega_0 x_l^2 - \frac{1}{2}\omega_0 c_4 x_l^4 + \frac{1}{6}\omega_0 c_6 x_l^6\right] - \frac{1}{2}\omega_0 d_2 \sum_{<l,l'>} x_l x_l' , \tag{25}$$

and Ω_f is the total phonon free energy,

$$e^{-\beta\,\Phi_f} = \int Dx\ e^{-\beta\,\Phi_f(x)} . \tag{26}$$

In the path integral notation, the internal energy of crystal E_f ($\equiv \left\langle\left\langle H_f\right\rangle\right\rangle$ is represented as

$$\Omega_f = \int Dx\ e^{-\beta\left[\Phi_f(x)-\Phi_f\right]}\left[\omega_0\sum_l\left(x_l^2 - \frac{3}{2}c_4 x_l^4 + \frac{2}{3}c_6 x_l^6\right) - \omega_0 d_2 \sum_{<l,l'>} x_l x_l'\right] , \tag{27}$$

from which the heat capacity can be derived as

$$C_f^V = \left(\frac{\partial E_f}{\partial T}\right)_V . \tag{28}$$

The Green's function of momentum space is given by,

$$G_q(\tau) = \frac{1}{N}\sum_{l,l''} e^{iq(l-l'')}\ G_{ll''}(\tau) , \tag{29}$$

which is connected with the phonon spectral function $A_q(\omega)$ through (Bonča & Gubernatis, 1996)

$$G_q(\tau) = -\int_0^\infty \frac{\cosh\left[\left(\frac{1}{2}\beta - \tau\right)\omega\right]}{\sinh\left(\frac{1}{2}\beta\omega\right)} A_q(\omega) d\omega \,. \tag{30}$$

Solving this integral equation is a notoriously ill-posed numerical problem because of the highly singular nature of the kernel. In order to analytically continue the imaginary time data into real frequency information, specialized methods are developed, such as maximum entropy method (Skilling, 1984) and least squares fitting method (Yamazaki et al., 2003). Ji (Ji et al., 2004) develops an iterative fitting approach to derive the electron spectral function, which gives a rapid and stable convergence of the spectrum without using any prior knowledge or artificial parameter. In the present study, however, the phonon spectral function does not yield a specific sum rule like the case of electron, the iterative fitting method cannot be applied directly. For this sake, we have modified this method by a renormalized iteration algorithm (Ji et al., 2009). Details of the standard and renormalized iterative fitting methods can be found in Appendices 6.2 and 6.3, respectively.

3. Numerical results and discussion

3.1 Optical responses

Based on the path integral formalisms, the imaginary time Green's function can be readily calculated via a standard quantum Monte Carlo simulation (Ji et al., 2004). Our numerical calculation is performed on a 10×10×10 cubic lattice with a periodic boundary condition. The imaginary time is discretized into 10-20 infinitesimal slices. As already noticed for the analytic continuation (Gubernatis et al., 1991), if the imaginary time Green's function is noisy, the uncertainty involved in the inverse transform might be very large, and the spectral function cannot be determined uniquely. In order to obtain accurate data from quantum Monte Carlo simulation, a hybrid algorithm (Ji et al., 2004) has been implemented in our calculation. This method is elaborated in the Appendix 6.1. Besides, we pick out each Monte Carlo sample after 100-200 steps to reduce the correlation between adjacent configurations. The Monte Carlo data are divided into 5-10 sets, from which the 95% confidence interval is estimated through 10,000 resampled set averages by the percentile bootstrap method. We found that about 1,000,000 Monte Carlo configurations are sufficient to get well converged spectral functions and real time dynamic quantities.

In the numerical calculation, the phonon frequency ω_0 is assumed to be 20 meV (Zhong et al., 1994), the inter-site coupling constant d_2 is fixed at a value of 0.032, whereas c_4 and c_6 are selected to make the on-site U_0 a symmetric triple-well potential. As shown in Fig. 5, this triple-well structure is featured by five potential extrema located at x_a, $\pm x_b$ and $\pm x_c$, where

$$x_a = 0 \,, \tag{31}$$

$$x_b = \sqrt{\frac{c_4 - \sqrt{c_4^2 - c_6}}{c_6}} \,, \tag{32}$$

$$x_c = \sqrt{\frac{c_4 + \sqrt{c_4^2 - c_6}}{c_6}} \,, \tag{33}$$

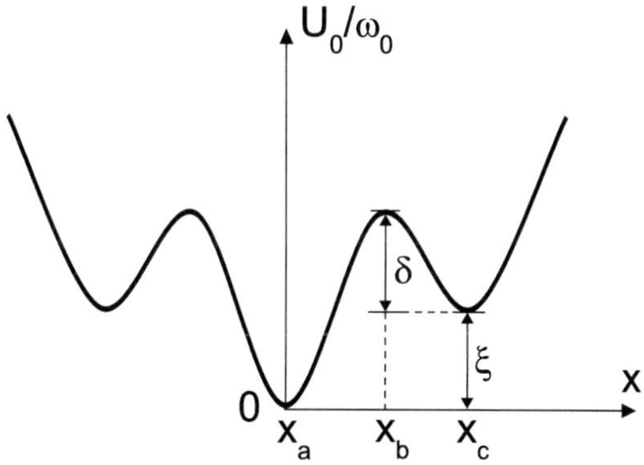

Fig. 5. On-site potential U_0 for the modified Krumhansl-Schrieffer model in the unit of ω_0. The coordinates of potential extrema are denoted by x_a, x_b, and x_c. δ and ξ are two parameters adopted to characterize this potential.

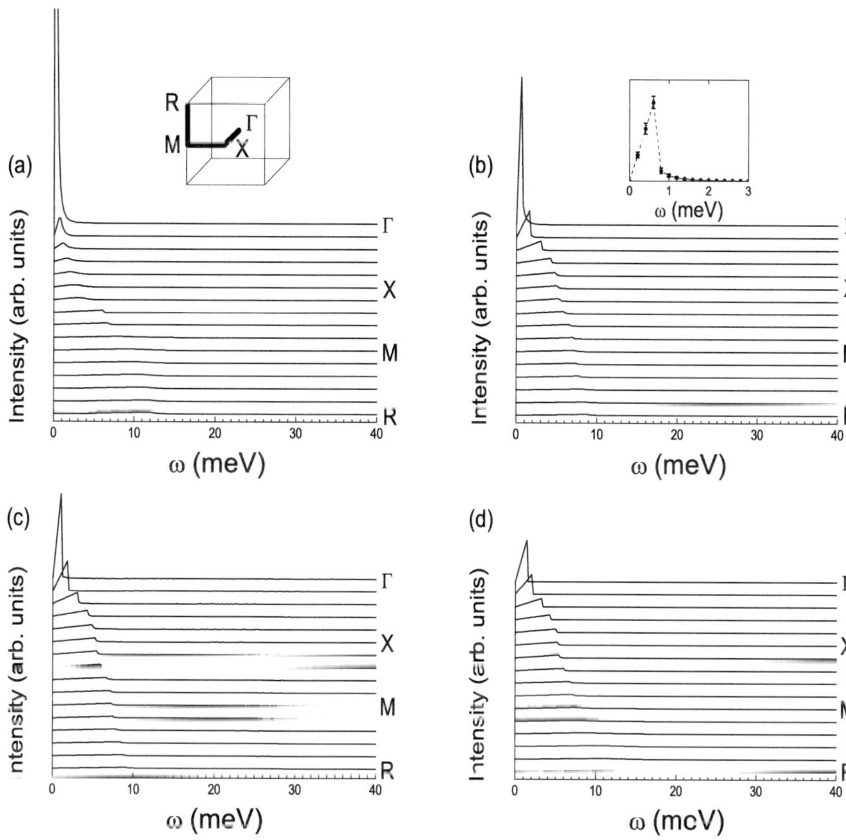

Fig. 6. Phonon spectral function along the line ΓXMR of Brillouin zone in the paraelectric phase at various temperatures: (a) $1.001T_c$, (b) $1.012T_c$, (c) $1.059T_c$, and (d) $1.176T_c$, where T_c=386 K. The inset of panel (a) shows the Brillouin zone with high symmetry lines. The inset of panel (b) represents the spectrum at Γ point when T=1.012 T_c.

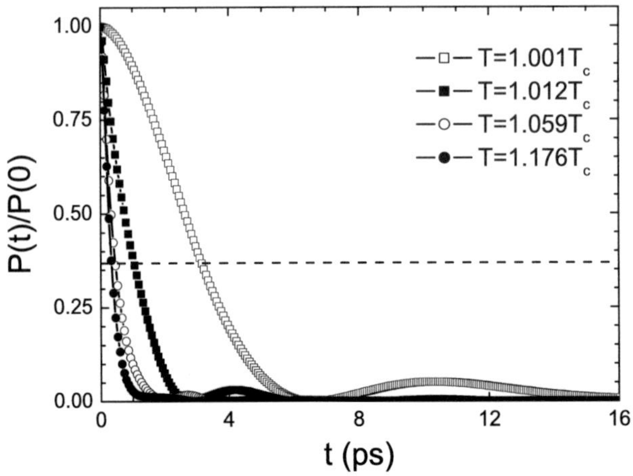

Fig. 7. Normalized speckle scattering probability as a function of time for paraelectric $BaTiO_3$, at various temperatures. Horizontal dashed line denotes $P(t)=P(0)/e$.

In Figs. 6 and 7, we show the optical responses of crystal, where c_4=2.0132x10^{-2} and c6=3.2595×10^{-4} are used. Fig. 6 presents the phonon spectral functions in the paraelectric phase at different temperatures: (a) T=1.001T_c, (b) T=1.012T_c, (c) T=1.059T_c and (d) T=1.176T_c, where T_c=386 K. In each panel, the spectra are arranged with wave vectors along the ΓXMR direction of Brillouin zone [see in the inset of panel (a)], and ω refers to energy. In the inset of panel (b), the spectrum at Γ point for T=1.012T_c is plotted. Since the spectra are symmetric with respect to the origin ω=0, here we only show the positive part of them. In Fig. 6, when the temperature decreases towards T_c, as already well-known for the displacive type phase transition, the energy of phonon peak is gradually softened. In addition, a so-called central peak, corresponding to the low energy excitation of ferroelectric cluster, appears at the Γ point. The collective excitation represented by this sharp resonant peak is nothing but the photo-created ferroelectric cluster. On decreasing temperature, spontaneous polarization is developed locally as a dipole fluctuation in the paraelectric phase. This fluctuation can stabilize the photo-created ferroelectric cluster, leading to a dramatically enhanced peak intensity near T_c.

The appearance of sharp peak at Γ point nearby T_c signifies a long life-time of the photo-created ferroelectric clusters after irradiation. Thus, near T_c, they are more likely to be probed by subsequent laser pulse, resulting in a high intensity of speckle pattern. Keeping this in mind, we move on to the results of scattering probability. In Fig. 7, we show the variation of normalized probability $P(t)/P(0)$ as a function of t (time interval between the pump and probe photons). Temperatures for these curves correspond to those in the panels (a)-(d) of Fig. 6, respectively. In this figure, $P(t)/P(0)$ declines exponentially, showing that the speckle correlation decreases with t increases as a result of the photorelaxation of ferroelectric cluster. When t is long enough, the crystal returns to the equilibrium paraelectric state. In addition, as shown in the figure, the relaxation rate bears a temperature dependence. On approaching T_c, the duration for return is prolonged, indicative of a critical slowing down of the relaxation. This is because with the decrease of temperature, the fluctuation of local polarization is enhanced, and a long range correlation between dipole moments is to be established as well, making the relaxation of photo-created clusters slower and slower.

3.2 Critical slowing down of photorelaxation

In order to quantitatively depict the critical slowing down, we introduce a relaxation time t_r to estimate the time scale of relaxation, which is the time for $P(t)$ to be reduced by a factor of e from $P(0)$. In Fig. 7, $P(t)=P(0)/e$ is plotted by a horizontal dashed line. Correspondingly, t_r is the abscissa of the intersection point of relaxation curve and this dashed line. In Fig. 8, the relaxation time for various local potential U_0 is presented at $T>T_c$. Here we adopt two legible parameters, δ and ξ, to describe the potential wells and barriers for U_0 (see Fig. 5), which are defined by

$$\delta = \frac{U_0(x_b) - U_0(x_c)}{\omega_0}, \tag{34}$$

$$\xi = \frac{U_0(x_c)}{\omega_0}. \tag{35}$$

Provided δ and ξ, c_4 and c_6 can be derived in terms of Eqs. (32)-(35). The values of c_4 and c_6 for the calculation of Fig. 8 are listed in Table 1, where we set ξ=3.061 and change δ from 4.239 to 4.639. The leftmost point on each curve denotes the t_r at just above T_c, which is a temperature determined from the singular point of C_f^V according to Eq. (28).

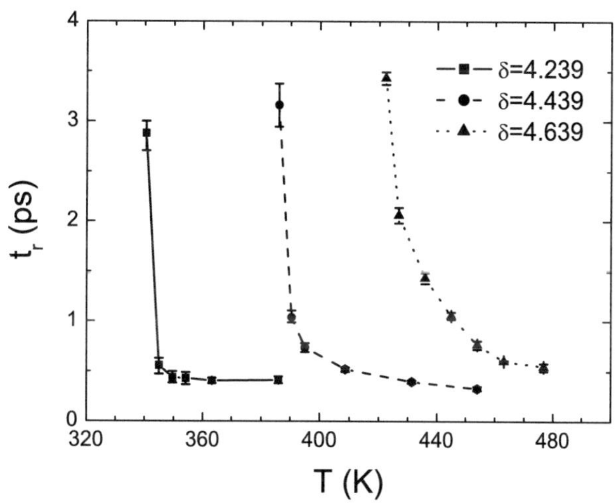

Fig. 8. Temperature dependence of relaxation time t_r for various δ when $T>T_c$, where ξ is fixed at 4.439.

As revealed by the NMR experiment (Zalar et al., 2003), the paraelectric-ferroelectric phase transition of $BaTiO_3$ has both displacive and order-disorder components in its mechanism. Short range dipole fluctuation arises in the paraelectric phase near T_c as a precursor of the order disorder transition, and condenses into long range ferroelectric ordering below T_c. Thus, in the present study, the relaxation of photo-created cluster is also subject to the dynamics of this dipole fluctuation and yields a temperature dependence. As illustrated by the three curves in Fig. 8, if a ferroelectric cluster is created at a temperature close to T_c, relaxation of this cluster is slow because of a rather strong dipole fluctuation, which holds the cluster in the metastable ferroelectric state from going back to the paraelectric one. Away from T_c, t_r decreases considerably for the dipole fluctuation is highly suppressed. This behaviour is nothing but the critical slowing down of photorelaxation.

c_4	c_6	δ	ξ	T_c (K)
2.0696×10^{-2}	3.4521×10^{-4}	4.239	3.061	340
2.0132×10^{-2}	3.2593×10^{-4}	4.439	3.061	386
1.9596×10^{-2}	3.0814×10^{-4}	4.639	3.061	422

Table 1. Parameters adopted for calculation of Fig. 8.

In Fig. 8, it can also be seen that with the increase of δ, T_c moves to the high temperature side so as to overcome a higher potential barrier between the ferroelectric and paraelectric phases. Furthermore, the evolution of t_r becomes gentle as well, implying a gradual weakening of dipole fluctuation at high temperature region.

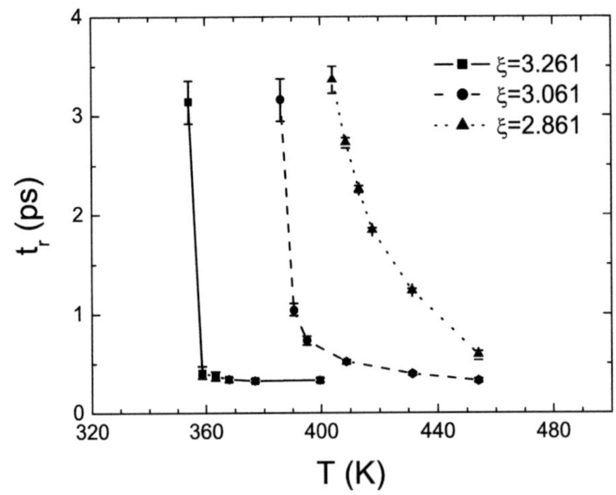

Fig. 9. Temperature dependence of relaxation time t_r for various ξ when $T>T_c$, where δ is fixed at 3.061.

In Fig. 9, we show the temperature dependence of t_r for different ξ when $T>T_c$, where δ is fixed at 4.439. The values of parameters for this calculation are given in Table 2. When ξ changes from 3.261 to 2.861, as shown in Fig. 9, T_c gradually increases. This is because with the decrease of ξ, the ferroelectric state at x_c (refer to Fig. 5) becomes more stable and can survive even larger thermal fluctuation. In a manner analogous to Fig. 6, the evolution of t_r also displays a sharp decline at low temperature, and becomes more and more smooth as temperature increases.

c_4	c_6	δ	ξ	T_c (K)
1.9626×10^{-2}	3.1070×10^{-4}	4.439	3.261	354
2.0132×10^{-2}	3.2593×10^{-4}	4.439	3.061	386
2.0663×10^{-2}	3.4223×10^{-4}	4.439	2.861	404

Table 2. Parameters adopted for calculation of Fig. 9.

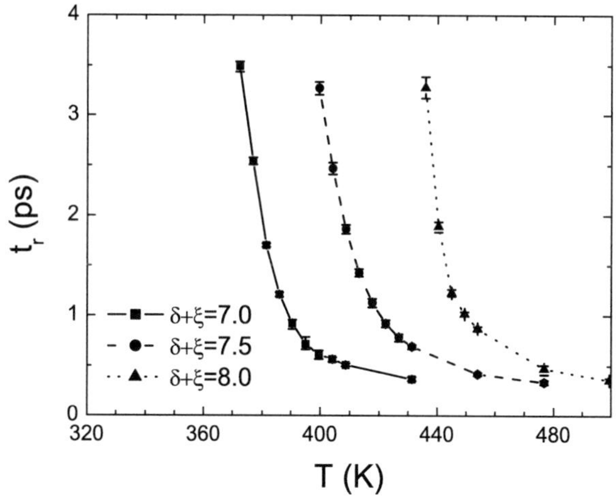

Fig. 10. Temperature dependence of relaxation time t_r for various barrier height $\delta+\xi$ when $T>T_c$, where $\delta/\xi=1.5$ is assumed.

In Fig. 10, we plot the temperature dependence of t_r for different barrier heights, i.e., $\delta+\xi$ varies from 7.0 to 8.0, while the ratio of δ/ξ is fixed at 1.5. Parameters for this calculation are provided in Table 3. As already discussed with Figs. 8 and 9, a larger δ tends to increases T_c, but a higher ξ applies an opposite effect on T_c. Combining these two effects, in Fig. 10, one finds that T_c increases if both δ and ξ are enhanced, indicating that in this case, the change of δ plays a more significant role than that of ξ. Meanwhile, in contrast to Figs. 8 and 9, all the three curves in Fig. 10 present smooth crossovers on decreasing temperature towards T_c, signifying that the dipole fluctuation can be promoted by lowering ξ even the temperature is decreased.

c_4	c_6	δ	ξ	$\delta+\xi$	T_c (K)
2.1557×10^{-2}	3.7309×10^{-4}	4.200	2.800	7.000	372
2.0122×10^{-2}	3.2505×10^{-4}	4.500	3.000	7.500	400
1.8860×10^{-2}	2.8557×10^{-4}	4.800	3.200	8.000	436

Table 3. Parameters adopted for calculation of Fig. 10.

In Namikawa's experiment, the wavelength of soft x-ray laser is 160 Å, hence the photo-created polarization cluster is of nano scale. However, it should be noted that relaxation process of such a nano-sized cluster is actually beyond our present quantum Monte Carlo simulations because of the size limitation of our model. For small-scale dipole fluctuations, the relaxation becomes relatively fast. This is the primary reason why the experimentally measured relaxation time can reach about 30 picoseconds, being several times longer than our calculated results. In spite of the difference, our calculation has well clarified the critical dynamics of $BaTiO_3$ and the origin of speckle variation.

4. Conclusion

We carry out a theoretical investigation to clarify the dynamic property of photo-created ferroelectric cluster observed in the paraelectric $BaTiO_3$ as a real time correlation of speckle pattern between two soft x-ray laser pulses. The density matrix is calculated by a perturbative expansion up to the fourth order terms, so as to characterize the time dependence of scattering probability. The cluster-associated phonon softening as well as central peak effects are well reproduced in the phonon spectral function via a quantum Monte Carlo simulation. We show that the time dependence of speckle pattern is determined by the relaxation dynamics of photo-created ferroelectric cluster, which is manifested as a central peak in the phonon spectral function. The photorelaxation of ferroelectric cluster is featured by a critical slowing down on decreasing the temperature. Near the T_c, cluster excitation is stablized by the strong dipole fluctuation, correspondingly the relaxation becomes slow. While, at higher temperature, dipole fluctuation is suppressed, ending up with a quicker relaxation of cluster. Our simulation also illustrates that the critical slowing down and dipole fluctuation are subject to the chemical environment of crystal.

5. Acknowledgments

We are indebted to Professor K. Namikawa and Professor H. Zheng for their valuable discussions. This work is supported by the Ministry of Education, Culture, Sports, Science and Technology of Japan, the peta-computing project, and Grant-in-Aid for Scientific Research (S), Contract No. 19001002, 2007.

6. Appendices

6.1 Hybrid Monte Carlo method

The Metropolis algorithm is a widely used sampling rule for the Monte Carlo simulation. It randomly generates a new configuration from the previous one and accept it with the probability min[1,exp(-ΔE)], where ΔE is the energy difference between the initial and final configurations. In the present case of interacting many-body system, however, the Metropolis algorithm works only slowly, because we have to sum up a great number of paths and the phonon coordinate covers a large spatial extent. Nevertheless, we find the so-called hybrid Monte Carlo algorithm (Duane et al., 1987) is very efficient. By introducing an extra degree of freedom, it adequately combines the heat bath scheme with the Metropolis algorithm so as to achieve a rapid and ergodic evolution through the configuration space.
In order to incorporate this hybrid Monte Carlo algorithm to our path-integral theory, we first discretize the imaginary time τ, dividing β into L segments with interval Δ. Then, from Eqs. (24) and (26) we write the partition function (≡Z) as,

$$Z = e^{-\beta\Phi_f} = \int Dx \exp\left\{-\int_0^\beta d\tau \Omega_f[x(\tau)]\right\}$$

$$= \int Dx \exp\left\{-\Delta \sum_{i=1}^{L} \Omega_f(x_i)\right\}. \tag{36}$$

Then we introduce a "pseudo time" t, and velocity field $\dot{x}$, which is the derivative of x with respect to this pseudo time. Then, we define a new pseudo crystal with its partition function (≡Z_P) given by

$$Z_P = Z \int D\dot{x} \exp\left(-\sum_{i,l} \frac{\dot{x}_{i,l}^2}{2}\right)$$

$$= \int Dx D\dot{x} e^{-H_P(x,\dot{x},t)} , \tag{37}$$

where i and l refer to imaginary time τ and real lattice site, respectively, and

$$H_P(x,\dot{x},t) = \sum_{i,l}\left[\frac{\dot{x}_{i,l}^2}{2} + \frac{1}{2\omega_0\Delta}(x_{i,l} - x_{i-1,l})^2 + \frac{\Delta}{2}\omega_0 x_{i,l}^2 - \frac{\Delta}{2}\omega_0 c_4 x_{i,l}^4 + \frac{\Delta}{6}\omega_0 c_6 x_{i,l}^6\right]$$

$$-\frac{\Delta}{2}\omega_0 d_2 \sum_{i,<l,l'>} x_{i,l} x_{i,l'} . \tag{38}$$

Because the Gaussian noise in Eq. (37) can be integrated out of the partition function Z_P, the introduction of the auxiliary degree of freedom actually gives no effect on the thermal average involving x, and the new configurations for Monte Carlo sampling thus can be determined by the Hamilton's principle

$$\delta \int_{t_1}^{t_2} L_P(x,\dot{x},t)dt = 0 , \tag{39}$$

where t_1 and t_2 are the initial and final pseudo times, respectively, and L_P is the Lagrangian of the pseudo crystal defined as

$$L_P(x,\dot{x},t) = \sum_{i,l} x_{i,l}^2 - H_P(x,\dot{x},t) . \tag{40}$$

Expanding Eq. (39) with respect to δx and $\delta\dot{x}$, we obtain the ordinary Hamilton's equation

$$\ddot{x}_{i,l} = -\frac{\partial H_P(x,\dot{x},t)}{\partial x_{i,l}} . \tag{41}$$

To generate the new configurations, the following scheme of molecular dynamics is cycled. For a given set of x, we first update $\dot{x}$ field using a heat bath, i.e., set each $\dot{x}_{i,l}$ to be a random number of Maxwell-Boltzmann distribution. Next we propagate x and $\dot{x}$ forward along pseudo time t according to the Hamilton's equation (41). To this end, we divide the time interval into N_t molecular-dynamics steps, which is 20 ~ 40 in this work. Compute $\dot{x}$ at $\frac{1}{2}\delta t$, $\frac{1}{2}\delta t$, $\frac{5}{2}\delta t$, ..., $(N_t - \frac{1}{2})\delta t$, $N_t\delta t$ [$\delta t \equiv (t_2 - t_1)/N_t$] and x at δt, $2\delta t$, $3\delta t$, ..., $N_t\delta t$ by the "leap-frog" algorithm (Allen & Tildesley, 1989). Then this evolved configuration is accepted according to the Metropolis algorithm.

6.2 Iterative fitting method

With the Green's function calculated by the QMC simulation, the phonon spectral function $A(\omega)$ can be numerically attained by the analytic continuation of Eq. (30). Before we proceed to calculate the phonon spectral function, in this section, we first address how to derive the spectral function of electron. We shall see, the electron-oriented iterative fitting algorithm developed here is the cornstone for a more general approach towards solving the problem of spectrum reconstruction, like that related with phonon.

In the case of electron, the spectral function $A(\omega)$ is connected to the electronic Green's function $G(\tau)$ by an equation similar to Eq. (30),

$$G(\tau) = -\int_{-\infty}^{\infty} \frac{e^{-\tau\omega}}{1+e^{-\beta\omega}} A(\omega) d\omega \,. \tag{42}$$

In terms of Eq. (42), the numerical calculation starts from the following equation,

$$G_i = \sum_j K_{ij} A_j \Delta\omega \,. \tag{43}$$

where i and j denote the discretized imaginary time τ and frequency ω, respectively, and

$$K_{ij} = -\frac{e^{-\tau_i \omega_j}}{1+e^{-\beta\omega_j}} \tag{44}$$

is the integral kernel. In order to perform this analytic continuation efficiently, we have developed an iterative fitting method (Ji et al., 2004), which can reproduce the normalized and positive-definite spectra self-consistently. Our algorithm is based on the sum-rule of electronic spectrum (Mahan, 1990)

$$\sum_j A_j \Delta\omega = 1 \,, \tag{45}$$

which suggests that the spectral function can be rewritten in an iterative form as

$$A_j^{(N)} = \frac{n_j^{(N)}}{N\Delta\omega} \,, \tag{46}$$

where $n_j^{(N)}$ is the bin counter for the j-th spectral segment $A_j^{(N)}$ and records the times that the j-th bin being used during the previous N iterative steps. The sum rule of Eq. (45) is thus fulfilled by

$$N = \sum_j n_j^{(N)} \,. \tag{47}$$

The positivity of $A_j^{(N)}$ is also guaranteed because $n_j^{(N)}$ is nonnegative.
To reproduce the spectral function, we start from a flat spectrum, and then repeat the following scheme:
(1) First we calculate the Green's function $G_i^{(N)}$ by the current spectral function $A_j^{(N)}$ as,

$$G_i^{(N)} = \sum_j K_{ij} A_j^{(N)} \Delta\omega \,. \tag{48}$$

(2) Then measure the distance [$\equiv\chi^{(N)}$] between $G_i^{(N)}$ and the true G_i (QMC results) as

$$\left[\chi^{(N)}\right]^2 = \sum_i \left[G_i^{(N)} - G_i\right]^2 \,, \tag{49}$$

which finally should be minimized by the present iteration.
(3) Make a trial spectral function $A_j^{(N+1)}$ of the next step $N+1$, whose $n_j^{(N+1)}$ is different from $n_j^{(N)}$ by only one, and only in a randomly selected bin j_0, as

$$n_j^{(N+1)} = n_j^{(N)} + \delta_{jj_0} \,. \tag{50}$$

By this change, we have new $A_j^{(N+1)}$, $G_i^{(N+1)}$ and $\left[\chi^{(N+1)}\right]^2$ for this trial step.
(4) Calculate the difference S as

$$S = \left[\chi^{(N)}\right]^2 - \left[\chi^{(N+1)}\right]^2 + \sum_i \left[G_i^{(N+1)} - G_i^{(N)}\right]^2 . \tag{51}$$

(5) If $S>0$, we accept the change Eq. (50), and start a new cycle from (1). Otherwise, reject this change Eq. (50) and return to (3).

Here let us explain the meaning of S of Eq. (51). In Eq. (49), $\chi^{(N)}$ gives the distance from the true $\mathbf{G}$ to the trial $\mathbf{G}^{(N)}$, in a hyperspace spanned by various $\mathbf{G}$'s, $\mathbf{G}^{(N)}$'s and $\mathbf{G}^{(N+1)}$'s, as schematically shown in Fig. 11. The equi-distant sphere is also symbolically denoted by S_1 in this figure. If $\chi^{(N)}>\chi^{(N+1)}$, or $\mathbf{G}^{(N+1)}$ is in this sphere S_1, we, of course, have to accept this change Eq. (50) unconditionally, since the distance becomes shorter than before.

However, $\chi^{(N)}$ may not be a simple function of this change ($\equiv\mathbf{\Delta}$),

$$\mathbf{\Delta} \equiv \mathbf{G}^{(N+1)} - \mathbf{G}^{(N)} , \tag{52}$$

but will be a nonlinear and complicated function. In some cases, $\chi^{(N)}$ will be in a local minimum with respect to the change $\mathbf{\Delta}$, and this change cannot make the distance shorter. For this reason, in Eq. (51), we introduce a "relaxation effect" through the third term, which avoids the search for $\chi_{\min}^{(N)}$ being trapped in such a local minimum. Since the projection of $\mathbf{\Delta}$ on the vector $\mathbf{G}^{(N)}$-$\mathbf{G}$ is just $-S/2\chi^{(N)}$, we find that the acceptable region for the trial $\mathbf{G}^{(N+1)}$ consists of two sections, i.e., i and ii (see in Fig. 11), both of which correspond to $S>0$. In section i, $\mathbf{G}^{(N+1)}$ is unconditionally accepted, since $\chi^{(N+1)}$ truly decreases. While in section ii, the acceptance is conditional, since it does not give a smaller $\chi^{(N+1)}$ but only make $\mathbf{G}^{(N+1)}$ moves toward $\mathbf{G}$ along the $\mathbf{G}$-$\mathbf{G}^{(N)}$ line. By taking this conditional acceptance, the minimization is relaxed, and correspondingly an uphill search for a more global minimum can be continued.

With the algorithm described above, the spectral function can be readily recovered from the Green's function. A few hundred iterations are sufficient to construct the convergent spectra, but with very rough contours. For a naturally refined line shape, much more iterations are necessary before no appreciable change can be observed any more.

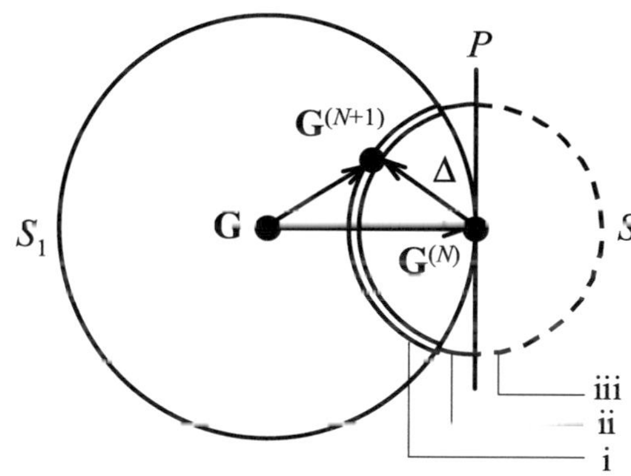

Fig. 11. Schematic interpretation for the recipe of analytic continuation. The hypersphere S_1 is centered at $\mathbf{G}$ with radius $\chi^{(N)}$, and S_2 is centered at $\mathbf{G}^{(N)}$ with radius $|\Delta|$. The hyperplane P is perpendicular to $\mathbf{G}^{(N)}$-$\mathbf{G}$ and bisects S_2. The surface of S_2 is thus composed of three sections: i (unconditional acceptance), ii (conditional acceptance) and iii (rejection).

6.3 Renormalization iterative fitting method

The iterative fitting method relies its convergence on the sum-rule of electronic spectrum, which conserves the total spectral weight through the iteration process. In a same manner, the Green's function and spectral function can be introduced to describe a bosonic system. It can be shown that in the case of boson, Eq. (42) should be reformulated as

$$G(\tau) = -\int_{-\infty}^{\infty} \frac{e^{-\tau\omega}}{1-e^{-\beta\omega}} A(\omega) d\omega \,, \tag{53}$$

and the sum rule also holds for the bosonic spectrum $A(\omega)$. Hence we can apply the iterative fitting method to calculate it.

However, for the phonon spectrum, there is not a rule that conserves the total spectral sum like that of electron. This is because the phonon Green's function is defined in the coordinate representation rather the particle number representation [see Eqs. (17) and (20)]. Such a definition leads to general properties of phonon spectrum as manifested by

$$\omega A(\omega) \geq 0 \,, \tag{54}$$

$$A(-\omega) = -A(\omega) \,. \tag{55}$$

Obviously, the phonon spectrum is anti-symmetric and the total sum of spectral weight should be zero,

$$\int_{-\infty}^{\infty} A(\omega) d\omega = 0 \,. \tag{56}$$

Therefore, the standard iterative fitting method does not work for phonon.

In order to get the phonon spectral function $A(\omega)$, we need to confine our calculation in the region $0<\omega<\infty$, and introduce a renormalized spectral function $\widetilde{A}(\omega)$ as

$$\tilde{A}(\omega) \equiv -\frac{1}{G(\beta)} \coth\left(\frac{1}{2}\beta\omega\right) A(\omega) \,. \tag{57}$$

Because of the anti-symmetry of $A(\omega)$, the integral in Eq. (53) can be rewritten as

$$G(\tau) = \int_{0}^{\infty} \frac{1}{\cosh\left(\frac{1}{2}\beta\omega\right)} \cosh\left[\left(\frac{1}{2}\beta - \tau\right)\omega\right] G(\beta)\tilde{A}(\omega) d\omega \,. \tag{58}$$

It is easy to see the renormalized spectral function is positive if $\omega>0$, and satisfies a sum rule,

$$\int_{0}^{\infty} \widetilde{A}(\omega) d\omega = 1 \,, \tag{59}$$

which allow us to solve the integral equation of Eq. (58) by the iterative fitting approach in the region $0<\omega<\infty$. Once $\widetilde{A}(\omega)$ is reproduced, the phonon spectral function $A(\omega)$ can be obtained from Eq. (57).

7. References

Allen, M.P. & Tildesley, D.J. (1989). *(1987). Computer simulation of liquids*, Clarendon, ISBN 0198556454, Oxford

Aubry, S. (1975). A unified approach to the interpretation of displacive and order–disorder systems. I. Thermodynamical aspect, *Journal of Chemical Physics*, 62, 3217-3229, ISSN 0021-9606

Bonča, J. & Gubernatis, J.E. (1996). Real-time dynamics from imaginary-time quantum Monte Carlo simulations: Tests on oscillator chains, *Physical Review B*, 53, 6504-6513, ISSN 1098-0121

Doniach, S. & Sondheimer, E.H. (1998). *Green's Functions for Solid State Physicists*, Imperial College Press, ISBN 1860940781, London

Duane, S.; Kennedy, A.D.; Pendleton, B.J. & Roweth, D. (1987). Hybrid Monte Carlo, *Physics Letters B*, 195, 216-222, ISSN 0370-2693

Goodman, J. W. (2007). *Speckle Phenomena in Optics: Theory and Applications*, Roberts & Company, ISBN 0974707791, Greenwood Village

Gubernatis, J.E.; Jarrell, M.; Silver, R.N. & Sivia, D.S. (1991). Quantum Monte Carlo simulations and maximum entropy: dynamics from imaginary-time data, *Physical Review B*, 44, 6011-6029, ISSN 1098-0121

Hamazaki, S.-I.; Shimizu, F.; Kojima, S. & Takashige, M. (1995). AFM observation of 90° domains of $BaTiO_3$ butterfly crystals, *Journal of the Physical Society of Japan*, 64, 3660-3663, ISSN 0031-9015

Harada, J.; Axe, J.D. & Shirane, G. (1971). Neutron-scattering study of soft modes in cubic $BaTiO_3$, *Physical Review B*, 4, 155-162, ISSN 1098-0121

Ji, K.; Zheng, H. & Nasu, K. (2004). Path-integral theory for evolution of momentum-specified photoemission spectra from broad Gaussian to two-headed Lorentzian due to electron-phonon coupling, *Physical Review B*, 70, 085110(1-9), ISSN 1098-0121

Ji, K.; Namikawa, K; Zheng, H. & Nasu, K. (2009). Quantum Monte Carlo study on speckle variation due to photorelaxation of ferroelectric clusters in paraelectric barium titanate, *Physical Review B*, 79, 144304(1-8), ISSN 1098-0121

Khare, A. & Saxena, A. (2008). Domain wall and periodic solutions of a coupled φ^6 model, *Journal of Mathematical Physics*, 49, 063301(1-18), ISSN 0022-2488

Kishimoto, M.; Namikawa, K.; Sukegawa, K.; Yamatani, H; Hasegawa, N. & Tanaka M. (2010). Intensity correlation measurement system by picosecond single shot soft x-ray laser, *Review of Scientific Instruments*, 81, 013905(1-5), ISSN 0034-6748

Krausz, F. & Ivanov, M. (2009). Attosecond physics, *Review of Modern Physics*, 82, 162-234, ISSN 0034-6861

Krumhansl, J.A. & Schrieffer, J.R. (1975). Dynamics and statistical mechanics of a one-dimensional model Hamiltonian for structural phase transitions, *Physical Review B*, 11, 3535-3545, ISSN 1098-0121

Mahan, G.D. (1990). *Many-Particle Physics*, 2nd ed., Plenum, ISBN 1441933395, New York

May, M. (1977). Information inferred from the observation of speckles. *Journal of Physics E: Scientific Instruments*, 10, 849-864, ISSN 0022-3735

Migoni, R.; Bauer, D. & Bilz, H. (1976). Origin of Raman scattering and ferroelectricity in oxidic perovskites, *Physical Review Letter*, 37, 1155-1158, ISSN 0031-9007

Morris, J.R. & Gooding, R.J. (1990). Exactly solvable heterophase fluctuations at a vibrational-entropy-driven first-order phase transition, *Physical Review Letter*, 65, 1769-1772, ISSN 0031-9007

Mulvihill, W.L.; Uchino, K.; Li, Z. & Cao, W. (1996). In-situ observation of the domain configurations during the phase transitions in barium titanate, *Philosophical Magazine B*, 74, 25-36, ISSN 1364-2812

Namikawa, K.; Kishimoto, M.; Nasu, K.; Matsushita, E.; Tai, R.Z.; Sukegawa, K.; Yamatani, H.; Hasegawa, H.; Nishikino, M.; Tanaka, M. & Nagashima, K. (2009). Direct Observation of the Critical Relaxation of Polarization Clusters in $BaTiO_3$ Using a Pulsed X-Ray Laser Technique, *Physical Review Letter*, 103, 197401(1-4), ISSN 0031-9007

Nasu, K. (1994). Resonance enhancement of inelastic x-ray scatterings induced by strong visible lights, *Journal of the Physical Society of Japan*, 63, 2416-2427, ISSN 0031-9015

Pang, G.K.H. & Baba-Kishi, K.Z. (1998). Characterization of butterfly single crystals of $BaTiO_3$ by atomic force, optical and scanning electron microscopy techniques, *Journal of Physics D: Applied Physics*, 31, 2846-2853, ISSN 0022-3727

Polla, D.L. & Francis, L.F. (1998). Processing and characterization of piezoelectric materials and integration into microelectromechincal systems, *Annual Review of Materials Science*, 28, 563-597, ISSN 0084-6600

Savkin, V.V.; Rubtsov, A.N. & Janssen, T. (2002). Quantum discrete φ^4 model at finite temperatures, *Physical Review B*, 65, 214103(1-12), ISSN 1098-0121

Schneider, T. & Stoll, E. (1978). Molecular-dynamics study of a three-dimensional one-component model for distortive phase transitions, *Physical Review B*, 17, 1302-1322, ISSN 1098-0121

Skilling, J. & Bryan, R.K. (1984). Maximum entropy image reconstruction: general algorithm, *Monthly notices of the royal astronomical society*, 211, 111-124, ISSN 0035-8711

Tai, R.Z.; Namikawa, K.; Kishimoto, M.; Tanaka, M.; Sukegawa, K.; Hasegawa, N.; Kawachi, T.; Kado, M.; Lu, P.; Nagashima, K.; Daido, H.; Maruyama, H.; Sawada, A.; Ando, M. & Kato, Y. (2002). Picosecond snapshot of the speckles from ferroelectric $BaTiO_3$ by means of x-ray lasers, *Physical Review Letter*, 89, 257602(1-4), ISSN 0031-9007

Tai, R.Z.; Namikawa, K.; Sawada, A.; Kishimoto, M.; Tanaka, M.; Lu, P.; Nagashima, K.; Maruyama, H. & Ando, M. (2004). Picosecond view of microscopic-scale polarization clusters in paraelectric $BaTiO_3$, *Physical Review Letter*, 93, 087601(1-4), ISSN 0031-9007

Takagi, M. & Ishidate, T. (2000). Anomalous birefringence of cubic $BaTiO_3$, *Solid State Communications*, 113, 423-426, ISSN 0038-1098

Völkel, G. & Müller, K.A. (2007). Order-disorder phenomena in the low-temperature phase of $BaTiO_3$, *Physical Review B*, 76, 094105(1-8), ISSN 1098-0121

Yamada, Y.; Shirane, G. & Linz, A. (1969). Study of critical fluctuations in $BaTiO_3$ by neutron scattering, *Physical Review*, 177, 848-857, ISSN 0031-899X

Yamazaki, M.; Tomita, N. & Nasu, K. (2003). Intermediately correlated many-electron systems studied by quantum Monte Carlo method, *Journal of the Physical Society of Japan*, 72, 611-617, ISSN 0031-9015

Yan, R.; Guo, Z.; Tai, R.; Xu, H.; Zhao, X.; Lin, D.; Li, X. & Luo, H. (2008). Observation of long range correlation dynamics in $BaTiO_3$ near T_C by photon correlation spectroscopy, *Applied Physics Letters*, 93, 192908(1-3), ISSN 0031-9007

Yin, Q.R.; Zeng, H.R.; Yu, H.F. & Li, G.R. (2006). Near-field acoustic and piezoresponse microscopy of domain structures in ferroelectric material, *Journal of Materials Science*, 41, 259-270, ISSN 0022-2461

Zalar, B.; Laguta, V.V. & Blinc, R. (2003). NMR evidence for the coexistence of order-disorder and displacive components in barium titanate, *Physical Review Letter*, 90, 037601(1-4), ISSN 0031-9007

Zhong, W.; King-Smith, R.D. & Vanderbilt, D. (1994). Giant LO-TO splittings in perovskite ferroelectrics, *Physical Review Letter*, 72, 3618-3662, ISSN 0031-9007

Two-Photon Polymerization of Inorganic-Organic Hybrid Polymers as Scalable Technology Using Ultra-Short Laser Pulses

Houbertz, Ruth[1], Steenhusen, Sönke[1],
Stichel, Thomas[1,3], and Sextl, Gerhard[1,2]
[1]Fraunhofer ISC, Neunerplatz 2, 97082 Würzburg,
[2]Chemische Technologie der Materialsynthese, Julius-Maximilians-Universität, Röntgenring 11, 97070 Würzburg,
[3]Nano-Optics & Biophotonics Group, Physikalisches Institut, Lehrstuhl für Experimentelle Physik V, Julius-Maximilians-Universität, Am Hubland, 97074 Würzburg, Germany

1. Introduction

For more than five decades, laser light properties have been taken advantage of, mostly using first order interactions between photons and materials. Well-established methods such as laser welding (Sieben & Brunnecker, 2009; El-Bataghy, 1997; Song et al., 2009), laser ablation (Conforti et al., 2007; Baudach et al., 2000; Spyridaki et al., 2003; Blanchet et al., 1993), and direct laser writing (Schweizer et al., 2008; Koechlin et al., 2009) were improved concomitantly with the laser quality (e.g. beam profile) or the laser energy (e.g. short laser pulses). The highly localized application of energy has led to more and more advanced rapid prototyping techniques such as laser sintering or stereolithography (Stampfl et al., 2004; Simchi, 2006).

Multi-photon polymerization (nPP, n = 2,3,...), in particular two-photon polymerization (2PP), has attracted continuously increasing attention after its first demonstration in 1997 as a method for the fabrication of three-dimensional structures with feature sizes far beyond the diffraction limit of the applied laser wavelength (Maruo et al., 1997; Tanaka et al., 2002). Contrary to conventional UV and (stereo-)lithography, 2PP is a truly three-dimensional laser writing technique. The underlying physical phenomenon of two-photon absorption (TPA) was theoretically postulated by Maria Göppert-Mayer in 1931 (Göppert-Mayer, 1931). It took, however, several decades to finally prove this effect due to the lack of highly irradiant light sources.

The invention of ultrafast lasers had led to many different applications. Particularly, the interaction of lasers with polymer surfaces and polymer or glass bulk samples is of high technological interest. The adaptation of polymers to laser beam characteristics and vice versa is very challenging from the scientific as well as from the technological point of view. Besides the implementation of two-photon absorption microscopy (Denk et al., 1990; Denk & Svoboda, 1997) and spectroscopy (Asaka et al., 1998; Yamaguchi & Tahara, 2003), TPA is particularly used for the three-dimensional micromachining of a large variety of materials. Two- or multi-photon absorption processes are particular examples, in which focused ultrafast laser pulses are employed to trigger reactions in either glasses or photosensitive polymers (Cheng et al., 2003; Wong et al., 2006; Deubel et al., 2004). This technique uses the fact that TPA or nPA can only occur in the focal volume of the laser irradiation. Thus, the triggered reactions are strongly confined to the focal region, and the production of 3D microstructures is performed in one production step just by moving the focal volume in 3D through the materials. Among the applications of two-photon absorption so far are the fabrication of waveguides (Nolte et al., 2004; Houbertz et al., 2008a), of photonic crystal structures (Serbin et al., 2004; Houbertz et al., 2008b; Li et al., 2008), of phase masks (Jia et al., 2007), and the production of three-dimensional structures for medical applications such as scaffolds for regenerative medicine (Doraiswamy et al., 2005; Claeyssens et al., 2009; Beyer et al., 2010). Due to the possibility of creating *arbitrary* 3D micro- and macrostructures which are basically determined by the (shape/size of the) focal light intensity distribution of the focusing optics, other future applications will be conceivable which manifest themselves in the fact that the use of TPA processes for micromachining, microscopy, or spectroscopy is a continuously increasing field.

This chapter is organized as follows. First, general aspects of the material class of inorganic-organic hybrid polymers such as ORMOCER®s[1] will be described in order to give a brief overview about the materials and the accessible material parameters, followed by a brief description of selected materials for the TPA process. Then, the experimental setup which was used to fabricate small- and large-scale structures is described in detail. This section is followed by the characterization of two-photon absorption cross-sections of selected photoinitiators for the TPA processing. Finally, selected examples are given, ranging from the determination of voxel sizes in dependence of the laser parameters to high-resolution small-scale as well as large-scale structures (from sub-100 nm to the cm regime) fabricated by TPA. Future aspects of technology development will be addressed as well.

2. Inorganic-organic hybrid polymers

2.1 General material aspects

In order to combine the advantageous properties of inorganic and organic materials, different synthesis strategies are reported to produce inorganic-organic materials, ranging from physically blending organic materials with inorganic compounds via multi-step syntheses to one-step synthesis (Wojcik & Klein, 1995; Haas & Wolter, 1999). Nanostructured materials such as inorganic-organic hybrid polymers (ORMOCER®s) are intensively investigated due to their wide range of physical and chemical properties, synergistically resulting from their hybrid structure. The material concept of ORMOCER®s comprises the combination of material properties of organic polymers such as toughness, functionalization, and processing at low temperatures, with those of glass-like materials like hardness, chemical and thermal stability, and transparency on a molecular level (Haas, 2000). This concept allows one to achieve material properties which are not simply accessible with composite or polymer materials and, simultaneously, to apply various technologies to fabricate structures and elements on a different size scale. Due to the fact

[1]Registered trademark of the Fraunhofer-Gesellschaft für Angewandte Forschung e.V.

that the material properties of ORMOCER®s can be tailored as required by the respective applications, a large variety of different structures for applications in optics, ranging from the nm to the cm regime, in microelectronics, or in micro- and biomedicine is enabled. As the range of applications has been increased, the requirements on the materials and their processing have become more challenging, and the fabrication of microstructures with nm precision is a prerequisite for optical elements. Not only due to their excellent optical and electrical properties, but also due to their enhanced thermal, mechanical, and chemical stability and, simultaneously, their simple processability, they overcome the restrictions of purely organic polymers for most applications. In addition, functionalization is used to create suitable binding sites, particularly for biomolecules and cells for enabling micro- and biomedical applications.

Contrary to class-I hybrids, the material class of ORMOCER® consists of materials, where the inorganic and organic components are covalently bonded to each other (Sanchez et al., 2005). Their synthesis is carried out via catalytically controlled hydrolysis and (poly)condensation reactions using differently functionalized alkoxysilanes. Upon synthesis, an organically modified inorganic-oxidic network is established on a molecular level (Haas & Wolter, 2001), whereas also hetero-elements such as, for example zirconium, aluminum or titanium can be introduced into the network for further tuning of the material properties (Haas & Wolter, 1999; Helmerich et al., 1994; Declerck, 2010). The resulting hybrid materials are storage-stable resins with negative resist behavior, i.e. an organic cross-linking can be initiated via photochemical and/or thermal reactions.

The organic groups are employed as network modifiers, connecting groups, and also organically polymerizable moieties (cf., Figure 1). The latter are chosen with respect to application requirements. For example, for patterning via UV or laser lithography, (oligo-) methacrylate, acrylate, or styryl moieties are favored. Epoxy groups are preferred for thermal cross-linking in screen-printing processes. Alkyl or aryl groups as non-reactive groups can be connected to the $-[Si-O]_n-$ network, and they also influence the material properties. An increase of their amount within the hybrid polymer can reduce the degree of polymerization due to sterical reasons, thus resulting in a reduced density in the coated layers. This directly influences the optical or dielectric properties such as the refractive index or the dielectric permittivity.

By varying the type and/or amount of alkoxysilane precursors and, in addition, the synthesis conditions (for example, the reaction temperature, the type and amount of catalyst, and/or the type and amount of solvent), custom-designed hybrid materials with a well-defined parameter profile can be provided for a large variety of applications. Among those are ORMOCER®s for microsystems which are applied, for example, as optical interconnects or waveguides (Streppel et al., 2002; Uhlig et al., 2006), in microoptics (Bräuer et al., 2001; Houbertz et al., 2006), in electro-optical applications (Robertsson et al., 1998; Houbertz et al., 2008a), as dielectric layers (Haas et al., 2006), and as passivation materials for the encapsulation of microelectronic devices and components (Houbertz et al., 2001; 2003a; 2003b).

The processing of an ORMOCER® typically consists of two steps. First, an organically modified $-[Si-O]_n-$ network is established via chemical syntheses, resulting in a pre-polymer sol (resin). Second, the organic cross-linking is initiated photochemically and/or thermally which finally results in the formation of an inorganic-organic hybrid network. For that, UV and/or thermal initiators have to be introduced into the pre-polymer resin.

In Figure 1, a schematic sketch of selected multifunctional precursors which are employed for material synthesis is shown. In addition, some material properties being influenced by the individual precursors are summarized as well. As mentioned above, the hybrid polymers' material properties can be adjusted in wide ranges with respect to application via photochemical and / or thermal reactions. It has to be mentioned, however, that a change in one material property also might influence other material properties, and a comprehensive know-how is necessary to control this in a well-defined manner.

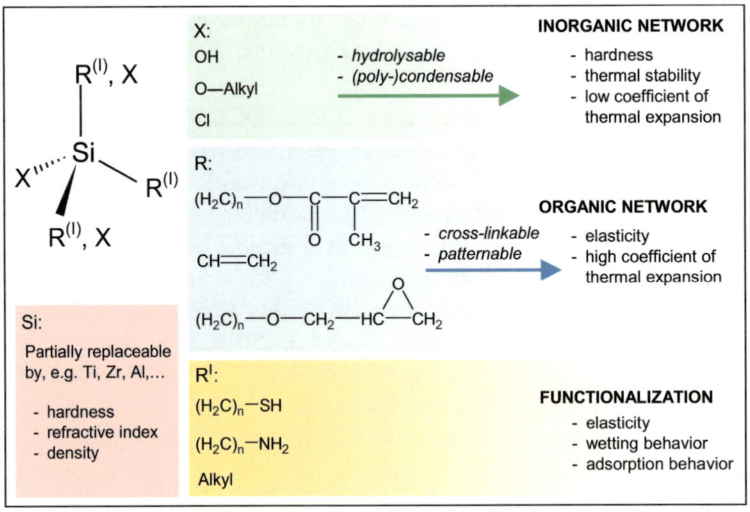

Fig. 1. Multifunctional precursors for ORMOCER® synthesis and their influence on resulting material properties.

For optical applications such as single- or multi-mode waveguides for the near infrared (NIR), the amount of Si-OH groups has to be as low as possible in order to reduce the materials' optical absorption losses which are basically determined by the amount of silanol groups due to 2nd harmonic (overtone) absorption at around 1438 nm (Rousseau & Boutevin, 1992). To the same extent, the materials' SiO_x content which intrinsically has a low optical loss is increased. It should be mentioned, however, that the higher the inorganic content, the more brittle the material will be, and the resulting layer thicknesses will also be reduced compared to a more organic material. The reduction of silanol groups can be performed, for example, by reducing them in the material by silylating agents, or by special synthesis conditions which are reported elsewhere (Buestrich et al., 2001). The best suited material for waveguide application should have no detectable -OH signal in multi-nuclei magnetic resonance (NMR) spectra. In addition, the amount of aliphatic -CH which is known to absorb around 1310 and 1550 nm was reduced in some ORMOCER®s to a minimum by increasing the amount of aromatic organic substituents. Scattering from the organically functionalized inorganic oligomers (polysilsesquioxanes) is not very likely, because their structural size, which was determined by small angle X-ray scattering (SAXS) measurements, is in the range from 1 to 3 nm for ORMOCER®s applied in optics. The structural features of the hybrid polymers are dependent on the precursors which are used for synthesis. For example, by NMR spectroscopy it can be shown that di-alkoxysilanes yield chain or ring polymers, while tri-alkoxysilanes can result in three-dimensional networks.

The mechanical properties of ORMOCER®s can be also varied over a wide range (Haas & Wolter, 1999). For example, the Young's modulus and the mechanical and thermal stability can be increased by increasing the inorganic content in the hybrid network. This simultaneously leads to a reduction of the coefficient of thermal expansion (CTE). Solvents which are produced upon syntheses are usually removed from the hybrid polymer resins in order to allow either solvent-free processing, or to replace them by other solvents or reactive diluters. This provides highest flexibility in thin-film processing, because the materials' viscosity can be simply adjusted to the process requirements. The storage stability with and without photoinitiators was characterized for selected ORMOCER® materials to be more than two years at room-temperature. For a prolonged storage, the materials should be kept at 4 or -18 °C, respectively.

2.2 Selected materials for TPA processes

For the TPA experiments, three different hybrid polymer modifications were chosen with respect to their reaction reactivity which significantly differs for each material. These ORMOCER®s are typically applied in optics and for biomedical applications. As reactive moieties, either acrylate, (oligo-)methacrylate, or styryl groups were chosen. The photochemical reactivity of acrylate groups in the TPA process is usually much higher than the one of methacrylate or styryl groups, thus resulting in higher reaction rate.
The materials were modified such that their response to the femtosecond laser pulses is very high in order to achieve a high densification of the laser light-exposed areas. Several chromophore and initiator systems, respectively, were used to radically initiate the organic cross-linking by TPA. Their concentration in the ORMOCER® materials was set to be between 0.05 and 3 wt.-%. However, the two-photon absorption cross-section σ_2 of these chromophores and initiators was reported in the literature to differ significantly, being as low as approximately 10^{-55} cm^{4}s (10^{-5} GM) for 2-benzyl-2-dimethylamino-4'-morpholinobutyrophenone (Serbin et al., 2003), or as high as $1250 \cdot 10^{-50}$ cm^{4}s (1250 GM) (Cumpston et al., 1999; Schafer et al., 2004). All selected materials are transparent for the laser wavelength, showing absorption losses of about 0.02 to 0.04 dB/cm.

3. Multi-photon patterning

3.1 General aspects

Multi-photon polymerization (nPP) processes are used to realize arbitrary three-dimensional structures without any complicated processing sequences by multi-photon absorption (nPA) in various materials such as, for example photosensitive materials. Typically, two photon absorption processes are considered due to their higher process probability compared to higher order processes, and they are used to trigger the materials' organic polymerization by photochemical reaction with ultra-short laser pulses (Maruo et al., 1997; Tanaka et al., 2002; Houbertz et al., 2007; 2008a; 2008b; Jia et al., 2007; Li et al., 2008). Particularly, the 3D patterning of hybrid polymers such as ORMOCER®s is of high technological interest for application in optics or biomedicine.
Due to the threshold behavior and the non-linear nature of the nPP process, structures far beyond the diffraction limit can be achieved by choosing a suitable combination of pulse energy, number of applied pulses, material, and initiator system (Tanaka et al., 2002; Steenhusen et al., 2010a; 2010b). This technique uses the fact that TPA only occurs in the

focal volume of the laser irradiation, where the photon density is locally high enough to initiate TPA, while the entire material has to be transparent for the laser wavelength. Thus, the organic cross-linking of the material is strongly confined to the focal region, enabling the computer-assisted fabrication of 3D microstructures in only one production step just by moving the focal volume through the ORMOCER® material in x, y, and z direction.

3.2 Experimental setup

In Figure 2, a principal sketch of the experimental setup for nPP patterning is shown which is used for small- and large-scale patterning of ORMOCER®s. The setup consists of a laser and a second harmonic generation (SHG) system, an acousto-optical modulator (AOM), and three air-bearing motion stages for the 3D movement of the sample and the optics, respectively. While most setups employ Ti:sapphire laser systems operating at 780 to 800 nm for the processing of polymer or hybrid polymer materials (Tanaka et al., 2002; Houbertz et al., 2008a; 2008b; Doraiswamy et al., 2005; Claeyssens et al., 2009; Schizas et al., 2010; Gittard et al., 2010), the laser system in the present setup emits at a fundamental wavelength of 1030 nm. This is converted to visible light for the 2PP process by SHG. Three-photon polymerization (3PP) processes are initiated at the fundamental wavelength, i.e. without using SHG.

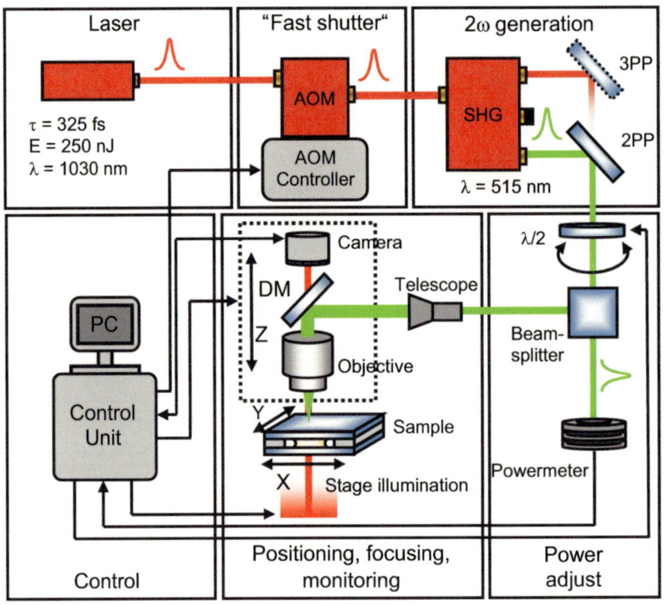

Fig. 2. Experimental setup for multi-photon polymerization (nPP) for the fabrication of 3D structures. DM: dichroic mirror, AOM: acousto-optic modulator.

The laser is a diode-pumped low-cost Ytterbium-based femtosecond laser system. The pulse duration is about 325 fs at 1030 nm (determined for 2.3 W). The lasers' repetition rate is 10.1 MHz, and the maximum pulse energy is 230 nJ. A combination of polarizing beam splitter and half-wave plate mounted on a computer-controlled rotary stage (Aerotech ADR 160) allows a well-defined control of the pulse energy. The acousto-optical modulator (AOM) which is also computer-controlled is used for switching the laser light on and off. All reported average powers were determined below the focusing microscope objectives.

For the 2PP patterning, the 515 nm (SHG) irradiation is expanded by a factor of three before being focused into the ORMOCER®. The focusing is carried out using a microscope objective with a numerical aperture (N.A.) of 1.4 for the fabrication of small-scale structures (up to approximately 100 µm in height) (Steenhusen et al., 2010a), and with either a Nikon CFI Super Plan Fluor ELWD 40XC objective with an NA of 0.60 or a Nikon CFI Super Plan Fluor ELWD 20XC objective with an NA of 0.45. Both of the latter have long working distances of about 3.8 mm and 8.2 mm, respectively, for the creation of large-scale structures (up to 1 cm) (Stichel et al., 2010). For the 3D movement of sample and optics, an Aerotech ABL-1000 three axis system is used with a travel distance of 10 cm x 15 cm x 15 cm in x, y, and z direction, respectively. The setup allows scanning speeds up to 300 mm/s which is useful for the fabrication of larger structures, and a precision of a few nanometers for the generation of high-resolution structures. The patterning can be monitored in-situ by illuminating the sample with red light from the backside, and detecting the transmission with a CCD camera placed behind a dichroic mirror (DM).

The hybrid polymer resin is dispensed on glass substrates. Dependent on the required structure sizes, the patterning is carried out using different exposure strategies which are depicted in Figure 3. For small-scale structures, two standard setups are used. In one setup, the material is sandwiched between two glass substrates, separated by a distance holder (approximately 100 µm in thickness; cf., Figure 3 (a)). In the second setup, the material is deposited as a droplet, being located head first on a glass substrate (Figure 3 (b)). Subsequent cross-linking in order to generate the complete 3D structure is then performed

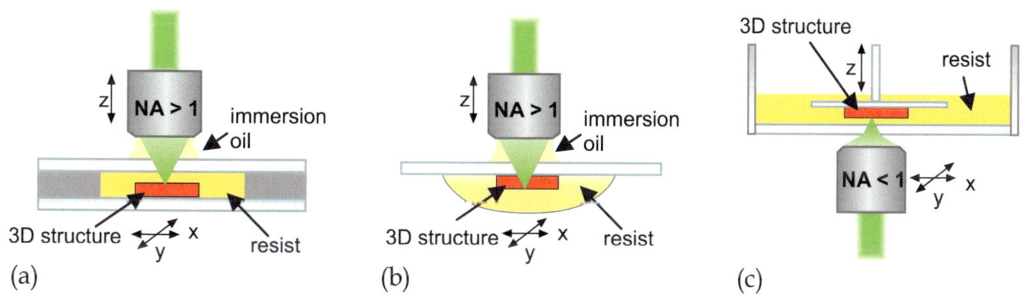

Fig. 3. Exposure strategies. For high-resolution patterning, the material is placed (a) between two glass plates separated by a distance holder, or placed (b) as droplet head first. (c) For large-scale fabrication, an inverted setup is used with a material bath, a sample holder, and an objective scanning the x-y-plane below the bath.

by moving the focus through the resin. In both cases, the structure is then built up starting at the glass substrate's surface. However, in the first case, patterning is carried out by moving the pulsed laser beam up through the hybrid polymer's volume, while for the second setup the patterning is performed through the volume of the already cross-linked material. It has to be mentioned that beam distortion might occur for the setup displayed in Figure 3 (b) due to the fact that the light has to travel trough the already cross-linked material, having a higher refractive index than the liquid resist. Finally, for very large-scale structures, the setup is inverted, using a bath as material reservoir, a sample holder movable in the z direction, and an x-y-scanning objective. The sample is moved upwards which enables one to build structures whose heights are not limited by the working distance of the employed objective anymore.

Contrary to other lithography methods such as conventional UV or laser lithography, the 2PP process can be carried out also in the volume of the materials. Thus, principally two types of processes can be performed which are schematically depicted in Figure 4. On the one hand, the patterning process can be carried out, initiating a cross-linking of the materials' photochemically active organic moieties, followed by a subsequent removal of the non-exposed parts in a suitable developer solvent, analogously to classical processing schemes (solvent-based processing, Figure 4 (a)). On the other hand, structural features can be written inside the hybrid resins' volume, while the surrounding material is not removed by a developer, and which can be differently treated in further processing steps (solvent-free processing, Figure 4 (b)) (Houbertz, 2007; Houbertz et al., 2008). Examples for both routes will be given in chapter 3.4.

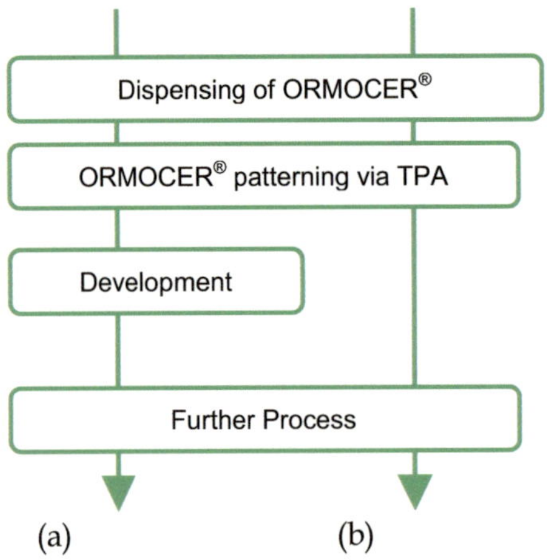

Fig. 4. Process flow diagram. (a) Solvent-based, and (b) solvent-free processing.

3.3 Absorption cross-sections

In order to characterize the non-linear properties of the photoinitiators which are introduced into the selected ORMOCER® materials, a z-scan experiment was set up which determines the transmitted laser light signal passing a quartz glass cuvette filled with initiator/solvent solution. It is widely accepted that the higher the initiators' absorption cross-sections, the more efficient should be the TPA process. It should be mentioned, however, that not only the absorption cross-section of the employed photoinitiators determine the process efficiency, but also the surrounding matrix, i.e. ORMOCER® material and solvents, if included. The latter two might have a much stronger influence on the process efficiency than the absorption cross-section of the initiators itself (Houbertz et al.; 2010).

There are two effects which can occur, namely non-linear absorption and non-linear refraction, and which were theoretically described in literature (Sheikbahae et al., 1989; Sheikbahae et al., 1990). Non linear refraction causes a phase shift of the incident laser beam depending on the irradiance, because the materials' refractive index also varies with irradiance. This is the reason why the incoming beam in a z-scan experiment is focused and defocused, if the cuvette is scanned across the focal region. Due to the fact that nPA is dependent on the irradiance at the sample, the transmitted signal will have a dip for the sample being directly located in the focus. From the magnitude of this dip, the TPA cross-section σ_2 can be calculated (van Stryland & Sheikbahae, 1998).

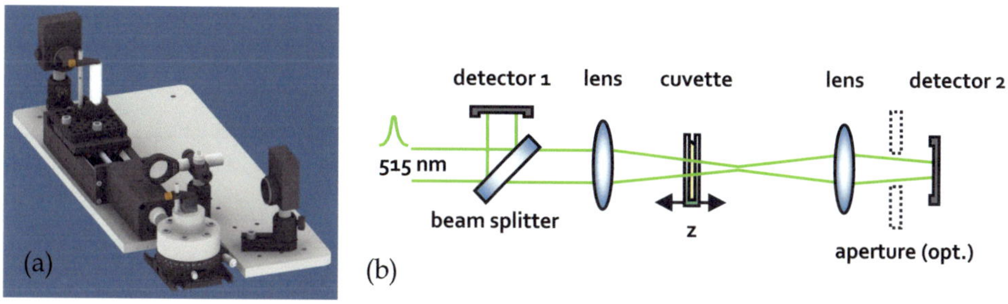

Fig. 5. Experimental setup for the z scan. (a) 3D view, and (b) schematic sketch of the setup.

In Figure 5, a 3D presentation of the experimental setup for the z-scan experiment as well as a schematic sketch is shown which was used to determine the TPA absorption cross-sections of the photoinitiators. As laser wavelength, 515 nm pulses were used which were split with a beam splitter for providing reference and transmission signal. The transmitted beam is then focused using a lens, and detected with a 9.8 mm photodiode (detector 2). Optionally, a variable aperture and an additional lens are located in front of the photodiode. The photoinitiators were dissolved in methylisobutylketone in a 1 mm quartz glass cuvette which is moved along the beam propagation path with a 75 mm linear travel stage. The signal-to-noise ratio was improved by recording multiple scans for each measurement.

For the determination of the absorption cross-sections, non-linear refraction should be neglectable. This can be achieved by an open-aperture scan, where the transmitted signal only depends on the non-linear absorption which is dominated by TPA. The fraction of non-linear refraction can be determined by using the aperture in front of the detecting photodiode, resulting in additional signals in the transmission curve (Sheikbahae et al., 1989).

Aside a high absorption cross-section of the photoinitiators, a high chemical reactivity of the hybrid resins is required. A first insight into their reactivity in selected hybrid polymer systems was deducted from photo-DSC (photo-differential scanning calorimetry) measurements of the ORMOCER®/initiator formulations. It has to be mentioned, however, that the underlying reaction is initiated in a classical one-photon process which already gives a good measure of the reaction enthalpy, and thus of the materials' cross-linking behavior upon UV light exposure. From these measurements, two different commercially available UV initiators were chosen, henceforth labeled as Ini1 and Ini2 (BASF), respectively, as well as a specially developed photoinitiator, labeled as Ini3 (Seidl & Liska, 2007).

In order to prove whether non-linear absorption and/or non-linear refraction are taking place, the magnitude of the absorption dip was determined in dependence of the excitation power. The result is shown in Figure 6 (a). For pure two-photon absorption, a linear power dependence with no offset is expected from the theory (equation (1)). For the exclusion of non-linear refraction, the transmission measurements were repeated with an additional lens and aperture placed in front of detector 2 (cf., Figure 5). If there is no influence on the transmission signal upon opening and closing the aperture, the detector area is large enough, and defocusing attributed to non-linear refraction can be neglected. In Figure 6 (b), a representative z-scan transmission curve is shown. The curve was recorded using a solution of Ini3 and MIBK at an average laser power of 243 mW.

According to the theory (van Stryland & Sheikbahae, 1989), the change in the transmission is given by

$$\Delta T(z) = \frac{\alpha_2 I_0}{2\sqrt{2}} \cdot \frac{1-\exp(-\alpha_1 L)}{\alpha_1} \cdot \frac{1}{1+z^2/z_R^2} \quad , \tag{1}$$

with α_1 and α_2 as linear and non-linear absorption coefficients, respectively, and z as the cuvette position. z_R is the Rayleigh length, and L is the sample thickness. The intensity I_0 is proportional to the average laser power.

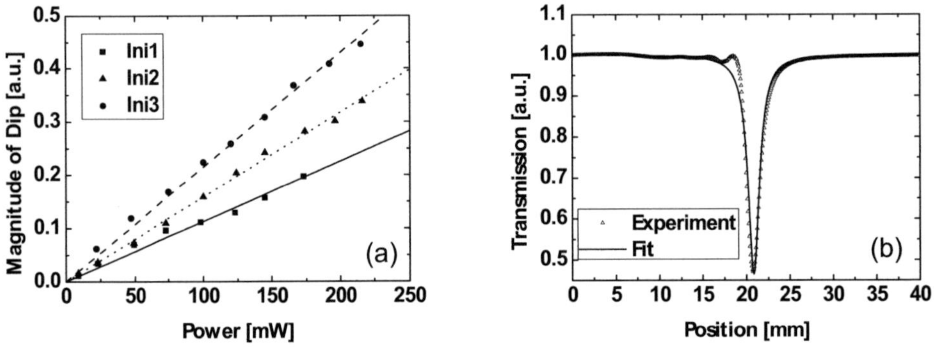

Fig. 6. z-scan results. (a) Magnitude of the transmission signal dip as a function of excitation power for three different photoinitiators. (b) Open aperture trace for Ini3 (for better illustration, only every fifth point is shown).

Due to a negligible linear absorption coefficient, the second term in equation (1) can be approximated by the sample thickness L, leading to a more simplified expression. The non-linear absorption coefficient α_2 can be correlated to the TPA cross-section σ_2 using the photon energy and the density of the initiator molecules in the cuvette. In addition, information on the beam waist w_0 is necessary for the determination of the incident on-axis irradiation I_0. This was determined with a home-built USB camera beam profiler which was scanned along the beam path. A beam waist of about 16 μm was found for the underlying focusing conditions, i.e. the thin sample approximation $z_R > L$ is valid (Sheikbahae et al., 1989).

The TPA cross-sections can be better determined from the slopes of the curves in Figure 6(a), which yield better statistics, because more measurements contribute to the determination of σ_2. From the data it was calculated that Ini3 has the highest absorption cross-section, and thus the highest TPA efficiency, followed by Ini2, while Ini1 has the lowest absorption cross-section which is about a factor of 10 lower than published for the same initiator by Schafer et al. (Schafer et al., 2004). The quantitative results are summarized in Table 1.

Initiator / Cross-section	Ini1	Ini2	Ini3
σ_2 (m^4s)	$(6.7 \pm 0.4) \cdot 10^{-59}$	$(1.4 \pm 0.3) \cdot 10^{-58}$	$(3.2 \pm 0.2) \cdot 10^{-56}$
σ_2 (GM)	0.7	1.4	320
σ_2 (relative to Ini1)	1 ± 0.1	2.1 ± 0.4	472 ± 32

Tab. 1. Calculated TPA cross-sections for Ini1, Ini2, and Ini3. The error bars were determined by identifying the minimum and maximum slope found for each photoinitiator.

There are several possible explanations for the difference in σ_2. The presence of non-linear refraction which significantly influences the TPA cross-section results towards higher values, and cannot be excluded in the data of Schafer et al. (Schafer et al., 2004) due to the fact that no details are given in their publication. Finally, the determination of the beam waist w_0 is difficult and a significant source of error in the determination of σ_2. This is related to the quadratic dependence of I_0 on w_0, i.e. only slight deviations in w_0 will significantly impact the value of σ_2. Therefore, Table 1 also gives relative absorption cross-sections (normalized to Ini1) in order to allow a better comparison of the different photoinitiators.

3.4 TPA patterning

3.4.1 TPA-written arbitrary 3D structures

The most impressing way of demonstrating the possibilities of TPA processing is to write computer-generated, arbitrary 3D structures which demonstrate the ability of scaling up structures from the µm to the cm scale. In order to show the power and the beauty of the technology, we have produced various 3D microstructures using differently functionalized ORMOCER® materials with two commercially available initiators (Ini1 and Ini2, alternatively). Figure 7 shows examples of arbitrary 3D structures which were fabricated in an acrylate and a methacrylate-functionalized ORMOCER®, henceforth labeled as OC-V and OC-I, respectively.

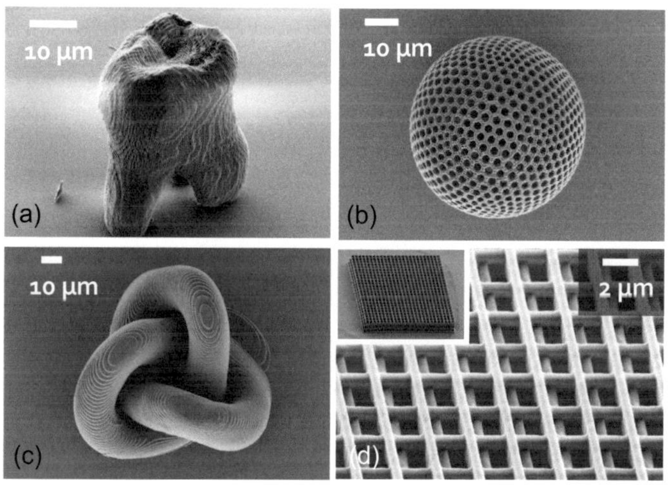

Fig. 7. Selected 3D structures, fabricated by 2PP for different ORMOCER® formulations. (a) Tooth created in OC-I/Ini1 [average power: 500 µW, dimensions: (32 x 37 x 55) µm³], (b) Hollow ball after (Hart,2009) written in OC-V/Ini2 [average power: 34 µW, diameter: 75 µm, hatch distance: 500 nm], (c) Knot after (Wei,2010) created in OC-V/Ini2 [average power: 105 µW, dimensions: (90 x 90 x 50) µm³]. (d) Photonic crystal structure after (Steenhusen, 2008) written in OC-V/Ini1 [average power: 48 µW, period of 2 µm, dimensions: (50 x 50 x 6) µm³]. The writing speeds were (a), (c) 50, (b) 100, and (d) 60 µm/s. All materials were formulated with 3 wt.-% photoinitiator except for (c) which includes only 1 wt.-% initiator.

3.4.2 Voxel size determination

While these types of structures typically inspire end-users, only little is known about the cross-linking behavior of hybrid polymers in this process due to the fact that many effects

influence the reaction kinetics. The minimum achievable feature sizes are related to different effects, which occur simultaneously in the 2PP experiment, influencing each other and which finally will determine the voxel size. Among them are the diffusion of initiators and oxygen molecules, the polarity of the ORMOCER® matrix or traces of solvents, and the process efficiency of the photoinitiator, only to mention some. It could be shown by Monte Carlo simulations that initiator molecules spread into free space after being excited by one or several laser pulses. According to this diffusion of initiator radicals, the voxel is enlarged significantly, because polymerization can be triggered outside the focal volume (Steenhusen, 2008). Oxygen which is present in each material is known to act as radical scavenger, i.e. upon formation of initiating radicals by (laser) light irradiation the initiator's triplet states will, for example be quenched, thus reducing the amount of initiating radicals in the resin (see, e.g. Studer et al., 2003). Although it is widely accepted that the TPA efficiency of the photoinitiators plays a major role in the initiation of the cross-linking, the matrix materials which contain these initiators as well as the propagation of chain growth and termination reactions also have significant impact on the reaction kinetics (Houbertz et al., 2010).
Thus, the voxel dimensions are not only dependent on the technical equipment such as optics used for patterning. Figure 8 shows a schematic of the different interaction volumes which influence the minimum voxel dimensions in TPA-initiated cross-linking experiments, impacting the resulting feature sizes significantly.
The technical interaction volume (red in Figure 8) is principally determined by the employed optics, by the stability of the laser, and by the stability and accuracy of the positioning system. From a technical point of view, this can be optimized by using specially adapted optics (Fuchs et al., 2006), by stabilizing the laser source, and by employing highly accurate positioning stages, mounted on suitable damping systems. The chemical interaction volume (green in Figure 8), however, is much more complicated to minimize, because this is dependent on many different factors such as, for example by the reaction kinetics of the material formulation and, consequently, on the laser-light initiated propagation and termination reactions in the hybrid resin, as already described above. In addition to them, the reaction rate is also influenced by the diffusion of radicals and radical scavengers in the liquid resin (Steenhusen, 2008; Struder et al., 2003).

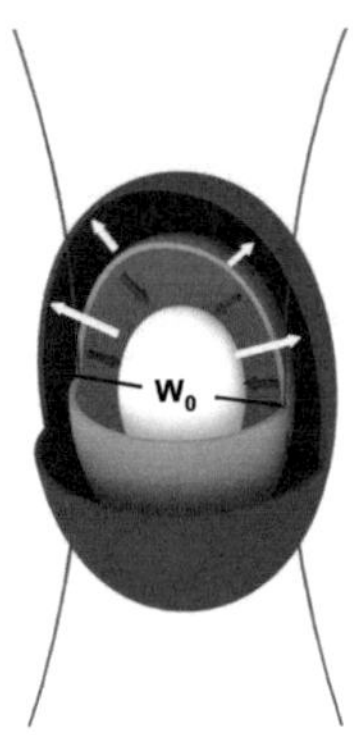

Fig. 8. Schematics of the different interaction volumes, influencing the achievable voxel sizes in a 2PP experiment: technical (gray ellipsoid) and chemical (black ellipsoid) interaction volume. The threshold behavior determines the third interaction volume (white ellipsoid).

The third effect, i.e. the threshold behavior (blue in Figure 8; Tanaka et al., 2002) of the reaction, could principally lead to infinitesimal small voxel sizes. However, aside the exposure dose (determined by the average power, the number of pulses, and the writing speed), the threshold behavior is also dependent on the minimum initiator (i.e. the threshold) concentration necessary to start the chemical reaction. This, however, is not really known, and thus not as well-defined as the laser parameters.

In order to gather information on the 2PP process for a given material formulation, voxel arrays were written using the ascending scan method which is described elsewhere (Sun et al., 2002). In Figure 9, a voxel array is shown which was written using a constant average power of 164 μW. From the left to the right, the exposure time was varied in 2.5 ms intervals, and the height of the laser focus was varied in intervals of 0.25 μm from the top to the bottom of the array. The voxel pitch was set to 2 μm. It has to be mentioned, however, that the degree of cross-linking also has to be considered which will be discussed in the next section.

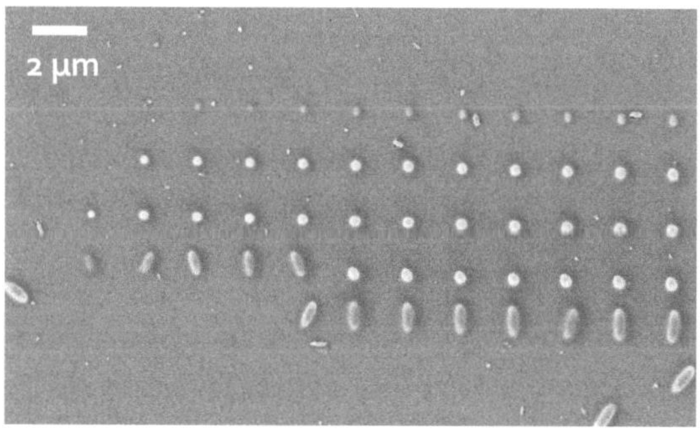

Fig. 9. Typical voxel field created in OC-V with the ascending scan method (Steenhusen et al., 2010a).

Contrary to 2PP experiments previously reported (Kawata et al., 2001; Serbin et al., 2003), the pulse energy for initiating a photochemical reaction is much lower in the present case, being only about 5 to 50 pJ, under the assumption that the focusing condition and the writing speed are comparable in the experiments. There are two possible reasons for this which will be briefly summarized in the following. First of all, the literature data were created using a central wavelength of 800 nm which is about 30 % higher than the wavelength used for our experiments. The overlap of the initiators' maximum in linear extinction coefficient with the laser spectrum significantly determines the process efficiency (Houbertz et al., 2006). For the chosen initiators, this overlap is much more pronounced at 515 than at 800 nm. In addition, a specially designed acrylate-based ORMOCER® system was used for the experiments which usually has a much higher reaction rate than, for example methacrylate-based materials (Odian, 1981).

3.4.3 Investigation on voxel sizes

In order to account for a well-defined fabrication of 3D functional structures for application, an understanding of the underlying polymerization processes initiated by the laser light/material interaction is necessary. By Serbin et al. (Serbin et al., 2003), a simple model which can be used in a first approximation for estimating the voxel diameter d was proposed, where d is given by

$$d(t,F_0) = w_0 \sqrt{\ln\left(\frac{\overset{*}{\sigma_2} F_0^2 t \nu \tau}{\ln(\rho_0/\rho_0 - \rho_{th})}\right)}. \qquad (2)$$

However, the beam waist w_0, the effective TPA cross-section σ_2*, and the threshold radical concentration ρ_{th} for the initiation of the 2PP process which are needed for the calculation of the voxel diameter are not known. The initial photoinitiator concentration is given by ρ_0, F_0 describes the incident photon flux, and t, ν, and τ are the temporal parameters exposure time, repetition rate, and pulse duration, respectively.
In order to investigate the 2PP process at 515 nm, exactly the same material formulation as reported by Serbin et al. was used to create voxel arrays (Steenhusen et al., 2010a). The average laser powers at which voxels could be fabricated were three orders of magnitude lower than reported in (Serbin et al., 2003), i.e. in the μW instead of the mW regime. From the data evaluation assuming the same threshold radical density of 0.25 wt.-%, a TPA cross-section was determined which is four orders of magnitude higher than the one given by Serbin et al.. These differences in the 2PP process are attributed to the higher overlap of the laser spectrum with the initiators' extinction spectrum, because the chemical composition in both experiments is the same.

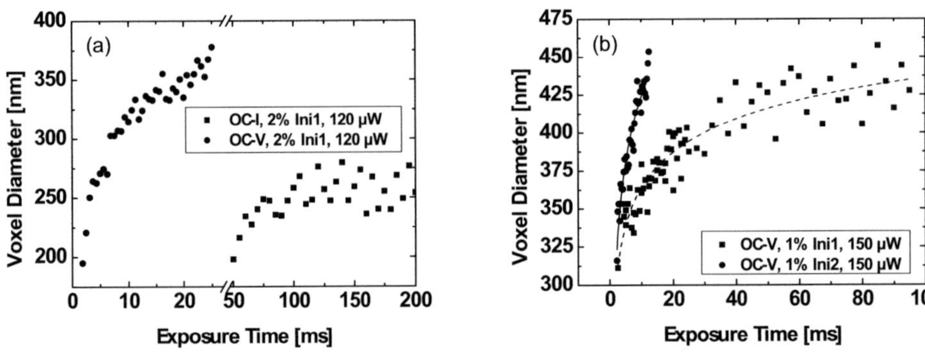

Fig. 10. (a) Voxel size dependence on the applied exposure time for OC-I and OC-V, both formulated with 2 wt.-% Ini1 at an average laser power of 120 μW. (b) Impact of the initiator on the voxel size of OC-V, formulated with 1 wt.-% of Ini1 and Ini2 (Steenhusen et al., 2010a).

In order to demonstrate the different reactivity of various ORMOCER® material systems, voxel arrays were written using OC-I and OC-V, both formulated with 2 wt.-% Ini1, and the resulting voxel diameters were evaluated. In Figure 10 (a), the voxel diameters determined from voxel arrays generated in acrylate-based (OC-V) and the methacrylate-based (OC-I) ORMOCER®s are compared. Obvious from the data is that OC-V requires a significantly shorter exposure time (and thus exposure dose) in order to produce a voxel equivalent in size of the ones fabricated in OC-I which is related to the different reaction rates of acrylate and methacrylate groups (Odian, 1981). A comparison of the TPA cross-sections, however, cannot be performed, since the threshold concentrations will significantly differ due to the different cross-linkable moieties. In addition, the materials have different polarity as well as different oxygen sensitivity.

The voxel diameters of OC-V in dependence of the exposure time are compared. The material was formulated with two different photoinitiators of the same concentration (1 wt.-% Ini1 and Ini2, respectively, at an average laser power of 150 µW). The results are shown in Figure 10 (b). For the formulation of OC-V with Ini2, the voxel diameter increases much steeper than for the same material formulated with Ini1, i.e. Ini2 is much more efficient. From the fits using the model of (Serbin et al., 2003), the TPA cross-section of Ini2 is approximately two times larger than the one of Ini1, which is in good agreement to the z-scan data (cf., Table 1; Steenhusen et al., 2010a). A more comprehensive study will be published elsewhere.

The effect of different initiator concentrations on the voxel formation was also investigated for OC-V at a given average laser power and varying the exposure times which is reported elsewhere (Steenhusen et al., 2010a). Beside other findings, it was observed that the dependency of the voxel sizes on the initiator concentration is not linear. From investigations on the cross-linking behavior and the resulting refractive indices in dependence of the UV initiator concentrations which were carried out by one-photon processes (classical UV exposure), it was concluded that different initiator concentrations lead to different inorganic-organic hybrid networks in the final layer (Houbertz et al., 2004; Fodermeyer, 2009; Landgraf, 2010).

Finally, the extraordinary performance of Ini3 should be underlined by the fact that voxel sizes comparable to the ones fabricated using Ini1 and Ini2 in a given ORMOCER® material system were achieved with an about 200 times lower initial initiator concentration of Ini3 than of Ini1 or Ini2.

An investigation of the voxel diameter in dependence of the exposure time at different average laser powers has revealed that the higher the laser power, the larger the voxel diameters will be (Steenhusen et al., 2010a). The determined TPA cross-section σ_2 by using equation (2), however, are about two times larger than derived from the z-scan experiments, which can be attributed to the fact that the assumed threshold concentration of 0.25 wt.-% is too high. Additional experiments with conventional UV exposure which were carried out to support this statement have revealed that the organic cross-linking can be initiated for initiator concentrations being as low as 0.01 wt.-% (Landgraf, 2010). However, although the model proposed by (Serbin et al., 2003) yields a reasonable starting point for theoretically determining the TPA cross-sections, it lacks of some important effects such as the diffusion initiator radicals or molecular oxygen.

As mentioned above, the minimum voxel sizes which can be fabricated are dependent on many different parameters, among which the chemical and the threshold behavior are the most difficult to quantify. In the following, some results will be presented for sub-100 nm patterning, and they will be discussed with respect to the degree of organic cross-linking.

The typical minimum feature sizes reported for several years were about 100 nm ("resolution limit"). Recently, several groups have reported sub-100 nm resolution using various polymer materials, where minimum feature sizes down to 40 nm were achieved, some of them using the stimulated emission depletion (STED) approach (Li et al., 2009; Andrew et al., 2009, Haske et al., 2007). In Figure 11, a representative image of a voxel, fabricated in a styryl-based ORMOCER®, formulated with 2 wt. % Ini1 is shown. The patterning was carried out at an average laser power of 65 µW and an exposure time of 100 ms, with no further optimization of the technical equipment, yielding a voxel diameter of about 90 nm. Features as small as about 75 nm can be routinely achieved, and the data will routinely achieved, and these data be published elsewhere.

From conventional UV lithography in dependence on the processing parameters it is known that the organic cross-linking is very sensitive to the process conditions. If these are not suitably chosen or adapted, part of the material will not be cross-linked, and will be removed in the development step. This then results, for example in lower layer thicknesses or smaller structures than adjusted. The same effects can be observed in 2PP experiments, since the underlying process is a laser light-induced organic cross-linking, i.e. if the 2PP parameters are not optimized with respect to the reaction kinetics of the material, smaller structures consequently will result. It has to be mentioned, however, that there is a trade-off between threshold effect and cross-linking by reducing the photon dose. By driving the threshold effect, smaller structures will definitely occur which, however, might not be as well cross-linked as voxels being fabricated with a higher photon dose and/or initiator concentration, i.e. the resulting voxels will be less stable, and further reduction in size by the development step might thus occur. This needs to be investigated in more detail.

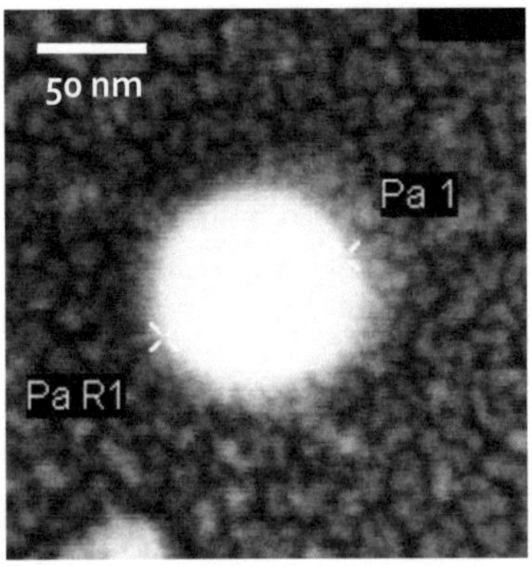

Fig. 11. Sub-100 nm voxel (diameter: 90 nm), fabricated by 2PP in a styryl-based ORMOCER® material, formulated with 2 wt.-% Ini1.

We therefore have started to investigate the degree of organic cross-linking of ORMOCER® materials which were processed by 2PP by high-resolution μ-Raman spectroscopy. In Figure 12, typical μ-Raman spectra are displayed as well as the degree of organic cross-linking of OC-I formulated with 1 wt.-% Ini1 in dependence on the average laser power. As μ-Raman sample, squares of 10 μm x 10 μm were written with a velocity of 100 μm/s and a hatch distance of 0.1 μm. In Figure 12 (a), two μ-Raman spectra are displayed for a different cross-linking state of OC-I. At about 1648 cm^{-1}, the C=C bond resulting from the methacrylate groups which decreases in intensity the more cross-linked the material is can be seen. As internal reference, the C=C bond of the diphenylsilane precursor at 1569 cm^{-1} was used. The calculation of the degree of cross-linking was performed as reported in (Houbertz et al., 2004), and the first result is shown in Figure 12 (b). Analogously to the results from ORMOCER® layers which were prepared by conventional UV lithography, the degree of cross-linking increases continuously until saturation for the given process conditions. However, almost the same magnitude in organic cross-linking is achieved in saturation by TPA processing as for classical UV exposure. A more comprehensive study on the TPA-initiated organic cross-linking will be published elsewhere.

Additionally to the 2PP experiments, first patterning by 3PP using the fundamental wavelength of 1030 nm was performed which was straightforward when considering the

extinction spectra of the initiators (Steenhusen et al., 2010a). From the spectra it can be concluded that no TPA processes will occur, because there is no absorption of the initiator at 515 nm. Excitation with three photons, is likely depending on the three-photon absorption cross-sections which have to be evaluated for the different systems from z-scan experiments at 1030 nm. The latter is still under investigation. A resulting voxel array written using OC-1 with 2 wt.-% Ini1 and an exposure time of 200 ms is displayed in Figure 13. A photonic crystal structure written by 3PP can be found in (Steenhusen et al., 2010b).
The average laser power was with 5.2 to 5.7 mW about three orders of magnitude higher than for the respective TPA process at 515 nm, indicating a higher order non-linear process, being related to the lower efficiency of the 3PA process compared to the TPA process.

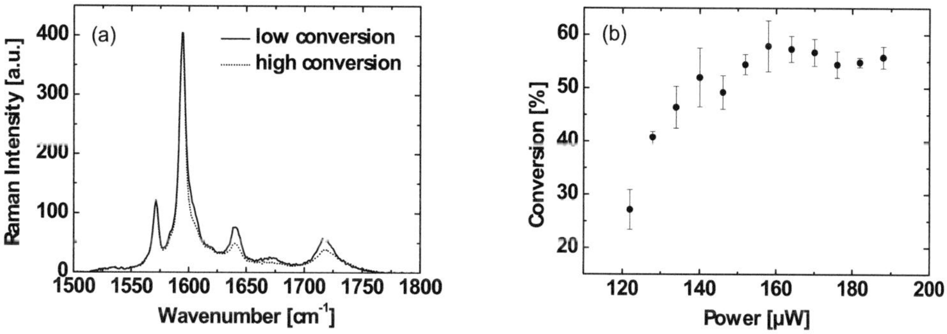

Fig. 12. Cross-linking investigations of OC-1/1 wt.-% Ini1. (a) Selected µ-Raman spectra, and (b) degree of organic cross-linking in dependence on the average laser power.

By evaluating the voxel size, it can be seen that features being only the seventh part of the fundamental wavelength are achieved. The voxel pitch was set to 2 µm, and the smallest voxel in these data has a diameter as low as 155 nm which is far beyond the diffraction limit. However, also for these data the degree of organic cross-linking needs further investigation in order to give final proof for real sub-diffraction limit structures.

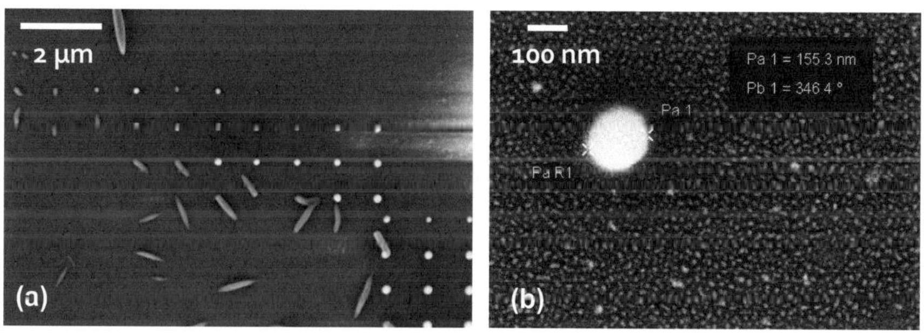

Fig. 13. (a) Voxel array (pitch 2 µm) written by 3PP in OC-1/2 wt.-% Ini1, and (b) zoom into (a), displaying an individual voxel of about 155 nm in diameter (i.e., a feature size of λ/7).

The data yield a proof of concept for 3PP experiments at 1030 nm. By varying the exposure parameters, a tremendous potential for further decreasing the feature sizes is seen. A more comprehensive study of 3PP processes at 1030 nm including z-scan experiments is presently carried out, and will be published elsewhere.

3.4.4 Large-scale TPA patterning

Up to now, most patterning results making use of TPA processes are restricted to smaller scale structures, where typically structures of view hundreds of µm in size were reported (Ostendorf & Chichkov, 2006). The restriction in structure dimensions is mainly related to limitations of the working distance of the high-NA focussing optics and to long fabrication times. Instead of the focussing objective with an NA of 1.4 which is used for high-resolution patterning, for large-scale fabrication this objective was replaced either by a microscopy objective with an NA of 0.60 or with an NA of 0.45, characterized by long working distances (cf., section 2.2). In addition, they offer a correction collar enabling an adaptation to different cover glass thicknesses ranging between 0 and 2 mm in order to reduce spherical aberration, resulting from a refractive index mismatch of air, glass substrate, and ORMOCER® resin, leading to blurring of the focal light distribution. Due to fact that the refractive index mismatch of glass and resin is very small compared to their difference to the refractive index of air, the spacer thickness can be included into the corrective adjustments. Nevertheless, this correction of the spherical aberration is only valid for a distinct penetration depth of the focal spot into the resin, and thus inhomogeneous patterning results can be observed during processing with the common sandwich configuration (cf., Figure 3 (a)) and varying the penetration depth by vertically objective movement (Stichel et al., 2010).
In order to demonstrate the full potential of the TPA technology, the experimental setup for the TPA patterning was modified (cf., Figure 3 (c)) in order to allow the fabrication of high resolution large-scale structures with structure heights being not limited by the objective's working distance. These structures might be employed, for example as scaffolds for regenerative or biomedicine (see also section 3.4.5). In Figure 14, two examples for the 3D fabrication of arbitrary 3D large-scale structures by 2PP in an acrylate-based ORMOCER® (OC-V/2 wt.-% Ini2) are shown.

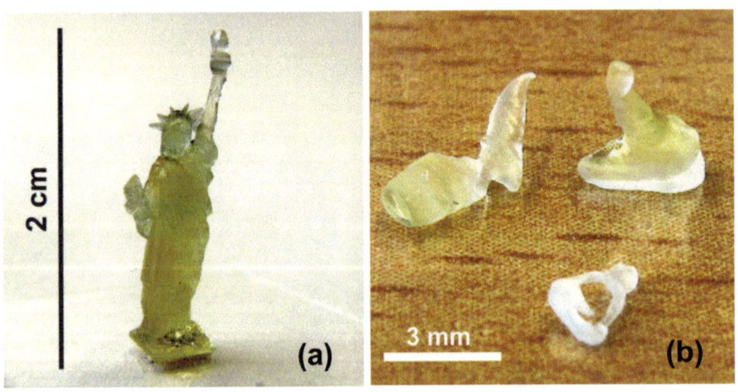

Fig. 14. Examples of large-scale structures fabricated by 2PP in OC-V, formulated with 2 wt.-% Ini2. (a) Statue of liberty, and (b) human ossicles in life-size.

3.4.5 Application examples

Finally, in this section two application examples will be given, one for optics and the other one for biomedical applications.

Due to the fact that the selected ORMOCER® materials exhibit particularly low absorption losses at data and telecom wavelengths (850, 1310, and 1550 nm) (Houbertz et al., 2003b), the employment of TPA for the fabrication of highly sophisticated optical designs would be advantageous, since this process can also be carried out on pre-configured substrates, already containing opto-electronic elements such as laser- or photodiodes, vertical cavity surface emitting lasers (VCSEL), or microlenses.

Two-photon absorption (TPA) processing was used for the fabrication of multimode waveguide (WG) using just one individual ORMOCER® material which was specially designed for the process. This reduces the process steps significantly, and only two to three process steps need to be performed in order to create the waveguide (Houbertz, 2007; Houbertz et al., 2008). The ORMOCER® material was coated onto a pre-configured printed-circuit board (PCB) substrate, where laser source (transmitter) and photo-diode (receiver) were already mounted. As laser source for this application, a femtosecond laser (fundamental wavelength λ = 800 nm, pulse durations between 130 and 150 fs) was employed, and focused about 80 to 250 µm deep into the ORMOCER® layer without using a cover glass. This depth is just dependent on the position of the optoelectronic devices' active surfaces. The patterning by TPA then results in solid polymerized structures embedded in the non-exposed resin. The waveguide is then finally obtained by thermally treating the samples for 2 h at 200 °C in a nitrogen atmosphere. This particularly avoids any solvent-based processing (cf., Figure 4). Dependent on the chosen optoelectronic elements, data transfer rates as high a 7 Gbit/s at a bit error ratio of about 10^{-9} were routinely achieved.

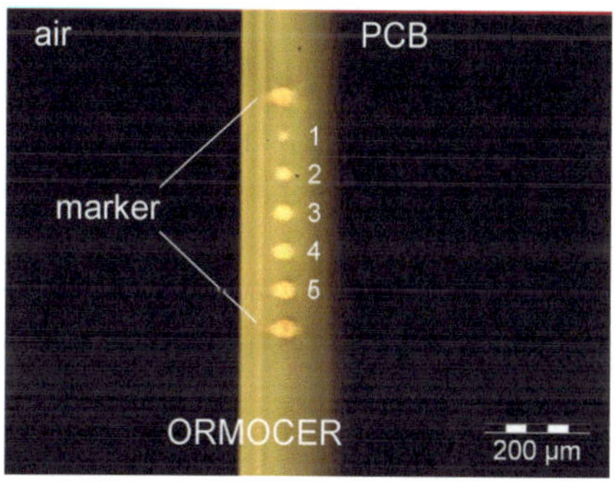

Fig. 15. TPA-WG fabricated by 2PP in a specially designed acrylate-based ORMOCER® after (Houbertz et al., 2008a). As laser source, a Ti:sapphire laser was used operating at 800 nm.

Another application example is related to the field of regenerative or biomedicine which attracts increasing attention. For example, micro-needle fabrication for drug delivery or the realization of scaffold structures using 2PP was already demonstrated (Doraiswamy et al., 2005; Narayan et al., 2005; Ostendorf & Chichkov, 2006). Scaffolds for medical applications provide 3D structures with well-defined shapes with an interconnecting pore structure in the range of a few up to several hundreds of µm, thus mimicking the properties of extracellular matrices. Such artificial matrices should support 3D cell formation, cell proliferation, and differentiation in order to create neo-tissue or grafts from autologous cell cultures.

The large-scale fabrication of biomedical scaffold structures with dimensions in the mm-range still remains very challenging from a technical and a materials' point of view. Most commercial rapid prototyping techniques cannot provide sufficiently small structure sizes of a few μm in order to produce highly-porous scaffolds. Thus, 2PP with tailored material systems is a promising technology for this application, because it allows a real 3D fabrication at high resolution and a free design of the structures. In Figure 16, various highly-porous scaffolds fabricated by 2PP in OC-V/Ini2 are shown. An objective with a NA of 0.45 in the inverted configuration (cf., Figure 3 (c)) was used. It has to be mentioned, however, that the processing times are not yet optimized, and the equipment is continuously modified in order to reduce the necessary production time. This will be published elsewhere.

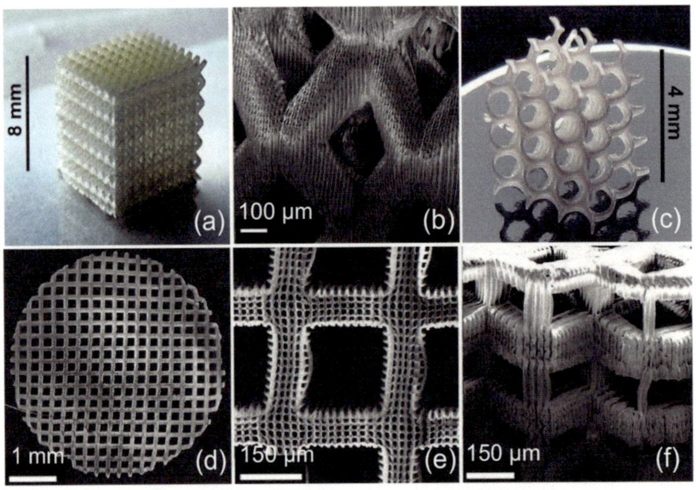

Fig. 16. Scaffolds fabricated by 2PP in OC-V/2 wt.-% Ini2. (a,b) Scaffold after a design from (Phoenix), redesigned by the authors, (c) scaffold after (Hart, 2009), and (d-f) scaffold with cubically designed pores of about 180 μm size (hatch distance 20 μm).

4. Conclusions

We have demonstrated the use of visible and infrared laser pulses for the fabrication of several types of sub-diffraction limit micro- and nanostructures by 2PP and 3PP. The data demonstrate that a well-defined fabrication of arbitrary structures can routinely be performed. The TPA cross-sections of some photoinitiators were characterized by the z-scan method, and were correlated to voxel size studies for differently functionalized hybrid polymers. First results on the fabrication of sub-100 nm features with a specially tailored hybrid material were presented as well. An investigation of the degree of organic cross-linking of patterns written by 2PP has yielded that it is comparable to the one achieved on conventionally UV-exposed ORMOCER® layers. First 3PP results demonstrate a significant circumvention of the diffraction limit, resulting in feature sizes of only λ/7 even without any optimization of process and material. Large-scale scaffolds up to the cm regime were fabricated using a modified TPA setup which has a huge potential for biomedical and tissue engineering applications. The scaffolds were fabricated with interconnecting pores, and a pore density as high as 90 % can be crated by using hybrid polymer materials due to their excellent mechanical stability. However, further work concerning the TPA cross-sections of initiators and voxel formation for differently functionalized materials including investigations on the organic cross-linking by spectroscopic methods and mechanical stability investigations are presently carried out.

5. Acknowledgements

We thank Carola Cronauer and Adelheid Martin for their excellent support in materials synthesis, formulation, and characterization. Financial support from the Deutsche Forschungsgemeinschaft (grant: HO 2475/3-1), and from the Fraunhofer-Gesellschaft für Angewandte Forschung e.V. (Challenge Programme) is gratefully acknowledged. One of us (R.H.) would like to express special thanks to the friends who supported writing of this manuscript by providing continuous inspiration, particularly J.W. All colleagues who have contributed to our work by supporting us with discussions and other support are greatly appreciated.

6. References

Andrew, T. L.; Tsai, H. J. & Menon, R. (2009), Confining Light to Deep Subwavelength Dimensions to Enable Optical Nanopatterning, *Science* 324 (5929), 917-921

Asaka, S.; Nakanishi, S.; Itoh, H.; Kamada, M.; Watanabe, M.; Arimoto, O.; Fujiwara, S.; Tsujibayashi, T. & Itoh, M. (1998), Time-gated photon counting method for two-photon spectroscopy using synchrotron radiation and laser, *Review of Scientific Instruments* 69 (5), 1931-1933

BASF, initiators from Ciba-Geigy, now part of BASF

Baudach, S.; Bonse, J.; Krüger, J. & Kautek, W. (2000), Ultrashort pulse laser ablation of polycarbonate and polymethylmethacrylate, *Applied Surface Science* 154-155 (1-4), 555-560

Beyer, M.; Stichel, T.; Houbertz, R. & Sextl, G. (2010), Novel biocompatible hybrid polymers processed by two-photon polymerization (unpublished data)

Blanchet, G. B.; Fincher Jr., C. R.; Jackson, C. L.; Shah, S. I. & Gardner, K.H. (1993), Laser Ablation and the Production of Polymer Films, *Science* 29 (262). (5134), 719 - 721

Bowman, C. N. & Kloxin, C. J. (2008), Toward an Enhanced Understanding and Implementation of Photopolymerization Reactions, *AIChE Journal* 54 (11), 2775 - 2795

Bräuer, A.; Dannberg, P.; Mann, G. & Popall, M. (2001), Precise Polymer Micro-Optical Systems, *MRS Bulletin* 26, 519-522

Buestrich, R.; Kahlenberg, F.; Popall, M.; Dannberg, P.; Müller-Fiedler, R. & Rösch, O. (2001), ORMOCER®s for Optical Interconnection Technology, *Journal of Sol-Gel Science and Technology* 20, 181 - 186

Cheng, Y.; Sugioka, K.; Msuda, M.; Toyoda, K.; Kawachi, M.; Shihoyama, K. & Midorikawa, K. (2003), 3D microstructuring inside Foturan glass by femtosecond laser, *RIKEN Review*, No. 50, 101-106

Claeyssens, F.; Hasan, E. A.; Gaidukeviciute, A.; Achilleos, D. S.; Ranella, A.; Reinhardt, C.; Ovsianikov, A.; Xiao, S.; Fotakis, C.; Vamvakaki, M.; Chichkov, B. N. & Farsari, M. (2009), Three-dimensional biodegradable structures fabricated by two-photon polymerization, *Langmuir* 25 (5), 3219–3223

Conforti, P. F.; Prasad, M. & Garrison, B.J. (2007), Simulations of Laser Ablation of Poly(methyl methacrylate): Fluence versus Number of Photons, *Journal of Physical Chemistry C* 111 (32), 12024–12030

Cumpston, B. H.; Ananthavel, S. P.; Barlow, S.; Dyer, D. L.; Ehrlich, J. E.; Erskine, L. L.; Heikal, A. A.; Kuebler, S. M.; Lee, I. Y. S.; McCord-Maughon, D.; Qin, J.; Rockel, R.; Rumi, M.; Wu, X.-L.; Marder, S. R. & Perry, J. W. (1999), Two-photon polymerization initiators for three-dimensional optical data storage and microfabrication, *Nature* 398, 51–54

Declerck, P. (2010), Synthesis and technological processing of hybrid inorganic-organic materials for photonic applications, *Dissertation*, Julius-Maximilian University Würzburg

Denk, W.; Strickler, H. & Webb, W. (1990), 2-Photon Laser Scanning Fluorescence Microscopy, *Science* 6, 248 (4951), 73-76

Denk, W. & Svoboda, K. (1997), Photon upmanship: why multiphoton imaging is more than a gimmick, *Neuron* 18 (3), 351–357

Deubel, M.; v. Freymann, G.; Wegener, M.; Pereira, S.; Busch, K. & Soukoulis, C. M. (2004), Direct laser writing of three-dimensional photonic-crystal templates for telecommunications, *Nature Materials* 3, 444-447

Doraiswamy, A.; Platz, T.; Narayan, R. J.; Chichkov, B.; Ovsianikov, A.; Houbertz, R.; Modi, R.; Auyeung, R. & Chrisey, D. B. (2005), Biocompatibility of CAD/CAM ORMOCER® polymer scaffold structures, *Materials Research Society Symposium Proceedings* 845, AA2.4.1

El-Batahgy, A. M. (1997), Effect of laser welding parameters on fusion zone shape and solidification structure of austenitic stainless steels, *Materials Letters* 32 (2-3), 55-163

Fodermeyer, V. (2009), Spektroskopische Untersuchung zur lichtinduzierten Vernetzung von Hybridpolymeren, *Diploma Thesis*, Julius-Maximilian University, Würzburg

Fuchs, U.; Zeitner, U. D. & Tünnermann, A. (2006), Hybrid optics for focusing ultrashort laser pulses, *Optics Letters* 31 (10), 1516-1518

Gittard, S. D.; Ovsianikov, A.; Akar, H.; Chichkov, B.; Monteiro-Riviere, N. A.; Stafslien, S.; Chisholm, B.; Shin, C. C.; Shih, C. M.; Lin, S. J.; Su, Y. Y. & Narayan, R. J. (2010), Two Photon Polymerization-Micromolding of Polyethylene Glycol-Gentamicin Sulfate Microneedles, *Advanced Engineering Materials* 12 (4), B77-B82

Göppert-Mayer, M. (1931), Über Elementarakte mit zwei Quantensprüngen, *Annalen der Physik* 401 (3), 273–294

Haas, K.-H. & Wolter, H. (1999), Synthesis, properties and applications of inorganic-organic copolymers (ORMOCER®s), *Current Opinion in Solid State and Materials Science* 4, 571–580

Haas, K.-H. (2000), Hybrid Inorganic-Organic Polymers Based on Organically Modified Si-Alkoxides, *Advanced Engineered Materials* 2, 571 - 582

Haas, K.-H. & Wolter, H. (2001), Properties of Polymer-Inorganic Composites, *Encyclopedia of Materials: Science and Technology* Oxford, Elsevier: 7584-7594

Haas, U.; Haase, A.; Satzinger, V.; Pichler, H.; Leising, G.; Jakopic, G.; Stadlober, B.; Houbertz, R.; Domann, G. & Schmitt, A. (2006), Hybrid polymers as tunable and directly-patternable gate dielectrics in organic thin-film transistors, *Physical Review B* 73, 235339.1-235339.7

Hart, G.W (2009), http ://www.georgehart.com/rp/rp.html

Haske, W.; Chen, V. W.; Hales, J. M.; Dong, W. T.; Barlow, S.; Marder, S. R. & Perry, J. W. (2007), 65 nm feature sizes using visible wavelength 3-D multiphoton lithography, *Optics Express* 15 (6), 3426-3436

Helmerich, A.; Raether, F.; Peter, D. & Bertagnolli, H. (1994), Structural studies on an ORMOCER® system containing zirconium, *Journal of Materials Science* 29, 1388 - 1393

Houbertz, R.; Froehlich, L.; Schulz, J. & Popall, M. (2001), Inorganic-Organic Hybrid Materials (ORMOCER®s) for Multilayer Technology - Passivation and Dielectric Behavior, *Materials Research Society Symposium Proceedings* 665, 321 - 326

Houbertz, R.; Schulz, J.; Froehlich, L.; Domann, G. & Popall, M. (2003a), Inorganic-organic hybrid materials for polymer electronic applications, *Materials Research Society Symposium Proceedings* 769, 239-244

Houbertz, R.; Domann, G.; Cronauer, C.; Schmitt, A.; Martin, H.; Park, J.-U.; Fröhlich, L.; Buestrich, R.; Popall, M.; Streppel, U.; Dannberg, P.; Wächter, C. & Bräuer, A. (2003b), Inorganic-organic hybrid materials for application in optical devices, *Thin Solid Films* 442, 194–200

Houbertz, R.; Domann, G.; Schulz, J.; Olsowski, B.; Fröhlich, L. & Kim, W. S. (2004), Impact of photoinitiators on the photopolymerization and the optical properties of inorganic-organic hybrid polymers, *Applied Physics Letters* 84 (7), 1105–1107

Houbertz, R., Wolter, H.; Dannberg, P.; Serbin, J. & Uhlig, S. (2006), Advanced packaging materials for optical applications: bridging the gap between nm-size structures and large-area panel processing, *Proceedings of SPIE* 6126, 612605.1-612605.13, doi: 10.1117/12.660140

Houbertz, R. (2007), http://www.fraunhofer.de/ueber-fraunhofer/wissenschaftliche-exzellenz/fraunhofer-wissenschaftspreise/archiv/

Houbertz, R.; Satzinger, V.; Schmid, V.; Leeb, W. & Langer, G. (2008a), Optoelectronic printed circuit board: 3D structures written by two-photon absorption, *Proceedings SPIE* 7053, 70530B.1 -70530B.13

Houbertz, R.; Declerck, P.; Passinger, S.; Ovsianikov, A.; Serbin, J. & Chichkov, B. N. (2008b), Investigations on the generation of photonic crystals using two-photon polymerization (2PP) of inorganic-organic hybrid polymers with ultra-short laser pulses, In: *Nanophotonic Materials: Photonic Crystals, Plasmonics, and Metamaterials*, R.W. Wehrspohn, H.S. Kitzerow, & K. Busch (Eds.) , 97-114, Wiley VCH, 978-3-527-40858-0, Weinheim, 97 - 113

Houbertz, R.; Steenhusen, S.; Beyer, M. & Sextl, G. (2010) (unpublished data)

Jia, B. H.; Serbin, J.; Kim, H.; Lee, B.; Li, J. F. & Gu, M. (2007), Use of two-photon polymerization for continuous gray-level encoding of diffractive optical elements, *Applied Physics Letters* 90, 073503.1-073503.3

Kawata, S.; Sun, H. B.; Tanaka, T. & Takada, K. (2001), Finer features for functional microdevices, *Nature* 412 (6848), 697–698

Koechlin, M.; Poberaj, G. & Günter, P. (2009), High resolution laser lithography system based on two-dimensional acousto-optic deflection, *Review of Scientific Instruments* 80, 085105

Landgraf, F. (2010), Spektroskopische Untersuchung zur UV-induzierten Vernetzung von Hybridpolymeren, Diploma Thesis, FH Regensburg

Li, J.; Jia, B. & Gu, M. (2008), Engineering stop gaps of inorganic-organic polymeric 3D woodpile photonic crystals with post-thermal treatment, *Optics Express* 16 (24), 20073-20080

Li, L. J.; Gattass, R. R.; Gershgoren, E.; Hwang, H. & Fourkas, J. T. (2009), Achieving lambda/20 Resolution by One-Color Initiation and Deactivation of Polymerization, *Science* 324 (5929), 910-913

Maruo, S.; Nakamura, O. & Kawata, S. (1997), Three-dimensional microfabrication with two-photon-absorbed photopolymerization, *Optics Letters* 22 (2), 132-134

Narayan, R. J.; Jin, C.; Doraiswamy, A.; Mihailescu, I. N.; Jelinek, M.; Ovsianikov, A.; Chichkov, B. N. & Chrisey, D. B. (2005), Laser Processing of Advanced Bioceramics, *Advanced Engineered Materials* 7 (12), 1083-1098

Nolte, S.; Will, M.; Burghoff, J. & Tünnermann, A. (2004), Ultrafast laser processing: new options for three-dimensional photonic structures, *Journal of Modern Optics* 51 (16-18), 2533-2542

Odian, G. (1981), *Principles of polymerization*, Wiley-Interscience, New York, 2nd edition

Ostendorf, A. & Chichkov, B.N. (2006), Two-Photon Polymerization: A new Approach to Micromachining, *Photonics Spectra*, October Issue

Phoenix GmbH & Co. KG, Gröbenzell, Germany

Robertsson, M. E.; Hagel, O. J.; Gustafsson, G.; Dabek, A.; Popall, M.; Cergel, L.; Wennekers, P.; Kiely, P.; Lebby, M. & Lindhal, T. (1998), O/e-MCM Packaging with New, Patternable Dielectric and Optical Materials, *Proceedings 48th Electronic Component Technology Conference* (Seattle, Washington, USA), IEEE Catalogue No. 98CH36206, 1413-1421

Rousseau, A. & Boutevin, B. (1992), Synthesis of low absorption halogenated polymers for POF, *Proceedings of the Plastic Optical Fibers Conference*, 33–37, Paris, June 22-23, 1992

Sanchez, C.; Julián, B. ; Belleville, Ph. & Popall, M. (2005), Applications of hybrid organic–inorganic nanocomposites, *Journal Materials Chemistry* 15, 3559–3592

Schafer, K. J.; Hales, J. M.; Balu, M.; Belfield, K. D.; Van Stryland, E. W. & Hagan, D. J. (2004), Two photon absorbtion cross-sections of common photoinitiators, *Journal Photochemistry and Photobiology* A162, 497–502

Schizas, C.; Melissinaki, V.; Gaidukeviciute, A.; Reinhardt, C.; Ohrt, C.; Dedoussis, V.; Chichkov, B. N.; Fotakis, C.; Farsari, M. & Karalekas, D. (2010), On the design and fabrication by two-photon polymerization of a readily assembled micro-valve, *International Journal of Advanced Manufacturing Technology* 48 (5-8), 435-441

Schweizer, Th.; Neumeister, A.; Guo, Q.; Wohlleben, W.; Leyrer, R. J.; Kling, R. & Ostendorf, A. (2008), Generation of Photonic Crystal End Faces Using Laser Microfabrication, *JLMN - Journal of Laser Micro/Nanoengineering* 3 (3), 141-146

Seidl, B. & R. Liska (2007), Mechanistic investigations on a diynone type photoinitiator, *Macromolecular Chemistry and Physics* 208 (1), 44-54

Serbin, J.; Egbert, A.; Ostendorf, A.; Chichkov, B. N.; Houbertz, R.; Domann, G.; Schulz, J.; Cronauer, C.; Fröhlich, L. & Popall, M. (2003), Femtosecond laser-induced two-photon polymerization of inorganic–organic hybrid materials for applications in photonics, *Optics Letters* 28, 301-303

Serbin, J.; Ovsianikov, A. & Chichkov, B. N. (2004), Fabrication of woodpile structures by two-photon polymerization and investigation of their optical properties, *Optics Express* 12 (21), 5221–5228

Sheikbahae, M.; Said, A. A. & Vanstryland, E. W. (1989), High-sensitivity, single-beam n_2 measurements, *Optics Letters* 14 (17), 955–957

Sheikbahae, M.; Said, A. A.; Wei, T. H.; Hagan, D. J. & Vanstryland, E. W. (1990), Sensitive measurement of optical nonlinearities using a single beam, *IEEE Journal of Quantum Electronics* 26 (4), 760–769

Sieben, M. & Brunnecker, F. (2009), Welding plastic with lasers, *Nature Photonics* 3, 270–272

Simchi, A. (2006), Direct laser sintering of metal powders: Mechanism, kinetics and microstructural features, *Materials Science and Engineering A* 428, 148–158

Song, J. H.; O'Brien, P. & Peters, F. H. (2009), Optimal laser welding assembly sequences for butterfly laser module packages, *Optics Express* 17 (19), 16406-16414

Spyridaki, M.; Koudoumas, E.; Tzanetakis, P.; Fotakis, C.; Stoian, R.; Rosenfeld, A. & Hertel, I. V. (2003), Temporal pulse manipulation and ion generation in ultrafast laser ablation of silicon, *Applied Physics Letters* 83 (7), 1474–1476

Stampfl, J.; Fouad, H.; Seidler, S.; Liska, R.; Schwager, F.; Woesz, A. & Fratzl, P. (2004), Fabrication and moulding of cellular materials by rapid prototyping, *Int. J. Materials and Product Technology* 21 (4), 285-296

Steenhusen, S. (2008), Mikrostrukturierung von Hybridpolymeren mit Zwei-Photonen-Absorption, *Diploma Thesis*, Julius-Maximilian University Würzburg.

Steenhusen, S.; Stichel, T.; Houbertz, R. & Sextl, G. (2010a), Multi-photon polymerization of inorganic-organic hybrid polymers using visible or IR ultrafast laser pulses for optical or optoelectronic devices, *Proceedings of SPIE* 7591, 759114.1–759114.12

Steenhusen, S.; Houbertz, R. & Sextl, G. (2010b), 3D sub-diffraction limit patterning of hybrid polymers with visible and infrared laser pulses, *Proceedings Laser Precision Micromachining LPM 2010*, Stuttgart, June 7-10, 2010

Stichel, T.; Houbertz, R.; Hecht, B. & Sextl, G. (2010), Two-photon polymerization as method for the fabrication of large scale biomedical scaffold applications, *Proceedings Laser Precision Micromachining* LPM 2010, Stuttgart, June 7-10, 2010

Streppel, U.; Dannberg, P.; Wächter, Ch.; Bräuer, A.; Fröhlich, L.; Houbertz, R. & Popall, M. (2002), New wafer-scale fabrication method for stacked optical waveguide interconnects and 3D micro-optic structures using photoresponsive (inorganic-organic hybrid) polymers, *Optical Materials* 21, 475 –483

Studer, K.; Decker, Ch.; Beck, E. & Schwalm, R. (2003), Overcoming oxygen inhibition in UV-curing of acrylate coatings by carbon dioxide inerting, Part I, *Progress in Organic Coatings* 48, 92–100

Sun, H. B.; T. Tanaka & Kawata, S. (2002), Three-dimensional focal spots related to two-photon excitation, *Applied Physics Letters* 80 (20), 3673-3675

Tanaka, T.; Sun, H. B. & Kawata, S. (2002), Rapid sub-diffraction-limit laser micro/nanopro-cessing in a threshold material system, *Applied Physics Letter* 80 (2), 312-314

Uhlig, S.; Fröhlich, L.; Chen, M.; Arndt-Staufenbiel, N.; Lang, G.; Schröder, H.; Houbertz, R.; Popall, M. & Robertsson, M. (2006), Polymer optical interconnects - a scalable large-area panel processing approach, *IEEE Transactions Advanced Packaging* 29, 158–170

Wojcik, A. B. & Klein, L. C. (1995), Transparent Inorganic/Organic Copolymers by the Sol-Gel Process: Copolymers of Tetraethyl Orthosilicate (TEOS), Vinyl Triethoxysilane (VTES) and (Meth)arcylate Monomers, *Journal of Sol-Gel Science and Technology* 4, 57–66

Van Stryland, E. W. & Sheikbahae, M. (1998), Z-scan measurements of optical nonlinearities, In: *Characterization Techniques and Tabulations for Organic Nonlinear Materials*, Kuzyk, M. and Dirk, C. (eds.), 655–692, Marcel Dekker Inc., 978-0824799687 New York

Wei, F. (2010), http ://www.eng.nus.edu.sg/LCEL/RP/u21/wwwroot/stl library.htm

Wong, S.; Deubel, M.; Pérez-Willard, P.; John, S.; Ozin, G. A.; Wegener, M. & von Freymann, G. (2006), Direct Laser Writing of Three-Dimensional Photonic Crystals with a Complete Photonic Bandgap in Chalcogenide Glasses, *Advanced Materials* 18 (3), 265-269

Yamaguchi, S. & Tahara, T. (2003), Two-photon absorption spectrum of all-trans retinal, *Chemical Physics Letters* 276 (1-2), 237-243

26

Several Diffractive Optical Elements Fabricated by Femtosecond Laser Pulses Writing Directly

Zhongyi Guo[1,2], Lingling Ran[2,3], Shiliang Qu[1,2] and Shutian Liu[1]
[1]Department of Physics, Harbin Institute of Technology, Harbin, 150001,
[2]Department of optoelectronic science, Harbin Institute of Technology at Weihai, Weihai 264209,
[3]College of Electronic Engineering, Heilongjiang University, Harbin 150080, China

1. Introduction

With the developments of the laser technology, femtosecond laser technology is emerging as one of the useful microfabrication tools in recent years for both microfabrication and micro-machining of various multi-functional structures in dielectric materials through multi-photon absorption because of its high-quality and damage-free processing. Many high-quality material processing techniques have been achieved to date by using femtosecond laser pulses with the methods of holographic fabrication [1-8] and direct writing [9-16], such as micro-gratings [1-4], photonic crystals [5-8], waveguide [9] and diffractive optical elements (DOE) [10-16].

In this chapter, we have reported to fabricate several diffractive optical elements (DOEs) on the surface of the metal film or inside transparent silica glass by femtosecond laser pulses writing directly. Firstly, we introduce a method for holographic data storage with the aid of computer-generated hologram (CGH) on the metal film (Au) by femtosecond laser pulses writing directly. Both the simulated and the experimentally restructured object wave show high fidelity to the original object. Then, we introduce a novel method for generating the optical vortex (OV) by fabricating the computer generated hologram (CGH) of the OV inside glass using femtosecond laser directly writing. And the superpositions of the photon orbital angular momentum (OAM) have also been obtained by using a combined computer generated hologram (CCGH). We also give a concrete explanation to the superpositions of the photon OAM. Lastly, we have fabricated volume grating inside silica glass induced by a tightly focused femtosecond laser pulses for improving the first order of the diffractive efficiency. Experimental results show the first order diffractive efficiency (FODE) of the fabricated gratings is depending on the energy of the pulses and the scanning velocity of the laser pulses greatly, and the highest FODE reaches to 30% nearly. The diffraction pattern of the fabricated grating is also numerically simulated and analyzed by using a two dimensional FDTD method and Fresnel Diffraction. The numerical simulated results proved our prediction on the formation of the volume grating is correct which agree well with our experimental results.

2. Realizing optical storage by method of computer-generated hologram

Because computer-generated holograms (CGHs) can produce wavefronts with any desired amplitude and phase distributions, they have yielded many applications since Lohmann *et al.* [17, 18] firstly demonstrated it several decades ago, such as optical interconnection [19], spatial filtering [20], three-dimensional display [21, 22], and holographic optical manipulation [23]. Two steps are needed for the production of a Fourier hologram. The first step is to calculate the complex amplitude of the virtual or physical object wave at the hologram plane. The second step involves encoding and production of a transparency.

Here, we introduce a method for holographic data storage with the aid of CGH on the metal film by femtosecond laser pulses writing directly. Firstly, the letter "E" consisted of 64×64 pixels was selected as the object image depicting in Fig. 1 (a), which was sampled and Fourier transformed by a computer to obtain the discrete complex amplitude distribution. Then, the discrete complex amplitude distribution was encoded by the detour phase method as depicted in Fig. 1 (b), in which the width of the rectangular aperture was set to the half width of the cell; the height of the rectangular aperture was proportional to the modulus of the complex amplitude; and the phase of the complex amplitude was expressed with the distance between the center of the aperture to the center of the cell. The concrete resulted encoded CGH could be found in Fig. 1 (c).

The resulted CGH could be directly written and recorded on the metal film ablated selectively by femtosecond laser pulses with right pulse energy. The experimental setup for the fabrication of the metal film is shown in Fig. 2. A regeneratively amplified Ti:sapphire laser system (Coherent. Co.) was used, which delivered pulses with a duration of 120fs (FWHM), with a center wavelength at 800nm and a repetition rate of 1kHz. The femtosecond laser pulses with proper energy turned by a ND (neutral density) filter is focused on the surface of the metal film with thickness of 130nm deposited on a silica glass substrate by a $50\times$ microscope objective (NA 0.80); the micro-stage with resolution of 0.1 μm could be controlled by the computer; a shutter system was used to control the ablating area on the surface of the metal film selectively. And the process of the fabrication can be observed by a CCD camera in real-time.

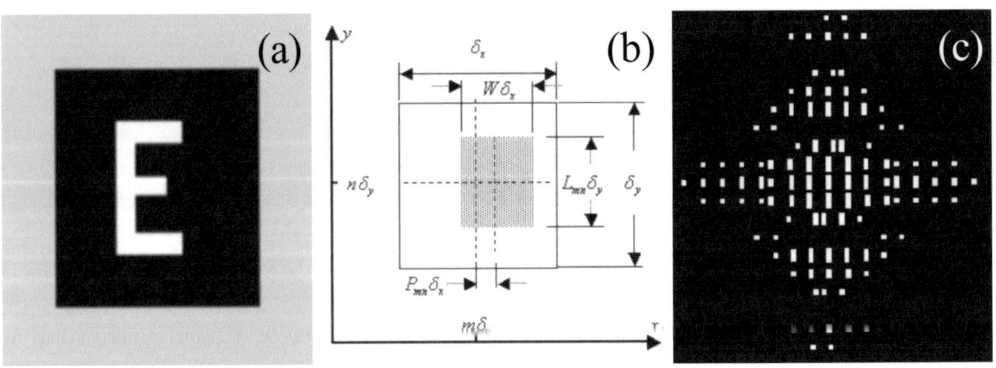

Fig. 1 (a) The object image, (b) The sketch for encoding by the detour phase method, $W = \frac{1}{2}$, L_{mn} and P_{mn} was proportional to the modulus and the phase of the complex amplitude in the cell (m, n) respectively. (c) The calculated encoded CGH of the object image.

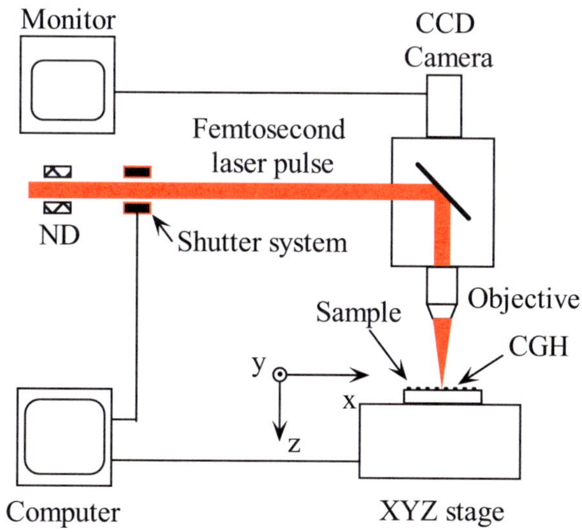

Fig. 2. Experimental scheme of holographic storage on the metal film by femtosecond laser pulses

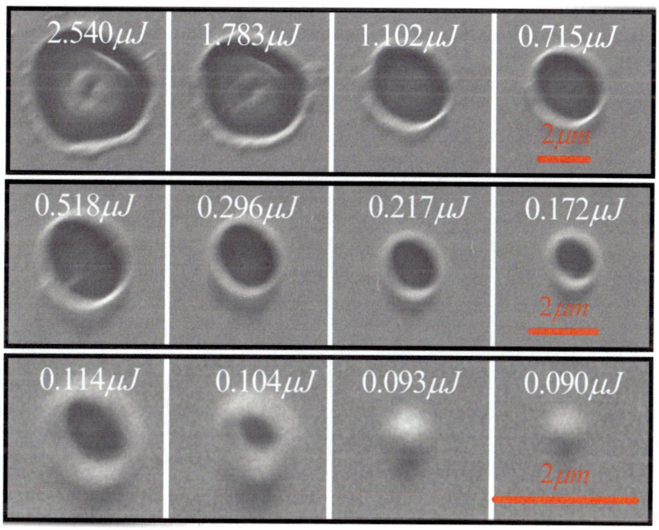

Fig. 3. The hole structures on metal film deposited on glass substrate fabricated by a single femtosecond laser pulse with different pulse energy, all the scale bars are 2 μm .

Before writing the CGH on the metal film, firstly we need decide the diameters and quality of the ablative spots on the metal film in order to achieve well-defined patterns of CGH. In general, the minimum achievable structure size in laser-processing is determined by the diffraction limit of the optical system and is of the order of the radiation wavelength. However, it is different for the femtosecond laser system, because if we choose the peak laser fluence slightly above the threshold value, only the central part of the beam can modify the material and it becomes possible to produce subwavelength structures [24-25]. The ablated microhole structures could be found in Fig. 3 by scanning electron microscope

(SEM). When the laser pulse energy is set to 2.54μJ, the diameter of the ablated hole for the metal film is about 4.26μm, and there exist an ablated tiny hole in the substrate. With the decrease of the pulse energy to 1.783μJ, the diameter of the ablated hole for the metal film is also decreased to 4.03μm. When the pulse energy was changed to 1.1μJ and 0.52μJ, the diameter of the ablated hole can reach to 2.6μm and 2.2μm respectively, and the ablated hole in the substrate vanishes either. Although there are not obvious ablated crater on the glass substrate, the diameter is somewhat bigger for fabricating CGH experiments. On the other hand, when the pulse energy was changed to 0.30μJ, the diameter of the ablated hole could reach to 1.1μm. With energy of the pulse decreasing to 0.217μJ and 0.172μJ, the diameters of the fabricated holes decrease to 800 nm and 600 nm, respectively. When the pulse energy decreases to 0.114μJ and 0.104μJ, the diameter of the fabricated holes decrease to 240 nm and 136 nm respectively, which is less than one third of the wavelength 800 nm and out of the diffraction limit. Especially, when the pulse energy decreases to 0.093μJ and 0.090μJ, there are no ablated holes for the metal film but a nanobump, which could be explained by the T-T model (Two temperatures model) [25]. Although the smaller pulse energy can attain smaller size of the holes and more accurate fabrication in theory, it is difficult to control because of the fluctuations of the energy of the pulses and the thickness of the metal film. Thus, to get a good property of the structured CGH, we choose the pulses energy as 0.30μJ in our experiment.

To record the desired CGH on the metal film, the sample was mounted on a computer-controlled XYZ translation micro-stage with a resolution of 0.1μm and moved step by step for being ablated selectively by the focused pulses according to the CGH pattern shown in Fig. 1 (c). When the area is black in the hologram, the shutter system will be closed for not being irradiated, while on the contrary, the shutter system will be open for irradiating. The

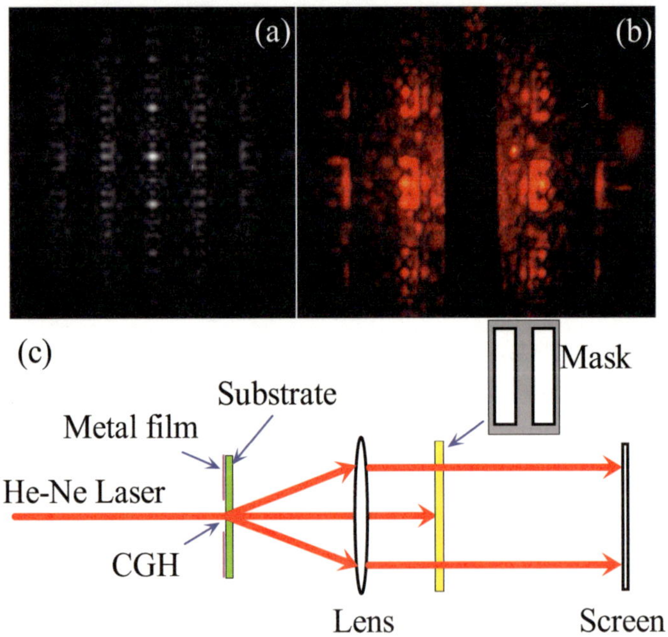

Fig. 4. (a) The simulation of the reconstruction from the CGH; (b) The experimental result; (c) The experimental scheme for the reconstruction of the CGH.

irradiated dots became "transparent" by ablation, while the unexposed dots remained "opaque". The simulated result of reconstruction from the fabricated CGH is given in Fig. 4 (a). In the reconstruction of the CGH, a collimated He-Ne laser beam was used to be incident normally to the CGH on the metal film, as depicted in Fig. 4 (c). The diffraction pattern can be observed in Fig. 4 (b). The letter "E" appears in the +1st order of the diffraction field, while the conjugated image appears in the -1st order of the diffraction field. The result in experiment shows high fidelity to the simulation of the reconstruction. The diffraction efficiency was also measured to be 4.68% by a power meter at the wavelength of 632.8 nm. Here, we defined the diffraction efficiency as a ratio of the intensity of first-order diffraction to that of incident beam.

3. Generating optical vortex

Optical vortex (OV) has been paid considerable attentions in the past two decades because of their special characteristics and potential applications [26-28]. OV has been described as a topological point defect (also known as a dislocation) on wavefront and manifest as a "null" within a light beam because the phase at the defect point is undetermined. OV has been applied in many fields, such as optical trapping [23, 29], optical manipulation for MEMS [23], and optical vortex coronagraph [30]. Several methods, such as mode-converters [31], phase mask [32], and computer-generated holograms [13, 28] (CGH), may be used to embed OV into a "background" beam, such as a Gaussian laser beam.

Here, we generate the OV by fabricating the CGHs of the OV inside glass using a near infrared 800 nm femtosecond laser directly writing. The continued and pulsed OV beams have also been reconstructed with both a collimated He-Ne laser beam and the femtosecond laser beam incident to the fabricated CGH, and the first order of the diffraction efficiency could reach to 3.2% nearly.

Nye and Berry [26] have analyzed the phase dislocation within a monochromatic wave in detail in 1974. They have also analyzed a number of optical wavefront dislocations including screw dislocations, edge dislocations and mixed screw-edge dislocations. A monochromatic beam propagating in the z-direction and containing a single vortex transversely centered at the origin (r = 0) can be expressed by the scalar envelope function:

$$u(r,\theta,z) = A_m(r,z)\exp(im\theta)\exp[i\Phi_m(r,z)] \quad (1)$$

Where $u(r,\theta,z)$ is the optical field expressed in cylindrical coordinates with the optical axis aligned along the z axis, $\exp(im\theta)$ is the characteristic expression of the optical vortex, m is a signed integer called the topological charge, Φ_m is the phase.

To construct a CGH of an OV, we numerically calculate the interferogram of two waves: a planar reference wave and an object wave containing the desired optical vortex. For simplicity, we choose the object wave to be a point vortex of unit charge on an infinite background field of amplitude C_o

$$E_o = C_o \exp(im\theta) \quad (2)$$

and a reference wave of amplitude C_r, whose wavevector lies in the (x, z) plane, subtending the optical axis, z, at the angle φ, can be expressed as

$$E_r = C_r \exp(-i2\pi x/\Lambda) \quad (3)$$

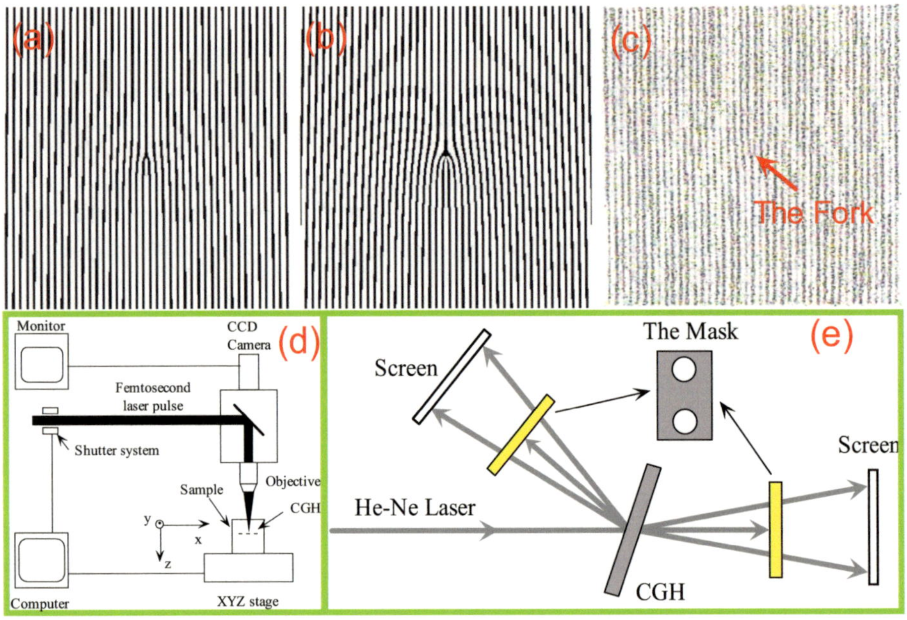

Fig. 5. The CGH interfering between a vortex beam and a planar beam for different topological charge, (a) $m=1$ and (b) $m=3$. (c) Top view of the fabricated CGH. (d) (e) The experimental setup scheme for the fabrication and reconstruction of the CGH respectively.

Where $\Lambda = \lambda/\sin\varphi$ is the spatial period of the plane wave in the transverse plane. The interferogram is given by the intensity of the interfering waves

$$I_{z=0}(x,\theta) = \left|E_o + E_r\right|^2_{z=0} = 2C_o\left[1+\cos(2\pi x/\Lambda + m\theta)\right] \tag{4}$$

Where we set $C_r = C_0$ to achieve unity contrast $((I_{max} - I_{min})/I_{max} = 1)$. The resulting interferogram, depicted in Fig. 5 (a) and (b), resembles a sinusoidal intensity diffraction grating. The pattern contains almost parallel lines with a bifurcation at the vortex core.

To record the desired CGH, silica glasses with four planes being polished have been selected as the sample and mounted on a computer-controlled XYZ translation micro-stage with $0.1\mu m$ resolution, which were moved step by step and irradiated by the focused femtosecond laser pulses by a microscope objective lens with a numerical aperture of 0.45 (20X, Nikon.Co.) with proper pulse energy according to the hologram pattern (Fig. 5 (a)) controlled by the PC. The schematic setup is shown in Fig. 5 (d). The irradiated dots would become "black" (opaque) because of microexplosion induced by femtosecond laser pulses inside silica glass while the unirradiated dots remained "white" (transparent). The top view of the fabricated hologram is presented in Fig. 5 (c) according to the CGH depicted in Fig. 5 (a). In general, it is difficult to fabricate microstructure on the surface of the silica glass. The silica glass is transparent at the wavelength of 800 nm because the light frequency is out of the linear absorption range of the silica glass. However, the intensity at focus point would be approximate to 100 TW/cm^2. So, high energy fluence within the focal volume would quickly ionize the silicate glass by the combined action of avalanche and multi-photon processes [13, 33].

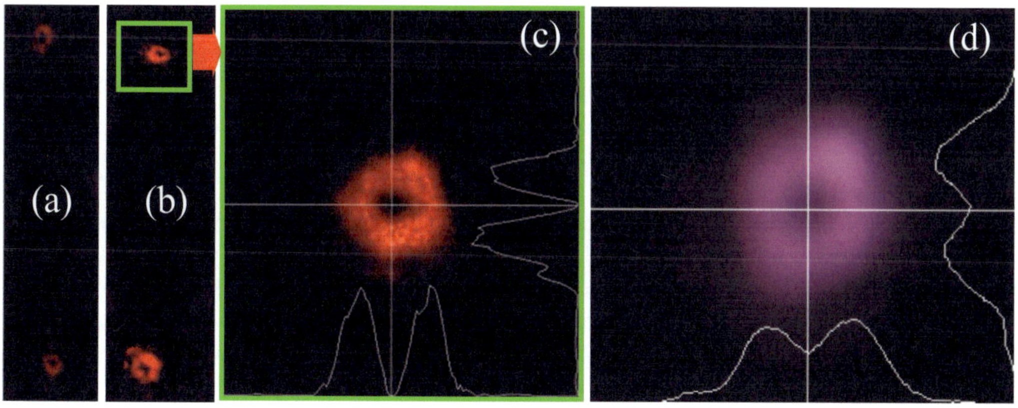

Fig. 6. Diffraction patterns with reconstruction of the CGH shown in Fig. 2. (a) Reflected pattern, (b) Transmitted pattern and (c) The enlarged version of the first-order diffraction of the Transmitted pattern. (d) Reconstructed optical vortex by a femtosecond pulsed laser.

To reconstruct the optical vortex beam from the fabricated CGH, a collimated He-Ne laser beam was incident on the CGH inclined at 30° against the sample, and both the transmission and reflection pattern grating can be realized. In order to get an excellent image, we lay two masks in the diffraction beam path as shown in Fig. 5 (d). Then, the output image could be taken by a digital camera on the screen. Both of the transmission and reflection patterns are shown in Fig. 6 (a) and (b). One vortex of the first-order diffraction in the transmission pattern was taken out for getting an intuitionistic image of the optical vortex in the transverse plane (Fig. 6 (c)). From the result depicted in Fig. 6 (c), we can see that the optical vortex was reconstructed with high fidelity. We have also reconstructed the optical vortex with the femtosecond laser beam with the wave length of 800nm as shown in Fig. 6 (d), which have also shown high fidelity to the optical vortex and been used as a irradiated source for special structures [34].

We have also measured the diffraction efficiency by a power meter at the wavelength of 632.8 nm. Here, we defined the diffraction efficiency as a ratio of the intensity of first-order diffraction to that of incident beam. The efficiency of transmission beams was 3.44%, and that of reflection beams was 1.35%. So, the total efficiency was about 4.79%.

4. Realizing the superpositions of the photon OAM by CCGH

As predicted [35] and observed [36], OV beams carry an orbital angular momentum (OAM) distinct from the intrinsic angular momentum of photons associated with their polarizations. This external angular momentum of the photon states is coming from the helical phase structure on the optical wavefront. The OAM of photon states is the reason why they have been suggested for optical data storage [37] and gearing micromachines, such as optical tweezers [23, 29] and optical manipulation for MEMS [23]. The superposition of OAM of photons by Mach-Zehnder interferometer [38-39] has also attracted great attentions because of their potential applications in quantum computation and quantum information processing.

Ordinary beams carry only "spin angular momentum", encoded in the polarization of light. All possible spin states can be constructed with just two polarization states (vertical and horizontal, or clockwise and counterclockwise). In 1992, however, Allen [35] showed that OV beams carry a discrete OAM of $m\hbar$ units per photon along their propagation direction because of having an azimuthal phase dependence of the form $\exp(-\mathrm{i}m\theta)$. For the light with OAM, the energy spirals around the beam axis. Therefore there would be OAM transferring to materials while interacting with materials, which can make the particles in the beam spin around the beam axis.

Here, the superpositions of the photon OAM has also been obtained by using a combined computer generated hologram (CCGH) which is fabricated inside the polished silica glass with a near infrared 800 nm femtosecond laser. The CCGH is combined by two special binary CGHs of OV which are recorded by the interference of OV beams with a planar beam simulating on the computer and directly written and recorded inside the glass by femtosecond laser pulses induced microexplosion because of multiphoton absorption. The OV beams with new topological charges can be reconstructed in the diffraction field with a collimated He-Ne laser beam incidence to the fabricated CCGH normally, which reveals the superposition of OAM of photons successfully.

From above, we can obtain two CGHs by choosing topological charges $m=1$ and $m=3$ with the same Λ (depicted in Fig. 5 (a) and (b)) for getting the same diffracting angle of the diffracting beams in the same diffracting order. In order to get the CCGH, two CGHs were combined with an orthogonal style by keeping one invariable and the other one rotated to vertical direction (depicted in Fig. 7 (a)).

After getting the binary CCGH, we directly fabricated it inside the silica glass sample with four planes being polished by femtosecond laser pulses writing directly. The same femtosecond laser system was used as the irradiated source, and the schematic setup was

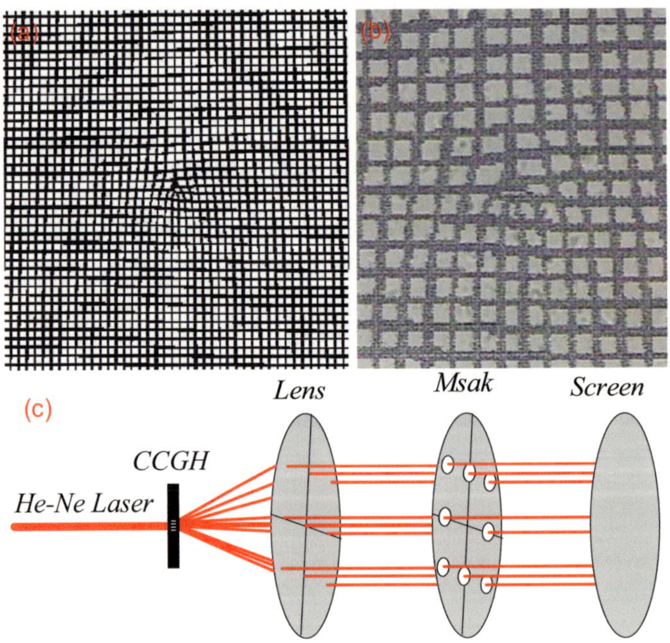

Fig. 7. (a) The CCGH combined with the two CGHs of the optical vortex with topological charges of m=1, m=3 respectively, (b) The top view of the central part of the fabricated CCGH, (c) The experimental setup scheme for the reconstruction of the CCGH.

depicted in Fig. 5 (d). The used pulse energy could be controlled by the ND filter, and the shutter system could control the switch of the femtosecond beam.
To record the desired CCGH, the sample was moved step by step and irradiated selectively by the focused pulses according to the hologram pattern controlled by the PC. When the area is blank in the hologram, the shutter system controlled by the PC will be closed. The irradiated dots became opaque and shown as the "black region" while the unirradiated dots remained transparent as the "white region". The size of the resulted CCGH is 256×256, and every pixel in the recorded CCGH is controlled as $2\,\mu m$. So the scale of the recorded CCGH is $512\mu m \times 512\mu m$ in the silica glass. The top view of the central part of the fabricated hologram under optical microscope is presented in Fig. 7 (b) according to the designed CCGH depicted in Fig. 7 (a).

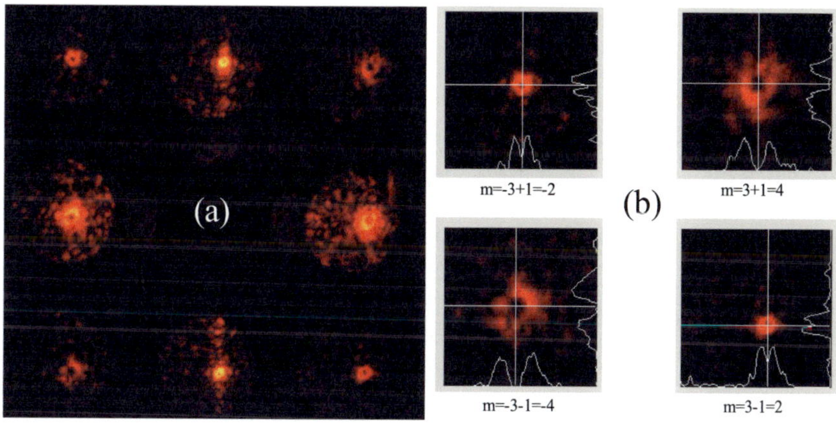

Fig. 8. (a) The diffraction pattern of the fabricated CCGH, (b) The analytical results for the four superposed modes

To reconstruct the optical vortex beams from the fabricated CCGH, a collimated He-Ne laser beam was been incident normally on the CCGH. In order to get an excellent diffraction result, we place a mask with a special aperture array in the diffraction beam path to block out the zero-order diffraction as shown in Fig. 7 (c). Then, the diffraction pattern depicted in Fig. 8 (a) could be taken by a digital camera on the screen. The OV beams with topological charges $m = \pm 1$ and $m = \pm 3$ should be generated in the first order of the diffraction field respectively. However, the OV beams with $m = \pm 2$ and $m = \pm 4$ can also be attained in the first order diffraction modes. We also give an analytical result for the OV beams with new topological charges as shown in Fig. 8 (b), which shows high fidelity to OV beams.
The OVs with new topological charges can be attributed to the superposition of the topological charges between ± 1 orders of the two CGHs. We give a sketch for the explanation of diffraction modes as shown in Fig. 8. We can see that the topological charge is $m = 4$ in the position of the cross between +1 order for CGH with $m = 1$ and +1 order for CGH with $m = 3$ from the Fig. 9, while the topological charge is $m = -2$ in the position of the cross between +1 order for CGH with $m = 1$ and −1 order for CGH with $m = 3$.
In order to confirm the theoretical explanation about the superposition of the topological charges, we use another CCGH made by two CGHs with $m = 1$ depicted in Fig. 5(a), and the diffraction pattern is shown in the Fig. 5(b). From the Fig. 5(b), we can find that the

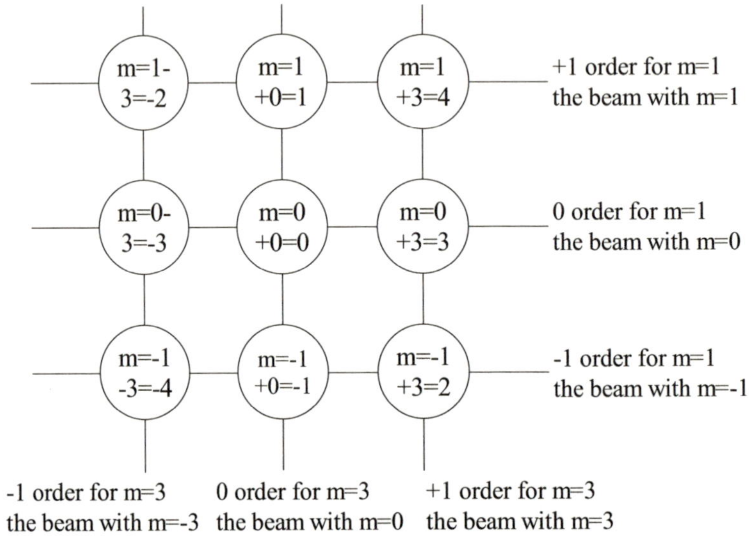

Fig. 9. The sketch for explanation of diffraction mode

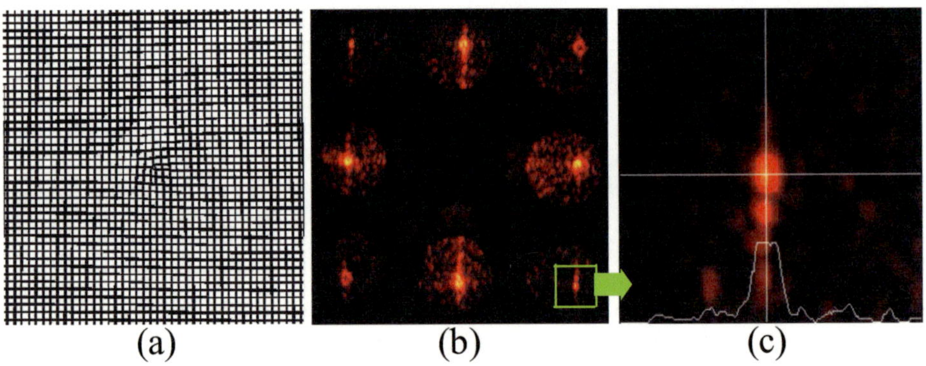

Fig. 10. (a) The CCGH combined by two CGHs with m=1, m=1, (b) The diffraction pattern for the CCGH showed in (a), (c) The enlarged version for the superposed gaussian mode with m=0

diffraction pattern in the position of the cross between the +1 order and the −1 order has become to a Gauss mode with the topological charge of $m=0$, and the analytical profile of the superposed beam is depicted in Fig. 5(c). It can be seen that our explanation about the superposition of the topological charges is reasonable.

We also measured the diffraction efficiency by a power meter at the wavelength of 632.8 nm. The efficiencies of diffraction beam with $m=\pm 1$ and $m=\pm 3$ were 2.2% respectively, and those of diffraction beam with $m=\pm 2$ and $m=\pm 4$ were 1.1% respectively.

The CCGH method would be an effective way to realize the superposition of the photon OAM. It would be significant for the micromachining, optical tweezer, quantum computation and quantum information processing to superpose the diffraction beams with the CCGH combined by two CGHs with the same spatial period Λ. Because it is much easier to generate the superposition of the OAM of the photons compared to by the methods of Mach-Zehnder interferometer [38, 39].

5. Self-assembling volume grating in silica glass

From above experimental results, we can know that the first diffraction efficiencies of the fabricated DOEs are comparatively small. It would be very significant if the diffraction efficiency of these diffractive optical elements could be increased. According to the grating parameter $Q = 2\pi\lambda d / (n\Lambda^2)$ defined by Kogelnik [11, 12, 16, 40], where Λ is the period of the gratings, d is the thickness of the diffraction grating, n is the refractive index, and λ is the wavelength of the incident beam, we can enlarge the thickness or reduce the period of the diffraction grating for obtaining a larger diffraction efficiency. Because of the self-focusing and self-defocusing effects, there would be a line of self-assembly periodic nanovoids generated spontaneously along the propagation direction of the laser beam, when a femtosecond laser beam is focused into transparent glass with proper pulse energy. That is to say the entire length of the void structure and the shapes of the generated voids could be controlled by varying the laser parameters. So far, there have been reported to form a submicrometer-sized void array in borosilicate glass [41], $SrTiO_3$ crystal [42], Al_2O_3 crystal [43], CaF2 crystals [44], fused silica glass [45] and so on.

Here, we fabricated volume grating inside silica glass induced by a tightly focused femtosecond laser pulses. The first order diffractive efficiency (FODE) of the fabricated gratings is depending on the energy of the pulses and the scanning velocity of the laser greatly, and the highest FODE reaches to 30% nearly. The diffraction pattern of the fabricated grating is also numerically simulated and analyzed by using a two dimensional FDTD method and Fresnel Diffraction. The numerical simulated results proved our prediction on the formation of the volume grating is correct which agree well with our experiment results.

5.1 Experimental results

The schematic illustration of the experimental setup is shown in Fig. 11 (a), which is similar with Fig. 5 (d). The commercially available polishing fused silica glass with size of 8mm×5mm×4mm was mounted on the micro-stage. The laser beam with a diameter of 6mm in Gaussian profile was focused inside the glass sample by a 50X objective lens with numerical aperture of 0.80. And the femtosecond laser beam is focused to 100 μm beneath the surface of the sample. The energy of the pulse can be turned by a ND (neutral density) filter. A shutter system was used to control the number of the deposited pulses selectively.

As depicted in Fig. 11 (b-d) , line style structures induced by femtosecond laser pulses consisting of a group of voids and a filamentation could be observed along the incident direction (arrow direction) inside samples. It is observed with the increase of the number of the pulses, the length and the diameter of the induced structure are becoming larger and larger and the void group is becoming more regular (see Fig. 11 (b)). Similarly, with the increase of the pulse energy, the length of the line structures induced by both single pulse (see Fig. 11 (c)) and the double pulses (see Fig. 11 (d)) are also becoming longer, which implies with the decrease of the scanning speed (more laser pulses irradiating in a certain spot) and the increase of the pulse energy, the length of the fabricated line structure would be increased. It is notable to mention that when the pulse energy is decreased to 3.80μJ for the single pulse interacting, the voids almost disappear, with a long sharp filamentation left, as is depicted in Fig. 11. (c). The formation of the filamentation in the silica glass sample is one of the most fundamental nonlinear optical phenomena and results from the dynamic balance between self-focusing arising from an increase in the refractive index and self-defocusing arising from diffraction or plasma formation [46-47].

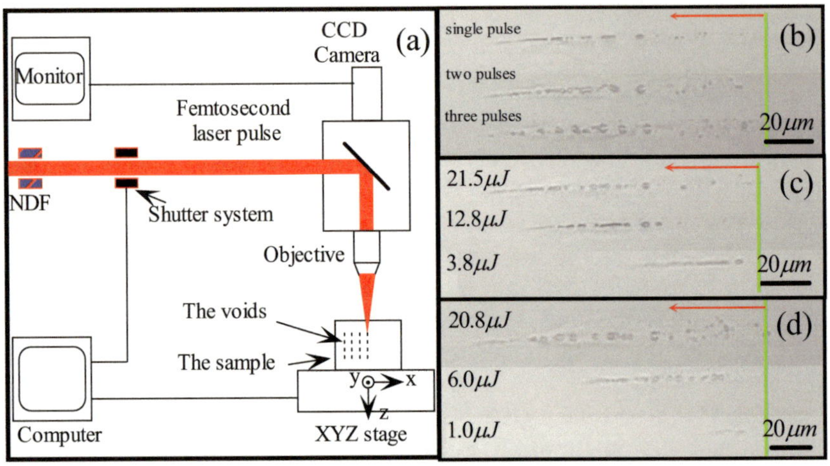

Fig. 11. (a) The sketch of the experimental setup for the generation of periodic voids, Side view of the voids array induced in silica glass by femtosecond laser pulses, (b) Different number of with pulse energy of 20μJ, (c) Single pulse with different energies, (d) Double pulses with different energies. The scale bars in the Figure are 20μm. The arrows show the direction of the pulses propagation, and the lop lines express the focus place

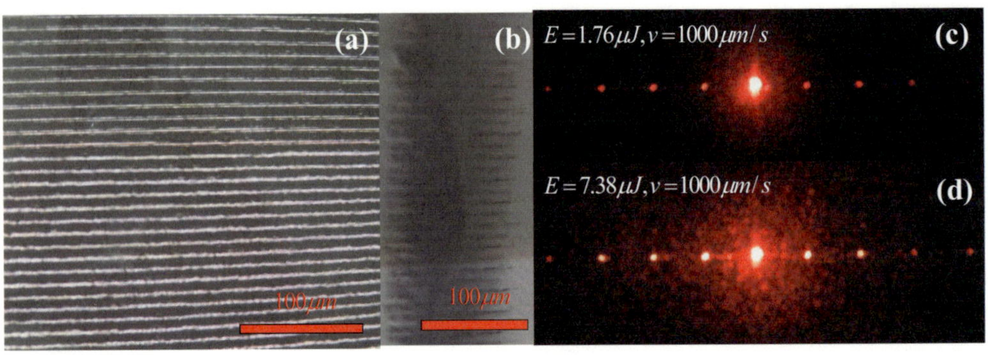

Fig. 12. the fabricated grating: (a) top view, (b) side view; diffraction pattern of the grating obtained under different conditions: (c) E=1.76μJ, v=1000$\mu m/s$, (d) E=7.38μJ, v=1000$\mu m/s$

A series of grating could be fabricated in the silica glass sample by moving the computer-controlled XYZ translation micro-stage at different scanning speeds and different pulse energies. One of the typical fabricated grating is shown in Fig. 12 (a) and (b) with the pulse energy and the moving speed of the stage of E=3.520μJ and v=500$\mu m/s$ respectively. The period of the fabricated grating is set to 20μm. From Fig. 12 (b), we can observe that the depth of the grating is reaching to 120μm nearly, which is six times of the periods, so we called it volume gratings. The diffraction pattern of the fabricated grating could be obtained by a collimated He-Ne laser beam normally incident on the fabricated grating, and two typical diffraction patterns are depicted in Fig. 12 (c) and (d) by keeping the scanning speed at v=1000$\mu m/s$ but different pulse energy E=1.176μJ and E=7.380μJ respectively. It is apparent that the first diffraction efficiency of the fabricated grating by E=7.380μJ is larger than that of E=1.176μJ, which shows the first diffraction efficiency of the fabricated grating is depending upon the pulse energies greatly.

We also measure the transmittance efficiency and diffraction efficiency of the fabricated gratings by laying a power meter at the wavelength of 632.8nm behind the sample for detecting the diffractive He-Ne laser. Here, we defined the transmittance efficiency (TE) to be the ratio between the intensity of whole transmittance and that of incident beam. The variation tendency of the transmittance efficiency with the different pulse of energy and scanning velocity is shown in Fig. 13 (a). We found that the transmittance efficiency decreases for higher pulse energy and slower scanning speed. It can be explained by the fact that the absorption and scatterance are enhanced because of stronger microexplosion and the more generation of color centers. We also defined the first order diffraction efficiency (FODE) and normalized first order diffraction efficiency (NFODE) to be the ratio of the intensity of the first-order diffraction to the incident intensity and the transmitted intensity respectively. And the corresponding variation tendencies are shown in Fig. 13 (b) and (c). With the increase of the pulse energy, the FODE would enlarge to a maximum firstly, and then decrease gradually depicted in Fig. 13 (b). However, NFODE nearly increases all the time as shown in Fig. 13 (c) expect for situations with too large energies.

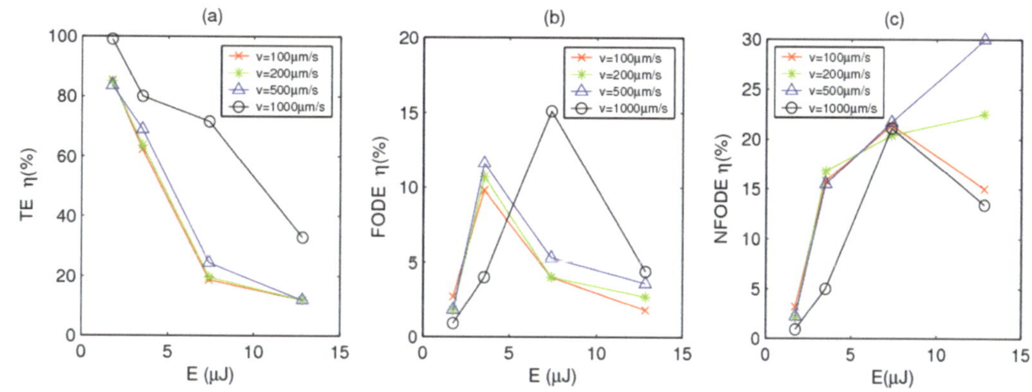

Fig. 13. (a) Variational tendency of the transmittance efficiency(TE) with the different pulse of energy and scanning velocity, (b) Variational tendency of the first order diffraction efficiency(FODE) with the different pulse of energy and scanning velocity, (c) Variational tendency of the normalized first order diffraction efficiency(NFODE) with the different pulse of energy and scanning velocity.

Now we give a physical description of the phenomenon observed above: With the increase of the pulse energy, the depth of the fabricated gratings would be increased, so does the first order diffraction according to Kogelnik's equation [13]. However, the absorption and scatterance are also enhanced at the same time. When the former is dominated, the FODE would be increasing; while the posterior is dominated, the FODE would be decreasing, as shown in Fig. 13 (b). On the other hand, if we neglect the influence of the absorption and scatterance, the NFODE should be always increasing as is depicted in Fig. 13 (c). For different scanning speed at the pulse energy of 1.760μJ, it is observed that the FODE decreases with increasing scanning speed, and the maximum is only 3.2% for v=100$\mu m/s$.

However, when the pulse energy increase to $3.520\mu J$, the maximum of the FODE reaches to 11.6% for v=$500\mu m/s$, and the FODE for circumstance of v=$100\mu m/s$ and v=$200\mu m/s$ are comparatively small compared to v=$500\mu m/s$. For the pulse energy of $7.380\mu J$, the biggest diffraction efficiency reaches to 15.1% for v=$1000\mu m/s$, but the others decrease to a lower level compared to that of E=$3.520\mu J$. Going on increasing the pulse energy to $12.83\mu J$, all of the FODE decrease to a relative lower level compared to E=$7.380\mu J$.

We deem that when the pulse energy is $1.760\mu J$, the induced microexplosion and the darkening region is comparatively small. When the scanning speed is slow, the fabricated grating is thicker than that for large scanning speed, so the FODE is larger. While the pulse energy reaches to $3.520\mu J$, the depth of the gratings would be bigger, so the FODE would become bigger than the former. The microexplosion and the darkening region would also increase, and the slower the bigger, so the FODE for v=$100\mu m/s$ is smaller than that for v=$500\mu m/s$ and v=$200\mu m/s$. For v=$1000\mu m/s$, the resulted grating is not so compacted compared with other scanning speed, so the FODE is smaller. However, when the pulse energy reaches to $7.380\mu J$, the resulted gratings could be compacted enough, so the FODE reaches maximum too. But for v=$500\mu m/s$, v=$200\mu m/s$, and v=$100\mu m/s$, the FODE is decreasing at E=$7.380\mu J$ because of the enhancement of the microexplosion and the darkening region. For pulse energy of $12.83\mu J$, resulted gratings is too compacted, and the absorption and scatterance dominate, so all of the FODE is decreased to a lower level. Moreover, the order of the FODE is in a reverse order compared to that of E=$1.760\mu J$, just because of the different dominant mechanism for lower and higher energy.

5.2 Simulated results

In order to verify our explanation on experimental results, we simulate the diffraction process of the induced gratings without regard to the optical absorption by the combined methods of two dimensional FDTD and Fresnel diffraction as shown in Fig. 14 (a-c). The scale of the simulation districts is chosen to be $h_x = 200\mu m$ and $h_z = 80\mu m$, and assume the sample is uniform along the direction of Y axes. The grating period is set as $d_x = 20\mu m$ according to our experimental reality, and the refractive indexes of silica glass and the filamentations are $n_s = 1.5$ and $n_f = n_0 + \delta n$ respectively, where δn is the refractive index change (RIC) induced by the laser. The refractive index of the background (air) is $n_b = 1.0$. According to our experimental results as depicted in Fig. 11 (b-d), the filamentations could be simplified as composite regions combined by a rectangle and followed by a triangle (see Fig. 14 (b)), and the width of the rectangle region is d_{rect}. In our simulation, the sample is divided into four regions labelled A, B, C and D respectively, and the length (along Z axes) of region A, B are set as $h_A = 5\mu m$ and $h_B = 10\mu m$ respectively. We could change the length of region C (h_C), the refractive index variation δn and the width of the rectangle region d_{rect}, so as to match different conditions in our experiment. We use a Gauss form continuous source in front of the sample with a twist $\omega_0 = 40\mu m$ and wavelength $\lambda_0 = 0.8\mu m$. The position of the source waist is marked with green line in Fig. 14 (b).

To obtain the far field diffraction, we extract the amplitude and phase distribution of the electric field in the plane (x,y) behind the silicate sample to be $E(x) = A(x)e^{i\phi(x)}$, and assume the scale of the sample in the third dimension to be constant. Then the diffraction pattern of the far field in the (x',y') plane can be solved using the Fresnel Diffraction Integral as below:

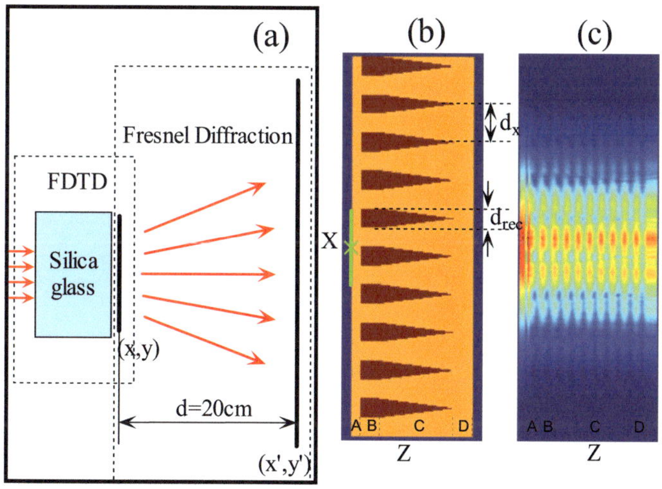

Fig. 14. (a) A simple sketch of our numerical method, (b) A typical example of the distribution of the amplitude of the electric field inside the sample. (c) Distribution of the amplitude of the Electric filed in the silicate glass sample.

$$E(x')=\frac{C}{j\lambda}\int_{-\infty}^{+\infty}E(x)\frac{\exp(ikr)}{r}\frac{\cos\theta+1}{2}dx \tag{5}$$

Where $r=\sqrt{d^2+(x-x')^2}$, $\theta=\arccos(d/r)$, k is the wave-number, C is an constant introduced by the integration along the third dimension. The distance between the two planes is fixed to be $d=20\text{cm}$, which is in consistent with our experiment. And we observe the distribution of the electric field in the (x',y') plane along the x' axis.

The numerical simulated results are shown in Fig. 15. After more than 20000 iterations, we could attain stationary distribution of the electric amplitude inside the silicate glass sample as depicted in Fig. 14 (c) (at $h_C=40\mu\text{m},\delta n=0.002,d_{rec}=10\mu\text{m}$). The far field diffraction patterns of the given volume grating at which $d_{rec}=10\mu\text{m},\delta n=0.005$ with different depths($h_c=10,20,40\mu m$) are depicted in Fig. 15 (a), which shows that with the increase of the depth of the grating, the diffraction efficiency is increasing, which verify our former physical explanation and agree with our experimental results very well as depicted in Fig. 13 (c).

With the increase of the pulse energy or decrease of the scanning speed, the fabricated lines' width (d_{rec}) of the induced grating would increase obviously as depicted in Fig. 11 (b-d). So we also give a numerical simulation on the far field diffraction patterns of the volume grating by keeping $h_c=20\mu\text{m}$ and $\delta n=0.005$ but with different line widths ($d_{rec}=15,10,5\mu m$) are depicted in Fig. 15 (b), which shows that with the increase of the line width (d_{rec}), the first order diffraction efficiency is increasing. In other words, increasing the incident pulse energy (leading to the increase of the line width and the depth of the induced grating) would enhance the FODE if we can take no account of the absorption.

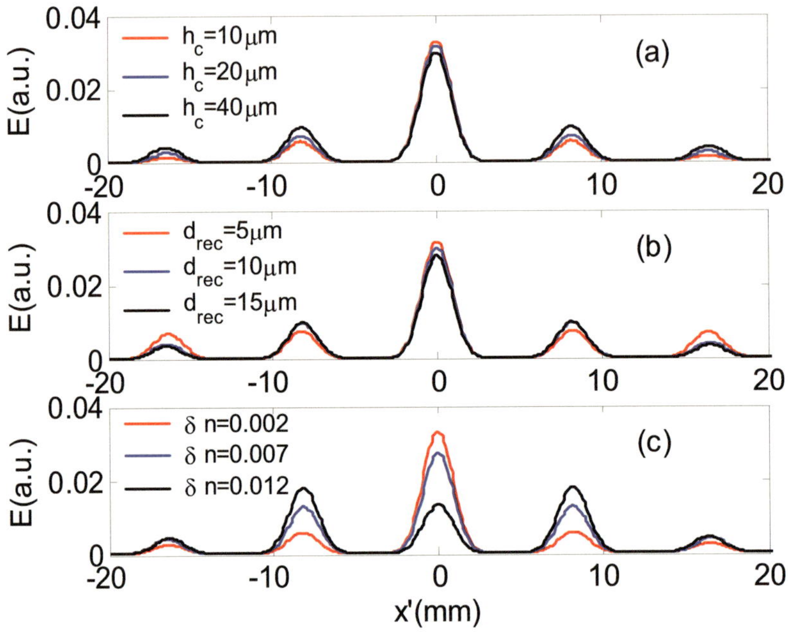

Fig. 15. The far field diffraction of the distribution of Electric filed along the x' direction, (a) $h_{rec} = 10\mu m, \delta n = 0.002$ and the three curves correspond to different depths of the induced gratings: $h_c = 10$(red), 20(blue) and 40 μm (black) respectively; (b) $h_c = 10\mu m, \delta n = 0.002$ and the three curves correspond to different width of the filaments: $d_{rec} = 15$(red), 10(blue) and 5 μm (black) respectively; (c) $d_{rec} = 10\mu m, h_c = 40\mu m$ and the three curves correspond to different variations of the refractive index: $\delta n = 0.007$ (red), 0.005 (blue) and 0.012(black) respectively.

We also simulated FODEs depending on the different RIC ($\delta n = 0.002, 0.007, .0012$) by remaining $d_{rec} = 10\mu m, h_c = 40\mu m$. From the simulated result as shown in Fig. 15 (c), we can observe that the RIC would affect the FODE strongly. Especially, for $\delta n = 0.012$, the FODE is higher than the zero order diffraction efficiency (ZODE) for the normally incident beam. In fact, in our simulation if we further enlarge the RIC, the ZODE would be minished correspondingly. It is possible that the ZODE minish to zero absolutely at a certain RIC that relates to the depth of the grating (h_c). That is to say the energy of the zero order diffraction could transfer to the first and higher order diffraction. Therefore, if we can accurately control the RIC and h_c inside the glass sample induced by femtosecond laser, we could control the diffraction pattern very well.

6. Conclusion

In conclusion, optical information has been stored on the metal film by femtosecond laser pulses with the aid of the computer-generated hologram (CGH) encoded by the detour phase method. Furthermore, the object wave has been reconstructed by using a collimated He-Ne laser beam in a proper way, and the result shows a high fidelity to the original object.

The CGH of the OV has been fabricated inside glass by the femtosecond laser pulses. And we realized the restructured optical vortex beam of both the transmission and reflection pattern with high fidelity using a collimated He-Ne laser beam. We have realized the superposition of the OAM of photons with CCGH of OVs. The CCGH was formed by combining two CGHs of OVs in a special way and directly written inside glass by femtosecond laser pulses induced microexplosion. The OV beams with different photon OAM can be generated from different topological charges carried by the diffraction beams from the CCGH. Theoretical analysis and experimental demonstrations have been given to explain the generations of the superposition of the OAM. The efficiencies of the first order diffraction waves have been measured to be about 6.6% in total.

In order to enhance the first order diffraction efficiency, we have researched the way to fabricating the volume grating inside silica glass by tightly focused femtosecond laser pulses. The femtosecond pulses induced multiple microexplosion and the self-focusing and self-defocusing effects in the transparent materials have been considered as the reasons for the formation of volume grating. The first order diffraction efficiency of the induced grating is measured by a collimated He-Ne laser and the highest NFODE reaches to 30% nearly. We also have constructed a grating mode and attained the simulated results by using a two dimensional FDTD method and Fresnel Diffraction. The numerical simulated results verified our physical viewpoint on the formation of the volume grating is correct which agree well with our experiment results.

7. Acknowledgment

This work was supported by the National Science Foundation of China (NSFC: 10904027), the China Postdoctoral Science Foundation (AUGA41001348) and the Heilongjiang Province Postdoctoral Science Foundation (AUGA1100074), and development program for outstanding young teachers in Harbin Institute of Technology, HITQNJS. 2009. 033.

8. Reference

[1] S. Qu, J. Qiu, C. Zhao, X. Jiang, H. Zeng, C. Zhu, and K. Hirao, "Metal nanoparticle precipitation in periodic arrays in Au_2O-doped glass by two interfered femtosecond laser pulses," *Appl. Phys. Lett.* 84, 2046-2048, (2005).

[2] Y. Han, S. Qu, Q. Wang, Z. Guo, and X. Chen, Controllable grating fabrication by three interfering replicas of single femtosecond laser pulse, Chinese physics B, 18, 5331, (2009).

[3] S. Qu, C. Zhao, Q. Zhao, J. Qiu, C. Zhu, and K. Hirao, One-off writing of multimicrogratings on glass by two interfered femtosecond laser pulses, Opt. Lett. 29, 2058 (2004).

[4] Z. Guo, S. Qu, L. Ran, and S. Liu, Modulation grating achieved by two interfered femtosecond laser pulses on the surface of the silica glass, Appl. Surf. Sci., 253, 8581, (2007).

[5] T. Kondo, S. Matsuo, S. Juodkazis, V. Mizeikis, and H. Misawa, "Multiphoton fabrication of periodic structures by multibeam interference of femtosecond pulses," *Appl.Phys.Lett.* 82 (2003) 2758-2760.

[6] Z. Guo, L. Ran, Y. Han, S. Qu, and S. Liu, "The formation of novel two-dimensional periodic microstructures by a single shot of three interfered femtosecond laser pulses on the surface of the silica glass", *Opt. Lett.* 33 (2008) 2383.
[7] Z. Guo, S. Qu, Y. Han, and S. Liu, Multi-photon fabrication of two-dimensional periodic structure by three interfered femtosecond laser pulses on the surface of the silica glass, Optics Communications, 280, 23-26, (2007).
[8] J. Si, Z. Meng, S. Kanehira, J. Qiu, B. Hua and K. Hirao, "Multiphoton-induced periodic microstructures inside bulk azodye-doped polymers by multibeam laser interference", *Chem. Phys. Lett.* 399 (2004) 276-279.
[9] C. Schaffer, A. Brodeur, J. García, and E. Mazur, "Micromachining bulk glass by use of femtosecond laser pulses with nanojoule energy," Opt. Lett. 26, 93-95 (2001)
[10] Y. Li, Y. Dou, R. An, H. Yang, and Q. Gong,, "Permanent computer-generated holograms embedded in glass by femtosecond laser pulses," *Opt. Express,* 13 (2005) 2433-2438.
[11] Q. Zhao, J. Qiu, X. Jiang, E. Dai, C. Zhou, and C. Zhu, "Direct writing computer-generated holograms on metal film by an infrared femtosecond laser," *Opt. Express,* 13 (2005) 2089-2092.
[12] Z. Guo, S. Qu, Z. Sun, and S. Liu, "Superposition of orbit angular momentum of photons by combined computer-generated hologram fabricated in silica glass with femtosecond laser pulses", *Chin. Phys. B.* 17 (2008) 4199.
[13] Z. Guo, S. Qu, S. Liu, Generating optical vortex with computer-generated hologram fabricated inside glass by femtosecond laser pulses, Optics Communications, 273, 2007, 286-289.
[14] Z. Guo, H. Wang, Z. Liu, S. Qu, J. Dai, and S. Liu, Realization of holographic storage on metal film by femtosecond laser pulses micromachining, Journal of Nonlinear Optical Physics and Materials, 18(4), 2009, 617-623.
[15] Z. Guo, W. Ding, S. Qu, J. Dai, and S. Liu, Self-assembled volume grating in silica glass induced by a tightly focused femtosecond laser pulses, Journal of Nonlinear Optical Physics and Materials, 18(4), 2009, 625-632.
[16] K. Zhou, Z. Guo, W. Ding, and S. Liu, Analysis on volume grating induced by femotosecond laser pulses, Opt. Express 18(13), 2010, 13640-13646.
[17] A. W. Lohmann and D. P. Paris, "Binary Fraunhofer holograms generated by computer," *Appl. Opt.* 5 (1967) 1739.
[18] B. R. Brown and A. W. Lohmann, "Computer-generated binary holograms," *IBM J. Res. Develop.* 13 (1969) 160.
[19] M. R. Feldman and C. C. Guest, "Computer generated holographic optical elements for optical interconnection of very large scale integrated circuits," *Appl. Opt.* 26 (1987) 4377.
[20] B. R. Brown and A. W. Lohmann, "Complex spatial filtering with binary masks," *Appl. Opt.* 5 (1966) 967.
[21] O. Bryngdahl and F. Wyrowski, "Digital holography - computer-generated holograms" in *Progress in Optics,* E. Wolf, ed. (North-Holland, Amsterdam, 1990), 28, pp.1–86.
[22] D. Abookasis and J. Rosen, "Computer-generated holograms of three-dimensional objects synthesized from their multiple angular viewpoints," *J. Opt. Soc. Am. A,* 20 (2003) 1537.
[23] D. G. Grier, "A revolution in optical manipulation," *Nature* 424 (2003) 810.

[24] Y. Li，W. Watanabe, K. Yamada, T. Shinagawa, K. Itoh, J. Nishii, and Y. Jiang "Holographic fabrication of multiple layers of grating inside soda–lime glass with femtosecond laser pulses". *Appl.Phys.Lett,* 80 (2002) 1508.
[25] J. Koch, F. Korte, T. Bauer, C. Fallnich, A. Ostendorf, B. Chichkov. Nanotexturing of Gold Films by Femtosecond Laser-induced Melt Dynamics. Appl. Phys. A. 2005, 81:325–328.
[26] J. F. Nye and M. V. Berry, 1974 Proc. R. Soc. A. 336 165-190.
[27] G. A. Turnbull, D. A. Robertson, G. M. Smith, L. Allen, and M. Padgett, 1996 Opt. Comm. 127 183-188.
[28] V. Y. Bazhenov, M. V. Vasnetsov, and M. S. Soskin, 1990 Pis'ma Zh. Eksp. Teor. Fiz. 52, 1037-1039 [1990 JETP Lett. 52 429-431].
[29] M. Dienerowitz, M. Mazilu, P. Reece, T. Krauss, and K. Dholakia, "Optical vortex trap for resonant confinement of metal nanoparticles," Opt. Express 16, 4991-4999 (2008).
[30] G. Foo, D. Palacios, and G. Swartzlander, Jr., "Optical vortex coronagraph," Opt. Lett. 30, 3308-3310 (2005)
[31] M. Beijersbergen, L. Allen, H. van der Veen and J. Woerdman, Astigmatic laser mode converters and transfer of orbital angular momentum, *Opt. Commun.* 96 (1993), 123.
[32] G. Turnbull, D. Robertson, G. Smith, L. Allen and M. Padgett, The generation of free-space Laguerre-Gaussian modes at millimetre-wave frequencies by use of a spiral phaseplate, *Opt. Commun.* 127 (1996), 183.
[33] S. Juodkazis, H. Misawa, T. Hashimoto, E. Gamaly, and B. Luther-Davies, Laser-induced microexplosion confined in a bulk of silica: Formation of nanovoids, *Appl. Phys. Lett.* **88** (2006), 201909
[34] L. Ran, and S. Qu, Self-assembled volume vortex grating induced by femtosecond laser pulses in glass, *Current Applied Physics*, 9, 2009, 1210-1212.
[35] L. Allen, M. Beijersbergen, Orbital angular momentum of light and the transformation of Laguerre-Gaussian laser modes, Phys. Rev. A, 45, 1992, 8185-8189.
[36] H. He, M. Friese, and N. Heckenberg, Direct Observation of Transfer of Angular Momentum to Absorptive Particles from a Laser Beam with a Phase Singularity, Phys. Rev. Lett. 75, 1995, 826-829.
[37] G. Gibson, J. Courtial, M. Padgett, M. Vasnetsov, V. Pas'ko, S. Barnett, and S. Franke-Arnold, "Free-space information transfer using light beams carrying orbital angular momentum," Opt. Express 12, 5448-5456 (2004).
[38] J. Leach, and J. Courtial, Interferometric Methods to Measure Orbital and Spin, or the Total Angular Momentum of a Single Photon, Phys. Rev. Lett. 92, 2004, 013601.
[39] W. Jiang, Q. Chen, Y. Zhang, and G. Guo, Computation of topological charges of optical vortices via nondegenerate four-wave mixing, Phys. Rev. A, 74, 2006, 043811.
[40] H. Kogelnik, "Coupled wave theory for thick hologram gratings," *Bell Syst. Tech. J.* 48 (1969) 2909-2947.
[41] S. Kanehira, J. H. Si, J. R. Qiu, K. Fujita, and K. Hirao, Periodic nanovoid structures via femtosecond laser irradiation, *Nano Lett.* 5 (2005) 1591.
[42] Juan Song, Xinshun Wang, Jian Xu, Haiyi Sun, Zhizhan Xu, and Jianrong Qiu, "Microstructures induced in the bulk of SrTiO3 crystal by a femtosecond laser," *Opt. Express.* 15 (2007) 2341-2347.

[43] Xiao Hu, Juan Song, Qinling Zhou, Luyun Yang, Xiaodan Wang, Congshan Zhu, and Jianrong Qiu, "Self-formation of void array in Al_2O_3 crystal by femtosecond laser irradiation," *Chin. Opt. Lett.* 6 (2008) 388-390.
[44] X. Hu , B. Qian, P. Zhang, X. Wang, L. Su, J. Qiu, C. Zhu, Self-organized microvoid array perpendicular to the femtosecond laser beam in CaF_2 crystals, Laser Phys Lett, 5(5) (2008) 394.
[45] E. Toratani, M. Kamata, and M. Obara, Self-fabrication of void array in fused silica by femtosecond laser processing, *Appl. Phys. Lett.* 87 (2005) 171103.
[46] J. Song, X. Wang, X. Hu, Y. Dai, J. Qiu, Y. Cheng and Z. Xu, Formation mechanism of self-organized voids in dielectrics induced by tightly focused femtosecond laser pulses, *Appl. Phys. Lett.* 84 (2008) 092904.
[47] Brodeur, F. Ilkov and S. Chin, Beam filamentation and the white light continuum divergence, *Opt. Comm.* 96 (1996) 193-198.

27

Sub-Wavelength Patterning of Self-Assembled Organic Monolayers via Nonlinear Processing with Femtosecond Laser Pulses

Nils Hartmann
Center for Nanointegration Duisburg-Essen (CeNIDE)
Universität Duisburg-Essen
Germany

1. Introduction

Self-assembled monolayers (SAMs) are ultrathin organic coatings, which allow one to precisely tailor surface properties, such as the wettability, the biocompatibility and the chemical reactivity and resistance. For this reason, SAMs are widely used as ultrathin resists and functional coatings in numerous micro- and nanofabrication schemes (Love et al., 2005; Onclin et al, 2005; Buriak, 2002). Femtosecond (fs) laser processing of SAMs here offers a variety of particularly promising perspectives (Hartmann et al., 2008; Franzka et al., 2010; Mathieu et al., 2010). On one hand, fs-laser processing is a maskless noncontact technique with a high flexibility, both, in two and three dimensions. Also, processing can be carried out at high pressures, in liquids and on - or even in - dielectric materials and rapid patterning is feasible over wafer scale areas. In addition, fs-lasers allow for nonlinear processing and hence provide a means for sub-wavelength patterning. On the other hand, because of their ultrathin nature, SAMs can be processed with single laser pulses, which ensures short processing times. The monomolecular thickness also allows for well-defined irradiation and burr-free patterning of the coating and avoids bubble and particle formation. All off which are problems commonly encountered in direct ablative fs-laser processing. Moreover, varying the chemical structure of the monolayer provides a means to tailor cross sections, incubation effects and resist properties and to enable chemical patterning. Surprisingly, though, despite these promising perspectives, fs-laser processing of organic monolayers still is largely unexplored (Chang et al., 2005; Chang et al., 2006; Kirkwood et al., 2007; Álvarez et al., 2008). Moreover, most of these contributions focus on micropatterning experiments (Chang et al., 2006; Kirkwood et al., 2007; Álvarez et al., 2008). This chapter reviews recent efforts in sub-wavelength patterning of organic monolayers via single pulse fs-laser processing including silane-based monolayers on quartz glass (Hartmann et al., 2008) and surface-oxidized silicon substrates (Franzka et al., 2010) and thiol-based monolayers on gold-coated silicon substrates (Mathieu et al., 2010). These results demonstrate the general versatility of fs-lasers in nonlinear processing of SAMs on dielectric, semiconducting and metallic platforms and emphasize the particular prospects of this approach as a tool for rapid micro- and nanofabrication.

2. Self-assembled monolayers

Fig. 1 displays the general structure of SAMs (Mathieu & Hartmann, 2010). Generally, these ultrathin films consist of densely packed elongated molecules, which stand up on top of the substrate's surface. The film thickness typically is 1-2 nm. Each molecule can be divided into three functional parts: i) the head group, which couples the molecules to the surface and defines the type of the SAM. Most prominent are thiol-based SAMs on gold and other coinage metals (Love et al., 2005), silane-based SAMs on oxidic or surface-oxidized substrates (Onclin et al., 2005) and organic monolayers from olefinic precursor molecules on semiconducting platforms (Buriak, 2002). ii) the backbone, which in the simplest case represents an aliphatic hydrocarbon chain, and iii) the end group, e. g. simple functional groups such as amine or carboxyl groups or more complex chemical or biochemical functionalities. Upon varying the chemical structure the chemical reactivity and resistance of SAMs can be tailored. This explains the wide-spread use of SAMs as ultrathin resists and functional coatings (Love et al., 2005; Onclin et al., 2005; Buriak, 2002). In particular, via the end group additional components can be attached to the surface. In conjunction with patterning routines this provides a facile means to fabricate chemical templates with domains of defined size, shape and chemical termination. Such platforms allow one to selectively couple nanoscopic building blocks in predefined surface areas in order to build up functional surface architectures (Woodson & Liu, 2007).

A common procedure for the preparation of SAMs considers immersion of clean substrates into millimolar solutions of respective precursor molecules. With time the molecules couple to the substrate surface via the head group, stand up on the surface and form a densely packed monolayer as depicted in Fig. 1. Monolayer growth is self-limiting, as the end groups usually are chosen as to not react with the head groups. The quality of the monolayer in terms of packing density and order, however, strongly depends on the type of the SAM and the detailed preparation conditions (Love et al., 2005; Onclin et al., 2005; Buriak, 2002).

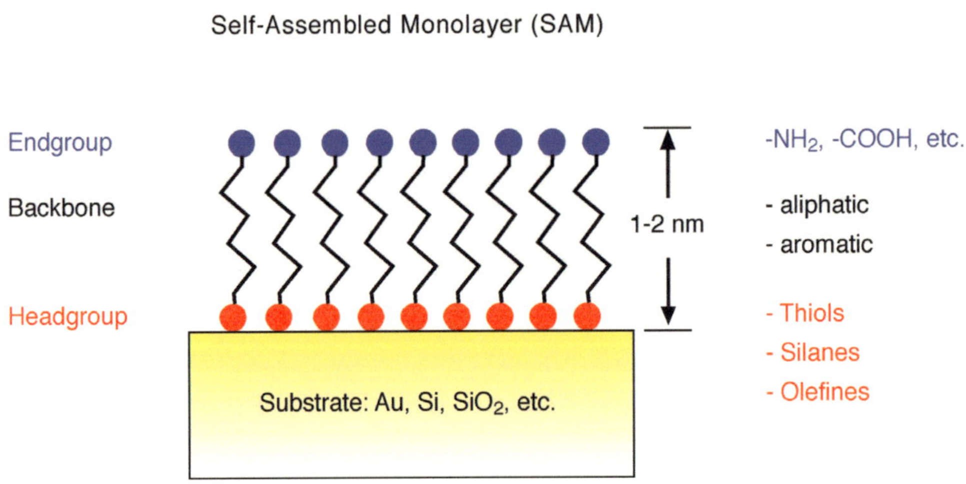

Fig. 1. Schematic diagram of the structure of SAMs. The thickness typically is 1-2 nm. Three functional parts of the adsorbed molecules are distinguished: the head group, the backbone and the end group. Adapted from Mathieu & Hartmann, 2010. © IOP.

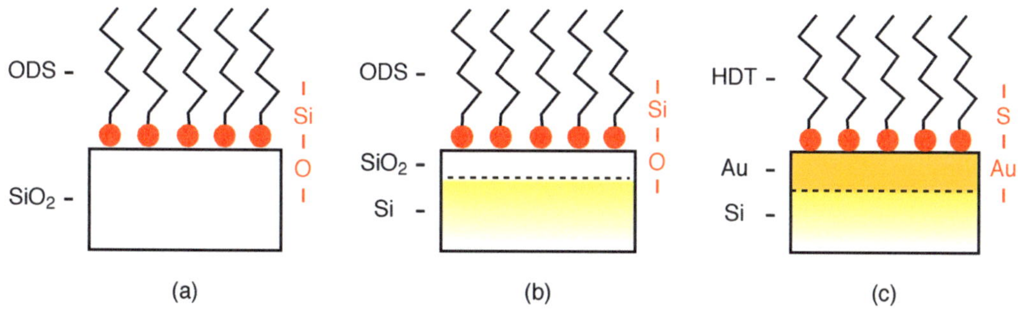

Fig. 2. Schematic presentations of the SAM/substrate combinations considered in this review: a) ODS monolayers on quartz glass, b) ODS monolayers on surface-oxidized silicon substrates, c) HDT monolayers on Au-coated silicon substrates. The surface bond of the molecules is shown in red on the right-hand side of each scheme.

In the work reviewed here, alkylsiloxane SAMs and alkanethiol SAMs are addressed. As substrates quartz glass, surface-oxidized silicon and Au-coated silicon samples are used. Surface-oxidized silicon samples expose a native oxide layer with a thickness of 1-2 nm on top. Au-coated silicon samples are silicon samples with a 5 nm Ti adhesion layer and a 40 nm Au layer on top. Detailed coating procedures are given in the literature (Hartmann et al., 2008; Franzka et al., 2010; Mathieu et al., 2010). Briefly, prior to coating, all substrates are cleaned in hot piranha solution (a mixture of sulfuric acid and hydrogen peroxide). Octadecylsiloxane (ODS) monolayers then are formed upon immersion of the substrates into a millimolar solution of octadecyltrichlorosilane in toluene. For coating with hexadecanthiol (HDT) monolayers a millimolar solution of HDT in ethanol is used. Schematic presentations of the distinct SAM/substrate combinations are shown in Fig. 2.

3. Nonlinear femtosecond laser processing

Laser processing is widely recognized as a facile and versatile means for direct patterning of SAMs (Mathieu & Hartmann, 2010). If nanopatterning is targeted, the optical diffraction limit, however, poses a significant constraint (Bäuerle, 2000). Even if highly focusing optics with a numerical aperture *NA* close to one is used the minimum structure size d_{min} usually is not much smaller than the wavelength λ of the laser light:

$$d_{\min} \approx \lambda / NA \tag{1}$$

A common means to extend the lateral resolution of laser techniques into the sub-wavelength range takes advantage of nonlinear effects (Bäuerle, 2000; Koch et al., 2005a; Ali et al., 2008; Chong et al., 2010). If one considers the complexity of photoexcited processes at surfaces, the pool of nonlinearities, of course, is large (Zhou et al., 1991; Richter & Cavanagh, 1992; Zimmermann & Ho, 1995; Mathieu & Hartmann, 2010). In general, direct and indirect excitation mechanisms can be distinguished. Direct mechanisms are based on an immediate excitation of the adsorbed molecules within the monolayer. Examples include photoexcitation via single or multiphoton processes and field-induced processes. In contrast, indirect mechanisms start with excitation processes that are initiated in the substrate. Absorption of laser light in the substrate at first generates excited charge carriers, such as hot electrons. On one hand, these electrons may interact with the adsorbed molecules building up the monolayer and initiate reactions. Examples are photochemical or photoelectrochemical reaction pathways. On the other hand, the excited electrons eventually scatter inelastically with the substrate lattice, which inevitably results in a certain temperature rise at the surface. This provides the basis for photothermal reaction pathways. All the aforementioned processes proceed on material-specific time, length and energy scales, which vary over several orders of magnitude. This makes laser patterning a rich and complex process (Bäuerle, 2000). For this reason, nonlinear laser processing of organic monolayers, at first, necessitates a proper choice of all laser parameters in order to trigger the desired surface reactions and avoid damage of the substrate (Mathieu & Hartmann, 2010). Fig. 3 depicts some of the essential energetic parameters: the photon energy E_P, the linear photodissociation threshold E_D of the monolayer and, if applicable, the band gap E_B of the substrate. Commonly, in nonlinear laser processing photon energies well below the photodissociation threshold are chosen. Femtosecond lasers here offer some particularly promising perspectives (Hartmann et al., 2008; Franzka et al., 2010; Mathieu et al., 2010). Processing of organic monolayers on dielectric materials with $E_B > E_P$ allows one to exploit multiphoton absorption processes. On semiconductors and metals with $E_B \leq E_P$, indirect processes, e. g. multiple electronic excitations or photothermal reactions can be initiated. All these processes introduce strong nonlinearities and ensure sub-wavelength resolution. Note, although fs-lasers commonly are used in order to minimize thermal impact, photothermal pathways could open up shortly after local irradiation with fs-laser pulses.

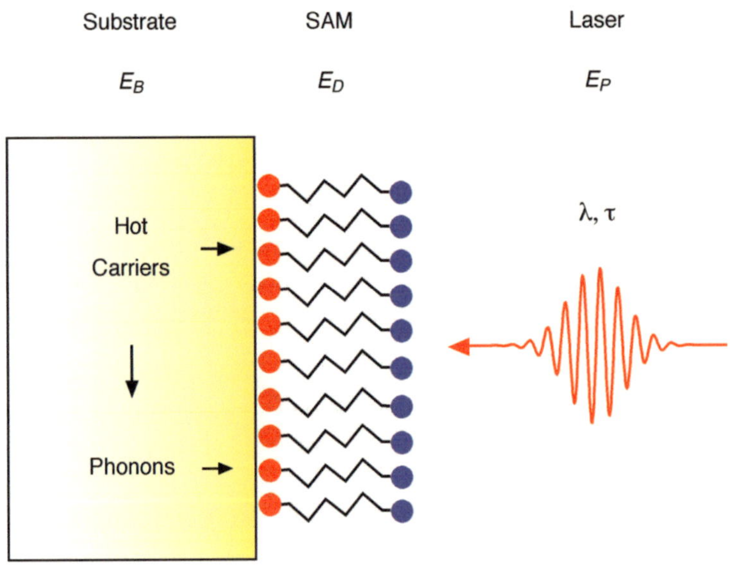

Fig. 3. Energetic constraints and photoexcited processes in fs-laser processing. Adapted from Mathieu & Hartmann, 2010. © IOP.

Laser processing experiments described in this work have been carried out at the Laser Zentrum Hannover (LZH) using a commercial Ti:Sapphire oscillator-amplifier system (Femtopower compact Pro, Femtolasers Produktions GmbH) at ambient conditions. A detailed description of the experimental setup is given in the literature (Koch et al., 2006). The laser system provides Gaussian laser pulses with a wavelength and a pulse length of $\lambda = 800$ nm and $\tau < 30$ fs, respectively. A Schwarzschild microscope objective with $NA = 0.5$

is used to focus the laser pulses onto the substrates. The 1/e focal spot diameter $d_{1/e}$ somewhat varies depending on the specific objective and the optical adjustment of the experimental setup. Respective values in each experiment are obtained upon fitting the experimental data as outlined below.

Typically, single-pulse processing at distinct pulse energies E is carried out. Fluences F are calculated via (Hartmann et al., 2008):

$$F = 4E/\pi d_{1/e}^2 \tag{2}$$

For characterization of the patterned samples atomic force microscopy (AFM) is used. AFM images are recorded in contact mode with standard cantilevers. The diameter d of the structures usually depends on the laser fluence as follows (Hartmann et al., 2008):

$$d = d_{1/e}\sqrt{\ln(F/F_{th})} \tag{3}$$

where F_{th} is a critical threshold for a given process, e. g. for monolayer decomposition or substrate ablation. For analysis, the experimental data is fitted on the basis of Eq. (3). This yields the corresponding critical threshold value F_{th} and the 1/e spot diameter $d_{1/e}$.

3.1 Alkylsiloxane monolayers on quartz glass

Femtosecond laser patterning of alkylsiloxane monolayers on quartz glass at $\lambda = 800$ nm allows one to exploit multiphoton absorption processes (Hartmann et al., 2008). In particular, photochemical patterning of these coatings usually is carried out at $\lambda < 200$ nm (Sugimura et al., 2000). The linear photodissociation energy E_D is expected to be around 6 eV. Quartz glass, in turn, exhibits a band gap E_B of 9 eV. Hence, at a photon energy of $E_P = 1.6$ eV considered here, nonlinear processing appears feasible. Moreover, in view of these energetic constraints, it has been anticipated, that sub-wavelength patterning of the coating can be carried out well below the ablation threshold of the substrate.

Fig. 4 shows AFM data from experiments, in which coated samples are patterned with single laser pulses at distinct fluences. Local irradiation results in circular spots of varying morphology and size. At high laser fluences, two regions can be distinguished (Fig. 4a). In the inner region ablation of the substrate is evident. In particular, here, depths reach deep into the bulk (Fig. 4c). In contrast, depths in the outer region remain below 2 nm indicative for a decomposition of the organic monolayer. Note, the outer and the inner region can also be clearly distinguished in the friction contrast (Fig. 4b). In the following the diameters of these regions, where monolayer decomposition and substrate ablation are observed, are denoted as d_{SAM} and d_{quartz}, respectively (Fig. 4d).

At lower laser fluences, no substrate ablation is observed (Figs. 4e-h). Decomposition of the monolayer, however, takes place down to very low fluences. In Fig. 5a a plot of the diameters d_{SAM} and d_{quartz} over the full range of laser fluences considered here is displayed. A fit of the data on the basis of Eq. (3) yields critical threshold values for monolayer decomposition $F_{th}^{SAM} = 3.1$ J/cm² and substrate ablation $F_{th}^{quartz} = 1.2$ J/cm². This points to a fairly large parameter range for selective processing of the monolayers. Moreover, in this regime sub-wavelength patterning below $\lambda/3$ is possible. In particular, as shown in Fig. 5b, at $d_{1/e} = 1.8$ µm structures with a width of 250 nm and below are fabricated. Note, these structures also come with sharp edges.

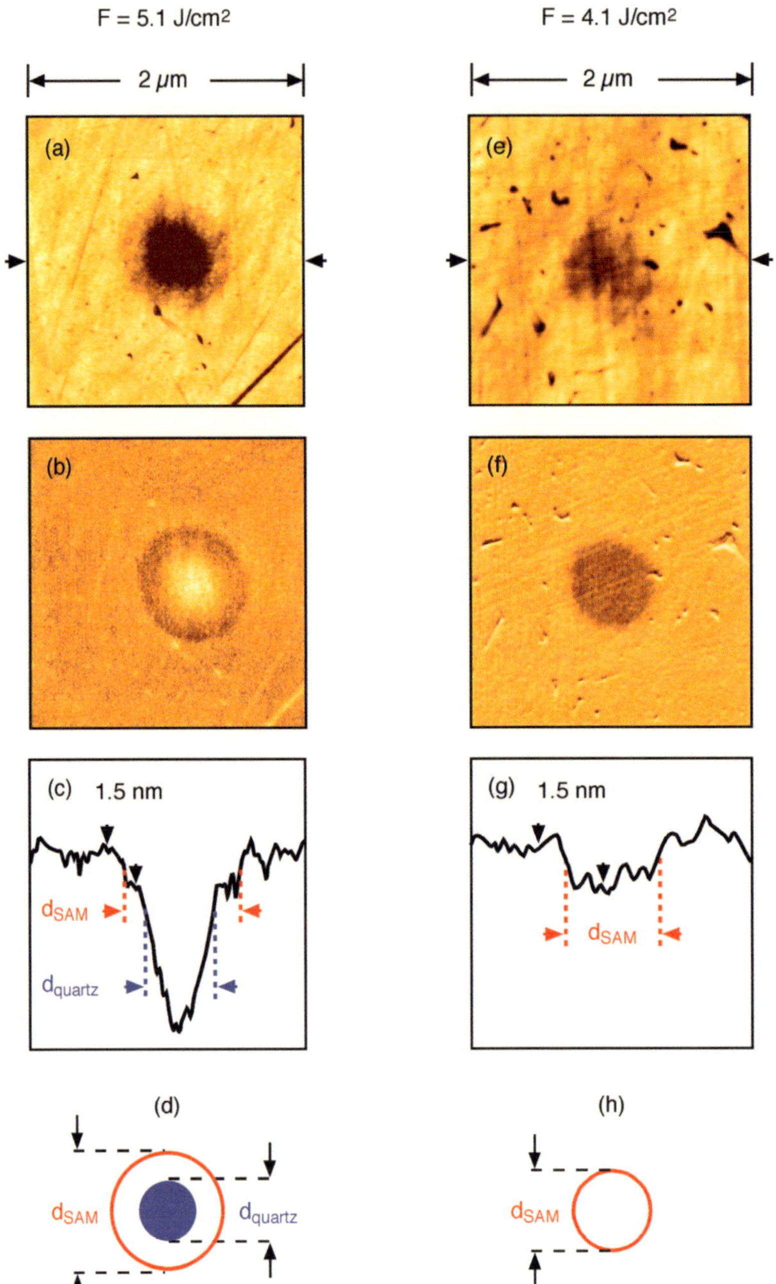

Fig. 4. AFM data from patterning experiments of ODS monolayers on quartz glass with single laser pulses at λ = 800 nm, τ < 30 fs and distinct fluences F: a) - d) F = 5.1 J/cm^2 and e) - h) F = 4.1 J/cm^2. The data display: a) and e) the topography, b) and f) the friction contrast, c) and g) height profiles. d) and h) are schematic presentations denoting the diameters d_{SAM} and d_{quartz}, where monolayer decomposition and substrate ablation are observed, respectively. Adapted from Hartmann et al., 2008. © AIP.

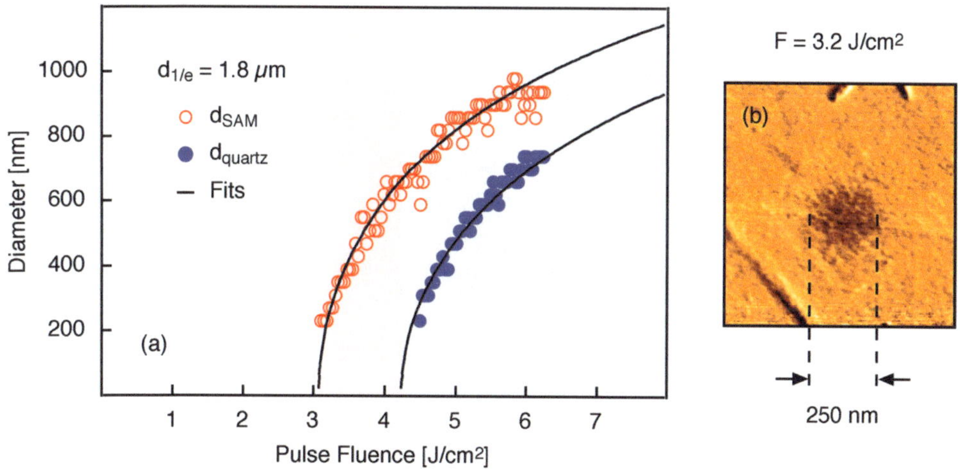

Fig. 5. AFM data from patterning experiments of ODS monolayers on quartz glass with single laser pulses at λ = 800 nm, τ < 30 fs: a) dependence of the diameters d_{SAM} and d_{quartz} on the laser pulse fluence. Lines are fits on the basis of Eq. (3). b) friction contrast image of a sub-wavelength structure at F = 3.2 J/cm². Adapted from Hartmann et al., 2008. © AIP.

This opens up an avenue towards laser fabrication of transparent templates with chemical structures down into the sub-100 nm regime. Such chemical templates represent promising platforms for biotechnological applications, e. g. DNA chips or other biosensor arrays. Processing at higher fluences, also provides a facile route towards combined chemical/topographic structures, e. g. for microfluidic applications. With these results and perspectives, quartz substrates represent an ideal platform for fs-laser processing of silane-based monolayers.

Noteworthy, only few techniques allow for direct nanopatterning of silane-based monolayers on dielectric substrates, such as quartz or glass. Photolithographic techniques, of course, can be used. Sub-100 nm patterning, e. g. using EUV lithography, though, remains challenging. Also, constructive techniques such as dip pen nanolithography and micro contact printing are complicated because of the intricate silane chemistry and surface charging impedes direct patterning with electron beam lithography.

3.2 Alkylsiloxane monolayers on surface-oxidized silicon

For comparison, fs-laser patterning experiments with alkylsiloxane monolayers on surface-oxidized silicon substrates are carried out (Franzka et al., 2010). With a linear photodissociation threshold E_D of 6 eV and a band gap E_B of 9 eV, both the organic monolayer and the surface silicon oxide film are highly transparent in the near-infrared regime. In contrast to quartz glass, however, the band gap of silicon is E_B = 1.1 eV. Hence, at a photon energy of E_P = 1.6 eV laser absorption essentially takes place in the silicon substrate. Monolayer decomposition appears possible via indirect processes. At sufficiently high fluences, multiple electronic excitations could to take place. In addition, photothermal pathways could open up even at comparatively low fluences. All these processes introduce strong nonlinearities and hence provide a means for patterning with sub-wavelength resolution.

AFM images from single-pulse patterning experiments at distinct fluences are shown in Fig. 6a. In contrast to the experiments with coated quartz glass substrates, local irradiation

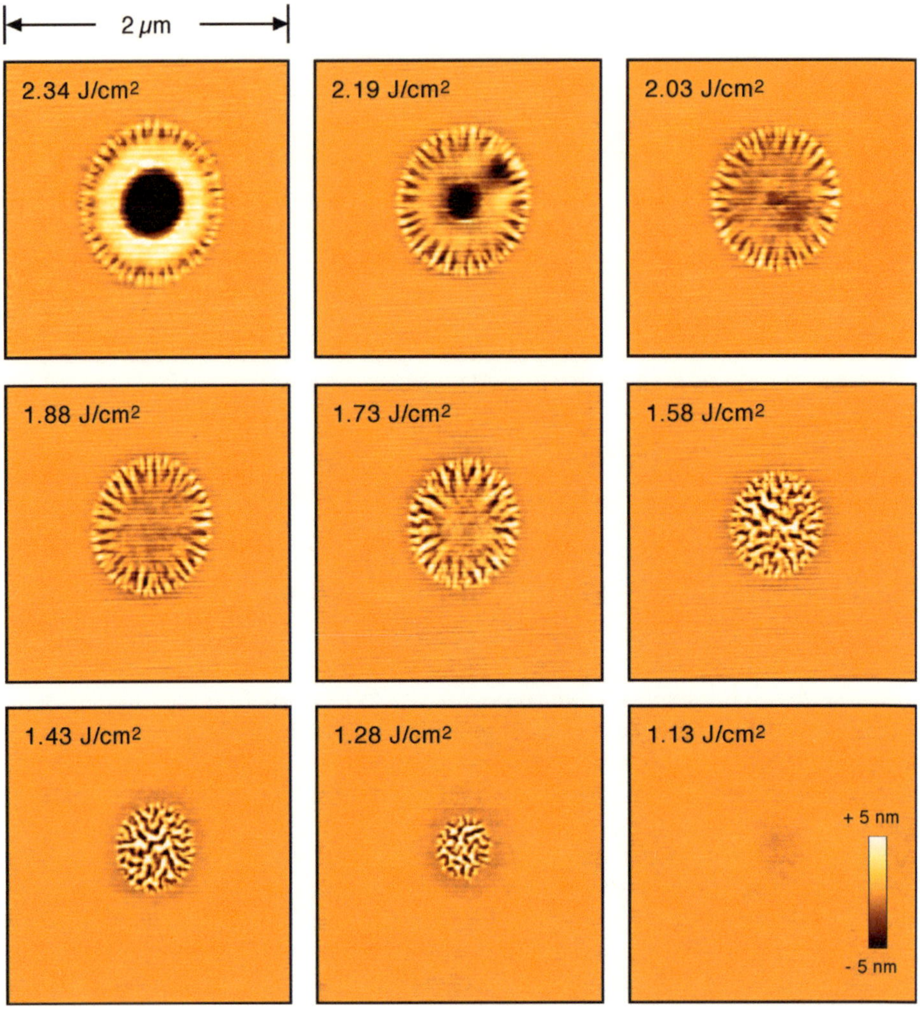

(a)

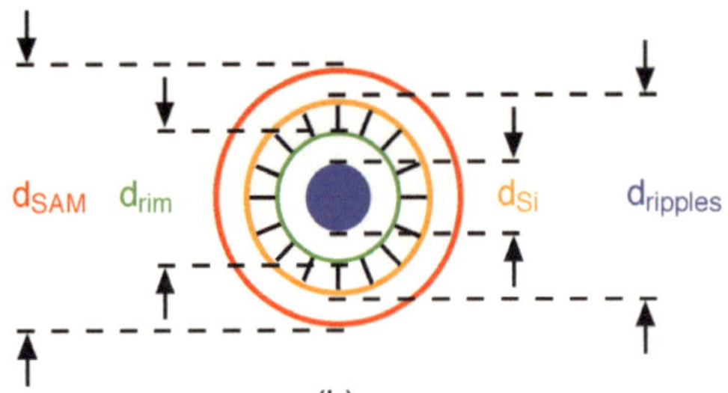

(b)

Fig. 6. a) AFM data from patterning experiments of ODS monolayers on surface-oxidized silicon substrates with single laser pulses at λ = 800 nm, τ < 30 fs and distinct fluences as indicated in the frames. b) schematic presentation depicting the diameters d_{SAM}, $d_{ripples}$, d_{rim} and d_{Si}. Adapted from Franzka et al., 2010. © AIP.

here results in a particularly rich complexity of surface morphology. At high fluences, at least four regions can be distinguished. In the centre region ablation of the substrate is evident (region 1). Radially outwards follow concentric areas where rim and ripples formation is observed (region 2 and 3, respectively) and a faint boundary area (region 4). If the fluence is reduced more and more, at first the depth of the hole in the centre decreases and the rim structure flattens. Finally, the inner regions disappear one after another and only a faint surface spot remains visible where monolayer decomposition has set in. In the following the diameters of the four regions are depicted as d_{Si}, d_{rim}, $d_{ripples}$ and d_{SAM}, respectively (Fig. 6b). In Fig. 7a a plot of these diameters over the laser fluence is displayed.

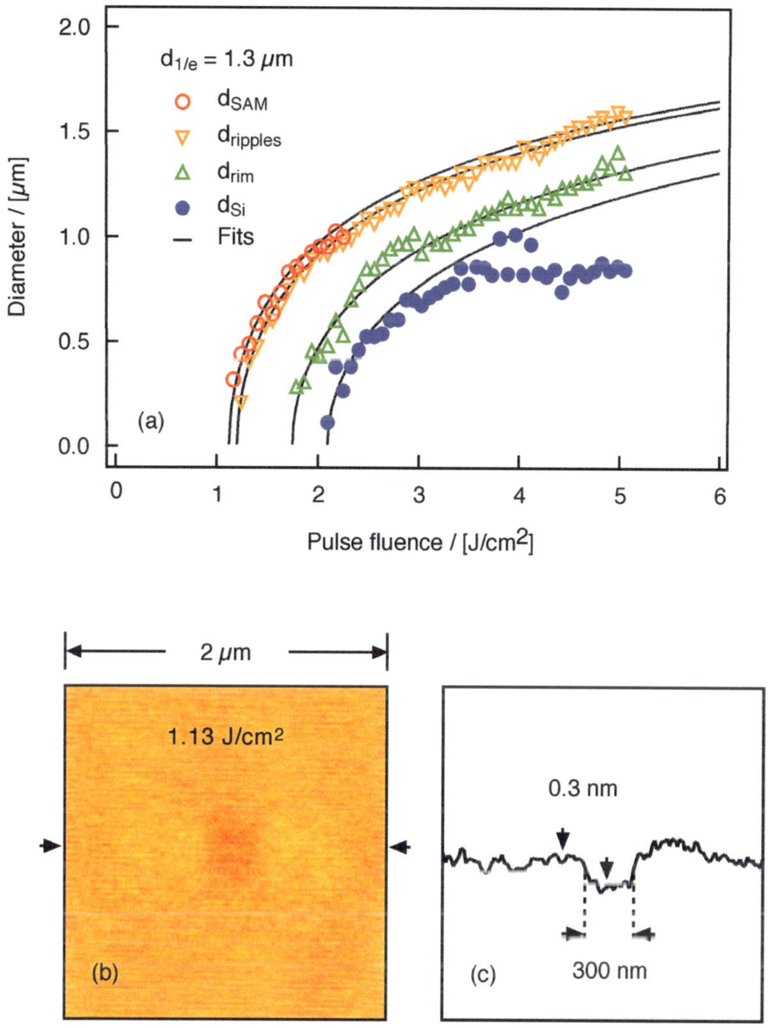

Fig. 7. AFM data from patterning experiments of ODS monolayers on surface-oxidized silicon substrates with single laser pulses at λ = 800 nm, τ < 30 fs: a) dependence of the diameters d_{hole}, d_{rim}, $d_{ripples}$ and d_{SAM} on the laser pulse fluence. Lines are fits on the basis of Eq. (3). b) and c) sub-wavelength structure at F = 1.13 J/cm²: b) topography and c) height profile across the structure in Fig. 7b. Adapted from Franzka et al., 2010. © AIP.

Noteworthy, despite a 1/e laser spot diameter of $d_{1/e}$ = 1.3 µm, minimum structure sizes for selective processing of the organic monolayer are about 300 nm. In particular, sub-wavelength patterning close to λ/3 is feasible. As shown in Figs. 7b and 7c, however, height profiles reveal depths of only 0.3 nm, that is, monolayer decomposition is largely incomplete. Moreover, the parameter range for selective fs-laser processing of alkylsiloxane monolayers on surface-oxidized silicon substrates is fairly narrow. Corresponding fits on the basis of Eq. (3) yield critical thresholds for substrate ablation, rim and ripples formation, and monolayer decomposition of F_{th}^{Si} = 2.1 J/cm², F_{th}^{rim} = 1.8 J/cm², $F_{th}^{ripples}$ = 1.2 J/cm² and F_{th}^{SAM} = 1.1 J/cm², respectively.
Alkylsiloxane monolayers, of course, exhibit an exceptional high thermal and photochemical stability (Onclin et al., 2005). Selective fs-laser processing of more sensitive organosiloxane monolayers is expected to be feasible in a significantly larger parameter range. Monolayers with tailored chromophors, here, are particularly promising (Onclin et al., 2005). In a recent contribution by Jonas, Kreiter and coworkers, for example, fs-laser patterning of silane-based SAMs with photoprotected carboxylic ester functionalities has been addressed (Álvarez et al. 2008). In addition, fs-laser processing of organic monolayers on oxide-free silicon appears promising (Buriak et al., 2002; Klingebiel; 2010). In particular, the Si/SiO_2 interface exhibits a valence band offset of around 4 eV. Hence in case of surface-oxidized samples, excited electrons in the substrate need higher energies in order to reach the organic coating on the surface. This could well block reactions, e. g. via multiple electronic excitations. SAMs on oxide-free silicon, in turn, are directly coupled to the semiconducting substrate and hence, generally, are more sensitive.

3.3 Alkanethiol monolayers on Au-coated silicon

If the fabrication of chemical templates is targeted, selective processing of the monolayers is a key requirement, e. g. in order to ensure precise vertical alignment of nanoscopic building blocks in subsequent deposition processes (Mathieu & Hartmann, 2010). SAMs, in turn, are also used as ultrathin resists. Generally, this makes lower demands on the selectivity of the process. In this respect, alkanthiol SAMs have been demonstrated to provide particularly promising perspectives in fs-laser processing as shown in Fig. 8 (Mathieu et al., 2010).
Similar to alkylsiloxane monolayers, alkanthiol monolayers are highly transparent at λ = 800 nm. Photochemical patterning of such SAMs usually is carried out at wavelengths in the deep ultraviolet range (Ryan et al., 2004). The linear photodissociation threshold E_D is close to 5 eV. At E_P = 1.6 eV, Au, in turn, exhibits a 1/e penetration depths of 14 nm. Hence, fs-laser processing of alkanthiol monolayers is expected to proceed via indirect mechanisms, which start with respective excitations in the substrate (cf. section 3.2).
Figs. 8a-c shows typical AFM images and corresponding height profiles from single-pulse patterning experiment at high fluences. As evident from this data, circular structures with two regions can be distinguished. In the outer region decomposition of the monolayer takes place. In particular, depths are 1-2 nm equivalent to the thickness of HDT SAMs on Au. In the inner region the formation of a fine tip structure indicative for substrate melting is visible (Koch et al.; 2005b). In agreement with these results, both regions exhibit a distinct friction contrast in comparison to the surrounding areas. Note, melting structures as those in Fig. 8a, of course, are undesirable when the fabrication of chemical templates is addressed. At lower fluences, melting structures are not observed anymore. The parameter window, for selective monolayer processing, though, is rather narrow. The melting structures, in turn, do

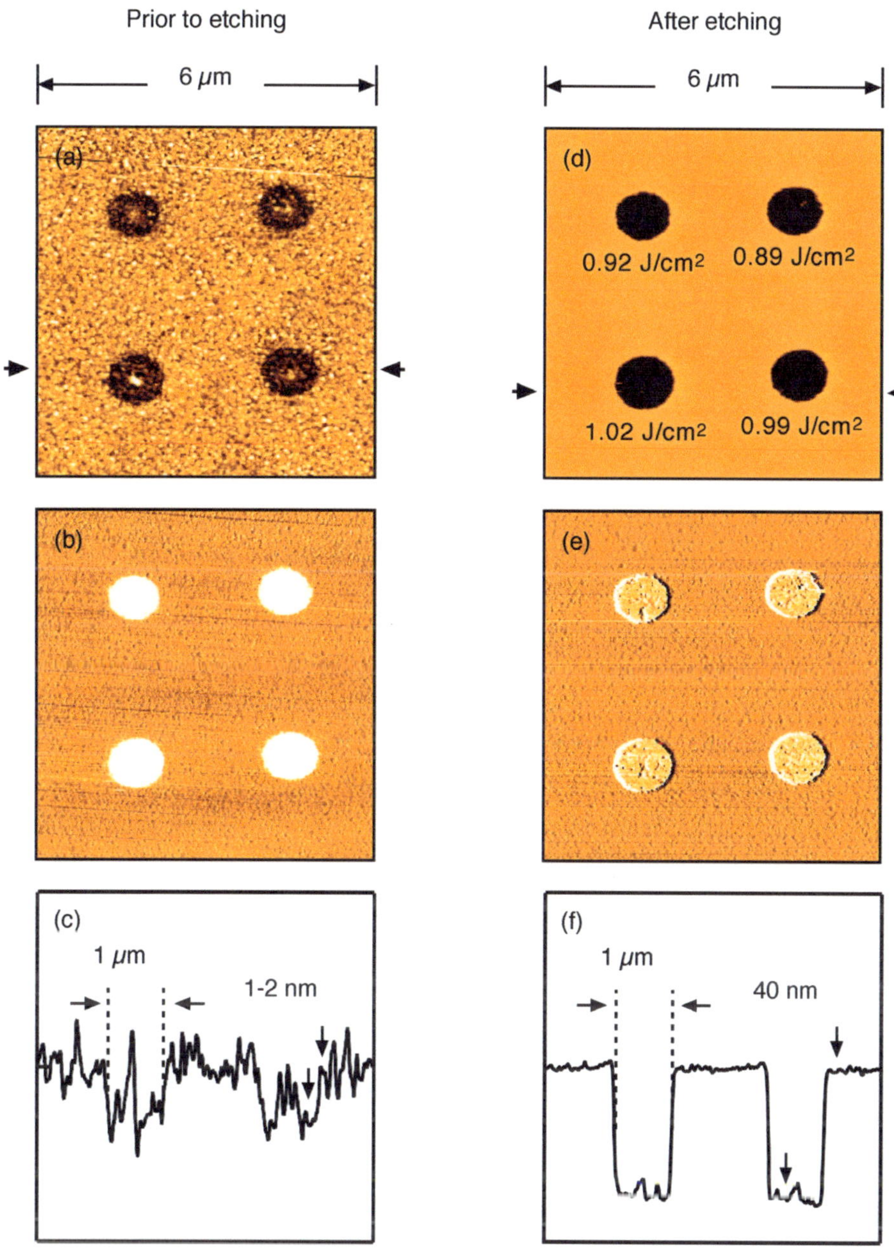

Fig. 8. AFM data from patterning experiments of HDT SAMs on Au-coated silicon samples with single laser pulses at $\lambda = 800$ nm, $\tau < 30$ fs. a)-c) and d)-e) show the same surface area prior to etching and after etching in ferri-/ferrocyanide solution, respectively. Fluences at the distinct positions as indicated in d) are $F = 0.89$ J/cm^2, $F = 0.92$ J/cm^2, $F = 0.99$ J/cm^2 and $F = 1.02$ J/cm^2. a) and d) show the topography, b) and e) the friction contrast and c) and f) display height profiles at the positions marked by black arrows in a) and b). Black arrows in c) and f) indicate the width and the depth of the structures prior to etching and after etching, respectively. Adapted from Mathieu et al., 2010. © Springer.

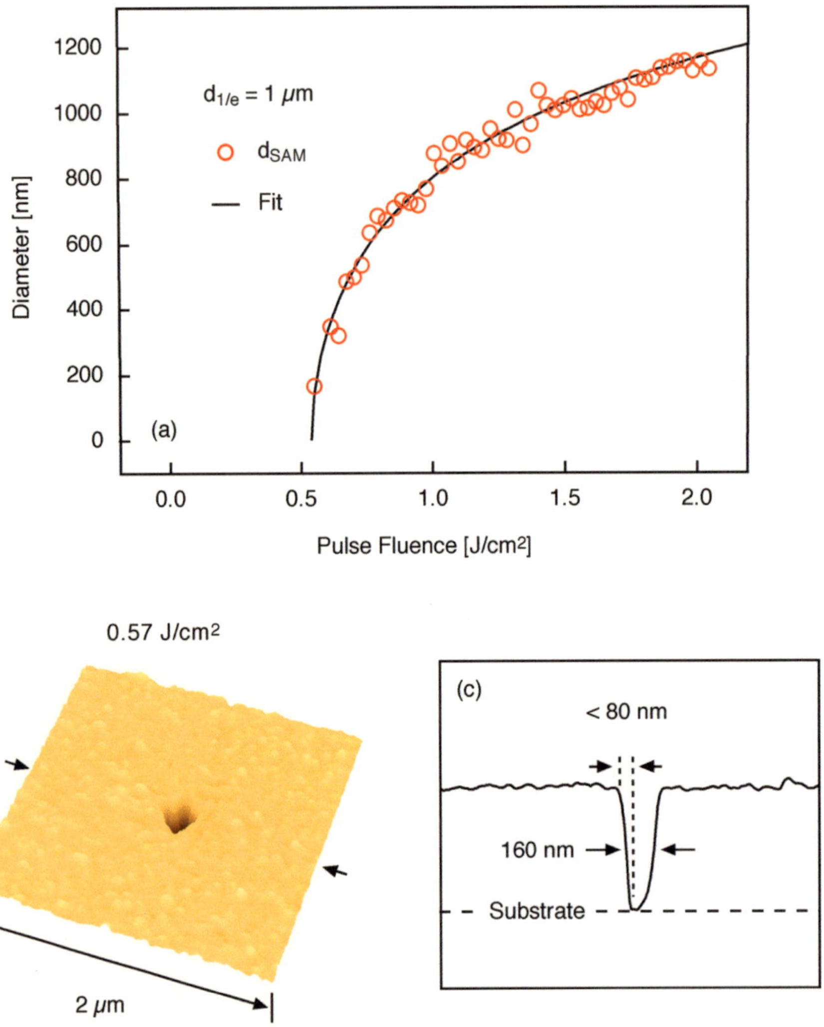

Fig. 9. AFM data from patterning experiments of HDT monolayers on Au-coated silicon substrates with single laser pulses at $\lambda = 800$ nm, $\tau < 30$ fs: a) dependence of the diameters d_{SAM} on the laser pulse fluence. The line is a fit on the basis of Eq. (3). b) and c) sub-wavelength structure at $F = 0.57$ J/cm^2. b) shows the topography and c) displays a height profile across the structure in Fig. 9b. Adapted from Mathieu et al., 2010. © Springer.

not affect subsequent etching steps. Figs. 8d-f, for example, display the same structures as in Figs. 8a-c after etching in a ferri-/ferrocyanide solution (Xia et al., 1995). In the course of this process the 40 nm thick Au film in the bare surface areas is completely dissolved, whereas the SAM in the surrounding areas represents an effective resist layer. Hence, in this way the patterns are transferred into the Au film. No difference in etching is observed between the outer and inner region that are present after laser processing. This results in an extremely large processing window for pattern transfer.

From AFM height profiles the diameters of the etched structures at half-depth are obtained. As evident from Fig. 8, these diameters essentially correspond to the diameters d_{SAM} where monolayer decomposition is initiated. In Fig. 9a a plot of d_{SAM} over the laser fluence is displayed. A fit on the basis of Eq. (3) yields a critical threshold for monolayer decomposition F_{th}^{SAM} = 0.53 J/cm². As expected from the observations discussed above, this value is close to the ablation threshold for thin Au films F_{th}^{Au} = 0.55 J/cm² reported in the literature (Kirkwood et al., 2007). Despite a 1/e spot diameter of $d_{1/e}$ = 1 µm, however, minimum structure sizes after pattern transfer are 160 nm. In particular, sub-wavelength patterning down to λ/5 is feasible. Respective AFM data are shown in Figs. 9b and 9c. Noteworthy, the full width of the step edge of these structures is 80 nm. This value is not corrected for tip size effects, that is, the actual width is expected to be even smaller. Altogether, these results demonstrate the prospects of alkanthiol monolayers as high-resolution resists in rapid nonlinear fs-laser processing.

4. Summary, conclusions and future perspectives

In conclusion, recent progress in single-pulse fs-laser processing of organic monolayers is reviewed. Processing with a micrometer-sized laser spot at λ = 800 nm and τ < 30 fs allows for sub-wavelength patterning. Generally, though, processing strongly depends on the particular chemical and electronic structure of the SAM/substrate system (Hartmann et al., 2008; Franzka et al., 2010; Mathieu et al., 2010). Tab. 1 summarizes the results discussed here. Processing of alkylsiloxane monolayers on quartz glass allows one to exploit multiphoton absorption processes (Hartmann et al., 2008). Selective patterning can be carried out in a fairly large range of laser pulse fluences with sub-wavelength resolution of λ/3 and below. This provides promising perspectives in fabrication of nanostructured chemical templates on transparent platforms, e. g. for biotechnological and microfluidic applications. Processing of alkylsiloxane monolayers on surface-oxidized silicon substrates is expected to follow an indirect mechanism, e. g. via multiple electronic excitations or photothermally induced reactions (Franzka et al., 2010). In contrast to the experiments on coated quartz samples, the parameter range for selective processing of alkylsiloxane monolayers on surface-oxidized silicon substrates is very narrow only. Moreover, monolayer decomposition here is largely incomplete. Hence, more sensitive monolayers are desired in order to exploit this approach on silicon substrates. In addition, fs-laser processing of organic monolayers on oxide-free silicon appears promising (Klingebiel; 2010). Similarly to alkylsiloxane monolayers on surface-oxidized silicon, processing of alkanethiol monolayers on Au-coated silicon is expected to proceed via indirect mechanims. The parameter range for selective processing, again, is rather narrow only. These SAMs, however, show great promise as high-resolution resists in rapid fs-laser processing of thin Au films. In particular, via etching the patterns can be transfered into the Au film. Minimum structures exhibit a width of λ/5.

Overall, these results demonstrate the general versatility of fs-lasers in nonlinear processing of SAMs on dielectric, semiconducting and metallic platforms. Considering tighter focusing optics and shorter wavelength sub-100-nm patterning appears feasible. In addition to its high lateral resolution, the procedure provides a powerful combination of other features, i.e., it is based on a noncontact maskless process and allows for rapid large-area patterning at ambient conditions.

Multiple-pulse patterning experiments also show that incubation effects generally are negligible warranting precise fabrication of complex patterns. In view of these results and perspectives, fs-laser processing constitutes a powerful tool for large-area micro- and nanopatterning of self-assembled organic monolayers.

SAM / Substrate	E_D	E_B	$d_{1/e}$	d_{min}	λ/n	Critical Thresholds F_{th}	Processing window
ODS on quartz	6 eV	9 eV	1.8 μm	250 nm	$< \lambda/3$	$F_{th}^{SAM} = 3.1$ J/cm² $F_{th}^{quartz} = 4.2$ J/cm²	Patterning: 3.1 - 4.2 J/cm²
ODS on SiO_2/Si	6 eV	1.1 eV	1.3 μm	300 nm	$\approx \lambda/3$	$F_{th}^{SAM} = 1.1$ J/cm² $F_{th}^{ripples} = 1.2$ J/cm² $F_{th}^{rim} = 1.8$ J/cm² $F_{th}^{Si} = 2.1$ J/cm²	Patterning: 1.1 - 1.2 J/cm²
HDT on Au/Si	5 eV	-	1.0 μm	160 nm	$\lambda/5$	$F_{th}^{SAM} = 0.53$ J/cm² $F_{th}^{Au} = 0.55$ J/cm²	Patterning: 0.53 - 0.55 J/cm² Pattern Transfer: 1.1 - 2.0 J/cm²

Tab. 1. Summary of the experimental results on fs-laser processing of organic monolayers with single laser pulses at $\lambda = 800$ nm, equivalent to $E_P = 1.6$ eV, $\tau < 30$ fs. For focusing, optics with $NA = 0.5$ are used (Hartmann et al., 2008; Franzka et al., 2010; Mathieu et al., 2010).

On a broader view, the results reported here nicely complement previous studies focusing on sub-wavelength patterning of SAMs via nonlinear processing with continuous-wave lasers exploiting photothermal processes (Balgar et al., 2006; Dahlhaus et al., 2006; Mathieu et al., 2009; Klingebiel et al., 2010; Mathieu & Hartmann, 2010).

5. Acknowledgements

Financial support by the Deutsche Forschungsgemeinschaft (DFG, Grant HA-2769/3-1), the BASF Coatings GmbH and the European Union and the Ministry of Economic Affairs and Energy of the State North Rhine-Westphalia in Germany (NanoEnergieTechnikZentrum, NETZ, Objective 2 Programme: European Regional Development Fund, ERDF) is gratefully acknowledged. It is a pleasure to thank my coworkers and collaborators, who contributed to the work presented here: Steffen Franzka and Mareike Mathieu at the University of Duisburg-Essen, Andreas Ostendorf at the Ruhr University Bochum and Jürgen Koch and Boris Chichkov at the Laser Zentrum Hannover, where all laser experiments have been carried out. I'm also grateful to Eckart Hasselbrink for his continuing support.

6. References

Ali, M .; Wagner, T.; Shakoor, M. & Molian, P.A. (2008). Review of laser nanomachining, J. Laser Appl., Vol. 20, pp. 169-184.

Álvarez, M.; Best, A.; Pradhan-Kadam, S.; Koynov, K.; Jonas, U. & Kreiter, M. (2008). Single-photon and two-photon induced photocleavage for monolayers of an alkyltriethoxysilane with a photoprotected carboxylic ester, Adv. Mater., Vol. 20, pp. 4563-4567.

Balgar, T.; Franzka, S.; Hartmann, N. (2006). Laser-assisted decomposition of alkylsiloxane monolayers at ambient conditions: rapid patterning below the diffraction limit, Appl. Phys. A, Vol. 82, pp. 689-695.

Bäuerle, D. (2000). Laser processing and chemistry, Springer, Berlin.

Buriak, J.M. (2002). Organometallic chemistry on silicon and germanium surfaces, Chem. Rev., Vol. 102, pp 1271–1308.

Chang, W.; Choi, M.; Kim, J.; Cho, S. & Whang, K. (2005). Sub-micron scale patterning using femtosecond laser and self-assembled monolayers interaction, Appl. Surf. Sci., Vol. 240, pp. 296-304.

Chang, W.; Kim, J.; Cho, S. & Whang, K. (2006). Femtosecond-laser-coupled near-field scanning optical microscopy patterning using self-assembled monolayers, Jpn. J. Appl. Phys., Vol. 45, pp. 2082-2086

Chong, T.C.; Hong, M.H. & Shi, L.P. (2010). Laser precision engineering: from microfabrication to nanoprocessing, Laser and Photonics Rev., Vol. 4, pp 123-143.

Dahlhaus, D.; Franzka, S.; Hasselbrink, E.; Hartmann, N. (2006). 1D nanofabrication with a micrometer-sized laser spot, Nano Letters, Vol. 6, pp. 2358-2361.

Franzka, S.; Koch, J.; Chichkov, B.N. & Hartmann, N. (2010). Nonlinear femtosecond laser processing of alkylsiloxane monolayers on surface-oxidized silicon substrates, J. Vac. Sci. Technol. A, Vol. 28, pp. 814-817.

Hartmann, N.; Franzka, S.; Koch, J.; Ostendorf, A. & Chichkov, B.N. (2008). Subwavelength patterning of alkylsiloxane monolayers via nonlinear processing with single femtosecond laser pulses, Appl. Phys. Lett., Vol. 92, pp. 223111-1-223111-3.

Kirkwood, S.E.; Shadnam, M.R.; Amirfazli, A. & Fedosejevs, R. (2007). Mechanism for femtosecond laser pulse patterning of self-assembled monolayers on gold- coated substrates, J. Phys.: Conf. Ser., Vol. 59, pp. 428-431.

Klingebiel, B.; Scheres, L.; Franzka, S.; Zuilhof, H. & Hartmann, N. (2010). Photothermal micro- and nanopatterning of organic/silicon interfaces, Langmuir, Vol. 26, pp. 6826-6831.

Koch, J.; Fadeeva, E.; Engelbrecht, M.; Ruffert, C.; Gatzen, H.H.; Ostendorf, A. & Chichkov, B.N. (2006). Maskless nonlinear lithography with femtosecond laser pulses, Appl. Phys. A, Vol. 82, pp. 23-26.

Koch, J.; Korte, F.; Fallnich, C.; Ostendorf, A. & Chichkov, B.N. (2005a), Direct-write sub-wavelength structuring with femtosecond laser pulses, Opt. Eng., Vol. 44, pp. 051103-1-051103-5.

Koch, J.; Korte, F.; Bauer, T.; Fallnich, C.; Ostendorf, A. & Chichkov, B.N. (2005b), Nanotexturing of gold films by femtosecond laser-induced melt dynamics, Appl. Phys. A, Vol. 81, pp. 325-328.

Love, J.C.; Estroff, L.A.; Kriebel, J.K.; Nuzzo, R.G. & Whitesides, G.M. (2005). Self-assembled monolayers of thiolates on metals as a form of nanotechnology, Chem. Rev., Vol. 105, pp. 1103-1170.

Mathieu, M.; Franzka, S.; Koch, J.; Chickov, B.N. & Hartmann, N. (2010). Self-assembled organic monolayers as high-resolution resists in rapid nonlinear processing with single femtosecond laser pulses, Appl. Phys. A, Vol. 101, pp. 461-466.

Mathieu, M. & Hartmann, N. (2010) New J. Phys., in press

Mathieu, M.; Schunk, D.; Franzka, S.; Mayer, C.; Hasselbrink, E.; Hartmann, N. (2009), Direct laser patterning of soft matter: photothermal processing of supported phospholipid multilayers with nanoscale precision, Small, Vol. 5, pp. 2099-2104.

Onclin, S.; Ravoo, B.J. & Reinhoudt, D.N. (2005). Engineering silicon oxide surfaces using self-assembled monolayers, Angew. Chem. Int. Ed., Vol. 44, pp. 6282-6304.

Richter, L.J. & Cavanagh, R.R. (1992). Mechanistic studies of photoinduced reactions at semiconductor surfaces, Prog. Surf. Sci., Vol. 39, pp. 155-226.

Ryan, D.; Parviz, B.A.; Linder, V.; Semetey, V.; Sia, S.K.; Su, J.; Mrksich, M. & Whitesides, G.M. (2004). Patterning multiple aligned self-assembled monolayers using light, Langmuir, Vol. 20, pp. 9080-9088.

Sugimura, H.; Ushiyama, K.; Hozumi, A. & Takai, O. (2000). Micropatterning of alkyl- and fluoroalkylsilane self-assembled monolayers using vacuum ultraviolet light, Langmuir, Vol. 16, pp. 885-888.

Woodson, M. & Liu, J. (2007). Functional nanostructures from surface chemistry patterning, Phys. Chem. Chem. Phys., Vol. 9, pp. 207-225.

Xia, Y.; Zhao, X.M.; Kim, E. & Whitesides, G.M. (1995). A selective etching solution for use with patterned self-assembled monolayers of alkanethiolates on gold, Chem. Mater., Vol. 7, pp. 2332-2337.

Zhou, X.L.; Zhu, X.Y. & White, J.M. (1991). Photochemistry at adsorbate/metal interfaces, Surf. Sci. Rep., Vol. 13, pp. 73-220.

Zimmermann, F.M. & Ho, W. (1995). State resolved studies of photochemical dynamics at surfaces, Surf. Sci. Rep., Vol. 22, pp. 127-247.

28

Applications of Short Laser Pulses

S. Mehdi Sharifi and Abdossamad Talebpour
Dena Green Inc.
Canada

1. Introduction

Even though the phrase "short pulse" has been in widespread use for more than two decades, there has been no indication of serious attempts to provide a clear limit on pulse duration beyond which the pulse is no more considered to be short. Unless this ambiguity is resolved devising applications peculiar to the short pulses will be prohibitively difficult as it has been the case over past years despite significant advances in untangling the phenomena arising from the interaction of laser pulses with matter. The parameter space determining the outcome of such interactions, which include wavelength, pulse duration, peak intensity, polarization state, density of the matter, and the energy levels of the molecular constituents of the matter, is too enormous to allow an efficient design appropriate for providing performances levels superior to what can be achieved using traditional techniques. Moreover, the parameter space becomes more complex when two or more temporally overlapping pulses of different wavelengths are desired to be applied for providing more flexibility. Thus, we are well justified to devote a whole section for the definition of "short pulse" in the context of the potential applications that are expected to emerge in coming years. Then, two application areas, which we consider to be promising despite being scarcely discussed in academic literature, will be discussed in some details. These include laser assisted high specific mass spectroscopy and flat panel display dicing.

2. Relevancy of applications and definition of "short pulse"

In the introduction we used the phrase "relevant application" without clearly specifying why the adjective relevant needed to be used in the first place. Unfortunately, there are many claims in the literature on the applications of short laser pulses while few, if any, of these claims live up to markets' expectations and soon sink into oblivion. Therefore, we believe that the relevant applications are much fewer than the large list one can compile by gathering those suggested by the researchers in universities. We support our assertion by providing as an example related to one of our recent experiences. Someone approached us for advice on a flat transparent thin sheet of inert polymer having mono-dispersed pores for application in health science to remove bacteria from biological fluids. The pores were desired to have diameters of 0.4 μm with a pore density of about 10^8 pores per cm^2. Obviously, such pores cannot be drilled using commercially available high repetition rate lasers since the pore diameter is smaller than their wavelength. However, recalling the published work on drilling to sub-wavelength diameters we suggested using the high repetition 100 fs Ti:Sapphire laser pulses as a suitable and feasible method. To our dismay, few days later, we learnt of commercially available polycarbonate membrane filters with well characterized characteristics having list prices as low as 0.2 $/cm^2$. Unfortunately, achieving similar levels of perfection and cost does not appear to be practical with short laser pulses; the number of the pores is too large to have a wide area filter to be manufactured in a relatively short time. Thus, as usual the merits of chemistry based fabrication techniques are easily manifested and the economical realities will dictate their selection.

We are aware that cases similar to the aforementioned example abound in technology transfer offices of the universities. The announcement of a new application may generate excitement in research circles and provide a new hot topic for the researchers looking for satisfaction by addressing real world problems and justifying large funds that are spend for conducting application oriented researches. Then, at the first glimpse of seemingly new observation patents are files with the hope that soon they will be licensed by some high tech firm. This approach suffers from the lack of locating the invention on a detailed development landscape. Such a map, often, is prepared at the lunch of development phase at the established industries by evaluating the survival potential of the invention encounter with the harsh realities of the markets.

In table 1 we provide a list of topics which were considered as breakthroughs for a while. For each item the most plausible reason for its final demise is briefly mentioned. The list is

The suggested application of short pulses	**The reason for failing the relevancy criterion**
Using the short laser pulses as a replacement to the mechanical drills in dentistry to avoid the transfer of AIDS virus from one patient to the other.	The early overwhelming AIDS scare subsided when the clinical reports did not support the prevalence of viral infections through dentistry tools. In addition, a user friendly laser system was not introduced into the market, perhaps, due to the difficulties in justifying the development costs.
Writing FBG's	The dotcom crash of late nineties nearly dried up the market for the FBGs. Remaining small market is very well served by the installations utilizing UV lasers equipped with phase masks, which enable the manufacturing of high performance FBGs.
Pattern carving on wooden structure without leaving behind burn traces.	The short pulses are not able to efficiently carve out deep enough as required by the industry. It is not clear if there is any demand for shallow and colourless patterns.
Using filaments generated by the short pulses for remote sensing of target molecules in atmosphere.	Efficient and selective detection has not been validated by data collected in the field.
Using filaments generated for lightening control.	Reviewing the published results does not support the claims that the clouds can indeed be discharged by filaments.
Continuum free spectroscopy of gases	Despite providing clean spectra with very low continuum noise, the method lacked enough specifity required for discrimination of molecules having similar structures.

Table 1. A short list of suggested applications for short pulses that could not live up to their initial promises.

very short and only presents few items that the author had been personally involved. In order to do justice to those who initiated these topics, we should add that applying the relevancy criterion, i.e. examining the proposed application with scrutiny, is not easy at the time of announcements due to the lack of required feedbacks from the intended end users.
Returning to the challenge of defining "short pulse", we select the multiphoton/tunnelling ionization of Xe interacting with 200 fs Ti:Sappire laser pulses reported by Larochelle et al (1998). They measured the ion yield versus peak intensity curve, the laser was focused using f#100 optics into an UHV chamber having a background pressure of 2 x10−9 Torr. Ion species were separated with a time-of-flight mass spectrometer. Ion curve was produced by combining a series of intensity scans, each having a different fill pressure in the interaction chamber. The experimentally measured ion yield versus intensity curves were compared with that predicted by the PPT model. The results are presented in figure 1.

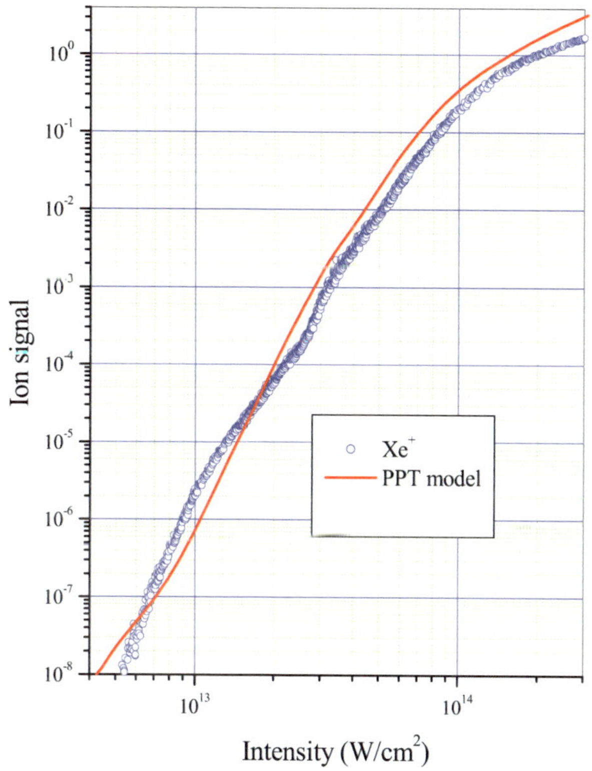

Fig. 1. Ion versus peak intensity plot for Xe interacting with linearly polarized laser pulses from a Ti:Sapphire laser (800 nm). Each datum corresponds to a three point average. The theoretical ion yield is calculated using PPT model.

The derivation of the PPT model proceeds as following (Talebpour 1998a). By finding the exact solution of the time dependent Schrodinger equation for a potential $V = \sqrt{2E_i}\delta(\mathbf{r})$ (E_i is the magnitude of the ionization potential of the atom), the probability current $\mathbf{J}(\mathbf{r}, t)$ is calculated. Then, the total ionization rate from short range potential for linear polarization, W, is found from

$$W=\left(\frac{\omega}{\pi}\right)\lim_{x\to\infty}\int_0^{2\pi/\omega}\int_{-\infty}^{\infty}\int_{-\infty}^{\infty}J_x(r,t)\,dydzdt$$

where ω is the frequency of the laser, which is assumed to be polarized in the direction of the x axis. The effect of the ionic potential, which behaves like Z/r (Z is the charge of atomic core) at a long distance from the nucleus, is calculated through first order correction on the semi-classical action. Accordingly, the PPT model simply disregards the excited energy levels of the atom. The overall fit between the model and the measured ion signal is amazingly good. However, the measured signal exhibits unexpected slope changes at intensities less than 3×10^{13} W/cm^2. These are believed to indicate the involvement of multiphoton resonances of some excited energy levels with the ground state followed by Lambda type trapping. The mechanism of the lambda type trapping is schematically presented in figure 2. At the rising part of the pulse (a) the excited state (with two neighboring levels 1 and 2 which are degenerate within the laser bandwidth) are not in multiphoton resonance with the ground state. The electron is ionized through multiphoton coupling with the continuum. As the intensity of the pulse is increased the excited state and the continuum are shifted in energy due to the Stark shift. At the peak of the pulse (b) the excited states go into multiphoton resonance with the ground state and most of the population is transferred from the ground state to the excited states 1. Coupling starts instantaneously and continues on through the rest of the pulse. As the intensity starts to decrease (c), the two states are coupled through continuum and the population is trapped in a coherent superposition of the two states.

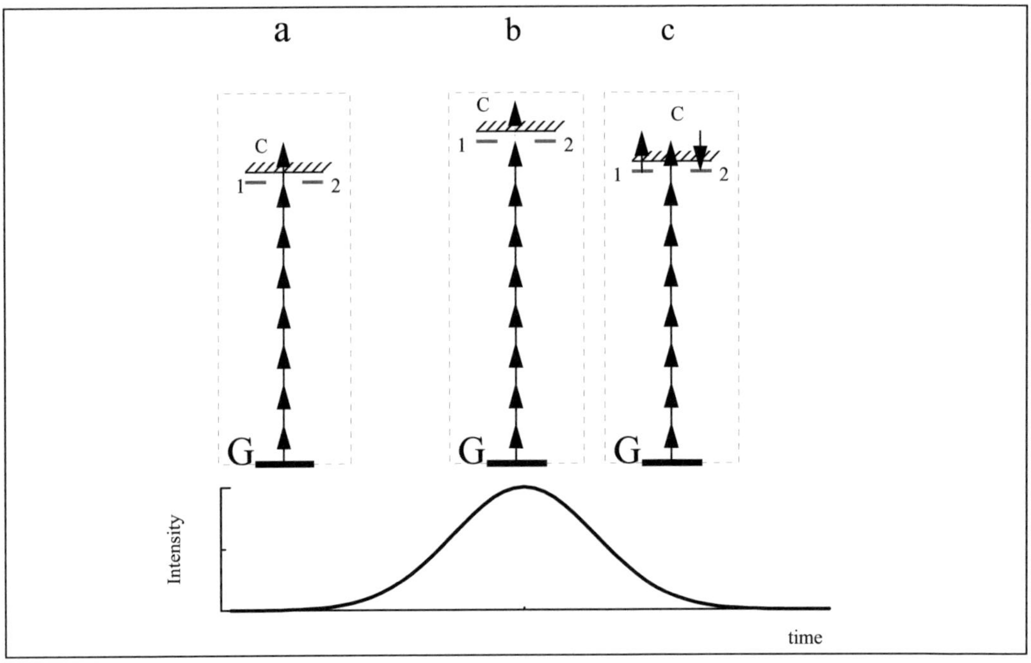

Fig. 2. The schematic presentation of lambda type population trapping. G is the ground state of the atom. 1 and 2 are two degenerate excited states. After the population is transfered to the states due to multiphoton resonance, these states are coupled through continuum c and the population is trapped in the superposition of these states.

If the multiphoton resonance had not been occured at the peak of the pulse, the population transfer to the excited state would not be appreciable. Within the dressed atom picture, the ground state dressed by m photons and the resonant state undergo an avoided crossing at the resonance intensity I_r. The minimum distance, V_m, at the avoided crossing is proportional to the generalized Rabi frequency $\Omega(t) = \Omega_m I(t)^{m/2}$ coupling the two states. According to Story *et al* (1994), the probability of remaining in the ground state, P_g, is given by $P_g = \exp\left(-\frac{2\pi V_m^2}{dW/dt}\right)$, where W is the time-dependent energy difference between the two dressed states. In interaction with a short pulse, if the dynamic resonance is reached in the rising or the falling part of the pulse, the population practically remains in the ground state and the effect of multiphoton resonances may be neglected. However, if the states go onto resonance at the peak of the pulse, where dW/dt is zero, then the excited state is populated. After being populated, since the ionization potential of the excited state is small, it is expected that the electron will be instantly ionized or subsequent interaction with field stabilize the atom against ionization.

Now we offer our definition of short pulse for the cases where the density of the interacting matter is very low.

A pulse interacting with a dilute gas is considered to be short if the rule of excited atomic or molecular levels in the outcome of the interaction phenomenon under consideration can be ignored.

According to our definition, in the case of the multiphoton ionization of Xe (corresponding to the data of figure 1) the pulses are in the brink of being considered short. The data of Larochelle *et al* (1998) indicate that ion versus peak intensity curves of rare gas atoms with high ionization potentials, such as Ar, do not have noticeable slope changes. Therefore, for the multiphoton ionization of Ar interacting with Ti:Sapphire laser pulses 200 fs pulse duration is short.

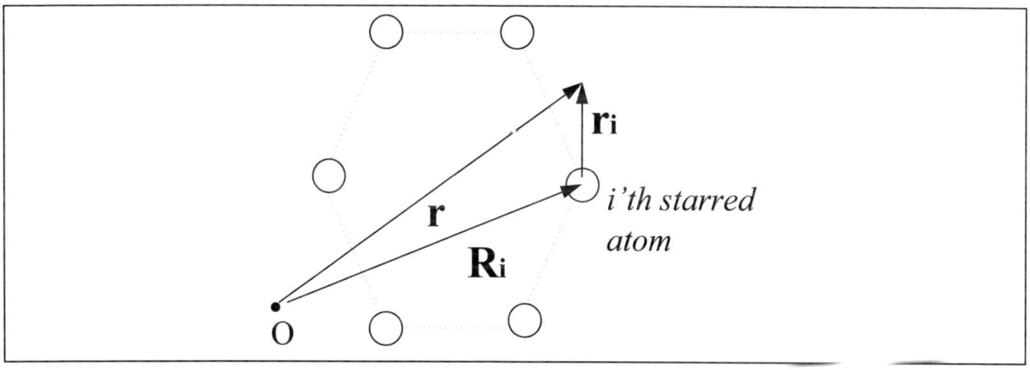

Fig. 3. The schematic diagram of an unsaturated hydrocarbon. The starred atoms are presented by circles. Note that the molecule contains other atoms like H which in Huckel's approximation are neglected. O is the origin of coordinate system.

The hydrocarbon molecules, whose interactions with lasers may be more important, have ionization potentials that are less than the ionization potential of Xe. However, a pulse with duration of 200 fs can be safely taken as short for the following reason. We consider an unsaturated hydrocarbon, such as benzene, which is schematically presented in figure 3. As it is well known from organic chemistry, using Huckel`s theory, the normalized wavefunction of the active electron in the molecule can be approximately represented as

$$\Psi(r)=\sum_i c_i \psi_i(r_i)$$

where $c_i\psi_i(\mathbf{r}_i)$ is the contribution of the p_z electron wavefunction of the i'th carbon atom (starred atom in the nomenclature of organic chemistry) situated at point $\mathbf{R}_i$ and c_i's are constants ($\sum|c_i|^2=1$), determined by variationally minimizing the total energy of the molecule. The variable $\mathbf{r}_i$ is the position vector in a coordinate system with origin at $\mathbf{R}_i$. A schematic definition of different variable are presented in fig. 3. Talebpour *et al* (1998b) have argued that the ionization amplitude can be thought as being a linear superposition of the ionization amplitudes of individual stared atoms. The MPI process in an unsaturated hydrocarbon might be considered as the emission of electron waves from multiple centers. At any observation point in space these waves interfere and, depending on their relative phase, give rise to constructive or destructive interference. The result is that the ionization of the unsaturated hydrocarbons are noticeably suppressed and the ionization process instead of occurring the pure multiphoton channel, by absorbing multiple photons, proceed through tunnel ionization for which the intermediate resonances play no rule in the electron detachment process. In order to verify our claim, we have presented the ion signal versus peak intensity of benzene in figure 4. Clearly there are no observable kinks, reminiscent of the involvement of the excited energy levels in the process, on the ion curve. Therefore, the pulses with wavelength of 800 nm and duration of 200 fs that interact with unsaturated hydrocarbons are short based on our definition.

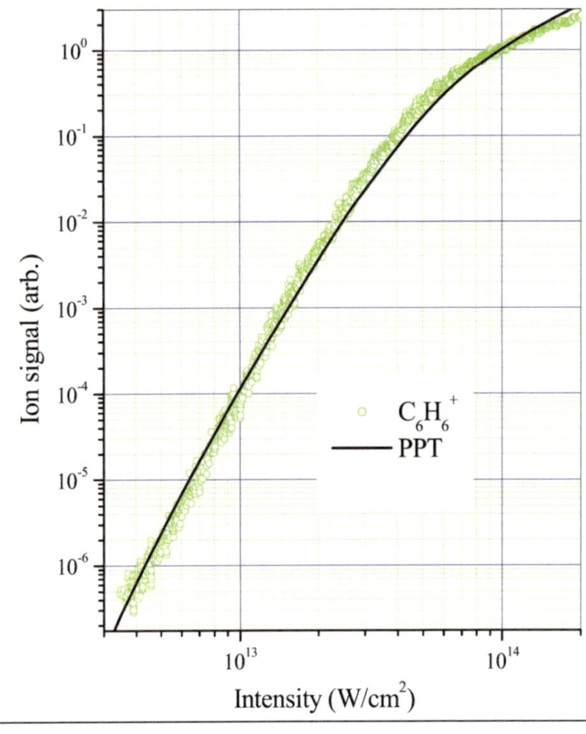

Fig. 4. Ion versus peak intensity plot for benzene interacting with linearly polarized laser pulses from a Ti:Sapphire laser (800 nm). Each datum corresponds to a three point average. The theoretical ion yield is calculated using PPT model.

In the case of interaction with dens media another definition for short pulse may appear to be necessary. Following ionization the electron is being accelerated by inverse Bremsstrahlung and can collide with the nearby molecules and generate new electrons through collisions. If the pulse duration is long, the newly ionized electrons can be accelerated and eventually avalanche or cascade ionization follows. Once the density of the electrons reached a critical value breakdown occurs and high density plasma is created which has no memory of the laser pulse. So, the criterion for the shortness of a pulse in dense media is as the following:

A pulse interacting with a dense matter is considered to be short if during the interaction the threshold for the avalanche ionization is not reached.

At the first glance this definition may appear to be too limiting. Fortunately, due to the delicately balanced behavior of the pulses in dense media the threshold does not reached easily. The phenomenon responsible for the balance is the well known filamentation process during the propagation of strong laser pulses in dense media. The vast amount of studies have resulted in the emergence of a qualitative understanding, which describes the phenomenon as being the manifestation of a self-tuning between two competing processes; the nonlinear self focusing of the pulses and the multiphoton ionization. The electrons are exclusively created through the MPI in the leading edge of the pulse. As time passes, the increase in the energy of the pulse cannot lead to an appreciable increase in the density of the electrons, since the defocusing effects caused by the plasma and by the high order nonlinearities decrease the local intensity and thus stop the generation of more electrons. Talebpour *et al* (2000a) have estimated that, in the case of propagation in atmospheric pressure D_2 gas the plasma density does not reach above 10^{-5} times the gas density. This is the reason why the strength of the light contributed by the continuum of radiation is so low in the visible spectrum of air monitored from the side of filament. Moreover the characteristic atomic lines resulting from the breakdown induced fragmentation of air molecules are missing. These statements can be easily verified by referring to figure 5, which presents the spectrum of the atmospheric pressure air interacting with 200 fs, 800 nm laser pulses.

3. Examples of potential applications for the short laser pulses

In this section we will discus two areas that short pulses may provide relevant applications. These are highly specific mass spectroscopy of molecules and flat panel display dicing. In the first case the basic physics has been very well developed by the contributions from multitude of research groups all over the world. What remain are intensive development efforts preferably supported by industry to resolve the issues related to robust instrumentations. We are aware that the second case is already being introduced in electronic industry particularly in south Korea.

In what follows the scientific background of the fragmentation of molecules, as related to the mass spectroscopy will be reviewed with sufficient detail. The second application will be briefly described by referring to a recent PCT patent application by Kamata et al (2008). Due to the lack of information on marketing front, we are not able to construct a solid case to unambiguously demonstrate the relevance of the suggested applications areas. The future events will prove whether or not our speculations are sound.

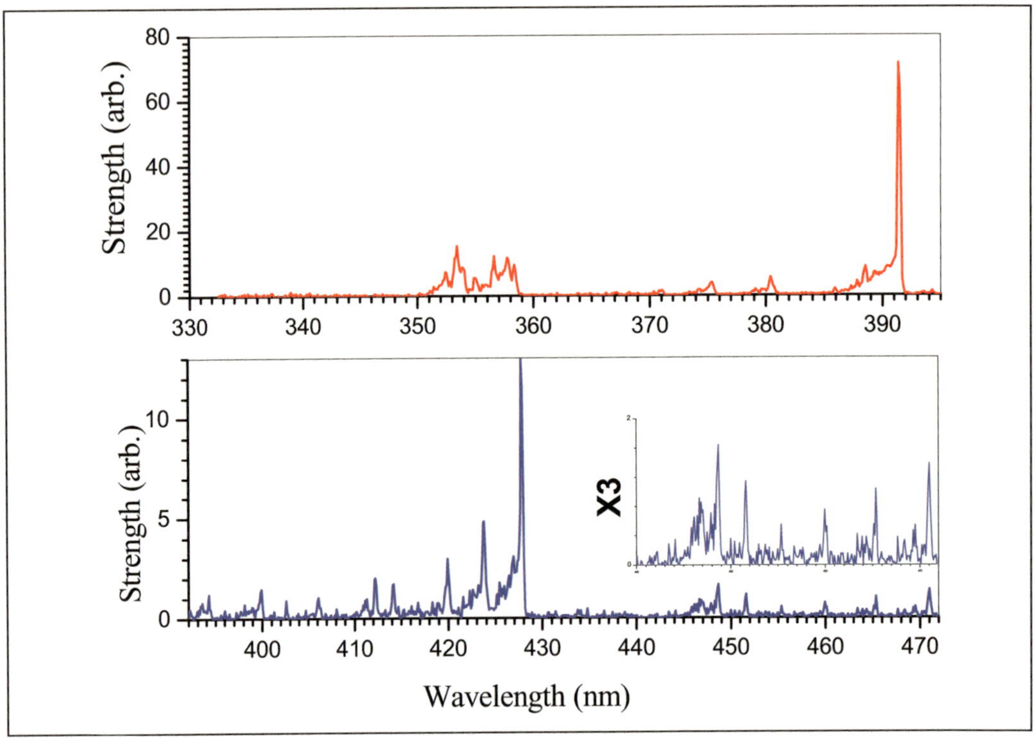

Fig. 5. The visible spectrum of air interacting with linearly polarized Ti:Sapphire pulses having a duration of 200 fs.

3.1 Potential application of short laser pulses as an ion source for high performance mass spectroscopy

3.1.1 Single pulse

The discussions of the previous section indicated that according to the criteria put forward by our definitions, pulses with durations of less than 200 fs are considered to be short. We do not set any restriction on the laser wavelength as the wavelengths of the commercially available lasers and their second harmonics are more or less in the 400-1200 nm range. It is true that all multiphoton process are wavelength dependent but such dependencies are not expected to result in failing the two aforementioned "shortness" characteristics of the pulses. Therefore, in the present subsection, except one case, the discussions will be restricted to the case of 200 fs pulses delivered by a Ti:Sapphire laser that has a wavelength of 800 nm.

We start by reviewing the mass spectra of Cyclohexane that have been prepared at some selected laser intensities (see figure 6). The experiment employed a commercial laser system that consists of a Ti:Sapphire oscillator followed by a regenerative and two multi-pass Ti:Sapphire amplifiers that can deliver pulses with energies of up to 100 mJ. The pulse duration is 200 fs (FWHM) and the central wavelength λ is 800 nm. The laser pulses were focused using *f*/100 optics (1 m focal length lens and a beam diameter of 1 cm) into an ultra-high vacuum chamber having a background pressure of 2 x10^{-9} Torr. The liquid was introduced into the vacuum chamber through a needle valve and the ions resulting from the dissociative ionization of the target molecule are extracted into a time-of-flight (TOF) mass

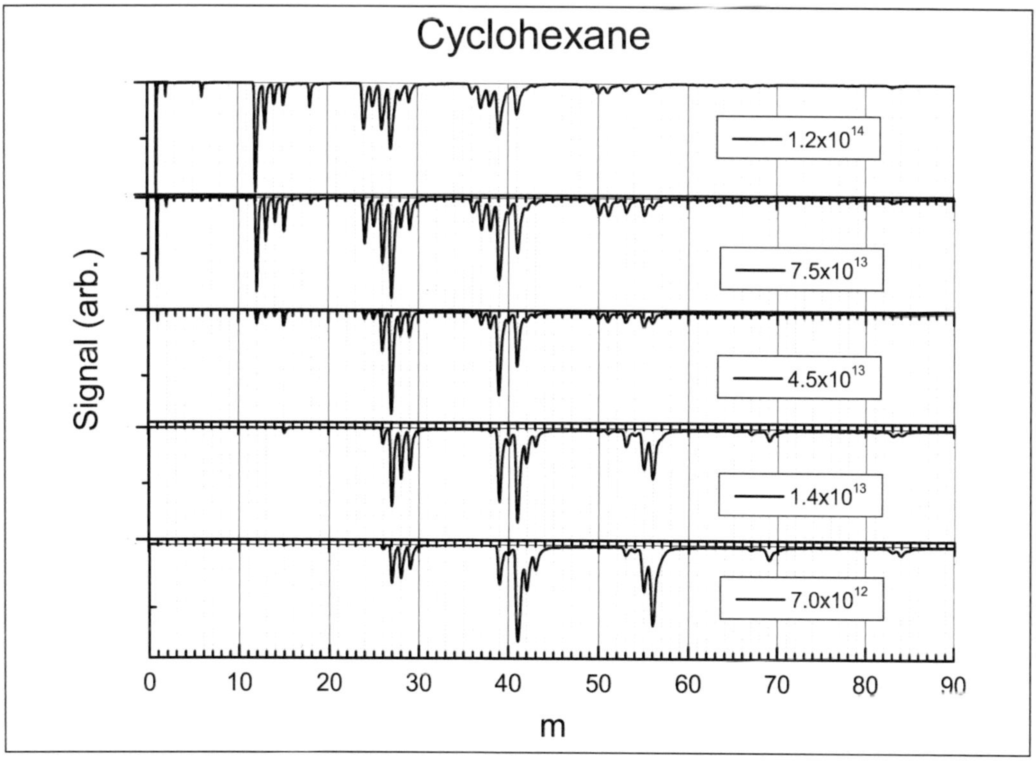

Fig. 6. The mass spectra of Cyclohexane interacing with linearly polarized Ti :Sapphire pulses having a duration of 200 fs.

spectrometer having a 60 cm long drift tube and detected with an electron multiplier in conjunction with a digital storage oscilloscope. The large f-number focusing lens was used to maximize the focal volume of the interaction region so as to obtain ion signals from very low ionization rates. The gas pressure in the interaction chamber was controlled by a precision and set at 10^{-7} Torr. The laser intensity was determined by measuring the laser energy, the pulse width and the focal spot size. The energy measurement was taken from a 4% reflection off a wedged beam-splitter situated after the interaction region and was detected with a pyroelectric energy meter. The pulse width was measured by a second-order autocorrelator. To measure the spot size, the beam was attenuated by successive reflections from high-quality glass plates. The beam was then focused by the 100 cm focal length lens and the image of the focused area was analyzed by a microscopic objective with an associated CCD camera. The beam had a nice Gaussian profile and a roundness of 0.95. The focal spot diameter (at $1/e^2$) was found to be 100 μm. Using this information the peak laser intensity was calculated and recorded on the corresponding spectrum.
When the molecule was exposed to 1 ns pulses from a CO_2 laser the mass specrum only possessed H^+ and C^+ peaks. Thus, the main characteristic peculiar to the mass spectra obtained using short pulses is the richness of the spectrum, as any peak expected from fragmentation of the molecule through known dissociation channels can be observed by varying the laser intensity in a 10x intensity interval. In this respect the short pulse ion source is superior to the well established electron beam based ion sourses, a fact that can be easily verified by referring to figure 7.

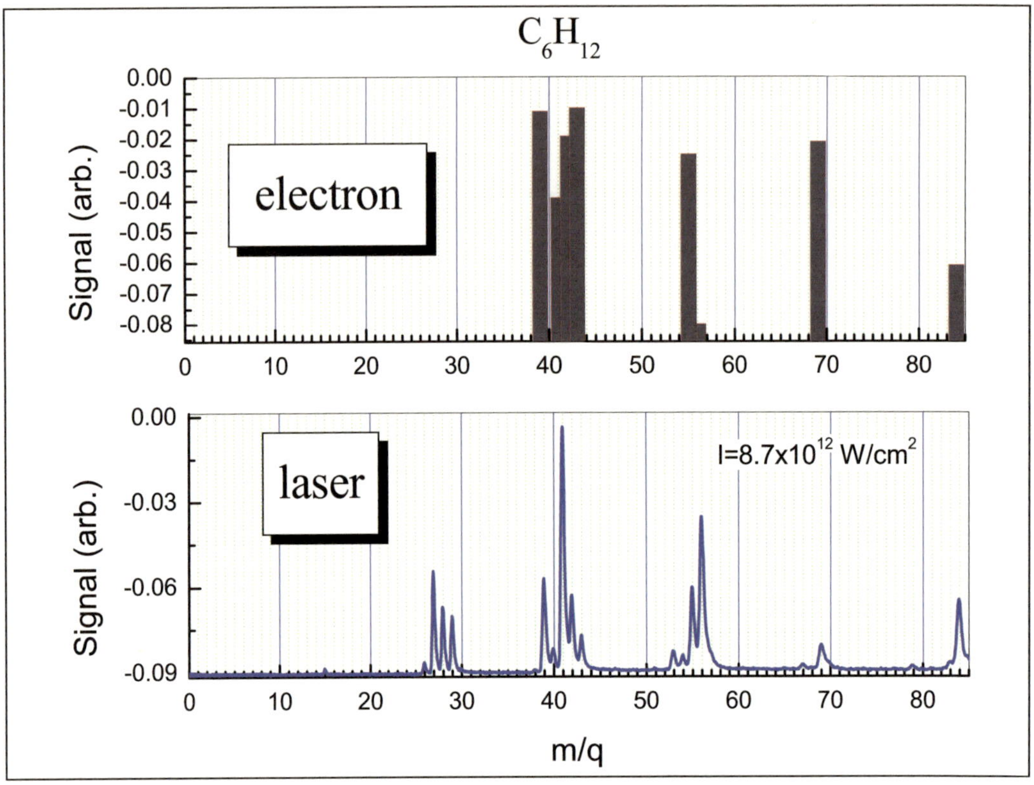

Fig. 7. The mass spectra of Cyclohexane interacing with linearly polarized Ti :Sapphire pulses having a duration of 200 fs.

The mechanism responsible for the fragmentation of polyatomic molecules interacting with short laser pulses was developed by Talebpour *et al* (2000b). Their model describes the short pulse induced fragmentation of molecules as happening in three steps:

a. An electron ionizes from either the outer or one of the inner orbitals of the molecule this generates a molecular ion. If the ionized electron had originated from an inner orbital the ion will be left in ro-vibrational levels of an excited electronic state;
b. The excited ion makes a rapid radiationless transition to the high-lying ro-vibrational levels of a lower electronicstate; and
c. Then the ion dissociates to different fragments through various fragmentation channels. 100 μm.

The importance of the short duaration of the pulse is that due very high intensity the ionization steps occurs rapidly without involving the intermediate excited states. Then, the pulse seizes before appreciably perturbing the geometry of the excited ion. So, the task of the pulse is rapidly ionizing the molecule by coupling the ground state of an electron to the continuum and then leaving the resulting ion alone to follow the path to fragmentation. A long duration pulse stays too long and heavily involves the intermediate states in the ionization process. In addition, there is the possibility that the molecule undergo dissociation before being ionized. Moreover, the pulse survies so long that the fragments interact with it and undergo their own fragmentation route. This explains why in the case of using 1 ns CO_2 laser pulses the molecule dissociates down to its constituent elements.

Now we are in a position to address the main aspect of the short pulse induced molecular fragmentation, i.e. the higher specifity compared to the other ion sources, through experimental studies. The intentention is to show that the selectivity provided by a short pulse based source is superior to that expected when using the conventional electron ionization based sources. The distinction becomes important when the simultaneous analysis of mixtures containing multiple isomers is required. Let us take, as an example, the mass spectra of butane isomers studied by Bennett and Cook (200). Visual inspection of the reported results indicates that the differences between the mass spectra of different isomers are not sufficient for accurately quantifying the partial concentrations of the isomers in a mixture. Any source capable of increasing the difference would improve the accuracy by which the concentration of each isomer is determined. In order to verify if the short pulse based ion source provides enhanced performance Sharifi et al (2008) studied the fragmentation of 1-butene and cis-2-butene, whose structures are presented in figure 8) interacting with femtosecond Ti:Sapphire laser pulses . The resulting fragmentation patterns were compared for the two isomers and it was shown that their difference, as quantified by the spectral similarity index (SI) , was more pronounced than the case of using 100 eV electrons as ionization/dissociation source. The experimental setup and conditions were similar to what was mentioned above in connection to the mass spectra of Cyclohexane.

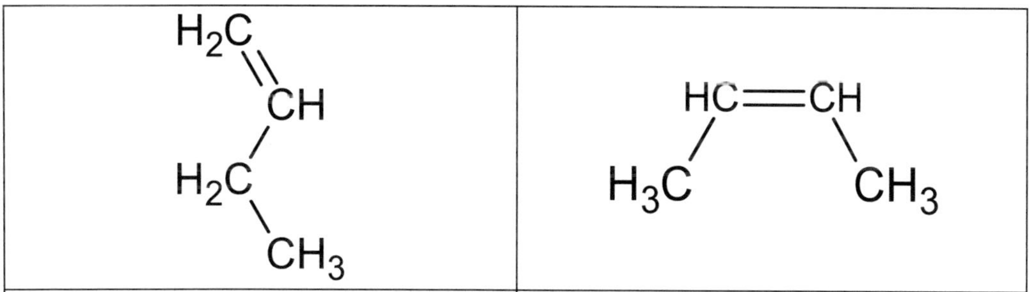

1-Butene	Cis-2-Butene

Fig. 8. The structure of 1-Butene and Cis-2-Butene.

The mass spectra of 1-butene and cis-2-butene were recorded at two different peak laser intensities and presented in figures 9 and 10. In both figures the signal corresponding to singly charged ion of cis-2-butene, at the mass to charge ratio of m/q=56, has been scaled to an arbitrary value of 100 and the peaks corresponding to other ions have been accordingly scaled. As it is observed at both peak intensities the difference in the spectra of the isomers are more pronounced than the cases of using electron impact ionization of reference.

In order to quantify the performance of the new ion source in terms of determining the composition of a gas containing the two isomers, we calculate the similarity index (SI), which is defined as,

$$SI \equiv \left(\sum_{m/q} \sqrt{I_x I_y} \right)^2 \Big/ \left(\sum_{m/q} I_x \sum_{m/q} I_y \right)$$

where x and y denote the two isomers, m and q are the mass and charges of a given ion, and I_x and I_y are the relative abundances of peaks for equal value of m/q on 1-butene or cis-2-

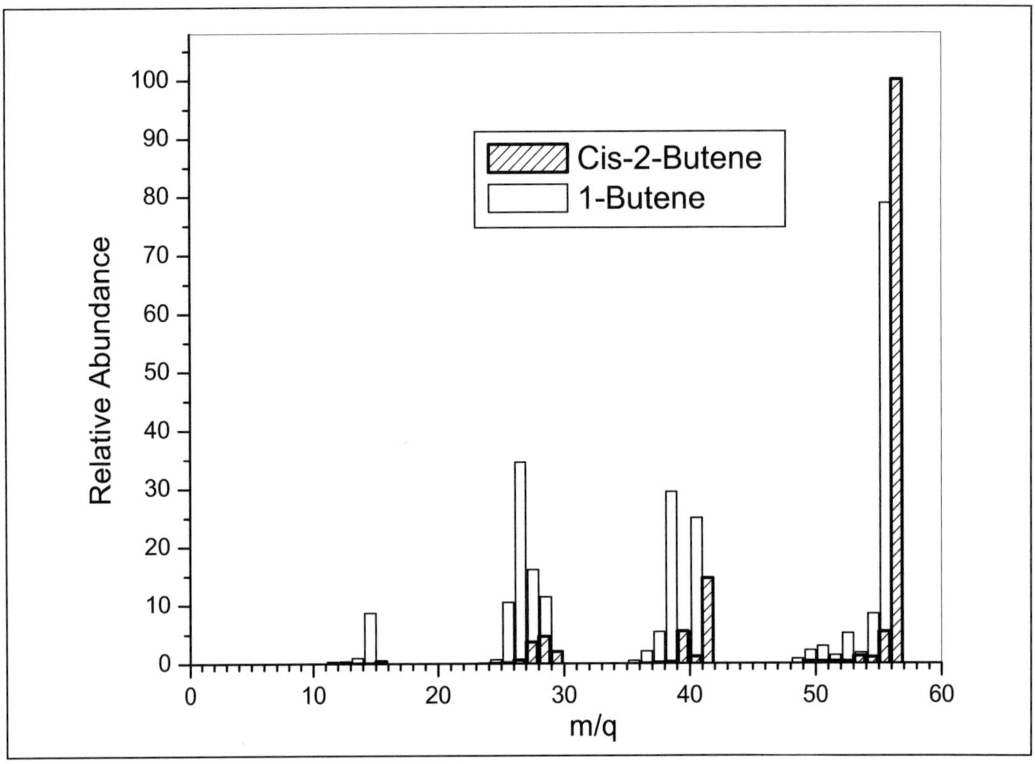

Fig. 9. Comparison of mass spectra for Cis-2-Butene and 1-Butene prepared by exposing pure gases to linearly polarized short laser pulses focused to peak intensity of $2x10^{13}$ W/cm^2 (SI=0.793).

butene mass spectra. The SI values were calculated to be SI=0.793 and SI=0.918 for the spectra presented in figures 2 and 3, respectively. The corresponding value reported in reference, in the case of using electron source for ionization, is SI=0.9957. This observation readily indicates that the dissociative ionization by ultra-fast laser pulses can potentially provide a high performance ion source for applications in mass spectroscopy. In addition, the dependence of the SI on the laser peak intensity could provide a practical optimization tool. We elaborate on these observations in the following section.

The peak laser intensity is the main parameter that can be used to optimize the dissociative multiphoton ionization as a source for high performance mass spectroscopy. This is due to the action of two well established phenomena; increase in ionization potential resulting from AC Stark shift and the field induced barrier suppression. Let us denote the appearance potential of a particular fragment by AP, which is defined as the minimum energy needed to form the fragment from a neutral molecule in the ground state. AP can be expressed as $AP = E_i + \Delta E$, where E_i is one of the ionization potentials (from the inner valence orbital) and ΔE is the minimum energy above that ionization potential needed to form the ion. At the peak laser intensity, I_0, E_i increases by the so-called ponderomotive potential $Up_{(ev)} = 9.33 \times 10^{-14} I_{0(W/cm^2)} \lambda^2{}_{(\mu m)}$, where λ is the wavelength of laser. The ponderomotive potential is the average energy of the oscillatory motion of a free electron in an electromagnetic field and explicitly appears in analytical formula for the multiphoton

ionization of atoms with applying the strong field approximation. In contrast to the ionization potential, ΔE decreases by an amount δE due to field induced barrier suppression. The magnitude of δE is estimated to be given by $\delta E \approx 0.5FR$, where F is the strength of the laser field and R is the size of the parent ion. For example at a peak intensity of $2x10^{13}$ W/cm^2 and for an ion with R=2 atomic unit, δE is calculated to be 0.62 eV.
The aforementioned arguments can be applied to qualitatively explain the dependence of SI on the peak laser intensity. For this purpose we define partial SI by excluding the signal of the parent ion from the summations in equation 1. This quantity is calculated to be SI=0.868 and SI=0.967 at low and high intensities, respectively. The relatively elevated value of the partial SI indicates that at higher intensities the fragmentation patterns of the two isomers are more similar. This can be explained by noting that the quantity δE at higher intensity (6.5x10^{13} W/cm^2) is higher than its value at the low intensity (2x10^{13} W/cm^2) by a factor of 1.8. Consequently, most of the parent ions resulting from the multiphoton/tunnel ionization of inner valence electrons will indiscriminately dissociate at higher intensities resulting in a mass spectrum that is independent of the differences in the geometries of the isomers. One might conclude that lower intensities are preferable for giving rise to smaller partial SI. However, we should remember that the ion signal decrease rapidly with lowering the intensity and the potential gain due to reduced SI might be compromised by the detection limit of the mass spectrometer. Therefore, a balance between (partial) SI and the magnitude of ion signal would lead to optimum ion source.

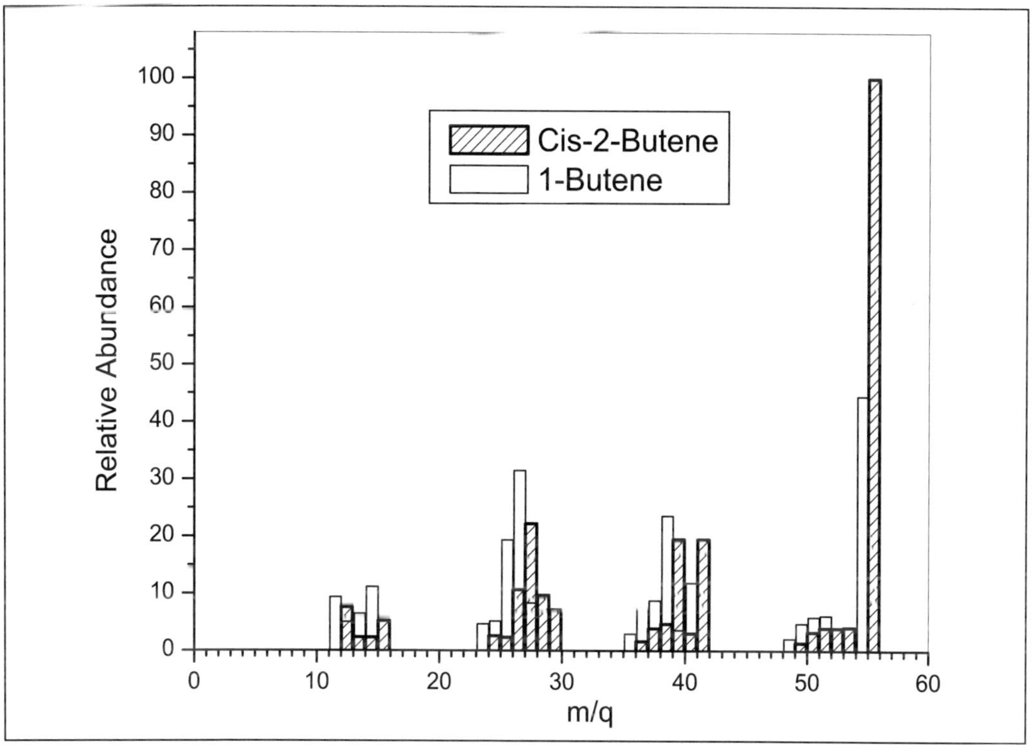

Fig. 10. Comparison of mass spectra for Cis-2-Butene and 1-Butene prepared by exposing pure gases to linearly polarized short laser pulses focused to peak intensity of 6.5x10^{13} W/cm^2 (SI=0.793).

Another parameter that can be used to optimize the short pulse based ion source is the laser wavelength. We have noticed that the fragmentation pattern is wavelength dependent therefore, in theory, by changing the wavelength we may enhance the SI of two isomers. Unfortunately, tuning wavelength is not easy and requires expensive setup, which often is bulky and is not able to supply stable output. This difficulty can be overcome by using our dual wavelength setup that by varying the angle between the polarization vectors of ω and 2ω fields results in an interaction outcome equivalent to what was expected from using a single frequency, ω', in the range ω < ω' <2ω. This finding will be discussed in the next section.

3.1.2 Dichroic pulses

In a recent publication (Sharifi *et al* (2010)) we studied the interaction of a two-color field by taking the angle between the polarization vectors of the two fields as the control parameter. The experimental setup is presented in figure 11. A Ti:Sapphire laser system provides a ten hertz train of 800-nm pulses having a full-width-at-half-maximum time duration of 150 fs. A KDP crystal of thickness 2 mm, and phase-matched for the second-harmonic generation,

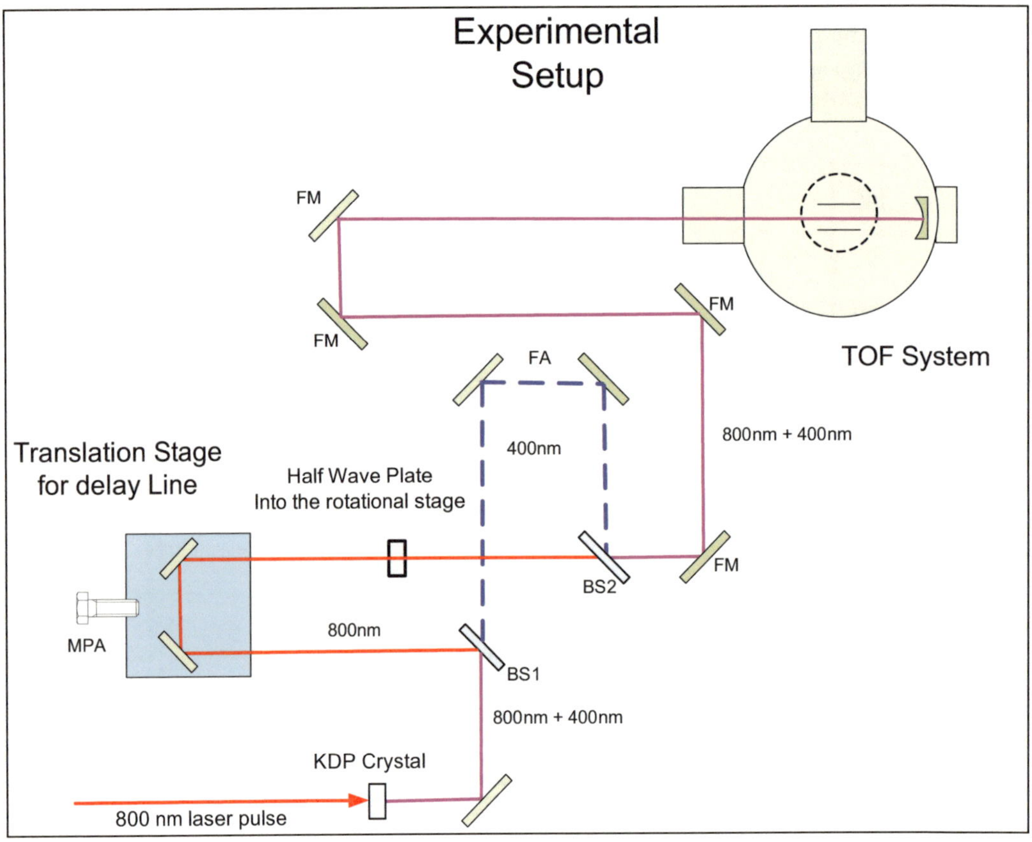

Fig. 11. Comparison Figure 1- Schematic diagram of the experimental setup for two-color multiphoton process study where, BS1: Harmonic separator HR@800 nm, AR/HT@400 nm; BS2: Harmonic separator, HR@400 nm, AR/HT@800 nm; MPA: Micrometer piezoelectric actuator; FA: Fixed arm ~400-nm arm; FM: Folding mirror.

was used to produce the 400-nm pulses whose characteristics will be explained later in this section. The fundamental and second-harmonic beams emerging from the doubling crystal are passed into a Michelson interferometer with a dichroic beam splitter. The IR arm of the interferometer includes a wave-plate oriented such that the angle between the polarization vectors of the two beams, θ, can be varied in the range of 0 to π/2. The laser pulses from both the arms of the set-up were collinearly focused using a 10-cm focal length concave mirror in the vacuum chamber.

The total field causing the interaction is given by $E(t)=E_\omega\cos(\omega t)+E_{2\omega}\cos(\theta)\cos(2\omega t)$. By defining two variables $\alpha=\left(1+E_\omega/E_{2\omega}\cos(\theta)\right)^{-1}$ and $F=E_\omega+E_{2\omega}\cos(\theta)$ the field can be expressed as; $E(t)=F\left((1-\alpha)\cos(\omega t)+\alpha\cos(2\omega t)\right)$. We have shown that, under the condition $I_{2\omega}<<I_\omega$, the rate of electron tunneling in this field can be calculated by applying the following transformations,

$$I_\omega\to I_\omega+I_{2\omega}\cos^2(\theta)+2\sqrt{I_\omega I_{2\omega}}\cos(\theta)\quad \omega\to\omega\sqrt{1+3\alpha}$$

Thus, the two color field is equivalent to a single laser whose wavelength can be adjusted by warying the angle between the polarization vectors of the two fields. This effect can be very useful for the mass spectroscopy application as we mentioned in the previous subsection.

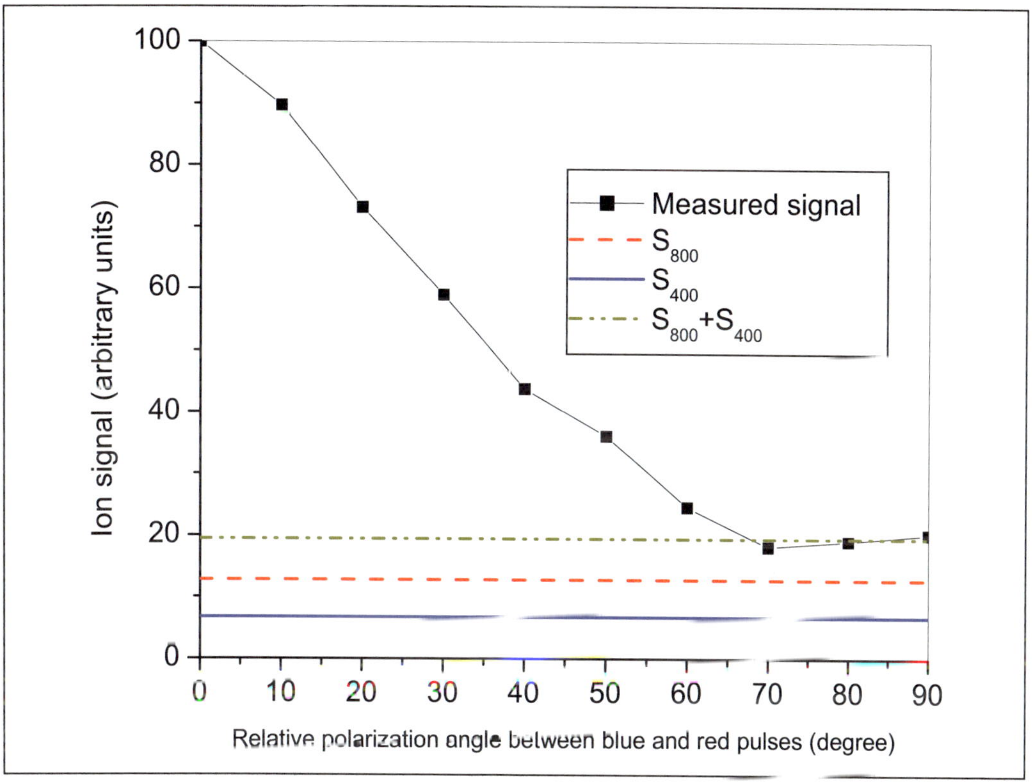

Fig. 12. The dependence of Xe^{++} signal on θ which is the angle between the polarization vectors of ω and 2ω pulses. The "measured signal" plot corresponds to the case when both pulses are present. The dashed and solid horizontal lines indicate the ion signals when either of the ω or 2ω pulses is present alone.

The key finding from the tests using the setup of figure 11, as related to the potential use for generating permanent index changes in transparent solids, is furnished by the following experimental result. the ionization of Xe was studied by setting the intensities of ω and 2ω pulses such that $I_{\omega} = 7 \times 10^{13}$ W/cm² and $I_{2\omega} = 7 \times 10^{12}$ W/cm², respectively. In figure 12, the signal of doubly charged ion is presented as a function of the angle between the polarizations of the two fields, θ. In a purely classical picture, it is expected that the rescattered electron (ionized and rescattered by the action of ω field) to be deflected by the action of 2ω field. Accordingly, the electron-ion collision probability and the resulting double ionization rate depends on the angle θ, because the double ionization process at our working intensity is dominated by the non-sequential channel.

3.2 Flat panel display dicing

Flat panel dicing using short laser pulses is, in our opinion, is a relevant application due to the fact that as the glass substrates become thinner it becomes more difficult to improve the process yield using conventional diamond blade dicing techniques. Using short pulses dicing speeds of over 400 mm/s has been successfully demonstrated on non-alkali glass and borosilicate glass, using a 50 kHz, 5W high-power femtosecond laser. The working principal of the technique is presented in figure 13, which has been adapted from Kamata *et al* (2008).

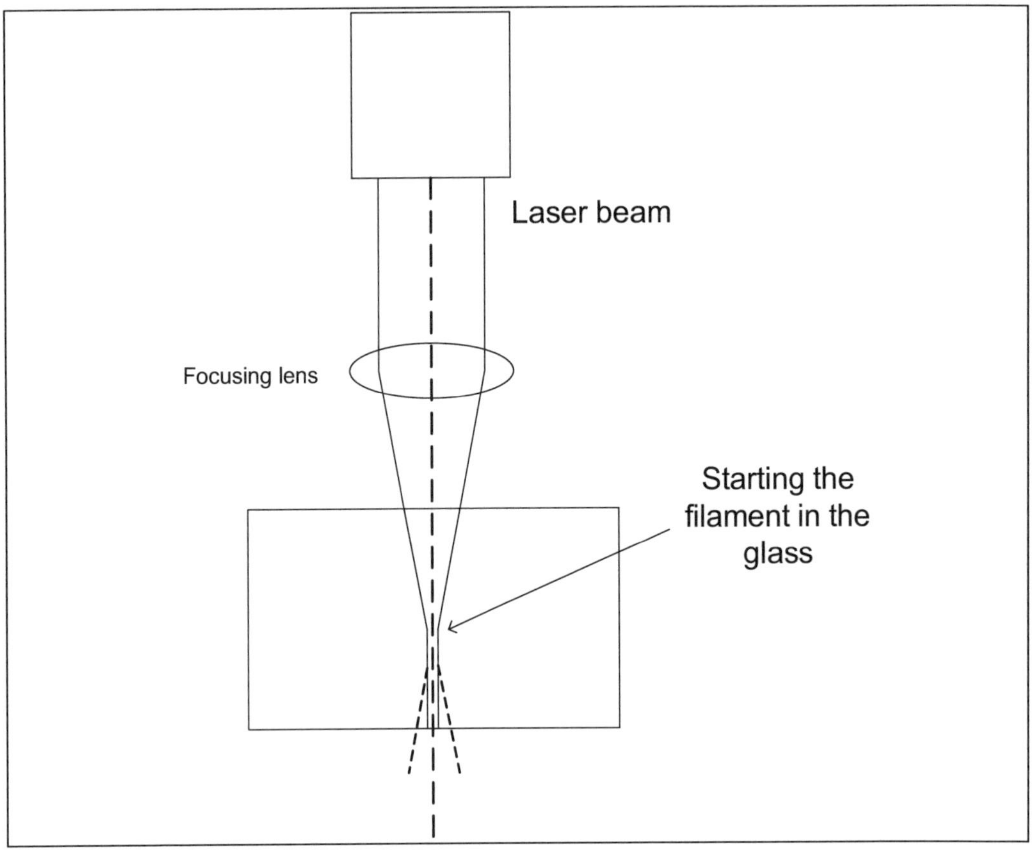

Fig. 13. Schematic of short laser pulse dicing technique.

The short pulse laser beam having a wavelength to which the work is transparent is directed to the front surface of the work toward the back surface and focused. A filament in the light beam traveling direction from the beam waist is formed by the auto-focusing action due to the laser beam propagation in the work is formed. The substance in the filament is decomposed by the laser beam and can be discharged from the back surface, and a cavity is formed in the channel. While forming the cavity, the laser beam is scanned, a machined surface is formed, and thereafter the work can be cut with a weak bending stress.

4. Conclusion

Over past two decades, robust lasers capable of providing pulses with sub-picosecond durations have been developed. This has resulted in extensive research activities attempting to find applications based on light-matter interactions involving, predominantly, electronic couplings. Such interactions minimize the, potentially undesirable, effects of transitions between the ro-vibrational states of the molecules, which may result in heating and uncontrolled disintegration of molecules. Though, even a single wide-spread application has not materialized yet, but nuclei of the potential applications have already formed. We discussed the fact that not all the suggested applications are relevant, in terms of addressing the market needs in cost effective manner, and will be adopted by the high technology companies. However, there are unexplored areas with the potential of offering viable solutions to the real life problems. We identified and discussed two areas with some details. These include the use of ultrafast pulses as an ion source for mass spectroscopy and flat panel display dicing.

5. References

Bennett, K. H.; Cook, K. D. (2000). Simultaneous analysis of butene isomer mixtures using process mass spectrometry, *J. Am.Soc. Mass Spectrum,* 11 1079-1085

Kamata, M.; Sumyoshi, T.; Tsujikaula, S., & Sekita, H. (2008). Laser machining method, laser cutting method, and method for dividing structure having multilayer board, *PCT Application,* WO/2008/126742

Larochelle, S.; Talebpour, A. & Chin, S. L. (1998). Coulomb effect in multiphoton ionization of rare gas atoms, *J. Phys. B,* 31 1215-1224

Sharifi, S. M.; Talebpour, A. & Chin, S. L. (2008). Ultra-fast laser pulses provide an ion source for highly selective mass spectroscopy, Appl. *Phys. B* 91 579-581

Sharifi, S. M.; Talebpour, A.; Yang, J. & Chin, S. L. (2010). Quasi-static tunnelling and multiphoton processes in the ionization of Ar and Xe using intense femtosecond laser pulses, *J. Phys. B,* 43 155601

Story, J. C.; Ducan, D. I. & Gallagher, T. F. (1994). Landau-Zener treatment of intensity-tuned multiphoton resonances of potassium, *Phys Rev A* 50 1607-1617

Talebpour, A.; (1998a). Ph.D. Thesis. Universite Laval.

Talebpour, A.; Larochelle, S. & Chin, S. L. (1998b). Multiphoton ionization of unsaturated hydrocarbons, *J. Phys. B,* 31 2769-2776

Talebpour, A.; Abdolfattah, M. & Chin, S. L. (2000a). Focusing limits of intense ultrafast laser pulses in a high pressure gas: road to new spectroscopic source, *Opt. Comm.*, 183 479-484

Talebpour, A.; Bandrauk, A. D.; Vijayalakshmi, K. & Chin, S. L. (2000b). Dissociative ionization of benzene in intense ultra-fast laser Pulses"., J. Phys. B, 33 4615-4626

29

Ultrashort Laser Pulses Applications

Ricardo Elgul Samad, Lilia Coronato Courrol,
Sonia Licia Baldochi and Nilson Dias Vieira Junior
Instituto de Pesquisas Energéticas e Nucleares – IPEN-CNEN/SP
Brazil

1. Introduction

Ultrashort laser pulses are considered to be pulses of electromagnetic radiation whose duration is shorter than the thermal vibration period of molecules, around tens of picoseconds (10^{-12} s). Pulses with durations of a few picoseconds were already produced in the 1960's (DiDomenico et al., 1966), shortly after the laser invention, using the mode-locking technique (Hargrove et al., 1964; Haus, 2000). In the next decade, refinements on this pulse generation scheme, and the use of bulky dye lasers with large emission bandwidths, shortened the pulses to the hundreds of femtoseconds (10^{-15} s) timescale (Diels et al., 1978; Shank & Ippen, 1974). In the 1980's, pulses with durations below 10 femtoseconds were generated from dye lasers (Fork et al., 1987; Knox et al., 1985), however the applications had to wait for the Ti:Sapphire Kerr-Lens mode-locked laser (Brabec et al., 1992; Spence et al., 1991) and the Chirped Pulse Amplification (CPA) technique (Strickland & Mourou, 1985) to really spread out. The large Kerr effect and broad emission bandwidth available in the Ti:Sapphire (Moulton, 1986), and the diode pumped solid state lasers (Keller, 1994; Keller, 2010; Scheps, 2002) that became available around this time, greatly simplified the setup needed to generate ultrashort pulses, and promptly replaced the dye lasers for this purpose. Finally, the invention of the CPA technique in 1985, allowed the generation of high intensity ultrashort pulses in all-solid state laser systems, and disseminated these laser systems due to its simplicity of operation when compared to the preceding systems, stability and relatively low cost. The CPA technique (Diels & Rudolph, 2006; Maine et al., 1988; Mourou et al., 1998) consists in generating ultrashort pulses with nanojoules of energy in a main oscillator laser, then temporally stretching these pulses by chirping its frequency (dispersion control) (Rullière, 1998) to decrease its power and intensity, allowing its amplification up to more than a million times without damaging the amplification chain components; after the amplification, the pulses are compressed to durations close to their original ones and directed to the applications. When the CPA technique became prominent, the pulses stretching and compression were mainly done with diffraction gratings (Fork et al., 1984; Martinez, 1987a; Martinez, 1987b; Treacy, 1969), although nowadays other techniques are disseminated, such as the use of Chirped Mirrors (Nisoli et al., 1997, Szipocs et al., 1994), fiber stretchers (Zhou et al., 2005) and prisms and grisms compressors (Chauhan et al., 2010).

As the years went by, scientific and technological developments led to the dissemination of ultrashort pulses systems based in other mode-locking schemes such as SESAMs (Semiconductor Saturable Absorber Mirrors) (Keller, 2010) and gain media including chromium, ytterbium and neodymium doped crystals (Diels & Rudolph, 2006), neodymium doped glasses (Badziak et al., 1997), erbium doped fibers (Krauss et al., 2010), Optical Parametric CPAs (OPCPA) (Dubietis et al., 2006) and picosecond semiconductor lasers (Koda et al., 2010). The availability of systems with varying characteristics in many universities and research laboratories resulted in numerous ultrashort laser pulses applications in many areas. The great variety of ultrashort pulses laser systems available nowadays, both commercially and under development in laboratories around the world, is capable of generating pulses with durations in the range from a few femtoseconds to hundreds of femtoseconds, and picoseconds for the semiconductor lasers, wavelengths ranging from the ultraviolet spectrum (Reiter et al., 2010) up to 1.5 μm, energies extending from nanojoules to thousands of Joules (Stoeckl et al., 2006), terawatt peak powers in repetition rates up to few kilohertz (Bagnoud & Salin, 2000), and reaching tens of MHz at lower powers (Malinowski et al., 2004; Naumov et al., 2005). Powers over one petawatt (10^{15} W) are generated in national laboratories (Chambaret et al., 2007; Gaul et al., 2010; Habara et al., 2010; Musgrave et al., 2007; Perry & Mourou, 1994; Tajima & Mourou, 2002), and intensities up to 10^{22} W/cm^2 (Bahk et al., 2005) were obtained.

Laser oscillators cannot generate pulses shorter than a few femtoseconds due to the wavelengths and bandwidths necessary for this, in the ultraviolet spectrum. Nevertheless, the use of High Harmonic Generation techniques allow the creation of pulses down to a few hundreds of attoseconds (10^{-18} s) (Sansone et al., 2006).

The applications of ultrashort laser pulses derive mainly from two characteristics of such pulses: firstly, their very short duration, which can be used to induce and measure ultrafast phenomena with temporal resolution of femtoseconds (and more recently with hundreds of attoseconds), besides minimizing heat transfer to the target. Secondly, the huge electric and magnetic fields associated to the pulses that can surpass the ones that bind electrons to atoms, resulting in large ionizations that lead to material modifications; the free electrons generated can also be accelerated by the electromagnetic field to high energies, even into the relativistic regime.

The following sections will explore ultrashort laser pulses applications, roughly going from lower to higher energy (power, intensity) ones. Not all applications will be covered due to space limitations, and references will be given for those wanting a deeper understanding on the various subjects.

2. Time resolved measurements

To measure an event, a faster phenomenon is needed, and the extremely short duration of ultrashort pulses make them ideal for measuring processes that occur in the picosecond and femtosecond time scales. Moreover, ultrashort pulses can be used to induce a series of fast phenomena in various kinds of samples with the advantage of shortening the transient behavior resulting from the optical excitation, so the consequences of this perturbation are free to develop in time. The shorter the pulse, the faster the transient behavior vanishes and the excitation evolution can be measured. Ultrashort pulses also have the benefit of minimal heat transfer to the sample, not exciting vibrational modes that can introduce noise or even mask the signal under measurement. The sample can be a dielectric material, whose absorption coefficient is changed due to saturation of this absorption, a semiconductor in which the charge carriers created in the conduction band by the ultrashort pulse change the material properties (Yariv & Yeh, 2007), a metal whose magnetic characteristics are altered by the ultrashort pulse (Beaurepaire et al., 1996), or even an atom that has its electronic properties modified by the light absorption (Drescher et al., 2002). Electronic process in biological molecules can also be stimulated (Li et al., 2010; Schreier et al., 2007).

When a pulse can be used to induce a phenomenon and to measure its temporal evolution, a pump-probe technique should be used. In this technique (Diels & Rudolph, 2006; Rullière, 1998) two replicas of an ultrashort pulse are created using a beam splitter, one with most of the original pulse energy (pump pulse), and the other one with a only small fraction of that energy (probe pulse), as represented on Fig. 1. Both pulses are focused upon the sample under study overlapping spatially at the focus, and one of the pulses (usually the pump) goes first through a delay line, in which a relative delay, Δt, can be adjusted between the pump and the probe pulses. At $t = 0$ the pump pulse excites the sample, and at the time $t = \Delta t$ the weaker pulse goes through the sample and has one or few of its characteristics (intensity, polarization, phase, temporal duration) modified, probing the excitation. Repeating the measurement varying the delay (Δt) from negative to positive values, determines the temporal dynamics of the excitation.

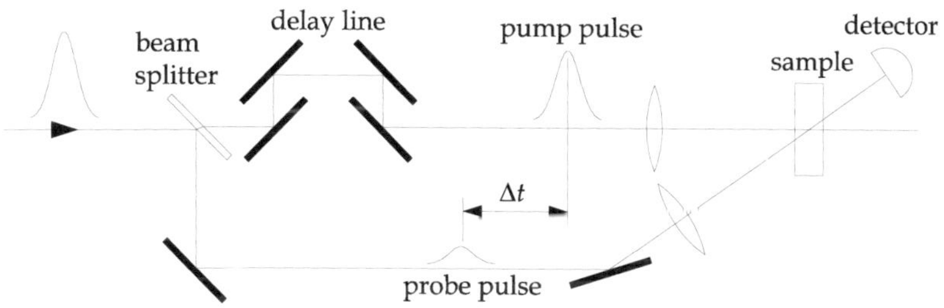

Fig. 1. Scheme of a pump-probe experiment

As an example, consider a two-level sample and the transmitted intensity given by the Beer law (Koechner, 2006; Rullière, 1998):

$$I(\omega,\Delta t) = I_0(\omega)\exp\left[-\sigma(\omega)n(\Delta t)l\right] \tag{1}$$

in which I_0 and I are the incident and transmitted intensities, respectively, $\sigma(\omega)$ is the absorption cross section at frequency ω, l is the sample length, and $n(\Delta t)$ is the ground level population density. An intense resonant pump pulse at $t = 0$ strongly promotes electrons to the upper level, decreasing the ground level population density $n(t = 0)$. If the weaker probe pulse hits the sample at this time $\Delta t = 0$, the sample transmittance I/I_0 will be larger than its nominal value, meaning that the absorption coefficient $\sigma\,n$ has decreased. As the transmittance measurements are repeated at increasing Δt values, the electrons return to the ground level, and consequently the sample absorption increases. Plotting the transmittance values dependence on Δt leads to the determination of the upper level lifetime. This technique is used when the lifetimes involved are shorter than tens or hundreds of picoseconds, to which solid state detectors do not respond and have low sensitivity. Moreover, the pump or the probe pulse can go through a medium that generates the second or third harmonic, or even a supercontinuum before impinging on the sample, and in this case the effect measured reveals the influence of a frequency on another, providing data to understand its mechanism.

Variations of the technique described can be used to determine the temporal dynamics of many processes in the femtosecond to picosecond time scale. Femtosecond pulses are known to induce electron heating in metals, and their relaxation, which occurs around 1 ps, can be measured by monitoring the metal reflectivity (Eesley, 1983; Schoenlein et al., 1987) and transmittance temporal evolutions in thin films (Sun et al., 1994). Variations of this technique enable the determination of the characteristic times of the microscopic interactions controlling basic metallic properties like electron transport and superconductivity (Brorson et al., 1990).

Many ultrafast phenomena occur in semiconductors (Rossi & Kuhn, 2002), mainly from the creation of charge carriers in the conduction and valence bands, either by single or multiphotonic absorption of ultrashort pulses (Yariv, 1989). The semiconductor is brought to states far from the thermodinamical equilibrium, and the dynamics of these states relaxations or interaction with electromagnetic radiations can be measured by various pump-probe techniques, with temporal resolutions below 10 fs. Besides transmission and reflection pump-probes measurements, other ones that investigate the carriers evolution bound to the bands, even in coherent states (Leo et al., 1991), can be performed (Axt & Kuhn, 2004).

The dynamics of fast magnetic effects also can be studied using pump-probe techniques. A possible way to do this is to use a polarized pump pulse to optically perturb the magnetization vector of a sample in a magnetic field. After the perturbation, the magnetization precesses in the external field to return to its initial state, modulating the material optical transmission and reflectivity of a polarized probe pulse, allowing the reconstruction of the trajectory of the magnetization vector in 3D space (Andrade et al., 2006; Vomir et al., 2005). Ultrafast spin dynamic can also be determined with similar techniques (Beaurepaire et al., 1996).

In the last decade, the robustness and reliability of ultrashort pulse systems made them valuable tools to pump probe studies of chemical (Zewail, 2000) and biological ultrafast phenomena. Pump-probe techniques have been used to investigate protein dynamics (Pal et al., 2002), intracellular processes such as the dynamics of electrons in photosynthesis (Brixner et al., 2005), and also in photoinitiated DNA degradation mechanisms (Schreier et al., 2007), demonstrating that ultrafast phenomena are essential to life.

Also in the last years, the availability of attosecond (10^{-18} s) pulses for pump-probe techniques allowed the observation of electronic processes inside atoms (Drescher et al., 2002; Mauritsson et al., 2010; Uiberacker et al., 2007) with temporal resolution in the hundreds of attoseconds. This time-scale is comparable to the revolution time of the electron in the Bohr atom, making possible the experimental investigation of fundamental aspects of quantum mechanics.

Summing it up, pump-probe techniques can use ultrashort pulses, the fastest events ever created by man, to measure and determine the evolution of ultrafast processes in many kinds of materials, molecules or even in internal states of atoms, opening many areas of research and development that a few decades ago were impenetrable to probing. In this context, we expect in the near future to witness the use of attoseconds pulses to probe the dynamics of electrons bound to atoms and investigate fundamental aspects of their wavefunctions.

3. Ultrashort pulses nonlinear optics

When light, an electromagnetic oscillating field, impinges on matter its electrons are subjected to the Lorentz force (Jackson, 1999):

$$\mathbf{F} = e\left(\mathbf{E} + \mathbf{v} \times \mathbf{B}\right) \quad (2)$$

where e is the electron charge, **E** and **B** the electric and magnetic fields, and **v** is the electron velocity. Usually the magnetic force is disregarded because the magnetic field amplitude B is given by the electric field amplitude divided by the speed of light (E/c), and the speed v is small. In this approximation, the electron can be considered to oscillate driven solely by the electric field, and the material response depends on how the electrons trajectories evolve in time. In metals, the electrons are free to follow the excitation field, absorbing its energy in a thin superficial layer and re-emitting it, not allowing electrical fields in its interior. If the electromagnetic field oscillating frequency is above a specific frequency to which the metal free electrons are unable to follow, called the plasma frequency (Jackson, 1999), the metal becomes transparent to the radiation. In a dielectric, the electrons are bound to the atoms by restorative forces, and when an electron is displaced from its equilibrium position by the electric field, an oscillating dipole momentum $\mathbf{p}(\mathbf{r},t)$ is induced, therefore emitting electromagnetic radiation at its oscillating frequency (Jackson, 1999). The sum of all dipoles is the material total polarization, **P**, whose spatial components in the frequency domain (Butcher & Cotter, 1990; Shen, 1984) can be written as an expansion in powers of the electric field with coefficients $\chi^{(n)}$, called the electrical susceptibilities:

$$\begin{aligned} P_i = &\ \chi_{ij}^{(1)}\left(\omega';\omega_1\right)E_j\left(\omega_1\right) \\ &+ \chi_{ijk}^{(2)}\left(\omega'';\omega_1,\omega_2\right)E_j\left(\omega_1\right)E_k\left(\omega_2\right) + \chi_{ijkl}^{(3)}\left(\omega''';\omega_1,\omega_2,\omega_3\right)E_j\left(\omega_1\right)E_k\left(\omega_2\right)E_l\left(\omega_3\right) + \ldots \end{aligned} \quad (3)$$

In expression (3) the right term on the first line represents the linear polarization, given by the product of the instantaneous electrical field component $E_j(\omega_1)$ oscillating at the frequency ω_1 and the linear susceptibility tensor $\chi_{ij}^{(1)}(\omega';\omega_1)$, in which the indices indicate the spatial coordinates $(i, j = x, y, z)$. In this susceptibility notation, the polarization oscillation frequency is the one at the left of the semicolon, and it is the sum of the electrical fields frequencies at the right. These frequencies can be either positive or negative, and in this case ($\omega_n<0$), the corresponding electrical field complex conjugate is used (Butcher & Cotter, 1990). When the exciting field amplitude is small and the electron oscillation is harmonic, the polarization is proportional to the field and oscillates at its frequency ($\omega'=\omega_1$). The arising effects, such as the refractive index and the absorption, are linear on the field and do not depend on its intensity (Shen, 1984).

As the light intensity and the electrical field amplitude increase, the electrons oscillations became anharmonic and higher orders of the expansion have to be added to the total polarization (Butcher & Cotter, 1990; Shen, 1984), as exemplified by the second and third orders nonlinear polarizations on the bottom line of expression (3). Each nonlinear polarization is the product of a nonlinear susceptibility tensor and various exciting fields. As long as the electrons stay bound to the atoms, this description can be used, and many nonlinear phenomena arise from the different nonlinear susceptibilities spatial components and frequencies combinations. As the nonlinear polarizations grow, the total polarization oscillates in various frequencies simultaneously, and consequently the emission will be the superposition of monochromatic fields.

Regarding the mechanisms that generate nonlinear optics phenomena, the main ones are electronic redistribution and molecular reorientation (electronic configurations changes), and thermal and electrostriction processes that modify the material density (Boyd, 2008). All these processes can be described by the formalism presented, but only the phenomena resulting from electronic redistribution will be discussed here once the other processes are slow and do not respond to ultrashort pulses.

Ultrashort pulses are prone to generate nonlinear effects due to the high powers and intensities readily attainable. Even a few nanojoules of energy in a 50 femtoseconds pulse, straightforwardly generated by ultrafast lasers, produce tens of kilowatts of peak power, and can be easily focused in a spot with a radius of 20 μm, resulting in intensities on the order of GW/cm^2. From the Poynting vector it can be shown (Born & Wolf, 1999) that the electric field amplitude, E, is related to the intensity, I, by:

$$E(\text{V/cm}) = 27.45\sqrt{\frac{I(\text{W/cm}^2)}{n_0}} \tag{4}$$

where n_0 is the refractive index of the medium. Intensities of GW/cm^2 correspond to electric fields of about 10^6 V/cm, which is on the order of 1% of the electric field binding electrons to molecules, meaning that at these intensities the contribution of the higher order electric field terms of expression (3) start to be relevant to the polarization. Higher intensities that increase the nonlinear effects can be obtained from CPA systems, extended cavity lasers and cavity dump techniques.

The first susceptibility to generate nonlinear effects is the second order one, $\chi^{(2)}$. If one of the fields is zero, the susceptibility $\chi^{(2)}(\omega; 0, \omega)$ describes the Pockels effect (Fowles, 1989), in which a static (DC) electric field applied to a medium rotates the polarization of a field at the optical frequency ω going through this medium. Choosing adequately the medium length and the DC field amplitude, it is possible to rotate by 90 degrees the polarization of an ultrashort pulse at ω, and a polarizer placed after the medium will transmit or block the pulse depending if the DC field is on or off. This arrangement, called a Pockels Cell, is commonly used as a fast optical switch in ultrashort laser pulse systems.

The second order susceptibility component $\chi^{(2)}(\omega; \pm\omega_1, \pm\omega_2)$ also describes a class of phenomena, called sum-frequency generation, in which two photons from different laser beams are mixed to create a third one (Boyd, 2008). As already explained, when a negative frequency is used, the complex conjugate corresponding field is used in the polarization expression, so in the particular case that ω_1 is positive and ω_2 is negative, the second order polarization is:

$$\begin{aligned} P_i^{(2)} &= \chi_{ijk}^{(2)}(\omega_1-\omega_2;\omega_1,-\omega_2)E_{0j}(\mathbf{r},t)e^{-i\omega_1 t}\left[E_{0k}(\mathbf{r},t)e^{-i\omega_2 t}\right]^* \\ &= \chi_{ijk}^{(2)}(\omega_1-\omega_2;\omega_1,-\omega_2)E_{0j}(\mathbf{r},t)E_{0k}^*(\mathbf{r},t)e^{-i(\omega_1-\omega_2)t} \end{aligned} \tag{5}$$

where $E_0(\mathbf{r},t)$ is the electric field temporal and spatial envelope (Jackson, 1999). The negative frequency $-\omega_2$ means that this field carries a phase of π relative to ω_1, so the polarization oscillates at the difference frequency, as evidenced in equation (5), and a photon is emitted with energy $\hbar(\omega_1\text{-}\omega_2)$. The exciting photons are not resonant with the transparent medium, and virtual electronic levels are involved (Yariv, 1989). Note that when two photons, at frequencies ω_1 and ω_2 interact with a nonzero second order susceptibility, both the frequencies $\omega_1\text{-}\omega_2$ and $\omega_2\text{-}\omega_1$ are simultaneously generated, and a negative values means π added to the phase. An immediate application of sum-frequency generation for ultrashort

pulses is the Optical Parametric Amplification (OPA) (Yariv & Yeh, 2007), in which a powerful pump beam at ω_p transfers energy to two beams at lower frequencies ω_i and ω_s, with $\omega_p = \omega_i + \omega_s$, where i and s stand for idler and signal, respectively. The susceptibility describing it is $\chi^{(2)}(\omega_s; \omega_i, \omega_p)$, determining how the energy moves back and forth between the three beams, and fine-tuning the phase matching conditions in the gain medium makes it possible to control the resulting amount of energy at each frequency, so a weak idler beam can be used to induce energy transfer from the pump to generate a powerful signal beam. This effect is used to extend ultrashort pulses to wavelengths differing from the main oscillator one, increasing the wavelength range of ultrashort pulses available for spectroscopic and other applications (Laenen et al., 1993). The OPA also can be used to amplify ultrashort pulses into the petawatt regime by the Optical Parametric CPA (OPCPA) technique (Dubietis et al., 2006). The pump beam and the ultrashort pulse to be amplified interact in a gain medium, generating an idler one; adjusting the phase matching conditions the idler beam intensity can be minimized and the signal beam amplified (Cerullo & De Silvestri, 2003). The gain medium is not resonant with either ω_p, ω_i or ω_s, so no energy is absorbed by it, and thermal problems that are known to limit the amplification performance (Koechner, 2006) are minimized; moreover, with the appropriate parameters selection, the gain medium bandwidth can be extended to all the gain medium transparency window.
A special case of the sum-frequency generation, called Second Harmonic Generation (SHG) or frequency doubling, occurs in the degenerate case when $\omega_1 = \omega_2 = \omega$ (Boyd, 2008). In this situation, governed by $\chi^{(2)}(2\omega; \omega, \omega)$, in which only one laser beam is used, the polarization oscillates at twice the exciting field frequency, emitting light with half of the excitation wavelength. The SHG can also be understood as the absorption of two photons with energy $\hbar\omega$ followed by the emission of a photon with energy $2\hbar\omega$, since energy has to be conserved. Homogeneous media, like gases or glasses, are not capable of SHG. This happens because the second order nonlinear polarization is $P^{(2)} = \chi^{(2)}E \cdot E$, and in a homogeneous surrounding, an inversion of the electrical field must result in an inversion of the polarization: $-P^{(2)} = \chi^{(2)}(-E) \cdot (-E) = \chi^{(2)}E \cdot E$. This is only possible if $\chi^{(2)} = 0$. The SHG signal increases quadratically with the excitation intensity, and generates ultrashort pulses at new wavelengths that can be used in experiments such as the pump-probe ones already mentioned in the previous section.
The SHG can be used to measure the temporal duration of ultrashort pulses by a technique known as second harmonic autocorrelation. As mentioned in the beginning of section 2, to measure an event, a faster phenomenon is needed, and since that there are no man-made events shorter than an ultrashort pulse to measure it, the pulse itself has to be used to determine its own duration. This is made possible by the use of a Michelson interferometer (Träger, 2007) and a SHG crystal placed at the crossing of the two beams, as shown in Fig. 2. The pulse to be measured is split in two copies by the beam splitter (BS), and each copy travels along a different arm. The mirror M in the delay line can be moved back and forth in order to control the temporal superposition of the pulse replicas at the SHG crystal, and when these arrive at the crystal delayed by more than the pulse duration there is no SHG; as the mirror is moved and the pulses overlap in time, a SHG beam proportional to the time superposition (autocorrelation signal) is produced at the bisecting angle defined by the two beams oscillating at ω, once momentum has to be conserved. The autocorrelation signal reaches its maximum when the delay between the pulses is zero. This technique generates a symmetrical autocorrelation trace that is proportional to the delay line position, and a slow

detector can be used to measure the second harmonic signal generated by a train of pulses. The deconvolution of the autocorrelation signal, taking into account that a displacement of 1 μm corresponds to 3.3 fs, determines the pulse duration (Trebino, 2000).

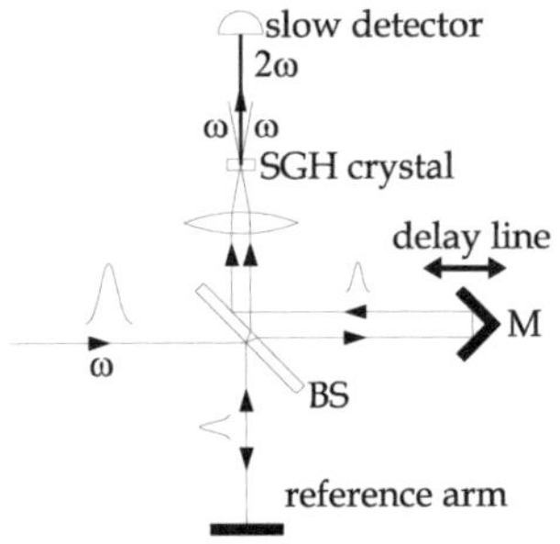

Fig. 2. Second order autocorrelation scheme. The thicker beam originating at the SHG crystal is the autocorrelation signal.

The second order autocorrelation is the most used technique to determine the duration of ultrashort pulses, although it presents some problems, such as ambiguities in the pulse temporal shape and no retrieval of any information about the pulse phase (Trebino, 2000). In order to overcame these problems and to completely determine the pulse temporal and spectral shapes and phases without ambiguities, other techniques, such as Frequency Resolved Optical Gating (FROG) (Kane & Trebino, 1993; Trebino, 2000) and Spectral Phase Interferometry for Direct Electric-field Reconstruction (SPIDER) (Gallmann et al., 1999) were created, allowing the measurement of pulses below 5 femtoseconds. Variations of these methods, in which an ultrashort pulse whose duration is know is used to generate the second harmonic interacting with a unknown pulse can be easily implemented, and are called cross-correlation techniques (Trebino, 2000). The deconvolution of the SHG signal with knowledge of one of the pulses determines the unknown one.
The two photon induced fluorescence is a variation of the SHG technique in which two photons are absorbed to a resonant level of a molecule, called fluorophore, and a more energetic photon is emitted (Peticolas et al., 1963). Although the process, differently from the previous one described, is resonant with two photons, it can also be described by the $\chi^{(2)}(2\omega; \omega, \omega)$ of the fluorophore. This effect can be used in a technique called two-photon scanning microscopy (Denk et al., 1990; Zipfel et al., 2003), in which a fluorophore that is transparent to the ultrashort pulses, but exhibits two-photon absorption for the same pulses, is used to image microscopic samples, mainly biological ones (Konig, 2000; Xu et al., 1996). A strongly focused infrared laser beam is used, and the two-photon fluorescence is confined to the focus region due to the nonlinear dependence of the emission on the excitation intensity. The laser beam is scanned across the sample, mapping the fluorophores with 3D high-resolution (Muller et al., 1998). Besides using naturally occurring fluorophores, molecules can be engineered to have high values of $\chi^{(2)}(2\omega; \omega, \omega)$ (Albota et al., 1998) and to bind to specific proteins so structures can be imaged unambiguously. The high precision obtained by the nonlinear effect at the vicinity of the focus spot can also be used to manufacture microstructures using two-photon polymerization. In this process, instead of using a fluorophore, a monomer that photopolimerizes under ultraviolet light and is transparent to the infrared is illuminated by an infrared laser. In this material the laser beam can be focused in deep regions, and the polymerization is initiated by two-photon absorption in a small region surrounding the beam focus, enabling the fabrication of three-dimensional structures

(Cumpston et al., 1999; Maruo et al., 1997). Both the two-photon microscopy and two-photon polymerization take advantage of the transparency of the medium for the fundamental wavelength, which allows the laser propagation inside the medium with minimal losses, and the nonthermal character of the ultrashort pulses interaction, that does not heat the samples, avoiding thermal effects that can modify the samples properties.

Moving to the third order susceptibility, some phenomena are extensions of the second order ones, such as the four wave mixing governed by $\chi^{(3)}(\omega; \pm\omega_1, \pm\omega_2, \pm\omega_3)$ (Träger, 2007), in which three photons at frequencies ω_1, ω_2 and ω_3 interact to generate a fourth one. Depending on the particular frequencies involved several phenomena as Coherent Anti-Stokes Raman Scattering (CARS) (Zumbusch et al., 1999), Degenerate Four Wave Mixing (DFWM) (Joo & Albrecht, 1993), Optical Phase Conjugation (Yau et al., 1996), Stimulated Brillouin Scattering, among others (Yariv & Yeh, 2007), can be generated. In the degenerate case for a single beam the third harmonic, described by $\chi^{(3)}(3\omega; \omega, \omega, \omega)$, is generated, and some applications are extensions of the second order ones (Träger, 2007).

Another class of phenomena emerge when the effects arising from the third order susceptibility $\chi^{(3)}(\omega; \omega, \omega, -\omega)$ come into play for a single beam. Under these conditions the third-order polarization is

$$P_i^{(3)} = \chi_{ijjj}^{(3)}\left(\omega;\omega,\omega,-\omega\right)E_{0j}(t)e^{-i\omega t}\, E_{0j}(t)e^{-i\omega t}\left[E_{0j}(t)e^{-i\omega t}\right]^{*} = \chi_{ijjj}^{(3)}\left(\omega;\omega,\omega,-\omega\right)e^{-i\omega t}E_{0j}(t)I_0(t) \quad (6)$$

where $I_0(t) = E_{0j}(t)\cdot[E_{0j}(t)]^*$ is the beam intensity. Expression (6) shows that the polarization oscillates at the same frequency of the exciting beam and it is proportional to its intensity, and this implies (Butcher & Cotter, 1990; Shen, 1984) that the total refractive index of the material, n, must have a dependence on the pulse intensity according to:

$$n(\mathbf{r},t) = n_0 + n_2 I(\mathbf{r},t) \quad (7)$$

where n_0 is the linear (usual) refractive index of the material, I is the pulse intensity and n_2 is the nonlinear refractive index given by $n_2 = \alpha\, \mathrm{Re}[\chi^{(3)}(\omega; \omega, \omega, -\omega)]$, where the constant α depends on the unit systems considered (Butcher & Cotter, 1990). Inserting equation (7) in the expression of a plane wave $E_0\exp[i(knx-\omega t)]$ propagating in the x direction results in:

$$E(\mathbf{r},t) = E_0(t)\exp[i(kn_0x - \omega t)]\exp[ikn_2 I(\mathbf{r},t)x] \quad (8)$$

where $k = 2\pi/\lambda$ is the plane wave wavenumber. The second exponential represents the nonlinear effects that can result from the spatial and temporal distributions of the intensity.

Considering a laser beam with a spatial Gaussian intensity distribution (Yariv, 1989) propagating in the x direction, the intensity reaches its maximum in the optical axis, and decays exponentially in the plane perpendicular to the propagation, and the spatial profile of the total refractive index follow this distribution. If n_2 is positive, the beam center goes through a medium with a longer optical path than its borders, similarly to a convergent lens, and the beam is focused. This effect is known either as Kerr effect or self focusing (or self-defocusing if n_2 is negative) (Shen, 1984), and is a self-effect in which the beam characteristics determine its propagation through a nonlinear medium. The self-focusing effect is used in many ultrashort lasers as the Kerr-Lens Mode-Locking mechanism (Haus et al., 1992) responsible to decrease the resonator losses for shorter pulses (Koechner, 2006), providing a robust and stable way to passively generate the shortest pulses possible directly from laser oscillators.

Based on the self-focusing effect, a simple technique was devised to measure the value of n_2, and consequently of $\chi^{(3)}(\omega; \omega, \omega, -\omega)$. This technique, called Z-Scan (Sheik-Bahae et al., 1990), consists in scanning a nonlinear sample across the waist of a focused beam, measuring the beam transmittance through an iris in the far-field as a function of its position. For a sample with positive n_2 self-focusing occurs, and when the sample is before the beam waist the tighter focusing results in a higher divergence, decreasing the transmittance through the iris; conversely, when the sample is after the waist, the self focusing reduces the beam divergence and the transmittance through the iris increases. These effects produce a peak-valley curve of the transmittance dependence on the sample position, and measuring the peak-valley signal variation immediately determines the third order nonlinear susceptibility. The Z-Scan technique quickly became widely used due to its simplicity, good sensitivity and the capability of measuring the electronic nonlinearities of solid and liquid materials. Many variations were introduced in the Z-Scan technique in order to increase its sensitivity (Kershaw, 1995), to resolve the time-scales of the self-focusing process with a pump-probe measurement (Ma et al., 1991), to measure dispersion of the nonlinearity (Balu et al., 2004) and to consolidate it as the most used technique to measure odd-orders nonlinearities (Zhan et al., 2002) in crystals, glasses, polymers and solutions among other kinds of samples. Alternatively, performing a Z-Scan measurement without the iris with all the beam energy impinging on a detector, the nonlinear absorption, corresponding to the imaginary part of $\chi^{(3)}(\omega; \omega, \omega, -\omega)$ is measured (Sheik-Bahae et al., 1990), providing more information about the electronic processes occurring inside the sample under study.

Taking into account now the temporal aspects of expression (8), consider a temporally symmetric pulse centered at the temporal origin with a temporal distribution $I(t)$ that can be expanded around $t = 0$. The expansion can be written as $I(t) = I_0 + \beta t$, with $\beta = \partial I(t)/\partial t$, and its substitution in expression (8) results in:

$$E(\mathbf{r},t) = E_0(t)\exp[ik(n_0 + n_2 I_0)x]\exp[-i(\omega - kn_2\beta x)t] \quad (9)$$

As can be seen in the second exponential of expression (9), the pulse carrier frequency is shifted by $kn_2\beta x$, meaning that before the pulse peak, lower frequencies are generated, and after it higher frequencies are created. This Phenomena is called Self-Phase Modulation (SPM) (Yariv, 1989) and is responsible for broadening the spectrum inside laser resonators for the generation of ultrashort pulses (Koechner, 2006). Without SPM the gain media would not generate enough gain to maintain the bandwidth needed, by Fourier Transform, to generate ultrashort pulses. In some specially designed resonators it is nowadays possible to generate octave-spanning spectra that support pulses shorter than 5 fs (Ell et al., 2001), and these ultra-broadband pulses are used to synthesize optical combs (Holzwarth et al., 2000; Ye & Cundiff, 2005). These optical combs are frequency synthesizers with uncertainties bellow 10^{-15} that are being used as time standards replacing atomic clocks (Takamoto et al., 2005), and to absolutely measure optical frequencies (Reichert et al., 1999; Udem et al., 2001). These combs are also used to stabilize the phase between the pulse carrier frequency and its envelope to a few milirads, corresponding to tens of attoseconds (Telle et al., 1999).

The SPM is also one of the main mechanism, along with wave mixing and harmonics generation, responsible by the generation of white light supercontinnums (Alfano, 2006) that can support shorter than 5 fs pulses with energies at the few milijoule level (Bohman et al., 2010). These supercontinnums extend the ultrashort pulses to new wavelengths, expanding the range of applications such as pump-probe ones, keeping the ultrafast aspect of the interaction with matter.

There are many other nonlinear phenomena acting in the time interval ranging from a few picoseconds down to the attosecond region, and only a few important ones were described here. Whole books were written on this subject (Diels & Rudolph, 2006; Hannaford, 2005; Rullière, 1998; Träger, 2007) and the reader interested in a deeper understanding is encouraged to resort to these references.

4. Ionization by ultrashort pulses

When the intensity of ultrashort pulses reaching the electrons in a material is high enough to exceed the electric field binding the electrons, ionization takes place. Depending on the density of free electrons generated in solid samples, ablation can occur on the material surface, or other phenomena can happen inside the sample.

Ultrashort pulse laser ablation of solids is due to an electron avalanche induced breakdown process (Bloembergen, 1974; Du et al., 1994) that occurs when seed electrons are accelerated in the laser field, exponentially generating free electrons by collisions. The breakdown takes place when the plasma originated by the avalanche electrons reaches a critical density and transfers energy to lattice ions, which expand away from the surface after the pulse has finished. In metals, the seed electrons are always present (conduction band free electrons), and in dielectrics and semiconductors they are excited from the valence to the conduction band by the pulse leading edge, either by multiphotonic ionization (Kautek et al., 1996; Perry et al., 1999) or by tunneling induced by the laser field (Keldysh, 1965; Lenzner et al., 1998). Although the seed electrons have dissimilar origins in different classes of materials, a metallization occurs in dielectrics and semiconductors after they are produced, and the avalanche evolves deterministically in time (Bass & Fradin, 1973; Du et al., 1994; Joglekar et al., 2003) in the same way in all solids, that behave like metals (Gamaly et al., 2002; Nolte et al., 1997). These mechanisms confer a nonselective characteristic to the ultrashort pulse ablation, and the intensity ablation threshold of a material, I_{th}, is the only parameter relevant to the etching process. The established method (Liu, 1982) to determine the ultrashort pulses ablation threshold of a given material consists in ablating this material with a TEM_{00} Gaussian beam at various intensities, and then measuring each ablation area diameter. From these data the ablation threshold is determined. A few years ago members of our group analytically described (Samad & Vieira, 2006) and introduced an alternative simpler method to measure the ultrashort pulses ablation threshold of solid samples, based in the very precise definition of the etching region resulting from the nonlinear character of the ultrashort pulse ablation together with the almost inexistent lateral heat diffusion. The method consists in moving the sample diagonally across the waist of a focused ultrashort pulses laser beam, etching the profile shown in Fig. 3 in its surface.

It can be easily shown that the maximum transversal dimension of the etched profile, ρ_{max}, is related to the ablation threshold, I_{th}, by:

$$\rho_{\max} = 0.342\sqrt{\frac{P_0}{I_{th}}} \tag{10}$$

where P_0 is the ultrashort pulse power, readily measurable in a laser laboratory. The values of ablation threshold measured with this technique agree with the ones given by the traditional method (Freitas et al., 2010; Samad et al., 2008).

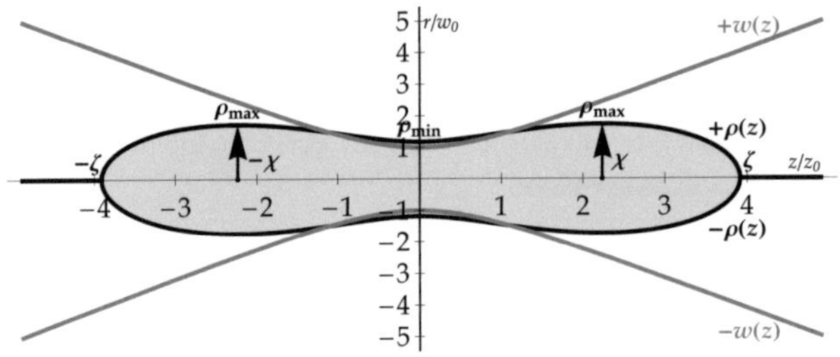

Fig. 3. Ablation profile etched in the surface of a sample by the diagonal scan method. The laser beam size is indicated by the curves labeled $\pm w(z)$, and the horizontal and vertical axes are normalized by the beam confocal parameter z_0 and beamwaist w_0, respectively.

Once the ablation threshold is known, ultrashort pulses can be used in a variety of materials ranging from transparent dielectrics (Gattass & Mazur, 2008) to metals (Dumitru et al., 2002; Itoh et al., 2006; Ke et al., 2005; Shirk & Molian, 1998), semiconductors (Nayak et al., 2007) and polymers (Baudach et al., 2000), to drill holes and to machine the material in a controlled way to modify its surface (Zoubir et al., 2003), create microstructures (Kruger & Kautek, 1999; Sugioka et al., 2005) and microchannels (Gomez et al., 2005). Due to the nonthermal character of the ablation, the heat affected zone is minimized, and can be brought to be almost nonexistent, allowing the use of ultrafast pulses to perform surgeries (Vogel et al., 2005; Vogel & Venugopalan, 2003) like neuron severing without damaging neighboring living tissues (Chung et al., 2006) and corneal cutting as high precision scalpels (Juhasz et al., 1999), among other application in biological samples and tissues (Braun et al., 2008).

At lower intensities than those needed for ablation, the free electrons density does not reach the value needed to produce the Coulombic explosion, but the localized excess of free electrons can be enough to promote modifications in the material bulk such as permanent changes in its refractive index (Davis et al., 1996) and the creation of defects like color centers (Courrol et al., 2004). We suggest that the defects creation mechanism begins with electrons being teared out from anions off the structure by multiphotonic or impact ionization, leaving a neutral atom behind. This neutral atom is no longer held in its position by the surrounding electric field, so it can be moved from its site by a collision with an electron oscillating in the laser field, and when this occurs, an electron can be trapped in the potential of the now empty position, creating a color center (Courrol et al., 2004; Gellermann, 1991). Alternatively, electrons can be trapped in the UV or valence bands, modifying electronic properties of the material, and consequently changing its refractive index. These processes and similar ones happen in glasses (Courrol et al., 2008), crystals (Martynovich et al., 2008; Orlando et al., 2010), and polymers (Samad et al., 2010), and when these defects are created with spatial control inside the material, structures such as waveguides (Nolte et al., 2003), diffraction gratings (Hirao & Miura, 1998) and photonic devices (Florea & Winick, 2003; Minoshima et al., 2001) can be manufactured by the ultrashort pulses. Once again, all these processes benefit from the minimized heat generation that preserves the material properties in the surroundings of the created structures.

5. Plasma, high harmonic generation and attosecond pulses

When ultrashort pulses interact with a material at intensities higher than those needed for ablation, the excess energy delivered by the pulses accelerates electrons, produce higher ionization states and excites the ions to upper energy levels, forming a light emitting plasma (Gibbon & Forster, 1996) whose properties are essentially controlled by the laser pulse characteristics.

Ultrashort pulses lasting 30 fs with modest energies under 1 mJ can be easily focused on sample surfaces to intensities above 10^{15} W/cm^2, generating plasmas whose emission is used to determine the elements of the sample in a technique called femtosecond Laser Induced Breakdown Spectroscopy (fs-LIBS) (Le Drogoff et al., 2001). The traditional LIBS technique (Cremers & Radziemski, 2006) relies on the characteristic light emission of each element, being used as an analytical tool for qualitative and quantitative analysis of the sample composition, and is performed with nanosecond pulses that promote ablation through thermal processes and then excite the ejected atoms by multiphotonic absorption. The thermal character of the ablation creates a dependence of the LIBS signal on the type of material under study, demanding the determination of many calibration curves. In fs-LIBS the nonselective character of the ultrashort pulses ablation makes the technique almost insensitive to the material under study, so less calibration curves must be created, and almost any solid material can be studied, including biological (Baudelet et al., 2006; Samek et al., 2006; Santos et al., 2008) and archeological samples, and works of art in order to determine the materials used by their artists (Svanberg, 2008). Because the plasma is formed by focused optical radiation, LIBS can also be used to interrogate samples remotely by stand-off analysis.

The plasma formation capability of ultrashort pulses can also be used in conjunction with LIDAR systems (Weitkamp, 2005) to investigate the atmosphere composition. For this, terawatt pulses are sent into the atmosphere with negative dispersion, and are compressed by the atmosphere positive dispersion until self-focusing occurs, leading to an abrupt intensity increase that leads to breakdown and plasma formation. The plasma introduces a defocusing effect that, under the right conditions, balance the self-focusing, channeling the plasma into filaments that can propagate through hundreds of meters (Kasparian et al., 2003), placing an intense light source in the sky. The plasma emission is collected by a ground based telescope, providing information about the atmospheric composition and transmittance. More recently, this plasma formation processes have been under study to induce water condensation in the atmosphere as a rain starting mechanism (Rohwetter et al., 2010), and to control lightning (Kasparian et al., 2010).

When plasmas are generated in gases at low pressures, the ionized electrons can be accelerated to high energies before colliding with ions. This mechanism is known as three-step model, in which the ultrashort pulse initially ionizes an atom by either tunneling or multiphotonic absorption (Kautek et al., 1996; Keldysh, 1965; Miyazaki & Takada, 1995), then the electron is accelerated away by the oscillating electric field that brings it back to collide with the ion, emitting x-rays as its kinetic energy is converted to electromagnetic radiation (Daido, 2002). Depending on the experimental details the x-rays generated can be generated in a coherent way, originating a beam with laser characteristics. These x-rays obey selection rules that determine the wavelengths generated, that can be understood as high harmonics of the fundamental ultrashort pulse (L'Huillier & Balcou, 1993; Lewenstein et al., 1994). Among other applications, these x-ray pulses can be used for high resolution imaging (Chao et al., 2005) in the water window (Chang et al., 1997; Gibson et al., 2003; Spielmann et al., 1997) of biological systems, and also of biomolecules and proteins (Neutze et al., 2000).

Assuming a pulse with a Gaussian temporal profile without generality loss, and referring to equation (3), it can be seen that when an harmonic is generated by the n-order polarization, the harmonic pulse is shortened by a factor $n^{1/2}$ (Shen, 1984). If the harmonic order is sufficiently high, the harmonic pulse can be generated with durations below a hundred attoseconds (Krausz & Ivanov, 2009; Tang & Chen, 2010) in the x-ray spectral region. This is the timescale of the electrons movements in atomic and molecular orbitals (Corkum & Krausz, 2007), and these pulses can be used in pump-probe experiments to directly observe electron tunneling in the dynamics of ionization (Kling & Vrakking, 2008; Uiberacker et al., 2007) and to control ionization processes (Johnsson et al., 2007). To deal with pulses in the attosecond time scale, new experimental tools and techniques are being developed to control and measure these pulses, like x-ray photoemission, cross correlation of light and x-rays (Hentschel et al., 2001). Today is possible to obtain information on the amplitude and phase of electronic wavefunctions using attosecond pulses.

6. Relativistic optics and high field science

When intensities on the order of 10^{18} W/cm^2 are reached and the corresponding electric field, given by expression (4), is an order of magnitude greater than the electric field in the Bohr atom, relativistic nonlinear effects start to come into play. At these intensities, the electron quivering motion on an ultrashort pulse reaches relativistic velocities, and the magnetic term on the Lorentz force, equation (2), has to be taken into account since v/c approaches the unity and the magnetic force is comparable to the electric one. In this situation the magnetic force pushes the electron in the forward direction, along with the Poynting vector. A direct consequence of this is wakefield generation, a relativistic optical rectification, in which the longitudinal field effects could be as large as the transverse ones (Umstadter et al., 1996). The electrons accelerated in this configuration are originated from a plasma, and create a strong static field that pull the ions left behind, accelerating them. Protons can be accelerated to energies over a GeV, in what are being called laser driven particle accelerators (Bulanov et al., 2010; Maksimchuk et al., 2000).

At these high intensities, relativistic phenomena analogous to nonlinear ones appear, as relativistic focusing in which the focusing is due to the relativistic electron mass increase (Monot et al., 1995), relativistic transparency, nonlinear modulation and multiple harmonic generation, and strong coupling to matter and other fields also take place (Mourou et al., 2006).

At very high intensities the light pressure can be used to inertially confine targets that will reach densities and temperatures high enough to obtain fast ignition in a laser driven atomic fusion process (Moses et al., 2009). The densities and temperatures expected are almost an order of magnitude higher than those at the center of the sun, allowing the accomplishment of experimental astrophysics.

Nowadays, many laboratories are working to increase the power generated by ultrashort lasers, and ways to focus it to ultimately reach intensities close to 10^{29} W/cm^2, the Schwinger limit (Bulanov et al., 2003; Schwinger, 1951), at which electron positron pairs can be created directly from the quantum vacuum as a consequence of its polarization by the electromagnetic field.

7. Conclusions

Applications of ultrashort pulses were presented and discussed here. Some basic optical mechanisms that permeate many applications were outlined to create a basis for the reader to expand his knowledge on the field of ultrashort pulses. Although the discussions of the many applications are far from complete once almost each topic presented here can be expanded to fill a complete book, updated references in the form of books, reviews and cornerstone papers were given, and the reader is encouraged to read them to obtain a deeper comprehension on the subjacent mechanisms and resulting phenomena.

8. References

Albota, M.; Beljonne, D.; Bredas, J. L.; Ehrlich, J. E.; Fu, J. Y.; Heikal, A. A.; Hess, S. E.; Kogej, T.; Levin, M. D.; Marder, S. R.; McCord-Maughon, D.; Perry, J. W.; Rockel, H.; Rumi, M.; Subramaniam, C.; Webb, W. W.; Wu, X. L. & Xu, C. (1998). Design of organic molecules with large two-photon absorption cross sections. *Science*, 281, 5383, 1653-1656, ISSN: 0036-8075

Alfano, R. R. (2006). *The supercontinuum laser source : fundamentals with updated references*, Springer, ISBN: 0387245049 (acid-free paper), New York

Andrade, L. H. F.; Laraoui, A.; Vomir, M.; Muller, D.; Stoquert, J. P.; Estournes, C.; Beaurepaire, E. & Bigot, J. Y. (2006). Damped precession of the magnetization vector of superparamagnetic nanoparticles excited by femtosecond optical pulses. *Phys. Rev. Lett.*, 97, 12, 127401, ISSN: 0031-9007

Axt, V. M. & Kuhn, T. (2004). Femtosecond spectroscopy in semiconductors: a key to coherences, correlations and quantum kinetics. *Rep. Prog. Phys.*, 67, 4, 433-512, ISSN: 0034-4885

Badziak, J.; Chizhov, S. A.; Kozlov, A. A.; Makowski, J.; Paduch, M.; Tomaszewski, K.; Vankov, A. B. & Yashin, V. E. (1997). Picosecond, terawatt, all-Nd:glass CPA laser system. *Opt. Commun.*, 134, 1-6, 495-502, ISSN: 0030-4018

Bagnoud, V. & Salin, F. (2000). Amplifying laser pulses to the terawatt level at a 1-kilohertz repetition rate. *Appl. Phys. B-Las. Opt.*, 70, S165-S170, ISSN: 0946-2171

Bahk, S. W.; Rousseau, P.; Planchon, T. A.; Chvykov, V.; Kalintchenko, G.; Maksimchuk, A.; Mourou, G. & Yanovsky, V. (2005). Characterization of focal field formed by a large numerical aperture paraboloidal mirror and generation of ultra-high intensity (10^{22} W/cm^2). *Appl. Phys. B-Las. Opt.*, 80, 7, 823-832, ISSN: 0946-2171

Balu, M.; Hales, J.; Hagan, D. & Van Stryland, E. (2004). White-light continuum Z-scan technique for nonlinear materials characterization. *Opt. Express*, 12, 16, 3820-3826,

Bass, M. & Fradin, D. W. (1973). Surface and Bulk Laser-Damage Statistics and the Identification of Intrinsic Breakdown Processes. *IEEE J. Quantum Elec.*, QE 9, 9, 890-896, ISSN: 0018-9197

Baudach, S.; Bonse, J.; Kruger, J. & Kautek, W. (2000). Ultrashort pulse laser ablation of polycarbonate and polymethylmethacrylate. *Appl. Surf. Sci.*, 154, 555-560, ISSN: 0169-4332

Baudelet, M.; Guyon, L.; Yu, J.; Wolf, J. P.; Amodeo, T.; Frejafon, E. & Laloi, P. (2006). Femtosecond time-resolved laser-induced breakdown spectroscopy for detection and identification of bacteria: A comparison to the nanosecond regime. *J. Appl. Phys.*, 99, 8, ISSN: 0021-8979

Beaurepaire, E.; Merle, J. C.; Daunois, A. & Bigot, J. Y. (1996). Ultrafast spin dynamics in ferromagnetic nickel. *Phys. Rev. Lett.*, 76, 22, 4250-4253, ISSN: 0031-9007

Bloembergen, N. (1974). Laser-Induced Electric Breakdown in Solids. *IEEE J. Quantum Elec.*, QE10, 3, 375-386, ISSN: 0018-9197

Bohman, S.; Suda, A.; Kanai, T.; Yamaguchi, S. & Midorikawa, K. (2010). Generation of 5.0 fs, 5.0 mJ pulses at 1 kHz using hollow-fiber pulse compression. *Opt. Lett.*, 35, 11, 1887-1889, ISSN: 0146-9592

Born, M. & Wolf, E. (1999). *Principles of optics : electromagnetic theory of propagation, interference and diffraction of light*, Cambridge University Press, ISBN: 0-521-642221 (hardback), Cambridge [England] ; New York

Boyd, R. W. (2008). *Nonlinear optics*, Academic Press, ISBN: 9780123694706, Amsterdam ; Boston

Brabec, T.; Spielmann, C.; Curley, P. F. & Krausz, F. (1992). Kerr lens mode locking. *Opt. Lett.*, 17, 18, 1292-1294,

Braun, M.; Gilch, P. & Zinth, W. (2008). *Ultrashort laser pulses in biology and medicine*, Springer, ISBN: 9783540735656 (alk. paper) 3540735658 (alk. paper), Berlin ; New York

Brixner, T.; Stenger, J.; Vaswani, H. M.; Cho, M.; Blankenship, R. E. & Fleming, G. R. (2005). Two-dimensional spectroscopy of electronic couplings in photosynthesis. *Nature*, 434, 7033, 625-628, ISSN: 0028-0836

Brorson, S. D.; Kazeroonian, A.; Moodera, J. S.; Face, D. W.; Cheng, T. K.; Ippen, E. P.; Dresselhaus, M. S. & Dresselhaus, G. (1990). Femtosecond room-temperature measurement of the electron-phonon coupling constant gamma in metallic superconductors. *Phys. Rev. Lett.*, 64, 18, 2172, ISSN: 0031-9007

Bulanov, S. S.; Bychenkov, V. Y.; Chvykov, V.; Kalinchenko, G.; Litzenberg, D. W.; Matsuoka, T.; Thomas, A. G. R.; Willingale, L.; Yanovsky, V.; Krushelnick, K. & Maksimchuk, A. (2010). Generation of GeV protons from 1 PW laser interaction with near critical density targets. *Phys. Plasmas*, 17, 4, ISSN: 1070-664X

Bulanov, S. V.; Esirkepov, T. & Tajima, T. (2003). Light intensification towards the Schwinger limit. *Phys. Rev. Lett.*, 91, 8, ISSN: 0031-9007

Butcher, P. N. & Cotter, D. (1990). *The elements of nonlinear optics*, Cambridge University Press, ISBN: 0521341833, Cambridge ; New York

Cerullo, G. & De Silvestri, S. (2003). Ultrafast optical parametric amplifiers. *Rev. Sci. Instrum.*, 74, 1, 1-18, ISSN: 0034-6748

Chambaret, J. P.; Canova, F.; Lopez-Martens, R.; Cheriaux, G.; Mourou, G.; Cotel, A.; Le Blanc, C.; Druon, F.; Georges, P.; Forget, N.; Ple, F.; Pittman, M. & Ieee (2007). *ILE 25PW single laser beamline: the French step for the European Extreme Light Infrastructure (ELI)*, Ieee, ISBN: 978-1-4244-3590-6, New York

Chang, Z. H.; Rundquist, A.; Wang, H. W.; Murnane, M. M. & Kapteyn, H. C. (1997). Generation of coherent soft X rays at 2.7 nm using high harmonics. *Phys. Rev. Lett.*, 79, 16, 2967-2970, ISSN: 0031-9007

Chao, W. L.; Harteneck, B. D.; Liddle, J. A.; Anderson, E. H. & Attwood, D. T. (2005). Soft X-ray microscopy at a spatial resolution better than 15nm. *Nature*, 435, 7046, 1210-1213, ISSN: 0028-0836

Chauhan, V.; Bowlan, P.; Cohen, J. & Trebino, R. (2010). Single-diffraction-grating and grism pulse compressors. *J. Opt. Soc. Am. B*, 27, 4, 619-624, ISSN: 0740-3224

Chung, S. H.; Clark, D. A.; Gabel, C. V.; Mazur, E. & Samuel, A. D. T. (2006). The role of the AFD neuron in C-elegans thermotaxis analyzed using femtosecond laser ablation. *BMC Neurosci.*, 7, 30, ISSN: 1471-2202

Corkum, P. B. & Krausz, F. (2007). Attosecond science. *Nat. Phys.*, 3, 6, 381-387, ISSN: 1745-2473

Courrol, L. C.; Messaddeq, Y.; Messaddeq, S. H.; Ribeiro, S. J. L.; Samad, R. E.; de Freitas, A. Z. & Vieira Jr, N. D. (2008). Production of defects in ZBLAN, ZBLAN:Tm^{3+} and ZBLAN:Cr^{3+} glasses by ultra-short pulses laser interaction. *J. Phys. Chem. Sol.*, 69, 1, 55-59, ISSN: 0022-3697

Courrol, L. C.; Samad, R. E.; Gomes, L.; Ranieri, I. M.; Baldochi, S. L.; de Freitas, A. Z. & Vieira, N. D. (2004). Color center production by femtosecond pulse laser irradiation in LiF crystals. *Opt. Expr.*, 12, 2, 288-293, ISSN: 1094-4087

Cremers, D. A. & Radziemski, L. J. (2006). *Handbook of laser-induced breakdown spectroscopy*, John Wiley, ISBN: 0470092998 (cloth alk. paper), Chichester, England ; Hoboken, NJ

Cumpston, B. H.; Ananthavel, S. P.; Barlow, S.; Dyer, D. L.; Ehrlich, J. E.; Erskine, L. L.; Heikal, A. A.; Kuebler, S. M.; Lee, I. Y. S.; McCord-Maughon, D.; Qin, J. Q.; Rockel, H.; Rumi, M.; Wu, X. L.; Marder, S. R. & Perry, J. W. (1999). Two-photon polymerization initiators for three-dimensional optical data storage and microfabrication. *Nature*, 398, 6722, 51-54, ISSN: 0028-0836

Daido, H. (2002). Review of soft x-ray laser researches and developments. *Rep. Prog. Phys.*, 65, 10, 1513-1576, ISSN: 0034-4885

Davis, K. M.; Miura, K.; Sugimoto, N. & Hirao, K. (1996). Writing waveguides in glass with a femtosecond laser. *Opt. Lett.*, 21, 21, 1729-1731, ISSN: 0146-9592

Denk, W.; Strickler, J. H. & Webb, W. W. (1990). Two-photon laser scanning fluorescence microscopy. *Science*, 248, 4951, 73-76, ISSN: 0036-8075

DiDomenico, M.; Geusic, J. E.; Marcos, H. M. & Smith, R. G. (1966). Generation of ultrashort optical pulses by mode lockin the YAlG:Nd laser. *Appl. Phys. Lett.*, 8, 7, 180-183, ISSN: 0003-6951

Diels, J.-C. & Rudolph, W. (2006). *Ultrashort laser pulse phenomena : fundamentals, techniques, and applications on a femtosecond time scale*, Academic Press, ISBN: 9780122154935 (hbk.), Burlington, MA

Diels, J. C.; Van Stryland, E. & Benedict, G. (1978). Generation and measurement of 200 femtosecond optical pulses. *Opt. Commun.*, 25, 1, 93-96, ISSN: 0030-4018

Drescher, M.; Hentschel, M.; Kienberger, R.; Uiberacker, M.; Yakovlev, V.; Scrinzi, A.; Westerwalbesloh, T.; Kleineberg, U.; Heinzmann, U. & Krausz, F. (2002). Time-resolved atomic inner-shell spectroscopy. *Nature*, 419, 6909, 803-807, ISSN: 0028-0836

Du, D.; Liu, X.; Korn, G.; Squier, J. & Mourou, G. (1994). Laser-induced breakdown by impact ionization in $Si0_2$ with pulse widths from 7 ns to 150 fs. *Appl. Phys. Lett.*, 64, 23, 3071-3073, ISSN: 0003-6951

Dubietis, A.; Butkus, R. & Piskarskas, A. P. (2006). Trends in chirped pulse optical parametric amplification. *IEEE J. Sel. Top. Quantum Elec.*, 12, 2, 163-172, ISSN: 1077-260X

Dumitru, G.; Romano, V.; Weber, H. P.; Sentis, M. & Marine, W. (2002). Femtosecond ablation of ultrahard materials. *Appl. Phys. A-Mat. Sci. Proc.*, 74, 6, 729-739,

Eesley, G. L. (1983). Observation of Nonequilibrium Electron Heating in Copper. *Phys. Rev. Lett.*, 51, 23, 2140-2143, ISSN: 0031-9007

Ell, R.; Morgner, U.; Kartner, F. X.; Fujimoto, J. G.; Ippen, E. P.; Scheuer, V.; Angelow, G.; Tschudi, T.; Lederer, M. J.; Boiko, A. & Luther-Davies, B. (2001). Generation of 5-fs pulses and octave-spanning spectra directly from a Ti : sapphire laser. *Opt. Lett.*, 26, 6, 373-375, ISSN: 0146-9592

Florea, C. & Winick, K. A. (2003). Fabrication and characterization of photonic devices directly written in glass using femtosecond laser pulses. *Journal of Lightwave Technology*, 21, 1, 246-253, ISSN: 0733-8724

Fork, R. L.; Cruz, C. H. B.; Becker, P. C. & Shank, C. V. (1987). Compression of optical pulses to six femtoseconds by using cubic phase compensation. *Opt. Lett.*, 12, 7, 483-485, ISSN: 0146-9592

Fork, R. L.; Martinez, O. E. & Gordon, J. P. (1984). Negative dispersion using pairs of prisms. *Opt. Lett.*, 9, 5, 150-152, ISSN: 0146-9592

Fowles, G. R. (1989). *Introduction to modern optics*, Dover Publications, ISBN: 0486659577, New York

Freitas, A. Z.; Freschi, L. R.; Samad, R. E.; Zezell, D. M.; Gouw-Soares, S. C. & Vieira, N. D. (2010). Determination of Ablation threshold for composite resins and amalgam Irradiated with femtosecond laser pulses. *Las. Phys. Lett.*, 7, 3, 236-241, ISSN: 1612-202X

Gallmann, L.; Sutter, D. H.; Matuschek, N.; Steinmeyer, G.; Keller, U.; Iaconis, C. & Walmsley, I. A. (1999). Characterization of sub-6-fs optical pulses with spectral phase interferometry for direct electric-field reconstruction. *Opt. Lett.*, 24, 18, 1314-1316, ISSN: 0146-9592

Gamaly, E. G.; Rode, A. V.; Luther-Davies, B. & Tikhonchuk, V. T. (2002). Ablation of solids by femtosecond lasers: Ablation mechanism and ablation thresholds for metals and dielectrics. *Phys. Plasmas*, 9, 3, 949-957, ISSN: 1070-664X

Gattass, R. R. & Mazur, E. (2008). Femtosecond laser micromachining in transparent materials. *Nat. Photonics*, 2, 4, 219-225, ISSN: 1749-4885

Gaul, E. W.; Martinez, M.; Blakeney, J.; Jochmann, A.; Ringuette, M.; Hammond, D.; Borger, T.; Escamilla, R.; Douglas, S.; Henderson, W.; Dyer, G.; Erlandson, A.; Cross, R.; Caird, J.; Ebbers, C. & Ditmire, T. (2010). Demonstration of a 1.1 petawatt laser based on a hybrid optical parametric chirped pulse amplification/mixed Nd:glass amplifier. *Appl. Opt.*, 49, 9, 1676-1681, ISSN: 0003-6935

Gellermann, W. (1991). Color center lasers. *J. Phys. Chem. Sol.*, 52, 1, 249-297, ISSN: 0022-3697

Gibbon, P. & Forster, E. (1996). Short-pulse laser-plasma interactions. *Plasma Phys. Control. Fusion*, 38, 6, 769-793, ISSN: 0741-3335

Gibson, E. A.; Paul, A.; Wagner, N.; Tobey, R.; Gaudiosi, D.; Backus, S.; Christov, I. P.; Aquila, A.; Gullikson, E. M.; Attwood, D. T.; Murnane, M. M. & Kapteyn, H. C. (2003). Coherent soft x-ray generation in the water window with quasi-phase matching. *Science*, 302, 5642, 95-98, ISSN: 0036-8075

Gomez, D.; Tekniker, F.; Goenaga, I.; Lizuain, I. & Ozaita, M. (2005). Femtosecond laser ablation for microfluidics. *Optical Engineering*, 44, 5, ISSN: 0091-3286

Habara, H.; Xu, G.; Jitsuno, T.; Kodama, R.; Suzuki, K.; Sawai, K.; Kondo, K.; Miyanaga, N.; Tanaka, K. A.; Mima, K.; Rushford, M. C.; Britten, J. A. & Barty, C. P. J. (2010). Pulse compression and beam focusing with segmented diffraction gratings in a high-power chirped-pulse amplification glass laser system. *Opt. Lett.*, 35, 11, 1783-1785,

Hannaford, P. (2005). *Femtosecond laser spectroscopy*, Springer, ISBN: 0387232931 (hd. bd.), New York, NY

Hargrove, L. E.; Fork, R. L. & Pollack, M. A. (1964). Locking of He-Ne Laser Modes Induced by Synchronous Intracavity Modulation. *Appl. Phys. Lett.*, 5, 1, 4-5, ISSN: 0003-6951

Haus, H. A. (2000). Mode-locking of lasers. *IEEE J. Sel. Top. Quantum Elec.*, 6, 6, 1173-1185, ISSN: 1077-260X

Haus, H. A.; Fujimoto, J. G. & Ippen, E. P. (1992). Analytic theory of additive pulse and Kerr lens mode locking. *IEEE J. Quantum Elec.*, 28, 10, 2086-2096, ISSN: 0018-9197

Hentschel, M.; Kienberger, R.; Spielmann, C.; Reider, G. A.; Milosevic, N.; Brabec, T.; Corkum, P.; Heinzmann, U.; Drescher, M. & Krausz, F. (2001). Attosecond metrology. *Nature*, 414, 6863, 509-513, ISSN: 0028-0836

Hirao, K. & Miura, K. (1998). Writing waveguides and gratings in silica and related materials by a femtosecond laser. *J. Non-Cryst. Sol.*, 239, 1-3, 91-95, ISSN: 0022-3093

Holzwarth, R.; Udem, T.; Hansch, T. W.; Knight, J. C.; Wadsworth, W. J. & Russell, P. S. J. (2000). Optical frequency synthesizer for precision spectroscopy. *Phys. Rev. Lett.*, 85, 11, 2264-2267, ISSN: 0031-9007

Itoh, K.; Watanabe, W.; Nolte, S. & Schaffer, C. B. (2006). Ultrafast processes for bulk modification of transparent materials. *Mrs Bulletin*, 31, 8, 620-625, ISSN: 0883-7694

Jackson, J. D. (1999). *Classical electrodynamics*, Wiley, ISBN: 0 471 30932-X, New York

Joglekar, A. P.; Liu, H.; Spooner, G. J.; Meyhofer, E.; Mourou, G. & Hunt, A. J. (2003). A study of the deterministic character of optical damage by femtosecond laser pulses and applications to nanomachining. *Appl. Phys. B-Las. Opt.*, 77, 1, 25-30, ISSN: 0946-2171

Johnsson, P.; Mauritsson, J.; Remetter, T.; L'Huillier, A. & Schafer, K. J. (2007). Attosecond control of ionization by wave-packet interference. *Phys. Rev. Lett.*, 99, 23, ISSN: 0031-9007

Joo, T. & Albrecht, A. C. (1993). Electronic dephasing studies of molecules in solution at room temperature by femtosecond degenerate four wave mixing. *Chem. Phys.*, 176, 1, 233-247, ISSN: 0301-0104

Juhasz, T.; Frieder, H.; Kurtz, R. M.; Horvath, C.; Bille, J. F. & Mourou, G. (1999). Corneal refractive surgery with femtosecond lasers. *IEEE J. Sel. Top. Quantum Elec.*, 5, 4, 902-910, ISSN: 1077-260X

Kane, D. J. & Trebino, R. (1993). Characterization of arbitrary femtosecond pulses using frequency-resolved optical gating. *IEEE J. Quantum Elec.*, 29, 2, 571-579, ISSN: 0018-9197

Kasparian, J.; Rodriguez, M.; Mejean, G.; Yu, J.; Salmon, E.; Wille, H.; Bourayou, R.; Frey, S.; Andre, Y. B.; Mysyrowicz, A.; Sauerbrey, R.; Wolf, J. P. & Woste, L. (2003). White-light filaments for atmospheric analysis. *Science*, 301, 5629, 61-64, ISSN: 0036-8075

Kasparian, J.; Wöste, L. & Wolf, J.-P. (2010). Laser-Based Weather Control. *Opt. Photon. News*, 21, 7, 22-27,

Kautek, W.; Kruger, J.; Lenzner, M.; Sartania, S.; Spielmann, C. & Krausz, F. (1996). Laser ablation of dielectrics with pulse durations between 20 fs and 3 ps. *Appl. Phys. Lett.*, 69, 21, 3146-3148, ISSN: 0003-6951

Ke, K.; Hasselbrink, E. F. & Hunt, A. J. (2005). Rapidly prototyped three-dimensional nanofluidic channel networks in glass substrates. *Analytical Chemistry*, 77, 16, 5083-5088, ISSN: 0003-2700

Keldysh, L. V. (1965). Ionization in the Field of a Strong Electromagnetic Wave. *Sov. Phys. JETP-USSR*, 20, 5, 1307-1314, ISSN: 0038-5646

Keller, U. (1994). Ultrafast all-solid-state laser technology *Appl. Phys. B-Las. Opt.*, 58, 5, 347-363, ISSN: 0946-2171

Keller, U. (2010). Ultrafast solid-state laser oscillators: a success story for the last 20 years with no end in sight. *Appl. Phys. B-Las. Opt.*, 100, 1, 15-28, ISSN: 0946-2171

Kershaw, S. V. (1995). Analysis of the EZ Scan Measurement Technique *J. Mod. Opt.*, 42, 7, 1361-1366, ISSN: 0950-0340

Kling, M. F. & Vrakking, M. J. J. (2008). Attosecond electron dynamics. *Annu. Rev. Phys. Chem.*, 59, 463-492, ISSN: 0066-426X

Knox, W. H.; Fork, R. L.; Downer, M. C.; Stolen, R. H.; Shank, C. V. & Valdmanis, J. A. (1985). Optical pulse compression to 8 fs at a 5-kHz repetition rate. *Appl. Phys. Lett.*, 46, 12, 1120-1121, ISSN: 0003-6951

Koda, R.; Oki, T.; Miyajima, T.; Watanabe, H.; Kuramoto, M.; Ikeda, M. & Yokoyama, H. (2010). 100 W peak-power 1 GHz repetition picoseconds optical pulse generation using blue-violet GaInN diode laser mode-locked oscillator and optical amplifier. *Appl. Phys. Lett.*, 97, 2, 021101-021103,

Koechner, W. (2006). *Solid-state laser engineering*, Springer, ISBN: 038729094X, New York

Konig, K. (2000). Multiphoton microscopy in life sciences. *J. Microsc. - Oxford*, 200, 83-104, ISSN: 0022-2720

Krauss, G.; Lohss, S.; Hanke, T.; Sell, A.; Eggert, S.; Huber, R. & Leitenstorfer, A. (2010). Synthesis of a single cycle of light with compact erbium-doped fibre technology. *Nat. Photonics*, 4, 1, 33-36, ISSN: 1749-4885

Krausz, F. & Ivanov, M. (2009). Attosecond physics. *Rev. Mod. Phys.*, 81, 1, 163-234, ISSN: 0034-6861

Kruger, J. & Kautek, W. (1999). The femtosecond pulse laser: a new tool for micromachining. *Las. Phys.*, 9, 1, 30-40, ISSN: 1054-660X

L'Huillier, A. & Balcou, P. (1993). High-order harmonic generation in rare gases with a 1-ps 1053-nm laser. *Phys. Rev. Lett.*, 70, 6, 774,

Laenen, R.; Wolfrum, K.; Seilmeier, A. & Laubereau, A. (1993). Parametric generation of femtosecond and picosecond pulses for spectroscopic applications. *J. Opt. Soc. Am. B*, 10, 11, 2151-2161,

Le Drogoff, B.; Margot, J.; Chaker, M.; Sabsabi, M.; Barthelemy, O.; Johnston, T. W.; Laville, S.; Vidal, F. & von Kaenel, Y. (2001). Temporal characterization of femtosecond laser pulses induced plasma for spectrochemical analysis of aluminum alloys. *Spectrochim. Acta B*, 56, 6, 987-1002, ISSN: 0584-8547

Lenzner, M.; Kruger, J.; Sartania, S.; Cheng, Z.; Spielmann, C.; Mourou, G.; Kautek, W. & Krausz, F. (1998). Femtosecond optical breakdown in dielectrics. *Phys. Rev. Lett.*, 80, 18, 4076-4079, ISSN: 0031-9007

Leo, K.; Shah, J.; Gobel, E. O.; Damen, T. C.; Schmittrink, S.; Schafer, W. & Kohler, K. (1991). Coherent oscillations of a wave packet in a semiconductor double-quantum-well structure. *Phys. Rev. Lett.*, 66, 2, 201-204, ISSN: 0031-9007

Lewenstein, M.; Balcou, P.; Ivanov, M. Y.; Lhuillier, A. & Corkum, P. B. (1994). Theory of high-harmonic generation by low-frequency laser fields. *Phys. Rev. A*, 49, 3, 2117-2132, ISSN: 1050-2947

Li, J.; Liu, Z.; Tan, C.; Guo, X.; Wang, L.; Sancar, A. & Zhong, D. (2010). Dynamics and mechanism of repair of ultraviolet-induced (6-4) photoproduct by photolyase. *Nature*, advance online publication, ISSN: 1476-4687

Liu, J. M. (1982). Simple technique for measurements of pulsed Gaussian-beam spot sizes. *Opt. Lett.*, 7, 5, 196-198, ISSN: 0146-9592

Ma, H.; Gomes, A. S. L. & Dearaujo, C. B. (1991). Measurements of nondegenerate optical nonlinearity using a two-color single beam method. *Appl. Phys. Lett.*, 59, 21, 2666-2668, ISSN: 0003-6951

Maine, P.; Strickland, D.; Bado, P.; Pessot, M. & Mourou, G. (1988). Generation of ultrahigh peak power pulses by chirped pulse amplification. *IEEE J. Quantum Elec.*, 24, 2, 398-403, ISSN: 0018-9197

Maksimchuk, A.; Gu, S.; Flippo, K.; Umstadter, D. & Bychenkov, V. Y. (2000). Forward ion acceleration in thin films driven by a high-intensity laser. *Phys. Rev. Lett.*, 84, 18, 4108-4111, ISSN: 0031-9007

Malinowski, A.; Piper, A.; Price, J. H. V.; Furusawa, K.; Jeong, Y.; Nilsson, J. & Richardson, D. J. (2004). Ultrashort-pulse Yb3+-fiber-based laser and amplifier system producing > 25-W average power. *Opt. Lett.*, 29, 17, 2073-2075, ISSN: 0146-9592

Martınez, O. E. (1987a). 3000 times grating compressor with positive group velocity dispersion: Application to fiber compensation in 1.3-1.6 μm region. *IEEE J. Quantum Elec.*, 23, 1, 59-64, ISSN: 0018-9197

Martinez, O. E. (1987b). Design of high-power ultrashort pulse amplifiers by expansion and recompression. *IEEE J. Quantum Elec.*, 23, 8, 1385-1387, ISSN: 0018-9197

Martynovich, E. F.; Glazunov, D. S.; Grigorova, A. A.; Starchenko, A. A.; Kirpichnikov, A. V.; Trunov, V. I.; Merzlyakov, M. A.; Petrov, V. V. & Pestryakov, E. V. (2008). Highly nonlinear fundamental mechanisms of excitation and coloring of wide-gap crystals by intense femtosecond laser pulses. *Optics and Spectroscopy*, 105, 3, 348-351, ISSN: 0030-400X

Maruo, S.; Nakamura, O. & Kawata, S. (1997). Three-dimensional microfabrication with two-photon-absorbed photopolymerization. *Opt. Lett.*, 22, 2, 132-134, ISSN: 0146-9592

Mauritsson, J.; Remetter, T.; Swoboda, M.; Klunder, K.; L'Huillier, A.; Schafer, K. J.; Ghafur, O.; Kelkensberg, F.; Siu, W.; Johnsson, P.; Vrakking, M. J. J.; Znakovskaya, I.; Uphues, T.; Zherebtsov, S.; Kling, M. F.; Lepine, F.; Benedetti, E.; Ferrari, F.; Sansone, G. & Nisoli, M. (2010). Attosecond Electron Spectroscopy Using a Novel Interferometric Pump-Probe Technique. *Phys. Rev. Lett.*, 105, 5, ISSN: 0031-9007

Minoshima, K.; Kowalevicz, A. M.; Hartl, I.; Ippen, E. P. & Fujimoto, J. G. (2001). Photonic device fabrication in glass by use of nonlinear materials processing with a femtosecond laser oscillator. *Opt. Lett.*, 26, 19, 1516-1518, ISSN: 0146-9592

Miyazaki, K. & Takada, H. (1995). High-order harmonic generation in the tunneling regime. *Phys. Rev. A*, 52, 4, 3007,

Monot, P.; Auguste, T.; Gibbon, P.; Jakober, F.; Mainfray, G.; Dulieu, A.; Louis-Jacquet, M.; Malka, G. & Miquel, J. L. (1995). Experimental Demonstration of Relativistic Self-

Channeling of a Multiterawatt Laser Pulse in an Underdense Plasma. *Phys. Rev. Lett.*, 74, 15, 2953,

Moses, E. I.; Boyd, R. N.; Remington, B. A.; Keane, C. J. & Al-Ayat, R. (2009). The National Ignition Facility: Ushering in a new age for high energy density science. *Phys. Plasmas*, 16, 4, ISSN: 1070-664X

Moulton, P. F. (1986). Spectroscopic and laser characteristics of Ti:Al_2O_3. *J. Opt. Soc. Am. B*, 3, 1, 125-133, ISSN: 0740-3224

Mourou, G. A.; Barty, C. P. J. & Perry, M. D. (1998). Ultrahigh-intensity lasers: Physics of the extreme on a tabletop. *Phys. Today*, 51, 1, 22-28, ISSN: 0031-9228

Mourou, G. A.; Tajima, T. & Bulanov, S. V. (2006). Optics in the relativistic regime. *Rev. Mod. Phys.*, 78, 2, 309-371, ISSN: 0034-6861

Muller, M.; Squier, J.; Wilson, K. R. & Brakenhoff, G. J. (1998). 3D microscopy of transparent objects using third-harmonic generation. *J. Microsc. - Oxford*, 191, 266-274, ISSN: 0022-2720

Musgrave, I. O.; Hernandez-Gomez, C.; Canny, D.; Collier, J. & Heathcote, R. (2007). Minimization of the impact of a broad bandwidth high-gain nonlinear preamplifier to the amplified spontaneous emission pedestal of the Vulcan petawatt laser facility. *Appl. Opt.*, 46, 28, 6978-6983,

Naumov, S.; Fernandez, A.; Graf, R.; Dombi, P.; Krausz, F. & Apolonski, A. (2005). Approaching the microjoule frontier with femtosecond laser oscillators. *New J. Phys.*, 7, ISSN: 1367-2630

Nayak, B. K.; Gupta, M. C. & Kolasinski, K. W. (2007). Spontaneous formation of nanospiked microstructures in germanium by femtosecond laser irradiation. *Nanotechnology*, 18, 19, ISSN: 0957-4484

Neutze, R.; Wouts, R.; van der Spoel, D.; Weckert, E. & Hajdu, J. (2000). Potential for biomolecular imaging with femtosecond X-ray pulses. *Nature*, 406, 6797, 752-757, ISSN: 0028-0836

Nisoli, M.; DeSilvestri, S.; Svelto, O.; Szipocs, R.; Ferencz, K.; Spielmann, C.; Sartania, S. & Krausz, F. (1997). Compression of high-energy laser pulses below 5 fs. *Opt. Lett.*, 22, 8, 522-524, ISSN: 0146-9592

Nolte, S.; Momma, C.; Jacobs, H.; Tunnermann, A.; Chichkov, B. N.; Wellegehausen, B. & Welling, H. (1997). Ablation of metals by ultrashort laser pulses. *J. Opt. Soc. Am. B*, 14, 10, 2716-2722, ISSN: 0740-3224

Nolte, S.; Will, M.; Burghoff, J. & Tuennermann, A. (2003). Femtosecond waveguide writing: a new avenue to three-dimensional integrated optics. *Appl. Phys. A-Mat. Sci. Proc.*, 77, 1, 109-111, ISSN: 0947-8396

Orlando, S.; Langford, S. C. & Dickinson, J. T. (2010). Generation of color centers in alkali halide single crystals using ultrafast laser pulses. *J. Optoelectron. Adv. Mater.*, 12, 3, 707-710, ISSN: 1454-4164

Pal, S. K.; Peon, J.; Bagchi, B. & Zewail, A. H. (2002). Biological water: Femtosecond dynamics of macromolecular hydration. *Journal of Physical Chemistry B*, 106, 48, 12376-12395, ISSN: 1520-6106

Perry, M. D. & Mourou, G. (1994). Terawatt to Petawatt Subpicosecond Lasers *Science*, 264, 5161, 917-924, ISSN: 0036-8075

Perry, M. D.; Stuart, B. C.; Banks, P. S.; Feit, M. D.; Yanovsky, V. & Rubenchik, A. M. (1999). Ultrashort-pulse laser machining of dielectric materials. *J. Appl. Phys.*, 85, 9, 6803-6810, ISSN: 0021-8979

Peticolas, W. L.; Rieckhoff, K. E. & Goldsborough, J. P. (1963). Double Photon Excitation in organic crystals. *Phys. Rev. Lett.*, 10, 2, 43-&, ISSN: 0031-9007

Reichert, J.; Holzwarth, R.; Udem, T. & Hansch, T. W. (1999). Measuring the frequency of light with mode-locked lasers. *Opt. Commun.*, 172, 1-6, 59-68, ISSN: 0030-4018

Reiter, F.; Graf, U.; Schultze, M.; Schweinberger, W.; Schröder, H.; Karpowicz, N.; Azzeer, A. M.; Kienberger, R.; Krausz, F. & Goulielmakis, E. (2010). Generation of sub-3 fs pulses in the deep ultraviolet. *Opt. Lett.*, 35, 13, 2248-2250,

Rohwetter, P.; Kasparian, J.; Stelmaszczyk, K.; Hao, Z. Q.; Henin, S.; Lascoux, N.; Nakaema, W. M.; Petit, Y.; Queisser, M.; Salame, R.; Salmon, E.; Woste, L. & Wolf, J. P. (2010). Laser-induced water condensation in air. *Nat. Photonics*, 4, 7, 451-456, ISSN: 1749-4885

Rossi, F. & Kuhn, T. (2002). Theory of ultrafast phenomena in photoexcited semiconductors. *Rev. Mod. Phys.*, 74, 3, 895-950, ISSN: 0034-6861

Rullière, C. (1998). *Femtosecond laser pulses : principles and experiments*, Springer, ISBN: 3-540-63663-3 (acid-free paper), Berlin ; New York

Samad, R. E.; Baldochi, S. L. & Vieira Jr, N. D. (2008). Diagonal scan measurement of Cr:LiSAF 20 ps ablation threshold. *Appl. Opt.*, 47, 7, 920-924, ISSN: 0003-6935

Samad, R. E.; Courrol, L. C.; Lugão, A. B.; de Freitas, A. Z. & Vieira Junior, N. D. (2010). Production of color centers in PMMA by ultrashort laser pulses. *Rad. Phys. Chem.*, 79, 3, 355-357, ISSN: 0969-806X

Samad, R. E. & Vieira, N. D. (2006). Geometrical method for determining the surface damage threshold for femtosecond laser pulses. *Las. Phys.*, 16, 2, 336-339, ISSN: 1054-660X

Samek, O.; Lambert, J.; Hergenroder, R.; Liska, M.; Kaiser, J.; Novotny, K. & Kukhlevsky, S. (2006). Femtosecond laser spectrochemical analysis of plant samples. *Las. Phys. Lett.*, 3, 1, 21-25, ISSN: 1612-2011

Sansone, G.; Benedetti, E.; Calegari, F.; Vozzi, C.; Avaldi, L.; Flammini, R.; Poletto, L.; Villoresi, P.; Altucci, C.; Velotta, R.; Stagira, S.; De Silvestri, S. & Nisoli, M. (2006). Isolated single-cycle attosecond pulses. *Science*, 314, 5798, 443-446, ISSN: 0036-8075

Santos, D.; Samad, R. E.; Trevizan, L. C.; de Freitas, A. Z.; Vieira, N. D. & Krug, F. J. (2008). Evaluation of Femtosecond Laser-Induced Breakdown Spectroscopy for Analysis of Animal Tissues. *Appl. Spectrosc.*, 62, 10, 1137-1143, ISSN: 0003-7028

Scheps, R. (2002). *Introduction to laser diode-pumped solid state lasers*, SPIE Press, ISBN: 0819442747 (pbk.), Bellingham, Wash.

Schoenlein, R. W.; Lin, W. Z.; Fujimoto, J. G. & Eesley, G. L. (1987). Femtosecond studies of nonequilibrium electronic processes in metals. *Phys. Rev. Lett.*, 58, 16, 1680-1683, ISSN: 0031-9007

Schreier, W. J.; Schrader, T. E.; Koller, F. O.; Gilch, P.; Crespo-Hernandez, C. E.; Swaminathan, V. N.; Carell, T.; Zinth, W. & Kohler, B. (2007). Thymine dimerization in DNA is an ultrafast photoreaction. *Science*, 315, 5812, 625-629, ISSN: 0036-8075

Schwinger, J. (1951). On Gauge Invariance and Vacuum Polarization. *Phys. Rev.*, 82, 5, 664,

Shank, C. V. & Ippen, E. P. (1974). Subpicosecond kilowatt pulses from a mode-locked cw dye laser. *Appl. Phys. Lett.*, 24, 8, 373-375, ISSN: 0003-6951

Sheik-Bahae, M.; Said, A. A.; Wei, T. H.; Hagan, D. J. & Van Stryland, E. W. (1990). Sensitive measurement of optical nonlinearities using a single beam. *IEEE J. Quantum Elec.*, 26, 4, 760-769, ISSN: 0018-9197

Shen, Y. R. (1984). *The principles of nonlinear optics*, J. Wiley, ISBN: 0471889989, New York

Shirk, M. D. & Molian, P. A. (1998). A review of ultrashort pulsed laser ablation of materials. *Journal of Laser Applications*, 10, 1, 18-28, ISSN: 1042-346X

Spence, D. E.; Kean, P. N. & Sibbett, W. (1991). 60-fsec pulse generation from a self-mode-locked Ti:sapphire laser. *Opt. Lett.*, 16, 1, 42-44, ISSN: 0146-9592

Spielmann, C.; Burnett, N. H.; Sartania, S.; Koppitsch, R.; Schnurer, M.; Kan, C.; Lenzner, M.; Wobrauschek, P. & Krausz, F. (1997). Generation of Coherent X-rays in the Water Window Using 5-Femtosecond Laser Pulses. *Science*, 278, 5338, 661-664,

Stoeckl, C.; Delettrez, J. A.; Kelly, J. H.; Kessler, T. J.; Kruschwitz, B. E.; Loucks, S. J.; McCrory, R. L.; Meyerhofer, D. D.; Maywar, D. N.; Morse, S. F. B.; Myatt, J.; Rigatti, A. L.; Waxer, L. J.; Zuegel, J. D. & Stephens, R. B. (2006). High-energy petawatt project at the University of Rochester's Laboratory for Laser Energetics. *Fusion Sci. Technol.*, 49, 3, 367-373, ISSN: 1536-1055

Strickland, D. & Mourou, G. (1985). Compression of amplified chirped optical pulses. *Opt. Commun.*, 56, 3, 219-221, ISSN: 0030-4018

Sugioka, K.; Cheng, Y. & Midorikawa, K. (2005). Three-dimensional micromachining of glass using femtosecond laser for lab-on-a-chip device manufacture. *Appl. Phys. A-Mat. Sci. Proc.*, 81, 1, 1-10, ISSN: 0947-8396

Sun, C. K.; Vallee, F.; Acioli, L. H.; Ippen, E. P. & Fujimoto, J. G. (1994). Femtosecond-tunable measurement of electron thermalization in gold. *Phys. Rev. B*, 50, 20, 15337-15348, ISSN: 0163-1829

Svanberg, S. (2008). Laser based diagnostics - From cultural heritage to human health. *Appl. Phys. B-Las. Opt.*, 92, 3, 351-358, ISSN: 0946-2171

Szipocs, R.; Ferencz, K.; Spielmann, C. & Krausz, F. (1994). Chirped multilayer coatings for broadband dispersion control in femtosecond lasers. *Opt. Lett.*, 19, 3, 201-203, ISSN: 0146-9592

Tajima, T. & Mourou, G. (2002). Zettawatt-exawatt lasers and their applications in ultrastrong-field physics. *Phys. Rev. Spec. Top.-Ac.*, 5, 3, ISSN: 1098-4402

Takamoto, M.; Hong, F. L.; Higashi, R. & Katori, H. (2005). An optical lattice clock. *Nature*, 435, 7040, 321-324, ISSN: 0028-0836

Tang, S. S. & Chen, X. F. (2010). Generation of isolated sub-100-as pulses with 30-fs lasers. *Phys. Rev. A*, 82, 1, ISSN: 1050-2947

Telle, H. R.; Steinmeyer, G.; Dunlop, A. E.; Stenger, J.; Sutter, D. H. & Keller, U. (1999). Carrier-envelope offset phase control: A novel concept for absolute optical frequency measurement and ultrashort pulse generation. *Appl. Phys. B-Las. Opt.*, 69, 4, 327-332, ISSN: 0946-2171

Träger, F. (2007). *Springer handbook of lasers and optics*, Springer, ISBN: 9780387955797, New York

Treacy, E. B. (1969). Optical Pulse Compression With Diffraction Gratings. *IEEE J. Quantum Elec.*, QE 5, 9, 454-458, ISSN: 0018-9197

Trebino, R. (2000). *Frequency-resolved optical gating : the measurement of ultrashort laser pulses*, Kluwer Academic, ISBN: 1402070667 (alk. paper), Boston

Udem, T.; Diddams, S. A.; Vogel, K. R.; Oates, C. W.; Curtis, E. A.; Lee, W. D.; Itano, W. M.; Drullinger, R. E.; Bergquist, J. C. & Hollberg, L. (2001). Absolute frequency measurements of the Hg+ and Ca optical clock transitions with a femtosecond laser. *Phys. Rev. Lett.*, 86, 22, 4996-4999, ISSN: 0031-9007

Uiberacker, M.; Uphues, T.; Schultze, M.; Verhoef, A. J.; Yakovlev, V.; Kling, M. F.; Rauschenberger, J.; Kabachnik, N. M.; Schroder, H.; Lezius, M.; Kompa, K. L.; Muller, H. G.; Vrakking, M. J. J.; Hendel, S.; Kleineberg, U.; Heinzmann, U.; Drescher, M. & Krausz, F. (2007). Attosecond real-time observation of electron tunnelling in atoms. *Nature*, 446, 7136, 627-632,

Umstadter, D.; Chen, S. Y.; Maksimchuk, A.; Mourou, G. & Wagner, R. (1996). Nonlinear optics in relativistic plasmas and laser wake field acceleration of electrons. *Science*, 273, 5274, 472-475, ISSN: 0036-8075

Vogel, A.; Noack, J.; Huttman, G. & Paltauf, G. (2005). Mechanisms of femtosecond laser nanosurgery of cells and tissues. *Appl. Phys. B-Las. Opt.*, 81, 8, 1015-1047, ISSN: 0946-2171

Vogel, A. & Venugopalan, V. (2003). Mechanisms of pulsed laser ablation of biological tissues. *Chem. Rev.*, 103, 2, 577-644, ISSN: 0009-2665

Vomir, M.; Andrade, L. H. F.; Guidoni, L.; Beaurepaire, E. & Bigot, J. Y. (2005). Real space trajectory of the ultrafast magnetization dynamics in ferromagnetic metals. *Phys. Rev. Lett.*, 94, 23, 237601, ISSN: 0031-9007

Weitkamp, C. (2005). *Lidar : range-resolved optical remote sensing of the atmosphere*, Springer Science+Business Media, ISBN: 0387400753 (acid-free paper), New York

Xu, C.; Zipfel, W.; Shear, J. B.; Williams, R. M. & Webb, W. W. (1996). Multiphoton fluorescence excitation: New spectral windows for biological nonlinear microscopy. *Proc. Natl. Acad. Sci. USA*, 93, 20, 10763-10768, ISSN: 0027-8424

Yariv, A. (1989). *Quantum electronics*, Wiley, ISBN: 0471609978, New York

Yariv, A. & Yeh, P. (2007). *Photonics : optical electronics in modern communications*, Oxford University Press, ISBN: 0195179463, New York

Yau, H.-F.; Wang, P.-J.; Pan, E.-Y. & Chen, J. (1996). Self-pumped phase conjugation with femtosecond pulses by use of BaTiO3. *Opt. Lett.*, 21, 15, 1168-1170,

Ye, J. & Cundiff, S. T. (2005). *Femtosecond optical frequency comb : principle, operation, and applications*, Springer, ISBN: 0387237909 (hd.bd.), New York, NY

Zewail, A. H. (2000). Femtochemistry: Atomic-scale dynamics of the chemical bond. *J. Phys. Chem. A*, 104, 24, 5660-5694, ISSN: 1089-5639

Zhan, C. L., Zhang, D. Q.; Zhu, D. B.; Wang, D. Y.; Li, Y. J.; Li, D. H.; Lu, Z. Z.; Zhao, L. Z. & Nie, Y. X. (2002). Third- and fifth-order optical nonlinearities in a new stilbazolium derivative. *J. Opt. Soc. Am. B*, 19, 3, 369-375, ISSN: 0740-3224

Zhou, S.; Kuznetsova, L.; Chong, A. & Wise, F. W. (2005). Compensation of nonlinear phase shifts with third-order dispersion in short-pulse fiber amplifiers. *Opt. Expr.*, 13, 13, 4869-4877, ISSN: 1094-4087

Zipfel, W. R.; Williams, R. M.; Christie, R.; Nikitin, A. Y.; Hyman, B. T. & Webb, W. W. (2003). Live tissue intrinsic emission microscopy using multiphoton-excited native fluorescence and second harmonic generation. *Proc. Natl. Acad. Sci. USA*, 100, 12, 7075-7080, ISSN: 0027-8424

Zoubir, A.; Shah, L.; Richardson, K. & Richardson, M. (2003). Practical uses of femtosecond laser micro-materials processing. *Appl. Phys. A-Mat. Sci. Proc.*, 77, 2, 311-315, ISSN: 0947-8396

Zumbusch, A.; Holtom, G. R. & Xie, X. S. (1999). Three-dimensional vibrational imaging by coherent anti-Stokes Raman scattering. *Phys. Rev. Lett.*, 82, 20, 4142-4145, ISSN: 0031-9007